Advances in Cryogenic Engineering

VOLUME 41, PART A

A Continuation Order Plan is available for this series. A continuation order will bring delivery of each new volume immediately upon publication. Volumes are billed only upon actual shipment. For further information please contact the publisher.

A Cryogenic Engineering Conference Publication

Advances in Cryogenic Engineering

VOLUME 41, PART A

Edited by
Peter Kittel
NASA–Ames Research Center
Moffett Field, California

PLENUM PRESS · NEW YORK and LONDON

The Library of Congress cataloged the first volume of this title as follows:

Advances in cryogenic engineering. v. 1–
New York, Cryogenic Engineering Conference; distributed by Plenum Press, 1960–
v. illus., diagrs. 26 cm.
Vols. 1– are reprints of the Proceedings of the Cryogenic Engineering Conference, 1954–
Editor: 1960– K. D. Timmerhaus.

1. Low temperature engineering—Congresses. I. Timmerhaus, K. D., ed. II. Cryogenic Engineering Conference

TP490.A3 660.29368 57-35598

Proceedings of the 1995 Cryogenic Engineering Conference,
held July 17–21, 1995, in Columbus, Ohio

ISBN 0-306-45300-2

10 9 8 7 6 5 4 3 2 1

Printed in the United States of America

CONTENTS

PART A

Cryobiology

Heat and Mass Transfer: General

Heat and Mass Transfer: Boiling

Heat and Mass Transfer: Helium II

Magnet: Design and Performance

Magnet: Technology and Applications

Current Leads

Large Scale Systems

Large Hadron Collider - LHC

TeV Electron Superconducting Linear Accelerator - TESLA

Mechanisms and Machinery

PART B

Cryofuels

Minesweeping

Space Applications

Cryocoolers: General

Cryocoolers: Pulse Tube

Cryocoolers: Other Regenerative

Instrumentation and Control

Safety

Miscellaneous Techniques and Applications

INDICES

FOREWORD

The Hyatt Regency Hotel, Columbus, Ohio was the venue for the 1995 Cryogenic Engineering Conference. The meeting was held jointly with the International Cryogenic Materials Conference. Jim Peeples, of CVI, Inc., was conference chairman. Columbus is the home of the Battelle Memorial Institute, a pioneer in cryogenic materials development; the home of CVI, Inc., and Lake Shore Cryotronics, Inc., two leading manufacturers of cryogenic equipment; and it is the home of Ohio State University, where research on liquid helium has long been conducted.

The program consisted of 315 CEC papers, nearly the same number as for CEC-91. This was the second largest number of papers ever submitted to the CEC. Of these, 252 papers are published here, in Volume 41 of **Advances in Cryogenic Engineering**. Once again the volume is published in two books.

This volume includes a number of photographs taken during the awards lunch on July 20, 1995. Photographs have often been taken during the conferences, but they have never been used. The pictures are of the awardees, the conference chairs, and the organizers. They are distributed through out the books on pages that would otherwise have been blank. The pictures can be found on the following pages: 28, 232, 334, 536, 640, 826, 990, 1032, 1202, 1462, 1682, 1888, and 1994.

Another first for the 1995 CEC was a set of awards for the best posters. Poster presentations are often viewed as being inferior to oral sessions. However, posters offer a more intimate forum for exchanging ideas. It also offers a less restrictive medium for expressing those ideas. To encourage authors to make full use of the medium, awards of $50 were given to about 10% of the posters. The selection was based on the aesthetics of the presentation as judged by a panel of nontechnical people.

Peter Kittel
Ames Research Center
Moffett Field, CA
May, 1996

SAMUEL C. COLLINS AWARD

The Samuel C. Collins Award for outstanding contributions to cryogenic technology was presented to Dr. Ronald W. Fast at the 1995 Cryogenic Engineering Conference.

The Collins Award was established in 1965 by the Cryogenic Engineering Conference to recognize outstanding contributions to cryogenic technology. The award is named in the honor of the late Samuel C. Collins, Professor of Mechanical Engineering at Massachusetts Institute of Technology, who is responsible for invention of the modern helium Liquefier. In addition to Dr. Collins, who received the award in 1965, this award has been presented to Professor Klaus D. Timmerhaus (1967), Dr. Edward F. Hammel (1973), Dr. Howard O. McMahon (1979), Professor Joseph L. Smith (1983), Professor William Gifford and Dr. Bascom Birmingham (1991), and Professor Roger W. Boom (1993). Since its establishment 30 years ago, the Collins Award has been presented only eight times.

Dr. Ron Fast has had a long and distinguished career in cryogenic engineering, primarily in the development of the high field magnets and related cryogenic systems, which have played and will continue to play a large role in high-energy physics, fusion and materials studies. He has also played a major role in the development of the Cryogenic Engineering Conference.

Ron received his B.S. in Physics from Washington and Lee College in 1956 and his M.S. and Ph.D. from the University of Virginia in 1958 and 1960, respectively. He spent five years at the Midwestern Universities Research Association in Madison, Wisconsin, two years at the University of Wisconsin at Madison and twenty-five years at the Fermi National Accelerator Laboratory in Batavia, Illinois, retiring in 1994. During this time his interests ranged widely over conventional and superconducting beam-line magnets, analysis magnets for high-energy particle experiments, refrigerated liquid H_2 targets for experiments, liquid helium refrigerators and cryogenic systems for superconducting magnets and multi-layer insulation. His work resulted in over 45 papers. In 1981, he and his co-authors were awarded the Russell B. Scott Memorial Award for best research paper presented at the Cryogenic Engineering Conference in 1979.

Ron was especially appreciated for his talents in mentoring young engineers, leading and guiding, but encouraging them to develop their own styles and approaches. He was also deeply concerned with safety, a vital matter as any experienced cryogenic engineer knows.

He has also played a major role in the progress of the Cryogenic Engineering Conference. from a small specialized meeting to the major conference of today. Ron spent twenty years on the CEC Board of Directors, thirteen as editor of the proceedings, *Advances in Cryogenic Engineering*, and ten years as secretary of the CEC Board and its registered agent. He worked tirelessly and efficiently to make the conference the smoothly functioning entity it is today. Those who worked with him learned to appreciate his wide and detailed knowledge of the CEC organization. During his years of service the conference and the proceedings have grown by 50% and have attained world-wide recognition.

Peter V. Mason
Jet Propulsion Laboratory
Pasadena, CA
July, 1995

RUSSELL B. SCOTT MEMORIAL AWARDS

The Russell B. Scott Memorial Awards honor the first head of the Cryogenic Engineering Laboratory of the Boulder Laboratories of the National Bureau of Standards, now the National Institute of Standards and Technology. Mr. Scott was the founder of the Cryogenic Engineering Conference, the first of which was held in 1954 in Boulder. He was author of the book: **Cryogenic Engineering**. He retired in 1965 after 37 years at NBS and died in 1967.

The objectives of the Scott Memorial Awards are to provide an incentive for high quality papers at the Cryogenic Engineering Conference and to provide recognition to authors who, in the judgement of the CEC board, presented the best papers at the proceeding conference. The award consists of a certificate and an honorarium.

The 1995 Scott Awards, for the best papers delivered at the 1993 CEC in the Albuquerque Convention Center, Albuquerque, New Mexico, and published in Vol. 39 of **Advances in Cryogenic Engineering**, were presented at the 1995 Awards luncheon in the Hyatt Regency Hotel, Columbus, Ohio.

Best Research Paper

K.A. Gschneider Jr., H. Takeya, J.O. Moorman, V.K. Pecharsky, S.K. Malik, and C.B. Zimm
for their paper
"New Magnetic Refrigeration Materials for the Liquefaction of Hydrogen"

Best Applications Paper

K.M. Schaubel, C.B. Baxi, G.L. Cambell, G.J. Laughon, K.A. Mahdevi, M.M. Menon, C.C. Marariou, J.P. Smith, and M.J. Schaffer
for their paper
"The Development of an In-Vessel Cryopump System for the DIII-D Tokamak"

STUDENT AWARDS

Student awards are presented by the CEC board to encourage the participation of students in the conference and to provide an incentive for students to write high quality papers. The awards are based of papers submitted prior to the conference.

The following students received
Travel Assistance Awards:

Adrian Corless	University of Victoria, Victoria, B.C., Canada
Krishna Iyengar	Florida State University - Tallahassee
Catherine Pinsky	University of California - Los Angles
Vlad Zarnescu	Pennsylvania State University

Student Meritorious Paper Awards
were not presented in 1995

Erratum

In Vol. 39 the affiliation of one of the Travel Assistance Awardees was incorrect. Vlad Zarnescu was from the Civil Engineering Institute, Bucharest, Romania. We apologize for this error.

1995 CRYOGENIC ENGINEERING CONFERENCE BOARD

J.B. Peeples Chairman	CVI, Inc. Columbus, OH
J.P. Kelley Program Chairman	Babcock and Wilcox Lynchburg, VA
J.A. Barclay	University of Victoria Victoria, B.C., Canada
J.R. Bergen	Lake Shore Cryotronics, Inc. Westerville, OH
B.W. Birmingham	Birmingham Associates Boulder, CO
E.W. Collings	Battelle Memorial Institute Columbus, OH
P. Gifford	Cryomech, Inc. Syracuse, NY
G.F. Green	Naval Surface Warfare Center Annapolis, MD
G.R. Karr	University of Alabama-Huntsville Huntsville, AL
P. Kittel	NASA-Ames Research Center Moffett Field, CA
W.A. Little	Stanford University Stanford, CA
P.V. Mason	Jet Propulsion Laboratory Pasadena, CA
W.F. Stewart	Las Cruces, NM
K.D. Timmerhaus	University of Colorado Boulder, CO
A.R. Urbach	Ball Cryogenics Boulder, CO
R.J. Witt	University of Wisconsin Madison, WI

Awards Committee

P.V. Mason, Chairman — W.F. Stewart — A.R. Urbach

ACKNOWLEDGMENTS

The Cryogenic Engineering Conference Board wishes to thank the Hosts,

Battelle Memorial Institute,
CVI, Inc.,
and
Lake Shore Cryotronics, Inc.,

of the 1995 Cryogenic Engineering Conference.
These organizations greatly contributed to the success of the conference.

Producing the **Advances in Cryogenic Engineering** requires the dedicated effort of many people. Of particular imporatnce were those who took time from their busy schedules to review papers. Without their assistance the process would have taken even longer.

ENGINEERING ASPECTS OF CRYOBIOLOGY

L. Wolfinbarger, Jr[1,2]., V. Sutherland[2], L. Braendle[2], and G. Sutherland[2]

[1]LifeNet Transplant Services
5809 Ward Court, Virginia Beach, VA 23455 USA
[2]Center for Biotechnology
Old Dominion University, Norfolk, VA 23529 USA

ABSTRACT

Cryobiology deals primarily with the art of freezing aqueous solutions. These solutions may reside inside or outside of living cells and the transition of this solution from the liquid phase into the solid phase constitutes freezing. For purposes of this presentation, the cryobiology of aqueous solutions has been divided into sections with the understanding that they are not distinct from each other. In addition, emphasis is placed on selected cells/tissues and on solution of problems encountered in freezing, storage, and thawing of these selected cells/tissues, rather than attempt to provide an all inclusive review of the subject of cryobiology,. Hence, the title "Engineering Aspects of Cryobiology" is intended to illustrate how problems encountered in the freezing (cryopreservation) of cells and tissues have been resolved through the engineering of devices used in the various steps of their cryopreservation. This presentation will deal with **freezing** (with focus on containers, controlled-rate freezers, and temperature monitoring), **storage** (with focus on mechanical versus liquid nitrogen, thermal/mechanical perturbations, and shipping), and **thawing** (with focus on rates of thawing, maintenance of sterility, and dilutional removal of cryoprotectants).

INTRODUCTION

Great advances have been made in the cryopreservation of living cells since the early works of Smith and Parkes during the 1950s. These advances have included development of sophisticated computer-controlled freezing equipment, characterization of a variety of cryoprotective agents, and establishment of major research and development programs in cryobiology within academia and industry. Most of this research and technologic development has focused on the preservation of dispersed cell suspensions and it has successfully enlarged our understanding of the basic principles of cryopreservation of biologic systems.

Complex tissues, however, are different from simple cell suspensions and require more sophisticated procedures for cryopreservation. Consequently, research in this area has lagged behind that of freezing of cell suspensions. Complex tissues include tissues that are essentially acellular, or contain only a minimal number of cells in a complex protein/polysaccharide matrix, to tissues that are primarily populations of different cell types. Examples of the former tissues include heart valves and ligaments/tendons whereas examples of the latter include veins/arteries and pancreatic tissues or parts thereof.

A typical mammalian cell contains low molecular weight solutes at concentrations equivalent to 0.15 M sodium chloride. Consequently, their freezing point is about minus 0.6 °C. Most cells, do not freeze internally unless cooled to minus 10 °C or lower. This supercooling effect means that the cells are not in thermodynamic equilibrium with the external solution, and eventually this equilibrium must be restored[1]. The preponderance of research to date suggests that this disequilibrium can be restored by one of

Advances in Cryogenic Engineering, Vol. 41
Edited by P. Kittel, Plenum Press, New York, 1996

several mechanisms, including the formation of intracellular ice crystals either independently within the cell or by intrusion from the extracellular environment. Considering that ice crystals grow by accretion at their surfaces, it is improbable that ice crystals "grow" through a hydrophobic domain such as the plasma membrane. Rather, for transmembrane growth of an extracellular ice crystal, the crystal would need to pass through a hydrophilic domain such as an aggregation of phase-separated cell surface glycoproteins (or similar structures) expected to occur at reduced temperatures[2].

Alternatively, the disequilibrium may be restored by movement of intracellular water across the plasma membrane to add to the growing ice crystals forming outside the cell. Because the movement of water molecules across the plasma membrane occurs more rapidly than solute movement[3] the cells become progressively more dehydrated as a function of extracellular ice crystal formation until the eutectic point is achieved. The eutectic point is defined as the temperature at which both solvent and solute solidify into an amorphous glass.

For single cells, it is possible to calculate the movement of solvent and solutes across the plasma membrane[4] as a function of temperature and thus to predict "accurately" the time required to effectively dehydrate the cell without the formation of intracellular ice. Figure 1 illustrates the alternative scenerios possible with single cells frozen at varying rates ranging from "rapid" to "slow". Definition of "slow" and "rapid" varies with the individual, but slow freezing rates are generally considered to range between 1 and 10 °C per minute. Rapid freezing rates are typically in excess of 100 °C per minute. At exceedingly rapid rates, cells may be frozen by a process refered to as vitrification. Vitrification of cells and tissues consists of formation of an amorphous glass without the formation of crystalline ice[5,6]. With very limited exceptions, vitrification of cells and tissues has not been successfully achieved. As illustrated in figure 1, slow freezing of cells results in the formation of extracellular ice which results in an increase in solute concentrations in the unfrozen solution. This increase in solute concentration results in cellular dehydration, i.e. cell shrinkage. Since the probability of ice crystal formation is proportional to the volume of water available in a given environment, it is more probable that ice crystal nucleation will occur extracellularly. With cell shrinkage, the probability of intracellular ice is reduced in proportion with the reduction in cell water lessening the likelyhood of cellular damage due to intracellular ice. When cells are frozen rapidly, formation of ice is more

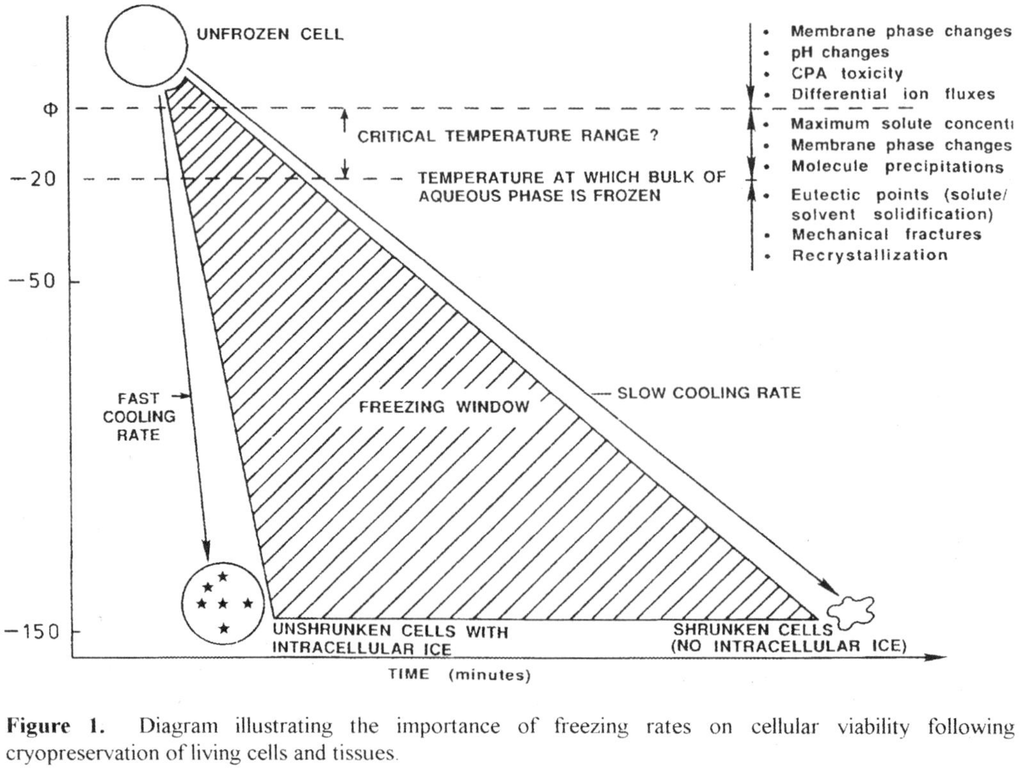

Figure 1. Diagram illustrating the importance of freezing rates on cellular viability following cryopreservation of living cells and tissues.

rapid, and ice crystals are small[7], allowing less time for cell dehydration to occur, thereby increasing the likelyhood of intracellular ice formation. From the perspective of cell damage due to increased solute toxicity, however, rapid freezing rates result in shorter exposure times of cells to higher solute concentrations, and their associated complications to cellular viability, with the result that cell damage is reduced. For slowly frozen cells, high solute concentrations frequently result in shifts in pH[8], salt mediated denaturation/dissociation of cell proteins[9], cell shrinkage with attendent loss of plasma membrane (as small vesicles)[10] and general cell damage. Thus, the optimal freezing rate varies with cell type and suspending solution and the cryobiologist must compromise between damage to cellular viability due to ice crystal formation and and that due to solute toxicity. For most cells, this optimum freezing rate appears to be between negative 1 and negative 5 ° C per minute.

Cryopreservation of tissues presents additional complications other than those associated with cellular viability. When complex tissues of any volume are cryopreserved they may be expected to reside in both the crystalline ice and the amorphous glass phases of the frozen volume. As cell suspensions are cryopreserved they distribute into the unfrozen volumes as crystalline ice forms. The cells become increasingly concentrated as the ice occupies more and more of the volume being frozen leading to problems in cell fusion and loss in cellular viability. The majority of the cells are eventually localized in the amorphous glass phase of the frozen volume as small dehyrated structures[11]. For large tissues, structural integrity can be compromised which may be more damaging than reductions in cellular viability. Crystalline ice and amorphous glass possess different coefficients of expansion/contraction and sudden and dramatic changes in temperature can lead to the formation of stress fractures in the frozen volume[11,12]. Equivalent stess fractures readily occur in ice cubes suddenly plunged into a warm drink. In addition, gases dissolved in the aqueous solution will be excluded from the crystalline ice, appearing as discrete bubbles at the growing front of the ice[13]. Cell damage associated with contact of these bubbles with cells has been previously described[14], however for purposes of this discussion the presence of these gas bubbles, at the approximate interface between the crystalline ice and amorphous glass phases, poses the additional problem of expanding and contracting with temperature changes contributing to formation of stress fractures, particularly during the warming phase associated with thawing. Rapid temperature fluctuations, such as associated with immersion of cryopreserved tissues into liquid nitrogen, with rapid thawing, or temperature fluctuations during storage, have been demonstrated to contribute to surface and full thickness fractures in cryopreserved heart valves and arteries[15,16]. This issue will be returned to in later sections of this article.

FREEZING OF CELL SUSPENSIONS AND TISSUES

Cell suspensions

Cells and tissues are cryopreserved in a wide variety of containers. Suspensions of cells are most frequently frozen in plastic, typically polyethylene or polypropylene, ampules holding volumes of one (1) to five (5) mls. Cell suspensions, at densities of 1 to 2 X 10^6 cells/ml, are typically suspended into the appropriate medium supplemented with 10 to 40 % fetal calf serum (FCS) and 10 % dimethylsulfoxide (DMSO) or 15 % glycerol. The cell suspension is then frozen using controlled rate freezers, for example, FTS Sytems, Inc. model TPC-44 or BioCool II freezers[17], in which ampules of cells are slowly lowered (using a screw drive mechanism) through progressively colder temperatures in the chamber or immersed in an alcohol bath that is progressively cooled. These types of systems are frequently accompanied by programmers which allow continuous recording of chamber (bath) temperatures.

Cheaper systems for freezing of cell suspensions include the time-honored styrofoam container into which the ampules containing cells are placed. The styrofoam box with cells is typically placed in a refrigerator for 1 hour, transfered to a minus 20 ° C freezer for 1 hour, then to minus 80 ° C for 24 hours or until ultimate storage in some form of ultralow (below minus 130 ° C) freezer. This system has proven to be quite effective, providing for as much as 95% cell viability following thawing. However, this sytem requires careful validation of position(s) in the styrofoam container and dimensions of the container. The author's laboratory has had considerable success for many years freezing cells in 2 ml ampules placed in the outermost series of slots present in the ever-available styrofoam containers used to hold disposable plastic cuvettes (Fisher Scientific, FISHERbrand cat. no. 14-385-942). Using this system, a total of 36 ampules can be frozen per container and provided the elastic-band securing the container does not break, the cells can be stored long-term in these same containers.

Laboratories growing small quantities of cells typically store cells in Dewar flasks in the liquid nitrogen (minus 196 ° C) or in the vapor phase of the liquid nitrogen (below minus 130 ° C). Storage of small ampules directly in liquid nitrogen is generally discouraged due to the potential for "explosion" of the ampule during thawing if liquid nitrogen leaked into the container during storage. Freezing of cell suspensions is a fairly stable technology now and improvements are directed to applications to specific cell types where variable freezing rate programs (with holds at necessary intervals) are available to maximize viability of specific cell lines. Innovations in freezing solutions will ultimately permit vitrification of cell suspensions without dramatic changes in cellular viability and function.

Complex tissues

Methodologies for freezing of tissues were emperically derived from the methodologies used for freezing of cell suspensions. Tissues are typically suspended in freezing media supplemented with 10 to 20 % FCS and 10 to 15 % DMSO for times determined to allow for equilibration of the cryoprotectants between the tissue and solution. For the cryopreservation of human heart valves, LifeNet Tranplant Services utilizes RPMI 1640 tissue culture media supplemented with DMSO and freezes the valve in a total volume of 100 mls. For an excellent review of this process the reader is refered to Lange and Hopkins[18]. Issues relevant to the cryopreservation of complex tissues include temperature sensitivities of the various populations of cells present in the tissue, the cryoprotectant (and it's concentration), freezing rate(s), the volume to surface area of the freezing container, the heat transfer characteristics of the container, and the orientation of the container in relation to the flow of the liquid nitrogen vapors in the controlled rate freezer. A very specific problem which has plagued the authors for years deals with the issue of temperature monitoring during the freezing process. Figure 2 illustrates a typical temperature profile of the freezing chamber and tissue during controlled-rate freezing of a heart valve (approximately 7 ml volume in a total volume of 100 mls) using a CryoMed Model 1010 Controlled Rate Freezer[19]. The controlled rate freezer utilizes two temperature probes to measure chamber and sample temperature and data is sent to an X-Y strip chart recorder. The temperature of the sample being frozen is controlled by the temperature of the chamber which is programmed into the onboard programmer (computer) in a series of step commands describing

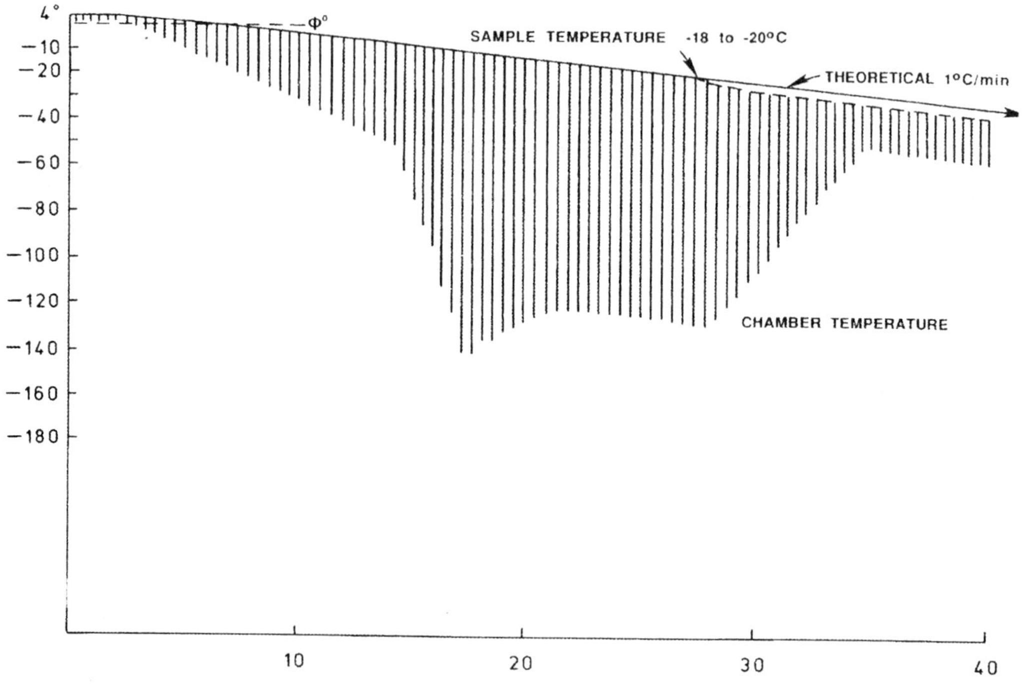

Figure 2. Diagramic representation of a typical freezing curve associated with cryopreservation of a human heart valve using a Cryomed Controlled-Rate freezer.

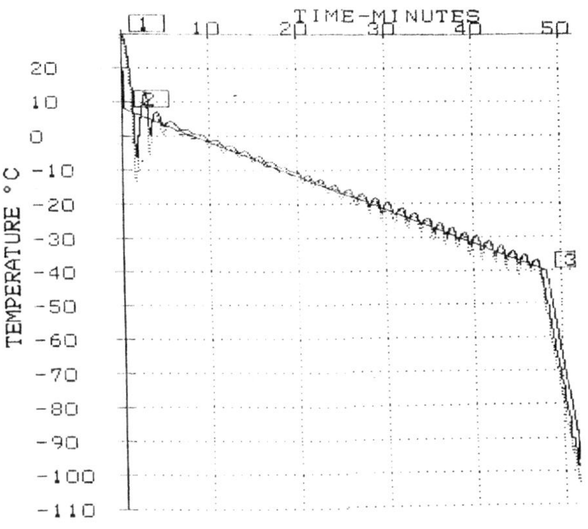

Figure 3. Diagramic representation of a typical freezing curve associated with the cryopreservation of a porcine heart valve using a Planar Controlled-Rate freezer.

desired chamber temperatures, rates of changes in chamber temperature and times at some specified temperature (holds).

Should the sample temperature deviate from that desired, the operator must immediately shift to manual control and either increase or decrease the chamber temperature to bring the sample back into the desired temperature range, at which time the automatic programming mode can be reentered.

Figure 3 illustrates a typical temperature profile of the freezing chamber and tissue during controlled-rate freezing of a heart valve using a Planar Controlled-Rate Freezer (Cryo 10/16, series 3) provided by T.S. Scientific[20]. Again, this controlled rate freezer uses two temperature probes to measure chamber and sample temperature and the data is sent to either a built in X-Y recorder or to software which permits data capture onto a personal computer. One essential difference in these two controlled rate

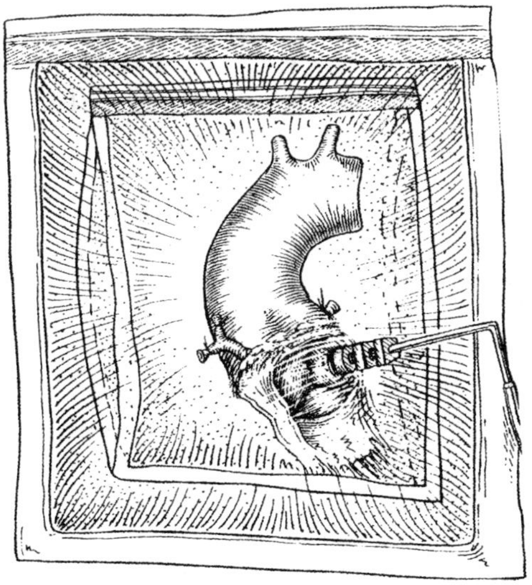

Figure 4. Illustration of a human heart valve contained in a typical double-bagged freezing pouch with a temperature probe inserted through a special sealing device.

freezers is the matter of programming of a particular freezing curve. With the CryoMed system, programming describes the chamber temperature necessary to achieve a specified sample temperature. With the Planar system, programming (version 1.04 and 1.08) involves setting the desired rate of change in sample temperature or specific sample temperature over some defined time interval and the system determines, using the PID controller, the chamber temperature necessary to obtain that parameter.

Both controlled-rate freezing systems however suffer from the problem that they employ temperature probes inserted directly into or mounted onto the surface of a "witness" container system used to predict the temperature(s) of the tissue(s) being cryopreserved. Relatively inexpensive radiation thermocouples are commercially available, but most can not measure temperatures as low as 0 ° C. Figure 4 illustrates the type of temperature probe system typically employed to monitor the temperature of control ("witness") pouches which are used to predict the temperatures of similar pouches containing heart valves which will ultimately be used in some clinical application. Although such temperature probes are accurate and reliable, they have the obvious disadvantage of serving as heat sinks, facilitating transfer of heat from the interior of the "control" pouch to the flow of liquid nitrogen vapor in the freezing chamber of the controlled-rate freezer. Heart valves in freezing pouches that lack this temperature probe lack this heat sink and their temperatures and pattern of ice crystal formation will be different from the control pouches. Using this witness temperature sensing system, temperatures frequently cited as representing the temperatures of tissues being controlled-rate frozen do not necessarily accurately portray the temperatures of tissues lacking the temperature probe. It has been suggested that surface mounted temperature probes should be used to monitor the temperature of each tissue being cryopreserved. To do so will require controlled-rate freezers with multiple sample temperature monitoring systems and validation that the presence of a temperature probe attached to the surface of the container will not introduce structural anomolies which will compromise the integrity of that container. An additional complication in monitoring the temperatures of tissues being cryopreserved lies in the need for sterility of the tissue. Most cryopreserved tissues which are used clinically are frozen in a double-pouch system where the outermost surface is not sterile and a insulating barrier of air is present between the outer and inner pouchs. Thus, attachment of a temperature probe to the surface of the outer pouch, for example the HyCal Zig-Zag tm thermocouple, would still not be in intimate contact with the sterile pouch containing the aqueous solution being frozen. This complication has prompted efforts at freezing tissues without the outer pouch and sterily adding the frozen tissue into a sterile outer pouch, adding a sterile-nonfreezing solution into the space separating the inner and outer pouches, and using witness samples. The obvious solution to this problem involves the use of 1.) external long-range temperature sensors (radiation thermometry) or, 2.) the use of disposible radiotelemetry temperature sensors in the solutions being frozen. Radiation equilibrium thermocouples of a design based on one such sensor developed at LaRC for the LZEEBE program in the 1970's or a UV fluorescence technique based on the commercial LUXTRON sensor, could be used to determine exterior sample bag temperatures (Jag J. Singh, NASA, personal communication). Such thermocouples would have to be repackaged in vacuum tight enclosures, with an infrared window which would not permit condensation of water on its exterior surface, if they were to be incorporated into a commercially available controlled-rate freezer. The LUXTRON device would require either a permanently installed sensing disc, i.e. two color thermographic phosphor, applied to the outside of the sample package and a UV transmitting fiber optic probe to illuminate the sensing disk (and pick up emitted radiation from the disc) or a probe designed to measure interior witness sample or production sample temperatures. Use of such systems are currently limited by their lower temperature

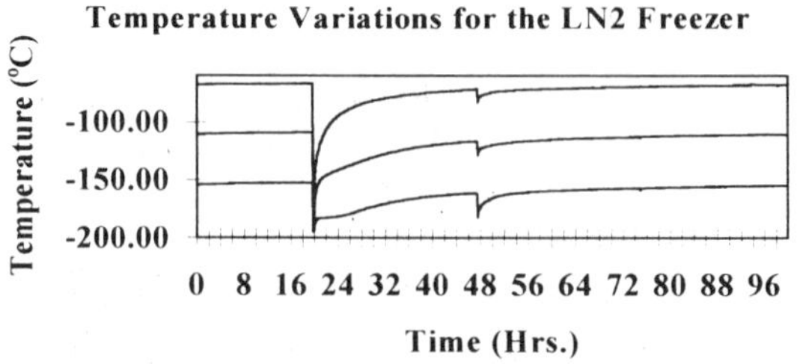

Figure 5. Temperature profiles at the top, middle, and bottom of a liquid nitrogen storage tank without a racking system to mitigate the temperature gradients present in the storage chamber. The chamber was approximately three fourths full of boxed tissue samples and was not opened during the testing interval. Note the automatic filling event.

ranges, typically minus 30 ° C, and their use at the lower temperatures of minus 80 to minus 100 ° C to which most samples are frozen will require modification. Alternatively, most samples being frozen reach their eutectic point somewhere around minus 22 to 25 ° C and below this temperature only minimal heat is liberated as a result of ice crystal formation. Below the eutectic point of a particular sample being cryopreserved, the freezing process becomes more manageable by todays controlled rate freezers, since the chamber temperature typically needs to be only a few degrees below the sample temperature to effect a routine freeze. It may be possible to use a monitoring device such as the LUXTRON device to measure temperatures to below that at which the samples are soldified, at which time alternative monitoring devices could be employed.. Such systems must however be sufficiently inexpensive so as to not add dramatically to the costs of the tissue being cryopreserved.

STORAGE OF CRYOPRESERVED CELLS/TISSUES

As described earlier, in research laboratories most cryopreserved cell suspensions are typically stored in Dewar flasks of varying design and construction (depending on the particular commercial source). Flasks with narrow mouths offer the advantage of restriction of loss of nitrogen, i.e. reduced heat gain and lower overall liquid nitrogen consumption, however the number of samples that may be stored is less than in flasks with a wide mouth. Wide mouth Dewars result in larger liquid nitrogen consumption rates and must be recharged more often. In laboratory operations where larger numbers of cell lines, or "cell banks" of specific cell lines are maintained in large numbers, cell suspensions are typically stored in racking systems designed to facilitate identification and correct retrieval of cells in either liquid nitrogen or mechanical freezers. Although mammalian cells may frequently be stored for short periods of time at minus 80 ° C, longe-term maintenance of a viable cell population requires storage of cells below minus 130 ° C. With the exception of innovations in flask design, materials used in construction, and development of materials which absorb and hold the liquid nitrogen restricting the potential for spillage of liquid-liquid nitrogen, it might be suggested that Dewar flask storage systems for cell suspensions is a mature technology.

The freezers used in storage of cryopreserved tissues are quite similar in design to freezers used in storage of cell suspensions. However, the storage chamber is typically of greater capacity to accomodate the large volume of packaging associated with tissue storage. These storage systems are typically of two types, liquid nitrogen or mechanical. Liquid nitrogen storage systems are designed to hold liquid nitrogen at a depth of approximately 5 to 8 inches such that when "full", the liquid nitrogen is just at the bottom of the stainless steel or aluminum rack present in the storage tank. Use of these types of storage systems, without some additional racking system to mitigate excessive temperature gradients should be with considerable care. Figure 5 illustrates the temperatures at the top, middle, and bottom of an "off the shelf" liquid nitrogen storage chamber. Temperatures were recorded at 5 minute intervals (using temperature probes and software provided by Medical Equipment and Maintenance Company[21]) and the data shown in figure 5 were selected to show how the temperatures at each level fluctuated during a routine refill (where refill was automatic at 5 inches of liquid nitrogen and refill was to the 9 inch level).

In an "unracked" system, the average temperatures were minus 72.40 +/-10.96° C at the top level, minus 115.46 +/-10.00 at the middle level, and minus 158.75 +/-8.00 ° C at the lower level. Thus, the system presents a temperature gradient, top to bottom (distance of approximately 19 to 20 inches), of 86.36 ° C, and only the bottom tier for storage provides a temperature below minus 130 ° C. During a routine refill of the system with liquid nitrogen, this system experiences extreme temperature variance, with the severity of the change proportional to the distance from the liquid nitrogen. Samples which might be stored in the upper part of the chamber may be expected to experience temperature fluctuations of as much as 130 ° C (note the dramatic changes in temperatures at approximately 240 minutes in figure 5) over the 4 to 5 minutes required for a refill. Note also that these temperature fluctuations actually occur twice, once during refill (very rapid changes) and once just after the refill event as the system reequilibrates (a slower process). Samples stored at the bottom of the chamber experience similar, although less extensive, temperature fluctuations of approximately 44 ° C. This latter temperature fluctuation is functionally equivalent to plunging of tissues stored at approximately minus 150 ° C into liquid nitrogen as described by Adam, et. al. in 1990[15] where it was demonstrated that such rapid temperature fluctuations contributed to microfractures on the surfaces of cryopreserved heart valves.

By adding some form of "racking" system to the systems currently provided by commercial sources, the undesirable temperature gradients resulting from an unracked system can be mitigated to some

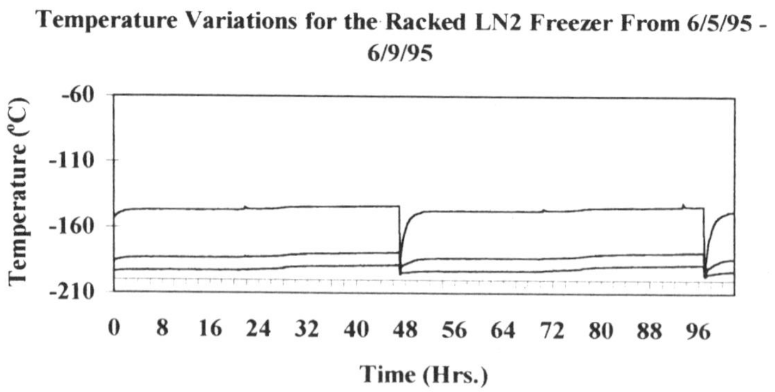

Figure 6. Temperature profiles at the top, middle, and bottom of a liquid nitrogen storage system containing an aluminum racking designed to mitigate the temperature gradients present in the storage chamber. The chamber was approximately three fourths full of boxed tissue samples and was not opened during the study interval. Note two automatic filling intervals.

extent. Such racking systems may include, but not be limited to, attachment of vertical aluminum rods around the perimeter of the platform at the bottom of the chamber. Alternatively, a racking system can be manufactured which not only "sits" in the liquid nitrogen, but also serves to hold the samples above the liquid phase. Such a racking system has been developed and is in use at LifeNet Transplant Services. By adding temperature probes at the top, middle, and bottom of the storage chamber in a liquid nitrogen storage system containing the racking system described we have obtained the temperature profiles illustrated in figure 6.

In the racked liquid nitrogen storage chamber, temperature gradients from top to bottom are reduced to approximately 44 ° C (over the 19 to 20 inches) with average temperatures of minus 147.02 +/- 4.69 ° C at the top, minus 182.00 +/- 2.19 °C at the middle, and minus 191.43 +/- 1.94 ° C at the bottom of the racked chamber. Of importance in this storage system is the observation that samples stored near the top of the chamber would still be below the minus 130 ° C necessary to restrict the process of recrystallization. Figure 6 also illustrates two refilling events where the storage tank is refilled using the automatic mode. It is obvious that samples stored in the upper part of the chamber will experience dramatic temperature fluctuations, approximately 40 ° C, during a refill event whereas samples stored in the lowest part of the chamber (but not in the liquid nitrogen) will experience the least dramatic fluctuations in temperature (approximately 8 ° C).

A variety of mechanical ultra-low "freezers" are commercially available for the storage of cells and tissues below minus 130 ° C. Figure 7 illustrates temperature profiles within a Harris Ultra-Low chest

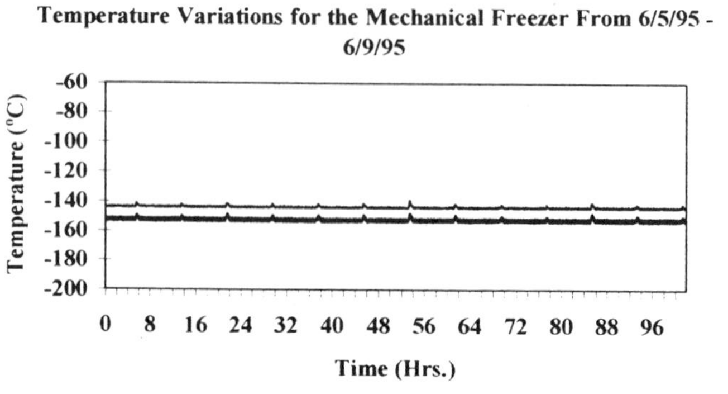

Figure 7. Temperature profiles at the top, middle, and bottom of a Harris Ultra-low Temperature mechanical freezer. The chamber was approximately three fourths full of boxed tissues and was not opened during the study interval.

freezer (Revco/Lindberg[22]) (Model XLT-7LS-150CBA), taken at the bottom, middle, and top of the storage chamber. Temperature gradients from top to bottom of this unit averaged 8.85 ° C with average temperatures of minus 152.82 +/- 0.96 ° C at the bottom, minus 151.64 +/-0.68 ° C at the middle, and minus 143.97 +/- 0.64 ° C at the top of the storage chamber (for a preselected temperature of minus 150 ° C). As illustrated in figure 7, minor temperature fluctuations occur in the mechanical system due to the thermostatic controller. These minor oscillations averaged 32.5 minutes with an average cooling period of 22.5 minutes until the temperature reached minus 154 +/- 0.2 ° C, at which time the unit warmed at an average rate of 0.21 degrees per minute for an average of 9.6 minutes until a temperature of minus 152 +/- 0.2 ° C, at which time the cycle repeats. Longer cycles appeared at 7.92 hour intervals which included an average 16.7 minute heating time (with a heating rate of 3.9 degrees per minute). This latter event was associated with a warming phase to volatilize refrigerants in the system (personal communication, Mike Brown at Revco/Lindberg). It is obvious that unlike the liquid nitrogen storage systems, the mechanical freezer described does not present dramatic temperature fluctuation at any position within the storage chamber. The temperature gradient within the chamber is minimal, compare 8.85 ° C for the mechanical system with 86.36 ° C and 44.40 ° C for the unracked and racked liquid nitrogen storage chambers, respectively, and extremely stable.

A concern of long-term storage of tissues at ultra-low temperatures is the potential for formation of fractures, both "minor" surface abrasions and "major" full-thickness fractures, in the tissues. In the late 1980s, the authors were involved in an incident in which a liquid nitrogen storage system, located in a hospital, failed to shut off following a normal refill cycle. The tank overflowed immersing the clinically usable heart valves. The incident was not considered a problem by the hospital personnel until the surgeons began to implant the tissues and observed a higher incidence of full-thickness fractures in the tissues. This incident prompted a study of whether or not the immersion contributed to the fractures in the tissues and the results of this study were eventually published[15,16]. It was concluded that it was not necessarily the immersion of the valves into the liquid nitrogen, but rather it was the sudden and dramatic changes (fluctuations) in temperature which were a contributing factor in the formation of fractures. Large pieces of tissues frozen such that the aqueous phase (presumably including the tissues themselves) solifies into both a crystalline ice and amorphous glass phase will presumably lie at some angle to the interface between these two phases. With dramatic changes in temperature during storage and/or thawing (see later section), these two phases will experience differential contraction/expansion generating stress in the solidified matrix which may be released by the formation of fractures, probably within the plane of the interface of the two phases. If this plane crosses the surface of the cryopreserved tissue, the fracture may penetrate into or through the tissue creating either surface or full-thickness fractures. Kroener and Luyet[23,24] reported that 97 % glycerol solutions contracted by 2.8 mm^3 /gm for every 10 degrees above the glass transition temperature (Tg) and by 0.4 mm^3/gm for every 10 degrees below Tg. As temperatures in these solutions approached 158 K (minus 115 ° C) cracks developed in the specimens and their readings became "irratic". They suggested that the formation and disappearance of cracks depended on "the interaction of several factors, in particular, the mechanical properties of the material, the concentration of solute in it, the temperature gradients through it, the overall temperature, and the rate of change of temperature". At cooling rates on the order of 1 to 3 K/minute (with temperature gradients of 6.5 to 8.5 K/cm) cracks appeared. When higher cooling rates and larger temperature gradients were attempted, the cracks appeared "earlier". When the solutions were warmed, the cracks disappeared, but only at temperatures above that of the glass transition.

Because of the potential for fracture (crack) formation in cryopreserved tissues stored for long-terms, it is also important to consider the issue of shipping of tissues to points distant from the place they are cryopreserved and stored. For example, heart valves (and other cryopreserved tissues) are frequently shipped to a hospital across town or across the Country where they may be transfered into a hospital storage system or used immediately in some surgical procedure. In instances where the tissue is not used, it may be necessary to ship the material back to its point of origin for placement back into the central storage system. This aspect of "storage" brings up the issue of the temperature to be used during shipping and how to validate that the temperature of the shipper did not allow for significant "thawing", where thawing can mean an increase in temperature above the minus 130 ° C value at which recrystallization may occur. Commercial air carriers will not permit transport of containers containing liquid-liquid nitrogen because of safety issues associated with spillage of the liquid nitrogen during transport. Some groups continue to ship cryopreserved tissues "on dry ice" at minus 70 ° C whereas other groups ship tissues on dry ice supplemented with liquid nitrogen as a nonspillable slurry. Companies have now developed cryoshippers which are capable of maintaining ultralow temperatures (below minus 130 ° C) for extended periods of time (typically 1 to 2 weeks) (see Brubaker, this issue, for a thorough discussion on cryoshipper construction and temperature maintenance). The issue to be considered with regards to shipping is the occurence of dramatic temperature transitions, for example from minus 150 ^{o} C to minus 70 ^{o} C and back to minus 150 ^{o} C. Do these temperature fluctuations/transitions contribute to fractures in tissue (or loss in cellular viability)?

"Cracks" in cryopreserved tissues are not a serious problem with the freezing protocols in use by most tissue cryopreserving groups. Groups like Cryolife, LifeNet, and the American Red Cross cryopreserve large numbers of human heart valves (and other human tissues) each year and the incidence of fractures in these tissues are exceedingly rare. However, it is an issue to be resolved and the use of ultra-low storage systems which permit dramatic temperature fluctuations can contribute to the problem. As more and more tissues, and eventually organs, are cryopreserved for use in clinical applications, this issue may become more pronounced demanding resolution.

THAWING OF CRYOPRESERVED TISSUES

For the many research groups working with cryopreserved cells, thawing is generally not a serious issue. The frozen cells in small ampules are rapidly thawed by simply immersing the containers in a warm (typically 37 ^{o} C) water baths followed by transfer of the cells into fresh media and addition to culture flasks. Cell viabilities are typically high (80 to 95 %) and the occurrence of fractures in the solified matrix contributes only minimally to damage to the cell population-comparatively few cells will be damaged along any formed stress fractures. For cell suspensions, osmotic imbalances, cryoprotectant toxicity, and cell fusions are more important than fractures in the frozen matrix. However for tissues, fractures formed during thawing can be an important issue. With rapid thawing the system experiences the same generation of stress, and the potential for stress fracture formation, as during storage. Conversely, it might be expected that all cryopreserved tissues would be thawed by "slow" methods. However, most of these tissues, particularly heart valves, will be used in clinical applications and a surgeon, with a patient on "by-pass" (life-support machines) and the patients own heart valve removed, is not inclined to wait more than a few minutes. Thus the issue of thawing rates is not always simple, but must meet the needs and demands of the situation.

A variety of thawing protocols for cryopreserved tissues are currently in use. Typically, the cryopreserved tissue in a double-bag system is placed into a warm/sterile water bath at 37 to 45 ^{o} C and gently moved within the water bath to facilitate heat transfer-thawing. Maintaining the temperature of the water bath serves to enhance thawing rates, but may result in a thawed tissue where the temperature of the cryoprotectant solution exceeds 10 ^{o} C, where toxicity of the solution may contribute to cell "death" in the tissue. Care must be taken in handling tissues being thawed in that residual ice frequently remains within the tissues after most of the solution has thawed and mechanical pressure can score and damage the tissue as it is pushed against ice crystals. Temperature monitoring near the interior lumen of a heart valve during thawing of a heart valve cryopreserved in a total volume of 100 mls indicates that thawing can be accomplished in as little as 3 to 5 minutes. Thawing of the cryopreserved tissue does not occur uniformly using this current methodology and efforts need to be directed at finding better ways to more uniformly thaw such tissues. The double bag system so necessary to maintenance of sterility of the inner pouch presents an insulating barrier between the inner and outer pouches slowing heat transfer to those sides of the inner pouch not in physical contact with the outer pouch. Few operating rooms contain a mechanical water bath where water temperature can be maintained and sterile basins with added sterile warm water can contribute to variable thawing rates, i.e. unreliable thawing, unless care is taken to monitor water temperature. Microwave warming of cryopreserved tissues might seem to be an obvious way to thaw cryopreserved tissues, however the possible deleterious effect that this radiation may have on the tissues must be evaluated. Alternatively, sterile chemical warming solutions might be injected into the space between the inner and outer pouches where the volumes could be calculated based on the temperature and volume of frozen solution/tissue to be thawed. Means of determining completion of the thawing process might include the aforementioned remote temperature sensors, for example the LUXTRON device with a permanently installed sensing disc, however, these devices would need to be inexpensive and useable in an operating room.

It should be obvious that this article has dealt with a narrow area of the field of cryobiology, namely the cryopreservation of cells and tissues. The intent of the article was to illustrate a part of this field with the objective of stimulating groups and individuals outside of cryobiology to bring their expertise to solution of problems. The subjects of cryosurgery, where the primary intent is to selectively kill tissues, vitrification, where the primary intent is to freeze cells and tissues without the formation of ice crystals, and cryopreservation of biologically important molecules are left untouched for other individuals to cover at later meetings of this group. The authors entered the realm of cryogenics some years ago in an effort to resolve issues of immediate concern to the cryopreservation of human heart valves. The need for collaboration between individuals and groups such as those present at the Cryogenics Engineering Conference and cryobiologists is immediate and important. Hopefully, topics raised in this article will serve to stimulate such collaborations.

ACKNOWLEDGMENT

Appreciation is extended to the members of the Cryogenics Engineering Conference and particularly Pat Kelley for providing the opportunity to attend the Conference. For those desiring assistance in networking or for additional information contact FXFJ51a@prodigy.com on e-mail or call Dr. Lloyd Wolfinbarger, Jr. directly at 804-683-3617, facsimile: 804-683-5283.

The references cited have been chosen to facilitate review of the field of cryobiology through important books and individuals with whom the author has had some form of professional interaction. The list of these books and individuals is certainly not complete.

REFERENCES

1. L. Wolfinbarger, Jr. and R. Hopkins, Biology of heart valve cryopreservation, In: "Cardiac reconstructions with allograft valves", R. Hopkins, ed., Springer-Verlag, New York (1989) p. 21.

2. E. Asahina, Frost injury in living cells, Nature 196:445 (1962).

3. P. Mazur, Fundamental cryobiology and the preservation of organs by freezing. In: "Organ preservation for transplantation", A.M. Karow and D.E. Pegg, ed., Marcel Dekker, New York (1981) p 143.

4. J. Farrant, General observations on cell preservation. In: "Low temperature preservation in medicine and biology", M.J Ashwood-Smith & J. Farrant, ed., Pitman, London (1980) p.1.

5. G. M. Fahy, D.R. MacFarlane, C.A. Angell, and H.T. Merryman, Vitrification as an approach to cryopreservation. Cryobiology 21:407 (1984).

6. C. A. Angell and H. Sanapti, Crystallisation and vitrification in cryoprotected aqueous systems. In: "The biophysics of organ cryopreservation", D.E. Pegg and A.M. Karow, ed., Plenum Press, New York (1987) p. 147.

7. C. Korber and G. Rau, Ice crystal growth in aqueous solutions. In: "The biophysics of organ cryopreservation", D.E. Pegg and A.M. Karow, ed., Plenum Press, New York (1987) p. 173.

8. L. van der Berg and D. Rose, Effects of freezing on the pH and composition of sodium and potassium phosphate solutions: the reciprocal system KH_2PO_4-Na_2HPO_4-H_2O. Archives Biochemistry & Biophysics 81:319 (1959).

9. J. H. Crowe, L.M. Crowe and M. W. Deamer, Hydration dependent phase changes in a biological membrane, In: "Biophysics of water" F.Franks and S. Mathias, eds., John Wiley & Sons, New York (1982) p. 295.

10. M. J. Ashwood-Smith, G. J. Morris, R. Fowler, T.C. Appleton, and R. Ashorn, Physical factors are involved in the destruction of embryos and oocytes during freezing and thawing procedures. Human Reproduction. 3:795 (1988).

11. D. E. Pegg, The effect of cell concentration on the recovery of human erythrocytes after freezing and thawing in the presence of glycerol. Cryobiology 18:221 (1981).

12. B. Rubinsky, E. G. Cravalho, and B. Mikic, Thermal stresses in frozen organs, Cryobiology 17:66 (1980).

13. P. L. Steponkus and M. F. Dowgert, Gas bubble formation during intracellular ice formation, Cryo-Letters 2:42 (1981).

14. S. Tondorf, J. J. McGrath, and C. R. Olien, On the adhesive interaction between ice and cell-size liposomes, Cryo-Letters 8:322 (1987).

15. M. Adam, J. F. Hu, P. Lange, and L. Wolfinbarger, Jr. The effect of liquid nitrogen submersion on cryopreserved human heart valves, Cryobiology 27:605 (1990).

16. L. Wolfinbarger, Jr., M. Adam, P. Lange, and J.F. Hu, Microfractures in cryopreserved heart valves: valve submersion in liquid nitrogen revisited. In: "Applications in Cryogenic Technology", J. P. Kelly, ed., Plenum Press, New York (1991) p. 227.

17. FTS Systems, Inc., P.O. Box 158 (Route 209), Stone Ridge, New York 12484

18. P. L. Lange and R. Hopkins, Allograft valve banking: Techniques and technology, In: "Cardiac reconstructions with allograft valves", R. Hopkins, ed., Springer-Verlag, New York (1989) p. 37.

19. Cryomed, P.O. Box 636, New Baltimore, Michigan 48047

20. TS Scientific, P. O. Box 198, Perkasie, Pennsylvania 18944

21. Medical Equipment and Maintenance Company, Inc., Washington, D.C. and Baltimore, Maryland

22. Revco/Lindberg: A Unit of General Signal, 275 Aiken Road, Asheville, North Carolina 28804

23. C. Kroener and B. Luyet, Discontinuous change in expansion coefficient at the glass transition temperature in aqueous solutions of glycerol, Biodynamica 10:41 (1966).

24. C. Kroener and B. Luyet. Formation of cracks during the vitrification of glycerol solutions and disappearance of the cracks during rewarming, Biodynamica 10:47 (1966).

POSSIBLE APPLICATIONS OF DIRECTIONAL SOLIDIFICATION TECHNIQUES IN CYOBIOLOGY

I. Heschel, C. Lückge, M. Rödder, C. Garberding, and G. Rau

Helmholtz-Institute for Biomedical Engineering
at the Aachen University of Technology
Pauwelsstr. 20, D-52074 Aachen, Germany

ABSTRACT

During freezing of cell suspensions, the cooling rate achieved generally varies markedly throughout the bulk sample. As the cooling rate strongly affects cell survival, this may lead to a variation of cell survival with position in the sample. To avoid these problems, in many basic investigations microscopic sample volumes are used. But even in this case effects like the undercooling of the aqueous samples occur that prevent the study of cells under well defined thermal conditions. Thus the directional solidification of microscopic biological samples according to the Bridgman-technique was introduced to cryobiology. This technique allows the highly defined freezing of small samples and the independent control of the two fundamental solidification parameters, ice front velocity, v, and temperature gradient, G. The cooling rate B is simply the product of the two parameters, v and G, and can easily be adjusted. Applications of the Bridgman-technique are the study of the crystallization morphology (e. g. measurement of the solute pile up and the breakdown time, dendrite spacings, tip radius, electrical potentials, etc.) and the study of the interactions between growing ice crystals and biological cells (e. g. intracellular ice formation, encapsulation or rejection of cells). But although the advantages are obvious for microscopic samples, the Bridgman-technique cannot be adapted to the cryopreservation of bulk systems. In this case an alternative approach, the directional solidification according to the power-down technique can be applied. The aqueous sample is enclosed between two blocks which provide the temperature control. After a constant temperature gradient is established between the two blocks both block temperatures are lowered simultaneously with identical cooling rates. The ice front then grows with a constant velocity from the cold to the warm block. Freezing simulations and experimental results show that the power-down technique offers a great potential for future studies. Possible applications are the cryopreservation of delicate cell types which require uniform cooling conditions (e. g. heart muscle cells, oocytes), basic investigations on the mechanisms of cell and tissue damage during freezing (cryosurgery), and the processing of frozen samples with a defined and uniform microstructure (morphology studies in bulk samples, improved freeze-drying protocols, and processing of biomaterials with defined pore sizes).

INTRODUCTION

It is well established that the recovery of cells following freezing is strongly dependent on the cooling rate B during freezing. The addition of cryoprotectants prior to freezing can increase survival, and for each cell type there is an optimal combination of cryoprotectant composition and cooling rate which produces maximum cell survival after thawing. For this reason, basic research in cryobiology often uses small sample volumes, so that the whole sample experiences a uniform cooling rate. For biological and medical applications, larger sample volumes must be preserved. The variation in the cooling rate as a function of position in the sample then becomes important, and the position-dependent recovery rates can vary enormously. As the optimal cooling rate is experienced by only a small fraction of the sample, the overall recovery rate can be severely reduced.

Figure 1 shows typical temperature-time curves during freezing in a cylindrical brass container (inner diameter d = 24 mm) of a physiological salt solution (c = 0.9 wt% NaCl in H_2O - used as a simple model for the suspension of biological cells). The filled container was immersed vertically in liquid nitrogen (T = -196 °C). The fact that the temperature reduction is not a constant function of time means that we need a sensible working definition of the cooling rate. We define the cooling rate B here as the average temperature rate of change over the temperature range (T = -0.5 °C ... -30 °C), the range where most cell damage occurs. When this definition is applied to the data in Figure 1, it can be seen that the cooling rate increases by a factor of 4.7 from the cylinder wall (B_W = 14 K/min) to the middle of the probe.

Extensive numerical studies were carried out to investigate the variation in cooling rates in aqueous salt solutions, as a function of the nominal cooling rate and the container geometry.[1,2] For this purpose non-stationary freezing was numerically simulated using finite volume analysis. To confirm the numerical solutions, and the validity of the assumptions used, the results were compared to measured curves (see figure 1). The numerical calculations were based on the initial experimental conditions (i.e. geometry, temperature, NaCl concentration) and the measured transient container wall temperature was used as the boundary condition. Further studies with suspended biological cells were performed in order to verify that cooling rate variations indeed lead to variations of cell survival.[3] In these experiments red blood cells were cryopreserved in a hydroxyethyl starch solution according to an established protocol for small samples (40 ml)[4]. Working with a large sample thickness of 12 mm, the cooling rate generally varied within the samples. By dividing the frozen samples into two segments the cell

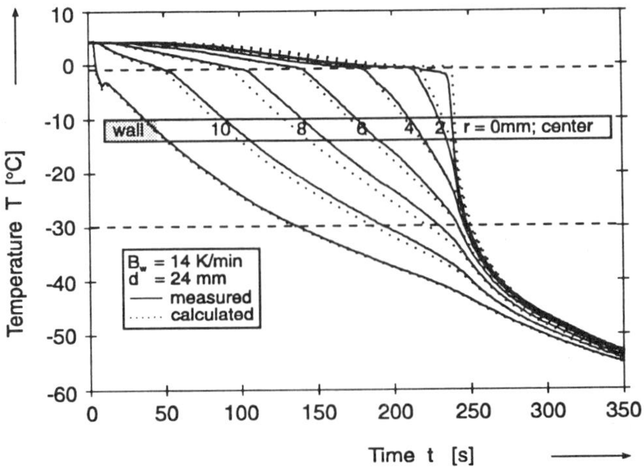

Figure 1. Comparison of measured and calculated temperature-time curves during freezing of a physiological salt solution (c = 0.9 wt% NaCl in H_2O) in a cylindrical brass container (inner diameter d = 24 mm) immersed in liquid nitrogen.

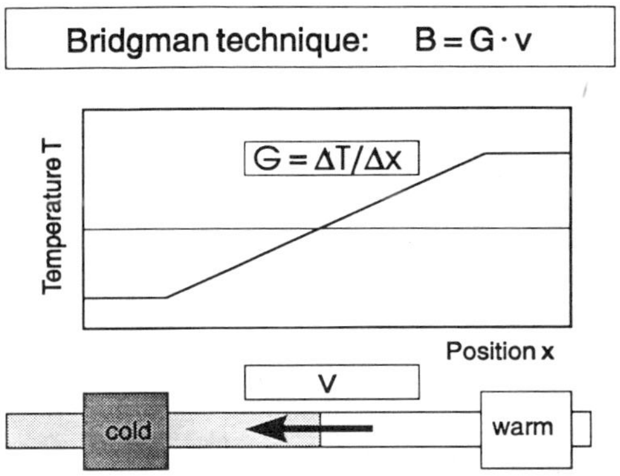

Figure 2. Bridgman technique for the directional solidification of microscopic aqueous samples. The samples are often enclosed in flat glass capillaries with rectangular cross section (e.g. 8). These capillaries are pushed with a constant preselected velocity v from a warm into a cold block (preselected individual block temperatures are controlled). Under optimum conditions, the temperature gradient G within the sample is constant and thus the entire sample can be frozen with the same cooling rate B = G·v.

survival in both segments could be determined individually. The maximal difference of the survival rate between both segments was found for an applied cooling rate of 120 °C/min (S_1 = 71 %, S_2 = 14 %) due to the strong decrease of cooling rate from 120 °C/min to 15 °C/min within the extended sample.

CONTROLLED FREEZING BY DIRECTIONAL SOLIDIFICATION

As shown above, simple freezing procedures often fail to provide a homogeneous, position-independent cooling rate throughout the sample needed for an optimum cryopreservation. To improve the uniformity of the cooling rate during freezing directional solidification techniques may be used.[5] Directional solidification processes permit the separate control of the velocity of the growing solid phase v and the temperature gradient $G = \Delta T/\Delta x$ at the phase front. The local cooling rate B which is generally used to characterize the cryopreservation process in cryobiology is simply given by the product $B = G \cdot v$. Directional solidification techniques are mainly used in materials science ond solidification processing in order to produce large single crystals (e. g. silicon crystals for semiconductors) or super alloys with a very uniform microstructure (e. g. for gas-turbine blades). In the following, the prospects of directional freezing devices in cryobiology according to the Bridgman technique and the Power Down technique are discussed.

Bridgman Technique for Microscopic Samples

Although stages for the freezing of microscopic samples with a defined cooling rate are widely used in cryobiology[6], several restrictions must be taken into account: The aqueous samples generally supercool prior to the nucleation of ice crystals. This supercooling is a stochastic event and thus not predictable. After nucleation the ice crystals grow with an increased velocity into the supercooled region. Therefore neither the ice front velocity nor the temperature field during supercooling are predictable. Furthermore, many investigations require the exact adjustment of a constant ice front velocity and a constant temperature gradient at the ice front. For this reason Rubinsky[7] developed a cryomicroscope for the directional solidification of biological samples according to the Bridgman technique. In the meantime, various modifications of the Bridgman freezing stages were performed in order to achieve a better process control. A typical Bridgman freezing stage is shown schematically in figure 2.

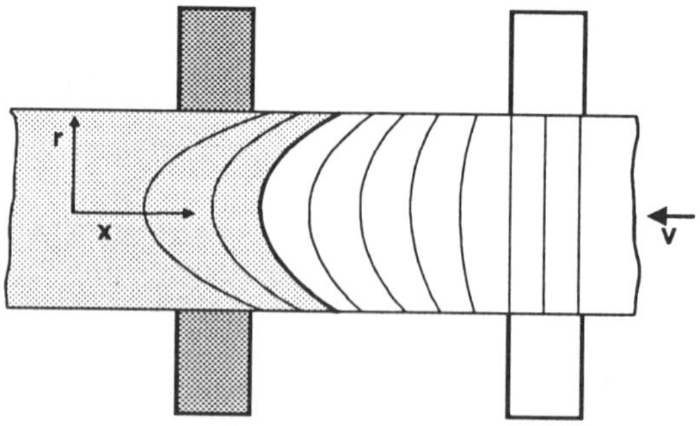

Figure 3. The Bridgman technique is not applicable for the directional solidification of macroscopic aqueous samples with high pushing velocites. In this case additional temperature gradients occur that cause curved isotherms and thus position dependent cooling rates (illustrated for a cylindrical sample).

Since the Bridgman technique allows the individual control of the temperature gradient G and the ice front velocity v (the absolute value is identical to the pushing velocity) it has further advantages. The cooling rate B which is defined as $B = G \cdot v$ can be realized in various ways. Thus several investigators studied whether the realization of a certain cooling rate by different combinations of G and v also has an influence on cell survival.[8,9] Other applications of this technique include the study of the breakdown times of planar ice fronts in aqueous solutions and the characterization of the developing dendritic solidification structures,[10] the nucleation of gas bubbles during freezing, the interaction of cells and particles with the growing ice fronts[11] and the morphological changes during directional solidification of tissue slices[12]. Our current investigations are concerned with the intracellular ice formation in isolated heart muscle cells.

The application of the Bridgman technique for the directional solidification of macroscopic biological samples is only possible for very small pushing velocities.[13] For higher pushing velocities the temperature field becomes two-dimensional (cylindrical sample, $T = f(x,r)$) or three-dimensional (parallelepiped, $T= f(x,y,z)$). The reason is the larger extent of the macroscopic (bulk) samples in the directions perpendicular to the applied temperature gradient. Thus additional temperature gradients in the r-direction (cylinder) or y,z-direction (parallelepiped) occur and therefore the cooling rate becomes position dependent (c. f. figure 3).

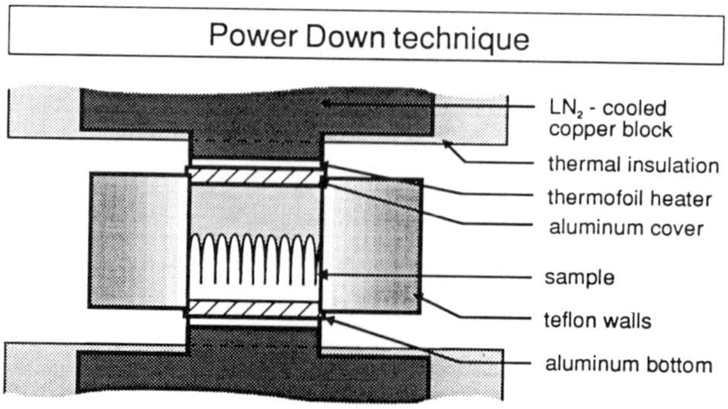

Figure 4. Schematic cross section of the power-down freezing apparatus (not true to scale). The aqueous sample is enclosed between two identical blocks which provide the temperature control. The controlled freezing procedure is described in figure 5.

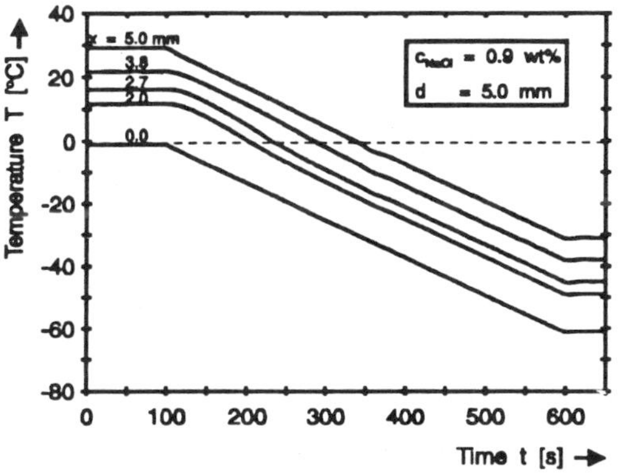

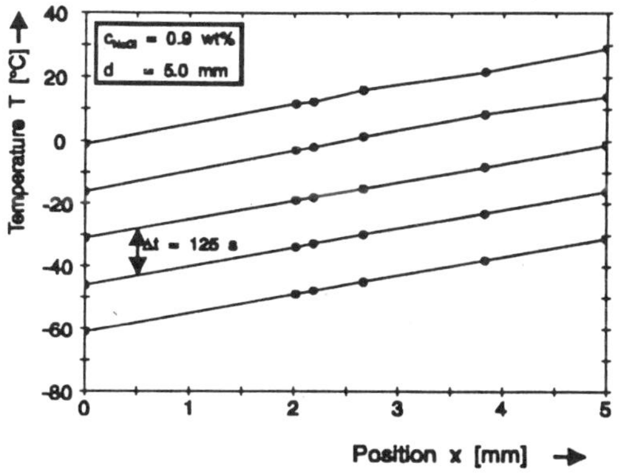

Figure 5. Measured temperature curves during freezing of a physiological salt solution within the Power Down freezing apparatus. After initially building up the desired temperature gradient G = 6 K/mm between lower (x = 0 mm) and upper block (x = 5 mm), the temperatures of the two blocks are simultaneously lowered with identical cooling rates of B = 7.2 K/min. The left diagram shows that the entire sample freezes uniform at the desired cooling rate. The right diagram demonstrates that the temperature gradient can also be kept constant.

Power Down Technique for Macroscopic Samples

As the Bridgman technique is in general not appropriate for the directional solidification of macroscopic aqueous samples an alternative technique must be applied. In case of bulk samples the Power Down technique is very promising. This technique is mainly used for the casting of superalloys.[14] The principle of the Power Down Technique is to solidify a fixed sample (no pushing device) between two adjacent blocks that are temperature controlled. Thus, after building up an inital gradient between both blocks, the directional solidification is performed by lowering both block temperatures with the same cooling rate. Since electric heating devices are generally used to control the temperature of the blocks, the term "power down" denotes the required reduction of the power to solidify the samples. Figure 4 shows a schematic cross section of a Power Down freezing apparatus that was developed for our basic investigations. The aqueous sample (dimensions: 5 mm • 25.4 mm • 25.4 mm) is surrounded by a square isolating teflon container. Above and below the sample are two identical blocks which provide the temperature control. Each block consists of an aluminum plate with a heating element, which is in thermal contact with a copper block continuously cooled with liquid nitrogen. The temperature is measured using a thermocouple embedded in the aluminum plate, and the temperature is controlled using the heating element and a PID controller. The aluminum plate is capable of achieving cooling rates of up to about 300 K/min.

The controlled freezing procedure first requires that a constant temperature gradient $G = (T_W - T_C)/d$ is established between the two blocks (see figure 5, t = 0 s, T_W = 29 °C, T_C = -1 °C, d = 5 mm, G = 6 K/mm). Then the temperatures of the two blocks are simultaneously lowered with identical cooling rates (in figure 5 with B = 7.2 K/min), thus keeping the gradient G constant. The ice front then grows with an approximately constant velocity v from the lower (cold) to the upper (warm) block. The finger-like (dendritic) structure of the ice front, briefly mentioned in figure 4, is characteristic for the directional solidification of aqueous solutions. The measured temperature-time curves in figure 5 show that under these conditions, the entire sample freezes uniform at the desired cooling rate.

When the cooling rate is increased, some position-dependent variation in the cooling rate can be seen even using this technique, although this variation is much smaller than for conventional techniques. The reason is that the temperature gradient can only be kept constant if the rate of enthalpy change is insignificant in comparison to the applied heat flow across the sample. Therefore a version of the freezing simulation was developed which allows the determination of the maximum acceptable cooling rate B_{max} using the power-down technique. The definition of homogeneity used for the program, is that the difference between the desired cooling rate and the cooling rate at any position in the sample cannot exceed ±25%. The important parameters of temperature gradient, sample thickness and desired cooling rate, can be expressed in dimensionless form, although the solutions are difficult to interpret. To make interpretation easier, the results shown in figure 6 use the fixed sample thickness of the power-down freezing apparatus (d = 5 mm) as an example. For a range of temperature differences between the two blocks, the variation in cooling rates in the sample was determined for a particular desired cooling rate, and this rate was then iterated until the maximum error of 25% in the cooling rate was found. This procedure defined B_{max} for that temperature difference.

The results show that increasing the temperature difference leads to a significant increase in the maximum achievable homogeneous cooling rate. For example, a temperature difference between the upper and lower block of $\Delta T = 37$ °C in comparison to classical freezing without any temperature difference ($\Delta T = 0$ °C) results in an order of magnitude increase in the acceptable cooling rate. Against a further increase of the temperature difference which would have a positive effect on the maximum cooling rate, there is of course the restriction that the biological samples cannot be heated above physiological temperatures.

The power-down freezing apparatus was used to study the dendritic structure of ice in aqueous salt solutions. The freezing parameters, temperature gradient G and ice front velocity v, were varied over a wide range of values (2 K/mm $\leq G \leq$ 12 K/mm, 20 µm/s $\leq v \leq$ 100 µm/s), and their influence on the dendritic tip separation λ_1 measured[15]. The equation developed to describe dendritic growth in molten alloys[5] $\lambda_1 = k \cdot G^{-1/2} \cdot v^{-1/4}$ could also be used to characterize the growth of ice in aqueous solutions. The results are in good agreement with experiments carried out using microscopic aqueous samples[10].

Future uses of the power-down technique in basic research include the possibility to study systematically the increasing solute concentrations in the aqueous phase during freezing, and to study the development of electrical potentials during freezing. In the field of process engineering, the technique could provide new insights into freeze-drying - using the property that the dendritic separation can be held constant, the effect of dendritic separation on the diffusion of water vapor in the sublimating sample can be systematically studied for the first time. For the cryopreservation of biomolecules and suspended cells the technique could be used when high activities or recovery rates after thawing are critical. For the preservation of tissues (e.g. human heart valves), it has the advantage that the thermal stresses which occur in standard freezing protocols, and can cause structural damage, are significantly reduced so that mechanical damage can be minimised.

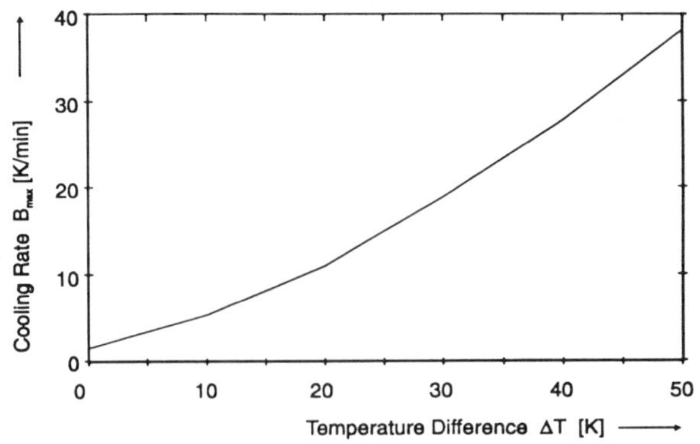

Figure 6. Numerically calculated cooling rates B_{max} as a function of the applied temperature difference between both blocks of the Power Down freezing apparatus (sample thickness d = 5 mm). For B_{max} the variation of cooling rate within the sample reaches 25%. Therefore, large temperature differences are advantageous for the homogeneity of Power Down freezing.

CONCLUSIONS

As shown above, there are many possible applications of directional solidification techniques in cryobiology. The two presented devices for the directional solidification of microscopic and macroscopic biological samples can quite easily be established in other research groups as well. Thus, the authors would like to encourage others to make use of directional solidification in basic and applied cryobiological research.

ACKNOWLEDGMENT

The support of the DFG to present this and four other papers on the Cryogenic Engineering Conference CEC 95 is gratefully acknowledged. Many thanks to Peter Schwindke who developed most of the electronic devices required for the accurate measurement and control of temperature, pressure, etc.

REFERENCES

1. U. Hartmann, B. Nunner, Ch. Körber, G. Rau, Where should the cooling rate be determined in an extended freezing sample, *Cryobiology* 28: 115-130 (1991).
2. I. Heschel, F. Schneider, Ch. Körber, A numerical study of the cooling rate variation within cylindrical samples, *Cryobiology* 28: 586 (1991).
3. C. Lückge. "Thermal Events during Freezing of Biological Cell Suspensions and their Influence on Cell Survival", diploma thesis, FH Aachen, Germany (1992).
4. A. Sputtek, G. Rau, Kryokonservierung von Humanerythrozyten mit Hydroxyethylstärke (HES) - Teil 1: Verfahrensbeschreibung, *Infusionstherapie* 19: 269-275 (1992).
5. W. Kurz, D. J. Fisher. "Fundamentals of Solidification", 3rd ed. Trans Tech Publications, Aedermannsdorf, Switzerland (1989).
6. Ch. Körber, S. Englich, P. Schwindke, M. W. Scheiwe, G. Rau, A. Hubel, E. G. Cravalho, Low temperature light microscopy and its application to study freezing in aqueous solutions and biological cell suspensions, *J. Microscopy* 141: 263-276 (1986).
7. B. Rubinsky, M. Ikeda, A cryomicroscope using directional solidification for the controlled freezing of biomaterials, *Cryobiology* 22: 55-68 (1985)
8. J. Beckmann, Ch. Körber, G. Rau, A. Hubel, E. G. Cravalho, Redefining cooling rate in terms of ice front velocity and thermal gradient: first evidence of relevance to freezing injury of lymphocytes, *Cryobiology* 27: 279-287 (1990)
9. A. Hubel, E. G. Cravalho, B. Nunner, Ch. Körber, Survival of directionally solidified B-lymphoblasts under various crystal growth conditions, *Cryobiology* 29: 183-198 (1992)
10. B. Nunner, Ch. Körber, G. Rau, Characterization of the ice front morphology developing during the directional solidification of aqueous solutions, submitted to *J. Crystal Growth*
11. Ch. Körber,Phenomena at the advancing ice -liquid interface: solutes, particles and biological cells, *Q. Rev. Biophys.* 21: 229-298 (1988)
12. B. Rubinsky, C. Y. Lee, J. Bastacky, G. Onik, The process of freezing and the mechanism of damage during hepatic cryosurgery, *Cryobiology* 27: 85-97 (1990)
13. W. Kurz, P. R. Sahm, Gerichtet erstarrte eutektische Werkstoffe, in: "Reine und angewandte Metallkunde in Einzeldarstzellungen, Vol. 25", W. Köster ed., Springer, Berlin (1975)
14. J. S. Erickson, C. P. Sullivan, F. L. Versnyder, Modern processing methods and investment casting of the superalloy family, in: "High-Temperature Materials in Gas Turbines", P. R. Sahm and M. O. Speidel eds., Elsevier, Amsterdam (1974)
15. I. Heschel, M. Rödder, C. Lückge, B. Nunner, G. Rau, A power-down device for the directional solidification of macroscopic samples, *Cryobiology* 29: 767 (1992)

CRYOGENIC HEAT TRANSFER OPTIMIZATION OF THE FREEZING FRONT GENERATED BY A CRYOPROBE

V. Zarnescu,[1] and F. Chiriac

Cryogenics Division, Heat & Mass Transfer and Applied Thermodynamics Department, Civil Engineering Institute
Bucharest, Romania
[1] Also, Department of Mechanical Engineering
The Pennsylvania State University
University Park, PA 16802

ABSTRACT

This paper is concerned with the analysis and optimization technique of the freezing front generated by a cryoprobe. The method, which is based on the placing of a flat cryoprobe against a body of tissue studies the resulting freezing front and optimizes the heat transfer, minimizing also the unnecessary freezing of the tissue. The optimization procedure is illustrated using a flexible model that starts from a one dimensional simplified situation and then is extended to a three-dimensional model that accounts for the density change of the tissue and the movement upon the readings of thermocouples placed at various depths in the tissue. The model uses an iterative procedure that continues until the optimal value is reached and the unnecessary freezing is minimized.The model can be used in a number of different ways, for example to study the system behaviour over the entire range of acceptable parameters, to test the possible effects of density change or to obtain simple estimates of the system behaviour in specific conditions. Using data fitting techniques, optimized values were obtained over the range from body temperature (310 K) to that of liquid nitrogen probe (73 K). For the model considered here, the optimization technique shows that the optimal value for density ration was found to be 1.085 and the minimization of the unnecessary freezing is reachable using an extensive approach that covers the entire range of parameters.

INTRODUCTION

In recent years, the microcryogenic technology for medicine and biology has greatly developed. Advances in cryosurgical techniques are connected with overcoming of theoretical, experimental and technological problems. The cryoprobes and their specifical functioning conditions require special consideration. The cooling performance of the cryoprobe needs to be controlled dynamically in order to cope with biological reaction and temperature immunity. Close monitoring is required in order to minimize, and if possible to eliminate the unnecessary freezing which in many cases limits the cryosurgical technique application. By its meaning, cryosurgery destroys by freezing and thawing unwanted cells. It is performed by bringing the region to be destroyed into contact with a cold source, ordinarily cooled by liquid nitrogen. While the objective is to kill certain cells, it is equally important that the treatment not be lethal to neighboring healthy cells. Therefore, the extent of freezing front generated by a cryoprobe and lethality conditions are of primary interest (Fig.1).

Elimination of detrimental freezing or at least minimization of the unnecessary destruction of healthy cells clearly requires more than visual monitoring or human decision. The purpose of this paper is to introduce and illustrate the optimization technique of the freezing front generated by a cryoprobe. Unlike other optimization techniques (Spendly et al,1962;Clegg et al, 1988; Keanini et al ,1992) that are concerned with the number and the size of the cryoprobes used in the procedure, we focus our attention on the surgical and non-surgical parameters and the specific boundary conditions. The optimization technique is illustrated using a flexible model because clinical applications will likely require that probe locations and orientations be optimized too.

MATHEMATICAL PROBLEM

Considering the density change of the tissue and the movement upon the readings of thermocouples placed at various depths in the tissue, we develop a mathematical model that is able to test the possible effects of density change with temperature and the difference between the solid and liquid densities (ρ_s, ρ_l). Since the freezing front is moving with speed v, the heat flux consists of the sum of the conductive and convective fluxes:

$$q_s(x,t) = -k_s T_x(x,t) + \rho_s c_s[T(x,t) - T_m]v(t) \qquad (1)$$

where the latter represents the sensible heat crossing a unit area per unit time due to a movement of the solid with speed v(t). Thus, the heat conduction equation in the solid takes the form:

$$\rho_s c_s[T_t + v(t)T_x] = k_s T_{xx} \qquad (2)$$

We assume $\rho_l \leq \rho_s$, c_L, c_s, k_L, k_s, L, T_m are all constants of positive value. So, in the solid region we have:

$$T_t + v(t)T_x = \alpha_s T_{xx} \qquad (3)$$

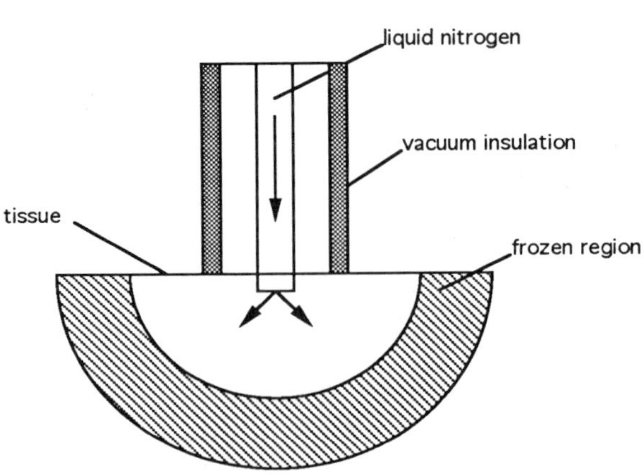

Fig.1 The freezing front generated by a cryoprobe.

The additional equation needed to determine the unknown v(t) comes from the requirement that mass be conserved. Thus, in order to find T(x,t), X(t) and v(t) we have:

$$v(t) = [1 - \rho_l/\rho_S]X'(t) \qquad ,t > 0 \tag{4}$$

$$T(X(t),t) = T_m \qquad ,t > 0 \tag{5}$$

The interface condition can be written as:

$$\rho_l L X'(t) = -k_L T_x(X(t)^-,t) + k_S T_x(X(t)^+,t) \qquad ,t > 0 \tag{6}$$

Using (4) we can obtain a similarity solution with the similarity variable as $\xi = (x/t)^{1/2}$ and the notation $\mu = \rho_l/\rho_S$. We find the unknown as :

$$X(t) = 2\lambda(\alpha_L/t)^{1/2} \tag{7a}$$

$$v(t) = \lambda(1-\mu)(\alpha_L/t)^{1/2} = v\lambda(1 - \mu)(\alpha_S/t)^{1/2} \tag{7b}$$

$$T(x,t) = T_L - (T_L - T_m)\frac{\operatorname{erf}\left[x/2\sqrt{\alpha_l t}\right]}{\operatorname{erf}\lambda} \quad , \quad 0 \le x \le x(t) \;, t > 0 \text{ (liquid)} \tag{7c}$$

$$T(x,t) = T_S + (T_m - T_S)\frac{\operatorname{erf}\left[x/2\sqrt{\alpha_s t} - (1-\mu)v\lambda\right]}{\operatorname{erfc}(\mu v\lambda)} \quad , \quad X(t) \le x \;, t > 0 \text{ (solid)} \tag{7d}$$

where λ is the root of the transcendental equation:

$$\frac{St_L}{\lambda e^{\lambda^2}\operatorname{erf}\lambda} - \frac{St_S}{(\mu v\lambda)e^{(\mu v\lambda)^2}\operatorname{erfc}(\mu v\lambda)} = \sqrt{\pi} \tag{8}$$

and

$$St_L = \frac{c_L(T_L - T_m)}{L} \;, \quad St_S = \frac{c_S(T_m - T_S)}{L} \;, \quad v = \sqrt{\frac{\alpha_l}{\alpha_S}} \;, \quad \mu = \frac{\rho_l}{\rho_s} \tag{9}$$

The root λ depends only on the parameters St_L, St_S and μv. Solving numerically for a range of μ's we found that the dependence of λ on μ is linear with ralatively small slope, typically changing by ± 2% when μ changes by ± 25% (Fig. 2a and 2b). The difference between the densities for our case was approximately 9% and the optimal value for μ was found to be 1.085.

THERMAL ANALYSIS

Since a significant percentage of the biological tissue is water, the thermal properties of the tissue will be determined to a large extent by those of water and ice. Over the range from body temperature (310 K) to that in liquid nitrogen cryoprobe, the conductivity k,

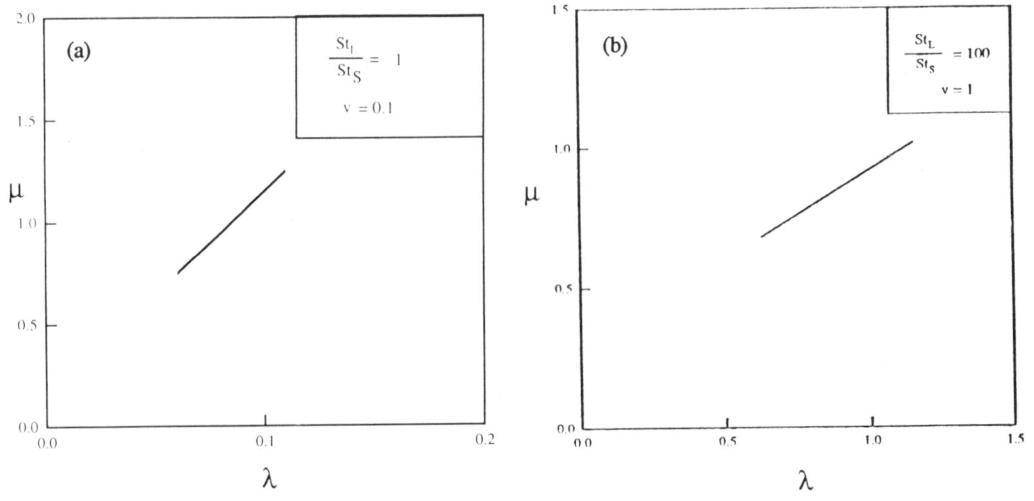

Fig.2 The dependence of the root λ with μ for different values for v and St_L/St_S ratio.

specific heat c, density ρ and thermal diffusivity α vary strongly. Consequently, we need to make the following assumptions:

1) constant thermophysical parameters ρ_S, ρ_l, c_S, c_L, k_S,k_L, α_S,α_L, some of them adapted from [3].

2) uniform initial tissue temperature T_L.

Using standard data fitting techniques we have obtained expressions of the above parameters temperature dependence over the range 73 K - 310 K:

$$c = 7.15 \times 10^{-3}\, T + 0.137 \quad (\text{kJ/kgK}) \quad \text{for} \quad T \leq 273 \text{ K (ice)}$$
$$c = 4.1866 \quad (\text{kJ/kgK}) \quad \text{for} \quad T \geq 273 \text{ K (water)}$$

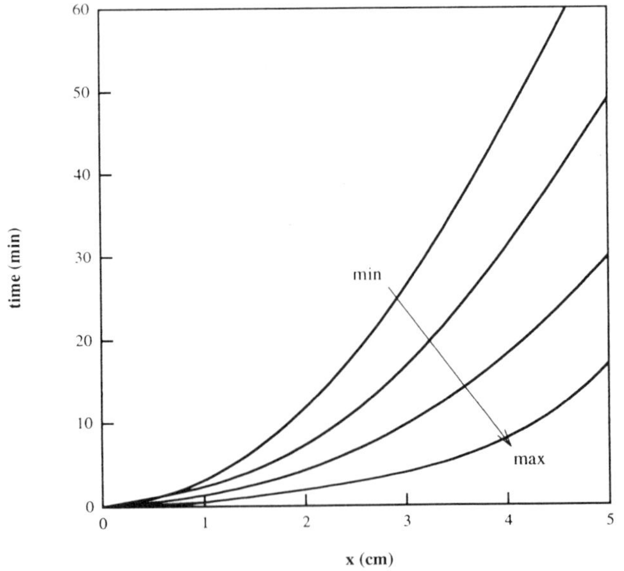

Fig.3 Time-dependent evolution of the freezing front generated by a cryoprobe.

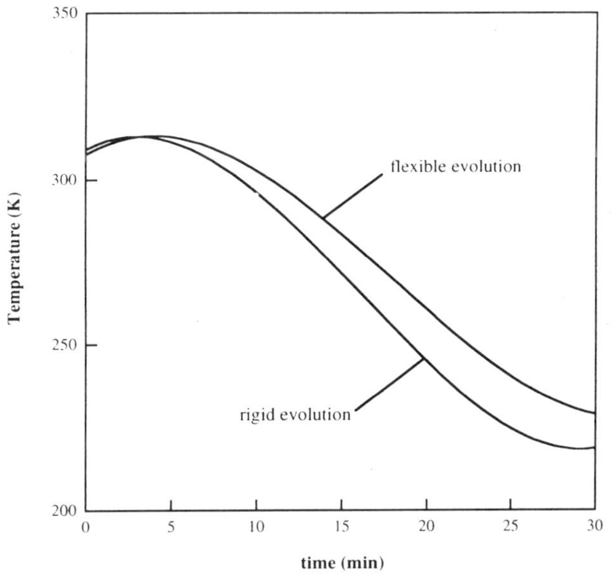

Fig.4 Rigid and flexible freezing front evolutions in time.

$k = 2.23 \times 10^{-3} + 5.973 \times 10^{-6}(273 - T)^{1.156}$ for $T \leq 273$ K
$k = 1.0185 \times 10^{-4} + 1.693 \times 10^{-6}T$ for $T \geq 273$ K

$\rho = 920$ (kg/m^3) for $T \leq 273$ K
$\rho = 1000$ (kg/m^3) for $T \geq 273$ K and $L = 333.73$ (kJ/kg)

Using the minimum, average and maximum values of c and k we obtain the extreme values for the interface as:

$$0.587\, t^{1/2} \leq X(t) \leq 2.018\, t^{1/2} \tag{10}$$

where t is the time in minutes and X is the depth in centimeters. The average value for the penetration depth of the freezing front is $X(t) = 1.302\, t^{1/2}$. The evolution of the freezing front penetration is shown in Fig.3.

Usually, the cryosurgical procedure targets a specific point or a specific control volume within the tissue. The entire monitoring procedure is directed toward that particular point. But, in practical cases, due to freezing-induced expansion the point of interest will freeze at a different location. In order to determine to what extent the temperature readings are affected by the expansion, we consider the location $x_0(t)$ at time t of the target point (which moves with velocity v(t) until it freezes). By integration of (4), we find:

$$x_0(t) = x_0(0) + (1 - \rho_s/\rho_l)X(t)\ , \quad 0 \leq t \leq t_{freezing} \tag{11}$$

So, the displacement of the location of the freezing point can be expressed by:

$$x_0(t_{freezing}) = (\rho_l/\rho_s)x_0(0) \tag{12}$$

and

$$t_{freezing} = [(\rho_l/\rho_s)(x_0(0)/2\lambda)]^2/\alpha_s \tag{13}$$

For a rigid evolution of the freezing front, the temperature profile can be written as:

$$T(x_0,t) = T_L - (T_L - T_m)\frac{\operatorname{erf}\left[\dfrac{x_0}{2\sqrt{\alpha_L t}} - \left(1 - \dfrac{\rho_s}{\rho_l}\right)\lambda\sqrt{\dfrac{\alpha_S}{\alpha_L}}\right]}{\operatorname{erfc}\left(\dfrac{\rho_s}{\rho_l}\lambda\sqrt{\dfrac{\alpha_S}{\alpha_L}}\right)} \tag{14}$$

whereas, for a flexible evolution we will have:

$$T(x_0,t) = T_L - (T_L - T_m)\frac{\operatorname{erfc}\left[\dfrac{x_0}{2\sqrt{\alpha_L t}}\right]}{\operatorname{erfc}\left(\dfrac{\rho_s}{\rho_l}\lambda\sqrt{\dfrac{\alpha_S}{\alpha_L}}\right)} \tag{15}$$

It is observed (Fig.4) that the freezing front location displacement has a considerable effect on the temperature profiles and therefore the density change can not be ignored.

THREE-DIMENSIONAL EXTENSION OF THE MODEL

The original freezing front penetration into the tissue is an exact solution for the freezing of a homogenous body, semi-infinite in extent. Three-dimensional heat transfer effects can be described mathematically by a system of partial differential equations that have to satisfy certain boundary conditions. They represent the situation when a small area on the surface of a semi-infinite medium is cooled. The solution depends on the shape of the cooled area (in our case circular). Assuming the dimensionless temperature $\theta = (T - T_\infty)/(T_S - T_\infty)$, the solution for a circular area of radius R is:

$$\theta_0 = 1 - x/(R^2 + x^2)^{1/2} \quad \text{at} \quad r = 0 \tag{16}$$

$$\theta = 0.5 - x(2/\pi)^{-1}[(x/2)^2 + R^2]^{-1/2}B\{R[(x/2)^2 + R^2]^{-1/2}\} \quad \text{at} \quad r = R \tag{17}$$

where B is the complete elliptical integral of the first kind. Numerical solutions were generated for constant temperature and constant surface heat flux boundary conditions. The results consistently showed the freezing front forming more rapidly under a constant temperature boundary condition than under constant heat flux. As the steady state condition is approached, the ice-front radius for a constant surface heat flux approaches and finally becomes equal to that for the constant temperature condition. Hence at the start of freezing and at the end of it the two solutions are equal as would be expected. In the case of negligible thermal capacity of the frozen phase we have:

$$\theta = R(x-1)/x(R-1) \tag{18}$$

where θ is the dimensionless temperature.

The variation of the heat flux into the cryoprobe with time for the constant temperature case was determined by writing Fourier's equation at the cryoprobe surface. Differentiating the frozen phase temperature profile and substituting into Fourier's equation yields:

$$q = [\ k_f(T_S - T_L)/r_0](\theta - 1)(x/x - 1) \tag{19}$$

where q is the heat flux into cryoprobe surface; T_L is the initial tissue temperature; T_S is the surface temperature and $x = R/r_0$.

The two solutions are equal at the limits of time interval as expected but the freezing front is forming more rapidly under a constant temperature boundary condition.

CONCLUSIONS

An optimization procedure for the freezing front generated by a cryoprobe has been carried out based on a flexible mathematical model. Different aspects of studying sysyem behaviour, testing possible effects of density change and influence of boundary conditions have been pointed out. For the present model, the optimal value of density ratio was found to be 1.085. The accuracy of this optimization technique can be further improved by considering effects of the internal geometry of the body, the existence of heat sources resistant to temperature changes like the presence of a major blood vessel situated in the neighborhood of the lesion. For this, a better knowledge of thermal properties of the bodies to be subjected to cryosurgery is needed. However , together with a successful procedure, the final goal is the efficient minimization of the unnecessary destruction of healthy tissue during cryosurgery.

REFERENCES

1. W.Spendly, G.R. Hext and F.R.Himsworth, "Sequential Application of Simplex Designs in Optimization and Evolutionary Operation.", *Technometrics* 4:441 (1962).
2. S.T. Clegg, C.T.Liauh and R.B. Roemer, "Selecting the Optimal Amount of Regularization to reconstruct Three-dimensional Temperature Fields from Limited Noisy Temperature Measurements during Hyperthermia" in: "Comp.Methods in Bioengineering", R.L.Spilker and B.R. Simon eds., ASME, New York (1988), pp.280-291.
3. R. Barron, "Cryogenic Systems.", McGraw-Hill, New York (1985).
4. R. Barron, "Heat Transfer Problems in Cryosurgery." *J.of Cryosurgery,* 1:4:316-325 (1968).
5. R.G. Keanini, B. Rubinsky, "Optimization of Multiprobe Cryosurgery.", *J. of Heat Transfer,* 114:796-801 (1992).
6. J.B. Bischoff, J. Bastacky and B. Rubinsky, "An Analytical Study of Cryosurgery in the Lung.", *ASME J. of Biomech. Eng.*, 114:59-68 (1992).
7. H. Budman, A. Shitzer and S. Del Giudice, "Investigation of Temperature Fields around Embedded Cryoprobes.", *ASME J. of Biomech. Eng.*, 108:42-48 (1986).
8. V. Zarnescu and F. Chiriac, "Heat Transfer and Distribution of Heat Flux Density in Microcryogenic Installations used in Medicine.", presented at the CEC/ICMC Conf., Albuquerque, NM (1993).
9. V. Zarnescu and F. Chiriac, "Applied Cryogenics in Medicine and Biology." in "Proc. of 23rd Symp. of AIIR", Timisoara, Romania, 4:440-446 (1991).

1995 CEC/ICMC Conference: The 1995 CEC Student Travel Assistance awardees: (from left to right) Krishna Iyengar, Catherine Pinsky, Adrian J. Corless, and Vlad Zarnescu.

ON A LUMPED HEAT TRANSFER MODEL OF FREEZING PROCESS OF CRYOPRESERVATION MEDIA

D.Y. Gao[1,2], J. Liu[1,3], F. Kleinhans[1,3], A. Ungan[2], J.K. Critser[1,4,5]

[1] Cryobiology Research Institute, Methodist Hospital of Indiana
Indianapolis, IN 46202, USA
[2] Department of Mechanical Engineering, Purdue University
Indianapolis, IN 46202, USA
[3] Department of Physics, Indiana University-Purdue University
Indianapolis, IN 46202, USA
[4] Departments of Physiology & Biophysics and Obstetrics and Gynecology, Indiana University School of Medicine
Indianapolis, IN 46202, USA
[5] Department of Veterinary Clinical Sciences, Purdue University
School of Veterinary Medicine
West Lafayette, IN 47907, USA

ABSTRACT

In cryopreservation, biological cells are usually suspended in and cooled together with a cryopreservation medium. A lumped heat transfer model and an apparent specific heat form of energy equation were developed to formulate the cooling process of the cryopreservation media. Using glycerol as a model cryopreotective agent, the freezing process of ternary cryopreservation media (glycerol-NaCl-water) in cylindrical plastic straws were studied numerically by computer simulation. Effects of four important factors on the temperature history in the media during the cooling process were investigated. These factors include: glycerol concentration, supercooling temperature of the media, ambient constant cooling rates, and radius of the straw. Results from computer simulation showed that these factors significantly influenced the thermal history in the ternary media, especially during the water phase transition period. It was also revealed that actual cooling rate in the media varied as a function of time and the maximum rate was much higher than the ambient constant cooling rate.

Advances in Cryogenic Engineering, Vol. 41
Edited by P. Kittel, Plenum Press, New York, 1996

INTRODUCTION

Using cryoprotective agents (CPAs, e.g. glycerol, dimethysulfoxide, ethylene glycol, etc.) cryopreservation of biological cells and tissues has been widely utilized in many scientific research and clinical applications. However, there is an apparent contradiction between the concept of preservation and the experimental finding that a part of cells are injured by the cryopreservation procedure itself. It is generally accepted that the cooling rate of a cryopreserved sample is a critical factor affecting the cryosurvival of the cells. As cells are cooled to about -5°C, both the cells and surrounding medium usually remain unfrozen and supercooled. Between -5°C and about -15°C, ice forms in the external medium but the cell contents remain unfrozen and supercooled, presumably because the plasma membrane blocks the growth of ice crystals into the cytoplasm. The supercooled water in the cells has, by definition, a higher chemical potential than that of water in the partially frozen extracellular solution, and thus water flows out of the cell osmotically and freezes externally. The subsequent physical events in the cell depend on cooling rate. If the cell is cooled <u>too rapidly</u>, intracellular water is not lost fast enough to maintain equilibrium; the cells becomes increasingly supercooled, eventually lethal intracellular ice formation (IIF) occurs. If the cell is cooled <u>too slowly</u>, the severe cell shrinkage (due to dehydration) and a long time exposure of the cell to highly concentrated electrolytes (due to the precipitation of liquid water as ice) before the eutectic is achieved can also damage the cells[1]. Since 1963, Mazur[2] first and then Pitt et al.[3], Toner et al.[4] and Mondrew and McGann[5], have developed different theoretical models and equations to describe the cell response to the environmental change during the cooling process. Using these models, the optimal constant cooling rates have been predicted to prevent the cryoinjury to the certain cell types. However, these models do not describe how to achieve the predicted optimal cooling rate in a biological sample. Direct control of the local cooling rates in a cell suspension or a tissue is technically difficult and sometimes impossible because of the limitation of putting a temperature sensor into a sample (e.g. a tissue or organ). Practically it is feasible and relatively easy to control the sample's ambient temperature change or cooling rate. In the present study, a mathematical formulation describing the relationship between the actual temperature change in a sample and the ambient temperature change is developed using a lumped heat transfer model and an apparent specific heat form of energy equation, which allows one to design a ambient cooling program to ensure actual optimal cooling rate in a sample.

RESEARCH MODEL AND MATHEMATICAL FORMULATION

In cryopreservation, biological cells are first suspended in a cryopreservation medium (e.g. an isotonic saline solution containing one or several cryoprotective agents e.g. glycerol, dimethylsulfoxide, propylene glycol, etc.). The cell suspension will then be placed in a container (e.g. a thin-wall plastic straw or bag) and cooled in a cooling chamber. The coolants in the cooling chamber are usually chilled air or the liquid nitrogen vapor which are released to the chamber through electromagnetic valves. Using a computer program the cooling rate in the chamber (i.e. the ambient cooling rate of the sample(s)) can be controlled by opening or closing the valves. Based on heat transfer principle, such a system (i.e. a cell suspension in a container) with a convective heat transfer boundary condition may be considered as a " lump" whose interior temperatures are nearly uniform if the following condition is satisfied[6] :

$$\text{Biot number, } Bi = h_{\infty}(V/A_w)/k \leq 0.05$$

where h_∞ is heat transfer coefficient of the fluid (e.g. air or liquid nitrogen) in the cooling chamber, V is the system volume, A_w is the surface area of the system, and k is thermal conductivity of the system. The above equation indicates that to ensure the "lumpability", a system should have a small ratio of volume to surface area and/or its internal "conductive resistance" should be much smaller than its external convective resistance. Mathematical formulation for energy conservation of a lumped system can be made as follows[6]:

$$h_\infty A_w\left[T_\infty(t) - T(t)\right] = m\frac{d\left(C_{pa}T\right)}{dt} \tag{1}$$

$$C_{pa}(T) = (1-\omega_1-\omega_2)C_{pt} + \omega_2\cdot C_{pc} + \omega_1\left\{\left[1-f(T)\right]C_{pu} + f(T)C_{pi} - \frac{df(T)}{dT}L_f(T)\right\} \tag{2}$$

From Kirchoff's Equation:

$$\left(\frac{\partial L_f}{\partial T}\right)_p = C_{pu} - C_{pi}$$

$$L_f(T) = \left(C_{pu} - C_{pi}\right)T + L_f^o(T = 0^\circ C) \tag{3}$$

where T = temperature of the system, t = time, T_∞ = ambient temperature, m = total mass of the system (cell suspension and the container), C_{pa} = apparent specific heat of the system, ω_1 = water mass fraction of the system, ω_2 = cryoprotective agent mass fraction of the system, C_{pt} = specific heat of the container material, C_{pu} = specific heat of liquid water, C_{pi} = specific heat of ice, $f(T)$ = mass fraction of frozen water (ice) over the total water mass as a function of T (NOTE: $f(T) \equiv 0$ if $T >$ freezing temperature of a solution.), $L_f(T)$ = latent heat of water solidification.

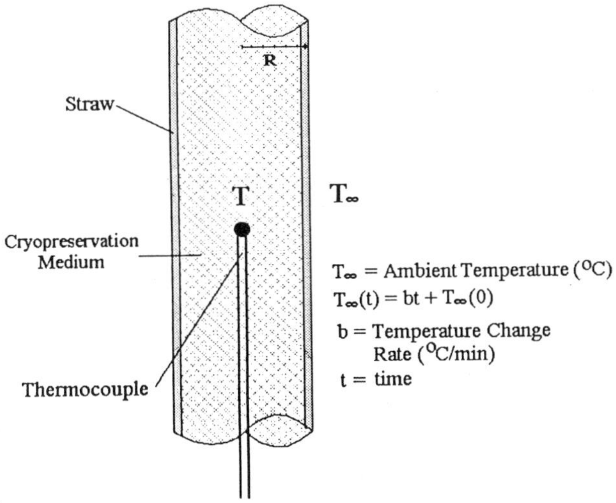

Figure 1. A schematic diagram showing the straw-medium model system with a convective heat transfer boundary condition. T = temperature of the model system, T_∞= ambient temperature.

In the present model study, thin-wall plastic cylindrical straw/tube is considered as a container; glycerol, as a cryoprotective agent; the ternary solution, water-NaCl-glycerol, as a cryopreservation medium. As shown in Figure 1, the straw-solution system is placed in a cooling chamber with a convective boundary condition. The ambient temperature of the straw-solution system, T_∞ varies as a linear function of time. The frozen fraction of water, f(T) in the water-NaCl-glycerol solution during freezing process can be derived from the analytical equations describing the phase diagrams of the ternary solution[7]:

$$f(T) = \left[1 + \frac{M_s}{M_{H_2O}}\right] - \left(\frac{M_s}{M_{H_2O}}\right) \cdot \frac{2}{a + (a^2 - 0.047T)^{0.5}} \tag{4}$$

$$a = \frac{1}{-1.6 - 1.27R - 0.25R^2} \tag{5}$$

$$R = \frac{M_s}{M_{NaCl}} \tag{6}$$

where Ms/M_{H_2O} = the ratio of glycerol mass to water mass in a water-NaCl-glycerol medium, and M_S/M_{NaCl} = the ratio of glycerol mass to NaCl mass in a water-NaCl-glycerol medium. The ambient temperature (T_∞) is a linear function of t:

$$T_\infty = Bt + T_\infty(0) \tag{7}$$

where B is constant ambient cooling rate, $T_\infty(0)$ is initial ambient temperature.

NUMERICAL CALCULATION

Equations (1) to (7) were calculated numerically using a differential equation solver, MLAB (Civilized Software, Inc., Bethesda, MD). The following assumptions and conditions were made in the calculation:

(a) The specific heat values of water, ice, glycerol, as well as straw material (polystyrene) were assumed to be constant (Table 1).

(b) Initial temperature $T(0) = T_\infty(0) = 5$ °C.

(c) NaCl concentration = 0.15M.

(d) The forced convective heat transfer coefficient (h_∞) for gases varies ranging from 10 to 300 $W/m^2 K$ (White, 1988)[6]. In this study, h_∞ is assumed to be 100 $W/m^2 K$.

Effects of following four factors on the actual temperature history or cooling rate in the water-NaCl-glycerol system were investigated by computer simulation:

(1) Constant ambient cooling rate: ranging from 1 to 1000°C/min.;

(2) Glycerol concentration: ranging from 0 to 2 M;

(3) Supercooling temperature (defined as an actual ice nucleation temperature in a solution during a cooling process, which is lower than the real freezing temperature of the solution): ranging from -5 to -15°C.

(4) Radius of the thin-wall (0.1 mm in thickness) cylindrical straw: ranging from 1 to 2 mm, which are typical sizes for the straws used to cryopreserve mammalian sperm, oocytes and embryos[8].

Table 1. Physical properties of the materials

Materials	Cp (J/kg°C)	Density (g/ml)	Molecular Weight
water	4217	1	18
ice	1945	0.91	18
glycerol	2377	1.26	92.1
polystyrene	1210	0.8	-
NaCl	-	2.16	58.4
	$L_f(0) = 333.84$ kJ/kg (water), $V/A_w = R/2$		
	$k_{ice} = 2.0$ W/m°C, $h_\infty = 100$ W/m^2°C		

The lumped model restriction for the cylindrical sample can be estimated by the following equation:

$$Bi = h_\infty \frac{V}{A_w k} \cong \frac{h_\infty R}{2k} \leq 0.05 \tag{8}$$

where R is the radius of the straw. Using k value of ice (2 W/mK) as a first approximation for the thermal conductivity of the straw-solution system, it is easy to prove that the system is " lumpable" (i.e. Equation (8) is satisfied) for the conditions used in the present model studies (Table 1).

RESULTS

Based on the lumped heat transfer model and numerical calculation, effects of the four factors (i.e. ambient cooling rate, glycerol concentration, supercooling temperature, and cylindrical straw radius) on the actual temperature history or cooling rate (i.e. the slope of the temperature history curves) in the water-NaCl-glycerol solutions were investigated and are shown in Figures 2 to 5. Figure 2 shows that when water starts to freeze at -10°C (an assumed supercooling temperature), the latent heat released from water solidification makes an increase (rebound) of sample temperature to achieve the freezing temperature (-1.8°C) of the 1 M glycerol solution. The temperature of the sample is kepted at near freezing temperatures for a short period of time (maintaining time) and then is rapidly lowered to catch up the ambient low temperature. The maintaining time is greatly decreased with an increase of ambient cooling rate, and the actual maximum cooling rate (i.e. the maximum slope of each temperature history curve) of the sample after the rebound can be much higher than the ambient cooling rate. It is also revealed that the maximum cooling rate is increased with a decrease of supercooling temperature (T_{super}) (Figure 3) and a decrease of glycerol concentration (Figure 4), as well as an decreased in the sample size (radius of the straw) (Figure 5). From Figure 2-5, it is clear that the actual cooling rate in the solution is close to the ambient cooling rate only before and after water solidification.

DISCUSSION

A lumped heat transfer model and corresponding mathematical formulation of the freezing process of biological cells or tissues suspended in cryopreservation media were developed. Using water-NaCl-glycerol as a representative cryopreservation medium, results from the numerical calculation indicates that actual cooling rate of the medium during the cooling process can be greatly different from ambient cooling rate, especially during the phase transition of water in the solution, which is fundamentally important to be understood

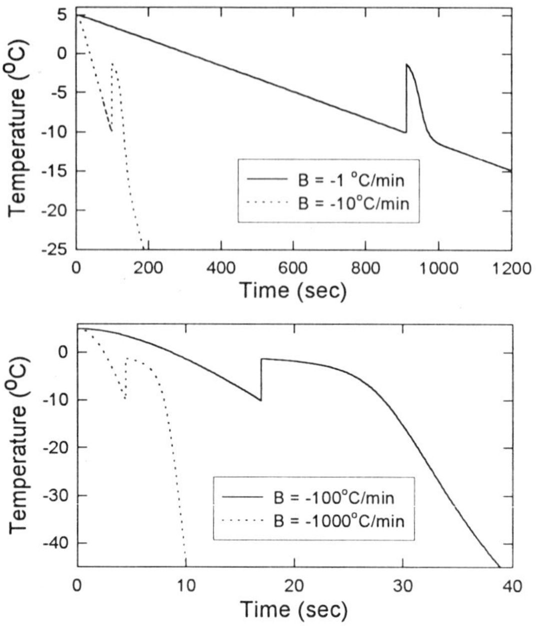

Figure 2. Calculated temperature histories in the straw-medium system with ambient cooling rate (B) as a parameter. (Glycerol concentration = 1M, supercooling temperature = -10°C, straw radius = 1 mm).

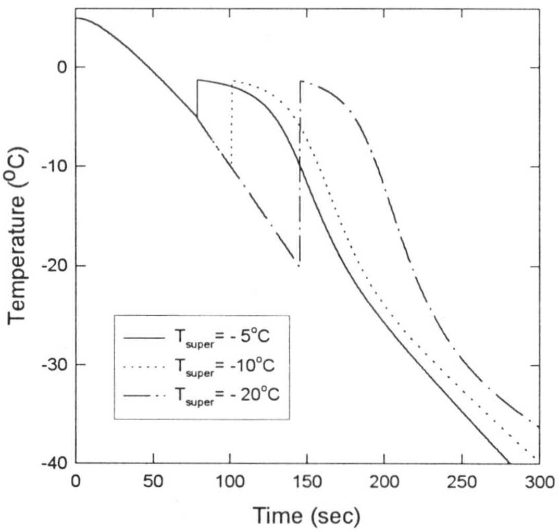

Figure 3. Calculated temperature histories in the straw-medium system with the supercooling temperature (T_{super}) as a parameter. (Glycerol concentration = 1M, ambient cooling rate = -10°C/min, straw radius = 1 mm).

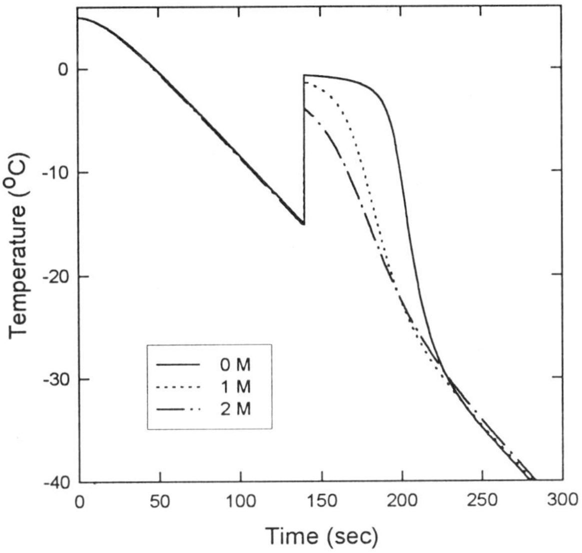

Figure 4. Calculated temperature histories in the straw-medium system with the glycerol concentration as a parameter. (Supercooling temperature = -15°C, ambient cooling rate = -10°C/min, straw radius = 1 mm).

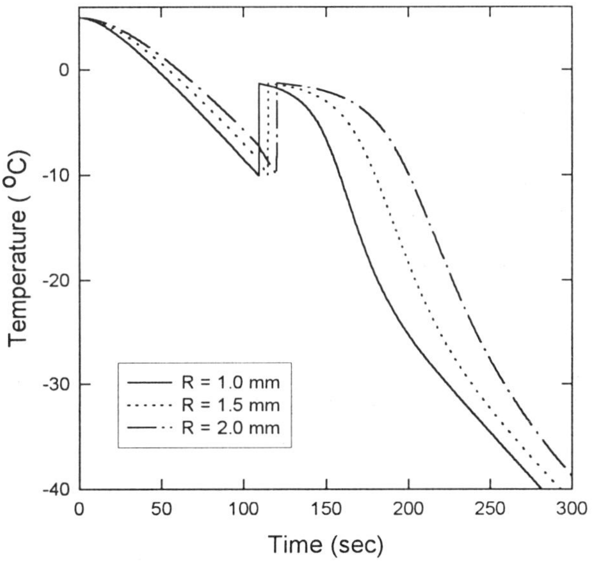

Figure 5. Calculated temperature histories in the straw-medium system with the straw radius (R) as a parameter. (Supercooling temperature = -10°C, ambient cooling rate = -10°C/min, glycerol concentration = 1 M).

for the cell cryopreservation because the actual cooling rate is crucial for the cryoinjury to the cells.

The glycerol concentration, ambient cooling rate, straw size and supercooling temperature significantly influence the temperature change or actual cooling rate of the solution. For a given constant ambient cooling rate, a high glycerol concentration, a relatively large size (radius) of the straw , or a higher supercooling temperature (i.e. a small degree of supercooling) can reduce the maximum cooling rate in the medium.

If a cryopreservation system is lumpable (i.e. $Bi \leq 0.05$), the major advantages of the developed lumped heat transfer model are:

(1) This model avoids difficulties associated with the complicated mathematical formation describing the microscopic growth of dendrite tip of the ice crystals and the redistribution of the solute concentration in a solution during freezing processes.

(2) The relationship between the temperature (T(t)) or cooling rate (dT/dt) in a cryopreservation medium and the ambient temperature ($T_{\infty}(t)$) or cooling rate (dT_{∞}/dt) is simply described by the equation (1)-(3). Given a predicted optimal cooling program (T) or cooling rate (dT/dt) for a specific cell type[2,3,4,5], one can readily derive and design a corresponding ambient cooling program (T_{∞}) which guarantees the actual cooling rate of biological sample is the optimal one.

(3) The current model and mathematical formulation can be applied not only to the cooling process but also to the warming process.

From equation (1)-(3), it is noted that because of the non-linearity of function f(T) (frozen fraction of water at temperature T), a non-linear cooling program (non-constant cooling rate) is required to ensure a constant, actual cooling rate in the solution-straw system. Finally, it should be mentioned that the theoretical results obtained from the lumped model for the freezing process of water-NaCl-glycerol system in straws agreed well with a series of recent experimental results .

REFERENCES

1. P. Mazur, Freezing of living cells: mechanism and implication, *Amer. J. Physiol.* 247:C125(1984),

2. P. Mazur, Kinetics of water loss from cells at subzero temperatures and the likelihood of intracellular freezing, *J. Gen. Physiol.* 47:347(1963),

3. R.E. Pitt and P.L. Steponkus, Quantitative analysis of the probability of intracellular ice formation during freezing of isolated protoplasts, *Cryobiology*, 26:44(1989).

4. M. Toner and E.G. Cravalho, Thermodynamics and kinetics of intracellular ice formation during freezing of biological cells, *J. of Applied Physics,* 67:1582(1990).

5. K. Muldrew and L.E. McGann, Mechanisms of intracellular ice formation, *Biophysics Journal,* 57:525 (1990).

6. F.M. White, "Heat and Mass Transfer", Addison-Wesley series in Mechanical Engineering. (1988)

7. D.E. Pegg, Simple equations for obtaining melting points and eutectic temperatures for the ternary system glycerol/sodium chloride/water., *Cryo-Letters,* 4:259(1983).

8. P.F. Watson, The preservation of semen in mammals, in: "Oxford Reviews of Reproductive Biology", L.A. Finn, ed, Vol.1, (1979), p.283.

SOME ENGINEERING CONSIDERATIONS IN DESIGNING A MULTIPROBE CRYOSURGICAL DEVICE USING LIQUID NITROGEN

Zhao Hua Chang

Cryomedical Sciences, Inc.
Rockville, MD 20850, USA

ABSTRACT

Visceral cryosurgery requires the use of multiple cryoprobes with refrigerative capacities greater than any conventional single cryoprobe device of the prior arts can provide. The present work reveals and discusses various engineering problems associated with the design of a multiprobe cryosurgical device utilizing liquid nitrogen including basic requirements, various applicable heat-transfer-enhancement methodologies, and design of the LN_2 distribution manifold. The disposal of liquid nitrogen and its clinical implications are also analyzed.

INTRODUCTION

The initial interest in cryosurgery was stimulated by the development of a cryosurgical system utilizing liquid nitrogen[1]. Due to the inability to internally monitor the cryosurgical process and the slow progress in the technical advancement of cryosurgical instrumentation, cryosurgery has been largely limited to the treatment of diseases in readily accessible areas of the body [2-3]. When visceral cryosurgery is indeed performed, on a highly selected basis, open incisions are often necessary to expose the tumors. This inevitably subjects the patient to additional surgical risk and complicates the surgical process. Depending upon the size and number of the lesions, in many cases, several cryosurgical instruments are required to operate simultaneously.

As it is becoming clear that the cryosurgical process can be monitored via real-time ultrasound imaging and magnetic resonance imaging (MRI) with precision,[4-5] the development of a sophisticated cryosurgical device becomes increasingly imminent. The recent development of a high-performance multiprobe cryosurgical system has rendered it possible to perform many minimally invasive visceral cryosurgical procedures, including percutaneous transperineal cryoablation of the prostate, laparoscopic cryosurgery of the liver, and removal of brain tumors, etc.[6-11]. Clearly, many intrinsic technical problems are involved in designing a multiprobe cryosurgical system that are not obviously associated with a single-probe cryosurgical device. The present work is intended to establish some fundamental technical requirements and reveal the unique problems associated with developing a multiprobe cryosurgical device. Appropriate solutions to the technical problems are also discussed.

BASIC TECHNICAL REQUIREMENT AND HEAT-TRANSFER ENHANCEMENT

Traditionally, the basic technical requirement of cryosurgery has been one probe attached to one surgical unit. While one probe has been satisfactory for many surface-lesion treatments, it has been realized that multiple probes are often needed to compensate for the insufficient freezing capacity of the device or shorten the surgical duration when large and/or multiple lesions are involved. Multiple probes are also frequently used for the purpose of controlling the geometric shape of the freezing zone and thereby minimizing the potential for freezing healthy tissue surrounding a lesion. In many cases, this has been done by simultaneously operating several individual cryosurgical units. The presurgical planning process will inevitably be affected by the availability of the cryosurgical units. There are many variables, uncertainties and potential human errors associated with operating the units that may negatively affect the ultimate clinical outcome.

A multiprobe cryosurgical system offers a surgeon the flexibility to choose the number and type of cryoprobes prior to or during the cryosurgical process. One immediate problem in designing a multiprobe cryosurgical system is the determination of how many cryoprobe channels should be designed into the system to meet the specific surgical requirements or to ensure a full readiness of surgical hardware for the surgery. Clinical experience indicates that most of the open liver cryosurgical procedures require either simultaneous or sequential use of 1-3 cryoprobes of the same or different types (trocar, flat, and paddle probes etc.); laparoscopic liver cryosurgery typically requires 1-2 long-shafted trocar cryoprobes; brain cryosurgery typically requires only one slender cryoprobe; and percutaneous cryoprostatectomy requires the simultaneous use of 5 slender trocar cryoprobes. While it is always possible that more probes might be desirable for some special surgical cases, five probe channels have proven to be satisfactory for most. Whenever technically feasible, it is recommended that 1-3 additional ports should be designed into the system as an option for any special clinical application.

For a multiprobe system, each probe port should be capable of independent control to afford a surgeon the flexibility of designing a specific protocol of cryoprobe application, including spatial arrangement of cryoprobes, the sequence of cryoprobe application and the refrigerative capacity for each cryoprobe. With an appropriate protocol, the tumors may be "sculpted" with the desired precision.[12] Figure 1 illustrates a typical protocol of cryoprobe application recorded during clinical cryoprostatectomy. Figure 1(a) depicts the relative spatial positions of five 3-mm cryoprobes in a prostate of approximately 33 grams. Figure 1(b) dictates the temperature profiles inside the gland (T_1) and at the periphery of the gland (T_2), in correspondence to the protocol of cryoprobe application in terms of the sequence and the temperature set-point of each cryoprobe as represented in Figure 1(c). All these actions of cryoprobe application were taken by the surgeon, based upon the ice-ball growth inside the prostate as monitored by transrectal real-time ultrasound imaging. Depending upon personal preference, the temperature profile at a preselected location may also be a determinant in designing the protocol of cryoprobe application.[10] For more sensitive cryosurgery, such as the safe removal of brain tumors or other tumors in the vicinity of vital organs, a conservative and cautious approach is often preferred. In this case, the refrigerative capacity needs to be delivered to the target in a controlled manner.

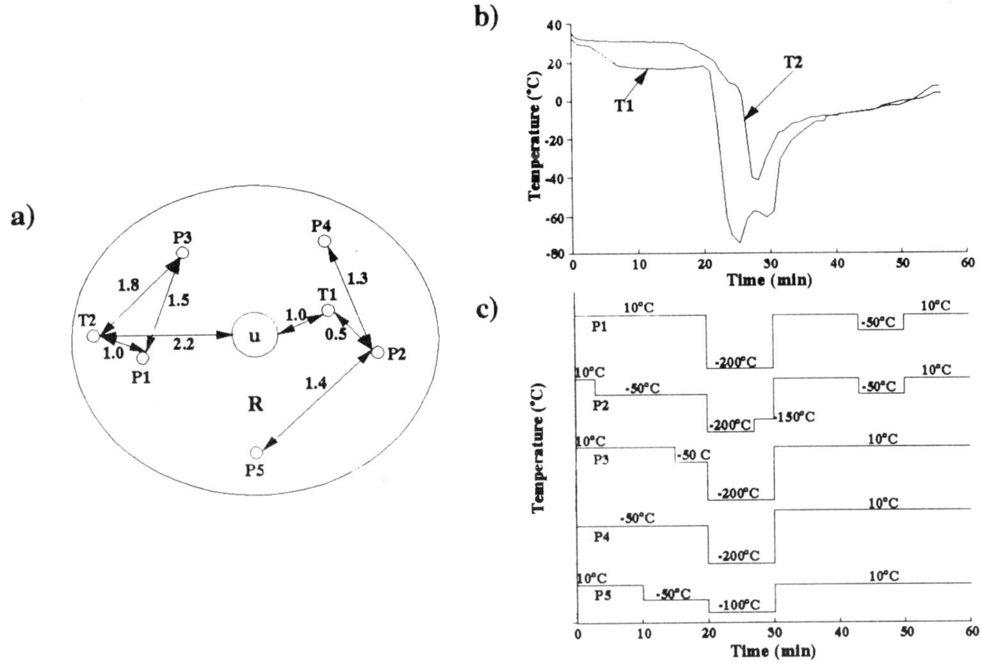

Figure 1. a) Schematic representation of cryoprobe and thermocouple positioning in a typical prostate cryosurgery. P_1...P_5, and T_1 and T_2 represent each individual 3-mm cryoprobe and thermocouples, respectively. U and R stand for the urethra and rectum. The numbers beside each line denote the distance in centimeters. b) Temperature profiles inside the gland (T_1) and at the periphery of the gland (T_2). c) Protocol of cryoprobe application. Notice that P4 was first applied at a "sticking" mode (-50°C), followed by P2, P5, and P3. P1 was sent directly to maximum freezing together with 4 other probes previously at "sticking" modes. P1-4 were set to maximum freezing for 10 minutes with the exception that P2 was set to -150°C momentarily prior to a "thaw mode." P5 was set to -100°C to avoid freezing the rectum. Both P1 and P2 were brought back to the "sticking mode" for a short period of time during the "thaw mode." The recordings of the procedure were conducted at Allegheny General Hospital, Pittsburgh, PA. The AccuProbe® System (Cryomedical Sciences, Inc.) was used for the procedure.

A controlled refrigerative capacity may be achieved either by adjusting the flow rate to the probe or by adjusting the voltage or current of an electrical heater installed internally in the probe tip. In the latter case, the control of the refrigerative capacity is immediate and accurate but requires a complex design of the probe body. The flow control can be obtained by directly presetting the percentage of liquid-nitrogen flow to the probe through either proportional (continuous flow) or on/off (intermittent or pulsed flow) control of an in-line cryogenic solenoid. For practical considerations, the on/off cryogenic solenoid should be chosen. The flow through the solenoid is controlled through a feedback signal from a thermocouple installed in the probe body. In addition to the set-point control of refrigerative capacity for each probe channel, the system should also provide an active means for facilitating the removal of the cryoprobe from the tumor after freezing. Traditionally, this typically implies an electrical heater at the probe tip. Most recently, dry nitrogen gas is used during the "thawing mode."[7, 13] In either case, the thawing function for each cryoprobe channel must be capable of being conducted independently.

Visceral cryosurgery requires a more intensive and elongated freezing than surface freezing. It is not uncommon in a prostate cryosurgery, for instance, that five probes are simultaneously set to maximum freezing for over 10 minutes especially for a relatively large gland. When multiple freeze/thaw cycles are conducted, as may be preferred by some cryosurgeons, an additional 10 minutes may be needed. In total, the operational time for the system could be as long as 20 minutes with simultaneous operation of all five probes (or 100 probe•minutes). Assuming that the liquid nitrogen flow rate through each probe is typically 0.6

liters/min., the total nitrogen consumption amounts to 60 liters per cryosurgical procedure. This is also a minimum requirement for the liquid nitrogen supply dewar of a universal cryosurgical system that is applicable to most cryosurgical procedures. For designated use in less aggressive cancer treatment, fewer probes may be needed, and the total operation time could be much shorter than 100 probe•minutes. Therefore, the liquid nitrogen supply dewar can be significantly smaller than 60 liters. In either case, an accurate level display of the remaining liquid nitrogen in the dewar is required.

From a functional point of view, the multiprobe cryosurgical system must be capable of delivering sufficient refrigerative capacity to the cryoprobes. With a large-diameter cryoprobe, such as a 9.5-mm trocar cryoprobe, the internal flow paths are large enough to allow effective venting of the gas generated in the conduit during the cool-down period or during cryosurgical operation when a thermal load is imposed on the cryoprobe tip. A satisfactory freezing capacity can be conveniently achieved because of high flow rate and large heat-transfer area at the tip. The heat-transfer efficiency can be further enhanced when the liquid nitrogen is subcooled. Figure 2 (a) illustrates the temperature profile of a 9.5-mm cryoprobe corresponding to different degrees of subcooling (no subcooling, subcooled to -196°C, -203°C and 208°C). The supply pressure of liquid nitrogen was maintained at 3.7 ATM (corresponding to a saturation point of approximately -183 °C). The test was conducted on a test apparatus as previously described[14] and in a clear medium (23 °C) containing 1.5% gelatin and 98.5% water to approximate biological tissue in terms of thermophysical properties.[15] As expected, when liquid nitrogen is subcooled, the temperature, as measured on the external surface of the probe tip, drops much faster than when no subcooling is applied. The ice-ball volume in relation to freezing time is represented in Figure 2(b). In general, the higher the degree of subcooling, the better the performance of the probe. The cryoprobe design is illustrated in Figure 2(b) (insert).

With respect to a slender cryoprobe, due to its high flow resistance, it is extremely difficult to deliver boiling cryogen to the cryoprobe tip and still maintain a liquid state, a condition for maximizing heat-transfer efficiency at the probe tip. Previous studies indicated that subcooling of liquid nitrogen together with gas venting through holes along the liquid-nitrogen delivery tube effectively enhance the liquid flow to the tip of a cryoprobe as small as 3 mm in diameter and as long as 200 mm when connected to a composite transport hose approximately 2 meters in length.[14, 16-17] In fact, experiments indicated that with sufficient subcooling, relatively high supply pressure and a vent hole of approximately 0.7 mm on the supply tube, the 3-mm cryoprobe can be made as long as 700 mm and still retain adequate refrigerative capacity.[18]

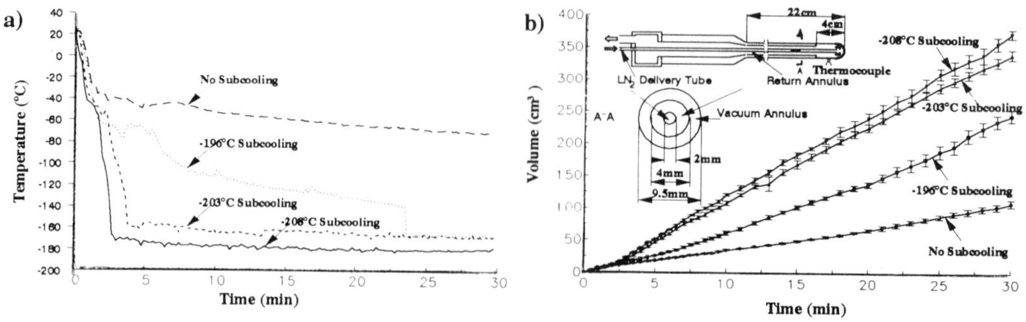

Figure 2. a) The temperature profile at the external surface of the probe tip in 23°C gel under different test conditions: no subcooling, subcooling to -196 °C, subcooling to approximately -203 °C, and subcooling to approximately -208 °C. b) The size of the ice ball in relation to the freezing time (n=3); the size in volume (S) is calculated based upon the measurement of ice-ball diameter (D) and length (L) $\{S = (0.52 D^2 - 0.71) L\}$ with an error less than 2%. Depicted in Figure (b) (insert) is a schematic representation of cryoprobe design with major dimensions; other than the probe tip made of pure copper, the probe body was constructed of 304 stainless steel.

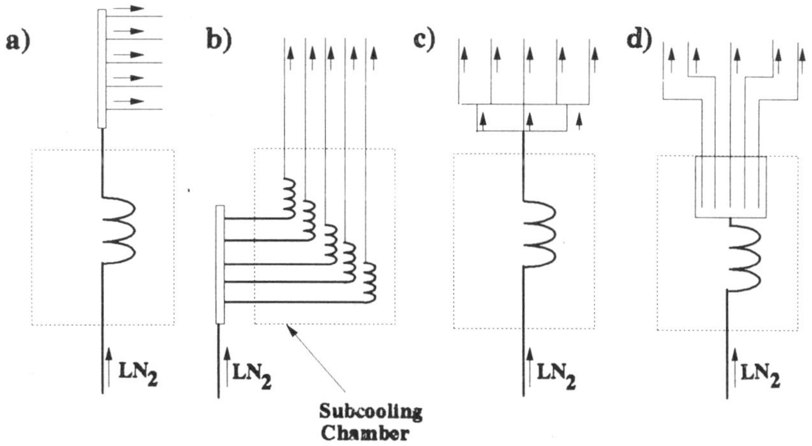

Figure 3. Typical patterns of LN_2 distribution manifold. a) Linear manifold located after subcooling; b) Linear manifold prior to subcooling; c) Tree manifold; and d) Cylindrical manifold.

To enhance its flexibility to be used for various cryosurgical procedures, the system should also be adaptable to various types of cryoprobes suitable for different cryosurgical procedures. This may be accomplished via detachable universal adapters between the system and the liquid-nitrogen transport hose or between the transport hose and the probe body, whichever is judged to be more technically feasible, economical, reliable, safer and convenient to use. Aside from the refrigerative-capacity requirement for each individual probe, all probe channels ought to deliver a uniform and an equivalent refrigerative capacity to the target in a predictable manner. The uniformity of performance among each channel depends, to a great extent, upon the design of the manifold to distribute liquid-nitrogen flow (as discussed below). Compared to conventional cryosurgical devices, special attention needs to be paid to the predictability and consistency of the probe performance. With improved liquid flow as a result of the combined effect of a relatively high supply pressure, subcooling and gas-venting mechanism, the cryoprobe performance may become predictable provided only one type of cryoprobe is used or only one single cryoprobe of any type is attached to the system. In fact, it has been observed that when multiple cryoprobes are attached to the system, the performance of each individual cryoprobe becomes increasingly degraded when more cryoprobes are simultaneously set to operation.[14] The performance could become even more uncertain when cryoprobes are set to different functional modes or different refrigerative capacities (such as different percentages of liquid flow or different temperature set- points). The problem may be alleviated, if not completely resolved, if the system is appropriately designed so that the liquid-nitrogen delivery capacity for each cryoprobe channel is always at maximum irrespective of the type, number and functional modes of the cryoprobe. Equally important is the cryoprobe design. Ideally, various design parameters should be optimized in a way such that the flow resistance along the flow path, and consequently the liquid-nitrogen flow rate, is equivalent for all types of cryoprobes.

DESIGN OF LN_2 FLOW DISTRIBUTION MANIFOLD

Unlike a single-cryoprobe device in which liquid nitrogen flows directly from the supply dewar to the cryoprobe, a multiprobe cryosurgical device requires a manifold to divert liquid-nitrogen flow. The structural design of the manifold, as well as its location, plays a primary role in uniformly distributing the liquid flow to each probe channel and thereby affects the uniformity of performance of the cryoprobes attached to different channels.

Figure 3 illustrates 4 typical manifold designs. In Design A, a linear type of manifold is located after the subcooling chamber. The liquid flow is introduced from one end of the manifold and then diverted to each probe channel. The linear type of manifold is characterized by its design simplicity but has the inherent problem of uneven distribution of liquid flow to each channel. Should all probe channels be opened simultaneously, more liquid nitrogen will be diverted to the channel closer to the inlet of the manifold. When only one probe channel is open, the gas generated in the pipeline upstream of the manifold or caused by direct heat input from its surroundings may "choke" the probe. Design B is similar to Design A in structure but is positioned prior to the subcooling chamber. One advantage of Design B is that any gas trapped in the manifold may be at least partially condensed through the heat exchanger in the subcooling chamber before reaching the probe channels. The disadvantages of Design B include uneven flow distribution and additional thermal mass of multiple individual heat exchangers in the subcooling chamber. An additional thermal mass in the subcooling chamber will impose an undesirable work load on the subcooling chamber and thereby potentially affect the size of the subcooling chamber and the selection of a vacuum pump (if a vacuum pump is used for subcooling). Design C is a tree type of manifold which may be positioned either prior to or after the subcooling chamber. Liquid nitrogen is introduced into the major branch of the manifold through two smaller branches at two different locations. More sub-branches in different arrangements may be introduced to further improve the uniformity of the flow distribution. Design C still suffers from a major problem, the trap of gases generated from the upstream pipelines and the bulky manifold itself. Dimensionally, the inner cross-sectional area of the linear manifold in either Design A or B or the major branch in Design C must be at least equivalent to the sum of the inner cross-sectional areas of the pipelines leading to the probe ports. Design D is a cylindrical type of manifold vertically positioned inside the subcooling chamber. Liquid nitrogen is introduced from the bottom of the manifold to each probe channel through vertically positioned tubes. The trapped gas, if not completely condensed in the subcooling chamber, will tend to accumulate in the ullage space of the manifold and flow to probe channel(s) after the flow conduit downstream of the manifold is cooled down by the initial flow of liquid nitrogen. Previous study has suggested that a transient pulsed flow of gas will not affect the temperature profile on the eternal surface of the probe tip due to the thermal inertia of the probe-tip materials.[14] The manifold may be completely submerged inside the liquid bath or suspended in the subcooling chamber. In contrast to Designs A, B and C, no additional thermal insulation is needed for Design D. Safety precautions, however, need to be exercised when designing and constructing a cylindrical type of manifold because of its dimensions and the pressure it sustains. Our preliminary results have indicated that Design D yields the best results in terms of uniformity of liquid-nitrogen delivery to each probe port. It is worthwhile to mention that all in-line cryogenic solenoids (or any other types of valves) should be located after the distribution manifold and the subcooling chamber. To install the valves prior to the manifold may cause accidental discharge of liquid nitrogen to the idle cryoprobes and thereby potentially result in unwanted tissue injuries.

DISPOSAL OF RETURN LIQUID NITROGEN AND CLINICAL IMPLICATIONS

In a conventional single-probe device, the nitrogen typically returns from the probe in gas state and may be disposed of in the operating room (OR) without any major safety concerns. In a highly effective multiprobe cryosurgical device, the liquid nitrogen is injected to each probe at a high flow rate (such as 0.6 liters/min.). A major portion of the liquid nitrogen will exit from each cryoprobe in a liquid state. To discharge return cryogen directly into the OR will create a disconcerting, unsightly, uncomfortable and potentially harmful clinical environment in the form of cold condensate and run-off of liquid nitrogen on to the floor. To alleviate the problem, the return nitrogen may be totally gassified through either a mechanical or electrical gas

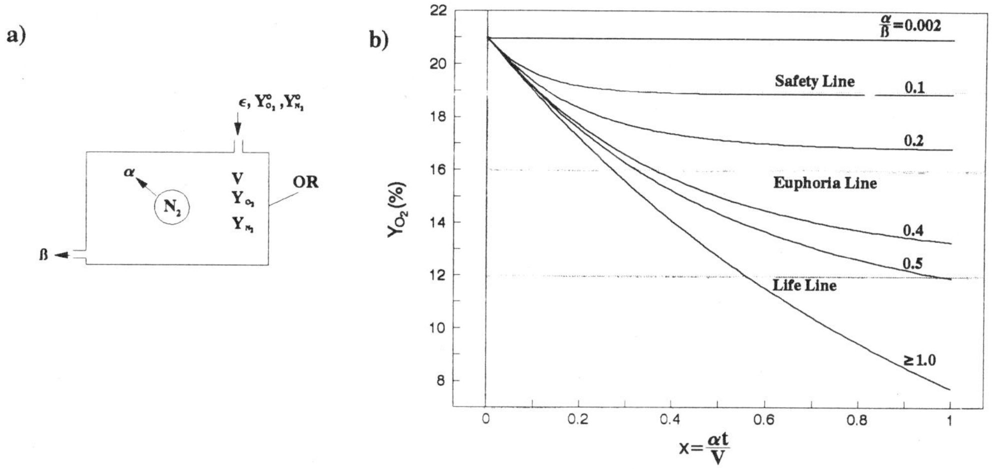

Figure 4. a): Schematic of physical modeling; Y_{N2}- volumetric concentration of nitrogen in air (%); Y_{O2}- volumetric concentration of oxygen in air (%); α - rate of LN_2 vaporization from the cryosurgical system (liters/min.); ϵ - rate of fresh air delivery to OR(liters/min.); β - rate of used air removal out of OR (liters/min.); V- Volume of OR space (liters); t - time (min.); $Y_{N2}{}^{o}$ - volumetric concentration of nitrogen in fresh air (71%). b): Oxygen level (% volume) in relation to vaporization rate (α, liters/min.) of a cryosurgical system, the volume of an OR (V, liters) and operation period (t, minutes). β denotes the rate of air removal out of OR (liters/min.).

vaporizer; still, the non-confined release of nitrogen gas in a vast amount from a multiprobe cryosurgical device may have the potential to create an oxygen-deficient environment in which exposed personnel could suffer extreme damage. Inhalation of high concentrations of nitrogen gas may cause anoxia, dizziness, nausea, vomiting, unconsciousness and even death. The following is a theoretical analysis of any clinical implications of nitrogen disposal with respect to its effect on oxygen level in an OR environment.

According to the U.S. government standards for hospital construction,[19] the minimum space requirements for a general operating room and cystoscopic room are, respectively, 33.5 square meters (360 square feet) and 23.2 square meters (250 square feet), at a height of 3.1 meters (10 feet). While most of the cryoprocedures are performed in a general OR, it is observed that some procedures of cryoprostatectomy may actually be performed in a cystoscopic room. To be conservative, we will include the cystoscopic room as an OR for cryosurgery in the following analysis. In either case, a positive pressure in relation to adjacent areas is balanced slightly above the atmosphere through the ventilation system. The air is required to be delivered to an OR at or near the ceiling and the return air to be removed near floor level. The minimum requirement for fresh air exchange rate for a recirculating air system is 5 times (out of a total of 25) per hour.

To simplify the analysis, the air is assumed to be composed of oxygen and nitrogen gases only. It is further assumed that the pressure and temperature in the OR remain constant throughout the cryosurgical process and that the air is thoroughly mixed throughout the space. Taking the entire OR space as the control volume and applying the principles of the conservation of species (Figure 4a), it follows,

$$Y_{N2} + Y_{O2} = 1 \qquad (1)$$

$$\epsilon = \beta - \alpha \qquad (2)$$

$$V dY_{N2} = \alpha\, dt + (\beta - \alpha)\, Y_{N2}{}^{o} dt - \beta Y_{N2}\, dt \qquad (\text{if } \beta > \alpha) \qquad (3)$$

$$V dY_{N2} = \alpha\, dt - \beta Y_{N2} dt - (\alpha - \beta) Y_{N2}\, dt \qquad (\text{if } \beta < \alpha) \qquad (4)$$

Eq. (4) suggests that when the vaporization rate (α) is higher than the removal capacity (β) of the ventilation system, the hospital ventilation system will be disturbed in order to maintain a constant pressure, i.e., no fresh air will be delivered to the affected OR, and the nitrogen-concentrated air will tend to leak to the adjacent area either through a doorway or any other type of channel.

Solving Eq. (1)-(4) with an initial condition of $Y_{O2}(t=0) = Y_{O2}^{o}$, yields,

$$Y_{O2}/Y_{O2}^{o} = (1 - \alpha\ \beta^{-1}) + \alpha\ \beta^{-1}\ \mathrm{EXP}(-V^{-1}\ \beta\ t) \qquad (\text{if } \beta > \alpha) \tag{5}$$

and

$$Y_{O2}/Y_{O2}^{o} = \mathrm{EXP}(-V^{-1}\ \alpha\ t) \qquad (\text{if } \beta < \alpha) \tag{6}$$

Assuming that the fresh air is composed of 21% oxygen and 79% nitrogen, Eq. (5) and Eq. (6) may be computed and are presented in Figure 4(b). Shown in the figure are also three critical lines, each representing a threshold for the specified physiological response to oxygen deficiency.[20] When the oxygen level is above the "safety line" (19% O_2), O_2 deficiency needs not be a particular concern. Below this line, however, special precautions should be taken. When the "euphoria line" (16% O_2) is crossed, an individual becomes incapable of correctly judging the situation because of the onset of such symptoms as loss of coordination, fatigue, lassitude and sleepiness. When the oxygen drops below the "life line" (12%), unconsciousness becomes immediate with virtually no warning.

When completely vaporized, one liter of liquid nitrogen will expand to 681 liters of gas at 15 °C.[21] At a typical liquid flow rate of 0.6 liters/min., the vaporization rate (α) will be approximately 2,043 liters/min. (72 CFM) with 5 cryoprobes simultaneously set to operation. In a cryosurgical OR with a total volume of approximately 70,825 liters (2,500 cubic feet), the used air is exchanged at a rate of 5,902 liters/min. (208 CFM) , equivalent to 5 fresh air changes per hour. For a five-probe cryosurgical system, it follows that as the operation continues the oxygen level will drop following the curve marked as 0.4 in Figure 5. It takes only 4 minutes for the oxygen level to cross the "safety line" and approximately 13 minutes to cross the "euphoria line." When operation proceeds beyond 37 minutes, as may occur in some cryoprostatectomy procedures, the "life line" is crossed and the oxygen level is below 12%. Should the ventilation system be ever accidently shut down in an open OR, it would only take 9 and 20 minutes to cross the "euphoria line" and the "life line," respectively. The results clearly indicate that a direct disposal of return nitrogen to the OR may create severe oxygen deficiency and potentially result in life-threatening consequences.

For safety as well as economy, it is absolutely necessary and desirable to effectively recover the return nitrogen from the probes so that only a minimal amount of nitrogen, the portion of nitrogen being normally gassified along the flow conduit, is vented into the OR. Experiments have indicated that with a full recovery of liquid nitrogen returning from the cryoprobes as previously described,[7] the gas is discharged into the atmosphere at a rate of approximately 150 liters/min. (5 CFM) when 5 cryoprobes, with a liquid flow rate of 0.6 liters/min. each, are simultaneously set to maximum freezing. As the operation continues, the oxygen level will slightly decrease, following the curve marked as 0.02 in Figure 5. It follows that the oxygen level will stay above the "safe line" regardless of the duration of the cryosurgical operation.

In a foreseeable situation, when no ventilation system is provided to the cryosurgical OR and the OR is airtight from its adjacent areas, the mean oxygen level in the OR with a presumed homogenous air composition should follow the following relationship:

$$Y_{O2}/Y_{O2}^{o} = (1 + V^{-1}\ \alpha\ t)^{-1} \tag{7}$$

With the above assumptions, in terms of OR dimensions and the liquid-nitrogen flow rate of each cryoprobe, a multiprobe cryosurgical system (with a return LN_2 recovery mechanism) could operate at maximum capacity over 250 probe•minutes before the oxygen level would drop below the "safety line." Similarly, it would take approximately 12.5 probe•hours for the oxygen level to cross the "euphoria line." The foregoing analysis has been based upon the assumption that the OR meets the minimum requirement for hospital construction.[19] The results generally suggest that the recovery of return liquid nitrogen will ensure a multiprobe cryosurgical system a great safety margin of operation in a clinical environment in the context of oxygen level. In no circumstance, however, should a multiprobe system be recommended for use in a structurally confined small area without ample air exchange.

ACKNOWLEDGEMENTS

The author wishes to thank Kathy Parrotte for her assistance in preparing the manuscript and Dr. HeWu Ma for preparing the graphs. The timely technical support of Bryan Trotman and Roy Cheeks is also greatly appreciated. Special thanks are also extended to Dr. Chaomin Zhang from Technische Universiteit Delft, the Netherlands, for his constructive input into the analysis of oxygen level in a clinical environment.

REFERENCES

1. I.S. Cooper. Cryogenic surgery: a new method of destruction or extirpation of benign or malignant tissues. *New Engl J Med.* 268:743(1963).
2. A.A. Gage. Cryotherapy for cancer, in: "Cryosurgery," R.W. Rand, A.P. Rinfret, H. von Leden, eds., Charles C. Thomas, Springfield, Ill (1968), p. 376.
3. A.A. Gage. Current progress in cryosurgery. *Cryobiology.* 25:483(1988).
4. G.M. Onik, C. Cobbs, J.K. Cohen, J. Zabkar, B. Porterfiel. US characteristics of frozen prostate. *Radiology.* 168:629(1988).
5. R. Matsumoto, K. Oshio, F.A. Jolesz. Monitoring of laser and freezing-induced ablation in the liver with T1-weighted MR imaging. *J Magn Reson Imaging.* 2:555(1992).
6. Z.H. Chang, JJ Finkelstein, J.G. Baust. Optimization of cryosurgical instrumentation for use in minimally invasive prostate surgery, in: "Recent advances in cryogenic engineering," J.P. Kelley, J. Goodman, eds. American Society of Mechanical Engineers, New York (1993), p. 45.
7. Z.H. Chang, JJ Finkelstein, H.W. Ma, J.G. Baust. Development of a high-performance multiprobe cryosurgical device. *Biomed Instrum Technol.* 28:383(1994).
8. G.M. Onik, J.K. Cohen, G.D. Reyes, B. Rubinsky, Z.H. Chang, and J.G. Baust. Percutaneous radical cryosurgical ablation of the prostate under transrectal ultrasound guidance. *Cancer.* 72:1291(1993).
9. J.C. Maroon, G. Onik, M.R. Quigley, J.E. Bailes, J.E. Wilberger, and J.S. Kennerdell. Cryosurgery revisited for the removal and destruction of brain, spinal and orbital tumors. *Neurol Res.* 14:294(1992).
10. F. Lee, D.K. Bahn, T.A. McHugh, G.M. Onik, F.T. Lee, Jr. How I do it: techniques of US-guided percutaneous cryoablation of prostate cancer. *Radiology.* 192:769(1994).
11. J.K. Cohen, G. Onik, R. Miller.Treatment of localized carcinoma of the prostate using percutaneous transperineal cryosurgical ablation: a preliminary report. *J Urol.* 149:232A(1993).
12. J.G. Baust and Z.H. Chang. Cryosurgery underlying mechanisms of damage and new concepts in cryosurgical instrumentation, in: "Cryosurgery" (in press).
13. A.V. Dilley, D.Y. Dy, A. Warlters, S. Copeland, D.L. Gillies. Laboratory and animal model evaluation of the CryoTech LCS 2000 in hepatic cryotherapy. *Cryobiology.* 30:74(1993).
14. Z.H. Chang. Various heat-transfer problems in designing a multiprobe LN_2 cryosurgical system, in: "Advances in heat and mass transfer in biological systems,".L.J. Hayes, R.B. Roemer, eds., American Society of Mechanical Engineers, New York (1994), p. 47.
15. C.P. Lentz. Thermal conductivity of meats, fats, gelatin gels and ice. *Food Technol.* 15:243(1961).
16. J.G. Baust, Z.H. Chang, JJ Finkelstein. Cryosurgical instrument with vent holes and method using same. U.S. Patent 5,254,116 (1993).
17. Z.H. Chang, L.Pottoroff, J.G. Baust. Cryogenic transport hose. U.S Patent 5,400,602 (1995).
18. Z.H. Chang, H.W. Ma, I.A. Gelman, and J.G. Baust. Engineering limitations in designing a slender, LN_2 cryoprobe. Proc of the Int Inst Refrig (in press).

19. U.S. Department of Health & Human Services. "Minimum Requirements of Construction and Equipment for Hospital and Medical Facilities," (1987).
20. Liquid Cylinder Users Manual - Safety Bulletin, Minnesota Valley Engineering Cryogenic Services, Inc.
21. B.A. Hand. "Cryogenic Engineering," Academic Press, Orlando (1986).

PHASE DIAGRAM PREDICTION FOR SOLUTIONS OF INTEREST IN CRYOBIOLOGY

C.V. Studholme and L.E. McGann

University of Alberta
Edmonton, Alberta, T6G 2R7, CANADA

ABSTRACT

Simulation of osmotic responses in cells and tissues at low temperatures requires descriptions of the composition of the extracellular solution and the osmotic responses of cells to the changing environment. The principles of irreversible thermodynamics have been used to describe osmotic water and solute flow across plasma membranes as the temperature and environment change. Insufficient experimental data is available for the empirical descriptions of phase diagrams now used to describe the composition of the extracellular solution. In order to describe the general phase behavior in binary mixtures of interest in cryobiology, a set of equations were derived from the Gibb's free energy equation, using changes in enthalpy and entropy upon unmixing and freezing. The parameters required are: the molecular weights, molecular volumes, van der Waals gas constant, b, applied to liquids, and a solubility parameter, δ. Where data was available, the excess enthalpy of mixing was used to determine δ, otherwise parameters were derived from least-squares fitting to experimental phase diagram data from the literature. Predictions fitted experimental phase diagram data very well (within 0.1K in some cases) from 0°C down to the eutectic point for solutions with various cryoprotectants. These equations, based on thermodynamics, provide an analytical tool for simulation and optimization of procedures used in cryopreservation, and can be generalized to mixtures with an arbitrary number of components.

INTRODUCTION

The freezing point of an arbitrary physiological solution is of importance to cryobiologists attempting to model cellular and tissue responses in non-ideal environments. The freezing point of a solution is depressed as a result of decreasing vapour pressure as solutes are added. A phase diagram represents the freezing point of solutions over a wide range of concentrations, and is usually determined by a set of experimental measurements of melting temperatures.

Advances in Cryogenic Engineering, Vol. 41
Edited by P. Kittel, Plenum Press, New York, 1996

The only general mathematical theory describing these phase diagrams is Raoult's Law for ideal dilute solutions which describes the vapour pressure of the solution as a linear function of solute concentration.[1] This results in a linear relation between freezing point and solute concentration which is only valid for very dilute solutions. Raoult's Law is inappropriate for solutions with solute concentrations of interest in cryobiology, since large deviations from linearity are observed at these concentrations.[2,3]

Others have used arbitrary mathematical expressions to describe the phase behaviour of specific aqueous solutions.[4] These equations are completely empirical in nature, having no physical basis for their derivation. Such equations normally describe the phase diagram curves for a few selected solutes, and cannot be generalized to arbitrary solutions containing an arbitrary number of solutes. This approach still requires experimental measurements of phase diagram information over the entire range of concentration to calculate the empirical parameters required in the equations. As the number of solutes present in a solution increases, experimental measurements of phase diagram information becomes impractical, and a general mathematical theory for predicting this phase behaviour becomes important.

We developed an approach to predicting the phase behaviour of solutions based entirely on thermodynamics. Thermodynamics is, by definition, the study of equilibrium states, and the phase change during freezing is a reversible, hence equilibrium, process. As a result of this equilibrium, one important thermodynamic state variable remains constant -- the Gibb's free energy. Hence, the total derivative of the Gibb's free energy is identically zero. This is the basis for development of a thermodynamic phase diagram theory.

THEORY

Since the process of freezing is an equilibrium process that occurs at a specific temperature depending on solute concentration, the Gibb's free energy of the system during this state change must remain constant. Free energy is given by

$$F = E - TS \,, \tag{1}$$

where F is the free energy, E is the energy (or enthalpy) of the system, T is the temperature, and S is the entropy.[1] Since F is constant, the differential of F should be identically zero. By formally differentiating Eq. (1) we get

$$dF = dE - TdS - SdT \,, \tag{2}$$

which can be simplified slightly by noting that freezing is an isothermal process, and thus $dT=0$. Furthermore, imposing our equilibrium condition, $dF=0$, we can write

$$dE - TdS = 0$$

or,

$$T = \frac{dE}{dS} \,. \tag{3}$$

Eq. (3) is an expression of the freezing temperature of a solution. We now need to obtain expressions for dE and dS.

Since E is the energy of the system, dE is the change in energy that occurs during freezing. Water releases energy, the latent heat of fusion, during the ice formation process and this energy will be one component of dE. The other component of dE is a result of the unmixing necessary for a phase change to occur. Here we assume that the phase change is the result of one component freezing out of the solution in a pure form. Hence, the unmixing occurs as this component separates from the solution and solidifies. During any mixing process it is generally the case that some energy is absorbed or released, usually called the excess energy of mixing or the heat of mixing. The unmixing process during freezing must therefore be accompanied by an absorption or release of energy, constituting the remaining component of the quantity dE.

We can present an analogous argument for dS. Freezing, being an ordering process, results in a decrease in entropy which constitutes one of the two components of dS. The other component comes from the unmixing that must occur since, as above, the mixing of distinct solutes generally results in a change in entropy. Specifically, mixing results in an increase in entropy since the mixture will be in a less ordered state than the separate components. Therefore, the unmixing process must result in a decrease in entropy and this decrease represents the second component of dS.

Consider a binary mixture consisting of n_1 moles of solvent and n_2 moles of solute. Since the solvent freezes in a pure form, to freeze dn_1 moles we must first remove dn_1 moles of the solvent from the solution and then solidify it. Removing this infinitesimal quantity from solution will cause changes in both the excess energy and the excess entropy of the solution, which we can label $-dE_M$ and $-dS_M$, respectively. Here the subscript M indicates mixture and the minus sign is included because we are removing substance from the mixture. Once the substance has been removed from solution, it can then undergo the necessary change of state. Let dE_F and dS_F be the changes in energy and entropy, respectively, that result from the process of fusion. We can now write the total changes in energy and entropy as follows:

$$dE = dE_F - dE_M, \tag{4a}$$

$$dS = dS_F - dS_M, \tag{4b}$$

which can be substituted into Eq. (3) to give,

$$T = \frac{dE_F - dE_M}{dS_F - dS_M} \tag{5}$$

Since the energy and entropy released during freezing is directly proportional to the amount of substance being solidified,

$$dE_F = \overline{E}_F \cdot dn_1, \tag{6a}$$

$$dS_F = \overline{S}_F \cdot dn_1, \tag{6b}$$

where $\overline{E}_F$ and $\overline{S}_F$ are the specific heat and specific entropy of fusion for the substance being frozen; both are constant. This only leaves dE_M and dS_M as unknown quantities that must be examined in greater detail.

In 1947, J. H. Hildebrand expressed the entropy of mixing for a binary solution whose molecules differ in size as

$$\frac{\Delta S^M}{R} = n_1 \ln \frac{V - n_1 b_1 - n_2 b_2}{n_1(v_1 - b_1)} + n_2 \ln \frac{V - n_1 b_1 - n_2 b_2}{n_2(v_2 - b_2)} \tag{7}$$

where ΔS^M indicates the excess entropy, what we label S_M, V is the total volume of the mixture, v_1 and v_2 are the molecular volumes of each component, and b_1 and b_2 are geometrical volumes or intrinsic van der Waals volumes of each component.[5]

The partial of S_M with respect to n_1, holding n_2 constant, can be expressed from Eq. (7)

as

$$\left(\frac{\partial S_M}{\partial n_1}\right)_{n_2} = R\left[\ln\frac{n_1(v_1-b_1)+n_2(v_2-b_2)}{n_1(v_1-b_1)} + \frac{n_2(v_1-b_1)-n_2(v_2-b_2)}{n_1(v_1-b_1)+n_2(v_2-b_2)}\right], \tag{8}$$

where we have assumed that the volume of the solution is additive, that is

$$V = n_1 v_1 + n_2 v_2 , \tag{9}$$

and thus

$$V - n_1 b_1 - n_2 b_2 = n_1(v_1 - b_1) + n_2(v_2 - b_2). \tag{10}$$

For a similar expression including E_M, we look to a book by Hildebrand, Prausnitz and Scott titled *Regular and Related Solutions* where an expression known as the Scatchard-Hildebrand equation is given.[6] This equation results from work done by Scatchard[7] in 1931 and describes the energy of mixing of two liquids as

$$\Delta E^M = (n_1 v_1 + n_2 v_2)\phi_1\phi_2(\delta_1{}^2 + \delta_2{}^2 - 2\delta_{12}), \tag{11}$$

where ΔE^M represents the excess energy of mixing, what we label E_M, ϕ_i represents the volume fraction of each component, and δ_i is defined to be a *solubility parameter*. Here δ_1 and δ_2 describe the interactions between like components in the solution while δ_{12} describes the interactions between unlike components. Practically, however, we cannot find all these parameters independently, therefore we define a composite solubility parameter

$$\delta = \delta_1^2 + \delta_2^2 - 2\delta_{12} \tag{12}$$

and write

$$E_M = \frac{n_1 v_1 n_2 v_2}{n_1 v_1 + n_2 v_2}\delta , \tag{13}$$

where we have also made substitutions for the volume fractions, ϕ. It should be noted here that $\delta > 0$ implies an endothermic mixing process, while $\delta < 0$ implies exothermic mixing. Again, we need the partial of this energy of mixing with respect to n_1, holding n_2 constant,

$$\left(\frac{\partial E_M}{\partial n_1}\right)_{n_2} = \frac{v_1 n_2{}^2 v_2{}^2}{(n_1 v_1 + n_2 v_2)^2}\delta . \tag{14}$$

Now, knowing Eq. (8) and Eq. (14) we can write

$$dE_M = \left(\frac{\partial E_M}{\partial n_1}\right)_{n_2} dn_1 , \tag{15a}$$

$$dS_M = \left(\frac{\partial S_M}{\partial n_1}\right)_{n_2} dn_1 , \tag{15b}$$

which, along with Eq. (6a) and Eq. (6b), can be substituted into Eq. (5) to get

$$T = \frac{\overline{E}_F \cdot dn_1 - \left(\frac{\partial E_M}{\partial n_1}\right)_{n_2} dn_1}{\overline{S}_F \cdot dn_1 - \left(\frac{\partial S_M}{\partial n_1}\right)_{n_2} dn_1} . \tag{16}$$

Cancellation of the dn_1's is a requirement since the freezing point of a solution is not expected to depend on the amount of solution being frozen. Thus, simplifying Eq. (16) gives

$$T = \frac{\overline{E}_F - \left(\frac{\partial E_M}{\partial n_1} \right)_{n_2}}{\overline{S}_F - \left(\frac{\partial S_M}{\partial n_1} \right)_{n_2}}. \tag{17}$$

By substituting Eq. (8) and Eq. (14) into Eq. (17), we get an equation for the freezing point of an arbitrary binary mixture as

$$T = \frac{\overline{E}_F - \frac{v_1 n_2^2 v_2^2}{(n_1 v_1 + n_2 v_2)^2}\delta}{\overline{S}_F - R\left[\ln \frac{n_1(v_1 - b_1) + n_2(v_2 - b_2)}{n_1(v_1 - b_1)} + \frac{n_2(v_1 - b_1) - n_2(v_2 - b_2)}{n_1(v_1 - b_1) + n_2(v_2 - b_2)} \right]}, \tag{18}$$

which depends on n_1 and n_2 along with the constants $\overline{E}_F$, $\overline{S}_F$, v_1, v_2, b_1, b_2 and δ. To show that this expression does not depend on the total amount of solution, $n_1 + n_2$, we make a transformation from n_1, n_2 to x_1, x_2, the mole fractions. This transformation is

$$x_1 = \frac{n_1}{n_1 + n_2}, \quad x_2 = \frac{n_2}{n_1 + n_2}, \tag{19}$$

and results in

$$T = \frac{\overline{E}_F - \frac{v_1 x_2^2 v_2^2}{(x_1 v_1 + x_2 v_2)^2}\delta}{\overline{S}_F - R\left[\ln \frac{x_1(v_1 - b_1) + x_2(v_2 - b_2)}{x_1(v_1 - b_1)} + \frac{x_2(v_1 - b_1) - x_2(v_2 - b_2)}{x_1(v_1 - b_1) + x_2(v_2 - b_2)} \right]}. \tag{20}$$

In this expression v_1 and v_2 are the molecular volumes of each component, which are usually well known. However, although δ_i may be found in tables for some compounds, the interaction term δ_{12} is generally not known, and therefore we will use the technique of least squares fitting to find δ. Finally, b_1 and b_2 are the intrinsic van der Waals volumes for each of the components and although this constant is known for several real gasses, it is difficult to obtain for most liquids. Therefore it will have to be found by other means, a few of which are proposed by Hildebrand.[1]

METHODS AND RESULTS

Values for the constants b_1, b_2 and δ are required to predict the phase behaviour of an arbitrary two component solution. In the absence of experimental values for these constants obtained from other sources, an effective method of determining these values is by fitting Eq. (20) to known phase information. With three parameters to fit, at least 4 data points are required, which can be obtained either experimentally or from previously published tables. A problem with this method pertains to the calculation of b_1. It is found that b_1 and b_2 are not entirely independent, and therefore, cannot be simultaneously fitted for. Furthermore, since our solutions of interest all include water as a primary component, there should be only one value for b_1 common to all the mixtures. Therefore, we must find b_1 independent of b_2 and then use curve fitting techniques to find b_2 and δ.

Determining b_1

Hildebrand proposes a few techniques to find this *geometrical volume*,[6] but due to limitations on the amount of experimental data to be found in the literature, only one technique was applicable. From a set of steam tables,[8] the value of b for water was determined by a linear extrapolation of the molar volume of water at constant pressure as a function of temperature to absolute zero temperature. The pressure chosen was 2000 kPa where the molar volume data in the range of 50°C to 210°C was nearly linear and thus an extrapolation was possible. By this method, the value of b_1 was determined to be 11.44 ml/mol.

Determining δ

Since Eq. (13) represents the excess enthalpy of mixing and does not contain the parameter b, information from the literature on the excess enthalpy of mixing can be used to calculate δ. Data on the excess enthalpy of mixing for four of the binary mixtures considered here (ethylene glycol, methanol, propylene glycol, and dimethyl sulfoxide, each as a binary mixture with water)[9] was fitted with Eq. (13) to determine values for δ. Figure 1 shows the data and fitted curves.

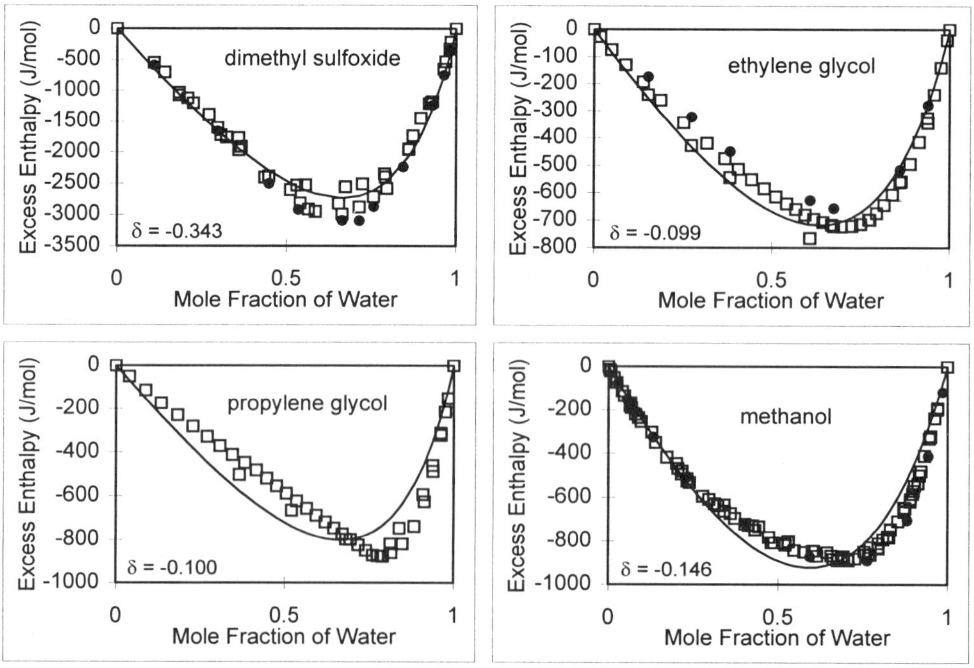

Figure 1. Excess enthalpy is shown as a function of water concentration. The solid line is the fitted curve using Eq. (13) with the indicated value of δ. All data is for 25°C except the closed circles which are for 35°C and were not used in the fitting, but reported here for comparison.

Determining b_2

With these values for δ, Eq. (20) was fitted to phase diagram data[2,3] for the single parameter b_2. For the cases where no excess enthalpy data was available, Eq. (20) was fitted for both b_2 and δ. In either case, Figure 2 shows this phase diagram data along with the fitted curves, while Table 1 summarizes the fitted parameters which are used to predict the phase behavior of these aqueous solutions.

CONCLUSIONS

This phase diagram theory, based on a phase change in one component of a binary solution and the unmixing necessary for the freezing of that component, predicts the freezing points of aqueous solutions of various solutes over a wide range of concentrations. This implies that these two factors are important in the thermodynamics of freezing and thawing.

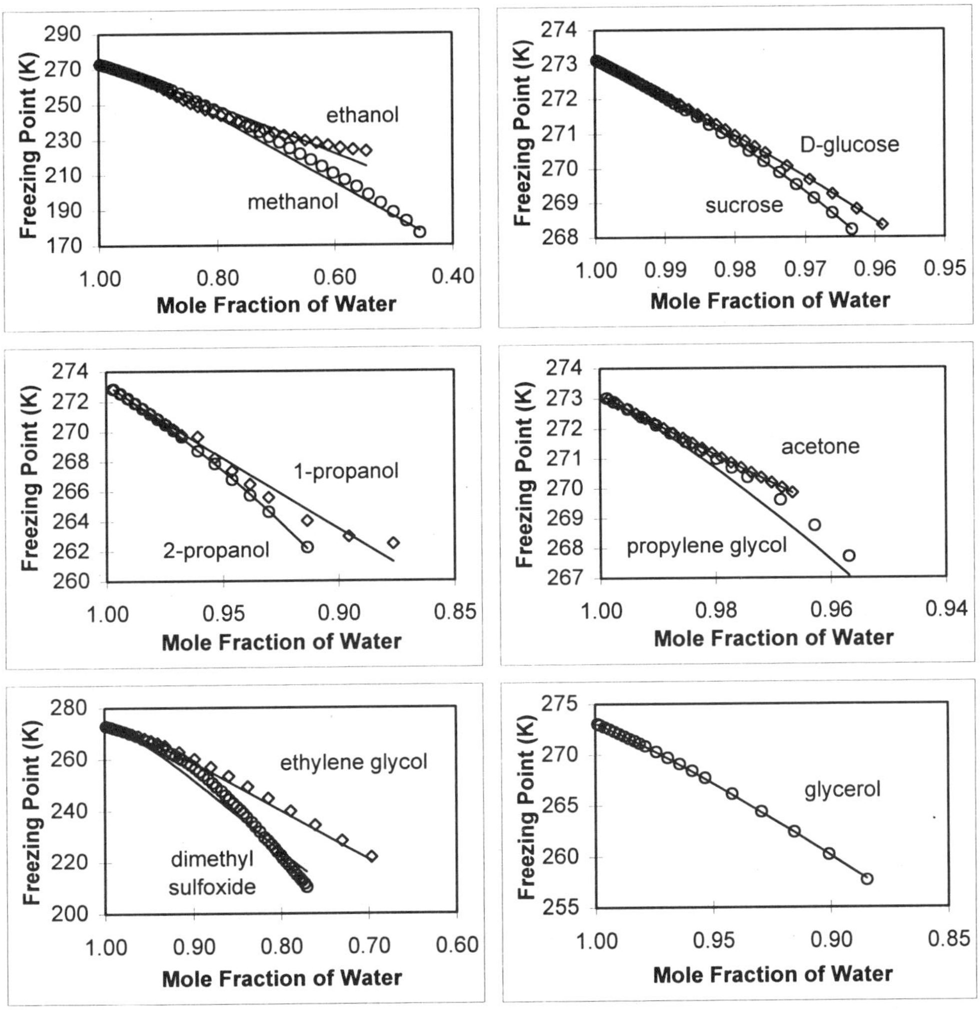

Figure 2. Phase diagrams from the literature were fitted to Eq. (20). The ethylene glycol, methanol, propylene glycol, and dimethyl sulfoxide curves were fitted for the single parameter b_2, while all the other curves were fitted for both parameters δ and b_2.

Table 1. Summary of parameters required in Eq. (20)

Solute	molecular weight g/mol	v ml/mol	b ml/mol	δ kJ/ml*
water	18.02	18.02	11.44	
ethylene glycol	62.07	54.63	48.05	-0.099
propylene glycol	76.09	71.66	67.18	-0.100
methanol	32.03	38.05	31.48	-0.146
ethanol	46.07	54.84	48.26	-0.021
1-propanol	60.09	70.46	63.89	0.011
2-propanol	60.09	70.23	77.92	0.005
glycerol	92.09	71.70	48.23	-0.004
D-glucose	180.16	113.99	123.84	0.002
surcose	342.30	214.12	232.74	0.001
acetone	58.05	67.59	60.95	0.015
dimethyl sulfoxide	78.13	70.94	52.49	-0.343

* negative values indicate exothermic mixing

Application of the theory requires three parameters describing the solution -- a solubility parameter, δ, and the intrinsic van der Walls volume of each solute, b_1 and b_2.

Current approaches to the cryopreservation of organs for transplantation use high concentrations of various solutes in aqueous solutions at low temperatures. This theory, although developed for binary solutions, can be generalized to mixtures with an arbitrary number of components, and used in strategies to model tissue responses and to optimize conditions for cryopreservation.

ACKNOWLEDGMENT

Supported by the Medical Research Council and the Natural Science and Engineering Research Council.

REFERENCES

1. P. W. Atkins, "Physical Chemistry Fourth Edition," W. H. Freeman and Company, New York (1990).
2. R. C. Weast, "CRC Handbook of Chemistry and Physics 63rd Edition," CRC Press, Inc., Boca Raton (1983).
3. D. H. Rasmussen, and A. P. MacKenzie, Phase Diagram for the System Water-Dimethylsulphoxide, *Nature.* 220:1315 (1968).
4. D. E. Pegg, Equations For Obtaining Melting Points And Eutectic Temperatures For The Ternary System Dimethyl Sulphoxide/Sodium Chloride/Water, *Cryo-Letters.* 7:387 (1986).
5. J. H. Hildebrand, The Entropy of Solution of Molecules of Different Size, *J. Chem. Phys.* 15:8:225 (1947).
6. J. H. Hildebrand, J. M. Prausnitz, and R. L. Scott, "Regular and Related Solutions," Van Nostrand Reinhold Company, New York (1970).
7. G. Scatchard, Equilibria in Non-Electrolyte Solutions in Relation to the Vapor Pressures and Densities of the Components, *Chem. Rev.* 8:321 (1931).
8. L. Haar, J. S. Gallagher, and G. S. Kell, "NBS/NRC Steam Tables," Hemisphere Publishing Corporation, Washington (1984).
9. C. Christensen, J. Gmehling, P. Rasmussen, U. Weidlich, "Heats of Mixing Data Collection," Deutsche Gesellschaft für Chemisches Apparatewesen, Frankfurt (1984).

STANDARDIZED IN VITRO MEASUREMENT OF HEAT FLUX AND DIAMETER OF FROZEN REGION FOR CRYOSURGICAL PROBES

R. Speetzen, M. Proske, I. Heschel, and G. Rau

Helmholtz-Institute for Biomedical Engineering
at the Aachen University of Technology
52074 Aachen, Germany

ABSTRACT

Cryosurgery has become an economic and easy to use technique for the surgical therapy of tumours in many fields of medicine. The volume of tissue that can be frozen by locally removing heat from the body with a thin cryosurgical probe is limited. To exactly predict the extent of the frozen region and for an efficient control of the therapy it is necessary to know the power and capacities of cryoprobes.

We present devices for the measurement of the transient ice ball diameter and the heat flux that is removed during the in vitro operation of a certain cryoprobe under various controlled thermal conditions. A segmented metal volume of nearly spherical inner shape is filled with a mixture of water and gelatine to avoid convection within the volume. The device is placed into a temperature controlled water bath to keep the temperature of the outer metallic surface constant. The transient temperature field of the inner water volume is recorded with an array of thermocouples located radially at various distances from the probes tip. The position of the advancing ice front is recorded continuously by a one-dimensional ultrasound signal that is evaluated by a computer. The heat flux that can be removed from the tissue during the steady state operation of a cryoprobe is measured quantitatively with a compensation heating device in an insulated stirred volume of water of computer-controlled temperature. Standardized conditions enable us to compare the power and therefore the cooling effects of cryosurgical probes. We suggest to use similar devices to quantify future improvements in cryosurgical technique more precisely.

INTRODUCTION

Cryosurgery is a powerful technique for the therapy of several kinds of diseases by the local application of subzero temperatures to living tissue in situ. Many investigations on the cryosurgical technique and the cryobiological effects have been published, but the current knowledge of the basic mechanisms of tissue and cell damage during cryosurgical operations is still limited.[1,2,3] Even though many investigations are concerned with specific technology[4] and medical applications[5], cryosurgical technique still has to be improved in several ways[1,2,3,4,5].

In addition to the expert knowledge and experiences of surgeons with different cryoprobes, objective data are required about their technical characteristics. To achieve comparable technical data, it is necessary to find a way of measurement under well-defined (thermal) conditions. For this purpose we have built two devices for the:

- standardized measurement of the frozen region, and
- standardized measurement of the heat flux.

The first system allows for a quantitative measurement of the transient advancing ice region around a cryoprobe in a controlled environment that mimics in vivo operation. This implies transient thermal conditions at the tip of the cryoprobe. With the second system we then establish almost constant thermal boundary conditions at the tip, which allows for the quantitative in vitro measurement of the heat flux removable by the cryoprobe under controlled conditions. The proposed conditions to be established for the two devices should be standardized in order to get comparable results for different types of cryoprobes.

STANDARDIZED MEASUREMENT OF THE FROZEN REGION AND TEMPERATURE FIELD

Growth and extent (and shape) of the advancing ice front around the probe's tip are of major importance for the effect of cryosurgical therapy[1,2]. Thus they are relevant to quantitatively determine surgically important performance characteristics of cryosurgical probes.

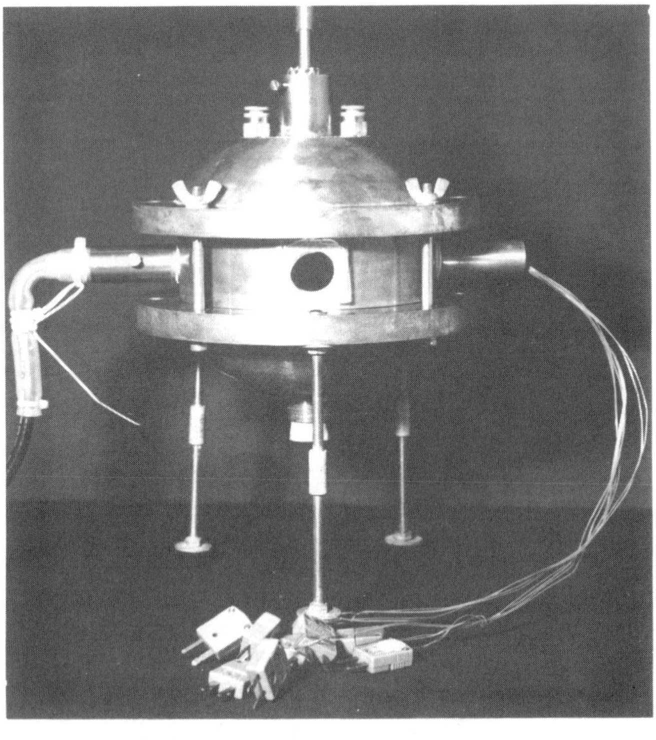

Figure 1. Sphere with inner Ø=100 mm for the measurements of the ice front position and temperature fields during the operation of a cryosurgical probe. Three segments form the aluminium sphere. The ultrasound probe is attached on the left hand side, the frame with thermocouple wires within the sphere on the right hand side, a glass window for optical observation on the front, and the cryoprobe itself is run through the top of the upper segment. The completely mounted sphere is filled with the test medium. The complete unit is submerged into a temperature controlled water bath to generate a homogeneous temperature of the metallic sphere.

The transient ice front position around the cryoprobe's tip and the corresponding temperature field are measured in a spherical volume shown in figure 1.

Considering ice regions of basically spherical shape a sphere is the most reasonable form to apply thermal boundary conditions (temperature of the inner surface) without influencing the ice growth. The cryoprobe is attached from the top and its tip is fixed in the centre of the spherical volume. The sphere is filled with a special medium (water and 3% w/w starch) to avoid influences due to convection. The temperature of the sphere's inner surface is kept constant at 37 °C by a temperature controlled water bath into which the completely mounted device is submerged. A nearly uniform temperature of 37±0,5°C in the test volume is achieved.

Cu-Const-thermocouples (TC) are used to measure the temperature field with a resolution of 0.2 °C. They are placed radially at various distances from the probe's tip with a spatial resolution of 2 mm. The frame with the thermocouple wires can slightly be moved in order to exactly position the wires relative to the cryoprobe. The ice front position (or cryoprobe surface, or TC wire) is determined with a resolution of 0.1 mm by a one-dimensional (1D) ultrasound (US) signal focused on the cryoprobe's tip. The time period between sending of the US-signal and receiving of the corresponding echo from the ice/liquid-interface is registered by a computer equipped with additional hardware. With the known velocity of sound of the test medium the distance from the ice front to the ultrasound receiver can be calculated. Ultrasound measurements can be performed from different directions relative to the cryoprobe which is useful for non-symmetric probes and ice shapes. In addition, optical observation of the growing ice front is possible through a transparent window. Shape and extent of the ice ball can be evaluated optically by photographs or videotapes (see figure 3).

An example of a test measurement can be examined in figure 2.

We tested this system with cryoprobes of relatively small power to prove its sensitivity. Test measurements were performed with N_2O-driven (Joule-Thomson-effect), flexible cryoprobes of 2.2 mm and 2.4 mm diameter respectively and 900 mm overall length (ERBOKRYO CA, ERBE GmbH, Tübingen, Germany). These particular probes are not equipped with an

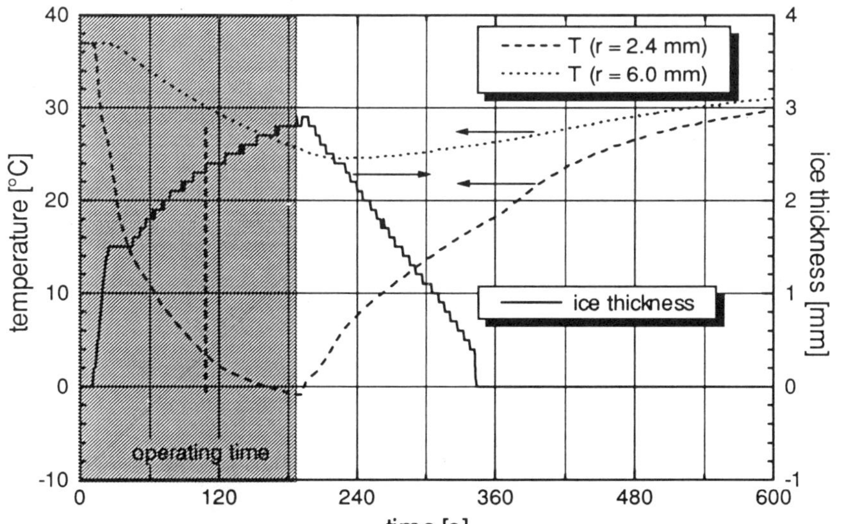

Figure 2. Ice thickness and temperatures around the cryoprobe's tip (N_2O, 2.4 mm, flexible, gas pressure 42 bar absolute). Test medium was water with 3% w/w starch at an initial temperature of 37.2 °C. The probe was operated for 185.5 s only. The dashed vertical line marks the moment when the ice front had reached the position of the nearest thermocouple (2.4 mm ice thickness). The thermocouples' positions are measured from the cryoprobe's surface. The short step after about 25 s is due to an irregular change in the probes' performance which for these particular probes was observed very often for a few seconds at the beginning of the freezing process. For the negative values of ice thickness see text.

internal heating device. Figure 2 shows an example of the advancing ice front during operation of the 2.4 mm-cryoprobe in the test medium.

After 3 min operation time, an ice thickness of 2.9 mm is achieved which corresponds to an ice ball diameter of 8.2 mm. Under the chosen conditions a maximum diameter of 10 mm can be reached without a significant pressure drop within the gas supply system. These in vitro results compare to in vivo experiences for different cryoprobes by Homasson.[6,7]

With the corresponding temperatures at two distances from the centre point further details of the system can be observed. During operation, the inner thermocouple becomes enclosed by the frozen volume. At $t \approx 110$ s its position should match the ice front position determined by US (marked in figure 2 with a dashed vertical line) and at this time it should show equilibrium freezing temperature. For the chosen medium, the equilibrium freezing temperature does not differ significantly from the value 0.0 °C for pure water (verified with a cryoscopic osmometer, OSMOMAT 030, gonotec GmbH, Berlin, Germany), whereas the thermocouple shows a temperature of 3 °C. This error is due to heat conduction along the

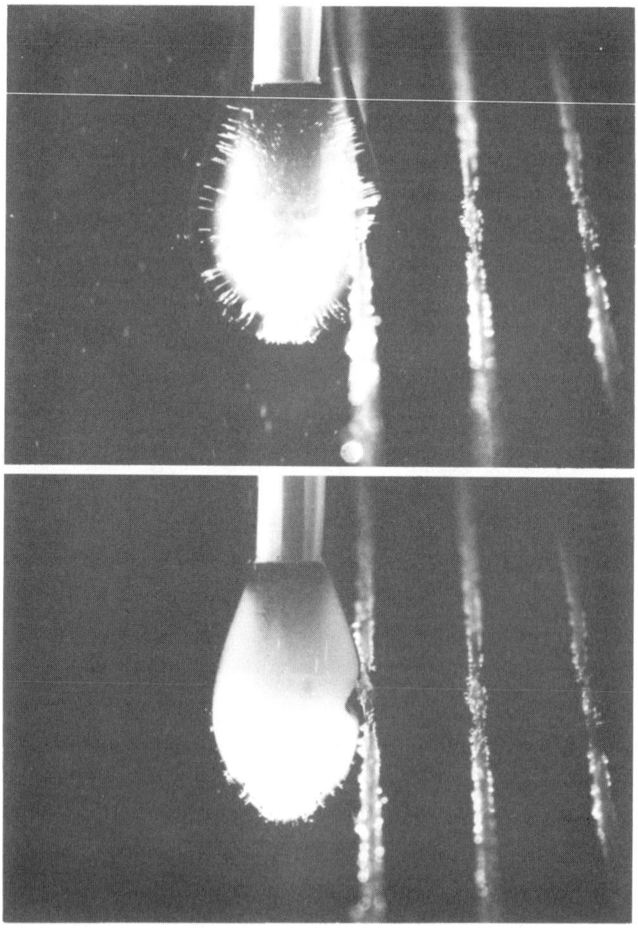

Figure 3. Photographs of the ice "ball" around the cryoprobe's tip (N_2O, probe-Ø = 2.4 mm) taken during freezing after about 3 min (top) and during thawing (bottom) with illumination from below. Test medium in this case was pure water to guarantee transparency. Consequently, convection processes lead to a slightly reduced extension of the ice compared to the starch mixture. The influence of heat conduction along the thermocouple wire can be seen clearly at the contact zone of the inner wire with the melting ice ball. During freezing the effect is much less pronounced due to the relatively great heat flux removed by the cryoprobe.

thermocouple wire (Ø 0.1 mm) itself, it can be diminished with an additional insulating plastic tube around the copper wire. Thinner wires are disadvantageous, because they might be displaced or even ruptured by mechanical interaction with the ice front. Since this largest error occurs at the outermost of the enclosed thermocouples only, temperature measurements are possible with better accuracy within the larger ice volume around more powerful (e.g. liquid nitrogen-driven) probes.

The second thermocouple shows a slightly delayed reaction - relative to the first - on the influence of the cryoprobe during freezing ($t \approx 25$ s) and thawing ($t \approx 220$ s) even though it is not enclosed by the ice volume itself. This delay is due to the heat conduction processes within the medium which are slow compared to convection processes. The main influence of convection should have led to much quicker and much more pronounced reactions of the second thermocouple, which has been sufficiently diminished by the chosen test medium.

Attention has to be paid to the calculation of the ice front position by means of the test medium's velocity of sound. This velocity decreases from 1590 m/s for 40 °C to 1470 m/s for 0 °C (own measurements) and therefore has to be compensated according to the measured temperature distribution in the liquid region. Without this compensation we observed errors for the ice thickness of up to 0.4 mm which decreases with increasing temperature of the medium.

The slow and uniform temperature increase for both thermocouples during thawing is a further evidence that convection does not play a significant role and heat transfer is mainly caused by conduction processes.

Photographs of the ice volume during freezing and thawing around a 2.4 mm cryoprobe in pure water are shown in figure 3. These photographs allow for a clear view on the extent and shape (symmetry) of the ice formed around the cryoprobe even though convection processes have to be considered. In this case, the assumption of a spherical shape can not exactly be confirmed. For this particular probe it can furthermore be shown that a considerable amount of heat is removed not only through the copper tip but also through the connected plastic tube. Videotapes of the moving ice front could be used to optically measure and verify its velocity and extent. These information can be of significance in medical practice and will be even more important for more powerful cryoprobes.

The error due to heat conduction along the thermocouple wire - corresponding to the temperature measurements in figure 2 - can be seen on the bottom of figure 3: melting of the ice is accelerated in the vicinity of the wire. This influence can not be observed during freezing because of the relatively large heat flux removed by the cryoprobe. This influence of the thermocouple wires has to be considered during temperature measurements to control cryosurgical therapy.

STANDARDIZED MEASUREMENT OF THE HEAT FLUX

To quantitatively measure the heat flux that can be removed in vitro by a cryosurgical probe under controlled thermal conditions, it is necessary to establish constant conditions at the heat exchanging tip of the cryoprobe for a sufficiently long period of measuring time. This can not be achieved with the device described above because it was designed for an advancing ice ball with changing thermal conditions around the probe's tip. An overall heat balance would allow for the calculation of the extracted heat flux but the immense heat capacity of the complete system would result in a large time lag until stationary conditions are achieved and corresponding calculation inaccuracies. Thus, for the heat flux measurement we use a specially designed second device consisting mainly of a water bath with a compensation heating device as it is shown in figure 4.

We use pure water (150 ml) as heat transferring medium. With this liquid we can easily test various cryoprobes of any width and any shape under equal thermal boundary conditions, even if these in vitro conditions do not perfectly match in vivo conditions. The water is enclosed by a thermally insulated Dewar vessel and it is stirred mechanically to get a uniform temperature throughout the sample volume. To reduce this stirred volume and to insulate it from the environment, parts of the empty vessel are filled with additional insulation material. The cryoprobe is fixed in the water volume by surrounding plastic tubes. The water is cooled by operating the cryoprobe. The water temperature is measured by a thermocouple and registered by a computer. With this data, the power of the electrical heating device is controlled with a resolution of 0.25 W in order to keep the water's temperature at a constant, preselected value. For our first measurements a bath temperature of 37 °C was chosen to achieve good correspondence to conditions in vivo. After equilibrium conditions had been attained (after total operating time Δt_{total}) the transient heat flux P_{probe} removed by the cryoprobe was compensated by the power supplied by the heating device $P_{heating}$. A heat loss to the environment ($P_{loss}\approx 0.7$ W) is taken into consideration. This value is evaluated by measuring the required heating power without operation of the cryoprobe. All data are stored to file for later processing.

Even for a small amount of water it is difficult to achieve equilibrium conditions within a sufficiently short period of time (Δt_{total} up to 15 min). This does not only lead to a considerable gas/liquid consumption but can also cause a remarkable pressure drop in the gas/liquid supply system. This causes - in case of gas-driven probes as described above - a drop of the cryoprobe's performance since it is not working under optimal conditions. The pressure drop would consequently cause difficulties in interpreting the results. For a longer operation time of the cryoprobe an additional problem of non-constant thermal conditions around the tip can occur due to ice formation outside the probe or changed heat transfer rates within the probe. These influences should be minimized in order to achieve a standardized and comparable way of heat flux measurement. We therefore reduced the operating time of the cryoprobes by the following integration method: the cryoprobe is operated for a short, preselected period of time Δt_{probe} (about 1 min) only. Subsequently the controller is used to establish initial conditions - in terms of compensation heating power - again. The transient electrical power $P_{heating}(t)$ is recorded continuously for the complete measuring time Δt_{total}. Considering the heat loss P_{loss}

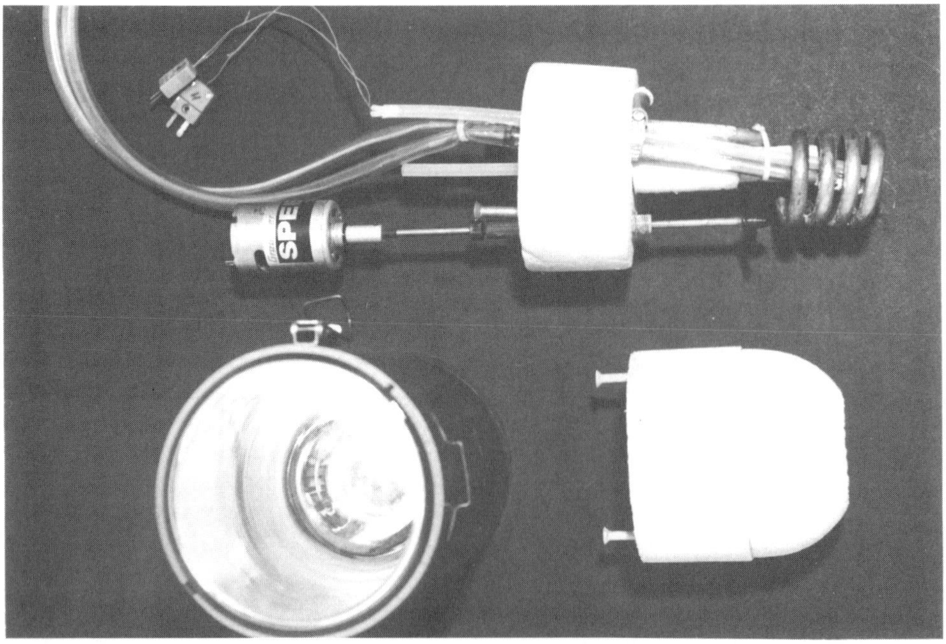

Figure 4. Dismounted device for the heat flux measurements with Dewar vessel (left, inner Ø=100 mm), additional insulation material (right), stirring device with propeller, thermocouples, and electrical heating device (top). Various cryoprobes can be attached and fixed by centred plastic tubes of variable diameter and length.

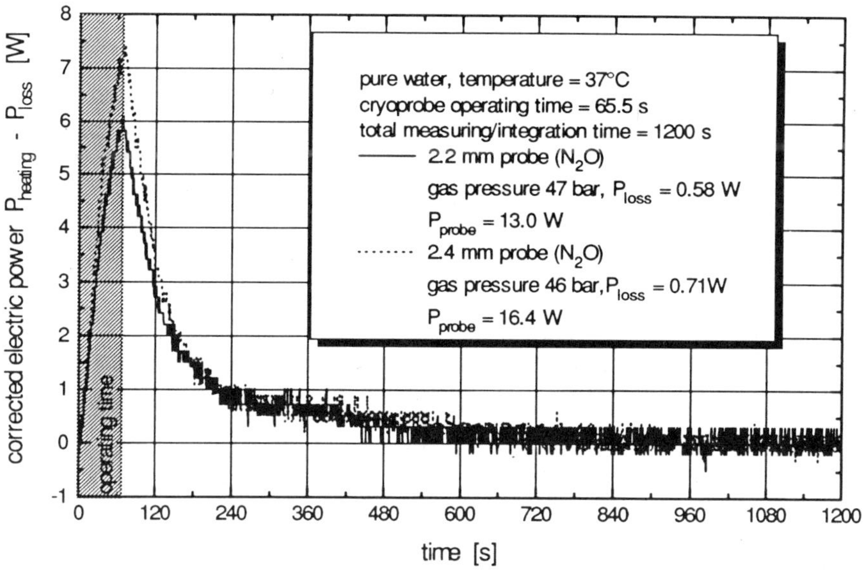

Figure 5. Corrected compensation heat flux [$P_{heating}(t)$ - P_{loss}] in terms of electric power applied by the heating device as a function of time. The cryoprobes (N_2O, 2.2 mm and 2.4 mm, flexible) were operated from time = 0 s for a period of time Δt_{probe} = 65.5 s only.

to be constant, the energy removed by the operating cryoprobe ($P_{probe} \cdot \Delta t_{probe}$) is equal to the energy supplied by the heating device for a longer period of time. The applied heating power is corrected for the heat losses by subtracting a constant value P_{loss}. The mean power removed by the cryoprobe P_{probe} can then be calculated from the following equation (1):

$$P_{probe} = \frac{1}{\Delta t_{probe}} \cdot \int_0^{\Delta t_{total}} [P_{heating}(t) - P_{loss}] \, dt \, . \tag{1}$$

Test measurements were performed with cryoprobes of relatively small power described above to verify the sensitivity of the system (N_2O-driven (JTE), flexible, 2.2 mm and 2.4 mm diameter). Examples of the test measurements are shown in figure 5.

Assuming that the thermal boundary conditions around the tip are approximately constant for the relatively short operating time (≈1 min) the calculated P_{probe} should give reasonable and comparable data for the performance characteristics of cryosurgical probes. With the described integration method, we determined mean heat fluxes of 13.0±1.0 W (2,2 mm) and 16.4±1.0 W (2,4 mm) respectively during operation of the mentioned cryoprobes under the described conditions (Δt_{total} = 1200 s). These results were confirmed in five experiments. Measurements with cryoprobes that were continuously operated - steady-state conditions with variations around the mean value of heating power not more than the resolution (0.25 W) for several minutes - gave comparable results. Differences were within an error interval of about 10% of the mean value. More powerful probes, e.g. those working with liquid nitrogen, can be analyzed with higher relative accuracy due to the less significant heat losses.

Nevertheless, it has to be noted that several influences have to be considered. Different probe operating times Δt_{probe} (1-3 min) as well as different heat losses P_{loss} (0.6-0.8 W) can lead to differences of the calculated heat fluxes of up to 10%. Total measuring time Δt_{total} is generally of minor importance, provided that this time period is sufficiently long for the system to reach initial conditions again (20-30 min). The stirring device has to be operated in a preselected, constant mode because different forced convection conditions at the tip will induce varying heat transfer coefficients. Induced stirring forces (rotating propeller, figure 4) should be sufficient but not much stronger than required to avoid severe temperature variations occurring at the thermocouple's tip. If the heat exchanging surface of a cryoprobe is not limited to the tip itself - as in case of the tested, non insulated gas-driven probes (see figure 3) - attention has to be paid to the exact length (surface) of the cryoprobe submerged into the stirred water volume (measured variations for the tested cryoprobes are up to 10%).

DISCUSSION AND CONCLUSIONS

The devices for in vitro measurement of the removable heat flux and the ice front and temperature field around cryoprobes give reasonable quantitative results for the applied thermal boundary conditions. The flexibility of the system allows for tests of cryoprobes in a great variety of power regimes and geometric shapes. Direct optical observation gives further information of the performance characteristics of cryoprobes. With the presented devices we want to propose a set of standard conditions - in terms of basic device design, media, temperatures, operating times, and calculation modes as described above - which should be uniformly applied for the measurement of cryosurgical probes. The sensitivity of the devices is shown to be sufficient to achieve data for different cryoprobes that can easily be compared even for the tested cryoprobes of relatively low power as presented above.

The data obtained under in vitro conditions certainly are not directly applicable to in vivo conditions during real cryosurgical therapy because tissue structure and thermal characteristics, crystallisation processes, varying influences of the blood flow (vasoconstrictive effect), and other influences were not taken into consideration. But the results of these in vitro measurements, combined with cryobiological data (cooling rate for lethal cell and tissue destruction), serve as a cryosurgical basis to estimate more exactly the extent and shape of the maximum region that can be treated successfully.

Further important data to be evaluated and different thermal boundary conditions to be applied to the systems for the quantitative measurements of cryoprobes have to be discussed with surgeons experienced in cryosurgery in order to ensure significance and accuracy of the in vitro measurements.

ACKNOWLEDGEMENT

The supply of the cryosurgical equipment ERBOKRYO CA and the mentioned cryoprobes by ERBE GmbH, Tübingen, Germany is gratefully acknowledged.

REFERENCES

1. P.J. Le Pivert, Basic considerations of the cryolesion, in: "Handbook of Cryosurgery," R.J. Ablin, ed., Marcel Dekker, New York (1980), p. 15
2. B. Rubinsky, The process of freezing and the mechanism of damage during cryosurgery, in: "Cryotherapy in Chest Medicine," J.-P. Homasson and N.J. Bell, eds., Springer Verlag, Paris (1992), p. 7
3. B. Rubinsky, G. Onik, Cryosurgery: advances in the application of low temperatures to medicine, *Int. J. Refrig.* 14:190-199 (1991)
4. R. D. Orpwood, Biophysical and engineering aspects of cryosurgery, *Phys. Med. Biol.* 26:555-575 (1981)
5. A. A. Gage, Cryosurgery in the treatment of cancer, *Surg., Gyn. & Obst.* 174:73-92 (1992)
6. J.-P. Homasson, Tracheo-bronchial endoscopic cryotherapy, in: "Cryotherapy in Chest Medicine," J.-P. Homasson and N.J. Bell, eds., Springer Verlag, Paris (1992), p. 35
7. J.-P. Homasson, J.-P. Thiery, M. Angebault, L. Ovtracht, O. Maiwand, The operation and efficacy of cryosurgical, nitrous oxide-driven cryoprobe, *Cryobiology* 31:290-304 (1994)

TEMPERATURE CONTROL OF A MICROSCOPE FREEZING STAGE FOR THE CRYOPRESERVATION OF LIVING CELLS : COMPARISON OF TWO DIFFERENT CONCEPTS

Th. Gross, Ch. Roling, I. Heschel, P. Schwindke, G. Rau

Helmholtz-Institute for Biomedical Engineering
at the RWTH Aachen, Pauwelsstr. 20, 52074 Aachen, Germany

ABSTRACT

Procedures for the cryopreservation of living cells are often developed in cryomicroscope freezing stages. In a cryomicroscope the temperature of the sample has to be controlled with a precision of $\Delta T \leq 1°C$ within the range from the physiological temperature (37°C) and the storage temperature (< -80°C). At the beginning of an optimization process for a new cell species the survival rates can be very low (<1%). Therefore the area of observation must be large (here: 20 x 15 mm^2) to freeze and thaw a large number of cells (here: > 1000) under uniform conditions. A freezing stage was developed to cryopreserve isolated heart muscle cells from adult rats. Two chambers form the stage. One is made of biocompatible materials and contains the cells. The second one is used to control the temperature of the first chamber by a nitrogen gas flow. The chambers are separated by a sapphire disc (thickness 1 mm), which generates a homogeneous temperature field and couples the two chambers thermally. The temperature is measured by a thermocouple on the disc and recorded on a small computer (PC). Cooling rates up to -15°C/min and heating rates up to 60°C/min are realised by the nitrogen gas flow through the stage. Two different concepts to control the temperature of the stage were tested. First, one single gas flow is cooled to -180°C in liquid nitrogen and then heated by a PID controlled electrical fan (compensation principle). The second concept is working with two alternative (180°C /-180°C) gas flows. Four valves to switch the gas flows are PC controlled (two-point controller). Higher precision and quicker response times were achieved with the second concept.

INTRODUCTION

Cryopreserved suspended living cells are used for the therapy of several human diseases[1]. The transfusion during elective operations or the transfusion of stem cells after chemotherapy are examples of clinical applications of suspended cryopreserved cells[2]. But not only in therapy cryopreserved cells are required. The cell injury during the cold storage of organs was also investigated. For organ banking a cryopreservation procedure has to be developed. But before whole organs can be cryopreserved the damaging processes during freezing and thawing need to be investigated with single suspended cells[3].

Isolated cardiomyocytes are cells which are object of interest in cardiophysiological research and pharmacological development. For the development of new drugs the contractility upon electrical stimulation serves as a good parameter of the pharmacological action. For electrophysiological investigations with the patch clamp technique and the measurement of the Ca^{++}-metabolism with fluorescence markers in microscope only single cells are required for the experiments. An isolation of a rat heart gives estimated a few million cells, which are viable in vitro only for a few hours[4]. Until now the cryopreservation of isolated cardiomyocytes was not successful. Only spontaneously cells contracting cells were found after thawing[5,6] and a rate higher than one spontaneous contraction per minute is a sign of damage[7]. So successful cryopreservation of suspended cardiomyocytes in small fractions of an isolation (a few millilitres in suspension with 10^4-10^5 cells) could make laboratory logistics easier, because cardiomyocytes of one isolation could be used whenever they are desired.

The initial steps to develop a cryopreservation process for a new cell type (e. g. cardiomyocytes) are often performed in cryomicroscopes[8]. A stage to control the sample temperature and the cooling and warming rates of the sample is mounted on a normal optical microscope. In cryomicroscopes the cells can be observed during freezing and thawing and therefore the main damaging processes can directly be investigated:

- intracellular ice formation (IIF) at high cooling rates can be observed by a darkening of the cells and
- osmotic shrinking at low cooling rates by changes in the visible cell area.

Furthermore, the viability of thawed cells can to a certain extent be tested in a cryomicroscope by observing their morphology, measuring their stimulated contraction or other cell-typical reactions.

To achieve significant results, accurate control of the temperature and the cooling and warming rates is necessary. A procedure for the cryopreservation of cardiomyocytes was developed in a specially designed cryomicroscope[9]. For this cryomicroscope two different concepts of temperature regulation were tested.

MATERIAL AND METHODS

Cryomicroscope

The highest temperature which is tolerated by biological systems is normally below 42°C. The temperature below which transport processes (e. g. diffusion and changes in crystal structure) are negligible is the glass transition temperature (for pure water T_g = -136°C)[10]. Cell suspensions contain proteins and glucose which increase the glass transition temperature. Furthermore, cryoprotectants like dimethyl sulfoxide (Me_2SO) or glycerol which are required to reduce cell damage during freezing and thawing also increase the glass transition temperature. Therefore the temperatures which have to be achieved for safe storage are higher than the glass transition temperature of pure water and -136°C is certainly sufficiently low. Most damages during cooling depend on the cell membrane permeability for water and the nucleation temperature of the medium. The developed cryomicroscope is especially designed to cryopreserve cardiomyocytes. In cardiomyocytes intracellular ice is generally formed at cooling rates higher than -15° C/min. We observed in other investigations that the formation temperatures of IIF in cardiomyocytes with Me_2SO-concentrations up to 20% (w/w) are always above -30° C[11]. Therefore a minimal temperature of -40°C is sufficient for investigations on the

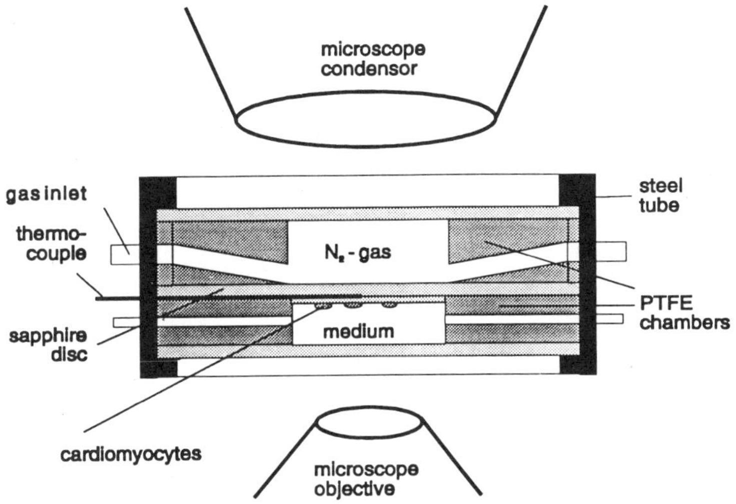

Figure 1. Schematic view of a microscope freezing stage with two chambers; the upper for a temperature controlled gas to regulate the temperature of the cardiomyocytes in the lower chamber; chambers connected by a sapphire disc.

injury of cardiomyocytes during cooling and warming. The damaging mechanisms which might occur at lower temperatures are much slower and thus of minor importance.

The cryomicroscope is formed by an inverse optical microscope (Zeiss IM 35), the freezing stage and a device for the gas temperature measurement and control.

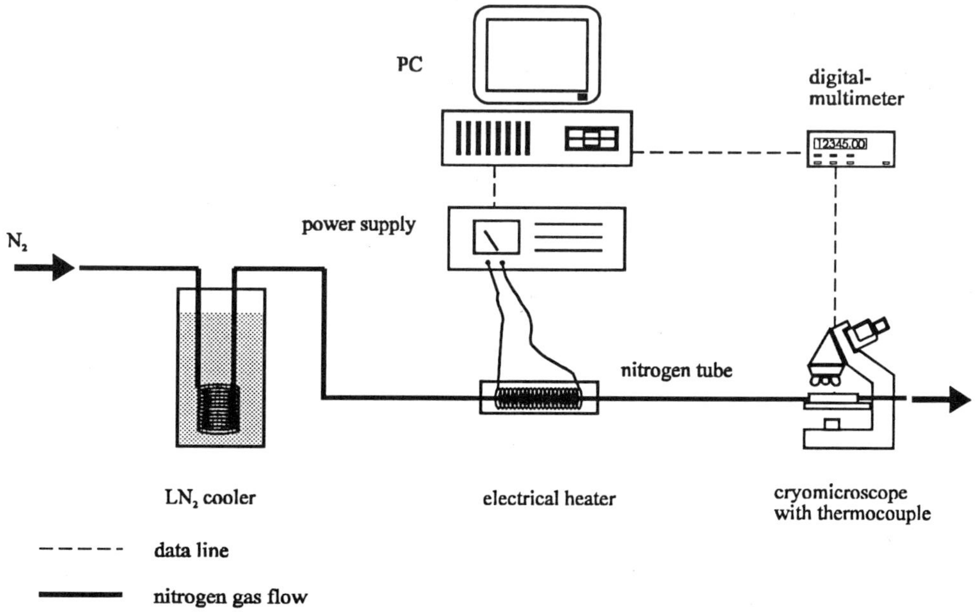

Figure 2. Schematic view of a PID regulation of a cryomicroscope by compensation heating of a cold nitrogen flow.

Freezing stage

The new freezing stage (fig. 1) was developed with the following specifications :

- controlled cooling and warming rates from 35°C to -40°C,
- exchange of the cell medium during the process at temperatures above 0°C, because toxicity depends on the metabolism which is reduced at lower temperatures[12] and
- electrical stimulation at 35°C before and after cryopreservation for viability testing.

For cryopreservation, a two chamber freezing stage for a cryomicroscope was developed. The diameter of the cylindrical steel tube which builds the skeleton of the freezing stage is 50 mm. The lower chamber is formed by two sapphire discs and a PTFE ring with a rectangular hole in its centre (20 x 15 mm^2). This chamber contains the cells. The cardiomyocytes are attached to a prepared cover slip. The cells are able to contract upon electrical stimulation and stay on the cover slip during the exchange of the cell medium. This cover slip is pressed against the sapphire disc which separates the lower from the upper chamber. The upper chamber is formed by two sapphire discs and a PTFE ring in between. Through this upper chamber nitrogen gas flows with a maximal mass flow of 2 g/s. This nitrogen gas is temperature controlled, it regulates the temperature at the cell location. Two different methods of gas temperature control are investigated:

- PID regulated compensated heating of a precooled gas flow with an electrical fan (fig. 2).
- Two point switching of a cold and a hot gas flow (fig. 3)

All materials in the sample are biocompatible (PTFE, sapphire and platinum). Both devices are programmed to the required cooling and warming rates which were used for the cryopreservation of cardiomyocytes:

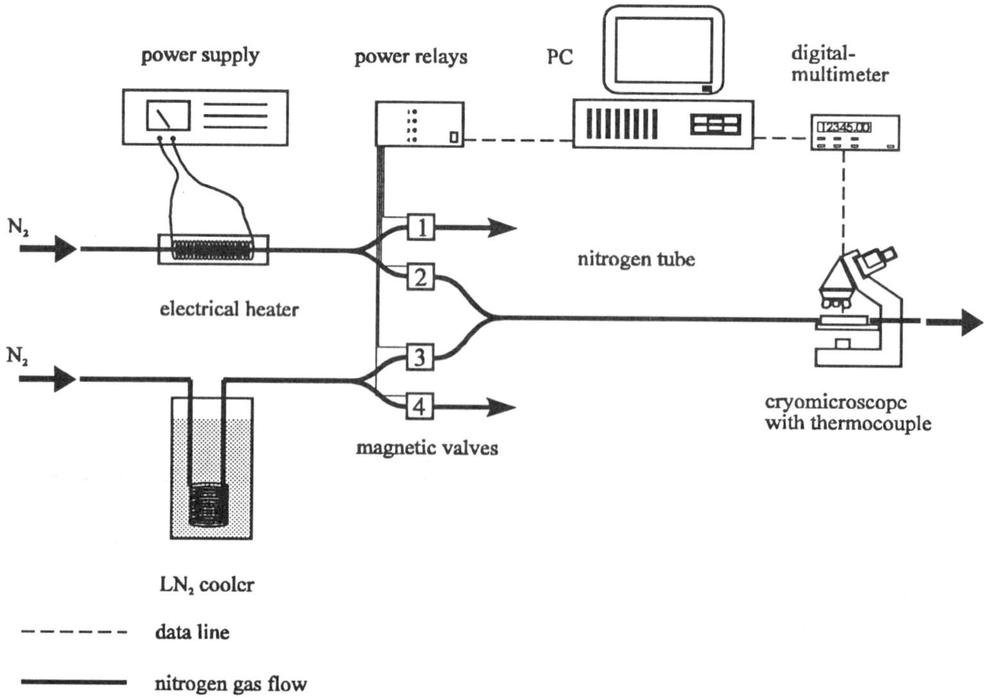

Figure 3. Schematic drawing of the two point regulation: cold and hot gas flows alternating through the freezing stage are controlled by the PC and magnetic valves.

- cooling from 35°C to -40°C with a maximum cooling rate of -15°C/min,
- storage at -40°C for 5 min,
- warming from -40° to 5°C with a maximum warming rate of 60°C/min, and
- warming from 5°C to 35°C with a moderate warming rate of 9°C/min.

Temperature measurement

The temperature is measured with a Cu/CuNi (copper/constantan) thermocouple. It is placed between the cover slip and the middle sapphire disc in a thin layer of cell medium. Therefore the distance between the cardiomyocytes, which are attached to the cover slip, and the thermocouple is about 300 μm. As a reference junction for the thermocouple voltage an ice water bath is used. The voltage of the thermocouple is recorded via an IEEE controlled digital multimeter (Keithley 199/1992) on a personal computer (PC). The sampling-rate is 7,3 Hz and temperature resolution 0.2°C[13].

PID - regulation by a compensation heating device

The gas temperature was cooled down to -190°C by submerging a coiled copper tube into liquid nitrogen. The cold gas flows through an electrical heat exchanger with a maximum power of 500 W. The power was PID controlled by a PC program and DC power supply. All tubes are insulated against thermal loss.

It is very difficult to find out the optimal parameter of the PID regulation. Therefore we followed the rules, which were suggested by Chien[14] and Ziegler[15]. In the first case a temperature jump from -20°C to 0°C is programmed and the reaction of the system is measured. Therefore the characteristic parameters, time lag, compensation time and proportional control factor can be determined. This parameter set for the regulation was suggested by Chien. We referred to the term "jump mode" in this case. The other set of parameters was determined by oscillations of the set-temperature with amplitude of 8°C and an a mean value of 0°C with different frequencies from 0.00075 Hz to 0.4 Hz. With the rules of Ziegler we determined the frequency where the system becomes unstable. Ziegler suggests PID parameters which are derived from this critical frequency. We called this mode the oscillation mode in figure 4.

Two point regulation by switching between a cold and a hot gas flow

The other concept of temperature regulation is the two point regulation of two different gas flows shown in figure 3:

- a hot flow which is produced by nitrogen flowing through an electrical heater of 500 W power reaching a temperature of 180°C, and
- a cold flow which is produced by a gas flow through a coiled copper tube submerged into liquid nitrogen.

Both the warm and the cold gas flow have a constant temperature, because the individual gas flows are kept constant. Four magnetic valves switch between the flows according to the temperature requirements. If the temperature is below the set temperature then only the magnetic-valves 2 and 4 are opened. Hot gas flows through the cryomicroscope and cold gas flows through the bypass into the environment. If the temperature is above the set temperature then valves 1 and 3 are opened. Cold gas flows through the freezing stage and hot gas flows into the environment. The maximal switching rate is limited by the sampling rate of the temperature measurement.

RESULTS

In fig. 4 the results of the different regulation concepts are shown. The solid line is the programmed temperature profile. The others show the measured temperatures by use of:

1. PID controlled system with a set of parameters in the oscillation mode (---- line),
2. PID controlled system with a set of parameters in the jump mode (--··-- line) and
3. Two point regulation (····· line).

The reproducibility of each experiment is better than $\Delta T < 2°C$ at all moments. During the cooling period at ≅ -16°C the influence of the latent heat can be seen after the ice crystal nucleation of the culture medium with 15% (w/w) of Me_2SO. The three curves do not vary significantly during cooling and storage at -40°C. But during the heating the PID regulation in the oscillation mode shows an overshooting of 7°C at 5°C and 4°C at 35°C. All temperatures above 37°C act like a fever to cardiomyocytes and are not tolerable. For the other parameter set (jump mode) the PID response times are not quick enough for the heating rates of 60°C/min between -40°C and 5°C. Therefore the parameter variation to investigate the influence of the warming rate is restricted to lower warming rates. Only the two point concept works precisely and fast enough. The differences between the set temperatures and the real temperatures are less than 1°C except during initial nucleation of the supercooled liquid.

Furthermore, the realisation of this regulation concept is cheaper, because a simple transformer without computer control can be used for the power supply.

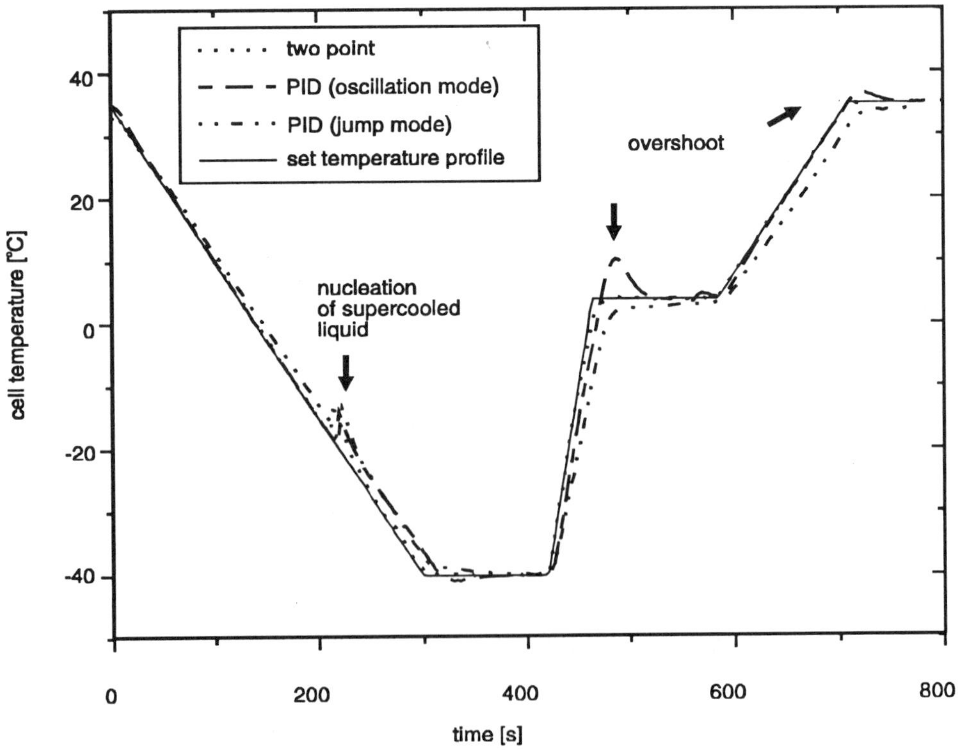

Figure 4. Typical temperature curves for the PID - temperature control with two different regulation parameters and for the two point temperature control. Comparison between two modes of PID regulation with the two point regulation for a basic cryopreservation protocol for cardiomyocytes .

CONCLUSIONS

This work shows that for a specific cryomicroscope a simple two point regulation of the gas temperature has a higher precision than a PID-regulated compensated system. The optimal parameters of a PID regulation are very difficult to find therefore two parameter settings are tested which are described in literature. Perhaps a further optimization of the PID parameters in the regulation of the heater could reach a similar accuracy as the simple two point regulation, but this procedure would require a lot of time.

The two point regulation works precisely with a accuracy of almost less than 1°C and fast enough for heating rates up to 60°C/min. It is simpler, require parameter setting and is cheaper, too, because the power supply can be a transformer without computer control. Therefore, we use this system for the development of a cryopreservation protocol for cardiomyocytes. The mean reason for the quicker response time of the two point controller is that no thermal lags occur. After opening the valves, the desired gas flow temperature is immediately achieved. In case of the PID controlled system an additional time delay is caused by the thermal capacity of the electrical heater. So even if the parameter settings for the PID controller are optimized, the two point controlled set-up has an additional principal advantage.

The successful preservation protocol shall be applied to a freezing device of 5 ml volume.

ACKNOWLEDGEMENT

This research work is partially supported by the Deutsche Forschungsgemeinschaft (DFG) grant no.: Ko 1054/3-1 and Ra 355/20-2. Thanks to the members of the Institute for Physiology of the RWTH Aachen for the helpful discussions.

REFERENCES

1. T. H. Meryman, Cryopreservation of blood cells and other tissues: current status, Haematologia 15: 337-350 (1982)
2. A. Sputtek and Ch. Körber, "Cryopreservation of red blood cells, platelets, lymphocytes, and stem cells", in Clinical Applications of Cryobiology, B. J. Fuller, and B. W. W. Grout (editors), CRC press Boca Raton (1991), pp. 95-149
3. A. M. Karkow, "Organ preservation" in Clinical Applications of Cryobiology, B. J. Fuller and B. W. W. Grout, CRC press Boca Raton 1991, pp 191-219
4. H. Kammermeier and H. Rose; Are isolated cardiomyocytes a suitable experimental model in all lines of investigation in basic cardiology; Basic research in Cardiology 62: 208-315 (1987)
5. J. O. Bustamente and D. Jachimowicz, Cryopreservation of human heart cells; Cryobiology 25: 394-408 (1988)
6. G. M. Alink, C. C. Verheul and F. G. Offerjins; Isolation and low temperature preservation of adult rat heart cells, J. Cryobiology 14: 399-408 (1977)
7. M. C. Capogrossi, M. D. Stern, H. A. Spurgeon and E. G. Lakatta; Spontaneous Ca^{2+} release from the sarcoplasmic reticulum limits Ca^{2+}-dependent twitch potentiation in individual cardiac myocytes, J. Gen. Physiol. 91: 133-155 (1988)
8. M. W. Scheiwe and G. Rau, Thermally defined cryomicroscopy and some applications on human leukocytes; J. Microscopy 126: 29-44 (1982)
9. Th. Gross, S. Sommerfeld, I. Heschel, G. Rau and H. Kammermeier, Cryopreservation of isolated cardiomyocytes from adult rats, a cryomicroscopical analysis of the freezing process; Pflügers Archiv, in print
10. D. R. Uhlmann, A kinetic treatment of glass formation, J. Non-Cryst. Solids 7, pp 337-348 (1972)
11. Th. Gross, S. Stegmeier, I. Heschel and G. Rau, Intracellular ice formation in isolated heart muscle cells of rats investigated with a bridgman-type cryomicroscope; Cryo Letters 16: 60 (1995)
12. S. Arrhenius; Über die Reaktionsgeschwindigkeit bei der Inversion von Rohrzucker durch Säuren; Z. Physik. Chem. 4: 226-248 (1889)

13. I. Heschel, M. Rödder, C. Lückge, B. Nunner, G. Rau, A power-down device for directional solidification of macroscopic samples, Cryobiology 29: 767 (1992)
14. K. Chien, et. al. On the automatic control of generalised passive systems; Trans ASME, 74: 175-185 (1952)
15. J. Ziegler and N. Nicols, Optimum setting for automatic controllers; Trans ASME, 64: 759-769 (1942)

GLASS TRANSITION CURVES OF AQUEOUS SOLUTIONS FOR THE FREEZE-DRYING OF BIOMATERIALS

G. Spieles, I. Heschel, and G. Rau

Helmholtz-Institute for Biomedical Engineering
at the Aachen University of Technology
Pauwelsstr. 20, D-52074 Aachen, Germany

ABSTRACT

Freeze-drying is a common process for the long-term preservation of biological material, such as therapeutical and diagnostical proteins. The maintenance of a glassy state is generally believed to guarantee long-term stability of the freeze-dried material. In order to yield high recovery rates, it is common practice to add stabilizers to the biomaterial. In this study, a large number of carbohydrate and macromolecular stabilizers are analyzed with respect to their glass transition temperatures. Aqueous solutions of these stabilizers were prepared and subjected to the freeze-drying process. It was found that the glass transition temperatures of these solutions increase during the drying process. They also tend to increase with increasing molecular weight of the stabilizer. In contrast, the addition of salts leads to a depression of the glass transition temperature. The resulting glass transition curves are described using a thermodynamic model. A rationale is presented to allow for the selection of stabilizers which have favorable properties with respect to glass formation.

INTRODUCTION

Freeze-drying is a common process for the long-term preservation of biochemicals which are unstable in aqueous solution. The freeze-drying process includes a series of unit-processes such as freezing, primary drying and secondary drying. The goal of the process is to protect a biochemical compound from undesired deterioration and to transform it into a shelf-stable and easy-to-use product.[1]

Aqueous solutions undergo a phase separation during freezing as the formation of ice crystals leads to an increase in concentration of all dissolved substances. Those substances which do not form eutectics will undergo a glass transition during freezing. This transition occurs at a characteristic temperature (T_g') which is strongly correlated with the water content of the glassy phase and the cooling rate. If a sufficiently low vacuum is applied to the frozen material, all ice crystals can be removed by sublimation, leaving behind the concentrated, glassy phase (primary drying). The water fraction which is incorporated in the glassy material is partially removed during the secondary drying step. Water removal from the glassy phase usually increases the glass transition temperature (T_g) of the remaining material.

The stability during processing and storage of the dried material depends largely on the storage temperature (T) in relation to the glass transition temperature. Knowledge of the glass

Advances in Cryogenic Engineering, Vol. 41
Edited by P. Kittel, Plenum Press, New York, 1996

Table 1. Review of recent publications on the freeze-drying of biomaterials

reference	freeze-dried material	stabilizer
3	hemoglobin	carbohydrates, macromolecules
4	lactic acid bacteria	polyethylene glycol, dextran
5	lactic acid bacteria	polyethylene glycol, dextran
6	phosphofructokinase	trehalose, maltose
7	cyclophosphamide	dextran, polyvinylpyrrolidone
8	catalase	various saccharides

transition curve is, thus, essential for the successful design of a freeze-drying process as well as the selection of safe storage temperatures.[2]

Numerous papers describe the application of freeze-drying processes to various biological materials (Table 1). From these works, it can be concluded that the addition of stabilizers generally enhances the performance of the freeze-drying process. The products show higher viability which can be maintained over longer periods of time. Furthermore, some applications require the addition of electrolytes to adjust the pH or to maintain osmotic balances. In order to characterize the effect of these substances on the glass transition temperatures of freeze-dried aqueous solutions, glass transition curves were recorded for a variety of common stabilizers. These curves relate the glass transition temperature to the water content of the solutions.

MATERIALS AND METHODS

Chemicals

Based on a survey of the commonly employed stabilizers in freeze-drying the glass transition curves for two classes of chemicals were studied: mono-, di-, and oligosaccharides and polymers. As for the saccharides, a homologous series of malto-oligomers was analyzed to allow an estimation of the effect of molecular weight. In addition, sucrose was studied because it is among the most easily available stabilizers. The polymers were chosen to be naturally occurring, modified and synthetic. For polymeric PVP, the influence of salt (NaCl) on the glass transition was studied (Table 2).

All substances were taken as supplied (i.e. without further purification) with two exceptions: hydroxyethyl starch and gelatin were dialyzed against demineralized water for 48 h and freeze-dried to yield a porous solid without lower molecular weight impurities (e.g., salts, oligosaccharides).

Aqueous solutions were prepared from the all substances (aqua ad iniect., Delta-Pharma, #62010). These solutions had initial total solids mass fractions of 50 – 65%. This rather high concentration of dissolved solids was chosen to avoid unnecessary losses of mass (water)

Table 2. List of analyzed stabilizers

classification	stabilizer	M_w [Da]	supplier
1. mono-, di-, and oligosaccharides	glucose	180	Serva, anal. grade
	maltose	360	Serva, res. grade
	sucrose	360	Serva, anal. grade
	maltotriose	504	Serva, pure
	maltopentaose	829	Serva, res. grade
	maltoheptaose	1,153	Serva, res. grade
2. polymers	dextran F70	≈ 64,000	Serva, pure
	ficoll 70	≈ 70,000	Serva
	gelatin	30,000	Biotest
	hydroxyethyl starch	200,000	Leopold
	polyethylene glycol	≈ 35,000	Merck, zur Synth.
	polyvinylpyrrolidone K15	≈ 10,000	Fluka, pract.
	polyvinylpyrrolidone K15	≈ 11,000	Serva, pract.
	polyvinylpyrrolidone K30	≈ 40,000	Fluka, purum pharm.
3. polymer & salt	polyvinylpyrrolidone K15 & NaCl	≈ 11,000 & 58	Serva, pract. & Merck, z.A.

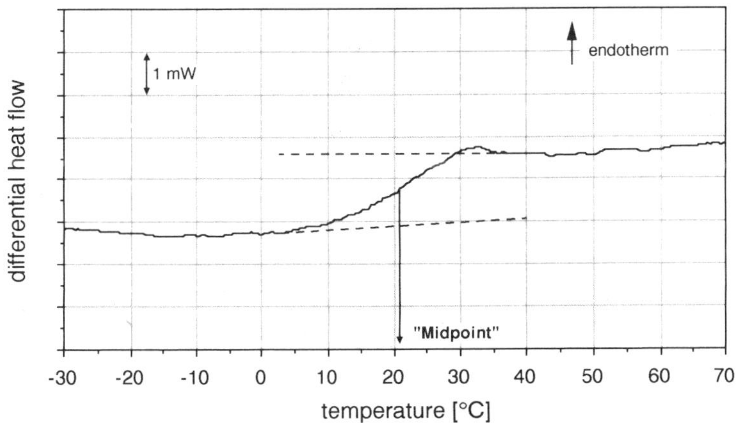

Figure 1. Typical DSC thermogram of a glass transition.

during the following drying step. Also, for the purposes of freeze-drying, the area of low water concentrations is of special interest.

Freeze-Drying

For each solution, 10 aluminum pans (Perkin-Elmer, #BO14-3021) of known weight were filled with 20 – 80 mg of solution. The aluminum pans were placed on a sample holder, capable of holding up to 11 pans. Ten of the 11 pans contained the solution to be analyzed, and an additional pan was attached to a thermocouple which monitored the pan temperature. The sample holder, with the pans, was cooled in liquid nitrogen and placed in a vacuum chamber. A circulating coolant regulated the temperature during vacuum drying. Starting with low temperatures, the sample temperature was raised as drying proceeded. Partially dried samples were removed from the vacuum chamber periodically to cover the concentration range from the initial solution composition to the totally dried material.

Differential Scanning Calorimetry

After weighing and calculating the concentration of the partially dried sample, the pans were hermetically sealed and monitored for their glass transition temperature using differential scanning calorimetry (Perkin-Elmer, DSC-4). An empty aluminum pan was used as the reference. Since glass transitions are accompanied by an incremental change in the specific heat capacity, they become manifest in a step change of the differential heat flow.

All samples were subjected to two heating scans (20 and 10 K/min) and two cooling scans (-20 and -10 K/min) through a temperature interval of 200 K.[9] Thermograms from the slow heating scan (10 K/min) were evaluated where a clear signal was detectable; otherwise the fast heating scan was used. The characteristic incremental change of the specific heat was analyzed using commercial software (Perkin-Elmer, TADS). The midpoint temperature was taken as the glass transition temperature (Figure 1).

Table 3. Publications of the glass transition of pure water

reference	measurement technique	T_{g1} [K]	Δc_{p1} [J/g K]
11	extrapolation of glass transition curves for	127 ± 4	–
12	various aqueous solutions	136	–
13		136	–
14		138 ± 2	–
15	change in heat capacity during warming	139	–
16	after vitrification	135	1.94
17		136 ± 1	–

MODELLING

A thermodynamic model was used to numerically describe the experimental glass transition temperatures.[10] It assumes a constant entropy at the glass transition and disregards the mixing entropies. Based upon these assumptions, a simple formula can be derived for a binary mixture. The glass transition temperature (T_g) of a mixture with components (1) and (2) is expressed by the mass fractions of the individual components (w_1 and w_2), the glass transition temperature of the pure components (T_{g1} and T_{g2}), and the incremental change in the specific heat at the glass transition of the pure components (Δc_{p1} and Δc_{p2}).

$$\ln T_g = \frac{w_1 \Delta c_{p1} \ln T_{g1} + w_2 \Delta c_{p2} \ln T_{g2}}{w_1 \Delta c_{p1} + w_2 \Delta c_{p2}} \quad (1)$$

In the present study, all solutions were made up from water and a stabilizer with the exception of the third group of substances (Table 2), which were ternary mixtures of water, PVP, and NaCl. However, all solutions were treated as binary mixtures with water as component (1) and all dissolved substances (i.e., saccharide, polymer, or PVP & NaCl) as component (2). This treatment was also justified for the ternary mixtures as DSC thermograms showed no indication of a phase separation, such as crystallization of pure or eutectic NaCl.

Since Eq. (1) requires the knowledge of the glass transition temperature and the incremental change in the specific heat at the glass transition of pure water the values T_{g1} = 136 K and Δc_{p1} = 1.94 J/g K were taken from a survey of the relevant vitrification literature (Table 3). The equivalent properties of the dissolved stabilizers (T_{g2} and Δc_{p2}) are considered as free parameters which are to be determined by numerical adaptation to the experimental data. The numerically determined values will be referenced to as T_{gS} and Δc_{pS} in order to distinguish them from the model properties.

RESULTS

Figure 2 displays three glass transition curves for the examined aqueous PVP solutions in the concentration range studied. The resulting numerical parameters are listed in Table 4 for all analyzed substances. They should be understood as the starting point of a collection of material properties which will allow the fast and simple calculation of glass transition curves. The accuracy for the determination of glass transition temperatures was less than 2°C.

All glass transition curves exhibit some characteristic features:

- glass transition temperatures increase during water removal by freeze-drying
- glass transition curves can be extended to the glass transition temperature of pure water
- glass transition curves can be appropriately described by Eq. (1)

The effect of molecular weight on the properties of the stabilizers is most expressed for the saccharides which exhibit a significant increase in the glass transition temperature and decrease in the incremental change of the specific heat capacity with increasing molecular

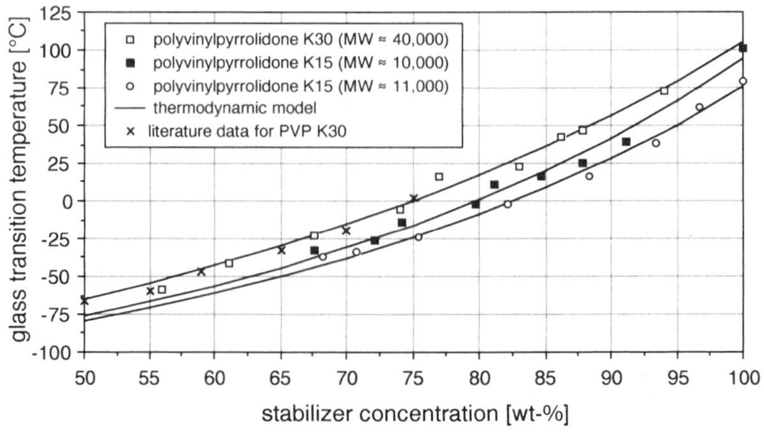

Figure 2. Glass transition curves for three PVP modifications.

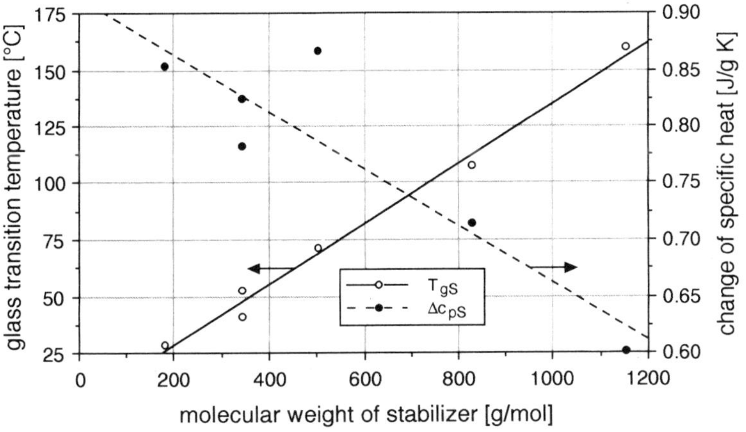

Figure 3. Variation of glass transition properties with molecular weight of saccharides.

weight (Figure 3). For the polymeric substances, the increase in glass transition temperature with molecular weight was still observed for PVP K30 if compared to PVP K15 (Figure 2), but at a far more moderate pace. However, the variation between the two PVP K15 modifications might be attributed to different molecular specifications other than the molecular weight. Of all polymeric stabilizers, gelatin exhibited the highest glass transition temperature.

For some stabilizers, it was not possible to determine glass transitions. Hydroxyethyl starch did not express a pronounced change of the specific heat capacity although glassy and brittle by appearance. This might be due to the inhomogeneity of hydroxyethylated starch molecules, varying in molecular weight and degree of substitution. We speculate that hydroxyethyl starch exhibits a continuous glass transition over a broad range of temperatures which makes it undetectable by DSC. In addition, thermograms for dextran and polyethylene glycol indicated crystallization of the stabilizer.

The addition of NaCl to aqueous PVP solutions results in a significant reduction of the incremental change of the specific heat capacity (Figure 4). Although the glass transition temperature of the dry material does show a slight increase, the reduced Δc_p of the dried PVP-NaCl mixtures results in a lower glass transition temperature at more elevated water concentrations.

The glass transition curve for PVP K30 compares well with previously reported data on the glass transition of PVP.[18] Some data are also available for dry saccharides which, in general, shows good agreement with our own experimental results and Eq. (2).[19,20]

Table 4. Material properties of analyzed stabilizers

class	stabilizer	T_{gS} [K]	Δc_{pS} [J/g K]
1.	glucose	301.8	0.854
	maltose	314.2	0.782
	sucrose	325.8	0.824
	maltotriose	344.7	0.866
	maltopentaose	381.0	0.714
	maltoheptaose	432.7	0.601
2.	dextran F70	–	–
	ficoll 70	366.2	0.770
	gelatin	401.7	0.932
	hydroxyethyl starch	–	–
	polyethylen glycol	–	–
	polyvinylpyrrolidone K15 (10,000 Da)	368.2	1.146
	polyvinylpyrrolidone K15 (11,000 Da)	349.3	1.160
	polyvinylpyrrolidone K30	378.6	1.381
3.	polyvinylpyrrolidone K15 (11,000 Da) & NaCl		
	m_{NaCl}/m_{PVP} = 0.0156	362.3	1.060
	0.0837	372.3	0.862
	0.3146	371.5	0.668

CONCLUSION

Most of the aqueous stabilizer solutions which were analyzed tended to form glasses when freeze-dried. The glass transition curves can be measured by applying DSC techniques to partially freeze-dried material. All curves could be thermodynamically modeled. It was found that glass transition temperatures increase with higher stabilizer concentration, higher molecular weight of the stabilizer, and lower salt content.

These findings have implications for the design of freeze-drying processes. In order to restrict losses of biological activity during the freeze-drying process and subsequent storage the temperature of the drying material should be kept below the glass transition temperature at any time. This requires primary drying to be performed at or below T_g'. During secondary drying, the temperature can be raised according to the glass transition temperature of the drying material. At this point, it becomes obvious that the stabilizer exerts a great influence on the process design. Stabilizers with high glass transition temperatures seem to be specially suited since they allow higher product temperatures and, as a result, faster drying.

Secondary drying can be terminated when the dried material's glass transition temperature has reached or surpassed the desired storage temperature. A safety margin might apply to allow for unexpected product treatment, such as warming by direct sunlight. Given a desired storage temperature, the glass transition curve indicates the amount of water which has to be removed and, therefore, determines the end-point of freeze-drying. Again, a high glass transition curve seems advantageous as it can reduce the total drying time. Storage temperature should always be kept below the glass transition temperature.

From these process design aspects, great care has to be taken in choosing a stabilizer for the protection of labile biochemicals in freeze-drying. In general, polymeric stabilizers are to be preferred, as they exhibit high glass transition curves. Low molecular weight saccharides should be avoided and, from all analyzed substances, pure glucose does not seem to be a promising stabilizer for products which are to be stored at 25°C. The salt content has to be limited and carefully controlled since the presence of NaCl in an aqueous PVP solution reduces the glass transition curve and, therefore, requires lower processing and storage temperatures. Any variation (e.g., due to previous separation and purification processes) will have an impact on the glass transition curve and, therefore, on the success of a subsequent cryopreservation or freeze-drying process.

An extension of these studies to further common stabilizers and salts is encouraged to build up a more comprehensive data collection. It is expected that the presented findings and the thermodynamic model can be expanded to more complex solutions which comprise biological material such as enzymes or suspended cells. Further measurement techniques, such as dynamic mechanical analysis (DMA) and electron spin resonance (ESR), should be applied to gain further insight into the meaning of the glass transition in terms of molecular mobility of water, stabilizers, and biological material.

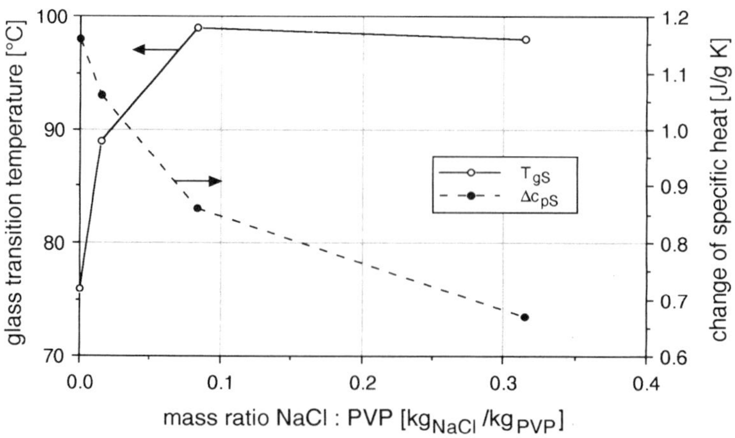

Figure 4. Variation of glass transition properties of PVP with NaCl content.

REFERENCES

1. F. Franks, Freeze-drying: from empiricism to predictability. *Cryo-Letters* 11:93 (1990).
2. M. J. Pikal, Freeze-Drying of Proteins, in: ACS Symp. Ser. Vol. 567, Washington, D.C. (1994), p. 120.
3. P. Labrude, B. Chaillot, F. Bonneaux, and C. Vigneron, Freeze-drying of haemoglobin in the presence of carbohydrates. *J. Pharm. Pharmacol.* 32:588 (1980) and P. Labrude, F. Bonneaux, and C. Vigneron, An attempt to freeze-dry haemoglobin in the presence of macromolecules. *J. Pharm. Pharmacol.* 33:115 (1981).
4. G. F. de Valdéz, G. S. de Giori, A. P. de Ruiz Holdago, and G. Oliver, Comparative study of the efficiency of some additives in protecting lactic acid bacteria against freeze-drying. *Cryobiology* 20:560 (1983).
5. T. Tsvetkov and R. Brankova, Viability of micrococci and lactobacilli upon freezing and freeze-drying in the presence of different cryoprotectants. *Cryobiology* 20:318 (1983).
6. J. F. Carpenter, L. M. Crowe, and J. H. Crowe, Stabilization of phosphofructokinase with sugars during freeze-drying: Characterization of enhanced protection in the presence of divalent cations. *Biochim. Biophys. Acta* 923:109 (1987).
7. T. R. Kovalcik and J. K. Guillory, Stability of cyclophosphamide in lyophilized cakes. II. Urea, polyvinylpyrrolidone, and dextran as excipients. *J. Parenter. Sci. Technol.* 42:165 (1988).
8. K. Tanaka, T. Takeda, and K. Miyajima, Cryoprotective effect of saccharides on denaturation of catalase by freeze-drying. *Chem. Pharm. Bull. (Tokyo)* 39:1091 (1991).
9. J. P. Wolanczyk, Differential scanning calorimetry analysis of glass transitions. *Cryo-Letters* 10:73 (1989).
10. P. R. Couchman and F. E. Karasz, A classical thermodynamic discussion of the effect of composition on glass–transition temperatures. *Macromolecules* 11:117 (1978).
11. I. Yannas, Vitrification temperature of water. *Science* 160:298 (1968).
12. D. H. Rasmussen and A. P. MacKenzie, The glass transition in amorphous water. Application of the measurements to problems arising in cryobiology. *J. Phys. Chem.* 75:967 (1971).
13. C. A. Angell, J. M. Sare, and E. J. Sare, Glass transition temperatures for simple molecular liquids and their binary solutions. *J. Phys. Chem.* 82:2622 (1978).
14. C. A. Angell, Supercooled water. in "Water – A Comprehensive Treatise" F. Franks, ed., Plenum Press, New York (1982), p. 1.
15. J. A. McMillan, and S. C. Los, Vitreous ice: irreversible transformations during warm-up. *Nature* 206:806 (1965).
16. M. Sugisaki, H. Suga, S. Seki, Calorimetric study of the glassy state. IV. heat capacities of glassy water and cubic ice. *Bull. Chem. Soc. Japan* 41:2591 (1968).
17. G. P. Johari, A. Hallbrucker, and E. Mayer, The glass-liquid transition of hyperquenched water. *Nature* 330:552 (1987).
18. A. P. MacKenzie and D. H. Rasmussen, Interactions in the water–polyvinylpyrrolidone system at low temperatures. in: "Water Structure at the Water–Polymer Interface" H. H. G. Jellinek, ed., Plenum Publishing, New York (1972), p. 146.
19. P. D. Orford, R. Parker, S. G. Ring, and A. C. Smith, Effect of water as a diluent on the glass transition behaviour of malto-oligosaccharides, amylose and amylopectin. *Int. J. Macromol.* 11:91 (1989).
20. L. Slade and H. Levine, Non-equilibrium behavior of small carbohydrate–water systems. *Pure & Appl. Chem.* 60:1841 (1988) and L. Slade and H. Levine, Beyond water activity: Recent advances based on an alternative approach to the assessment of food quality and safety. *Crit. Rev. Food Sci. Nutr.* 30:115 (1991).

DESIGN OF A HIGH-Tc SQUID BASED HEART SCANNER COOLED BY SMALL STIRLING COOLERS

H.J.M. ter Brake, H.A. de Boer, W.A.M. Aarnink, H.J. Holland, P.C. Bruins, P.J. van den Bosch and H. Rogalla

University of Twente, Faculty of Applied Physics
P.O.Box 217, 7500 AE Enschede, The Netherlands

ABSTRACT

A heart scanner based on high-T_c SQUIDs is under development. It should be used in standard clinical environments without magnetically shielded room. Because the scanner should be simple to use, the SQUIDs will be cooled by small cryocoolers thus realizing a turn-key apparatus. The aimed field resolution is 0.05 pT/√Hz, requiring an extremely stable temperature (fluctuations in the mK-range and drift below 0.1 mK/s). The mechanical cooler interference is reduced by incorporating two coolers and operating them in counter phase. The magnetic cooler interference is reduced by positioning the coolers and the SQUIDs in a coplanar arrangement, and by separating the SQUIDs from the cold tips with a solid conducting thermal interface. A design is presented in which a temperature of 55 K is expected with a cool-down time of less than 1 hour.

INTRODUCTION

Superconducting QUantum Interference Devices (SQUIDs) are the most sensitive magnetic flux-to-voltage converting sensors. At present, their main application lies in the field of biomagnetic research where multichannel low-T_c dc-SQUID based magnetometer systems are used. These systems are usually cooled by liquid helium and operated in magnetically shielded rooms to obtain an extremely low-noise measuring environment. They are expensive, require helium refills, and cannot be transported in a simple manner.

Because of the higher operating temperature, a much more flexible magnetometer system can be realized with high-T_c SQUIDs. The workhorse of current high-T_c superconducting electronics, $YBa_2Cu_3O_7$, with its critical temperature (T_c) of 92 K opened the possibility to operate SQUIDs in liquid nitrogen at 77 K. Also small-scale turn-key cryocoolers, originally developed for cooling infrared sensors, are available and are very reliable [1]. The advantages of cryocoolers compared to a liquid-nitrogen cryostat are the lower operating temperature (which gives a potentially better SQUID performance [2]), the turn-key operation (no refills required), and the possibility to operate the system in all directions.

In this paper the design of a heart scanner is presented that can be equipped with up to 25 SQUIDs cooled by two small cryocoolers. In the next section the heart signal and the required sensor resolution are considered, followed by a discussion on the operating temperature and the temperature stability. After that, the specific coolers that are used in this project are described. Their magnetic and mechanical interferences are dealt with in a separate section. Then, the thermodynamical aspects are considered and the paper is concluded with the actual design of the heart scanner.

HEART SIGNAL AND SENSOR RESOLUTION

The QRS complex in the magnetic heart signal (magnetocardiogram, MCG) of an healthy adult is typically 10-100 pT peak-to-peak. The spectral information of the heart signal is mainly contained in the frequency band 0.5 Hz to 40 Hz. However, in order to detect anomalies in the MCG, higher frequencies should also be accessable (up to 100 Hz). High-T_c SQUIDs can be used to adequately record MCGs provided their noise levels are 0.05 pT/√Hz or below.

In the heart scanner project we use high-T_c SQUIDs with bicrystal grain boundary junctions. Different coupling schemes are studied: monolithic flux-transformer coupled, flip-chip coupled and directly coupled (or inductively shunted) [3]. A directly coupled SQUID shunted with a single pick-up loop requires only one layer of high-T_c material and is, therefore, relatively simple to manufacture. So far, we have made magnetometers with a 8 mm x 8 mm sensing loop and a noise level of roughly 0.3 pT/√Hz. To arrive at a field resolution of 0.1 pT/√Hz we have redesigned the magnetometer; The SQUID inductance is optimized as well as the line width of the pick-up loop. Furthermore, the whole substrate area of 10 mm x 10 mm is used. As a next step larger substrates of 20 mm x 20 mm can be used to lower the intrinsic field noise to 0.05 pT/√Hz.

OPERATING TEMPERATURE AND TEMPERATURE STABILITY

The higher operating temperature of high-T_c compared to low-T_c systems is a big advantage with respect to flexibility and costs. As an option, liquid nitrogen boiling at 77 K can be used instead of liquid helium (nitrogen is one order of magnitude cheaper and has a factor of 50 higher volumetric heat of evaporation). However, a lower operating temperature gives a potentially better field resolution [2]. Such a lower operating temperature can be realized with small-scale cryocoolers, as is indicated in the following sections.

The required temperature stability of high-T_c SQUIDs is determined by the variation of the London penetration depth. The higher the operating temperature and the higher the background magnetic field, the more stringent is the required stability. To reduce various noise contributions, we plan to compensate the earth's magnetic field by 40 dB [4]. In that case temperature fluctuations should be limited to 0.3 mK_{pp} when operating the inductively shunted SQUIDs at 77 K (assuming a band width of 100 Hz) [5]. Because in a liquid nitrogen bath temperature fluctuations of a few tenths of a K may be expected [6], thermal buffering via relatively large heat capacities will be required. In our design cryocoolers will be used to cool the SQUIDs and a temperature in the range of 55 K is attainable. Then, the stability requirements are roughly one order of magnitude less severe (2 mK_{pp}) [5]. A stable temperature can be realized by actively controlling the cryocooler input power [7].

Figure 1. Signaal Usfa cryocoolers; referring to the compressors from left to right: types 7057, 7058 and 7056; resp. cooling powers 0.75 W/ 1.5 W/ 0.5 W; compressor size typ. 12 cm. (Courtesy of Signaal Usfa).

COOLERS TO BE USED IN THE HEART SCANNER

Small coolers have been developed to cool infrared sensors mainly for military applications (Stirling-type as well as Linde-Hampson type) [1]. These coolers are available with cooling powers ranging from typically 0.2 W to 2 W, specified at 80 K with an environmental temperature of 23 °C. Among other companies, Stirling-type coolers are manufactured by Signaal Usfa (Eindhoven NL), see figure 1. In these coolers, schematically depicted in figure 2, two compressor pistons are driven by linear motors. Motion of a piston is generated by feeding an alternating current through a coil connected to the piston in a permanent magnetic field. Thus a pressure wave is obtained in the working gas (helium). A displacer in the cold head moves in such a way that the expansion of the gas mainly takes place at the end of the cold finger, whereas compression occurs in the space between the compressor pistons and the displacer. The mass of the compressor piston/coil assembly, in combination with the mechanical spring and gas forces, creates a mass-spring system that is tuned to the drive frequency of 50 Hz. Two dual opposed pistons are used in a single compressor in order to balance the axial forces, thus reducing the mechanical vibrations. The Signaal Usfa coolers were described more extensively elsewhere [8].

For the heart scanner we plan to use a set of two coolers (type 7058). Firstly, because two coolers can be mounted back-to-back so that also the axial forces of the displacer units can be balanced and secondly, to have sufficient cooling power.

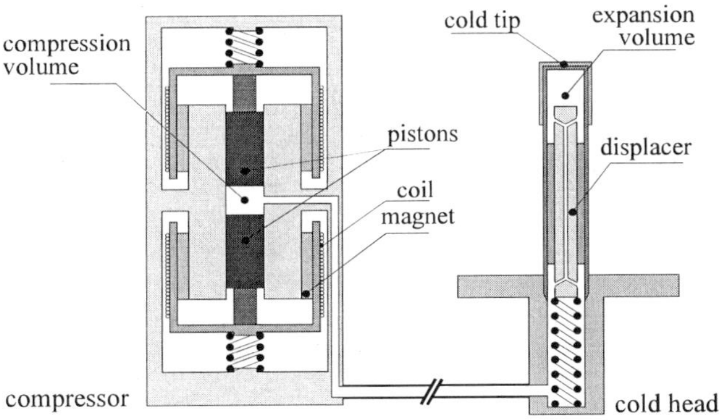

Figure 2. Schematic representation of the compressor and displacer units (Courtesy of Signaal Usfa).

Table 1. Noise experiments with SQUID mounted directly on cold tip of 7058 cooler

Configuration	50 Hz		10 Hz	
on tip, cooler on (with as well as without μ-metal)	0.5	μT_{RMS}	0.5-1	nT/√Hz
on tip, with power connections, cooler off	5	nT_{RMS}	5-10	pT/√Hz
on tip, cooler off, μ-metal around compressor	0.1	nT_{RMS}	5-10	pT/√Hz
on tip, no power connections	1	pT_{RMS}	1	pT/√Hz
in magnetically shielded flow cryostat	2	pT_{RMS}	0.7	pT/√Hz

COOLER INTERFERENCE

When a cryocooler is applied for cooling a SQUID magnetometer it will generate noise in different ways. Firstly, magnetic interference: direct magnetic noise due to the compressor currents or to the vibration of magnetic cooler components; Secondly, mechanical interference: noise due to vibrations of the SQUID magnetometer in the environmental field; And finally, noise due to the fact that metal is used in the vicinity of the SQUID pick-up loops: magnetic noise arising from the thermal motion of electrons in the metal and distortion of environmental fields via eddy-currents creating relatively strong gradients. In this paper we pay attention to the magnetic and mechanical interference only. The thermal noise can be evaluated by means of refs. 9 and 10, whereas the field distortion can be approximated with the worst case in which the metallic components are considered to be superconductive. Then the effect can be estimated via ref. 11.

Direct cooling of a high-T_c SQUID on the cold tip of a small-scale cryocooler was claimed to be possible by Khare and Chaudhari from IBM in 1994 [12]. They used a directly coupled bicrystal SQUID with a relatively poor performance: effective sensing area 0.013 mm^2, field resolution 75 pT/√Hz at 1 Hz and 10 pT/√Hz at 10 Hz. This SQUID was attached to the tip of a small integral-type Stirling cooler (Inframetrics Inc, 0.15 W at 77 K). The SQUID was shielded from the external magnetic field and from the noise of the cryocooler motor by two layers of μ-metal. In this configuration Khare and Chaudhari measured no increase in the field noise in the SQUID (compared to measurements in a liquid nitrogen bath) in the frequency band of 1 Hz to 10 Hz (no data were presented for higher frequencies). However, the fact that the cooler did not contribute to the noise can be attributed to the high intrinsic noise level of the SQUID and to the two-layer μ-metal magnetic shielding. This small-size shielding is only applicable if the measuring object can be placed close to the SQUID and inside the shielding (e.g. magnetic measurements on small materials). For the majority of applications (e.g. biomagnetism, non-destructive testing of relatively large materials, magnetotellurics) this type of small-size shielding cannot be applied. Therefore, we performed similar experiments in the magnetically shielded room of the Biomagnetic Centre Twente with a high-T_c SQUID directly mounted on the tip of a Signaal Usfa 7058 cooler.

The SQUID that we used for our test experiments was one of the first that were manufactured and had a noise level of 0.7 pT/√Hz at about 77 K, as measured in a magnetically shielded flow cryostat. The 50 Hz noise was 0.5 pT_{RMS}. The results of the test experiments are summarized in table 1. In all the measurements we applied bias-current modulation to reduce the low-frequency noise of the high-T_c SQUID [3]. With the SQUID attached to the cold head and the cryocooler running, the noise appeared to have increased dramatically: 0.5 -1 nT/√Hz at 10 Hz, at a stable temperature of about 60 K. At 50 Hz the noise was 0.5 μT_{RMS}. A μ-metal shield around the compressor did not have any reducing effect on the noise. Because in earlier experiments we measured a shielding factor of 40 for this shield [13], we concluded that the noise did not originate from the compressor. To investigate the effect of the connecting leads on the noise (50 Hz and RF pick-up), we turned down the power supply so that the cooler stopped operating. Despite the SQUID slowly warming up, the noise could be measured (an increase in low-frequency noise occurs due to the temperature drift). The noise in this case was roughly 40 dB lower than with the cooler running. Also the μ-metal appeared to shield the remaining 50 Hz noise effectively. This noise, therefore, is caused by currents running in the compressor coils that are due to 50 Hz noise in the power supply and pick-up in the leads. The "white" noise level is high compared to that measured in the flow cryostat because of the temperature drift, and because of RF interference coupled into the shielded room via the power lines [14]. Finally, also the leads were disconnected from the cryocooler. Then, the noise decreased by a further factor of 5, roughly to the noise level measured in the flow cryostat.

These experiments indicate that the noise coupled into the SQUID is dominated by contributions originating from the cold head. The largest part is most probably due to slightly magnetic regenerator material inside the displacer that oscillates with a 3 mm amplitude, relatively close to the SQUID. From these results we conclude that the SQUIDs cannot be directly attached to the cold tip of the cooler. A separation in space with a thermal interface is necessary, and special precautions have to be taken to reduce the magnetic and the mechanical cooler interference.

Reduction of magnetic cryocooler interference

The magnetic field generated by the 50-Hz compressor currents can be described as dipolar with the magnetic moment located in the centre of the compressor, pointing along its axis. The magnetic moment of the Signaal Usfa 7058 compressor is about 0.5 Am^2_{RMS} [13]. Because the field is dipolar it is possible to position the coolers with respect to the SQUID magnetometers in such a way that minimum pick-up is established. SQUID magnetometers are sensitive to magnetic fields which have a component normal to the plane of their sensing coils. Therefore, we should position the coolers in such a way that their resulting magnetic field at the SQUIDs lies in the plane of the sensing coils. We expect that we can realize this coplanar alignment within an accuracy of 1°. Furthermore, the 50-Hz magnetic moments of the two compressors should point in opposite directions. In this case, with a distance of 15 cm between the centres of the compressors and a distance of 0.5 m to the SQUIDs a maximum normal component of 6 nT_{RMS} is generated. We consider this as acceptable because it is below the 50 Hz noise level present in clinical environments (10 - 50 nT_{RMS}) [4].

Another contribution to the 50 Hz noise arises from movements of magnetic cooler components. This concerns the compressor magnet and the slightly magnetic displacer inside the cold head. However, we have estimated that in our design this contribution is far below the acceptable level [15].

Reduction of mechanical cryocooler interference

As mentioned before, each compressor comprises two pistons that move in opposite directions to balance the axial forces. In standard coolers the two compressor coils are connected in parallel. A small imbalance remains resulting in vibration peaks of about 1 N corresponding to accelerations of roughly 1 m/s^2. As a specialty our cryocoolers were manufactured with separate current connections for the two compressor coils. By tuning the compressor currents we were able to improve the balance and thus to reduce the 50 Hz peak by a factor of 50, whereas also the 100 Hz and the 150 Hz peaks were reduced. These experiments were performed in a test set-up in which the compressor was mounted on thin flexible plates allowing relatively large displacements [15].

As a further result, the tuning and thus the balance of the axial forces in the compressor appeared to be very stable in time, and also independent of the cold-tip temperature. Therefore, we plan to use the following concept for the heart scanner. Each of the two compressors is first characterized in the flexible test set-up. The optimum ratio of the amplitudes and the optimum phase difference of the compressor-coil currents are determined. After that, both coolers (compressors and cold heads) are rigidly mounted on a supporting structure that also carries the SQUIDs. The compressor-coil currents are controlled by means of a Digital Signal Processor board according to the predetermined conditions. In this configuration the remaining vibrations of the supporting structure are predominately caused by the imbalance in the axial forces of the two separate displacers. These vibrations will be detected and fed back to the computer. Then, also these vibrations can be minimized by automatically tuning the amplitudes and phases of the currents in one compressor with respect to those in the other.

THERMODYNAMICAL ASPECTS OF THE HEART SCANNER

Because of the magnetic interference the SQUIDs have to be separated from the compressors by 0.5 m (then the distance to the cold heads is 0.3 m). Therefore, a thermal interface is required. For a larger cooler we have utilized a gas-circulating interface [16], but for the smaller coolers a solid heat conducting strip is much simpler. To minimize the mechanical coupling between the cold tips and the SQUIDs this conducting interface should at least partly be flexible. The SQUIDs will be mounted on a plate that is connected to the conducting interface. This plate should have a good thermal conductivity combined with a small electrical conductivity (necessary to limit Johnson noise). To simulate the thermodynamics of this configuration, a model was developed containing the following parameters:

1 - the cooling power of the 7058 cryocooler as a function of the cold-tip temperature,
2 - the heat load from the environment,
3 - the thermal interface: dimensions, heat conductivity and heat capacity,
4 - the SQUID plate: dimensions, heat conductivity and heat capacity.

Ad.1. The cooling power depends on the input power and the heat contact of the compressor and the displacer housings to the environment (to carry off the heat of compression). We measured a cooling power Q_e that linearly relates to the temperature as $Q_e = aT\text{-}b$ with a = 0.04 W/K and b = 1.73 W. In these measurements the input power was 53 W and the cold head housing temperature 50 °C.
Ad.2. The cold part of the heart scanner is thermally insulated from the environment by placing it in a vacuum box. Furthermore, it will be wrapped in multi-layer insulation (MLI). The heat load from the environment consists of heat flow through spacers and through wires connecting the SQUIDs to the read-out electronics, and heat flow through the MLI blanket. The former flow is expressed via standard heat-conductivity integrals, the latter is determined via ref. 17.
Ad.3 and 4. The dimensions and the material properties of the thermal interface and the mounting plate are variables in the model and are subject to optimization.

The model we developed is based on the implicit Crank-Nicolson differencing scheme and uses the Newton-Raphson method for linearizing the radiative heat flow [18]. It was validated in a setup in which a 7 mm diameter copper rod of 30 cm in length was connected to a single cold head via a 13 cm long copper flexible link (cross-sectional area 20 mm^2). These test experiments are described elsewhere [19].

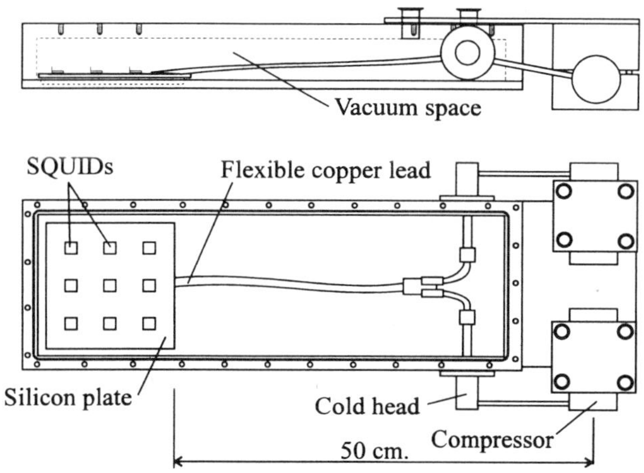

Figure 3. Heart scanner design (depicted with 9 SQUIDs).

HEART SCANNER DESIGN

In the actual design both coolers are incorporated. The cold heads are connected to the SQUID plate via two copper flexible links (each with a cross-sectional area of 15 mm^2 and a length of 30 cm). The SQUID plate is cut from an 8 inch silicon wafer and measures 15 cm x 15 cm x 0.7 mm. Together with the thermal links the plate will be wrapped in about 15 layers MLI and placed in a more or less rectangular non-metallic vacuum box. The coolers are mounted on this box in a rigid manner as discussed before. The SQUIDs will be glued on to the silicon plate and the electric connections to the outside world will be made with manganine high-resistance wire (to reduce the thermal load). This design is schematically depicted in figure 3. A cool-down curve for this heart scanner design, when equipped with 25 SQUIDs, is presented in figure 4. In this simulation no extra heat capacity was taken into account for SQUID packaging. In this case a SQUID temperature below 55 K can be realized in roughly 50 minutes.

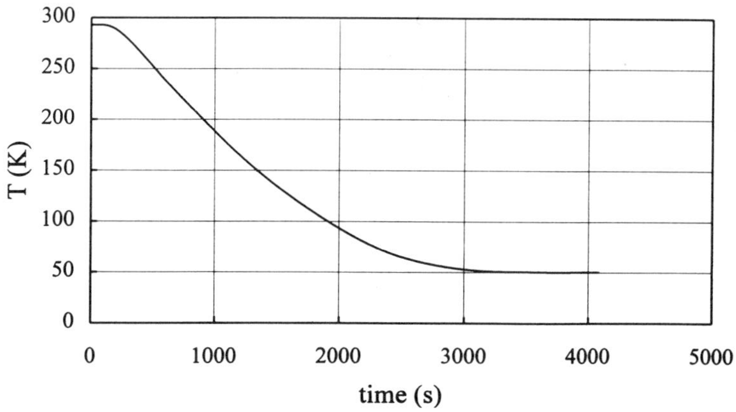

Figure 4. Cool-down curve of the heart scanner (with 25 SQUIDs). Temperature of the SQUID plate versus time.

ACKNOWLEDGEMENTS

This work is financially supported by the Dutch Research Program on High-T_c Superconductors (NWO/NOP-HT_c) and by Philips Research Laboratories Hamburg in Germany.

REFERENCES

1. G. Walker and E.R. Bingham, "Low Capacity Cryogenic Refrigerators" Oxford University Press (1994).
2. H. Rogalla, Superconducting electronics, *Cryogenics* 34 ICEC:25 (1994).
3. J.W.M. Hilgenkamp, High-Tc dc SQUID magnetometers, Ph.D. thesis, University of Twente, Enschede (1995).
4. H.J.M. ter Brake et al., Design of a high-Tc SQUID based heart scanner, European Conf. on Appl. Supercond. EUCAS '95, Edinburgh 3 - 6 July 1995.
5. W.A.M. Aarnink et al., Temperature dependence of the effective pick-up area of high-Tc dc SQUIDs; implications for stabilization of the SQUID operating temperature, submitted to *Appl. Physics Lett.*
6. R.G. Scurlock, private communication.
7. R.R. Clappier and R.J. Kline-Schoder, *Adv. Cryogenic Eng.* 39B:1177 (1994).
8. D. Verbeek et al., Performance of the Signaal Usfa Stirling cooling engines, Proc. 7th Int. Cryocooler Conf. Santa Fe 1992 (1993) p. 728.
9. T. Varpula and T Poutanen, *J. Appl. Phys.* 55:4015 (1984).
10. N. Kasai et al., *Cryogenics* 33:175 (1993).
11. J.A. Overweg and M.J. Walter-Peters, *Cryogenics* 18:529 (1978).
12. N. Khare and P. Chaudhari, *Appl. Phys. Lett.* 65:2353 (1994).
13. H.J.M. ter Brake et al., *Adv. Cryogenic. Eng.* 39:1287 (1994).
14. R.H. Koch et al., *Appl. Phys. Lett.* 65:100 (1994).
15. J.F.C. Verberne et al., Reduction of the vibrations generated by Stirling cryocoolers for cooling a high-Tc SQUID magnetometer, in: "Cryocoolers 8", proc. 8th Int. Cryocooler Conf. Vail USA 1994, R.G. Ross ed., Plenum Press New York (1995) p. 465.
16. P.J. van den Bosch et al., *Cryogenics* 35:109 (1995).
17. H.J.M. ter Brake and J. Flokstra, Computer aided cryostat design at Twente: recent developments, proc. 12th Int. Cryogenic Eng. Conf. , Southampton 1988 p.88.
18. W.H. Press et al., "Numerical Recipes", Cambridge University Press (1986).
19. H.J.M. ter Brake et al., The application of small Stirling-type cryocoolers in a heart scanner based on superconducting magnetic sensors, 19th Int. Congress of Refrigeration, The Hague The Netherlands, August 20-25 1995.

PREDICTION OF THE CHARACTERISTICS OF BIOLOGICAL TISSUE BEING DESTROYED WITH A LOCAL CRYOAFFECTION

Yu.P.Filippov

Joint Institute for Nuclear Research
141980 Dubna, Russia

ABSTRACT

The report presents the technique, calculation results and diagrams which allow to determine the interrelations between such parameters of cryogenic action as the time period, rate of cooling of the frozen phase, the relation between the depth of proposed necrosis and the depth of the frozen tissue, and the size of the cryoprobe and its surface temperature. The results of simulation are in good agreement with the data obtained during cryotherapy applications carried out at a zoological clinic.

INTRODUCTION

In medical practice biological tissue can be irreversibly destroyed in a localized area by freezing. For such a process the simulating temperature profiles help to predict the results of cryoaffection and obtain necessary information to choose a cryoprobe.

Of particular importance while freezing of biological tissue for medical purposes are the following[1, 2]:

- the period, t_n, of cryogenic action to reach temperature T_n for necessary depth R_n of tissue;
- the rate of cooling or the rate of temperature change $V = (T_f - T_n)/\Delta t$ which influences the probability of the cell destruction;
- the relative depth of necrosis, R_n/R_f, that is the correlation between the depth of proposed necrosis R_n and the depth of frozen tissue R_f, as shown in Figure 1.

In this report it is considered that parameters T_n and R_n are those where the probability of the tissue destruction is extremely high. Quantities of T_f and T_n depend on the kind of tissue. While freezing they can change, usually they are within the T_f range from 268 K to 273 K and T_n from 243 to 253 K.[1] Of course, values V, $R_n(t_n)$ and $R_f(t)$ depend on the surface temperature T_c of the cryoprobe, its size R_c and shape; here t is a current time.

TECHNIQUE OF CALCULATION

In principle, one can analyze the heat regime of the biological tissue solving numerically a system of equations for the frozen phase and tissue. For this purpose it is necessary to know the metabolic heat generation rate, the heat generation rate due to the blood flow and the thermal properties of the both phases. However, to obtain these data is difficult enough for the tissue: as shown in Reference 3, the value of the heat generation rate due to the blood flow can change while freezing by several times and significantly depends on T_c and R_c.

A solution of the task may be easier if to describe the properties of the tissue with an integral parameter which characterizes processes of the unfrozen tissue. It was suggested [2] using alteration of the radius of the frozen phase $R_f(t)$ as the boundary condition to simulate the heat regime of the frozen tissue. The way to determine dependence $R_f(t)$ and temperature profiles both in the frozen zone and tissue, was described in Reference 2 for the case of hemispheric cooling surface. Certain experimental data are presented in Reference 2 as well. These data have been obtained for the muscle femoral tissue (in vivo) of dogs during cryotherapy applications carried out at R_c = 3.5 mm, R_c = 6 mm and T_c = 88, 133 and 173 K at the Moscow Center of Oncology, Zoological Division. The cryogenic system reduced the needed temperature up to $T_c \geq$ 88 K in the interval from 5 to 8 sec. This time is much less than the total time of cryoaffection. The way to determine the temperature distribution in the frozen phase at the known law of the frozen tissue boundary movement $R_f(t)$ and its properties (density ρ, specific heat capacity c(T) and thermal conductivity $\lambda(T)$), is presented in Reference 4. Comparison of the results of the T(R,t) calculations with the experimental data[3] has shown that the maximum deviations did not exceed 4 or 5 K (2.5 or 3 %) for initial periods when t values are less than 8 sec. and $\simeq$1.5 K for t>30 sec. In this case values c(T) and $\lambda(T)$ were determined as $c(T) = c_0 + aT$ and $\lambda(T) = \lambda_0 - bT$ basing on the experimental data for ice. If it is not done (λ = const), the errors may be as large as 20 %. The comparison has also shown that for practical application one can assume a steady state temperature distribution in the frozen tissue at any given moment if $\lambda = \lambda_0 - bT$.

The technique of approximate analytical determination of $T[R_c, R_f, R, T_f, T_c, \lambda(T)]$ and $t[R_c, R_f, T_c, T_f, \lambda(T)]$ is presented in Reference 2; here value R is a coordinate. This technique allows to calculate the characteristics V, $R_n(t)$, $R_f(t)$ for those values of R_c and T_c at which experiments have not been carried out.

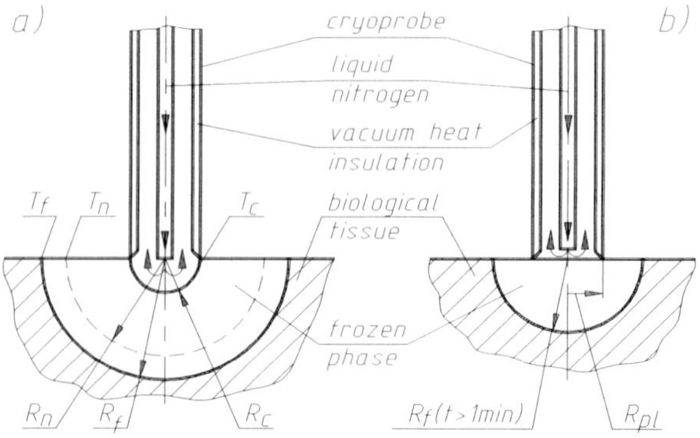

Figure 1. The scheme of the cryoprobe and its position in the tissue during cryoaffection.

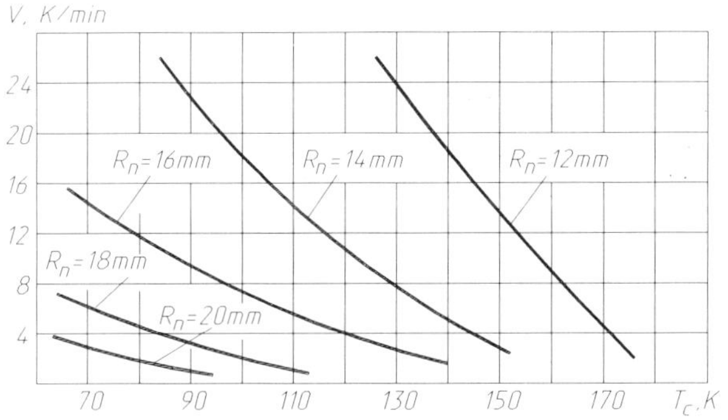

Figure 2. The dependence of cooling rate of the frozen tissue versus temperature of cryoprobe surface, R_c=6.0 mm.

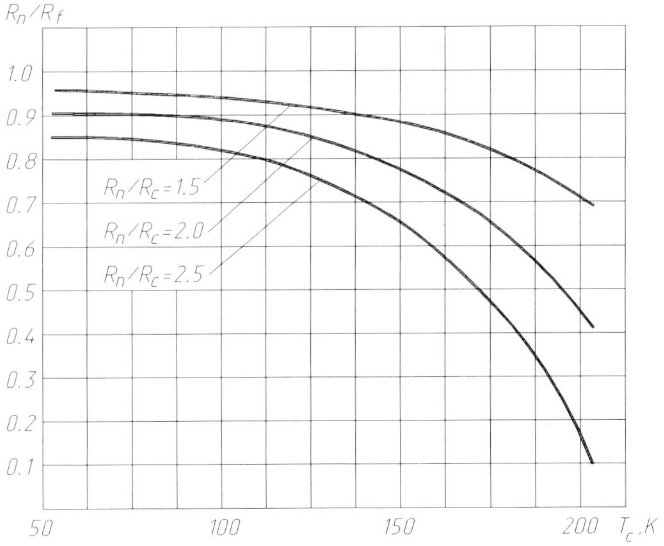

Figure 3. The relative depth of necrosis on the cryoprobe surface temperature.

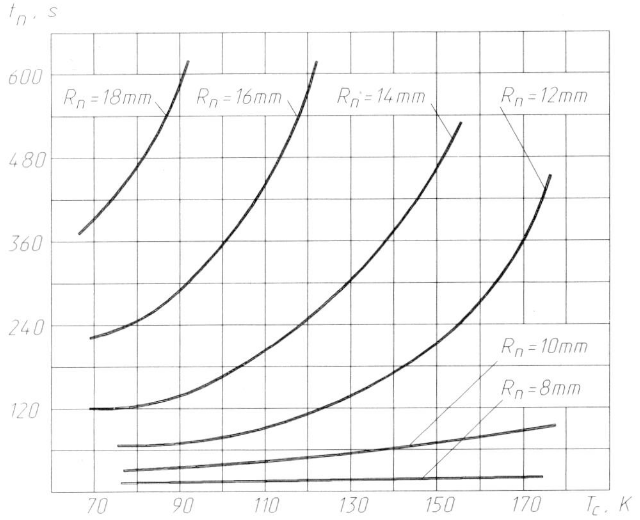

Figure 4. The period of cryogenic action to reach temperature T_n for depth R_n versus the cryoprobe surface temperature, R_c=6.0 mm.

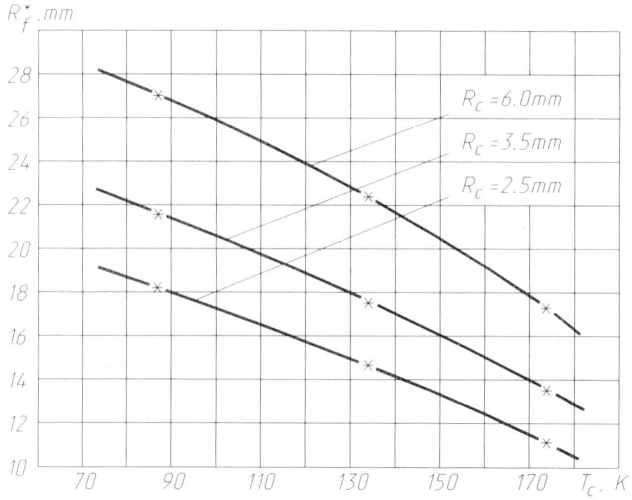

Figure 5. The steady state depth of frozen tissue versus the cryoprobe surface temperature.

CHOICE OF CRYOPROBE TEMPERATURES AND SIZES

The choice of cryoprobe temperatures for the certain case can be performed basing on such characteristics as $V(T_c, R_c, R_n)$, $R_n/R_f(T_c, R_n/R_c)$ and $t_n(T_c, R_n, R_c)$ which may be determined with approximate technique mentioned above for the muscle type of tissue in vivo (see Figures 2, 3, 4).

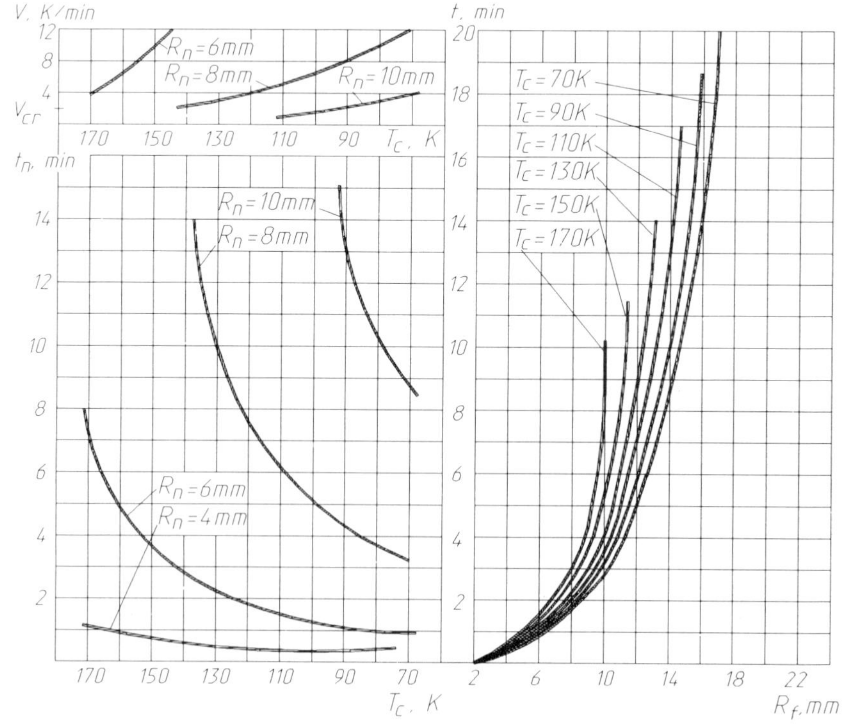

Figure 6. Diagram to predict the results of cryoaffection for the hemispheric cooling surface with radius R_c=2 mm.

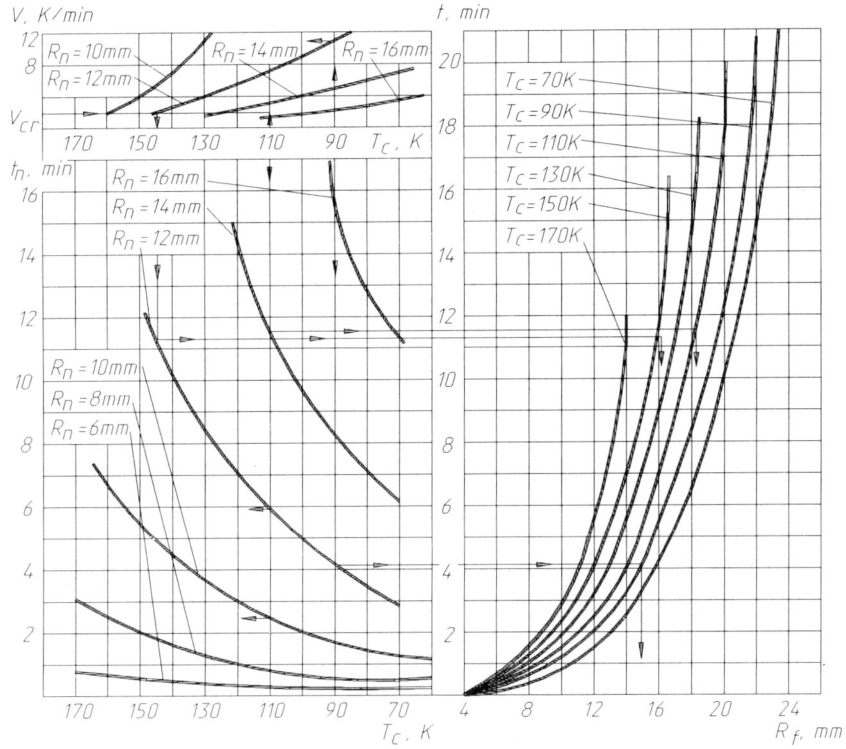

Figure 7. The same as for Figure 6, R_c=4 mm.

Curves V(T_c) are shown in Figure 2. One can see that V values are approximately proportional to T_c decreasing; and the usage of cryogenic temperatures, $T_c \leq 120$ K, allows one to provide comparatively rapid cooling, V≥2 K/min , at relatively large distances, $R_n/R_c \geq 2.5$.

Figure 3 illustrates curves $R_n/R_f(T_c)$ for different relative depths R_n/R_c of necrosis. The data analysis reveals that cryogenic temperatures allow to decrease significantly value $\Delta R = R_f - R_n$ near the area of supposed necrosis. For example, the value of ΔR is about 6.5 mm for R_n=12 mm, $R_n/R_c = 2.0$ and T_c=175 K. In its turn, ΔR is about 1.3 mm for T_c=75 K. One can stress that good results for comparatively small sizes of lesion, for example, $\Delta R \simeq 2$ mm only for R_n=9 mm at $R_n/R_c = 1.5$, may be provided if to use relatively high temperatures – T_c=175 K.

T_c decreasing cuts down appreciably the interval of t_n to reach the temperature of T_n for the R_n depth, particularly, it is essential for large values of R_n (see Figure 4).

Figure 5 allows to estimate the steady state depths R_f^* which can be reached at different temperatures. The values of R_f^* marked with the stars have been obtained experimentally for the muscle femoral tissue of dogs. The cryosurgery system was supplied with the control valve to regulate the value of T_c. One can note that T_c decreasing in the range of $T_c \leq 120$ K, provides a slower growth of R_f^* with respect to the range of $T_c > 120$ K.

Thus, satisfactory regimes of cooling can be provided by T_c from 170 to 200 K for relatively small depths of necrosis ($R_n/R_c \leq 1.5$). It seems reasonable to use nitrogen temperatures (T_c from 77 K to 90 K) in order to reach the necessary heat regimes within comparatively large sizes ($R_n/R_c \geq 2.5$) of the frozen tissue.

DIAGRAMS TO PREDICT THE RESULTS OF A LOCAL CRYOAFFECTION

Before cryosurgery procedures it is necessary to choose the key parameters of cryoaffection: the shape and size of the cryoprobe cooling surface, its temperature, and to estimate interval t_n. These parameters have to provide the given characteristics of the tissue heat regimes, $T_n \leq T_{cr}$ and $V \geq V_{cr}$, which allow to destroy the cells in area R_n with a high probability and minimum possible destruction of tissue near R_n. In this report the values of T_{cr} and V_{cr} are chosen as 253 K and 2 K/min [5], correspondingly. Of course, they depend on the type of the tissue and can be revised.

The diagrams to predict the results of local freezing are shown in Figures 6, 7, 8. They are composed for hemispheric cooling surfaces, R_c=2, 4, 6 mm, its temperatures, T_c from 70 K to 170 K, and the tissue (in vivo) like the muscle one mentioned above. Each diagram consists of three plots: plot I, cooling rate (T_c, R_n); plot II, time of freezing (T_c, R_n); plot III, depth of frozen tissue (t, T_c). The technique to estimate the results for a plate circle cooling surface at the same T_c is described below.

Let us consider two examples. First, let it be R_c=4 mm and T_c=110 K, and the task is to estimate capabilities of these key parameters. One can see (Figure 7, plot I) that the values of $V > V_{cr} = 2K/\min$ may be reached at $R_n \leq 15$ mm. The intervals to reach T_n=253 K at R_n=14, 12 and 10 mm are $t_n \simeq 11.5, 5.8$ and 2.5 minutes, correspondingly (plot II). Plot III helps to determine, for example, that for R_n=14 mm

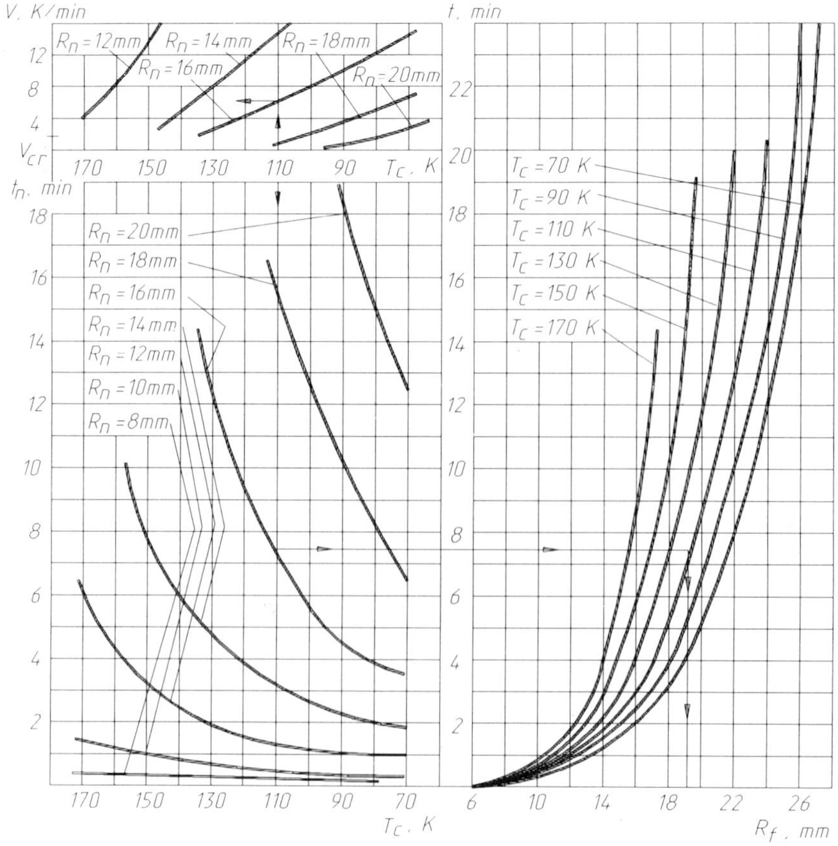

Figure 8. The same as for Figure 6, R_c=6 mm.

values R_f and $\Delta R = R_f - R_n$ are $\simeq$ 18.2 and 4.2 mm. In order to provide an acceptable heat regime of the tissue at $R_n \geq 16$ mm and T_c=110 K, it is necessary to use another cryoprobe: with R_c=6 mm. Figure 8 shows that in this case the values of V, t_n, R_f and ΔR are $\simeq$6 K/min , 7.4 min., 19.2 mm and 3.2 mm.

Second, let us suppose that it is necessary to choose R_c and T_c in such a manner to provide an acceptable heat regime for R_n=12 mm at $R_c = 4$ mm. Figure 7, plot I, shows that it requires $T_c \simeq 145K$ to reach $V_{cr} = 2K/\min$; in this case $t_n \simeq 11.3\,\text{min}$ (plot II), the values, R_f and ΔR, are 16.2 and 4.2 mm. To cut down t_n, we need to decrease temperature T_c. Thus, at $T_c = 90K$ the values of V and t_n are $\simeq$ 11 K/min (it can increase a probability of cells destruction with respect to the previous case) and $t_n \simeq 4.2$ min. In its turn, the values of R_f and ΔR are approximately equal to 15 and 3.0 mm.

As for the cryoprobes with the plate circle cooling surfaces (see Figure 1.*b*), one can stress the following. Experiments have shown that the shape of frozen zones look like hemispheric for the intervals of cryoaffection more than 1 minute. The data analysis showed that the heat regime characteristics are approximately the same for both – hemispheric and plate circle cooling surfaces if $R_c^{eq} \simeq kR_{pl}$, where R_c^{eq} is the radius of the equivalent hemispheric probe, R_{pl} is the radius of the plate circle probe and k is coefficient, $k \simeq 2^{-1/2}$. Thus, the cryoprobes with the hemispheric surface R_c=4 mm and the plate circle one $R_{pl} \simeq \sqrt{2} \cdot R_c \simeq 5.6$ mm, can yield the similar results. Such a procedure allows to use the diagrams mentioned above for the cases with the plate circle cooling surfaces as the first approximation if $t_n > 1$ min.

CONCLUSIONS

The used method allows one to reduce significantly the amount of experimental data required to simulate the heat regime of a biological tissue in vivo during freezing.

One can use the diagrams shown in Figures 6, 7 and 8 in order to predict the results of a local cryoaffection destroying a biological tissue in vivo with the cryoprobes of hemispheric and circle plate cooling surfaces at its radii R_c from 2 mm to 6 mm and temperatures T_c from 70 K to 170 K. The results are in a satisfactory agreement with the experimental data obtained for a muscle tissue in vivo whose metabolic heat generation, the heat generation due to the blood flow and thermal properties are similar to the ones of the muscle femoral tissue of dogs. For another type of a tissue one can obtain the corresponding results using the proposed method if the steady-state depths of the frozen tissue are known.

REFERENCES

1. R.Rand, A.Rinfert, and H.Leden, Cryosurgery, Springfield, 1968.
2. M.Boyarsky and Yu.Filippov, Calculation of Heat Regime of Biological Tissue during Freezing, *Cryobiology*, 16:492 (1979).
3. Yu.Filippov and A.Vasilkov, Simulation Technique to Calculate a Heat Regime of Biological Tissue while Local Freezing, *Engineering Physical Journal,* Vol.36, 6:1100, (1979), Minsk, in Russian.
4. Yu.Filippov, Removal of Intensive Heat Flux with Liquid Nitrogen Jet Impingement, in: "Advances in Cryogenic Engineering," Vol. 37A, Plenum Press, New York, (1991), p. 189.
5. A.Paches, editor, Cryogenic Method to Treat the Swelling of Head and Neck, Medicine, Moscow, (1978), p. 24, in Russian.

SUPPLEMENTAL MULTILAYER INSULATION RESEARCH FACILITY

P. J. Dempsey and R. J. Stochl

National Aeronautics and Space Administration
Lewis Research Center
Cleveland, Ohio, 44135, U.S.A.

ABSTRACT

The Supplemental Multilayer Insulation Research Facility (SMIRF) provides a small scale test bed for conducting cryogenic experiments in a vacuum environment. The facility vacuum system is capable of simulating a Space Shuttle launch pressure profile as well as providing a steady space vacuum environment of 1.3×10^{-4} N/m^2(1×10^{-6} torr). Warm side boundary temperatures can be maintained constant between 111 K(200 R) and 361 K(650 R) using a temperature controlled shroud. The shroud can also simulate a typical lunar day-night temperature profile. The test hardware consists of a cryogenic calorimeter supported by the lid of the vacuum chamber. A 0.45 m^3(120 gal) vacuum jacketed storage/supply tank is available for conditioning the cryogen prior to use in the calorimeter. The facility was initially designed to evaluate the thermal performance of insulation systems for long-term storage in space. The facility has recently been used to evaluate the performance of various new insulation systems for LH_2 and LN_2 ground storage dewars.

INTRODUCTION

SMIRF is a unique facility for ground based liquid hydrogen, liquid nitrogen, and vacuum experimentation. Tests that require vacuum and cryogenic capabilities of all types can be performed in this facility. The facility was designed to evaluate thermal performance of cryogenic insulation systems for long-term storage in space. The facility can simulate the pressure history of a shuttle launch, the pressure and temperature history on the lunar surface, and the pressure and temperature of space. SMIRF has most recently been used to evaluate various ground based insulation systems to be used by a commercial dewar manufacturer. This paper describes the SMIRF test facility and its capabilities.

DESCRIPTION OF FACILITY

General

The test facility (figure 1) is located in building 204 of the South 40 area of NASA Lewis Research Center. The steel frame building consists of a control room, shop, test area, and a basement under the test area. The building is surrounded on three sides by 4.6 m (15 ft) earthen mounds. The control room is separated from the shop by a 0.4 m (1.3 ft) thick concrete wall. The shop in turn is separated from the test area by a 20.3 cm (8 in.) thick concrete block wall. The test area is 31 ft high and surrounded by thirteen 3.0 m by 3.7 m (10 ft by 12 ft) panel lift doors. A crane with a 1814 kg(4000 lb) 6 m (20 ft) lift capacity is located in the test area.

Vacuum System

A 304 stainless steel cylindrical vacuum chamber is located at ground level in the test area. The chamber is approximately 250 cm(98.3 in.) high by 181 cm(71.3 in.) in diameter with a dished lid and bottom (figure 2). A mezzanine level allows access to the chamber and lid. Three diffusion pumps and 5 mechanical vacuum pumps provide the required vacuum capability. Four mechanical pumps are used as roughing pumps to evacuate the chamber. The fifth pump is used as a backing pump for the diffusion pumps. The exhaust line from the pumps is outside of the building and 4.6 m(15 ft) above the roofline. The total capacity of the roughing pumps is 33.4 m^3/min(1180 ft^3/min).

Each of the mechanical vacuum pumps are isolated from the vacuum line by a remote operated vacuum gate valve. The vacuum system is configured such that one pump can be used to vacuum purge the test article and LH_2 supply lines. All four of the vacuum system pumps can be used to pump down the vacuum chamber. Thermocouple vacuum gages are installed on the inlet to the vacuum pumps to measure inlet pressure. A thermocouple is installed in the pump oil reservoir to measure oil temperature. Flow switches verify cooling water is supplied to the pumps.

A water chiller provides cooling water for the pumps. The water chiller is a self-contained unit consisting of a reservoir chiller and pump. The reservoir has a capacity of 0.053 m^3(14 gal) and contains a 50/50 water glycol solution. The chiller has a heat rejection capability of 53 kW.

Three NRC diffusion pumps are attached directly to the vacuum chamber. The displacement of each pump is 4000 liters/sec. Each diffusion pumps is isolated from the vacuum chamber by a 25.4 cm(10 in.) gate valve.

The diffusion pumps are equipped with cryotraps. The LN_2 filled cryotraps trap (by freezing) any oil from the diffusion pumps which may back stream into the vacuum chamber. The LN_2 level in the cryotrap is controlled by a closed loop level controller.

The three diffusion pumps are backed up by a mechanical vacuum pump located in the basement of the test area. A GN_2 gas ballast valve is used on the pump to prevent contamination

Figure 1.—Supplemental multilayer insulation research facility.

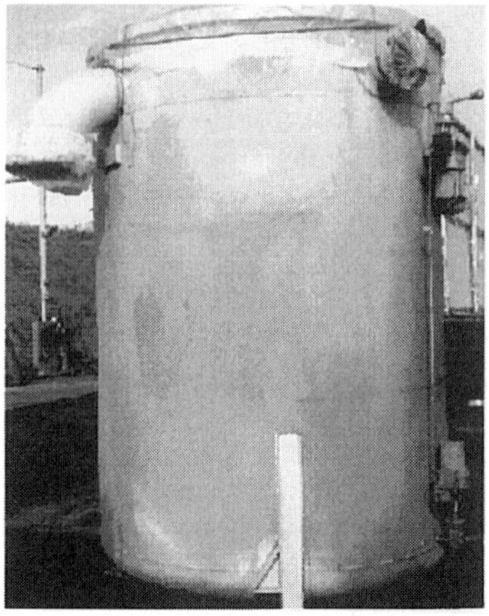

Figure 2.—Vacuum chamber.

of the oil reservoir. A thermocouple vacuum gage is installed on the inlet to the vacuum pump to measure inlet pressure.

Each of the diffusion pumps can be isolated from the roughing pumps. In the event of a failure of the roughing pump it is possible to switch to one of the other three mechanical pumps to back the diffusion pumps.

Flow switches on the diffusion pumps and mechanical backing pump verify sufficient cooling water is provided to the pumps. Thermocouples on the diffusion pump heaters indicate if the heaters malfunction. A thermocouple is also installed in the oil reservoir of the backing pump. If any of the temperature sensors or flow switches are outside the pump operating range, the pumps will automatically shutdown.

Shroud System

A temperature controlled shroud is located inside the vacuum chamber between the vacuum chamber wall and the research article. The purpose of the shroud is to provide a known radiative background temperature for the research article. The shroud can be controlled between 111 K (200 R) and 361 K(650 R). The shroud control system is also capable of producing a temperature profile over this range to simulate a lunar day. Temperature ramping capabilities are at least 0.17 K per minute over the entire range. Shroud temperature top to bottom can be controlled within ±3.0 K over the entire operating range.

The shroud is made of aluminum finned tubes that are attached to an upper and lower manifold. The shroud is shown on figure 3. The emissivity of the inner and outer surfaces of the shroud is less than 0.1. Two removable shields (one end attached to the shroud) are installed between the shroud and the test article. The shields are used to shield the test article from heat sources other than the shroud. The shroud is designed for operation at 1.3×10^{-4} N/m^2 (1×10^{-6} torr).

The temperature of the shroud is maintained by recirculating thermally conditioned GN_2 through the finned tubes. The shroud temperature is controlled at a desired value by a programmable controller which compares the input value from a thermocouple located on the shroud with the desired set temperature. Based on this temperature measurement, the GN_2 will either be heated or cooled. Heating is achieved by a 5kW electric immersion heater. The incoloy heater is enclosed in a stainless steel housing. The heater transfers heat by forced convection directly to the gas stream. Cooling is achieved by an LN_2 to GN_2 heat exchanger. A 14.2 m^3/min(500 ft^3/min) blower with a brushless motor is used to recirculate N_2. The blower is water cooled using the chilled water system.

Cryogenic Supply System

The cryogenic supply system consists of a roadable dewar, vacuum jacketed transfer lines, vent systems for the dewars, and a 0.45 m^3(120 gal) vacuum jacketed supply tank. The dewars are connected to the supply tank by a vacuum jacketed transfer line. Vacuum jacketed control valves control the flow to the supply tank. Dewar connections are made using bayonet fittings in the field. Each section of the vacuum jacketed piping is equipped with a combination pump-out, vacuum gaging, and pressure relief port.

A 0.45 m^3(120 gal) supply tank is used to supply cryogens to the test article. The tank is a double walled vacuum jacketed pressure vessel made of 304 stainless steel. The inside diameter is 0.6 m(2 ft) and the height is 2.4 m(8 ft). Reflective multilayer gettering materials are installed in the vacuum space. From the supply tank the fluid can enter the chamber through any or all of four separate fill lines.

Cryogenic Vent System

There are three vent lines available from the chamber, one 5.0 cm(2 in.) and two 2.5 cm(1 in.) lines. All three lines feed into a 7.6 cm.(3 in.) main vent which discharges 4.6 m (15 ft) above the roof. A separate test article vent path is required when evaluating insulation system thermal performance (boil-off). This is to prevent pressure disturbances in the other tanks from affecting the pressure controls system of the test tank.

Liquid Nitrogen System

The liquid nitrogen system consists of a 8 m^3(2100 gal) roadable dewar (N-65), flexible vacuum jacketed transfer lines and a vent system for the dewar. The dewar is equipped with a burst disc that relieves at 363 KPa(38 psig). The liquid nitrogen dewar is located on the west side of building 204. The main LN_2 vent discharges into a stainless steel drum with a level sensor installed in the top. If the level sensor indicates liquid nitrogen, the main LN_2 supply valve is automatically closed. This is to prevent a LN_2 spill in the area.

LN_2 is provided to the cryotraps on the diffusion pump inlet valves. The LN_2 level in the cryotrap is controlled by a closed loop level controller. LN_2 is also used in the shroud temperature control system.

Gaseous Helium System

GHe is used at SMIRF to purge the LH_2 system and the vacuum chamber. The facility is supplied by a 16,649 KPa(2400 psig), 1982 m^3(70,000 ft^3) tube trailer. GHe pressure is regulated to 170 KPa(10 psig) using two pressure regulators in series. If pressure of the main supply falls below 3549 KPa(500 psig), the supply automatically switches to two auxiliary 3.0 m^3(106 ft^3), 14,581 KPa(2100 psig) tubers.

Gaseous Nitrogen System

GN_2 is used for valve operator pressure, for the gas ballast supply, and for purging vent lines. GN_2 is also used to purge electrical boxes which contain components that are not rated for Class I, Division 2, Group B service per the NEC.

The main GN_2 supply is from a 8.8 m^3(310 ft^3), 16,649 KPa(2400 psig) storage tuber. If the supply pressure from this main tuber falls below 3549 KPa(500 psig), supply automatically switches to two auxiliary 3.0 m^3(106 ft^3), 14,581 KPa(2100 psig) tubers. The supply pressure is regulated down to 963 KPa(125 psig), 274 KPa(25 psig), and 170 KPa(10 psig) for the various operational usages.

Water System

A closed loop water chiller, operating at 289 K(520 R) supplies cooling water to the vacuum pumps, diffusion pumps, and shroud blower. The water chiller consists of a reservoir, a pump, and a heat pump type device for rejection of heat. The reservoir is filled with a 50/50 mixture of glycol and water to prevent freezing during winter months.

The chiller is equipped with temperature switches that shut down the compressor if the water temperature is less than 278 K(501 R) and shut down the entire unit if the water temperature exceeds 304 K(548 R). Pressure switches shut down the compressor and discharge fans if refrigeration pressure is less than 136 KPa(5 psig) and shut down the entire unit if refrigeration pressure is greater than 2790 KPa(390 psig). A pressure switch on the compressor shuts down the compressor and discharge fans if pressure is less than 163 KPa(9 psig).

Electrical Systems

All electrical components in the building are rated for Class I, Division 2, Group B service per the NEC or enclosed in nitrogen purged cabinets. Purge pressure is verified using a pressure switch that alarms if the pressure of a cabinet is less than 0.037 KPa(0.15 in. of H_2O).

Figure 3.—Temperature controlled shroud.

Figure 4.—Calorimeter.

Instrumentation cables run through the test area wall to signal conditioning in the shop area, then through an underground tray to the control room. Control cables run from the test area to the control room through the underground trays. Power cables run through the test area wall and control room wall. All penetrations are sealed at the test area eliminating any paths for the hydrogen to travel from the test area to the control room.

Hydrogen detectors are installed in the test area ceiling and basement. The H_2 detectors are set to alarm at 20% and 40% of the lower flammable limit (LFL) of H_2 in air. The hydrogen detectors alarm in the control room at 20%(LFL), in both the control room and fire station at 40%(LFL). Smoke detectors are installed in the control room and shop area. The smoke detectors alarm in both the control room and fire station when smoke is detected.

SMIRF operates with a Gould Modicon 984 programmable logic controller. The controller is used to control facility interlocks, permissives, and shutdowns.

Calorimeter

The calorimeter is used to provide a constant heat sink temperature during the evaluation of thermal performance of insulation systems. The boil-off from the calorimeter is used to determine the heat flux through the insulation.

The calorimeter, shown in figure 4, consists of three separate 76.2 cm(30 in.) diameter tanks. The center, or measure tank, is located between two guard tanks. The purpose of the guard tanks is to eliminate heat flow into the measure tank through its two ends. The tanks are constructed of 1.3 cm(0.5 in.) thick 1100 alloy aluminum.

The measure and guard tanks are supported through their centers by a stainless steel tube. During a test this tube is filled with the same cryogen that is in the tanks. This is to prevent conduction or radiation from reaching the measure tank. Insulation systems are placed and supported between two G-10 rings near the ends of each guard tank.

Calorimeter Tank Pressure Control

The determination of insulation thermal performance (heat flux) by measuring the boil-off from the measure tank requires precise pressure control of the measure tank. The tank pressure, 114 KPa(16.5 psia), is controlled by regulating boil-off flow. The measure tank control system consists of four parallel pneumatic control valves, a controller, a 6.8 KPa(1 psid) pressure transducer, and a constant pressure reference. The control system is sized to measure flows from 0.00056 m^3/hr to 90.6 m^3/hr(0.02 ft^3/hr to 3200 ft^3/hr) and maintain measure tank pressure to within $\pm$0.14 KPa($\pm$ 0.02 psia). The measure tank boil-off is measured by one or more of a series of five flowmeters.

The two guard tanks have to be maintained at a slightly higher temperature and therefore higher pressure of 117 KPa(17.0 psia) to prevent the condensation of the measure tank boil-off as it passes through the upper guard tank. This is achieved through another set of pneumatic control valves, controller, and a 34 KPa(5 psid) pressure transducer referenced to the same constant pressure volume as the measure tank. This guard tank pressure can be maintained to within $\pm$2.1 KPa($\pm$ 0.3 psia).

Accomplishments/Plans

To date this facility has been used to evaluate the performance of seven different insulation systems. Three were evaluated using LH_2 in the calorimeter and four used LN_2. All tests were successful.

Future plans for this facility include; a program to evaluate the venting characteristics of MLI insulation systems during rapid pressure decays, developing concepts to minimize heat transfer through insulation penetrations such as tank feed and vent lines and tank support members, measuring the thermal permeability of composite material, and demonstrating a propellant liquefaction and storage system in a Mars environment.

VARIABLE DENSITY MLI TEST RESULTS

R.J. Stochl,[1] P.J. Dempsey,[1] K.R. Leonard,[2] and G.E. McIntosh[2]

[1]NASA Lewis Research Center
Cleveland, OH

[2]Cryogenic Technical Services, Inc.
Boulder, CO

INTRODUCTION

A previous CEC paper reviewed MLI fundamentals and discussed a layer by layer computational method set up to take advantage of variable density separators to optimize performance. The work has advanced both conceptually and experimentally since 1993 and appears to be nearing commercial feasibility. In this paper we update and focus the MLI review, discuss current variable configurations, and compare experimental and predicted Q/A values.

BACKGROUND

We are reminded first that "floating shields" represent the ideal MLI with no intervening spacer material. In this case, neglecting gas conduction, heat transfer is solely by radiation and, for constant surface emissivities, varies with $1/(n+1)$ where n is the number of radiation shields. However, the floating shield concept is not generally feasible because some kind of spacer between reflectors is required to prevent thermal shorts. Spacing has been accomplished by wrinkling or dimpling aluminized Mylar® and by inserting a wide variety of separators between foils and metallized films. The function of the separators is to prevent the reflectors from shorting while creating a minimum solid conduction heat path.

A frequent misconception is that more reflectors always equate to lower heat transfer. This is only true if the spacing between the reflectors is kept reasonably constant or <u>increased</u> toward the low temperature surface. To illustrate this point, a simple calculation of combined radiation and solid conduction (Gas conduction was neglected.) was made for 0 to 100 reflectors between 20 and 300 K with 15 g/m^2 sheet glass paper separator. A shown in Figure 1, Q/A does decrease continuously with the number of layers although there is evidently a point of diminishing returns. It is important to note that the separator glass paper was assumed to have a constant thickness equivalent to 23.6 layers per cm.

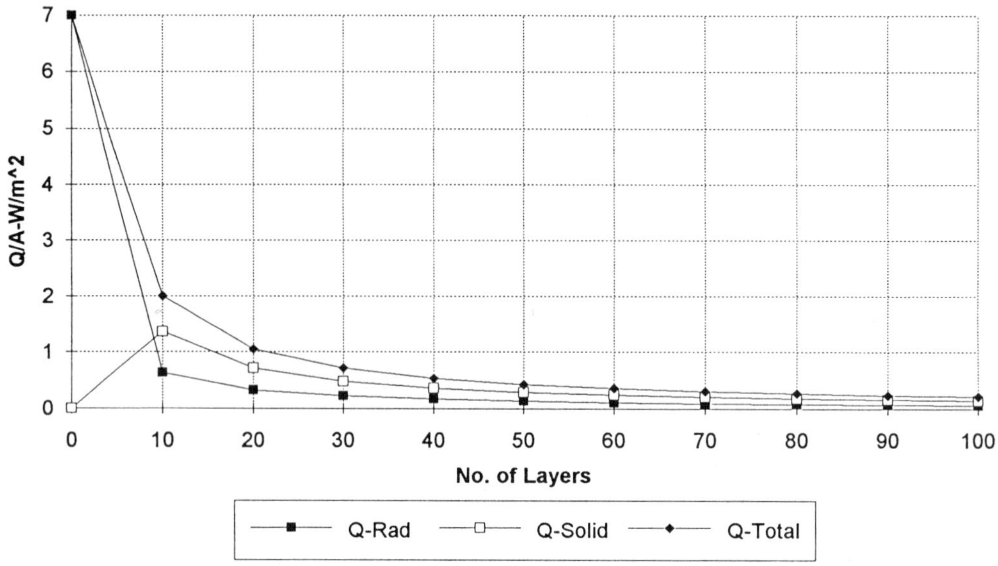

Figure 1. MLI Q/A vs. Layers and Thickness.

More commonly, a larger number of reflectors also results in a higher layer density which may actually increase heat transfer. A second example was prepared to illustrate this case using the same materials, temperature range, and emissivity (3 %) as above except that the total thickness was restricted to 0.0152 m (0.6 inch). As shown in Figure 2, the results of both sets of curves are the same out to 30 reflectors. However, between 30 and 40 layers, the glass paper separator starts to be compressed and solid conduction begins to increase. At 50 layers total Q/A has increased and continues to do so thereafter.

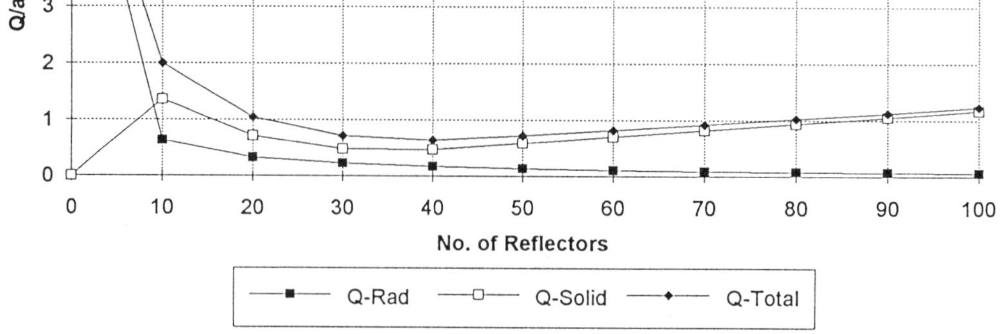

Figure 2. Q/A versus Layer Density.

Table 1. MLI design guidelines

1. "Floating shields" are superior to practical MLI with separators.
2. Q/A is continuously reduced with an increasing number of reflectors if the layer density does not increase.
3. There is an optimum layer density for each material combination and temperature range.
4. Optimum layer density decreases for lower temperature ranges.
5. For "warm" temperatures below 150 K, the outer MLI layer should be a reflector which does not touch the warm surface.

Although it remains a practical problem, the relation of Q/A to MLI density has been known for many years. Excellent publications[2,3&4] on the subject are referenced. It has also been generally known that the layer density should decrease toward the cold surface. Lockheed personnel treated this problem particularly well in Reference 4. In fact, Figure 3 consists of layer density values taken from a series of minimum Q/A values presented in Figure 6 of the referenced paper. It is clear from Figures 2 and 3 that excess layer density can lead to higher than expected MLI heat leak and that the minimum layer density decreases at lower temperature ranges. These conclusions are summarized in Table I.

VARIABLE DENSITY MATRIX

Typical MLI arrangements consist of a single separator layer between each reflector. This is quite effective at higher temperatures where radiation is the dominant mode

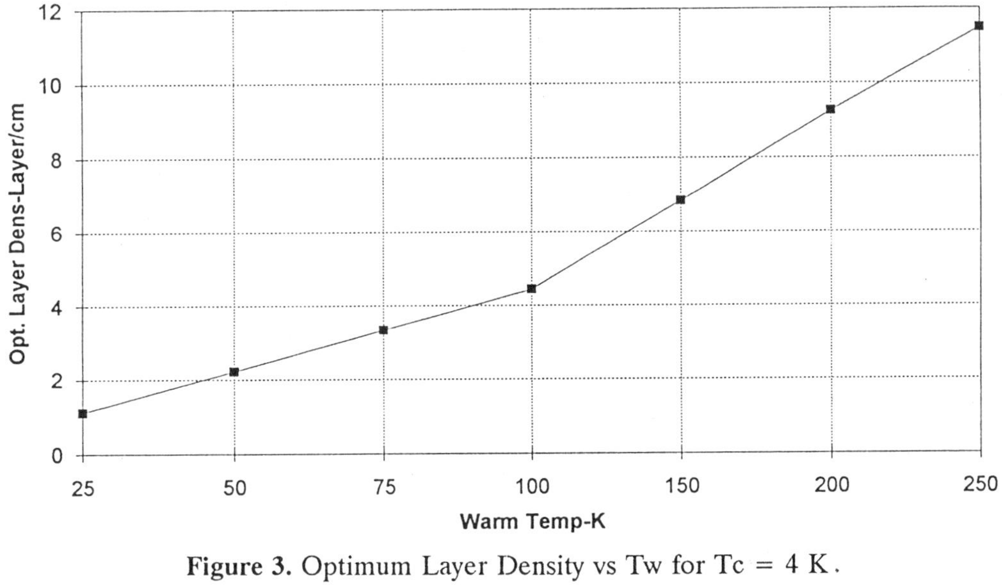

Figure 3. Optimum Layer Density vs Tw for Tc = 4 K.

of heat transfer but increasingly less efficient at lower temperatures where solid conduction is most important. In fact, standard one to one MLI is frequently worse than high vacuum and low emissivity surfaces between 77 and 4.2 K. A concept[1] to optimize MLI at each temperature level consists of using thin spacers at higher temperatures and thicker material in the coldest layers. This variable density concept and its related computer optimization technique makes it possible to calculate spacer thickness and density so that solid conduction more nearly matches radiation heat transfer through the coldest layers.

Study of variable density MLI applications brought two significant factors to light:

1. Variable density applications are more flexible if combinations of two or more thicknesses of material are used.

2. With the floating shield concept in mind, it becomes obvious that complete separator coverage is undesirable if reflector shorting can be prevented or minimized.

Thus the variable density matrix evolved. One configuration is illustrated in Figure 4. As shown, the matrix consists of longitudinal and cross strips of separator material which may be woven or laid up like railroad tracks and ties. In addition to the material itself, all of the dimensional parameters may be varied. With glass paper material, we have found it convenient to use a single thickness, X_1, of the longitudinal strips and vary thickness of the cross strips to achieve variable density. For the test reported herein, W = 30 mm, Z = Y = 152.4 mm, X_1 was about 0.2 mm (30 g/m^2), and X_2 was 0.2, 0.3, and about 0.5 mm (30, 45, and 84 g/m^2). This matrix covered about 39 % of the surface area and its density as actually applied to the NASA calorimeter ranged from 12.8 to 24.8 kg/m^3.

VARIABLE DENSITY MATRIX TEST

A quantity of 1.372 m (54 inch) wide matrix material was made up for testing on the NASA guarded cylindrical calorimeter described elsewhere. For reasons not recorded at the time, it was decided to put on 24 layers of thick matrix, 22 intermediate thickness wraps, and 10 layers of thin material on the outside. Wrapping details are given in Table 2.

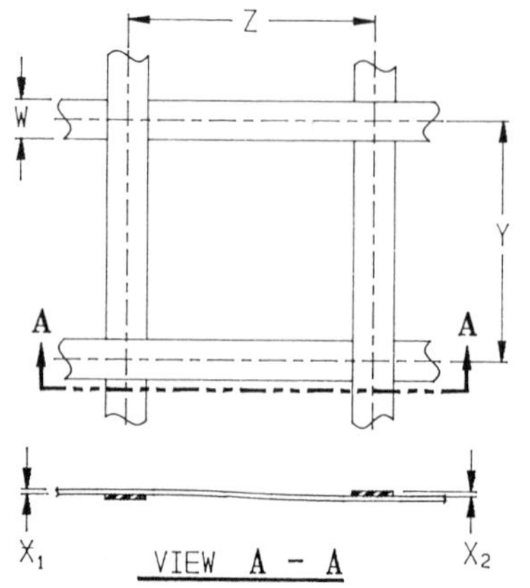

Figure 4 Variable density matrix concepts.

Table 2. Calorimeter matrix insulation wrapping detail

Layup	Layers	Actual Thickness	Density	Rel. Dens.
Thick	24	27 mm (1 1/16 in)	17.702 kg/m^3	0.0076
Medium	22	25.4 (1 in)	12.788	0.0055
Thin	10	4.76 (3/16 in)	24.797	0.0107

Double aluminized Mylar® was used as the reflector material and it was wound on the 30 inch diameter calorimeter as a continuous sheet. The outer wrap was Mylar so that the warm shroud would "see" a non-contacting reflective surface. 25 Type "E" (Chromel Constantan) thermocouples were installed as the insulation was wrapped. 13 of the TC's were located on the MLI center line to measure the radial temperature gradient. The remaining 12 were located on layers 11, 26, and 50 (Measuring from the cold surface.) to check lateral temperature gradients over the outer edges of the test cylinder and the adjacent edges of the upper and lower guard chambers. The thermocouple leads were tempered for about 48 inches to assure reasonable accuracy and minimal thermal impact on the test results. Thermocouple readings were automatically monitored during the heat leak test and, as reported below, they made a major contribution to the accuracy of the results.

After installation and evacuation, the test and guard volumes were filled with liquid hydrogen and the thermal control shroud temperature was set at 310.94 K (100 F). The volumes were topped off twice during the first two days of cooldown and again on the fourth night. By the fifth day, heat leak readings were essentially stable at a Q/A value approximately 50 % higher than 0.213 W/m^2 predicted by the layer by layer computational method previously[1] reported. Fortunately, just before shut down, Ms. Dempsey and her crew noted that the over-driven diodes being used as liquid level sensors might be influencing the results. When the diodes were turned off, the boiloff rate immediately dropped by a third to a final value of Q/A = 0.212 W/m^2.

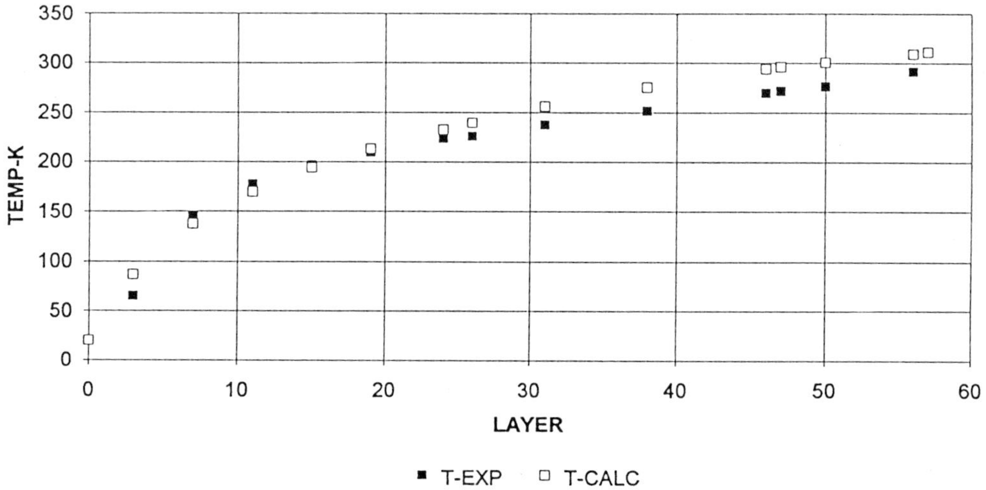

Figure 5. Matrix MLI Temperature Distribution.

DISCUSSION OF TEST RESULTS

At the end of the variable density matrix MLI test it appeared that the test and predicted Q/A values were almost identical. However, as shown in Figure 5, the experimental temperature distribution indicated a large jump from the outer reflector to the shroud in contrast to the smooth calculated curve. Since radiative view factors and emissivities were suspected, the calorimeter was removed from the test enclosure and an emissometer was used to check the viewing surfaces. These values and the radiation configuration are shown in Figure 6. An extensive view factor analysis[5] conducted by R. Stochl indicated that all of the emissivities were higher than anticipated and that the G-10 surfaces at each end of the insulation "spool" were most troublesome. (In subsequent tests, these surfaces have been covered with reflector material and the temperature jump has become much smaller.)

Adjusting the computational model to match the data proved easier than first feared. Q/A and the outer reflector temperature were known experimentally and ΔT between the hypothetical shroud and MLI surface needed to be about 2 K to make the first layer heat balance work. Therefore, the shroud was assigned a temperature of 293 K which resulted in calculated outer surface temperatures of 291.1 to 291.3 k depending on the surface emissivities used. The remaining concern was that of the actual emissivity value down in the MLI wrap as compared to the value of 0.045 measured on the outer surface. Three calculations were made for the 56 layer configuration over the range from 293 to 20.27 K:

Average Emissivity	Q/A
0.03	0.1655 W/m2
0.0375	0.1943
0.045	0.2232

An average emissivity of slightly over 4 % is required to calculate the measured Q/A of 0.212 W/m^2 . This is not to say that the computational method is strictly accurate but it may well be in the range of ± 10 %.

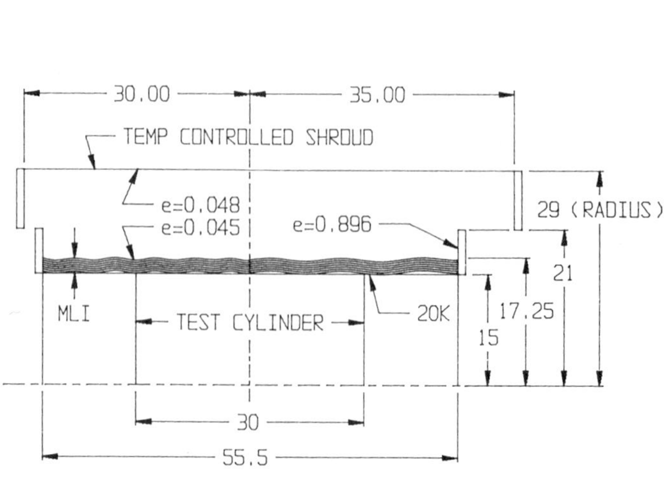

Figure 6 Calorimeter radiation configuration.

CONCLUSIONS

Tests on variable density matrix MLI made with glass paper spacers show that this concept can be utilized to produce material with heat leaks only about 50 % higher than floating shields. Refinements in the computer optimization technique indicate that MLI performance can be predicted to reasonable accuracy although further confirmation tests are needed for different materials. A recently fabricated "loom" is capable of making 8 ft wide matrix material and installation on a commercial dewar is planned.

ACKNOWLEDGEMENTS

The authors acknowledge the assistance of Joseph D. Gaby, Robert J. Stochl, and Paula J. Dempsey at NASA Lewis Research Center and project support under Contract NAS3-27009.

REFERENCES

1. G.E. McIntosh, Layer by layer MLI calculation using a separated mode equation, in: "Advances in Cryo. Eng.," Vol. 39B, Plenum Press, New York (1994).
2. Union Carbide Corp., "Linde Super Insulations-Manual for Licensees," (1967).
3. E.M.W. Leung et al, Techniques for reducing radiation heat transfer between 77 and 4.2 K, in: "Advances in Cryo. Eng.," Vol. 25, Plenum Press, New York (1980).
4. I.E.Spradley et al. Experimental studies of MLI systems at very low boundary temperatures, in: "Advances in Cryo. Eng.," Vol 35A, Plenum Press, New York (1990).
5. D.C. Hamilton & W.R. Morgan, "Radiant Interchange Configuration Factors," NACA TN 2836 (1952).

THE EFFECTS OF VIBRATIONS ON THE PERFORMANCE OF PEARLITE VACUUM INSULATION

L. Xu[1], A.Z. Gu[1], R.S. Wang[1], and S.M. Li[2]

[1]Low Temperature Center, Shanghai Jiao Tong University
Shanghai 200030, P.R. China
[2]Cryogenic Lab. Zhejiang University
Hangzhou, 310027, P.R. China

ABSTRACT

The vacuum-powders insulation is the commonest cryogenic insulation. Adding metal powders improves the performance of the insulation, but it is concerned that metal powders sink resulting from vibration, shock and thermal cycling in the using process decreasing the performance of the insulation. In this paper it was obtained that the above effect was eliminated if the diameter of metal powder reached a critical value by theoretical analyses, calculations and experimental result. A method was proposed that extended this insulation application and improved the performance of insulation.

Heat Transfer Analysis

The heat transfer processes of metal-pearlite vacuum insulation are intricate. This has been discussed by theoretical analysis and experimental study in many works. 1-3 It has been pointed out that the heat transfer occurs in the following ways:

a. heat conduction through a single particle
b. heat transfer among the contact particles
c. convective heat transfer of gases in the space among particles
d. radiative heat transfer among the particles

The calculation of heat transfer of this material is very complicated under atmosphere pressure, gas and solid conduction composes approximately 90 percent of total heat flux losses in this type of insulation. But the gas heat transfer decreases greatly under vacuum condition, the effective thermal conductivity is $K_\alpha = K_t + K_c + K_r$

Where K_t is thermal conductivity of solid and gases

K_c is thermal conductivity between particles

K_r is thermal conductivity of radiation

The type of conduction-radiation heat transfer was also studied by R.Viskanta[3] and C.L.Tien[4].

By theoretical analysis and calculations, K_t is found to be

$$K_t = K_g \left[\frac{5.8(1-m)^2}{R} \left(\frac{1}{R} \ln \frac{K_m}{K_g} - 1 - \frac{R}{2} \right) + 1 \right] \tag{1}$$

Where K_g is gas thermal conductivity, K_m is thermal conductivity of the particle, m is porosity, and $R = 1 - \frac{K_g}{K_m}$.

K_c is given by

$$K_c = \frac{3.37(1-m)^{1.33} K_m P^{0.33}}{E^{0.33}} \tag{2}$$

Where P is contact force of two particles,E is Young's modules K_m is thermal conductivity of particle, and other notes are same as those in Equation (1). The thermal conductivity of aerogel is gotten by Equation (2). The computed values nearly equals that obtained by experiment.

Effect of Aluminum Powder Content Percentage α on K_r

K_r is given by

$$K_r = 7.54 \times 10^{-8}(T_1 + T_2)(T_1^2 + T_2^2)/\beta \tag{3}$$

When $T_1 = 77K$ and $T_2 = 278.15K$, $K_r = 2.236 W/m \cdot K$ is obtained. Of course, it is complex to calculate above values.

A series of tests was conducted to evaluate the effect of AL powder content percentage on K_r. The experimental results are shown in Fig.1.

In Fig.1, the curve shows the dependence of K_r on α. While α is less than 10 percent (weight ratio), the radiation heat transfer K_r is decreased sharply while increasing the amount of Al powder. As the amount of Al powder is further increased, the K_r-α curve becomes progressively level. So the improvement is limited to decrease radiative transfer capacity only by means of increasing the amount of Al powder.

Effect of Al Powder Content Percentage α on K_c

Addition of Al powder can lower the radiation heat conductivity, but it also increases the solid heat conductivity. This is because the heat conductivity of Al powder particle is big, and

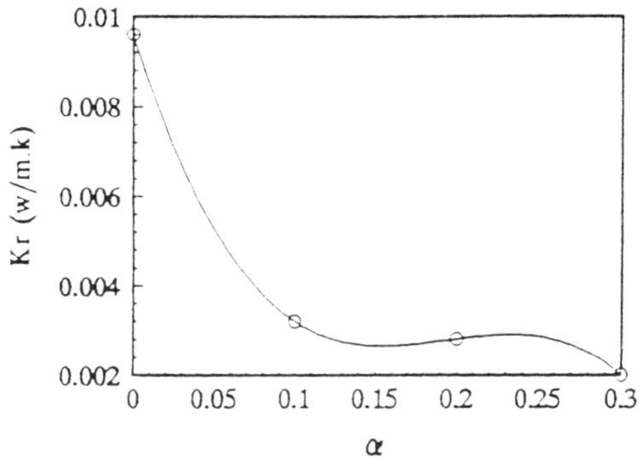

Figure 1. Curve of Radiation conductivity Kr vs. Al Powder Content Percentage α .

the contact resistance of particles decreases when pearlite surface adsorbs Al powder. The value of K_c can be gotten from analytical equations[1,3]. In engineering utilization process, it is difficult to obtain pearlite with the proper particle size and uniform density, so the calculation obtained by theoretical formula of above materials is not precise. The K_c values according to different of Al powder was gotten by experimental measurement. The result are shown in Figure 2.

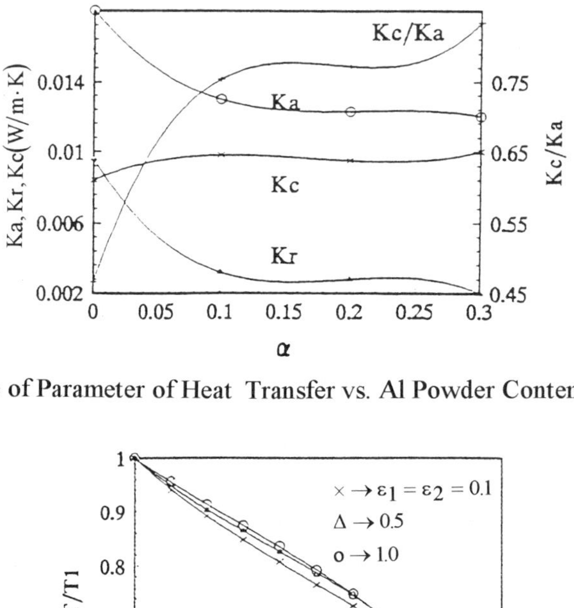

Figure 2. Curve of Parameter of Heat Transfer vs. Al Powder Content percentage α.

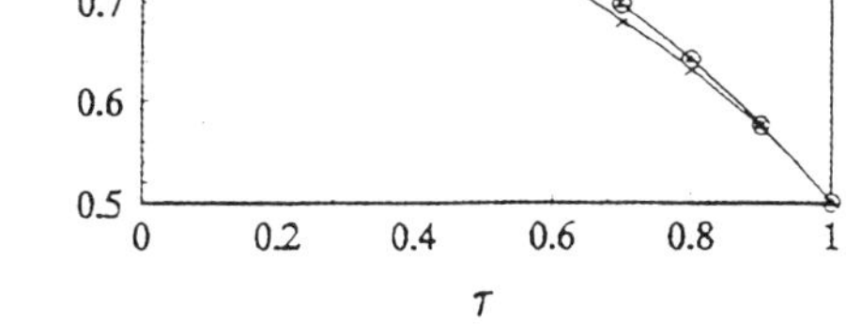

Figure 3. Effect of Surface Emissivity on Temperature Distribution with $N = 10$, $\omega_o = 0.5$, and $\tau_o = 1.0$.

It can be seen from above result that the K_c take up 46.6 percent of total thermal conductivity K_α without adding Al powder; when the amount of Al powder reaches 10 percent, K_c is 75.4 percent of K_α; when α reaches 30 percent, K_c us 83.3 percent. This shows that, with the addition of Al powder, the heat of radiation decreases sharply and the solid heat transfer becomes the leading factor. But when the content of Al powder is less than 30 percent, the increasing tendency of the solid heat transfer is not apparent along with the increasing Al powder content. The Solid heat transfer of this type of insulation mainly comes from the solid heat transfer of pure pearlite. So in order to further reduce its solid heat transfer, the K_c of pearlite should be decreased only.

Effect of Surface Emissivity and Diffuse Transmission Factor on Heat Transfer

Many numerical solutions have been derived for conduction-radiation heat transfer problem, and some useful conclusion of this type of problem has been obtained by using double heat flux model in recent years. The influence of emissivity on temperature fields is not great when optical thickness $\tau \rightarrow 1$, the emissivity only acts on the particle layer of the interfaces. As can be seen from Fig.3, the influence of emissivity of particles on heat transfer is not great.

Fig.4 shows the effect of diffuse transmission factor on heat transfer.

Ordinary medium is neither pure scattering medium ($\omega=1$) nor the no scattering medium ($\omega=0$) , the value of ω is between 0 and 1. Commonly, the bigger the percentage of heat transfer radiation, the bigger the effect of ω on heat transfer. The bigger the relative temperature difference, the smaller the effect of ω on heat transfer. For ordinary pearlite insulation, the effect of ω on radiation is very small compared to absorption coefficient. Thus it can be seen, the effect of the pearlite insulation with absorption capacity, scattering capacity and emission capacity on radiative transfer is mainly embodied in attenuation coefficient β . The absorption coefficient has a dominant effect on β .

Therefore, in order to improve the performance of evaluated pearlite insulation, the value of β should be increased while trying to decrease the solid heat transfer of pearlite itself by adding metal powder which has a larger absorption coefficient.

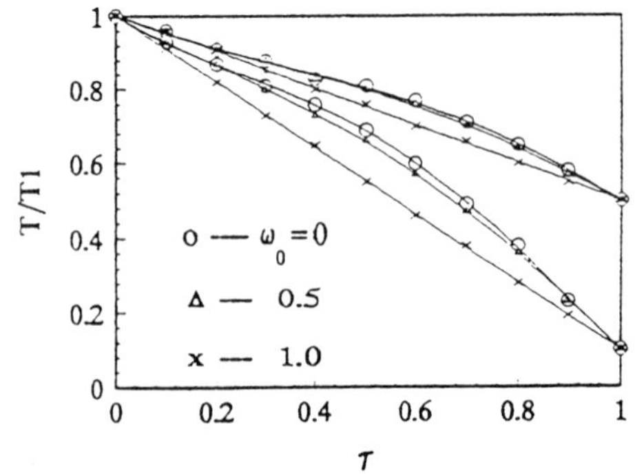

Figure 4. Effect of Albedo ω_0 on Temperature Distribution with $N = 10$ and $\tau_0 = \varepsilon_1 = \varepsilon_2 = 1.0$.

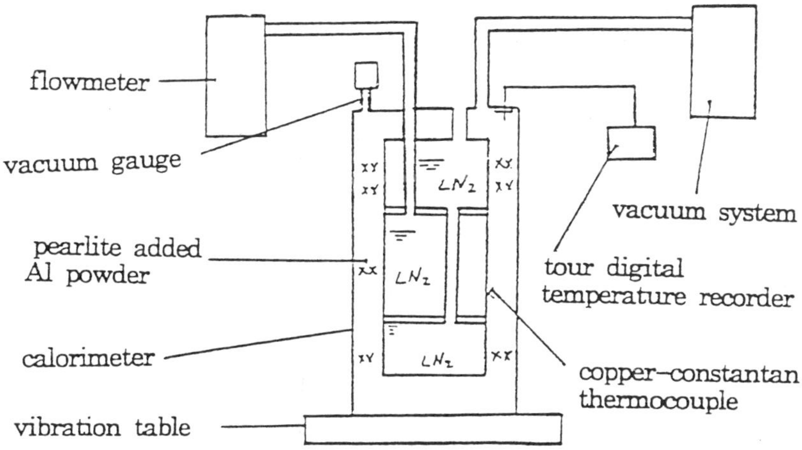

Figure 5. Illustrative Diagram of Test Equipment.

ACTING FORCE BETWEEN PARTICLES

Pearlite and metallic powder are both particles. At first, let's consider the properties of these two materials. The main ingredients of pearlite are S_iO_2 and $F_{e2}O_3$, and $S_iO_2$takes up the greatest percentage. The molecules of pearlite are polar molecules and have strong polarity. The particles made up of polar molecules also have strong polarity. At the same time, Aluminum is a king of active metal, its electronic structure is $3S^13P^1$, its crystal structure is face-centered cubic structure, and it is easy to form Al_2O_3 in the open air. The ionic radius of Aluminum is 0.57 A, its atomic radius is 1.431 A, and its main absorption spectrum lines are 2367.1 μ , 2373.4 μ , 2575.4 μ , 3082.2 μ , 3944.0 μ , 3961.5 μ and 3092.7 μ . The particle of Al powder is also polar molecule bunch. So the acting force between pearlite particle and Al powder particle are mainly orientation force, induction force and dispersion force. The size of pearlite particle is small and its volume weight is also small. Although the volume weight of Al powder is bigger than that of pearlite particle, the diameter of the Al powder we chose is absolutely smaller than 44 μ . It is so small that the van der Waals forces among them are substantial. It is proved by calculation that the van der Waals forces among Al powder are bigger than their gravity by as much as a factor of 6. The van der Waals forces is enough to counteract the gravity and the Al powder will not sink in the conditions of vibration, shock and heat cycling. Certainly, when Al powder is added to pearlite, not only is the Al powder adsorbed on the surface of pearlite but also some Al powder penetrate into the interior hole among pearlite powders and each other is connected with van der Waals forces.

RESULT OF EXPERIMENTS

A series of experiments was conducted to analyze the performance of this insulation material and to test the effect of conditions in use on its performance.

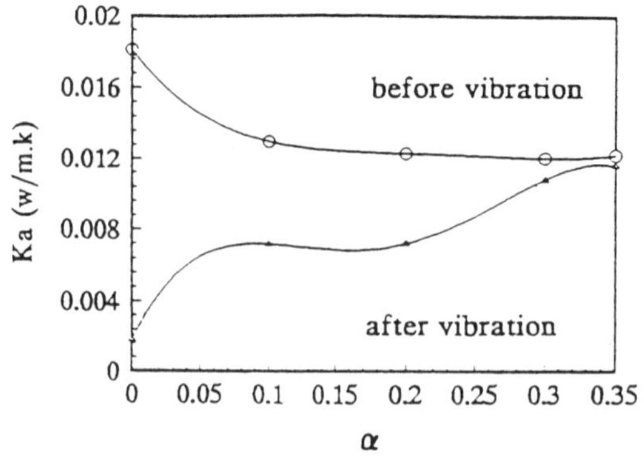

Figure 6. Curve of Effective Thermal Conductivity Ka vs. α Before and After Vibrations.

Vibration Test of Pearlite Added Al Powder

The pearlite insulation material mixed with different amount of Al powder was put in the cylindrical calorimeter and the calorimeter was put on the electron vibration table. Fig.5 schematically shows the illustrative diagram of test equipment.

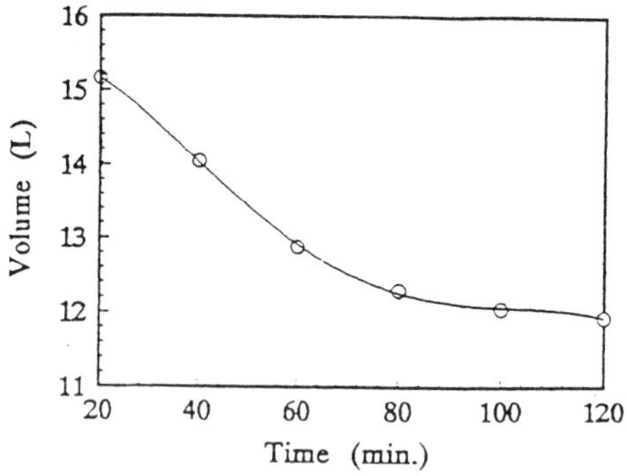

Figure 7. Volume of Insulation Layer vs. Time of Vibration.

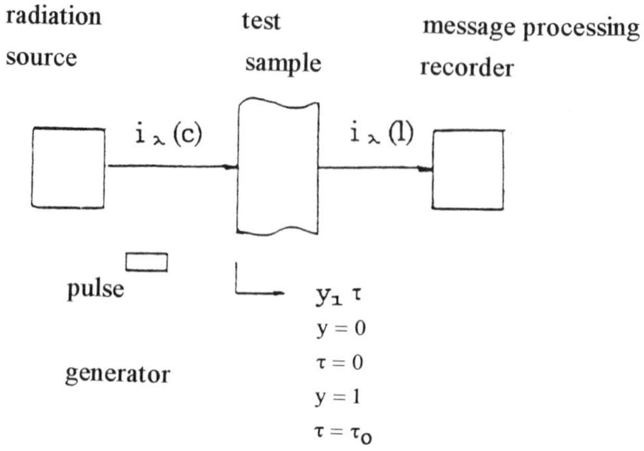

Figure 8. Illustrative Diagram for Radiation Penetration.

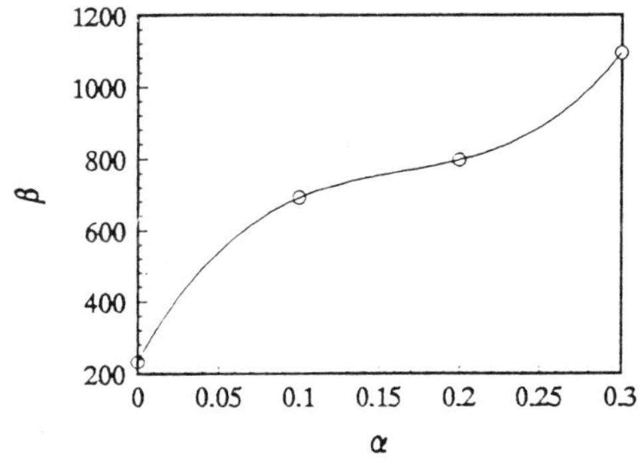

Figure 9. Curve of β vs. α .

Then we chose different frequency of 10 Hz, 30 Hz, 125 Hz and 500 Hz, with vibration acceleration of 5 g, and duration of 30 minutes for each condition. Experiments were done in these vibration conditions, the test results are shown in Figure 6.

Test of The Volume Change in The Vibration Process

The results are shown in Figure 7.

Test of Radiation Attenuation Coefficient

The illustrative diagram of the test is shown in figure 8. We used the infrared monochromatic optical spectrometer (Perkin-Elmer model 983 G). Its measurement range of wave length is 2.5-20 μ . The test sample piece is flimsy and its thickness is 1 mm. Transmissivity was measured on the test table. The dependence curve of value A and wave numbers $1/\lambda$ was printed by a printer and was integrated to the curve's area. The measured results are shown in Figure 9.

CONCLUSION

1. The insulation performance of evacuated pearlite mixed with Al powder does not decrease under vibration. Moreover it improves . This is because the volume weight of the insulation material increases from 105 Kg/m^3 to 148 Kg/m^3 under vibration and reaches the optimum value of this insulation.
2. If evacuated pearlite is mixed with Al powder of proper particle size, sinking of Al powder under vibration, which leads to heat short circuit and decrease of insulation performance will not occur. It is because molecules of pearlite are polar molecules and have strong polarity. Aluminum is a kind of active metal. So strong adsorption is formed between the two particles because of orientation force, induction force and dispersion force. The force of the adsorption is larger than the weight of Al powder if the diameter of Al powders is smaller than 44 μ . Thus sinking of Al powder will be avoided.

3. The attenuation coefficient increases with the increase of Al powder added, but not in direct proportion. So we could not improve the insulation performance only by this means.
4. The volume of evaluated pearlite will decrease under vibration in our test conditions. This should be noticed in engineering applications.
5. In cryogenic insulation, because of the volume change of inner chamber in using process, the volume of insulative space will change too. These two factors make the thickness of upper insulation thinner. It should be considered in engineering design.

REFERENCE

1. R.Viskanta, "Heat Transfer by Conduction and Radiation in Absorbing and Scattering Materials", Journal of Heat Transfer, Feb., 1965.
2. Xu Lie, Fang Rongshen, and Ma Qingfang, "Technology of Insulation", 1990.7.
3. R.Viskanta and R.J.Grosh, "Heat Transfer by Simultaneous Cinduction and Radiation in Absorbing Medium", Journal of Heat Transfer, Feb., 1963.
4. G.R.Cunnington and C.L.Tien, "Heat Transfer in Microshere Cryogenic Insulation", Adv.in Cry.Eng, V.181972.
5. S.T.Stoy, "Cryogenic Insulation Development", Adv.in Cry.Eng., V5, 1959.
6. Hu Lianfang and Lin Lihe, "Test for Insulation of Evacuated Powder Added Al Powder", Cryogenics, No.4, 1985. (China).

THERMAL CONDUCTIVITY OF POLYURETHANE FOAM AT LIQUID HYDROGEN TEMPERATURE REGION

M.Yamaguchi, K.Takahashi,and T.Ohmori

Ishikawajima-Harima Heavy Industries Co., Ltd.
Yokohama, Kanagawa, 235, Japan

ABSTRACT

Hydrogen is a promising, clean energy source which never produces CO_2. Liquid hydrogen (LH_2) is considered one of the large scale hydrogen storage methods. LH_2 must be thermally insulated from the ambient temperature region and must be stored with minimum evaporation loss. It is difficult to apply the usual cryogenic thermal insulation technique for a large scale LH_2 storage tank with moderate construction cost. We have been studying applications of polyurethane foam (PUF) to the thermal insulation material of the LH_2 storage tank. In order to design the storage tank, thermal conductivity data at LH_2 temperature region is required, but unfortunately no thermal conductivity data of PUF is available. For the above reason, we have developed a test facility and have measured thermal conductivity of PUF. This test facility has some design features: (1) thermal conductivity is measured by the steady cylindrical method, (2) PUF is cooled by a GM-refrigerator, and (3) measurement can be done at any point between LH_2 and ambient temperature and under or over atmospheric pressure. Some samples which have different density have been measured, and characteristics of thermal conductivity for temperature and density have been examined.

INTRODUCTION

Hydrogen is considered as a promising energy source in the 21st century because of its excellent characteristics. CO_2 is never produced when hydrogen is burned, which prevents acid rain and warming effects on the earth. Moreover, we don't have to worry about exhaustion of the energy source like fossil fuel, because hydrogen can be produced from water.

There are many media of hydrogen storage, but liquid hydrogen is the most efficient storage medium. Nevertheless, a liquid hydrogen storage tank is required to insulate the

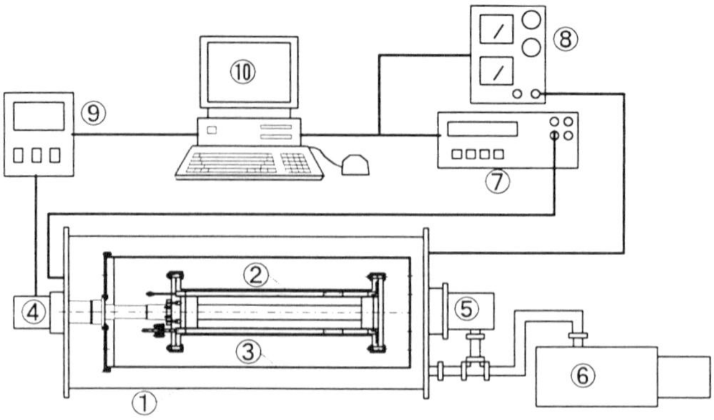

Figure 1. Experimental Apparatus.

inside of the tank from heat leak in order to reduce evaporation loss. Thermal insulation techniques for a small scale liquid hydrogen storage tank have been established, but techniques for a large scale storage tank have not been developed yet. We have been considering the application of polyurethane foam (PUF) as thermal insulation material, which is commonly used in LNG storage tanks.

Thermal conductivity data of insulation material is important for the design of the thermal insulation structure of the storage tank. PUF data is available above the liquid nitrogen temperature region, but no one has reported at the liquid hydrogen temperature region.

In order to prepare the thermal conductivity data, we have developed a test facility and have measured the thermal conductivity of PUF.

EXPERIMENTAL APPARATUS

The experimental apparatus, which is shown in Figure 1, consists of a test section, vacuum chamber, vacuum pump, data logger and computer.

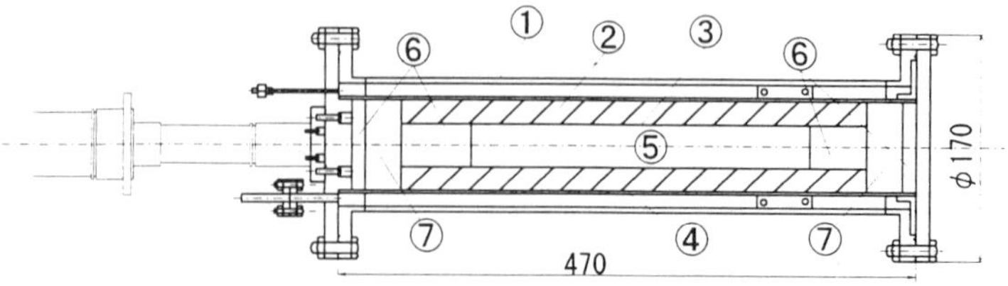

Figure 2. Test section.

The test section is horizontally set in the vacuum chamber, so that the influence of natural convection in the test section is negligible. The test section is also a sealed vessel, and pressure in it can be set at any value. Details of test section are shown in Figure 2. The heater is made of Nichrome foil with 0.1mm thickness and 300mm length, which is stuck on the inner surface of PUF, and joule heated by DC power supply. Aluminum powder is used to fill the inside of the heater, so that uniform temperature is realized on the heater surface. At both ends of the PUF, another PUF and fiber glass insulator is set in order to reduce heat loss. Cooling plates are made of copper. Several plates with 0.2mm thickness enclose and cool the test sample of PUF. The temperature of the cooling plates is controlled by a GM-refrigerator and held at any temperature between 20K and 300K. The inner and outer surface temperature of the PUF is measured by a Celnox sensor (Lake Shore Co., Ltd.) . This sensor is a kind of resistance thermometer and has a wide temperature measuring range (4 ~ 350K). Four sensors are stuck on each surface. A radiation shield is set in the vacuum chamber and connected to the first stage of the GM-refrigerator. The influence of radiation from outside the test section can be reduced with this shield.

MEASURING METHOD

The fundamental measuring method is prescribed in Japan Industrial Standard (JIS) A1412. The authors' measuring method also conformed to this standard. There are several kinds of thermal conductivity measuring methods. In this experiment, the steady cylindrical measuring method is adopted. Thermal conductivity is calculated from the power of the Nichrome foil heater, the temperature difference of the inner and outer surface of the PUF, and using equation (1)

$$\lambda = \frac{Q}{2\pi L}\frac{1}{\left(T_{in} - T_{out}\right)}\ell n\left(r_{out}/r_{in}\right) \qquad (1)$$

Temperature is measured by the 4-wire method. Resistance of the sensor is measured, and temperature is calculated from resistance-temperature curve which is already known. The power of the heater is calculated with the amount of supply current and the voltage drop between both ends.

Setting the temperature of the refrigerator and power of the heater is controlled by the computer. When the indicated value of the thermometer become constant (temperature changes less than 0.1K in 15 minutes), data is recorded automatically on floppy disk.

When data recording finishes at a temperature setting, signals setting the temperature and power of the heater are sent from the computer, and measurement at the next temperature mode starts. In this way, all instruments are controlled by the computer, so that automatic measurement can be performed during 24 hours a day. The algorithm of automatic measurement is shown in Figure 3.

Pressure in the test section is 1atm at ambient temperature, vacuum state is realized at the lower temperature region. This vacuum condition is established by the test sample being sealed in the test section at 1atm and ambient temperature.

TEST SAMPLES

Specifications for test samples of PUF are shown in Table 1. The samples were foamed in a cylindrical mold, and outer skins were shaved in order to adjust their

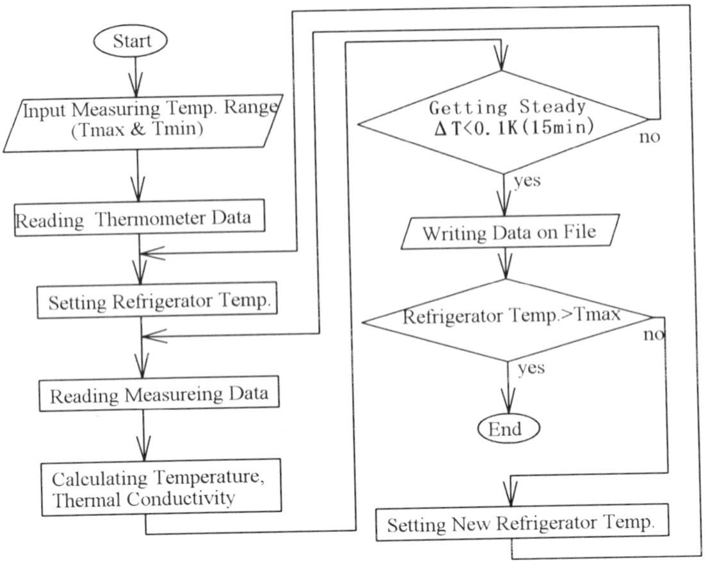

Figure 3. Algorithm for automatic measurement.

Table 1. Specification of test sample

Density [kg/m^3]	Length [mm]	Outer diameter [mm]	Inner diameter [mm]	Heated length [mm]
199	400	62	30	300
253				

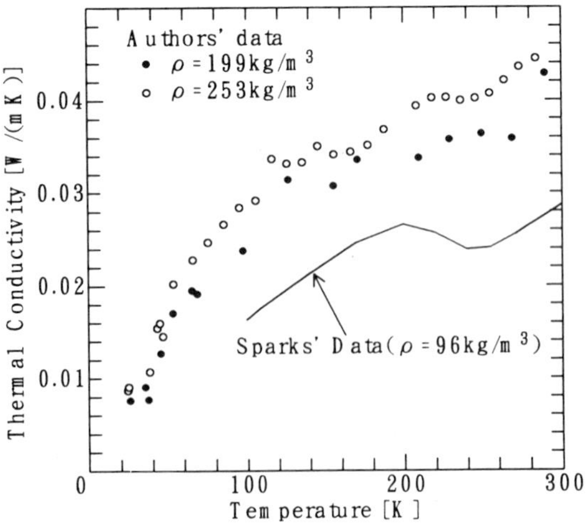

Figure 4. Experimental results.

dimension. The blowing agent of the samples is hydrochlorofluorocarbon 141b (HCFC-141b, CH_3CCl_2F). The use of chlorofluorocarbons like CFC-11 (CCl_3F), which had been commonly used as blowing agents, was forbidden because they break down the ozone layer of the earth. For this reason, alternative refrigerants like HCFC-141b have become major blowing agents.

EXPERIMENTAL RESULTS

Measured thermal conductivity as a function of temperature is shown in Figure 4. Data which was measured by Sparks et. al.[1)] is plotted together in the figure.

Thermal conductivity abruptly increases with temperature under 100K, and its increase become gentle above 100K. The authors' data is 80% greater than Sparks', however, so it is difficult to compare the two data set because of the large difference of density between the samples.

The maximum point which is observed at around 200K in Sparks' data is caused by condensation of gas in cells of PUF. In the authors' data, clear maximum point is not found. But, looking at the curve, one can see a slight trend similar to that in the Sparks' data. Several reasons for this feature of the authors' thermal conductivity curve can be considered. The first reason is that the authors' samples have much higher density than Sparks' samples. In the case of PUF with high density, over all thermal conductivity is more strongly influenced by solid thermal conductivity than that of the gas in the cells. The second reason is difference of component of the gas in the cells. In the Sparks' samples, the blowing agent is CFC-11. On the other hand, HCFC-141b was used in the authors' samples. Difference of the properties of the blowing agents (especially thermal conductivity and saturation temperature) bring these trend of the thermal conductivity. The third reason is effect of air. The authors' samples have thin thickness (15mm), and it is easy for air to enter into the cells. Properties of air influence on over all thermal conductivity of the PUF.

Gas in the cells pass into a state of solidity in the liquid hydrogen temperature region, and conduction of structure only dominates overall thermal conductivity. Therefore, thermal conductivity at liquid hydrogen temperature region is 1/5 that of 280K.

Data of ρ=253kg/m^3 is 10~20% greater than that of ρ=199kg/m^3. The ratio of the structure of the PUF with ρ=253kg/m^3 is more than that of ρ=199kg/m^3, which contributes to the increase of thermal conductivity.

CONCLUSION

A test facility was developed and thermal conductivity of PUF was measured at the liquid hydrogen temperature region using a GM-refrigerator. Thermal conductivity data was obtained between 20K and 290K for the PUF with two kinds of density, 199kg/m^3 and 253kg/m^3.

Some experiments will be carried out under different conditions in order to complete thermal conductivity data bases. These will be used for the development of technology for large scale liquid hydrogen storage.

NOMENCLATURE

ρ : Density of PUF
λ : Thermal conductivity

Q : Power of heater
L : Heating length of PUF
T_{in} : Inner surface temperature of PUF
T_{out} : Outer surface temperature of PUF
r_{in} : Inner radius of PUF
r_{out} : Outer radius of PUF

REFERENCE

1. L.L.Sparks and J.M.Arvidson. "P.B. Rep. PB85-10094", (1984)

THERMAL CONDUCTIVITY OF VACUUM PRESSURE IMPREGNATED Cu AND Al STABILIZED NbTi SUPERCONDUCTING COILS

J.D. Walters, T.L. Cooper, and T. Fikse

Annapolis Detachment
Carderock Division, Naval Surface Warfare Center
Annapolis, Maryland 21402

ABSTRACT

Superconducting magnets conductively cooled by cryocoolers are cooled by thermal bus systems that do not cover the entire magnet as does liquid helium in an immersion cooled magnet. Therefore, internal temperature profiles of conductively cooled magnets may be significantly different than immersion cooled magnets. Understanding the differences in the effective thermal conductivities of superconducting magnet composites becomes necessary in order to determine the magnets steady state and dynamic thermal performance. For this reason, the effective axial and radial thermal conductivities of vacuum pressure impregnated Cu and Al clad NbTi superconducting magnet composites have been measured, and the results are presented and discussed in this paper.

INTRODUCTION

The Annapolis Detachment of the Naval Surface Warfare Center for the past two years has been developing technologies in support of a superconducting Advanced Lightweight Influence Sweep System(ALISS) in order to clear magnetic influence sea mines.[1] Operational logistics need to be minimized for this Naval application. Therefore, the magnet is not cooled with liquid cryogens, but by one or more cryocoolers. All aspects of magnet system design have been addressed within this development phase including component interfaces, robust cold to warm supports, lightweight cryostat, novel magnet protection schemes, lightweight magnets, and stable low loss conductors.[2-5]

A conductively cooled NbTi magnet is no longer constrained to operate at 4.2 K. Thus, knowing the effective thermal conductivities of superconducting magnet composites is necessary to determine the magnets steady state and dynamic thermal performance at other than liquid helium temperatures. For this reason, the effective axial and radial thermal conductivities of vacuum pressure impregnated Cu and Al-Cu clad NbTi

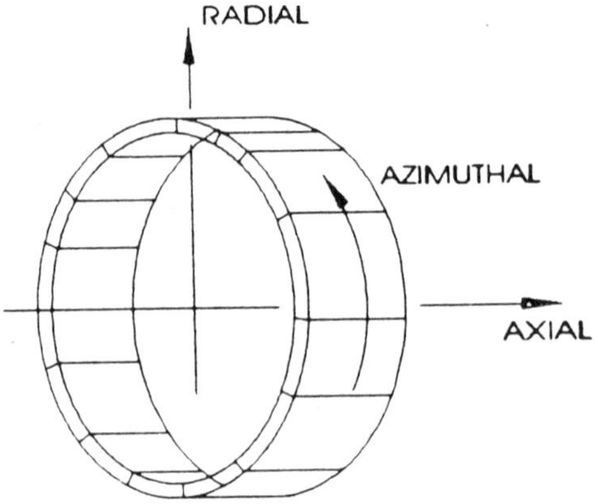

Figure 1. A solenoidal magnet defining the axial, radial, and azimuthal thermal conductivity directions.

superconducting magnet composites have been measured. These experimentally obtained conductivities make it possible to estimate thermal gradients within the magnet for different thermal bus concepts, to aid in quench initiation and propagation analyses, and examine thermal gradients in magnets during rapid cool down.

MAGNET SAMPLES

The superconducting magnets of present interest for the ALISS program are solenoidal magnets. The thermal conductivity components of the magnet are defined as axial, radial, and azimuthal as shown in Fig. 1. The axial and radial components of the effective thermal conductivity have been measured. The azimuthal conductivity was not because the measurements are not easily made within the present measurement apparatus. Additionally, the azimuthal conductivity of the magnet is dominated by the metal stabilizer in the conductor, and this property is more readily calculated than the axial or radial conductivities.

Magnet samples are sectioned from 13.3 cm(5.25 in.) diameter test coils fabricated and used routinely at NSWC, Annapolis for thermal energy to quench studies. These magnets are fabricated as shown in Fig. 2a by laying 3 layers of 0.008 cm(0.003 in.) fiberglass cloth over the winding fixture. The 0.56 mm by 1.15 mm wire used in this work is laid in place. Next a layer of 0.008 cm(0.003 in.) fiberglass cloth is laid down. This layering process of fiberglass cloth and conductor is repeated until the number of layers desired is achieved. Finally, fiberglass is laid over the windings for additional strength.

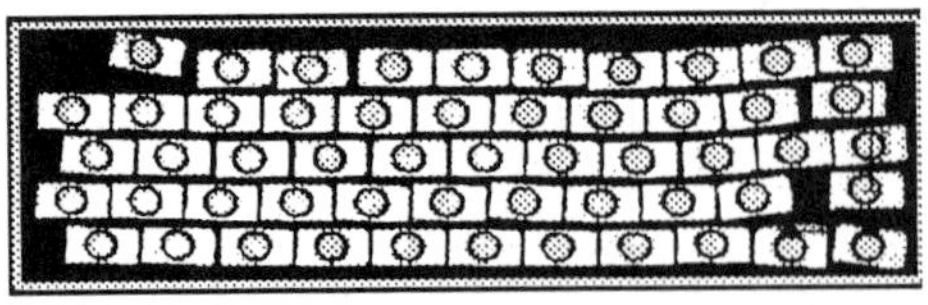

Figure 2a. Cross sectional view of a magnet sample used in this project. The conductor in this photo is an Al-Cu/NbTi superconductor. The circular core is approximately 1:1 Cu to NbTi. The core is surrounded by 3 parts of high purity Al. Fiberglass cloth and epoxy fill between conductor layers.

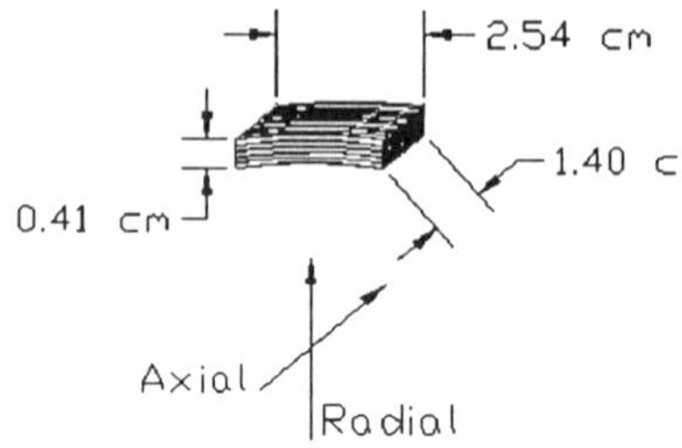

Figure 2b. Sketch of superconducting magnet sample with dimensions used to make conductivity measurements. The axial and radial orientations are shown for reference.

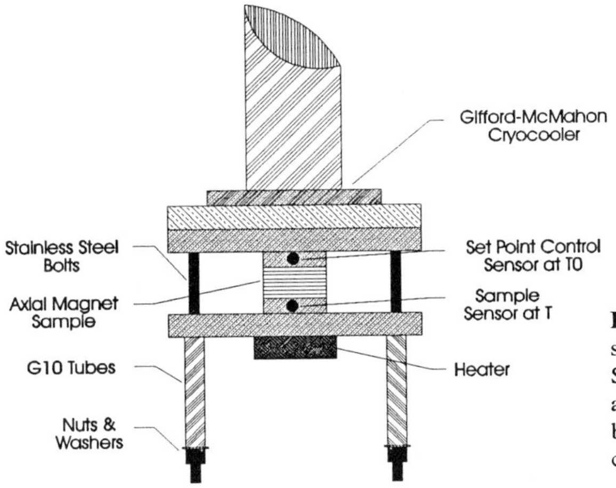

Figure 3. Test fixture for holding magnet samples to measure thermal conductivity. Shown are the temperature sensors, steel bolt and G10 tube re-entrant clamps, copper fixture backing plates, sample mounting pads, and cryocooler.

Then the magnets are vacuum pressure impregnated with epoxy and cured. The magnet samples discussed here have 5 conductor layers and 11 turns per layer. The conductor fill factor is approximately 75%.

For the measurements made here, the outer layers of glass cloth and epoxy were removed after curing. These layers were removed because the remaining magnet sample is more nearly representative of a sample from the interior of a large or full scale coil. The final dimensions of the magnet samples are measured individually, and are on average 2.54 cm x 0.41 cm x 1.40 cm as shown in Fig. 2b. Two sample types were prepared. The first being a Cu/NbTi core with coextruded high purity aluminum in ratios of 1:1:3, and the second being a Cu/NbTi wire in ratios of 4:1.

APPARATUS

The basic thermal conductivity measurement apparatus used to make the measurements discussed here is described in detail elsewhere.[6] The essentials of the apparatus include a two stage Gifford-McMahon cryocooler with a low pass thermal filter affixed to the second stage. The filter is used to remove the temperature swings inherent in the operation of the cryocooler. Sample fixtures are bolted to the bottom of the thermal filter.

The magnet sample fixture is shown schematically in Fig. 3. The same test fixture is used for both axial and radial measurements. The magnet samples are clamped between two copper plates which are 8.89 cm x 8.89 cm x 0.95 cm (3.50 in. x 3.50 in. x 0.38 in.). The upper plate is bolted to the thermal filter. Two sets of copper pads are soldered onto the plates. The fixture for the axial measurements has two flat pads 0.63 cm(0.25 in.) high between which the magnet sample is clamped. For radial measurements, two pads were machined to match the inner and outer radii of the magnet samples, respectively. A thin layer of Cry-Con grease is applied between both sets of pads and the magnet samples for axial and radial measurements in order to reduce interface thermal resistance. The degree to which the copper grease interface may have affected measurements is discussed below.

The magnet samples are clamped in the test fixture by stainless steel bolts and thin walled G10 tubes also shown in Fig. 3. The bolts and tubes are arranged in a re-entrant configuration to make the heat leak path between the two sides of the sample as long as possible. Clamping is achieved using UNC 10x24 stainless bolts with an effective heat transfer length of 9.04 cm(3.56 in.). These bolts pass from the top plate down through the G-10 cylindrical tubes which have an effective length of 4.55 cm(1.79 in.) and wall thickness of 0.056 cm(0.022 in.). The G10 tubes pass back up to recessed cut outs in the underside of the fixture's bottom plate. The bolts and tubes are held in place by nuts at the end of the bolts where the G10 begins.

Parasitic heat leak through this re-entrant arrangement has been estimated as if the bolts and tubes are a continuous piece. The added contact resistances of the bolt at the top plate, bolt to nut interface, nut to washer, washer to G10, and G10 to bottom plate were not incorporated. The heat leak due to all four sets of re-entrant clamps are 0.22×10^{-3} W to 6.0×10^{-3} W for temperature differences between the plates of 4.5 - 7.0 K and 40 - 63 K, respectively. The heat leak through the clamps was less than 1 % of the total applied heater power over all temperature ranges. Therefore, this heat leak was not included in the conductivity calculations.

Two ruthenium oxide temperature sensors were screwed to the appropriate fixture pads with Cry-Con grease between the sensor bobbins and the pads to enhance thermal contact. A 25 Ohm cartridge heater mounted in a 2.54 cm x 3.81 cm heat sink is fixed to the underside of the bottom copper plate.

A measurement is made by holding the top plate temperature constant using a Conductus LTC10-G temperature controller, providing a known input to the heater, and measuring the temperature difference developed across the magnet samples. Temperatures were recorded every 30 s. Once thermal equilibrium was achieved, data was acquired for approximately 10 min. Zero heat input measurements were always taken first at each set point in order to determine the offset calibration between the two temperature sensors. This offset was subsequently subtracted from each set of temperature differences. In this way, the bottom sensor is normalized relative to the fixed point sensor. This allowed the measurements to rely on the sensors sensitivity for temperature differences rather than the absolute calibration uncertainty. This approach has been validated using a gold-chromel 3 junction thermopile as discussed elsewhere.[6]

In the present test configuration, a copper and grease interface exists between magnet samples and temperature sensors. Although the Cry-Con grease was applied in as thin a layer as possible in order to minimize thermal resistance, a temperature difference occurred due to this interface. In searching the literature, the thermal conductivity of Cry-Con was available up to approximately 10 K, while much data is available for silicone based vacuum greases.[7,8] A curve describing the lowest thermal conductivity of vacuum greases as a function of temperature was scaled to match the available low temperature Cry-Con data and used to extrapolate Cry-Con to higher temperatures. The extrapolated curve was positioned slightly below the Cry-Con curve so that the extrapolation would produce underestimated thermal conductivities. The resulting calculated temperature differences developed across the grease interfaces are therefore upper bounds. This method of estimation was used separately at the upper and lower temperature interfaces for both axial and radial measurements. Assuming a 0.0025 cm thick greased layer, the maximum temperature gradients developed across the grease interfaces occurred at the lowest temperatures and were estimated to be 5% radially and 1% axially of the total measured temperature difference across the magnet samples. Each temperature difference due to a grease interface is subtracted from the total measured temperature difference in order to isolate the temperature difference developed across only the magnets.

DATA

Temperature differences developed across samples as functions of heater input have been measured for both the Al-Cu/NbTi and Cu/NbTi magnet samples. Axial and radial data for Al-Cu/NbTi samples were taken at set point temperatures of 5.5 K and 7.0 K, and 4.5 K and 6.0 K, respectively. These data were adequate to determine the conductivity within the practical operating range of a NbTi magnet. Axial data for the Cu/NbTi sample

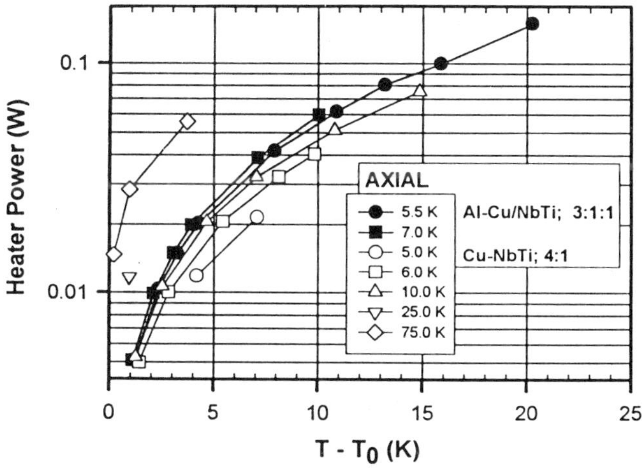

Figure 4. Heater input power as a function of temperature difference developed across axially oriented magnet samples.

were taken at set point temperatures of 5.0 K, 6.0 K, 10.0 K, 25.0 K, and 75.0 K. Radial data was acquired at 5.0 K, 6.0 K, 8.0 K, 10.0 K, 20.0 K, and 40.0 K. The results of all axial and radial measurements are summarized in Figs. 4 and 5, respectively.

RESULTS

The thermal conductivities have been calculated using the approximation

$$k(T) \propto \left(\frac{l}{A}\right)\frac{\dot{Q}}{T - T_0}, \tag{1}$$

where l, A, $\dot{Q}$, T, and T_0 are the sample length and cross sectional area in direction of heat flow, power dissipated in heater, and sample and set point temperatures, respectively. Thermal conductivities calculated for Al-Cu/NbTi and Cu/NbTi magnet samples are summarized in Table 1.

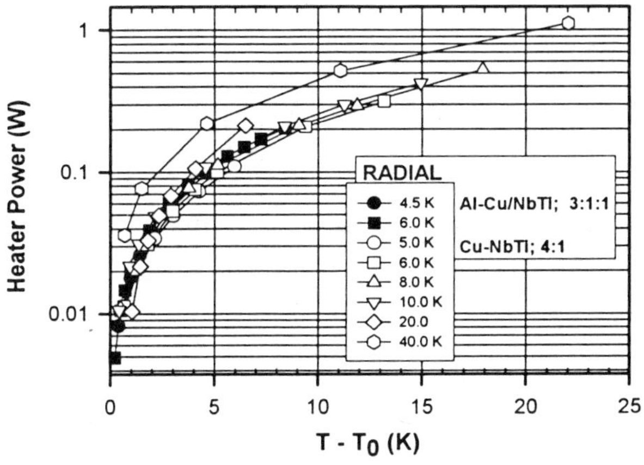

Figure 5. Heater input power as a function of temperature difference developed across radially oriented magnet samples.

Table 1. Thermal Conductivities of Magnet Composites

	T (K)	k (mW/cm-K)		T (K)	k (mW/cm-K)
Conductor	**Radial**	**Direction**		**Axial**	**Direction**
Al-Cu/NbTi 3:1:1	4.5	2.45		5.5	6.52
	6.0	2.85		7.0	7.06
Cu/NbTi 4:1	5.0	2.01		5.0	4.34
	6.0	2.23		6.0	5.19
	8.0	2.61		10.0	6.17
	10.0	2.97		25.0	18.5
	20.0	2.93		75.0	32.9
	40.0	6.39			

It is difficult to gauge the role epoxy and metallic matrix materials play in determining the conductivities of the magnets by examining the numbers in Table 1. To make this comparison clearer, the thermal conductivities of magnet constituents including epoxy, copper with RRR =100, and aluminum with RRR=500 are plotted in Fig. 6 along with the thermal conductivities from Table 1.[9] Data is also shown for limited measurements that were made for a 1.8:1 Cu/NbTi magnet composite.[10] It is clear from Fig. 6 that over all temperatures the conductivities of the Al-Cu/NbTi and Cu/NbTi conductors are both dominated by the epoxy fill within the magnet.

DISCUSSION

It is interesting to note that at 5.0 K the ratio of the thermal conductivities between Al-Cu and Cu stabilized magnet composites is approximately 1.5. Whereas the relative

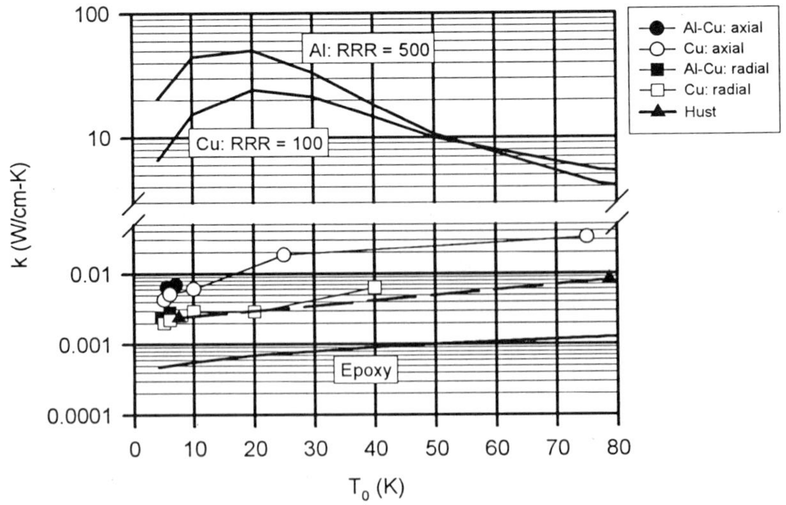

Figure 6. Summary of axial and radial magnet composite thermal conductivities for Al-Cu/NbTi and Cu/NbTi superconducting wires. Also shown for comparison are the thermal conductivities of epoxy, Al(RRR=500), Cu(RRR=100), and data acquired by Hust for a 1.8:1 Cu/NbTi superconducting magnet composite.

thermal energy to quench of these two coil types is approximately 3.[11] It might have been expected that the magnet composite conductivities would provide a stronger correlation with the energy to quench measurements. The difference may reside in that the thermal conductivity measurements presented here are the averaged thermal conductivities of the magnet composite constituents, and the thermal energy to quench is governed more strongly by the local conductivity in combination with the heat capacity of the metallic matrix material. The magnet composite conductivities may therefore be most useful in helping to describe normal zone propagation through magnets once a quench has initiated. The ratio of axial to radial thermal conductivities for both magnet samples is 2.3, and may be suggestive of the relative roles these directional conductivities play in quench propagation.

It was mentioned earlier in this paper that close attention must be paid to proper cooling distribution through a conductively cooled magnet. Magnet designs which have the cooling source a substantial distance from portions of the magnet may cause temperature gradients to develop in the magnet. For example, consider a NbTi solenoid built at NSWC which will be operated at 5.0 K that has an inner radius, outer radius, and length of 61.21 cm, 68.43 cm, and 15.24 cm, respectively. If the magnet were cooled solely through a metallic bobbin on the inner diameter of the magnet, and assuming a heat leak of 500 mW, radial temperature differences of approximately 0.20 K and 0.31 K would develop across Al-Cu/NbTi and Cu/NbTi magnets, respectively. For the same coil, if all 500 mW were applied over an end face of the magnet, axial temperature differences of 0.42 K and 0.62 K would develop for the Al-Cu/NbTi and Cu/NbTi magnets, respectively. These axial and radial temperature gradients must be accounted for in magnet designs because the warmest spot on the magnet defines its operating temperature. If the magnet is to operate at 5.0 K with a 1.0 K temperature margin and a 0.6 K gradient exists across the magnet, then the cryocooler needs to operate at 3.4 K. Currently, a single Gifford-McMahon cooler cannot produce 500 mW at 3.4 K. Situations such as in this latter example can be avoided with appropriate magnet designs, but this simple example serves to illustrate that potential axial and radial temperature gradients must be considered in designing conductively cooled superconducting magnet systems.

CONCLUSION

The thermal conductivities of two vacuum pressure impregnated magnet composites have been measured in the axial and radial directions. The first composite consisted of an 3:1:1 Al-Cu/NbTi superconducting wire, and the second a 4:1 Cu/NbTi wire. The conductors were identical in size and electrical designs. Internal temperature profiles of conductively cooled magnets may be significantly different than magnets cooled by immersion in liquid helium. Thermal properties of materials not needed before at temperatures above liquid helium temperatures, now seem necessary to support the design of conductively cooled magnets which operate above 4.2 K. Therefore, understanding the differences in the effective thermal conductivities of superconducting magnet composites becomes necessary in order to determine the magnets steady state and dynamic thermal performance.

ACKNOWLEDGMENT

This work has been conducted as a portion of the Advanced Lightweight Influence Sweep System project sponsored by the Naval Project Management Office 407.

REFERENCES

1. E.M. Golda, J.D. Walters, and G.F. Green, Applications of Superconductivity to Very Shallow Water Mine Sweeping, Naval Engineers Journal, May (1992).
2. E.M. Golda, T.H. Fikse, J. Tannenbaum, Development of a superconducting magnet for an advanced lightweight influence sweep system, To be presented at the 1995 Cryogenic Engineering Conference, Columbus, Ohio.
3. M. Heiberger, A. Langhorn, T.W. Hurn, D.G. Morris, J.D. Walters, L.F. Aprigliano, D.J. Waltman, and G.F. Green, A lightweight rugged conduction-cooled NbTi superconducting magnet for U.S. Navy mine sweeping applications, To be presented at the 1995 Cryogenic Engineering Conference, Columbus, Ohio.
4. E.M. Golda, J. Tannenbaum, and R.P. Reed, Characterization of Unidirectional Graphite-Fiber -Reinforced Composite Straps, To be presented at the 1995 Cryogenic Engineering Conference, Columbus, Ohio.
5. T.L. Cooper, J.D. Walters, and T.H. Fikse, Thermal conductance of heat transfer interfaces for conductively cooled superconducting magnets, To be presented at the 1995 Cryogenic Engineering Conference, Columbus, Ohio.
6. J.D. Walters, T.H. Fikse, and T.L. Cooper, Variable temperature thermal conductivity measurements using a Gifford-McMahon cryocooler, Cryocoolers 8, Proceedings of the International Cryocooler Conference, Ed. R.G. Ross, Jr., pp. 823-833 (Plenum Press, New York 1995).
7. Thermal conductivity of Cry-Con grease was obtained from the following: M.M. Kreitman and J.T. Callahan, Thermal conductivity of apiezon-N grease at liquid helium temperatures, Cryogenics April (1970).
8. Thermal conductivity of silicone greases was obtained from NIST report presented for "Miniature Refrigerators for Cryogenic Sensors and Cold Electronics", UCLA extension Short Course Program, coordinators R. Radebaugh and G Walker, (1991).
9. J.G. Hust, Low temperature thermal conductivity measurements on longitudinal and transverse sections of superconducting coil, Cryogenics (1975).
10. Copper and Aluminum Thermal Conductivities were obtained using Cryocomp Software V. 2.0, Cryodata,Inc. by Eckels Engineering.
11. Thermal Energy to quench information obtained from the following: T.H. Fikse, D.J. Waltman, and M.J. Superczinski, Mechanical Shock Stability of Model Superconducting Magnets, Report Number CARDIVNSWC-TR-81-94/16, April (1994).

THE THERMAL CONDUCTIVITY OF SOME MODERN ENGINEERING MATERIALS USED IN HIGH BANDWIDTH I/O SYSTEMS AT CRYOGENIC TEMPERATURES

D. L. Patelzick, E. L. Hershberg, T. J. Hendricks

System Integration Laboratory
Technology Group - Video and Networking Division
Tektronix Inc.
POB 500, M/S 50-470
Beaverton, OR 97077

ABSTRACT

Superconducting digital microelectronics require multi-gigahertz bandwidth performance of the signal I/O system, including multi-channel flat flexible I/O cables. The limited information on the thermal conductivity of modern engineering materials available to the designer of low temperature, high performance systems and devices, makes it difficult to optimize a design or even to know if it will perform as specified. This is particularly true of the newer dielectrics for which bulk thermal conductivity data do not exist. This paper presents the thermal conductivity measurements we have made of various metals and dielectrics commonly used for high performance, low loss flexible I/O cables and presents the design tradeoffs that these materials offer the designer. We also describe the relatively low cost procedure we are using to make thermal conductivity measurements of metals and dielectrics between 8 K and 100 K that satisfy design engineering requirements. A discussion of the value of thermal conductance vs. bulk thermal conductivity is also presented.

INTRODUCTION

Tektronix is a major participant in the nation's premier programs for applying digital superconducting microelectronics to the switching of multi-gigahertz data streams. These programs include the development of enabling technologies, without which the program goals cannot be met. One of the most significant issues in using low temperature superconducting digital microelectronics is the transfer of the digital data from room temperature to cryogenic temperatures and back at multi-gigahertz frequencies[1]. This requires the development of multi-channel, flat, flexible cables with very low thermal losses from room temperate to 4.5 kelvin that are also capable of transferring data at multi-gigahertz bandwidths. There is a fundamental need to push competing design constraints to the absolute limit of potential performance so that the thermal load to the cryosystem is minimized while the electrical performance is maximized. These competing demands set the stage for a classic conflict between the ideal and the practical[2].

MULTI-CHANNEL FLAT FLEXIBLE I/O CABLES

The general design of an I/O cable for communicating with low temperature superconducting microelectronics is shown in Figure 1[3]. In this example there are seventy signal lines in a one inch wide cable. These are controlled impedance lines of the microstrip design in which there is a ground plane, a signal line and an intervening dielectric spacer. This design requires precise control of the line width, the thickness of the dielectric material and the uniformity of the dielectric properties. Furthermore, to maintain signal integrity over the length of the cable, very high quality copper must be used. Signal dissipation losses must be at an absolute minimum. Heat is conducted along these cables from the room temperature end that passes through the vacuum envelope, to the cryogenic end that is attached to the superconducting microelectronics operating at 4.5 K. Cooling these cables is a challenging design task. The ground plane can be connected directly to a heat sink, but the signal lines cannot be touched. The only path for heat from a signal line to a cold stage is through the dielectric material separating the signal line from the ground plane. Thus the designer of these critical components must be successful in removing all the conducted heat from the signal lines through a thermal insulator.

The electronics components industry has been successful in developing materials that meet the demanding electrical performance requirements of these I/O cables[4,5]. Their success is a direct result of their ability to produce very high quality copper foils along with dielectric materials with highly uniform dielectric properties. The results of their efforts meets the electrical performance requirements but further complicates the thermal design. Only copper can provide the required electrical performance, but the materials are of such high quality that the thermal conductivity is exceptionally high. Rolled annealed and electro-deposited copper are the two varieties available. Whereas both of these materials would be expected to have compromised thermal performance due to their processing and handling, we have found the opposite to be the case.

Our experience with the design of multi-channel I/O cables for low thermal load cryogenic applications convinced us that the only way for us to be successful was to obtain primary data on the candidate materials, and design the cryosystem, thermal management system and the I/O cables specifically for the materials chosen. There are many new materials in use today, most of which have never been characterized at cryogenic temperatures. The exceptional activity in the characterization of engineering materials that took place in the 1950's and 1960's has given way to an almost total lack of

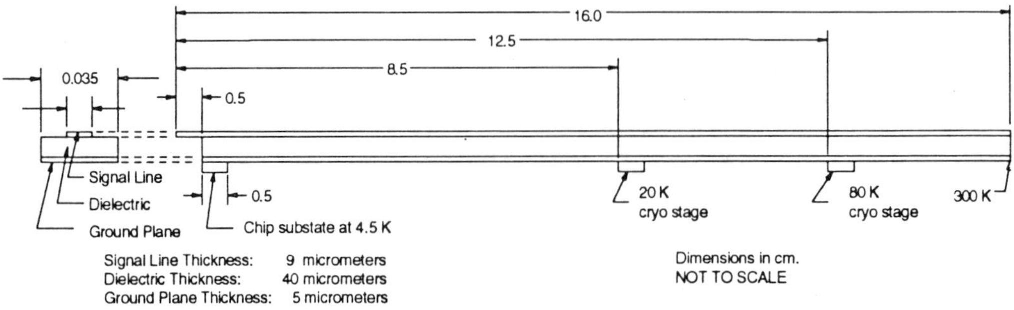

Figure 1 - Geometry of a typical signal/dielectric/ground section.

characterization. Thermal integral tables are of limited value when design factors of safety are below two. In our case the designs required an almost exact prediction of performance with essentially no factor of safety. This is primarily due to the limited cooling power of 4.5 K closed cycle refrigerators and the need to keep the size and input power of the cryosystem as low as possible. Anyone who attempts to design systems such as these and does not use primary data for the thermal analysis is courting disaster.

MEASURING THERMAL CONDUCTIVITY OF METALS AND DIELECTRICS AT CRYOGENIC TEMPERATURES

We have developed the capability to measure the thermal conductivity of metals and dielectrics at cryogenic temperatures. At 20 K, the conductivity of copper can vary by a factor of two from sample to sample, and can be a factor of ten higher than the room temperature value. Dielectrics used for flexible I/O cables can also display thermal conductivity differences of a factor of two. This imprecision makes any attempt at precise thermal design futile. The method we are using makes a direct measurement of the thermal conductivity of a material sample, is reproducible and uses closed cycle cryosystems operating in an industrial laboratory setting. A full set of measurements on one material can be completed in a day, and with sample preparation time and data analysis time factored in, the entire process of characterizing a material sample takes less than three working days. Our conductivity data ranges over six orders of magnitude, from .02 mW/cm-K to 40 W/cm-K, for temperatures of 8 K to 80 K.

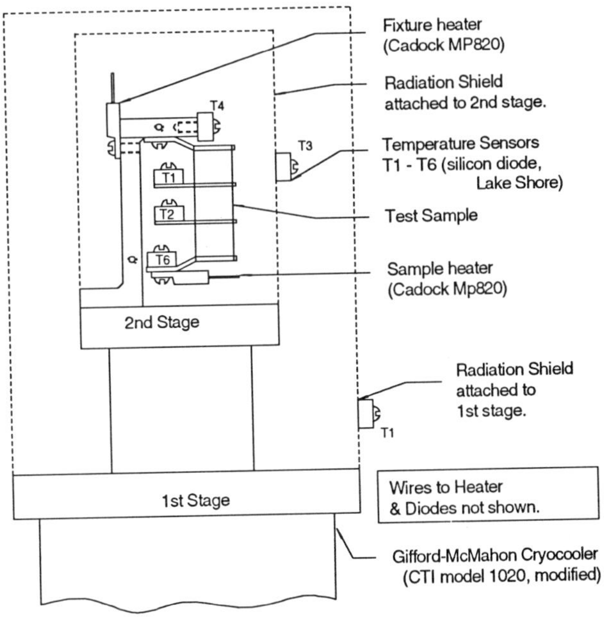

Figure 2 - Experimental Apparatus in Schematic form.

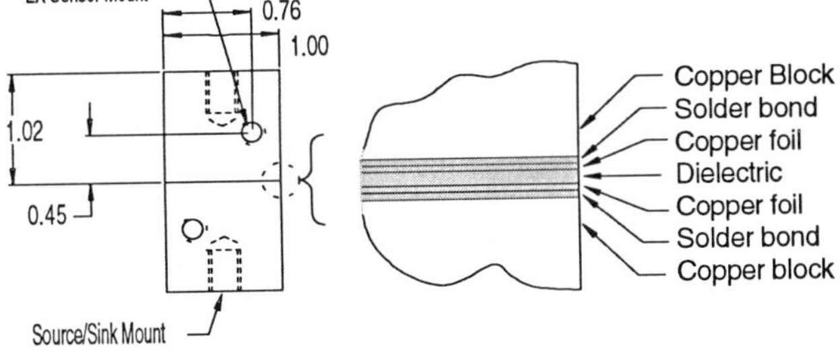

Figure 3- Dielectric Characterizaton Sample.

Our apparatus is shown schematically in Figure 2. It consists of a high capacity Gifford-McMahon refrigerator produced by CTI Cryogenics[6], that has been modified to provide a no-load minimum temperature of 4.6 K. It has sufficient cooling power on the second stage to allow the collection of thermal conductivity data down to 8 K. This system is mounted in a rectangular vacuum chamber (not shown) with removable sides that allow easy access to the cold finger for assembling and instrumenting the measurement apparatus. Instrumentation includes Lake Shore silicon diode temperature sensors, Lake Shore Cryotronic temperature measurement electronics, Tektronix programmable power supplies and multi-meters, and a computer based data acquisition system. Details of the samples and the configuration used for measuring thermal conductivity in dielectrics and metals are show in Figure 3 and Figure 4.

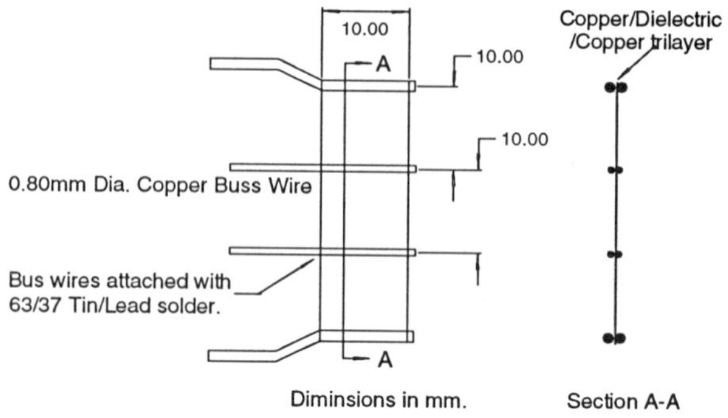

Figure 4 - Metals Characterization Sample.

Sizing the Test Sample

At cryogenic temperatures many materials have non-linear thermal conductivities. In order to reduce errors caused by non-linearity, temperature regulation on the sample should be maintained according to[7]

$$\frac{\Delta T}{T} \leq 0.1 \tag{1}$$

The cryocooler used for these measurements produces a fixture temperature of 8.0 K with no power dissipation in the heaters. Dissipating any power in the sample or sample fixture holding increases the sample temperature. With these system limitations and the desire to have data close to 10 K, it becomes desirable to keep the sample heater power at less than 50 mW. This maintains the mean sample temperature at 10 K for the first measurement point. At a sample temperature of approximately 10 K, the following things happen.

$$\Delta T \approx 1\text{K}, \; Q \approx 0.05 \text{ W}$$

$$\text{and } \frac{\Delta T}{Q} \approx 20 \text{ (K / W)} \tag{2}$$

Fourier's law defines the thermal conductivity as

$$k = \frac{Q}{\Delta T}\frac{l}{wt} \text{ (W / cm - K)} \tag{3}$$

where Q is the heat flow in watts, w and t are the sample cross section width and thickness, respectively. ΔT is the temperature difference between two points defined by heat path length l. This makes it desirable to arrange the geometry term defined by l, w, & t as follows

$$\frac{l}{wt} \approx 20k \tag{4}$$

An initial estimate of the thermal conductivity of the material in the range of interest is therefore required. A totally unknown material requires a preliminary experiment in order to have an optimized sample geometry.

An important requirement is uniformity of heat flux through the sample test length l. This can be ensured by positioning the heat source and sink the same distance l from the hot and cold sensors on the test sample.

Sample Geometry and Other Issues

Metal and robust dielectric samples allow using the thermal analog of an electrical four point measurement as shown by the equivalent circuits in Figure 5a and 5b. That measurement reduces the effect of the contact resistance at the current source and sink and at the voltage measurement points. The thermal four point measurement equivalent eliminates any concern for the contact impedance R_c at the heat source and sink as can be seen by inspecting the circuit. These impedances have no effect on the difference between T_2 and T_3, but it is wise to make them small so the mean sample temperature does not become too large. The electrical wires for the temperature sensors (R_s) provide a very high thermal impedance compared to the sample and contact impedance. This is accomplished by using appropriately long, small gage phosphor bronze wires for the temperature sensor leads. The desire is to have Q_h equal Q_{ts} and for most metal samples this is true. For some dielectric samples, heat from the diode temperature sensors may be a sizable fraction of the sample heater power. This requires including the diode heat dissipation in the thermal conductivity calculations. Sensor heat dissipations can be calculated given the voltage drops at the diode and the diode excitation current.

Thin dielectric materials pose a different problem. A four point measurement is difficult because the samples have low conductivity and are mechanically thin and flimsy. The solution is to sandwich the dielectric (a trilayer of Cu-dielectric-Cu) between two blocks of copper as shown in Figure 3. The corresponding modified four point schematic is shown in Figure 5b. The bond between the copper block and the copper film on the dielectric is indium-tin solder. The thermal impedance of the copper block, solder and copper film should be significantly smaller than the thermal impedance of the dielectric. Therefore the difference between T_2 and T'_2 is small and the difference between T'_2 and T'_3 is relatively

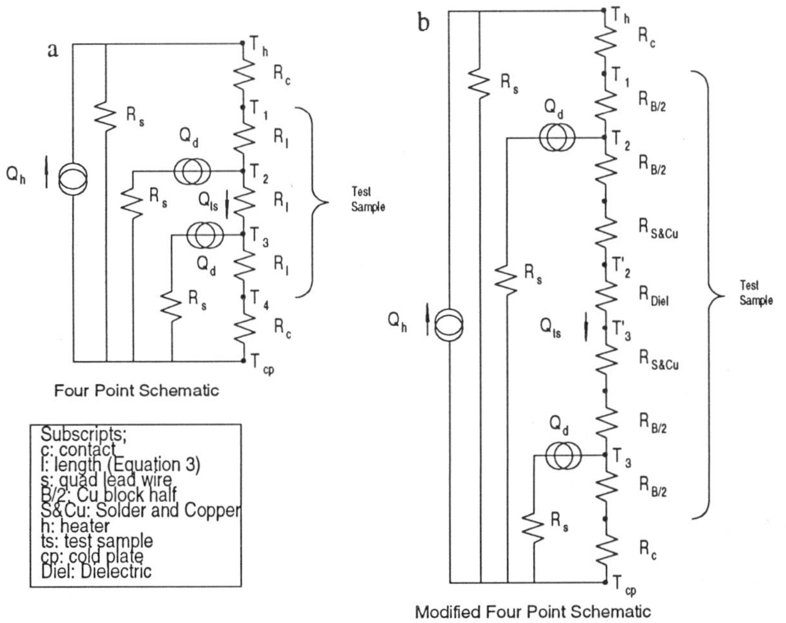

Figure 5 a & b - Four Point Schematic and Modified Four Point Schematic.

large. This works with materials compliant enough to withstand the dimensional change of copper from room temperature to cryogenic temperatures.

Thermometry Issues

The requirements for a small temperature difference between the two test points on the sample requires accurate thermometers. Our laboratory has been using silicon diodes from Lake Shore Cryotronics successfully for several years. For these measurements the DT-470 series calibrated in the range of interest provide ±20 mK accuracy at 10 K and ±55 mK accuracy at 50 K[7]. The error produced by two thermometers for a ΔT of 1 K would be 4% at 10 K and 2.2 % at 50 K. We do not use a scanning type thermometer because we have found that having the diodes always energized promotes temperature stability and accuracy for the samples with a low thermal conductivity.

Thermal Conductivity vs. Thermal Conductance

The measurements we are making on the metal samples are true conductivity measurements since we are making a four point measurement and there are no intervening materials. The measurements made on foils ususally consist of foil bonded to a dielectric material, but the heat path is parallel to the plane of the material. Since the thermal conductivity of the dielectric is at least three orders of magnitude lower than the copper thermal conductivity, the thermal path is confined primarily to the metal, thus yielding an accurate measurement of the thermal conductivity of the thin metallic foil, along the plane of the foil. However, the measurements made on the dielectric samples are really a thermal conductance measurement. A good example of what we mean by this is provided by the dielectric material used in the flexible cables. The manufacturer supplies the dielectric with copper foil bonded to both sides. These foils are typically only a few micrometers thick, and this is the only way to handle them. Photolithographic processes are used to form the signal lines on one side while the unprocessed foil on the other side forms the ground plane. When making the thermal conductivity measurement of the dielectric, the thermal path from the heater to the cold finger attachment is from the signal line material, through the dielectric, and through the ground plane material, and includes the adhesive layers that bond the copper foils to the dielectric. Thus we are actually measuring the thermal conductance of the cable material, not the thermal conductivity of the bulk dielectric. Since the signal lines can only be cooled through the dielectric, the measurement of the conductance of the thermal path is the correct measurement to make, and is more important than the conductivities of the individual bulk materials. The measurements we are making also take into account the contact resistances between the dissimilar materials, along with the fact that thin materials frequently have different

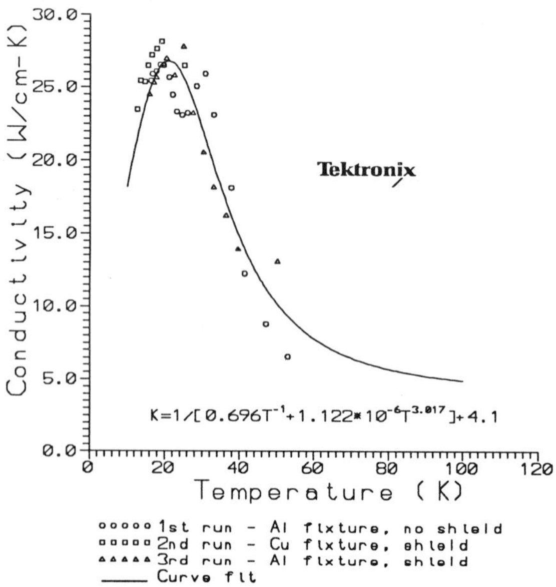

Figure 6 - Rolled Annealed Copper (W. L. Gore).

morphologies than bulk materials. In fact, this is the best possible measurement to make for our application.

THE THERMAL CONDUCTIVITY OF SOME ENGINEERING MATERIALS AT CRYOGENIC TEMPERATURES

Copper Foils

A copper foil test sample consisting of a copper-dielectric-copper trilayer was tested several times while we were refining our technique. Data were taken at different times with different fixtures and procedure changes, such as the addition of more sophisticated radiation shielding. Figure 6 shows the resulting data. The data for copper were fit to a curve of the form

$$k(T) = \frac{1}{\left(\frac{\alpha}{T} + \beta T^{\gamma}\right)} + 4.1 \qquad (5)$$

where the variables α, β & γ are fit to the calculated value of k and the sample mean temperature T.

We obtained samples of foil laminates for making high bandwidth flexible cables. These included coppers of different thickness and processing histories and two different dielectric materials. The thermal conductivities of the copper foils of three of the samples are shown in Figure 7. The most obvious characteristic of these foils is the remarkable thermal properties they display. A sample of the best quality of electronics grade OFHC copper available on the commercial market is shown on Figure 7 for comparison. The other important conclusion drawn from the results presented is that there is a factor of two difference in the range of thermal conductivities in the foils we obtained. This was very valuable information for the cryosystem and I/O cable designer. It had a significant impact on our choice of materials and the design goals we were able to attain.

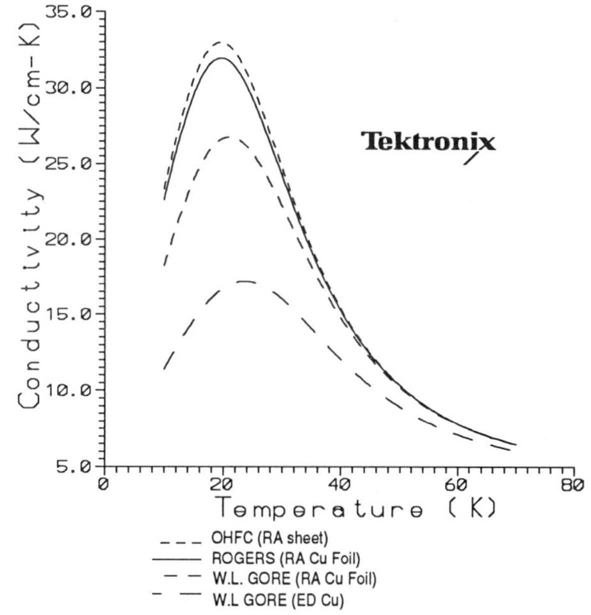

Figure 7 - Thermal conductivity at cryogenic temperatures of various copper foils used in high bandwith I/O cables.

Dielectrics

The two dielectric materials available to us have almost identical dielectric properties but are made of very different materials. One is a polyimide material and the other is a ceramic filled teflon. Our measurements, presented in Figure 8, show that the thermal conductances of these two laminates differ by a factor of two in the temperature range most critical to the thermal design of the cryosystem. Since the signal lines on the I/O cable can only be cooled through the dielectric, the best design calls for the most conductive dielectric. Judicious design can take advantage of the thermal conductance of the cable material to limit the cable length while maximizing the total number of signal lines that can be cooled by a closed cycle cryocooler of fixed cooling capacity.

SUMMARY AND CONCLUSIONS

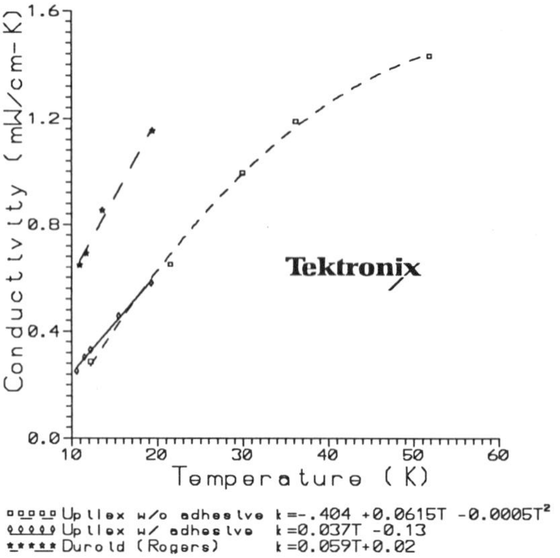

Figure 8 - Three copper clad dielectric conductivities. These curves are only valid in the range shown.

There is a substantial opportunity cost whenever an industrial laboratory engages in making material property measurements as part of regular operations. Each measurement is expensive and displaces other activities. Tektronix entered the realm of thermal conductivity measurements of metals and dielectrics to solve a specific problem: design and deliver high bandwidth multi-channel I/O cables with extremely low thermal conductance at cryogenic temperatures. The results of this effort are compelling. The thermal conductivity of modern copper foils for flat flexible cables is remarkably high, yet there are substantial differences among the choices. Furthermore, the dielectric materials key to the electrical performance of these cables did not exist when the highly active period of material characterizations took place during the early days of the nuclear and space programs. We have found as much as a factor of two difference in the thermal conductance of materials with similar dielectric properties. We could not have met the goals of our programs without the information obtained from these efforts. The investment Tektronix made in obtaining primary data on material properties for the I/O project paid off handsomely.

REFERENCES

1. B. Murdock, M. Bruns, K Godshalk, E. Hershberg, Cryosystems Design for Superconductive Digital Electonics, Government Microcircuit Applications Conference, 1992.
2. T.J. Hendricks, M.W. Bruns, E.L. Hershberg, Thermal Transport & Electrical Dissipation in Ultra-Low Thermal Load, Multi-Gigahertz I/O Cables For Superconducting Microelectronics, Advances in Cryogenic Engineering, Vol 41, 1996.
3. G.N.Ellison, E.L. Hershberg, D.L. Patelzick,The Cryocooler Implications of Flexible, Multi-Channel I/O Cables for Low Temperature Superconducting Microelectronics, Advances in Cryogenic Engineering, Vol 39, 1994.
4. Rogers Corporation, Microwaves and Circuit Materials Divison, Chandler, AZ.
5. W.L. Gore & Associates, Newark ,DE.
6. CTI Cryogenics, Mansfield, Massachusetts.
7 Seminar notes, Martin Wybourne, Materials Science Institute, University of Oregon, December 9, 1992.
8. Lake Shore Cryotronics, Inc., Temperature Measurement and Control, 1995.

GAS - GAP THERMAL SWITCH FOR PRECOOLING OF CRYOCOOLER - COOLED SUPERCONDUCTING MAGNETS

G. R. Chandratilleke, Y. Ohtani, H. Hatakeyama, T. Kuriyama, and H. Nakagome

R &D Center, Toshiba Corporation
4-1 Ukishima-cho, Kawasaki-ku, Kawasaki, 210, Japan

ABSTRACT

We have developed a heat switch for use with superconducting magnets that are cooled by two - stage GM cryocoolers. In developing cryocooler-cooled 6T and 10T superconducting magnets, which weigh some tens of kilograms, we have found that shortening cooldown times of this type of magnets is of vital importance for their wide-spread use. To shorten the magnet cooldown time, a gas-filled thermal switch having a gas-gap configuration was developed; the gap between heat transfer cylinders inside the switch is designed to be 1 mm and is filled with nitrogen. The switch thermally bridges the first and second stages of the cryocooler above 70 K when the switch is "ON." It de-couples the two stages below 50 K when the switch is "OFF." It was observed that in the temperature range 70 to 300 K, the switch provides a 80 W refrigeration capacity at the second refrigeration stage, which is about four times higher capacity than the second stage alone can provide. Using a copper block of 40 kg, we demonstrated that the cooldown time from 300 K to 4 K was only a quarter of the time required without the switch.

INTRODUCTION

Presently, development of superconducting magnets that do not require liquid helium as the coolant has become a trend, because these cryocooler-cooled magnets can be 'push-button' operated and, moreover, they do not require user skills for handling cryogens. In developing 6T and 10T cryocooler - cooled superconducting magnets[1,2], we found that reduction of cooldown times of these magnets is vital for their wide-spread use. Longer cooldown times are a consequence of the superconducting coil being connected (as in Figure 1) to the second refrigeration stage of the cryocooler, which has a low refrigeration capacity over a broad range of cryogenic temperatures.

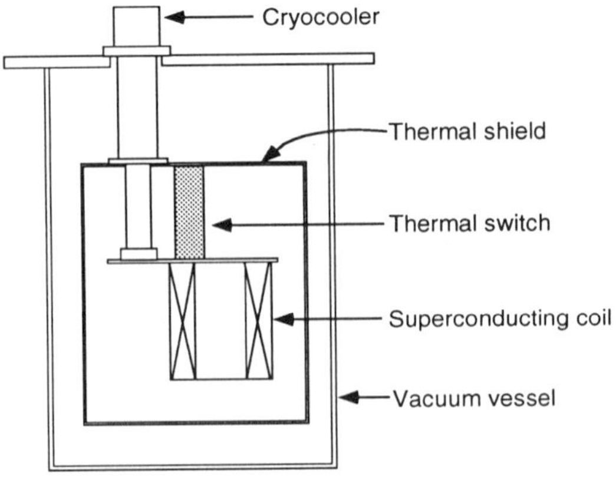

Figure 1. Utilizing a thermal switch in a superconducting magnet cooled by a two stage cryocooler.

In order to reduce cooldown time of cryocooler-cooled superconducting magnets, we developed a heat switch, which utilizes the refrigeration capacity of the first stage of the cryocooler for pre-cooling the superconducting coil as shown in Figure 1. The refrigeration capacity of a GM cryocooler first stage is about four times higher than that of the second stage in the temperature range 70 to 300 K, as shown in Figure 2. In this range of temperatures, the thermal switch will bridge the superconducting coil to the first stage of the cryocooler obtaining a higher cooling rate than the second stage alone can provide. When the coil temperature drops below the lowest temperature in the above range, the switch thermally disconnects the coil from the cryocooler's first stage letting the cryocooler's second stage do the rest of the cooling.

To date various types of cryogenic switches have been developed to perform different duties[3,4,5]. In the present case, the thermal switch is required to have a low thermal resistance over the 70 to 300 K temperature range. Consequently, a heat-pipe type thermal switch[5], which offers a low resistance only in a narrow band of temperatures, was not regarded as a candidate for the present heat switch.

THERMAL SWITCH CONSTRUCTION

The thermal switch we used has a gas - gap type construction, and has basically the same configuration as reported elsewhere[3,4]. However, there is one basic difference from

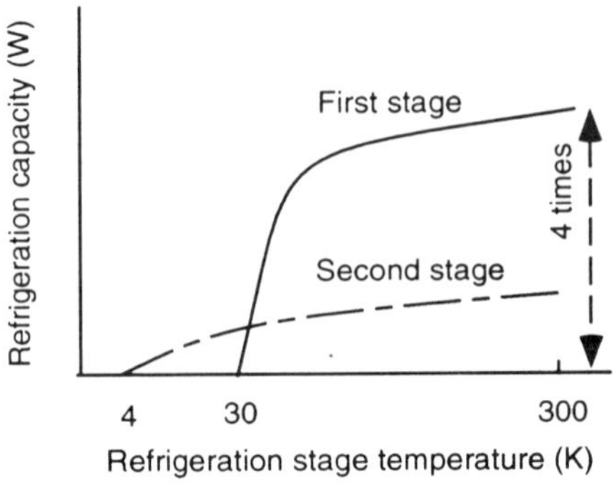

Figure 2. Schematic representation of refrigeration capacities of a GM-cryocooler.

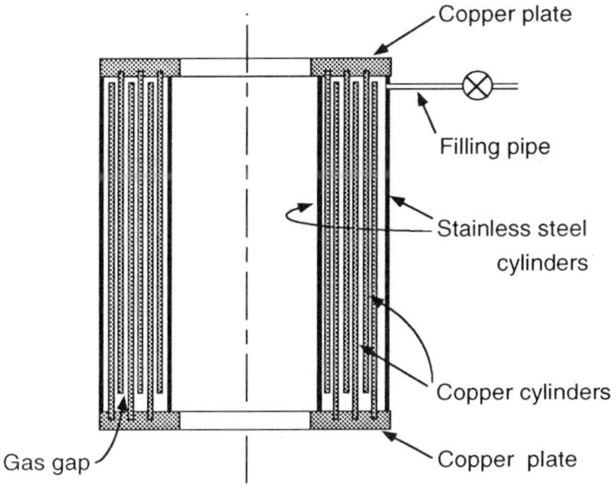

Figure 3. Construction of the thermal switch.

other gas-gap type switches: the present switch has a cryogenic gas enclosed in it as in heat-pipe type switches[5] and therefore there is no need for evacuating the gas in the switch to turn it OFF.

Figure 3 shows schematically the construction of the switch. To make it an integral part of the cryocooler, the switch has a hollow construction and is fitted between the first and second stages of a cryocooler. Overall dimensions of the switch are: 60 and 120 mm in inner and outer diameters and 212 mm in height. The switch is capped at top and bottom with copper plates and is bounded on the inside and outside by thin stainless steel pipes. The top capping copper plate is connected to the first refrigeration stage and the bottom one to the second refrigeration stage of the cryocooler. For heat transfer to take place between these two copper plates, inside the switch are several concentric copper cylinders, each having 1 mm thickness and 200 mm height and each being connected alternately to the copper plates. The gap between adjacent cylinders is designed to be 1 mm. The switch is filled with nitrogen gas at 0.01 MPa pressure at room temperature.

The filled gas thermally bridges the first stage of the cryocooler to the second stage, which in turn is coupled to a superconducting coil. In this set-up, when the cryocooler functions, it will utilize the refrigeration power of the first stage for pre-cooling the coil. In this state, the switch is turned ON. When the temperature of the first stage of cryocooler falls below the freezing point of the gas in the switch, the gas condenses on the copper cylinders connected to that stage, creating a vacuum inside the switch. The vacuum formed thermally de-couples the second stage from the first stage, and only the refrigeration power of the second stage is then available for cooling the coil. In this state, the switch is turned OFF.

DESIGN CONSIDERATIONS

In order to ensure that heat conduction mode is the dominant mode of heat transfer between copper cylinders in the switch, the following conditions[6] should be satisfied:

$$\frac{H}{w} >> 1 \text{ and } Gr_w \leq 1400\left(\frac{H}{w}\right)^{0.389}$$

where H is the height of the concentric cylinder, w the gap between adjacent cylinders and Gr_w the Grashoff number using w as characteristic length. These conditions were satisfied by the present switch and imply that convection is prevented within the switch.

The switch should have a low thermal resistance throughout the temperature range in which it is turned ON, so that the temperature difference between the first and second stages of the cryocooler is minimum. This condition will ensure that the temperature of the second stage is close to that of the first stage when the switch is turned OFF. The thermal resistance, defined by $w/\lambda A$, can be reduced when the gap w is decreased for given λ and A, which represent respectively the thermal conductivity of the gas and the total heat transfer area of the copper cylinders.

The minimum value of w is set up by mechanical limitations as well as by the thickness of the condensed gas layer on the surfaces of copper cylinders connected to the first cryocooler stage. Note that even though, liquefied nitrogen can drip[3] onto the base of the switch (i.e. the second stage), the drops quickly evaporate as a result of the second stage having a higher temperature than that of the first stage. It is thus assumed that gas condenses on the copper cylinders connected to the first stage. The thickness of this condensed gas layer can be roughly obtained by $w\rho_g/\rho_s$, where ρ_g and ρ_s are respectively the density of gas at room temperature and that of the solidified gas. Using ρ_g=1.25 kg/m^3 at room temperature and ρ_s=946 kg/m^3 at the triple point (63.14 K), the solid layer thickness is seen to be 0.0013 times the gap w, and therefore, gas solidification is not a critical issue in the design of the switch.

EXPERIMENTAL APPARATUS

The objective of the present experiment was to study the gain on cooldown time for a given superconducting coil when a heat switch is used with a two stage GM refrigerator. Figure 4 shows a schematic diagram of the apparatus, where instead of the coil, a 40 kg copper block has been used: The 6T cryocooler-cooled magnet that we have developed has roughly the same weight. Temperatures were measured at three locations on the block: bottom, middle and the top of the block. Also measured are the temperatures at the first and second stages of the cryocooler. Pressure inside the switch was measured using a pressure

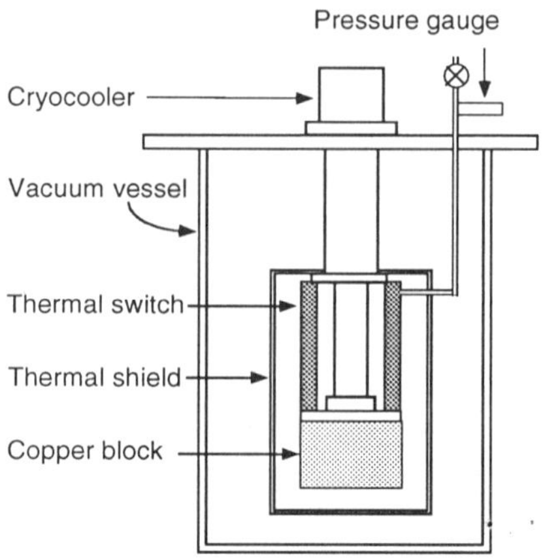

Figure 4. Schematic diagram of experimental apparatus

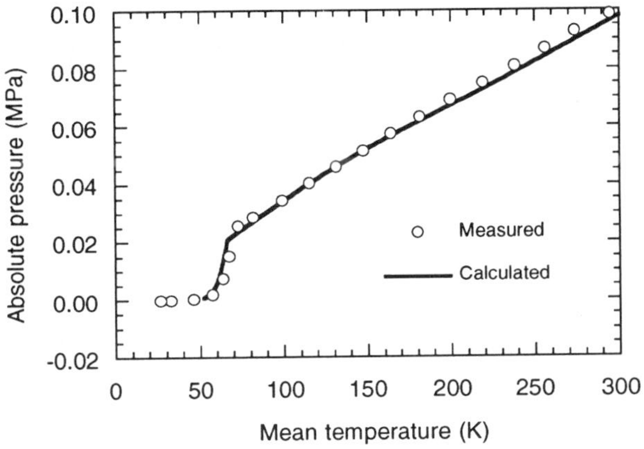

Figure 5. Pressure variation in the heat switch as cooling proceeds.

transducer. The switch is filled with one atmosphere of nitrogen gas at room temperature through the pipe connected to the switch and the valve is then closed to trap the gas inside the switch. Cooldown experiments were performed for two conditions: with the switch filled with nitrogen gas, and with the switch evacuated.

EXPERIMENTAL RESULTS

Figure 5 shows the pressure inside the switch as a function of the mean temperature of the gas, which was obtained by averaging the first and second stage temperatures. As the cryocooler is started, the first stage temperature starts dropping more rapidly than that of the second stage and the mean temperature also drops as a result. This drop in the temperature causes the pressure inside the switch to drop. As seen from the figure, at about 70 K, the drop in pressure becomes sharp, indicating that gaseous nitrogen inside the switch has become liquefied at that temperature. As the temperature drops further, the pressure approaches zero as a result of complete solidification of liquid nitrogen which occurs at about 50 K. The relationship between pressure and temperature can be reasonably well predicted by assuming that density of nitrogen remains roughly constant during the cooling

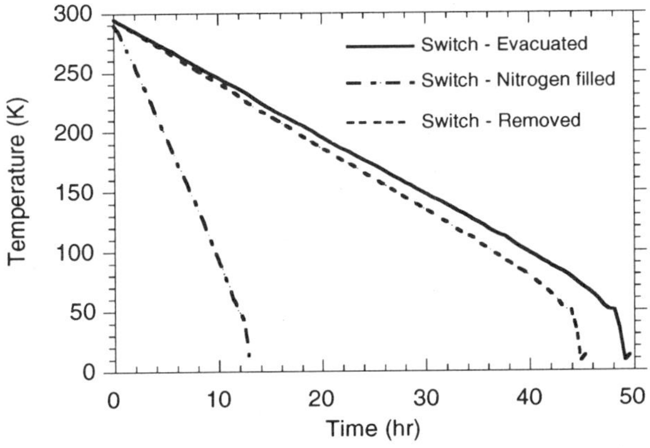

Figure 6. Comparison of the cooldown times of 40 kg mass under three states: switch filled with nitrogen, switch evacuated and switch removed.

process; the solid line in the figure was obtained from the constant density contour passing through the room temperature condition on a temperature-entropy chart.

Cooling curves for the 40 kg copper block are shown in Figure 6 for two states of the switch: nitrogen-filled and evacuated states of the switch. In the case of nitrogen-filled state, the switch turns off by itself when nitrogen solidifies. In the case of evacuated state, the switch is evacuated at room temperature so that it does not function, but is only an added mass to the 40 kg copper block. In both cases, the mass finally reached nearly an identical temperature of 4 K. At 4 K, the switch is estimated to have a 500 K/W thermal resistance compared to a 0.4 K/W when the switch is ON.

A correction is now made for the mass of the switch so that it is possible for us to compare the cooldown times with and without the switch. To do this, it is assumed that the refrigeration capacity $\dot{q}$ of the second stage is only a function of temperature, irrespective of the mass that is attached to that stage. Under this assumption, Eq. (1) defines the cooling rate $\dot{q}$ at a given temperature for the cooling curves corresponding to the masses, $(m+m_s)$ and m, where m represents the mass of copper block and m_s that of the switch.

$$\dot{q} = (m + m_s)\ C \left(\frac{dT}{dt}\right)_w = m\ C \left(\frac{dT}{dt}\right)_{wo} \tag{1}$$

Here, C represents the specific heat of copper, (dT/dt) the rate of change of temperature and the suffixes w and wo represent the two states of cooling, with and without the switch.

Simplification and integration of Eq. (1) result in the following relationship for the coordinates of cooling curves for the masses, $(m+m_s)$ and m at a given temperature.

$$t_{wo} = \frac{m}{m + m_s} t_w \tag{2}$$

Here, suffixes w and wo represent the two states of cooling, with and without the switch.

Based on the correction defined by Eq. (2), the cooling curve for the 40 kg copper block when the heat switch is not present is shown in Figure 6 by the dashed line. If we now consider cooldown from room temperature, 300 K, to the switch's turn-off temperature, 50 K, it can be seen from Figure 6 that the heat switch makes it possible to cool the 40 kg mass from 300 K to 50 K in 12 hours, which is only one quarter of the time required by the cryocooler without the switch.

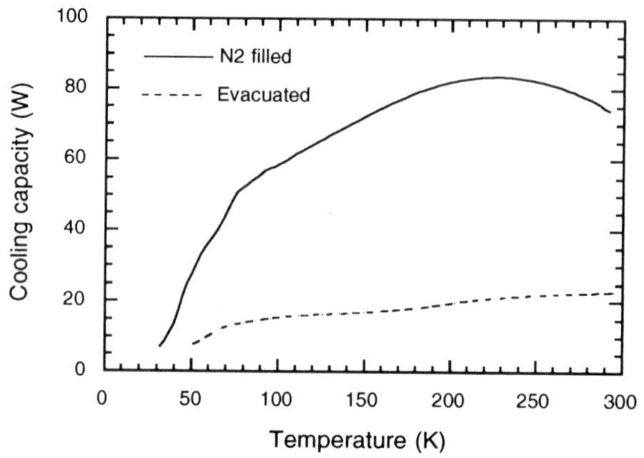

Figure 7. Comparison of cooling rates, when the switch functions and does not function.

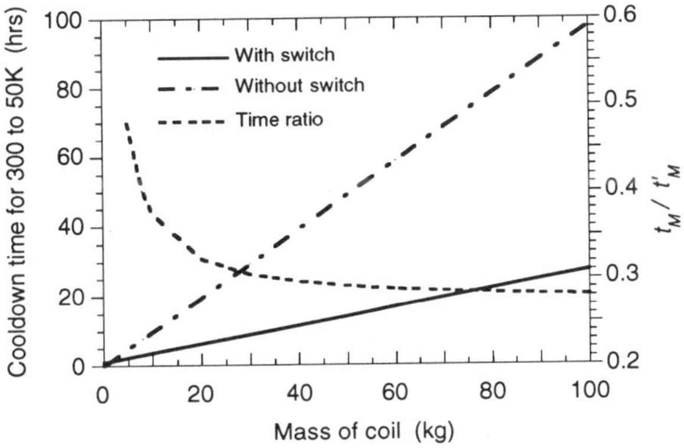

Figure 8. Comparison of cooldown times for the cases of with and without the switch in the range of temperatures, 300 K to 50 K, and ratio of the cooldown times.

The cooling rate $\dot{q}$ available at the second stage is obtained from Eq. (3), and is shown in Figure 7 as a function of temperature for the two states of the switch: nitrogen-filled and evacuated states of the switch. Here, the symbols carry the same meaning as in Eq. (1).

$$\dot{q} = m\, C(T)\, \frac{dT}{dt} \tag{3}$$

Comparison of the two curves shows that the switch has made it possible to obtain on the average a 80 W cooling capacity in the temperature range 300 to 70 K, in contrast to 20 W obtained when the switch is not functioning. The cooling rate is thus increased by about four times owing to the heat switch. Note that Eq. (3) has not been corrected for the mass of switch m_s and that the cooling rate increment is not affected by this correction because the cooling rate can be thought of as a function only of temperature, as in the derivation of Eq. (1).

In order to predict the cooldown times for different masses, it is assumed that the cooling rate available at the second stage is only a function of the temperature and is independent of the mass attached to the stage, whether the switch is present or not. Under these conditions, Eq. (2) is applicable. The cooldown times for a coil of mass M, with or without the switch can be obtained in terms of respective cooldown times for a coil of mass m as follows:

$$t_M = \frac{M + m_s}{m + m_s} t_m \tag{4}$$

$$t'_M = \frac{M}{m} t'_m \tag{5}$$

where, t_m indicates the cooldown time for a mass m when the switch is present, and the primed t_m indicates the cooldown time of mass m when the switch is absent.
Thus the ratio of cooldown times is given by:

$$\frac{t_M}{t'_M} = \frac{(M + m_s)m}{(m + m_s)M} \frac{t_m}{t'_m} \tag{6}$$

Figure 8 plots as a function of the cooling mass M, the cooldown times defined by Eq. (4) and Eq. (5), which respectively represent cooling with or without the switch; by the experiment on a mass of roughly 40 kg, t_m and t'_m are both known from Figure 6. The figure also plots the cooling time ratio defined by Eq. (6) as a function of the cooling mass.

It is seen from this curve that the cooldown time ratio is asymptotic when the cooling mass is large, implying that mass of the switch m_s becomes negligible in comparison with the cooling mass M.

CONCLUSIONS

Utilization of the present heat switch has made it possible to obtain a 80 W refrigeration capacity in the temperature range 70 to 300 K at the second refrigeration stage. This is a four times higher capacity than in the absence of the heat switch. Consequently, the heat switch is expected to shorten the cooldown times drastically. For a 40 kg mass, the cooldown time from 300 K to 4 K was 12 hours, which is only about a quarter of the time required in the absence of the switch. By using the experimental results for a given mass, it is possible to estimate the cooldown times for other masses.

REFERENCES

1. T. Kuriyama, et al., Cryocooler directly cooled 6 T NbTi superconducting magnet system with 180 mm room temperature bore, *Cryogenics ICEC Supplement* 34:643 (1994).
2. Y. Ohtani, et al., Cryocooler-cooled 10 T superconducting magnet system with 100 mm room temperature bore, *CEC Paper* TH-B3-1 (1995)
3. R. P. Bywaters and R. A. Griffin, A gas-gap thermal switch for cryogenic applications, *Cryogenics* 13:344 (1973).
4. T. Nast, G. Bell and C. Barnes, Development of gas gap cryogenic thermal switch, in: "Advances in Cryogenic Engineering, Vol. 27 ," Plenum Press, New York (1981), p. 1117.
5. J. Yamamoto, Performance of a gas - filled thermal switches, in: "Advances in Cryogenic Engineering, Vol. 25 ," Plenum Press, New York (1982), p. 261.
6. S. Ostrach, in: "Advances in Heat Transfer, Vol. 8," J. P. Hartnett and F. I., Jr. Thomas, ed., Academic Press, New York (1972), p. 161.

NITROGEN HEAT PIPE FOR CRYOCOOLER THERMAL SHUNT

F. C. Prenger[1], D. D. Hill[1], D. E. Daney[1],

M. A. Daugherty[1], G. F. Green[2], and E.W. Roth[2]

[1]Los Alamos National Laboratory
Los Alamos, NM, 87545, USA
[2]Naval Surface Warfare Center
Annapolis, MD, 21402, USA

ABSTRACT

A nitrogen heat pipe was designed, built and tested for the purpose of providing a thermal shunt between the two stages of a Gifford-McMahan (GM) cryocooler during cooldown. The nitrogen heat pipe has an operating temperature range between 63 and 123 K. While the heat pipe is in this temperature range during the system cooldown, it acts as a thermal shunt between the first and second stage of the cryocooler. The heat pipe increases the heat transfer to the first stage of the cryocooler, thereby reducing the cooldown time of the system. When the heat pipe temperature drops below the triple point, the nitrogen working fluid freezes, effectively stopping the heat pipe operation. A small heat leak between cryocooler stages remains because of axial conduction along the heat pipe wall. As long as the heat pipe remains below 63 K, the heat pipe remains inactive. Heat pipe performance limits were measured and the optimum fluid charge was determined.

INTRODUCTION

Cooldown times for components attached to the second stage of two-stage GM cryocoolers are significantly influenced by the thermal capacitance of the components and by the parasitic heat loads on the cryocooler. For systems containing superconducting magnets with large thermal capacitance, the cooldown times can be considerable. These components are attached thermally and mechanically to the second stage of the cryocooler because, ultimately, this is where the lowest temperatures are attained. However, because the cryocooler's second stage has a significantly lower refrigeration capacity than its first stage, cooldown times can be reduced by temporarily shifting part of the heat load to the cryocooler's first stage where capacity is greater. A gravity-assisted, nitrogen heat pipe is well suited for this task.

The heat pipe is installed between the two stages of the cryocooler with the evaporator attached to the second stage and the condenser attached to the first stage. When the heat pipe is between about 123 and 63 K during cooldown, it acts as a thermal shunt connecting the first and second stages together. The heat removed from the load, which is mechanically attached to the second stage, is then shared by the first and second stages of the cryocooler. As the second stage temperature continues to decrease below 63 K, the triple point of nitrogen, the heat pipe becomes inoperable as the working fluid freezes, and the two cryocooler stages become thermally disconnected. There remains a parasitic heat leak along the heat pipe wall which can be made small by proper selection of the heat pipe material and geometry. At temperatures above the critical point (123 K), the heat pipe is also inoperable because of the absence of a two phase interface within the heat pipe. The freezing and thawing process is reversible and does not degrade the operation of the heat pipe.

DESIGN

The heat pipe is gravity-assisted, whereby the condenser is assumed to always be above the evaporator, and should be sized to handle the maximum first stage capacity of the cryocooler. In this way the heat pipe cannot be cooled below the triple point by the cryocooler during cooldown. Only when the second stage approaches the nitrogen triple point will the heat pipe begin to freeze and "shutoff". The simplest and most economical configuration for this heat pipe is a smooth, thin-wall, stainless steel tube joining copper evaporator and condenser sections. The stainless steel tube minimizes conduction heat leak between the cryocooler stages when the heat pipe is in the off mode and provides adequate strength to contain the heat pipe internal pressure at 300 K. The latter requirement results from the pressure increase between the maximum operating temperature of 123 K and the 300 K ambient temperature. The copper condenser and evaporator sections minimize thermal gradients at the heat transfer surfaces.

For a smooth-wall, gravity-assisted heat pipe, the capacity is primarily determined by the heat pipe cross-sectional area[1]. Figure 1 shows the calculated heat pipe capacity for different tube sizes as a function of the heat pipe temperature at the evaporator exit. Under most operating conditions the heat pipe vapor temperature is nearly isothermal and specifying a particular location for the heat pipe temperature is unnecessary. The 9.5 mm (3/8") OD x 0.7 mm (0.028") wall tube was selected to insure adequate capacity for the heat pipe. The maximum capacity of the cryocooler first stage in the temperature range 63 to 123 K was estimated at 80 W.

The heat pipe tube wall thickness was determined by the heat pipe internal pressure and the tube wall stress[2].Figure 2 shows the allowable internal pressure for each tube OD considered as a function of wall thickness. An allowable tensile stress in the tube wall of 68.9 MPa (10,000 psi) was assumed with a factor of safety of 2.0. Also shown for reference is the heat pipe internal pressure at 300 K, which is the maximum pressure anticipated. From the figure the minimum wall thickness that does not exceed one half of the allowable stress for a 9.5 mm (3/8") OD tube is 0.7 mm (0.028"). This minimum value was selected to minimize axial conduction along the tube. The calculated axial conduction

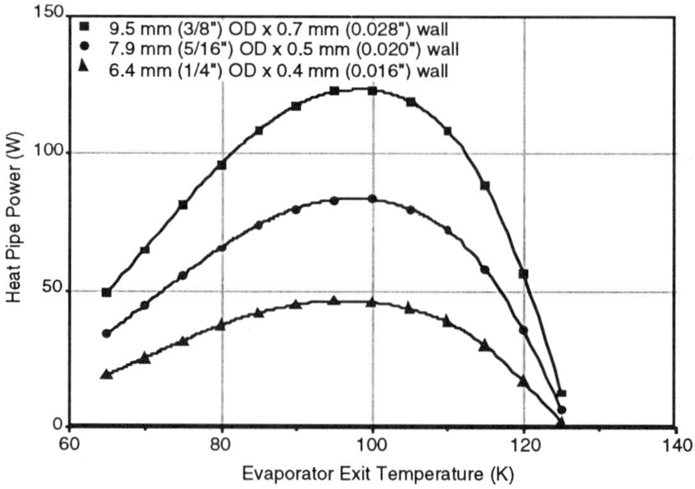

Figure 1. Calculated heat pipe performance as a function of operating temperature and tube size.

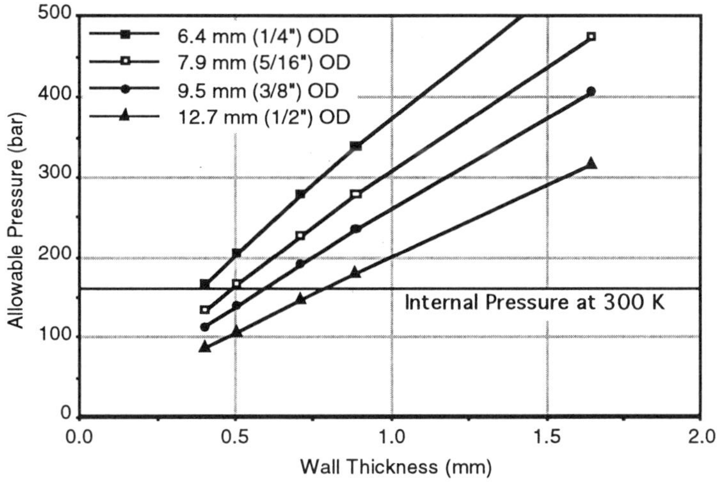

Figure 2. Allowable pressure for heat pipe tube wall as a function of OD and wall thickness. N_2 temperature is 80 K, fill fraction is 0.2, tube length is 127 mm (5.0"), and allowable stress is 68.9 MPa (10,000 psi) (FS = 2.0).

as a function of tube size at steady state operating temperatures (4.2 to 80 K) is shown in figure 3. For the 9.5 mm (3/8") OD x 0.7 mm (0.028") wall tube selected, the axial heat leak is approximately 50 mW.

Because of limited surface area in the heat pipe condenser, the thickness of the liquid condensate film on the inside wall of the heat pipe could be unnecessarily large. This film in the condenser has a thermal resistance and creates an unwanted temperature drop in the heat pipe. Increasing the internal diameter of the condenser reduces the condensate film thickness because the surface area is increased. This leads to a reduction in thermal resistance and a smaller temperature drop across the liquid film. The temperature difference across the film was calculated as a function of condenser length at an operating temperature of 80 K and is shown in figure 4. Results are shown for the nominal heat load of 60 W and, because the heat pipe is capable of higher performance, for a heat load of 100 W. The active condenser length is approximately 32 mm, which at the nominal heat load gives a film temperature drop of 9 K.

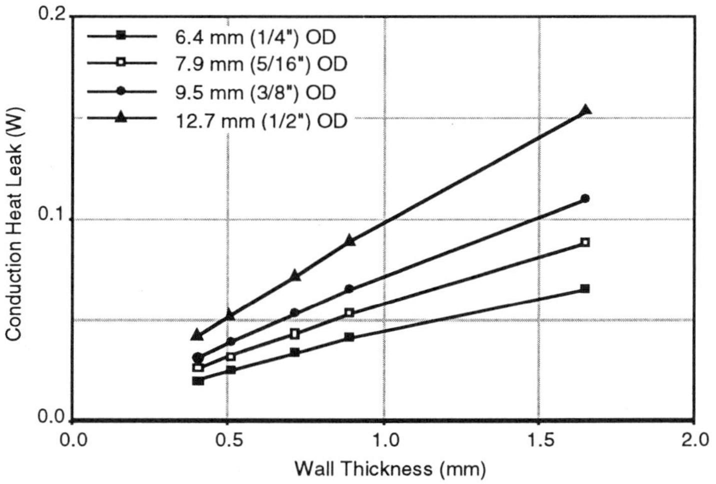

Figure 3. Calculated conduction heat leak along the heat pipe wall with boundary temperatures of 4.2 and 80 K.

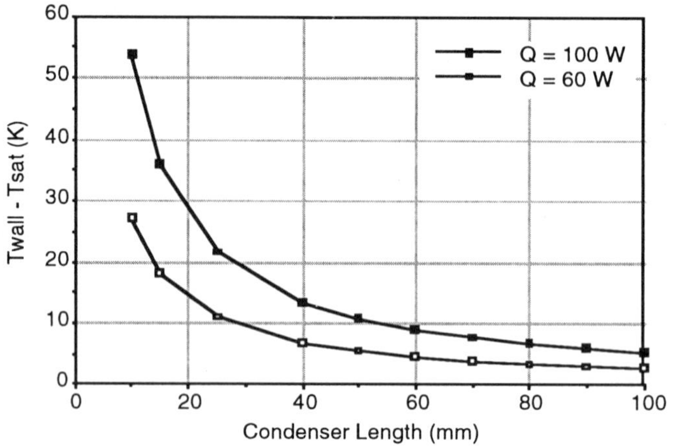

Figure 4. Calculated temperature drop across the liquid film in the heat pipe condenser.

A schematic of the heat pipe is shown in figure 5. The copper evaporator was externally threaded to mate with an internally threaded boss on the cryocooler second stage. Thermal contact was enhanced by mechanically loading the threaded joint with a locknut. The condenser extended above the first stage of the cryocooler. The smooth external surface of the condenser was inserted into a socket that was fitted with flexible louvers and was mounted on the first stage of the cryocooler. The louvers had an interference fit and

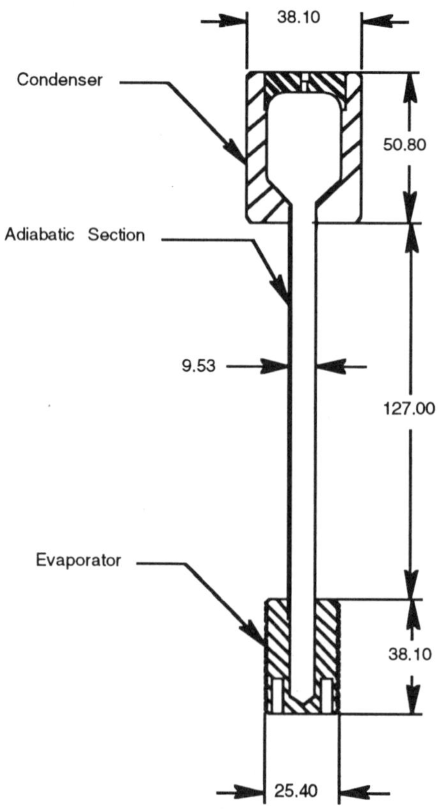

Figure 5. Shunt heat pipe configuration (dimensions in mm).

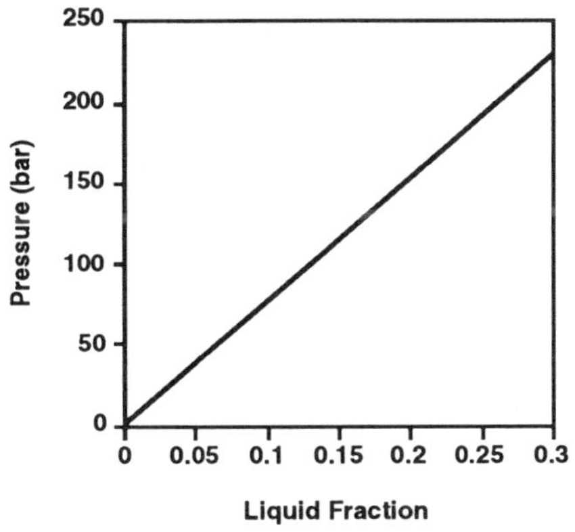

Figure 6. Relationship between liquid fraction in the heat pipe at 75 K and charge pressure at 300 K.

maintained contact with the heat pipe condenser surface. This joint accommodated relative movement between the heat pipe and the cryocooler.

FILL CHARGE

The heat pipe was evacuated and charged with nitrogen gas at 300 K. Figure 6 shows the relationship between charge pressure and liquid fraction in the heat pipe at 75 K. The heat pipe performance is sensitive to the amount of liquid in the evaporator. If there is insufficient liquid the evaporator will not be isothermal. In a smooth wall, gravity assisted heat pipe the evaporator is cooled from above by the descending liquid film and from below by splashing from the liquid pool[3]. Initially, the heat pipe was charged to 41 atm (600 psia) giving a liquid ratio at 75 K of 0.05. This corresponded to a liquid height at no load of 25 mm. Normally, for a constant diameter heat pipe, this would be a sufficient quantity of liquid for satisfactory operation, however, because of the abnormally large volume of the condenser the heat pipe underperformed at this fill charge (4.1 MPa or 600 psia). The fill charge was then increased to compensate for the holdup of liquid in the oversized condenser. Improved performance resulted.

TESTING

The heat pipe was tested in a vacuum insulated cryostat. The evaporator was electrically heated with a circumferentially wrapped Ni-Cr alloy wire. The heat pipe condenser

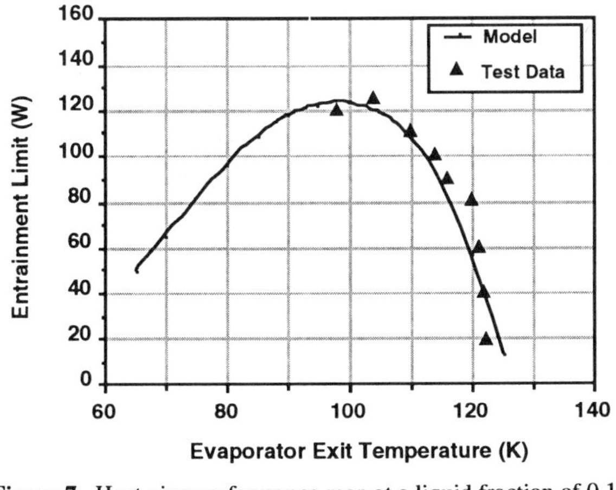

Figure 7. Heat pipe performance map at a liquid fraction of 0.19.

temperature was controlled by clamping the heat pipe condenser to a variable temperature nitrogen bath. A vacuum pump and throttle valve were used to control the pressure of the nitrogen bath and provided means to vary the bath temperature between 63 and 80 K. Because temperatures higher than 80 K were not required, pressurization of the bath was not necessary. With the variable temperature bath acting as heat sink and the electric heater controlling the input power to the heat pipe, the heat pipe operating temperature could be varied independent of the heat pipe power. Using this technique, a performance map of maximum power versus operating temperature of the heat pipe was generated.

Silicon diode thermometers were used to measure heat pipe wall temperatures and were attached to the heat pipe at the evaporator end, on the adiabatic section near the evaporator exit and on the condenser 20 mm from the condenser inlet. All instrumentation and heater leads were thermally anchored to the nitrogen bath and the thermometer leads were thermally tempered at the measurement points as well. Phosphor-bronze wire was used for the thermometer leads to further minimize heat leak and reduce measurement errors.

The heat pipe performance limit is defined as the maximum power that the heat pipe can transmit under nearly isothermal conditions. Exceeding this power limit results in an interruption of the circulation of the working fluid and subsequent dryout of the evaporator region. With an evaporator no longer supplied with liquid, the heat pipe cannot continue to transfer heat by internal fluid circulation and the axial temperature gradient increases as axial conduction becomes the primary heat transfer mode. For the smooth wall gravity assisted heat pipe, the entrainment limit is reached when flooding or entrainment of the liquid layer occurs[4].

The heat pipe was charged with nitrogen gas to 134 bar (1950 psia) at 300 K. This fill charge corresponded to a liquid fraction of 0.19 at 75 K. Figure 7 shows the measured performance data for the heat pipe compared with the predicted performance using the LANL heat pipe model[6]. The operating temperature range extended to 123 K, near the critical temperature, as predicted by the model. Heat pipe performance limits below 95 K were not measured because of limitations of the test apparatus. The nitrogen bath was capable of reaching 63 K but the thermal resistance between the heat pipe condenser and the nitrogen bath was larger than expected. Also, the heat pipe capacity was higher than expected and at a heat load of 100 W the heat pipe temperature was 95 K. To measure performance limits below 95 K would require decreasing the thermal resistance to the heat sink.

SYSTEM COOLDOWN TESTS

After performance testing was completed, the heat pipe was shipped to the Naval Surface Warfare Center (NSWC), Carderock Division for system cooldown tests. The heat pipe was installed between the first and second stages of a Balzers model UCH-130 cryocooler equipped with a neodymium second stage regenerator. The total weight of the second stage cold mass (copper disc and NbTi coil) was 4.1 kg.

The heat pipe was connected to copper plates on the first and second stages of the refrigerator. The external threads of the evaporator were mated with an internally threaded plate mounted to the refrigerator's second stage. To enhance thermal contact, the threaded joint was packed with Apiezon "N" grease and a nut was installed to increase the joint contact pressure. The condenser end was inserted into a socket on a copper plate mounted to the first stage of the cryocooler. The socket contained flexible louvers (Multi-Lam™) of silver plated beryllium-copper that maintained an interference fit but would allow some relative motion due to thermal contraction or during installation. In practice, the Multi-Lam connection was not as compliant as expected and was difficult to assemble. Significant temperature differences developed across this interface during cooldown of the magnet, which hindered performance.

The goal of the first test was to determine the cooldown time from room temperature to 5 K for the system without the heat pipe installed. The cooldown time was approximately 3.9 h for this case as shown in figure 8. After the heat pipe was installed, the test was repeated resulting in a cooldown time of 3.5 h. During this test the heat pipe's performance was impaired because of overcooling of the condenser and freezing of the working fluid while the evaporator remained above the critical temperature. While some heat was transferred by the heat pipe during the cooldown, freezing of the working fluid in the condenser most likely reduced the fluid inventory reducing heat pipe performance. As a

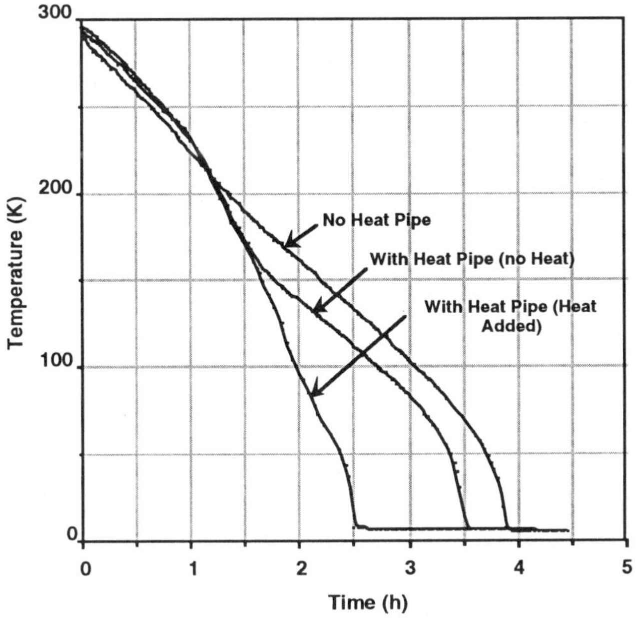

Figure 8. System cooldown performance with and without shunt heat pipe installed.

result, the heat pipe operated for only a short time. A third test was run during which supplemental heat was added to the heat pipe condenser for a few minutes during the cooldown to prevent overcooling of the condenser. This increased the effectiveness of the heat pipe and shortened the cooldown time to 2.5 h. Cooldown temperature profiles measured using the heat pipe with and without heat addition are also shown in figure 8. The inability of the heat pipe to increase the heat load on the first stage of the cryocooler at temperatures above about 125 K resulted in the observed overcooling of the heat pipe condenser. Addition of a second heat pipe with a higher operating temperature (ethane) would eliminate the overcooling problem and further reduce the cooldown time of the system.

CONCLUSIONS

This heat pipe, with a liquid fraction of 0.19 at 75 K, achieved a 30 % higher transport power than predicted; however, the liquid fraction needed to achieve this performance was also larger than expected. This larger liquid inventory is required because the liquid tends to accumulate in the oversized condenser, which is required to reduce the film thickness and, therefore, the temperature drop. The heat pipe operating temperature should be between 63 and 123 K. Although not tested at the lower end of this range, the LANL heat pipe computer code was used to determine the heat pipe performance.

As expected, heat pipe performance was sensitive to the liquid inventory within the pipe. Filling the heat pipe at liquid fractions below 0.15 resulted in a reduced operating temperature range with the maximum operating temperature reduced substantially below the nitrogen critical temperature. The maximum operating temperature of the heat pipe at a liquid fraction of 0.05 was 106 K. Larger liquid fill fractions increased the maximum operating temperature.

Upon cooldown of the cryocooler to below the nitrogen triple point, 63 K, the heat pipe working fluid freezes and heat transfer along the heat pipe wall is by conduction only. This heat leak was calculated to be 50 mW with heat pipe end temperatures of 4.2 and 80 K.

System cooldown tests at the NSWC showed a 30 percent decrease in cooldown time when using the nitrogen heat pipe. Because of the narrow operating temperature range of the nitrogen heat pipe, some supplemental heating was required to maintain the heat pipe condenser above the nitrogen triple point during the initial cooldown phase when the heat pipe was not yet fully operational. This problem could be eliminated with the addition of a second heat pipe with a higher operating temperature (such as ethane). By increasing the heat load to the cryocooler at a higher temperature, subcooling of the nitrogen heat pipe would be eliminated, and the additional capacity of the second heat pipe would further reduce the cooldown time.

ACKNOWLEDGMENTS

This work was sponsored by the Carderock Division of the Naval Surface Warfare Center as part of the U. S. Navy program on power applications of superconductivity.

REFERENCES

1. F. C. Prenger, J. E. Kemme, M. Groll and T. Spendel, "Performance Limits for Smooth-Wall, Gravity Assisted Heat Pipes", AIAA 18th Thermophysics Conference, Montreal, Canada, June 1983.
2. E. P. Popov, Mechanics of Materials, Prentice-Hall, Inc., Englewood Cliffs, NJ
3. F. C. Prenger and J. E. Kemme, "Performance Limits of Gravity-Assist Heat Pipes with Simple Wick Structures", 4th International Heat Pipe Conference, London, England, September, 1981.
4. D. D. Hill, F. C. Prenger, and J. A. Barclay, "Performance Limits in Gravity-Assisted Hydrogen Heat Pipes", ASME National Heat Transfer Conference, Houston, Texas, July 1988.
5 F. C. Prenger, "Liquid Limits in Gravity-Assisted Heat Pipes", 6th International Heat Pipe Conference, Grenoble, France, May 1987.
6. K. A. Woloshun et. al., "HTPIPE: A Steady-State Heat Pipe Analysis Program", Los Alamos National Laboratory internal report no. LA-11324-M, Nov 1988.

NUMERICAL SIMULATION OF THERMAL STRATIFICATION IN LIQUID HYDROGEN

Z. Tanyun, H Zhongping, and Shimo Li*

Cryogenic Lab. , Zhejiang University
Hangzhou, 310027 P. R. China

ABSTRACT

This paper presents the results of numerical simulations of thermal stratification in liquid hydrogen as a cryogenic fluid. We used the vorticity-stream function method and correlated the surface temperature with the heat flux, fluid level and time. In addition, the simulation results are compared with experimental data.
Key words: cryogenic, thermal stratification, ullage volume

INTRODUCTION

With the rapid development of the astronautic technologies since the 1960s, the United States, Europe, Japan, former Soviet Union and China have launched thousands of satellites, space shuttles and cosmic probes using high-specific thrust hydrogen-oxygen engines. Among application technologies for liquid hydrogen and oxygen, thermal stratification, especially in liquid hydrogen, has been an important research subject.

Previous investigations have shown,[1,2,3] that in sealed liquid hydrogen vessels, varied thermal stratification exists whether the vessels are on the ground or in flight. This phenomenon can cause gaseous corrosion in transport pumps, or inhibit the injection of hydrogen into engines when severe stratification has occurred, resulting in premature ignition failure and unsuccessful launch.

There has been considerable amount of reserch on thermal stratification in cryogenic liquids. The methods can be categorized in the following classes: 1. boundary layer integration based on turbulent boundary layer theory for natural convection; 2. heat transport; 3. finite element, and 4. statistical method. We point out that most researchers restricted their studies to thermal stratification alone and ignored the influence of thermal stratification on the gas pressure in the vessels, and the influence of ullage volume on thermal stratification.

---

* Corresponding author

A SIMULATION MODEL FOR THERMAL STRATIFICATION IN LIQUID HYDROGEN

Figure 1 illustrates the numerical simulation model for thermal stratification. Assuming thermodynamic equilibrium between the gas and the liquid phases, only the liquid phase is simulated since the evaporation at the free surface and the latent heat is relatively small and can be neglected. Since the viscosity of the liquid hydrogen is small, the dissipation and expansion work terms are neglected in the energy equation. Boussinesq approximation is also used, i. e. only the temperature dependence of density is kept in calculating buoyancy force. According to reference[4], the hydrodynamic equations are

$$\frac{\partial u}{\partial x} + \frac{\partial v}{\partial y} = 0 \tag{1}$$

$$\frac{\partial u}{\partial t} + u\frac{\partial u}{\partial x} + v\frac{\partial u}{\partial y} = -\frac{1}{\rho}\frac{\partial P}{\partial x} + \gamma\left(\frac{\partial^2 u}{\partial x^2} + \frac{\partial^2 u}{\partial y^2}\right) \tag{2}$$

$$\frac{\partial v}{\partial t} + u\frac{\partial v}{\partial x} + v\frac{\partial u}{\partial y} = -\frac{1}{\rho}\frac{\partial P}{\partial y} + \gamma\left(\frac{\partial^2 v}{\partial x^2} + \frac{\partial^2 v}{\partial y^2}\right) \tag{3}$$

$$\frac{\partial T}{\partial t} + u\frac{\partial T}{\partial x} + v\frac{\partial T}{\partial y} = \alpha\left(\frac{\partial^2 T}{\partial x^2} + \frac{\partial^2 T}{\partial y^2}\right) \tag{4}$$

For 2-D problems, a stream function ψ is introduced for convenience

$$u = \frac{\partial \psi}{\partial y}, \quad v = -\frac{\partial \psi}{\partial x} \tag{5}$$

This condition leads to automatic satisfaction of eq. (1). To eliminate P, a vorticity ω is introduced:

$$\omega = \frac{\partial v}{\partial x} - \frac{\partial u}{\partial y} \tag{6}$$

which transformed the momentum eqs. (2) and (3) to a vorticity transport equation

$$\frac{\partial \omega}{\partial t} + u\frac{\partial \omega}{\partial x} + v\frac{\partial \omega}{\partial y} = \gamma\left(\frac{\partial^2 \omega}{\partial x^2} + \frac{\partial^2 \omega}{\partial y^2}\right) + g\beta\left(\frac{\partial T}{\partial x}\right) \tag{7}$$

After proper non-dimensionalization, eqs. (1)-(6) become

$$u = \frac{\partial \psi}{\partial y}, \quad v = -\frac{\partial \psi}{\partial x} \tag{8}$$

$$\omega = -\left(\frac{\partial^2 \psi}{\partial x^2} + \frac{\partial^2 \psi}{\partial y^2}\right) \tag{9}$$

$$\frac{\partial \omega}{\partial \tau} + u\frac{\partial \omega}{\partial x} + v\frac{\partial \omega}{\partial y} = GrPr^2\frac{\partial \theta}{\partial x} + Pr\left(\frac{\partial^2 \omega}{\partial x^2} + \frac{\partial^2 \omega}{\partial y^2}\right) \tag{10}$$

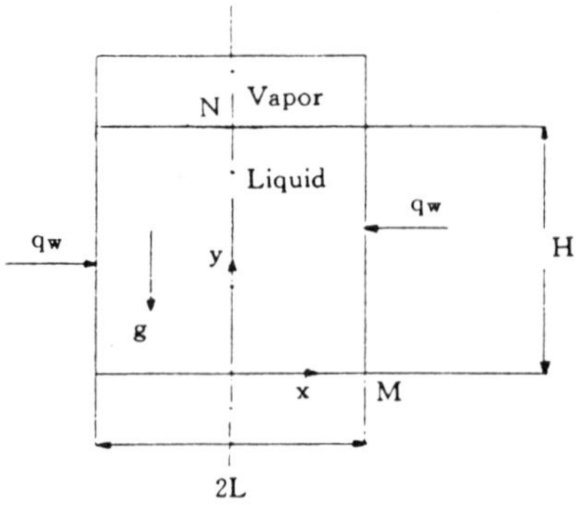

Fig. 1 Cartesian Coordinates.

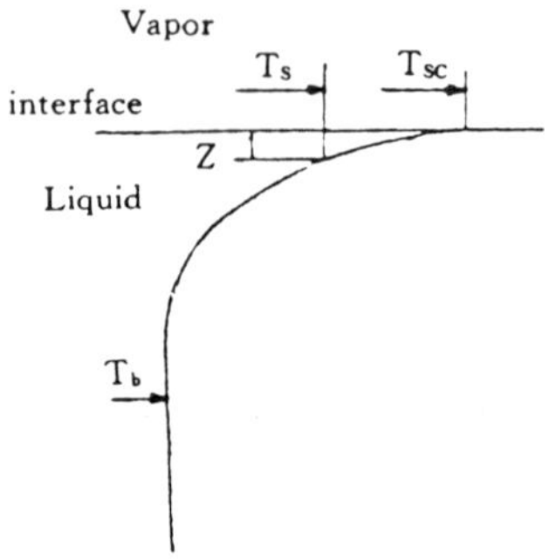

Fig. 2 Change in interface temperature.

$$\frac{\partial\theta}{\partial\tau} + u\frac{\partial\theta}{\partial x} + v\frac{\partial\theta}{\partial y} = \frac{\partial^2\theta}{\partial x^2} + \frac{\partial^2\theta}{\partial y^2} \tag{11}$$

To solve eqs. (8)-(11) numerically, the equations are discretized, transforming the nonlinear partial differential equations to linear algebraic equations[5]. Because of the larger transverse velocity gradient and temperature gradient in the natural convection boundary layer adjacent to the sidewalls, finer meshes should be employed, while there is no need for fine meshes far away from the walls. Since it is hard to use non-uniform meshes, two transformation functions p(x) and q(y) are introduced, which is equivalent to using finer meshes near the sidewalls:

$$p(x) = \frac{S_x}{2}\left\{1 + tg\left[\frac{\pi}{2}\left(\frac{2x}{S} - 1\right)\varepsilon\right] \Big/ tg\left(\frac{\pi}{2}\varepsilon\right)\right\} \tag{12}$$

$$q(y) = \frac{S_y}{2}\left\{1 + tg\left[\frac{\pi}{2}\left(\frac{2y}{S} - 1\right)\varepsilon\right] \Big/ tg\left(\frac{\pi}{2}\varepsilon\right)\right\} \tag{13}$$

where ε is a transformation coefficient ($0<\varepsilon<1$), its value adjusted for specific configurations. If ε is chosen too large, there are too few grid points in the middle and the degree of inaccuracy increases. On the other hand, if ε is chosen too small there is little effect upon transformation. We chose $\varepsilon=0.8$. $S_x=L/H$, $S_y=H/H=1$ are dimensionless coefficients.

Assuming

$$A_x = \frac{\partial p}{\partial x}, \quad A_y = \frac{\partial q}{\partial y}, \quad B_x = \frac{\partial Ax}{\partial x}, \quad B_y = \frac{\partial Ay}{\partial y}$$

eqs. (8)-(11) become

$$u = A_y\frac{\partial\psi}{\partial q}, \quad v = -A_y\frac{\partial\psi}{\partial p} \tag{14}$$

$$\omega = -\left(A_x^2\frac{\partial^2\psi}{\partial p^2} + A_y^2\frac{\partial^2\psi}{\partial q^2} + B_x\frac{\partial\psi}{\partial p} + B_y\frac{\partial\psi}{\partial q}\right) \tag{15}$$

$$\frac{\partial\omega}{\partial\tau} + A_x u\frac{\partial\omega}{\partial p} + A_y v\frac{\partial\omega}{\partial q} = GrPr^2A_x\frac{\partial\theta}{\partial p} + Pr\left(A_x^2\frac{\partial^2\omega}{\partial p^2} + A_y^2\frac{\partial^2\omega}{\partial q^2} + B_x\frac{\partial\omega}{\partial p} + B_y\frac{\partial\omega}{\partial q}\right) \tag{16}$$

$$\frac{\partial\theta}{\partial\tau} + A_x u\frac{\partial\theta}{\partial p} + A_y v\frac{\partial\theta}{\partial q} = A_x^2\frac{\partial^2\theta}{\partial p^2} + A_y^2\frac{\partial^2\theta}{\partial q^2} + B_x\frac{\partial\theta}{\partial p} + B_y\frac{\partial\theta}{\partial q} \tag{17}$$

upon transformation.

NUMERICAL METHOD

The small viscosity of the cryogenic liquids, especially liquid hydrogen, causes significant stability problem for numerical calculation since typical Rayleigh number can reach 10^{13}. This requires special care during discretization. For instance, alternating direction implicit method (ADI) is used for solving energy and vorticity transport equations. As a result, the discretization is accurate in time and space to the second order, and the coefficient matrix is tri-diagonal, thus the equations can be solved with tri-diagonal matrix algorithm (TDMA). The nonlinear advection term is discretized with the second order upwind scheme, making the coefficient of the equations positive-definite, thus allowing physical solutions. This stabilization scheme also makes the linear problem unconditionally stable and allows for larger time steps. Successive over-relaxation method (SOR) is used for solving Poisson type equations. Following the above procedure for discretization of the vorticity equation

(16) and energy equation (17), we obtain two sets of equations for x and y directions[6]:

$$\begin{cases} RR^n_{i,j}\omega^{n+1/2}_{i-1,j} + SS^n_{i,j}\omega^{n+1/2}_{i,j} + TT^n_{i,j}\omega^{n+1/2}_{i+1,j} = UU^n_{i,j} & (18a) \\ RR^{n+1/2}_{i,j}\omega^{n+1}_{i,j-1} + SS^{n+1/2}_{i,j}\omega^{n+1}_{i,j} + TT^{n+1/2}_{i,j}\omega^{n+1}_{i,j+1} = UU^{n+1/2}_{i,j} & (18b) \end{cases}$$

and

$$\begin{cases} R^n_{i,j}\theta^{n+1/2}_{i-1,j} + S^n_{i,j}\theta^{n+1/2}_{i,j} + T^n_{i,j}\theta^{n+1/2}_{i+1,j} = U^n_{i,j} & (19a) \\ R^{n+1/2}_{i,j}\theta^{n+1}_{i,j+1} + S^{n+1/2}_{i,j}\theta^{n+1}_{i,j} + T^{n+1/2}_{i,j}\theta^{n+1}_{i,j+1} = U^{n+1/2}_{i,j} & (19b) \end{cases}$$

Since the heat transport into the liquid hydrogen vessel is unevenly distributed, part of it is used for raising the liquid temperature and the rest is used for vaporizing the surface liquid to maintain the thermodynamic equilibrium in the ullage volume. Its value is determined by

$$Q_v = L_v|(\partial\rho/\partial T)_s|(T_{sc} - T_b)V_u \qquad (20)$$

The thickness of the surface vaporization layer is

$$Z = Q_v/(L_v A_c \rho) = |(\partial\rho/\partial T)_s|(T_{sc} - T_b)V_u(A_c\rho) \qquad (21)$$

This free surface is adiabatic in computation, leading to a surface temperature T_{sc} that is greater than the actual surface temperature T_s. This discrepancy can be corrected knowing the temperature gradient in the surface layer, as is shown in Fig. 2.

COMPUTATION RESULTS AND DISCUSSION

Thermal stratification in liquid hydrogen is calculated with the above method, whereas the experimental data are taken from literatures[3]. The internal vessel height $h=2.49$m, radius $L=0.51$m, liquid hydrogen height $H=1.98$m, sidewall heat flux $q_w=262.6\,W/m^2$, initial temperature $T_b=20.156$K. The time step is 1s, relaxation factor $\omega_0=1.8$. Beacuse of the axisymmetry of the container, only half of the area needs to be computed. The initial conditions are:

$$\theta_{i,j} = \omega_{i,j} = \psi_{i,j} = u_{i,j} = v_{i,j} = 0$$

The boundary conditions are:
At the free liquid surface:

$$\theta_{i,N} = \theta_{i,N-1}; \quad \omega_{i,N} = \psi_{i,N} = v_{i,N} = 0; \quad u_{i,N} = u_{i,N-1}$$

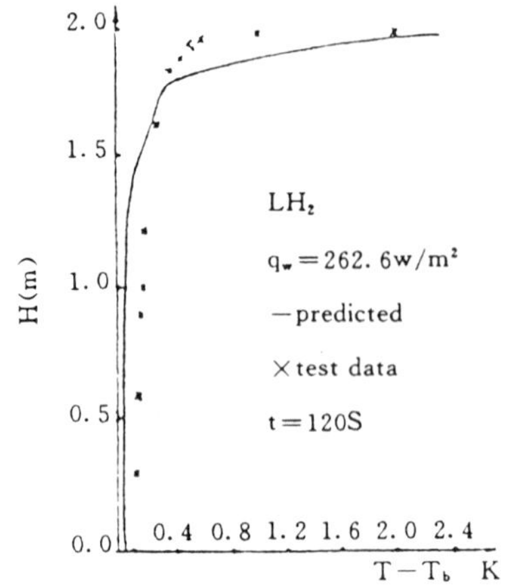

Fig. 3 tamperature vs. height.

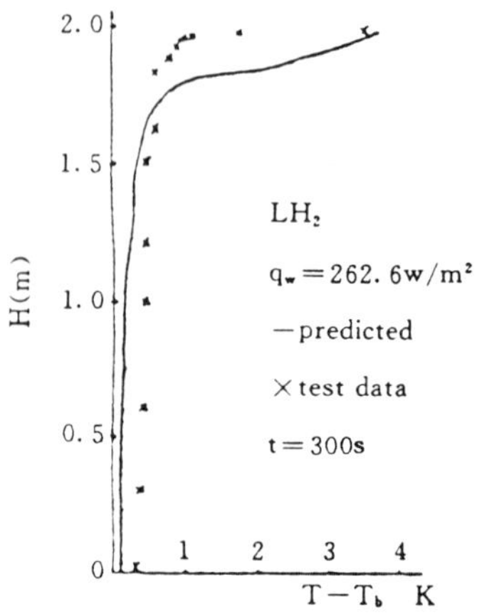

Fig. 4 temperature vs. height.

At the center of the container:

$$\theta_{1,j} = \theta_{2,j}; \quad \omega_{1,j} = \psi_{1,j} = u_{1,j} = 0; \quad v_{i,j} = v_{2,j}$$

At the bottom:

$$\theta_{i,1} = \theta_{i,2}; \quad \omega_{i,1} = -2\psi_{i,2}Ay(1)/\Delta q^2; \quad \psi_{i,1} = u_{i,1} = v_{i,1} = 0$$

At the sidewall:

$$\theta_{M,j} = 1; \quad \omega_{M,j} = -2\psi_{M-1,j}Ax(M)/\Delta p^2; \quad \psi_{M,j} = u_{M,j} = v_{M,j} = 0$$

The isotherms at two different times are shown in Figure 3 amd 4. It is obvious that in the upper area, especially near the boundary layer, the axial temperature gradient is larger while it is small in the lower area. On the other hand, the transverse temperature gradient is large in the lower area and small in the upper area. The transverse temperature gradient is small in areas far away from the boundary layer, which is similar to the characteristics of the thermal stratification distribution. This is because the liquid near the walls flows upward through the boundary layer during the early stage, the liquid near the boundary layer flows in to replenish the upward moving warm liquid. The warmer liquid flowing toward the upper surface stays at the top, forcing colder liquid downward and eventually establishing vertical thermal stratification.

It can also be observed from the two isotherm figures that temperature inversion exists near the sidewall, i.e. the temperature first decreases in a small region away from the wall and then increases. This phenomenon can be understood in the following way:

(1) Since the Prandtl number Pr is greater than 1, or the viscous boundary layer is thicker than the thermal boundary layer, some cold fluid outside the thermal boundary layer is dragged to flow upward;

(2) vortex flow might exist near the intersection of the thermal stratification layers and the boundary layer, which could lead to upward flowing cold fluid;

(3) the warm fluid in the thermally stratified area flows downward near the center of the container and can have higher temperature than the cold fluid flowing upward in the viscous boundary layer.

It is also observed that the area of thermal inversion decreases as time goes by, which can be used to roughly determine the thickness of the thermally stratified layers and the degree of stratification.

The solid lines in the figures represent the computation results while the crosses are experimental data points. They appear to agree with each other. Overall, the computed values in the upper area tend to be larger while they are smaller in the lower area. This could be caused by the adiabatic conditions at the free surface and the bottom boundary respectively. The figures show that the convex ($d^2T/dy^2>0$) temperature distribution curves gradually become concave ($d^2T/dy^2<0$) as time goes on. This might be the result of the high Rayleigh number of the liquid hydrogen: it causes most of the fluid in the boundary layer flows into the thermally stratified layer with higher temperatures, thus accelerating the temperature increase in the stratified fluid and resulting in $d^2T/dy^2<0$.

The degree of the thermal stratification is closely related to the heat flux, and the gap size affects the surface temperature rather severely. In general, the surface temperature is a function of the heat flux, liquid filling factor (f) and time. To discover the order in the temperature distributions, the surface temperature is approximated using the results of a series of calculation. The approximation function is assumed to be

$$T_s = a + b(q_w ft) + ce^{d(q_w ft)} \tag{22}$$

The fitted coefficients are $a=20.9188$, $b=5.5677\times10^{-6}$, $c=-4.976\times10^{-5}$, and $d=1.35\times10^{-4}$. The experimental results and the computed values are shown in Figure 5. This result shows that while $q_wft<7\times10^4 J/m^2$, using equation (22) to approximate the surface temperature of the liquid hydrogen vessel is reasonable. The method employed in this paper to calculate unsteady two-dimensional natural convection flows, the vorticity function method, is rather practical. The results also verified that using second order upwind discretization for treating the nonlinear advection term can increase the accuracy for the solutions to the instantaneous states.

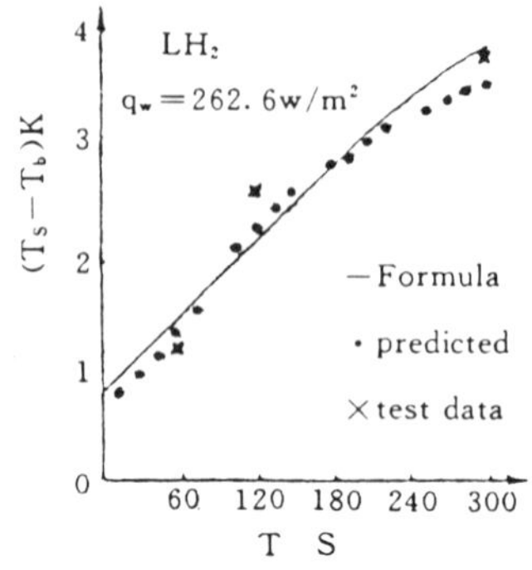

Fig. 5 Interface T vs time.

NOMENCLATURE

A,B transformation coefficient

A_c cross-sectional area, m^2

f ratio of liquid volume to total volume in a tank

Gr Grashof number $= g\beta H^4(T_w-T_b)/\nu^2$

H liquid height, m

k thermal conductivity, W/m-k

L tank radius, m

L_v average latent heat over the range of interest, J/kg

M number of spatial grid points in x-direction

Pr Prandtl number $= \nu/\alpha$

q_b heat flux through tank bottom, W/m^2

q_w heat flux through tank sidewall, W/m^2

R_a Rayleigh number = Gr • Pr

$RR_{ij}, SS_{ij}, TT_{ij}, UU_{ij}$ matrix coefficients

t time, s

T_b bulk liquid temperature, K

T_{sc} calculated surface temperature, K

T_s surface temperature, K

T_w tank sidewall temperature, K

u_o reference velocity, m/s

u x-component of the velocity, m/s

v y-component of the velocity, m/s

V_u tank ullage volume, m^3

x normal distance from center line, m

y axial distance from bottom, m

Greek Symbols

α thermal diffusivity, m^2/s

β coefficient of volumetric thermal expansion 1/K

θ dimensionless temperature $= (T-T_b)/(T_w-T_b)$

γ kinematic viscosity, m^2/s

ρ density, kg/m^3

τ dimensionless time(Fourier numebr) $= \alpha t/H^2$

ψ stream function, m^2/s

ω vorticity, s^{-1}

Subscripts

i,j position indices in the spatial grid

REFERENCES

1. J. A. Clark. "A Review of Pressurization, Stratification, and Interfacial Phenomena," Advances in Cryogenic Engineering, Vol. 10B(1965)
2. D. O. Barnett, T. W. Winstead, and L. S. Mcreynolds. "An investigation of liquid-hydrogen stratification in a large cylindrical tank of the saturn configuration," Advances in Cryogenic Engineering, Vol. 10B(1965)

3. J. W. Tatom, W. H. Brown, L. H. Knight, and E. F. Coxe. "Analysis of thermal stratification of liquid hydrogen in rocket propellant tanks," Advances in Crygenic Engineering, Vol. 9(1964)
4. Guo Kuanling, Wu Sican, "An Iteration Computation Method for 2-D Natural Convections," Engineering Thermal Physics Journal, 4(4) 1983
5. S. V. Patankar, Numerical Heat Transfer and Fluid Flow, McGraw-Hill, USA, 1980
6. Zhan Tanyun, "Numerical Simulation and Approximate Analysis of Thermal Stratification in Liquid Hydrogen," Master thesis, Cryogenic Engineering, Zhejiang University, 1989. 5

NUMERICAL SIMULATION OF THERMAL ACOUSTIC OSCILLATIONS IN A LIQUID HELIUM SYSTEM

Youfan Gu[1] and K. D. Timmerhaus[2]

[1]HPS Div. of MKS Instruments, Inc.
Boulder, CO 80301
[2]University of Colorado
Boulder, CO 80309

ABSTRACT

Thermal acoustic oscillations (TAOs) can generally be described by the general hydrodynamic equations for flow processes. Therefore, an understanding of the mechanisms associated with TAOs requires a full solution of these equations. However, the complexity of these equations make an analytical solution for TAOs essentially impossible. Accordingly, a numerical method has been employed to simulate the oscillation process.

A combination of the finite difference and finite element method has been found to be most effective in simulating the oscillation process. Simulated oscillation results match the experimental observations with reasonable accuracy. Detailed axial pressure distribution, axial mean gas velocity and radial temperature gradient in an oscillation tube have been obtained and are included in this study.

INTRODUCTION

The stability characteristics for TAOs have been analyzed by perturbating and linearizing the general hydrodynamic equations[1]. This method has been used to predict the stable and the unstable regions for TAOs and has proven to be relatively effective and accurate if the actual temperature profile along the length of the tube is incorporated in the analysis. However, the stability theory can not predict the oscillation pressure amplitude since the acoustic variables were assumed to vary exponentially with time. This implies that the pressure amplitude may become infinite in the unstable region. To obtain the pressure amplitude, two methods may be used. One involves the mass-spring method which considers the inertial force, driving force, spring force, viscous resistance and kinetic energy loss at the open end of the tube. The other method involves solving the general hydrodynamic equations numerically utilizing the proper boundary and initial conditions. This approach for investigating TAOs is presented in this paper.

Advances in Cryogenic Engineering, Vol. 41
Edited by P. Kittel, Plenum Press, New York, 1996

During a TAO process, thermal acoustic waves travel at a relatively high velocity approaching or even exceeding the local sonic velocity when the tube radius is relatively small. Movement of these waves occurs in two directions, parallel to the tube wall or in the x direction, and perpendicular to the tube wall or in the r direction. If TAOs occur in a tube with a length of one meter and a radius of five millimeters, the maximum time interval Δt which can be selected to obtain convergence of the numerical iterations is less than 5×10^{-8} seconds as determined from the numerical stability rule ($\Delta t \leq \min(\Delta x/a, \Delta r/a)$). Here a is the local sonic speed (~1000 m/s for helium at room temperature), Δx and Δr are the finite difference grid steps in the axial and radial directions (assumed to be one percent of the tube length and ten percent of the tube radius), respectively. These dimensions indicate that approximately twenty million iteration steps will be required just to simulate the TAOs occurring in one second. For each iteration step, thousands of calculations are required to obtain values for pressure, temperature, density, and velocity at each node, and any small numerical error that accumulates in the iterative calculation can prevent convergence of the solution. In addition, variation of the pressure amplitude with time makes it difficult mathematically to achieve convergence for such a dynamic oscillatory problem.

The boundary condition at the open end of the tube is difficult to establish since this affects both the pressure amplitude of the oscillation and the converging characteristics of the numerical iteration. Since, mass moves in and out of the open end of the oscillation tube during the oscillation process and varies with time, a very small amount of error in discreting the continuity equation can result in a major divergence problem.

Finally, both the diffusive and convective terms in the hydrodynamic equations are prominent when they are used to describe the TAO process since the oscillations of the gas in the tube are initiated by the large heat transfer that occurs during the TAO process coupled with the relatively large velocity gradient in the r direction. However, the principles for numerically solving the diffusion dominant equation and the convection dominant equation are quite different[2]. The latter also creates further difficulties with the proper discretization of the hydrodynamic equations. In order to obtain a stable numerical simulation, a combination of the finite difference method and the finite element method has been found to be most effective. Detailed numerical methods and typical simulation results are presented.

GENERAL EQUATIONS AND THEIR DISCRETIZATIONS

Numerical iteration of the general hydrodynamic equations requires that they are rewritten in discrete form. Unfortunately, truncation errors are introduced into these iterations because of discretization of the equations. Round-off error is inevitable due to the finite resolution of the computer. To obtain convergence of the numerical calculations requires consistency between the discretized equation and the original equation along with stabilization of the numerical iterations.

The general continuity equation is represented by

$$\frac{\partial \rho}{\partial t} + \frac{\partial(\rho u)}{\partial x} + \frac{1}{r}\frac{\partial(r\rho v)}{\partial r} = 0 \tag{1}$$

where ρ represents the density of the fluid, while u and v are the velocities of the fluid in the axial and radial directions, respectively. This equation represents the conservation of mass. Note that mass can move in both the r and the x directions. However, mass transfer in the r direction implies that the acoustic wave travels in both the x and the r directions which, as noted in the introduction of this study, results in too small a value for Δt during the numerical calculation. Therefore, a one-dimensional acoustic wave was assumed during the numerical simulation. This one-dimensional continuity equation is generally a convective dominate type of equation. If an ordinary FTCS (forward time, centered space) finite difference equation is used to solve this equation, no representative numerical solution will result since the FTCS scheme is generally unstable for a purely convective problem. Accordingly, the following three-level fully implicit scheme has been used during the numerical simulation:

$$\frac{3}{2}\left(\frac{\Delta\rho_{i,j}^{n+1}}{\Delta t}\right)-\frac{1}{2}\left(\frac{\Delta\rho_{i,j}^{n}}{\Delta t}\right)+\frac{1}{2}L_x\left[(\rho u)_{i,j}^{n+1}+(\rho u)_{i,j}^{n}\right]=0 \tag{2}$$

where $\Delta\rho_{i,j}^{n+1}=\rho_{i,j}^{n+1}-\rho_{i,j}^{n}$, and difference operator $L_x=(1/2\Delta x)(-1,0,1)$. An implicit scheme can provide a more stable solution for this discretized continuity equation. Because a one-dimensional acoustic wave was assumed, this system of implicit equations is tridiagonal and can be solved in the x direction by using the Thomas algorithm. Since no mass leaves the closed end of the tube, the boundary condition for ρu at the closed end is zero. However, at the open end of the tube where gas moves in and out of the tube periodically, the zero axial derivative, $\partial(\rho u)/\partial x=0$, is used as the boundary condition. This implies that the mass flow rate attains its maximum at the open end of the tube.

Since one-dimensional acoustic waves have been assumed, the equation of motion used to describe thermal acoustic oscillations in a tube becomes

$$\frac{\partial u}{\partial t}+u\frac{\partial u}{\partial x}=-\frac{1}{\rho}\frac{\partial p}{\partial x}+\nu\left[\frac{1}{r}\frac{\partial}{\partial r}\left(r\frac{\partial u}{\partial r}\right)+\frac{\partial^2 u}{\partial x^2}\right] \tag{3}$$

where p is the pressure and ν the kinematic viscosity of the oscillating fluid. A combination of finite element and finite difference methods has been used in the discretization of this motion equation. The finite element method is used since a more stable solution can usually be obtained by using the information from neighbor nodes. By using the finite element and finite difference discretization methods, the preceding equation can be discretized as

$$\begin{aligned}M_xM_r\left[(1+\beta)\frac{\Delta u_{i,j}^{n+1}}{\Delta t}-\beta\frac{\Delta u_{i,j}^{n}}{\Delta t}\right]+\frac{1}{2}M_r\left[u_{i,j}^{n+1}\frac{\partial u_{i,j}^{n+1}}{\partial x}+u_{i,j}^{n}\frac{\partial u_{i,j}^{n}}{\partial x}\right]=-\frac{1}{2\rho_{mi}}\left[\frac{\partial p_i^{n+1}}{\partial x}+\frac{\partial p_i^{n}}{\partial x}\right]\\+\frac{1}{2}\nu M_r\left[\frac{\partial^2 u_{i,j}^{n+1}}{\partial x^2}+\frac{\partial^2 u_{i,j}^{n}}{\partial x^2}\right]+\frac{1}{2}\nu M_x\left[\frac{1}{r_j}\left(\frac{\partial u_{i,j}^{n+1}}{\partial r}+\frac{\partial u_{i,j}^{n}}{\partial r}\right)+\left(\frac{\partial^2 u_{i,j}^{n+1}}{\partial r^2}+\frac{\partial^2 u_{i,j}^{n}}{\partial r^2}\right)\right]\end{aligned} \tag{4}$$

where β is a parameter controlling the degree of implicitness $(0\le\beta\le 1)$, mass operator $M_x=(\delta_x,(1-\delta_x),\delta_x)$, and $M_r=(\delta_r,(1-\delta_r),\delta_r)^T$. Normally, a value of 1/6 is selected for δ_x and δ_r. It is apparent that this specific combination of the finite element and finite difference method becomes a pure finite difference method when a value of zero is selected for δ_x and δ_r. It should be noted that the $(1/\rho)(\partial p/\partial x)$ term is very critical in the convergence of the numerical simulation since it represents the driving force for the oscillation of the gas in the tube. During the discretization process, a mean density ρ_{mi} is used since a small disturbance in the density can amplify the numerical error rather quickly thereby minimizing convergence of the numerical solution. Physically, this implies that the driving force across the small cross sectional area of the tube is a constant which is consistent with the assumption of a one-dimensional acoustic wave in the tube. By neglecting the second order terms in Eq. (4), this equation can be reformulated as a two-step split formulation, as

$$\left[M_x + \frac{\Delta t}{2(1+\beta)}\left(u_{i,j}^n L_x + \frac{\partial u_{i,j}^n}{\partial x} - \nu L_{xx}\right)\right]\Delta u_{i,j}^{n+1*} = M_x M_r \frac{\beta}{1+\beta}\Delta u_{i,j}^n$$
$$-\frac{\Delta t}{1+\beta}\left\{\frac{1}{\rho_{i,j}^n}\left[\frac{1}{2}\frac{\partial \Delta p_i^{n+1}}{\partial x} + \frac{\partial p_i^n}{\partial x}\right] - M_r\left(u_{i,j}^n L_x\right)u_{i,j}^n + \nu\left[M_x\left(\frac{1}{r_j}L_r + L_{rr}\right) + M_r L_{xx}\right]u_{i,j}^n\right\} \tag{5}$$

and

$$\left[M_r - \frac{\nu \Delta t}{2(1+\beta)}\left(\frac{1}{r_j}L_r + L_{rr}\right)\right]\Delta u_{i,j}^{n+1} = \Delta u_{i,j}^{n+1*} \tag{6}$$

where $L_r = (1/2\Delta r)(-1, 0, 1)$, $L_{rr} = (1/(\Delta r)^2)(1, -2, 1)$, and $L_{xx} = (1/(\Delta x)^2)(1, -2, 1)$. Upon splitting the implicit discretized equations, the latter can be solved independently in the axial and radial directions by using the Thomas algorithm.

For the one-dimensional acoustic wave in the oscillating tube, the equation describing the energy conservation can be written as

$$\left(\frac{C_p}{z} - 1\right)\left(\frac{\partial p}{\partial t} + u\frac{\partial p}{\partial x}\right) + \frac{C_p}{z}p\frac{\partial u}{\partial x} = k\left[\frac{1}{r}\frac{\partial}{\partial r}\left(r\frac{\partial T}{\partial r}\right) + \frac{\partial^2 T}{\partial x^2}\right] \tag{7}$$

where C_p is the heat capacity of the fluid at constant pressure, k the thermal conductivity, T the temperature, and z the coefficient of compressibility. If we integrate the energy equation in the r direction, the relation becomes

$$A'\frac{\partial p}{\partial t} + B'\frac{\partial p}{\partial x} + C'p = \frac{2k}{r_0}\left(\frac{\partial T}{\partial r}\right)_w + k\frac{\partial^2 T_w}{\partial x^2} \tag{8}$$

where the subscript w represents the tube wall, and

$$A' = \frac{C_p}{z} - 1; \quad B' = \frac{2}{r_0^2}\left(\frac{C_p}{z} - 1\right)\int_0^{r_0} urdr; \quad C' = \frac{2}{r_0^2}\frac{C_p}{z}\int_0^{r_0}\frac{\partial u}{\partial x}rdr \tag{9}$$

This equation can be discretized as

$$\left(M_x + \frac{1}{2}BL_x\right)\Delta p_i^{n+1} = \frac{\beta}{1+\beta}M_x\Delta p_i^n - (BL_x + CM_x)p_i^n + A\left[\frac{2k}{r_0}M_x L_r + kL_{xx}\right]T_{iw}^n \tag{10}$$

where

$$A = \frac{\Delta t}{A'(1+\beta) + 0.5C'\Delta t}; \qquad B = AB'; \qquad C = AC' \tag{11}$$

The equation of state for helium can generally be expressed as $p = z\rho RT$. For helium, the value of the compressibility factor is equal to one when the temperature of the gas is above about 20 K. When the temperature is below 20 K, the factor is calculated with the aid of the following regressed equation:

$$z = \exp\left[\frac{-1.625}{T}\left(\frac{T}{4.224}\right)^{2.3}\right] \tag{12}$$

Since thermal physical properties (thermal conductivity and viscosity) vary with temperature, a computer table has been established and these properties are obtained by interpolation within the table during the numerical calculation.

INITIAL AND BOUNDARY CONDITIONS

Previous work has shown that TAOs will not occur in a tube if the cold end is closed. This observation permits iteration of an analysis by assuming that the tube is already inserted in a dewar containing the cryogenic fluid. The gas inside the tube in this arrangement is in thermal equilibrium with the tube wall and the warm end of the tube is then closed when such thermal equilibrium is reached. Since the cold end of the tube is closed, the system is stable. However, if the cold end of the tube is suddenly opened, this system becomes unstable, and TAOs may be initiated because of the large longitudinal temperature gradient existing along the length of the tube. Therefore, mathematically, we can proceed with the following initial conditions; namely, the temperature of the gas at each location inside the tube at $t=0$ is the same as the temperature of the tube wall at these locations, implying a zero temperature gradient in the r direction. Further, the velocity of the gas in the tube is zero at all locations within the tube, and the pressure in the tube is assumed to be equal to the pressure in the dewar vessel.

The following boundary conditions are assumed when the cold end of the tube is suddenly opened: (1) velocity along the tube wall is zero (nonslip boundary); (2) in the center of the tube; $\partial p/\partial r = 0$ and $\partial T/\partial r = 0$ (3) at the closed warm end of the tube; $\partial p/\partial x = 0$ (4) at the open cold end of the tube; $\partial(\rho u)/\partial x = 0$ and (5) the pressure at the open end of the tube is a critical boundary condition which determines the pressure amplitude of the oscillation in the tube, and eventually the convergence of the numerical simulation. In fact, such a condition reflects the kinetic energy loss at the open end of the tube since the kinetic energy loss is proportional to ρu^2 at the open end of the tube. That is, $p = \Psi \rho u^2$. Relatively good results have been obtained when a value of 1.0 to 1.5 is used for Ψ.

NUMERICAL RESULTS

One of the numerical simulation results for TAOs in a helium system is presented in this section. This oscillation occurs in a tube with a length of 1.5 m and a radius of 2.48 mm. The temperature at the warm end of the tube is at 293.8 K, while the temperature at the cold end of the tube is 21 K. The tube length ratio, ξ, between the warm and cold sections is fixed at 1.05. The simulated pressure waves are shown in Fig. 1. Note that curve (b) is a continuation of curve (a) and curve (c) is a continuation of curve (b). It can be observed that the modulus of pressure amplitude varies exponentially with time at the beginning of the TAO process. This indicates that initially, there is minimal resistance for TAOs since the oscillation amplitudes are relatively small. Also, the viscous resistance between the fluid and the tube wall is small initially since the velocity is relatively small. However, the temperature gradient between the fluid and the tube wall increases rapidly, and results in a significant increase in the heat exchange between the fluid and the tube wall. In other words, large amounts of heat from the warm environment are transferred into the system. The latter causes an increase in the pressure amplitude. The steep rise in the pressure amplitude decreases with time as the amplitude becomes larger. This is because the kinetic energy loss at the open end of the tube is proportional to the square of the velocity. This resistance becomes significant when the velocity at the open end of the tube is relatively large, i.e., when the pressure amplitude is relatively large. In addition, the viscous resistance also increases with an increase in the gas velocity in the tube. Both of these parameters contribute to a finite pressure amplitude, which eventually, attains a fixed pressure amplitude with steady-state oscillations.

Figure 1 shows a maximum peak-to-peak oscillation pressure amplitude of about 38 torr at 1.6 seconds which is close to the experimental results of 47.5 torr obtained under the same operating conditions (experimental run #4-9C) as used in the simulation calculations. The time to attain steady state was about 1.5 seconds. The time step Δt used in the simulation was 1.5×10^{-5} second. The latter condition resulted in 1.1×10^{5} iterations, requiring nine hours of CPU time to simulate the 1.6 seconds to attain steady state.

The axial pressure distributions and the axial mean velocity in the oscillation tube during steady-state oscillation are shown in Figs. 2 and 3, respectively. Note that $x=0$ represents the cold open end of the tube while $x=L$ represents the warm closed end. At $t=0$ (arbitrarily selected when steady state is attained), the pressure in the oscillation tube is negative and the velocity is positive. Note that a positive velocity indicates that the gas is flowing into the tube while a negative velocity indicates that the gas is moving out of the tube. At this time, the pressure in the tube should increase with time as the cold gas flows into the tube. If we compare these results with those *30* ms later, it is apparent that both the

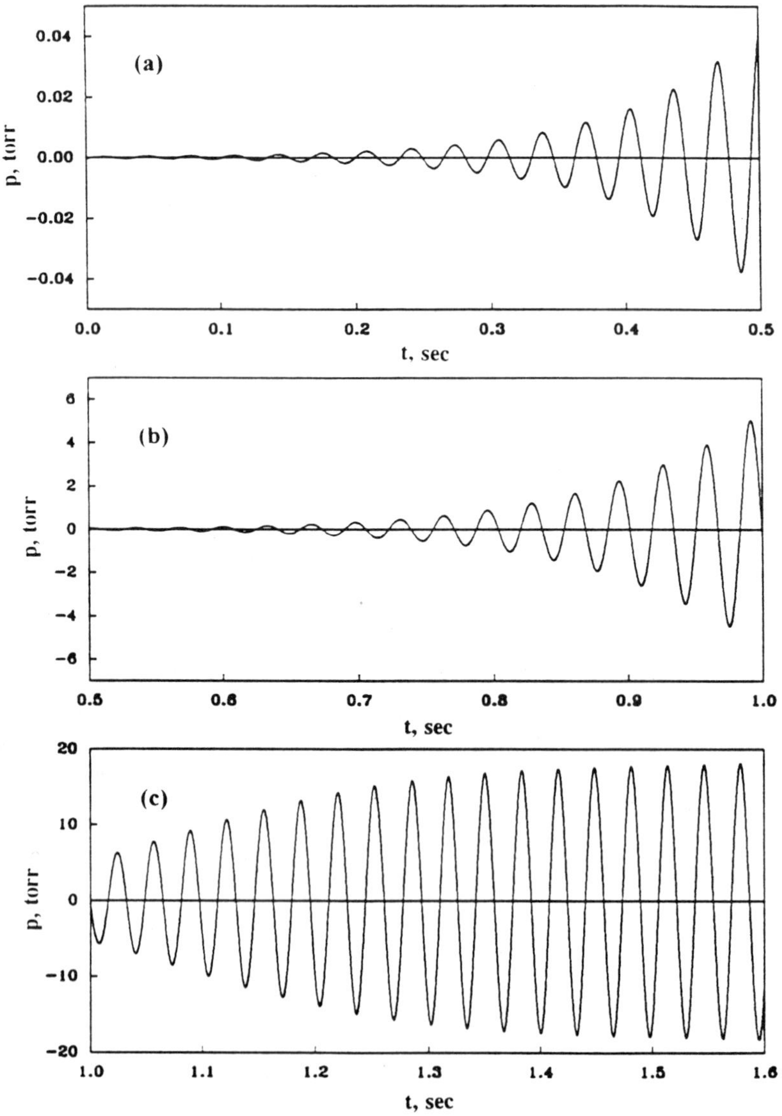

Figure 1. Numerical simulation of thermal acoustic oscillations in a liquid helium system ($L=1.5$ m, $r_0=2.48$ mm, $T_h=293.8$ K, $T_c=21$ K, $\xi=1.05$).

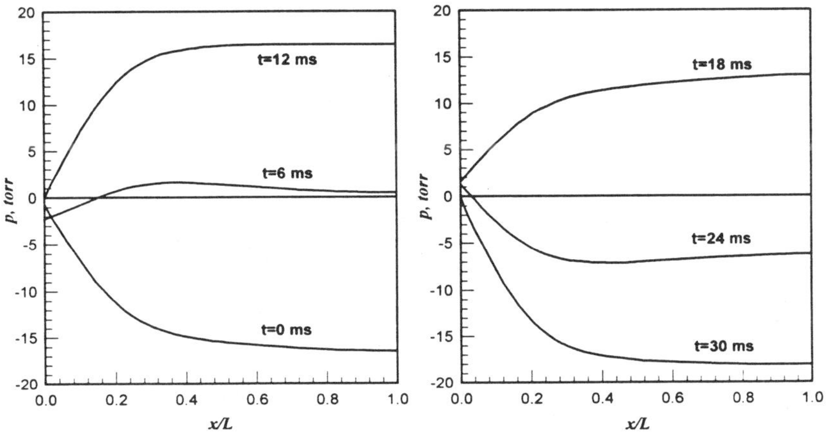

Figure 2. Axial pressure distributions in the tube at different times during TAOs.

pressure and the velocity are negative at $t=30$ ms; thus, the pressure should continue to decrease until a maximum negative pressure is attained.

These numerical results show that the axial pressure profiles as well as the velocity profiles are considerably different from those obtained when the oscillation occurs in a tube with a constant temperature along the warm and cold lengths of the tube. This is particularly true for the axial velocity profile. The axial velocity profile and the axial pressure profile are sinusoidal in a tube exhibiting a constant temperature along the length of the tube. However, when TAOs do occur in a tube with a large temperature gradient along the length of the tube, the pressure gradient in the warm section of the tube is almost zero, while a linear pressure gradient is present in the cold section of the tube. This condition occurs because the sonic speed is much higher in the warm section of the tube than in the cold section of the tube; thus, a much higher traveling wave speed can be observed in the warm section of the tube. In the case of the axial velocity, a velocity maximum can be observed in the tube at $x/L=0.6$; this is close to $x/L=0.525$ where the largest temperature gradient occurs in the tube when $\xi=1.05$. This is due to a lower mean value in the density and a larger pressure fluctuation in the warm section of the tube.

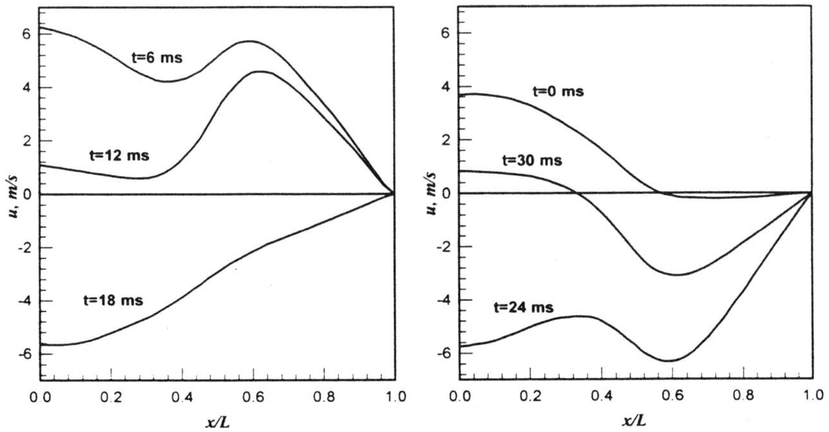

Figure 3. Axial mean gas velocity in the tube at different times during TAOs.

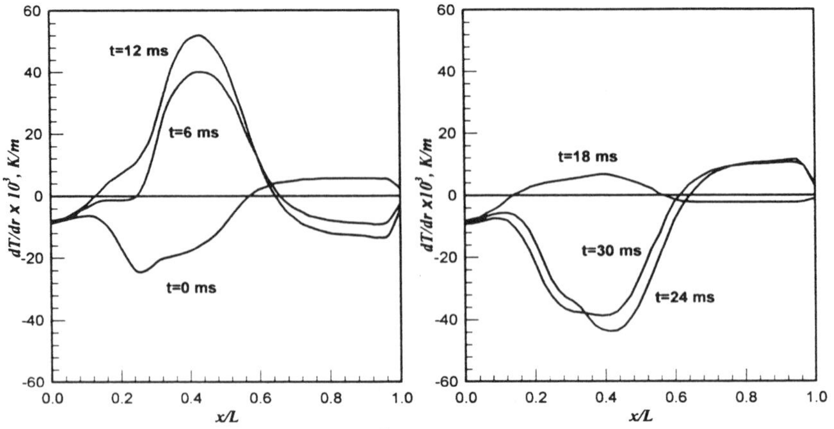

Figure 4. Radial temperature gradients at the tube wall and proceeding along the length of the tube during TAOs.

Both positive and negative values for the radial temperature gradient are observed during the oscillation cycle in Fig. 4. Note that a positive radial temperature gradient represents heat transfer from the tube to the fluid. The existence of a negative temperature gradient occurs because compression of the gas during oscillations can result in a temperature rise in the gas. However, the maximum positive temperature gradient is higher than the absolute value of the maximum negative temperature gradient, and this results in a significant net heat leak into the system.

CONCLUSIONS

The thermal acoustic oscillation process has been simulated numerically in this study. Stable discretized equations have been developed to minimize numerical error during the iteration process. From such simulation results, it becomes clear that a relatively steep temperature gradient along the length of the tube can result in a movement of the gas in the tube. This, in turn, induces a very large radial temperature gradient as well as a large axial gas velocity in the tube near the location of the largest axial temperature gradient. Rapid increases in the oscillation pressure amplitude and heat transfer are then observed in the tube, and eventually, sustained oscillations are initiated. However, a finite pressure amplitude is attained since the kinetic energy loss at the open end of the tube becomes dominant when the gas velocity (or pressure amplitude) is relatively high.

Axial velocity, pressure and radial temperature profiles in the tube during TAOs have been obtained from the simulation. A maximum radial temperature gradient occurs at the location where the maximum axial temperature gradient is observed and is much higher than the axial temperature gradient. Maximum quantity of heat is transferred to the gas at this location, and eventually, transferred into the cryogenic system due to the macroscopic movement of the gas in the tube.

ACKNOWLEDGMENT

This work was supported by NSF under grant CTS-9115040 and by NASA Lewis Research Center under research grant NAG3-1018.

REFERENCES

1. Gu, Y., Thermal Acoustic Oscillations in Cryogenic Systems, PhD Thesis, University of Colorado at Boulder, 1993.
2. Fletcher, C. A. J., Computational Techniques for Fluid Mechanics I & II, *Springer-Verlag, Berlin, Heidelberg, New York, Lodon, Paris, Tokyo*, 1991.

LOCALIZED HEAT TRANSFER TO VERTICAL FORCED FLOW TWO-PHASE HELIUM

J. Panek, X. Huang, and S. W. Van Sciver

National High Magnetic Field Laboratory
1800 E. Paul Dirac Dr., Tallahassee, FL 32306

ABSTRACT

Localized heat transfer measurements in vertical two-phase helium are reported. The test loop contains two short heat transfer sections made of 5 mm thick oxygen-free high conductivity (OFHC) copper discs. These test sections were installed in a U-shaped vertical flow loop driven by a single-stroke bellows pump. The surface temperature of each test section is measured with two germanium thermometers placed on different radial positions in each test section. With one test section placed on the downflow side and one on the upflow side of the loop, the effect of flow orientation on heat transfer characteristics in vertical two-phase helium flow is investigated. The study includes the effects of system pressure, mass flow rate, and geometry on the heat transfer coefficient, critical heat flux, and recovery heat flux.

INTRODUCTION

Many superconducting devices are cooled at their operating temperatures by convection of forced-flow normal helium liquid, He I. Detailed knowledge of heat transfer characteristics of He I is required to design and operate these devices reliably and safely. The general trend of the past has been to operate these devices at pressures slightly above 1 atmosphere. However, superconductors have higher performance at lower temperatures, and operation at lower saturated pressures is sometimes desired. Existing installations, such as the Tevatron at Fermilab, have been modified to run at lower saturated pressures.[1] Geometric and space constraints may dictate the use of downward flow cooling, which has not been as extensively studied as upward flow for He I. This study addresses those special considerations in forced-flow cooling of vertical channels.

Many works have studied the problem of convective flow boiling. Johannes and Steiner both report heat transfer coefficients in vertical tubes similar to the present experiment, described below.[2,3] Ogata and Sato, and Keilin et al. also report experimental results in a vertical tube.[4,5] These works all studied small diameter tubes of inner diameter of 1 to 3 mm. The present study differs in two important aspects: the channel is larger (6.35 mm) and downward-flowing convective boiling is studied.

EXPERIMENTAL APPARATUS

The experiment consists of a helium bath which contains a single stroke bellows pump, shown in Figure 1. The bellows is driven by a stepper motor, and has an effective

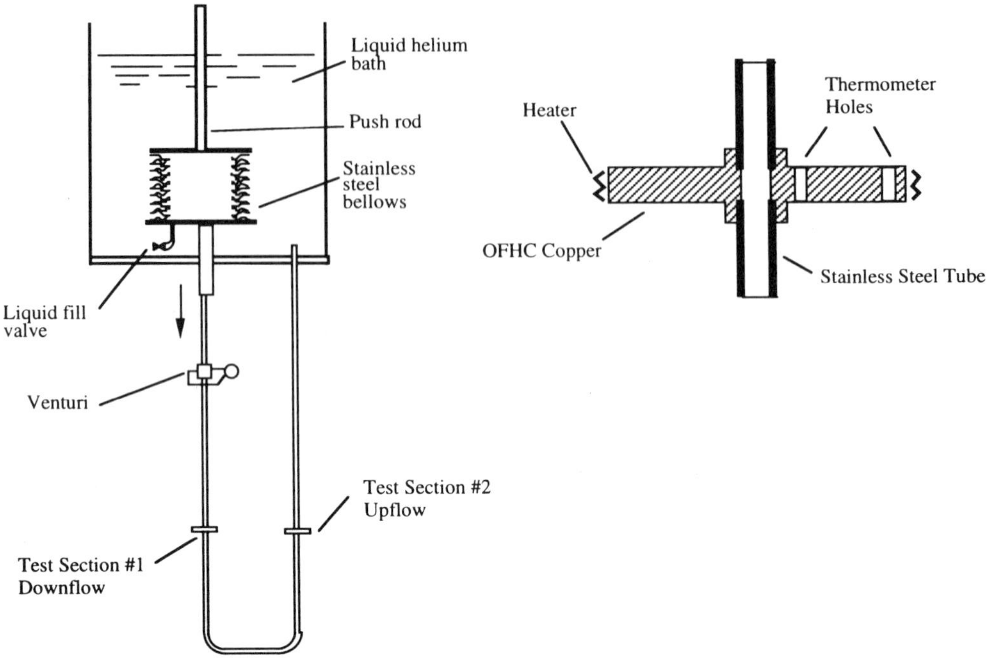

Figure 1. Experiment schematic and test section detail.

area of 181.3 cm^2 and stroke of 10 cm. Further details concerning the pump design and operation are in Reference 6. The test sections and a venturi flow meter are attached to the bottom of the bath. The entire cold mass is wrapped in multilayer insulation and housed inside a 35.5 cm diameter evacuated dewar with liquid nitrogen shield. Total heat leak from ambient is approximately 0.75 watt, with most of that going to the helium bath.

The test sections, made from OFHC copper, are also shown in Figure 1. They are identical to each other, with an inside diameter of 6.35 mm and length of 5.0 mm. A manganin wire heater is wrapped around the outer diameter of the test section, and two thermometer wells are located at radial distances of 6.0 mm and 23.0 mm. The thermometers

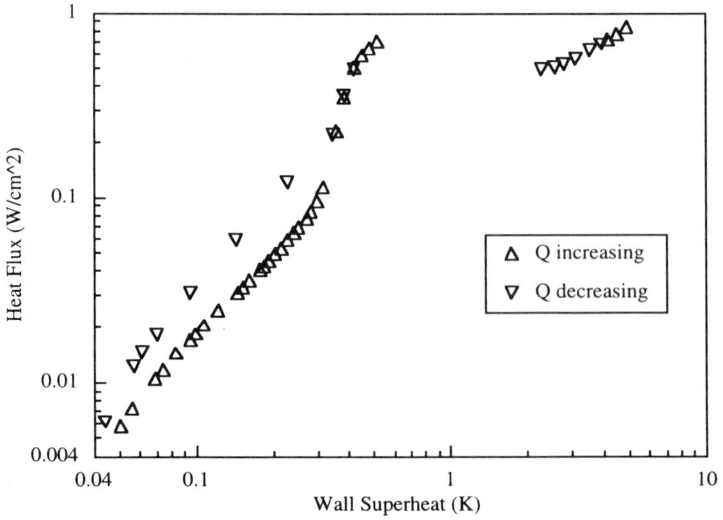

Figure 2. Boiling curve for P_{sat} = 1.0 atmosphere.

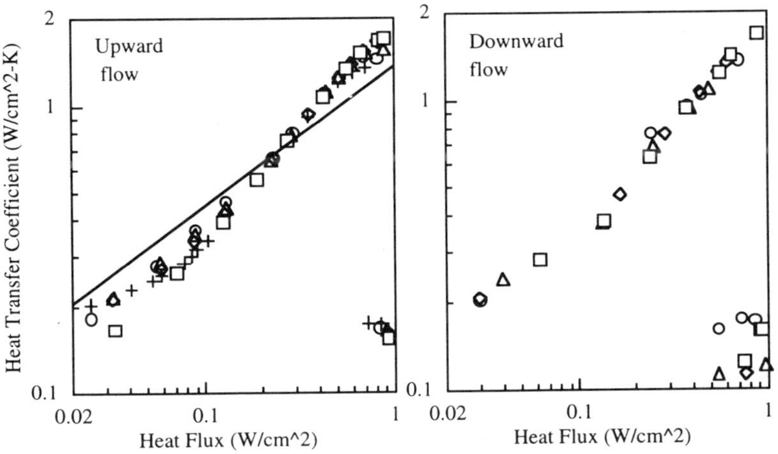

Figures 3 and 4. Heat transfer coefficient as a function of heat flux at P_{sat} = 1.0 atmosphere for different mass flow rates and flow directions. Cross, 0.0 g/s; circle, 0.25 g/s; triangle, 0.5 g/s; diamond, 0.75 g/s; square, 1.0 g/s. The solid line in Figure 3 is the correlation by Steiner, Eq. (1).

are calibrated germanium resistors, driven with 1 microamp of current. Care was taken to reduce thermal resistance between the thermometers and the copper test section, and the manganin lead wires were routed along isotherms whenever possible.

The experimental procedure begins by opening the fill valve, and slowly drawing up the bellows. The desired bellows velocity is programmed into the motor controller, and the bellows is driven downwards. The fill valve is then closed, and voltage is applied to one test section heater at a time. The venturi flow meter is monitored to ensure steady conditions and results are recorded from three precision voltmeters which measure voltage drops across the thermometers and heater. A warning light turns on when the bellows has reached 90% of its stroke, and the heater is turned off. This helps to prevent bubble accumulation inside the bellows when flow has stopped. The experiment is then ready for the next run.

RESULTS

A typical boiling curve is shown in Figure 2. Data for this run was taken with the fill valve closed to prevent induced flow, the bellows pump stationary, and with heat added to test section #2 only. The curve clearly shows the maximum heat flux, recovery heat flux, as well as nucleate and film boiling regimes. Smaller temperature differences in the nucleate boiling regime on the decreasing heat flux curve indicate the higher heat transfer coefficients present when the surface is fully educated with bubble nucleation sites.

Figures 3 and 4 illustrate the effect of mass flow rate on heat transfer coefficient for a fixed saturated pressure. Generally, no effect of mass flow rate is seen. Steiner[3] proposed the following correlation for nucleate boiling of He I flowing upward in vertical tubes:

$$\frac{\alpha(z)_{nb}}{\alpha_{0.3}} = C_F \left(\frac{\dot{q}}{\dot{q}_o}\right)^{0.7-0.13\times 10^{0.48p'}} \left[1.66p'^{0.45} + \left(2 + \frac{1}{1-p'^7}\right)p'^{3.7}\right]\left(\frac{d_o}{d}\right)^{0.4}\left(\frac{R_p}{R_{po}}\right)^{0.13} \qquad (1)$$

where $\alpha(z)_{nb}$ and $\alpha_{0.3}$ are the heat transfer coefficients for upward flow nucleate boiling and pool boiling at a reduced pressure of 0.3 (equal to 0.505 W/cm^2K), C_F is a coefficient equal to 0.57 for He I, $\dot{q}_o$ is the standard heat flux value of 1.0 W/m^2, p' is the reduced pressure, d is tube diameter (d_o = 1.0 cm), and R_p is surface roughness (R_{po} = 1 μm). This correlation is plotted as a solid line on Figure 3.

The effect of saturated temperature for a fixed flow rate of 0.5 g/s is shown in Figures 5 and 6. The four different temperatures correspond to the following four saturated

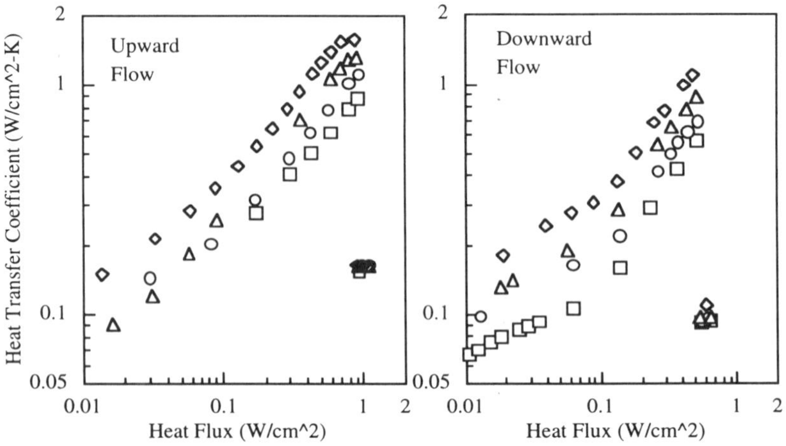

Figures 5 and 6. Heat transfer coefficient as a function of heat flux for different saturated temperatures at a fixed mass flow rate of 0.5 g/s and different flow directions. Square, 3.379 K; circle, 3.722 K; triangle, 3.993 K; diamond, 4.222 K.

pressures: 1.0, 0.8, 0.6, and 0.4 atmospheres. For both upward and downward flow, higher saturated temperatures result in higher heat transfer coefficients in the nucleate boiling regime. Little effect of saturation temperature is seen in the film boiling regime, but the downward flow has lower heat transfer coefficients than the upward flow.

The largest effects of flow direction with respect to gravity are observed in the critical heat flux transition from nucleate to film boiling, Figure 7. Low mass flow rates flowing downward result in significantly smaller critical heat fluxes than for upward flow. Similar behavior has been observed by Papell in liquid nitrogen flow boiling.[7] The recovery heat flux, Figure 8, is also smaller for downward flows at low mass flow rates.

CONCLUSION

Flow orientation, system saturated temperature, and mass flow rate can have significant effects on heat transfer in vertical channels cooled by He I. Critical and recovery

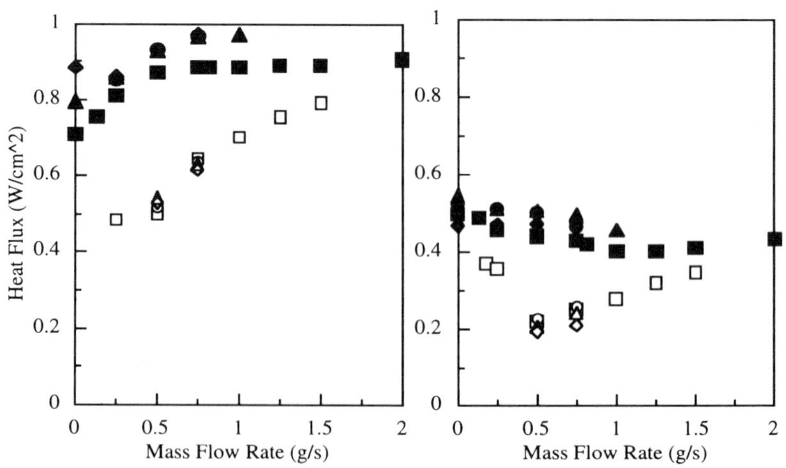

Figures 7 and 8. Critical and recovery heat fluxes as a function of mass flow rate for different saturated temperatures and flow directions. Square, 4.222 K; circle, 3.993 K; triangle, 3.722 K; diamond, 3.379 K. Open symbols are for downward flow, and dark symbols are for upward flow.

heat flux are also studied under various flow conditions. Further work could expand the envelope of mass flow rates, channel diameter, and system saturated pressures. Another avenue of investigation would investigate the effect of inlet quality on these parameters.

ACKNOWLEDGMENTS

The authors would like to thank Vincent Cochran, Hubie Payne, and Jermaine Taylor for their valuable technical support. This work is supported by the U. S. Department of Energy - Division of High Energy Physics under grant DOE-FG05-92ER40751.

REFERENCES

1. J. Theilacker. Upgrade of the Tevatron cryogenic system, in "Advances in Cryogenic Engineering," Vol. 39, Plenum Press, New York (1994), p. 517
2. C. Johannes. Studies of forced convection heat transfer to helium I, in "Advances in Cryogenic Engineering," Vol. 17, Plenum Press, New York (1971), p. 352
3. D. Steiner. Heat transfer during flow boiling of cryogenic fluids in vertical and horizontal tubes, *Cryogenics* May 1986
4. H. Ogata and S. Sato. Forced convection heat transfer to boiling helium in a tube, *Cryogenics* July 1974
5. V. E. Keilin, I. A. Kovalev, V. V. Likov, and M. M. Pozovonkov. Forced convection heat transfer to liquid helium I in the nucleate boiling region, *Cryogenics* March 1975
6. X. Huang. "Hydrodynamic study of one dimensional two-phase helium flow," Ph.D. Dissertation, University of Wisconsin-Madison, 1994.
7. S. S. Papell. Buoyancy effects on forced convective boiling, ASME 67-HT-63, 1967.

VACUUM ROUGHING PROCESS FOR LONG CONDUIT WITH LARGE CROSS SECTION

B. X. Zhang

Reliance Electric Company
Cleveland, Ohio 44117, USA

ABSTRACT

A discrete volume element scheme is developed for the analysis of vacuum roughing process of long conduits with large cross sections. Local conductances for such configuration become highly dependent upon the local vacuum level and the overall conductance of the conduit varies dynamically. Numerical schemes are developed to evaluate at each time iteration the overall conductance and hence to establish the pressure profile along the conduit. Time history of the pressure profile along the conduit is presented, and techniques used for computation of conductance due to local flow restrictions are discussed. Dynamic iterations yield time required to complete the roughing process.

INTRODUCTION

There are applications where long conduits are to be evacuated to sub-micron levels. Typical examples are vacuum jacketed superconducting magnet rings and cryogenic transfer lines for particle accelerators. The low conductance of the slender conduit makes it capable of sustaining substantial pressure gradient along the conduit. As a result, the local pressure at the pump-out port quickly reaches sub-micron level before the far end can be effectively evacuated. The goal of this study is to establish criteria for allowable interval between two neighboring pump-out ports to avoid infinite roughing process for a specific system. A specific system configuration (sector cryogenic transfer line for the Superconducting Super Collider) has been used for illustration purposes. The vacuum space of this system is the void enclosed by the vacuum jacket (0.5969 m I.D.) subtracting the space occupied by seven cryogenic circuit tubes, MLI blanket, and internal tube supporting system. The cross-sectional area of the vacuum space is approximately 0.211 m^2. The maximum length of a single transfer line is estimated at 80 meters. Depending on the site location, the vertical portion of the transfer line, which runs along the utility shaft, can be as long as over 70 meters. The total impedance along the entire conduit is substantially enhanced by the existence of the tube supporting system. Although pump-out ports are provided at the construction unit level (L = 12 m) for maintenance purpose, it is desirable not to pump at any port located in the vertical portion of the cryostat to avoid the inconvenience in equipment handling. A discrete volume element scheme is used to analyze the roughing process characteristics at various cross sections. Numerical solutions yield the pressure profile along the cryostat as a function of time during a vacuum roughing process.

SYSTEM MODELING

The basic equation generally used for the vacuum roughing process is the mass conservation equation:

$$-Vdp = Spdt - \dot{Q}_t dt \tag{1}$$

or

$$V(dp/dt) = \dot{Q}_t - Sp \tag{2}$$

which states that at any instance, the mass changing rate in a chamber is equal to the difference between the outgasing rate $\dot{Q}_t$ and pump-out rate Sp, where V is the chamber volume, S the pump speed, and p the pressure. Denoting the pump speed at the pump inlet as S_i, the pumping speed at the chamber can be calculated using the equation

$$S = \frac{S_i C_t}{S_i + C_t} \tag{3}$$

where C_t is the total conductance of the passage between the pump and the chamber. The ultimate achievable vacuum level can be estimated using the formula $p_u = \dot{Q}_t/S$. A simplified procedure is often used to estimate the total roughing time required for any specific vacuum level by assuming constant pumping speed and negligible outgasing. For an initial pressure p_o in the chamber, the mass conservation equation is integrated to yield pressure history in the chamber

$$p = p_o \exp[-t/(V/S)] \tag{4}$$

or the total roughing time required to reach a final pressure p

$$t = (V/S)\ln(p_o/p) \tag{5}$$

The conductance through each element is evaluated individually for viscous, intermediate, and molecular flow regimes. Use of CGS units yields the conductance in terms of cc/s. For viscous flow, the conductance through an annular cross section is given by the equation

$$C_v = \frac{\pi \bar{P}}{128\eta L}\left[D_o^4 - D_i^4 - \frac{\left(D_o^2 - D_i^2\right)^2}{\ln(D_o/D_i)}\right] \tag{6}$$

where η is the dynamic viscosity (η = 3.03×10^{-4} poise for air at standard condition), $\overline{P}$ average pressure, L aperture length, and D diameter.

For intermediate flow, Knudsen[1] suggested that an additional term C_2D^3 be used:

$$C_i = C_v + C_2 D^3 / L \tag{7}$$

$$C_2 = 3.81\times10^3\sqrt{\frac{T}{M}}\;\frac{1+1.1\times10^{-4}\sqrt{\frac{M}{T}}\frac{D\overline{P}}{\eta}}{1+1.36\times10^{-4}\sqrt{\frac{M}{T}}\frac{D\overline{P}}{\eta}} \tag{8}$$

where T is the absolute temperature and M molecular weight. This intermediate flow term is always added to the viscous term. Its significance is only manifested when the ratio of the inner diameter of the tube and the mean free path of the gas particle falls in the range $1 < D/\lambda < 100$, which corresponds approximately to a pressure range between 10^{-1} and 10^{-3} torr for a tube with an inner diameter of 0.0508 m. The mean free path is calculated by

$$\lambda = 2.33\times10^{-20}\,T/(\varsigma^2 p) \tag{9}$$

with molecule diameter ζ and pressure p in centimeter and torr respectively.

The molecular flow regime starts when the ratio D/λ falls below unity. Assuming the drift velocity of a molecule being proportional to its random velocity, Knudsen found that the conductance for molecular flow in long tube of constant cross section to be

$$C_m = \frac{8}{3\sqrt{\pi}}\left(\frac{2kT}{m}\right)^{1/2}\left(\frac{A^2}{BL}\right) = \frac{3.44\times10^4}{\sqrt{\pi}}\left(\frac{T}{M}\right)^{1/2}\left(\frac{A^2}{BL}\right) \tag{10}$$

in CGS units where k is Boltzmann constant, and m mass of a molecule. For an annular cross section, $A = \pi/4(D_o^2 - D_i^2)$, $B = \pi(D_o + D_i)$, and the conductance becomes

$$C_m = 3810.78\left(\frac{T}{M}\right)^{1/2}\frac{(D_o + D_i)(D_o - D_i)^2}{L}K \tag{11}$$

$$K = 1 + 0.481\frac{D_i}{D_o} - 2.013\left(\frac{D_i}{D_o}\right)^2 + 6.933\left(\frac{D_i}{D_o}\right)^3 - 9.798\left(\frac{D_i}{D_o}\right)^4 + 5.196\left(\frac{D_i}{D_o}\right)^5 \tag{12}$$

where correction factor K is approximated through curve fitting on the data from Roth[2]. Similarly, the correction factor for rectangular cross section is approximated by the following formula based on data from Guthrie and Wakerling[3]

$$K = 1.672 - 2.688X + 3.82X^2 + 4.245X^3 - 18.8X^4 + 1.994X^5 - 7.078X^6 \tag{13}$$

where $X \leq 1$ denotes the aspect ratio. The elbow piece which completes a 90° bend is treated as a section of straight module with an equivalent length L_e[4].

$$L_{ax} < L_e < (L_{ax} + 1.33\,D) \tag{14}$$

where L_{ax} is the length measured along the axis of the elbow, and D is the inner diameter of the vacuum jacket (0.5969 m).

A radial support of the cryostat consists of three plates as discussed elsewhere[5]. All annular passages for gas particles through the plates are treated as apertures. The total conductance of the radial support is calculated by adding the conductance for each plate:

$$\frac{1}{C}=\frac{1}{C_{4K}}+\frac{1}{C_{20K}}+\frac{1}{C_{80K}} \tag{15}$$

For each plate, the conductance is the sum of those for annular apertures that are in parallel:

$$C_{20K}=\sum_{j=1}^{6}C_j, \quad \text{and} \quad C_{80K}=\sum_{j=1}^{5}C_j \tag{16}$$

Three schemes are utilized to simulate the pump-down process. The impact of the relevant parameters on the overall performance of the pumping system is analyzed for all the schemes adopted.

A. Initial Estimate

An initial estimate on pumping capacity requirement is made based on the following assumptions and pump-down schemes:

- constant pump speed at all pressures
- cryostat considered as one large chamber
- negligible outgasing
- roughing limited to four hours
- maximum length limited to 80 meters
- initial and final pressures set at 760 and 0.001 torr respectively
- connection hose has a length of L_h = 1 m and an I.D. of 0.0508 m

The conductance of the hose in viscous flow is evaluated using the following formula

$$C=\frac{\pi\bar{p}D^4}{128\eta L_h} \tag{17}$$

which yields a value of $C = 2.05\times10^5$ cc/s. The pumping speed required at the chamber becomes $S = 1.587\times10^4$ cc/s using Eq. (3), and the pump speed is determined by

$$S_p=\frac{SC}{C-S} \tag{18}$$

as $S_p = 1.72\times10^4$cc/s. Leybold TRIVAC rotary vacuum pump Model D60B is selected. It has a speed of 45.9 CFM or 2.16×10^4 cc/s. Consequently, Eq. (5) yields the roughing time t = 2.94 hr. A turbomolecular pump TMP 1000 with 250 ISO-K or 15.24 cm (6") ASA flange is chosen to give a pumping speed of nitrogen at 1.15×10^6 cc/s in the molecular flow regime. It should be noted that this simplified approach under estimates the required roughing efforts.

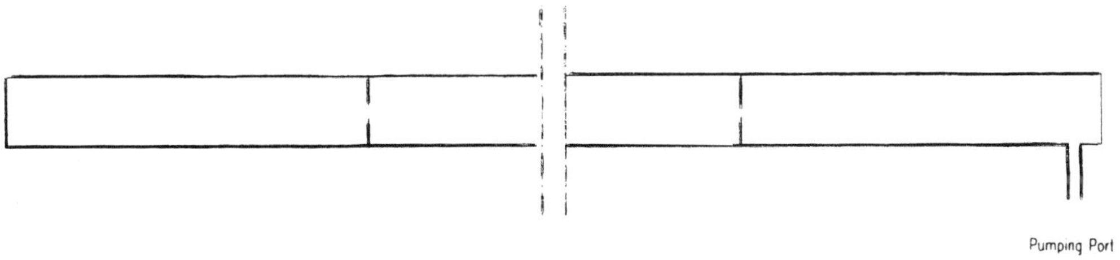

Figure 1. Schematic of discrete volume element.

B. Impact of Pressure on Pumping Speed

All assumptions in scheme A apply here except that the pressure effect is considered when calculating the actual pumping speed. An iteration process is involved to update the current pressure level.

C. Discrete Volume Element

This approach treats the vacuum space between two internal supports as a discrete volume element. The conductance through each support assembly is represented by an aperture. The system configuration is shown in Figure 1.

A straight transfer line module consists of four such volume elements, and the interconnect region is considered as a special element. Starting at the first element from the chamber inlet side, the pressure is calculated for a pre-selected time interval and the total conductance is updated before the next element is evaluated. The results from the calculation on the first volume element are then used for the calculation of the second element. This process is continued until the far end (mid-point when roughing is conducted at both ends) of the transfer is reached. After the calculation for each volume element, the same process is repeated for every downstream volume element to account for the gas influx from upstream. The iteration is terminated when a preset pressure convergence criterion is met. Numerically, for the *ith* volume element, Eq. (2) becomes

$$V_i(dp_i / dt) = \dot{Q}_i - (S_i p_i - S_{i+1} p_{i+1}) \tag{19}$$

The roughing process in the molecular flow regime is analyzed using similar approaches with conductances calculated using Eq. (11). Further more, the outgasing effect is also counted as it becomes predominant at low pressure levels and ultimately determines the dynamic equilibrium of the final state.

According to the experimental data from Franks[6], there will be approximately 25 kg water deposited on the MLI blankets for the longest transfer line cryostat. Generally, the water vapor outgasing rate is mainly a function of vacuum level and the residue water content. It also depends on the system configuration, which manifests in the molecular flow regime. The data published by Todd et al[7,8] show a time history of the outgasing rate for a specific system configuration. For the first 30 hours of roughing, these data can be approximated by the following equation

$$\dot{Q} = Ae^{[-1.952\ln(T/3600)-17.61]} \text{ torr l/s} \tag{20}$$

where A is the total area (two sides) of double aluminized mylar (DAM), and T is in seconds. The outgasing rate for the spun-bonded polyester spacer material is one order of

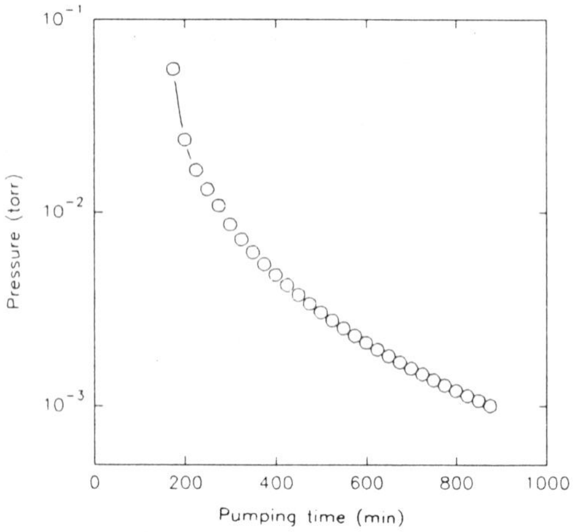

Figure 2. Pressure history for lumped-volume approach with a roughing pump and turbo pump combination.

magnitude lower than that of the DAM. In this analysis, the outgasing rate for DAM is multiplied by a factor of 1.5 to represent the total outgasing rate of the MLI, which is fairly conservative. A better representation of the outgasing rate requires its correlation with pumping speed and pressure history.

SIMULATION RESULTS

Lumped Volume Approach

The cryostat is considered as a large chamber and pressure distribution within its volume is uniform. Neglecting the outgasing effects and assuming constant pumping speed, the closed form solution requires a pumping time around 200 minutes to bring the pressure down below the micron level with the specified roughing pump alone. This is the only case when the variation of the pumping speed with vacuum level is not considered. Inclusion of the pumping speed variation brings the total time required up to approximately 220 minutes.

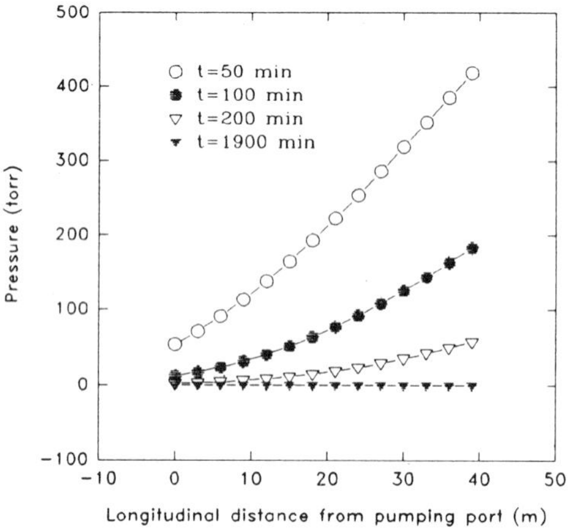

Figure 3. Time history of pressure profile along cryostat.

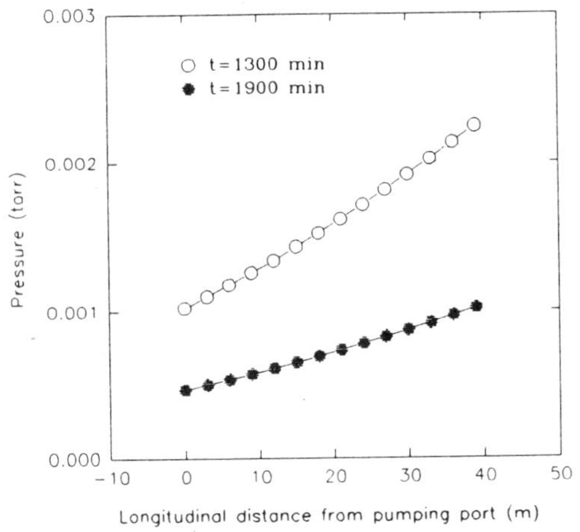

Figure 4. Pressure profiles when the near and far ends reach micron level.

If the turbomolecular pump is also engaged at the micron pressure level, the pumping time will be reduced to 179 minutes. When the water vapor outgasing from the MLI is considered, the pumping times required to reach the same vacuum level with and without the turbo pump become 848 and 1,400 minutes respectively. Figure 2. presents the pressure history in the lumped volume when outgasing is included. It is noted that approximately 70 % the roughing time is consumed to bring the vacuum level down from 10^{-2} to 10^{-3} torr.

Discrete Volume Element Approach

This approach treats the cryostat volume between two supports as a discrete element. The outgasing rate and the mass fluxes into and out this volume element are incorporated into the mass conservation equation which is solved iteratively for specified time intervals until a pre-selected tolerance for convergence is reached. The pressure profile along the cryostat is presented in Figure 3. as a function of roughing time.

It should be emphasized that due to the presence of the conductance through the support mechanism, large pressure gradients (relative to pressure level) exist along the cryostat. The pump-out port end (near end) of the cryostat has to be kept at sub-micron level in order to sustain reasonable pumping speeds at upstream locations. Evidently, this condition can only be met by using either a turbo pump or a cryopump. The curves in Figure 4. represent the pressure profiles when the near and far ends reach the micron level respectively.

The final state (one micron at the far end) is reached after the pumping equipment has operated at sub-micron level for a fairly long time.

DISCUSSIONS AND CONCLUSIONS

A discrete volume element model for vacuum roughing process is developed for long, large cross section conduits with apertures distributed at various intervals. A time iteration process is carried out to determine the throughput at each aperture. The pressure profile along the cryostat is computed as a function of roughing time. Conventional methods for roughing process estimation for such system configuration leads to substantial error. As the vacuum pump works at the micron and sub-micron level most of the time in order to bring the far end of the cryostat down to the micron level, a pump with a reasonable speed at such vacuum level is required, although a high speed at sub-micron level does not seem to help much while the conductance term is dominating.

The approximated outgasing rate appears to be the only uncertainty in the roughing process analysis. To improve the validity of the analysis, experiments need to be conducted to produce a better correlation between the outgasing rate, and the vacuum level and pumping time.

ACKNOWLEDGMENT

Research performed in part at the SSCL under DoE contract DE-AC35-89ER40486 with Universities Research Association, Inc.

REFERENCES

1. Knudsen, M., Ann. Phys. 31, 205; 32, 890; 33, 1435, (1910).
2. Roth, A., "Vacuum Technology", 3rd ed., North-Holland/American Elsevier, New York (1990).
3. Guthrie, A., and Wakerling, R.K., "Vacuum Equipment and Techniques", McGraw-Hill, New York (1949).
4. Xu, Tingwei and Wang, Kaiping, "The Relation between the Conductance of an Elbow and the Angle
   . between Tubes", Vacuum 32, p.655, (1982).
5. Zhang, B. and Stifle, K., "Preliminary Design of Sector Cryogenic Shaft Transfer Line System", IISSC, Supercollider 5, P. Hale ed., Plenum Press, New York (1994).
6. Franks, D., "Humidity Absorption Test on MLI", SSCL, Dallas, TX (1993).
7. Todd, R.J., Pate, D., and Welch, K.M., "Modeling of RHIC Insulating Vacuum for System Pumpdown Characteristics", Brookhaven National Laboratory, Upton, NY (1993).
8. Todd, R.J., Pate, D., Welch, K.M., "Outgasing Rate of Spunbonded Polyester and Dupont Double Aluminized Mylar", Brookhaven National Laboratory, Upton, NY (1992).
9. Welch, K.M., "The Pressure Profile in a Long Outgasing Vacuum Tube", Vacuum 23, p.271, (1973).

INFLUENCE OF HEATER ORIENTATION ON FLUCTUATIONS IN STEADY-STATE NUCLEATE BOILING

W.F. Osborne, Erik Bodegom, J.S. Semura, and L.C. Brodie

Department of Physics and
Environmental Sciences and Resources Program
Portland State University
Portland, OR 97207

ABSTRACT

In observations of steady-state nucleate boiling, fluctuations in the temperature and heat flux might initially appear to be completely random. However, it was shown that the fluctuations about the steady-state exhibit an average counterclockwise circulation when the heat flux is plotted versus the superheat temperature for a *vertically* mounted platinum wire in liquid nitrogen. An area associated with the average circulation was proposed as a numerical measure of stability for steady-state nucleate boiling. The mechanisms for the generation of these fluctuations are thought to be the feedback of the bubbles rising past the wire and the differential heating and cooling that this engenders. Data similar to the data on the vertical wire have been obtained using the same wire mounted horizontally. Although the counterclockwise circulation mentioned above is still seen, the measure of stability, as proposed earlier, is less useful for prediction of the transition to film boiling. This reduced sensitivity can be attributed to the fact that the possibility of feedback through the rising bubbles has been eliminated.

INTRODUCTION

Heat dissipation is a primary factor in the design of any temperature sensitive system and is often achieved by immersing an object in a cryogenic liquid. The boiling of the liquid allows the transfer of heat from the immersed object to the cryogen. Research involving heat flux and superheat temperatures have helped identify quantities such as the critical heat flux (CHF) for different materials.[1,2,3] Geometry, cryogen type, and it turns out, spatial orientation all play a role in determining the critical heat flux of a system. This investigation, along with previous research,[4,5] indicates that although the applied power is steady, the temperature of a heater/thermometer can undergo significant fluctuations. The conclusions of Roth et al.[4,5] were not totally clear-cut and it was thought that more investigation was needed. The intent of this research was threefold. First, we attempted to recreate previous results involving a vertically mounted platinum wire. That research had suggested a possible measure as a predictor for the transition from nucleate to film boiling. Second, we investigated the properties of a horizontally mounted wire and compared the results with the vertically mounted wire. Finally, we have clarified the use of the predictor in the transition to film boiling.

An apparatus was built to rotate the platinum wire from the vertical to a horizontal orientation without having to remove the wire from the liquid nitrogen bath. This small change shed new light on the heat flux near a platinum wire. The ability to rotate the wire revealed differences in temperature fluctuations between the horizontal and vertical orientations. For the same applied power, a horizontally oriented wire had significantly smaller temperature fluctuations than the same wire mounted vertically. This result was not unexpected, but the difference was striking. We suspected that the phenomenon of synchronization was at the root of the difference in temperature fluctuations between horizontal and vertical orientations.

Synchronization is the process whereby a bubble released on a lower portion of a vertically mounted wire rises past higher points on the wire. The bubble rising upwards causes a convective current in the liquid nitrogen near the wire. When this bubble passes near another bubble that is still attached to the wire, either the bubbles collide and form an even larger bubble or the attached bubble is pulled from its site. In either case, there is a change in wire temperature. This process continues as the bubbles move up the wire causing even more bubbles to detach from the wire. The result is a sweeping action separating most of the bubbles from the wire. As a result, there is a sharp decrease in the temperature of the wire. The process continues while new bubbles form around the wire until these bubbles are again swept away from the wire. This sweeping and reforming process appears to be the cause of large temperature fluctuations in the vertical case. It will be shown that synchronization accounts for the differences in the results between the horizontal and vertical orientation.

This investigation collects the temperature fluctuation data of a heater/thermometer. Steady-state power was applied to a platinum wire immersed in a liquid nitrogen at ambient pressure. Analysis of this data shows that the fluctuations about the average in the heat flux-temperature plane generally follow a counterclockwise trajectory. This behavior can be explained by considering the power balance equation.

When characterizing the heat transfer properties of a boiling liquid, a plot of the *steady-state* values of heat flux and superheat temperature is usually made. The two quantities of interest in this experiment were the *time-dependent* temperature and heat flux at the heater/liquid interface for a steady-state application of power to the heater. The heat flux per unit area ($\dot{q} / A$) into the liquid was calculated from conservation of energy considerations as

$$\dot{Q} \equiv \frac{\dot{q}}{A} = \frac{P}{A} - \frac{C}{A}\frac{dT}{dt} \tag{1}$$

where $\dot{q}$ is the heat flow, A is the surface area of the heater, P is the applied power which is held nearly constant in these experiments, C is the heat capacity of the platinum wire and dT/dt is the time derivative of the temperature.

Figure 1 shows a typical trajectory in the $\dot{Q}$ - ΔT plane for moderate power inputs under nucleate boiling conditions. Note that the average superheat temperature and the average heat flux have been subtracted out from the data to emphasize the fluctuations. The arrowheads are drawn to indicate the direction of circulation. Peak to peak fluctuations in the temperature and the heat flux approach 1 kelvin and 1 W/cm^2, respectively. The average values and the range of the fluctuations depend on the applied power. The qualitative physical picture that is consistent with this direction of the circulation is as follows.

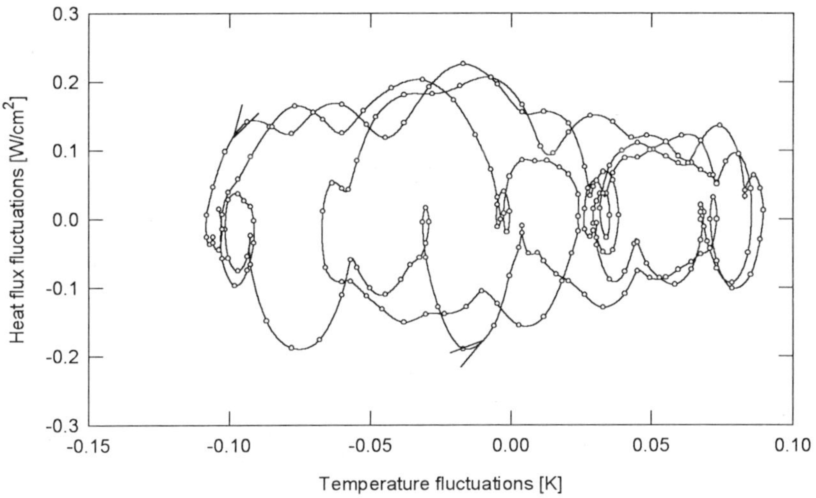

Figure 1. Fluctuations about the average in the temperature-heat flux plane.

The trajectory in the $\dot{Q}$ - ΔT plane of the state of the system is described roughly by an ellipse. *Individual* nucleation sites undergo the following "life" cycle. In this microscopic point of view, the bubble detaches from the heater surface and rises through the liquid. A new thermal boundary layer starts to form as the surface heats up. At some later time, and at a high enough superheat, a bubble begins to grow at the site. As the bubble continues to grow the heat of vaporization is the dominant factor in cooling at the solid interface. When the bubble increases in size it begins to insulate the wire, thus diminishing cooling by means of conduction or convection. Thus the overall rate of cooling begins to decrease with eventual increase in the temperature. Finally, when the bubble becomes large enough, it breaks off and leaves the surface because of buoyant forces. The liquid then rushes in, thereby initiating the process again. Investigations involving single nucleation sites in water appear to concur with this description.[6,7]

EXPERIMENTAL SETUP

The apparatus constructed to carry out this investigation consisted of the wire support assembly, the support rod and the lift arm. The wire support assembly contains the platinum wire/heater and is the part of the equipment that can be rotated from a horizontal position (Figure 2) to a vertical position (Figure 2 inset). The diameter of the platinum wire is 0.102 mm and it has an overall length of 7 cm. The effective length (distance between voltage leads) is 3.16 cm. The use of the platinum wire, as a resistance thermometer as well as a heater, minimizes errors when determining the heat flux and temperature when compared to a spatially separated thermometer and heater. The support rod is attached to the wire support assembly through a pivot point, allowing movement of the wire assembly. The rod is then used to lower the entire apparatus into the nitrogen bath and keep the current and voltage leads separated. The current leads between the support rod and the wire support assembly need to be flexible. We determined that wires with vinyl insulation were too rigid to allow for rotation of the wire assembly. Instead we used braided shielding which could conduct enough current and remain flexible in liquid nitrogen. The lift arm was connected to the right end of the wire support assembly through another pivot point. It is used to change the orientation of the platinum wire without having to remove the apparatus from the nitrogen bath. This allowed us to compare temperature fluctuations from both orientations while changing the minimum amount of parameters in the experiment. Thin leads were soldered to the platinum wire for voltage measurements. The small voltage fluctuation data from the platinum wire, which corresponds to temperature fluctuations, was amplified using a differential amplifier with a DC offset. The signal was then routed to a data acquisition board in a computer for data collection.

Data collection involved first placing the wire support assembly in the vertical position. The entire apparatus was then carefully lowered into a liquid nitrogen storage container with the liquid at ambient pressure. A storage container was used instead of a dewar to accommodate the width of the wire support assembly while situated in the horizontal position. The only size constraint with the storage dewar was the diameter of the filler neck (approximately 6.4 cm). The apparatus was constructed to fit through the filler neck while taking full advantage of the much larger diameter of the storage container's main body. The wire support assembly was then moved to the horizontal position and locked into place. A constant current was then set and recorded. The absolute voltage across the heater for this current is also recorded. The relative DC offset was adjusted so that the voltage fluctuations of the platinum wire would occur within the range of the data acquisition card. Voltage data were acquired over a 3 second time period, at 4000 samples per second, for a total of 12000 data points at that specific current level. The current level was then changed and the recording process was repeated. All of the horizontal data was collected starting at the lowest current level and then adjusting the level upwards incrementally until transition to film boiling occurred. Next, the wire support assembly was moved to the vertical position and locked into place. The procedure used for acquiring vertical position data is identical to the procedure previously described for the horizontally positioned wire.

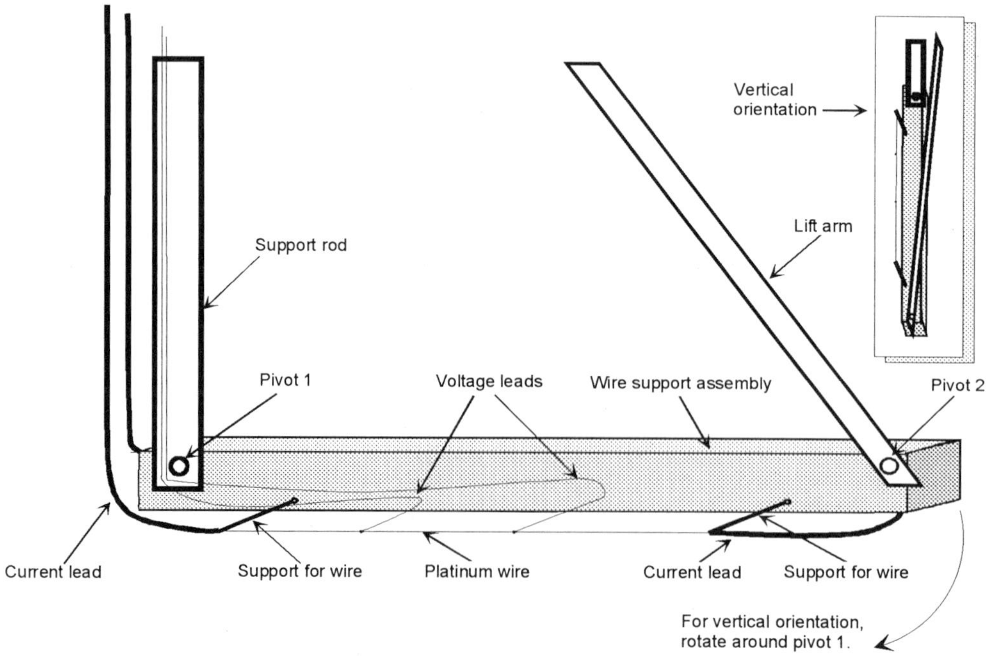

Figure 2. Schematic presentation of the platinum wire in the horizontal position. The inset shows the wire in the vertical position.

The thermal time constant of the platinum wire thermometer, τ, is given by

$$\tau = d^2 \frac{\rho c}{k} \tag{2}$$

where d is the diameter of the wire, ρ is the density of the wire, c is the specific heat of the wire and k is the thermal conductivity of the wire. This is roughly the time for the interior and surface temperatures of the wire to equalize. For the platinum wire in this investigation, τ was determined to be 60 μs. Since our observed temperature fluctuations were much slower than 60 μs and the sampling time was 250 μs, it is assumed that the temperature everywhere in a cross-sectional area of the wire is uniform. Therefore, the response time of the platinum thermometer is deemed adequate to record the fastest temperature fluctuations (around 300 Hz). However, it is not assumed that the temperature is necessarily constant between any two points along the length of the wire.

The heating current was controlled by an op-amp/transistor array configured to produce the desired constant current. This amounted to nearly a constant applied power (within 1%) since the overall resistance of the platinum wire changed insignificantly with the small temperature fluctuations in the experiments.

A typical data run consisted of setting the constant current source to a predetermined value such that nucleate boiling could be observed. Measurements of the applied current and the voltage across the heater/thermometer were performed by the data acquisition system installed in our laboratory PC. Values of the resistance and therefore of the temperature of the heater/thermometer were then digitally computed. This process was then repeated for increasing current levels until the transition from nucleate boiling to film boiling occurred. The time dependent heat flux was calculated as described below.

RESULTS AND DISCUSSION

The analysis of the experimental data for a run is shown in Figure 3. It shows the area associated with the two orientations of the same wire. The error bars were calculated by partitioning the data set at each current level into four parts and calculating the area and the standard deviation of the resulting sets. All the data reported in Figure 3 were obtained by first performing the measurements at all the selected current levels in the horizontal orientation and then switching to the vertical position. Similar results were obtained by taking the measurements in both orientations at every selected current. The vertical data are similar to previous observations.[4,5] It is seen that, again, upon approaching the critical heat flux in the vertical orientation, (CHF, approximately 9.5 W/cm^2) the area seems to level off. As was done previously, the curve through the vertical data represents a quartic fit, which seemed to adequately follow the experimental points.

Comparing this with the data for the horizontal orientation, several significant differences are observed. First, the CHF is higher (about 11 W/cm^2) for the horizontal orientation.[8] This can be explained by the fact that rising bubbles interfere with the efficient heat transfer in the vertical case and, in fact, limit the maximum heat flux that the vertical wire can support.

Second, Figure 3 shows a factor of three difference in the area between the two orientations. It should be pointed out that the *only* difference between the two data sets is the orientation. The wire is the *same*; it has the same nucleation sites, the same boiling sites, and the same imperfections. So the observed differences are due solely to the effect of the horizontal or vertical geometry. This can be understood by referring to the model for the generation of the area. In the vertical case different nucleation sites tend to be synchronized because of the feedback provided by the bubbles rising past the vertical wire. Thus the average temperature of the wire between the potential leads, all other things being equal, will tend to undergo larger fluctuations and the associated area will also be larger. Although, in the horizontal case adjacent nucleation sites tend to be somewhat synchronized, this synchronization is much more pronounced in the vertical case. So, for the first time, these measurements provide direct support for the proposed model.[4,5]

Third, the data for the horizontal orientation is well represented over the total range by a linear regression, in contrast to the vertical orientation where above about 7 W/cm^2 the relationship ceases to be linear. The linear relationship between area and the applied power can be understood based on the fact that the area is proportional not only to amplitude of the temperature fluctuations but also to the frequency of these fluctuations. The simplest way to see this is that frequency is related to the time for a bubble to grow to a size at which it detaches from the wire. This time is roughly inversely proportional to the applied power and thus the frequency increases linearly with the applied power. On the other hand, the magnitude of the temperature fluctuations is approximately constant because it is determined by the maximum size of the bubble and by the influx of the cold liquid. Thus, since the area is roughly equal to the product of this magnitude and the frequency, area is linear with applied power. In the vertical case, around 7 to 9 W/cm^2 the linear relationship between area and the applied power is lost because the rate of release of bubbles is so large that the synchronization decreases in significance. If somehow film boiling could be prevented, one would expect that at large applied powers, the area would drop down to the area for the horizontal wire. This loss of synchronization occurs over a certain range of applied powers, resulting in the generation of a broad peak. The data in this region is subject to perturbations extrinsic to the experiment per se (e.g., release of bubbles from the support structure, convective flow in LN_2, etc.). These perturbations have a large effect, as can be observed from the error bars in Figure 3.

CONCLUSIONS

This work has shown that for steady-state nucleate boiling in liquid nitrogen, the temperature fluctuations of the heater wire depend upon whether it is oriented vertically or horizontally. The temperature fluctuations were analyzed by calculating the area swept out

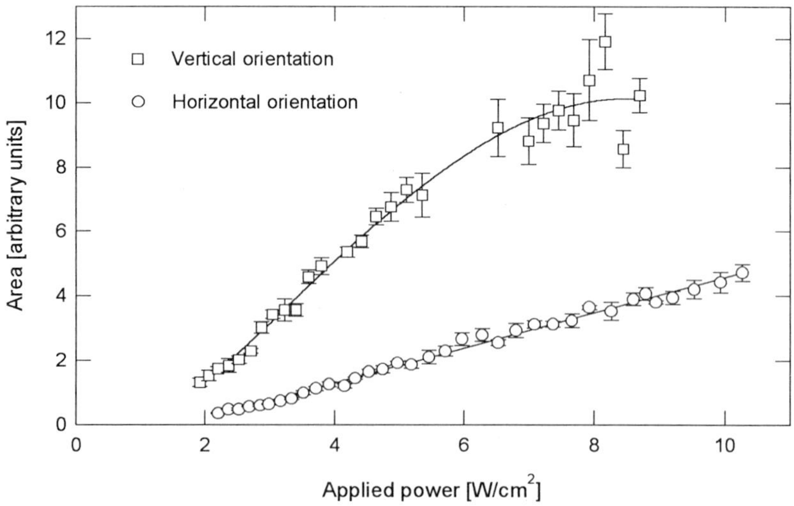

Figure 3. Circulation in the heat flux-temperature plane vs. the applied power.

in unit time by the counterclockwise closed circulation in the heat flux-superheat temperature $\dot{Q}$ - ΔT plane.

When plotted as a function of the applied heater power, the $\dot{Q}$ - ΔT area for the horizontal wire differs in three main respects from that for the vertical wire. The two most significant differences are shown in Figure 3. The peak which appears so clearly in the 6 to 9 watts/cm^2 region for the vertical wire is completely absent in the case of the horizontal wire. The $\dot{Q}$ - ΔT area for the horizontal wire is only about 1/3 that of the vertical wire at any particular heater power. In addition, it was found that the critical heat flux was less for the vertical wire.

It is probable that the two differences evident in Figure 3 arise from the same basic reason. In our previous paper,[4,5] we pointed out that there are probably many active nucleation sites on the heater wire surface between the two potential leads. Therefore, one would not expect the counterclockwise circulation in the $\dot{Q}$ - ΔT plane, which exists at any one nucleation site, to be manifest in our data unless there is a cooperative phenomenon occurring that tends to synchronize the activity at the various sites. The work reported here supports this idea. Since the bubbles leaving each site rise vertically through the liquid, they will have a much greater synchronizing effect for the vertical wire than for the horizontal wire. However, even for the vertical wire the synchronizing effect of the rising bubbles will tend to decrease as the applied heating power is increased because the frequency of bubble detachment increases. The maximum in the $\dot{Q}$ - ΔT area vs. applied power for the vertical orientation is the result.

The above discussion suggests further studies. For the vertical wire, the extent of the synchronization can be determined by measuring the area $\dot{Q}$ - ΔT for wires of increasing length. Such measurements might shed light on the speculation mentioned in the previous paper, that the peak might serve as a predictor of the transition from nucleate to film boiling. The possibility of generalizing the use of the area as a predictor in other configurations where cooperative phenomena are present is still an open question.

Other possibilities for future studies are suggested by this work. For example, by taking measurements on heating wires oriented at angles between the horizontal and vertical, it could be determined whether the transition between the horizontal and vertical heating wire is smooth or whether there is some intermediate orientation for which the cooperative phenomenon is a maximum. Also, it would be worthwhile to investigate the possibility of reducing the scatter in the higher power region of Figure 3 by increasing the sampling times. Finally, it is tempting to speculate on the applicability of the further development of the techniques in chaos theory to the analysis of the temperature fluctuations.

REFERENCES

1. D.N. Sinha, L.C. Brodie, J.S. Semura, and F.M. Young, Premature transition to stable film boiling initiated by power transients in liquid nitrogen, *Cryogenics* **19:225** (1979)

2. E.W. Roth, E. Bodegom, L.C. Brodie, and J.S. Semura, Suppression of the premature transition to film boiling in LN2 using forced convection in: *Advances in Cryogenic Engineering*, **Vol. 37A**, (1991), p. 197

3. M. Shiotsu K. Hata and A. Sakurai, Heterogeneous spontaneous nucleation temperature on solid surface in liquid nitrogen, in: *Advances in Cryogenic Engineering*, **Vol. 35A**, (1990) p.437

4. E.W Roth, Nucleation and heat transfer in liquid nitrogen, Ph.D. Dissertation, Portland State University, Portland, OR (1993)

5. E. W. Roth, J.S. Semura, L.C. Brodie, and E. Bodegom, Novel approach to the analysis of fluctuations in steady-state nucleate boiling, in: *Advances in Cryogenic Engineering*, **Vol. 39**, (1994), p. 1741

6. S.T. Hsu and F.W. Schmidt, Measured variations in local surface temperatures in pool boiling of water, *J. Heat Transfer, Trans. ASME*, (August 1961) p. 254

7. Thomas F. Rogers and Russel B. Mesler, An experimental study of surface cooling by bubbles during nucleate boiling of water, *A.I.Ch.E. Journal,* (September 1964) p. 656

8. C. Beduz, R.G. Scurlock, and A.J. Sousa, Angular dependence of boiling heat transfer mechanisms in liquid nitrogen, , in: *Advances in Cryogenic Engineering*, **Vol. 33**, (1988), p. 363

BOILING ENHANCEMENT BY MICROGROOVES

R.F. Barron and M.A. Baig

Mechanical and Industrial Engineering Department
Louisiana Tech University
Ruston, Louisiana 71272

ABSTRACT

Several techniques have been proposed to improve the boiling heat transfer process, including surfaces with re-entrant holes, porous sintered metal coatings, etc. The purpose of this work was to examine the effect of circumferential microgrooves, produced by diamond-bit machining, on the film boiling heat transfer from a horizontal cylinder immersed in liquid nitrogen. The depth of the grooves was varied from 25 to 75 micrometers, and the width of the grooves was varied from 25 to 50 micrometers. Auxiliary tests were run using a cylinder with a smooth (polished) surface and one run was conducted with a cylinder having both circumferential microgrooves and longitudinal microgrooves. The boiling heat transfer coefficients were measured using a transient cool-down method.

The results of the film boiling tests indicated that the use of microgrooves resulted in an improvement in the cool-down time and an enhancement of the boiling heat transfer coefficient by a factor of approximately 1.25. The cool-down time for the sample with no microgrooves was approximately 130 seconds; whereas, the sample having 50 μm deep by 25 μm wide microgrooves has a cool-down time of approximately 100 seconds. The improvement of the boiling heat transfer coefficient due to longitudinal microgrooves was insignificant, in comparison with that for the circumferential grooves.

INTRODUCTION

In many cryogenic systems, it is important to cool the system from ambient to the operating temperature as rapidly as possible. Several techniques have been proposed to improve the boiling heat transfer process, including surfaces with re-entrant holes, porous sintered metal coatings, etc. The purpose of this work was to examine the effects of circumferential microgrooves, produced by diamond-bit machining, on the film boiling heat transfer from a horizontal cylinder immersed in liquid nitrogen.

BACKGROUND LITERATURE

Nakayama[1] discussed the heat transfer enhancement in nucleate boiling for surfaces with internal cavities, such as tunnels and pores, using various fluids, including liquid nitrogen. He found that a reduction in wall superheat of 80% to 90% could be achieved. Webb[2,3] gave a discussion of the development of surface enhancement treatments, including porous coatings, for nucleate boiling. Vasiliev *et al.*[4] reported experimental results for liquid nitrogen in nucleate boiling on copper discs with a porous coating. They found that the critical heat flux (CHF) was the same for a porous coated surface and an uncoated surface.

Ashworth *et al.*[5] evaluated the boiling enhancement characteristics of a porous metallic coating for liquid nitrogen, liquid argon, liquid oxygen, and liquid R-12. They discussed the various regions of heat transfer enhancement for porous coatings, and their results were in agreement with those of Afgan *et al.*[6]

Ogata and Nakayama[7] presented both nucleate and film boiling data for liquid helium boiling on machined surfaces. The width of the machine grooves in their work ranged from 0.25 mm to 1.00 mm, and the groove depth was 1.00 mm. The heated surface was a flat plate, with the grooves oriented vertically in some samples and horizontally in other. For the macro-grooves, they found that the heat transfer enhancement in film boiling could be correlated in terms of the ratio of the machined surface area and the surface area of a corresponding smooth surface.

EXPERIMENTAL APPARATUS AND PROCEDURE

The test samples used in this study were cylindrical rods, approximately 25 mm in diameter and 100 mm long, as shown in Fig. 1. The material used for the rods was 1100F aluminum. A Type T (copper-constantan) thermocouple was installed in each end of the sample to measure the sample temperature as a function of time. The thermocouple output was directed to a Keithley data acquisition system for recording on a 3.5-in. diskette. The sample was supported by a wire frame in the horizontal position. The liquid nitrogen container was a styrofoam insulated (40 mm thick) container having a rectangular cross section, 250 mm by 800 mm. Details of the test set-up are given by Baig.[8]

The dimensions of the samples are given in Table 1. All samples had circumferential grooves machined with a diamond-point tool in the surface. One sample (#9) had both circumferential and longitudinal grooves in the surface.

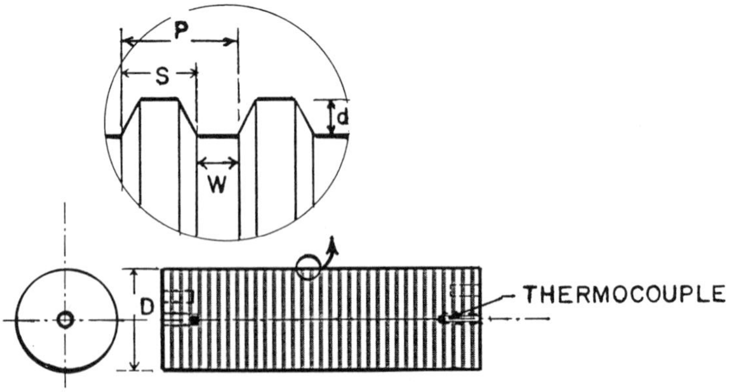

Figure 1. Test sample geometry. The outer diameter of the cylinder was approximately 25 mm (1.00 in.), and the length was approximately 100 mm (4.00 in.).

Table 1. Dimensions of the samples

SAMPLE NO.	DEPTH OF GROOVE d, μM	WIDTH OF GROOVE W, μM	AXIAL PITCH P, μM	ROD DIAMETER mm	AREA RATIO A/A_O
1	25	25	150	24.7	1.333
2	50	25	150	24.1	1.665
3	75	25	200	24.2	1.747
4	25	25	200	24.7	1.249
5	50	25	200	24.3	1.498
6	25	25	250	24.8	1.200
7	50	25	250	24.6	1.399
8	75	25	300	24.8	1.502
9*	25	50	200	24.4	
	20	50	400		

*Sample #9 had both circumferential grooves (first table entry) and longitudinal grooves (second table entry).

The film boiling heat transfer coefficient was measured using a transient technique. Experimental data were acquired twice for each sample to check repeatability, and little variation was observed between the two readings for a particular sample. The samples were suddenly immersed to depth of 75 mm in liquid nitrogen at atmospheric pressure (T_{sat} = 77.4 K). The depth of immersion was held constant during the cool-down process.

The data reduction expression for the heat transfer coefficient was:

$$h_c = \frac{mc}{A_0(T - T_{sat})} \frac{dT}{dt} \quad (1)$$

The mass of the specimen was measured with a precision balance, and the nominal surface area was calculated from the geometry of the sample. The temperature variation of the specific heat for the aluminum samples was considered by using a 4th-degree polynomial curve-fit, given by:

$$c = 3R\,[\,b_o + b_1\,(\theta_D/T) + b_2(\theta_D/T)^2 + b_3(\theta_D/T)^3 + b_4(\theta_D/T)^4\,] \quad (2)$$

where the numerical values for the constants were:

R = 0.30828 kJ/kg-K
θ = 385 K
b_o= +1.140565
b_1= -7.67846 x 10^{-2}
b_2= -5.35520 x 10^{-2}
b_3= +1.03192 x 10^{-2}
b_4= -5.64888 x 10^{-4}

The temperature variation with time was calculated by using a least-squares curve-fit to the cool-down data, in the form:

$$t = a_o \ln(T - T_{sat}) + a_2\,[\,\ln(T - T_{sat})\,]^2 + a_3\,[\,\ln(T - T_{sat})\,]^3 \quad (3)$$

The reciprocal of the temperature derivative was then calculated from Eq. (3).

$$\frac{dt}{dT} = \frac{a_1 + 2a_2 \ln(T - T_{sat}) + 3a_3\,[\,\ln(T - T_{sat})\,]^2}{(T - T_{sat})} \quad (4)$$

The values for the coefficients, a_o, ..., a_3, are given by Baig[8] for each experimental run.

The data reduction procedure assumes a uniform temperature throughout the cylinder at any instant of time. For a typical case (42 seconds into Run #6), the cylinder diameter D = 24.7 mm, the thermal conductivity of the aluminum k = 206 W/m-K, and the heat transfer coefficient h_c = 204 W/m^2-K. Using these values, the Biot number is:

$$Bi = \frac{h_c D}{4k} = 0.0061 << 0.1$$

Kreith and Bohn[9] suggest that the error introduced by assuming uniform temperature is negligible when the Biot number is less than 0.1. It was estimated that the actual difference between the center and surface temperature for the samples was less than 0.8 K.

RESULTS

A plot of the sample temperature vs. time during the cool-down is shown in Fig. 2. The initial cool-down period involved film boiling on the surface of the cylinder. When the cylinder temperature reached approximately 105 K, the boiling changed from film to nucleate boiling, and the cylinder temperature rapidly dropped to the temperature of the liquid nitrogen within a few seconds. Only the film boiling portion of the cool-down curve was considered in this research.

The Sherwood number is defined by:

$$Sh = \frac{g\rho_g(\rho_L - \rho_g)\, \lambda\, D^3}{\mu_g k_g(T_w - T_{sat})} \tag{5}$$

where:

$$\lambda = h_{fg} + 0.34\, c_g\, (T_w - T_{sat})$$

A plot of the Nusselt number, Nu = $h_c D/k$, as a function of the Sherwood number is shown in Fig. 3.

The film boiling data for the cylinder with no microgrooves could be correlated by the following expression, for Sh > 10^9:

$$Nu = h_c\, D/k_g = 0.165\, Sh^{1/3} \tag{6}$$

Frederking and Clark[10] obtained a similar expression for film boiling over a sphere immersed in liquid nitrogen.

The film boiling data for the cylinders with microgrooves could be correlated by the following relation, for Sh > 10^9:

where the quantity C_d is the Nikuradse drag coefficient[11], based on the depth of the microgroove,

$$Nu = 0.134\, (C_d)^{1/5}\, Sh^{0.40} \tag{7}$$

d, for 100 < D/d < 1000

$$C_d = [1.89 + 1.62 \log_{10}(D/d)]^{-2.5} \tag{8}$$

All fluid properties, except the heat of vaporization, are evaluated at the mean film temperature, $T_f = (T_w - T_{sat})/2$. The heat of vaporization is evaluated at the saturation temperature. From an

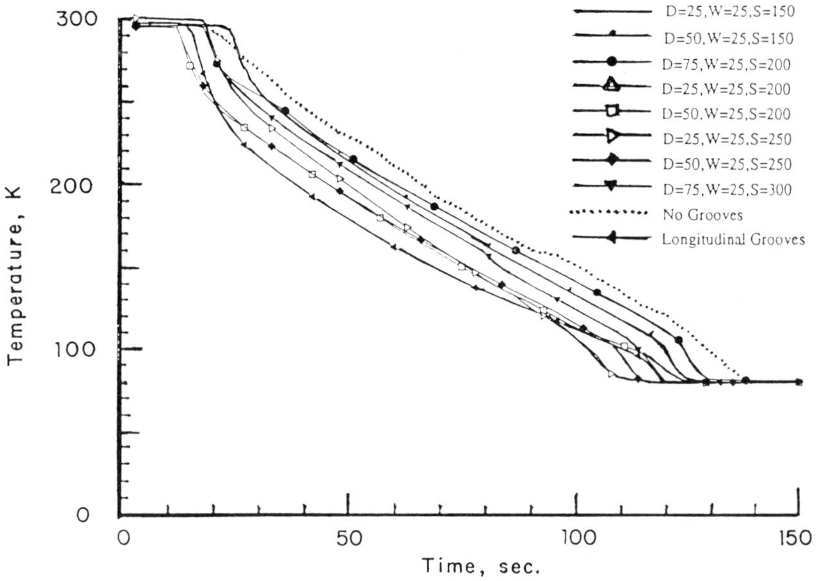

Figure 2. Plot of sample temperature as a function of time.

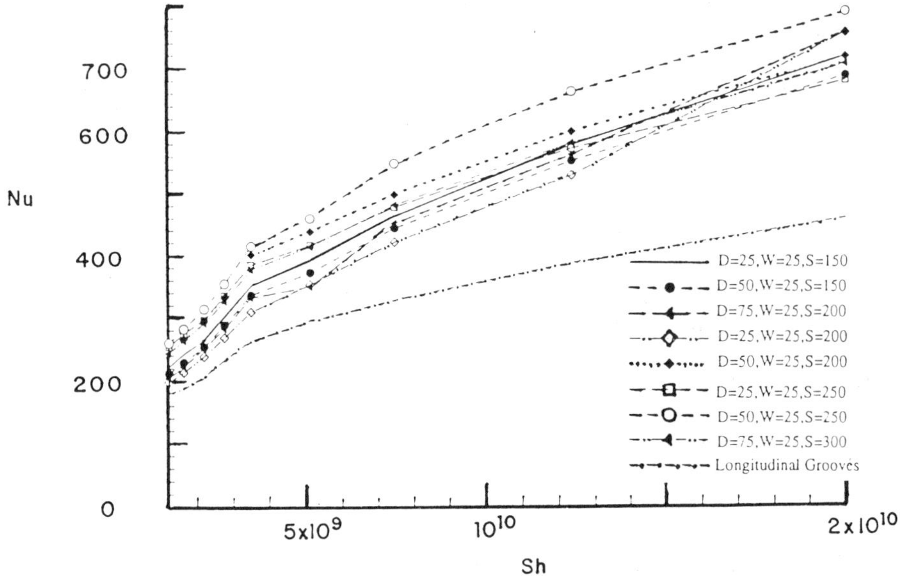

Figure 3. Plot of the Nusselt number as a function of the Sherwood number for the test samples.

uncertainty analysis[8], the uncertainty in the experimental values of the Nusselt number was estimated to be 10.1%.

The primary mechanism responsible for the heat transfer enhancement in film boiling over the cylinder with microgrooves was the roughness introduced by the grooves. The additional surface area resulting from the grooves did not correlate with the enhanced heat transfer, in the present study.

For a Sherwood number of 10^{10}, the estimated Nusselt number for a cylinder with no microgrooves was 355; whereas, the estimated Nusselt number for a 25 mm diameter cylinder with 50 μm depth microgrooves was 535. The enhancement in the boiling heat transfer due to the presence of the microgrooves was a factor of 1.50 or about a 50% increase over the case for a cylinder with no microgrooves.

For the cylinder with both circumferential and longitudinal microgrooves, the heat transfer coefficient during the initial 15 seconds of the run was approximately 20% larger than the corresponding heat transfer coefficient with circumferential grooves only. After approximately 45 seconds in the run, the heat transfer coefficients for the two cases were approximately the same (within 5%). This result would indicate that the improvement of the boiling heat transfer coefficient due to longitudinal microgrooves was not significant, in comparison with that for the circumferential grooves.

SUMMARY AND CONCLUSIONS

The purpose of this work was to examine the effect of circumferential microgrooves, produced by diamond-bit machining, on the film boiling heat transfer from a horizontal cylinder immersed in liquid nitrogen. The depth of the grooves was varied from 25 to 75 micrometers, and the width of the grooves was varied from 25 to 50 micrometers. The boiling heat transfer coefficients were measured using a transient cool-down method.

The results of the boiling tests indicated that the use of microgrooves resulted in an improvement in the cool-down time and an enhancement of the boiling heat transfer coefficient by a factor of approximately 1.25 to 1.50. The improvement of the boiling heat transfer coefficient due to the circumferential microgrooves was the result of the roughness effect of the cylinder surface produced by the grooves. The improvement of the boiling heat transfer coefficient due to longitudinal microgrooves was insignificant, in comparison with that for the circumferential grooves.

NOMENCLATURE

Symbol	Explanation
a_0, a_1, a_2	constants in eqn. (3)
A_0	nominal surface area of the sample, πDL, m^2
b_0, b_1, b_2	constants in eqn. (2)
$Bi = h_c D/4k$	Biot number
c	specific heat, J/kg-K
C_g	vapor specific heat, J/kg-K
C_d	drag coefficient, eqn. (8)
d	depth of the grooves, m
D	nominal diameter of the sample, m
g	local acceleration due to gravity, m/s^2
h_c	boiling heat transfer coefficient, W/m^2-K
h_{fg}	heat of vaporization, J/kg
k_g	vapor thermal conductivity, W/m-K
m	mass of the sample, kg

$Nu = h_cD/k_g$	Nusselt number
R	specific gas constant, J/kg-K
Sh	Sherwood number, see eqn. (5)
t	time, seconds
T	temperature, K
T_{sat}	fluid saturation temperature, K
T_ω	surface temperature, K
θ_D	Debye temperature, K
λ	"effective" heat of vaporization, J/kg-K
μ_g	vapor viscosity, Pa-s
ρ_g	vapor density, kg/m^3
ρ_L	liquid density, kg/m^3

REFERENCES

1. W. Nakayama, Dynamic model of enhanced boiling heat transfer in a porous surface, *ASME J. Heat Transfer* 102:445 (1980).
2. R.L. Webb, The evolution of enhanced surface geometries for nucleate boiling, *Heat Transfer Eng.* 2:46 (1981).
3. R.L. Webb, Nucleate boiling on porous coated surfaces, *Heat Transfer Eng.* 4:71 (1983).
4. L.L. Vasiliev, S.K. Vinokurov, and O.G. Rasin, Nitrogen boiling heat transfer in porous coating, in: "Heat and Mass Transfer in Refrigeration and Cryogenics," J. Bougard and N.H. Afgan, ed., Hemisphere Pub. Corp., New York (1987), p. 553.
5. S.P. Ashworth, C. Beduz, J. Mayne, A. Pasek, and R.G. Scurlock, Evaluation of a novel enhanced boiling surface in cryogenic fluids, in: "Advances in Cryogenic Engineering," vol. 35, Plenum Press, New York (1990), p. 429.
6. N.H. Afgan, L.A. Jovic, S.A. Kovalev, and V.A. Lenykor, Boiling heat transfer from surfaces with porous layers, *Int. J. Heat & Mass Trans.* 28:415 (1985).
7. H. Ogata and W. Nakayama, Heat transfer to boiling helium from machined and chemically treated copper surfaces, in: "Advances in Cryogenic Engineering," vol. 27, Plenum Press, New York (1982), p. 309.
8. M.A. Baig, "Thermal Cool-down Analysis of a Superconducting System With Microgrooves," *M.S. thesis,* Louisiana Tech University, Ruston, LA (1994).
9. F. Kreith and M.S. Bohn, "Principles of Heat Transfer," 5th ed., West Pub. Co., St. Paul, MN (1993), p. 120.
10. T.H.K. Frederking and J.A. Clark, Natural convection film boiling on a sphere, in: "Advances in Cryogenic Engineering," vol. 8, Plenum Press, New York, (1963), p. 501.
11. H. Schlichting, "Boundary Layer Theory," 6th ed., McGraw-Hill, New York, (1968), p. 612.

INCIPIENT BOILING SUPERHEATS AND CRITICAL HEAT FLUXES DUE TO INCREASING HEAT INPUTS IN SUBCOOLED He I AT VARIOUS PRESSURES

A. Sakurai, M. Shiotsu, and K. Hata

Institute of Atomic Energy, Kyoto University
Uji, Kyoto 611, Japan

ABSTRACT

Dynamic boiling heat transfer processes on a 0.2 mm diam. horizontal wire due to exponentially increasing heat inputs, $Q_0e^{t/\tau}$, with the periods, τ, from 0.2 ms to 25 s in subcooled He I at pressures of atmospheric, 142 kPa and 196 kPa were investigated to clarify the effect of liquid subcooling on the incipient boiling superheat and critical heat flux. The steady-state critical heat fluxes for the subcoolings ranging from 0 K to around 2.5 K at pressures of atmospheric, 142 kPa and 196 kPa were obtained by the heat input with a period of around 20 s; almost data obtained did not agree with the values derived from the existing correlations based on hydrodynamic instability model. An empirical correlation was given. The liquid temperature close to the test wire surface, T_{lw}, is given as the difference between the wire surface temperature and the temperature corresponding to the Kapitza resistance: the Kapitza conductance was estimated based on that experimentally obtained for the same test wire in He II. It was confirmed that the boiling incipience caused by a heat input over the range from quasi-steadily increasing one to rapidly increasing one occurs due to the heterogeneous spontaneous nucleation, HSN, at the T_{lw} lower than the corresponding theoretical value of homogeneous spontaneous nucleation temperature. The correlations for critical heat flux at which the transition to film boiling occurs due to the HSN in originally flooded cavities on the wire surface were given for the wide ranges of exponential periods, τ, pressures and subcoolings.

INTRODUCTION

Understanding and prediction of dynamic boiling heat transfer processes including boiling incipience and transition from non-boiling regime to film boiling due to the increasing heat inputs with a wide range of increasing rate in saturated and subcooled He I are important not only as the database for the stable and safe cooling design of large scale superconducting system but also as the general understanding for the boiling phenomena in all liquids; the mechanism of the phenomena for liquid He I has never been resolved in general.

The most important problem for the boiling incipience phenomenon on the solid surface in cryogenic liquids such as liquid He I and liquid nitrogen is that there are no active cavities entraining vapor and gases on a solid surface in the liquids for boiling incipience. Flint et al.[1] reported that they have found evidence that the superheat at boiling onset near a heater surface in He I is limited by the physics of homogeneous nucleation; the superheat measured agrees with the corresponding value derived from homogeneous spontaneous nucleation theory.

Recently, Sakurai et al.[2] have observed the boiling incipience on a horizontal test wire surface in saturated He I due to a quasi-steadily increasing heat input at the liquid temperature near the wire surface lower than the corresponding homogeneous spontaneous nucleation temperature at pressures from 5.465 kPa to 101.3 kPa. The liquid temperature was given as the difference between wire surface temperature and the temperature corresponding to the Kapitza resistance: the Kapitza conductance in He I was estimated based on the Kapitza conductance measured on the same test wire in subcooled He II for various liquid temperatures at atmospheric pressure. They assumed that the boiling incipience occurs due to the heterogeneous spontaneous nucleation, HSN, in originally flooded cavities on the wire surface.

Before the He I experiment, the experiments for transient heat transfer processes including boiling incipience and transition to film boiling due to exponential heat inputs in liquid nitrogen, ethanol, and water (which is pre-pressurized for a while by applying a high pressure before each experimental run) had been performed by Sakurai et al.[3] to clarify the boiling incipience and direct transition to film boiling due to the HSN in originally flooded cavities on a solid surface in the liquid: there is no contribution of active cavities in the liquids similar to that in liquid He I.

By using the theoretical model suggested by Kottowsky[4] for the initial boiling in liquid sodium, they presented the correlation which is able to evaluate the lower limit of HSN temperature (namely, steady-state HSN temperature in the liquids) due to a quasi-steadily increasing heat input: the model is based on the HSN in originally flooded conical cavities with an aperture angle on a solid surface which reduces the energy required for the formation of bubbles: two parameters including the aperture angle and liquid-solid contact angle in the theoretical model were determined by using the lower limit values of HSN temperatures experimentally obtained at various saturated pressures for liquid nitrogen, liquid helium, ethanol, and water pre-pressurized by a high pressure[3].

The main purposes of the present work are first to measure the steady-state critical heat fluxes for the wide ranges of subcoolings and pressures in He I, and to obtain a new correlation for the data obtained which is different from that based on hydrodynamic instability model, and second to clarify the effect of subcooling on the boiling incipience and the transitions from non-boiling regime to film boiling due to the heat inputs with various increasing rates from quasi-steadily to rapidly increasing ones for the subcoolings at the pressures.

EXPERIMENTAL APPARATUS AND METHOD

A vacuum insulated, liquid nitrogen shielded cylindrical stainless steel cryostat, 50 cm in inner diameter and 128 cm in height with view ports was used as a test vessel. The test wire heater is supported horizontally in the liquid with two helium gas cooled electrodes. The potential taps are located at each end of the heater. The pressure in the vessel is automatically controlled within $\pm$ 70 Pa of a specified value by a pressure control system consisting of a PID controller and a control valve actuated by a stepper-motor. A superconducting liquid level meter, with a 1 mm resolution, is installed in the vessel.

The wire test heater is 0.2 mm in diameter, D, and 64.3 mm in length, made of Pt-Co(0.5 Atomic %) alloy. The gradient of the electrical resistivity versus temperature curve for this alloy is positive for the entire experimental range. The electrical resistance versus temperature relation of the test heater was calibrated in the temperature-controlled vessel in liquid helium: the temperature control accuracy is 0.05 K for a temperature setting below 10 K, 0.1 K from 10 to 50 K, and 0.3 K up to 300 K.

The average temperature of the test wire was measured by resistance thermometry using a double bridge circuit which included the wire as a branch of the bridge. The output voltages of the bridge circuit together with the liquid temperature signal from the germanium diode in the vessel were amplified and recorded in a 4ch digital memory system. These voltages are simultaneously sampled at the shortest available time interval of 0.5 μs and stored in a 64 k word/channel memory. The stored data were then transferred to a digital computer. The average temperature of the wire, $T_a(t)$, was calculated from its electrical resistance with the aid of the previously calibrated temperature resistance relation. The heat generation rate in the wire, $Q(t)$, was determined from the voltage drops across the potential taps of the wire and across the standard resistance. The heat flux from the wire surface to the liquid, $q(t)$, is the difference between the heat generation rate per unit surface area, $DQ(t)/4$, and the rate of change of energy storage in the wire, $\rho_h c_h (D/4)(dT_a(t)/dt)$, obtained from the faired average temperature versus time curve. Temperature of the wire surface was calculated from the one-dimensional transient conduction equation for the wire having a time dependent heat generation rate $Q(t)$, with the heat flux $q(t)$ set as a boundary condition at the surface. The surface temperature was obtained numerically utilizing the finite difference method. However, even considering the experimental error, the difference between the average and the surface temperature was negligibly small for the wire tested. Experimental error is estimated to be 0.1 K in the surface temperature of the wire and 2 percent in the heat flux.

The heat fluxes, q, due to an exponential heat input are plotted against the surface liquid superheat, $\Delta T_{sat,l} (= T_{lw} - T_{sat})$, to make clear the dynamic boiling phenomenon in He I. The surface liquid temperature, T_{lw}, in liquid He I is evaluated by the following equation.

$$T_{lw} = T_w - (h_k)^{-1}_{HeI} q \tag{1}$$

where T_w is the surface temperature of the test wire, and $(h_k)^{-1}_{HeI} q$ is the temperature drop due to the Kapitza resistance, $(h_k)^{-1}_{HeI}$, for liquid He I. The Kapitza conductance for the test wire in He II was measured for the bulk liquid temperatures of 1.8, 2.0 and 2.1 K at atmospheric pressure. Following empirical equation for the Kapitza conductance on the test heater was obtained.

$$h_k = 2.0 \times 10^3 T_b^3 \tag{2}$$

By substituting T_b in Eq. (2) for T_{lw}, the Kapitza conductance in He I close the wire surface is given by,

$$(h_k)_{HeI} = 2 \times 10^3 T_{lw}^3 \tag{3}$$

The temperature drop, $\Delta T_k = (h_k)^{-1}_{HeI} q$, due to the Kapitza resistance at each heat flux was determined by an iterative calculation. Details appear elsewhere[5].

EXPERIMENTAL RESULTS AND DISCUSSION

Experimental Conditions

The steady-state critical heat fluxes for the subcoolings from zero to around 2.5 K at the pressures of atmospheric, 142 kPa and 196 kPa were measured on a 0.2 mm diameter horizontal wire. The dynamic boiling heat transfer from the wire surface due to an exponential heat input, $Q_0 e^{t/\tau}$, was measured for the period, τ, from 0.2 ms to 25 s, for the subcoolings up to around 2.5 K at the pressures from atmospheric to 196 kPa.

Steady-state Critical Heat Flux

The steady-state critical heat fluxes, $q_{cst,sub}$ for the subcooling, ΔT_{sub}, from zero up to around 2.5 K at pressures of 101.3, 142 and 196 kPa are shown on the graph of $q_{cst,sub}$ versus ΔT_{sub} in Fig. 1 with pressure as a parameter.

The q_{cr} values measured at saturated pressures from 5.465 to 142 kPa except the pressure of 196 kPa agree well with the values derived from the Kutateladze's correlation[6] with the coefficient of 0.18

$$q_{cst,sat} = 0.18 L \rho_v [\sigma g (\rho_l - \rho_v)/\rho_v^2]^{1/4} \tag{4}$$

Most of the data are within ±10% of the predicted values. This means that the mechanism of heat transfer crisis at $q_{cst,sat}$ under saturated condition is explained by

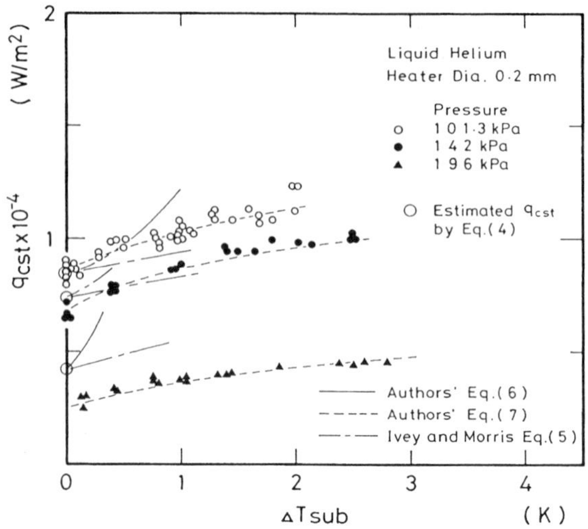

Fig. 1 Steady-state critical heat flux versus subcooling with pressure as a parameter.

the hydrodynamic instability model though there exist several different models based on hydrodynamic instability expressed by the correlation just same as the Kutateladze's one. However, the predicted value for a higher pressure becomes higher than the experimental data with the increase in pressure: it is about 65 % higher than the experimental data at 196 kPa.

On the other hand, there exist several $q_{cst,sub}$ correlations. The curves obtained from $q_{cst,sub}$ correlations given by Ivey and Morris[7],

$$q_{cst,sub} = q_{cst,sat}[1 + 0.1(\rho_v/\rho_l)^{1/4}(\rho_l c_{pl}\Delta T_{sub})/(L\rho_v)] \quad (5)$$

and that by Sakurai et al.[8]

$$q_{cst,sub} = q_{cst,sat}[1 + 0.87(\rho_l/\rho_v)^{0.69}(c_{pl}\Delta T_{sub}/L)^{1.5}] \quad (6)$$

with the $q_{cst,sat}$ values given by Eq.(1) are shown in the figure for comparison: these correlations were obtained by modifying the Kutateladze's one based on hydrodynamic instability. The values obtained from the latter correlation were found to be in good agreement with the experimental data on a horizontal cylinder in water for the subcoolings up to around 50 K at high pressures up to about 2 MPa as compared with the values obtained from the former correlation. However, both curves do not represent the experimental data in He I as shown in Fig. 1. This means that the the mechanism of heat transfer crisis at the $q_{cst,sub}$ is different from that due to the hydrodynamic instability suggested by Kutateladze[6] and Zuber[9]. The values of $q_{cst,sub}$ which are different from those derived from the correlation based on hydrodynamic instability were observed in water for high subcoolings at high pressures by Sakurai et al.[8] recently. They found that the heat transfer crisis at the critical heat flux for higher subcoolings at higher pressure would occur on fully developed nucleate boiling due to the explosive like HSN in originally flooded cavities: the heat flux is lower than that expected by the correlation based on the hydrodynamic model. It can be assumed that the heat transfer crisis at the $q_{cst,sub}$ would occur in subcooled He I due to the HSN in originally flooded cavities, though more study is required. Following empirical correlation for $q_{cst,sub}$ for high subcoolings at high pressures was given.

$$q_{cst,sub} = q^*_{cst,sat}[1 + 0.22(\rho_l/\rho_v)^{0.48}(c_{pl}\Delta T_{sub}/L)^{0.65}] \quad (7)$$

Best fitted values of $q^*_{cst,sat}$ for pressures of 101.3, 142, and 196 kPa are 8.4×10^3, 6.8×10^3, and 2.6×10^3 W/m^2, respectively. Although the values of $q^*_{cst,sat}$ are supposed as the saturated critical heat fluxes due to the HSN for the pressures, they are approximately the same as the corresponding experimental data seen in the figure.

Typical Transitions to Film Boiling due to Exponential Heat inputs

For Saturated Pressures of 5.465 and 101.3 kPa. Typical data of incipient boiling heat flux q_i and critical heat flux q_{cr} obtained for the periods τ, ranging from about 0.2 ms to 10 s at pressures of 5.465 and 101.3 kPa are shown in Fig. 2. The q_{cr} value for the corresponding value of τ increases, then decreases and again increases with the decrease in the period. They are clearly classified into three groups for the periods: first group of q_{cr} are those for the periods longer than about 150 ms and 500 ms, second ones are those for the periods shorter than about 4 ms and 2 ms, and third ones are those for the periods between 150 ms and 4 ms, and between about 500 ms and 2 ms, respectively.

Typical heat transfer processes for the periods of 10 s, 21.6 ms and 0.22 ms at the pressure of 5.465 kPa are shown on the graph of q versus $\Delta T_{sat.l}$ with the incipient boiling points and the points of transition to film boiling for all periods in Fig. 3 as typical ones. The q_{cr} values for the periods belong to the first, second, and third groups. The heat transfer process for $\tau = 10$ s is as follows. The initial boiling occurs at point A on the natural convection curve shown in the graph due to the heterogeneous spontaneous nucleation in originally flooded cavities as already assumed by Sakurai et al.[2]. After that the heat flux rapidly changes with the decrease of $\Delta T_{sat,l}$ through the process ABC to fully developed nucleate boiling FDNB and then heat transfer crisis occurs at the q_{cr} of point C. The decrease of $\Delta T_{sat,l}$ through the process AB is due to the activation of flooded cavities on the wire surface, and the FDNB curve CD is obtained by the gradual decrease of the heat input from the heat flux slightly lower than the q_{cr}. The q_{cr} for the first group occurs on the extension of FDNB curve as seen in the figure. The gradual increase of q_{cr} for the τ is roughly explained as due to the time lag of hydrodynamic instability which starts at steady state q_{cr}. The heat transfer process for the period of 0.22 ms which corresponds to the second group is as follows. The initial boiling occurs at point A' on the corresponding conduction curve due to the heterogeneous spontaneous nucleation, HSN, in originally flooded cavities. After that the heat flux rapidly increases with slight increase in $\Delta T_{sat,l}$ within around 0.05 K up to the point B' at which transition to film boiling occurs. The $\Delta T_{sat,l}$ at point B' is slightly lower than the homogeneous spontaneous nucleation superheat,

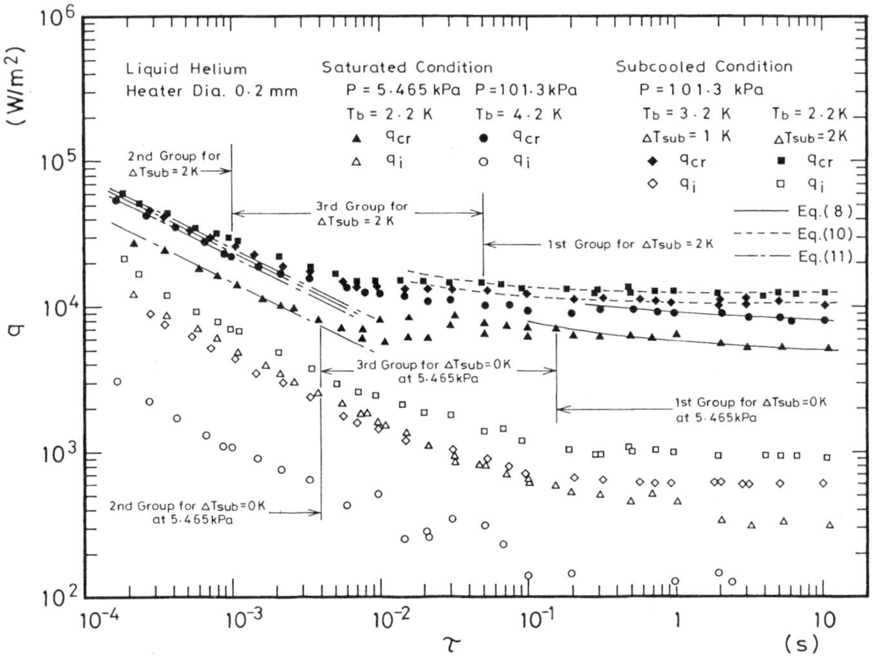

Fig. 2 q_i, and q_{cr} versus period under saturated and subcooled conditions at pressures.

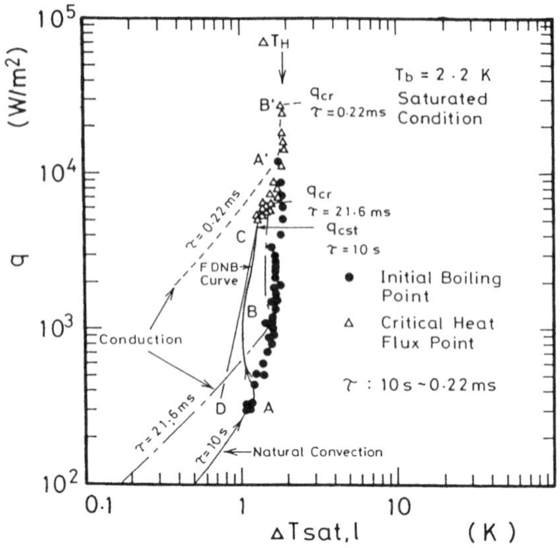

Fig. 3 Typical heat transfer processes under saturated condition at $T_B = 2.2$ K for the periods of 0.22 ms, 21.6 ms and 10 s together with the initial boiling and critical heat flux points for periods ranging from 10 s down to 0.22 ms.

ΔT_H, given by the theory. The transitions at A' and B' are called as the first and second kind ones. The first kind transition occurs for a short period at the q_i at which initial boiling occurs due to the HSN. The process including both transitions is called as semi-direct transition by the authors[3]. The semi-direct transition for a short period occurs at the pressures higher than around 1 MPa in liquid nitrogen though the direct transition occurs at lower pressures. The direct transition was not observed in He I for the pressures and periods tested here. The q_{cr} values belonging to the first and second groups at the pressures are expressed by the following correlations.

$$q_{cr} = q_{cst,sat}(1 + 0.21\tau^{-0.5}) \tag{8}$$

$$q_{cr} = [950\{(P/P_{cr})^{0.19} - 1.2(P/P_{cr})^3\}]\tau^{-0.5} \tag{9}$$

where P and P_{cr} are the system pressure and critical pressure. The curves obtained are shown in Fig. 2. The q_{cr} values of third group gradually decrease with the decrease in period from the maximum q_{cr} of the first group down to the minimum q_{cr} of second group. It is assumed that the heat transfer crisis at the q_{cr} in the third group occurs due to the HSN in originally flooded cavities on the test wire surface in the FDNB or insufficiently developed nucleate boiling at near the lower limit of HSN $\Delta T_{sat,l}$ because the activations of originally flooded cavities by the bubbles on the wire surface due to the HSN decrease with the decrease in the period for the increase in the increasing rate of $\Delta T_{sat,l}$.

For Subcoolings at Various Pressures. Typical data of q_i and q_{cr} versus τ for the subcoolings up to around 2.5 K at the pressures of 101.3 kPa and 196 kPa are shown for the periods ranging from 0.22 ms to 10 s in Figs. 2 and 4(a). The q_{cr} values normalized by the corresponding steady-state value $q_{cst,sub}$ are also shown in Fig. 4(b). The q_{cr} values for the periods for the subcooling very gradually increase with the decrease in periods up to around 120 % value of the steady-state one at the period of about 30 ms. As shown in Fig. 2, the increasing rate of q_{cr} for τ for saturated condition at atmospheric pressure is higher than that for the subcoolings of 1 and 2 K at the pressure. As mentioned before, it was assumed that the steady-state q_{cr} for the subcooling occurs due to the HSN in originally flooded cavities: the heat transfer crisis occurs at the critical heat flux lower than the value derived from the correlation based on hydrodynamic instability model. On the other hand, the HSN surface liquid superheat, $\Delta T_{sat,l}$ slightly increase with the decrease in period for the first group range; this would cause the increase of q_{cr} with the decrease in periods. The mechanisms of heat transfer crisis at the q_{cr} values for periods in the second and third group are the

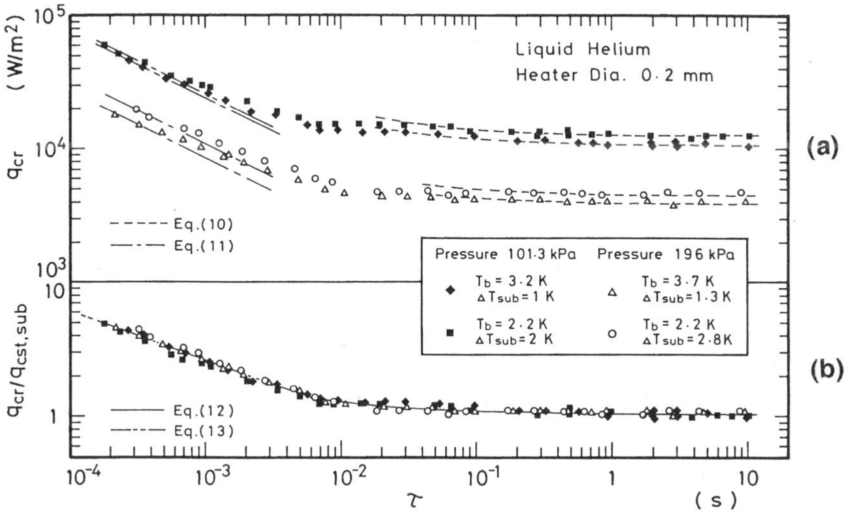

Fig. 4 (a) q_{cr} versus τ, and (b) $q_{cr}/q_{cst,sub}$ versus τ for various subcoolings at 101.3 and 196 kPa.

same as those for the saturation condition mentioned before. The q_{cr} values for periods in the first and 2nd groups are expressed by the following correlations.

$$q_{cr} = q_{cst,sub}[1 + 2.3 \times 10^{-2}\tau^{-0.7}] \tag{10}$$

$$q_{cr} = [950\{(P/P_{cr})^{0.19} - 1.2(P/P_{cr})^3\} + 45\Delta T_{sub}]\tau^{-0.5} \tag{11}$$

The $q_{cr}/q_{cst,sub}$ values are shown versus periods in Fig. 4(b). They are almost in agreement with each other independent of subcooling and pressure and approximately expressed by the two curves which are represented by the following correlations.

$$q_{cr}/q_{cst,sub} = [1 + 1.6 \times 10^{-2}\tau^{-0.6}] \tag{12}$$

$$q_{cr}/q_{cst,sub} = 0.22\tau^{-0.36} \tag{13}$$

The q_{cr} values for the third group would be included in the first group; the heat transfer crisis at the q_{cr} occurs due to the HSN in the FDNB or insufficiently developed nucleate boiling; the q_{cr} becomes lower than that estimated from the correlation based on the hydrodynamic instability model.

The typical heat transfer processes for the periods of 10.4 s and 4.92 ms for the subcooling of 2.0 K at atmospheric pressure are shown on the graph of q versus $\Delta T_{sat,l}$ with the incipient boiling points and the points of transitions to film boiling for the periods from 3.37 ms to 10 s in fig. 5. As seen in the figure, the $\Delta T_{sat,l}$ values at which initial boiling and transitions to film boiling occur for the periods from 3.37 ms to 10 s are lower than the theoretical liquid superheat of the homogeneous spontaneous nucleation, ΔT_H. The initial boiling for the period of 10.4 s occurs at the point A on natural convection regime and then the heat flux gradually increases up to the q_{cst} through FDNB regime. The $\Delta T_{sat,l}$ values corresponding to the heat transfer process near the q_{cst} are very close to the $\Delta T_{sat,l}$ values corresponding to the initial boiling points for the periods. It can be assumed that the heat transfer crisis at the q_{cst} occurs due to the HSN; the q_{cst} value is lower than that derived from the authors' correlation based on the hydrodynamic instability model as shown in Fig. 1. The initial boiling for the period of 4.92 ms occurs at point A' on conduction regime, and then the heat flux rapidly increases up to the heat flux at point B' at which the transition to film boiling occurs: those at the points of A' and B' correspond to the first and second kinds of semi-direct transitions to film boiling.

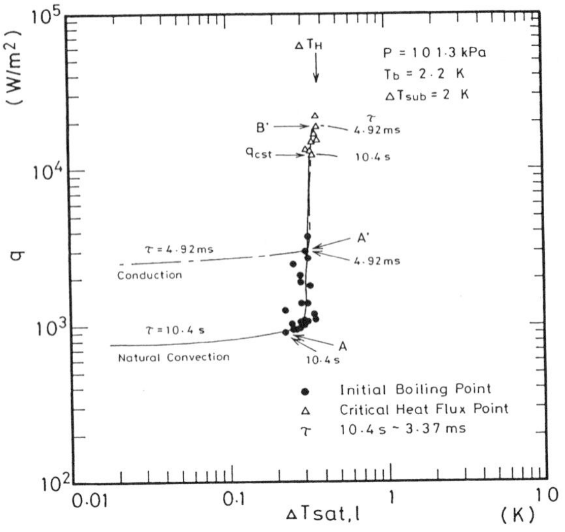

Fig. 5 Typical heat transfer processes for $\Delta T_{sub} = 2$ K at 101.3 kPa for the periods of 4.92 ms and 10.4 s together with the initial boiling and critical heat flux points for periods ranging from 10.4 s down to 3.37 ms.

CONCLUSIONS

1. It was confirmed that the boiling incipience on a wire surface caused by a heat input over the range from quasi-steadily increasing one to rapidly increasing one occurs at the surface liquid superheat lower than the corresponding theoretical value of homogeneous spontaneous nucleation not only for the saturated condition but also for subcooled condition.

2. The q_{cr} values for the subcoolings up to about 2.5 K at the pressures of 101.3, 142 and 196 kPa did not agree with the corresponding values derived from the existing correlation based on hydrodynamic instability model. It was assumed that the heat transfer crisis at the q_{cr} occurs due to the HSN. The empirical correlation was given.

3. The $q_{cr}/q_{cst,sub}$ values for the wide ranges of period, subcooling, and pressure were expressed by two curves on log $(q_{cr}/q_{cst,sub})$ versus log τ graph: the period at the intersection of the curves is around 7 ms. The correlations representing the curves were given.

REFERENCES

1. Flint, E., Van Cleve, J., Jenkins, L., and Guernsey, R., " Advances in Cryogenic Engineering," Vol.27, Plenum Press, New York, (1982), p.283.
2. Sakurai, A., Shiotsu, M., and Hata, K., " Advances in Cryogenic Engineering," Vol.39, Plenum Press, New York, (1990), p.1759.
3. Sakurai, A., Shiotsu, M., and Hata, K., in Instability in Two-Phase Flow Systems, J.H. Kim ed., ASME Publication, HTD-Vol.260 (1993) p.27.
4. Kottowsky, H.M., 1973, in "Progress in Heat and Mass Transfer," O.E. Dwyer ed., Vol.7, Pergamon Press, Oxford (1973) p.299.
5. Shiotsu, M., Hata, K., Takeuchi, Y., Hama, K., and Sakurai, A., Estimation of Kapitza Conductance Effect on Steady and Transient Boiling Heat Transfer in He I based on Kapitza Conductance Results in He II, in "Proc. of Internat. Symp. on Safety of Superconductors and Related Heat Transfer at Low Temperatures", August 1994, Fukuoka, Japan Paper No. 1A-3 (1994); to appear in *Cryogenics.*
6. Kutateladze, S.S, AEC-tr-3770, USAEC (1959).
7. Ivey, H.J. and Morris, D.J., UKAEA, AEEW-R 137 (1962).
8. Sakurai, A., Shiotsu, M., and Hata, K., Transition Phenomena from Non-Boiling Regime to Film Boiling on a Cylinder Surface in Highly Subcooled and Pressurized Water due to Increasing Heat Inputs, to appear in Phase Change Heat Transfer, ASME Internat. Mech. Eng. Congress and Expo., 1995.
9. Zuber, N., AECU-4439, USAEC (1959).

HEAT TRANSFER FROM ALUMINUM SURFACES TO POOL BOILING He I

Y. Huang and S. W. Van Sciver

National High Magnetic Field Laboratory
Florida State University
Tallahassee, FL 32306, USA

ABSTRACT

Heat transfer measurements between different aluminum surfaces and pool boiling He I at 4.21 K are reported. The samples are contained in a channel configuration similar to what might be found in a large bath cooled superconducting magnet. Results include heat transfer coefficients in the nucleate and film boiling regimes as well as values for the peak nucleate and minimum film boiling heat flux. The effects of sample orientation and surface condition are also studied.

INTRODUCTION

High purity aluminum has been considered as the stabilizer of the superconductor for large size superconducting magnet, such as the 30 MW SMES program for utility applications, to take advantage of its lower magnetoresistivity and lower density.[1] The proposed superconducting magnet will be operated in the pool boiling mode with liquid helium at 4.35 K, at a pressure slightly above atmospheric pressure. Although it is simple and easy to cool superconducting magnets this way, the superconducting magnets must be designed to be fully cryostable. This requirement means the Joule heating of stabilizer has to be transferred from its surface to the liquid helium if a portion of the conductor becomes normal. In order to design a cryostablized superconducting magnet, reliable boiling heat transfer data are needed.

Lyon[2] studied the orientation dependence of peak nucleate boiling heat fluxes in saturated liquid helium between a 9.91 mm diameter flat polished platinum surface and liquid helium and found out the effect of surface orientation on peak heat flux values is significant. Chen and Van Sciver[3] reported the experimental results on the angular variation of channel heat transfer in helium at 4.2 K and pointed out that the peak heat flux strongly depends on the channel width. They also noticed that for wide channels the peak heat flux should correlate with the open bath pool boiling results by Lyon.[2]

In this study, we conducted the boiling heat transfer study from four different aluminum surfaces in liquid helium at 4.21 K. The samples are contained in a channel configuration similar to that of a large bath cooled superconducting magnet. Results include heat transfer measurements in both the nucleate and film boiling regimes as well as values for the peak nucleate and minimum film boiling heat flux. The effects of sample surface orientation and condition on peak nucleate and minimum film boiling heat flux are also studied.

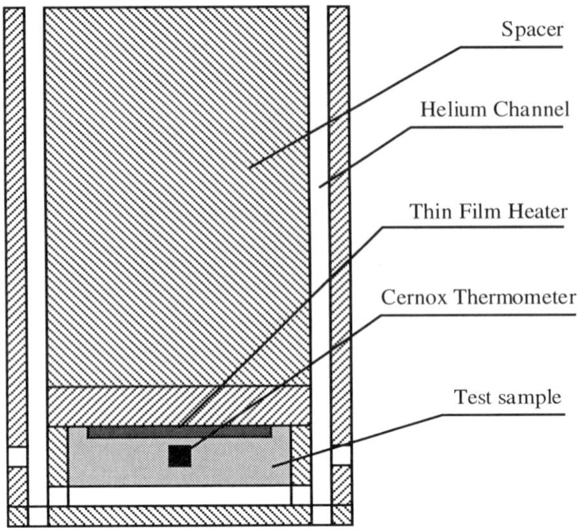

Figure 1. Experimental test element assembly.

EXPERIMENTAL APPARATUS

The test element for the heat transfer study is shown in Figure 1. The samples are contained in a channel configuration similar to that of a large bath cooled superconducting magnet. The test sample was placed in a holder made of G-10 material. After proper assembly, the gap between the heating surface and bottom cover is about 3 mm. The width of helium flow channel is 5 mm. Four small slots at the bottom cover allow the bath helium to flow in and replace the helium leaving heating surface during the experiment. There is also one 3×40 mm slot on each side cover of helium flow channel. The test sample is replaceable after each run.

The test sample and experiments run are summarized in Table 1. Sample #1 is made of high purity aluminum with one surface textured and has a nominal dimension of $11 \times 58 \times 76$ mm. The textured surface consists of the right square pyramid with the base size of 1.5×1.5 mm and 0.7 mm high. The nominal surface area of sample #1 is 58×76 mm. Sample #2 is also made of high purity aluminum but the surface is untreated smooth surface. The surface area of sample #2 in contact with liquid helium is also 58×76 mm as a comparison.

For samples #1 and #2, the temperature sensors are (Lake Shore) Cernox resistors which are sensitive to temperature up to about 40 K. The heaters are MINCO thin film type with 55 Ohms at helium temperature and can provide a maximum of 60 W steady state heat. The thermometer was first indium soldered into the slot at the center of the test sample and then covered with STYCAST to fill the space between them. The heater was then mounted with epoxy on the back of the test sample. After the sample element was properly assembled, only one surface was exposed to the liquid helium. Samples #1 and #2 are similar with assembly as shown in Figure 2.

Table 1. Test samples summary

No.	Sample surface	Surface orientation (degrees)
1.	Textured surface	0, 5, 10, 15, 30, 45, 60
2.	Smooth surface	0, 5, 10, 15, and 180
3.	Aluminum alloy jacket with internal helium cooling channels	0
4.	Aluminum alloy jacket only	0

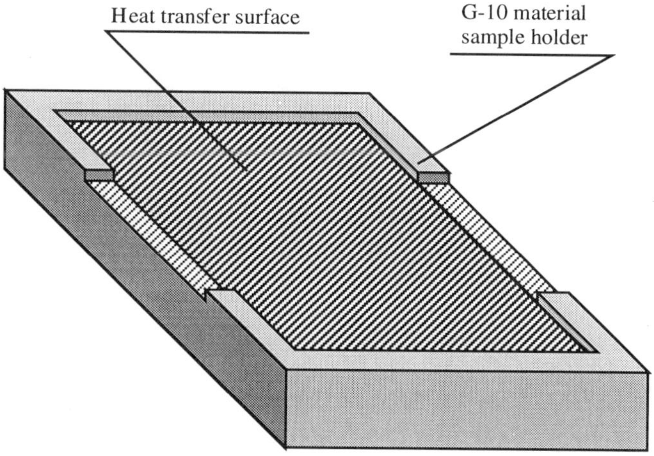

Figure 2 Test sample element after assembly of samples #1 and #2.

Sample #3 consists of high purity aluminum covered with aluminum alloy jacket and with internal helium cooling channels. It has a dimension of 18×67×80 mm. The sample has about 50% of its high purity aluminum surface area exposed to liquid helium. Sample #4 is the high purity aluminum also covered with aluminum alloy but without internal cooling channels. It has the same dimension as that of sample #3.

For test samples #3 and #4, four spacers are used and four surfaces are exposed to liquid helium. The test elements are similar and are shown in Figure 3. The gap on the top and bottom surfaces is about 3 mm and 5 mm for the side surfaces. The heaters are cartridge type with two heaters mounted in each sample. The gap between the heater and sample is filled with the STYCAST. The temperature sensor is a calibrated Carbon Glass thermometer. The thermometer was mounted into a 5 mm O.D. copper tube with STYCAST and with its end open. The thermometer was fit into the hole in the sample with CRYOCON® grease at the bottom and vacuum grease to fill the gap.

EXPERIMENTAL RESULTS

The measured experimental data in this study include the bath pressure of liquid helium, total heat applied to the sample and the temperature of the sample. The bath

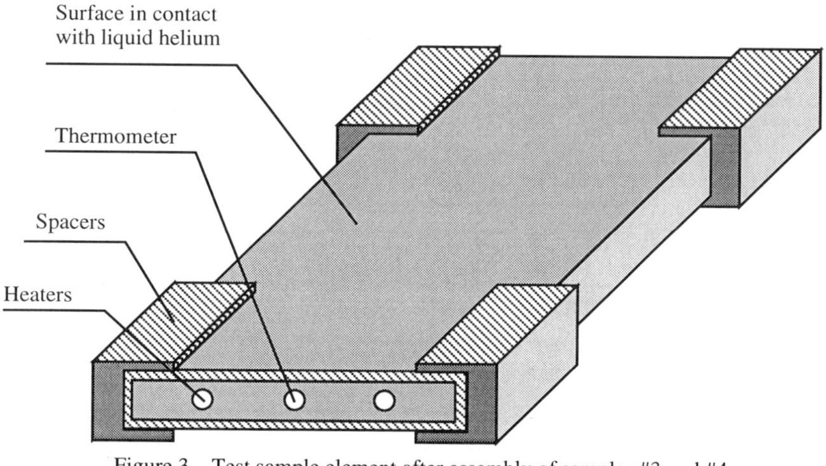

Figure 3. Test sample element after assembly of samples #3 and #4.

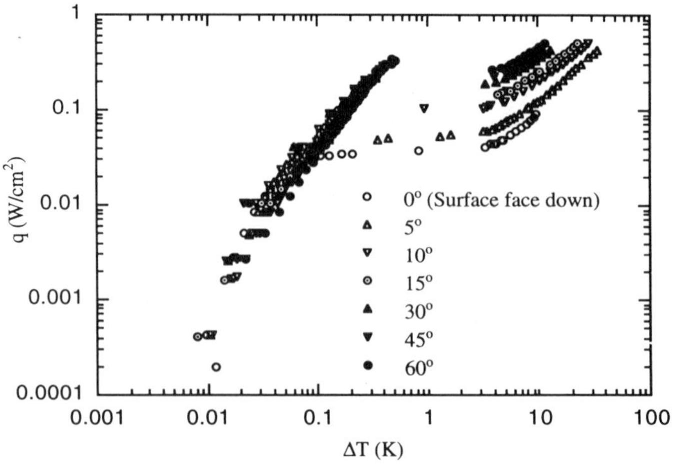

Figure 4. Heat flux vs. temperature difference for high purity aluminum with textured surface, sample #1, in liquid helium at 4.21 K with different inclination angles.

pressure is kept nearly constant during experiment so the bath temperature is constant at 4.21 K. The helium level is about one meter above the sample at the start of the experiment and will drop to 0.4m above the sample after the experiment. The heat applied is calculated from the measured current and voltage across the heater using four wire technique. The temperature difference is calculated from the measured temperature of the sample subtracted from the sample temperature before any heat was applied. The temperature variation before and after experiment is less than 5 mK and mainly caused by the helium level change during the experiment. The heat was applied by adjusting a constant current mode in steps of 10 mA at first and up to 50 mA at high heat flux. After the film boiling region, the heat is stepwise reduced to zero.

The boiling heat transfer measurements of the high purity aluminum sample with textured surface in liquid helium at 4.21 K for different inclination angles are shown in Figure 4. The experimental data show both nucleate boiling and film boiling regions. The transition occurs when the temperature difference is between 0.1 K and 0.5 K at different heat flux values depending on the heating surface inclination angles. The face down orientation gave the lowest heat transfer coefficient and transition from nucleate to film boiling starts at low heat flux and small temperature difference values. At small inclination angle, no hysteresis was observed for the textured surface sample while at large inclination

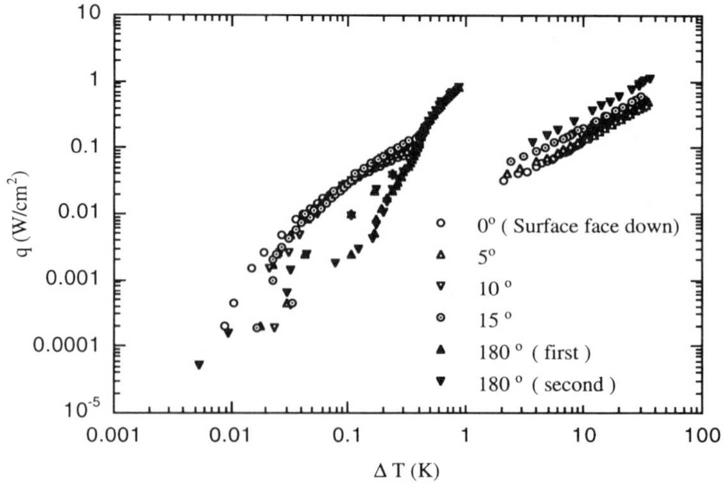

Figure 5. Heat flux vs. temperature difference for high purity aluminum with smooth surface, sample #2, in liquid helium at 4.21 K with different inclination angles.

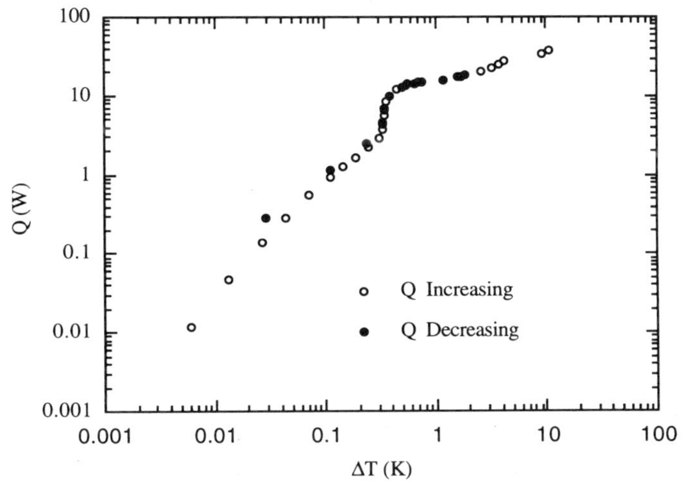

Figure 6. Heat flux vs. temperature difference for the sample with high strength aluminum jacket, sample #3, and internal helium cooling channels.

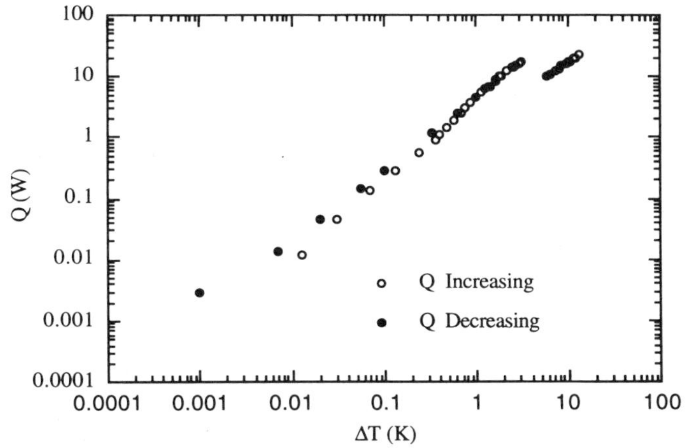

Figure 7. Heat flux vs. temperature difference for the sample high strength aluminum jacket, sample #4, without helium cooling channel.

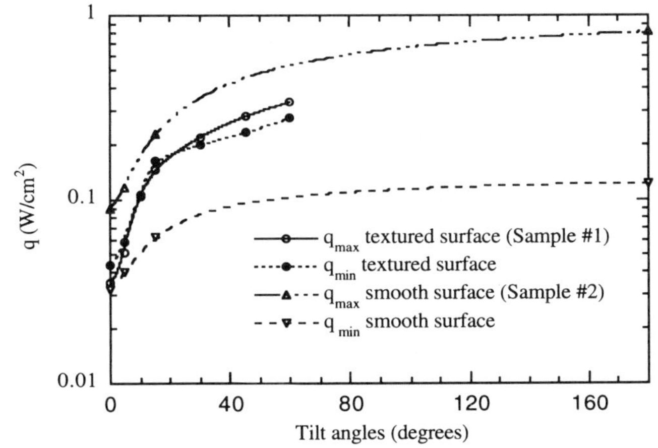

Figure 8. Peak nucleate boiling flux and minimum film boiling flux for the high purity aluminum samples in liquid helium vs. surface inclination.

angles, the hysteresis is significant. We are not sure if the hysteresis is an experimental artifact related or the surface orientation effects at this moment.

The boiling heat transfer measurements of the high purity aluminum smooth sample in liquid helium at 4.21 K with different inclination angles are shown in Figure 5. The transition from the nucleate to film boiling region starts at a temperature difference of 0.3 K to 0.9 K with varied heat flux values. With the face up surface orientation, the two experimental data sets match quite well in the nucleate boiling region for both heat power raising up and reducing down process. The results also show that the face down orientation is the worst position of heating surface from the heat transfer point of view.

Figure 6 shows the results of sample #3, which is the high purity aluminum covered with high strength aluminum jacket with internal helium cooling channels. Any heat applied to the sample will be transferred to liquid helium by two ways, one is through the high purity aluminum surface area in contact with helium by convection and the other through aluminum alloy jacket by conduction first and then to bath helium by convection. The configuration of the sample constructed is responsible for the Q vs. ΔT curve behavior like the one in Figure 6. At the beginning of the experiment, heat flux increases as the heat applied increases as well as the temperature of the sample. The heat transferred through aluminum alloy jacket is relatively small compared with that of high purity aluminum since the surface temperature difference between the aluminum alloy jacket and bath helium is quite small at low heat flux. When heat flux reached a certain point, nucleate boiling initiated on the surface of jacket due to relative high temperature difference. With the increase of heat flux further, film boiling initiated within the helium cooling channels and transition starts but not as sharp as that of the other samples due to the reason that heat can be conducted away through the aluminum alloy jacket to bath helium.

The experimental data of sample #4, which has a high strength aluminum jacket but is without internal helium cooling channels, is shown in Figure 7. The boiling heat transfer characteristics of this sample is similar to that of samples #1 and #2. The only difference is that sample #4 needs higher temperature difference to initiate the transition due to the low thermal conductivity of aluminum alloy. The hysteresis was observed for the sample #4 too. In order to have helium film boiling heat transfer start, a higher temperature of sample is needed even though the peak nucleate heat flux almost the same compared with other samples.

Figure 8 shows the peak nucleate boiling flux and minimum film boiling flux as a function of surface orientation with respect to face down. For the smooth surface sample, hysteresis occurred for all inclination angles. For the textured surface sample, however, the hysteresis happened only when the inclination angle is larger than 30 degrees.

CONCLUSIONS

Heat transfer measurements between four aluminum surfaces and pool boiling helium at 4.21 K are reported. The results show that the surface conditions will significantly affect the boiling heat transfer characteristics. The sample with aluminum alloy jacket shows that transition starts at larger temperature difference due to its lower thermal conductivity value. The surface inclination has a significant impact on its heat transfer behavior. The face down orientation gave the lowest peak nucleate heat flux for both samples #1 and #2. The smooth surface sample data show the hysteresis no matter what the inclination angles are while the textured surface shows the hysteresis only when the inclination angles are larger than 30 degree. The experimental data are repeatable and depends on the heat applied history.

REFERENCES

1. X. Huang, et al. 30 MW Babcock and Wilcox SMES Program for Utility Applications, IEEE Transaction on Applied Superconductivity, 5:428 (1995)
2. D. N. Lyon, Boiling heat transfer and peak nucleate boiling fluxes in saturated liquid helium between the λ and critical temperatures, Adv. Cryog. Eng. 10:371 (1965)
3. Z. Chen and S. W. Van Sciver, Channel heat transfer in He I----- Steady state orientation dependence, Adv. Cryog. Eng. 31:431 (1986)

HEAT TRANSFER FROM AN OXIDIZED LARGE COPPER SURFACE TO LIQUID HELIUM: DEPENDENCE ON SURFACE ORIENTATION AND TREATMENT

A. Iwamoto, T. Mito, K. Takahata, N. Yanagi, and J. Yamamoto

National Institute for Fusion Science
322-6 Oroshi, Toki, Gifu 509-52, Japan

ABSTRACT

Heat transfer of large copper plates (18 x 76 mm) in liquid helium has been measured as a function of orientation and treatment of the heat transfer surface. The results relate to applications of large scale superconductors. In order to clarify the influence of the area where the surface treatment peels off, we studied five types of heat transfer surface areas including: (a) 100% polished copper sample, (b) and (c) two 50% oxidized copper samples having different patterns of oxidation, (d) 75% oxidized copper sample, (e) 90% oxidized copper sample, and (f) 100% oxidized copper sample. We observed that the critical heat flux depends on the heat transfer surface orientation. The critical heat flux is a maximum at angles of 0° – 30° and decreases monotonically with increasing angles above 30°, where the angle is taken in reference to the horizontal axis. On the other hand, the minimum heat flux is less dependent on the surface orientation. More than 75% oxidation on the surface makes the critical heat flux increase. The minimum heat fluxes of the 50 and 90% oxidized Cu samples approximately agree with that of the 100% oxidized Cu sample. Experiments and calculations show that the critical and the minimum heat fluxes are a bilinear function of the fraction of oxidized surface area.

INTRODUCTION

Large scale superconducting magnets for fusion applications are planned and under construction. An example is the Large Helical Device (LHD) at NIFS[1,2]. For the cryogenic stabilization of pool-cooled superconducting magnets, heat transfer between the surface of a superconductor and liquid helium is very important. The cooling channel and the surface orientation, area, and treatment affect the heat transfer characteristics[3-8]. In the magnets, the surface orientation of the superconductor varies according to the winding position[1,2] and a surface treatment is applied to improve the heat transfer characteristics. On the other hand, since the conductor is channel-cooled, the stability of the superconducting magnet is degraded. Many studies measured the heat transfer from samples with small surface areas[3,4,8], for example, from wires[4,5]; however, the measured values cannot be used for stability analysis of large scale superconductors with large heat transfer surfaces. We have already reported the dependence of the heat transfer from a polished Cu surface on surface orientation and channel width[6], for comparison with the data of Lyon[8]. We found that the critical heat flux decreases for samples with larger heat transfer surfaces, and the channel degrades the heat transfer characteristics. The heat transfer characteristics from large, partially treated surfaces are required to estimate the stability of full scale superconductors exactly. The purpose of this study is to measure the heat transfer from a larger copper plate with surface orientations and treatments appropriate to the cryogenic stability analysis of large scale superconductors. We chose oxidation as the surface treatment because this treatment is applied to the superconductor of the helical coils for the LHD[9]. During the magnet winding process, the winding machine may strip off the surface treatment at some locations. The resulting damage may change the heat transfer characteristics and degrade the stability of the superconducting magnet. To clarify the influence of such damage and improve the accuracy of stability analysis, we measured the heat transfer of partially oxidized surfaces also.

EXPERIMENT

We measured the heat transfer from large copper surfaces to liquid helium to investigate the dependence on the heat transfer surface orientation and the area fraction of the surface treatment. The experiments were carried out using a glass Dewar with a viewing window permitting visual observation of the heat transfer surface and the behavior of helium vapors. During measurements, we kept the liquid helium constantly at atmospheric pressure. We used a copper plate 18 mm wide, 76 mm long, and 7.5 mm thick to simulate large scale superconductors, like those of the helical coils for LHD[1,2]. The samples were treated surfaces of polished Cu or oxidized Cu. The roughness of the polished Cu surface was less than 10 μm. A schematic illustration of a sample is shown in Fig. 1. Figure 2 shows the different surface treatments of the samples. In Fig. 2, samples (a), (b), (c), (d), (e), and (f) are a polished Cu sample, two types of 50% oxidized Cu sample, a 75% oxidized Cu sample, a 90% oxidized Cu sample, and a 100% oxidized Cu sample, respectively. Samples (b) and (c) were investigated to evaluate the effect of oxidized surface shape. The surface was heated uniformly by a thermofoil heater attached on the reverse side of the copper plate. The sample was covered with an FRP holder, except for the heat transfer surface. Temperature differences between the surface and the liquid helium were measured using AuFe-Chromel thermocouples. The thermocouples were planted 1 mm below the surface and at equal intervals of 17 mm in the longitudinal direction (Figs. 1 & 2). An apparatus to change the surface orientation is shown in Fig. 3. The surface orientation was rotated from 0°(horizontal, heat transfer surface upward), through 90°(vertical), to 180°(horizontal, downward) at 15° intervals. For sample (b), the terms "+ degrees" and "- degrees" are defined to clarify the orientation of the sample whether the oxidized part is up(+) or down(-) (Fig. 4). To measure the heat flux from the copper plate to the liquid helium precisely, we first determined the leakage flux from the FRP holder. It is confirmed that thermal leak from FRP to liquid helium hardly influences the measured value; at all angles of the surface orientation, the thermal leak was less than 5%.

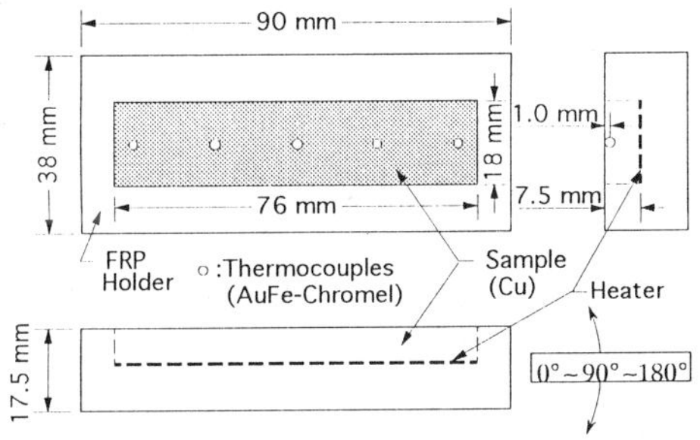

Figure 1. Schematic illustration of a sample.

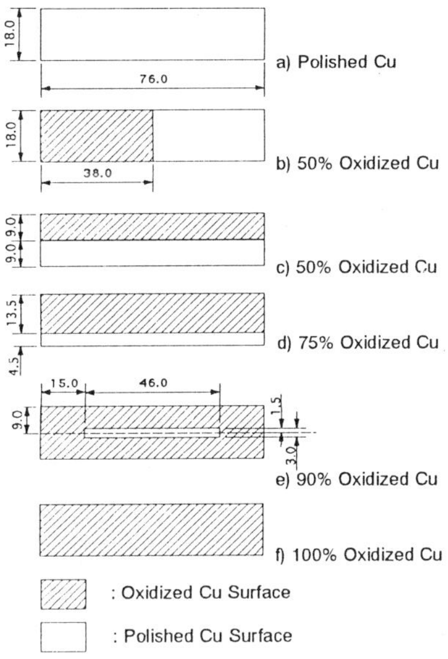

Figure 2. Various surface treatments.

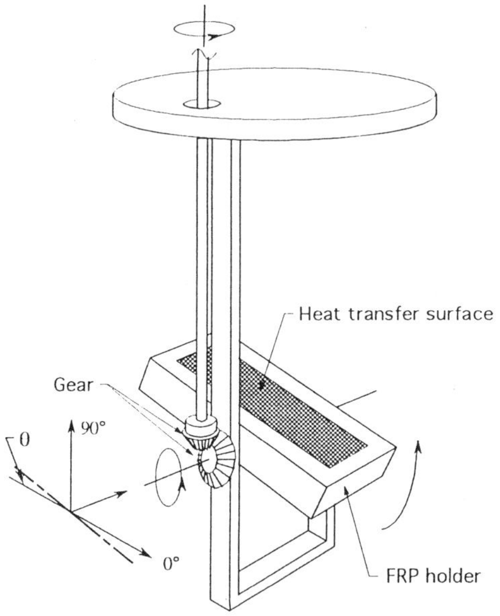

Figure 3. Apparatus to change the surface orientation.

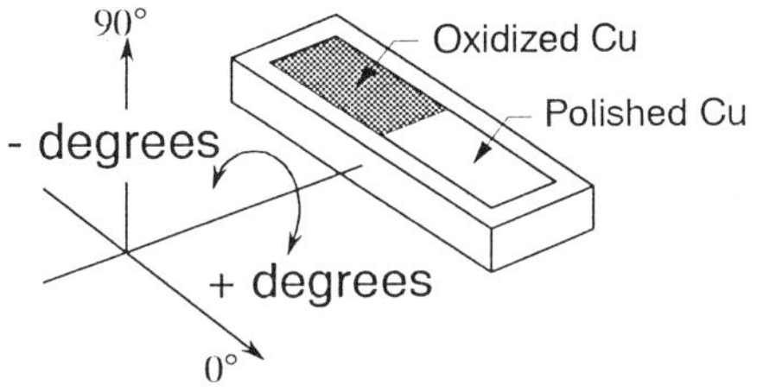

Figure 4. Definition of "+ degrees" and "- degrees" for the sample (b).

RESULTS AND DISCUSSION

Dependence of Heat Transfer on Surface Orientation

Figures 5(a)-(f) show heat flux versus surface temperature as a function of surface orientation. There is no difference in the heat transfer characteristics between + degrees and - degrees for sample (b). The gradient of the nucleate boiling curve does not depend on the surface orientation. The gradient of the film boiling curve depends on orientation, and the heat transfer coefficient reaches a maximum at 90°. During the transition from nucleate boiling to film boiling, we measured the time dependence of the surface temperature using thermocouples distributed longitudinally. No time delays and no positional temperature variations were observed for any of the samples or angles of orientation. Figure 6 shows the dependence of the critical heat flux (CHF) and the minimum heat flux (MHF) on the surface

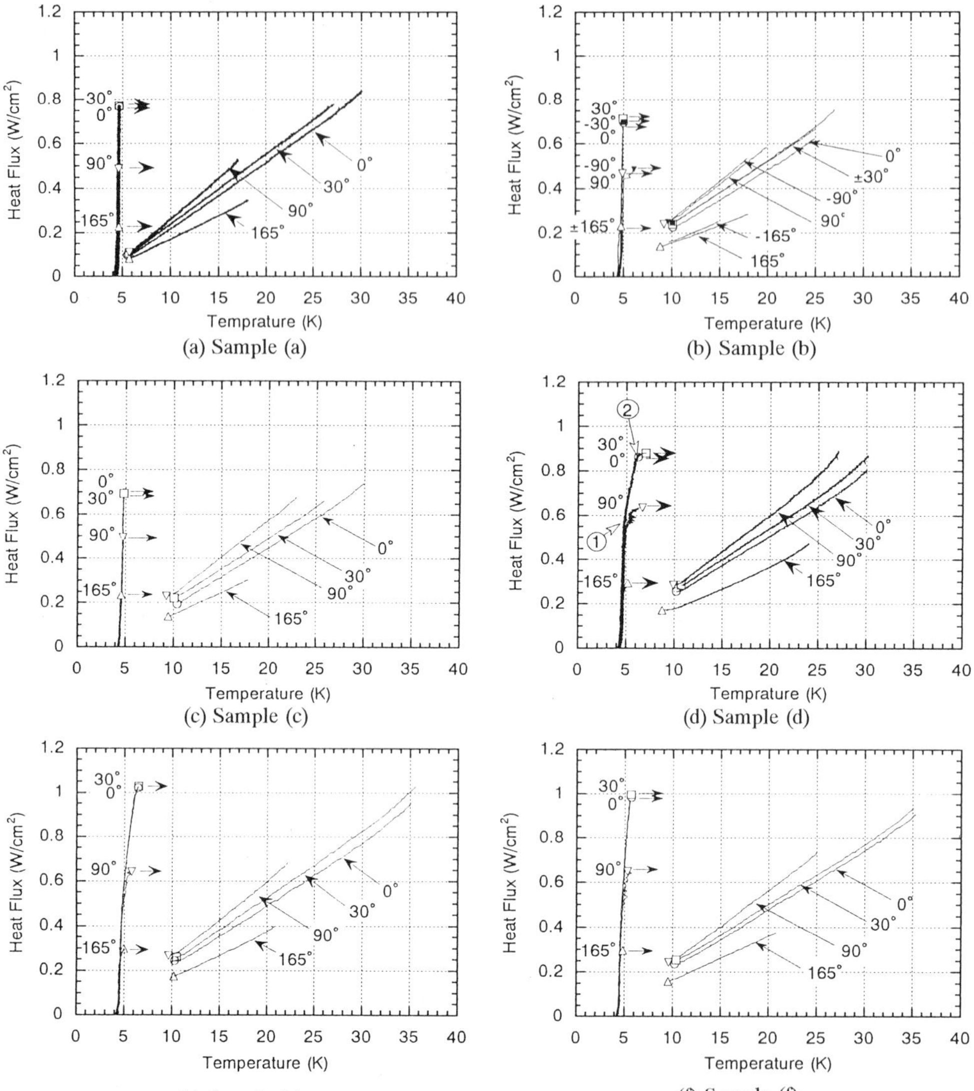

(a) Sample (a)

(b) Sample (b)

(c) Sample (c)

(d) Sample (d)

(e) Sample (e)

(f) Sample (f)

Figure 5. Heat flux versus surface temperature.

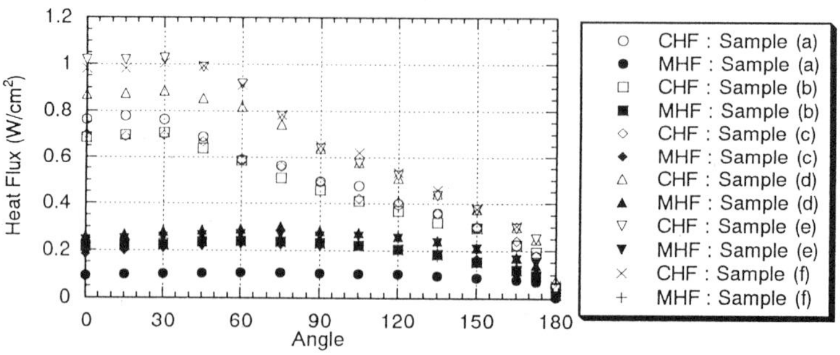

Figure 6. Dependence of the critical and minimum heat fluxes on surface orientation.

orientation of samples (a), (b), (c)(only at + degrees), (d), (e), and (f). We found that the critical heat flux strongly depends on surface orientation. At angles above 30°, the critical heat flux decreases as the surface orientation increases. A maximum is reached at angles between 0 and 30°. The convection around 30° drives the critical heat flux to its maximum. On the other hand, the minimum heat flux shows little dependence on orientation, except around 180°. At 90°, the heat transfer characteristics change and an inflection point appears on the curve of critical heat flux as a function of surface orientation. We presume that the convection of liquid helium causes the characteristics to change between 0°-90° and 90°-180°. As mentioned above, the heat transfer coefficient changes with increasing surface orientation. It is thought that the change of heat transfer coefficient is due to the different convection, and surface treatment can affect release of helium vapor from the surface.

Effect of Heat Transfer Surface Treatment

Experimental Results. As Figure 6 shows, oxidation of the entire heat transfer surface improves the critical and minimum heat fluxes compared with those of the entire polished surface. The minimum heat fluxes of samples (b) and (c) agree with that of the sample (f). On the other hand, the critical heat fluxes of the samples (b) and (c) almost agree with that of sample (a). It appears that the critical and minimum heat fluxes change discontinuously as a function of the area fraction of the oxidized surface. The critical heat flux suddenly changes at around 75% oxidization. Between 0 and 50% oxidation, the minimum heat flux does also. The oxidation drives the critical and minimum heat fluxes to improve and cannot affect the heat transfer coefficient both of nucleate and film boilings. No difference in heat transfer characteristics between (b) and (c) can be measured. Therefore equivalent surface treatment areas produce identical heat transfer characteristics. Changing the surface treatment cannot affect the heat transfer coefficient but does change the critical and minimum heat fluxes.

Calculation of the Boiling Curve for a Partially Oxidized Surface. First, consider the boiling curve of the 75% oxidized Cu sample. As shown in Fig. 6(d), the heat transfer curve has two gradients for the nucleate boiling state between 0° - 90°. Figure 7 shows the assumption used to calculate the heat transfer of partially oxidized samples. We assume that a transition state exists between the nucleate and film boiling states. The interpretation is as follows: At point 1, Fig. 7, the polished Cu surface area in sample (d) reaches a critical heat flux. After this, conditions for the polished Cu surface area enter the transition state, and heat which cannot transfer from the polished Cu surface area to liquid helium flows to the oxidized Cu surface area instead, so that the heat transfer from the oxidized area increases. Finally, at point 2, Fig. 7, the oxidized Cu surface area reaches its critical heat flux, and entire surface changes to the film boiling condition. Heat is transferred to the liquid helium with contributions from the polished and the oxidized regions proportional to their area fractions. Table 1 summarizes the heat transfer of the 100% polished and the 100% oxidized Cu samples. We approximate the heat transfer curve for each case as h(T)=aT+b, then calculate the total heat transfer for a partially oxidized sample using Table 1 and a rule of mixtures, Eq. (1):

$$h_{POS}(T) = h_{Oxidized}(T) \cdot x + h_{Polished}(T) \cdot (1 - x). \qquad (1)$$

In this expression, $h_{POS}(T)$, $h_{Oxidized}(T)$, and $h_{Polished}(T)$ are the heat fluxes of the partially oxidized, 100% oxidized, and 100% polished Cu samples at T K, and x is some area fraction of oxidation. The effect of thermal conductivity are neglected in such calculations. We calculate the heat transfer at 0° using Eq. (1). At point 1 (ΔT = 0.5 K), the heat fluxes of the polished and oxidized Cu surface areas are 0.75 (Critical heat flux) and 0.5 W/cm^2, respectively. At point 2 (ΔT = 1.0 K), the heat fluxes of the polished and oxidized Cu

Table 1. Summary of heat transfer characteristics (at bath temperature of 4.2 K)

Angle	Critical Heat Flux (W/cm^2) (at T_1 K)	Minimum Heat Flux (W/cm^2) (at T_2 K)	Boiling Curve Nucleate $T < T_1$		Transition $T_1 < T < T_2$		Film $T_2 < T$	
			a	b	a	b	a	b
Polished Cu Sample								
0°	0.766 (at 4.68 K)	0.0969 (at 5.66 K)	1.58	-6.65	-0.687	3.98	0.0294	-0.0694
30°	0.772 (at 4.66 K)	0.108 (at 5.70 K)	1.70	-7.12	-0.638	3.74	0.0310	-0.0691
90°	0.484 (at 4.60 K)	0.108 (at 5.71 K)	1.20	-5.06	-0.340	2.05	0.0368	-0.102
165°	0.404 (at 4.53 K)	0.103 (at 5.76 K)	0.745	-3.13	-0.145	0.878	0.0332	-0.0878
100% Oxidized Cu Sample								
0°	0.983 (at 5.71 K)	0.235 (at 10.2 K)	0.652	-2.74	-0.166	1.93	0.0257	-0.0276
30°	0.996 (at 5.65 K)	0.254 (at 10.4 K)	0.687	-2.89	-0.156	1.88	0.0263	-0.0201
90°	0.648 (at 5.33 K)	0.243 (at 9.58 K)	0.573	-2.41	-0.0951	1.16	0.0318	-0.0617
165°	0.297 (at 4.85 K)	0.159 (at 9.53 K)	0.451	-1.89	-0.0294	0.440	0.0191	-0.0225

surface areas are 0.2 and 0.98 W/cm^2 (Critical heat flux), respectively. The calculated results based on the assumption for points 1 and 2 are 0.56 and 0.79 W/cm^2, respectively. The measured values are 0.57 and 0.85 W/cm^2. Figure 8 shows that the calculated and measured heat transfer results for the 75% oxidized Cu sample at 0° almost agree. Therefore the existence of a transient state of boiling is supported by these results.

The dependence of the critical and minimum heat fluxes on surface treatment is now considered. Using the measured heat transfers of the polished and 100% oxidized Cu samples and Eq. (1), we calculated the critical and minimum heat fluxes at 0°, 30°, 90°, and 165° for surfaces with between 0 (polished Cu surface) and 100% oxidation under the assumption stated earlier. Figures 9(a)-(d) show the calculated and the measured results. The calculated critical heat flux begins to increase when surface oxidation exceeds 70% (at 0°), 75% (at 30°), 60% (at 90°), or 50% (at 165°). The measured critical heat flux increases at 75% of oxidization. An inflection point in the plot of critical heat flux occurs because the critical heat flux of the partially oxidized sample begins to increase when the heat transfer from the oxidized Cu surface area begins to compensate for the decrease of heat transfer caused by the onset of the transition state for the polished Cu surface area. On the other hand, the measured minimum heat flux degrades if the surface oxidation falls below 50%. Specifically, the calculated minimum heat flux begins to degrade below 15% (at 0°), 17% (at 30°), 28% (at 90°), or 40% (at 165°). The calculated critical and minimum heat fluxes are in

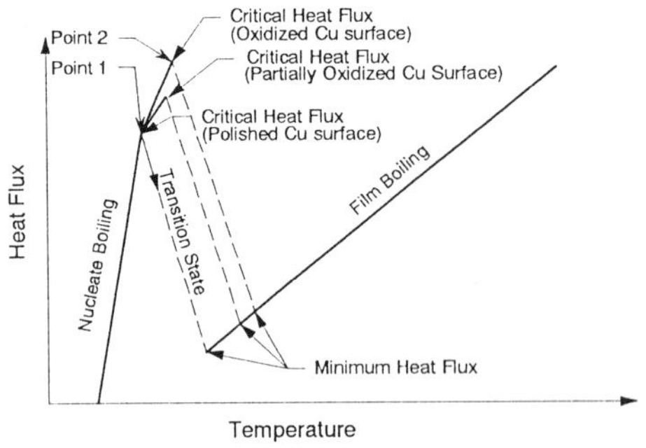

Figure 7. Assumption of heat transfer for calculation.

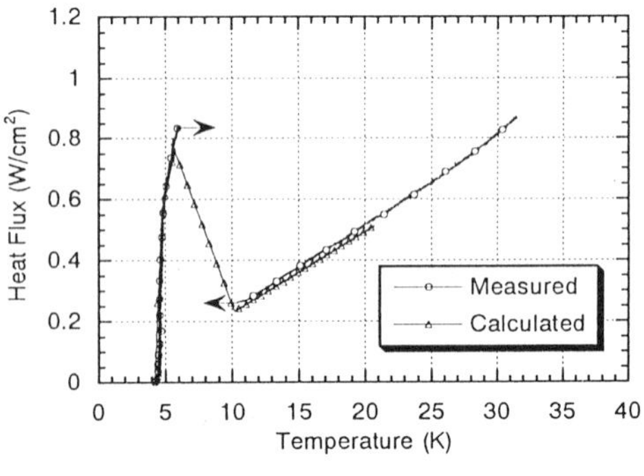

Figure 8. Calculated heat transfer for the 75% oxidized Cu sample at 0°.

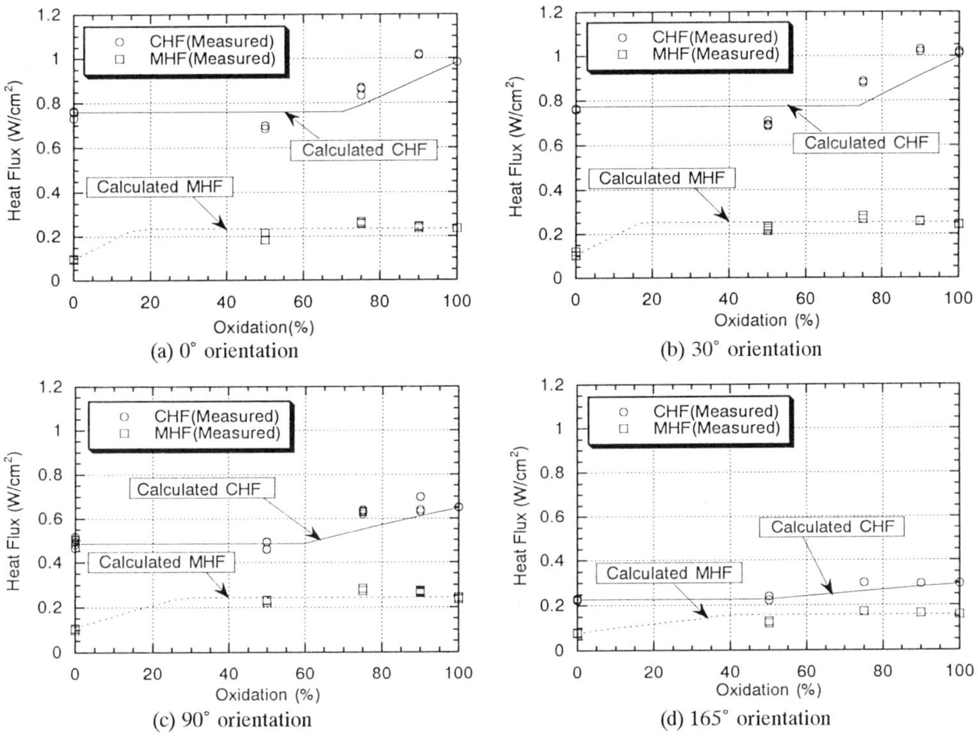

Figure 9. Calculated dependences of the critical and minimum heat fluxes on percentage of the oxidized area.

approximate agreement with the measured values. Therefore the dependence of critical and minimum heat fluxes on oxidation can be easily calculated from the measured heat transfer using the assumption shown in Fig. 7. Judging from these calculated and measured results, a transition state of boiling exists on the surface for specific conditions.

Standard for Superconducting Wire

The heat transfer from superconducting wires depends on the surface treatment. It is found that the surface treatment improves the heat transfer characteristics and magnet stability. According to the observed dependence of the heat transfer on percentage of surface area treated, some area of oxidation on the surface will improve the critical and the minimum heat fluxes compared to the polished Cu surface. We can adopt these results as the standard to judge effects of surface treatment damage on superconducting wires.

CONCLUSIONS

We investigated the dependence of heat transfer from a large copper plate to liquid helium on heat transfer surface orientation and treatment. The following conclusions are obtained:

The critical heat flux depends on the surface orientation and is a maximum at 0° - 30°. The minimum heat flux shows little orientation dependence. The heat transfer coefficient for nucleate boiling is not a function of orientation. The change of the surface orientation affects the heat transfer coefficient for film boiling and it reaches a maximum at 90°. It is thought that changing the orientation influences the convection for film boiling, and that this changes the heat transfer coefficient for film boiling.

The dependence of critical and minimum heat fluxes on the surface treatment is interesting. The critical heat flux begins to increase when the oxidized surface area reaches 75%, whereas the minimum heat flux degrades below 50%. These experimental results almost agree with calculated results for the dependence of the critical and minimum heat fluxes on surface treatment area. Experiments and calculations show that the critical and the minimum heat fluxes are bilinear functions of the area fraction of surface treatment. Surface treatment cannot alter the heat transfer coefficient.

Judging from the measured and calculated results, a transition state of boiling exists for specific surface conditions.

This investigation of heat transfer characteristics applies to large scale superconducting applications; the results can be used as a standard of judgment for damage to the surface treatment of superconducting wires, and to estimate the stability of superconducting magnets.

ACKNOWLEDGMENTS

The authors would like to thank Professor M. Shiotsu of Kyoto University for helpful suggestions and Drs. S. Imagawa, S. Satoh, A. Nishimura, S. Yamada, T. Satow and O. Motojima of the National Institute for Fusion Science for discussing the experimental results. Special thanks are due to Dr. R. Tobler, visiting professor at NIFS, for helpful discussions. Thanks are due to Mr. T. Tamaki and S. Suzuki of Hitachi, Ltd., and Mr. H. Nakazawa and H. Moriai of Hitachi Cable, Ltd., for providing the samples.

REFERENCES

1. O. Motojima, et al., Physics and engineering design studies on the Large Helical Device, *Fusion Engineering and Design,* 20:3 (1993).
2. T. Mito, et al., Short sample tests of aluminum-stabilized superconductors for Large Helical Device, *Fusion Engineering and Design,* 20:233 (1993).
3. G. R. Chandratilleke, S. Nishio, and H. Ohkubo, Pool boiling heat transfer to saturated liquid helium from coated surface, *Cryogenics* 29:588 (1989).
4. S. Sato, and H. Ogata, Review of boiling heat transfer to helium-Part 1, *Journal of the Cryogenic Society of Japan,* 12:2 (1977) (in Japanese).
5. M. Shiotsu, K. Hata, and A. Sakurai, Effects of diameter and system pressure on critical heat flux for horizontal cylinder in saturated liquid He I, *Cryogenics* 29:593 (1989).
6. A. Iwamoto, T. Mito, K. Takahata, N. Yanagi, and J. Yamamoto, Heat transfer of a large copper plate to liquid helium applicable to large scale superconductors, *Cryogenics* 34 ICEC Supplement:321 (1994).
7. Z. Chen and S.W.Van Sciver, Channel Heat Transfer in He I - Steady State Orientation Dependence, *Advances in Cryogenic Engineering* 31:431 (1986).
8. D. N. Lyon, Boiling heat transfer and peak nucleate boiling fluxes in saturated liquid helium between the λ and critical temperatures, *Advances in Cryogenic Engineering,* 10:371 (1965).
9. N. Yanagi, et al., Experimental observation of anomalous magneto-resistivity in 10-20 kA class aluminum-stabilized superconductors for the Large Helical Device, *Advances in Cryogenic Engineering(Materials),* 40:459 (1994).

STUDIES ON A HEAT EXCHANGER PRODUCING SUBCOOLED LIQUID HELIUM

Norihito Ohuchi and Kiyosumi Tsuchiya

KEK National Laboratory for High Energy Physics
1-1 Oho Tsukuba 305 Japan

ABSTRACT

A heat exchanger for cooling the TRISTAN final focusing superconducting quadrupole magnet was studied. This heat exchanger could produce subcooled liquid helium of 20 g/s and 0.16 MPa below 4.4 K with saturated liquid helium at 0.11 MPa. For the heat transfer of single phase flow in the heat exchanger, the calculations with the Dittus-Boelter and Kutateladze equations agreed well with the measurements. For the heat transfer during the condensation process, the Shah equation including vapor quality was applied. The calculated heat transfer energy was twice as large as the measured value and the vapor effect on the heat transfer coefficient was not as large as that calculated by the Shah equation.

INTRODUCTION

As a cooling method for a large superconducting magnet system, though forced-flow cooling of helium is mainly applied, subcooled liquid helium is also one possibility, except for supercritical helium. Producing subcooled liquid helium, however, involves a condensation process of which the thermal characteristics have not been studied enough. In order to understand the process, we studied a heat exchanger which can supply subcooled liquid helium of 20 g/s and 0.16 MPa at 4.4 K. In the experiment, we produced subcooled liquid helium and measured the helium temperature at the outlet while changing the heat exchanging surface area. In calculating the temperature profile along the heat exchanger, heat transfer coefficients of forced-flow helium in the condensation process were estimated with the Shah equation[1]. Earlier this equation was examined with R-11, R-12 and ethanol etc., but it was never applied to liquid helium. We tried to analyze the temperature profile along the heat exchanger with the Kutateladze[2], Dittus-Boelter and Shah equations.

EXPERIMENTAL SYSTEM

The heat exchanger of which the performance was estimated is shown schematically in Fig. 1. The heat exchanger was used in a prototype cryogenic system for TRISTAN final focusing insertion quadrupole magnets[3]. The heat exchanger was designed to produce subcooled liquid helium of 20 g/s and 0.16 MPa at 4.4 K. The heat exchanger was made of copper tubing 43 meters long and with inner and outer diameters of 17.05 mm and 19.05 mm, respectively. The material was phosphorized copper. The copper tubing was wound in a diameter of 140 mm at the first layer increasing to 300 mm at the outer layer. The total

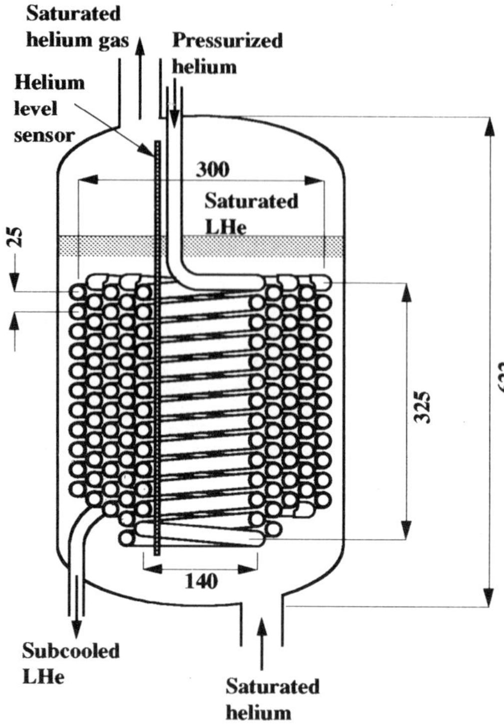

Figure 1. Schematic view of the studied heat exchanger.

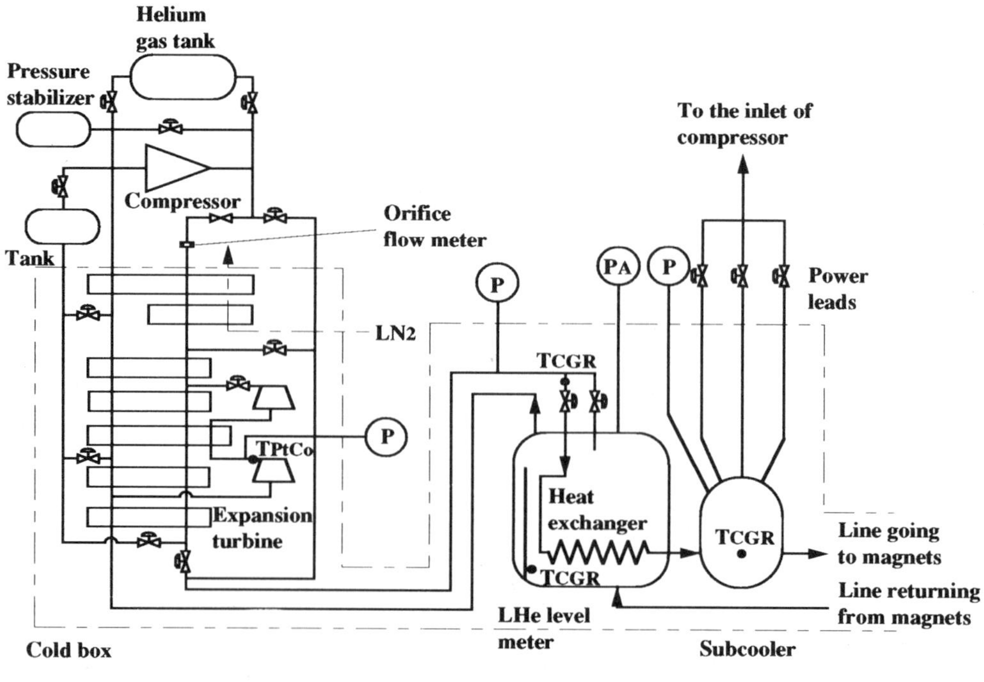

Figure 2. Flow diagram of the experimental system. T_{CGR} and T_{PtCo} represent the carbon glass resister thermometer and the platinum cobalt resister thermometer, respectively. P and P_A represent the pressure gauge and the absolute pressure gauge, respectively.

number of turns was 63. The heat exchanger was located in a helium vessel whose height and inner diameter were 622 mm and 350 mm, respectively. The pressurized helium inside the tubing was designed to be heat exchanged with the saturated liquid helium outside the tubing.

INSTRUMENT

The temperatures of the helium which passed through the tubing were measured by carbon-glass resister thermometers just before the inlet and at the outlet of the heat exchanger. The saturated helium temperature outside the tubing was measured by a carbon-glass resister thermometer too. These thermometers had already been calibrated by the supplier, and we estimated that the measurement error of the temperatures was within 30 mK by immersing these thermometers in saturated helium and comparing these measured temperatures and the saturated temperature, which was measured with a precise pressure gauge. The pressure of saturated helium in the vessel was measured by an absolute pressure gauge. This pressure gauge is accurate to 7.8×10^2 Pa within the full range of 4.1×10^5 Pa. The mass flow rate of helium through the heat exchanger was calculated by subtracting the mass flow rate through the expansion turbines from the total mass flow rate through the refrigerator. The total mass flow rate was measured by an orifice flow-meter at room temperature, and the flow rate through the turbine was calculated from the pressure and temperature of helium at the inlet of the second turbine. The temperature was measured by a platinum-cobalt resister thermometer. The mass flow rate of helium through turbines was measured by quickly closing the control valve of the turbine circuit. This measurement was carried out 9 times. The difference between the measured mass flow rate through the turbines and the calculated one was 4.46 %, on an average, of the measured mass flow rate, and the standard deviation was 8.45 %. Important sensors are shown in Fig. 2. The liquid helium level sensor was installed in the vessel and the wetted surface of the heat exchanger was calculated at that level.

MEASUREMENT RESULTS

Figure 3 shows the temperature changes of helium at the tubing outlet and in the vessel. The helium pressures inside the tubing were 0.1602 ~ 0.1612 MPa and the pressures in the vessel were 0.1098 ~ 0.1135 MPa. The lateral axis represents a wetted length of the tubing calculated by the liquid helium level sensor. Figure 4 shows the mass flow rate and an enthalpy of helium at the inlet of the heat exchanger which was calculated from the temperature and the pressure of helium just before the inlet valve. Since the helium is in a supercritical condition before it reaches the inlet valve, we can estimate the enthalpy of the helium at the inlet of the heat exchanger. The measured mass flow rate varied from 20 to 23 g/s. The Reynolds number of helium in the tubing was greater than 5×10^5, and the helium was in turbulent flow. The vapor qualities at the inlet were estimated from 35 % to 85 %. The helium condition at the inlet was dependent on the thermal condition of the cryogenic system, and the enthalpy and vapor quality changed with the liquid helium level in the vessel. The cryogenic system, however, was thermally stable when the measurement was carried out at each liquid helium level.

From Fig. 3, the helium temperature in the tubing at the outlet began to fall sharply at a length of 10 ~ 12 meters. At positions up to this point, the temperatures were constant and the liquid helium was considered to be in two phase flow. The average temperature and pressure in this region were 4.727 K and 0.1606 MPa, respectively. The saturated helium temperature of 0.1606 MPa was 4.752 K from a helium property[4]. From these results, it is considered that the systematic error was within 30 mK for this measurement. In this region, the heat transfer mechanism involves a condensation process. At positions beyond this point, the liquid helium was in single phase flow. The helium temperature at the outlet of the tubing was almost constant over 20 meters of the wetted length and the temperature reached around 4.39 K. For the 43 meters of wetted length, the helium temperature inside the tubing was 4.391 K at a helium temperature of 4.356K in the vessel. Figures 5 and 6 show the results for helium at a pressure of 0.15 MPa inside the tubing. The helium temperatures show the same changes as the results of 0.16 MPa. The temperature difference between the helium in the tubing and in the vessel was 40 mK for the wetted length of 43 meters.

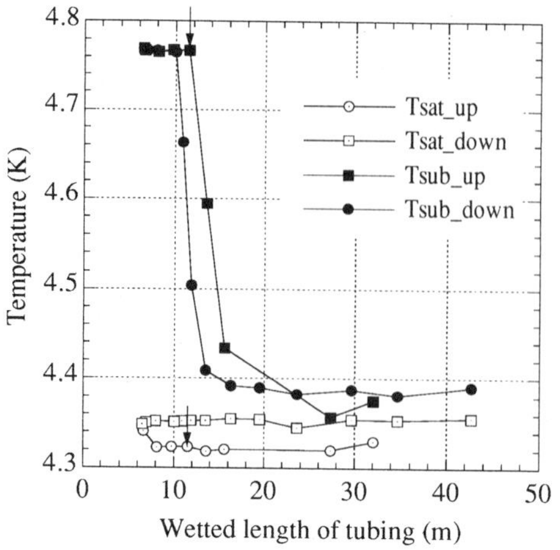

Figure 3. Temperature changes for 0.16 MPa helium in the tubing. T_{sat} and T_{sub} represent the temperature of saturated helium in the vessel and helium in the tubing, respectively. Up and down correspond to the conditions of increasing and decreasing liquid helium level in the vessel.

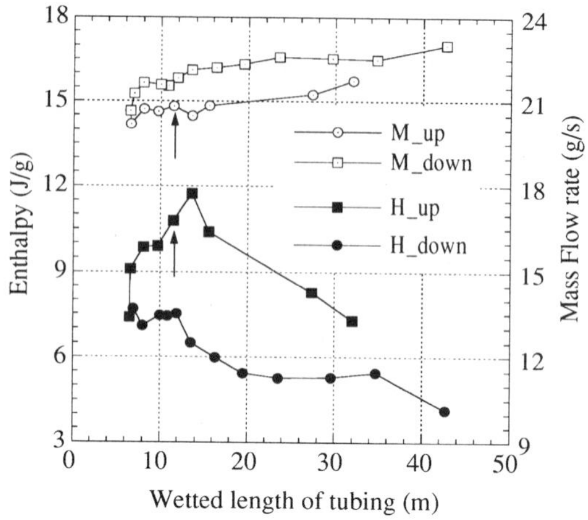

Figure 4. Enthalpy and mass flow rate changes for 0.16 MPa helium in the tubing. M and H represent the mass flow rate and the enthalpy of helium in the tubing, respectively.

NUMERICAL ANALYSIS

In order to calculate the temperature change of the helium inside the tubing, we used the following equations. The heat flux from the outer surface of the tubing to the liquid helium in the vessel was estimated by the Kutateladze equation

$$Q_{out} = A_{out} \times 4.87 \times 10^{-11} \times \left\{\frac{C_{pl}}{\varepsilon \times \rho_v}\right\}^{1.5} \times \frac{\lambda_l \times \rho_l^{1.282} \times P^{1.75}}{\sigma^{0.906} \times \eta_l^{0.626}} \times (T_w - T_{hv})^{2.5}$$

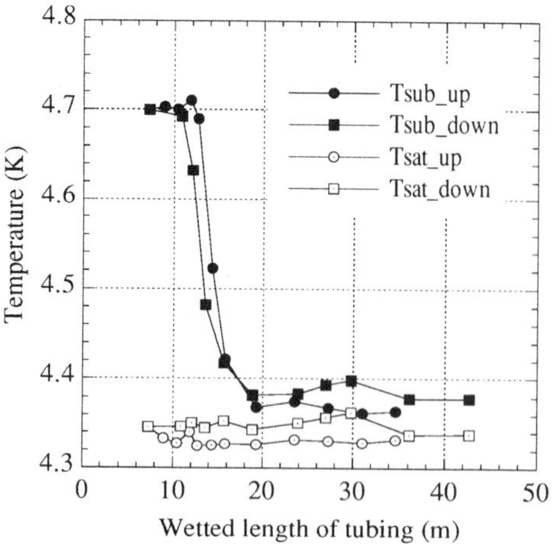

Figure 5. Temperature changes for 0.15 MPa helium in the tubing.

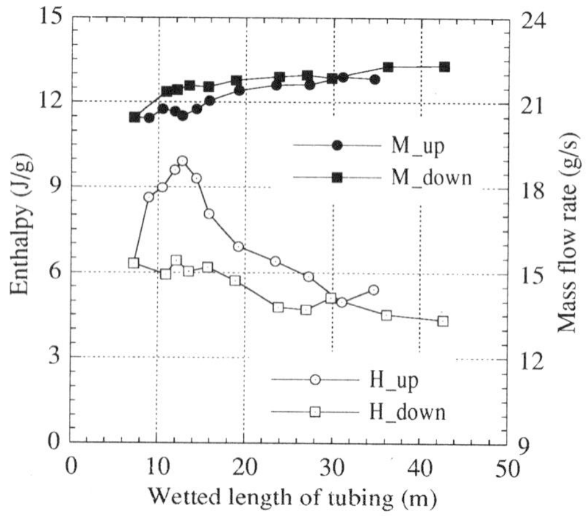

Figure 6. Enthalpy and mass flow rate changes for 0.15 MPa helium in the tubing.

The heat flux from the helium inside the tubing to the inner surface of the tubing was estimated by the following equations;

$$Q_{in} = A_{in} \times H_{in} \times (T_{ht} - T_w)$$

For liquid helium in a single phase, the forced-flow heat transfer coefficient in the tubing was defined by the Dittus-Boelter equation;

$$H_{in}(S) = 0.023 \times Re^{0.8} \times Pr^{0.4} \times \frac{\lambda_l}{D}$$

For the two phase helium during condensation, the heat transfer coefficient was defined by the Shah equation;

$$H_{in}(T) = H_{in}(S) \times \left[(1 - x)^{0.8} + \frac{3.8 \times x^{0.76} (1 - x)^{0.04}}{P_{rd}^{0.38}} \right]$$

On the heat transfer surface in the steady state, $Q_{out} = Q_{in}$. In the thermal model, we neglected the temperature distribution in the copper wall and the heat exchange at the dry wall. Figure 7 shows the calculated results. The calculated helium condition inside the tubing at the inlet was 0.1606 MPa, 20.8 g/s of mass flow rate and 10.77 J/g of enthalpy. The helium pressure outside the tubing was 0.1103 MPa. The temperatures were calculated from the saturated pressure. This condition corresponds to the experimental results which are marked by arrows in Figs. 3 and 4. The entire coil of copper tubing is assumed to be wetted by the saturated liquid helium. As shown in Fig. 7, the helium temperature inside the tubing began to decrease 5 meters from the tubing inlet. Experimental results show that the temperature began to decrease at around 11.5 meters. From a comparison between the calculations and the experimental results, the calculated heat flux is two times larger than the heat flux estimated by the experiment. The calculations show that the wall temperature of the copper tubing decreased in the two phase region in spite of the constant helium temperature in the tubing and in the vessel. These are due to the vapor effect on the heat transfer coefficient. In this calculation, the vapor quality at the tubing inlet was 77.3 % and the heat transfer coefficient was enhanced to 3.6 times larger than the heat transfer coefficient of the single phase helium by the Shah equation.

DISCUSSION

For calculating the heat transfer between the subcooled helium and the saturated helium via the copper tubing, we applied the Dittus-Boelter and Kutateladze equations. With the 0.15 MPa data, we estimated the difference between the calculations and the measurements. As shown in Figs. 5 and 6, the enthalpy of helium at the inlet from 7 to 20 meters along the wetted length is almost constant and around 6 J/g. In this case, the helium temperature in the tubing began to decrease at 10.9 meters along the wetted length and reached a constant temperature of 4.38 K at 18.9 meters. The temperature decrease and the exchanged heat energy within this length were 0.31 K and 39.6 J, respectively. In the calculations beyond the temperature decreasing point, the temperature decrease and the exchanged heat energy within 8 meters were 0.29 K and 44.7 J, respectively. The difference between the calculations and the measurements was around 13 %.

As shown in Fig. 7, the calculated length of the two phase region was almost half of the measured length. Comparing with the data from E. Sauvage-Bouter et al.[5], the flow pattern is estimated to be annular at the inlet and to change into intermittent billowing flow at 4 meters from the inlet. It is considered that the surface inside the tubing was wetted by the saturated liquid helium at 0.16 MPa. In order to estimate the heat transfer coefficient of the two phase region, the Dittus-Boelter equation was applied instead of the Shah equation. The helium was assumed to be single phase liquid. The calculated results are shown in Fig. 8.

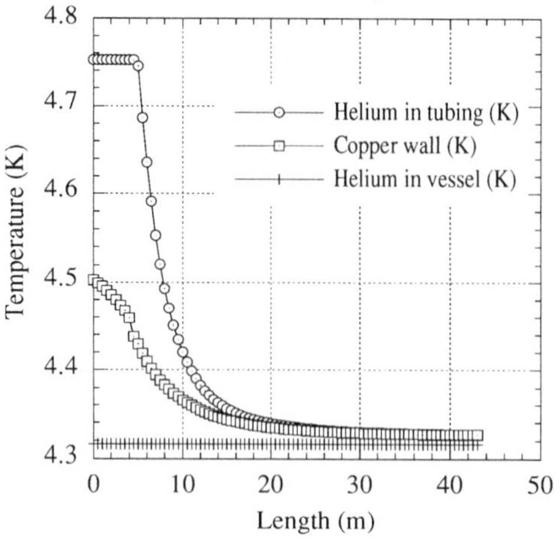

Figure 7. Calculated temperature changes with the Shah, Dittus-Boelter and Kutateladze equations.

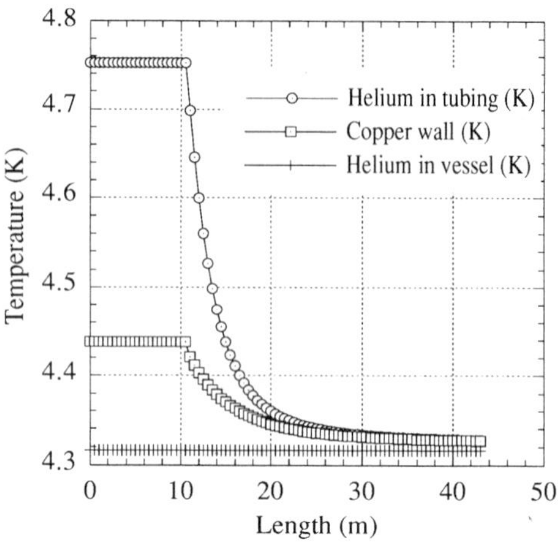

Figure 8. Calculated temperature changes with the Dittus-Boelter equation instead of the Shah equation.

The calculated length was almost the same as the experimental results. In this kind of heat exchanger with the condensation process, vapor can not flow faster than liquid because the crest touches the upper wall of the tubing in the case of intermittent flow of low vapor quality. It is then considered that the heat transfer coefficient is not enhanced as much as that calculated by the Shah equation.

CONCLUSION

A heat exchanger which can produce subcooled liquid helium was studied. It was made of phosphorized copper tubing 43 meters long and with inner and outer diameters of 17.05 mm and 19.05 mm, respectively. The results are as follows;

1. For helium conditions of which enthalpy was around 6 J/g at the inlet, the heat exchanger could produce subcooled liquid helium at 20 g/s, 0.16 MPa and below 4.4 K with saturated liquid helium of 0.11 MPa. In this case, the wetted length of 20 meters was enough to produce the subcooled liquid helium.

2. Heat transfer between the subcooled liquid helium and the saturated helium via copper tubing could be calculated with the Kutateladze and Dittus-Boelter equations. The calculations agreed well with the measurements.

3. In order to estimate the heat transfer during the condensation process, the Shah equation was applied but the measured heat transfer energy was half as much as the calculated energy. In this kind of heat exchanger, it is possible that the vapor effect on the heat transfer coefficient was not as much as that calculated by the Shah equation.

NOMENCLATURE

A_{out} = outer area of heat transfer
A_{in} = inner area of heat transfer
C_{pl} = specific heat at constant pressure of the liquid
D = inside diameter of the tubing
H_{in} = heat transfer coefficient inside the tubing; (s) represents a single phase, (T) represents a two phase
P = pressure of the boiling system
P_{rd} = reduced pressure
T_w = wall temperature of the tubing
T_{hv} = helium temperature in the vessel

T_{ht} = helium temperature inside the tubing
x = vapor quality
ε = latent heat of vaporization at saturation
ρ_v = density of the vapor
ρ_l = density of the liquid
λ_l = thermal conductivity of the liquid
σ = surface tension between the liquid and its own vapor
η_l = viscosity of the liquid
Re = Reynolds number
Pr = Prandtl number

REFERENCES

1. M. M. Shah, A General Correlation for Heat Transfer During Film Condensation Inside Pipe, Int. J. Heat Mass Transfer, Vol. 22, p547 (1979).
2. E. G. Brentari and R. V. Smith, Nucleate and film pool boiling design correlations for O2, N2, H2, and He, in: *Advances in Cryogenic Engineering*, Vol. 10, Plenum Press, New York (1965), p325.
3. K. Tsuchiya, N. Ohuchi, A. Terashima and K. Shinkai, Cryogenic system of the superconducting insertion quadrupole magnets for TRISTAN main ring, in: *Advances in Cryogenic Engineering*, Vol. 35, Plenum Press, New York (1990), p941.
4. PROPATH group, PROPATH: a program package for thermophysical property, Version 3.1 (1989).
5. E. Sauvage-Boutar, C. Meuris, J. Poivilliers and M. X. Francois, Observation of two-phase helium flows in a horizontal pipe, in: *Advances in Cryogenic Engineering*, Vol. 33, Plenum Press, New York (1988), p441.

1995 CEC/ICMC Conference: The Conference Chairs of the 1995 CEC/ICMC Conference: (from left to right) James B. Peeples, CEC, and E.W. (Ted) Collings, ICMC.

SPECIAL PHYSICAL PROBLEMS RELATED TO HEAT TRANSFER IN BATH OF SUPERFLUID HELIUM

Ruzhu Wang

Institute of Refrigeration and Cryogenics, Shanghai Jiao Tong University
Shanghai 200030, P. R. China

ABSTRACT

On the basis of transient heat transfer measurement of film boiling onset time in superfluid helium (HeII) bath, a criterion correlated to bath temperature and pressure for quantum turbulence onset after a rectangular heat pulse in HeII bath is shown. With the previously measured data of time dependent heat transfer, the boundary phenomena after the onset of film boiling are well defined. The bath temperature and pressure influences on heat transfer in HeII are studied.

INTRODUCTION

Superfluid helium HeII (including saturated superfluid helium HeIIs, and subcooled superfluid helium HeIIp) has developed a new way for the cooling of superconducting magnets because of its relatively low bath temperature and much better cooling property than that of the normal liquid helium HeI. As is known it is advantageous to use HeIIp as the cooling medium for superconducting magnets. Experimental researche and theoretical analyses concerning heat transfer in a bath of HeII have been performed. In my research[1-5] on transient and steady state heat transfer in HeIIp bath, I have noticed that several special physical problems related to heat transfer in a bath of HeII should be clarified, which may help to enrich the knowledge of superfluid physics.

The criterion of quantum turbulence onset, boundary phenomena after the onset of film boiling around a heater, temperature and pressure effects on heat transfer in HeII bath - the above special physical problems related to heat transfer in bath of HeII are studied in this paper.

CRITERION OF QUANTUM TURBULENCE ONSET

The previous studies of quantum turbulence onset after a heat pulse were limited to confined tubes containing saturated superfluid helium (HeIIs)[6-11] . However the recent application of subcooled superfluid helium (HeIIp, typically HeII at a pressure of 0.1 MPa) for cooling superconducting magnets, has induced intensive research of heat transfer (both of transient and steady state) in HeIIp. How the onset time of quantum turbulence after a heat pulse can be determined , and whether the quantum turbulence in HeII will be influenced by bath pressure - these two problems are not only related to the transient heat transfer processes, but also to superfluid physics itself. The answer must be based upon the experimental studies of transient heat transfer in a HeII bath.

With the help of experimental work of transient heat transfer in a HeII bath, a criterion of quantum turbulence onset after a heat pulse is developed, which shows both temperature and pressure influences.

Quantum Turbulence Onset Time after A Rectangular Heat Pulse

Superfluid helium possesses extremely high thermal conductivity, which can be explained by the mutual friction between superfluid and normal components. In HeII, two heat transfer regions - linear and Gorter-Mellink - can be observed, which depend on the formation of quantum vortex lines.

For real applications and physics itself, transient processes are of more interest, specially the time τ needed to initiate quantum turbulence after a heat pulse q ($q>q_c$, q_c is a critical value, however it is very small[10]). Vinen and Chase have had intensive studies on the transient heat transfer processes in HeIIs. They have done a lot of experiments in various capillaries with different sizes and configurations, and their results are well known in superfluid physics. Usually a correlation of τ and q can be expressed as[12]

$$\tau = \alpha q^{-3/2} \qquad (1)$$

The value α shows a criterion parameter and with this value it is possible to determine how long it is needed to turn the heat transfer in HeII from linear region to Gorter-Mellink region.

As is known, the criterion parameter of quantum turbulence onset is $\alpha = \alpha$(T, p, configuration, size) for heat transfer in HeII confined in a capillary tube, thus it is clear that the criterion parameter for HeII bath is $\alpha = \alpha$(T, p) .

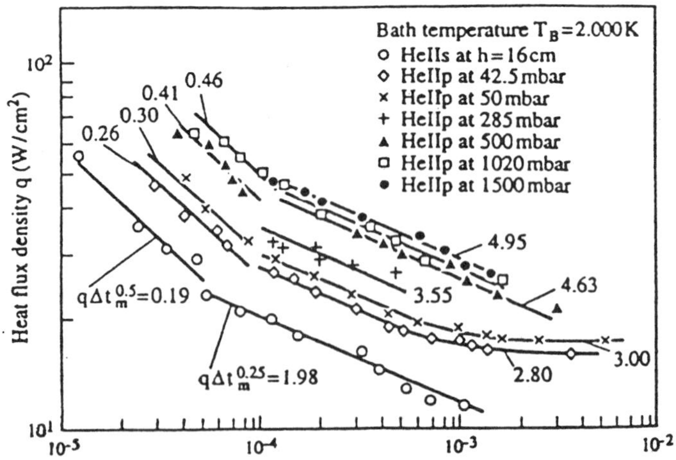

Figure 1. Film boiling onset time Δt_m in HeII at 2.0K with respect to various bath pressures. ○: HeIIs at a hydrostatic head of 16cm; ◇: 42.5 mbar; ×: 50.0 mbar; +: 285 mbar; ▲: 500 mbar; □: 1020 mbar; ▼: 1500 mbar.

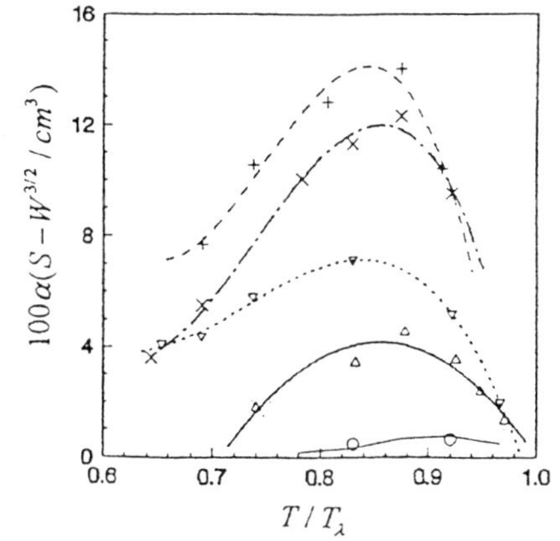

Figure 2. Temperature effect on the criterion parameter of quantum turbulence onset in HeII bath. +: HeIIs, ϕ 4.04mm capillary[11] ; × : HeIIs, ϕ 4.04--1.59mm circular channel[11] , ∇: HeIIs, 2.4 × 6.5mm rectangular channel[7] , Δ: HeIIp bath (p=0.1MPa); O :HeIIs bath

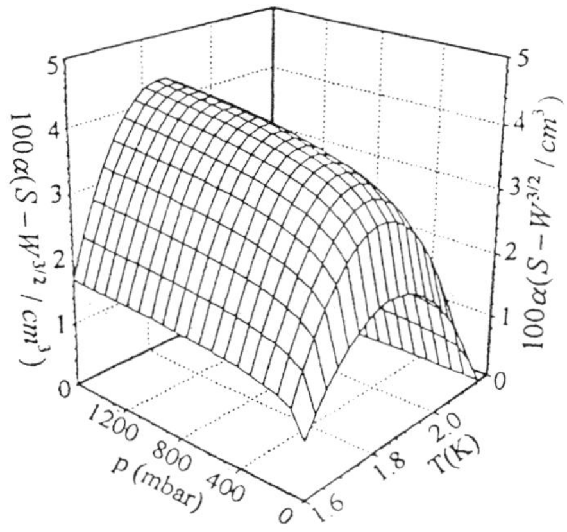

Figure 3. Temperature and pressure effect on the criterion parameter of quantum turbulence onset in a HeII bath.

Transient Heat Transfer in HeII Bath

The detailed experimental method (the experimental apparatus and measurement techniques) and results of transient heat transfer in HeII bath have been reported earlier[1], especially the study of the film boiling onset time Δt_m. It is found that the film boiling onset time Δt_m after a rectangular heat pulse q could be expressed as

$$\Delta t_m = \frac{K_{GM}}{q^4} \tag{2}$$

for Gorter-Mellink heat transfer region; and in the linear heat transfer region

$$\Delta t_m = \frac{K_L}{q^2} \tag{3}$$

Fig. 1 shows the experimental results of film boiling onset time Δt_m at 2.0 K and various pressures. It can be shown that in the log(q) vs. log(Δt_m) diagram, the relation between q and Δt_m behaves as two groups of lines with two different slopes for various pressures. These two groups represent two different heat transfer mechanisms: linear and Gorter-Mellink. The crossing points of the two sets of lines are just the transition points between linear and Gorter-Mellink heat transfer regions for various bath pressures. According to the transition points for various pressures, the criterion parameter value α can be determined by Eq.(1) in which the values q and $\tau = \Delta t_m$ are obtained from the transition point. It is clear that when HeII bath temperature is a constant, α is a function of pressure p, the correlation between α and p in a bath of 2.0 K is written in a unitless relation[4]

$$\frac{\alpha(T,p)}{\alpha(T,0.1MPa)}\Big|_{T=2.0K} = 1 + 0.2235\ln\frac{p}{p_{atm}} \tag{4}$$

If the HeII bath is kept at atmospheric pressure (p_{atm}=0.1MPa), the transient heat transfer experiments at different bath temperatures are shown in diagram[1,4] like Fig.1. Thus, for T ⩾ 2.0 K, two groups of log(q) vs. log(Δt_m) curves with two different slopes are presented due to the two different heat transfer mechanisms. However, for T<2.0K, the slopes of the two curve-groups will be changed. The lower the temperature, the more the curve behavior changes. It could be explained by the second sound effect on heat transfer in HeII[1]. Anyway the transition points between linear and G-M heat transfer regions could still be found. The criterion for quantum turbulence onset in HeIIp at atmosphere pressure (0.1MPa, or 1013mbar) is

$$\alpha(T,0.1MPa) = -136.7 + 329.8\frac{T}{T_\lambda} - 193.0(\frac{T}{T_\lambda})^2 \qquad (5)$$

Fig. 2 shows the criterion parameter with respect to the unitless temperature T/T_λ, the curve represents the calculated result of Eq.(5). Here T_λ is related to the HeII bath pressure, for HeII at 0.1MPa T_λ=2.163K; and for HeII at saturated pressure T_λ=2.172K. The criterion parameters of HeIIs confined in various configurations and sizes are also shown in Fig.2 in comparison. Clearly these criterion parameters have close behavior, but their values are strongly related to the configuration and size of the channels, and are pressure dependent for the HeII bath.

Criterion Parameter of Quantum Turbulence Onset in HeII Bath

The above experimental results and correlation show a strong temperature and pressure dependent relation for the criterion parameter. If the bath temperature and pressure affect the parameter independently, that means the two effects could be separated, and a relation Eq.(6) of α (T,p) can be expressed in a three-dimension diagram (Fig.3). For verification, a calculated curve of α (T,p) for HeIIs at $p=p_s$ is shown in Fig.2, which fits the experimental results well - the two experimental parameters α (T=1.8K, 2.0K) are determined by the transient heat transfer measurement of film boiling onset time in HeIIs at 1.8 K and 2.0K. This verification may show that the suggested formula, Eq.(6), can be used to determine the quantum turbulence onset criterion parameter α (T,P).

Through the above studies, the criterion parameter of quantum turbulence onset in HeII bath is

$$\alpha(T,p) = \alpha(T,0.1MPa) \times \frac{\alpha(T,p)}{\alpha(T,0.1MPa)}$$

or

$$\alpha(T,p) = (1 + 0.2235\ln\frac{p}{p_{atm}}) \times [-136.7 + 329.8\frac{T}{T_\lambda} - 193.0(\frac{T}{T_\lambda})^2] \qquad (6)$$

The experimental work and the above analyses show (1) The criterion parameter of quantum turbulence onset can be determined by transient heat transfer measurement of the film boiling onset time in HeII bath; (2) The criterion parameter of quantum turbulence onset is strongly related to the bath temperature and pressure, although the bath temperature and pressure affect the criterion parameter independently. The value of α in the HeII bath is much smaller than the value for HeII confined in small channels. This result verifies the prediction of Arp[13], which shows that the tangled quantum vortex lines are formed easier in a free HeII bath; (3) The formation of vortex tangled lines is affected by the bath pressure of HeII. Vortex lines are more difficult to form in a high bath pressure of HeII.

BOUNDARY PHENOMENA AFTER THE ONSET OF FILM BOILING

Steady state heat transfer in a HeIIp bath shows that the peak heat flux density of a HeIIp bath is strongly related to the complicated boundary phenomena around a heater, in which a gas helium (GHe) -helium I (HeI) - helium II (HeIIp) boundary exists. With careful considerations on the boundary phenomena GHe/HeI/HeIIp, a clear view of the steady state heat transfer mechanism can be shown. The study of the temperature rise process of a thin wire after applying a rectangular heat / current pulse can show more interesting results.

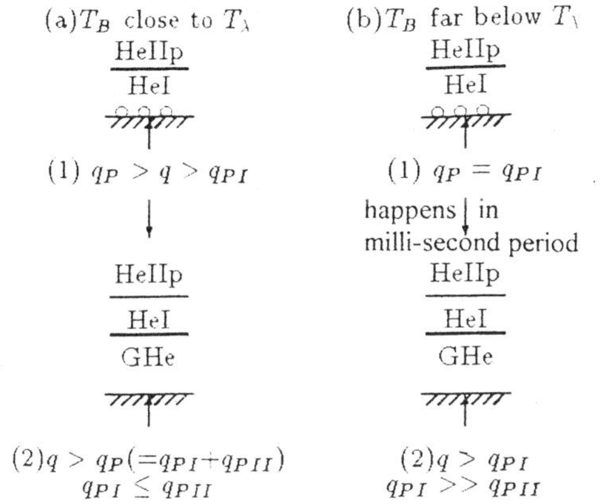

Figure 4. GHe,HeI and HeIIp around sample wire in the film boiling state in HeIIp bath.

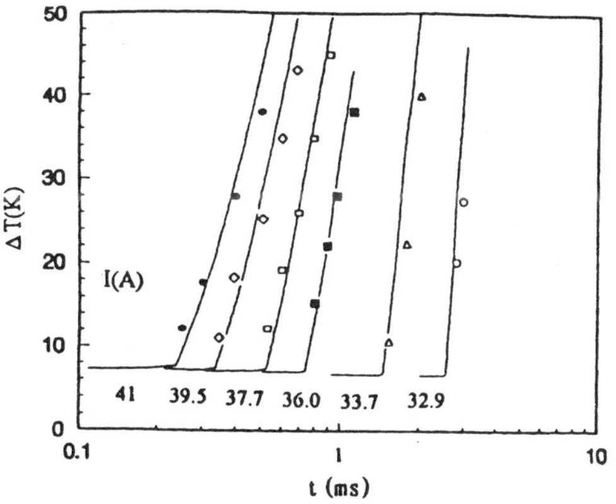

Figure 5. The temperature rise processes of the copper wire for various stepwise current pulses in HeIIp at 1.9K and 0.1MPa. Solid curves:the simulated results.

Regarding the boundary phenomena, we can model a wire in HeIIp subjected to heat, by the analyses of peak heat flux density of HeIIp [2]. Fig.4a shows the two steps in the formation of the triple-phase phenomenon when T_B is close to T_λ. Step 1 represents that which occurs when the heat flux increases to a certain value q_{PI}, the lambda transition starts and thus a HeI layer forms around the wire surface. A further increase in heat flux causes increased nucleate boiling in the HeI layer. Step 2 represents that which occurs when the heat flux reaches the value $q_{PI} + q_{PII}$, a vapor film forms spontaneously around the wire, and a triple phase is formed with GHe,HeI and HeIIp around the sample wire.These two steps were verified by Caspi and Frederking[14]. For my experiments the bath temperature is far below T_λ, and there are no stable points of steady state HeIIp/HeI two-phase transport as the formed HeI layer is unstable and it undergoes a spontaneous transformation to vapor. Thus this step transition is not well defined, as shown in Fig.4b. Anyway, in the analysis of the peak heat flux density of HeIIp bath, the two heat flux densities q_{PI} and q_{PII} could be treated separately. As is clear, q_{PI} is the heat flux density needed to initiate a HeI film around the sample wire, and q_{PII} is the heat flux density needed to initiate change of the respective HeI film into a vapor film around the sample wire- this is just the peak heat flux density of subcooled HeI$_\lambda$ (P=0.1Mpa, $T_B=T_\lambda$=2.163K).

By applying the Gorter-Mellink equation, for a cylindrical sample (radius r_0) q_{PI} is shown as

$$q_{PI} = [\frac{2\phi(r_0)}{r_0}\int_{T_B}^{T_\lambda}\frac{dT}{f(T)}]^{1/3} \tag{7}$$

Where $\phi(r_0)$ is radius-dependent correction parameter.For simplification, a relative critical heat flux density method may be applied. If we know $q_{PI}(T_0)$ ($T_0 < T_\lambda$), then $q_{PI}(T_B)$ at T_B is

$$q_{PI}(T_B) = q_{PI}(T_0)[\frac{\int_{T_B}^{T_\lambda}\frac{dT}{f(T)}}{\int_{T_0}^{T_\lambda}\frac{dT}{f(T)}}]^{1/3} \tag{8}$$

As to the peak heat flux density of subcooled HeI$_\lambda$, q_{PII} may be treated according to the theory for the peak heat flux density of a subcooled liquid[12], thus

$$q_{PII} = q_{sat}(1+\frac{a_h C_p \Delta T_{sub}}{L}) \quad (9)$$

here q_{sat} is the peak heat flux density of saturated HeI at 0.1 MPa, α_h is an empirical parameter found to be close to 1.75[12], $\Delta T_{sub} = T_s - T_\lambda$, T_s is the saturated temperature and L is the latent heat of vaporization of HeI at T_S. The solution for peak heat flux density is

$$q_P(T_B) = q_{PI}(T_0)\left[\frac{\int_{T_B}^{T_\lambda} \frac{dT}{f(T)}}{\int_{T_0}^{T_\lambda} \frac{dT}{f(T)}}\right]^{1/3} + q_{PII} \quad \text{in case of } T_B \text{ close to } T_\lambda \quad (10)$$

$$q_P(T_B) = q_P(T_0)\left[\frac{\int_{T_B}^{T_\lambda} \frac{dT}{f(T)}}{\int_{T_0}^{T_\lambda} \frac{dT}{f(T)}}\right]^{1/3} \quad \text{in case of } T_B \text{ far below } T_\lambda \quad (11)$$

The above theory has been verified with good accuracy[2] compared with the experiments.

By the transient heat transfer measurement, the temperature dependent curve for a certain heat or current pulse can be demonstrated in a simplified way. Here shows measurement results of time dependent temperature curves for a high purity copper wire in HeIIp (d=60μm, 99.999%) after subjected to various current (I) pulses . Before film boiling onset, heat transfer obeys Kapitza theory

$$q = \alpha_K(T_S^4 - T_B^4) \quad (12)$$

Where $\alpha_K = 180 W/m^2K^4$ is the Kapitza constant for a clean copper surface[15] (T_S=T, the sample wire temperature). After film boiling is initiated, the wire is covered by GHe, HeI and HeIIp layers in series, the applied heat power contributes to (1) temperature rise of the wire, (2) temperature rise of the GHe film and its growth and (3)transfer to HeI layer and HeIIp bath. But in this transient heat transfer process, most of the heat is contributed to items (1) and (2). The film growing process can be supposed to be initiated at thickness $\delta_0 = 10.3\mu m$[2,3] and the film growing can be calculated by the conducted heat through the GHe divided by latent heat of HeI. A simplified formula for the time dependent temperature rise process, $T-t$, after being subjected by a current pulse is[5]

$$I^2\rho_T\frac{1}{S} = CS\frac{dT}{dt} + d_{He}C_{He}S_{He}\frac{dT_{He}}{dt} + \lambda_{av}\frac{2\pi(T-4.2)}{\ln(1+\delta/r_0)} \quad (13)$$

where C and C_{He} are the specific heat of copper and GHe respectively, S and S_{He} are their cross sectional areas, d_{He} is density of GHe, ρ_T is resistivity of copper, λ_{av} is average thermal conductivity of GHe film, and the GHe average temperature is T_{He}=(T+4.2)/2. An example of the above theory is shown in Fig.5. The experimental results of the temperature rise process of the high purity copper wire subjected the various current pulses are also shown, good agreement is obtained.

TEMPERATURE AND PRESSURE EFFECTS ON HEAT TRANSFER IN HeII

Although temperature effects on HeII are well understood[2], pressure effects on HeII seem to be worth studying. For example, in the discussion of peak heat flux density of HeIIp, the pressure term is usually neglected, however the effective thermal conductivity of HeII is a function of both temperature and pressure[12]. For example, as far as HeIIs is concerned, and for a hydrostatic head not too great (i.e. in case of noiseless film boiling), the peak heat flux density is given as

$$q_P^3 = \frac{2(r_0+\delta)^2}{r_0^3}\cdot\frac{T}{\rho_v L}\cdot\frac{1}{f(T)}\cdot\rho g h \quad (14)$$

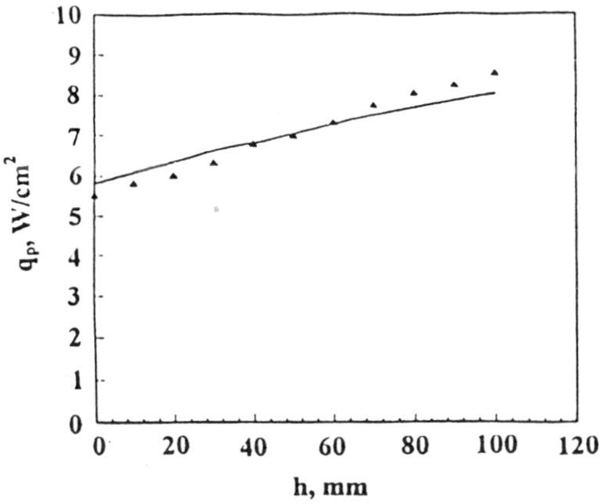

Figure 6. The peak heat flux density as a function of the immersion depth h. HeIIs bath at 2.0K(PtRh10 wire with a diameter 40μm. Δ -experimental result[16]).

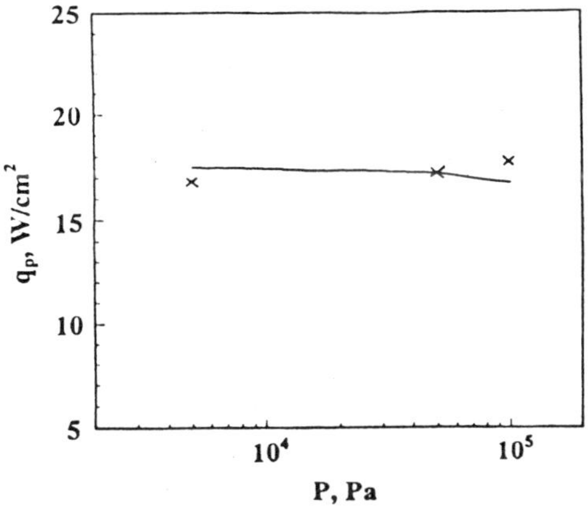

Figure 7. Pressure effect on the peak heat flux density of HeII bath at 2.0K ($P>P_\lambda$, ×: experimental result with a 51μm diameter RhFe wire).

where r_0, δ are the radius of the sample wire and the formed GHe thickness respectively, and ρgh is the hydrostatic head, ρ_v is the density of vapor, L is the latent heat of vaporization. The above equation may show that if the hydrostatic head is zero, then the q_P value is zero, which is in contradiction to experimental results both for nearly zero immersion depth ($h \to 0$) and zero gravity ($g \to 0$)[16] . The solution to the above problems is the introduction of pressure terms.

In HeIIs Bath

As is known, in a HeIIs bath, if film boiling is initiated, a GHe film will surround the heater. In the calculation of peak heat flux density q_P, a subcooling of $\Delta T_{\max}$ should be overcome which is caused by hydrostatic head $\Delta p_{\max} = \rho g h$ shown as

$$\Delta T_{\max} = \frac{T}{\rho_v L} \cdot \rho g h \tag{15}$$

Gradt[16] has noticed the limitation of Eq.(15) and Eq.(14) and suggested to introduce Vander-Waals pressure, however it is still impossible for real calculations. In reality, in a film boiling state, the Vander-Waals pressure can be determined by the Vander-Waals equation

$$(p + \frac{a}{v^2})(v - b) = RT \tag{16}$$

Where a and b are Vander-Waals constants for helium gas, and v is the specific volume of GHe. the Vander-Waals pressure is

$$\Delta p_A = \frac{a}{v^2} = a\rho_v^2 \tag{17}$$

thus the maximum subcooled pressure of the GHe film in the film boiling state is

$$\Delta p_{\max} = \rho g h + a\rho_v^2 \tag{18}$$

Thus a normalized equation to calculate the peak heat flux density q_P is

$$q_P^3 = \frac{2\Phi(r_0)}{r_0} \cdot \frac{T}{\rho_v L} \cdot \frac{1}{f(T)} \cdot (\rho g h + a\rho_v^2) \tag{19}$$

The above theory is valid for noiseless film boiling in HeIIs bath. If the experimental results of Gradt, et al.[16] are taken (Fig.6) , and $r_0 = 20\,\mu m$, T=2.0K, the evaluated parameters in Eq.(17,19) are $\Phi(r_0) = 0.011, \Delta p_A = 120\,Pa$. Thus if $h \to 0$, q_P=5.8 W/cm^2, which is close to the experimental result 5.5 W/cm^2 [16], and the theoretical results (the solid line) agree the experimental data well shown as Fig.6.

In HeIIp Bath

As is shown that in case of T_B not close to T_λ, Eq.(7,11) can be used to calculate the peak heat flux density of HeIIp at 0.1 Mpa, however it can be also applied to HeIIp at various bath pressures , as shown in the following:

$$q_P = [\frac{2\phi(r_0)}{r_0}\int_{T_B}^{T_\lambda}\frac{dT}{f(T)}]^{1/3} \tag{20}$$

where the pressure effect contributes mainly to the values of T_λ, f(T) and also $\phi(r_0)$. The above theory can be compared with experimental work. Steady state peak heat flux density q_p was measured earlier[2,17], in which a RhFe wire (r_0=25.5μm) was horizontally immersed in HeIIp bath. Fig.7 shows the q_P values with respect to various bath pressures, the solid line representing the calculated result of Eq.(20) for $\phi(r_0)$=0.0709 , thus it is seen that the theory agrees the experimental data well. The discrepancy is caused by the neglect of pressure effect on $\phi(r_0)$. Anyway the developed theory can be applied for data treatment.

DISCUSSIONS

Criterie of quantum turbulence onset, boundary phenomena, temperature and pressure effects are related to heat transfer in HeII baths. These problems are studied with the help of experimental work on time dependent heat transfer to a HeII bath. Other special physical problems related to heat transfer to a HeII bath such as linear thermal conduction and second sound effects will be discussed later.

ACKNOWLEDGMENT

This work is supported by the Chinese National Natural Science Foundation under contract No. 59406010. The author thanks Mr. Zhang Peng for some calculations.

REFERENCES

1. R. Wang, T. Gradt, U. Ruppert, et al. , *Cryogenics* 30 (suppl.): 360 (1990).
2. R. Wang, *Cryogenics* 34: 983 (1994).
3. R. Wang, *J. Phys. D: Appl. Phys.* 28:25 (1995).
4. R. Wang, Cryogenics 35: to be published (1995).
5. R. Wang, Chinese J. Low Temp. Phys. 15:300 (1993).
6. W. F. Vinen, *Proc. Roy. Soc. A* 240: 114 (1957).
7. W. F. Vinen, *Proc. Roy. Soc. A* 240: 128 (1957).
8. W. F. Vinen, *Proc. Roy. Soc. A* 242: 493 (1957).
9. W. F. Vinen, *Proc. Roy. Soc. A* 243: 400 (1957).
10. C. E. Chase, *Phys. Rev.* 127: 361 (1962).
11. C. E. Chase, *Phys. Rev.* 131:1898 (1963).
12. S.W. Van Sciver. " Helium Cryogenics ," Plenum Press, New York (1986)
13. V. Arp, *Cryogenics* 10: 96 (1970)
14. S. Caspi, T.H.K. Frederking, *Cryogenics* 19:513 (1979)
15. D. Gentile, W.V. Hassenzahl. " Advances in Cryogenic Engineering " Vol. 25:385 (1980)
16. T. Gradt, Z. Szücs, H.D. Denner, G. Klipping. " Advances in Croygenic Engineering " Vol. 31:499 (1986)
17. R. Wang. " Time Dependent Heat Transfer in Subcooled Superfluid Helium", Ph. D. Thesis, Shanghai Jiao Tong University, Shanghai (1990)

TRANSIENT HEAT TRANSFER FROM A HORIZONTAL WIRE IN SUBCOOLED He II AT ATMOSPHERIC PRESSURE FOR A WIDE RANGE OF WIRE DIAMETER

M. Shiotsu, K. Hata and A. Sakurai

Institute of Atomic Energy, Kyoto University
Uji, Kyoto 611, Japan

ABSTRACT

Transient heat transfer caused by large stepwise heat inputs to a horizontal wire in subcooled He II at atmospheric pressure was measured for the wire diameters of 0.08, 0.2, 0.5 and 0.7 mm. Liquid temperatures tested were 1.8, 1.9, 2.0 and 2.1 K. The steady-state critical heat flux which is dependent on liquid temperature and wire diameter was well expressed by the authors' correlation based on Gorter-Mellink equations. The lifetime of quasi-steady Kapitza conductance state, which corresponds to that of a certain point on the extrapolation of stead-state Kapitza conductance curve, was systematically measured on the wires: the quasi-steady state rapidly changes to film boiling regime at the end of the lifetime. The lifetime at an excess heat flux beyond steady-state critical heat flux, is longer for a larger diameter wire, though the steady-state critical heat flux is lower. Excessive heat flux beyond steady-state critical heat flux, when integrated from the time that the heat flux reaches the steady-state critical heat flux to the end of the quasi-steady state lifetime is almost constant and it is independent of the step heights and of the initial waveform during the rise time to the step height. A correlation for the average integrated value was given as a function of the wire radius and bulk liquid temperature. The lifetimes for the ideal step heat inputs (rise time = 0) are given from the correlation.

INTRODUCTION

Superfluid liquid helium, especially that under a subcooled condition at atmospheric pressure is expected as a coolant for large scale superconducting magnets for superconducting magnetic energy storage systems and experimental fusion facilities because of its excellent cooling properties. The study of steady-state and transient heat transfer on a solid surface in He II for various subcooling conditions is necessary to evaluate whether a He II cooled superconducting magnet will quench when the conductor experiences instantaneous thermal disturbances with peak heat fluxes far higher than the steady-state critical heat flux.

The steady-state and transient heat transfer in He II have been investigated experimentally for a one-dimensional heat flow system, namely, that for a metal heater positioned at the end of a He II filled constant cross section duct or tube whose sides are insulated and other end is connected to a He II tank, and two-dimensional heat flow system such as that for a test wire in a He II bath. The one dimensional steady-state heat transfer experiments[1-3], and transient heat transfer experiments[4,5] caused by step heat inputs have been analyzed well in terms of two-fluid hydrodynamics and theory of mutual friction. The two dimensional steady-state heat transfer experiments under saturated condition were performed by many workers[6-10]. Effects of liquid head and wire diameter on the critical heat flux were clarified by these studies. The two dimensional transient heat transfer experiments were performed for a cylinder by Van Sciver and Lee[9] and for a thin wire by Gradt et al.[11]. The existence of a time delay before the onset of film boiling was observed in one-dimensional and two-dimensional transient heat transfer phenomena caused by stepwise heat inputs.

The authors investigated steady-state[12-17] and transient[12-15] heat transfer from a horizontal wire in a He II bath caused by stepwise heat inputs. Experimental conditions included saturated vapor pressures on the liquid surface of the bath with a wide range of liquid temperatures and liquid heads, and subcooled conditions at atmospheric pressure for various liquid temperatures. They[13,14] derived an equation for the steady-state critical heat flux from a horizontal wire by modifying the solution obtained from the Gorter-Mellink equations. They[16,17] showed that their steady-state critical heat flux data for wire diameters ranging from 0.08 to 1.2 mm were well described by the correlation. The transient heat transfer processes under these saturated and subcooled conditions were such that the heat flux increased along the extrapolation of the steady-state Kapitza conductance curve, stopped increasing and resided on the extrapolated curve for a period of time designated as the lifetime, and then departed from the Kapitza conductance regime to film boiling.

The purpose of the present study is twofold. First is to obtain the experimental data of the lifetime for stepwise heat inputs to single horizontal wires of various diameters in subcooled He II at atmospheric pressure. Second is to derive the correlation of the lifetime for wide ranges of subcoolings and cylinder diameters.

EXPERIMENTAL APPARATUS AND METHOD

Details on the experimental apparatus were already reported[16,17]. The Claudet type cryostat used is 45 cm in inner diameter and 157 cm in height, with a liquid helium content of about 150 L. The lower part of the cryostat constitutes the He II compartment (He II bath with no free surface) with a volume of about 74 L. Four horizontal test wires are supported in the He II compartment.

Single horizontal test wires with the diameters of 0.08, 0.2, 0.5, 0.7 and 1.2 mm were used. The 0.08 mm diameter test wire was fabricated from Au-Mn (0.9 atomic %) alloy and those with the larger diameters were fabricated from Pt-Co (0.5 atomic %) alloy. The electrical resistivity versus temperature curve for these alloys has a positive gradient over the entire experimental range from liquid helium temperatures to several hundred degrees Kelvin; the latter alloy was used for thicker diameter wires because it has higher sensitivity in resistance thermometry. Two fine $30 \mu m$ diameter platinum wires were spot welded at around 13 mm from each end of the horizontal wires. The effective lengths of the test wires between the potential taps on which critical heat fluxes were measured were about 60 mm. The electrical resistance versus temperature relationship of each wire was calibrated in a temperature controlled vessel immersed in liquid helium (control accuracy is 0.05 K for temperature settings below 10 K, 0.1 K from 10 to 50 K, and 0.3 K up to 300 K).

The heating current to the test wire was supplied by a power amplifier. The power amplifier used for the wires with the diameter larger than 0.2 mm can supply a direct current of up to 300 A at a power level of 6 kW; the minimum increasing time for the current from 0 to the maximum value is around 250 μ s. Another power amplifier used for the 0.08 mm wire can supply 20 A maximum with the minimum increasing time of 60 μ s. The input signals of the power amplifiers are controlled by an analog computer so that the heat generation rate in the test wire agrees with a desired value.

The average temperature of the test wire was measured by resistance thermometry using a double bridge circuit which included the wire as a branch of the bridge. The output voltages of the bridge circuit were amplified and passed to a digital memory system. These voltages are simultaneously sampled at the shortest available time interval, 0.5 μ s, and stored in a 64 k word/channel memory. The stored data were transferred to a digital computer. The heat generation rate of the wire was determined from the heating current for the wire and the voltage difference between potential taps

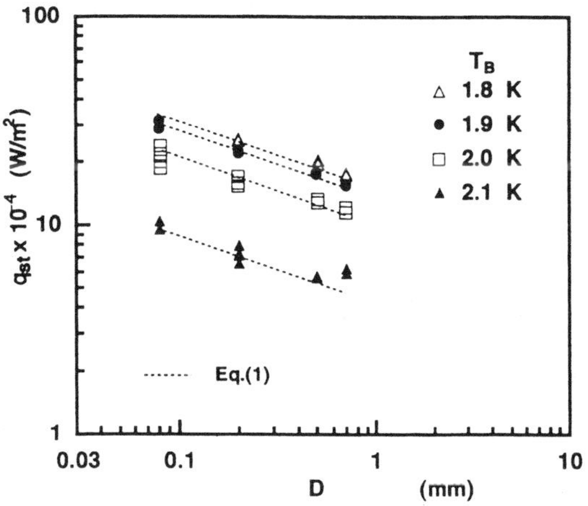

Fig. 1 Steady-state critical heat flux versus test wire diameter with liquid temperature as a parameter.

on the wire. The heat flux from the wire surface is the difference between the heat generation rate per unit surface area and the rate of change of energy storage in the wire obtained from the faired average temperature versus time curve. The wire surface temperature was obtained from the measured average temperature and heat generation rate by solving the conduction equation for the wire. However, when the experimental error was considered, the difference between the average and surface temperatures was negligibly small for the wire tested. Experimental error is estimated to be 0.1 K in the heater surface temperature and 2 percent in the heat flux.

RESULTS AND DISCUSSION

Steady-State Critical Heat Flux

The authors[13,14] have already presented the following correlation for the steady-state critical heat flux, q_{st}, on a horizontal test wire in subcooled He II (for $T_{sat}(P_L) > T_\lambda$).

$$q_{st} = K\left[\frac{2}{r_0}\int_{T_B}^{T_\lambda}\frac{1}{f(T)}dT\right]^{\frac{1}{3}} \tag{1}$$

where

$$f(T)^{-1} = g(T_\lambda)[T_R^{6.8}(1 - T_R^{6.8})]^3 \ ,$$
$$g(T_\lambda) = \rho^2 s_\lambda^4 T_\lambda^3 / A_\lambda \ , \qquad T_R = T/T_\lambda \ ,$$
$$s_\lambda = 1559 \quad \mathrm{J/(kg\ K)} \ , \qquad A_\lambda \simeq 1150 \quad \mathrm{m\ s/kg} \ .$$

where $T_{sat}(P_L)$ is the saturation tempearture for the pressure at the level of test wire; T_B and T_λ are bulk liquid temperature and λ temperature, respectively; and r_0 is the test wire radius. The modification coefficient K, in Eq.(1) was determined to be 0.58 by using the experimental data of q_{st} for various diameter wires. The exponent of 6.8 in the equation of $f(T)^{-1}$ was changed from the original value of 5.7 for one-dimensional heat transfer in He II; the bulk liquid temperature at which the maximum of $f(T)^{-1}$ occurs was shifted from 1.923 K to 1.96 K with this modification.

Figure 1 shows the q_{st} data versus test wire diameter, D, with bulk liquid temperature, T_B, as a parameter. The curve predicted from Eq.(1) for each T_B is also shown in the figure for comparison. The experimental data are well expressed by the correlation.

Transient Heat Transfer for Stepwise Heat Inputs

To realize experimentally the ideal step heat input without any rising time is difficult. The stepwise heat input with the maximum available increasing rate (minimum rise time) is called in this work the pseudo-step heat input. The rising time up to the step height for the pseudo-step heat input is around 60 μ s for the heat input

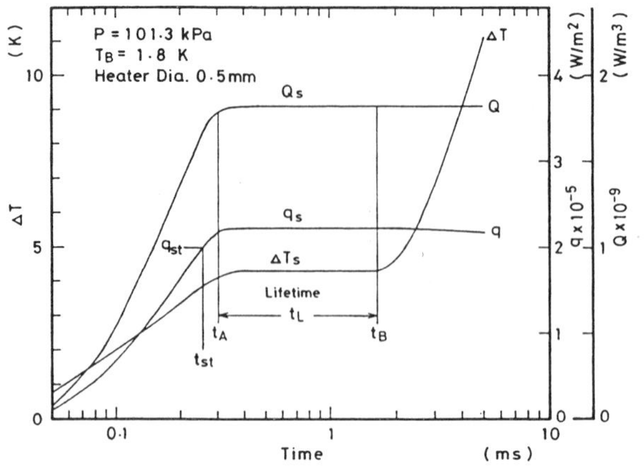

Fig. 2 Time dependencies of the heat flux and surface temperature difference for a pseudo-step heat input.

to the 0.08 mm wire, and around 250 μ s for the heat input to the thicker wires as mentioned above. An exponential-step heat input, which increases exponentially in time ($Q = Q_o e^{t/\tau}$) at first with the period τ of 0.5 ms and then keeps a constant value, was used to investigate the effect of the heat input waveform in its early stage.

Transient heat transfer coefficients caused by pseudo-step heat inputs with the heights larger than the values corresponding to the steady-state critical heat fluxes, q_{st}, were measured for 0.08, 0.2, 0.5 and 0.7 mm diameter wires for T_B of 1.8, 1.9, 2.0 and 2.1 K at atmospheric pressure.

Typical time traces of the pseudo-step heat input Q, heat flux q, and surface temperature difference $\Delta T (= T_w - T_B)$ are shown in Fig. 2. Initially, the heat input rapidly increases in time and then it takes a constant value, Q_s, after $t = t_A$. As shown in the figure, the surface temperature difference and the heat flux remain constant at ΔT_s and q_s, respectively, for a certain duration $(t_B - t_A)$, then they begin to increase and decrease, respectively, at $t = t_B$. The duration $t_L = t_B - t_A$ is defined as the lifetime of the quasi-steady state heat flux q_s. It was first found by the authors[12] that the transient heat flux for the stepwise heat input increases along the Kapitza conductance curve and the curve's extrapolation, and then reaches a quasi-steady-state $(\Delta T_s, q_s)$ on the extrapolated curve.

The lifetimes, t_L, of the quasi-steady-state heat fluxes due to the pseudo-step heat inputs on the 0.2 mm diameter wire for T_B of 1.8, 1.9, 2.0 and 2.1 K at atmospheric pressure are shown versus q_s in Fig. 3 as a typical example of the experimental data.

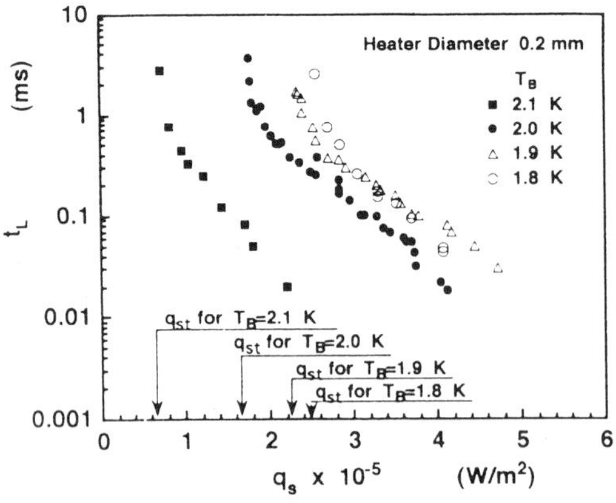

Fig. 3 Lifetime versus quasi-steady heat flux with liquid temperature as a parameter on 0.2 mm-diameter wire.

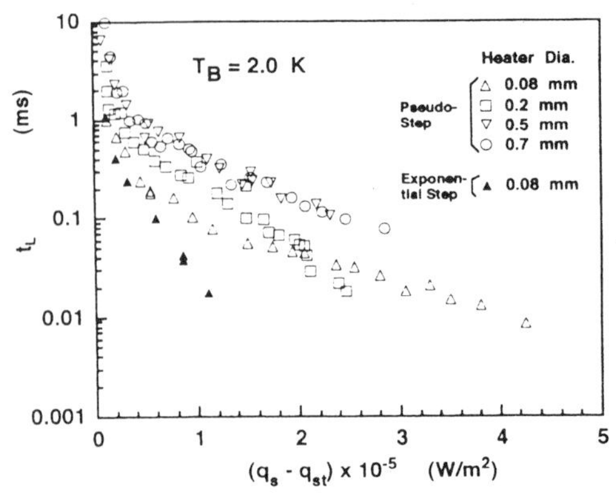

Fig. 4 Relation between the t_L and $(q_s - q_{st})$ for pseudo-step heat inputs to various diameter wires.

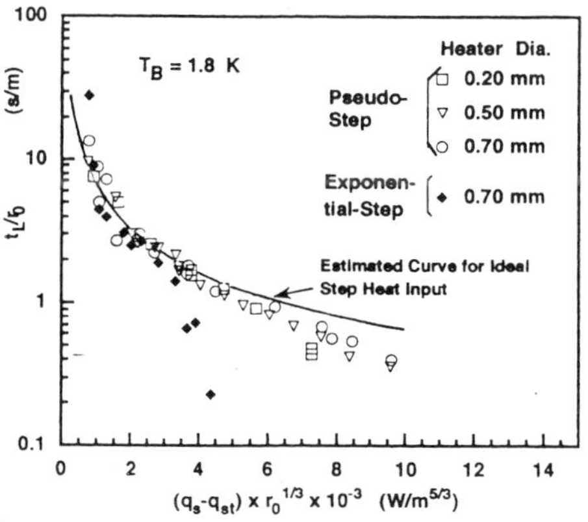

Fig. 5 Relation between the t_L/r_0 and $(q_s - q_{st})r_0^{1/3}$ for various diameter wires at $T_B = 1.8$ K.

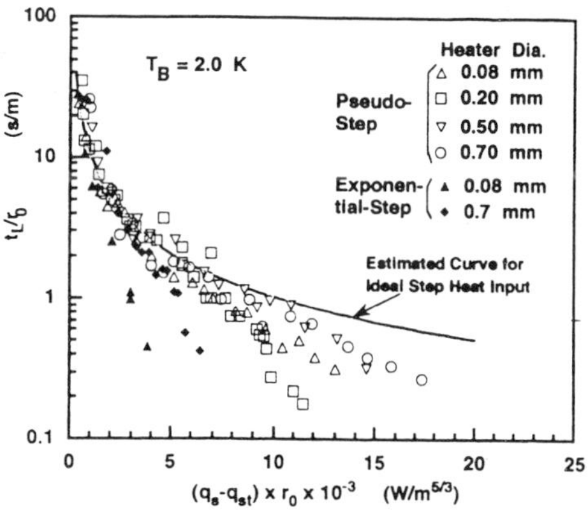

Fig. 6 Relation between the t_L/r_0 and $(q_s - q_{st})r_0^{1/3}$ for various diameter wires at $T_B = 2.0$ K.

The lifetime is infinite for a step height corresponding to a heat flux lower than the steady-state critical heat flux which is higher for lower bulk liquid temperature. With increasing the step height and the corresponding quasi-steady heat flux q_s from the steady-state critical heat flux, the lifetime of the quasi-steady state decreases. The decreasing rates for the T_B of 2.1 K and 1.8 K appear to be greater than those for 1.9 K and 2.0 K.

The values of t_L due to the pseudo-step heat inputs for the wires with the diameters of 0.08, 0.2, 0.5 and 0.7 mm are shown against excess heat flux beyond steady-state critical heat flux, $(q_s - q_{st})$, in Fig. 4. Bulk liquid temperature in this case is 2.0 K. As shown in the figure, the lifetime at an excess heat flux is longer for a larger diameter wire, though the steady-state critical heat flux is lower for a larger diameter as already shown in Fig. 1. The data of t_L on 0.08 mm-diameter wire for the exponential-step heat inputs are also shown in the figure in comparison with the data on the same wire for the pseudo-step heat inputs. The lifetime for the exponential-step heat input decreases almost in agreement with that for the pseudo-step heat input with the increase in excess heat flux up to around 5×10^4 W/m^2, and decreases at a much greater rate for the excess heat flux higher than the value. The effect of the heat input waveform in the early stage was observed significantly. It is assumed from this fact that the data of t_L for the pseudo-step heat inputs would also be affected by the initial waveform up to the step heights, especially in the high excess heat flux regime.

It was already shown in Fig. 1 that the steady-state critical heat flux, q_{st}, was well expressed by the authors' theoretical correlation given by Eq.(1): the q_{st} for a certain liquid subcooling is proportional to $r_0^{-1/3}$. Figures 5 and 6 show the relation of t_L/r_0 versus $(q_s - q_{st}) \times r_0^{1/3}$ on various diameter wires due to the pseudo-step heat inputs for T_B of 1.8 and 2.0 K, respectively. As seen from these figures, the values of t_L/r_0 on various diameter wires almost agree with each other and can be expressed as a single curve on this graph, although relatively large difference can be observed in the high excess heat flux regime for each wire where the effect of the heat input waveform in the early stage is significant.

The integrated values, $(1/r_0^{2/3}) \int_{t_{st}}^{t_B}(q - q_{st})dt$, obtained from each runs for the pseudo-step heat inputs to various diameter wires for the liquid temperatures of 1.8 and 2.0 K are shown on the graph of $(1/r_0^{2/3}) \int_{t_{st}}^{t_B}(q - q_{st})dt$ versus $(q_s - q_{st})$ in Figs. 7 and 8, respectively. At the time t_{st}, the increasing heat flux reaches the q_{st} , and at the time t_B, the lifetime for the value of q_s ends as shown in Fig.2. The integrated values for all the experimental runs obtained for each liquid temperature are almost constant and they are independent of the step heights and of the initial waveforms during the rise time to the step heights.

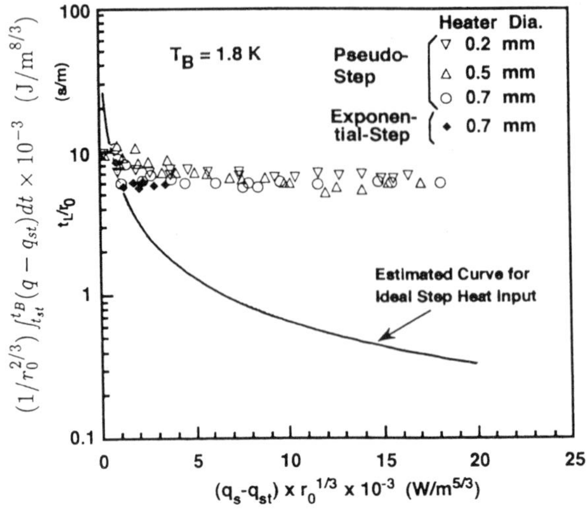

Fig. 7 Integrated values of excess heat fluxes beyond q_{st} until the end of lifetimes for pseudo-step and exponential-step heat inputs to various diameter wires at $T_B = 1.8$ K.

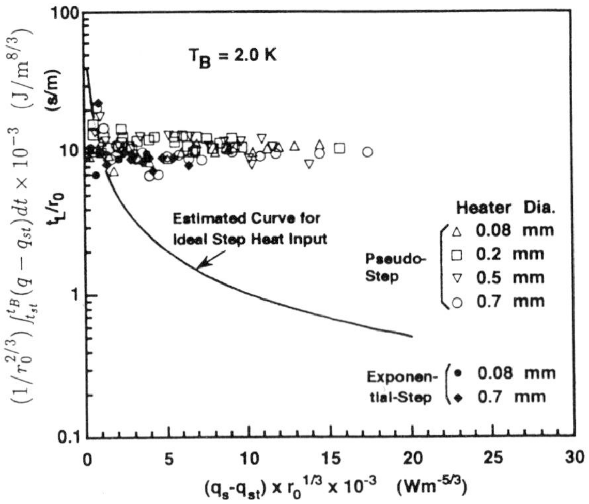

Fig. 8 Integrated values of excess heat fluxes beyond q_{st} until the end of lifetimes for pseudo-step and exponential-step heat inputs to various diameter wires at $T_B = 2.0$ K.

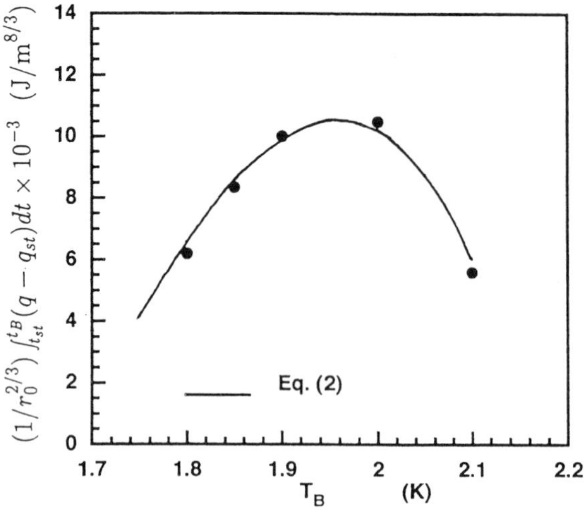

Fig. 9 Average of integrated values on various diameter wires for each liquid temperature.

The average of the integrated values for each bulk liquid temperature is shown in Fig. 9. Most of the experimental data for each bulk liquid temperature exist within ±20% of the average value. A curve representing the average value has a maximum at T_B of around 1.96 K which is in agreement with the value for which $f(T)^{-1}$ in Eq.(1) has a maximum as mentioned previously. The curve can be expressed by the following equation,

$$[(1/r_0^{2/3})\int_{t_{st}}^{t_B}(q - q_{st})dt]_{av.} = F(T_B) \qquad \mathrm{J/m^{8/3}} \tag{2}$$

where

$$F(T_B) = 9.7928 \times 10^5 - 1.8620 \times 10^6 T_B + 1.1437 \times 10^6 T_B^2 - 2.2748 \times 10^5 T_B^3$$

for $1.8 \leq T_B \leq 2.1$.

It is possible to evaluate the lifetimes due to stepwise heat inputs on various diameter wires from Eqs.(1) and (2). For instance, the lifetimes for the ideal step heat inputs which cannot be realized experimentally can be estimated as follows. By assuming that the integrated values, $\int_0^{t_B}(q - q_{st})dt$, for the ideal step heat inputs at a given liquid temperature are the same as the integrated values, $\int_{t_{st}}^{t_B}(q - q_{st})dt$, obtained for pseudo-step heat inputs with rising times up to Q_s for the same liquid temperature, the lifetimes of the quasi-steady heat fluxes, q_s, corresponding to the ideal step heights for each liquid temperature are given by

$$t_L = [1/(q_s - q_{st})][\int_{t_{st}}^{t_B}(q - q_{st})dt]_{av.} = [1/(q_s - q_{st})]r_0^{2/3}F(T_B) \tag{3}$$

The curves of (t_L/r_0) versus $(q_s - q_{st})r_0^{1/3}$ for the ideal step heat inputs derived from Eq.(3) for T_B of 1.8 and 2.0 K are shown in Figs. 7 and 8. The curves are also shown in Figs. 5 and 6 in comparison with the corresponding experimental data for pseudo-step heat inputs. The experimental values of t_L on each diameter wire for the pseudo-step heat inputs become lower than the evaluated values for the ideal step heat inputs for larger values of $(q_s - q_{st})$.

If one wishes to consider the impact on conductor stability of a transient heat pulse in a subcooled He II cooled superconducting magnet, the transient heat transfer caused by the ideal pulse heat inputs with the heights Q_s and the width t_w will be of interest; the transient heat transfer is in the quasi-steady state Kapitza conductance regime and a rapid rising of the surface temperature does not occur as long as the width of the pulsed heat input is equal to or shorter than the lifetime of the quasi-steady heat flux mentioned above.

CONCLUSIONS

Heat flux for a stepwise heat input increases along the Kapitza conductance curve and its extrapolation, attaining a quasi-steady state on the extrapolated curve. This quasi-steady state lasts a period of time which is defined as the lifetime for the quasi-steady state Kapitza conductance heat flux.

The lifetime for a pseudo-step heat input depends on the value of Q_s; it becomes shorter with increased values of Q_s. The relationship between the quasi-steady state heat flux and its lifetime was quantitatively clarified for the wire diameters of 0.08, 0.2, 0.5 and 0.7 mm for bulk liquid temperatures ranging from 1.8 to 2.1 K at atmospheric pressure.

The lifetime at an excess heat flux beyond steady-state critical heat flux, $(q_s - q_{st})$, is longer for a larger diameter wire, though the steady-state critical heat flux is lower for a larger diameter. The relations between the lifetime and $(q_s - q_{st})$ for various diameter wires almost agree with each other on t_L/r_0 versus $(q_s - q_{st})r_0^{1/3}$ graph.

The excessive heat flux beyond q_{st} when integrated from the time that the heat flux reaches the q_{st} to the end of the quasi-steady state lifetime is almost constant independent of the step height and of the initial heat input waveform up to the step height. The integrated value for a wire has a maximum at T_B of around 1.96 K, and it is larger for thicker wires being proportinal to $r_0^{2/3}$. The correlation for the average integrated value was given as a function of the wire radius and bulk liquid temperature.

The lifetimes of the quasi-steady state heat fluxes for the stepwise heat inputs can be estimated for various diameter wires from this correlation and the theoretical correlation of q_{st} already presented by the authors. The lifetimes for the ideal step heat input without any rising times up to the step heights are given from these correlations.

ACKNOWLEDGMENT

This research was supported by a Grant-in-Aid for Scientific Research from the Ministry of Education, Science and Culture, Japan.

REFERENCES

1. Bon Mardion, G., Claudet, G., Seyfert, P., Steady State Heat Transport in Superfluid Helium at 1bar, in: "Proceedings of 7th International Cryogenic Engineering Conference", IPC Science and Technology Press, London (1978), p.441.
2. Van Sciver, S.W., Kapitza Conductance of Aluminum and Heat Transport through Subcooled He II, *Cryogenics,* 18:521 (1978).
3. Lottin, J.C., He II Experimental Facilities at Saclay, "Advances in Cryogenic Engineering", Vol.27, Plenum Press, New York, (1982), p.431.
4. Van Sciver, S.W., Transient Heat transport in He II, *Cryogenics,* 19:385 (1979).
5. Sayfert, P., Lafferanderrie, J., and Claudet, G., Time Dependent Heat Transport in Subcooled Superfluid Helium, *Cryogenics,* 22:401 (1982).
6. Frederking, T.H.K., Wärmeübergang bei der Verdampfung der verflüssigten Gase Helium und Stickstoff, *Forschung,* 27:17 (1961).
7. Goodling, J.S. and Irey, R.K., Non-Boiling and Film Boiling Heat Transfer to a Saturated Bath of Liquid Helium, "Advances in Cryogenic Engineering", Vol.14, Plenum Press, New York (1969), p.159.
8. Haben, R.L., Madsen, R.A., Leonard, A.C., and Frederking, T.H.K., Breakdown of Superfluidity for Cylinders in Saturated Liquid Helium II, "A dvances in Cryogenic Engineering", Plenum Press, New York, (1972), p.323.
9. Van Sciver, S.W., and Lee, R.L., Heat Transfer from Circular Cylinders in He II, in: "Cryogenic Process and Equipment in Energy Systems", ASME Publication No. H00164 (1981), p.147.
10. Lemieux, G.P., and Leonard, A.C., Maximum and Minimum Heat Flux in Helium II for a 76.2-μ Diameter Horizontal Wire at Depths of Immersion up to 70 centimeters, "Advances in Cryogenic Engineering", Vol.13, Plenum Press, New York (1981), p.624.
11. Gradt, T., Wang, R., Ruppert, U., and Luders, K., Transient Heat Transfer to Superfluid Liquid, "Advances in Cryogenic Engineering",Vol. 35A, Plenum Press, New York (1989), p.117.
12. Shiotsu, M., Hata, K., and Sakurai, A., Transient Heat Transfer from a Horizontal wire in Superfluid Helium Caused by Exponential and Step Heat Inputs, in: "Superfluid Helium Heat Transfer", J.P. Kelly and W.J. Schneider ed., ASME Publication No. HTD-Vol.134.(1990), p.9.
13. Sakurai, A., Shiotsu, M., and Hata, K., Transient Heat Transfer for Large stepwise Heat Inputs to a Horizontal Wire in Saturated He II, "Advances in Cryogenic Engineering", Vol.37A, Plenum Press, New York (1992), p.25.
14. Shiotsu, M., Hata, K., and Sakurai, A., Transient Heat Transfer for Large Stepwise Heat Inputs to a Horizontal Wire in Subcooled He II, "Advances in Cryogenic Engineering", Vol.37A, Plenum Press, New York, (1992), p.37.
15. Shiotsu, M., Hata, K., and Sakurai, A., Steady and Unsteady Heat Transfer from a Horizontal Wire in a Pool of Subcooled He II, in "Heat Transfer and Superconducting Magnetic Energy Storage",J.P. Kelly and M. J. Supercznski ed., ASME Publication No. HTD-Vol.211 (1992), p.19.
16. Shiotsu, M., Hata, K., and Sakurai, A., Effect of Test Heater Diameter on Critical Heat Flux in He II, "Advances in Cryogenic Engineering", Vol.39, Plenum Press, New York (1994), p.1797.
17. Shiotsu, M., Hata, K., Takeuchi, Y., Hama, K., and Sakurai, A., Critical Heat Flux on Single Horizontal Wires in Superfluid Helium at Pressures, in: "Heat Transfer 1994 (Proc. of the 10th Internat. Heat Transf. Conf.)", G.F. Hewitt ed. Vol.5, Taylor and Francis, USA (1994), p.141.

VISUALIZATION STUDY OF HIGHLY TRANSIENT THERMO-FLUID DYNAMIC PHENOMENA IN He II

T. Iida, M. Murakami, T. Shimazaki,
H. Nagai and T. Furukawa

Institute of Engineering Mechanics
University of Tsukuba
Tsukuba 305 Japan

ABSTRACT

A laser holographic interferometer is applied to the study of the thermo-fluid dynamic phenomena in superfluid helium. A pressure wave accompanied by a thermal wave and a vapor wave by a thermal wave incidence to a liquid-vapor interface are investigated, all excited by pulse heating from a plane heater under the saturated vapor pressure condition. A new technique for interferometry - finite-fringe method - is introduced in order to observe the above phenomena not only visually but also quantitatively. It is understood from the visualized photos that a finite amplitude pressure wave of the first sound origin propagates with a speed of sound, corresponding to the local Mach number of unity in He II, in immediate response to the onset of heating, being ahead of a thermal shock wave. The density variation and propagating speed of the vapor wave with the variable heat flux from the heater are measured from the visualized photos. It is found that the two have their own limits with applied heat flux.

INTRODUCTION

The thermo-fluid dynamic phenomena in He II, a thermal shock wave and boiling, have been investigated by many researchers. It is well-known to us that a thermal shock wave transfers entropy in a form of wave in He II and that the diffusive process becomes remarkable due to the accumulation of the quantized vortices, thus leading to boiling. However, a pressure wave except those two phenomena is detected by some researchers[1-5], with different measurements. In their papers it is said that the pressure wave is generated by a formation of vapor layer adjacent to a heater, that is, boiling. If so, the onset of boiling becomes about one hundred times earlier than the rapid rise in temperature that is detected by a conventional thermometer. This is remarkably important to fabricate a cryogenic device as a superconducting magnet to which a high heat flux is applied.

We successfully visualized[6] a vapor wave from a liquid-vapor interface by the incidence of a thermal shock wave and presented it at the last CEC. A study on an evaporation is very difficult because it is practically hard to set a single material environment and to uniformly heat the single material. However, the He II-helium vapor environment offers such a single material environment and also the uniform heating by a thermal pulse. Therefore we have chance of studying the process of evaporation in such a cryogenic system, measuring the governing parameters which are related to the process.

In the present paper the experiments of the thermo-fluid dynamic phenomena, a pressure wave and a vapor wave, resulted from highly transient heat transfer in He II are conducted by using the laser holographic interferometer visualization. This technique gave us the successful visualization of thermal shock wave of three types, as presented last time. This time another interferometry method - finite-fringe - is also introduced in order to measure the density, temperature or pressure variation/profile and the propagating speed of a pressure wave and a vapor wave.

EXPERIMENTAL SET-UP AND PROCEDURE

Laser Holographic Interferometry

The arrangement of the optical elements for the laser holographic interferometer is sketched in Fig. 1. The density variation in He II appear in the form of interference fringes with the holography interferometer system. The variation in the refractive index of He II due to the density variation gives the variation in optical path length along which the laser light passes. In the present study, we applied both the finite-fringe method and infinite-fringe method complementarily. In the finite-fringe method, the parallel fringes are intentionally produced in a quiescent state by adjusting an optical element and then they are shifted according to the variation in the refractive index in question. In this method the interval between two adjoining fringes is finite in contrast to infinite interval in a quiescent state in the infinite-fringe method. The variation in the refractive index in a medium in question would cause the deflection of the fringes. The amount of shift of one reference fringe can directly represent a density profile of interest in a local area in the finite-fringe method, while each fringe in a reconstructed image of a hologram in the infinite-fringe method can be regarded as a density contour.

The density can be approximately related linearly to the refractive index, n, by

$$(n-1)/(n_0-1) = \rho/\rho_0 \tag{1}$$

where n_0 and ρ_0 are the refractive index and density at the initial condition, and n and ρ are the refractive index and density at the condition of interest. This results in

$$\Delta n = [(n_0-1)/\rho_0]\Delta\rho \tag{2}$$

Thus the density variation corresponding to one interference fringe is

$$\Delta\rho = \lambda\rho_0/[(n_0-1)L] \tag{3}$$

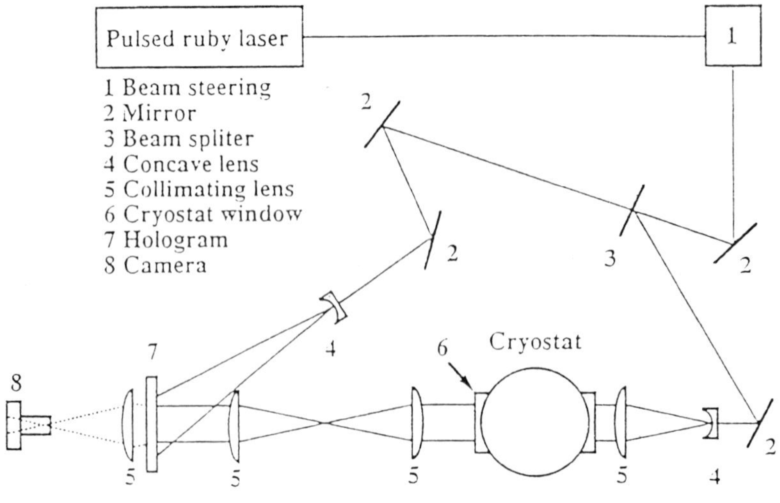

Figure 1. Arrangement of the optical elements for the laser holographic interferometer.

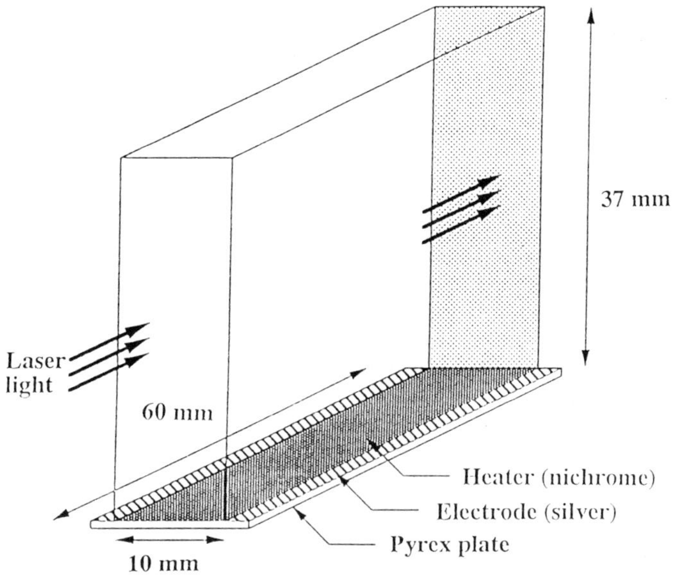

Figure 2. Sketch of the thermal shock tube installed in the cryostat.

where L is the optical path length across the shock tube and λ is the wavelength of the laser we used. The refractive index data of He II is given by Edwards[7].

Thermal Shock Tube and Thin Film Heater

A sketch of the thermal shock tube is shown in Fig.2, which is soaked in He II in the test section of the cryostat. It is a rectangular parallelepipeds surrounded with two optical glass plates, a ceiling, two wedge-shaped side walls made of Bakelite and a planar heater at the bottom. It is 10 mm wide, 60 mm deep between window glass plates and 37 mm high. The image of the thermal shock tube is of two dimensions, looking like a rectangular with 10 mm in width and 37 mm in height. The planar heater located at the bottom, which generates highly transient heat flux, is made of Nichrome thin-film (approx. 50 Å-thick) evaporated onto a Pyrex glass plate, whose area is 60 mm x 10 mm. The stripe electrodes on both sides are consisted of silver thin-film, being also evaporated on Nichrome film heating surface. The electric resistance of the heater is approximately 20 Ω. The heat flux released out of the heater is up to 65 W/cm^2. Electric current varying in a rectangular form in time is applied to the heater to generate a thermal pulse and a pressure wave for the observation.

Procedure of Experiment

It is of crucial importance to control the timing of flashes of laser light and an emission of wave from the heater. With the double exposure technique, the first flash of a double-pulsed laser is made for recording an initial quiescent state, say an initial He II temperature or density field, and then following the onset of heating the second flash is done to record a disturbed state due to a thermal pulse and a pressure wave in question. The wavelength of the double-pulsed ruby laser is 694.3 nm for recording hologram. The time interval between the onset of heating and the second flash of laser beam is termed t_H (μs), ranging from several hundred μs to several ms. A hologram information is recorded on a holographic plate, Agfa 10E75. For image reconstruction, a holographic plate is illuminated by continuum Ar-ion laser light with the wavelength 514.5 nm and a reconstructed holographic image showing a fringe field is photographed by a still camera.

EXPERIMENTAL RESULTS AND DISCUSSION

1. Finite-amplitude Pressure Wave

The photos of a finite-amplitude pressure, or first sound, wave propagating in He II are shown in Fig.3-a and b. These two images are recorded with the different methods, that is to say, the infinite- and finite-fringe methods. The He II temperature T (K), peak heat flux Q (W/cm^2) are all the same for two cases and the images are receded a little after the heater switched-off, at a delay times t_D (μs) from the beginning of heating. The fringes in Fig.3-a are considered to correspond to the isobars in this particular case because the temperature variation within the pressure wave is found to be quite small as shown later. The main pressure wave is considered to be almost half the image from the front to the heater, as marked in Fig.3-a and b. It is impossible that the intersection of fringes behind the main pressure wave, as seen in Fig.3-a, may be attributed to the edge effect caused at the heater-side wall corner. These figures can not tell us whether the finite-amplitude pressure wave is a compression or expansion wave. The finite-fringe method can solve this question as clearly shown in Fig.3-b. The interferogram exhibits a pressure wave and a thermal pulse, where the former approaches the top surface and the latter is still in the vicinity of the heater because of a slower propagation speed of the second sound by a factor than that of the pressure wave. The finite-fringes resulted from the thermal pulse are deflected toward the right, representing a positive density variation. Thus it is inferred that the pressure wave is a compression wave because the displacement of the finite-fringes are in the same direction as the thermal pulse.

In Fig.4, the temperature variation due to the compression wave is recorded by a superconducting thermometer. It is very small in amplitude, being within a few mK, and is negative. The latter fact is quite natural because the Maxwell relation

$$\Delta T = \Delta P(\beta TV/C_p) \tag{4}$$

where C_p is the heat capacity per unit volume of He II and β is the thermal expansion coefficient. The relation states that a positive pressure variation causes a negative temperature variation due to the negative thermal expansion coefficient of He II β, in this temperature range. It is observed in Fig.3-b that the compression wave has two rises, steep and gradual ones, in pressure. The very front becomes shock-like as seen from the extremely steep slope of fringes and the following has a more gentle slope with a smaller magnitude in variation.

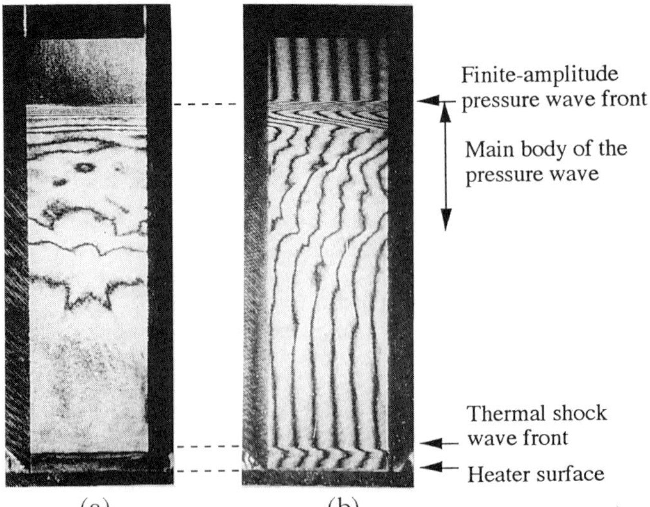

Figure 3. Finite-amplitude pressure wave at lower temperature (T=2.08 K).
(a) Q=55 W/cm^2, t_H=50 μs, t_D=140 μs (infinite-fringe interferogram).
(b) Q=55 W/cm^2, t_H=50 μs, t_D=140 μs (finite-fringe interferogram).

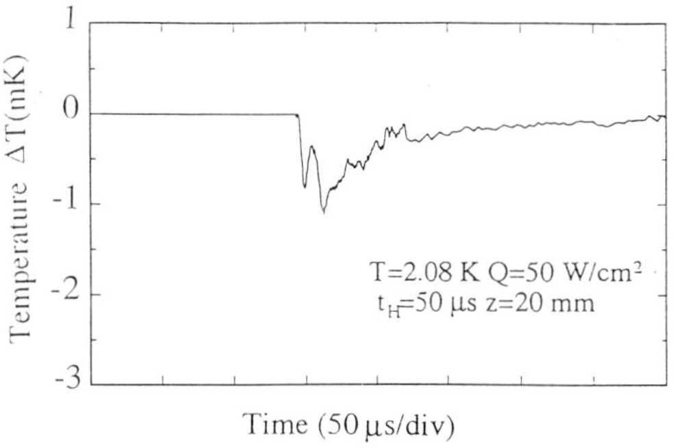

Figure 4. Time variation of a finite-amplitude pressure wave at higher temperature (detected with a superconducting temperature sensor).
T=2.08 K, Q=50 W/cm^2 and t_H=50 μs.

The double peak structure is also seen in the temperature profile shown in shown in Fig.3-b. One may think that the maximum pressure of the wave is beyond the lambda line by judging from the finite-fringe interferogram of the wave. As a matter of fact, the pressure corresponding to the temperature 2.08 K is 8.61 bar on the lambda line and so it is impossible. This is confirmed by measuring the pressure variation of the wave as follows. With the magnitude of the heat flux not only the pressure amplitude but also the profile of the finite-amplitude pressure wave changes. The relation between the heat flux and the maximum amplitude of the finite-amplitude pressure wave is investigated by using interferogram pictures, as shown in Fig.5. The density variation obtained from visualization is converted into the pressure variation from the initial pressure in He II in isothermal process because the temperature variation in the finite-amplitude pressure wave is very small as detected by the superconducting temperature sensor, shown in Fig.3-b. That is,

$$\Delta\rho = (\partial\rho/\partial P)_T \Delta P = \rho_0 K_T \Delta P \qquad (5)$$

where K_T is the isothermal compressibility of He II. We also use Eq.(3) and then we have the pressure variation corresponding to one interference fringe as follows,

$$\Delta P = \lambda/[(n_0-1)LK_T] \qquad (6)$$

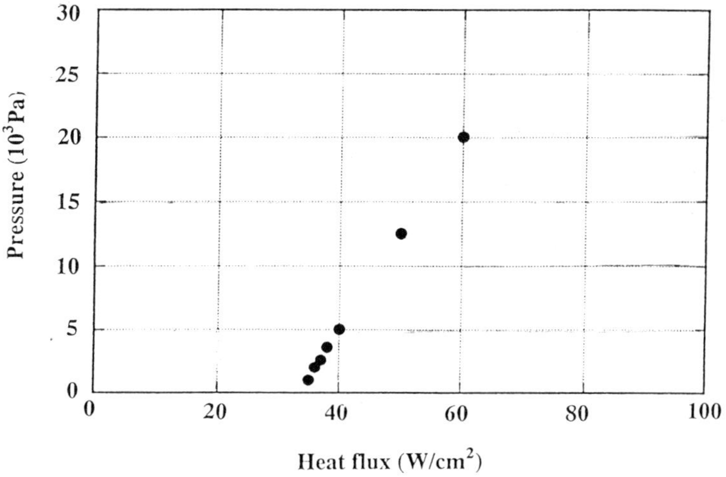

Figure 5. Relation between an applied heat flux and maximum pressure amplitude of a finite-amplitude pressure wave (T=1.74 K and t_H=50 μs).

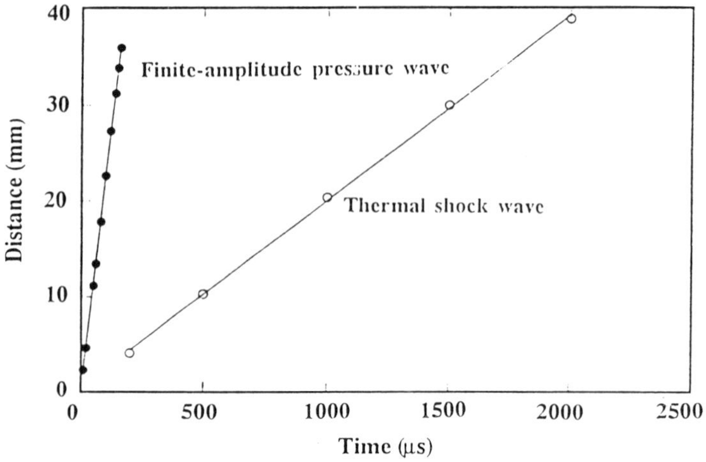

Figure 6. x-t diagram of a finite-amplitude pressure wave (T=2.08 K, Q=60 W/cm^2 , t_H=50 μs) and a thermal shock wave (T=1.74 K, Q=60 W/cm^2 , t_H=50 μs). The propagating speed of the finite-amplitude pressure wave is 224 m/s, while that of the thermal shock wave is 19.4 m/s ,in each condition.

It is found from Fig.5 that the maximum amplitude is linearly proportional to the heat flux applied to the heater.

An x-t diagram is drawn in Fig.6 by plotting the wave front during propagation in visualization pictures, from which the propagation speed is estimated to be 224 m/s, being equal to the speed of first sound in He II at 2.08 K. This diagram also indicates that the time of occurrence of the pressure wave is within as small as a few micro-seconds after the onset of heating. It is naturally considered from this fact that such a sudden occurrence of the compression can not be attributed to a rapid growth of a vapor layer above a heater. A macroscopic vapor layer is observed to be formed about a few ms after the onset of heating though it depends on Q.

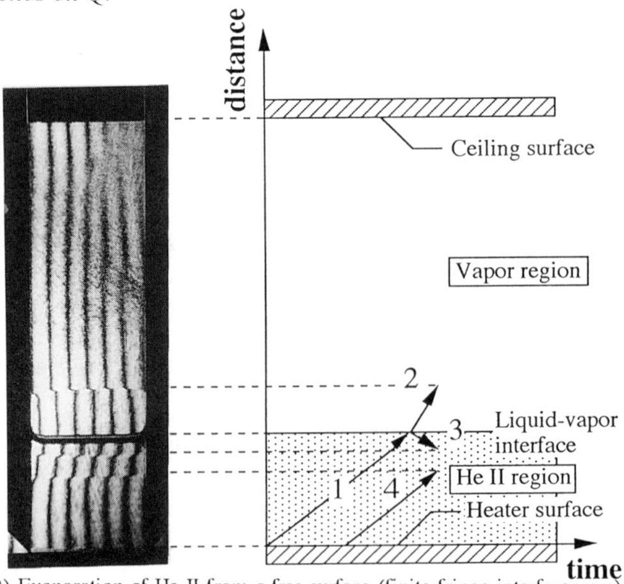

Figure 7. (left) Evaporation of He II from a free surface (finite-fringe interferogram) T=1.74 K, Q=25 W/cm^2 and t_H=200 μs.(right) The histories of 1-incident thermal shock wave front 2-vapor wave front 3-reflected thermal shock wave front 4-incident thermal shock wave tail.

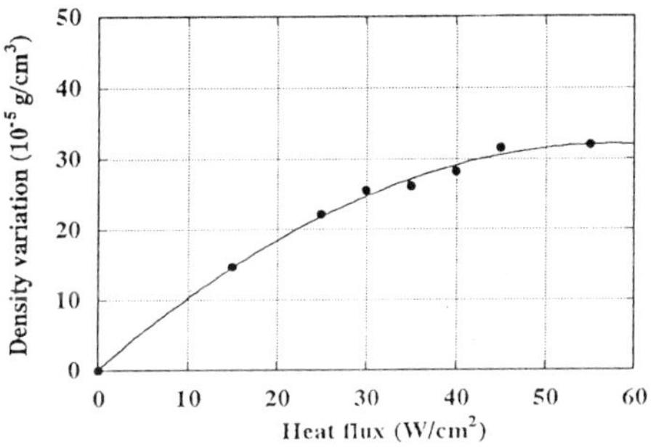

Figure 8. Relation between an applied heat flux and maximum density variation in a vapor wave.

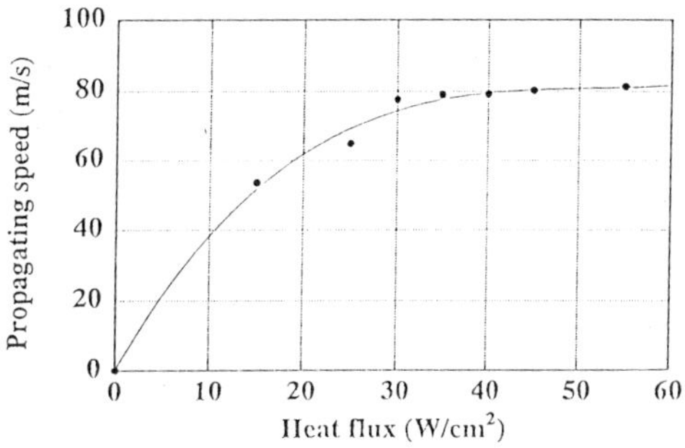

Figure 9. Relation between an applied heat flux and propagating speed of a vapor wave.

2. Evaporation from Free Surface

An evaporation from a He II free surface can be caused by the incidence of a frontal thermal shock wave to the free surface or liquid-vapor interface. This experiment can be conducted in a practically pure liquid-vapor system where nothing but He II and helium vapor exists in solid state at the temperature near the absolute zero. The He II environment also offers an advantage that the peak heat flux and heating duration are clearly defined because input thermal energy is transferred in a wave form, free from the interference of any diffusive processes. The visualization photo of a vapor wave taken with the finite-fringe method is shown in Fig.7. The history of the evaporation process is drawn aside. At the first stage a thermal shock wave is emitted from the heater. Then the thermal shock wave is propagating in He II, reaching the liquid-vapor interface and brings about an evaporation from the interface. At the next stage a vapor wave is formed and propagating in a vapor region of helium. The incident thermal shock wave also reflects on the interface and propagating downward to the heater. Note that a high heat flux with a long heating duration causes not only a thermal shock wave but also boiling. Thus the heating duration is reduced to 50 μs in order to prevent boiling so that all the applied heat flux is transferred by a thermal shock wave, giving rise to an evaporation in the following experiment.

The density variation in evaporation against the applied heat flux is measured, as shown in Fig.8. The density variation is estimated by using Eq.(3). The refractive index data of He vapor corresponding to the vapor temperature is given by Edwards[8]. The density variation increases with the heat flux. However, it gets close to the constant value as the heat flux goes high. This results leads to that the limiting heat flux exists in the evaporation. The relation between the applied heat flux and the propagating speed of a vapor wave is also investigated from the visualized photos, shown in Fig.9. The propagating speed tends to increase with increasing the applied heat flux. It also has the limit to the speed which is almost equal to the speed of sound, 77.61 m/s, at the temperature of vapor 1.74 K. These two facts seem to result in that the density variation of the wave with a heat flux determines the propagating speed.

CONCLUSIONS

1 The occurrence of a finite-amplitude pressure wave is found to be at most 10 μs after heating.

2 The pressure variation of a finite-amplitude pressure wave is measured with visualization, which is not beyond the lambda line.

3 The propagating speed and density variation of a vapor wave has a limit in spite of the increase of applied heat flux.

REFERENCES

1. A.Yu. Iznankin and L.P. Mezhov-Deglin, Transmission of short, strong thermal pulses through liquid helium, JETP Lett., 29(12) (1979) 693
2. A.Yu. Iznankin and L.P. Mezhov-Deglin, Shock waves in liquid helium, Sov. Phys. JETP 57(4) (1983) 801
3. B.A. Danil'chenko, M.O. Lutset, and V.N. Poroshin, Transient transfer processes for large heat pulses from a film heater to superfluid helium, Sov.J.Low Temp. Phys. 13(12) (1987) 730
4. B.A. Danil'chenko, M.O. Lutset, and V.N. Poroshin, Limit of transient heat absorption by superfluid helium for very large heat pulses, Cryogenics Vol. 29 (1989) 444
5. S.K. Nemirovskii and A.N. Tsoi, Transient thermal and hydrodynamic processes in superfluid helium, Cryogenics. Vol. 29 (1989) 985
6. T. Iida, T. Shimazaki and M. Murakami, Laser holographic interferometer visualization of a thermal shock wave in He II, Advances in Cryogenic Engineering Vol. 39 (1994) 1851
7. M.H. Edwards, Refractive index of He^4:LIQUID, Can. J. Phys. 36(1958) 884
8. M.H. Edwards, Refractive index of He^4:Saturated Vapor, Physical Review. 108(5) (1957) 1243

THERMO-FLUIDDYNAMIC ASPECT OF NOISY FILM BOILING PHENOMENA IN He II

M. Murakami, Y. Katsuki[1] and M. Yamaguchi

Institute of Engineering Mechanics, University of Tsukuba
Tsukuba 305 Japan
[1]Presently, Yokosuka Research Laboratory
Central Research Institute of Electric Power Industry
Yokosuka-city, Kanagawa 240-01 Japan

ABSTRACT

Noisy film boiling appears in He II in the case of large heat flux and large hydrostatic pressure. The thermo-fluiddynamic phenomena of noisy boiling on a planar heater are studied with the aid of Schlieren and shadowgraph visualization methods. For a pulsative heating, transient nucleate boiling first appears and then film boiling starts and, furthermore, may proceed to noisy boiling if the hydrostatic pressure is sufficiently large. The frequency at which a large vapor bubble repeats expansion-and-crush quasi- regularly is investigated from the visualization video images. Forced transition to noisy boiling is investigated by imposing small oscillating thermal disturbance superposed on constant heat flux. The effect of oscillating thermal disturbance is found to be so large at rather low frequencies between 0.05 and 100 Hz that the transition takes place even at the half value of the steady critical heat flux.

INTRODUCTION

There are two film boiling modes in He II, noisy boiling characterized by a loud acoustic noise generation and silent boiling. For larger hydrostatic pressures, noisy boiling exclusively appears. The critical hydrostatic pressure for the transition to noisy boiling becomes small as the temperature drops. Loud noise in a noisy boiling state results from violent unstable motion of large scale vapor bubbles[1]. Intense audible noise or mechanical vibration may be harmful to cryogenic systems. Furthermore, it is known the heat transfer coefficient of noisy boiling is smaller than that of silent boiling. The study of the fundamental mechanism of noisy boiling and its control is of crucial importance for such practical applications of He II as superconducting magnet cooling. Most experiments of noisy boiling have been conducted by using a thin wire heater[2,3,4] Only a few reports have been published on noisy boiling on a planar heater[1,5,6]. We have carried out experiments primarily by using visualization method for noisy boiling on a planar heater in open space He II in order to study the fundamental mechanism.

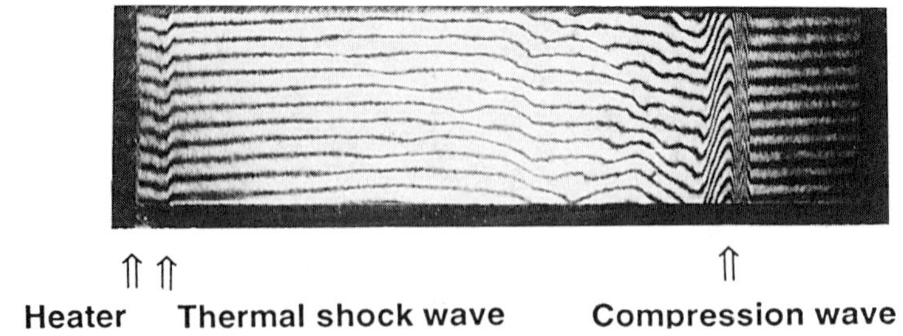

Fig. 1 Finite fringe interferogram of a compression wave and a 2nd sound (thermal shock) wave. T=2.08 K, q=60 W/cm^2, heating duration(t_H)=50 μs, taken at 140 μs after the onset of heating (=t_D).

EXPERIMENTAL SET-UP AND PROCEDURE

A summary of the whole experimental set-up is presented here, since the detail was given elsewhere[1]. The visualization study was carried out by using the Schlieren and the shadowgraph methods. The dark and bright of an image is proportional to the spatial gradient of the refractive index, or of the density, in the test section in the Schlieren method, and to the second derivative in the shadowgraph method. For the visualization of a He II-vapor interface region the shadowgraph method is usually more relevant than the Schlieren method because the variation in the density is too large in the interface region for the Schlieren method to be applied. The Schlieren method with an inherently far higher sensitivity cab be employed for the observation of the density fluctuation in bulk He II as a result of noise generation. Two kinds of planar heaters, a vertical transparent heater and a horizontal opaque one are used. The transparent heater consists of evaporated indium oxide thin film (approximately 1000 Å thick) on an optical glass substrate with dimensions of 2.5 cm by 2.5 cm. Detail phenomena in the vicinity of the heater is well visualized through this heater. The other heaters consist of evaporated Ni-Cr thin film on Pyrex substrate of various sizes. Experiments are conducted under the saturated vapor pressure or weakly subcooled condition with a hydrostatic head up to 40 cm. Heat is pulsatively generated from a planar heater. The Mg-spark light source with a spark duration of 3 μs fires at t_H (between 10 μs and 100 ms) after the onset of heating, when a visualization photograph is taken with a 35-mm camera with a shutter open in a dark room. For continuous image recording the high speed video camera is used.

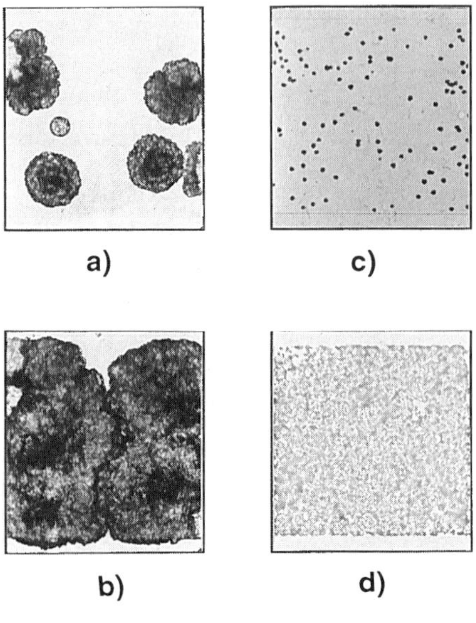

Fig. 2 Shadowgraph visualization pictures observed through a transparent heater in precursory nucleate boiling stage. Heater with dimensions of 2.5 cm by 2.5 cm is vertically mounted.
-a and -b : T=1.80 K, 7 W/cm^2.
-a ; t_H=t_D=5 ms, -b ; t_H=t_D=10 ms.
-c and -d : T=1.80 K, 30 W/cm^2.
-c ; t_H=t_D=100 μs, -d ; t_H=t_D=100 μs.

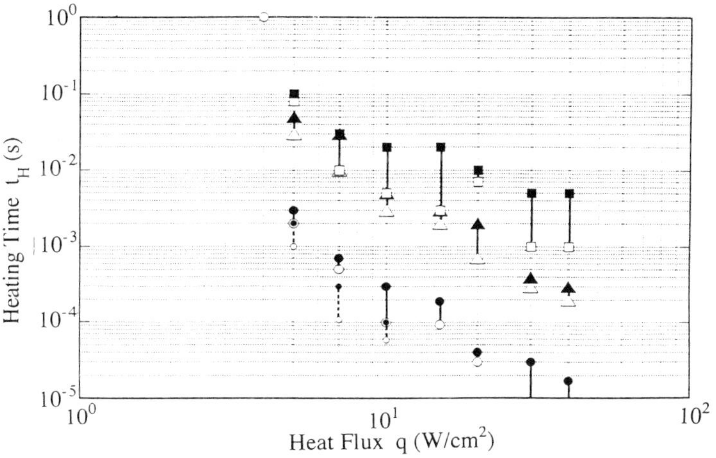

Fig. 3 Onset times of precursory nucleate boiling (○ - ●), film boiling (△ - ▲) and noisy boiling (□ - ■) at 1.80 K. The open marks indicate that a target boiling phenomenon is not visually confirmed, and the closed marks indicate that it is visually confirmed.

RESULTS AND DISCUSSION

Transient Boiling Process toward Noisy Film Boiling

Following the onset of a strong pulsative heating, a compression (first sound; 200 m/s) wave and a second sound wave (20 m/s) with a finite temperature amplitude which soon develops into a thermal shock wave are emitted[7] as shown in Fig. 1. These waves appear in immediate response to the onset of heating far ahead of the appearance of boiling. They must be a good signal for the detection of such a high density heating as SC magnet quench. Then, high density quantized vortices develop on the heater, which cause the formation of a thermal boundary layer; a rapid temperature rise of He II is caused due to the diffusive heat transfer through this layer. Finally boiling starts on the heater surface. Before appearing of film boiling a precursory transient nucleate boiling takes place as shown in Figs. 2, which are shadowgraph pictures observed through a transparent heater mounted vertically. It is seen from Figs. 2-a and -b that each site expands to result in overlapping with each other without any appreciable increase in site number, and finally film boiling begins. On the other hand, for high values of q as shown in Figs. 2-c and -d, the transition process to film boiling seems different from that for smaller q. The number density of boiling sites is far larger and this state turns into a film boiling after further increase in the site number. The critical value of q between the apparent two types of transient nucleate boiling states seems to be between 10 and 20 W/cm². Three onset times of initial precursory nucleate boiling, film boiling and noisy film

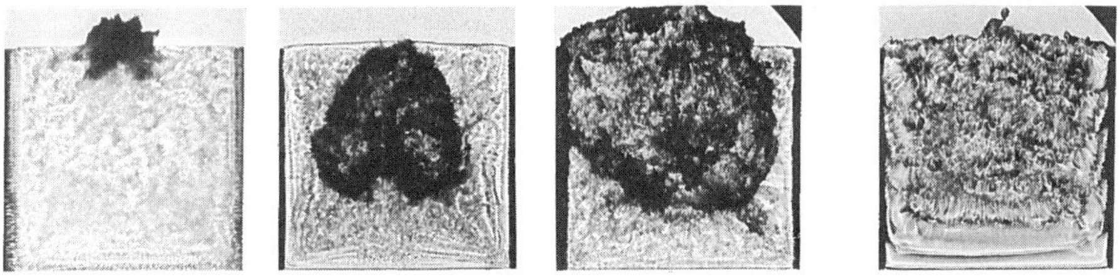

a) b) c) d)

Fig. 4 Typical boiling patterns visualized in the form of shadowgraph through a transparent heater mounted vertically. T=1.80 K, 5 W/cm². -a, -b, -c : noisy boiling, -d : silent boiling.

boiling are plotted against the heat flux q in Fig. 3, in which all the data are obtained from the visualization pictures. Precursory transient nucleate boiling appears far earlier than film boiling. A film boiling state is then reached for large q. If the hydrostatic pressure is sufficiently large, noisy boiling finally appears after the establishment of a film boiling state. Under small hydrostatic pressure condition silent film boiling still continues. The present result seems to agree qualitatively with some existing experimental results[8,9], though such an experimental condition as the heater geometry is different from the present one.

Noisy Film Boiling Phenomena

In noisy boiling case a single vapor bubble full of wrinkles with a size comparable to a heater is seen to repeat an expansion-crush process quasi-regularly above an irregular vapor layer (Figs. 4-a b,c). These are shadowgraphs taken through the transparent heater. The back ground vapor layer under a vapor bubble is seen to be fundamentally similar to that of silent boiling as seen from Fig. 4-d. Loud audible noise and mechanical vibration are generated in noisy boiling state. The characteristic difference between noisy and silent boilings are also visually understood from Figs. 5-a and b. These are side view pictures of boiling patterns on the horizontal heater visualized by the Schlieren method for higher sensitivity. A large vapor bubble is seen in Fig. 5-a in the case of noisy boiling. In the case of silent boiling only a vapor layer with small scale unsteady irregular fluttering is formed as seen in Fig. 5-b. Another clear distinction between the two boiling states is the density fluctuation in bulk He II appeared in connection with pressure disturbance radiation in the noisy case as seen in Fig. 5-a. In the transition state for intermediate magnitude of hydrostatic pressure, a silent boiling is intermittently interrupted by noisy boiling. It is found that such a mechanical disturbance as by knocking the cryostat may readily give rise to a transition from silent to noisy boilings, or even from a non-boiling state to noisy boiling. In the case of large hydrostatic pressure, once the transition occurs, noisy boiling continues until heater switched-off, while in the case of moderate hydrostatic pressure, noisy boiling can be excited only when external disturbance is being imposed. The transition to noisy boiling may be considered as a kind of noise-induced instability. The effect of heater orientation is visually examined for three orientations. The bubbles are found, in general, similar in both orientations, face up and face down, as shown in Fig. 6-a and b. This fact indicates that a vapor bubble is little affected by the buoyancy effect. In the experiments using a rectangular heater with an aspect ratio of 3, a long single bubble mostly appears as seen from Figs. 6-a and c. It seems that the aspect ratio of 3 is not large enough to induce a bubble instability to cause bubble split into several smaller ones.

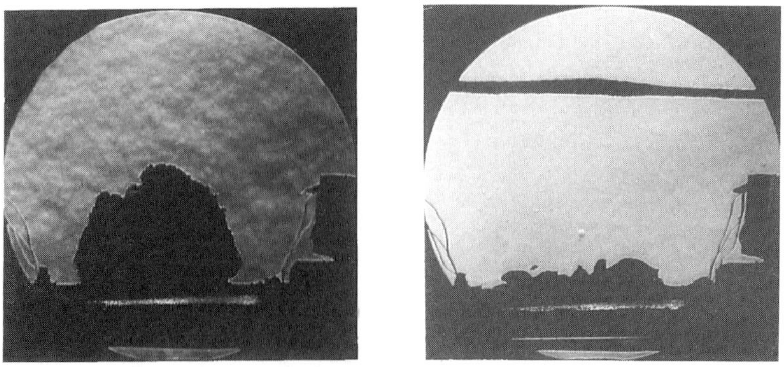

a) b)

Fig. 5 Typical side view pictures of noisy and silent boiling states on a horizontal heater (S=2 cm by 2 cm) visualized by Schlieren method. T=1.80 K, 5 W/cm^2. -a : noisy boiling, -b : silent boiling (the horizontal fat line seen upper half of the picture is a free surface)

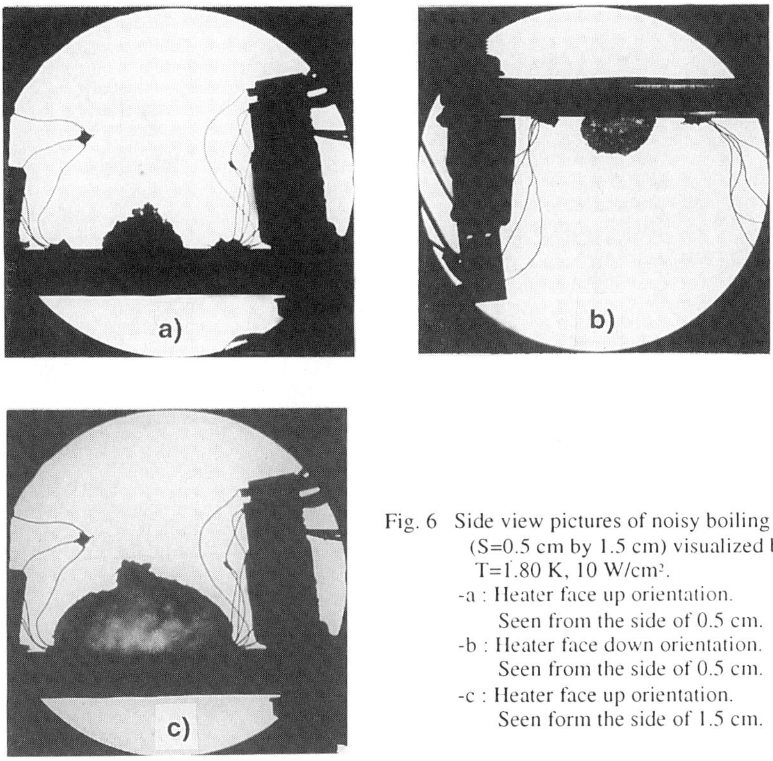

Fig. 6 Side view pictures of noisy boiling on a horizontal heater (S=0.5 cm by 1.5 cm) visualized by Schlieren method. T=1.80 K, 10 W/cm^2.
-a : Heater face up orientation. Seen from the side of 0.5 cm.
-b : Heater face down orientation. Seen from the side of 0.5 cm.
-c : Heater face up orientation. Seen form the side of 1.5 cm.

Frequency of Bubble Formation and Pressure Fluctuation

The frequency of expansion and crush of large vapor bubbles during noisy boiling is measured from the high-speed video image (720 pictures/s). The result given in Fig. 7 suggests that a smaller heater results in higher frequency because of smaller bubble size. Larger q results in lower frequency because of larger vapor bubbles. The pressure fluctuation is directly measured with a quartz cryogenic pressure transducer (PCB 106M122) mounted just above the heater in He II. The FFT analysis data indicates that the fundamental frequency agree with the vapor bubble formation frequency. In addition to the fundamental, a number of higher harmonics and white noise component are included in the pressure fluctuation.

Upper Critical Heat Flux for Noisy Boiling

It is found that boiling state does not turn into noisy boiling in spite of under high hydrostatic pressure condition when the heat flux q is too large. We designate the critical heat flux for noisy boiling appearance as the lower critical heat flux and that for noisy boiling disappearance as the upper one. The experimental result is plotted in Fig. 8, where noisy boiling occurs at the heat fluxes below closed marks and only silent boiling appears above the open marks. No film boiling phenomenon is observed for q smaller than about 4 W/cm^2 indicated by the open circles even for a long heating duration t_H=10 s, though the lower critical heat flux slightly depends on the surface condition of heater and on the bath temperature. The experimental results indicates that the upper critical heat flux depends on the heater size as well as on the hydrostatic pressure. These experimental facts indicate that noisy boiling characteristically appears in He II and relates to the heat transfer in the superfluid turbulent state around vapor bubbles. The disappearance of noisy boiling above the upper critical heat flux seems to indicate that the heater surface is in completely dried-out state. It can be

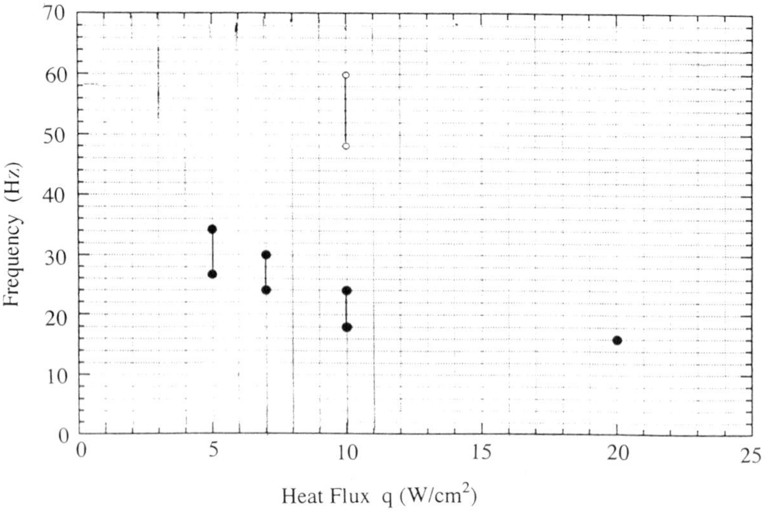

Fig. 7 Frequency of vapor bubble expansion and crush in noisy boiling. T=1.80 K.
● : vertical heater (S=2.5 cm by 2.5 cm) ○ : horizontal heater (S=0.5 cm by 0.5 cm).

concluded that noisy boiling appears at high hydrostatic pressure and in the heat flux range between the lower and the upper critical heat fluxes.

Forced Transition to Noisy Boiling

In the transition state, a natural transition from silent to noisy boilings and sometimes vice versa occurs without any external disturbance. The forced transition to noisy boiling, whichever it is from silent or even from non-boiling state to noisy boiling, is readily caused by giving a thermal disturbance or a mechanical disturbance for large hydrostatic pressure. The dependence of the transition on the frequency and the amplitude of thermal disturbance

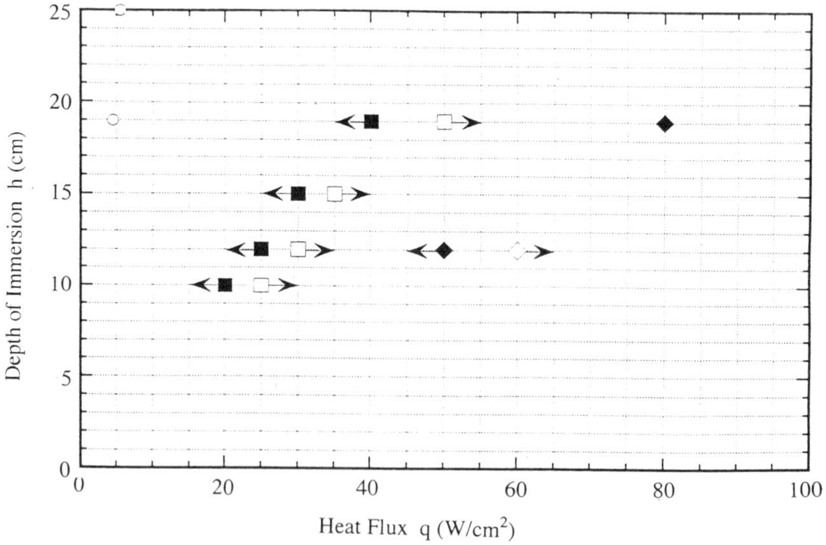

Fig. 8 Upper critical heat flux for noisy film boiling at 1.80 K. Noisy boiling occurs at the heat flux below closed marks and silent boiling occurs above open marks. Horizontal heater (■□ S=0.5 cm by 0.5 cm, ♦◇ S=0.5 cm by 0.5 cm).
○ : The lower critical heat flux for both heaters.

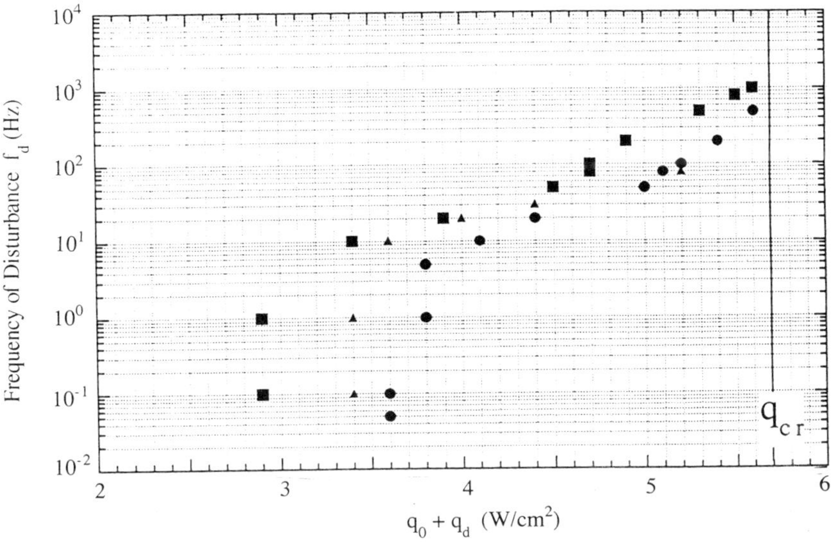

Fig. 9 Noisy boiling transition curves for oscillating heat input. Dependence of the critical frequency on $q_0 + q_d$. T=1.80 K, h=18 cm, horizontal heater (S=0.5 cm by 0.5 cm), q_{cr}=5.7 W/cm². ● : q_d=0.3 W/cm², ▲ : q_d=0.5 W/cm², ■ : q_d=1.0 W/cm²,

is experimentally investigated by imposing a thermal disturbance from a heater, which is a weak sinusoidal thermal disturbance with an amplitude of q_d and a frequency of f_d superposed on constant base heat flux q_0. The experimental result is shown in Fig. 9, where three experimental neutral curves are given for three cases of q_d, where at larger values of $q_0 + q_d$ marked here the transition to noisy boiling takes place. The steady critical heat flux q_{cr} is also given in this figure. It is quite evident that the transition takes place at considerably smaller value of the heat flux than q_{cr} in the case of oscillating heat flux. It is concluded that the disturbance with lower frequencies, between 100 Hz and 0.05 Hz, is extremely effective to induce the transition to noisy boiling. The curve for q_d = 0.3 W/cm² must be very close to the natural neutral curve for infinitesimally small disturbance. The transition occurs at lower $q_0 + q_d$ in the case of larger q_d, higher bath temperature and larger heater area. This transition enhancement by imposing oscillating thermal disturbance seems to disappear at high frequencies over 1 kHz. It is further found that noisy boiling continues until the heater switched-off in the case of large hydrostatic pressure, while for moderate values of it noisy boiling mode periodically appears corresponding to every peak of oscillating disturbance for lower frequencies than about 10 Hz. The data presented in Fig. 9 is not quite reproducible, though all the data obtained within a series of experimental runs is qualitatively consistent.

CONCLUDING REMARKS

1 In the case of noisy boiling, a large vapor bubble repeats expansion-and-crush, generating loud audible noise. Density fluctuation in bulk He II is also observed.
2 The size of bubble becomes larger for larger q and larger heater size. The larger bubbles induce lower frequency of bubble formation.
3 Noisy boiling mode appears only between the upper and lower critical heat fluxes.
4 The transition to noisy boiling is readily enhanced by imposing a thermal disturbance. The disturbance with lower frequencies is extremely effective for the transition to noisy boiling.

ACKNOWLEDGMENT

This work was partly supported by a Grant-in-Aid for Scientific Research from the Ministry of Education, Science and Culture.

REFERENCES

1 Katsuki, Y. and Murakami, M., Visualization study of film boiling onset and transition to a noisy film boiling in He II, Cryogenics, Vol. 35-7 (1995) in press.
2 Leonard, A. C., Helium II noisy film boiling and silent film boiling heat transfer coefficient values, Proc. ICEC 3 (1970) 109-114
3 Steed, R. C. and Irey, R. K., Correlation of the depth effect on film-boiling heat transfer in liquid Helium II, "Adv. Cryog. Eng." 15 (1970) 299-307
4 Ebright, F. L. and Irey, R. K., High-speed motion-picture studies of film boiling in liquid Helium II, "Adv. Cryog. Eng." 16 (1971) 386-392
5 Betts, K. R. and Leonard, A. C., Free convection film boiling from a flat, horizontal surface in saturated He II, "Adv. Cryog. Eng." 21 (1975) 282-292
6 Kobayashi, H. and Yasukochi, K., Maximum and minimum heat flux and temperature fluctuation in film-boiling states in superfluid Helium, Adv. Cryog. Eng. 25 (1980) 372-377
7 Murakami, M., Iida, T., Shimazaki, T., Katsuki, Y. and Sato, T.,The initial thermo-fluiddynamic processes of boiling phenomena in He II, to be published in Cryogenics (1995)
8 Van Sciver, S. W., Transient heat transport in He II, Cryogenics 19 (1979) 385-392.
9 Tsoi, A. N. and Lutset, M. O., Inzhenerno-Fizicheskii Zhurnal 51 N1 (1986) 5-9.

TEMPERATURE MEASUREMENT IN TRANSIENT HEAT TRANSPORT PHENOMENA THROUGH A THERMAL BOUNDARY LAYER WITH HIGH VORTEX DENSITY IN He II

T. Shimazaki, M. Murakami and T. Iida

Institute of Engineering Mechanics
University of Tsukuba
Tsukuba Ibaraki 305
Japan

ABSTRACT

Highly transient heat transport phenomena are investigated with a superconductive temperature sensor. The heat transport process in He II is categorized into three major processes, the second sound wave mode, the dissipative process through a high vortex density thermal boundary layer and the consumption for the phase transition of He II. The contribution of each process is evaluated by analyzing the measured temperature variation. The result is plotted in a map characterized by a parameter $q_p t_H^{1/2}$. Experimental data are also compared with the analytical solution obtained by Dresner. It is found that if the contribution of the diffusive process can be properly evaluated, the thermal behavior in the diffusive region fairly agree with the Dresner's approach in the highly transient case.

INTRODUCTION

One of the most important practical applications of He II is efficient cooling of superconducting magnets and infrared detectors in space. In these practical applications, the heat load to He II may exceed the critical heat flux for superfluid breakdown in many cases. For instance, in the case of superconducting magnet quench, strong pulsed heat load is fed to He II and dense quantized vortices are generated.

The quantized vortices accumulate in the vicinity of a heat source and form a thermal boundary layer. Understanding of the effect of tangled mass of quantized vortices to the transient heat transport phenomena and the thermal behavior of He II are indispensable for the practical applications. However, owing to the complexity of the development and decay of the quantized vortices, sufficient solution for these problems have not been obtained yet. Recently a number of works have been carried out the investigation of the transient heat transport in dense quantized vortex field by using pulsed heating[1,2,3]. In the present study, highly transient heat transport phenomena are experimentally investigated with a superconductive temperature sensor. The experiments are mainly arranged to investigate the heat transport phenomena when the applied heat flux is sufficiently large to induce quantized vortices and to form a thermal boundary layer in the vicinity of a heat source.

EXPERIMENTAL SET-UP

The experiments in the present study are carried out by using an experimental apparatus which consists of three essential parts, a Ni/Cr thin film heater, a cylindrical shock tube and a superconductive temperature sensor immersed in a Dewar containing He II under a saturated vapor pressure condition. Since detailed description of the apparatus was reported in reference 4, only the outline of it will be described here. Pulsed heat is generated by feeding a trapezoidally varying current pulse to the Ni/Cr thin film heater, 27 $mm \times 27$ mm, placed at the bottom of the cylindrical shock tube. The cylindrical shock tube having a height of 150 mm and an inner diameter of 25 mm prevents the temperature variation due to the pulsed heating from diffracting and maintains the one-dimensional character. The temperature variation is measured with a superconductive temperature sensor which is quite similar to the one originally developed at the Max-Planck-Institut[5,6]. The configuration of the sensor resembles a conventional hot wire probe for flow velocity measurement in aerodynamic experiments. The sensing element of the sensor is a superconductive thin film consisting of gold and tin, which are vacuum deposited on the outer surface of fine quartz fiber having a length of 2 mm and is 40 μm in diameter. The temperature sensor is supported by the sensor mount which is movable to move within the cylindrical shock tube along its axis. The distance from the heater to the sensor can be varied within 0.5 mm to 150 mm with 0.1 mm precision. The temperature variation is detected as a voltage drop along the sensing element of the superconductive temperature sensor. The measured signals are stored on floppy disks via a digital synchro-scope and a personal computer. Each successive measurement is performed after a rest time $t_R = 120$ s to avoid the uncertainty due to the initial vortex line density in the experimental field.

RESULTS AND DISCUSSION

Transient heat transport in He II

The total length of the trapezoidal heating ,t_H from a Ni/Cr heater to He II is varied 100 to 4000 μs with constant rise and fall times of 6 μs. The peak heat flux q_p is also varied from 5 to 40 W/cm^2.

Typical waveforms generated on various heating condition of q_p and t_H are shown in Figure 1. As long as the applied heat flux is below a certain critical value or the heating time is short, the measured waveform is free from the influence of the interaction with quantized vortices (waveform 1). The wave height and length are in proportion to those of the applied heat pulse. The wave height and length change during propagation only due to the non-linear features of thermo-fluiddynamics. Theoretically predicted height and length agree well with the experimental values.

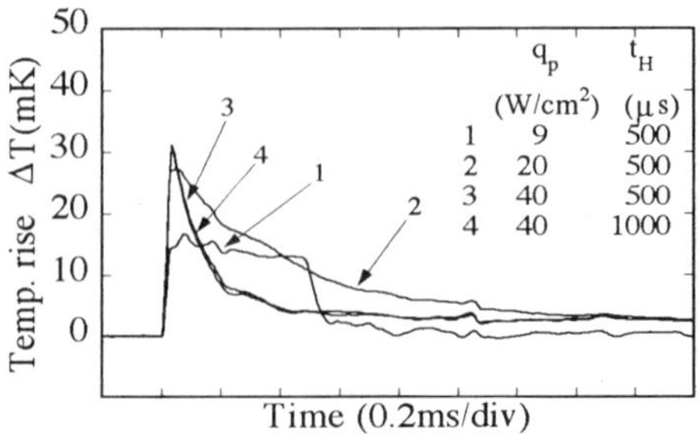

Fig. 1. Typical waveforms generated on the various heating conditions. Temperature sensor is fixed at a distance from the heater of z = 30 mm, $T_B = 1.70$ K.

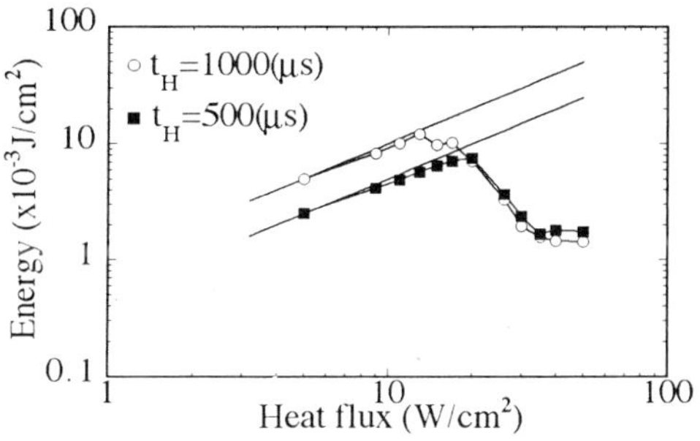

Fig. 2. Amount of heat transported by the second wave mode. Experimental data obtained for two different heating times are plotted. Solid lines represent the total amount of heat applied from a heater. $z = 30\ mm$, $T_B = 1.70\ K$.

On the other hand, when the applied heat flux exceeds a critical value, the superfluid breakdown occurs. The quantized vortices are generated and a thermal boundary layer is formed due to the accumulation of dense quantized vortices in the vicinity of a heater. The influence of the interaction with quantized vortexes on the waveform becomes noticeable i.e. dissipative effect becomes apparent (waveform 2). The measured waveform becomes triangular and the length of it becomes longer than that of the original trapezoidal heat pulse due to the dissipative effect. Further increase in heat flux brings more significant feature, the appearance of a limiting profile (waveforms 3 and 4). When applied heat flux is considerably large, quantized vortex development to a very high density occurs in quit a short time. Consequently the waveform is subjected to strong attenuation from the wave front and the length becomes considerably shorter than that of the original heat pulse from the heater. The shape of a limiting profile is determined only in the initial stage of heating and so the profile becomes insensitive to the length of heating time. Actually the waveform 3 generated on the condition of $t_H = 500\ \mu s$ and the waveform 4 generated on the condition of $t_H = 1000\ \mu s$ almost coincide with each other irrespectively of t_H.

The amount of heat transported by a second sound wave, defined as the second sound wave mode heat transfer is calculated by integrating the measured waveform as follows[4]

$$E_w = C_{20}\rho c_p \int_0^{t_w} \Delta T(t)dt \qquad (1)$$

where C_{20}, ρ and c_p are the propagation speed of second sound, the density and the heat capacity of He II, respectively. $\Delta T(t)$ is the temperature rise within a waveform. t_w is the length of a measured waveform defined as double the half-value width of the waveform. The integration is carried out from the arrival of the wave t = 0 to end of the wave t = t_w. Figure 2 shows the results. It is found that the applied heat is wholly transported by the second sound wave mode when the applied heat flux is small. The amount deviates downward as the heat flux still increases beyond

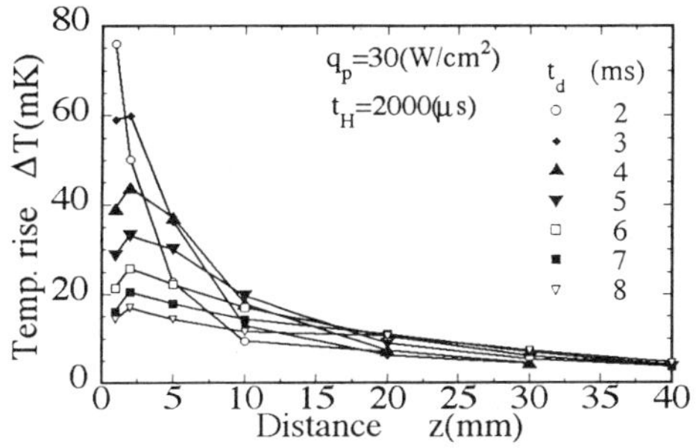

Fig. 3. Time variation of the temperature distribution near a heater. The time after the onset of heating t_d is taken as a parameter. $T_B = 1.70\ K$.

the superfluid breakdown. When waveform reaches a limiting profile, only limited amount of heat can be transported by the second sound wave mode. The rest of heat which is not transported by the second sound wave mode accumulates in the thermal boundary layer and causes a temperature rise in the layer. Afterward heat is transported outward from the heater by the thermal counterflow under influence of dense vortices whose relative velocity between a normal fluid component and the superfluid component exceeds a critical value for the superfluid breakdown. The time variation of spatial distributions of temperature near heater due to the counterflow is similar to that due to a diffusion. Then the diffusion-like heat transport process is designated as the diffusive process. Figure 3 shows the time variation of the temperature distribution near a heater. The high temperature region in a thermal boundary layer is clearly observed in the vicinity of a heater. Gradually the temperature in the region decreases and heat spreads outward from the vicinity of the heater with the laps of time. It should be noted that the temperature slightly drops in the very vicinity of the heater. The reason of the appearance of it is still an open question. But it must certainly relate to highly transient boiling phenomena on a heater surface. The amount of heat transported by the diffusive process is also calculated by analyzing the time variation of the temperature distribution near a heater. Contribution of the diffusive process to heat transport becomes more significant as the applied heat flux rises. Boiling in the thermal boundary layer is in fact observed when applied heat flux and heating time are sufficiently large and long.

On the basis of these results, we may divide the highly transient heat transport process in He II into three major processes for further investigation. The first is the second sound wave mode heat transport. It is the governing process in the case freefrom the superfluid breakdown. The second is the diffusive process in the thermal boundary layer. Finally, it is latent heat of evaporation of He II. The ratio of the amounts of the heat transported by these processes are evaluated and shown in Figures 4-a and b. Heat flux, q_p is varied from 5 to 40 W/cm^2. Heating time, t_H is selected to be 1, 2, 4 ms. The ordinate represents the ratio of the amount of heat transported by each process to total amount of heat input. The difference between Figures 4-a and b is the abscissa. The abscissa of Figure 4-a is simply an applied heat flux q_p but that of Figure 4-b is $q_p t_H^{1/2}$. It is seen in Figure 4-a, that two boundaries dividing three processes represented by closed and open symbols vary with the heating time, t_H. However, it is found that the variation of the boundaries can be almost eliminated if a variable $q_p t_H^{1/2}$ is taken as the abscissa as shown in Figure 4-b. Several curves are found to reduce to almost universal two curves irrespectively of t_H.

It may be concluded from the result of the Figure 4-b that the second sound wave mode is a dominant heat transfer mechanism when $q_p t_H^{1/2}$ is small. As $q_p t_H^{1/2}$ becomes large, contributions of the diffusive process and evaporation become significant instead

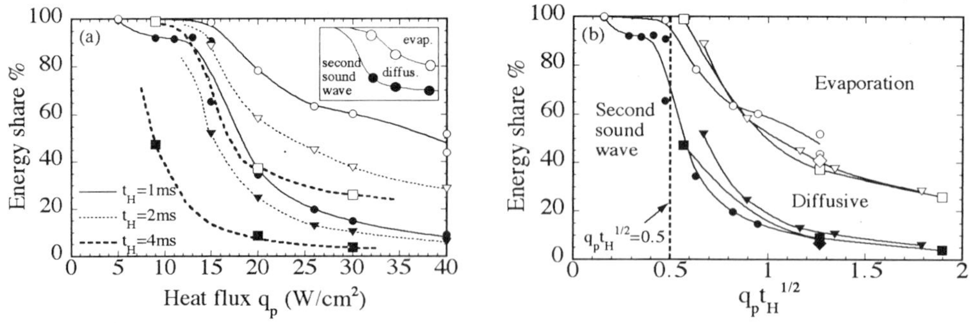

Fig. 4. (a) Ratio of the amount of energy transport by each process. Closed symbols indicate the boundary between the contribution of the second sound process and that of the diffusive process. Open symbols indicate the boundary between the contribution of the diffusive process and that of evaporation. These boundaries are schematically drawn in the inset of the figure. (●,○): $t_H = 1$ ms (▼,▽): $t_H = 2$ ms (■,□): $t_H = 4$ ms , $T_B = 1.70$ K. (b) Ratio of the amount of heat transported by each process. The difference from Figure 4-a is the is the abscissa of $q_p t_H^{1/2}$. A pair of data obtained as $T_B = 1.90$ K, $t_H = 1$ ms is also plotted for reference. (◆,◇)

of the second sound wave mode. A dotted line in the Figure 4-b represents the boil-up criterion obtained by Tsoi and Lutset[7,8]. They obtained an empirical relation for the boil-up criterion from numerous measurements as

$$q_p t_H^{1/2} = 0.5 \tag{2}$$

Figure 4-b shows the contribution of the evaporation becomes significant when $q_p t_H^{1/2}$ exceeds 0.5, which is consistent with the result by Tsoi and Lutset. It is also found from a lot of experimental results that the temperature dependence of the Figure 4-b is rather weak blow 2.0 K. A pair of data obtained at $T_B = 1.90\ K$, $t_H = 1\ ms$ is also added in the figure for comparison.

Comparison with the analytical solution

Experimental data of the diffusive process are compared with the analysis by Dresner[9]. He analytically investigated the time variation of an one-dimensional the temperature distribution near a hater on the basis of the Gorter-Mellink formulation and the energy conservation law after a pulsed heating from a plainer heat source. Heat transport in He II is governed by the following equation in the case of steady or quasi-steady state, where the Gorter-Mellink relation holds.

$$k\frac{\partial}{\partial z}\left(\frac{\partial T}{\partial z}\right)^{1/3} = s\frac{\partial T}{\partial t} \tag{3}$$

where k is a thermal conductance parameter s is the volumetric heat capacity. The Eq. (3) is solved with the boundary condition of pulsed heating from a planer heat source.

$$\int_{-\infty}^{+\infty} s\left(T - T_B\right) dz = Q \qquad t > 0, \quad T\left(z, 0\right) = T_B, \quad T\left(\infty, t\right) = T_B \tag{4}$$

The particular solution of Eq. (3) for the boundary conditions is

$$\frac{\left(kt/s\right)^{3/2}\left(T - T_B\right)}{\left(Q/s\right)^2} = \frac{4/3\sqrt{3}}{\left(x^4 + b^4\right)^{1/2}} \tag{5}$$

$$\begin{aligned} x &= z\left(Q/s\right)/\left(kt/s\right)^{-3/2} \\ b &= 2\left[\Gamma\left(1/4\right)\right]^2/3\sqrt{3\pi} = 2.855 \end{aligned}$$

According to the solution, it is seen that all experimental data of the temperature distribution such as Figure 3 should reduce to a single curve if the similarity variables $\alpha = (T - T_B)t^{3/2}$ and $\beta = z/t^{3/2}$ are used. Eq. (5) is rewritten in terms of the similarity variables is

$$\alpha = \frac{4/3\sqrt{3}k^{-3/2}s^{-1/2}Q^2}{\left(Q^4 s^2 k^{-6}\beta^4 + b^4\right)^{1/2}} \tag{6}$$

When β becomes large, an asymptotic solution to the equation (6) can be written as

$$\alpha = 4/3\sqrt{3}k^{3/2}s^{-3/2}\beta^{-2} \tag{7}$$

The curve becomes a line with a gradient of -2 in a logarithmic plot. When β becomes small, the opposite extreme case, equation (6) is approximately

$$\alpha = 4/3\sqrt{3}b^{-2}k^{3/2}s^{-1/2}Q^2 \tag{8}$$

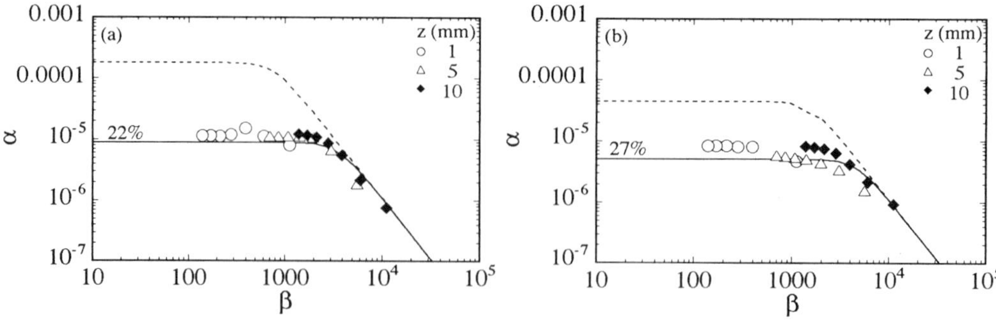

Fig. 5. (a) Comparison of the experimental data and analytical solutions. Experimental data obtained at the sensor location, z of 1, 5, 10 mm are plotted. The theoretical curves are represented by dotted and solid lines. The dotted line obtained by regarding the total input heat as Q. According to Figure 4-b, 22 % of the applied heat is transported by the diffusive process in this case. $T_B = 1.70\ K$, $q_p = 40\ W/cm^2$, $t_H = 2\ ms$. (b) Comparison of the experimental data and analytical solutions in the case of $q_p = 20\ W/cm^2$, the amount of heat transported by the diffusive process is 27 %. Other conditions are the same as those of (a).

α becomes constant if the bath temperature is kept constant. Q should be regarded as the amount of the heat transported by the diffusive process in a thermal boundary layer in the present experiment not as the total amount of heat fed from a heater. It is easy to note that the accurate estimation of Q is essential to make a comparison between the analytical solution and the experimental results. Figure 5-a shows the comparison of the experimental data and analytical solutions. The dotted line representing the theoretical curve is obtained by regarding the total heat input as Q which doesn't agree with the experimental data. In highly transient case, some part of applied heat is transported by the second sound wave and also is consumed for a latent heat of evaporation. The amount of heat transported by those processes need to be extracted from the total heat input to make a comparison between the experimental data of the diffusive process. According to the result in Figure 4-b, the amount of heat transported by the diffusive process is 22 % of the total amount. The theoretical curve described by a solid line is obtained by using the 22 % of the heat input as Q in equation (6) . The curve shows fair agreement with the experimental data. Figure 5-b shows the similar result in the case of $q_p = 20\ W/cm^2$. The solid curve represents fair agreement with the experimental data as far as the diffusive portion is correctly evaluated.

CONCLUSION

1.Three main heat transport processes, the second sound wave mode, the diffusive process and the evaporation are discussed in the transient heat transfer phenomena in He II.
2.The map which describes the contributions of three processes are obtained by using the parameter of $q_p t_H^{1/2}$ irrespectively of t_H.
3.It is found that if the contribution of the diffusive processes can be properly estimated in highly transient case, the thermal behavior in the diffusive region fairly agree with the Dresner's approach.

ACKNOWLEDGMENT

This research was partly supported by the Japan Society for the Promotion of Science.

REFERENCES

1. W. Fiszdon, M. v. Schwerdtner, G. Stamm and W. Poppe,Temperature overshoot due to quantum turbulence during the evolution of moderate heat pulse in He II, *J. Fluid Mech.* 212 : 663 (1990).
2. W. Fiszdon, S. K. Nemirovskii and M. v. Schwerdtner, Interaction of second sound acoustic waves with a counterflow-induced turbulent field, *Pysica B.* 168 : 93 (1991).
3. S. K. Nemirovskii and A. N. Tsoi, Transient thermal and hydrodynamic processes in superfluid helium, *Cryogenics* 29 : 985 (1989).
4. T. Shimazaki, M. Murakami and T. Iida, Second sound wave heat transfer, thermal boundary layer formation and boiling:highly transient heat transport phenomena in He II, *Cryogenics* in press.
5. H. Borner, T. Schmeling and D. W. Schmidt, Experimental investigations of fast gold-tin metal film second-sound detectors and their application, *J. Low Temp. Phys.* 50 : 405 (1983).
6. M. v. Schwerdtner, W. Poppe and D. W. Schmidt, Distortion of temperature signals in He II due to probe geometry, and a new improved probe, *Cryogenics* 29 : 132 (1989).
7. A. N. Tsoi and M. O. Lutset, *Inzhenerno-Fizicheskii N1* 51 : 5 (1986).
8. S. K. Nemirovskii and A. N. Tsoi, Transient thermal and hydrodynamic processes in superfluid helium, *Cryogenics* 29 : 985 (1989).
9. L. Dresner, Transient heat transfer in superfluid helium-part II, in: "Advances in Cryogenic Engineering 29, Plenum-press, London, p323 (1984).

TRANSITION TO TURBULENT STATE OF SUPERFLUID HELIUM FLOW THROUGH THERMOMECHANICAL PUMP ELEMENT

Hirotaka Nakai,[1] Nobuhiro Kimura,[1] Masahide Murakami,[2]
Tomiyoshi Haruyama,[1] and Akira Yamamoto[1]

[1] National Laboratory for High Energy Physics (KEK)
1-1 Oho, Tsukuba-shi, Ibaraki-ken 305 JAPAN

[2] Institute of Engineering Mechanics
University of Tsukuba
1-1 Tennohdai 1-chome, Tsukuba-shi, Ibaraki-ken 305 JAPAN

ABSTRACT

It has been found experimentally that there are two distinct flow regions in the flow characteristics of superfluid helium through the porous element of a thermomechanical pump (TMP). [1] One is the ideal flow region and the other is the turbulent one. However, it is not known yet how the flow turns turbulent. The concept of the critical heat input to TMP is introduced in this study to define the transition from the ideal flow region to the superfluid turbulent one. The temperature dependence and the geometrical dependence of the critical heat input are experimentally investigated and the results are discussed in this paper. This critical heat input gives one of the necessary information to design a TMP properly for the given specifications such as the required mass flow rate and the temperature of superfluid helium.

INTRODUCTION

A thermomechanical pump (TMP) is a device to drive superfluid helium in a cooling system. The working principle of the TMP is the thermomechanical effect of superfluid helium. The main parts of the pump are a porous element of a superleak grade and a heater which establishes the temperature difference across a porous element. Since the TMP has no moving part in it, this pump is highly reliable and often adopted to the cooling system of superfluid helium used in space. The mass flow rate of superfluid helium driven by the heat input from a heater is expressed on the basis of the conservation laws of mass and energy as:

$$\dot{M} = \frac{Q}{ST} , \qquad (1)$$

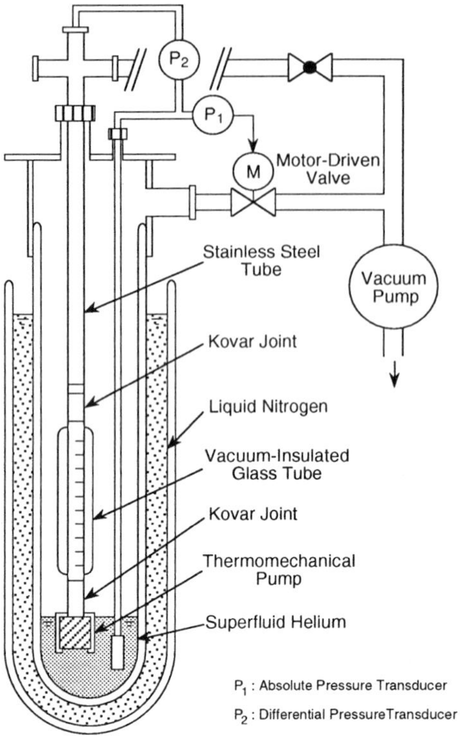

Figure 1. A schematic of experimental setup of thermomechanical pump.

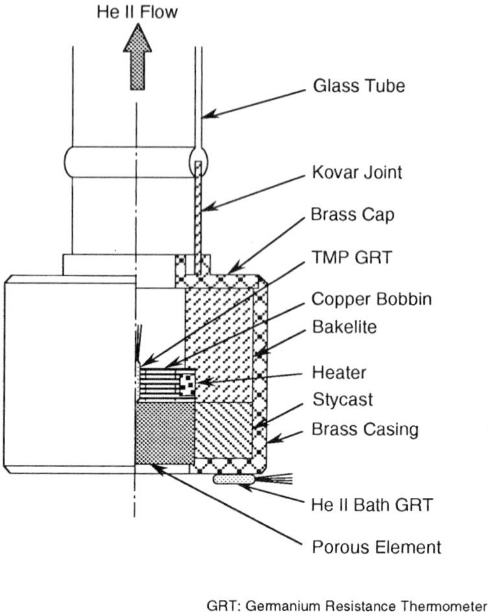

Figure 2. Cut-out view of thermomechanical pump section.

where $\dot{M}$ is the mass flow rate, Q the heat input, S the specific entropy of superfluid helium and T the temperature of superfluid helium, that is, the bath temperature. This is the theoretical prediction of the flow characteristics of superfluid helium through the porous element of the TMP. This equation suggests that the mass flow rate depends only on the superfluid helium temperature and the input heat, and not on any geometry of the porous element of the TMP.

The actual flow characteristics of superfluid helium through the porous element of the TMP is not so simple as predicted from Eq. (1). It has been observed that if large heat input is applied to the pump, the flow of superfluid helium turns so-called turbulent, and not Eq. (1) but the Gorter-Mellink mutual friction law[2] is applicable to describe the flow characteristics.[1] It is unknown yet, however, under which conditions the flow of superfluid helium through a

Table 1. Parameters of porous elements of thermomechanical pumps

TMP No.	Cross-sectional Area A (x 10^{-4} m^2)	Thickness L (x 10^{-3} m)	Permeability κ (x 10^{-15} m^2)	Porosity ε	Mean Pore Diameter d (x 10^{-6} m)
221	4.34	5.0	3.8	0.226	1.07
222	4.32	5.0	3.1	0.226	1.07
241	4.34	10	4.4	0.226	1.07
242	4.32	9.9	3.3	0.226	1.07
421	5.21	5.1	2.3	0.232	0.8
521	4.87	5.1	0.56	0.210	0.4
541	4.87	10	2.8	0.210	0.4

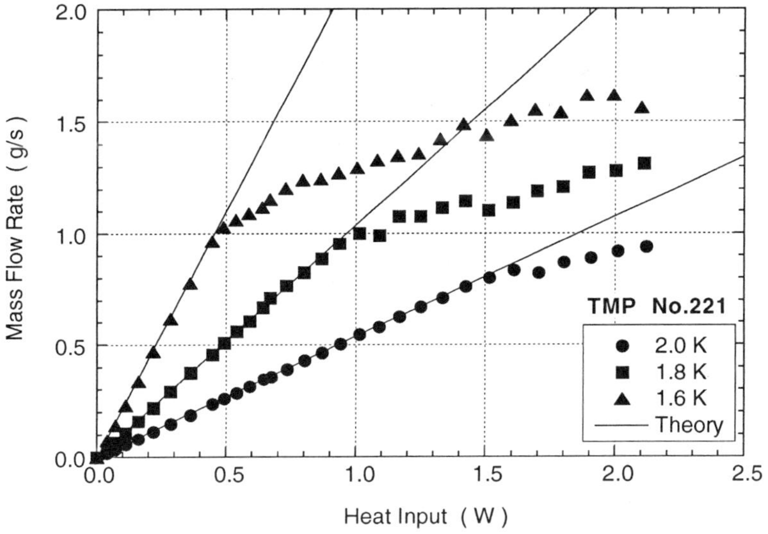

Figure 3. Mass flow rate vs. heat input at various temperatures.

porous element transits from the ideal flow region to the turbulent one. The objective of this study is to clarify experimentally the conditions of the transition of the superfluid helium flow. If the conditions of the transition can be found, the flow characteristics of the superfluid helium flow through the porous element of the TMP can be fully understood, and the design scheme of a TMP under given specifications can be established.

EXPERIMENTS

Commercially available ceramic alumina (Kurosaki Yogyo AR-9D and AR-10D) is selected as a porous element of the TMP. The geometrical parameters of porous elements used in this study are listed in Table 1. The permeability, κ, is measured with air at room temperature, as described in Reference 3, and the mean pore diameter is measured with a mercury porosimeter.

The experimental setup is shown schematically in Figure 1, and the TMP section is drawn in detail in Figure 2. Heat driving superfluid helium is generated by the heater placed at the downstream end of the porous element. The mass flow rate of superfluid helium is calculated from the ascending rate of the free surface of superfluid helium in the vacuum-insulated and graduated glass tube.

While the temperature and the pressure in superfluid helium bath are regulated to be constant, the heat input to the TMP is set successively to various values by changing the constant current to the heater. After establishing a steady state of the whole system at a certain heat input, measured are the temperatures and pressures across the porous element, the mass flow rate of superfluid helium through the element, and the voltage and current at the heater. To avoid the influence of the hydrostatic pressure of superfluid helium on the measurement, the measurement of these quantities is always initiated when the superfluid helium free surface in the glass tube reaches to the fixed level. The temperatures are measured with germanium resistance thermometers (Lake Sore Cryotronics, GR-200) excited by 1 μA of DC current. The pressures are measured with an absolute pressure transducer and a differential pressure transducer (MKS Instrument Type 122A and 223B).

RESULTS AND DISCUSSIONS

Flow Regions of Superfluid Helium

Typical experimental results of TMP No. 221 are plotted in Figure 3 for three different bath temperatures. Solid lines in the figure represent Eq. (1) at the temperatures. For small heat input, data points of the mass flow rate are just on the lines. This means that the flow of superfluid helium in this region is in the ideal state, which can be described by the theory. On the other hand, for large heat input, data points deviate from the line and seem to saturate at larger heat input.

Figure 4 shows the data sets of various TMPs at the same bath temperature, 1.8 K. It is clearly found that for small heat input, the data points coincide with each other and with the theory. There is no geometrical dependence in this flow region as expected from Eq. (1). For large heat input the difference among the data of the mass flow rate can be seen. Hence, the theory is no longer valid in this flow region. This region can be ascribed to the superfluid turbulence and be approximated by using the Gorter-Mellink mutual friction law.

Critical Heat Input

Even though it has been found that there are two flow regions of the superfluid helium flow though a porous element of the TMP, the condition of the transition from the ideal state to the turbulent one has not been researched. It is derived from the equation of motion of superfluid helium and the Gorter-Mellink mutual friction law that the temperature difference across the flow constriction is proportional to the cube of the heat input in the turbulent state of the superfluid helium flow. This relation is experimentally observed.[4]

In Figure 5 the temperature difference across a porous element is plotted against the cube of the heat input. The temperature difference is almost constant at very small values of Q in the ideal flow region because of a constant hydrostatic pressure, while the linear relation between the temperature difference and the cube of the heat input is readily seen in the turbulent region. From this figure, the critical heat input, Q_{cr}, which bounds the ideal flow region and the turbulent one, can be decided experimentally as shown by a dotted vertical line in the figure.

Dependence of Critical Heat Input

The critical heat input for TMP No. 221 is plotted against the temperature of the superfluid helium bath in Figure 6. The temperature dependence of the critical heat input is recognized, but it is not so straightforward to find how the critical heat input depends on the temperature. As a result of searching various temperature-dependent parameters, it is found that the temperature dependence of the critical heat input arises from the density of superfluid component, ρ_s, the specific entropy, S, and the temperature of superfluid helium, T. The critical heat input divided by the product of these quantities yields some constant value, as

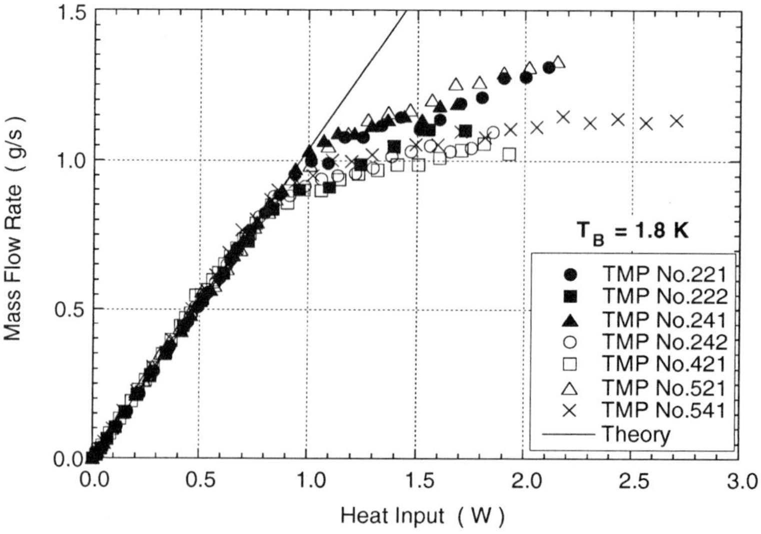

Figure 4. Mass flow rate vs. heat input of various TMPs.

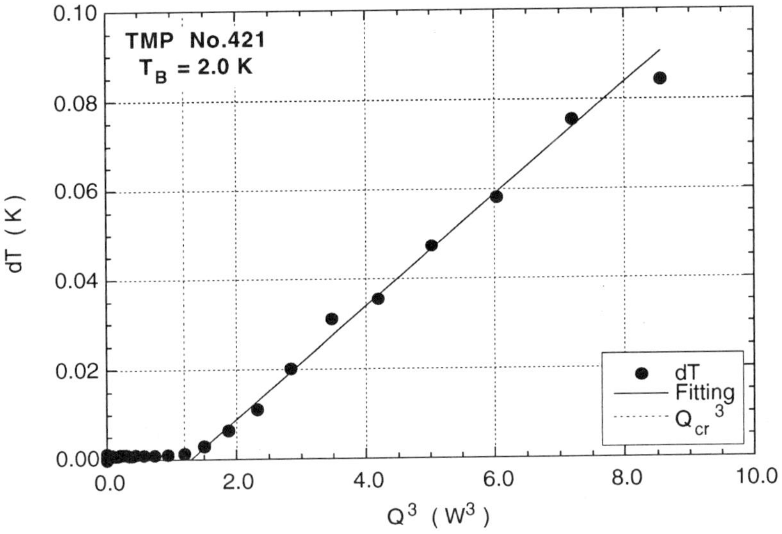

Figure 5. Temperature rise and the cube of heat input and the definition of critical heat input, Q_{cr}.

shown in Figure 7. From this figure the temperature dependence of the critical heat input is completely explained by these quantities.

The critical heat input should be reduced also by the effective cross-sectional area, εA, where ε is the porosity of a porous medium and A the geometrical cross-sectional area of a porous element, to evaluate the flow characteristics properly among the porous elements having different geometry. Moreover, the critical heat input may be considered to correspond to the critical velocity of the superfluid component, which defines the beginning of the superfluid turbulence. The critical velocity, $v_{s,c}$, is empirically expressed as[5]:

$$v_{s,c}\, d^{1/4} = 3.2 \times 10^{-3} \ (m^{5/4}/s)\ , \qquad (2)$$

where d is the inner diameter of a capillary tube, for example. This equation suggests that

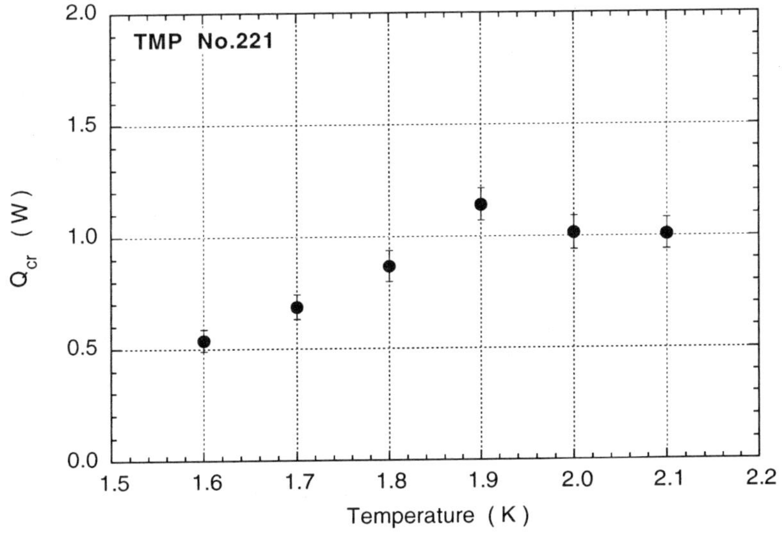

Figure 6. Temperature variation of critical heat input.

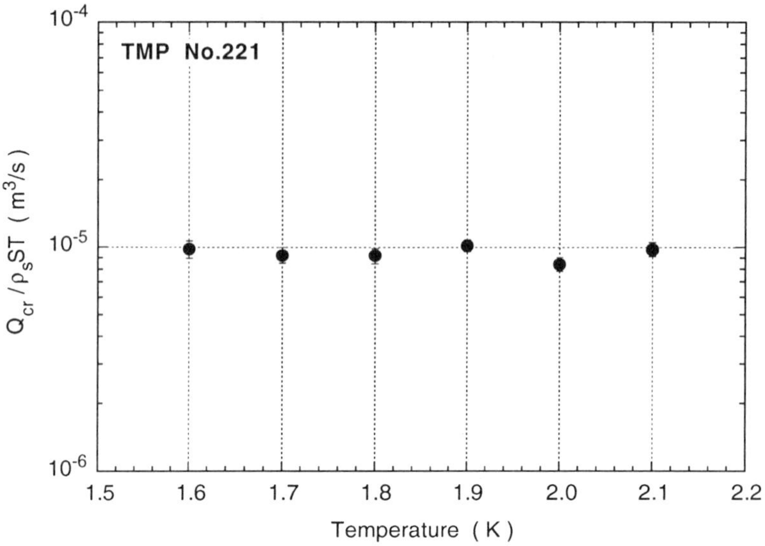

Figure 7. Reduced heat input.

the critical heat input is also reduced by the fourth root of the mean pore diameter of the porous medium. Hence, the geometrical dependence of the critical heat input can be expressed by the effective cross-sectional area and the fourth root of the mean pore diameter.

The critical heat input of various TMPs reduced by the temperature-dependent term, $1/(\rho_s ST)$, and the geometrical term, $d^{1/4}/(\varepsilon A)$, is plotted against the superfluid helium temperature in Figure 8. It is moreover derived from the Gorter-Mellink mutual friction law that the mass flow rate driven by a TMP is inversely proportional to the cubic root of the thickness of a porous element. It is found, however, that the introduction of the effect of the thickness makes the scattering of data worse than those in the figure. Since the difference in the geometrical parameters of the porous elements is small (by a factor of 2 or so at most) in this study and there may be some unknown factors such as the penetration of the epoxy resin into a porous element, then the geometrical dependence does not appear so clearly in the experimental data beyond the experimental errors and there is a scattering of data in the

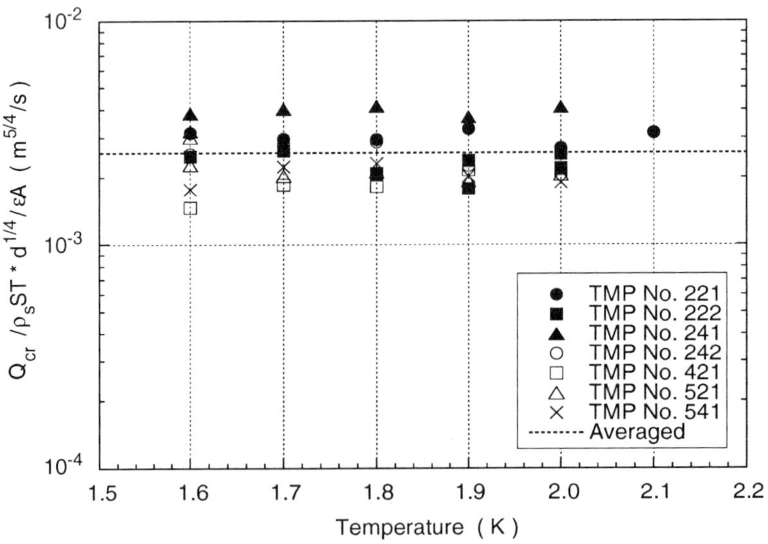

Figure 8. Reduced critical heat input of various TMPs.

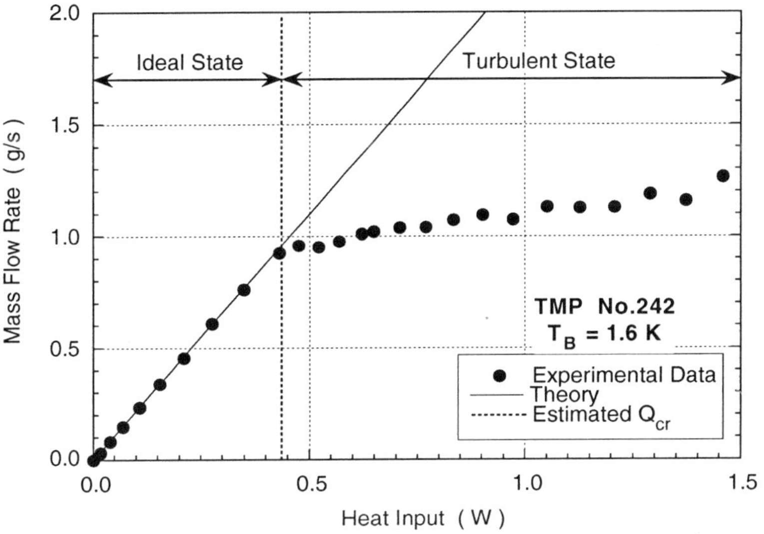

Figure 9. Critical heat input shown on flow characteristic curve.

figure. As a rough approximation, a constant value can be estimated from this figure. Using this constant expressed as:

$$(Q_{cr} \cdot d^{1/4}) / (\rho_s ST \cdot \varepsilon A) = \text{constant}\,, \tag{3}$$

the estimated critical heat input can be calculated, and this critical heat input is compared with the experimental data in Figure 9. The estimation of the critical heat input agrees fairly well with the experimental data for various porous elements. Hence, by using Eq. (3) and the given parameters as the required specifications for a TMP, the critical heat input can be calculated and then the flow characteristics of superfluid helium through the porous element of the TMP can be fully predicted.

CONCLUSIONS

The transition of the superfluid helium flow from the ideal state to the superfluid turbulent one can be characterized by the critical heat input which is determined experimentally. This critical heat input has the dependence on the temperature of superfluid helium and the geometry of the porous element. The temperature dependence can be explained in terms of the density of superfluid component, ρ_S, the specific entropy, S, and the temperature of superfluid helium, T, as $1/(\rho_S ST)$. On the other hand, the geometrical dependence of the critical heat input has the form of $d^{1/4}/(\varepsilon A)$ from the analogy with the critical velocity, where d is the mean pore diameter, ε the porosity of the porous medium and A the geometrical cross-sectional area of the porous element. This reduced critical heat input gives a useful information to design a TMP properly for the required specifications.

REFERENCES

1. M. Murakami and T. Hanyu, Hydrodynamic characteristics and limitation of He II flow through a porous element for a fountain effect pump, *Adv. in Cryog. Eng.* 37A:123 (1992).
2. C.J. Gorter and J.H. Mellink, On the irreversible process in liquid helium II, *Physica* 15:285 (1949).

3. H. Nakai, N. Kimura, T. Haruyama, M. Murakami and A. Yamamoto, Effect of macroscopic parameters of porous material on flow characteristics for superfluid helium thermomechanical pump, *Adv. in Cryog. Eng.* 39B:1805 (1994).
4. K.P. Martin and J.T. Tough, Evolution of superfluid turbulence in thermal counterflow, *Phys. Rev. B* 27:2788 (1983)
5. W.M. van Alphen, G.J. van Haasteren, R. de Bruyn Ouboter and K.W. Taconis, The dependence of the critical velocity of the superfluid on channel diameter and film thickness, *Physics Letters* 20:474 (1966).

THE ASPECT-RATIO EFFECT OF He II CHANNELS ON THE HEAT TRANSPORT CHARACTERISTICS

H. Kobayashi, Y. Akedo, and K. Kawakami

Atomic Energy Research Institute, Nihon University
1-8 Kanda, Surugadai, Chiyoda-ku, Tokyo 101, Japan

ABSTRACT

The critical heat flux q_λ at which normal-fluid (He I) sets in at the center of a heated surface distributed along the inner wall of a He II channel can be smaller than the value calculated from the theory of Gorter-Mellink heat conduction. The reduction in q_λ becomes noticeable as the ratio of channel gap d to channel length L increases. The effect may be explained by the presence of a temperature gradient across the channel gap, so that the temperature at the heated side reaches the λ-point at an earlier q_λ than the other side.

INTRODUCTION

In the past heat transport characteristics in long, narrow He II channels with distributed heat sources (see Figure 1b)[1,2,3] were regularly investigated within the framework of the Gorter-Mellink theory[4]. For such narrow channels the observed critical heat flux q_λ does not differ much from that predicted by theory. In the present work, we study the heat transport in channels with a relatively large aspect ratio of channel gap to channel length in order to investigate the applicability of the G-M theory to superconducting windings where the cooling channels have a geometry with a large aspect ratio, and resistive dissipation is distributed along the conductor walls as shown in Figure 1c.

Since a relatively good agreement of experiment and theory has been obtained by use of the equation for distributed heated surfaces in the case of small d/L channels or normal G-M channels (see Figure 1a), and since the discrepancy of q_λ depends on the aspect ratio d/L it is to be expected that some modification of the equation for distributed heat source should be sufficient to reconcile theory and experiment.

This aspect ratio effect becomes important in the estimation of the stability margin of superconducting magnets cooled with He II because here large aspect ratios d/L occur rather frequently in the channels of superconducting windings formed between the conductors and spacers.

THE DEFINITION OF CRITICAL HEAT FLUX

The heat flux $q(x)$ [W per channel cross-section] in the liquid adjacent to a heated surface at one side of a channel as shown in Figures 1b and 1c depends on the distance

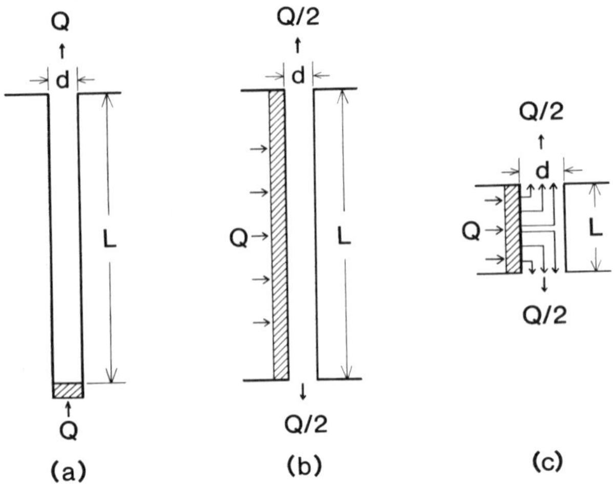

Figure 1. Sketches of various channels. (a) Normal Gorter-Mellink channel, (b) Long He II channel with a distributed heat source and a small aspect ratio and (c) a short channel with a large aspect ratio. Arrows in (c) indicate the heat flow

x along the distributed heater. Thus $q(x)$ becomes $Q \cdot x/A_c L$ where Q is the total heat dissipation from the heated surface and A_c is the cross-section of the channel. By integrating $f(T)^{-1}$ from T_b (bath temperature) to T_λ (λ−point) in the G-M equation $dT/dx = -f(T)q(x)^3$, and $q(x)$ from $x = 0$ to $L/2$(cf. Figure 2b), the critical heat flux q_λ per unit heated surface area becomes

$$q_\lambda = 4d \cdot L^{-4/3} \cdot \left(\int_{T_b}^{T_\lambda} f(T)^{-1} dT \right)^{1/3} \tag{1}$$

where $f(T)^{-1}$ is some kind of thermal conductivity[5]. Note that we define a heat flux per unit heated surface and not per unit channel cross-section since our discussion treats the stabilization of conductor surfaces. Furthermore, since the discrepancy between the experimental $q_\lambda(Exp.)$ and $q_\lambda(Cal.)$ as calculated from Eq.(1) depends on the aspect ratio[1], we define a correction factor γ and, assuming a linear relation between aspect ratios d/L and γ for small d/L, we set

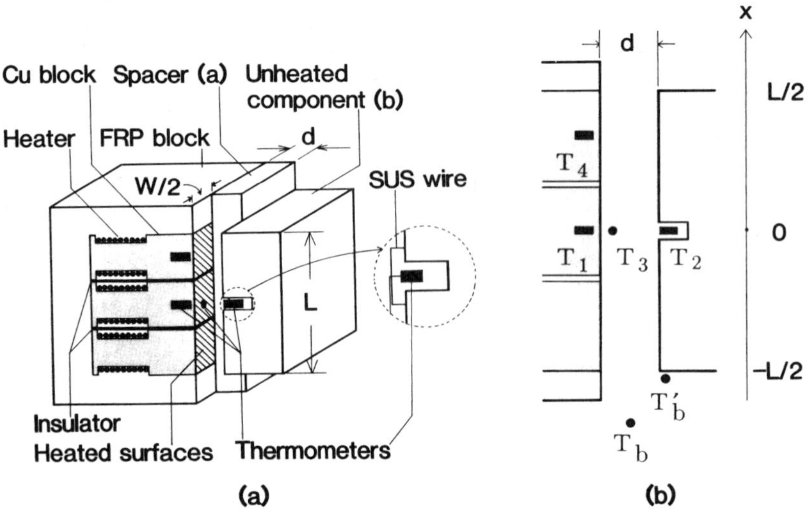

Figure 2. Arrangement of the test channel (a) and thermometers (b). The x-axis is taken along the channel length with the origin at the mid-point of the channel length.

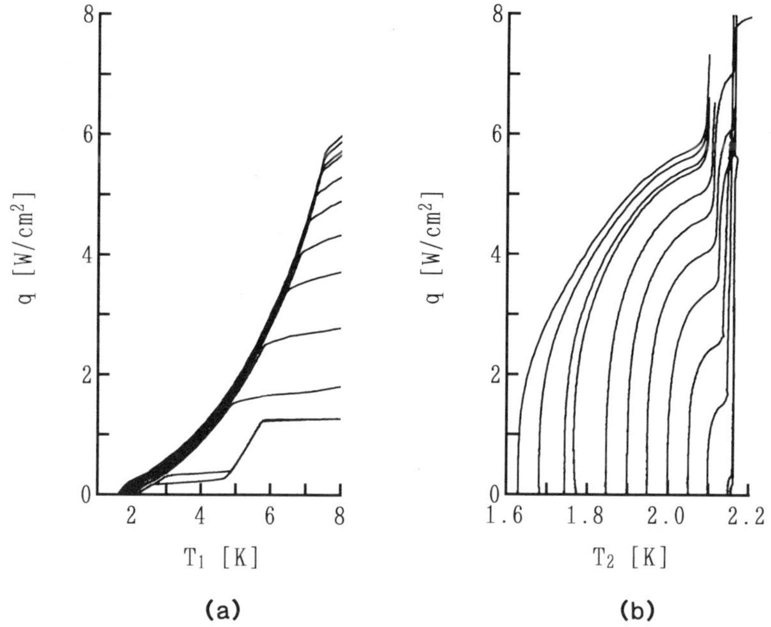

Figure 3. Temperature rise in T_1 and T_2 from different bath temperatures T_b as a function of heat flux q. The heat flux is defined as W/heated surface. $L = 8mm$, $d = 2mm$.

$$\gamma = \frac{q_\lambda(Exp.)}{q_\lambda(Cal.)} = a - b(\frac{d}{L}) \tag{2}$$

Thus, contrary to normal G-M channels, $q_\lambda(Exp.)$ is not directly proportional to the gap d. Similarly, the critical heat flux in terms of W per channel cross-section depends on the gap.

EXPERIMENTAL

For the measurement of q_λ as a function the aspect ratio d/L and other geometrical influences, we prepared a series of rectangular channels by using spacers of various thickness d (see Figure 2a, part (a)), and by varying the length L of part (b) from 8 mm to 25 mm. Figure 2a shows the rear half of the channel of variable length L, constant width $w = 8$ mm and variable gap d. The channels were fabricated from rectangular fiberglass epoxy blocks and three copper blocks with a total heated area of 8×8 $mm^2 \times 3$, thermally separated by mylar foils of 0.5 mm thickness. The sides of the channel were plugged with fiberglass epoxy spacers so that heat flows only up and down along the channel. The channel length L is adjusted by replacing the front block. The rectangular channel has dimensions $d \times L \times w$ with an heated area $L \times w$.

Joule heat produced in manganin wires (154.2 Ω each) wound around the bottoms of the copper blocks is totally delivered to the liquid through the exposed copper faces.

Germanium thermometers T_1 and T_4 are buried in two of the copper blocks. To measure the temperature of the liquid in the channel a germanium thermometer T_2 was suspended by thin stainless steel wires in a small hole in the epoxy block facing the copper surfaces in such a way that the thermometer is surrounded by liquid and does not touch the fiberglass block, see the inset of Figure 2a. In experiments including the measurement of the temperature gradient across the channel gap, another small germanium thermometer T_3 was inserted into the liquid at $x = 0$ and close to the heated surface. The bath temperature T_b was monitored both with a further germanium thermometer T_b and a small diode thermometer T_b' exactly at the exit point of the channel $x = -L/2$. Thermometer positioning is schematically shown in Figure 2b. The channel was immersed in atmospherically pressurized He II with the copper surfaces

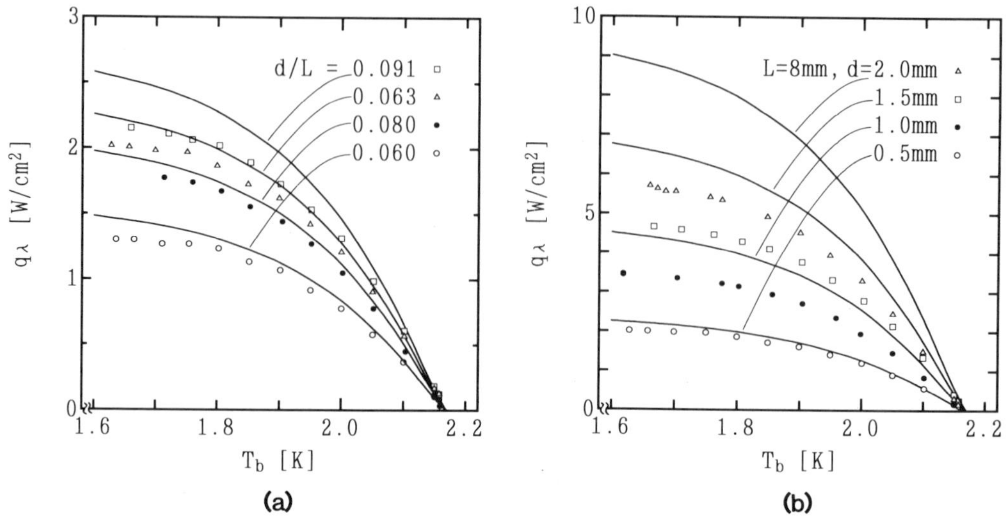

Figure 4. Critical heat flux q_λ as a function of bath temperature T_b, (a) with aspect ratio d/L and (b) with the channel gap d as parameters. Solid curves show the values calculated from Eq.(1).

arranged vertically. Since the refrigeration power of the cryostat is large enough to keep the bath temperature T_b constant within a few mK during a sweep of the heater current no dependence on sweep rate was observed over the range from 0.43 to 0.86 mA/s.

RESULTS AND DISCUSSION

Figures 3a and 3b show samples of the observed temperatures T_1 and T_2 as a function of q. We define the critical heat flux q_λ as the flux corresponding to the sudden turns in T_1, which also corresponds to the heat flux at which the temperature near the heated surface T_3 reaches T_λ (cf. Figure 6a).

The dependence of q_λ on T_b for different aspect ratios is shown in Figure 4a. In Figure 4b q_λ is plotted as a function of T_b with d as parameter. The solid curves in Figures 4a and 4b represent $q_\lambda(Cal.)$ calculated from Eq.(1). Agreement with experiment deteriorates with increasing aspect ratio d/L.

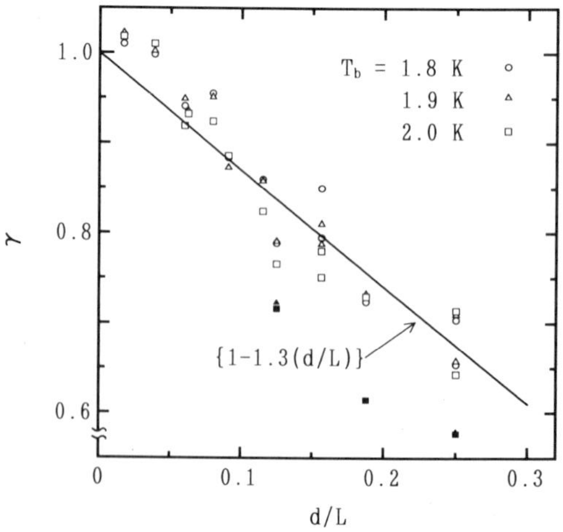

Figure 5. Reduction of q_λ as a function of the aspect ratio d/L. The solid line shows Eq.(2) with $a = 1$ and $b = 1.3$ as a fit to the open marks. Solid marks are measurements with $L = 8$ mm with all three heaters activated at the same time.

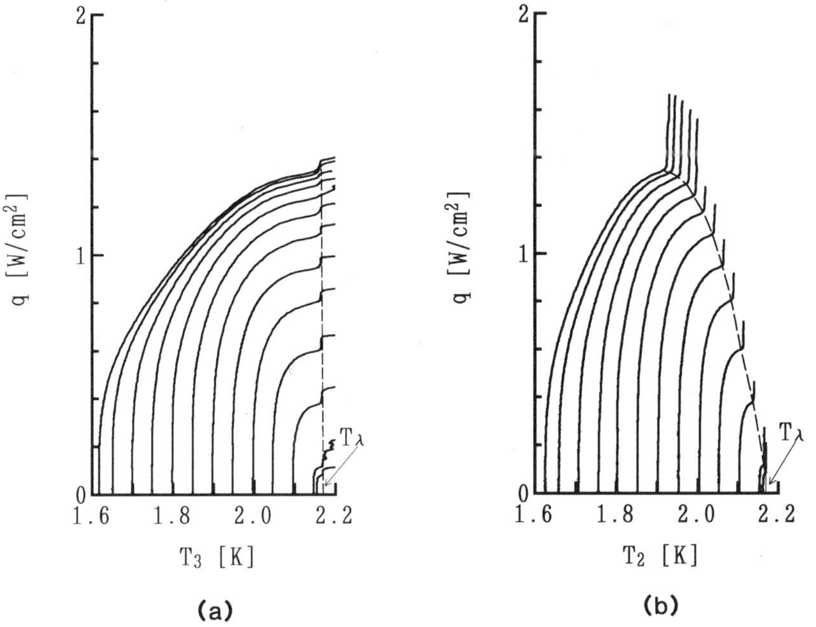

Figure 6. Temperature rise in the liquid at T_3 and T_2 from different bath temperatures T_b. The dashed lines are the traces of the sudden turning points in T_1 at q_λ (cf. Figure 3a). $L = 25mm$, $d = 1.5mm$.

Correction factor γ ($= q_\lambda(Exp.)/q_\lambda(Cal.)$) is displayed in Figure 5 as a function of the aspect ratio d/L. A manual linear fit to the open marks of Figure 5 with $a = 1$ and $b = 1.3$ is included and seems to indicate that constants a and b depend slightly on bath temperature. For $L = 8$ mm, b increases when the heat leak to the surroundings is blocked by activating all three heaters at the same time (solid marks in Figure 5).

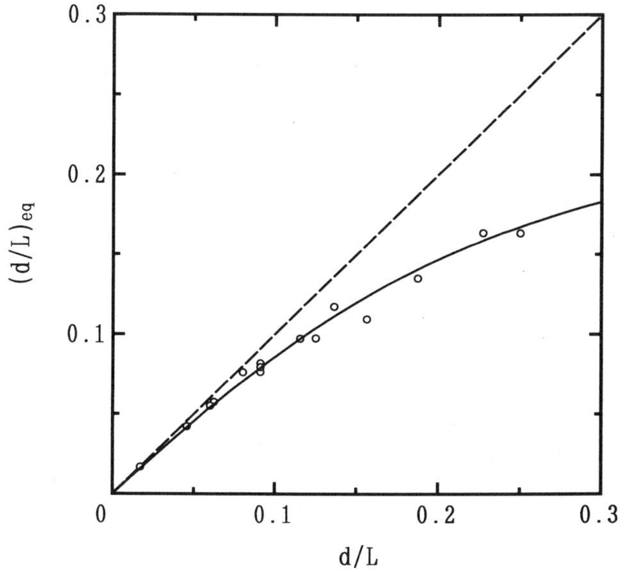

Figure 7. Equivalent aspect ratio $(d/L)_{eq}$ versus actual aspect ratio d/L. The solid curve is a smooth trace through the experimental values.

In the cooling channels of superconducting magnets with their large aspect ratio the geometrical effect may be significant. For a Large Helical Device (LHD, $L = 12.5$ mm and 18 mm, $d = 2$ mm and 3.5 mm)[6] for instance, q_λ will be about 25 % smaller than that estimated from Eq.(1).

To investigate the physical reasons behind the observed aspect-ratio dependence, the temperature distribution across the gap was measured with thermometers T_2 and T_3. Temperature T_2 is always smaller than T_λ even at q_λ, while the liquid temperature T_3 near the heated surface does reach T_λ, see Figure 6. The broken lines show the temperature T_2 corresponding to the sudden turns in T_1. Note that this was used as the definition of q_λ. It implies that stream lines along the length of the channel cease to be one-dimensional, and that the simple G-M theory would have to be extended to two-dimensional or more complex flow situations.

The aspect-ratio effect does not change whether heating is performed from two heaters attached to opposite walls of the channel (data not shown).

Even if q_λ were simply calculated along the two-way elongated stream lines of Figure 1c, which then should be taken as channels of a larger effective length, the correction is insufficient to obtain q_λ as a function of the aspect ratio.

The disagreement between experimental and calculated values rather seems to be due to the fact that the calculation assumes one-dimensional flow lines. Figure 7 shows how the measured discrepancy factor would translate into an "effective aspect ratio" if only Eq.(1) were to be used.

For relatively large d, the temperature T_2 would remain constant above q_λ in spite of an increased heat flux, even after film boiling sets in (see Figure 3b). It could be caused by superfluid flow at T_b along the unheated wall. Fluctuations of T_3 of a higher temperature than T_2 due to the turbulent flow of He I was observed above q_λ as shown in Figure 8. If the channel gap is small, the high temperature fluctuations of He I may spill over into T_2.

Above q_λ, the pressure rise will affect the flow situation even in our large shannels. However, we neglectes the pressure effect below q_λ which is significant in a micro-channel.

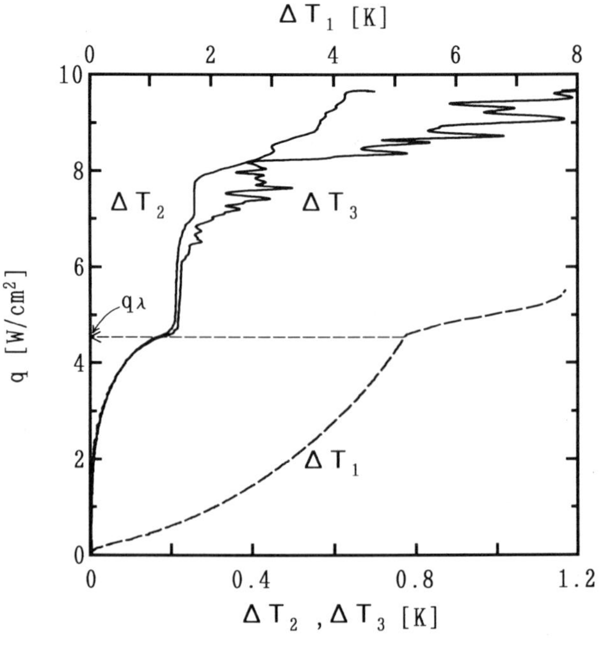

Figure 8. Temperature rise ΔT_1, ΔT_2, ΔT_3, as a function of heat flux q, $T_b = 1.9$ K, $L = 8$ mm and $d = 2$ mm.

ACKNOWLEDGMENTS

This work was supported partially by the National Institute for Fusion Science, Japan. The authors wish to thank Dr. L. Boesten for the kind reviewing of the manuscript.

REFERENCES

1. H. Kobayashi et al., Heat transport in branched channels filled with He II, Cryogenics, to be published.
2. Z. Chen and S. W. VanSciver, Channel heat transfer in He II steady state orientation dependence, Cryogenics 27:635 (1987).
3. T. Okamura, R. Suzuki, N. Seki and S. Kabashima, Heat transport in He II channel with phase transition, Cryogenics 34:187 (1994).
4. C. J. Gorter and J. H. Mellink, On the irreversible processes in liquid helium II Physica 15:285 (1949).
5. S. W. VanSciver, "Helium Cryogenics", Plenum Press, New York and London (1986).
6. T. Satow et al., Present status of design and manufacture of the superconducting magnets for the large helical device, IEEE Trans. on Applied Superconductivity 3:365(1993).

STEADY-STATE HEAT TRANSFER IN He II THROUGH POROUS SUPERCONDUCTING CABLE INSULATION

B. J. P. Baudouy*, F.-P. Juster, C. Meuris, L. Vieillard, and M. X. François

CEA, DAPNIA/STCM, CE-Saclay
Gif sur Yvette, 91191, France

ABSTRACT

The LHC program includes the study of thermal behavior of the superconducting cables wound in the dipole magnet cooled by superfluid helium (He II). Insulation of these superconducting cables forms the major thermal shield hindering the He II cooling. This is particularly a problem in magnets which are subjected to thermal loads. To investigate He II heat transfer processes an experimental model has been realized which creates a one-dimensional heat transfer in such media. Insulation is generally realized by wrapping around the superconducting cable a combination of different kind of Kapton® tapes, fiber-glass impregnated by epoxy resin or Kevlar® fiber tapes. Steady-state heat transfer in He II through these multi-layer porous slabs has been analyzed. Experimental results for a range of heat flux show the existence of different thermal regimes related to He II. It is shown that the parameters of importance are a global geometrical factor which could be considered as an equivalent "permeability" related to He II heat transfer, the transfer function $f(T)$ of He II and the thermal conductivity of the slab. We present and analyze results for different insulations as a function of the temperature.

INTRODUCTION

This work is a part of the general framework of the Large Hadron Collider study program and concerns thermal behavior of the dipoles' superconducting cables cooled by He II. It has been designed such that the whole superconducting coil of LHC will be maintained at a temperature around 1.9 K and will be immersed in a He II pressurized bath at 0.1 MPa[1]. Dipoles may be submitted to a permanent heat production up to 10^4 Wm^{-3} due to beam losses. In modeling the thermal behavior of a LHC dipole, it appeared that the cable's insulation, made of two distinct layers, is one of the critical parameters driving heat transfer. Indeed insulation of these superconducting cables forms the main thermal resistance to He II cooling[2,3]. Even if insulations are permeable to He II, current cooling of cables is insufficient. For a better understanding of heat transfer, thermal behavior of the insulation is approached through an experimental model in which characteristics can be modified. The model, rather than considering the modeling of entire heat transfer in a dipole, focuses on one-dimensional heat transport through an elementary insulation pattern. A first attempt of modeling steady-state heat transfer through this particular porous media is proposed, using the assumption of decoupled thermal paths between insulation and helium.

* Doctoral fellow CEA-Jeumont Industrie-CERN

Advances in Cryogenic Engineering, Vol. 41
Edited by P. Kittel, Plenum Press, New York, 1996

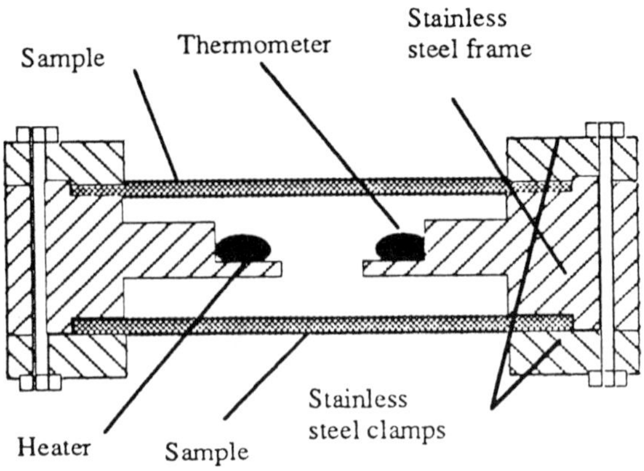

Figure 1. Schematic representation of the experiment.

EXPERIMENTAL DESCRIPTION

Experimental set-up

The main constituent of the apparatus is a stainless steel cylindrical support oriented vertically in the cryostat. Two insulation samples are attached on the sides of the cylindrical support. A schematic of the experimental apparatus is shown in Figure 1. The insulation samples have an external diameter of 80 mm and an internal diameter of 60 mm. They are fixed over 20 mm of their diameter by Stycast resin to the frame and tightened by two stainless steel clamps to avoid helium leak. A heater and a Allen-Bradley carbon resistor are located in the internal bath. In the temperature range of the experiment, this thermometer has a high sensitivity (10 Ω/mK at 2 K). A capillary tube of one meter length is wrapped around the support. It is insulated from the external bath by stycast resin. Instrumentation wires are located in the capillary tube. Heat through the capillary tube filled by instrumentation wires is negligible compared to the heat dissipated by the heater. The internal bath is naturally filled by He II through the insulation and through the capillary tube.

Experimental procedure

The measurement technique uses the thermal properties of the He II. Due to its high effective thermal conductivity, the internal and external temperature baths are uniform during steady-state measurement. In measuring the temperature of the two baths, temperature of the insulation walls can be reached through Kapitza boundary layer. The temperature difference of the two baths is measured as a function of heat flux. Two measurements are separated by 30 s which is higher than the establishment time of the steady state of He II and conductive heat transfer for a heat step. During tests, the external bath is regulated within 2 mK while heat is injected in the internal bath. Experimental sessions have been performed in a "double bath" cryostat which operates between 1.6 K to 2.15 K with a constant helium volume pressurized at atmospheric pressure. Before each experimental session, the thermometer is

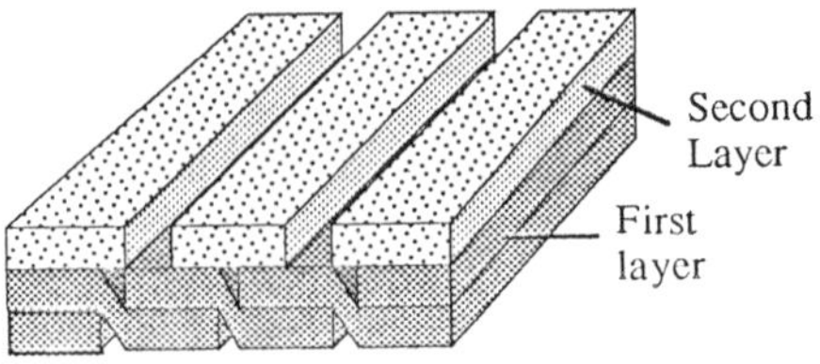

Figure 2. Schematic insulation pattern. First layers has a 50% overlap. The pattern of the second layer creates a gap.

Table 1. Samples characteristics

	Sample #1	Sample #2	Sample #3
First layer	Kapton 100 HN (25 μm x 11 mm) 50% over lapping	Kapton 100 HN (25 μm x 11 mm) 50% over lapping	Kapton 150 HN (37.5 μm x 11 mm) 50% over lapping
Second layer	Kapton XRCI (2x 38 μm x 12 mm) 2 mm spacing	Fiberglass prepreg (125 μm x 12 mm) alternated with dry glass-Kevlar woven. (120 μm x 10 mm) (No gap)	Kapton XRCI (2x 38 μm x 12 mm) alternated with dry glass- Kevlar woven (120 μm x 10 mm) (No gap)
Thickness after polymerization	0.132 mm	0.125 mm on prepreg 0.154 mm on glass-Kevlar	0.164 mm on Kapton 0.184 mm on glass-Kevlar
Total cross-sectional area (A_i)	2 x 2.827 10^{-3} m^2	2 x 2.827 10^{-3} m^2	2 x 5.026 10^{-3} m^2
Conductivity at 2K	0.0087 $Wm^{-1}K^{-1}$	0.0159 $Wm^{-1}K^{-1}$	0.0131 $Wm^{-1}K^{-1}$
Conductivity at 4K	0.0136 $Wm^{-1}K^{-1}$	0.0276 $Wm^{-1}K^{-1}$	0.0255 $Wm^{-1}K^{-1}$

calibrated in situ between 1.6 K to 2.15 K. Calibration is governed by computer and based on calibrated Germanium thermometers. This provides a precision around 1 mK.

CHARACTERISTICS OF SAMPLES

Patterns

In most of superconducting accelerator dipoles, cable's insulation is composed of two distinct layers. The first one answers to the need of electrical insulation while the second layer protects the first one and serves as a mechanical binding for the whole coil during its manipulation. Helium cooling channels are also formed between cables by wrapping them with a gap. Generally, the first layer is composed of Kapton tapes whereas for the second adhesive Kapton or fiberglass impregnated by resin tapes are used.

For these tests, Kapton tapes were used for the first layer. For the second layer, adhesive Kapton tapes, fiberglass impregnated with epoxy resin and dry mixed Kevlar and fiberglass weaving were used. A description of the samples is listed in Table 1. For these samples, the real 3D pattern of cable's insulation has been recreated in a 2D pattern. A representation of a typical insulation is shown in Figure 2. The preparation of the samples is exactly the same as the real insulation wrapped around a superconducting cable : same thickness, patterns and polymerization are used[2,3].

Thermal conductivity

Thermal conductivity of the samples is calculated from thermal conductivity, pattern and dimensions of the insulation foils. Conductivity of basic components as Kapton, fiber glass, epoxy resin and Kevlar fiber has been found in the literature at no lower than 2 K[4,5,6,7]. In the temperature range of investigation, values of conductivity have been linearly extrapolated. For sample #1, a series-parallel model has been considered to evaluate the conductivity. Particular attention has been given to samples #2 and #3. Models taking into account the pattern of fiber weaving and thermal contact resistance have been used for the evaluation of the conductivity of the fiberglass impregnated by epoxy resin[8]. During polymerization of sample #2, resin around fiberglass is able to flow out of the ribbon and cover the glass Kevlar. We included such a possibility in conductivity evaluation. Conductivity of the fiberglass Kevlar has been determined using a series model without taking into account the thermal contact resistance between the two components. Final results are listed in Table 1. Considering imprecision in the conductivity of components and models used, conductivity of an insulation cannot be evaluated more precisely than ± 20%.

EXPERIMENTAL RESULTS

Figure 3 shows typical data for heat flux versus temperature difference (ΔT) between the internal bath (T_i) and the external bath (T_b) for sample #1 at different external bath temperatures. Conductive heat transfer curves are also displayed in Figure 3. At a certain bath temperature, the strong non linear variation of temperature difference with respect to heat flux suggests that turbulent He II heat transfer influences the general heat transfer. Temperature dependence of the curves follows the temperature dependence of the turbulent He II heat transfer function $f(T)$ as it is described in Eq. 1. As $f(T)$ is maximum at 1.9 K, the curve for T_b=1.9 K presents a better heat transfer conductance than at other bath temperature. When T_i reaches T_λ, the heat flux reaches a maximum value (Q^*) where HeI appears in the internal bath around the heater. Above this maximum flux the measured temperature difference reaches a plateau increasing slightly with increasing Q. The plateau indicates that temperature reaches T_λ in the internal bath. Temperature of the internal bath is no longer uniform and the advantage of using the high effective thermal conductivity of He II is lost. Temperature of the inside wall is not obtainable easily. Results above Q^* are not presented in this paper.

ANALYSIS

Pure superfluid regime

Considering the problem of steady-state heat transfer with no net mass flow, He II internal convection, conductive heat transfer through insulation and Kapitza boundary are the major transport mechanisms. Above Q^*, helium channel start to be saturated and HeI heat transfer, He II-He I heat transfer and thermal constriction should be considered.

Equivalent thermal permeability. As we noted previously, turbulent He II heat transport has a large influence on the total heat transfer. We can expect that it is predominant at low heat flux, corresponding to small temperature difference within the sample, because conductive heat transfer is small. We define a pure superfluid regime where we make the assumption that turbulent He II dominates the total heat transfer and where conductive heat transfer and Kapitza thermal boundary layer can be negligible. With increasing Q (causing increasing ΔT), effect of conductive heat transfer increases. The pure He II regime ends at Q_o where conductive heat transfer is no longer negligible. Heat transfer in this regime is defined by the turbulent heat transport process where the relation between heat flux and temperature gradient is expressed by the Gorter-Mellink relation[9] as:

$$\vec{q}^{\,n} = -f(T)\vec{\nabla}T, \tag{1}$$

where n is found to be 3 theoretically but which has been experimentally determined to be between 3 and 4, especially when the temperature approaches T_λ. To consider Eq. (1) under that form, the linear term describing the viscous flow of non turbulent He II is neglected[9]. This assumption will be verified later. In defining a global geometric characteristic independent of temperature, Eq. (1), with n =3, can be rearranged to:

$$Q = \zeta\left[\int f(T)dT\right]^{1/3}, \tag{2}$$

where $\zeta = A/L^{1/3}$, A and L are the global equivalent cross-section and the global equivalent length of the helium channels. ζ can be considered as a global "thermal permeability" of the media related to He II heat transfer with a dimension of $m^{5/3}$.

With analogy to classical concept of permeability, the higher ζ is, the lower the gradient of potential is (pressure gradient in fluid dynamics and temperature gradient in heat transfer). Notice that this parameter is only related to He II heat transfer and represents the geometrical characteristic of the heat flow conductance through the media. It is useful to introduce an equivalent permeability per unit cross-sectional of insulation, $\zeta_u = \zeta/A_i$ ($m^{-1/3}$), to be able to

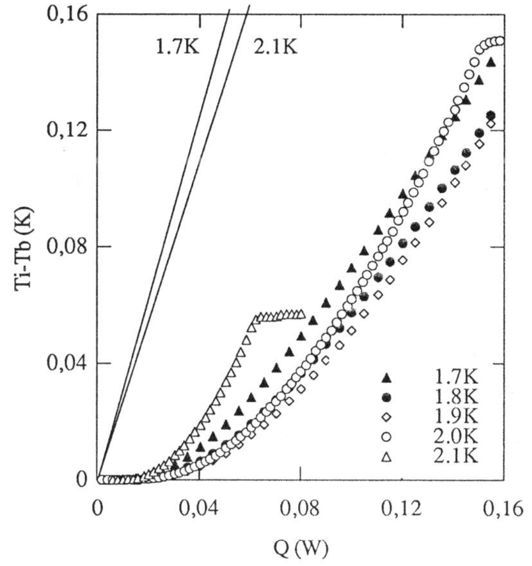

Figure 3. Measurements at different bath temperature for sample #1. Solid lines represent conductive heat transfer in the sample.

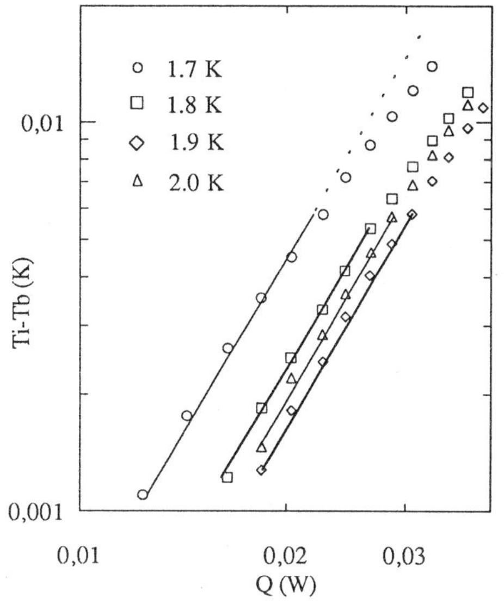

Figure 4. Comparison between measurements and fitted equations (solid lines) in the He II regime for sample #2 at different bath temperatures.

compare different samples. Values of A_i are presented in Table 1. The concept of permeability is only valid if for different bath temperatures the heat transfer curves can be fitted with the same permeability value in the low heat flux regime (He II regime). For the determination of ζ, $f(T)$ data are from a curve-fit to summary data[10].

Limit of the pure He II regime. The He II regime is no longer valid as Q reaches Q_o. Q_o is obviously function of conductive and superfluid heat transfer. The ratio of conductive to turbulent He II heat transfer in the low heat flux regime has to be low. In this regime (small dT), temperature dependence of $f(T)$ and $k(T)$ can be neglected, then this ratio is expressed by:

$$\chi = \frac{A_i}{e\zeta} \frac{k(T)}{f(T)^{1/3}} \Delta T^{2/3} \tag{3}$$

where k is the conductivity and e is the thickness of the media.

We have determined ζ and Q_o for $\chi \approx 0.1$. Data and fits for sample #2 are shown in Figure 4 for different bath temperatures. Results of permeability are displayed in Table 2; they are the mean values taken over the different fits at different temperatures. The variation from the mean value is around 5% for the three samples. Considering the experimental variation of $f(T)$ found in literature[10], assumption of a pure superfluid regime at low heat flux, where superfluid turbulence is developed, can be considered valid.

The length of He II channels can be estimated to be in order of 10 mm. A global cross-section of these channels can be calculated from ζ which is around 10^{-6} m^2. The equivalent hydraulic diameter of these channels can be estimated to be from 10 µm to 200 µm for

Table 2. Equivalent thermal permeability

	Sample #1	Sample #2	Sample #3
Global permeability ζ (m$^{5/3}$)	9.50 10^{-6}	6.61 10^{-6}	2.80 10^{-5}
Permeability per unit area ζ_u (m$^{-1/3}$)	1.68 10^{-3}	1.17 10^{-3}	2.79 10^{-3}

different aspect ratios. For sample #1 at 1.9 K, the smallest heat flux generating a measurable temperature difference is $5x10^5$ W/m^2. It is found in the literature[11] that the critical heat flux for development of turbulent transport at 1.9 K is 10^4 W/m^2 for a diameter of 10 μm and $2x10^3$ W/m^2 for a diameter of 200 μm. From these remarks, the He II regime can be considered as fully turbulent and the linear term can be negligible as assumed in Eq. (1). Similar remarks are valid for other samples at different temperatures.

Mixed regime

We have seen that in the low heat flux regime, the Gorter-Mellink expression describes the heat transfer. But at higher heat flux, as is shown in Figure 5, measurements deviate constantly from this expression with Q increasing. With increasing Q, relative effect of conductive heat transfer increases because temperature difference increases. So, above Q_O, assumption of the pure superfluid regime is not valid anymore. We define a "mixed regime" taking in account of turbulent He II and conduction through insulation andKapitza boundary. The mixed regime is defined for $Q_O \leq Q \leq Q*$. This regime is characterized by the condition of a non local thermal equilibrium between solid and helium. Indeed, at the lowest geometric level, a difference of temperature between the solid and the liquid exists locally due to Kapitza conductance. To construct the model, we assume that heat transfer in the solid and He II takes place in parallel so that there is no net heat transfer from the solid to He II. Thermal paths are decoupled. Then, in this regime, the total heat flux is given by:

$$Q = Q_{HeII} + Q_k = \zeta \left[\int_{T_b}^{T_i} f(T)dT \right]^{1/3} + \frac{A_i}{e} \int_{T_l}^{T_2} k(T)dT \tag{6}$$

where $$Q_k = \frac{A_i}{e} \int_{T_l}^{T_2} k(T)dT = 4\alpha A_i T_b^3 (T_l - T_b) = 4\alpha A_i T_i^3 (T_i - T_2) \tag{7}$$

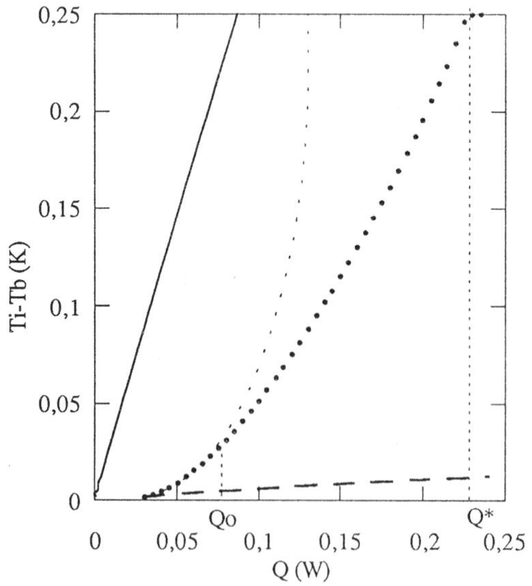

Figure 5. Heat transfer measurement for sample #1 at 1.9K. Conductive heat transfer (solid line), Turbulent He II heat transfer (small doted line) and Kapitza conductance (large doted line) are displayed.

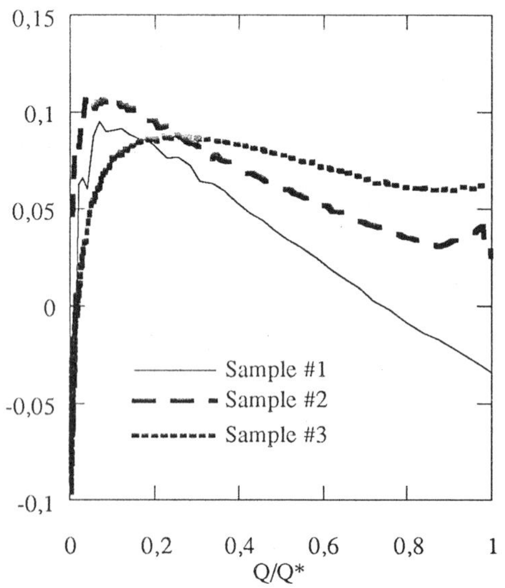

Figure 6. Relative error ((Qcalculation-Q)/Q) in the mixed region between calculation and measurements at 1.9 K for different samples.

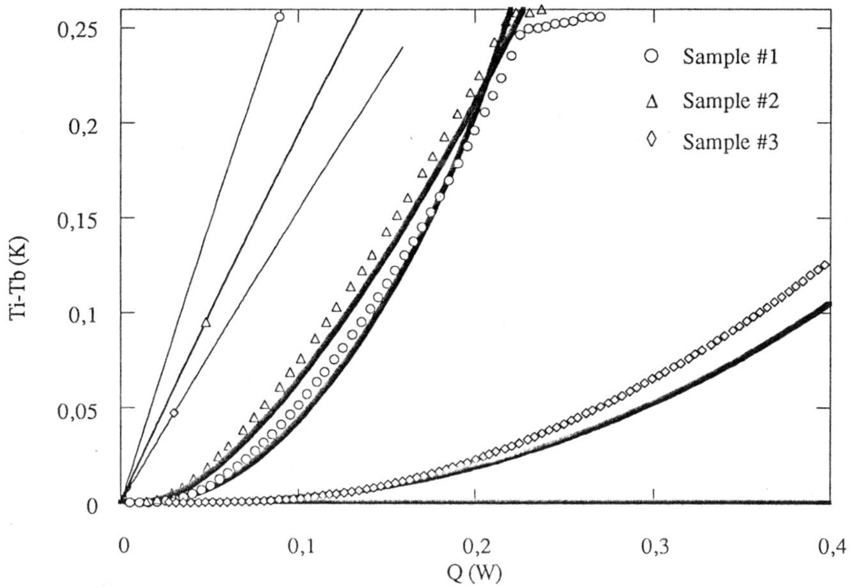

Figure 7. Comparison between measurement and calculation (thick solid line) at 1.9K for different samples. Conductive heat transfer is also displayed (thin solid line).

Q_k is the conductive heat flux. T_1 and T_2 are the wall temperatures of the insulation, results of the Kapitza thermal boundary layer given by the right hand side of Eq. (7). Expressions of the heat flux through the Kapitza boundary have been linearized because for all ranges of heat flux, the condition $\Delta T \ll T$ is satisfied. The term $4\alpha T^3$ is defined as the Kapitza conductance per unit cross-sectional[4]. For this model we have assumed that $Ai \gg A$. Figure 7 presents a comparison between calculation and measurements at 1.9 K for different samples.

Calculations are within ±10% (Figure 6) of the measurements which is a good agreement considering accuracy on the conductivity of the samples and the function $f(T)$. For all samples error increases until it decreases after reaching a maximum in the same region.

Sample #3 has a better cooling efficiency than the two others. It can be related to the second layer. The second layer of sample #3 is composed by a dry glass-Kevlar oven alternated with Kapton with polyimide glue whereas sample #2 has fiber impregnated with epoxy resin: the resin tends to decrease the permeability of the insulation (factor 2.4) by either reducing the number of channels or their cross-section whereas the polyimide glue does not flow. Sample #1 has better cooling than Sample #2 even if the conductivity of sample #2 is twice as high. The He II channels formed by the overlapped Kapton of the first layer end directly in the external bath. through the gap of 2 mm for the sample #1 second layer whereas for sample #2 second layer they end through the dry woven. Note that helium volume is the same in the 2 mm gap as in the dry woven.

CONCLUSION

Results show that heat transfer is influenced by He II heat transfer even for small helium volume inside insulation. At low heat flux, heat transfer is purely governed by He II and we have determined a thermal permeability of the media related to He II heat transfer and independent of temperature. Resin in the fiberglass of the second layer tends to decreases the permeability of the insulation by either reducing the number of channels or their cross-section. For higher heat flux, we proposed a first model which considers insulation and He II thermal paths in parallel. The permeability has a stronger dominance on the total heat transfer than the conductivity. Calculation agrees with measurement within ±10% but does not follow exactly qualitatively. One explanation could be that thermal paths are not decoupled. More investigations will be carried on to improve the model to take account of a non zero heat transfer coefficient between solid and He II.

ACKNOWLEDGEMENTS

The authors address warm thanks to Mrs. A.-M. Puech and Mr. R. Gaubert for samples preparation and measurements. This research is supported by CEA (Saclay, France), Jeumont-Industrie (Jeumont, France) and CERN (Geneva, Switzerland).

REFERENCES

1. LHC Study Group, "The Large Hadron Collider Accelerator Project" CERN/AC/93-03 (1993).
2. L. Burnod et *al.*, "Thermal modelling of the LHC dipoles functioning in superfluid helium". Proc. 13th Int. Conf. on Magnet Technology (1994).
3. B. Baudouy et *al.*, " Transfert thermique dans l'isolation de câbles supraconducteurs refroidis par hélium suprafluide", 4èmes Journées d'Aussois (1993).
4. C. Meuris, "Measurement of Kapitza conductance and thermal conductivity of Kapton®", CEA internal report (1991).
5. M. Jäckel *et al.*, "Thermal conductivity of modified epoxy resins with different cross-link densities at low temperature", *Cryogenics* 31:229 (1991).
6. D.J. Radcliffe and H.M. Rosenberg, "The thermal conductivity of glass-fibre and carbon-fibre/epoxy composites from 2 to 80K", *Cryogenics* 22:245 (1982).
7. B. Poulaer *et al*, "Temperature variation of the thermal conductivity of Kevlar", *Polymer Communications* 26:132 (1985).
8. J.M. Dorvaux, "Modélisation des conductivités thermiques dans les matériaux composites", Journée GUT/SFT (1985).
9. C. J. Gorter and J.H. Mellink, "On the irreversible processes in liquid helium II", *Physica* XV:285 (1949).
10. R. Srinivacan and A. Hofmann, "Investigation on cooling with forced flow of He II. Part 1", *Cryogenics* 25:641 (1985).
11. V. Arp, "Heat transport through helium II", *Cryogenics* 10:96 (1970).

THERMAL CONDUCTANCE OF HEAT TRANSFER INTERFACES FOR CONDUCTIVELY COOLED SUPERCONDUCTING MAGNETS

T. L. Cooper, J. D. Walters, T. H. Fikse

Naval Surface Warfare Center
Annapolis, Maryland, 21041, USA

ABSTRACT

Minimizing thermal resistances across interfaces is critical for efficient thermal performance of conductively cooled superconducting magnet systems. Thermal conductance measurements have been made for a flexible thermal coupling, designed to accommodate magnet-to-cryocooler and cryocooler-to-shield relative motion, and an interface incorporating Multilam designed as a sliding thermal connector for cryocoolers. Temperature changes were measured across each interface as a function of heat input. Thermal conductances have been calculated for each interface, and the impact of each interface on conductively cooled magnet systems will be discussed.

INTRODUCTION

The U. S. Navy is currently investigating the use of conductively cooled superconducting magnet systems for use in magnetic mine sweeping. These systems will be conductively cooled in order to eliminate the logistical burden accompanying the supply and storage of liquid helium. Closed cycle cryocoolers are to be used in place of liquid helium to cool superconducting magnets.

A cold sleeve has been designed as shown in Fig. 1 to allow for removal of a cryocooler from a magnet system without breaking the vacuum surrounding the magnet. Minimizing the thermal resistance of the materials soldered or clamped together within the cold sleeve is critical. Two interfaces have been evaluated for use in this cold sleeve. First, a flexible copper rope connection between the cold sleeve and the superconducting magnet has been fabricated as shown in Fig. 1. This flexible connection must accommodate relative motion between the cold sleeve and the magnet due to extreme shock and vibration the system may experience. Second, Multilam has been used in a sliding connection between the cooler first stage and the cold sleeve, also shown in Fig. 1. The thermal conductances of the Multilam and flexible coupling interfaces have been measured, and the results are presented herein.

MEASUREMENT APPARATUS

Conductance measurements were made using a Gifford-McMahon cryocooler in a variable temperature apparatus described in detail elsewhere.[1] Full scale interfaces were fabricated in so much as practical to better understand actual interface thermal operating characteristics. Measurements were taken which simulated heat transfer conditions thought to be those in full scale magnet systems.

Ruthenium-oxide (RO) temperature sensors were bolted with Cry-Con thermal grease to the top and bottom of test samples. Because small temperature differences were anticipated across the flexible coupling, temperature sensor readings were verified with a three-junction gold-chromel thermopile attached immediately adjacent to the RO sensors. Readings obtained from the ruthenium-oxide sensors and the thermopile were not significantly different. Verification of temperature differences developed across the Multilam interface was not necessary because the temperature differentials were large.

A control heater was used to maintain a constant temperature, on one side of each interface. A sample heater was located across the interface from the control heater on the sample being tested. Following initial cool down, a temperature set point was established by the control heater using a Conductus LTC10-G temperature controller. Control and sample temperature measurements were taken with no power applied to the sample heater to obtain the temperature offset between the RO sensors. Temperatures were recorded in a computer every 15 seconds for the flexible coupling and every 60 seconds for the Multilam. The system was assumed to be in equilibrium when temperature differentials did not change more than 5 mK during a five minute period for the flexible coupling and during a twenty minute period for the Multilam. Power dissipated in the sample heater was incrementally increased. The power input to the sample heater was obtained from independent current and voltage readings and was known to within 5%. The temperature differences developed across the interfaces were obtained by subtracting the set point temperature and the sample sensor offset from the temperature at the bottom of the sample.

Conductances were calculated using Eq. (1), the power dissipated in the sample heater, and the temperature differences developed across the interfaces.

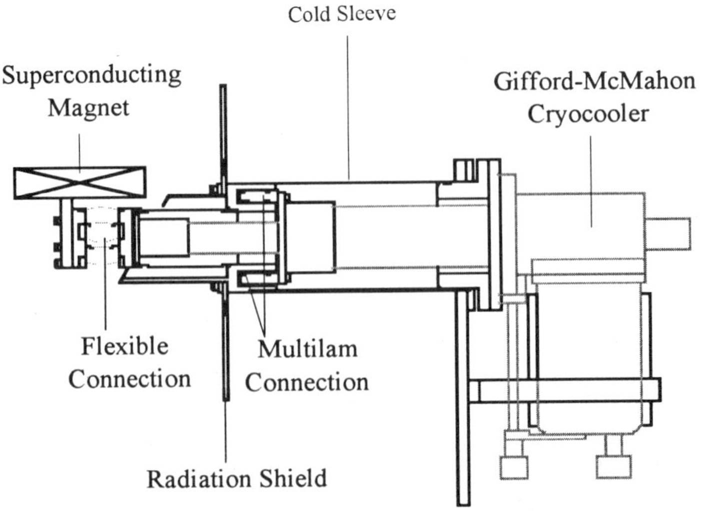

Figure 1. Cold sleeve design for isolation of magnet vacuum space from cryocooler for use in conductively cooled superconducting magnet system.

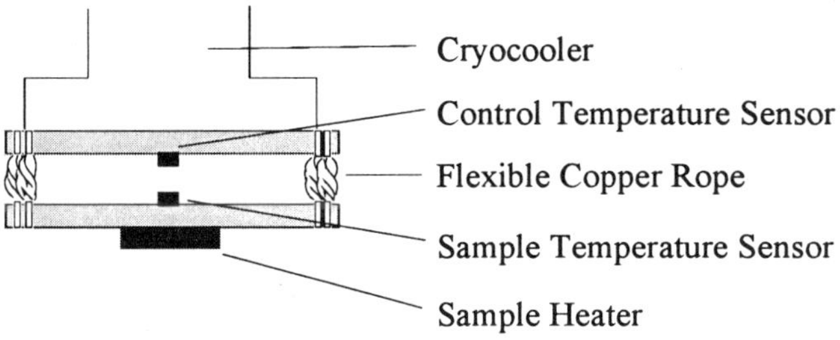

Figure 2. Sketch of flexible coupling interface test fixture showing sensor and sample heater locations.

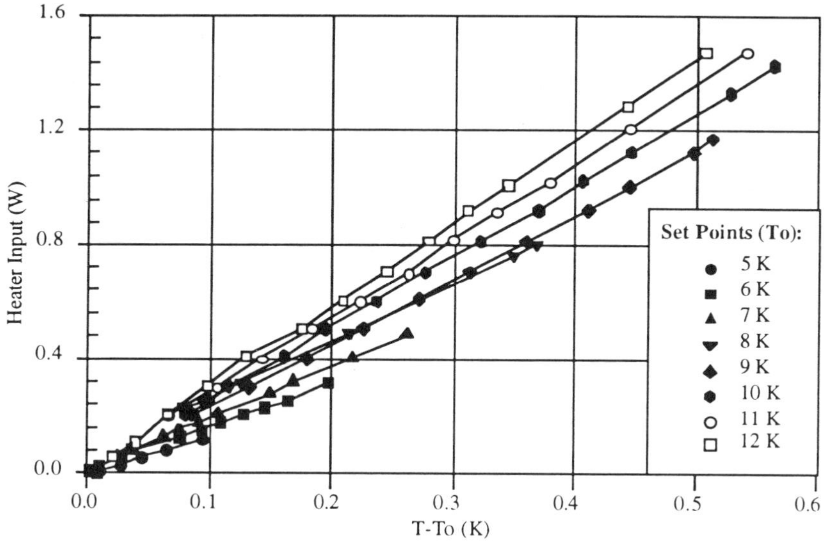

Figure 3. Heater input power as a function of temperature difference across the flexible coupling interface for various set point temperatures.

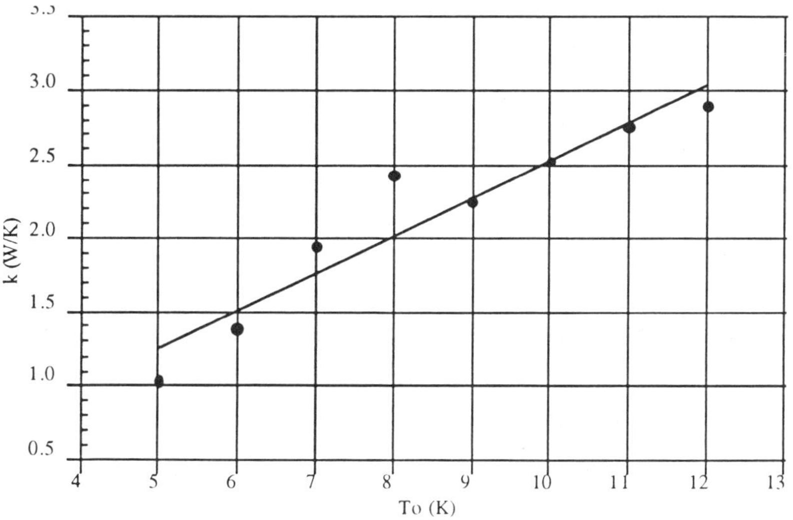

Figure 4. Conductance across the flexible coupling interface as a function of set point temperature.

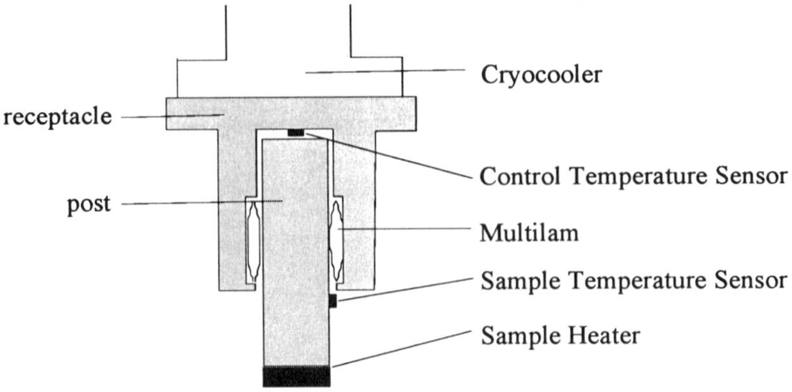

Figure 5. Sketch of Multilam interface test fixture with temperature sensor and sample heater locations shown.

$$\dot{Q} = \frac{dQ}{dt} = \frac{A}{L}\int_{T_0}^{T} k(T)\,dT, \tag{1}$$

where $\dot{Q}$, T, To, k, A, and L are the power input to the sample heater, sample temperature, controlled set point temperature, conductance, and sample cross sectional area and length, respectively. The detailed conductance calculations for the flexible coupling and for the Multilam interfaces are given below. Error estimates included uncertainty in current and voltage readings and in temperature sensor accuracy. Weighted averages of the root-mean-squared deviations in conductance were calculated and are generally within ±10%.

FLEXIBLE COUPLING

A flexible cold sleeve to magnet thermal mechanical coupling has been fabricated using eight 4.4 cm long segments of 7x150 strands of 0.00236 cm diameter copper. The residual resistance ratio of the copper rope is estimated to be 40. The ends of the rope segments were swaged down in diameter to obtain the highest possible copper fill factor. Finally, four sets of two swaged copper rope segments were soldered between the corners of two 0.9525 cm (0.375 in.) thick copper plates which are 8.89 cm (3.50 in.) on an edge. The key to this seemingly simple interface, as shown in Fig. 2, is to swage the ends of the rope segments to increase the copper density as high as possible and to maintain the high density during soldering.

The entire flexible coupling, including end plates, is considered to be the coldsleeve to magnet interface. Therefore, it is not necessary to consider temperature gradients that

Table 1. Curve fit parameters for low temperature Multilam conductance

To (K)	T (K)	a_1	a_2
5.5	5.5 ≤ T ≤ 10.3	1.019744E-03	1.241212E-04
6.0	6.0 ≤ T ≤ 10.9	1.184433E-03	1.105602E-04
7.0	7.0 ≤ T ≤ 13.8	1.358310E-03	1.190728E-04
8.0	8.0 ≤ T ≤ 17.9	1.615356E-03	1.145589E-04
9.0	9.0 ≤ T ≤ 18.2	1.900322E-03	1.094856E-04
11.0	11.0 ≤ T ≤ 23.3	2.316265E-03	1.034919E-04
12.0	12.0 ≤ T ≤ 24.8	2.689461E-03	0.996062E-04

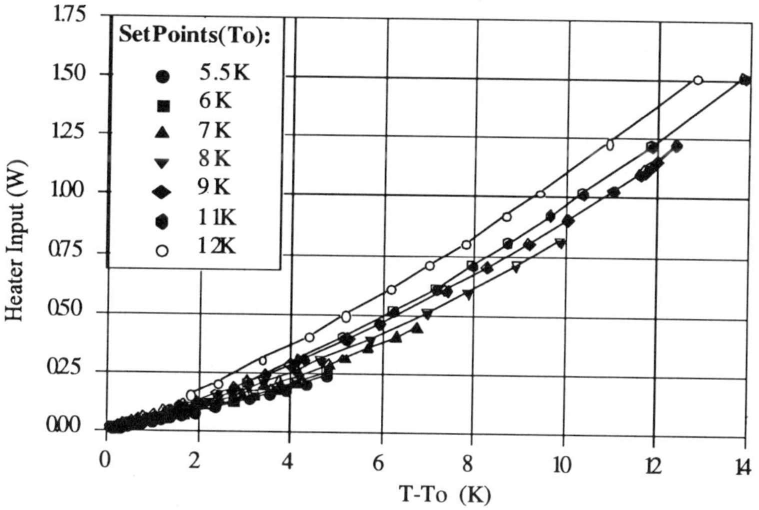

Figure 6. Heater input power as a function of temperature differential developed across the Multilam interface for data acquired below 25 K.

may develop across the copper endplates. The temperature differences developed across the flexible interface for given power inputs are summarized in Fig. 3.

The conductance of this flexible copper interface was calculated using Eq. (1), and the conductance of the coupling was found to be constant over the input powers used in these cases. The conductances of the copper interface as a function of set point temperature are summarized in Fig. 4.

MULTILAM

Type LAI/.5/0° Multilam is being evaluated as a portion of a sliding disconnect which would simplify cooler insertion into the cold sleeve. Moreover, the edges of the Multilam louvers aid in reestablishing good thermal contact by scraping oxide layers off the copper.

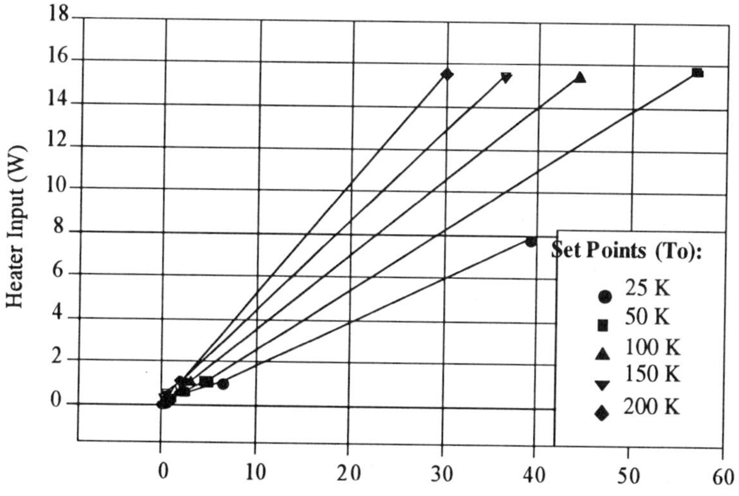

Figure 7. Heater input power as a function of temperature differences developed across the Multilam interface for set point temperatures greater than and equal to 25 K.

A subscale Multilam interface has been fabricated as shown in Fig. 5. A 6.0 cm long and 3.5 cm diameter cylindrical copper post, representative of a cryocooler, is inserted into a cylindrical copper receptacle which functions as a cold sleeve. The inner diameter of the receptacle was fabricated to manufacturer's specifications and lined with 33 louvers of Multilam. The control sensor was placed at the end of the receptacle, and the lead wires were brought out of the receptacle through a vertical channel in the post. The vertical channel removed 3 louvers of Multilam, so only 30 louvers were actually in contact with the post. The copper comprising the receptacle was made thicker than necessary so that the thermal resistance of the copper is small compared to that of the interface. The sample sensor was located on the outside of the post, just below the bottom of the receptacle, and a sample heater was placed at the bottom of the post. The results obtained are divided into those for set points below 25 K and for set points above 25 K and are summarized in Figs. 6 and 7, respectively. Two sets of measurements were made at 8 K with the test apparatus having been disassembled between tests. Upon comparison, the data agreed within 10%.

The low temperature conductances for Multilam varied strongly with set point temperatures and with heater input, as seen in Fig. 6. A second order curve fit was performed on the data comprising Fig. 6.[2] Low temperature conductance values were obtained using Eq. (1) as follows:

$$k(T) = \frac{A}{L}\frac{dQ}{dt} = \frac{A}{L}\frac{d}{dT}[a_0 + a_1(T\text{-}To) + a_2(T\text{-}To)^2], \qquad (2)$$

where a_0, a_1, and a_2 are the curve fit coefficients. Eq. (2) may be expressed in the following form:

$$k(T) = \frac{1}{N}[a_1 + 2a_2\,(T\text{-}To)], \qquad (3)$$

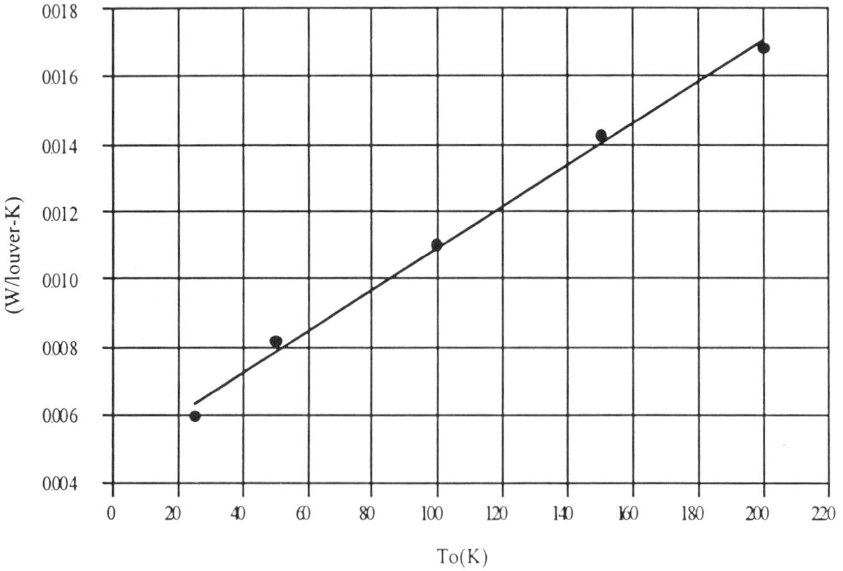

Figure 8. Conductance per louver of the Multilam interface as a function of set point temperatures at or above 25 K.

where the number of Multilam louvers, N, has been substituted for the geometric factor A/L. Parameters a_1 and a_2, for the conductance are summarized in Table 1.

The conductance of the Multilam interface at and above 25.0 K was determined to be constant for each set point over the range of heater inputs used. Therefore, only a simple average was necessary to obtain the interface conductances. These results are summarized in Fig. 8.

DISCUSSION AND SUMMARY

It is important to consider the impact of these conductance measurements on the design and performance of the cold sleeve shown in Fig. 1. The flexible copper interface performs acceptably for certain conditions anticipated for conduction cooled magnet systems. The impact of this interface on a cold sleeve design is straight forward. For example, if the magnet is designed to operate at 5.0 K with a 0.75 W heat leak, the cryocooler would need to operate at 4.4 K due to the thermal resistance of the interface. This interface can be augmented by adding more copper rope segments to further lower the thermal resistance. The thermal resistance of the present copper interface was intentionally kept large enough to be measurable.

The temperature change across a Multilam interface is likely to be substantial at the cryocooler's second stage. An interface with 200 louvers at 35 K with a 20 W heat leak would have a 15 K differential across the interface. More louvers can be incorporated into the design than were used here to lower the thermal resistance, but the number of louvers cannot be increased without bound for two practical reasons. First, it is usually beneficial to design magnet systems to be as compact as possible, and reducing the temperature increase by increasing the number of louvers takes space which may not by available. Second, a substantial force of approximately 0.09 N/louver (0.4 lb/louver) is required to engage the Multilam into its receptacle. Even for the small device discussed in this paper, a jacking arrangement was necessary to remove the post once it was engaged. The force of engagement and extraction will only increase with increasing numbers of louvers, and this force may quickly become impractically large.

CONCLUSION

Minimizing thermal resistances across interfaces is critical for efficient thermal performance of conductively cooled superconducting magnet systems. Thermal conductance measurements have been presented for a flexible thermal coupling designed to accommodate magnet-to-cryocooler and cryocooler-to-shield relative motion, and an interface incorporating Multilam designed as a sliding thermal connector for cryocoolers. Both of these interfaces were designed to be used in a cold sleeve which isolates the cryocooler from the magnet in a distinct vacuum space. Temperature changes were measured across each interface as a function of heat input. Results obtained for the flexible copper coupling are encouraging. The Multilam does not hold as much promise for low thermal resistance interfaces. However, additional Multilam interface designs are being considered because a sliding connection offers substantial benefits to cold sleeve design.

ACKNOWLEDGMENTS

The work presented in this paper was completed as a portion of the Advanced Lightweight Influence Sweep System project sponsored by the U.S. Navy Project Management Office 407.

REFERENCES

1. J. D. Walters, T. H. Fikse, and T. L. Cooper, Variable temperature thermal conductivity and conductance measurements using a Gifford-McMahon cryocooler, Cryocoolers 8: Proceedings of the International Cryocooler Conference, Ed. R. G. Ross, Jr., pp. 823-833 (Plenum Press, New York, 1995).
2. Curve Fit performed using SigmaPlot for Windows Version 2.0, Jandel Scientific, Corp.

THERMAL DESIGN AND ANALYSIS OF A CRYOGENLESS SUPERCONDUCTING MAGNET FOR INTERVENTIONAL MRI THERAPY

K. M. Obasih and M. T. Mruzek

General Electric Medical Systems
P. O. Box 414
Milwaukee, Wisconsin 53201 USA

ABSTRACT

The thermal design and analysis of a cryogenless open superconducting magnet used in interventional magnetic resonance therapy (MRT) is presented and discussed. The magnet consists of two halves, with an open space between the halves, for physician access to the patient. The magnet cartridge is designed to operate at 10K, while the thermal radiation shield is at 60K. The magnet's refrigeration is provided by two closed cycle Gifford-McMahon cryocoolers. Heat load management was a major design challenge, as was the necessity of incorporating robust thermal paths to minimize thermal conduction temperature gradients. Extensive 3D finite element analysis was performed to investigate the behavior of the magnet cartridge and thermal shield. The results of the analysis are compared to the actual thermal performance.

INTRODUCTION

Magnetic Resonance Imaging (MRI) has become a major diagnostic tool available to physicians during the last 15 years. More recently, imaging equipment manufacturers such as General Electric have begun to develop devices for performing real time interventional therapy.[1] In Magnetic Resonance Therapy (MRT) the attending physician has access to the patient to perform medical procedures *as the patient is being imaged.* As shown in Figure 1, the magnetic field for the imaging is provided by two concentric toroids spaced sufficiently apart to allow physician access between the halves. The .5 Tesla magnetic field is produced by superconducting coils wound with Nb_3Sn tape operating at 10K. An important facilitating feature of the tape is the opportunity to use thermal conduction to refrigerate the coils with dual cryocoolers, and hence eliminate the complicating factors of containing liquid helium in the unique geometry.

Two basic features of conduction cooled magnets have driven the MRT design effort: 1) The total heat load delivered to the cryocoolers determines their operating temperature and hence the temperature of the "sink", and 2) The thermal temperature

Table 1. Steady State and Transient Heat Load Summary for the MRT Magnet

HEAT LOAD SOURCE	STEADY STATE	TRANSIENT
Thermal Radiation	0.46	0.00
Suspension Members	0.156	0.00
Power Leads	0.50	0.06
Sleeve Conduction	0.174	0.00
Switch Quench	0.0	0.82
AC Losses	0.0	0.42
Thermal Shield Support	0.15	0.00
Residual Gas Conduction	0.07	0.00
Miscellaneous	0.238	0.14
Total Heat Loads	1.75 W	1.4 W

gradients developed in the conduction members and joints determines the temperature at the point on the magnet most distant from the source of cooling. It is especially important to manage the total heat load effectively because an excess will both raise the temperature of the sink and increase the temperature gradients. Because of the high sensitivity to excess heat loads we incorporated very robust thermal paths in the conduction circuit. We also minimized temperature differentials across thermal joints by using a knurled surface imbedding into a soft layer of cast indium on the mating piece.[2]

MAGNET AND CRYOSTAT DESIGN

The MRT magnet is refrigerated with dual Leybold cryocoolers which provide refrigeration at two temperature levels; each cryocooler nominally provides 4 Watts at 10K on the second stage and 40 Watts at 40K on the first stage. The second stages of the cryocoolers are thermally connected to the magnet cartridge, and the warmer first stages are connected to the thermal radiation shield enclosing the magnet.

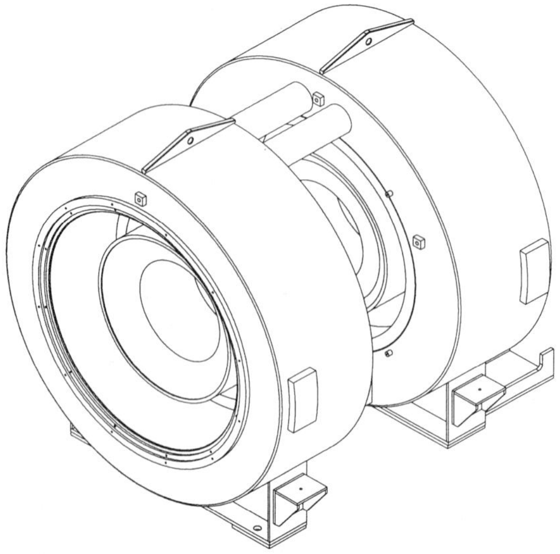

Figure 1. Isometric view of the MRT. The space between the toroidal halves allows physician access to the patient while imaging is in progress.

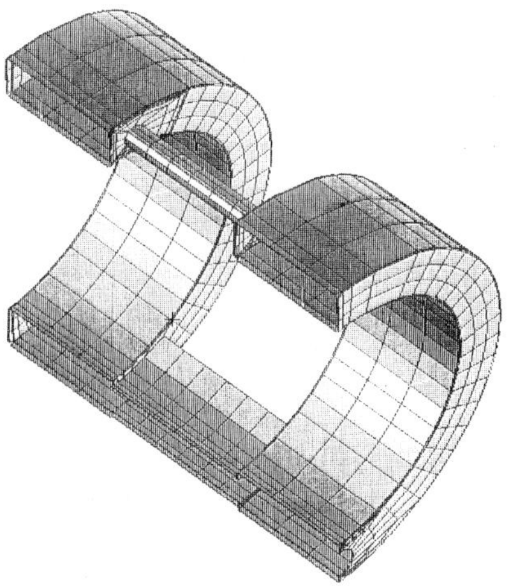

Figure 2. Finite element mesh used for analysis of the thermal radiation shield of the MRT.

The 2nd stage heat load summary for the MRT magnet is presented in Table 1. The steady state heat load of 1.75W on the 2nd stage and 53.0 Watts on the 1st stage represents normal magnet operating conditions, whereas the transient case represents charging or discharging of the magnet. The steady state heat load is dominated by thermal radiation and the conduction losses through the electrical current leads. The transient heat load is dominated by AC losses in the coils during ramping and the quenching of the persistent switch to 20K.

Dual cryocoolers were used on the MRT for several reasons. The primary reason is the increased cooling capacity. The design team felt a relatively large factor of safety was warranted for the initial clinical prototypes. Additionally, the dual coldheads provide the potential for servicing a defective or failing coldhead without ramping down the magnet.

The MRT magnet is designed to operate at a nominal current of 101 amperes. The current leads use High Temperature Superconductors (HTS) to reduce heat leak to the cold mass. We have found the HTS leads reduce the heat load by about a factor of two compared to optimized high purity copper wires. The HTS leads were also found to offer better protection against burnout, by withstanding higher first stage temperatures, such as might occur during cryocooler stoppages.

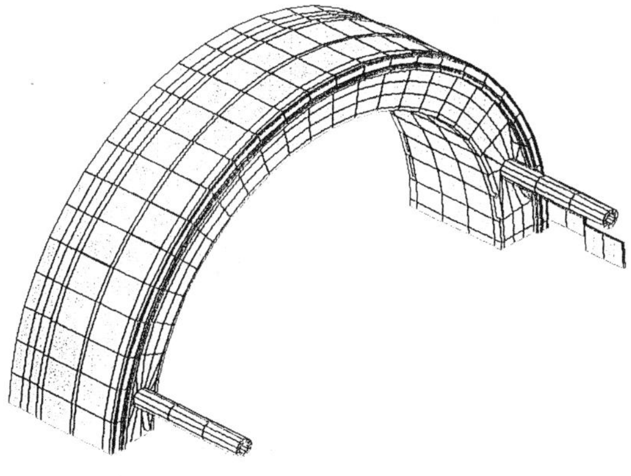

Figure 3. Finite element mesh for the magnet cartridge of the MRT.

One unique feature of conduction cooled magnets is the requirement to accurately monitor the temperature of the components along the thermal path. We used Ruthenium Oxide temperature sensors manufactured by Scientific Instruments for this purpose. The RuO_2 sensors are generally insensitive to magnetic fields up to 2 T, less expensive than Carbon Glass Resistors and more accurate compared to Diodes. We were also able to arrange for RuO_2 sensors manufactured to a common calibration curve, thereby eliminating the tedious tracking of individual devices and their locations.

Aside from the unique geometry, the cryostat of the MRT is relatively conventional. The vacuum vessel is constructed of nonmagnetic 316 Stainless Steel. The operating vacuum is specified at 1 X 10^{-6} Torr. Vacuum gettering is used to maintain the quality of the vacuum over the life of the product. Multilayer superinsulation is used between the 40K shield and the vacuum vessel. Emissivity control is applied to the entire surface of the magnet cartridge. The suspension system uses fiberglass epoxy straps in both the radial and axial directions.

The magnets 10 K thermal mass is substantial: approximately 2 metric tons. Computer simulations of cooldown using the cryocoolers indicated it would take approximately 3 months. To circumvent the cooldown delay, a heat exchanger is attached to the cold mass for the purpose of expediting cooldown using liquid nitrogen and liquid helium.

ANALYSIS

The MRT thermal analysis consists of two separate models. The first model was the solution of temperature distribution and heat load to the thermal radiation shield. The second model was for the temperature distribution in the magnet cartridge, whose boundary conditions depended on the results of the first model. The governing equation solved in both cases was the energy equation with temperature dependent material properties.

The finite element mesh for the radiation shield is shown in Figure 2. There were 12 elements in the axial direction, 18 elements in the hoop direction, and 3 elements in the radial direction. Symmetry allowed only half of the geometry to be modeled. All elements were brick elements.

The boundary conditions for the thermal radiation shield analysis consisted of specifying the heat sink temperature of the shield to be 50 K. All geometric surfaces at room temperature had a temperature specification of 300 K. An effective thermal conductivity was calculated for the multilayer insulation. Heat loads from suspension members to the shield were also modeled. The solution was performed using the commercial code ANSYS.

The finite element mesh for the magnet cartridge is shown in Figure 3. There were 9 elements in the axial direction, 26 elements in the hoop direction and 12 elements in the radial direction. As was the case in the model for the thermal radiation shield, symmetry allowed only half of the geometry to be modeled. Brick elements, Stif 70, were used. The preprocessing of both models was done using Patran 3 commercial code. Both models were translated to ANSYS code.

The various heat loads to the magnet cartridge were all computed using closed form methods, and are listed in Table 1. In the model, the radiative and gas conduction heat loads were added together and applied to all nodal points on the surface of the

Table 2. Predicted and Actual Thermal Shield Temperatures

	Predicted	Measured
Heat Sink	50 K	50.59 K
Shield Max	60.5 K	60.86 K

Table 3. Predicted and Actual Magnet Temperatures

	Predicted	Measured
Coil Maximum Temperature	9.90 K	9.95 K
Hoop Coil Temperature Gradient	0.23 K	0.28 K
Heat Sink Temperature	9.50K	9.65 K
Radial Magnet Cartridge Gradient	0.10 K	0.06 K

cartridge as nodal heat sources. The conductive heat loads were specified as element heat sources at those locations on the magnet cartridge at which they occur. For the transient case, the additional transient heat loads were superimposed on the steady state heat loads. The heat sink for the magnet cartridge was specified as 9.5K. The heat sink temperatures for the thermal shield and magnet cartridge were predetermined based on the heat load summary of Table 1, the cryocooler load maps and the calculated temperature gradients across the thermal joints in the penetration area.

RESULTS

The heat load summary and was used to determine the cooling capacity of the cryocoolers and required heat sink, as well as the maximum allowable temperature

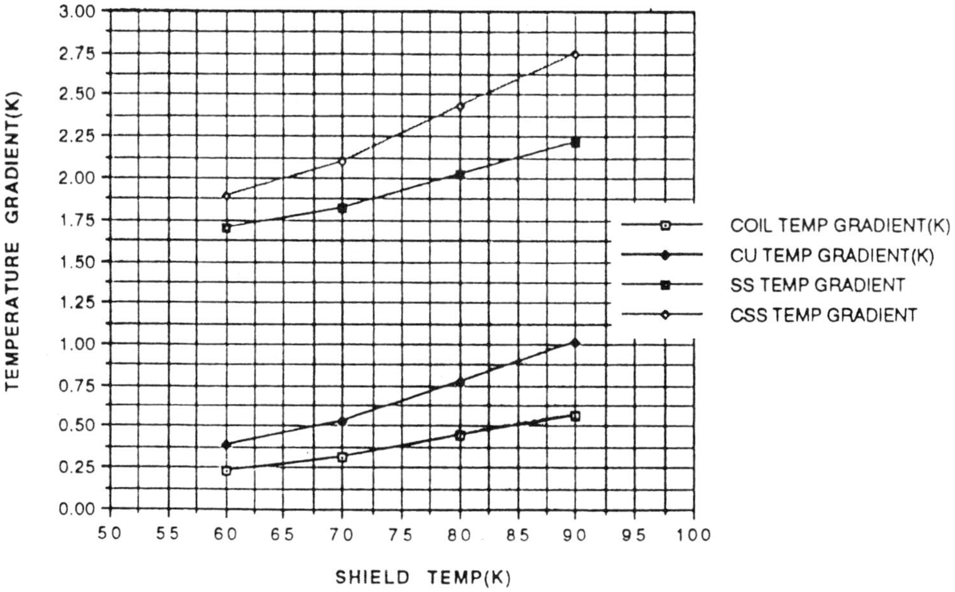

Figure 4. The effect of the thermal shield temperature on the temperature gradient across the MRT coil support structure .

gradient in the magnet. The predicted maximum temperature gradient in the shield is 10.5 K. The measured maximum temperature gradient in the shield was 9.84 K. A comparison of the predicted solution to the measured solution is shown in Table 2 for the thermal shield.

The predicted maximum temperature on the magnet cartridge coil was 9.90 K with a specified heat sink temperature of 9.5 K during steady state operation. The maximum predicted temperature gradient across the whole magnet cartridge structure for steady state was 1.75K, and 0.23 K for the coils. During transient operation, the predicted maximum temperature gradient across the coils was 0.334 K and a delta of 2.05 K across the whole magnet structure. In Table 3 the predicted and measured steady state temperatures are compared. The predicted and measured temperatures have reasonable agreement.

A sensitivity study was performed to determine the effect of various parameters on the magnet temperature. In Figure 4 a plot of the effect of thermal radiation shield temperature on magnet cartridge temperature gradient is plotted. Figure 5 shows the effect of heat sink temperature on the predicted temperature gradient across magnet system components.

CONCLUSION

The good agreement between calculated and actual temperature profiles suggests the heat load sources for the magnet cartridge and thermal shield were properly identified and quantified. The finite element prediction of temperature distribution on the thermal radiation shield and the magnet cartridge were in excellent agreement with measured values for steady state operations. Additionally, we found the finite element predictions of the transient case were also in excellent agreement with measured temperature values during the ramping of the magnet.

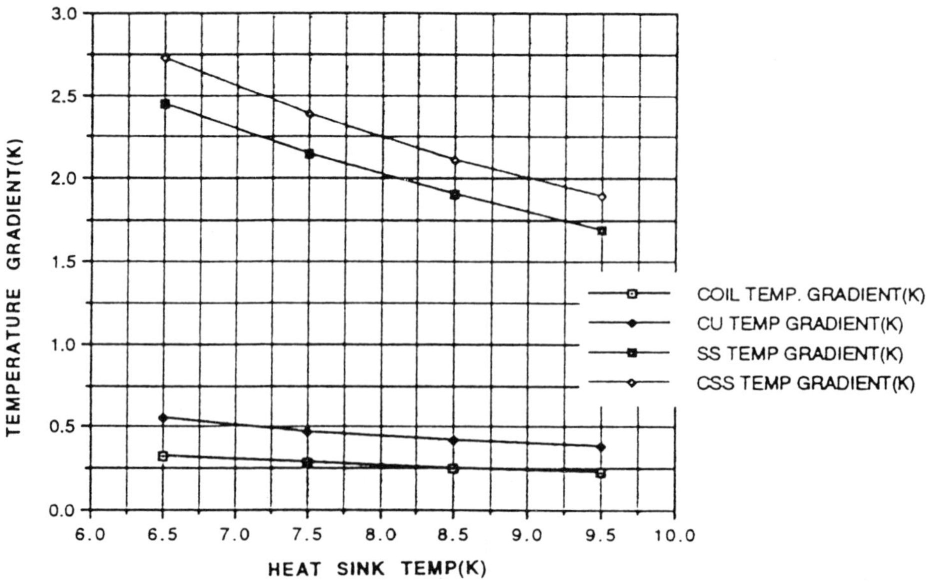

Figure 5. The temperature gradient across the various coil support structure components of the MRT.

ACKNOWLEDGMENT

This work was done when the authors were at the GE Magnet Plant in Florence, SC. We would like to thank all the team members and staff at the facility that made this work a success. In particular, we would like to acknowledge Dr. Timothy Havens, Mr. John Scaturro Jr., and the rest of the MRT Team.

REFERENCES

1. Superconductivity News, Monthly Technical Edition, January 26, 1994, Volume 6, Number 16, ISSN 0897-2427
2. United States Patent 5247800 by M. T. Mruzek, P. Eckels and C. Gouldsberry assigned to the General Electric Company. Issued September 28, 1993.

COOLDOWN ANALYSIS OF A CONDUCTION-COOLED SUPERCONDUCTING MAGNET FOR MAGNETIC RESONANCE THERAPY(MRT)

K. M. Obasih[1], and O. O. Ige[2]

[1]GE Medical Systems
P. O. Box 414
Milwaukee, Wisconsin 53201
[2]GE Medical Systems,
Magnets Systems
Florence, South Carolina 29501

ABSTRACT

A three dimensional finite element model of the cooldown process and the resulting thermal stresses for a cryogen-free superconducting magnet used in interventional magnetic resonance therapy is examined and evaluated. The magnet cartridge consists of two cylindrical halves connected together by four structural posts. The magnet cartridge is cooled primarily by conduction cooling. The cooldown heat load from the cartridge can be removed in two ways; with cryogens via a heat exchanger attached to the cartridge, or via a cryocooler coldheads operating as the heat sink. The model is used to predict the fastest allowable cooldown time without introducing unacceptable stresses in the magnet. The predicted cooldown temperatures are compared to the measured performance. The cooldown stresses are presented and discussed. The model was built in Patran 3 as the preprocessor and the solution was run in ANSYS.

INTRODUCTION

Magnetic resonance therapy(MRT) is the new real time MRI imaging device for operating rooms recently developed by General Electric. The magnetic field for imaging is provided by two doughnut shaped coils connected by four structural posts. The field strength of the magnet is 0.5 Tesla, which is produced by superconducting coils wound with Nb3Sn tape conductors operating at 10K. The magnet cartridge is cooled purely by conduction. The magnet is kept cold using two, two-stage cryocoolers, after cooldown. The first stage of the cryocooler is attached to the thermal radiation shield, while the second stage is attached to the magnet cartridge as the heat sink.

The process by which the magnet gets to its operating temperature, is a very important part of the magnet design and development. Economically it is essential to know the fastest allowable cooldown and warm-up rate for the magnet without introducing unacceptable thermal stresses in the magnet. Simulation of the cooldown and warm up of the MRT magnet was essential for understanding, control and improvement of the cryogenic and thermal design. These simulations provided information about cooldown rates, temperature gradients and their resulting thermal stresses. The cooldown results were used to define the cooldown and warm up procedures, as well as defining machine availability times in the cooldown area. In this paper, the cooldown analysis technique developed for the conduction cooled MRT superconducting magnet is presented.

ANALYSIS

The calculated enthalpy change between room temperature, 300K, and 10K for the magnet cartridge was 163.5 MJ. The enthalpy change calculated for the thermal shield between room temperature and the average shield operating temperature of 55K was 123.8 MJ. In this paper, only the cooldown analysis of the magnet cartridge is presented.

Figure 1 shows the finite element mesh of the magnet cartridge used in the cooldown analysis. The governing equation solved for the temperature distribution is the energy equation with temperature dependent material properties. Geometric and load symmetry allowed only one half of the magnet cartridge to be modeled as shown in figure 1. There were nine elements in the axial direction, 26 elements in the hoop direction and 12 elements in the radial direction.

Symmetric boundary conditions were imposed at the plane of symmetry, Z=0. The temperature dependent material properties were input in tabular format.

SOLUTION

All simulations were performed using ANSYS commercial code. In the analysis, cooldown was initiated by introducing lower temperature value at the heat sink and the solution marched with time. The solution time steps were selected to ensure numerical stability and accuracy. Two different simulations were performed in this study. The first simulation was for the case where the magnet was cooled down using cryocoolers only. In this case, only one heat sink value was provided at 9.5K. The second simulation was for the

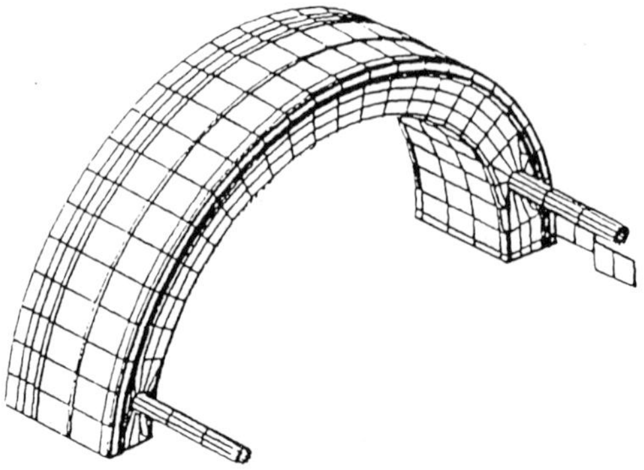

Figure 1. Finite element mesh of magnet cartridge.

Table 1. Typical cooldown solution load steps

Time (Hours)	Heat sink temperature (K)
0 - 24	250
24 - 48	200
48-72	150
72-96	100
96-144	9.5

case where cryogens were used in a heat exchanger attached to the magnet cartridge. In this case, several heat sinks were provided. The particular time during the cooldown period at which these heat sinks are introduced in the solution determines the cooldown rate. Table 1 above show typical load steps for cooldown simulation.

RESULTS

The results from the first simulation using cryocoolers predicted a 93 days cooldown time period. This predicted cooldown period was too long for all practical purposes. The predicted results was considered fairly accurate based on data obtained from previous conduction cooled magnet that took 18 days to cool down. This previous prototype magnet had only one fifth the enthalpy of the MRT.

The second simulation was for the case of using cryogens in a heat exchanger arrangement in the cooldown of the magnet cartridge. The predicted and measured cooldown temperature of the coil is shown in figure 2. below, while the predicted maximum temperature in the magnet cartridge and the measured value are presented in figure 3. It shows that cooldown can be achieved in 7-8 days. The temperature distribution for each load step was used in the computation of the resulting thermal stresses at the end of each load step.

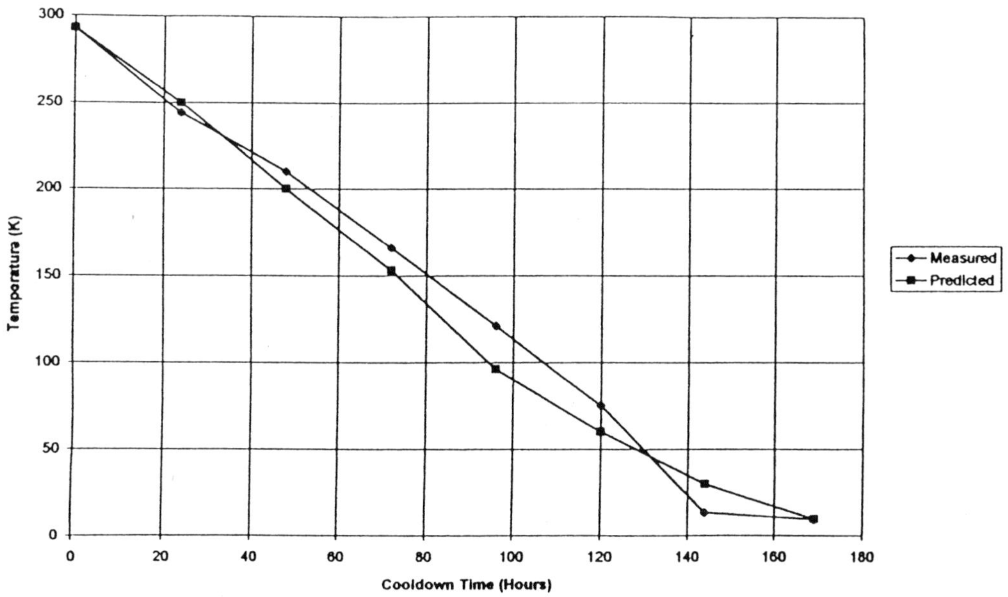

Figure 2. A comparison of predicted and measured temperature of the coil during cooldown.

Table 2. Cooldown equivalent stresses for magnet cartridge components

Time	Temp.		Equivalent	Stress	(ksi)
(hrs)	(K)	Prepreg	Copper	Epoxy	Stainless Steel
4.8	290/299	< 0.1	0.13	0.20	<0.1
9.6	280/295	<0.1	0.43	0.40	0.19
14.4	270/289	<0.1	0.78	0.60	0.36
24.0	250/272	0.22	1.54	0.97	1.0
28.8	240/262	0.34	1.94	1.14	1.51
48.0	200/222	1.08	3.80	1.81	4.64
52.8	190/212	1.33	4.33	1.98	5.69
72.0	150/172	2.60	6.61	2.64	10.96
76.8	138/161	3.03	7.81	2.83	12.77
91.2	102/127	4.62	10.57	3.46	19.41
96.0	90/114	5.23	11.32	3.67	21.96
100.8	74/100	5.82	11.21	4.25	25.32

Table 3. Cooldown component stresses for the MRT coil

Time	Temp	Radial	Hoop	Axial		Shear	(ksi)	Equiv.
(hrs)	(K)	(ksi)	(ksi)	(ksi)	XY	YZ	XZ	(ksi)
4.8	290/299	<0.1 / <0.1	<0.1 / <0.1	<0.1 / <0.1	<0.1	<0.1	<0.1	<0.1
9.6	280/295	<0.1 / <0.1	<0.1 / <0.1	<0.1 / <0.1	<0.1 / <0.1	<0.1	<0.1	0.1
14.4	270/289	<0.1 / <0.1	<0.1 / <0.17	<0.1 / <0.1	<0.1	<0.1	<0.1	0.19
24.0	250/272	<0.1 / 0.18	<0.1 / 0.36	<0.1 / 0.16	<0.1	<0.1	<0.1	0.38
28.8	240/262	<0.1/ 0.28	0.14/0.46	-0.1/ 0.26	<0.1	<0.1	<0.1	0.48
48.0	200/222	-0.18/ 0.84	0.43/0.98	-0.21/ 0.82	<0.1	<0.1	0.20	0.94
52.8	190/212	-0.22/ 1.03	0.49/1.19	-0.24/ 1.01	<0.1	<0.1	0.23	1.08
72.0	150/172	-0.44/ 1.98	0.71/2.27	-0.39/ 1.97	0.1	0.11	0.40	1.91
76.8	138/161	-0.51/ 2.31	0.84/2.64	-0.45/ 2.29	0.12	0.12	0.47	2.23
91.2	102/127	-0.80/ 3.50	1.10/4.00	-0.62/ 3.50	-0.17	0 18	0.66	3.34
96.0	90/114	-0.92/ 3.96	1.16/4.51	-0.68/ 3.97	-0.19	0.20	0.73	3.76
100.8	74/100	-1.12/ 4.56	1.48/5.16	-0.66/ 4.49	-0.23	0.23	0.81	4.35

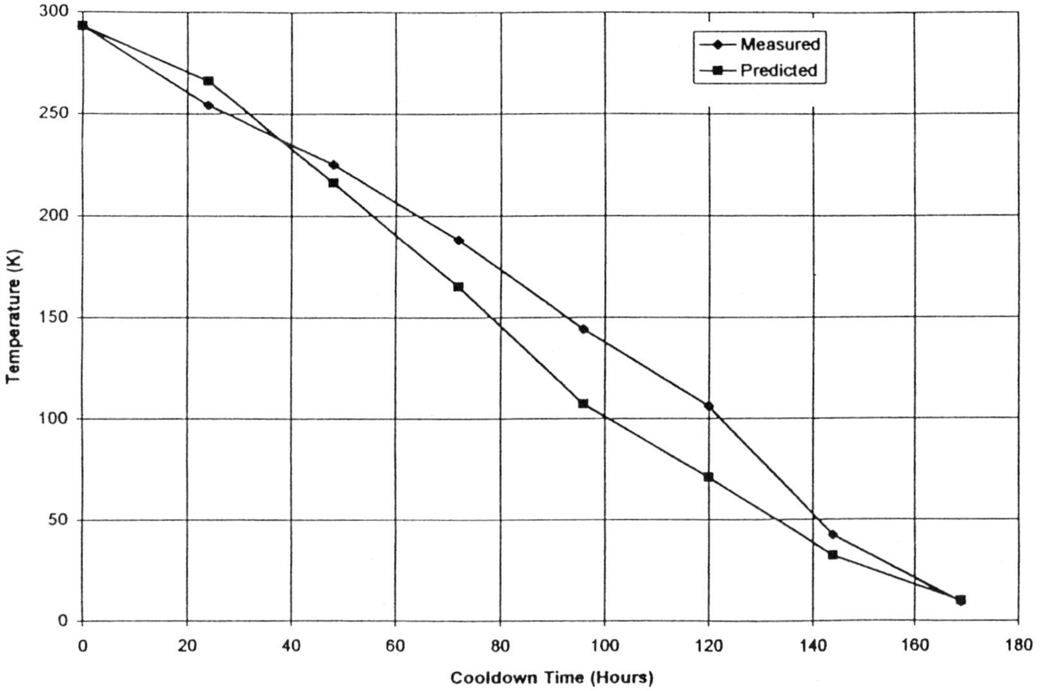

Figure 3. Maximum Temperature in magnet cartridge during cooldown.

The thermal stresses developed in the magnet cartridge during cooldown are presented in tables 2 and 3. The equivalent Von Misses stresses for the various materials that comprise the cartridge is presented in table 2. The stresses reported are all in ksi units. The evaluation of thermal stress was terminated when the temperature of the cartridge was between 74 - 100K since all the contractions after 77K are minimal. In table 3, the maximum component stresses are tabulated. All the above thermal stresses, especially for the coils, are within allowable limits.

CONCLUSION

A cooldown numerical model which properly predicts the cooldown characteristics of the MRT magnet cartridge is developed. The predicted cooldown temperatures are in good agreement with measured cooldown temperatures. The cooldown technique developed was easily put into practice and used in developing the cooldown procedure for the MRT type magnets. The model has also been used in further optimizing the cooldown time for the MRT magnet from 8 days to 4-5 days. This model was also used in the prediction of the warm up time for the magnet.

ACKNOWLEDGMENTS

We would like to thank the MRT team, and in particular Dr. Timothy Havens and Mr. John Scaturro Jr. for their support during the course of this project.

REFERENCES

1. J. P. Holman, "Heat Transfer", McGraw-Hill Book Company, New York (1976)
2. R. G. Budynas, "Advanced strength and Applied Stress Analysis", New York (1977)
3. Swanson Analysis Systems, Inc., "ANSYS4.4A users Manual, Houston, PA(1990)
4. PDA Engineering, Inc., PATRAN 3 Users Manual, Costa Mesa, CA(1994)

CRYOGEN-FREE SPLIT-PAIR SUPERCONDUCTING MAGNET WITH A ϕ 50 mm × 10 mm ROOM TEMPERATURE GAP

K. Watanabe,[1] S. Awaji,[1] J. Sakuraba,[2] K. Jikihara,[2] K. Watazawa,[2] T. Hasebe,[2] F. Hata,[2] C.K. Chong,[2] Y. Yamada,[2†] and M. Ishihara[2]

[1] Institute for Materials Research, Tohoku University
2-1-1 Katahira, Aoba-ku, Sendai 980-77, Japan
[2] R & D Center, Sumitomo Heavy Industries, Ltd.
63-30 Yukigaoka, Hiratsuka 254, Japan

ABSTRACT

A cryocooler-cooled split-pair superconducting magnet which will be a new functional system combined with important studies such as X-ray diffraction, neutron diffraction, and magneto-optics or opto-magnetism has been constructed. The split-pair NbTi superconducting magnet has a vertical clear bore of ϕ 68 mm and a horizontal gap of 58 mm, and is operated in a vacuum below 6 K using a GM-cryocooler with magnetic regenerator material $ErNi_{0.9}Co_{0.1}$. This magnet system is designed to generate a magnetic field of 5.0 T at the center of an experimental bore, and the maximum field is estimated to be 6.7 T at the coil winding.

INTRODUCTION

In the basic research of materials, it is a general practice to use X-ray or neutron diffraction for crystal structure analyses and phase transformation investigations. Currently, X-ray diffraction experiments under the special sample environment of low temperature, high temperature or high pressure are being vigorously carried out. However, an X-ray diffraction apparatus combined with a high field superconducting magnet has not yet been developed, and the high field research using X-ray diffraction remains unfamiliar.[1] If the phase transformation due to the change of magnetic moment or electron state in high fields can be directly observed by X-ray diffraction, the results offer the potential of revealing new informations.

High magnetic fields are usually generated using a superconducting magnet. Liquid helium is needed as the coolant for operating a superconducting magnet. Since liquid helium is easy to evaporate, a cryogenic dewar is designed to have a complicated cryostat structure with a liquid nitrogen reservoir, radiation shield, and vacuum thermal insulation in order to realize the desired liquid helium hold time for a long-term experiment. This results in a large and heavy cryostat. In addition, a high field superconducting magnet requires vapor cooled current leads to reduce the heat input into the helium reservoir. The vapor cooled current leads need an optimized lead length, whose length is never short. This causes a limitation on minimizing the size of the cryostat. Therefore, it is very difficult to

† present address: Faculty of Engineering, Tokai University, 1117 Kitakaname, Hiratsuka 259-12, Japan

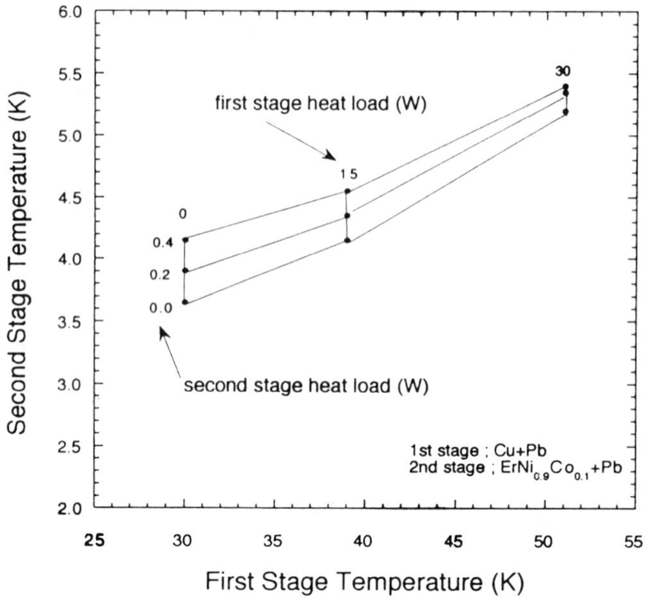

Figure 1. Refrigeration capacity of a 4 K GM-cryocooler.

design a high field superconducting magnet system combined with the X-ray diffraction apparatus.

This paper reports the development of a very compact split-pair superconducting magnet system without liquid helium and nitrogen which is available for an X-ray diffraction or a neutron diffraction experiment. A conduction-cooled superconducting magnet [2] is constructed using a 4 K Gifford McMahon (GM)-cryocooler and high temperature $Bi_2Sr_2Ca_2Cu_3O_{10}$ (Bi(2223)) current leads.

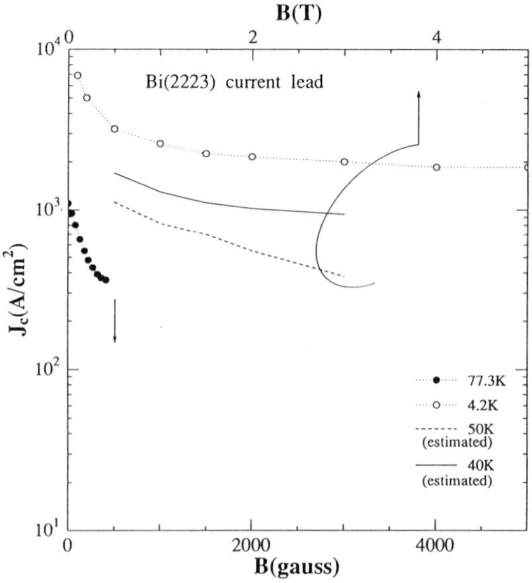

Figure 2. J_c-vs-B properties at temperatures ranging from 77 K to 4.2 K for the Bi(2223) current lead.

Table 1. Heat load estimation for a cryogen-free split-pair superconducting magnet

2nd stage	Bi(2223) current leads	0.089 W
	Joule heating	0.018 W
	heat radiation	0.005 W
	measuring wires	0.066 W
	ac loss etc.	0.020 W
	total	0.198 W
1st stage	Cu current leads	4.0 W
	Joule heating	0.4 W
	heat radiation	10.0 W
	measuring wires	0.3 W
	total	14.7 W

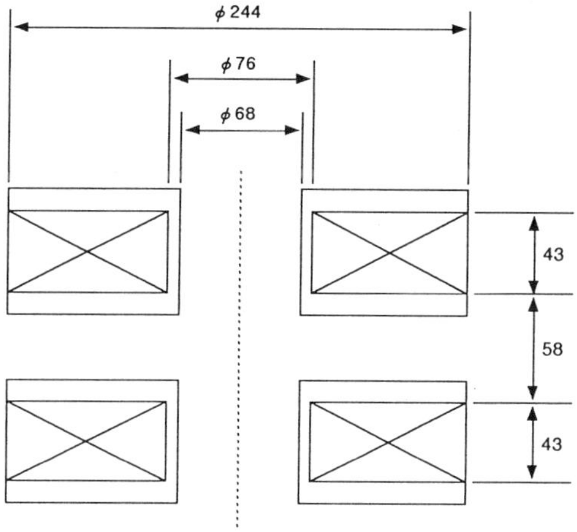

Figure 3. Coil parameters for two coils separated by a gap of 58 mm.

CRYOSTAT DESIGN AND CONSTRUCTION

Refrigeration Power of GM-Cryocooler

Recently developed GM-cryocoolers with a two stage cycle provide temperatures below 4 K.[3] This is because magnetic materials such as $ErNi_{0.9}Co_{0.1}$, which have much larger specific heat below 10 K than that of ordinarily used Pb, have just become available for a 2nd stage regenerator of the GM-cryocooler. Figure 1 shows the relation between refrigeration capacity and temperature for the GM-cryocooler with the hybrid regenerators of $ErNi_{0.9}Co_{0.1}$ and Pb for a 2nd stage and Cu and Pb for a 1st stage. The refrigeration power for the 2nd stage is about 0.4 W at 4.5 K under the heat load of 15 W at 39 K for the 1st stage.

Bi(2223) Current Leads

The high temperature $Bi_2Sr_2Ca_2Cu_3O_{10}$ current leads were fabricated by a powder sintering process, and were formed into a tubular shape with a 23 mm outer diameter, 20 mm inner diameter, and 190 mm long. Ag was coated along a 30 mm length at both ends for current terminals using a plasma spray method. Figure 2 shows the field dependence of the critical current density at 77 K and 4.2 K for the Bi(2223) current leads. The external field B is applied perpendicular to the transport current I. As this tubular lead has the strong c-axis oriented structure perpendicular to the surface, the critical current properties in B ⊥ I are inferior to those in B // I. The calculated leakage fields of the B ⊥ I component are 0.44 T at the cold end of the Bi(2223) lead and 0.14 T at the warm end, when the split-pair superconducting magnet generates 5 T at a center of a room temperature gap. The Bi(2223) leads are thermally anchored at the 1st stage, and as a result the warm end of Bi(2223) is expected to be cooled down to about 50 K. The estimated I_c-vs-B curves at 40 K and 50 K are also exhibited in Fig. 2. One notices that the Bi(2223) lead with an areal cross section of 1 cm^2 at temperatures below 50 K has more than the critical current of 1000 A in fields up to 0.5 T for B ⊥ I.

We have been testing the long-term performance of Bi(2223) leads which were used in a cryogen-free magnet system about 3 years ago.[2] Up to now, those leads experienced cooling time of 15,240 hours in a vacuum and thermal cycles of 17 times from room temperature to 10 K. The magnet operated at 10 K generates 4 T at 400 A. Experiments in continuous holding time at 370 A for 1,200 hours were performed twice. From these performance tests, the long life of the Bi(2223) lead was fully confirmed.

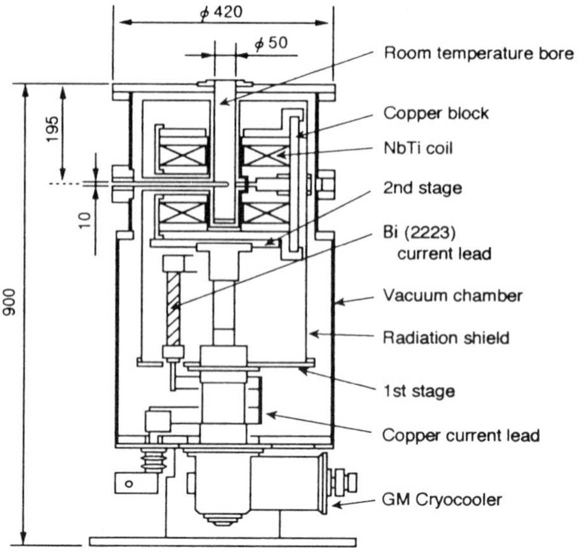

Figure 4. Outline of the cryocooler-cooled split-pair NbTi superconducting magnet using Bi(2223) current leads.

Heat Loads Evaluation

Heat input into the GM-cryocooler was estimated. Table 1 lists the heat load from each component of the cryogen-free split-pair superconducting magnet system.

The largest heat input of 10 W for the 1st stage is due to thermal radiation from room temperature. The current leads from room temperature to the 1st stage temperature were made with phosphorous dioxide copper, and their conduction heat load is estimated to be about 4 W. Since the total heat input for the 1st stage is about 15 W, the temperature of the

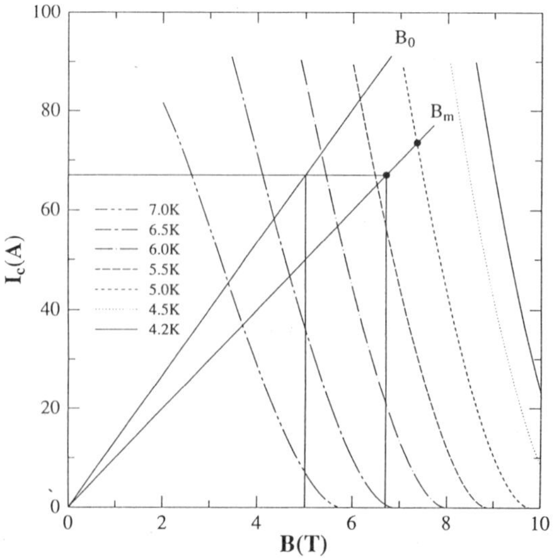

Figure 5. Load lines at a center of a gap (B_0) and at a winding (B_m). I_c-vs-B curves at various temperatures for the multifilamentary NbTi superconducting wire are calculated using a scaling law fitted to I_c data at 4.2 K

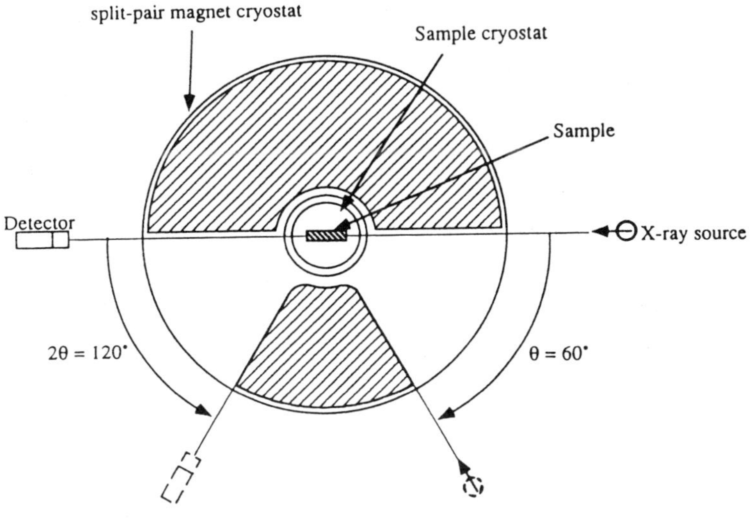

Figure 6. Plane configuration of X-ray source, detector, and sample.

1st stage is expected to be less than 50 K. Then, the heat input due to the Bi(2223) current leads from 50 K to 4 K can be reduced to only 0.089 W, using the thermal conductivity integral of 0.58 W/cm. These results confirm that the cryocooler-cooled split-pair superconducting magnet can be operated at 5 K, considering the temperature difference between the 2nd stage and the coil.

SPLIT-PAIR MAGNET DESIGN AND CONSTRUCTION

The split-pair coil was made employing a multifilamentary NbTi superconducting wire. The wire parameters are 0.65 mm diameter, 6 μm filament diameter, 13 mm twist pitch, and 1.8 : 1 Cu/NbTi ratio. Figure 3 shows a pair of coils separated by a gap. The individual coils consist of 244 mm outer diameter, 76 mm inner diameter, and 43 mm coil height, and two solenoids are set coaxially with a 58 mm gap. This split-pair coil is equipped in a cryostat with a room temperature bore of ϕ 50 mm and a room temperature gap of 10 mm as shown in Fig. 4. The split-pair superconducting magnet is designed to generate the field of 5 T at the center of the gap, when the current of 67 A is applied. The effective coil current density corresponds to 151 A/mm^2 at 67 A. The maximum field of about 6.7 T is applied at the coil winding. In case that a coil quench occurs at 5 T, the stored energy of 47 kJ is absorbed by the coil mass through a protection diode installed inside the cryostat. The temperature rise of the coil is estimated to be about 55 K maximum under the adiabatic condition.

PERFORMANCE OF THE MAGNET SYSTEM

The split-pair superconducting magnet was cooled down to 4.5 K, and the initial cooldown time was 44 hours from room temperature to 4.5 K. Figure 5 shows the load line together with the field dependence of I_c at temperature ranging from 4.2 K to 7.0 K for the NbTi superconducting wire, which are calculated from a scaling law using I_c data at 4.2 K. In the performance test, the cryocooler-cooled split-pair superconducting magnet quenched at 4.9 T at the coil temperature of 5.6 K, which corresponds to the maximum field of 6.6 T at the coil winding. When the coil was tested by immersion in liquid helium beforehand, it generated 5.0 T at 67 A without quench. The problem encountered is that there may be a small temperature gradient in a pair of coils separated by the gap, although the temperature difference between two coils can not be clearly observed by a carbon glass resistance thermometer attached in each coil. However, it seems that this quench point almost reaches the critical current at 5.6 K. If the coil is cooled down to 5.0 K while sweeping the field, the center field of 5.0 T at a gap will be available.

HIGH FIELD X-RAY DIFFRACTION APPARATUS

The high field X-ray diffraction instrument consists of a source of X-rays, a sample-positioning device, a detector for the diffracted beam, a cryocooler-cooled split-pair superconducting magnet system, and a cryocooler-cooled sample cryostat. The cryostat of the split-pair magnet system is made for use with a horizontal diffractometer, and has a room temperature horizontal gap as shown in Fig. 6. In this configuration, specimens can be scanned over a range of 2θ from -0° to +120°. The magnet cryostat combined with the sample cryostat is mounted on a xyz-goniometer to modify a position, and the X-ray source and detector are rotated to get θ-2θ diffraction data. The specimen is set into a cryocooler-cooled cryostat with beryllium windows,[4] and temperature control can be easily achieved from room temperature to 4 K. As the X-ray source, which will be located at a position of 400 mm apart from the magnet center, experiences the leakage field of about 350 G, a magnetic shield for the X-ray source will be needed to ensure the constant beam intensity.

For further applications of the newly developed cryocooler-cooled split-pair superconducting magnet, it is easy to extend this system for neutron diffraction investigations, if the cryostat is made of aluminum. This is because a room temperature gap is not necessary.

CONCLUSIONS

The cryocooler-cooled split-pair superconducting magnet without use of a supplemental cryogen such as liquid helium and nitrogen has been developed. The magnet system provides a room temperature experimental gap of ϕ 50 mm × 10 mm and generates 5 T at a center of the gap. An attempt will be made to combine a compact high field superconducting magnet with an X-ray diffraction apparatus.

ACKNOWLEDGMENTS

This work is partly supported by Toray Science Foundation, and Grant-in-Aids for Scientific Research from the Ministry of Education, Science and Culture, Japan. We are grateful to Y. Watanabe of Institute for Materials Research and A. Shibata of Rigaku Corp. for useful discussions of the X-ray diffraction equipment.

REFERENCES

1. K. Tajima, N. Suzuki; and K. Oguchi, Jpn. J. Appl. Phys. 29 : 986 (1990).
2. K. Watanabe, Y. Yamada, J. Sakuraba, F. Hata, C.K. Chong, T. Hasebe, and M. Ishihara, Jpn. J. Appl. Phys. 32 : L488 (1993).
3 S. Masuyama, H. Yamamoto, and Y. Matsubara, IEEE Trans. Appl. Supercond. 3 : 262 (1993).
4. K. Henriksen, F.K. Larsen, and S.E. Rasmussen, J. Appl. Cryst. 19 : 390 (1986).

TEMPERATURE AND PRESSURE RISE IN SUPERCRITICAL HELIUM DURING THE QUENCH OF INDIRECTLY COOLED SC COILS

R. M. Schöttler and H. W. Lorenzen

Lehrstuhl für Elektrische Maschinen und Geräte
Technische Universität München
D-80290 München, Germany

ABSTRACT

A 1 MJ energy storage device consisting of six toroidally arranged superconducting coils is under construction at TU München. Each coil is cooled indirectly by supercritical helium in two layers of cooling pipe. All cooling layers are connected in series. The heat input is removed by heat exchangers to keep the coolant's temperature rise small in normal operation.

The quench of a coil causes high heat input to the helium cooling system which leads to a fast pressure rise, blocks the helium flow and vaporizes the liquid helium pools. A computer program was developed to calculate the coils' temperature evolution and the helium's thermodynamic state during normal operation and quench.

The simulation results show that the proposed protection system can detect the quench within 0.01 s and that the coil's current will be nearly zero after 1 s due to heating and bypass devices. In the case of two coils quenching simultaneously, the peak pressure in the helium tube arises after 3.8 s and can be limited to 3 MPa by pressure relief valves. The maximum helium temperature reaches 70 K within 7 s.

INTRODUCTION

To have a testing plant for experiments with a superconducting magnetic energy storage (SMES) device working at the power network, Lehrstuhl für Elektrische Maschinen und Geräte currently realizes a toroidal SMES (1 MJ) with six sc coils fabricated from the LCT 10+1 NbTi superconductor.[1–3] It is optimized for fast charging/discharging cycles[4] lasting 1...5 s and is scheduled to be set into operation the end of 1995.

Although a sophisticated control unit should avoid quenching by monitoring all characteristic data the device must be able to withstand the stresses during such a fault without damage. Besides magnetic forces and thermal stresses the pressure rise in the cooling system due to the excessive heat input is of special interest. The results of computer simulations are presented in the following.

Advances in Cryogenic Engineering, Vol. 41
Edited by P. Kittel, Plenum Press, New York, 1996

COOLING SYSTEM

As presented in former publications[1–4] the sc coils are cooled indirectly by two cooling layers made of insulated copper pipe: one at the inner and one at the outer coil radius. Fig. 1 shows the cross–section of the winding region. Supercritical helium is piped through the cooling tube at approximately 0.6 MPa and 5 K. It is supplied with a mass flow rate of 5.7 g/s by a refrigeration plant, Linde Standard I, which has a rated refrigeration capacity of 60 W, equivalent to 16 L/h liquefaction capacity.

The coolant gets heat input by superconductor a.c. losses, heat radiation and conduction from the environment (LN_2 cooled thermal shield), and — in case of a quench — by ohmic heat generation in the coils. Even in normal operation this heat input can be up to 20 W per coil.

In contrast to cases using liquid helium (LHe) as coolant, where latent heat is used to absorb heat at constant temperature, supercritical helium warms up while removing heat from the coil. This is a significant disadvantage of supercritical helium as coolant for sc devices, where low temperatures are desired. However, supercritical helium was chosen for several reasons:

1. the refrigeration plant delivers it at a pressure high enough to force its circulation through the pipe;
2. the coolant remains in single phase state, so calculations are more reliable, especially the heat transfer is known better;
3. design and manufacture are simpler for indirectly cooled coils than for bath-cooled ones;
4. due to shorter and wider cooling pipes compared to internally cooled or cable-in-conduit superconductors, stationary pressure drop and pressure rise during coil quenching are lower.

The complete helium cooling circuit is shown in Fig. 2 and explained as follows: As the magnetic field is always highest at the innermost superconductor layer, this layer should have the lowest temperature to allow a high operating current and stored energy. Good cooling of this layer at given heat generation is achieved by supplying the cooling layer with:.

1. the full helium mass flow to keep the temperature rise small between inlet and outlet. So all cooling layers are connected in series to form an inner and outer cooling circuit;
2. cold coolant: This is done by heat exchangers which remove the heat input of the preceding coil to the LHe bath. So every coil is supplied with coolant only slightly warmer than the pool's temperature.

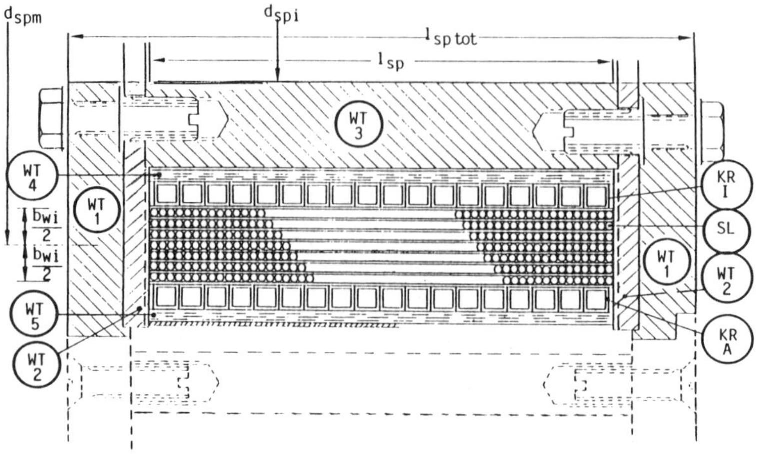

Fig. 1. Cross–section of the superconducting coils.[3] WT1–5, Winding support; SL, superconductor (7 layers, 55 windings each); KR, inner (I) and outer (A) cooling layer.

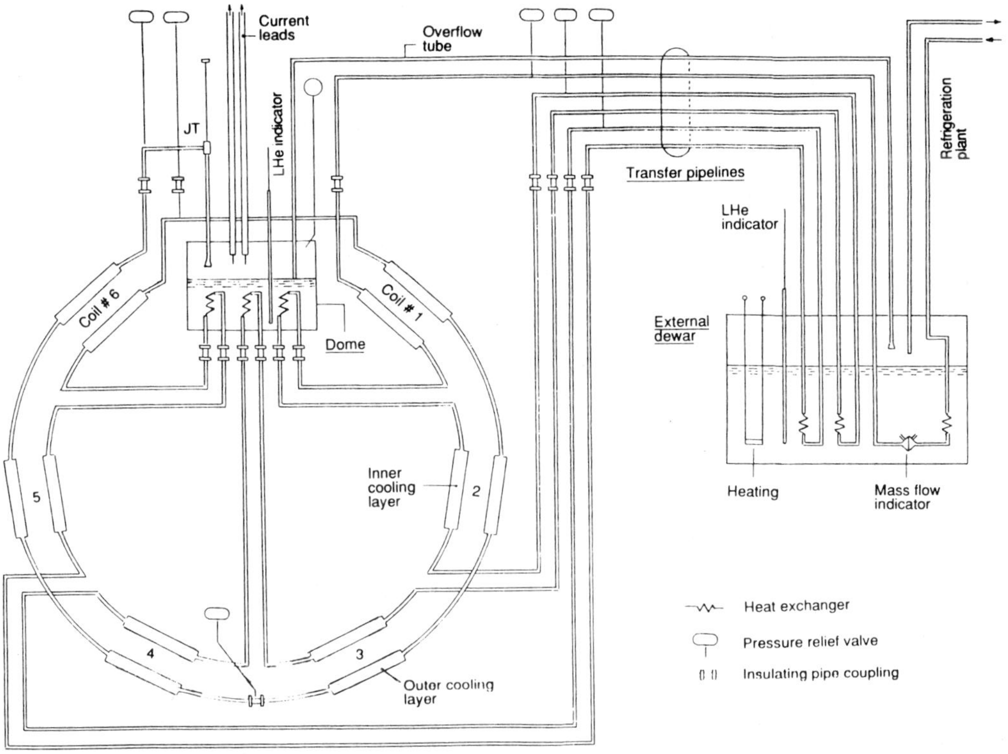

Fig. 2. Complete helium cooling system of the testing plant.[4] There are two cooling layers per coil (an inner and an outer) which are connected in series to form two circuits: The inner with the heat exchangers and the outer without. The heat exchangers are located in a small helium pool inside the dome and in a big external Dewar.

Due to the lower magnetic flux density and also heat input at the outer superconductor layers, the temperature is less critical here and no heat exchangers are needed. Leaving the last coil with about 0.55 MPa and 5.0...5.9 K, the helium is expanded through a Joule–Thomson (JT) valve where it liquefies partially. This will fill up the small pool in the dome. From this pool the current leads as well as three of the heat exchangers vaporize helium for cooling. Any excess of liquid and gaseous helium will flow through an overflow tube to a big external Dewar where the other three heat exchangers are located.

To protect the helium pipe from the high pressure during a quench, six relief valves are placed in the circuit. The locations were determined to limit the maximum pressure to 3 MPa even in the worst case[4] and are marked in Fig. 2.

SIMULATION MODELS

Numerical Calculation of Dynamic Helium Flow

The time and space dependent thermodynamic state of the helium in the cooling pipe is calculated numerically by evaluating the three nonlinear differential equations

$$\frac{\partial \rho}{\partial t} + \frac{\partial}{\partial x}(\rho u) = 0 \tag{1}$$

$$\rho \frac{\partial H}{\partial t} + \rho u \frac{\partial H}{\partial x} = \frac{\partial}{\partial x}\left(\frac{\lambda}{C_P}\frac{\partial H}{\partial x}\right) + u\frac{\partial P}{\partial x} + \frac{\partial P}{\partial t} + q + \frac{\rho \xi}{2 D_H} u^2 \left| u \right| \tag{2}$$

$$\rho \frac{\partial u}{\partial t} + \rho u \frac{\partial u}{\partial x} = \frac{\partial}{\partial x}\left(\eta \frac{\partial u}{\partial x}\right) + \frac{\partial P}{\partial x} + \frac{\rho \xi}{2 D_H} u\, |u| \; . \tag{3}$$

The equations represent the continuity of mass (1) and enthalpy (2) and the balance of momentum (3). The nomenclature for the equations can be found in Table 1.

Only one space coordinate x, along the length of the cooling tube, is considered. Variations over the tube's cross-section are represented by a wall and bulk temperature and an overall heat transfer coefficient α. The equations are already simplified by constant pipe dimensions and neglecting the influence of the gravitational force. For the cooling pipes the external heat input per unit volume q is calculated from the heat balance inside the coils. In the heat exchangers the heat transfer to the LHe is evaluated with the formulae for nucleate or film boiling in the pool[5] and heat conduction in the copper tube wall. The forced convection heat transfer coefficient α is calculated from the *Hausen* equation for the Nusselt number

$$\mathrm{Nu} = 0.037(\mathrm{Re}^{0.75} - 180)\mathrm{Pr}^{0.42}\left[1 + \left(\frac{D_H}{l_P}\right)^{2/3}\right], \qquad \mathrm{Re} > 2300 \tag{4a}$$

for turbulent flow, and in case of laminar flow

$$\mathrm{Nu} = 3.78 \; , \qquad \mathrm{Re} \le 2300 \; . \tag{4b}$$

The heat transfer coefficient

$$\alpha = \frac{\mathrm{Nu}\,\lambda}{D_H} \tag{5}$$

is slightly smaller than that determined by *Giarratano et al*[6] for small temperature differences. But the generally accepted Eq. (4) – (5) were chosen since the temperature difference between helium and tube wall can reach some ten Kelvin during a coil's quench.

The differential equations Eq. (1)–(3) are discretized and integrated using the iterative ISIMPLER–Algorithm published by *Patankar*[7] and improved for strongly compressible fluids by *Cornelissen*.[8] The element length is usually 2.5 m in the cooling layers and 0.2 m in the heat exchangers. A factor of 0.1...0.2 is used for underrelaxation to accelerate convergency. The iterative evaluation of the independent variables (P, H, u) and the helium properties is stopped when the relative change between two steps falls below a threshold value. Due to the fully implicit nature of this algorithm, the results are physically realistic for every time step Δt, as long as the boundary conditions (e. g. heat input q) do not change during Δt.

The simulation turned out to oscillate when the heat input q was calculated from the wall temperature in every iteration step. The reason is the slowly converging calcu-

Table 1. Nomenclature and units for the helium's thermodynamic state

Dependent variables:			Derived variables and constants:		
P	Pa	Pressure	T	K	Helium bulk temperature
H	$\mathrm{J\,kg^{-1}}$	Enthalpy	K	$\mathrm{s^2\,m^{-2}}$	Compressibility
u	$\mathrm{m\,s^{-1}}$	Velocity	$\dot{m}$	$\mathrm{kg\,s^{-1}}$	Mass flow rate
Coordinates:			Re	–	Reynold's number
x	m	Space	Pr	–	Prandtl number
t	s	Time	Nu	–	Nusselt number
			α	$\mathrm{W\,m^{-2}K^{-1}}$	Heat transfer coefficient
Material properties:			ξ	–	Fanning friction coefficient
ρ	$\mathrm{kg\,m^{-3}}$	Density	q	$\mathrm{W\,m^{-3}}$	Heat input to helium
λ	$\mathrm{W\,m^{-1}K^{-1}}$	Thermal conductivity	D_H	m	Pipe hydraulic diameter
C_P	$\mathrm{J\,kg^{-1}K^{-1}}$	Heat capacity	l_P	m	Pipe length
η	Pa s	Dynamic viscosity			

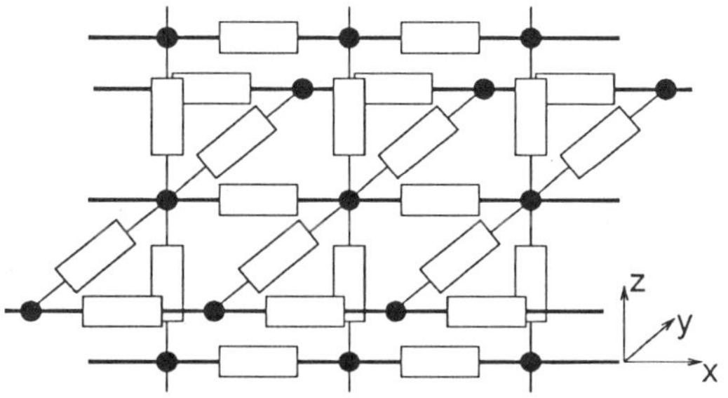

Fig. 3. Thermal model of the winding region with lumped masses and concentrated conductivities:
The thick lines indicate the superconductor, the thin ones the insulation material.

lation of the flow field u which strongly influences q by varying the transfer coefficient α. Stability was obtained by the determination of α from the temperatures and properties in the previous time step. As mentioned above, this leads to the necessity of small time steps. $\Delta t = 1 \cdot 10^{-3}$ s did well for the simulation of a coil quench and can be up to 0.2 s for slower developing thermal processes.

Calculation of the Temperature Evolution inside the Coils

To compute the temperature distribution and evolution of the winding region, the differential, three dimensional heat balance equation

$$\rho C_V(T)\frac{\partial T}{\partial t} = \frac{\partial}{\partial x}\lambda_x(T)\frac{\partial T}{\partial x} + \frac{\partial}{\partial y}\lambda_y(T)\frac{\partial T}{\partial y} + \frac{\partial}{\partial z}\lambda_z(T)\frac{\partial T}{\partial z} + P_V \tag{6}$$

is discretized ($\partial \rightarrow \Delta$). Each superconductor winding is subdivided into 25 lumped masses which are coupled to their six neighboring masses by thermal conductivities (Fig. 3). With its 7 layers and 55 windings per layer (Fig. 1) this gives a total of 9625 elements per coil.

The heat source per element P_V is calculated from the superconductor losses during normal operation[4,9,10] or the resistance loss, when the element has lost superconductivity. The element's heat capacity and the conductivity along the winding are derived from the superconductor properties, the conductivity between the windings and layers from the insulation material's properties (glassfibre reinforced plastic). Both vary strongly with temperature and are recalculated continuously. The temperature of the cooling layers serves as a boundary condition.

To reduce computation time and memory requirements, the algorithm uses the Alternate Direction Implicit scheme. As thermal capacities and conductivities are very low at cryogenic temperatures, the time step Δt must be chosen with care ($1 \cdot 10^{-3}$ s for quench simulations, $5 \cdot 10^{-3}$ s for cool down or transient effects).

Electrical System and Coupling

The seven thermal systems — helium and six coils — are coupled by the helium and tube wall temperature which are exchanged in every time step. Additionally, the proposed electrical and protection systems[1,2] are simulated. As shown in Fig. 4, each coil can be heated by discharging an external capacitor into the coil through a tap between the third and fourth sc layer. This will be done when a quench in this coil is detected either thermally or electrically.[1] As the resulting inductance is very small, a short pulse with high amplitude is obtained which makes the coil normalconducting nearly instantly by superconductor losses, field distortion and current amplification. This boosting avoids excess temperature differences inside the coil. To accelerate the current decline in the quenched coil, a bypass device (IBGT module) is activated. The SMES current will commutate partly from the coil to the bypass.

The simulation of the electrical system has regard to all self and mutual inductances and ohmic resistances of the coils, the characterisitics of the semiconductors and the behavior of the proposed power converter. The currently working version can handle the boosting and bypassing of one coil, so the two ohmic resistances of the quenching

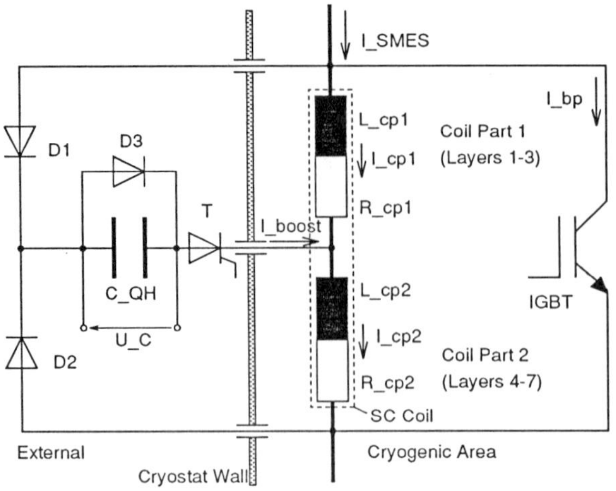

Fig. 4. Booster and bypass devices for one sc coil. A current pulse from the external capacitor accelerates the quench and thus reduces the temperature difference inside the coil. The SMES current partly passes the coil by the IBGT.

coil and eventually the sum of the other five coils are updated continuously from the coils' simulation.

The simulation programs are written in FORTRAN, C and C++ as independent modules. To reduce the execution time, the resulting eight tasks (helium, electrical and six coil systems) are distributed to different computers (workstations) by the PVM programming interface.[11]

SIMULATION OF TWO COILS QUENCHING SIMULTANEOUSLY

Starting with a stationary helium and coil state from cycle M,[4] an assumed fault in the SMES control unit keeps charging the device, exceeding the maximum safe operating current. At $t = 0.031$ s, the SMES current reaches the critical one in coil #6, causing a quench in the first sc layer. 4 ms later — due to slightly lower temperatures — coil #5 quenches, too. The protection system detects the quench in coil #6 at $t = 0.041$ s, when the temperature at the sensor reaches 20 K. Immediately, the SMES is switched to a dump resistor and the booster is fired. It increases the current in the outer sc layers, causing them to quench much earlier than without the pulse. The resulting current curves are shown in Fig. 5, the evolution of the coil resistances is drawn in Fig. 6.

With the developing resistance the ohmic losses heat up the coil and the helium. The maximum heat generation rate is 650 kW, the dissipated heat in total $\int Ri^2 dt$ is 180 kJ for coil #6 and 120 kJ for #5. The coils reach their temperature maximum of approx 140 K after 1 s.

The heat input to the helium in the cooling pipe causes a pressure rise which starts with 1.5 MPa/s in the quenching coils and spreads over the whole system with the speed of sound. The maximum pressure of 2.7 MPa is reached at $t = 3.8$ s (Fig. 7). The simulation is strongly influenced by the boundary conditions: Here $P =$ const was assumed for the cooling pipe front end ($x = 0$) and $\dot{m} =$ const for the Joule–Thomson

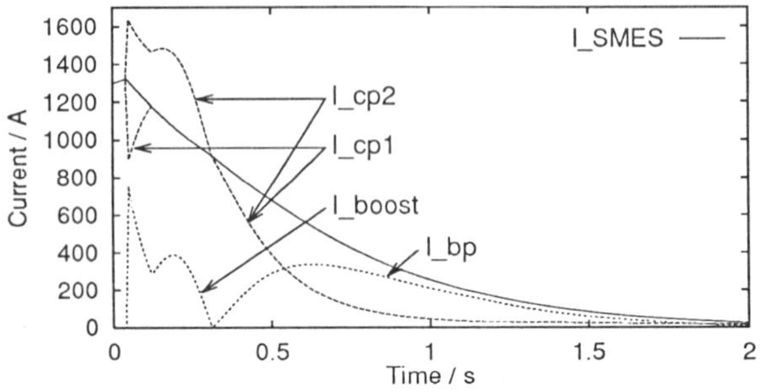

Fig. 5. Current curves during the quench of two sc coils.
Only coil #6 is heated and bypassed, coil #5 gets I_{SMES}.
The bypass stays idle until the booster current vanishes as the coil voltage remains negative.
The nomenclature refers to Fig. 4.

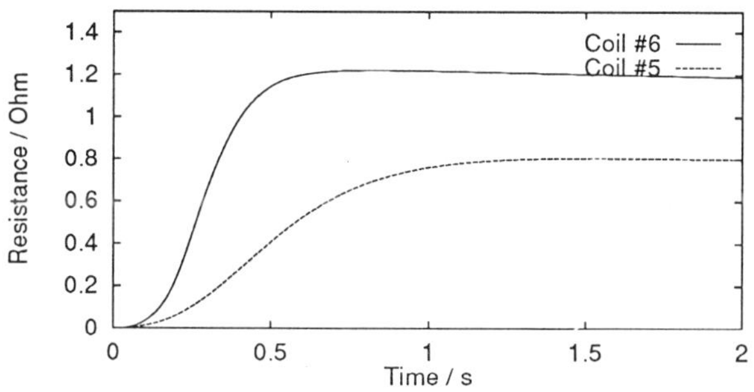

Fig. 6. Growth of ohmic resistance. Coil #6 quenches faster due to the booster current.

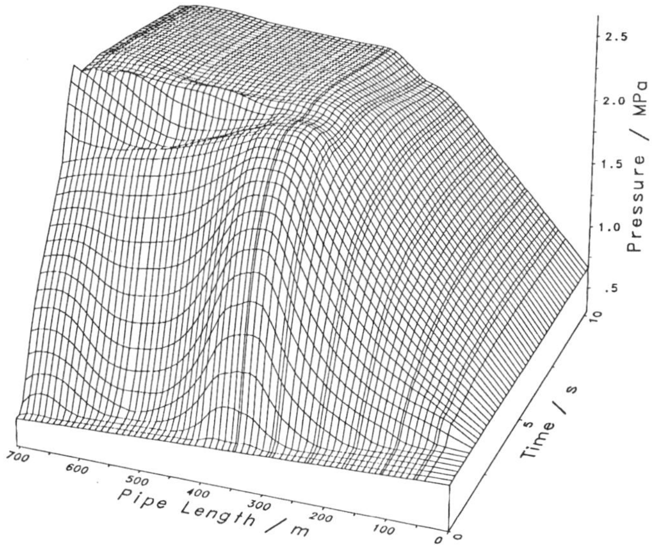

Fig. 7. Pressure rise in the cooling pipe (data thinned out for better visibility). The quenching coils are at 293 m$\leq x \leq$ 400 m (inner cooling layers) and 600 m$\leq x \leq$ 700 m (outer layers). The peak pressure is about 2.7 MPa, boundary conditions are explained in the text.

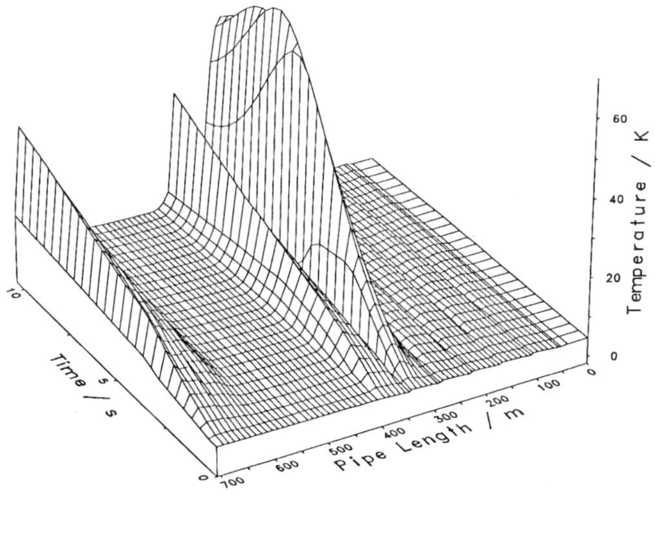

Fig. 8. Temperature rise in the cooling pipe (only coil data is plotted). As the helium velocity remains below 15 m/s, the heating up is restricted to the area of the quenched coils and spreads over the pipe length very slowly. The heat exchangers avoid quenches in neighboring coils by recooling the warm helium before it enters them.
The maximum temperature of 70 K arises after 7 s.

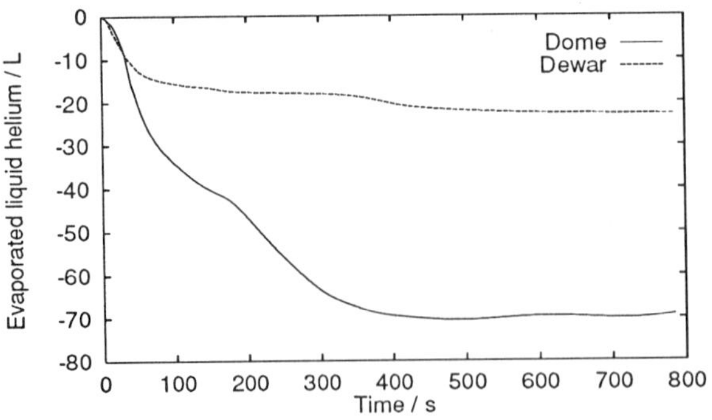

Fig. 9. Evaporation of liquid helium in the pools during recooling after the quench. Heat input by heat exchangers and the Joule–Thomson valve vaporizes LHe, slowly decreasing the reservoirs.

valve ($x = 700\,\mathrm{m}$). This is probably wrong for the real cooling system but the exact behavior is unknown until now and former simulations revealed that the value of the peak pressure is quite independent from the boundary pressure due to the relief valves. These valves open at 1.0 MPa and release a total of 0.5 kg helium (which is about 17 % of the initial mass in the pipe) in 45 s. Another 1.1 kg escape at the pipe's front end due to the assumed boundary condition.

The temperature rise in the helium is slow compared to that in the coils (Fig. 8): The maximum gradient is 14 K/s at the inner cooling layer of coil #5 which is neither boosted nor bypassed. The heat input is nearly negligible at its outer cooling layer as this superconductor layers quench very late. The situation in coil #6 is completely different: The boosting evens out the heat input to both cooling layers. This reduces the temperature gradient to about 5 K/s.

The liquid helium pools in the dome and Dewar get high heat inputs by the heat exchangers and the Joule–Thomson valve which blows warm helium — up to 70 K — to the liquid. The pools will vaporize slowly. As shown in Fig. 9 the collective heat input to the Dewar is equivalent to the evaporation of 22.9 L LHe and 70.6 L in the dome pool, respectively. If this pool contains less LHe, the cooling down after the quench will take somewhat longer than 13 min resulting from this simulation, as three heat exchangers are deactivated. After refilling the dome pool the device would be ready for operation again.

ACKNOWLEDGMENT

This study and the realization of the SMES project have been enabled by financial support of Deutsche Forschungsgemeinschaft (DFG).

REFERENCES

1. J.F. Kärner, H.W. Lorenzen, F. Rosenbauer, J. Schaller, and R.M. Schöttler, A protection system for small high power SMES with power semiconductors working at cryogenic temperature, to be published in *IEEE Trans Appl Supercond.* (May 1995).
2. U. Brammer, H.G. Herzog, J.F. Kärner, H.W. Lorenzen, J. Schaller, and R.M. Schöttler, A fast-acting 1 MJ superconducting magnetic energy storage, In *Proc ICEM, Paris* (3):373 (1994).
3. P.J. Birkner, U. Brammer, H.W. Lorenzen, J.F. Kärner, W. Rehm, J. Schaller, and R.M. Schöttler, Testing plant with small fast superconducting energy storage at TU München, *IEEE Trans Appl Supercond.* 3(1):207 (1993).
4. R.M. Schöttler, Cooling of small SMES system with forced flow supercritical helium, *Cryogenics* 34(10):825 (1994).
5. H. Frey and R.A. Haefer. "Tieftemperaturtechnologie," VDI–Verlag, Düsseldorf (1981).
6. P.J. Giarratano, V.D. Arp, and R.V. Smith, Forced convection heat transfer to supercritical helium, *Cryogenics* 11(10):385 (1971).
7. S.V. Patankar. "Numerical Heat Transfer and Fluid Flow," Hemisphere Publishing Corp., Washington, New York, London (1980).
8. M. Cornelissen. "The Thermal Stability of Superconductors Cooled by a Turbulent Flow of Supercritical Helium," Dutch Efficiency Bureau, Pijnacker (1984).

9. J.F. Kärner, H.W. Lorenzen, and A. Nitsche, Quench initialization and propagation in superconducting magnetic energy storage systems, in: "Advances in Cryogenic Engineering," 37(A):379 (1992).
10. P.J. Birkner. "Technical Aspects of Design and Operation of Superconducting Magnetic Energy Storage in Toroidal Form" (Thesis, in german), Technische Universität München (1994).
11. A. Geist, A. Beguelin, J. Dongarra, W. Jiang, R. Mancheck, and V. Sunderam. "PVM: Parallel Virtual Machine. A User's Guide and Tutorial," MIT Press, Cambridge, Massachusetts (1994).

1995 CEC/ICMC Conference: Ronald W. Fast (right), the winner of the 1995 Samuel C. Collins award, and Jean Fast.

FINITE ELEMENT CODE FOR QUENCH AND STABILITY ANALYSIS OF SUPERCONDUCTING MAGNETS COOLED BY He II

M. B. Gorbounov,[1] L. Bottura,[2] J. R. Miller,[1] S.W. Van Sciver[1]

[1]National High Magnetic Field Laboratory
Tallahassee, FL, 32306, USA

[2]The NET Team, Garching, FRG

ABSTRACT

The design of the protection system for the superconducting portion of the 45-T Hybrid Magnet System being constructed at the National High Magnetic Field Laboratory and the establishment of safe and reliable operating procedures are critical tasks. To support these efforts, the finite element code SARUMAN has been extended to include capability for analysis of the stability and quench in Cable-in-Conduit Conductors (CICC) cooled by pressurized superfluid He II. This modification consists of introducing a temperature gradient term in the energy equation for the counterflow mechanism and extension of the properties of helium and other materials of the magnet to the He II temperature region. Using this new version of the code, the stability of the conductors for the 45-T Hybrid have been analyzed against different thermal disturbances and under a variety of possible operating conditions, as well as the evolution of the quench in the conductor.

INTRODUCTION

Cooling large superconducting magnets to He II temperatures can have important benefits: 1) operation at the lower temperatures gives more margin with respect to the current sharing temperature, 2) within defined limits, the heat transport in He II is essentially passive, 3) surface heat transfer (limited only by Kapitza conductance) is both appreciable and predictable, and 4) the combination of high heat transport rate and large heat capacity of He II allows absorption of large transient heat pulses. These factors can have an important impact on the reliable and safe operation of such magnets and thereby are critical to their design. The design of the superconducting portion of the 45-T Hybrid Magnet combines CICC technology with He II cooling and depends critically on our ability to predict stability and quench in such a system. Therefore, the availability of sophisticated tools for analysis of these processes is essential.

Up to now, none of the existing codes for simulation of quench and stability have allowed reliable analysis in the pressurized He II regime. The finite element code SARUMAN, which was developed by the NET Team for the He I regime and tested and verified against the available experimental data, has been taken as a base and developed by extension and modification to produce a code capable of treating stability and quench of CICCs cooled by pressurized superfluid He II.

DESCRIPTION OF THE FINITE ELEMENT CODE

The most comprehensive description of all the details of the finite element approach for the transient processes in CICC used in SARUMAN code can be found elsewhere[1]. Here we concentrate on the latest modifications although for complete understanding some review of the overall approach is required. The SARUMAN code has been created in two modifications: strictly one dimensional model (1-D) and one dimensional including radial and longitudinal heat transfer between the adjacent parts of the conductor (3-D). The He II related extension has been done so far only for 1-D case.

The basic equations of the modified version of SARUMAN code

1. Fluid (He I and He II) equations:

- continuity and momentum equations

$$\frac{\partial\rho}{\partial t}+\frac{\partial(\rho v)}{\partial x}=0; \qquad \frac{\partial(pv)}{\partial t}+\frac{\partial(pv^2)}{\partial x}+\frac{\partial p}{\partial x}=2f\frac{\rho v|v|}{D_h} \tag{1}$$

- energy equation

$$\frac{\partial(\rho e)}{\partial t}+\frac{\partial(\rho ev)}{\partial x}+\frac{\partial(pv)}{\partial x}+\frac{\partial(q_{cf}+q_c)}{\partial x}=\frac{Q_{He}}{A_{He}}, \tag{2}$$

where

$$Q_{He}=hP_{st}\left(T_{st}-T_{He}\right)+hP_{jk}\left(T_{jk}-T_{He}\right), \qquad q_c=-K_{He}\frac{\partial T}{\partial x},$$

$$q_{cf}=-\left(F(T)\frac{\partial T}{\partial x}\right)^{1/3} \tag{3}$$

In these set of equations: A_{He} - cross-section of the helium in the cable, ρ - helium density, v - velocity, t - time, x - co-ordinate along the conductor, p - pressure, f - friction factor, T - temperature, $F(T)$ - effective heat conductivity function for counterflow mechanism in He II, h - heat transfer coefficient, e - total specific energy, Q_{He} - linear heat flux to helium, q_{cf} - counterflow mechanism heat flux, q_c - conductivity heat flux, P - perimeter of the heat exchange surface, K_{He} - thermal conductivity of helium; subscripts st and jk refer to strands and jacket.

The set of equations (1)-(3) implements the simplified approach to the He II which states that for practical purposes it can be approximately treated as a fluid with high effective thermal conductivity having a special form dependent on temperature and its gradient. This approach has been justified and used in many publications[3-6], although some of its limitations are discussed in others[6,7].

The initial conditions are arbitrary. The boundary conditions are represented by reservoir with given temperature and pressure (they can be variable in time) and/or closed end (symmetry). In application to the 45-T Hybrid magnet the first type of condition is used at one end of conductor and the second at the other end.

2. Heat conduction in the solid material (strands and jacket). No modifications have been done in this part of the code compared to the original version[1].

3. Helium properties The new version of SARUMAN extensively uses commercially available HEPAK code (Cryodata Inc.) for calculations of all properties in He II region, while in He I region it mostly uses less precise but at the same time less time consuming approximations of the original version of SARUMAN.

4. Solid materials properties. All properties of most of the solid materials (copper, stainless steel, G-10) are calculated using commercially available CRYOCOMP code (Cryodata Inc.). Unfortunately CRYOCOMP does not include properties of superconducting materials, so for that purpose the correlations from the original version of SARUMAN are used.

5. Flow coefficients. The correct calculation of friction factor and heat transfer coefficient is crucial for the stability and quench evolution simulation. The following formulae are used for flow coefficient calculations.

Friction factor. As in original version of SARUMAN the correlation of Hooper[8] based on the available experimental data for CICC conductors is used:

Heat transfer coefficient. There are several possible regimes of heat transfer between the helium and the solid part of the conductor.

A. *He II region.* In this region the heat exchange process is limited only by Kapitza conductance and is calculated using the following formulae[3]:

$$\begin{cases} h_k = 900 T_{He}^3 & for \quad \Delta T << T \\ h_k = 500\left(T_s^3 - T_{He}^3\right) \Big/ \left(T_s - T_{He}\right) & for \quad \Delta T \approx T \end{cases}$$

where T_s is temperature of the solid surface. In the code the transition from one formula to another takes place when the values calculated from the second exceed those from the first.

B. *He I region.* There are two different stages of heat transfer in this region:

a) Initial heat transfer mainly by heat diffusion during the formation of the boundary layer when the temperature profile changes from almost uniform in He II to its stable form. The heat transfer coefficient in this regime is estimated by the value calculated from the classical analytical solution for semi-infinite body with a step change in surface temperature:

$$h_t = \sqrt{\frac{K_{He} \rho C_{He}}{\pi t}} \tag{4}$$

where C_{He} is specific heat of helium. All properties are average in the boundary layer.

As existence of Kapitza conductance still limits heat exchange in He I region for the first instants of time the heat transfer coefficient should be calculated as equivalent for series of thermal resistances corresponding to h_t and h_k.

b) Developed heat transfer. At this stage the heat transfer coefficient is estimated from the steady state turbulent regime Dittus-Boelter's correlation with correction which takes into account the degradation of heat transfer for large temperature differences[9].

Mathematical model and solution

The equations (1)-(3) form the system of non-linear partial differential equations with strong hyperbolic character. This system can be written in a more compact way:

$$\frac{\partial \mathbf{u}}{\partial t}+\frac{\partial \mathbf{f}}{\partial x}=\mathbf{s} \qquad (5)$$

with

$$\mathbf{u}=\begin{pmatrix}\rho\\ \rho v\\ \rho e\end{pmatrix}, \qquad \mathbf{f}=\begin{pmatrix}\rho v\\ \rho v^2+p\\ (\rho e+p)v+q_{cf}+q_c\end{pmatrix}, \qquad \mathbf{s}=\begin{pmatrix}0\\ 2f\rho v|v|D_h^{-1}\\ Q_{He}A_{He}^{-1}\end{pmatrix},$$

where $\mathbf{u}$ is the vector of unknowns, $\mathbf{f}$ are convective fluxes and $\mathbf{s}$-vector of sources.

The original SARUMAN[1] code uses finite element method (FEM) both for the solution of the set of fluid equations and for the heat diffusion equations in solid part of the conductor, which are solved separately for each time step. First for solid part, then for fluid. One of the well known implicit algorithms (Euler-backward, Crank-Nicholson, Lees') is used for the solid part. Nothing new was introduced in this procedure in the modified version. For the fluid an explicit two-step form of the Taylor-Galerkin algorithm is used.

To preserve the logic of the code the same FEM technique is used in the modified version. The FEM has never been used before for analysis in He II region, the existing works[5,6] used finite difference method (FDM) in numerical approach for that region. The unique non-linear form of the He II heat flux term corresponding to the counter flow mechanism (3) represents a difficulty for FEM, as no linearization can be used in this case. The method implemented in the SARUMAN code uses average values of all fluxes in the elements, including counterflow term, for integrating of the full spatial Galerkin approximation at the second step of two-step algorithm. This procedure tends to need many elements for good accuracy, especially in the regions of fast changes.

Adaptive mesh

For the quench propagation calculations typical mesh size is of order of some mm, while the typical lengths are of order of 100 m or more. In the He II region close to the heat deposition the necessary mesh size should be less than 1 mm to calculate the correct value of the heat flux due to the counter flow mechanism. Therefore, to facilitate calculations we have used the adaptive meshing, which means variable mesh density depending on the position and time, is the most acceptable solution to the problem[2]. The concept includes the introduction of two types of nodes in the mesh: fixed (hard) nodes and movable (soft) nodes. The mesh density is changing by the relocation of soft nodes. Two different methods for mesh generating are used: mesh density approximately proportional to the temperature gradient in helium (most effective before formation of the normal zone in the conductor

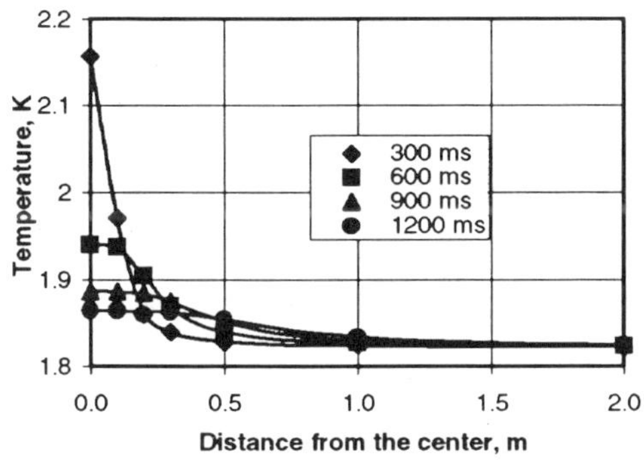

Figure 1. Comparison with Dresner's solution[4]

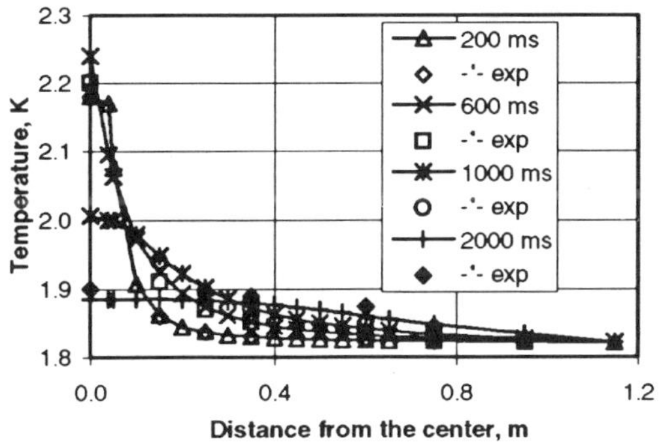

Figure 2. Comparison with experimental data[10].

while all helium is in He II region) and mesh density that approximately follows gaussian profile centered at the normal front location (for the quench propagation).

Comparison of the code with analytical solution in He II region.

To verify the treatment of the He II counterflow term in the FEM formulation discussed above a comparison was made with existing analytical solution for 1-D transient heat transfer in heat pulse problem[4]. This analytical solution is done for the following assumptions: 1) infinite channel, 2) perfect heat exchange between He II and solid wall, 3) constant He II effective conductivity $F^{1/3}$=10.4 W·cm$^{-5/3}$K$^{-1/3}$, 4) constant $(\rho C)_{He}$ = 0.410 J·cm^{-3}·K^{-1}. The test calculations were performed for a long (10.3 m) stainless steel tube with internal diameter (ID) of 6 mm initially filled with helium at 1.82 K and 0.1 MPa, and the heat pulse of 1 J applied to the center of the tube. The assumptions 2-4 were introduced into the code for the test calculations, although assumptions 3,4 are not very close to the reality. The symmetry of the problem was used to reduce the necessary time of calculations. For the initial condition in numerical calculations the temperature distribution from analytical solution at the moment when helium reenters He II region (at 300 ms) was used. The results are on fig. 1. The test showed the reliability of the used FEM procedure in He II.

Comparison of the code with experimental data in He II region.

In order to test the code with the real He II properties the comparison with available experimental data in He II region[10] was done. The calculations were performed for the same conditions as in the experiment - 2.3 m long and 6 mm ID stainless steel tube with 100 mm long copper section in the middle for heating. Initial temperature is 1.82 K along the whole tube and pressure is 0.1 MPa. The calculations were made for all experimental heat pulses of 0.92 J (20 ms), 1.84 J (40 ms) and 4.6 J (100 ms). As example the results for 1.84 J pulse are shown on fig. 2. It can be seen that the results are rather good. The most serious discrepancy is in the central point where the He II-He I-He II transition takes place, which most probably indicates some limitations for the use of formula (4).

APPLICATION TO THE 45T HYBRID PROJECT

The modified of the SARUMAN code was used for the analysis for the 45-T Hybrid magnet project in order to determine the margins of cable stability and parameters of the quench system.

There are two known types of disturbances capable of generating heat deposition to the superconducting cables of the coils: localized, uniform, sudden (less than 100 ms) disturbances, typically as a result of conductor motion (occurring perhaps over lengths comparable to a transposition length of cable, ~ 100 mm) (type I) and spatially distributed disturbances of more extended duration AC-loss transients (for example, as a result of a resistive-insert trip, whereby the insert current goes to zero and the superconducting outsert current surges upward, causing significant changes in the field pattern over large regions of the outsert windings, with a characteristic time of several hundred ms) (type II). Both may cause a quench in the coils. The innermost coil A (Nb_3Sn superconductor) is the most critical for the stability and quench considerations as it is situated in the region of the highest magnetic field. The details of the geometry of the coils can be found in Ref. [11].

Protection allowables must finally be related to stresses resulting from differential thermal expansion of the coil constituents or to stresses in the conductor jacket containing the helium driven to elevated pressures. In lieu of detailed, iterative analyses, it is easy to show simple estimates, substantiated by experience, that a maximum hot-spot temperature of 150 K or a maximum helium pressure inside the conductor jacket of approximately 100 MPa are quite safe.

Analysis of the Stability for 45-T Hybrid magnet

The stability of the coil A conductor (first layer) against type I disturbances has been studied in the range of all possible lengths for the times 100 ms, 10 ms and 1 ms. The importance of the following issues for the stability is investigated: the improvement of stability due to the existence of the counterflow mechanism in He II; the importance of the purity of copper (residual resistivity ratio - RRR); the change in stability due to the change in the jacket material from Incoloy 908 to the stainless steel 316LN.

Some of the results of the calculations are presented on fig. 3-5. Several important observations have been made: 1) The improvement in stability due to the counterflow mechanism is not very significant for the times of disturbances under consideration, the most substantial gain takes place for shorter heated lengths and longer times. To determine this effect a special calculation has been done for the system (1,2) with and without counterflow term (3). On fig. 5 Ecf is the stability margin for the first case while deltE is the difference between two cases. 2) The stability margin is independent upon the heated length for lengths more than 300 mm, it exceeds the level determined from the enthalpy difference in helium between operating temperature (Top=1.8 K) and temperature of lambda transition (Tlamb), mainly because of the existence of the initial stage of heat exchange in He I region determined by formula (4). This level is shown on the fig. 4 by the solid line. 3) For the most typical times Incoloy 908 (with estimated strain of 0.003%) has some but not very significant advantage against stainless steel (with estimated strain of -0.41%) as a jacket material. This advantage is increasing for shorter times, where the margin is determined

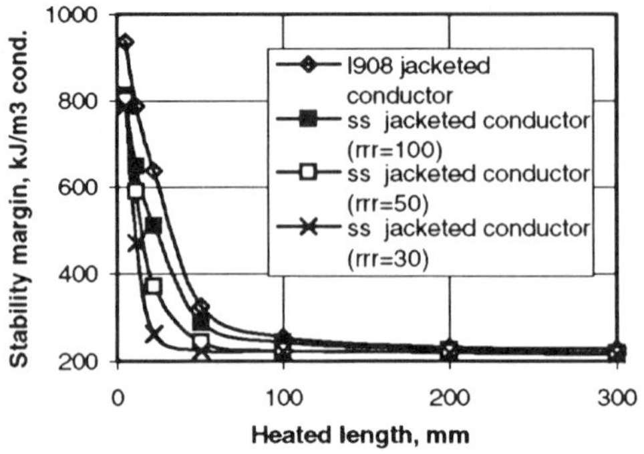

Figure 3. Stability curves for 10 ms heat load.

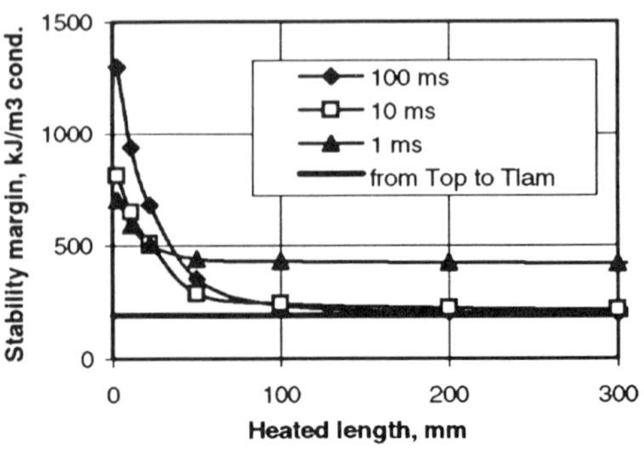

Figure 4. Stability for stainless steel jacketed conductor.

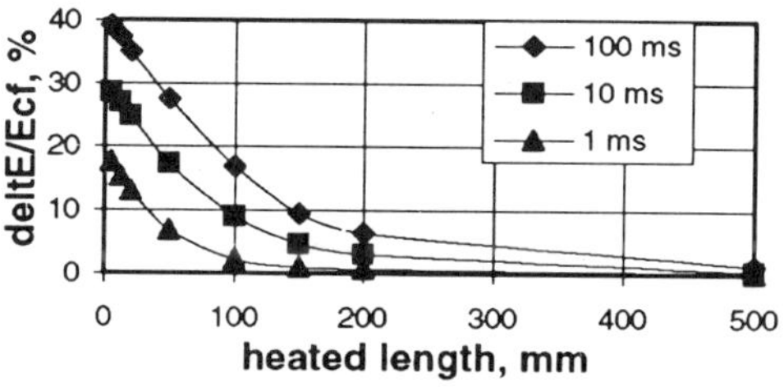

Figure 5. Relative gain in stability due to the counterflow mechanism in He II.

mostly by the current sharing temperature - Tcs (4.4 K with Incoloy jacket and 2.7 K with stainless steel jacket). Because of the better mechanical properties of stainless steel it is possible to increase the helium cross-section (50.97 mm^2 instead of 47.25 mm^2 with Incoloy) and to reduce the difference. 4) The stability of the conductor is not very sensitive to the purity of used copper material, except for the very short heated lengths, where the conduction through copper is essential. 5) The stability against the disturbances of type I is very high and much more than the estimated upper bound of ~ 20 kJ/m3 of conductor, based on the effective transverse modulus of the cable and the Lorentz force applied to it.

Analysis of the Quench for 45-T Hybrid magnet

When the deposited energy exceeds the limit of stability a quench takes place. The quench evolution in the first layer of the coil A was studied for both types of disturbances in order to determine the possible maximum temperature, maximum pressure and period of time for voltage to reach its threshold value (100 mV). For the type I disturbance the calculations were done for the most probable estimated disturbance time (10 ms) on the

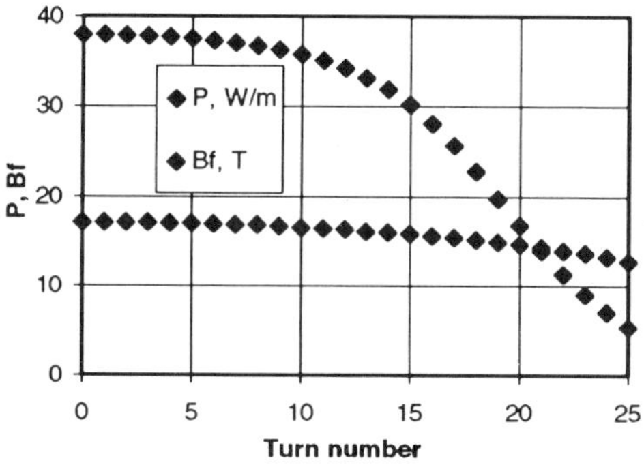

Figure 6. Heat load and magnetic field distribution.

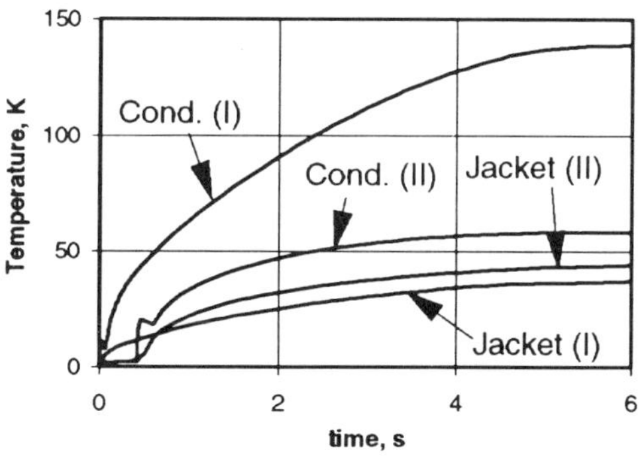

Figure 7. Temperature evolution during quench.

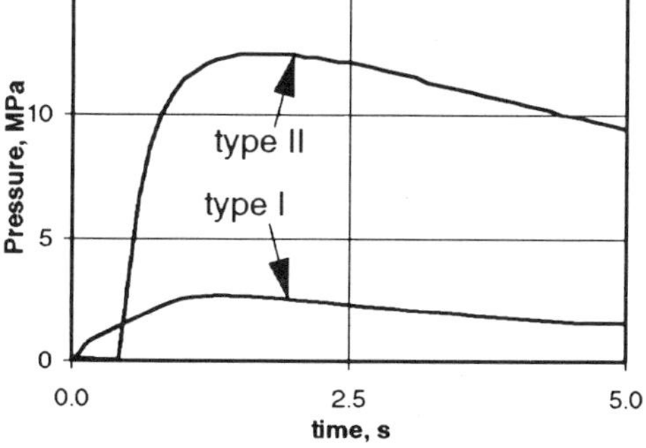

Figure 8. Pressure evolution during quench.

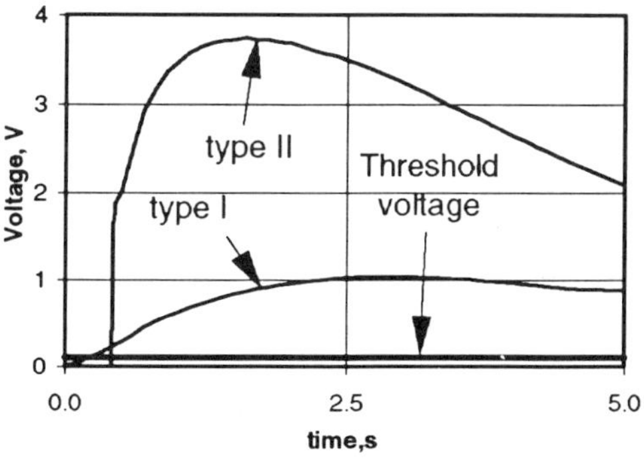

Figure 9. Voltage evolution during quench.

most probable estimated length (300 mm). The type II disturbance takes place in the whole length of the conductor (116.1 m). The distribution of final magnetic field and of the linear heat load that just causes the quench in the first layer of the coil A is shown on fig. 6. One turn corresponds to 2.28 m. There are the evolution curves of the cable and the jacket temperatures and of the pressure in the central point on figs. 7,8 for both types of disturbances. The evolution of voltages with time delay of 0.7 s and current dumping time constant of 4 s is shown on fig. 9.

CONCLUSIONS

1. A finite element code for analysis of stability and quench of CICCs has been extended to the He II regime.

2. The He II heat transport features included in the code have been benchmarked against experiment and analytic expressions with good results.

3. When used to calculate stability of the highest-field coil of the superconducting outsert for the 45-T Hybrid, the code predicts very high stability against sudden, localized disturbances (type I) and quite reasonable stability against broadly distributed disturbances of longer duration (type II).

4. In the event of a quench caused by either type of disturbance, the conductor appears safe from the perspective of two important protection criteria: hot-spot temperature of the conductor and maximum pressure in the partially confined helium. The maximum pressure resulting from the local (type I) quench is much lower, because the cold helium outside the quenched region is more easily expulsed, whereas the hot-spot temperature for this case is higher, in large part for the same reason.

ACKNOWLEDGMENT

The NHMFL is supported by the State of Florida and the National Science Foundation through NSF Cooperative Grant No. DMR 9016241

REFERENCES

1. L. Bottura. Quench analysis of superconducting magnets. A numerical study, NET report No 102, EUR-FU/XII/185/93
2. L. Bottura. A quench simulation model for the ITER magnets, NET report /R/0821/54/A
3. S.W. Van Sciver. "Helium Cryogenics," Plenum Press, New York (1986).
4. L. Dresner, Transient heat transfer in superfluid helium - part II, Adv. Cryog. Eng., 29:323 (1984).
5. S.W. Van Sciver et al. Thermal hydraulics of the 45-T hybrid superconducting magnet system, in: " Proc. Int. Symp. on Safety of Superconductors", Fukuoka, Japan (1994)
6. Y.F. Rao et al. Transient characteristics of He II forced flow heated at the center of a pipe line, in: " Proc. Int. Symp. on Safety of Superconductors", Fukuoka, Japan (1994)
7. S.K. Nemerovskii, et al. Unsteady heat transfer in He II with cylindrical geometry, Cryogenics, 34:733 (1994)
8. R.J. Hooper, Friction factor correlations for CICC, memo. FEDC-M-84-E/M-016, Fusion engineering design center, ORNL, (1984).
9. L.A. Yaskin et al. A correlation for heat transfer to supercritical helium in turbulent flow in small channels, Cryogenics, 17:549 (1977)
10. J.C. Lottin and S.W. Van Sciver, Heat transport mechanism in a 2.3 m long cooling loop containing He II, Proc. ICEC. 9:269 (1982), and additional unpublished data.
11. T. A. Painter et al. Progress in manufacture of the cable-in-conduit Nb3Sn outsert coils for the 45 Tesla Hybrid magnet, IEEE Tran. Magn., 30:2204 (1994)

DESIGN AND PROJECTED PERFORMANCE OF THE WESTINGHOUSE μSMES UNIT

O. R. Christianson, D. J. Hall, D. A. Hoecker, D. W. Scherbarth,
R. L. Swensrud, J. M. Toms, and D. T. Hackworth

Westinghouse Science & Technology Center
Pittsburgh, PA, 15235, USA

ABSTRACT

The Westinghouse design of a 5 MJ, 2 MW, 2 second discharge μSMES unit is described in this report. The magnet consists of concentric, potted coil modules, each with a support structure designed to limit eddy currents. It operates in a bath of liquid helium. The individual modules are wound with a superconducting Rutherford cable which operates at 1500 amperes in a peak field of 3.5 T. The coil modules are layer wound and potted with the support structure. The magnet is designed so that it will not quench during discharge and the subsequent charge. This performance is calculated by determining the losses due to hysteresis and eddy currents and the resulting temperature rise. The magnet current is compared to the temperature and field dependent critical current to determine that the superconductor has adequate stability margin to prevent a quench. Losses are reduced by using a Rutherford cable and by controlling the interstrand resistance. When fully charged, adequate margin ensures superconducting operation of the magnet where the conductor margin is found by comparing the operating point to the critical surface. The quench current and field are determined from the intersection of the magnet load line and the superconductor critical surface. Magnetic field calculations are performed with the finite element code MAFCO, and the superconductor critical surface is defined by an empirical relationship. Although no quenches are expected during normal operation, a quench detection circuit and protection resistor are provided so that the temperature rise, estimated from the MIITS generated at a selected discharge voltage, is small ensuring no magnet damage. Joint design, operation, and margin are also described in this paper.

INTRODUCTION

The superconductor in the Westinghouse 5 MJ, 2 MW, 2 second discharge μSMES unit is designed such that no quench will occur during discharge and subsequent charge of the magnet. When the μSMES unit is discharged at a constant power output, the magnet current drops decreasing the magnetic field and increasing the critical current. However,

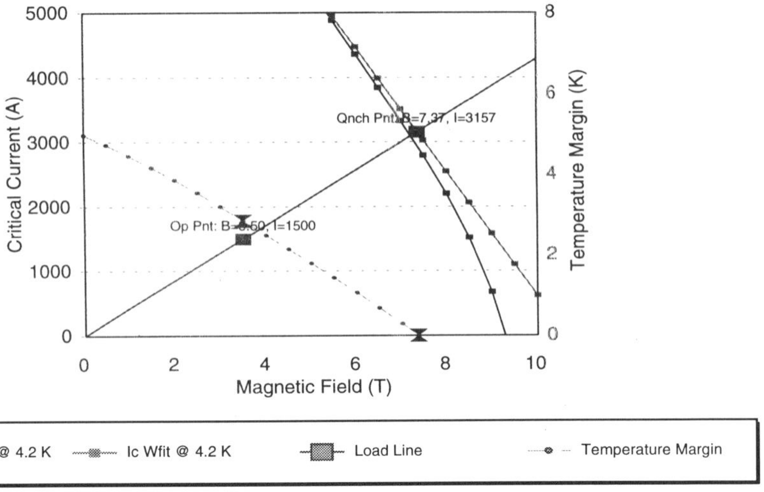

Figure 1. The load line, superconducting critical surface, and temperature margin of the μSMES unit. The superconductor operates at 48% along the load line with a temperature margin of 2.9 K.

changing the magnet current produces losses in the superconductor. These losses are due to eddy currents flowing between the strands, and hysteresis in the superconductor, and losses appear as heat increasing the temperature of the conductor. The increase in temperature lowers the critical current of the superconductor. Depending on the rate of decrease of current, the temperature rise in the conductor may reduce the critical current more quickly than the decrease in magnetic field increases the critical current. If the critical current decreases past the magnet current during the discharge a quench will result and the superconductor will turn normal. The μSMES conductor is designed so that the critical current of the superconductor is always greater than the magnet current. This results in a conductor with a large superconducting margin where the operating point is 48% along the μSMES unit load line, and the temperature margin is 2.9 K, see Figure 1. The potential for inelastic mechanical disturbances in the μSMES unit are reduced by a robust mechanical support structure so that they should not initiate a quench.

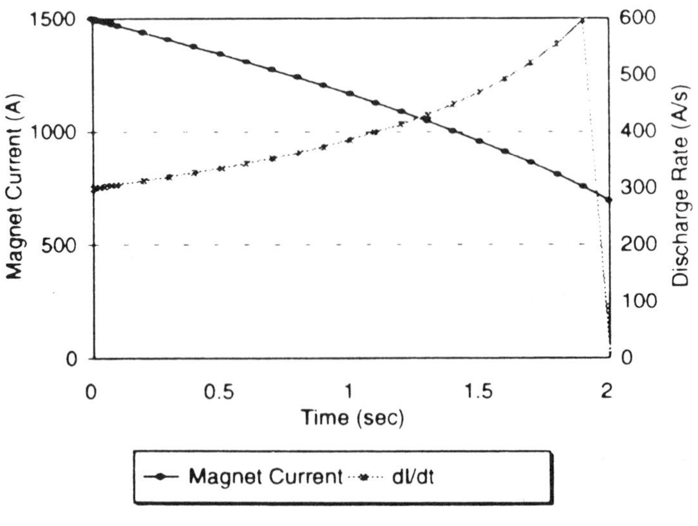

Figure 2. Magnet current and discharge rate.

Table 1. Main Parameters Selected by WSTC

Stored Energy	MJ	5
Usable Energy	MJ	4
Peak Power	MW	2
Pulse Length at Peak Power	sec	2
Charging Time	sec	60

The stability of the superconductor ensures that the conductor will not quench in normal operation. However if a quench were to occur the μSMES unit is designed such that there would be no damage to the magnet. The μSMES unit is actively protected by a quench protection circuit that dumps the magnetic stored energy to an external dump resistor in the unlikely event of a quench. The quench protection circuit is activated by an automatic quench detection circuit. The conductor is designed such that if a quench occurs, no damage will result from the temperature rise of only about 100 K at a quench voltage of 4200 volts. The conductor is also adequately insulated to ensure no voltage breakdown during a quench. The cable insulation is identical to that used for the FNAL and BNL 40mm SSC Dipole Magnets.

DISCHARGE AND CHARGE OF THE μSMES

The major system parameters of the μSMES system are shown in Table 1 and Table 2.

Theory, Methodology, and Calculations

The required discharge rate for the system to provide a constant 2 MW power output is shown in Figure 2. During discharge the magnetic field decreases allowing the critical current to increase. However, changing the magnet current produces losses in the superconductor.[1] The largest losses are due to eddy currents flowing between the strands, and hysteresis in the superconductor. (Other losses such as intrastrand eddy currents or dynamic resistance are less, but are also included in these calculations.) These losses appear as heat increasing the temperature of the conductor. The increase in temperature lowers the critical current of the superconductor. The conductor is designed such that the magnet current is less than the critical current of the superconductor throughout the discharge and subsequent charge of the μSMES unit.

The interstrand eddy current loss[2] is given by

Table 2. Conductor Specifications

Strand Diameter	mm	0.808
Actual Strands (exact qty rounded up)		15
Cu/SC		2.89
Conductor Length	m	6772
Filament Diameter	μm	6
Cable Transposition Length	mm	45
Cable Overall Size	mm	6.22 x 1.37
Interstrand Resistance	μΩ	200

$$W = \frac{N^2 h^2 L \dot{B}^2}{60R} \tag{1}$$

where W is Watts per unit length of conductor, h is the width of the Rutherford cable, L is the half braid pitch length, R is the crossover resistance of the braid, and N is the number of strands in the Rutherford cable. The hysteresis loss is

$$W_h = Q_h \frac{\lambda V}{\tau} \tag{2}$$

where λ is the percent of superconductor, V is the conductor volume, τ is the pulse length, and Q_h is given by

$$Q_h = \frac{1}{\mu_o} \int_{B_i}^{B_f} M dB \tag{3}$$

where the initial magnetic field B_i and the final magnetic field B_f reflect the change in the magnetic field at the conductor due to the discharge, and M is the magnetization given by

$$M = \frac{2}{3\pi} \mu_o J_c d_f \left(1 - \frac{J}{J_c}\right) \tag{4}$$

where d_f is the diameter of a superconducting filament, J is the current density at some instant in the discharge, and J_c is the critical current density at the filament's field and operating temperature. Losses for the WSTC proposed conductor are shown in Figure 3. The temperature rise during discharge where the conductor is treated adiabatically is

$$dT = \frac{Pt}{\left(C_{NbTi} V_{NbTi} + C_{Cu} V_{Cu} + C_{He} V_{He}\right)} \tag{5}$$

where the specific heat is a function of temperature. During charge there is cooling to the LHe bath limited by thermal conduction through the conductor insulation

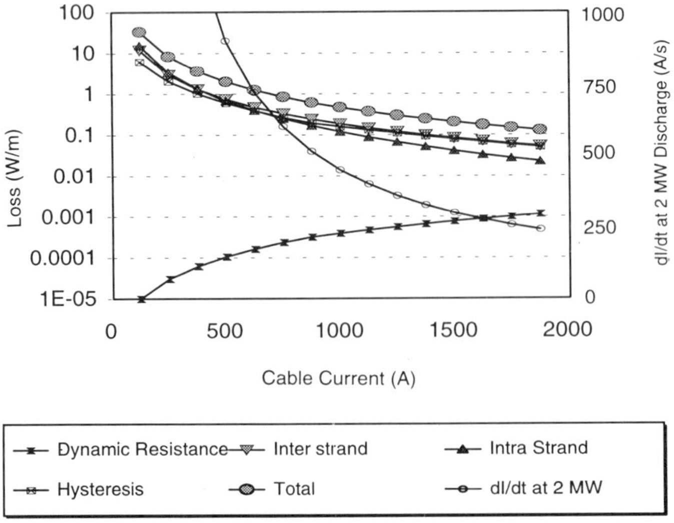

Figure 3. Instantaneous losses in the WSTC proposed conductor for the μSMES unit. The superconductor operating point is 1500 A at 3.5 T.

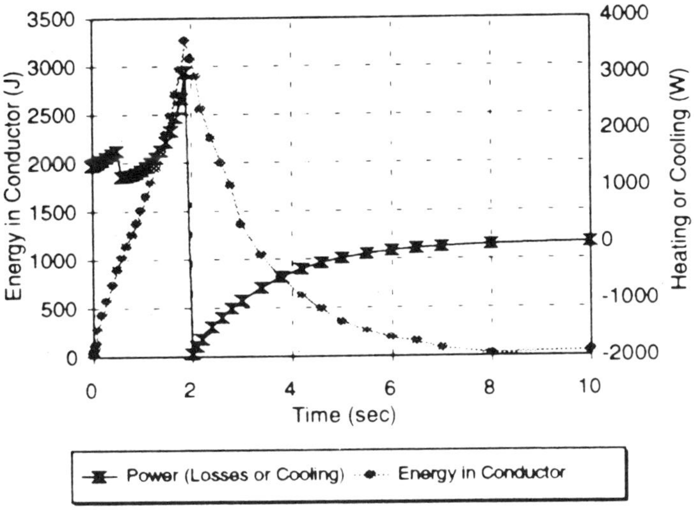

Figure 4. The heating and cooling of the conductor during discharge and subsequent charge.

$$q_{cooling} = kA\frac{\left(T_{conductor} - T_{bath}\right)}{\delta} \tag{6}$$

where δ is the insulation thickness. The critical surface of the superconductor as a function of magnetic field and temperature is given by an empirical correlation proposed by Morgan.[3] The load line of the magnet is given by the transfer function of 3.5T/1500 A.

Results

The loss calculations assume an interstrand cross over resistance of 200 μΩ, that is the conductor strands are insulated. This may be accomplished by Ebonol coating, chrome coating, niobium coating, or perhaps varnish. The instantaneous losses are shown in Figure 3. The cumulative energy deposited in the conductor is shown in Figure 4. The current carrying capacity of the superconductor, the magnet current, and the conductor temperature during a discharge and subsequent charge of the μSMES unit are shown in Figure 5. For insulated strands with cross over resistances greater than 175 μΩ, the conductor will not

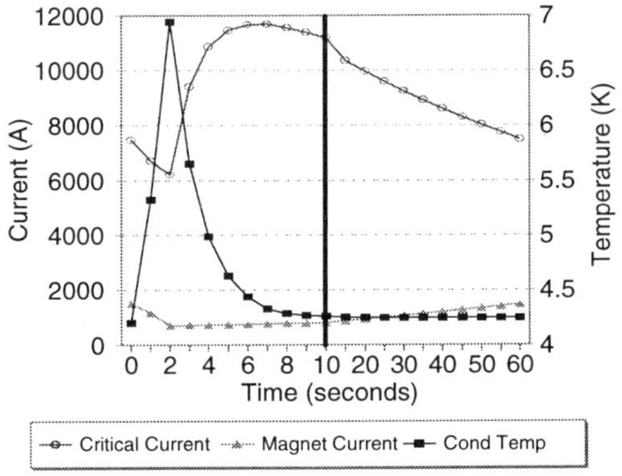

Figure 5. Discharge and charge of μSMES unit.

Table 3. Magnetic Field at the Coil Assembly Joints

Joint	Magnetic Field
Inner	2.442 T
Middle	2.262 T
Outer	2.027 T

quench during a discharge or charge. The conductor temperature has recovered within 10 seconds during the charging of the magnet.

The energy dissipated in the discharge is about 4 kJ, and must be handled by the cryogenic system. This energy is equivalent to vaporization of 1.5 liter of LHe. If the discharge occurs once a day, this represents an additional 5% load added to the steady state heat load.

μSMES JOINT DESIGN

Joints exist at the magnet leads and between the different coil assemblies. Since the joints are resistive, joule heating occurs raising the temperature of the joints. The joint location is also in a high magnetic field region. It is important that the superconductor in the joints remain in the superconducting state, otherwise a quench will be initiated and propagate into the coil assemblies. This analysis will consider joint resistance resulting in joule heating, and the joint cooling which is dependent upon orientation of the joint to determine any temperature rise and compare to the superconductor critical surface determined by the magnetic field.

The joint is a lap joint resembling an SSC splice. The resistance of typical joints made in SSC accelerator magnets is 10 nΩ. The joule heating at a transport current of 1500 A is then 0.02 W. Typical joints are approximately 8 cm long. This gives a joint surface area of 10 cm^2. Orienting the cable face vertical maximizes the surface cooling. The surface heat flux is about 0.002 W/cm^2. Comparing to the nucleate boiling heat flux of LHe indicates that the surface temperature increase will be small.

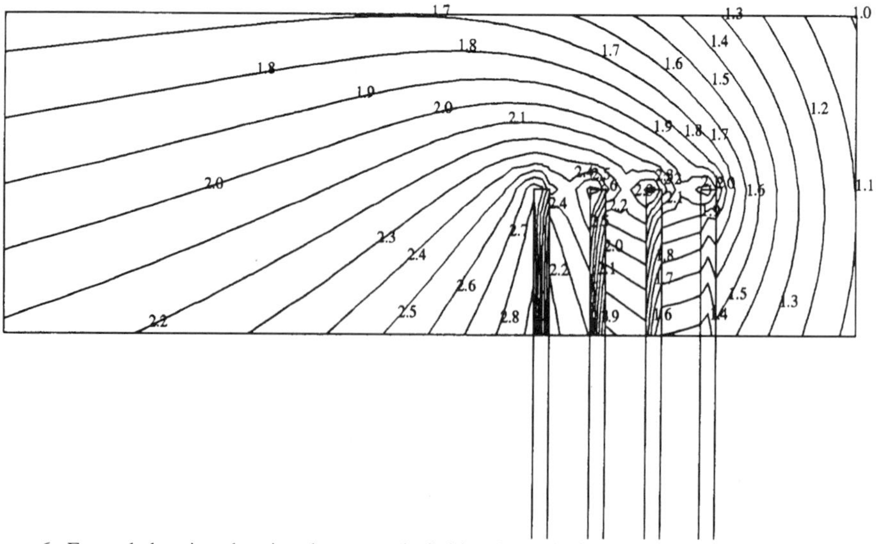

Figure 6. Expanded region showing the magnetic field at the end of the coil assemblies where the joints will be placed.

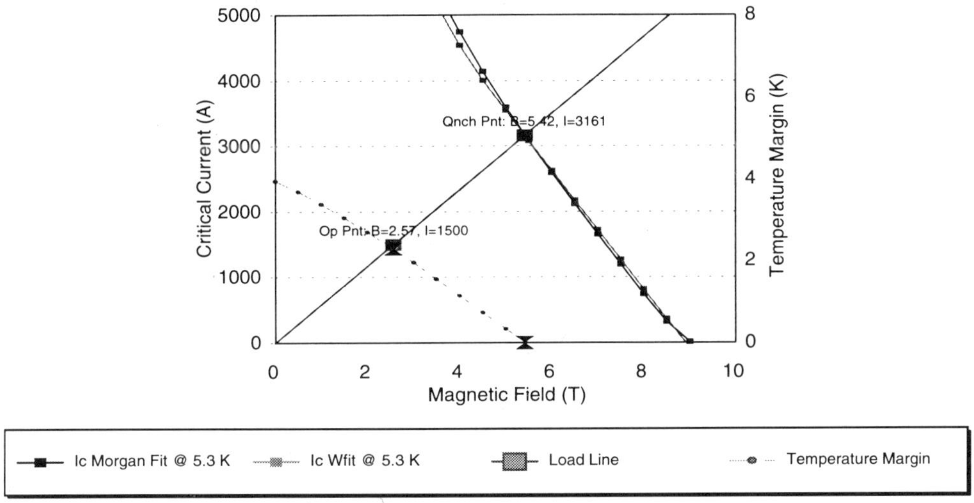

Figure 7. Operating point of the joint compared to the superconductor critical surface. The superconductor operates at 47% along the load line with a temperature margin of 2.3 K.

The joint will be insulated with 0.05 mm of Kapton. Kapton has a thermal conductivity of 10^{-4} W/cmK at 4.2 K. The temperature drop across the insulation is given by $dT = dx\frac{q}{kA}$, so that the temperature across the Kapton insulation is 1.03 K. This gives a conductor temperature of 5.25 K

The joints between coil assemblies will be located at the ends of the coil assemblies in a magnetic field of 2.44 T, see Figure 6. The self field of the cable is 0.13 T, therefore the maximum field in a joint is 2.57 T. Taking the joint operating point as 1500 A at 5.25 K and 2.572 T, one finds that the joint superconductor is 47% along the load line and has a temperature margin of 2.3 K, see Figure 7. This margin is adequate to insure stable operation of the joints.

QUENCH AND CU:SC RATIO

If a quench should occur, the resulting temperature rise is predicted using an adiabatic, one dimensional approximation of a conductor carrying a current. A simple heat balance is

$$\frac{I^2(t)\rho(T)}{A_{cu}}dt = AC(T)dT \tag{7}$$

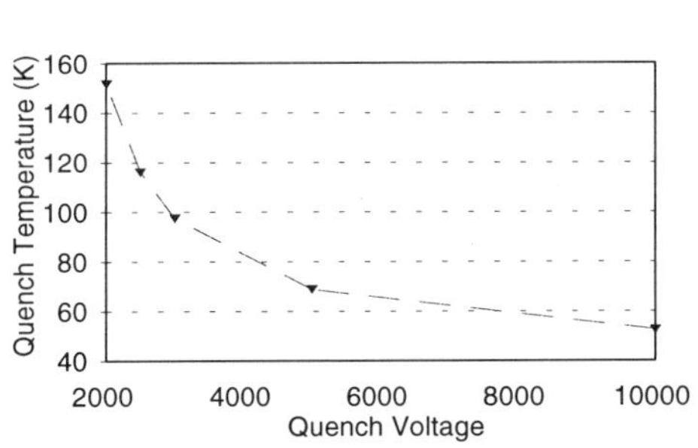

Figure 8. MIITS calculation of the quench temperature at different quench voltages.

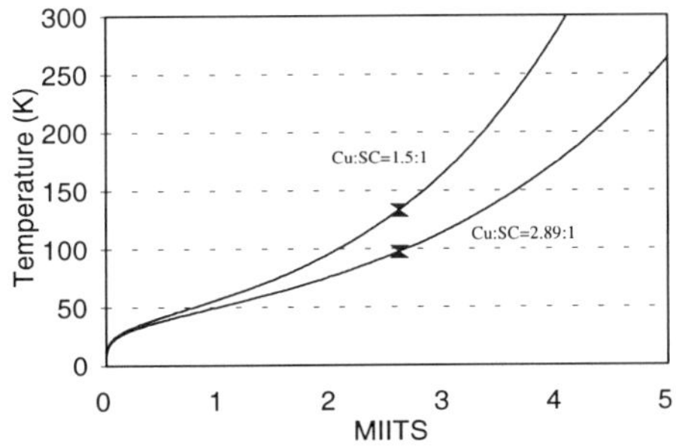

Figure 9. Quench temperature of the μSMES coil for Cu:SC ratios of 2.89 and 1.5.

Rearranging and integrating gives[4]

$$\int_0^t I^2 dt = A_{cu} A \int_{T_o}^{T} \frac{C(T)}{\rho(T)} dT \qquad (8)$$

The units are MIITS, 10^6 Amps^{2}-sec. The left hand integral is a function of the circuit parameters, and the right hand integral is a function of temperature. To simplify the problem, it is assumed that the current decay is entirely controlled by the dump resistor. The current decays exponentially with a time constant of L/R where L is the inductance of the magnet and R is the resistance of the dump resistor. Thus as time increases the temperature becomes the final temperature of the coil.

The specific heat of copper, iron, and the superconductor are obtained from the computer program Cryodata.[5] Values beyond 300 K are extrapolations from the Cryodata program. The resistance of copper as a function of field is derived from a Kohler plot with Fickett's empirically derived coefficients,[6] and the temperature dependence follows the equation[3]

$$\rho(T, RRR) = \frac{1.545E-8}{RRR} + 1E-8\left(\frac{2.32547E9}{T^5} + \frac{9.57137E5}{T^3} + \frac{1.62735E2}{T}\right)^{-1} \qquad (9)$$

Varying the dump resistance (and hence the peak quench voltage) results in different decay times and different maximum temperature rises. Choosing a quench voltage of 4200 V produces a quench temperature rise less than 100 K, see Figure 8.

The copper to superconductor ratio, Cu:SC, chosen for the µSMES unit is larger than values used in the SSC cable, 2.89 versus 1.5 (or 1.3 and 1.8). The Cu:SC ratio is chosen to set the quench temperature less than 100 K. Changing the Cu:SC ratio from 2.89 to 1.5 raises the quench temperature from 97 K to 133 K, an increase of 36 K, see Figure 9.

CONCLUSIONS

The Westinghouse design of a 5 MJ, 2 MW, 2 second discharge µSMES unit using a 1500 A LHe bath cooled Rutherford superconducting cable operated in a 3.5 T field is described. Adequate margin ensures superconducting operation of the magnet during discharge and the subsequent charge. If a quench should occur, the temperature rise estimated from the MIITS generated at a given voltage, is small ensuring no magnet damage. Joint design is also described in this paper.

REFERENCES

1. W. J. Carr, Jr., "AC Loss and Macroscopic Theory of Superconductors," Gordon and Breach, NY (1983).
2. G. Morgan, "Eddy Currents in Flat Metal-Filled Conducting Braids," JAP 44, No. 7, July 1973, p. 3319.
3. G. Morgan, "A Comparison of Two Analytic Forms for the Jc(B,T) Surface," SSC Tech. Note. No. 310-1 (SSC-MD-218), (January 6, 1989).
4. McAshan, M.S., "MIITS Integrals for Copper and for Nb-46.5% Ti," SSC-N-468, 1988.
5. Cryodata, P.O. Box 558, Niwot, CO 80544
6. Fickett, F.R., "A Preliminary Investigation of the Behavior of High Purity Copper in High Magnetic Fields," INCRA Project No. 186, June 1972.

FABRICATION AND TESTING OF 1-M-LONG SSC MODEL DIPOLE MAGNETS

K. Hosoyama,[1] K. Hara,[1] N. Higahi,[1] A. Kabe,[1] H. Kawamata,[1] Y. Kojima,[1] Y. Morita,[1] H. Nakai,[1] A. Terashima,[1] T. Takahashi,[2] T. Shintomi,[1] H. Hirabayashi,[1] and Y. Kimura[1]

[1]KEK-National Laboratory for High Energy Physics
Tsukuba, Ibaraki, Japan
[2]Hitachi Work, Hitachi, Ltd., Hitachi, Japan

ABSTRACT

A series of fifteen 5-cm-aperture, 1-m-long, SSC model dipole magnets with various types of cable have been designed and fabricated at the National Laboratory for High Energy Physics (KEK). These magnets have been quench tested in a 4.2-K vertical cryostat and a 1.8-K pressurized-superfluid cryostat. In order to improve the quench characteristics of the magnets, we have carried out modifications of the coil end and splice design. We will describe the basic features of the magnet design, the various fabrication steps of the magnets, and results of quench tests.

INTRODUCTION

The design and construction of a series of fifteen 5-cm aperture, 1-m-long SSC model dipole magnets, KEK#1 to #15, was started in May 1990 to establish construction techniques for full-scale dipole magnets as part of the cooperative R&D program between SSCL and KEK. Except for length, these magnets have the same baseline design and features as the full size magnet. We have paid much attention to the design of magnet end part and developed new-type end spacers,[1, 2] which have constant perimeter profiles to minimize internal stress of the cables at the coil end. Following the 1-m-long dipole magnet R&D program, the construction of a 13-m-long, full-scale dipole magnet was started in April 1993 at KEK. The first quench test of this magnet was performed successfully in February 1995 at KEK.

MAGNET DESIGN

A cross-sectional view of the cold mass is shown in Fig.1. The baseline design of the magnet has a cross section referred to as W6733[3], which was designed and extensively studied at Brookhaven National Laboratory (BNL) for SSC main ring dipole magnet. This coil has a two-layer cosine-theta coil geometry with 19 turns in the inner coil and 26 turns in the outer. The coil is mechanically constrained radially by 1.5mm thick, stainless-steel collars. The collared coil is covered by a horizontally-split iron yoke made of 1.5mm thick, low-carbon steel laminations to enhance the magnetic field by about 20 percent and further mechanical

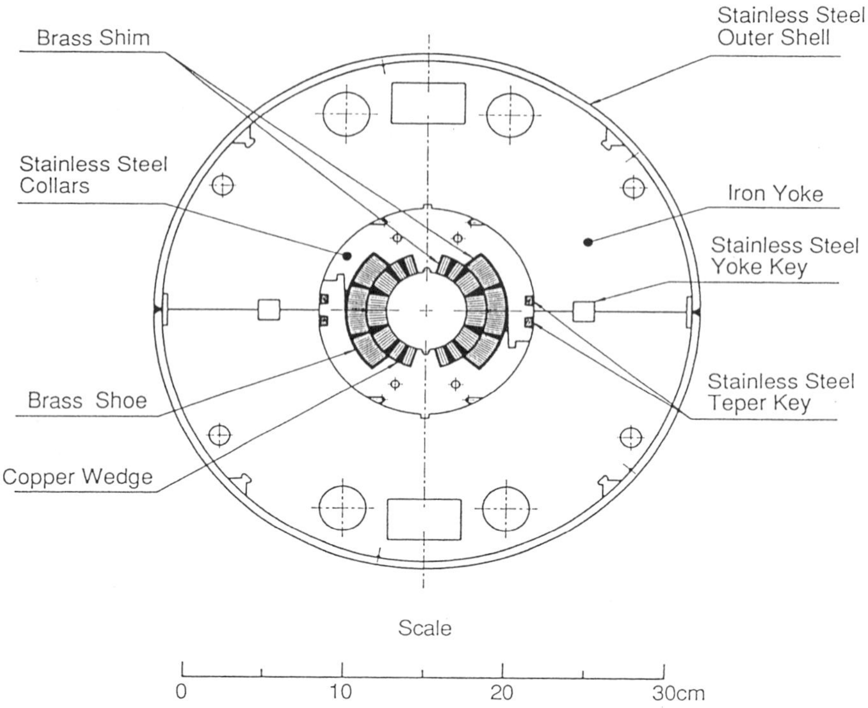

Figure 1. Cross sectiona of the cold mass of KEK 5 cm aperture 1 m-long dipole magnet.

stability. The yoke modules at the ends have stainless-steel voids to lower the magnetic field there. The main parameters of the magnets are listed in Table 1.

COIL AND SUPERCONDUCTING CABLES

The inner layer of the coil is wound using a 1.2°-keystone-angle cable of 30 strands (strand diameter 0.808 mm) and 3 copper wedges. The outer coil is wound using a 1.0° keystone-angle cable of 36 strands (strand diameter 0.648 mm) and 1 copper wedge. Table 2 shows the typical selected parameters of superconducting cables. These cables are insulated with 0.025-mm thick, double-half-lapped polyimide (Upilex-R®) tapes and a 0.05-mm thick B-stage epoxy-impregnated polyimide tape wrapped with a 1-mm gap.

All magnets except magnets KEK#3 and #5, which are fabricated with 2.5-μm filament

Table 1. Selected parameters of SSC model dipole magnets

Design current	6.5 kA
Central field	6.6 T
Overall magnet coil length	1.3 m
Coil inner radious	50 mm
Number of turns	
inner layer	19
outer layer	26
Collar thickness	27 mm
Yoke thickness	98 mm
Outer shell thickness	5 mm

Table 2. Cable parameters for SSC model dipole magnets

Cable Parameters	Inner	Outer
Number of wires	30	36
Mid-thickness (mm)	1.46	1.17
Width (mm)	12.13	11.7
Keystone angle (degree)	1.23	1.0
Cable pitch (mm)	94	86
Wire diameter (mm)	0.81	0.65
Wire twist pitch (mm)	12.8	12.8
Filament dia. (mm)	6 / 2.5	6 / 2.5
Cu / SC	1.3 / 1.5	1.8
RRR	160	160

conductors to meet the requirements for fast-ramp cycle of the SSC High Energy Booster (HEB), are wound from 6-μm-NbTi-filament cables. In the 2.5-μm-filament cables, a 0.5 wt % Cu-Mn matrix is used to eliminate the proximity effect coupling loss. In the magnets KEK#14 and #15, we used cables with 6-μm NbTi filaments and 0.5 wt % Cu-Mn matrix. All cables, except outer-coil cables for KEK#12 and #13, are annealed after cabling to improve the flexibility for coil winding. By this annealing, the RRR values of the cables are about 150 (RRR values are 45 for KEK#12, #13 outer-coil cables).

COIL END

We adopted a practical constant-perimeter end design,[1] developed at KEK to confine the conductors to a consistent shape and minimize the internal stress in the cables. The spacers are cut out from filament-wound G10 pipe usuing a numerically controlled milling machine. The axial motion of each coil end is restricted by a G10 shoe. This shoe is pressed on by four screws set in stainless-steel end plates. The end plates are connected to the outer shell of the iron yoke.

WINDING AND CURING

The inner coil is wound around a center post set on a mandrel of the rotary winding machine. During the coil winding, the cable tension is maintained constant, starting with 490 N and reduced by 9.8 N every two turns. After winding the coil is heated up to 90 ℃, and at this temperature, the epoxy resin on the surface of insulation tape enters the gel phase. At this stage, the coil ends are loaded by two screws from both ends to a nominal force of 13000 N. Then the temperature is increased up to 150 ℃ and the pressure are to nominal curing pressure 50 MPa. These conditions are maintained for 5 hours.

For the magnets KEK#1 to #9, the outer coils are wound on the cured inner coils and then cured under the same curing conditions of the inner coils. For the magnets KEK #10 to #15, inner and outer coils are wound and cured separately and then assembled into coil halves.

SPLICING AND ASSEMBLING

Inner and outer coils are connected electrically in series by what is called a ramp-splice. For the magnets KEK#1 to #9 before the outer coil winding, the outer-coil cable is soldered to the inner cables at the ramp-splice and placed in a straight groove at the outer key.[1] For the magnets KEK#10 to #15, inner conductor is spliced to the outer conductor and the ramp and splice are encased in a G10 holder.[4]

In order to reinforce the electrical insulation, the four quarter-coils are wrapped with 0.1 mm polyimide films and between the inner and outer coils 0.1mm thick G10 spacer are inserted. The coil halves are assembled around a stainless-steel beam tube. The outer surface

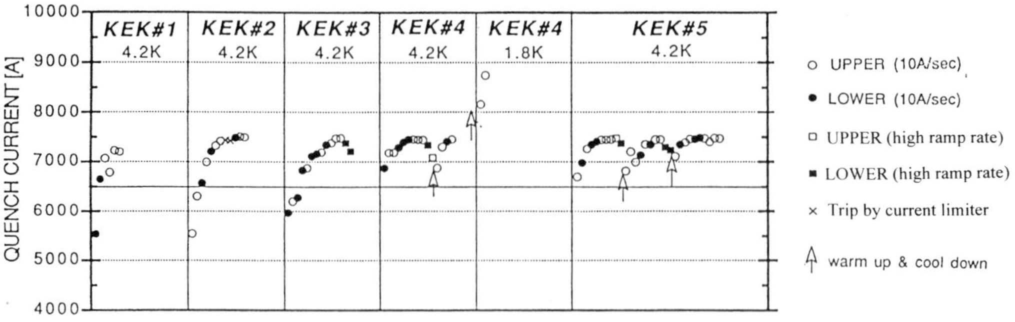

Figure 2. Quench histories of the magnets KEK#1 to #5.

of the assembled coil is covered by several layers of polyimide film to complete the ground insulation. Four stainless-steel strip heaters are set in the polyimide-insulation film wraps outside of the outer coils for quench protection. Two copper-plated stainless-steel spot heaters are also set between the upper and lower coils for an induced-quench test.[2] The coil assembly is finally covered by 0.5-mm thick layer of brass, called the collaring shoe, to protect the coil against the ripples of the collar laminations.

COLLARING AND YOKING

The entire coil is clamped firmly in place and pre-compressed by collar stacks constructed of 1.5mm thick stainless steel (YUS130S®) laminations. High-manganese-steel (KHMN) collar laminations are also used for recollaring magnets KEK#7' and KEK#15. Vertical cylinders are used to press collar packs vertically and horizontal cylinders press the stainless-steel tapered keys (3° taper per side) into the keyways. The collar stacks are pressed by vertical cylinders and then the four taper keys are inserted by the horizontal cylinder. During this process the vertical pressing force is reduced to keep the coil pre-stress constant and to prevent damaging the coil insulation by over stressing. A set of beam-type strain-gauge transducers installed in the collar pack are used to measure coil azimuthal prestress during magnet assembly and testing. During the collaring process the prestress in the straight section is controlled to be 90 MPa for the inner coil and 70 MPa for the outer coil by using brass shims between the collar poles and the coil.

An horizontally split yoke made of 1.5-mm thick, low-carbon-steel laminations is used to enhance the magnetic field and further mechanical stability. The 5-mm thick stainless-steel shells surrounding the yoke stacks are welded together under the pressurized condition in the yoking press.

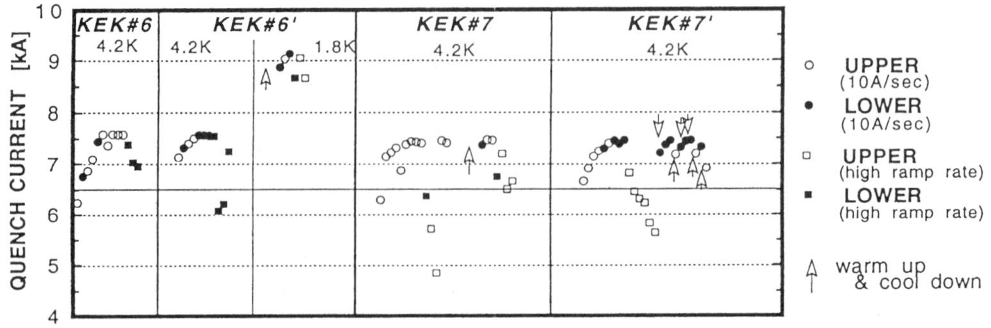

Figure 3. Quench histories of the magnets KEK#6 to #7'.

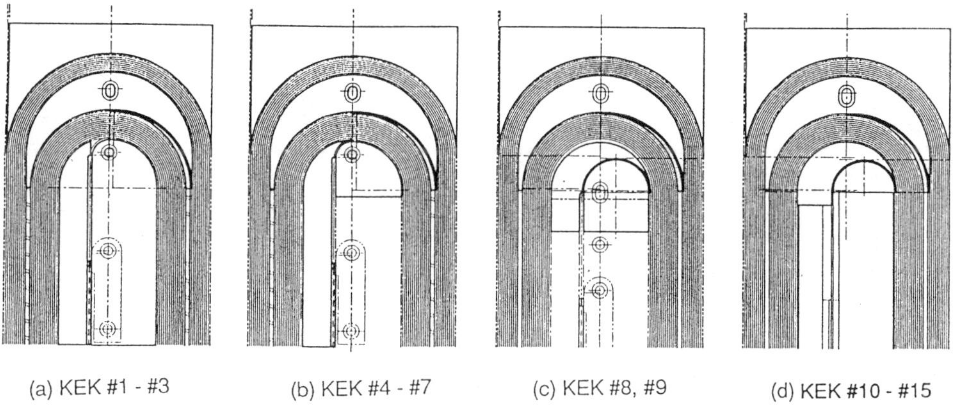

Figure 4. Variants of coil end designs for the magnet KEK#1 to #15.

QUENCH TEST

Cryogenic testing of the magnets was performed in vertical cryostats with both pool-boiling liquid helium at 4.2 K and pressurized superfluid helium at 1.8 K. The quench test was performed by ramping up the magnet current at about 10 A/s until the magnet quenched. In order to test the limits of the mechanical performance of the magnets under high-stress conditions excitation tests of the magnets KEK#4 and #6' were also performed in pressurized superfluid helium at 1.7 K where the quench current could exceed 9000A.

The inner coils and splice parts were instrumented with 50 voltage taps and 32 voltage signals were monitored and recorded by data-acquisition system to know the quench origins and quench-propagation velocities.

QUENCH TEST AND MODIFICATION OF DESIGN

The quench history plots of KEK#1 to #5 are shown in Fig.2. For the first three magnets, KEK#1, #2, and #3, the first quench currents are lower than 6000 A. A few training quenches were required to reach the design operating current of 6500 A, and several more quenches were required to reach the plateau at 7500A. These low-current quenches were caused by insufficient prestress of the splice part of the magnets. We designed asymmetrical

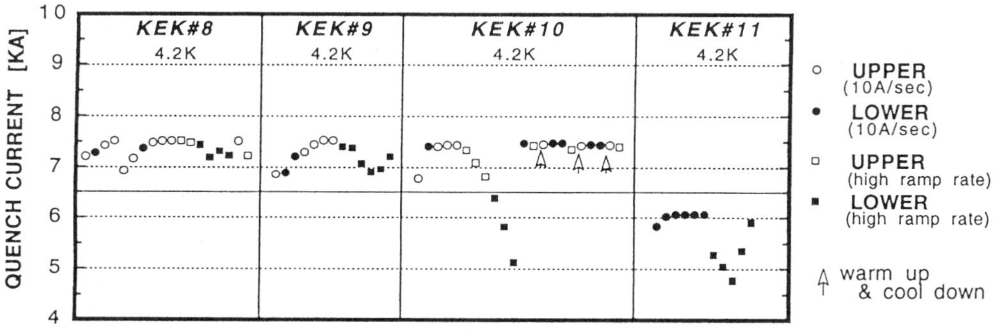

Figure 5. Quench histories of the magnets KEK#8 to #11.

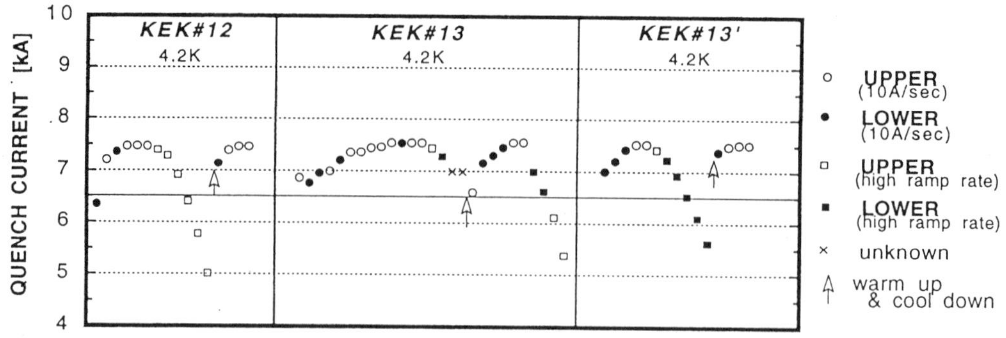

Figure 6. Quench histories of the magnets KEK#12 and #13.

collars, so-called "transition collars", which can press the splice part directly. The prestress on the splice part was also monitored with the same type strain-gauge transducer used at the straight part. The level was controlled to be about 80 MPa by insertion of a brass shim between the splice and the asymmetrical collars. We adopted this newly designed transition collar for all magnets after KEK#4. By this modification, we could eliminate the low-current quenches at the splice part, but still several more quenches were required to reach the plateau. For the magnets KEK#4, #5, and #6, low-current quenches occurred at the starting point of outer-coil winding. After the quench test of the magnet KEK#6, we disassembled this magnet and found the "pop-up" of the strand at the sharp bending corner of the first turn of outer coils. We soldered these parts (upper and inner coils) of the magnets KEK#6 and #7 to increase the mechanical stability of the cable and assembled magnets KEK#6' and #7 (KEK#6' is reassembled version of KEK#6). By this modification, the quench performance was improved as shown in a quench history plot of KEK#6' and #7 in Fig.3. But the high-ramp-rate quench currents decreased drastically due to eddy-current heating at the soldered parts.[4] In order to prevent the "pop-up", we made a new outer-coil splice design that has a smooth curve with constant-perimeter surface. We adopted this new design to the magnets KEK#8 and #9. Figure 4 shows various designs of the coil end part adopted in our magnets. In KEK#8 and #9 the bending corner of the first turn of outer coil was not mechanically stabilized by soldering to reduce the eddy current heating effect there. In the magnets KEK#10 to #15 we adopted the splice design developed by BNL[5]. The quench history of magnet KEK#10 is shown in Fig.5, the quench performance of these magnets were improved. After the first quench at 6800A this magnet reached the plateau. The first quench originated at the transition section between the splice and the pole turn of the outer coil. Subsequent quenches moved to the straight section of the inner pole turn after two quenches and remained there. The first quench of KEK#11 was about 6000A and quench currents stayed there without training The quench origins were the same place, on the pole turn of lower inner coil. By

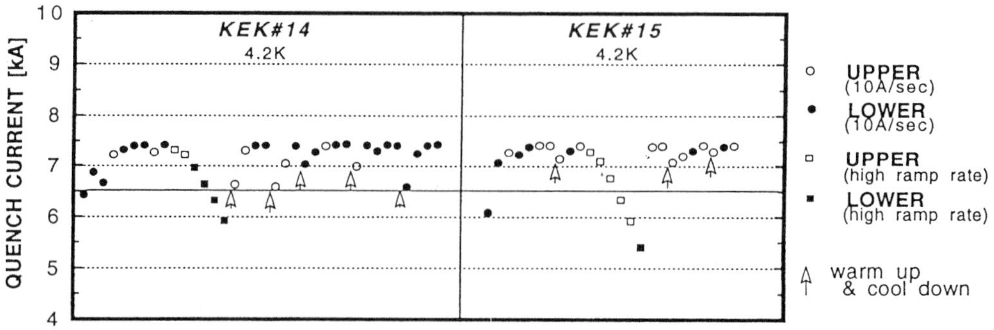

Figure 7. Quench history of the magnets KEK#14 and #15.

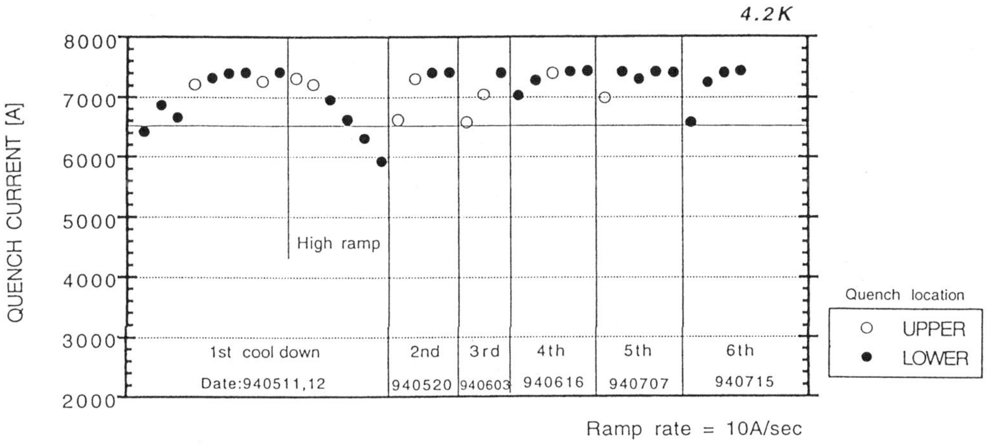

Figure 8. Quench history of the megnet KEK#14.

disassembling this magnet we found one of the pole shims of the inner coil had dropped inward by about 3 mm and the prestress of this part was insufficient. Figures 6 and 7 show the quench history of magnets KEK#12 to #15. For the magnet KEK#13, it took about 10 quenches to reach the plateau and the origins were in the splice part. We found that shims at the transition collar, which are used to control the prestress of the splice part, were not correctly adjusted.

Figure 8 shows the quench history of the magnet KEK#14. In this magnet we reduced the end force to 6000 N. The first quench current of this magnet is low, and it took about 10 quenches to reach the plateau. The first quenches after the thermal cycle in 2nd and 3rd cooldown tests were also low. After this test we increased the end force up to nominal values of 36000 N and performed the 4th quench test. This time the first-quench currents is improved, and after thermal cycle, the first-quench current remained at 7000 A. Before the 6th cooldown test we decreased the end force to 6000 N and the first-quench current was again very low.

CONCLUSION

A series of fifteen 5-cm aperture, 1-m long SSC model dipole magnets have been constructed and tested at KEK. In the first magnets of the series, quenches at low excitation current were found to originate at the splice part due to insufficient clamping force. We could overcome these quenches by increasing the azimuthal prestress to the splice part with a modified collar. The quench characteristics were further improved by adoption of a new splice design and the test results were very promising. We have established construction techniques of full scale magnets through the 1-m long-dipole magnet R&D program.

ACKNOWLEDGMENTS

The authors express their gratitude to Professor H. Sugawara for his continuous support and encouragement. They also wish to thank Dr. E.Willen and Dr. R.Coombes for their helpful advice and useful discussions.

REFERENCES

1. K. Hosoyama, et al., "Fabrication and test of 5-cm aperture 1-m long SSC collider dipole magnet," in : Supercollider 3, John Nonte, ed., Plenum Press, New York p615, (1991).

2. K. Hosoyama, et al., "A status report on the development of 5-cm aperture, 1- m long SSC dipole magnet at KEK" in : Supercollider 5, P. Hale, ed., Plenum Press, New York, p229 (1994)
3. R.C. Gupta, S.A. Kahn, and G.H. Morgan, "SSC 50 mm dipole cross section," in : Supercollider 3, John Nonte,ed., Plenum Press, New York and London, p587, (1991).
4. K. Hosoyama, et al., "Quench characteristics of 1-m-long SSC model dipole magnet," to be presented at CEC conference, Columbus (1995)
5. A. Devred, et al. ,"Status of 4-cm aperture. 17-m-long SSC dipole magnet R&D program at BNL part I: magnet assembly," in : Supercollider 3, John Nonte,ed., Plenum Press, New York and London, p549, (1991).

QUENCH CHARACTERISTICS OF 1-M-LONG SSC MODEL DIPOLE MAGNETS

K. Hosoyama,[1] A. Kabe,[1] K. Hara,[1] N. Higahi,[1] H. Kawamata,[1] Y. Kojima,[1] Y. Morita,[1] H. Nakai,[1] A. Terashima,[1] T. Takahashi,[2] T. Shintomi,[1] H. Hirabayashi,[1] and Y. Kimura[1]

[1]KEK-National Laboratory for High Energy Physics
Tsukuba, Ibaraki, Japan

[2]Hitachi Work, Hitachi, Ltd., Hitachi, Japan

ABSTRACT

A series of fifteen 5-cm-aperture, 1-m-long SSC model dipole magnets with various types of end design and cable have been designed and fabricated at National Laboratory for High Energy Physics (KEK).[1,2,3] The ramp-rate-dependent quench tests of the magnets KEK#3 to #15 were performed in a 4.2-K vertical cryostat. A ramp-rate-dependent test of the magnet KEK#6' was also performed in 1.7-K pressurized superfluid helium. Special ramp tests so called "heating" and "cooling" experiments[4] were also performed on the magnet KEK#10, as well as heat induced quench tests using the spot heaters installed in midplane of inner coils of the magnets KEK#7' and #15 and in the splice part of the magnet KEK#13Y.

INTRODUCTION

The baseline design of 5-cm-aperture, 1-m-long SSC model dipole magnets KEK#1 to #15 have coil cross section, referred to as W6733[5], which was designed and extensively studied at Brookhaven National Laboratory (BNL) for the SSC main ring dipole magnet. This coil has a two-layer cosine-theta coil geometry with 19 turns in the inner coil and 26 turns in the outer. All magnets except magnets KEK#3 and #5, which are fabricated with 2.5-μm-filament conductors to meet the requirements for fast-ramp cycle of the SSC High Energy Booster (HEB), are wound from cables with 6-μm NbTi filament. In the magnets KEK#14 and #15 we used the conductors with 6-μm NbTi filaments and with 0.5-wt-% Cu-Mn matrix. The coils were mechanically constrained radially by 1.5-mm thick stainless-steel collars. The collared coil was covered by a horizontally-split iron yoke made of 1.5-mm thick low-carbon-steel laminations to enhance the magnetic field by about 20 percent and further mechanical stability. The detailed structure of the magnet is described in reference 1 and 2.

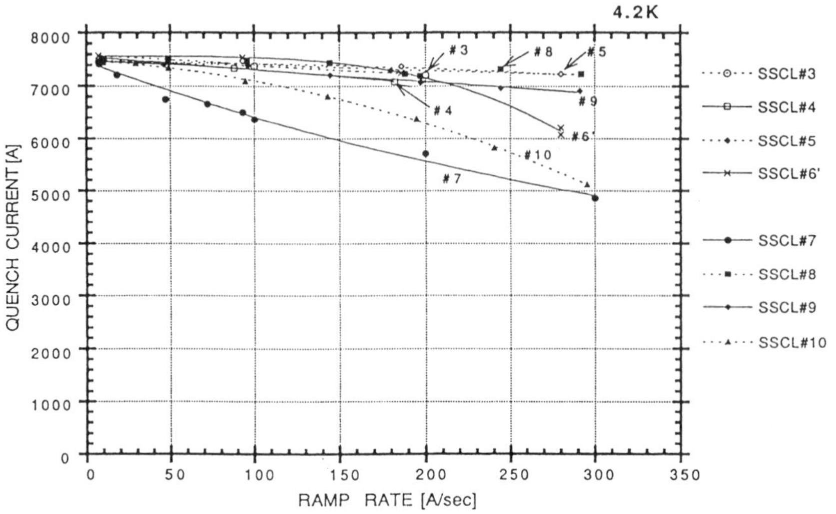

Figure 1. Quench test at varying ramp rates for the 1m dipole magnets KEK#3 to #10.

RAMP RATE DEPENDENCE STUDY

Cryogenic testing of the magnets was performed in vertical cryostats with pool-boiling liquid helium at 4.2K and pressurized superfluid helium at 1.8 K. After the quench current reached the plateau at a low ramp rate of 10 A/s, the ramp rate-dependent quench tests were performed at varying ramp rates up to 300 A/s. The test results for 11 magnets are summarized in Figs.1 and 2. The magnets KEK#3, #4, #5, #8, and #9 did not show larger ramp-rate dependence than the #6', #7, and #10 did. In the magnets KEK#6' and #7, quenches at low ramp rate occurred in straight section of the inner pole turn, close to the ramp section. Quenches at higher ramp rate started at the soldered part of the bending corner of the outer coil. In these magnets, this part was soldered using 96% tin- 4% silver solder to improve the mechanical stability.[3] The magnetic field at this point is low compared to the inner pole turn

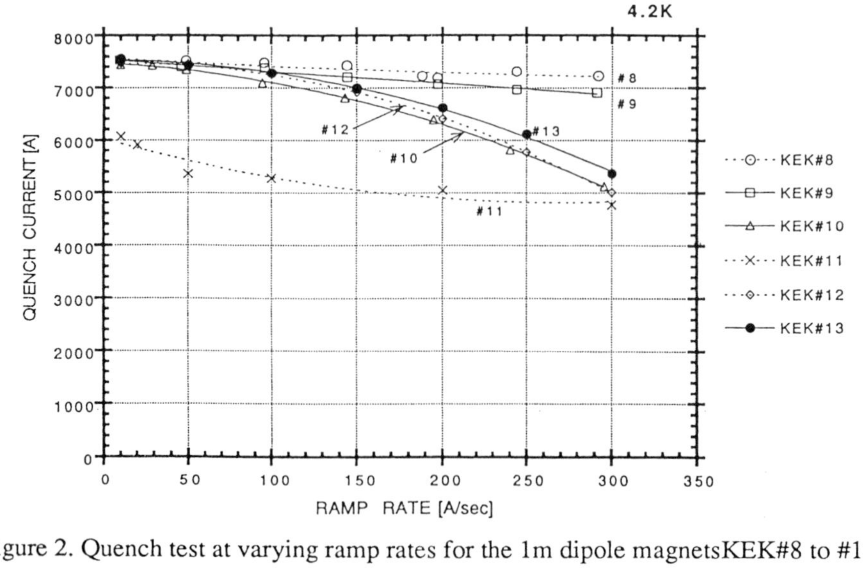

Figure 2. Quench test at varying ramp rates for the 1m dipole magnetsKEK#8 to #13.

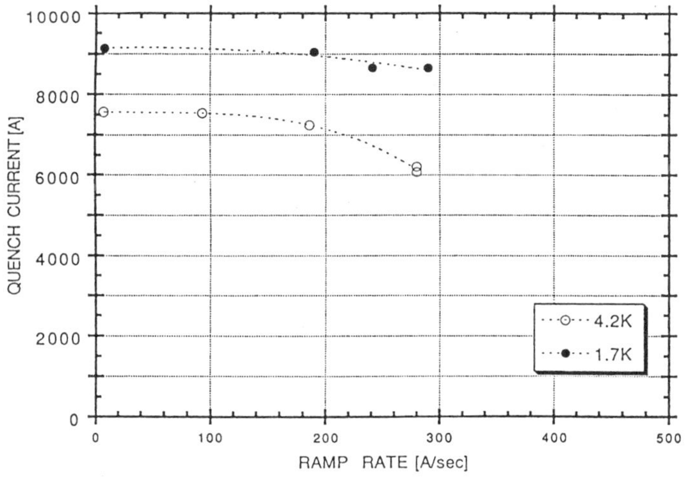

Figure 3. Quench test at varying ramp rates for the 1 m dipole magnet KEK#6' in 4.2 K and 1.7 K pressurized superfluid helium.

(about 60 % of central field), but unfortunately the field direction is nearly perpendicular to the flat face of the cable and the eddy-current loss of this part is very large. The magnet KEK#7 showed larger ramp-rate dependence of the quench current than KEK#6'. This corresponds to the fact that the mechanically weak points of KEK#7 were soldered so heavily that the eddy-current loss was large. Figure 3 shows the high-ramp rate dependence of the quench current of magnet KEK#6' in 4.2-K liquid helium and 1.7-K pressurized superfluid helium. In pressurized superfluid helium, ramp-rate dependence of the quench current was improved due to the higher cooling capacity of the superfluid helium. Figure 2 shows the ramp-rate dependence of magnets KEK#8 to #13. Magnets KEK#10 to #15, which has the same splice design as DAS207 and DAS208 constructed at BNL, showed rather large ramp rate dependence. Quenches at high ramp rate were inside the G10 box where the cable has been solderfilled to increase the mechanical stability. Figure 4 shows the ramp-rate dependence of the quench current of magnets KEK#13, #14, and #15. The ramp-rate quench behavior of these magnets are similar to that of magnet KEK#10.

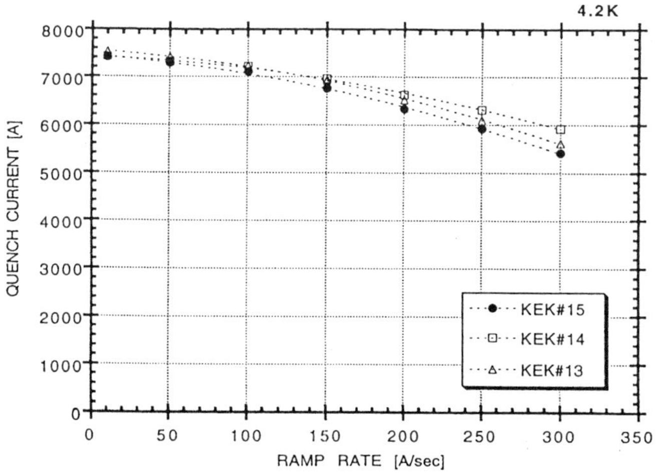

Figure 4. Quench test at varying ramp rates for the 1 m dipole magnets KEK#13 to #15.

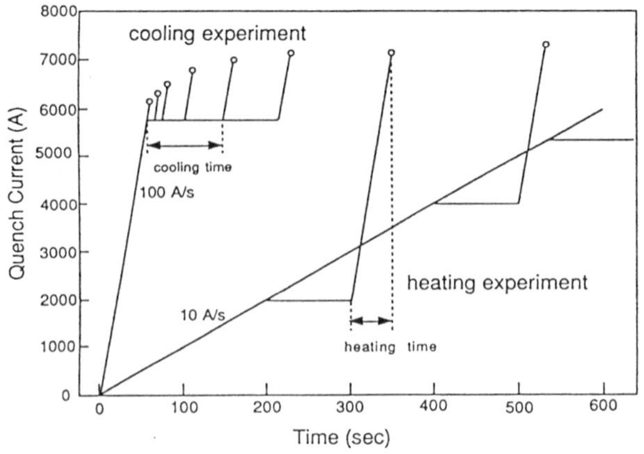

Figure 5. Special ramp pattern of "cooling" and "heating" experiments.

HEATING AND COOLING EXPERIMENTS

For understanding the quench behaviors of these magnets, special ramping patterns, so called "cooling" and "heating" experiments, were introduced at SSCL.[4] Figure 5 shows the conceptual procedures of these experiments. In the heating experiment, the magnet was ramped up to some current just below the quench current with ramp rate nominally 100 A/s. After a dwell period, it was then ramped up again to the quench. The magnet was heated up by eddy-current loss during the ramping up operation. During the dwell, there is no heating and the magnet cooled down. When the magnet was ramped up again, the magnet quenched, but at a higher current, because of the cooldown during the dwell period. The quench current can be plotted as a function of dwell time. In the heating experiment, the magnet was ramped up at low ramp rate 10 A/s to the starting current of high ramp rate and then held there for about 100 sec. After this dwell, the magnet was ramped rapidly up to the quench. The quench currents were plotted as a function of heating time. Figure 6 shows the results of the cooling experiment for the magnet KEK#10, where the current dwells at 5800 A, 6490 A and 6700 A and the ramp rate was 100 A/s. Figure 7 shows the results of heating experiment of the magnets KEK#10 where ramp rate are 100 A/s and 150 A/s. The quench currents decay in about 10 seconds.

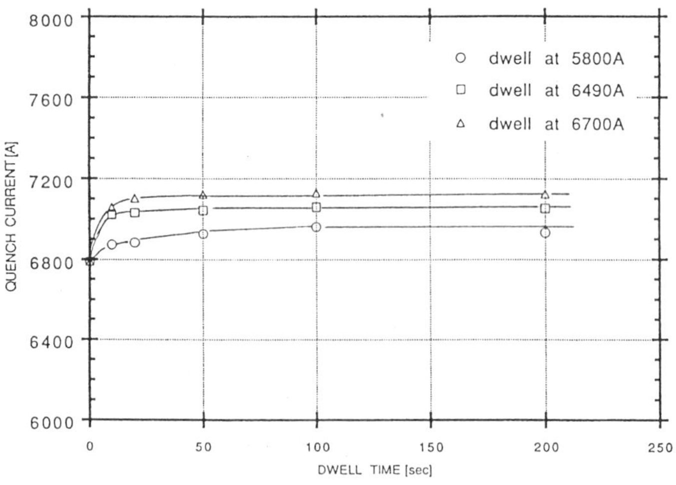

Figure 6. Result of "cooling" experiment for the 1 m dipole magnet KEK#10.

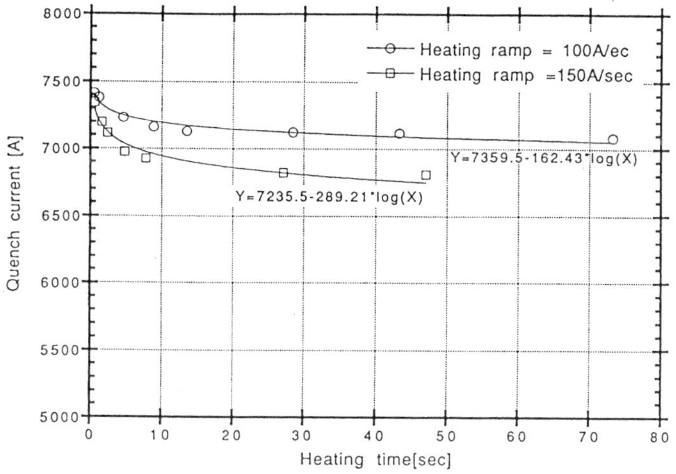

Figure 7. Result of "heating " experiment for the 1 m dipole magnet KEK#10.

HEAT INDUCED QUENCH TEST

Heat-induced quench tests were performed using spot heaters that were installed in the midplane of the inner coils of magnets KEK#7' and #15 and in the splice part of magnet KEK#13Y. Selected parameters of superconducting cable used in magnets KEK#7', #13Y, and #15 are shown in Table 1. Figure 8 (a) shows the electric spot heaters installed in magnets KEK#7' and #15. The heating element is the narrow part of a 0.05-mm thick stainless-steel strip. Both sides of the strip are copper plated to localize the heater power to the narrow heating area. The heating element is laminated between layers of 0.05-mm polyimide film. The inner coil is wrapped with 0.1-mm thick polyimide film. The cable in the coil is also

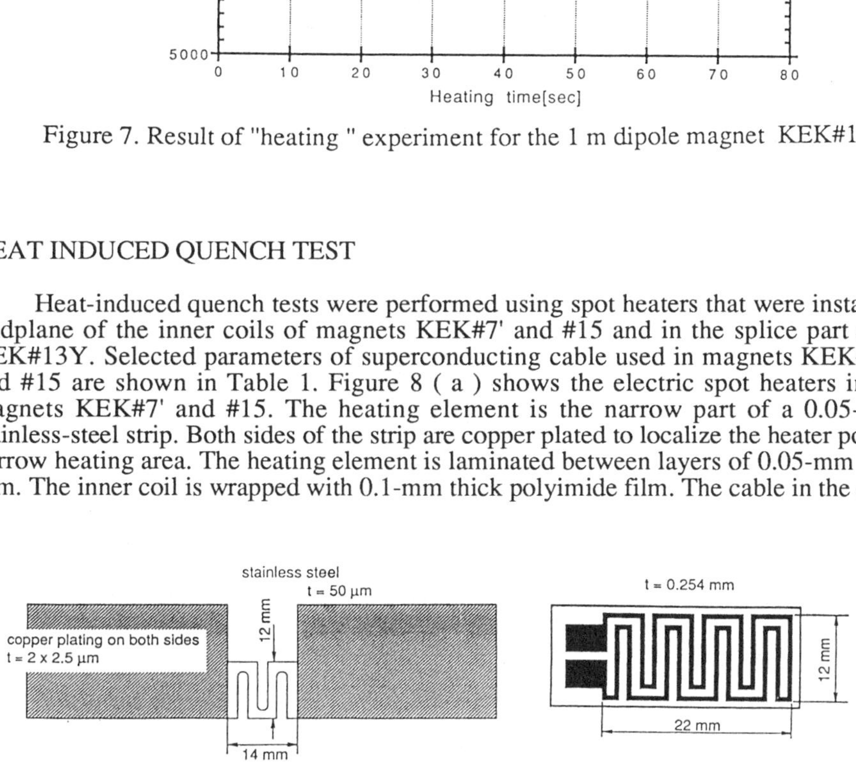

(a) heater installed in inner coil midplane of the magnet .

(b) heater installed in the splice part of the magnet KEK#13Y.

Figure 8. Spot heaters used in heat induced quench tests .

Table 1. Selected parameters of cables for the magnet KEK#7, #13Y, #15

Magnet	(Inner coil) Cu / SC	Filament dia. (μm)	RRR	(Outer coil) Cu / SC	Filament dia. (μm)	RRR
KEK#7	1.4	5.8	162	1.8	5.9	174
KEK#13Y	1.3	6.0	147	1.7	6.3	45
KEK#15	1.5*	6.2	160	1.8 *	6.3	178

* 0.5 % Cu-Mn Matrix + Cu Sheath

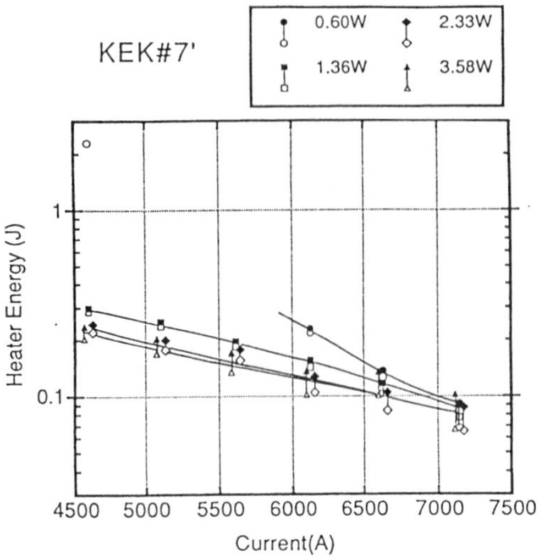

Figure 9. Induced quench energy for the 1m dipole magnet KEK#7'.

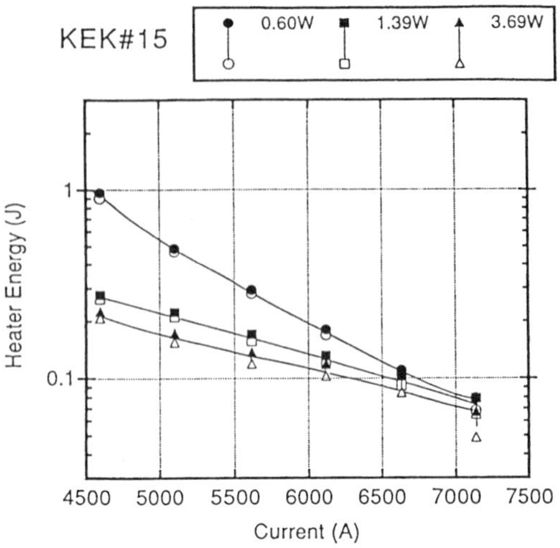

Figure 10. Induced quench energy for the 1 m dipole magnet KEK#15.

wrapped with one 0.05-mm and two 0.025-mm layers of polyimide film. The input energy to the spot heater is transfered to the cable through 5 layers of polyimide film (total thickness is 0.25 mm). Figure 8 (b) shows a thermofoil heater, flexible heating element consisting of an etched-foil resistive element laminated between layers of Kapton. This heater was set directly on the surface of the ramp section of the inner cable, which was soldered to increase mechanical stability. The splice part was double half overlapped with 0.025-mm polyimide film and then encased in a G10 holder. By keeping the magnetic current at constant values from 4500 A to 7000 A, the heater input energy is increased step by step up to the quench by increasing the heating time with constant heater current. In Figures 9, 10, 11, and 12, the maximum heater input energy at which the magnet did not quench and minimum heater input energy at which the magnet quenched are shown by white and black figures, respectively, as a function of the magnet current. Figure 9, 10 show induced quench energy of the magnets KEK#7' and #15 as a function of the magnet current where heating powers 0.6 W, 1.4 W,

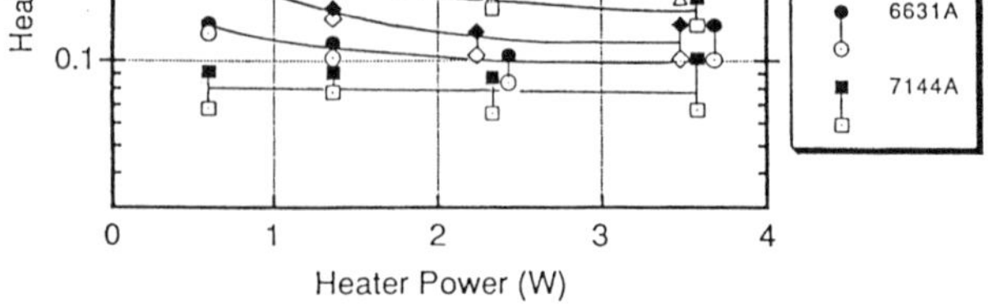

Figure 11. Heater power dependence of induced quench energy for the 1 m dipole magnet KEK#7'.

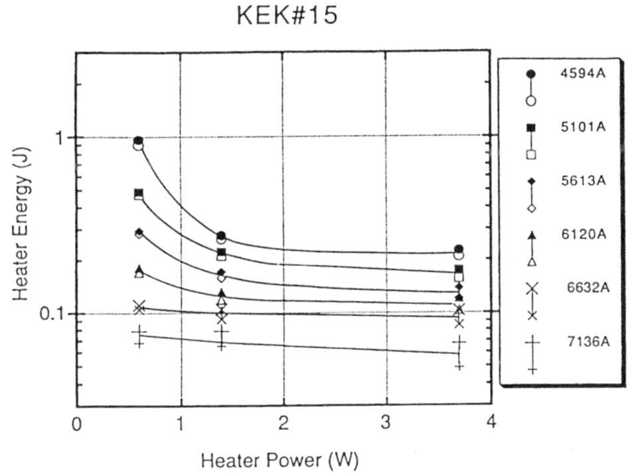

Figure 12. Heater power dependence of induced quench energy for the 1 m dipole magnet KEK#15.

2.3 W and 3.6 W are chosen as parameters. When the heating power was low, induced quench energy is high especially at low magnet current. We can also plot the induced quench energy of the magnets KEK#7' and #15 as a function of heater power as shown in Figure 11, 12 respectively. These figures shows that at the higher magnet current of 7000 A there is no heater-power dependence of induced quench current, but at lower magnet current of 4500 A, induced quench current increase rapidly by decreasing heater power. There is no difference in the induced quench energy between the magnets KEK#7' and #15 except at low heater power 0.6 w and low magnet current 4500A. Under these conditions the magnet KEK7' did not quench at the heater power of 2.3 Joule, the limit of the heater power supply. This difference may come from difference of longitudinal thermal conductivity in the cable due to the difference of matrix material. Figure 14 shows the induced quench energy of the magnet KEK#13Y, where heater power is 0.9 W. The quench heater energy of the magnet KEK#13Y is about one order lower than it of KEK#7' and #15 at high magnet current. If we take into account the following fact that : (a) in the magnet KEK#7' and #5 input heater power is

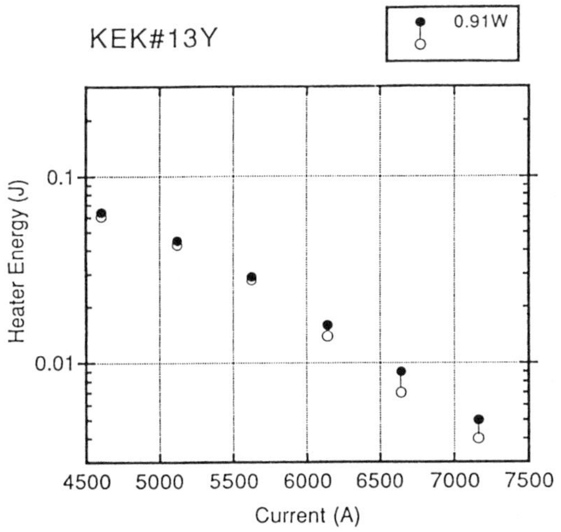

Figure 13. Heater power dependence of induced quench energy for the 1 m dipole magnet KEK#13Y.

transfer to both upper and lower coils, only a half of the input energy reached the conductor through the total of 0.25-mm polyimide film ; (b) in the magnet KEK#13Y the input heater is set directly on the soldered ramp section ; (c) the magnetic field at the ramp section is between 7.2 to 5.7 T and midplane of inner coil is 6.8 to 2.9 T. We consider this difference to be not so large.

CONCLUSION

We have performed the ramp rate quench tests for magnets KEK#3 to #15 with various types of end design and cables to investigate the stability at high-ramp-rate operation. It is very effective for the improvement of mechanical stability to solder the weak part of the cable, but the eddy-current heating at soldered part causes quenching at high-ramp-rate operation. We found it is very important for good performance at high-ramp-rate operation to improve the cooling at the splice ramp section. In 1-m long KEK magnets, there is no conductor dependence in the ramp-rate quench test. By "cooling" and "heating" with special ramping patterns, we found the time constant of cooling at splice part is about 10 seconds. The results of heat-induced quench tests using the spot heaters give us valuable information about the stability of the magnet against heat input to the coil.

ACKNOWLEDGMENTS

The authors express their gratitude to Professor H. Sugawara for his continuous support and encouragement. They also wish to thank Dr. E.Willen and Dr. R.Coombes for their helpful advice and useful discussions.

REFERENCES

1. K. Hosoyama et al., "Fabrication and test of 5-cm aperture 1-m long SSC collider dipole magnet," in Supercollider 3, John Nonte, ed., Plenum Press, New York p615, (1991).
2. K. Hosoyama et al., "A status report on the development of 5-cm aperture, 1- m long SSC dipole magnet at KEK" in : Supercollider 5, P. Hale, ed., Plenum Press, New York, p229 (1994).
3. K. Hosoyama, et al., "Fabrication and testing of 1-m-long SSC model dipole magnets," to be presented at CEC conference, Columbus (1995).
4. M. Wake "Special ramp tests of DCA312" in SSC MSD Technical Note, Jan. 29, 1993
5. R.C. Gupta, S.A. Kahn, and G.H. Morgan, "SSC 50 mm dipole cross section," in : Supercollider 3, John Nonte,ed., Plenum Press, New York and London, p587, (1991).

A SUPERCONDUCTING BENDING MAGNET SYSTEM FOR A COMPACT SYNCHROTRON LIGHT SOURCE

M. A. Green and A. A. Garren

E. O. Lawrence Berkeley National Laboratory
University of California
Berkeley, CA 94720

E. M. Leung and D. D. Madura

ADO Lockheed Martin
Rancho Bernardo, CA 92125

D. B. Cline, J. J. Kolonko and L. C. Schachinger

Particle Beam Lasers, Inc.
Northridge CA 91324

ABSTRACT

High intensity, high energy X-rays for use in protein crystallography, nano-machining and medical applications, such as non invasive coronary angiography, can be produced by a 1.2 to 1.5 GeV electron storage ring compact light source with 6 to 8 tesla superconducting bending magnets. Because the bending magnets are to be superconducting, the storage ring energy can be over factor of two lower than a conventional storage ring that delivers same photon energy. The ring, which has superconducting bending magnets, is smaller in circumference and has the advantage of having fewer particles in the ring for a given x ray source intensity. The proposed storage ring is a separated function accelerator ring with six superconducting bending magnet units. Conventional quadrupoles and correction elements would be located between the bending magnets. Because the synchrotron radiation is generated in the bend, the superconducting bending magnets must have a warm vacuum chamber for the electron beam. Variations of a superferric magnet design have been studied for this application. This report presents a superferric H magnet design that can produce good quality magnetic field in a region that is 50 mm high by 100 mm wide. This modified superferric H magnet design has saturated iron poles but the magnetic flux is returned from one pole to the other through an unsaturated iron return path. The dipole magnet required for a compact storage ring must be physically short (380 mm long), and the field must fall off rapidly at the ends of the magnet. This report describes a preliminary design for a pair of 6.894 tesla, thirty degree bending magnets in a common vacuum vessel for use in a 1.5 GeV compact storage ring light source.

DIPOLE REQUIREMENTS FOR A COMPACT STORAGE RING

Figure 1 shows a diagram of an industrial 1.2 to 1.5 GeV compact electron storage ring with superconducting bending magnets and room temperature quadrupoles and sextupoles. The light source ring design shown in Fig. 1 resulted from cost optimization studies that were combined beam dynamics studies[1,2] The required central induction of the dipole at a ring energy of 1.5 GeV is 6.894 T. The dipoles that are shown in Fig. 1 and that are presented in this report have a design central induction commensurate with 1.5 GeV storage ring energy. The peak critical energy of the x rays generated by the ring shown in Fig. 1 is 10.4 keV (at a beam energy of 1.5 GeV). Intense synchrotron x rays with an energy up to 40 keV are produced at six points around the ring. The ring is compact; its circumference is 26 meters. There is adequate room in the straight sections for rf cavities and a system to inject a 100 MeV electron beam into the ring. The ring shown in Fig. 1 has a design beam current of 20 mA. As a result, the synchrotron x rays are extracted from the downstream dipole of the pair. The x rays from the upstream dipole of the pair will be absorbed by the vacuum chamber passing through the dipoles. If the dipoles are separated (with focusing elements between them), the x ray beam intensity can be increased over an order of magnitude, but the ring circumference and cost will increase.

Each dipole in the pair shown in Fig. 1 has a magnetic length of 380 mm along the beam orbit. The dipoles are straight magnets so that the lattice defocusing comes from the edge of the dipole. The key to making a compact electron storage ring is the fabrication of short high field dipoles that have the end field characteristics similar to room temperature, copper and iron dipoles. Within the magnetic length magnet, the integrated field quality has to be very good. Beam dynamics studies suggest that the integrated field has to be good to one part in 10000 over a region that is ±20 mm wide around the electron beam.

Picture frame magnets have a very good field uniformity over the entire width of the magnet pole until the iron in the pole and return yoke saturates. The design for the dipole presented here is based on a modified picture frame dipole proposed by Pavel Vobly at INP Novosibirsk[3,4]. The Vobly dipole has saturated iron in the poles, but shield coils keep the magnetic flux in the pole until it can be returned by an unsaturated iron return yoke. As a result, the end field fall off is similar to that of conventional iron dominated dipoles. This is very important for a dipole with a magnetic length of only 0.38 meters.

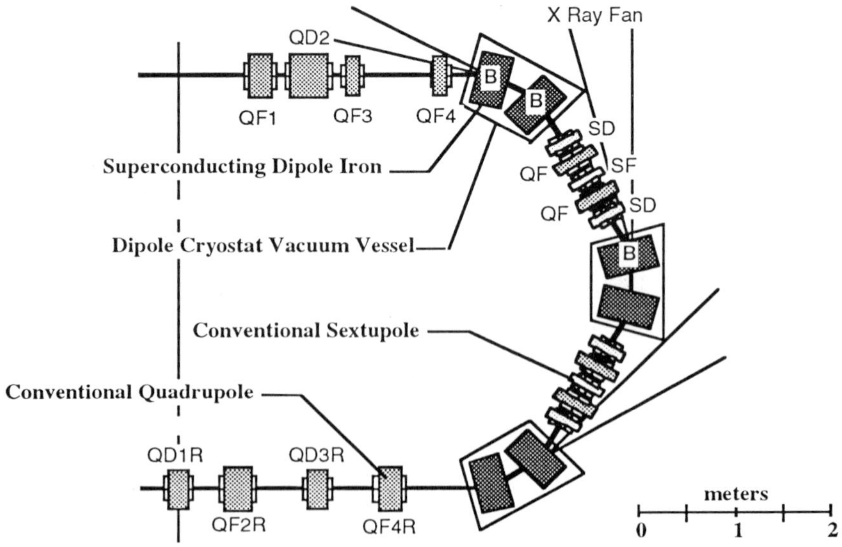

Figure. 1 Half of a 26 meter Circumference Compact Storage Ring Showing the Superconducting Dipoles, and the Conventional Quadrupoles and Sextupoles.

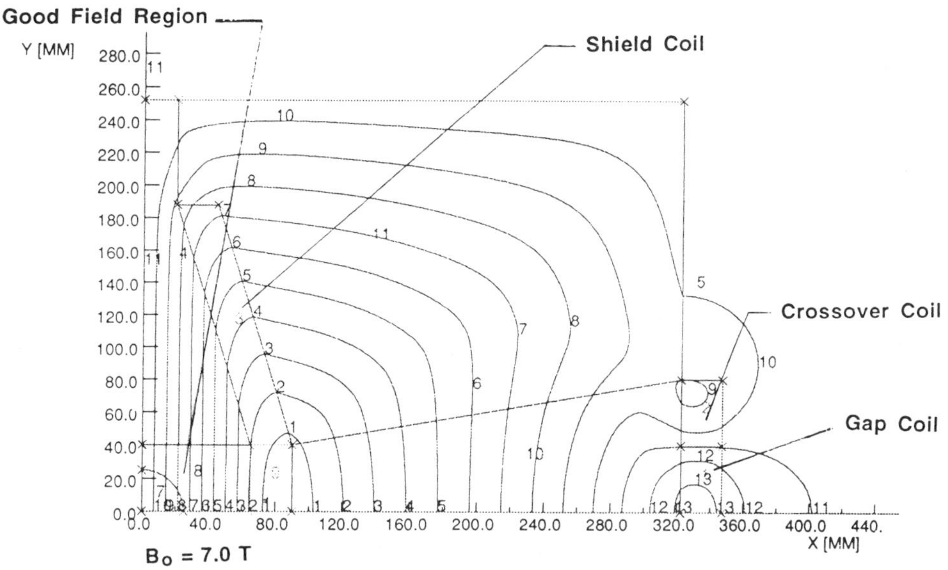

Figure 2 A Quarter Cross-section of the Compact Light Source Dipole Showing the Magnetic Flux Lines at a Central Induction of 7.0 T.

THE SUPERCONDUCTING DIPOLE

Figure 2 shows a quarter cross-section through the center (in the longitudinal direction) of a Vobly type of dipole designed to produce a central induction of 7.0 tesla. Fig. 2 shows the flux lines generated by the magnet when the field is uniform across the pole and the central induction is 7.0 tesla. The dipole shown in Fig. 2 has three different coils: 1) The gap coils will generate the magnetic field within the gap. The current in the gap coils will be linear with the magnetic induction within the gap. 2) The crossover coils will carry the current from one side of the gap to the other side of the gap. This coil, which is in series with the gap coil, acts in a way similar the end crossover coils in a conventional H dipole magnet with unsaturated iron poles. The gap and crossover coil currents cancel each other, away from the magnet gap. 3) The primary function of the shield coil is to keep flux from leaking out of the pole when its iron becomes saturated. The flux lines from gap stay perpendicular to the pole face even within the pole iron, when the current in the shield coil is set to the proper value. The shield coil current will be nearly zero when the iron in the pole is unsaturated. As the pole iron saturates, the current in the shield coil increases linearly as the central induction minus the saturation induction of the iron. The shield coil will be powered separately from the gap and crossover coils.

Figure 3 shows a three dimensional view of the three coils. The iron boundary is shown as dashed line. The arrows in Fig. 3 show the direction of current flow in each of the coils. From Fig. 3, one can see that coils are not complex in their construction. All of the bends occur in a single bend. Features such as turned up ends have been avoided. Figure 4 shows the iron assembled around the dipole coils

The dipole shown in Fig. 2 through 4 will generate a uniform field over a wide range of central inductions, provided that the current density in the shield coil system is correctly chosen with respect to the current density in the gap and crossover coil system. Fig. 2 shows that the flux density in the pole is the same as the flux density within the gap. The field in the magnet gap is very uniform. The current density in the shield coil is 166.6 MA per square meter while the current density in the gap and crossover coils is 270.0 MA per square meter. The excitation function for the shield coil can be modified by changing the height and slope angle of the shield coil[5].

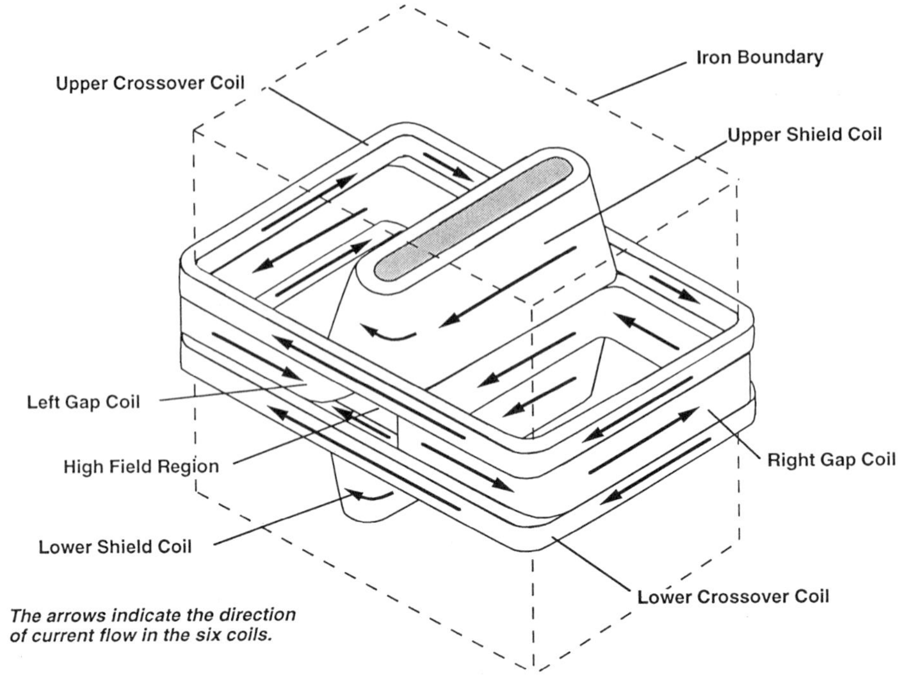

Figure 3 A Three Dimensional View of the Coils in a Compact Light Source Dipole

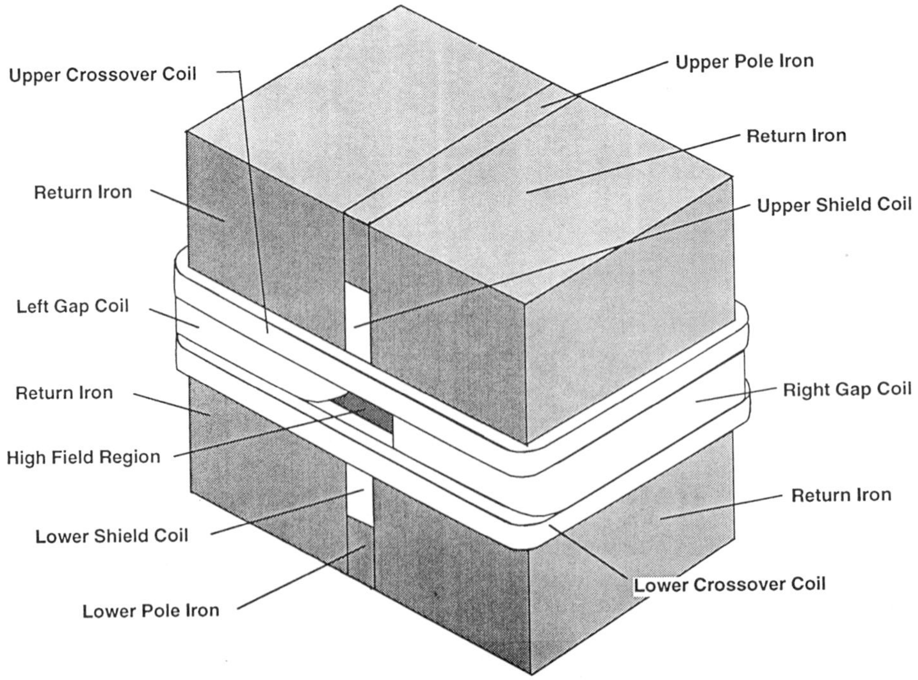

Figure 4 A Three Dimensional View of the Fully Assembled Light Source Dipole.

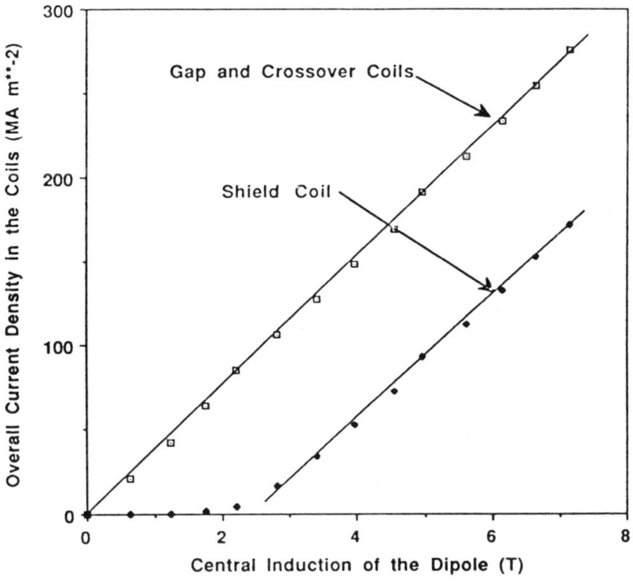

Figure 5 The Current Density in the Gap and Crossover Coils and the Shield Coils Needed to Achieve a Uniform Dipole Field as a Function of Central Induction.

Figure 5 shows the current density in the gap and crossover coils (The gap and crossover coils have the same current density.) and shield coils needed to produce a uniform dipole field as a function of the central induction of the dipole. One can adjust the current density in the shield coil with respect to the current density in the gap and crossover coils such that the sextupole generated in the gap is zero. If the gap coil height and slope angles are correctly selected the higher multipoles from decapole on up will be very small.

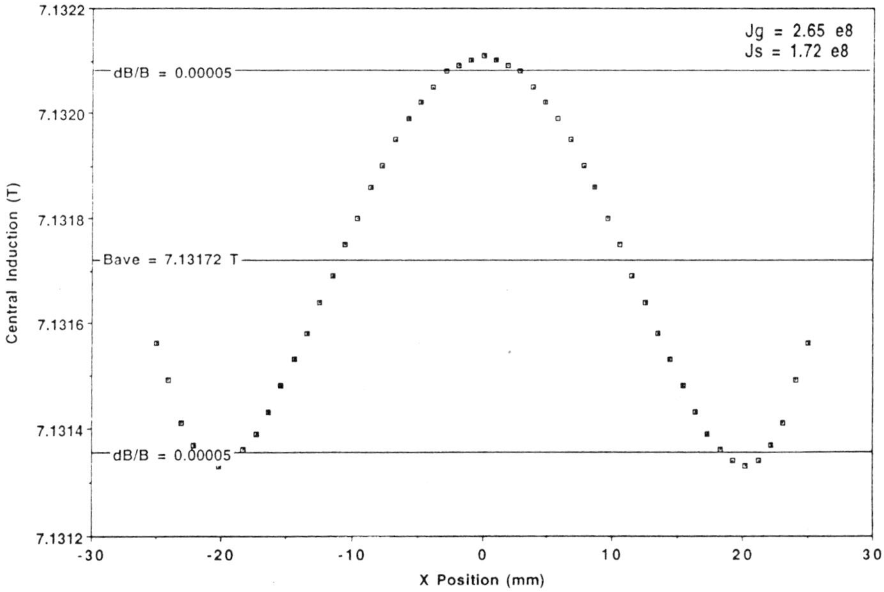

Figure 6 The Two Dimensional Magnetic Induction on the Beam Plane as a Function of X Distance along the Pole from the Magnet Center.
(The shield coil surrent is set for uniform Field.)

Table 1 Compact Light Source Superconducting Dipole Parameters

Number of Bending Units	6
Number of Dipoles per Bending Unit	2
Dipole bend Angle (degrees)	30.0
X Ray Fan Angle (degrees)	20.0
Required Integrated Induction (T m)	2.620
Design Induction at Center (T)	6.894
Magnet Cold Gap (mm)	80.0
Magnet Iron Pole Width (mm)	130.0
Nominal Pole Length (mm)	~376
Shield Coil Height (mm)	148.0
Coil Thickness (mm)	25.0
Stored Energy* (kJ)	~190
Peak Induction in Winding* (T)	~7.0
Iron Length (mm)	376.0
Iron Width (mm)	644.0
Iron Height (mm)	504.0
Estimated Cold Mass per Dipole (kg)	~1100

* at the design central induction for a 1.5 GeV storage ring

To check this, the two dimensional field uniformity was calculated at a central induction 7.132 T. A field uniformity of 0.6 parts in 10000 was calculated in a region that was ± 30 mm about the center of the dipole. (See Figure 6) This means that sextupole, decapole and 14 pole were very small. The even multipoles (quadrupole, etc) are zero by symmetry.

Table 1 presents the parameters for the 7 tesla dipole shown in Figs. 2, through 4. When the electron beam energy in the ring configuration shown in Fig. 1 is 1.5 GeV, the dipole central induction will be 6.894 T. The required integrated dipole for each of the twelve ring dipoles is about 2.62 T m If one bases the coil design shown in Figs. 2 and 3 on a cable made from SSC inner conductor (with a copper to superconductor ratio of 1.4),

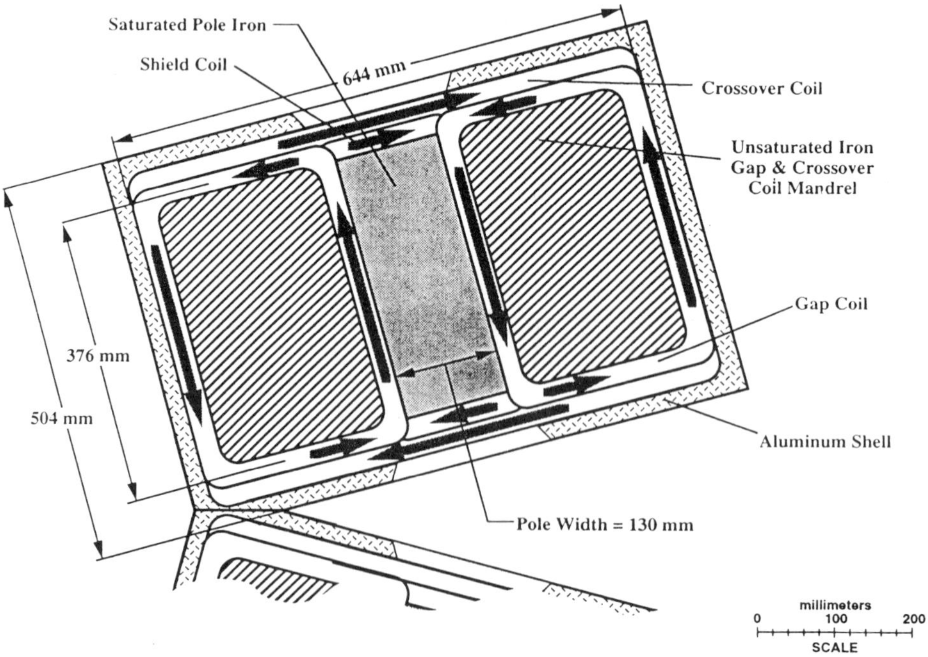

Figure 7 Dipole Cross-section in the Plane of the Electron Beam.

the dipole will operate at about 87% of its critical current along the load line (based on a metal packing fraction of 0.7 and an operating temperature of 4.4 K), when the dipole central induction is 6.9 T. It appears that the SSC inner superconductor can be cabled into conductor of almost any desired current for this application.

Figure 7 is a Cross-section of the dipole on the electron beam orbital plane. Fig. 7 shows that the cap and crossover coils can be wound as flat pancakes with small radius bends in the corners. The shield coils are more difficult. These coils can be wound in a stair like structure and still have the dipole meet its field uniformity requirements. The primary magnetic force acts in a direction to force the two gap coils apart. There is a strong force on the iron pole pieces to draw them together. Preliminary force calculations suggest that the magnetic forces can be carried by a 19 mm thick aluminum shell. The shell can also insure that the coils maintain contact with the iron during cool down and excitation

The cost of a short compact light source dipole is primarily a function of the pole width, not the gap between poles. Since most of the iron in the magnet is unsaturated, the overall iron height and width are functions of the pole width and the saturation induction of the iron. Most of the dipole cold mass in the dipole is iron. The Vobly type dipole is not very efficient in its use of superconductor. Unlike most superconducting dipoles, the cost of the superconductor is not a large part of the capital cost of this magnet.

CONCLUSION

A compact synchrotron with 6.9 tesla superconducting bending magnets appears to be feasible. A short high field dipole magnet is the key element for a compact synchrotron light source storage ring. The short high field bending magnet must have end field characteristics that are similar to short conventional low field magnets. The magnetic flux must be constrained in the poles, even when they are saturated while the magnet is running at high field. The modified picture frame magnet design presented in this report has this characteristic. The primary disadvantage of the magnet design presented in this report is that it requires two separate power supplies in order for the magnet to produce good field quality over a range of inductions. At low field, when there is almost no current in the shield coil, the shield coil current can be used to eliminate the remnant field sextupole due to magnetization of the iron and persistent currents in the superconductor. The Vobly dipole design concept, discussed here, appears to be viable for dipole magnets that have a length to gap ratio of greater than four.

ACKNOWLEDGMENTS

This work was performed with the support of the Director of the Office of Energy Research, Office of High Energy and Nuclear Physics, High Energy Physics Division, United States Department of Energy under contract number DE-FG03-94ER81826 and contract number DE-AC03-76SF00098.

REFERENCES

1. A. A. Garren, et al. "A 1.5 GeV Compact Light Source with Superconducting Bending Magnets," to be published in the Proceedings of the 1995 Particle Accelerator Conference, Dallas TX, USA, 1-5 May 1995
2. M. A . Green and A. A. Garren, "Optimization of Superconducting Bending Magnets for a 1.0 to 1.5 GeV Compact Light Source," to be published in the Proceedings of the 14th Magnet Technology Conference, Tampere, Finland, 11-16 June 1995
3. G. N. Kulipanov, N. A. Mezentsev, L. G. Morgonov, et al. "Development of a Superconducting Compact Storage Rings for Technical Purposes in the USSR," Rev. Sci. Instrum. 63 (1), p 731 (1992)

4. P. D. Vobly, private communication concerning compact synchrotron dipole magnet designs
5. M. A. Green, "Superconducting Dipole Magnets for Compact Light Source Storage Rings," Lawrence Berkeley Laboratory Internal Report LBID-2005, 9 Feb. 1994

HIGH GRADIENT, LARGE APERTURE QUADRUPOLES FOR THE NSCL SUPERCONDUCTING SPECTROMETER

B. Zhang, A. F. Zeller, J. DeKamp, P. Johnson, B. Sherrill, R. Swanson, R. Zink

National Superconducting Cyclotron Laboratory
Michigan State University
East Lansing, MI 48824, USA

ABSTRACT

The National Superconducting Cyclotron Laboratory (NSCL) is constructing a magnetic spectrometer, the S800, which will consist of two superconducting dipoles and two superconducting quadrupoles. One of the quadrupoles (Q1) has a 24 cm bore and a 34 cm length. The peak gradient of 21 T/m is produced by 2100 turns of 0.7 mm wire operating at 79 A. The maximum stored energy is 64 kJ. The second quadrupole (Q2) has a 42 cm bore and a 30 cm length. Using 0.85 mm wire with 2700 turns operating at 87 A it produces a 6.8 T/m gradient. The stored energy is 146 kJ. A superconducting sextupole is inserted in Q2, which uses 0.3 mm wire and operates at 2 A . All elements are assembled into the same cryostat.

INTRODUCTION

The NSCL's K1200 cyclotron is capable of providing high quality heavy ion beams with rigidities of 1.6 GeV/c. The highest energy beams will be 200 MeV/nucleon, or 1.28 GeV/c. In order to study the properties of single particle states populated in reactions at these energies, a magnetic spectrometer with an energy resolution E/ΔE of 10,000 is necessary. The spectrometer system chosen for the S800 is a QQDD configuration.[1] A strongly y focussing Q1 immediately after the target allows a large solid angle (20 msr) and an x focussing Q2 before the dipoles makes efficient use of the dipoles.

MAGNETIC DESIGN AND CONSTRUCTION

The quadrupoles Q1 and Q2 have high gradients or large apertures. This necessitates the use of superconducting magnets, since space is not available for longer quadrupoles. The choice of quadrupoles is limited to three basic types: a) cos(2θ) type, where the field is shaped by the coil, b) Panofsky type,[2] which is a square or rectangular current sheet approximation, and c) The NSCL beamline type,[3] where the iron pole tip primarily shapes the field. We have decided against cos(2θ) quadrupoles because of the high current operation and its helium requirements. We have also investigated superconducting Panofsky quadrupoles. A Panofsky design was not chosen due to the

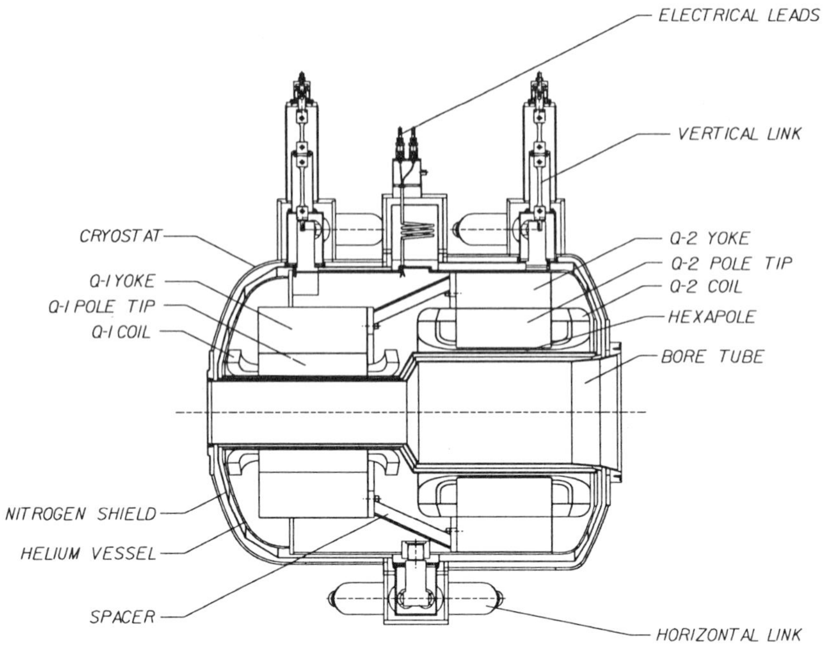

Fig. 1. View of the longitudinal section of Q1, Q2 and sextupole.

Table 1. The parameters of Q1 and Q2 magnets

Item	Q1	Q2
Peak gradient	21 T/m	6.8 T/m
Pole tip diameter	0.24 m	0.42 m
Length of iron	0.34 m	0.30 m
Outside diameter	0.68 m	0.88 m
Wire diameter	0.7 mm	0.85 mm
Cu:SC (wire spec.)	4.31 : 1	4.4 : 1
Ic @4.0T	240 A	—
Ic @3.5T	—	275 A
Turns per coil	2100	2700
Operation current	79 A	87 A
Stored energy	64 kJ	146 kJ
Inductance	23.9 H	38.8 H

Table 2. The parameters of the sextupole magnet

Item	Sextupole
Wire diameter	0.3 mm
Cu:SC (wire spec.)	6:1
Turns per coil	2900
Operation current	2 A
Stored energy	112 J
Inductance	56 H

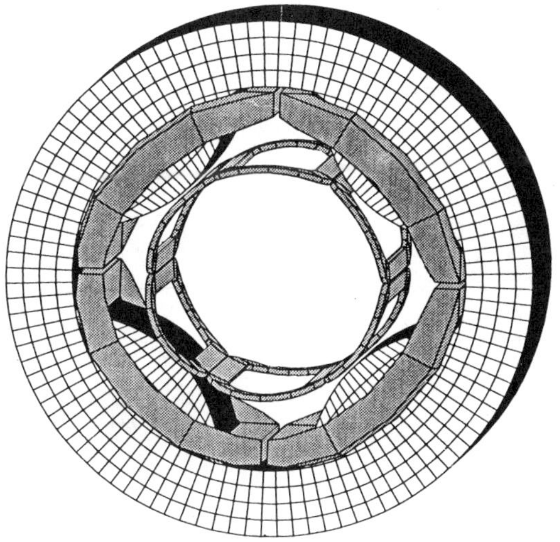

Fig. 2. The sextupole in Q2.

sensitivity of the field quality to errors in conductor placement. Our conclusions from these studies and from experience with our cold-iron, iron dominated superconducting beamline quads is that a larger version of the latter type is our choice for both Q1 and Q2 quadrupoles. For the S800 system, the dominant aberration at the focal plane is x/θ^2. We inserted a sextupole in Q2 to correct this aberration. Fig. 1 shows the details of the quadrupoles in their common cryostat.

We have studied the design of the S800 spectrometer quadrupole magnets in detail. The original design is based on two dimensional magnetic field calculations performed with POISSON.[4] We recently did the three dimensional calculation using the programs TOSCA and VF/OPERA.[5] We used six 20-node bricks and two straight bars to model Q1 and Q2 coils. The parameters for Q1 and Q2 magnets are given in Table 1. These magnets are similar to those used in the CEBAF Hall C spectrometer,[6] only ours are shorter.

For the sextupole we used constant perimeter end racetrack model. The parameters for the sextupole magnet are given in Table 2.

The sextupole is placed within Q2, shown in fig. 2. TOSCA was used to compute the total magnetic force on each coil. Since the fields produced near the straight sections are stronger than the fields produced near the ends of the coils, the forces acting on the straight sections are larger. The quadrupole coils will be shimmed against each other and radially against the yoke. The radial shimming is more important since the forces in this direction are higher. The forces per unit length acting on the straight section of Q1 and Q2 are 283 and 255 kN/m, respectively.

The forces acting on the straight section of the sextupole coil are more complicated. Depending on the location of the coil, the force can be radially inward or outward, or in an azimuthal direction. The maximum force on the straight section of the sextupole coil which is mounted onto the bore tube is 9.5 kN/m.

The random-wound coils are wet-wound with epoxy on a simple winding table. All of the required Q2 coils have been wound as shown inFig. 3. All 2700 turns of the 0.85 mm conductor are wound in three hours. Winding of Q1 coils is underway and the completed Q1 and Q2 will be tested in an open dewar. The sextupole coil winding fixture has been completed.

The low carbon iron assemblies were machined with numerically controlled machines. Fig. 4 shows Q2 pole tip and yoke steel. The magnet proved to be inexpensive and straightforward to build. Since the field is shaped primarily by the iron, conductor placement is less important. A high precision stainless steel spacer positions the magnets with respect to each other.

Fig. 3. A Q2 coil. The length of the straight section is 30 cm.

QUENCH CALCULATION

An evaluation of the quench behavior is needed for coil safety. A dump resistor is not used for quench protection because a high field and small coil size make the quench propagation fast. The energy spreads more uniformly throughout the coil. The higher stored energy requires larger wire and higher operating current for protection in case of a quench. The code QUENCH[7] has been used to evaluate the worst case scenario. Fig. 5 shows the internal voltage of Q1 increasing to 637 volts after quenching, which is still lower than the voltages of the NSCL standard beamline quads or the S800 beamline quads. Since the NSCL standard beamline quads have survived hundreds of quenches and the S800 beamline quads dozens of quenches without damage, no problem is anticipated for Q1 or Q2.

Since the sextupole coils are inside Q2, mutual inductance in the coils is significant and must be evaluated. The mutual inductance has been calculated to be 14.5 H. The

Fig. 4. Q2 pole tip and yoke steel. The pole tip diameter is 42 cm and outside diameter is 88 cm.

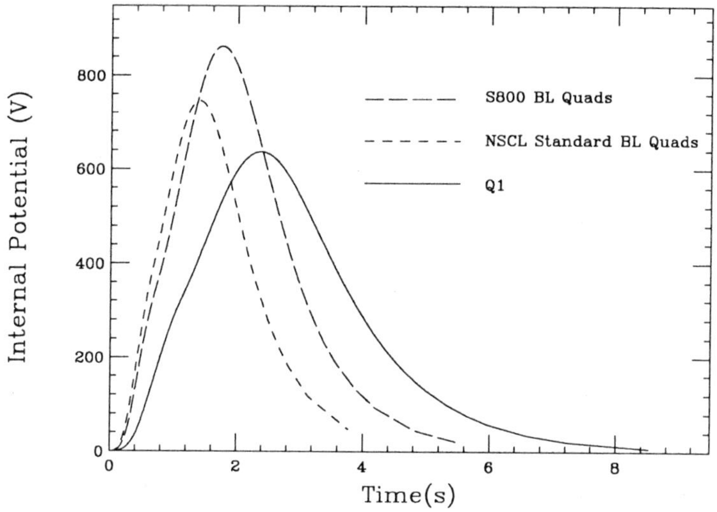

Fig. 5. The comparison of the internal voltages in Q1, NSCL standard beamline quads and S800 beamline quads after quenching.

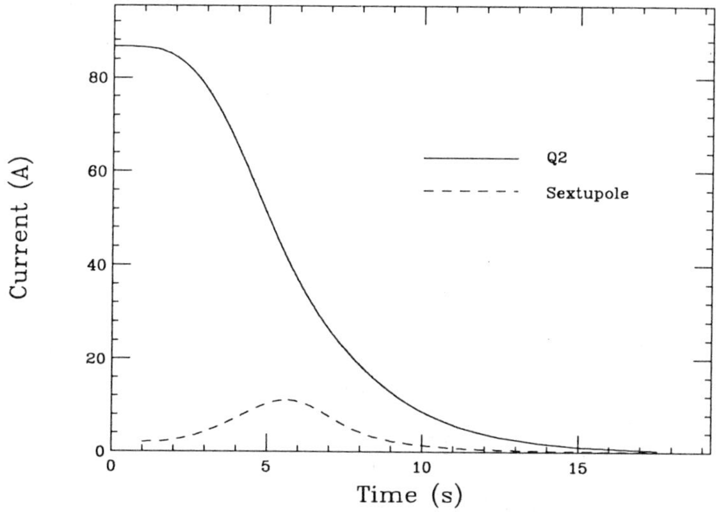

Fig. 6. Decay of currents in Q2 and sextupole after quenching.

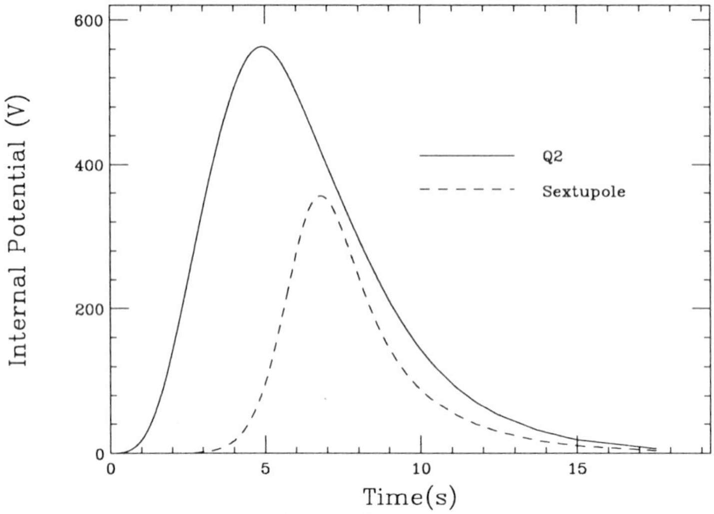

Fig. 7. Internal voltages in Q2 and sextupole after quenching.

worst case is Q2 quenching first, the energy from Q2 transferring to the sextupole, as seen in Fig. 6. The internal voltages Q2 and sextupole are shown in Fig. 7 and are not excessive.

FIELD MAPPER

To achieve the required energy and angular resolution for nuclear physics experiments, the magnetic field must be accurately known. A mapper consists of a gaussmeter and a dual axis motor controller. The three axis hall probe of the gaussmeter has been calibrated to absolute field with NMR measurements. The mapper was tested by using an S800 beamline quad, which is shown in Fig. 8.

Fig. 8. The field mapper for Q1, Q2 and the S800 beamline quads.

CONCLUSIONS

Two large aperture, cold iron quadrupoles have been designed and assembly is underway. The coil winding is almost complete and the coils are being prepared for installation into the iron assemblies. The complete magnets will be tested in an open dewar before being installed in a common cryostat. Two dimensional and three dimensional magnetostatic calculations were used to produce a design consistent with the required optics. A magnetic field mapper has been constructed to confirm this. The coils are calculated to be self protecting in case of a quench.

ACKNOWLEDGMENT

This work supported by the National Science Foundation under Grant No. Phy92-14992. The assistance of Dr. F. Marti in producing the three dimensional TOSCA calculations is appreciated.

REFERENCES

1. J. A. Nolen, A. F. Zeller, B. Sherrill, J. C. DeKamp and J. Yurkon, "A proposal for construction of the S800 spectrograph," Technical Report MSUCL_694, National Superconducting Cyclotron Lab, July, 1989.
2. L. N. Hand and W. H. Panofsky, "Magnetic quadrupole with rectangular aperture", Rev. Sci. Instr. 33, 927(1959).
3. A. F. Zeller et al., "Design and construction of a low current, cryogenically efficient beamline quadrupole", Proceedings of 10th International Conference on Cyclotron and Their Application, New York: IEEE Press, 1984, p 79-81.
4. R. F. Holsinger, Program POISSON (unpublished) 1979.
5. TOSCA version 6.3, Vector Field, Inc. 1700 N.Farnsworth Avenue, Aurora, IL 60505.
6. S. R. Lassiter et al., "Final design and construction progress for CEBAF's cold iron quadrupoles", IEEE Transactions on Applied Superconductivity, Vol.33, No.1, Match 1993, p 118-121.
7. M. N. Wilson, Computer Simulation of the Quenching of a Superconducting Magnet, RHEL/M 151 (1968)

CRYOGENIC TESTS OF THE g-2 SUPERCONDUCTING SOLENOID MAGNET SYSTEM

L. X. Jia, J. R. Cullen Jr., A. J. Esper, R. E. Meier, C. Pai, L. Snydstrup, and T. Tallerico

Brookhaven National Laboratory
Upton, NY, 11973

M. A. Green

E. O. Lawrence Berkeley National Laboratory
University of California
Berkeley, CA 94720

ABSTRACT

The g-2 muon storage ring magnet system consists of four large superconducting solenoids that are up to 15.1 m in diameter. The g-2 superconducting solenoids and a superconducting inflector dipole will be cooled using forced two-phase helium in tubes. The forced two-phase helium cooling will be provided from the J-T circuit of a refrigerator that is capable of delivering 625 W at 4.5 K. The two-phase helium flows from the refrigerator J-T circuit through a heat exchanger in a storage dewar that acts as a phase separator for helium returning from the magnets. The use of a heat exchanger in the storage dewar reduces the pressure drop in the magnet flow circuit, eliminates most two phase flow oscillations, and it permits the magnets to operate at variable thermal loads using the liquid in the storage dewar as a buffer. The g-2 magnet cooling system will consist of three parallel two-phase helium flow circuits that provide cooling to the following components; 1) the four large superconducting solenoids, 2) the current interconnects between the solenoids and the solenoid gas cooled electrical leads, and 3) the inflector dipole and its gas cooled electrical leads. This report describes a cryogenic test of the two 15.1 meter diameter superconducting solenoids using two-phase helium from a dewar. The report describes the cool down procedure for the 3.5 ton outer solenoid magnet system using liquid nitrogen and two-phase helium. Low current operation of the outer solenoids is discussed.

THE g-2 STORAGE RING MAGNET SYSTEM

An experiment to measure the value of g-2 for the muon has been under construction at the Brookhaven National Laboratory in Upton New York since 1988[1,2,3]. The g-2 experiment magnet system consists of three large superconducting solenoid magnets that form a storage ring, a superconducting inflector magnet for injecting the muon beam into the storage ring, and a string of conventional bending and quadrupole magnets that are used

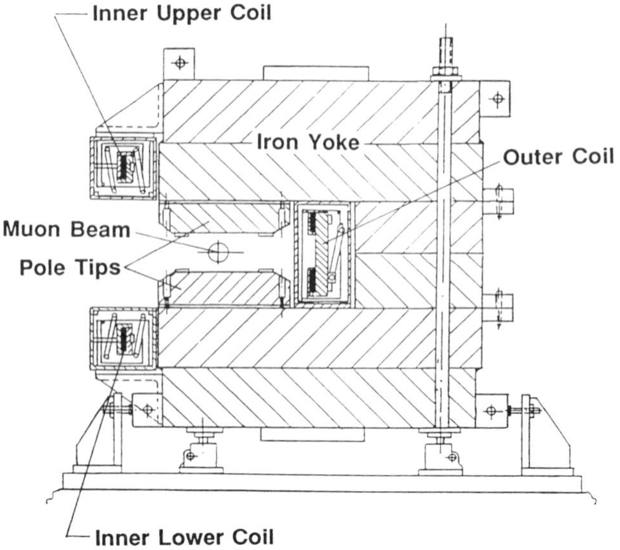

Fig. 1 A Cross-section of the g-2 Storage Ring Bending Magnet Showing the Location of the Three Large Superconducting Solenoid Magnets.

to carry a decaying pion beam from a fixed target at the Brookhaven 33 GeV Alternating Gradient Synchrotron (AGS) to the storage ring that is located in a building some distance from the AGS.

The storage ring consists three large indirectly cooled superconducting solenoids wound with an aluminum stabilized superconductor. There are two 24 turn inner solenoids that have a coil diameter of 13.4 meters and a single 48 turn outer solenoid that has a coil

Fig. 2 The g-2 Storage Ring (June 1995) before the Electron Detectors were Installed inside of the Storage Ring.

Table I Parameters of the g-2 Solenoid Magnets in the Iron Return Yoke

Iron Height (mm)	1570.
Iron Width (mm)	1394
Total Iron Mass for the Ring (metric tons)	681.6
Number of Solenoid Packages	3
Inner Solenoid Coil Radius (m)	6.70
Outer Solenoid Coil Radius (m)	7.55
Pole Width (mm)	560.0
Nominal Gap between the Poles (mm)	180.0
Nominal Beam Bend Radius (m)	7.112
Good Field Region Diameter (mm)	90.0
Dipole Central Induction (T)	1.451
Inner Solenoid Number of Turns (each)	24
Outer Solenoid Number of Turns	48
Coil Current* (A)	5300
Magnet Self Inductance (H)	0.39
Storage Ring Dipole Stored Energy* (MJ)	5.50
Total Solenoid Cold Mass (metric tons)	~6.2

* at the design central induction

diameter of 15.1 meters. The three solenoids are the largest diameter superconducting solenoids that have been built to date. The current in the two inner solenoids will be run in opposition to the current in the outer solenoid creating the storage ring dipole field. An iron return yoke in the shape of a C returns the magnetic flux. Separate iron pole pieces control the quality of the 1.451 tesla induction within the gap. The integrated induction around the storage ring must be good to 1 ppm within a 90 mm diameter region that contains the muon beam[4]. Focusing for the storage ring is supplied by electrostatic quadrupoles mounted within the storage ring beam vacuum chamber.

The open part of the C points toward the center of the storage ring so the electrons that result from the decay of the muons (muons at rest have a decay half life of about 0.06 ms) will spiral inward to the detectors for the experiment. The beam enters the storage ring from the outside through a hole in the C shaped iron return leg and outer solenoid between the two 24 turn coils that are mounted on the outer solenoid cold mandrel. A cross-section of the g-2 storage ring magnet showing the three solenoids is shown in Fig. 1. A photo of the nearly completed g-2 storage ring is shown in Fig. 2.

At the ring entrance, a superconducting DC inflector[5] will locally cancel the field of the main storage ring magnet so that the muon beam enters, as closely as possible, tangentially to the equilibrium orbit of the storage ring. The inflector magnet is designed to "eat" its own flux so that stray field from the inflector is greatly reduced. In addition the inflector has a superconducting shield so that virtually all of the remaining stray field from the inflector is eliminated. Thus, the inflector has no influence on the field in the gap.

Construction of the g-2 solenoids and their cryostats is complete. The outer solenoid has been mounted on the lower part of the iron yoke. The iron back leg of the C and the upper part of the iron yoke has been mounted around the outer solenoid cryogenic vacuum vessel. The two inner solenoids have been mounted on the C shaped iron yoke to form a continuos storage ring dipole as shown in Fig 1. The basic mechanical and electrical parameters of the three solenoid storage ring magnet system is given in Table 1.

CRYOGENIC TEST OF THE OUTER SOLENOID

Since the outer solenoid is to be buried within the C shaped iron yoke, a cryogenic test of the outer solenoid coil package was performed before the outer solenoid cryostat was buried. The refrigerator, which had been previously tested, was not available for cooling the outer solenoid down to liquid helium temperature. It was decided that the outer coil cold mass (about 3.5 metric tons) could be cooled down using cold nitrogen gas, then liquid nitrogen, and finally liquid helium. This cool down was done in the winter of 1995.

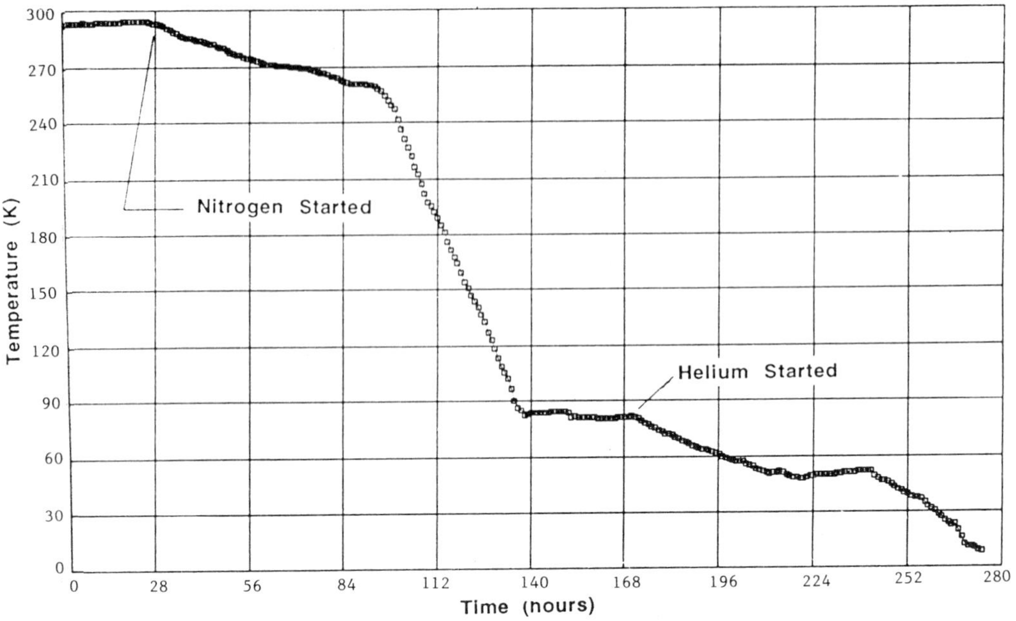

Fig. 3 The Average Temperature of the g-2 Outer Solenoid Mandrel as a Function of Time During a Cool Down Using Nitrogen and Liquid Helium from Dewars.

The liquid nitrogen and liquid helium was transferred from liquid dewars. The mandrel of the outer solenoid is cooled by cryogen that flows through an aluminum tube that is welded to the mandrel. The superconducting coil is indirectly cooled by conduction from the mandrel. The inside dimensions of this cooling tube is 11.1 mm by 30.1 mm. The effective diameter of an equivalent round tube is 20.6 mm, but the hydraulic diameter of this tube is only 16.2 mm. (The hydraulic diameter is always smaller than the equivalent

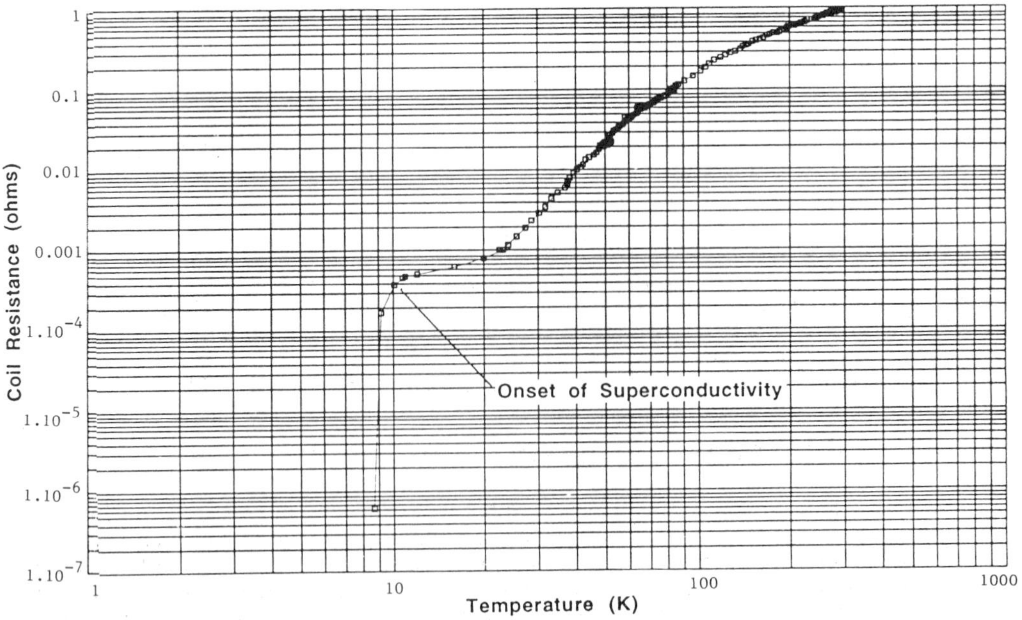

Fig. 4 The Outer Solenoid Coil Resistance as a Function of the Average Mandrel Temperature as Measured during a Liquid Nitrogen and Helium Cool Down.

round tube diameter for a rectangular flow tube.) The length of the tube attached to the mandrel is about 96 meters. The cooling tube makes two turns around the outer coil mandrel. There is good thermal conduction between the tube and between the entrance and exit of the tube. As a result, during the cool down, the lowest mandrel temperature is right at the cryogen entry point. The high temperature point on the mandrel was on the opposite side of the solenoid about 190 degree from the fluid entry point.

The nitrogen cooled shield has two cooling tubes in parallel. The nitrogen makes a single turn around the solenoid shield. Unlike the solenoid mandrel, the shield high temperature point is at the exit point of the nitrogen circuit (during the cool down). The reason for this is that the shield must be broken electrically to prevent circulating currents from flowing in the shield during a magnet quench. At the electrical break in the shield, the two pieces are attached through a small tab of fiber glass epoxy composite. This insulator is also a poor conductor of heat so the shield high temperature point occurs very close to the nitrogen exit point. The fact that the temperature distribution in the shield is different than the temperature distribution in the mandrel during the magnet cool down causes the screws that attach the shield to the mandrel to be stressed. It is likely that a number of these screws broke during the solenoid cool down despite trying to keep the shield temperature within 25 K of the mandrel temperature.

The outer coil was cooled from room temperature (294 K) to about 150 K using warm nitrogen gas mixed with liquid nitrogen boil off gas. At some point when the mandrel temperature was below 150 K, liquid nitrogen was used to cool the coil down to 80 K. It took about 115 hours to cool the outer coil package from 294 K to 80 K. (See Figure 3) The driving pressure during the liquid nitrogen phase of the cool down was less than 0.2 MPa (30 psi) The coil cool down from 80 K to the point where the outer coil became superconducting used liquid helium from dewars to provide the cooling. Through most of this phase of the cool down, the sensible heat of the helium was used along with the heat of vaporization. It took 135 hours to cool the coil from 80 K until the magnet was superconducting

The conductor used in the g-2 solenoid is about 19 parts ultra pure aluminum, 1 part copper and 1 part Nb-Ti. The stabilizing aluminum (99.999 percent pure) has a residual resistance ratio RRR (the resistance at 273 K divided by the resistance at 4 K) of 2000 to 2500 before the conductor was wound into the coil. At low temperature, the resistance of the conductor is strongly dominated by the aluminum. At room temperature the coil resistance was about 0.95 ohms. The resistance of the coil dropped about a factor of 10 as the coil cooled from 294 K to 80 K. The coil resistance dropped another factor of 220 as the coil cooled from 80 K to 10 K. The RRR of the ultra pure aluminum matrix material has remained above 2000 despite the cold work induced in the conductor during winding. Figure 4 shows the outer solenoid resistance as a function of the average mandrel temperature (as determined by several thermometers on the mandrel). As the average mandrel temperature dropped below 9.5 K, the outer solenoid coil became superconducting. The coil was fully superconducting when the average mandrel temperature reached 8.7 K. (The critical temperature of Nb-Ti is about 9.4 K.)

The lowest temperature achieved in the outer coil was about 7 K. At that point, the liquid helium supply was exhausted. All during the magnet cool down, the cryostat vacuum was monitored for leaks. The cryostat vacuum remained good even when there was liquid helium within the cooling circuit, as expected. The heat leak into the outer solenoid mandrel was measured to be a little over 50 watts. Since the cold mass supports carry about 3.5 metric ton and the cryostat area is about 70 square meters, this value is consistent with the heat leak calculations.

Since the outer solenoid cryostat is encased in the C shaped iron return yoke, the clearance between the cryostat vacuum vessel and the pole is an issue. The minimum clearance between the pole and the cryostat vacuum vessel (about 6.4 mm) limits the minimum temperature allowed for the outer solenoid cryostat vacuum vessel. This vacuum vessel temperature should always be kept above 260 K. Calculations show that the vacuum vessel wall can drop below this temperature if the vacuum is broken by helium gas at 0.1 MPa. An air leak into the vacuum vessel will result ina vacuum vessel wall temperature that is greater than 270 K Measurements of the vacuum vessel temperature when the vacuum was broken with helium appear to agree with the calculated temperatures for the wall. The calculations indicate that the vacuum wall temperature will never reach 260 K with helium in the vacuum space, provided the helium pressure in the cryostat vacuum vessel is kept below 3 kPa (about 23 torr).

CONCLUSION

Fabrication of the three large solenoids for the g-2 storage ring is complete. The three solenoids have been assembled into the iron return yoke. The outer solenoid magnet was successfully cooled down using liquid nitrogen and liquid helium. The outer solenoid became superconducting. The RRR of the conductor in the outer solenoid remained above 2000 despite the limited cold work that occurred in the matrix during winding. The refrigerator that will be used to cool down the experiment has been tested. It should deliver 625 watts when the compressor input flow is 100 grams per second at 1.8 MPa. A cool down of the three solenoids using the refrigerator will occur in the summer of 1995. The solenoid will be powered in the fall of 1995.

ACKNOWLEDGMENTS

This work was performed at the Lawrence Berkeley Laboratory and at Brookhaven National Laboratory with the support of the Director of the Office of Energy Research, Office of High Energy and Nuclear Physics, High Energy Physics Division, United States Department of Energy under contract number DE-AC02-76CH00016 and contract number DE-AC03-76SF00098.

REFERENCES

1. "Muon g-2 Design Report," Brookhaven National Laboratory, Upton NY 11973, AGS 821, April 1994
2. The g-2 Collaboration, "The g-2 Storage Ring Superconducting Magnet System," IEEE Transactions on Magnetics MAG-30, No. 4, p 2423 (1994)
3. G. Bunce et al., "The Large Superconducting Solenoids for the g-2 Ring," to be published in IEEE Transactions on Applied Superconductivity 5, No. 2 (1995)
4. G. T. Danby and J. W. Jackson, "Shimming Techniques for the Ultraprecise Muon g-2 Storage Ring at the AGS, IEEE Transactions on Magnetics MAG-30, No. 4, p 1710 (1994)
5. F. Krienen et al., "The Superconducting Inflector Dipole for the g-2 Muon Storage Ring," to be published in IEEE Transactions on Applied Superconductivity 5, No. 2 (1995)

MAGNETIC FIELD MEASUREMENT OF SUPERCONDUCTING DIPOLEMAGNETS WITH HARMONIC COIL AND HALL PROBE

Hirotaka Nakai, Atsushi Kabe, Yuuji Kojima, Kazufumi Hara, Kenji Hosoyama, Yoshiyuki Morita, Takakazu Shintomi, and Akio Terashima

National Laboratory for High Energy Physics (KEK)
1-1 Oho, Tsukuba-shi, Ibaraki-ken 305 JAPAN

ABSTRACT

Magnetic field measurements and field analyses of 1-m long superconducting dipole magnets fabricated at the National Laboratory for High Energy Physics (KEK) have been carried out using a harmonic coil with the bucking scheme.[1] Conditions of the data acquisition are optimized to achieve the accurate and efficient measurements. Not only in the steady state of the magnet excitation by constant currents, but also on the way the excite current increases until the magnet quenches the field measurements have been tried, and the results are discussed in this paper on the possibility of the "on-the-fly" measurement using a harmonic coil. Some results on the so-called remnant field of the magnets measured with a Hall probe are also described.

INTRODUCTION

Magnetic field measurements of magnets are essential to assure the field quality when the magnets are to be adopted in accelerators, for example. At KEK several 5-cm aperture, 1-m long superconducting magnets have been fabricated alternating their designs of the end part and the cable specifications for optimizing the magnet design.[2] Some of the magnets are measured their magnetic fields with a harmonic coil to evaluate their field quality, and other some are measured their remnant field after excitation of the magnet by using a probe with three Hall devices.

Magnetic field measurements with a harmonic coil and magnetic field analysis by the bucking and the Fast Fourier Transform (FFT) may be already established.[3] It was needed, however, to establish the actual method of the magnetic field measurement with a harmonic coil at KEK. As the first step, the field measurement system with a harmonic coil was constructed to measure the static magnetic field while the superconducting magnet was excited at some constant current. After assuring the system can measure the magnetic field and acquire the proper data for analyzing the higher order field components, several conditions of the data acquisition, such as the averaging the data and the integral time of the voltage measurement were varied to optimize the magnetic field measurement. This measurement system was applied also to measure the magnetic field while the excite current increased up to the value at which the magnet quenched. Of course, the harmonic coil is designed to measure the magnetic field in the steady state, but it is worth checking the possibility of the harmonic coil to measure the varying magnetic field.

On the other hand, at several laboratories where the magnetic field measurements are carried out for their magnets, the periodic remnant field along the magnet axis is observed.[4,5] It is also needed to measure the remnant field of our magnets if the phenomenon is very common in the superconducting magnets. To measure this weak remnant field, a probe with three Hall devices was designed and fabricated at KEK. With this probe the existence of the remnant field in our magnets was verified, and the variation of the field along the axis of the magnet was measured.

Some results of the magnetic field measurements with the harmonic coil and the Hall Probe carried out at KEK are described and discussed in this paper.

MEASUREMENTS

Harmonic Coil

The harmonic coil used in this study is shown schematically with some important dimensions in Figure 1. The harmonic coil consists of two bucking coils formed in the plane which includes the axis of rotation, and one tangential coil which does not include the axis of rotation and can detect the higher order components of the magnetic field. Numbers of turns of each coil are decided so that the output signals of the induced voltages in the coils have almost the same level. The bore of the superconducting magnet in a liquid helium vessel is occupied by a vacuum-insulated "warm finger" as shown in Figure 2. The harmonic coil is inserted into the warm finger, and the dry nitrogen gas is introduced into the warm finger to ensure the harmonic coil not to be cooled.

The harmonic coil is rotated by the drive unit located at the top flange of the cryostat at room temperature. Its rotational frequency is 0.25 Hz, and the rotary encoder outputs 1024 pulses per revolution. This pulse rate is so high that the integral time for voltage measurement becomes too short to avoid noises in data acquisition. Then we adopt a pulse divider which reduces the pulse rate to 128 pulses per revolution.

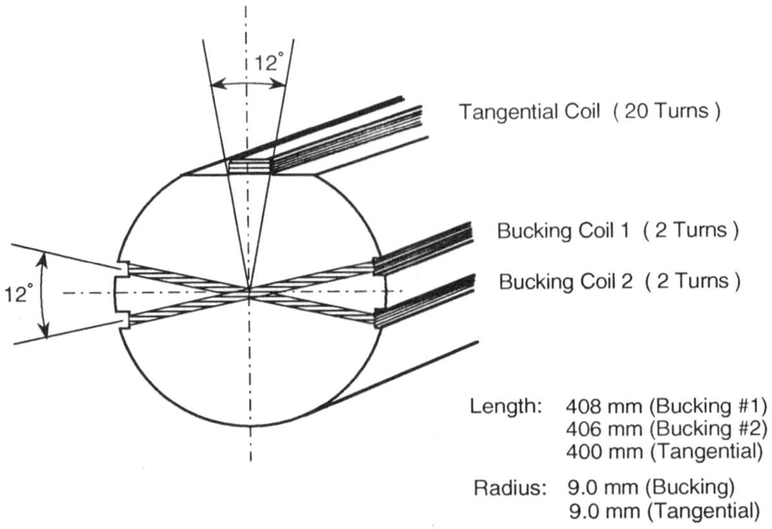

Figure 1. A harmonic coil consisting of one tangential coil and two bucking coils.

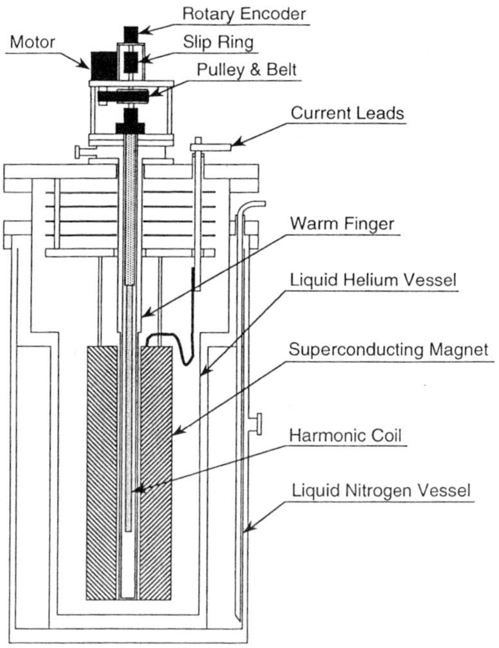

Figure 2. Setup of field measurement with a harminic coil.

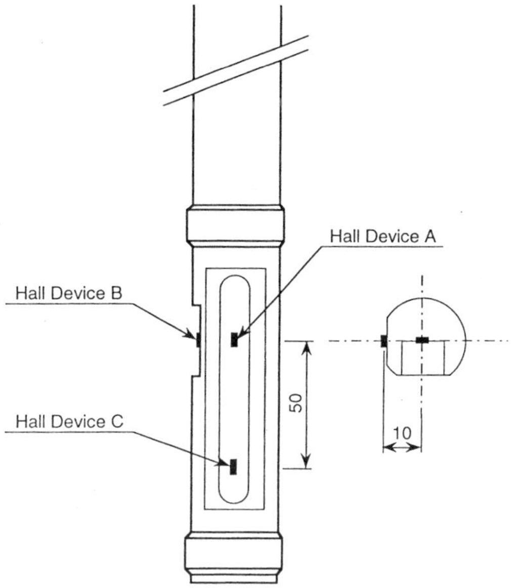

Figure 3. A Hall probe with three Hall devices.

Magnetic Field Measurement

An analyzing recorder (YOKOGAWA AR1100) is used to acquire the induced voltages in the three coils of the harmonic coil. This recorder has some filtering and averaging functions. We use 40 Hz low pass filtering functions for all voltage measurements, since the power line frequency is 50 Hz in the measuring site. The voltage measurements are triggered by pulses from the rotary encoder connected to the rotating axis of the harmonic coil as shown in Figure 2. To measure the magnetic field in the steady state, the excite current is set to a certain value. After establishing the current constant, the measurement is initiated. This procedure is repeated while the constant values of the excite current are varied successively.

Data of the induced voltages and of the shunt resistance voltage are stored in floppy disks and analyzed by the help of the bucking scheme and FFT according to the mathematical procedure described in Reference 1. When FFT is applied to the induced voltage, 512 points of data, which corresponds to 4 revolutions of the harmonic coil, are used.

Hall probe

The remnant field of a superconducting magnet seems to be very weak compared with the magnetic field generated while the magnet is excited. A Hall device (TOSHIBA THS102A) which is one of the magnetoelectric devices is selected to measure the remnant field. To achieve the efficient measurement three Hall devices are adhered to the G-10 FRP body as shown schematically in Figure 3. This probe is inserted into the warm finger located inside the bore of a superconducting magnet, as the same way of the magnetic field measurements with the harmonic coil. Each Hall device is excited by constant DC current of 10 mA. The Hall devices have been previously calibrated in the known magnetic field.

This Hall probe can be rotated and moved along the axis of the magnet with the drive unit on the top flange of the cryostat. The output voltages from the Hall devices are measured by the analyzing recorder and stored in floppy disks.

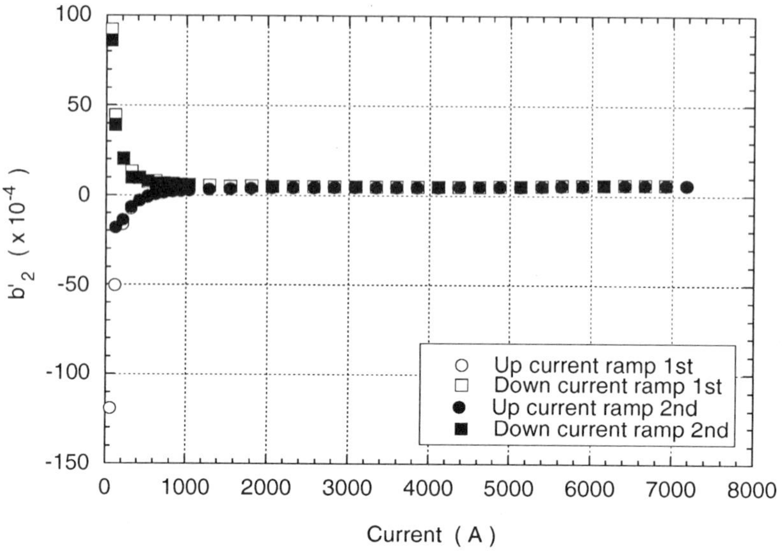

Figure 4-a. Current dependence of the sextupole component of KEK#14 dipole magnet. Sextupole component is expressed as a fractional coefficient.

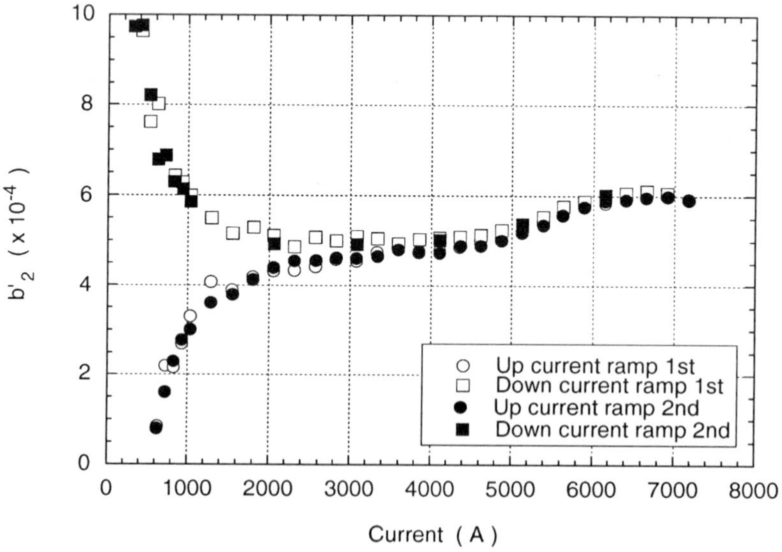

Figure 4-b. Close-up view of Figure 4-b.

RESULTS AND DISCUSSIONS

Magnetic Field Measurements

A dipole magnet generates not only the main dipole field but also sextupoles, decapoles and so on in effect. Hence the higher order component of interest is the sextupole in the magnetic field analysis for the dipole magnet. In Figures 4-a and 4-b shown are the normal component of the sextupole as typical results of the field measurement. The component is plotted against the excite current as the fractional field coefficients in the unit of 10^{-4}. It is readily seen the effect of the magnetization of the filament as a hysteresis feature at small excite currents. On the other hand, at high excite currents some variation of the component can be seen. This part appears as the iron yoke comes to saturate in the high magnetic field.

These figures show that this system for the magnetic field measurement works well and the analyzing scheme is correct.

"On-the-Fly" Measurements of the Magnetic Field

It is needed to measure the variation of the magnetic field when the magnet is excited by the current on ramp, and there is certain eddy current in the magnet which may change the magnetic field. The harmonic coil requires at least the time to make one revolution of rotation to pick up the magnetic field. Hence this coil seems to be inadequate theoretically to measure the magnetic field varying with time. If the ramp rate is not so fast that the time variation of the magnetic field is small, there is some possibility that even the harmonic coil can measure the field on the current ramp.

The results of the field measurements on ramp are shown in Figure 5, where results at various ramp rates are plotted with results at the steady constant currents as mentions above. To analyze the field on ramp, 256 points of data (2 revolutions of the harmonic coil) are used. A set of 128 points of data is also tried to analyze the magnetic field, but it is found that the data for only 1 revolution of the harmonic coil are inadequate to withdraw the physical result. On the other hand, 256 points of data mean that the all physical conditions are averaged over the time for 2 revolutions of the harmonic coil. Hence, this is the trade-off between the temporal and the magnetic resolutions.

The excite current changes while the field measurement is executed, and the problem is how the current is evaluated for the data acquired on the current ramp. The convenient and plausible way to evaluate the excite current is as follows. At first, the acquired data on ramp are analyzed by the same way as if they were measured at the constant excite current. Then we can get information on the field strength of the dipole field. By using this dipole field strength, the excite current on ramp is calculated from the transfer function which is determined experimentally from the results of the magnetic field measurements at constant excited currents.

In Figure 5, apparent difference between the data at steady state and those on the ramp can be seen clearly, and this means that the harmonic coil can detect the change of the field by the eddy current. However, results at ramp rates faster than 100 A/s have no longer the physical meanings, as shown by scattered data in the figure, because the current changes so large while the harmonic coil rotates.

Remnant Field Measurements

The remnant field measurement with the Hall probe was carried out for two different types of the magnet. One is the ordinary type magnet (KEK#6), which is made of the superconducting cable consisting of the filament whose diameter is 6 μm and the copper matrix. Another (KEK#5) is made of the cable consisting of the filament of 2.5 μm in diameter and the Cu+0.5wt%Mn matrix.

The results of the remnant field measurement for the magnets are shown in Figures 6 and 7. These measurements are made along the axes of the magnets. The excite current and the duration of the excitation are shown for each data. Full scale of the magnetic field corresponds to 10 Gauss. Every data is measured after quench of the magnet, otherwise the history of the excitation of the magnet are accumulated and the current-dependence of the remnant field is hidden.

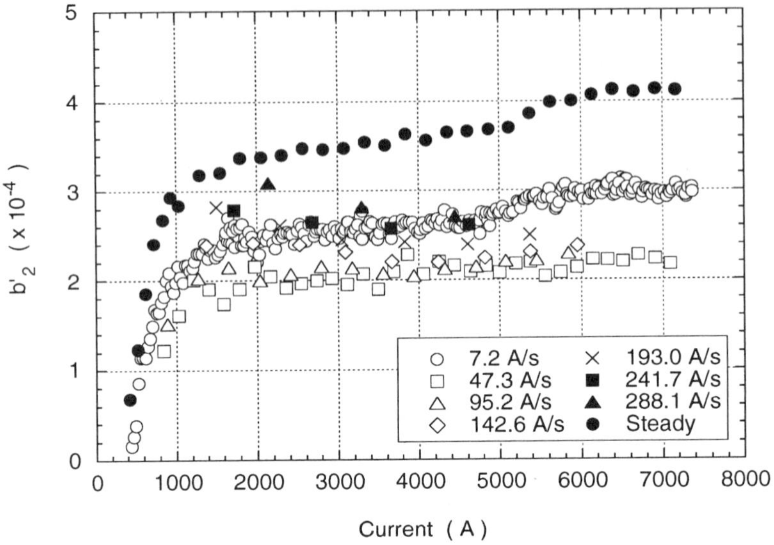

Figure 5. Ramp rate dependence of the sextupole component of KEK#13'.

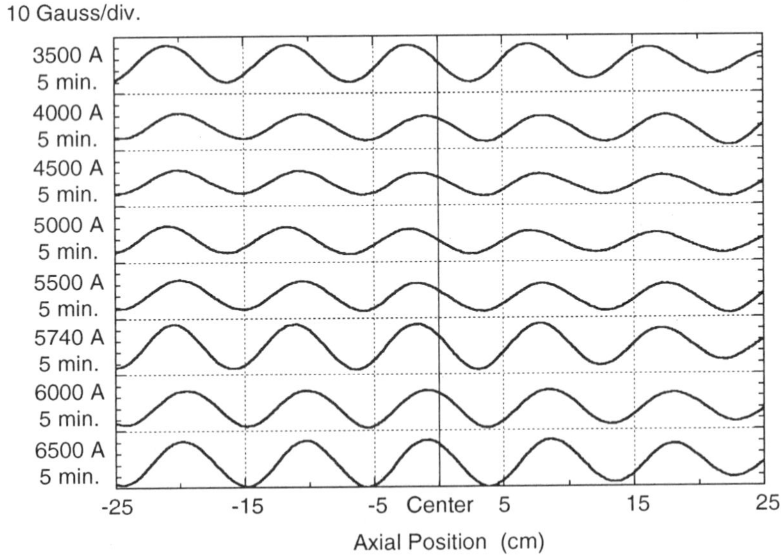

Figure 6. Current dependence of remnant field of KEK#6 dipole magnet.

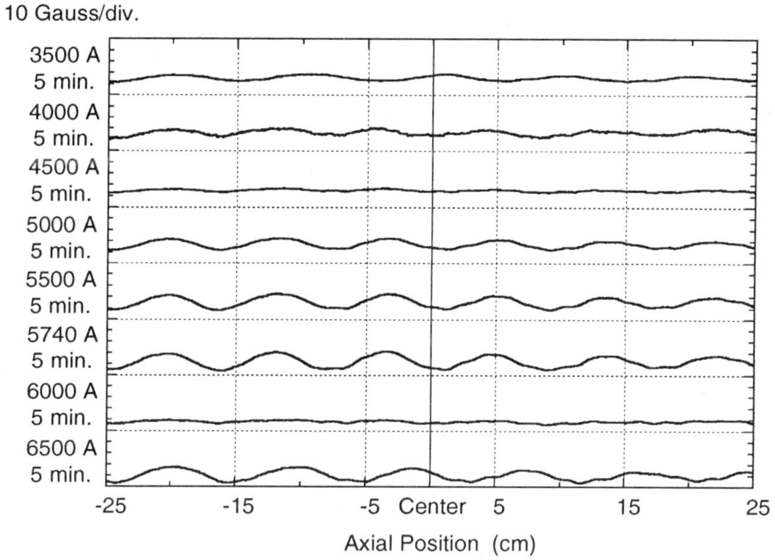

Figure 7. Current dependence of remnant field of KEK#5 dipole magnet.

KEK#6 magnet shows the clear periodic remnant field at every excite current, while data of KEK#5 are complicated. At some excite currents, the periodicity of the remnant field or the existence of the field itself cannot be seen clearly, while at other currents they can. The periodicity of the remnant field along the axis is exactly coincided with the transposition pitch length of the cable. On the other hand, the current-dependence of the field are not readily deduced even from the data of KEK#6. Anyway, the remnant field is the common phenomenon for superconducting magnets, and the Hall probe can be successfully used to detect the remnant field.

CONCLUSIONS

Magnetic field measurements with the harmonic coil and the Hall probe have been successfully carried out at KEK, and the measurement systems are verified to work well, and optimized under given conditions of the system. It is found that the harmonic coil can be used also in the "on-the-fly" measurements, even though there are some ambiguities of the measured data. Some problems of the measurement errors, such as the sagitta of the magnet and the bending of the harmonic coil, are, however, still left to be discussed.

The existence of the remnant field of the superconducting magnets is observed also at KEK by using the Hall probe. Characteristics of the remnant field are also not clear yet, and then the measurements should be continued further in various conditions as possible.

ACKNOWLEDGMENTS

One of the authors (H. N.) expresses his special acknowledgments to Dr. P. Wanderer, Brookhaven National Laboratory and to Dr. M. I. Green, Lawrence Berkeley Laboratory for their helpful information on the magnetic field measurement and analysis.

Mr. T. Takahashi, Hitachi Works, Hitachi, Ltd., Mr. H. Hattori, Kobe Shipyard & Machinery Works, Mitsubishi Heavy Industries, Ltd. and Mr. Y. Yuge, MEC Co. Ltd. are appreciated for their fabrication of some parts of the measuring apparatus and for their supports on the magnetic field measurements.

REFERENCES

1. E. Willen, P. Dahl and J. Herrera, Superconducting Magnets, *AIP Conf. Proc.* 153-2:1228 (1987).
2. K. Hosoyama, to be published in this proceedings.
3. J. Herrera, Basic Equations for Tangential Coil FFT Analysis, Magnet Testing Group Note MTG 225, Brookhaven National Laboratory (1982).
4. H. Brück et al., Observation of a Periodic Pattern in the Persistent-Current Fields of the Superconducting HERA Magnets, DESY-HERA 91-01 (1991).
5. A.K. Ghosh, K.E. Robins and W.B. Sampson, Axial Variations in the Magnetic Field of Superconducting Dipoles, *Supercollider* 4: 765 (1992).

NEW CALORIMETRIC AC LOSS MEASUREMENT TECHNIQUE INVOLVING SUPERFLUID HELUIM

B. J. P. Baudouy, K. Bartholomew, and S. W. Van Sciver

National High Magnetic Field Laboratory / Florida State University
Tallahassee, FL, 32306, USA

ABSTRACT

We have developed a new calorimetric AC loss measurement facility involving superfluid helium (He II). At present, the Test of AC Loss (TACL) facility performs AC loss measurements on Cable-in-Conduit Conductors (CICC) under development for the NHMFL 45 Tesla hybrid superconducting outsert magnet. TACL can handle large scale conductors up to one meter in length. Measurements utilize the exceptional high heat conductivity of He II, which provides an isothermal environment and is the dominant enthalpy in the system. The test conductors are placed in an independent cryostat containing He II which is inserted in a superconducting dipole magnet producing a transverse magnetic field up to 7 T. For a change of the magnetic field and associated AC loss, the temperature variation of the He II surrounding the conductor is measured and directly converted to enthalpy variation of the He II. This paper describes the measurement technique and compare its resolution to that of more conventional calorimetric AC loss measurements.

INTRODUCTION

The measurement of energy dissipation in a composite superconducting cable while it is subjected to a variation of an external field is indispensable information for magnet design. Basically, methods to measure AC losses can be divided in two groups: electromagnetic and calorimetric methods. The first method is based on the electromagnetic properties of the superconductor whereas calorimetric measurement can be performed by helium boiled off measurement (isothermal calorimetry) or by a direct measurement of the temperature increase of the conductor (adiabatic calorimetry). A third calorimetric method has been developed at the NHMFL. This new calorimetric measurement can be categorized as isothermal calorimetry and uses the high heat transfer capability of superfluid helium. It is based on temperature measurement of a saturated He II bath. Such a technique can allow measurement at high energy dissipation level up to 0.4 J/cm^3 of conductor for large scale conductors up to one meter in length. The technique was previously used to measure losses in CIC conductor developed for SMES applications[1] and Nb_3Sn and NbTi conductors under development for the 45 Tesla hybrid superconducting outsert magnets.[2,3]

MEASUREMENT FACILITY

Facility

TACL facility consists of a vertically oriented superconducting dipole magnet with an insert cryostat in its bore. The magnet is a 7 Tesla superconducting SSC model dipole magnet that operates at 4.2 K. A schematic representation of the assembly is shown in Figure 1. The magnet is protected by a quench detection system connected to various voltage taps and a set of two parallel breakers in series with the power supply. The breakers are also used to discharge the magnet into the dump resistor by isolating the magnet and the dump resistor from the power supply.

The insert cryostat consists of a double wall vacuum insulated He II container, which insulates the conductor from the surrounding 4.2 K liquid helium bath. The conductor is supported by a probe which can be removed from the cryostat even if the magnet is cold. The probe consists of a conductor holder attached to a conical plug through which passes the instrumentation wires. This component is used to separate the upper portion of the cryostat from the test section by a thermal isolation during the AC pulse. The conical plug can be opened, by a pneumatic linear actuator, to bring the conductor back to initial conditions after a heat pulse, Fig. 1. The upper portion of the cryostat is used as a reservoir of saturated He II for the test portion and is filled by the surrounding 4.2 K liquid helium through the fill valve.

Instrumentation

The measurement instrumentation within the inner cryostat consists of six Carbon Glass Resistance thermometers, 20 W power resistor and an Hall probe. One of the thermometers is located above the plug and the five others are taped on the conductor and the probe in the test section. The resistive heater is placed at the bottom of the instrumentation bus to provide assistance with He II volume measurement.

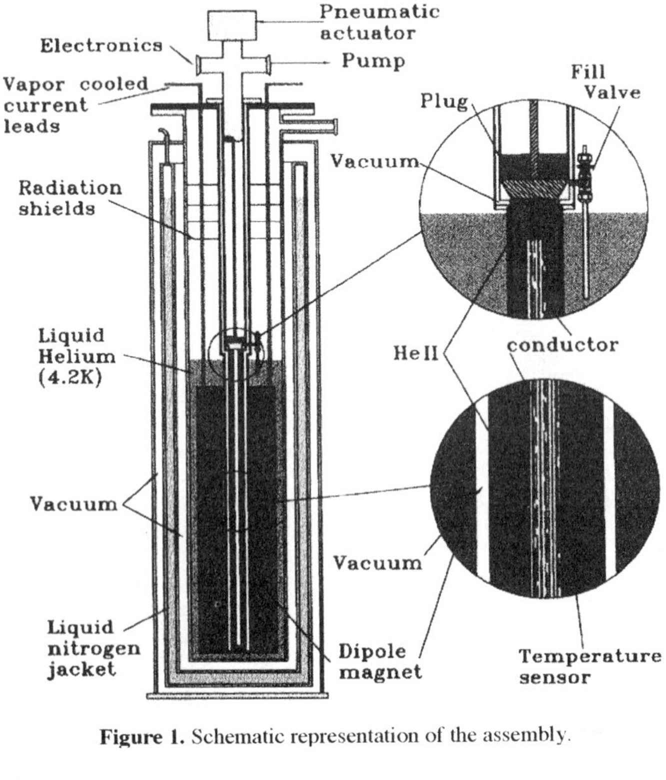

Figure 1. Schematic representation of the assembly.

EXPERIMENTAL TECHNIQUE

The technique consists of a measurement of the temperature rise in the closed volume of He II, caused by the energy dissipation by AC loss in the conductor volume. All measurements are carried out from 1.4 K to 1.8 K. When the desired initial temperature is obtained in the insert cryostat, the portion where the conductor is located, is insulated from the upper reservoir by closing the conical plug and the AC pulse is performed. The time required for one AC loss test is around one minute.

AC pulse is achieved by a rapid discharge, with time constants between about 0.2 s and 2 s, of the current through the dump resistor or by any controlled variation of the external field with a ramp rate (dB/dt) up to 1 T/s. Measurement of the temperature rise, which is converted to an enthalpy variation of the liquid helium, determines the energy deposited by the AC loss as:

$$E = \int_{T_i}^{T_f} \rho v C_v dT \tag{1}$$

where v is the closed helium volume surrounding the sample, ρ the density of helium, C_v the constant volume specific heat, and T_i and T_f the initial and final measured temperatures.

The helium volume contained in the insert cryostat is separately determined, following the same principle, by applying a known heat pulse from the power resistor. With a sufficient amount of energy, the reproducibility of these measurements is excellent, reducing the uncertainty in the measured volume under 2%.

RESOLUTION AND ACCURACY

Measurement and thermometry resolution

The He II calorimetric AC loss measurement technique has the potential for high resolution, which can be adjusted according to the level of loss to be measured. Under most circumstances the heat capacity of the sample is negligible, e.g. at 2 K the ratio of volumetric heat capacities of helium to copper is about 3000 decreasing to 1000 at 1.5 K. Therefore, the conductor heat capacity can be neglected provided the helium volume is at least 10% of the conductor volume. Further, the resolution and dynamic range of the measurement can be adjusted by varying the volume and initial temperature of the helium in contact with the conductor. Since the heat capacity of helium goes approximately as $T^{5.6}$ and most common temperature sensors have a sensitivity that increases with decreasing temperature, a decrease in temperature will significantly increase the resolution of the measurement. Clearly, the measurement must be carried out such that the final temperature does not exceed $T_\lambda = 2.17$ K. The maximum energy that can be absorbed during the AC pulse is limited by the allowable enthalpy change of the contained He II (about 0.4 J/cm^3 of He II between 1.5 K and T_λ).

The resolution of this technique can be estimated using the known properties of the various components in the system. For a given energy deposition, dE, the differential temperature change is given by

$$dT = \frac{dE}{\rho v C_v} \tag{2}$$

The volumetric heat capacity of He II may be approximated by $\rho C_v \approx 17\, T^{5.6}$ mJ/cm^{3}K.

For the temperature sensor to be used in the measurement, the specific sensitivity is defined as,

$$S = \frac{T}{dT}\frac{dV}{V} \tag{3}$$

where V is the voltage output from the resistive temperature sensor. Combining the above two expressions, we may write the energy deposition as,

$$dE = \frac{\rho v C_v T}{S}\frac{dV}{V} \tag{4}$$

In our facility, the temperature measurement is made using a carbon glass thermometer. For such a thermometer in the helium temperature range the specific sensitivity may be approximated by[4],

$$S \approx -\frac{10}{T^{0.8}} \tag{5}$$

which can be combined with the empirical expression for the volumetric heat capacity, to yield

$$dE \approx 1.7T^{7.4}v\frac{dV}{V} \tag{6}$$

For most experiments we use a constant current, so the term in parenthesis can be replaced by dR/R. Equation (6) can, in principle, be integrated over a finite change; however, since V (or R) is a complex function of T, the integration can not be done in closed form.

We can use Eq. (6) to estimate the resolution of the He II calorimetric AC loss measurement technique. To date, experiments performed in the TACL facility have helium volumes around one liter and conductor volumes of 30 cm^3. Thus, the large helium volume allows us to ignore the heat capacity of the conductor. Our digital data acquisition system has a maximum resolution (dV/V) of about +/- 0.05%, which corresponds to an absolute resolution of 17 mJ (0.57 mJ/cm^3 of conductor) at 1.5 K. This value, which is about 0.1% of a typical loss measurement on 45-T hybrid CIC Conductor[2,3], corresponds to about 0.1% of the range between 1.5 K and 2.17 K.

A comparison between resolution based on thermometry resolution, Eq. (6), and experimental resolution based on Eq. (2) can be made as a function of temperature. To use Eq. (2) as an experimental resolution, the noise level have been carefully recorded at different bath temperatures between 1.5 K to 2.1 K with the magnet power supply off. This temperature noise level (δT) has been introduced in Eq. (2) to calculate the experimental resolution, dE_{exp}, as

$$dE_{exp} \approx 17vT^{5.6}\delta T \tag{7}$$

Figure 2 displays the two different resolutions at different inital bath temperature for a one liter volume of helium. The experimental acquisition resolution is not equal to the resolution of the experiment and it is about a factor of 2 to 10 worse than predicted. The effect which could cause this difference is likely the acquisition system temperature noise level which has not been included in the experimental predicted resolution (dE).

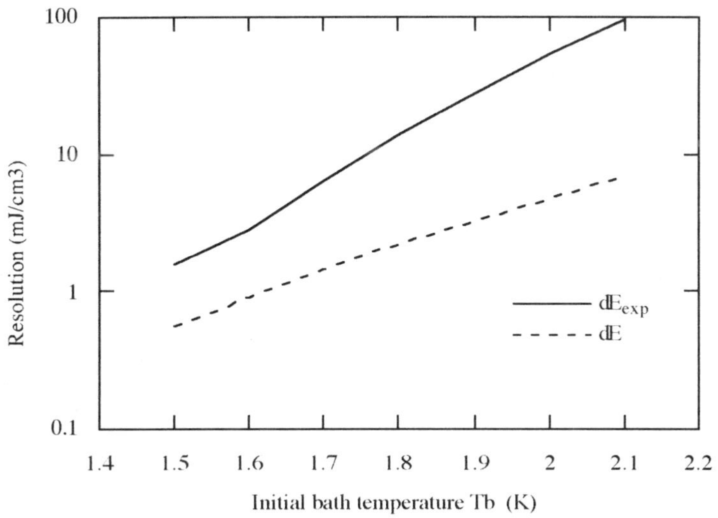

Figure 2. Comparison between resolutions for helium to conductor volume ratio of 33.

Note that if the helium volume is reduced by two orders of magnitude, to 10 ml, which will still give a helium to conductor volume ratio of 0.33, and a measurement resolution of 5.7 $\mu J/cm^3$ at 1.5 K, it would still correspond to about 0.1% of range. Further increase in resolution could be achieved by improving the temperature resolution or by operating at even lower initial temperature.

AC loss accuracy

The relative maximum error depends on the accuracy with which the helium volume is determined. In TACL, the helium volume is measured by the same technique, as the AC loss measurement. This measurement produces an overall error in the volume determination of about ± 2%. Note, however, that since the same volume appears in all measurements, the error introduced is systematic.

We define the relative maximal error on the energy as

$$\left[\frac{dE}{E}\right]_{max} = \sum_n \left|\frac{\partial E}{\partial n}\right| \frac{dn}{E} = \frac{dv}{E}\left|\frac{\partial E}{\partial v}\right| + \frac{dT}{E}\left|\frac{\partial E}{\partial T}\right| \tag{8}$$

Then, according to Eq.(2) and Eq. (8), the relative maximun error can be rewritten as

$$\left[\frac{dE}{E}\right]_{max} = \frac{dv}{v} + \frac{\rho C_v dT}{\int \rho C_v dT} \tag{9}$$

We present relative maximum error for a volume of helium of one liter. It is plotted AC loss term of Eq. (9) versus the temperature increase for different initial temperatures of the bath in Fig. 3. The dependence of the maxium error versus temperature can be approximated by the empirical law:

$$\left[\frac{dE}{E}\right]_{max} \approx \frac{dv}{v} + \frac{\alpha(T_b)}{T - T_b} \tag{10}$$

where $\alpha(T) \approx T^5$ and T_b is the initial bath temperature.

It is shown that the error decreases with respect to decreasing initial bath temperature and increasing variation of temperature. For large temperature variation around 0.1 K, the systematic error introduced by the determination of the helium volume is the main term of the error whatever the initial bath temperature is.

Measurements carried out with a smaller relative level of volume helium, around 10% of the conductor volume would improve the resolution of the experiment and reduce error in the determination of AC loss as well as the determination of the helium volume in the test section.

CALORIMETRIC MEASUREMENTS COMPARISON

Either isothermal or adiabatic calorimetric methods can be used to measure total AC losses. It is of interest to compare the relative accuracy of these methods.

Isothermal methods

The traditional isothermal method measures the gas flow rate induced by the sample loss. The most common technique is to measure the boil-off rate of helium with a flow-meter[5] or with a collection tube[6,7]. In the first method, the sample is placed in 4.2 K liquid helium in a bell jar connected to a gas flow-meter outside of the cryostat. The gas produced by the AC loss is conducted through a chimney towards the flow-meter. The flow-meter has to be calibrated previously by means of a heater. To achieve accurate measurements the gas flow through the flow-meter must be at standard temperature and pressure condition. Problems such as acoustics resonance in the chimney, mass flow rate change due to varying magnetic field, temperature fluctuation or pressure fluctuation from the bell jar to the flow-meter can introduce error. But the main disadvantages of this technique are a low resolution and a long thermal equilibrium. Time required for one AC loss measurement is typically around 30 minutes. To reduce the experimental time required some authors[4] use a heater to provide power to the liquid, at the background field, which is almost equivalent to the AC loss of the sample. When the AC loss test is performed the power is adjusted to have a constant gas flow. The variation of the power in the heater is equal to the AC loss in the sample. This method can reduced the time of the experiment and eliminate the problem of pressure and temperature fluctuation to the chimney. Relative error is said to be around 10% for this method.

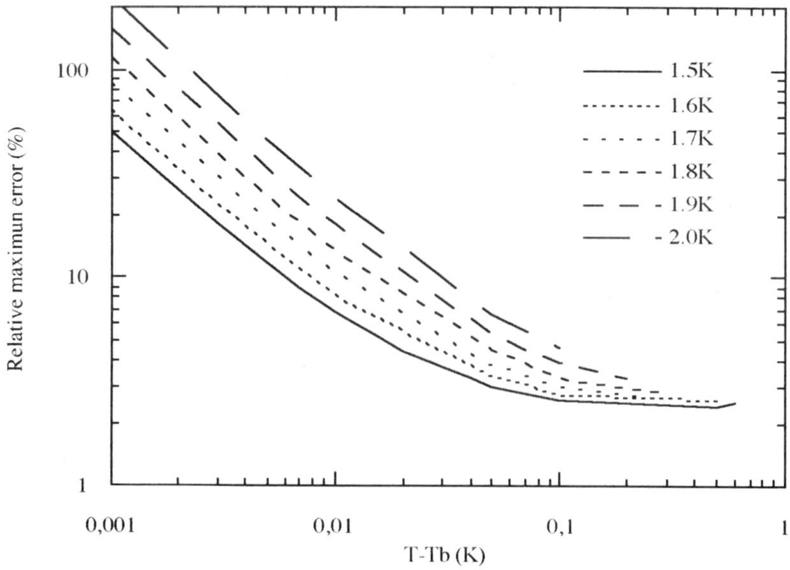

Figure 3. Calculated maximum error versus temperature variation for different initial bath temperature.

A second method which is similar to the flow-meter method consists in collecting and storing the gas in a closed chimney[6,7] filled by liquid helium. This method has the advantage of maintaining a thermal equilibrium during the measurement. The gas flow rate is determined by measuring the level of liquid helium in the chimney. Typically the accuracy of this method is around 3% of range of 3-170 mW. But a background dissipation up to 3 mW is reported. The maximum energy is limited by the volume of the chimney because only a finite amount of liquid can displace the liquid in the chimney. The other method[7] deduces the level of helium by two infrared LEDs. It is said that measurement can be made with an accuracy of 3% to 5% in the range of 1-100 W. The maximum energy measurable in this case is around 2 J dissipated in the liquid helium. This method also has a measuring time of order of 30 minutes.

Adiabatic methods

The energy dissipation can also be calculated from the temperature difference and related to heat capacities of the sample and the sample holder. Position of the thermometer and thermal contact resistance between the thermometer is critical to the experimental results. Schmidt[8,9] presented such a technique which consists in measuring the temperature of the sample with an AC loss applied. The measurement is an indirect measurement of AC loss. First, the technique must be calibrated by measuring of the temperature variation of the sample as a function of energy dissipated by a heater which is also in contact with the sample. Then the AC loss measurement is performed and the temperature variation is compared with the calibration made by the heater to deduce the AC loss. The sample is connected to the 4.2 K liquid reservoir by a determined thermal resistance, which can be changed to vary the resolution. The resolution presented is in the range of 1 $\mu J/cm^3$, note a error of order of 1% (100 J/m^3 of conductor) to around 3% (1J/m^3 of conductor). These results are valid if the heater calibration is made before each measurement[8]. The time required for a measurement is around few seconds[8].

This method cannot be applied to measure AC loss with transport current samples because the adiabatic condition needs to be conserved. It also is not suited for large scale and CIC conductors since large temperature variation could destroy the adiabatic condition.

He II calorimetric method

With a one liter helium volume, AC loss dissipation has been measured up to 150 J, with a precision of around 3% and with average resolution of 10 mJ/cm^3 at an initial temperature of 1.6 K. On the other hand, a 1 J energy dissipation is measured with an accuracy of 10% and with a resolution of 2 mJ/cm^3 at an initial temperature of 1.6 K[2,3]. Note that the maximum energy dissipation measurable is 0.4 mJ/cm^3 of He II. The main advantage of this method is that the time of an measurement is around one minute and the possibility exists to have various sample types with a length up to one meter. Resolution can be adapted for various samples by changing the volume of helium in the test portion and using different probes obtain results within few percents.

CONCLUSION

We present a new calorimetric AC loss measurement which is suited for high resolution measurements on large composite conductors which is capable to detect conductor dissipation in the range of the micro-Joule/cm^3. This technique is applied for measurement of AC loss on CIC conductor developed for the NHMFL 45 Tesla magnet.

ACKNOWLEDGEMENT

Research supported by the State of Florida and the National Science Foundation through NSF Cooperative Grant No. DMR 9016241. The authors would like to express appreciation to A. Bonito-Oliva, V. Cochran, J.R. Miller and S. Welton.

REFERENCES

1. S.W. Van Sciver, K. Bartholomew, A. Bonoto-Oliva and S, Welton, AC loss measurement of the SMES/CIC conductor, *Cryogenics* 34 ICEC supplement:615 (1994).
2. B. J. P. Baudouy, K. Bartholomew, J. Miller, S. W. Van Sciver and A. Bonito Oliva, AC Loss Measurement of the 45-T Hybrid/CIC Conductor, *IEEE Trans. on Appl. Super.* 5 no2:688 (1995).
3. A. Bonito Oliva, B. J. P. Baudouy, K. Bartholomew and J. Miller, S. W. Van Sciver, AC Loss in Superconducting Nb_3Sn and NbTi Conductor, *IEEE Trans. on Magn.*, proceedings of MT14, in press.
4. S.S. Courts et al., Cryogenic Thermometry-An Overview, in: "Application of Cryogenics Technology Vol. 10", Plenum Press, NY (1991) p. 55.
5. M. Takayasu *et al.*, Calorimetric Measurement of AC Loss in Nb3Sn superconductors, *Elsevier*.(MT11 proceedings 2:1033 (1989).
6. K. Kuroda, Modified Boil-off Method for Measuring A.C. Losses of Superconducting Composite, *Cryogenics* 26:566 (1986).
7. J. A. Eikelboom, Apparatus for calorimetric measurement of a.c. losses in superconductors, *Cryogenics* 31:363 (1991).
8. C. Schmidt, Calorimetric A.C. loss measurement in the microwatt range: a new simplified technique, *Cryogenics* 34: 3 (1994).
9. C. Schmidt, Measuring a.c. losses of superconductors, *Cryogenics* 25: 492 (1985).

AC LOSS MEASUREMENT OF SUPERCONDUCTING DIPOLE MAGNETS BY THE CALORIMETRIC METHOD

Y. Morita,[1] K. Hara,[1] N. Higashi,[1] K. Hosoyama,[1] A. Kabe,[1] H. Kawamata,[1] Y. Kojima,[1] H. Nakai,[1] T. Shintomi,[1] T. Takahashi,[2] and A. Terashima[1]

[1]National Laboratory for High Energy Physics (KEK)
1-1, Oho, Tsukuba, Ibaraki 305, Japan
[2]Hitachi Works, Hitachi, Ltd.
3-1-1, Saiwai, Hitachi, Ibaraki 317, Japan

ABSTRACT

AC losses of superconducting dipole magnets were measured by the calorimetric method. The magnets were model dipole magnets designed for the SSC. These were fabricated at KEK with 50-mm aperture and 1.3-m overall length. The magnet was set in a helium cryostat and cooled down to 1.8 K with 130 L of pressurized superfluid helium. Heat dissipated by the magnet during ramp cycles was measured by temperature rise of the superfluid helium. Heat leakage into the helium cryostat was 1.6 W and was subtracted from the measured heat to obtain AC loss of the magnet. An electrical measurement was carried out for calibration. Results of the two methods agreed within the experimental accuracy.

We present the helium cryostat and measurement system in detail, and discuss the results of AC loss measurement.

INTRODUCTION

In some accelerator magnets, it is required to operate in fast-ramp cycles. In the SSC project, the High Energy Booster magnets were this type. The nominal ramp rate was 60 A/s. In rapidly time-varying field, the magnet generates heat, that is the AC loss. The heat generated by the AC loss is transferred to liquid helium, then becomes a heat load to the refrigerator. Moreover, the AC loss increases the temperature of the magnet and can cause a quench at significantly lower current than the designed one.

The AC losses of the magnet are divided into two types. One is the hysteresis loss both in the iron yoke and the superconductor cable. Another is the eddy-current loss in the superconductor cable, that is, the inter-filament and inter-strand couplings. The inter-strand coupling loss becomes important in the AC loss at fast ramp rates. This type of loss is expected to be influenced by a magnet fabrication, such as, heat treatment or pressure applied in the cure cycle. Therefore it is important to measure the AC loss of the magnet.

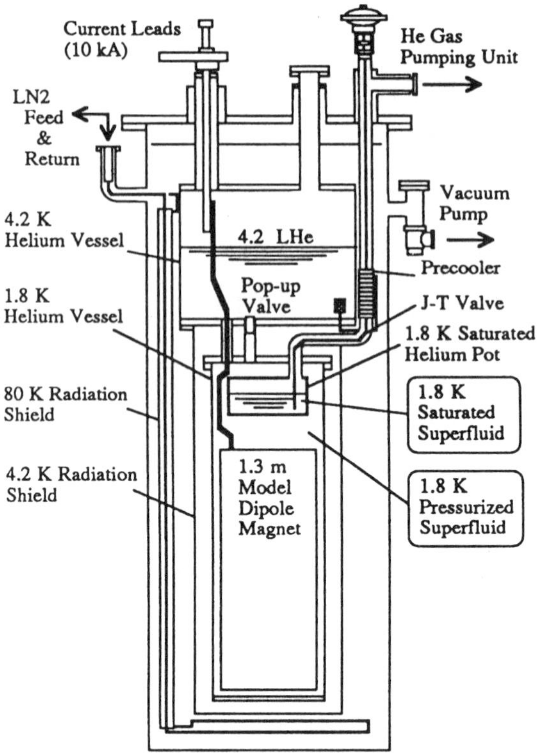

Figure 1. 1.8 K cryostat.

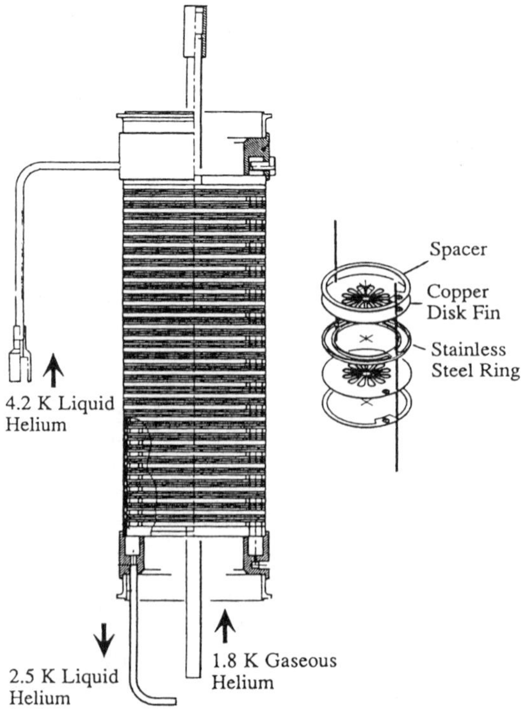

Figure 2. Pre-cooler.

We have fabricated 15 model dipole magnets for the SSC accelerator-magnet R&D.[1-3] The model dipole magnet is a two-layer and cosθ-type magnet and has a 1 m straight section, a 1.3 m overall length and a 50 mm aperture. Nominal operating current was 6500 A and nominal dipole field was 6.6 T. The superconductor cable was made of NbTi filamentary wires, which had filaments of 6 μm in diameter and Cu inter-filament matrix. To develop AC-resistant magnets, two magnets were wound with superconductor cable having filaments of 2.5 μm in diameter and Cu+0.5wt%Mn inter-filament matrix.

There are various measurement methods to measure the AC loss of the magnet. The electrical method, which is the direct measurement of the coil voltage and current, was applied at several laboratories.[4] The electrical method measures the stored energy of the magnet and the AC loss is much smaller than the stored energy. The current and voltage should be measured with 10^{-4} accuracy through proper handling of the signal. In our case, the magnet power supply had a large noise source, which may leave the measurement with insufficient accuracy. The calorimetric method using pressurized superfluid helium is another method to measure the AC loss of the magnet.[5,6] Heat generated during ramp cycles is detected by temperature rise of the superfluid. Although this measurement needs a cooling system from 4.2 K to 1.8 K, temperature measurement does not need great accuracy unlike the case of the electrical method. Furthermore, the calorimetric method is a direct measurement for the cryogenic heat load.

We have measured the AC loss of the model dipole magnets by the calorimetric method using pressurized superfluid helium. The electrical method was also applied for calibration. Results of the measurement are presented in this paper.

CALORIMETRIC METHOD

1.8K cryostat

We have fabricated a helium cryostat with a cooling system from 4.2 K to 1.8 K (1.8 K cryostat) for the calorimetric AC loss measurement. This cryostat is similar in principle to that of Claudet[7] and Bon Mardion.[8,9] The cryostat consists of a vacuum vessel, a 80 K radiation shield, a 4.2K helium vessel, a 4.2K radiation shield, a 1.8 K pressurized superfluid helium vessel, a pre-cooler, a J-T valve, and a 1.8K saturated superfluid heat exchanger. The model dipole magnet is set in the 1.8 K vessel. The 1.8 K cryostat is shown in Figure 1. The 4.2K helium vessel and 1.8 K helium vessel are connected by stainless-steel pipes, which provides paths for current leads, signal wires and liquid helium supply to the 1.8 K vessel. The pipes are packed with FRP to prevent heat leakage from the 4.2 K vessel.

The magnet was first cooled from room temperature to 80 K. Liquid nitrogen was fed in the 1.8 K heat exchanger, cooled pressurized gaseous helium and the magnet. The magnet was cooled from 300 K to 80 K for 5 days. After cooldown to 80 K, liquid nitrogen was extracted from the 1.8 K heat exchanger, then liquid helium was fed to the 4.2 K helium vessel. The liquid helium dropped to the 1.8 K vessel through one of the connection pipes (the feed pipe). Evaporated gaseous helium in the 1.8 K vessel returned through another pipe (the return pipe) to the 4.2 K vessel. It took 5 hours to fill the 1.8 K vessel with liquid helium.

After filling the helium, the feed and return pipes were shut with FRP valves (pop-up valves), which functioned as a safety valve during the magnet quench. The J-T valve opened and liquid helium expanded to the 1.8 K heat exchanger, then the magnet and helium of 130 L were cooled to 1.8 K. The 1.8K heat exchanger was a stainless-steel can of 300 mm diameter and 130 mm height with copper tube fins. The pre-cooler was made of a stack of copper fins, stainless-steel rings and spacers as shown in Figure 2. These components were nickel-brazed in vacuum. Liquid helium was fed through the path of sandwiched stainless-steel rings and copper fins while the evaporated gas was fed through the tear-drop shaped copper fins and cooled the liquid helium to 2.5 K. Evaporated gas was extracted by a rotary pump and mechanical booster combination.

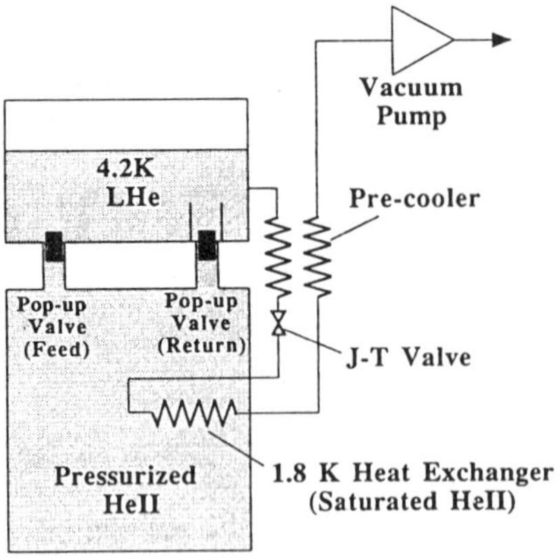

Figure 3. Schematic flow diagram.

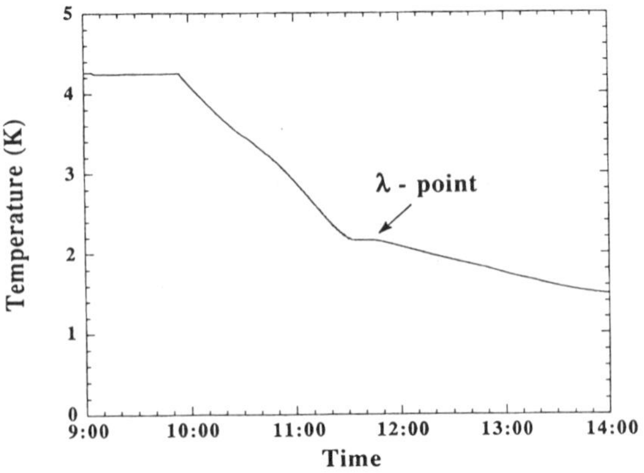

Figure 4. Cooling curve.

The pumping rate was 700 m^3/hr at 1.7 kPa. A schematic flow diagram is shown in Figure 3. The liquid helium reached λ-point within 90 minutes and was cooled to 1.8 K within 3 hours. A cooling curve is shown in Figure 4. The cooling power of the 1.8 K cryostat was 7.5 W at 1.8 K, and 3.5 W at 1.5 K. The heat leakage into the cryostat was around 1.6 W.

AC loss measurement

When the magnet temperature reached to 1.8 K, the J-T valve was closed and the 1.8 K heat exchanger was evacuated to extinguish the cooling power. A carbon glass resistor (CGR) thermometer, which was set in the superfluid helium bath, measured the temperature. It was calibrated from 44 K to 1.2 K. Constant DC current of 1μA was supplied to the CGR and the voltage was measured by a digital voltmeter. The magnet was ramped several cycles and temperature rise was measured. The standard current cycle was a ramp up from 500 to 5000 A and ramp down to 500 A with 5 second duration at the flat top. The ramp rates ranged from 100 to 400 A/s. A typical example of the temperature rise and ramp cycles is shown in Figure 5.

The AC loss was evaluated by the integration of the heat capacity of the superfluid helium;

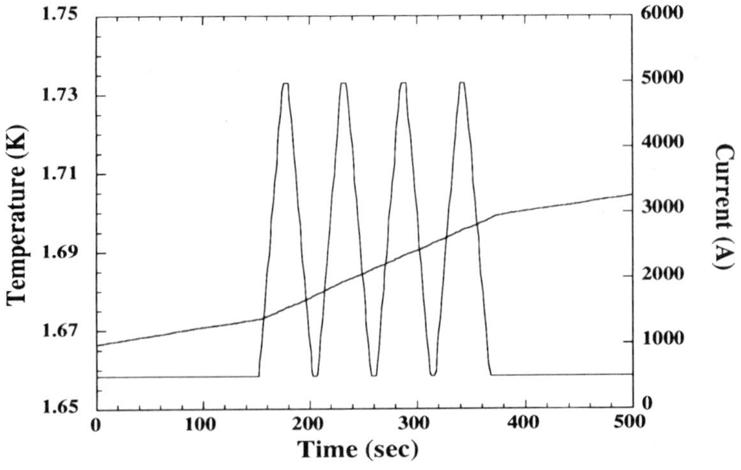

Figure 5. Temperature rise during ramp cycles.

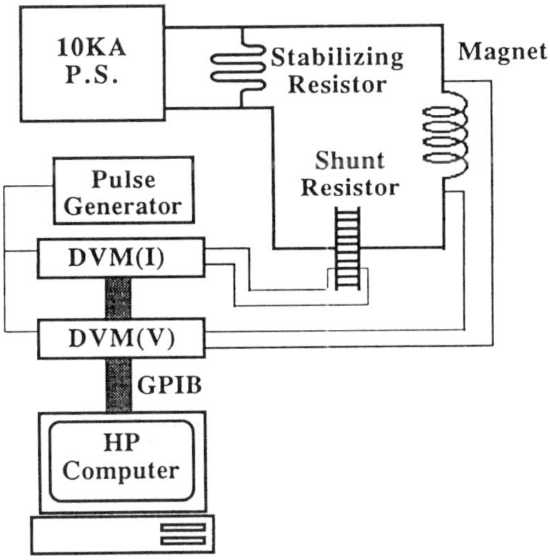

Figure 6. Electrical measurement system.

$$Q = \int m\, C_p\, dT\,,$$

where Q is the AC loss, m the mass, C_p the specific heat, and T the temperature of the superfluid helium. The background heat leakage of the cryostat was subtracted from the measured heat. The heat leakage was measured before and after the AC loss measurement. A heater, which was set at the bottom of the 1.8 K helium vessel, was used to calibrate the measurement system. The calibration was done from 1 to 6 W, which corresponded to the heat generation by the AC loss. The AC loss measurement was carried out in the temperature ranged from 1.5 K to 1.8 K. The temperature rise was measured to be about 20 mK.

ELECTRIC METHOD

AC loss of the magnet was evaluated by the integration of the coil voltage and current product over a ramp cycle;

Table 1. Summary of AC losses measured by the calorimetric method

Magnet	Filament diameter (μm)	Hysteresis (J/cycle)	Eddy Loss (J/cycle/(A/s))
KEK#3	2.5	79	0.21
KEK#4	6	147	0.22
KEK#6	6	124	0.32
KEK#9	6	127	0.21
KEK#10	6	124	0.25

Table 2. Summary of AC losses measured by the electrical method

Magnet	Hysteresis (J/cycle)	Eddy Loss (J/cycle/(A/s))
KEK#3	72	0.32
KEK#10	112	0.37

$$Q = \int I(t)\, V(t)\, dt\,,$$

where Q is the AC loss in a ramp cycle, I the current and V the voltage of the coil. This formula also counts the stored energy of the magnet, which was about 100 times greater than the AC loss. The coil voltage and the current must be measured with good accuracy and simultaneously. For this purpose, we used an integrating digital voltmeter (DVM) and measured the coil voltage directly. The coil current was determined by measuring the voltage across a shunt resistance using another DVM. Figure 6 shows the electrical measurement system. The voltage and the current were digitized by the DVM simultaneously with trigger signals from a pulse generator. The signal triggered the data acquisition every 0.2 sec (5 Hz). To reduce the large thyristor noise (600 Hz) of the 10 kA power supply and other commercial power-line noise (50 Hz), the DVM integrated the measuring voltage over 9 power-line cycles. To improve the response of the power supply, a stabilizing resistor (30 mΩ) was inserted parallel to the magnet. The digitized coil voltage and current were numerically integrated to obtain the AC loss of the magnet. The off-set voltage in the voltage measurement was assumed to be independent of time and was corrected to satisfy the completeness of the cycle;

$$\oint V(t)\, dt = 0\,.$$

RESULTS AND DISCUSSION

Calorimetric measurements were carried out for the magnet KEK#3 with 2.5-µm filaments and the magnets KEK#4, #6, #9 and #10 with 6-µm filaments. Figure 7 shows measured AC losses in Joule per ramp cycle as a function of ramp rate. The losses are linearly dependent of the ramp rate. In the previous measurement,[10] AC losses of KEK#3 were observed to be systematically large at low ramp rates where the losses should be small. So the loss of KEK#3 was re-measured. The losses of KEK#9 were measured three times in the different experiments. The reproducibility of the measurement was 3 %.

The losses are divided into the hysteresis losses in the superconductor cable and the iron yoke and the eddy-current losses of the inter-filament and inter-strand couplings. The extrapolated AC loss at zero ramp rate represents the hysteresis loss, while the slope represents the eddy-current loss. Summary of the loss is listed in Table 1. The magnets with 6-µm filaments (KEK#4, #6, #9, #10) show the hysteresis losses around 130 J/cycle, while the magnet with 2.5-µm filaments (KEK#3) shows the loss of 79 J/cycle. This difference can be understood as the filament diameter dependence of the hysteresis loss in the superconductor cable.

The eddy-current losses do not show significant difference from magnet to magnet. The loss of the inter-strand coupling depends on the strand to strand contact resistance. The contact resistance is expected to be influenced by the heat treatment and pressure applied in the curing. Our magnets were made in the same manner in cure cycle. This means the contact resistance is almost the same for all magnets and results in almost the same eddy-current losses.

Electrical AC loss measurements were carried out for the magnets KEK#3 and #10. Figure 8 shows the results of the electrical measurement together with the results of calorimetric measurement. Reproducibility of the electrical measurement was 10 %. The results of two methods agree well at low ramp rates. At higher ramp rates, discrepancy increases, however, it does not exceed the experimental reproducibility. The electrical measurement shows slightly higher ramp rate dependence. Summary of the losses measured by the electrical method is listed in Table 2.

Hysteresis loss in the superconductor filament with cylindrical geometry perpendicular to the field is proportionally dependent on the filament diameter.[11] The hysteresis loss is given by

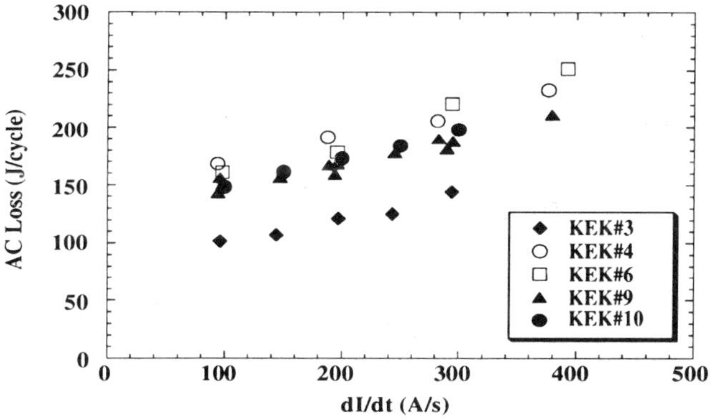

Figure 7. AC losses measured by the calorimetric method.

$$Q = \frac{8}{3\pi} d_f J_c(B_0) B_0 \ln(\frac{B_2 + B_0}{B_1 + B_0}),$$

where Q is the hysteresis loss, d_f the filament diameter, J_c the critical current density, B_1 the maximum field, B_2 the minimum field, and B_0 the characteristic field. Definition of B_0 is $J_c(B) = J_{c0}B_0/(B+B_0)$, where $J_{c0} = 2J_c(B_0)$. To clarify the filament dependence,

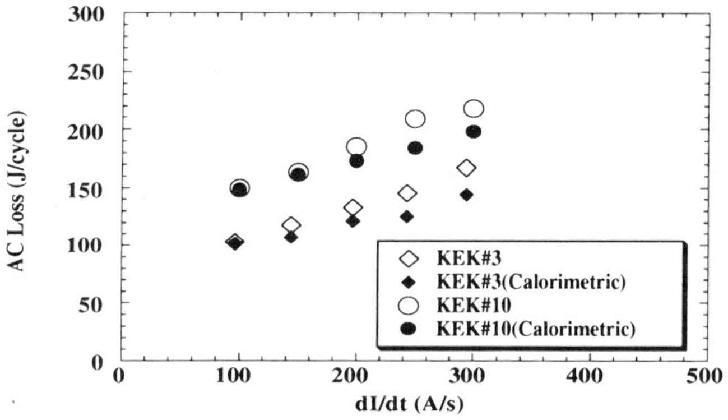

Figure 8. AC losses measured by the electrical method.

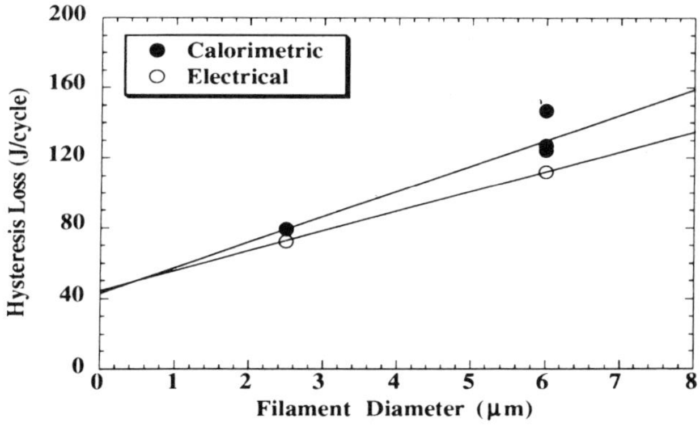

Figure 9. Hysteresis losses as a function of filament diameter.

hysteresis losses measured by the calorimetric and electrical methods are plotted in Figure 9 as a function of filament diameter. The volume fraction of the superconductor in the inner cable with 2.5-μm filaments is 0.4 while that with 6-μm filaments is 0.44 and the critical current density in the cable with 2.5-μm filaments is slightly different from that with 6-μm filaments, however, these differences are not greater than 10 %. If we assume the difference in the hysteresis losses is only due to the filament diameter, the extrapolated AC loss at zero diameter represents the hysteresis loss in the iron yoke. The hysteresis loss of the magnet measured by the calorimetric method can be divided into the loss in the iron yoke (42 J) and the loss in the superconductor (37 J for 2.5-μm filaments and 88 J for 6-μm filaments).

CONCLUSION

We have measured AC losses of the 1-m long model dipole magnets by the calorimetric method using pressurized superfluid helium at 1.8 K. The eddy-current losses were almost the same for all magnet. The hysteresis loss of the 2.5-μm filaments was significantly smaller than that of the 6-μm filaments. An electrical method was applied for the AC loss measurement. Results of the electrical measurement showed slightly higher ramp rate dependence, however, agreed with the calorimetric measurement within the experimental accuracy.

ACKNOWLEDGMENTS

Authors would like to express our gratitude to Professors H. Hirabayashi and Y. Kimura for their continuous support and encouragement in this work.

REFERENCES

1. K. Hosoyama et al., Fabrication and test of a 5-cm aperture, 1-m long SSC collider dipole magnet, in: "Supercollider 3," John Nonte, ed., Plenum Press, New York (1991), p. 615.
2. K. Hosoyama et al., A status report on the development of 5-cm aperture, 1-m long SSC dipole magnet at KEK, in: "Supercollider 5," Phyllis Hale, ed., Plenum Press, New York (1994), p. 229.
3. K. Hosoyama et al., Fabrication and testing of 1-m-long SSC model dipole magnets, to be presented at CEC, Columbus (1995).
4. S. Delchamps et al., AC loss measurement of SSC dipole magnets, in: "Proc. Workshop on AC Superconductivity, " Kiyosumi Tsuchiya, ed., KEK proceedings 92-14, Japan (1992), p. 19.
5. R. Aymar, et al., Test of a model coil of "TORE SUPRA," IEEE Trans. Magn., Vol. MAG-17, (1981), p. 38.
6. S. Caspi, The use of calorimetry in superfluid He II to measure losses in superconducting magnets, in:"Proc. 9th International Cryogenic Engineering Conference," Yasukochi and Nagano, ed., Butterworth & Co, Guilford, England (1982), p. 347.
7. G. Claudet, A. Lacaze, P. Roubeau and J. Verdier, The design and operation of a refrigerator system using superfluid helium, in: "Proc. 5th International Cryogenic Engineering Conference," K. Mendelssohn, ed., IPC Science and Technology Press, Guilford, England (1974), p. 265.
8. G. Bon Mardion, G. Claudet and J. C. Vallier, Superfluid helium bath for superconducting magnets, in: "Proc. 6th International Cryogenic Engineering Conference," K. Mendelssohn, ed., IPC Science and Technology Press, Guilford, England (1976), p. 159.
9. G. Bon Mardion, G. Claudet, P. Seyfert and J. C. Verdier, Helium II in low-temperature and Superconductive magnet engineering, in: "Advances in Cryogenic Engineering," Vol. 23, K. D. Timmerhaus, ed., Plenum Press, New York (1978), p. 358.
10. Y. Morita et al., Calorimetric AC loss measurement of 1.3-m model dipole magnet by using a 1.8 K cryostat, in: "Supercollider 5," Phyllis Hale, ed., Plenum Press, New York (1994), p. 679.
11. M. N. Wilson, "Superconducting Magnets," Clarendon Press, Oxford (1983).

HTS COIL AND JOINT DEVELOPMENT FOR A 5T NMR INSERT COIL

P.V. Shoaff, Jr.[1,2], J. Schwartz[1,2], S.W. Van Sciver[1,2], and H. W. Weijers[1]

[1]National High Magnetic Field Laboratory
Tallahassee, Florida, 32306-4005, USA
[2]Department of Mechanical Engineering, FAMU/FSU College
Tallahassee, Florida, 32306, USA

ABSTRACT

A HTS coil fabrication and joint development project was undertaken to develop the technology necessary for the production of a 5 T HTS insert coil for NMR applications. A series of minicoils are being fabricated to explore various construction options such as winding technique, structural support, and electrical insulation. The results from one of these coils are presented. Present conductor technology cannot produce the long length of conductor needed in the insert coil design, therefore, a means must be found to join several lengths of conductor together. Several joint prototypes are investigated.

INTRODUCTION

A major goal of the National High Magnetic Field Laboratory is to provide 1 GHz NMR facilities.[1] A preliminary design for such a facility includes a 25 T magnet which contains a set of outer LTS coils and a high temperature superconductor insert coil operated at low temperature. This design calls for the HTS coil to generate 5 T in a background field of 20T. It will require a high quality conductor in longer lengths than have been demonstrated to date. In addition, the size and field of the coil result in stresses in the windings which will require reinforcement of the conductor. The HTS coil fabrication and joint development project was undertaken to explore and develop the technology necessary for the production of the HTS insert coil.

EXPERIMENTAL SETUP

The superconducting properties of many high temperature superconductors are excellent for possible applications in high field magnets. Our project chose to use BSCCO 2212 conductor to produce a wind and react coil with the required delta B of 5 T. This decision places special requirements on the winding, reinforcement, and insulation of the coil. Currently, a series of minicoils are being produced to develop the fabrication techniques needed to create wind and react coils using BSCCO 2212 wires and tapes. This series will enable us to investigate various insulation and reinforcement options to determine which ones are best suited to meet the demands placed on the proposed insert coil. The minicoil studied in this paper is one of the first in the above mentioned series. It

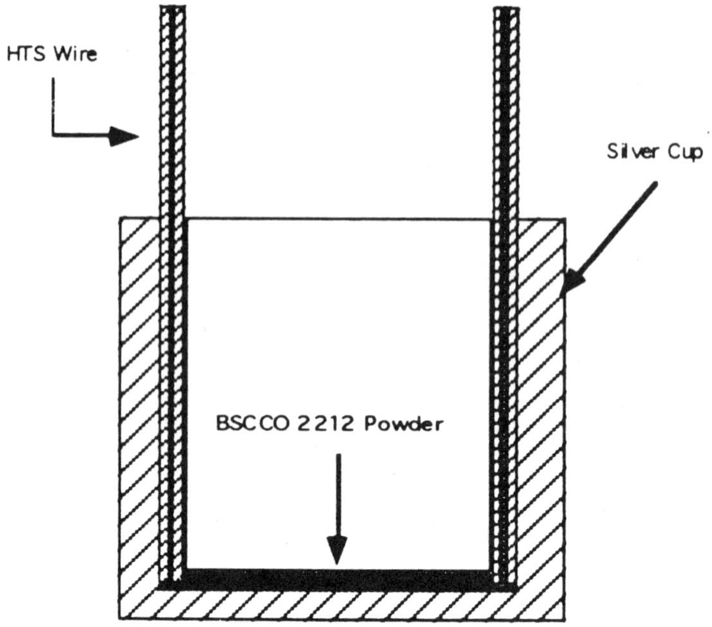

Figure 1. Schematic of a typical cup joint. The silver cup is 1.2 cm high with an I.D. of 0.6 cm and an O.D. of 1.0 cm.

is 2.0 cm high and has a diameter of 3.8 cm. This minicoil is wound using monocore BSCCO 2212 round wire conductor with a silver matrix on a stainless steel former. It is insulated with a quartz glass sleeve and reacted using the heat treatment developed for round wires at NHMFL.[2] Finally, the outer layer is painted with epoxy resin to provide reinforcement.

In addition to the investigations of coil fabrication, development has begun on superconducting joints using HTS materials. Initial coil design calls for the insert to be wound from several lengths of multifilament, powder in tube, BSCCO 2212 tapes or round wires joined together. The 5 T coil will require on the order of 10 km of conductor which makes the use of one length impractical. In order to meet these demands, the HTS joint development program studied the following joint prototypes. These prototypes utilize the partial melt BSCCO 2212 PIT reaction process to connect the current paths between superconducting cores of the free ends.[3] It is hoped that these techniques can be successfully upgraded to multifilament conductors.

The cup joint model was the first prototype to be investigated (See Figure 1). These joints are made by placing two green BSCCO 2212 monocore tapes into a silver 'cup', filling the cup with unreacted powder, pressing the powder into the cup against the tapes, and then running the whole assembly through the usual heat treatment for monocore tapes. The powder used is of the same stoichiometric composition as that used to produce the PIT conductor.[2]

The second joint type investigated is the wrap joint (See Figure 2). It is similar to the cup joint except it uses a thin silver film to wrap the joint package. The wrap joints are

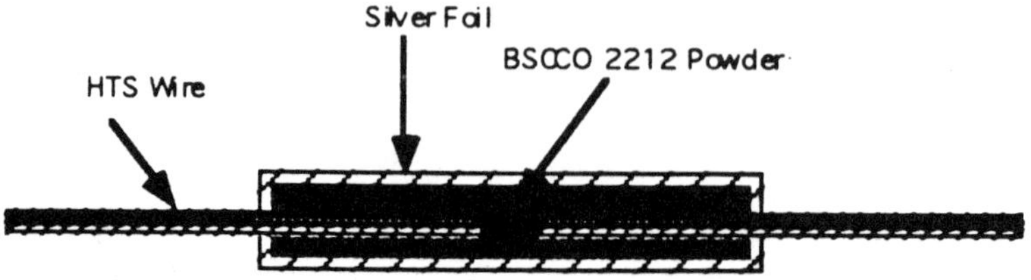

Figure 2. Schematic of a typical wrap joint. The joint body measures 2.5 X 0.2 cm.

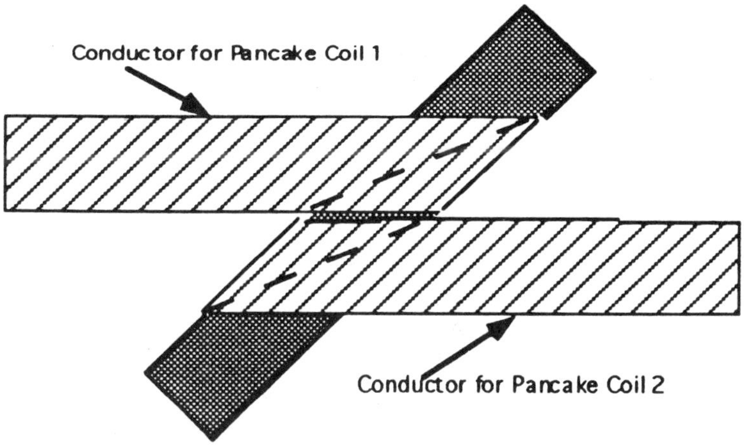

Figure 3. Top view schematic of a bend junction.

constructed by overlapping the free ends of the green conductor while laying them on a piece of 0.1 mm thick silver foil inside of a press mold. The tapes enter from opposite sides but this geometry can be easily modified. The free space around the conductor is filled with unreacted powder and the interior of the mold is compacted using a uni-axial press. When this compaction is completed, the assembly is removed from the mold, the silver foil is wrapped around the joint, and its edges are sealed with Leitsilver 200 silver paint. It is then reacted with the same heat treatment as the cup joints.

The final prototype is the bend junction (See Figure 3). It is not a true joint but a way of winding two adjacent pancakes from one piece of conductor. If a stacked pancake insert coil design is ultimately chosen, the bend junctions could possibly reduce the number of joints needed and, as a result, make coil production easier.

Four of these bend junctions were modeled by taking lengths of monocore tape 10 cm long, annealing them at 400° C for 30 minutes to soften the silver, and bending them twice around a form to produce parallel current paths on either side of the form. Note that in an application these current paths will be used to roll the adjacent pancakes. In this test the form is constructed from a stainless steel tube that is rolled flat to a thickness of 0.1 mm and a width of 5 mm. The assembly is mounted on a G-10 probe and each tape electrically connected in series.

For all samples, the four point measurement method is used to categorize the superconducting transport properties at 4.2K. The electric field criterion of 1 μV/cm is used to determine the critical current values. In the cases of both the wrap joint and the bend junction the current path is assumed to equal the tape length between voltage taps. Applying a standard criterion to characterize the cup joint is difficult because the current path cannot be easily defined. However, for the application as a 5T insert coil, the coil performance should be limited by the transport properties of the superconductor and not by the performance of the joint. Therefore, as long as the joint properties are at least equal to the bulk conductor's properties, we have a viable prototype. Thus, by taking the critical current value of a straight sample determined by a criterion of 1 μV/cm, we can characterize the cup joint by comparing the voltage produced across the cup joint at this current value to the limit imposed by the criterion.

RESULTS AND DISCUSSION

The Minicoil

The transport properties for the minicoils across both the coil as a whole and independent sections are illustrated below in Figure 4. It can be seen that the overall coil

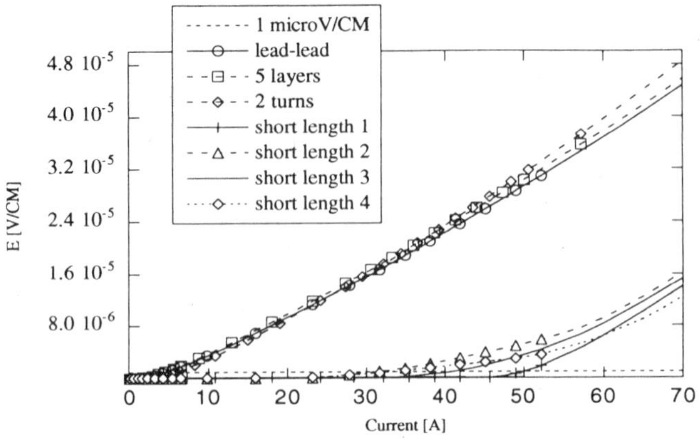

Figure 4. E-I curves across different sections of a HTS mini-coil.

performance is poor, with critical currents less than 10 A. However, some of the short sections exhibit properties in the same range as straight short samples of the same conductor batch. These short samples are heat treated with the minicoil and have critical currents of 50+ A. The exact mechanisms that cause these inhomogeneities are under investigation. The sections with strong properties and the mechanical stability of the coil indicate a promising start for our coil fabrication technology.

Cup Joints

The finished cup joints are bulky and awkward to handle. Using the stated criterion, the straight sample is found to have an Ic of 7.6 A (See Figure 5). At this current value the joint, which includes 2 cm of tape, shows a voltage of 1.5 μV. Thus, there appears to be a good chance of developing a joint that will not limit the performance of the coil beyond the restrictions imposed by the straight wire itself. Note also that the graph of the normal transition of the joint has several distinct regions suggesting an uneven distribution of superconducting properties. Therefore, the cup joint might be usable in the prescribed application if slight optimization will increase the superconducting transport current and sharpen the transition.

Wrap Joints

The wrap joints are less bulky than the cup joints and easier to fabricate. An example is seen in Figure 6 where the joint has excellent transport values relative to a straight

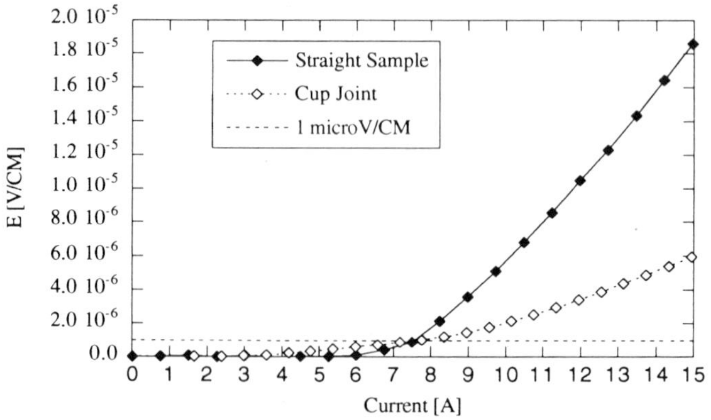

Figure 5. A comparison of the E-I curves measured across a cup joint and straight sample.

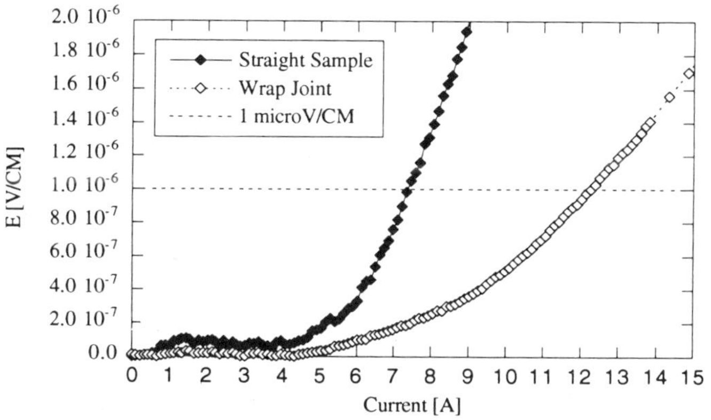

Figure 6. A comparison of the E-I curves measured across a wrap joint and a straight sample.

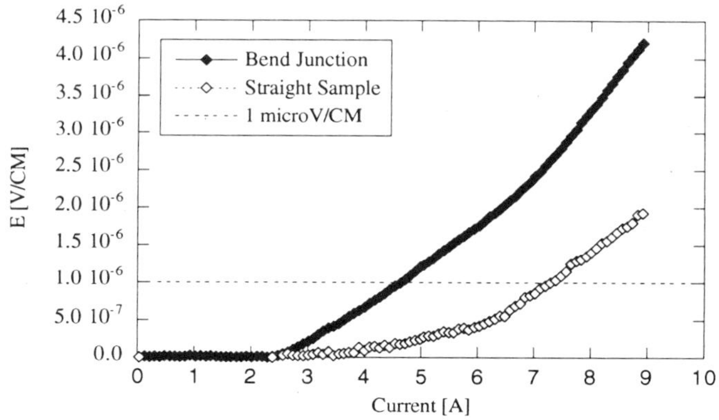

Figure 7. A comparison of typical E-I curves for bend junctions obtained in zero applied field.

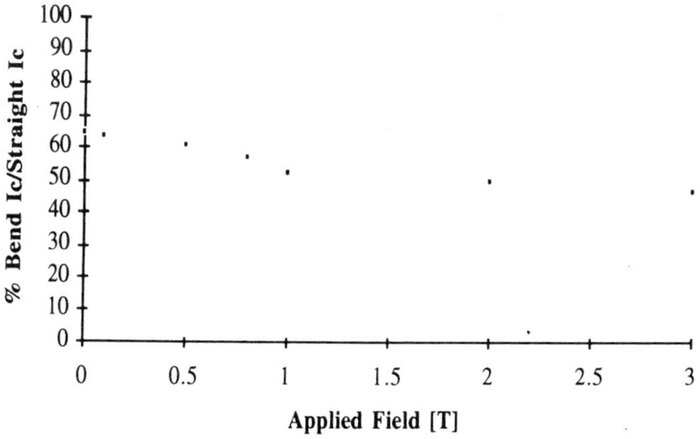

Figure 8. A comparison of Ic degradation between the straight sample and the bend junction. At zero applied field the bend junction performs at 65% of the straight sample capacity.

sample and a sharper transition than the cup joint. Using the 1μV/cm criterion, the wrap joint has an Ic of 12.3 A which outperforms the best equivalent straight sample Ic of 7.7 A. This appears to be the most promising option to date.

Bend Junctions

The application of the bend junctions will only work if the mechanical deformations required do not substantially reduce the transport properties along the whole coil by damaging the green conductor irreparably or by making the coil more sensitive to an external magnetic field. It is predicted that the bend junctions will not behave as well as the straight sections of the coil because, across the bends, the applied magnetic field will not be perpendicular to the c-axis of the tape grains. This study determines the degradation of the current carrying capabilities of the reacted tape as a result of the bending of the green conductor and the relative performance of bend junctions versus straight samples in an applied magnetic field. It is seen that the bending process causes 35% degradation in the performance of the conductor used in the bend junction. In zero applied field the bend junction has an Ic of 4.6 A while the Ic of the straight section measured 7.2 A (See Figure 7). The percentage of decrease in Ic due to an applied magnetic field is illustrated in Figure 8 for both the bend junction and a straight sample. It is evident that the difference between the bend junction Ic and the straight sample Ic increases as the applied field increases.

CONCLUSIONS

Some sections of the minicoil have transport properties in the same range as straight short samples of equivalent conductor, however, the overall coil performance is far below these values. The sections with strong properties and the mechanical stability of the coil are a promising start for our coil fabrication technology.

The cup joints are physically bulky but have encouraging transport properties. Optimization is needed to increase the superconducting transport values and sharpen the normal transition.

The wrap joints have the best transport properties to date. They are easier to fabricate and their geometry can be modified. The wrap joint appears to be the most promising prototype studied.

The presented bend junctions operate at only 65% of the capacity of the straight samples in zero applied field. Thus the bending of the green tapes strongly affects the performance of the reacted tape. It is also evident that the decrease in Ic of the bend junction is more field sensitive than the Ic of an equivalent straight sample. This is probably caused by the changing field orientation across the junction. The prospects for this type of joint are unclear.

ACKNOWLEDGMENTS

The authors wish to thank Jeorg Kessler and Wangshui Wei for their contributions and Yusuf Hascicek for providing invaluable advice and vital equipment. This work was supported by the National Science Foundation through the National High Magnetic Field Laboratory, Grant-No. 9016241.

REFERENCES

1. D. Markiewicz et. al., 25 T High Resolution NMR Program and Technology, Presented at the Fourteenth International Conference on Magnet Technology, Finland, June 1995, to be Published in IEEE Transactions on Magnetics, July 1996.

2 J Kessler et. al., Properties of Bi(2212) Short and Long Wires and Tapes, to be Presented at the ICMC, Columbus, Ohio, July 17-21, 1995.

3. R. D. Ray, II, "Alignment and Microstructural Evolution in Melt -Processed Ag-Clad $Bi_2Sr_2CaCu_2O_y$ High T_c Superconducting Tape," University of Wisconsin-Madison, USA, Ph.D. Thesis, (1993).

Bi-2212/Ag TAPE CONDUCTOR AND COIL DEVELOPMENT

G.W. Albert ,[1] R. Zhou,[2] E. M. Leung,[1] K.V. Salazar,[2]
T.G. Holesinger,[2] M.D. Daugherty,[2] and J. O. Willis[2]

[1]Lockheed Martin Advanced Development Operations
San Diego, California, 92128, USA
[2]Los Alamos National Laboratory
Los Alamos, New Mexico, 87975, USA

ABSTRACT

This paper presents recent progress by Los Alamos National Laboratory/ Superconductivity Technology Center (LANL/STC) and Lockheed Martin Corporation/ Advanced Development Operations (LMC/ADO) Energy and Power Systems group to jointly develop and test high temperature superconductor coils for practical applications such as motors, generators, magnetic bearings, superconducting magnetic energy storage (SMES), MAGLEV trains, or electrical inductors. Critical currents of 10^5 A/cm^2 have been achieved in a Bi-2212/Ag tape conductor using a patented Ag addition process to improve core morphology and uniformity. Results of testing to measure key conductor parameters are presented along with design and analysis details of a general-purpose multi-layer HTS coil.

INTRODUCTION

High temperature superconductors (HTS) are still plagued with problems which limit their usefulness in applications presently using copper and low temperature superconductors. The problems include relatively low mechanical strength, low available strain, low current density, flux pinning/weak links, difficulty in manufacturing long lengths, wind/react electrical insulation systems, and high cost, to name a few. Our efforts have been directed toward characterizing and understanding these problems and developing solutions for them. In the past five years researchers have made great leaps in increasing current densities by improving texture and uniformity of core materials. Cost of HTS conductors has also been consistently decreasing. HTS magnet applications are gradually becoming a reality.

CONDUCTOR DEVELOPMENT

Researchers at LANL have found that the addition of large Ag particles to the superconductive core of oxide powder in tube tapes results in better geometric uniformity during mechanical deformation thus resulting in thinner cores with concomitant better transport properties than can usually be achieved without resorting to multifilamentary conductors. We added 20 wt% of 99.99% pure silver powder with an average size of 38 μm to Bi-2212 powder from Seattle Specialty Ceramics, which had an average size of 4 μm, and thoroughly mixed them. The mixed powders were hand-packed into a Ag tube, sealed,

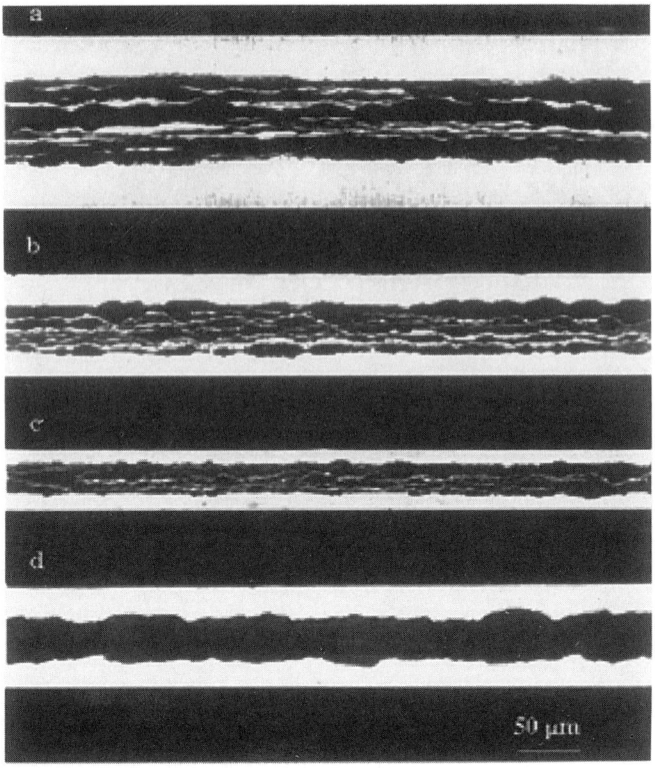

Figure 1. The longitudinal cross sections of as-rolled tapes containing Ag ribbons with core thicknesses of (a) 70 μm, (b) 40 μm, and (c) 23 μm; (d) has no ribbons and shows significantly more cracking and sausaging.

drawn to 2 mm diameter, then rolled to various thicknesses. The superconductor core morphology and cross sections were examined by optical microscopy.

The samples were heated in air at a rate of 10°C/hour to 880°C and held for 20 minutes. Then the samples were cooled to 845°C at 10°C/hour, from 845°C to 835°C over 24 hours, and finally furnace-cooled to room temperature. The phase assemblage was investigated by x-ray diffraction after mechanically removing the silver sheath. We found that the silver particles remained spherical during wire drawing due to the low (45%) packing density of the superconductor powder but they were deformed into long, thin ribbons during the rolling process. Silver powders with particle sizes from 3 to 200 μm were investigated; a size of 38 μm resulted in the best core uniformity and Ag deformation.

Figure 1 shows the longitudinal cross sections of as-rolled tapes containing Ag ribbons and one without for comparison. It can be seen that the silver particles are stretched into ribbons in the longitudinal direction. These silver ribbons not only improve the mechanical properties during deformation by eliminating the sausaging and cracking, etc., that are observed in the tape without Ag ribbons but also provide extra superconductor/Ag interfaces, which is believed to benefit the Bi-2212 phase formation and grain alignment. This is

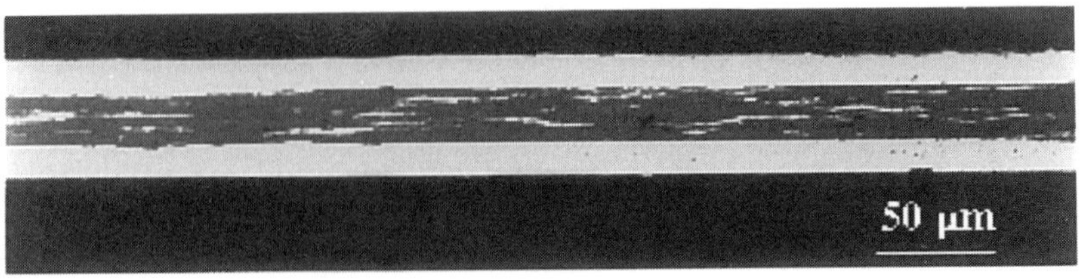

Figure 2. The transverse cross section of the 51 μm thick tape containing Ag ribbons and a 23 μm core.

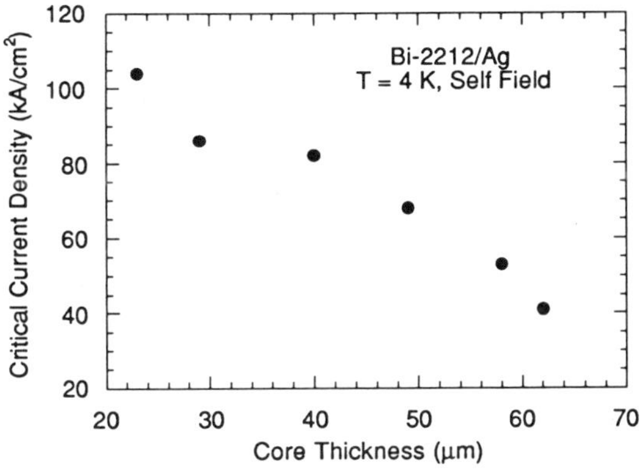

Figure 3. Critical current density J_c *vs.* superconductor core thickness at 4 K, self field.

supported by our test results. Because of the presence of Ag ribbons in the superconductor core, we could roll the tape as thin as 51 μm without necking, sausaging, or cracking, which is the best result we have had by various techniques (see Figure 2).

The silver ribbons are spheroidized at the end of the heat treatment. This phenomenon results from the diffusion of silver and occurs to minimize surface area. The ribbons remain after the melting of the Bi-2212 and consequent cooling to 845°C with some thickening but spheroidize during the cooling from 845°C to 835°C. X-ray diffraction patterns of such a fully processed tape with Ag ribbons showed no obvious secondary phases and a high degree of c-axis orientation.

TEST RESULTS

Critical Current

Transport critical current measurements were performed at 4 K and 26.2 K with a 1 μv/cm criterion. Figure 3 shows the critical current density J_c at 4 K and self field as a function of superconductor core thickness. As the core thickness decreases to about 23 μm, the critical current density exceeds 10^5 A/cm^2, which is about 60% higher than that of the tape without the Ag ribbons but processed identically. In Figure 4, J_c of the tape with a 29 μm core is plotted as a function of magnetic field at liquid neon temperature. At 1 T the critical current density of the tape is about 2.5×10^4 A/cm^2 with the field parallel to the tape (≈ab) plane.

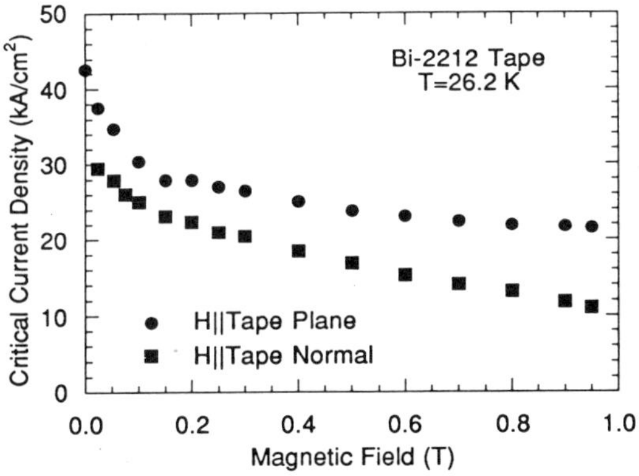

Figure 4. Critical current density J_c of the tape with a 29 μm core plotted as a function of magnetic field H at liquid neon temperature (26.2 K).

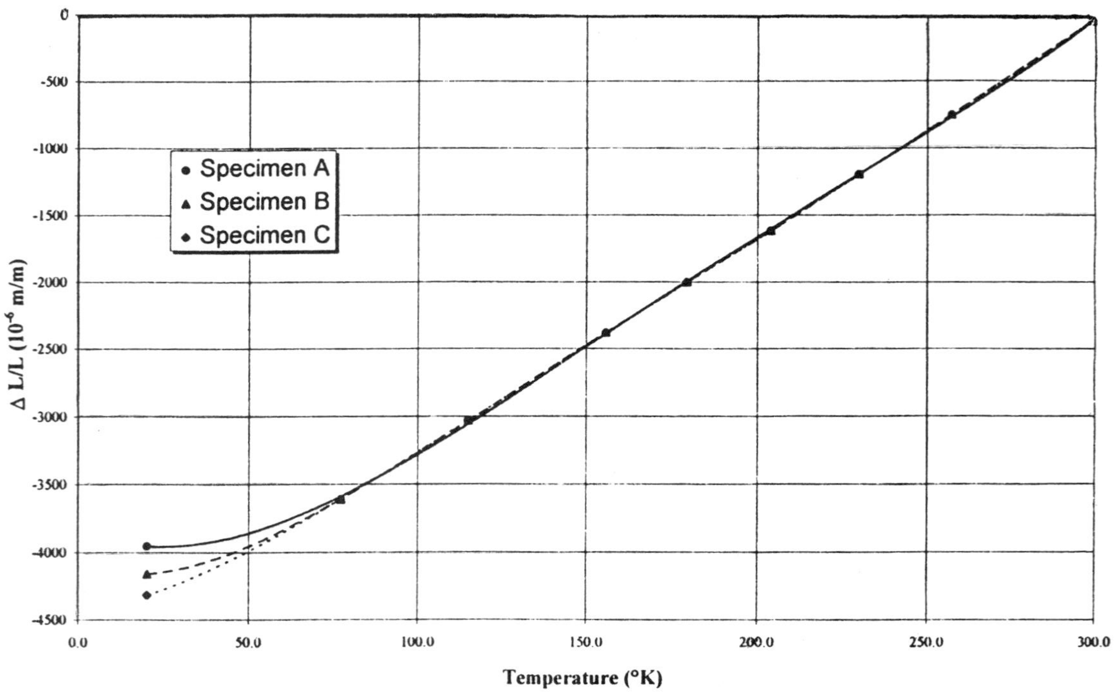

Figure 5. Linear thermal expansion versus temperature for tape specimens RZ694-A, B, and C.

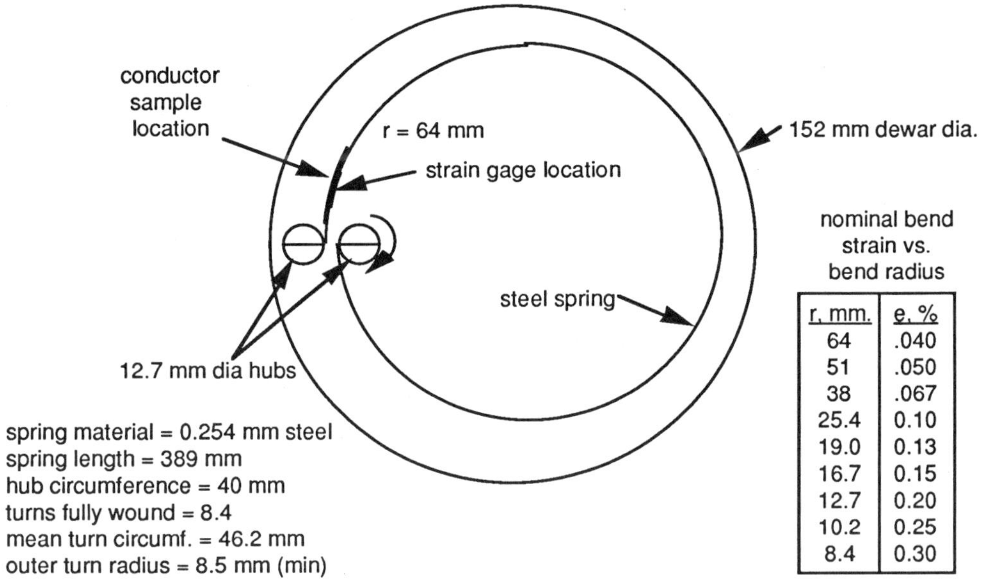

r, mm.	e, %
64	.040
51	.050
38	.067
25.4	0.10
19.0	0.13
16.7	0.15
12.7	0.20
10.2	0.25
8.4	0.30

Figure 6. Bending strain fixture, top view.

Thermal Expansion

Three 100 mm tape specimens of Bi-2212 were tested for thermal expansion from 20 K to 300 K using a tube dilatometer cooled with liquid hydrogen. The chart in Figure 5 reflects the average coefficient of linear thermal expansion (CTE) between 100 and 300 K of 16 µm/m•K. This value is essentially identical to published values for pure Ag.[1] Below 100 K we observe some variability in the data between specimens. This may be attributed to interaction between the Ag sheath in tension and BSCCO core material in compression at the lower temperatures. In general, the Ag was the dominant factor in the thermal expansion behavior of the tapes throughout most of the cryogenic temperature range.

PLANNED TESTING

Additional testing is planned to investigate tensile strength, modulus, and susceptibility of the tape conductors to bending strain. The fragile nature of the thin tapes make them difficult to handle and instrument for strain measurement. A special batch of 0.13 mm thick, 90 mm long Bi-2212 tape specimens have been prepared specifically for this purpose. After considering several alternatives for measuring conductor strain, we decided to use a single small thin strain gage (Measurements Group EA 13-031DE-350) on each specimen. The calculated effect of gage reinforcement would be around 5%. Tensile tests will be performed first on dummy tape specimens made of copper foil with similar dimensions to the tape conductor specimens to establish validity of the test method. A detailed test plan has been prepared for this task.

The results of the thermal expansion testing indicate that the annealed silver sheath is the dominant component along the linear axis of the tape conductor. Thus, it is very likely that the sheath will dominate the tensile strength and modulus as well. To maintain strain of the BSCCO core material below the typical damage threshold of 0.1%,[2] a conservative approach is to keep tensile loads below the yield point of the annealed silver. For a tape conductor with a silver-to-superconductor ratio of 3:1, maximum allowable tensile load is calculated as follows.

yield strength of silver at 300 K: $\sigma = 2.0\text{ e+7 Pa}$ (1)

cross section area of the silver sheath: $a = 3.0\text{ e-7 m}^2$ (2)

Table 1. Conductor and coil parameters

Parameter	Value
conductor width	3.2 mm
conductor thickness	0.13 mm
conductor length	27.9 m
I_C @ 4K, 0T	75 A
Ag/SC ratio	2:1
outside diameter	5.28 cm
inside diameter	2.64 cm
height	1.32 cm
turns/layer	56
layers	4
inductance	1.7 mH

maximum allowable load:
$$F = \sigma a \quad (3)$$
$$= (2.0\ e{+}7)(3.0\ e{-}7)$$
$$= 6\ N\ (1.35\ lbf)$$

This defines our upper limit for tensile loads such as handling and coil winding at ambient temperature. The tensile strength of silver increases by 50% at 77K, so our maximum conductor operating load should be limited to 9 N (30 MPa) including thermal and magnetic loads.

For measuring conductor degradation as a function of bending strain, a special fixture was designed for controlling conductor bending strain at the bottom of a cryogenic dewar is shown in Figure 6.

This setup is capable of varying tape specimen bend radii from r = 64 to 8.4 mm by rotating the inboard hub clockwise until the steel spring is tightly wound around it. This corresponds to a strain range of 0.04 to 0.3%. A strain gage fixed to the spring near the tape specimen could be used to monitor bending strain, hence bend radius, of the spring adjacent to the specimen. Values in the table of strain versus bend radius were calculated using the following equation:

$$e = \partial / r \quad (4)$$

where:
e = strain, $\Delta l / l$
∂ = offset from the neutral axis
r = bend radius at the neutral axis

To keep strain below the 0.1% limit, a monocore tape conductor with a core thickness of 0.05 mm must not be bent to a radius of less than 25 mm, assuming the neutral axis passes through the center of the core region, as shown in Figure 7.

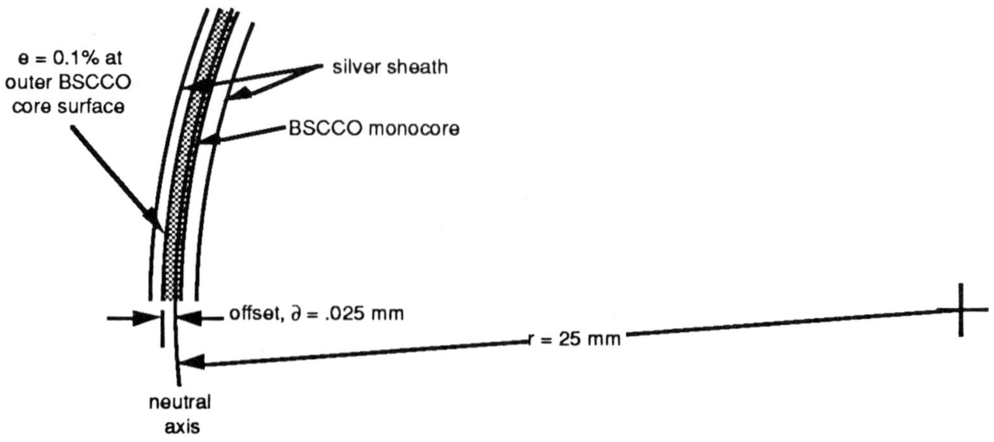

Figure 7. Tape conductor bend radius for 0.1% maximum strain .

Conductors with thicker core sections should be limited to even larger minimum bend radii according to Eq. (4).

COIL DEVELOPMENT

We have developed a coil design for a generic electrical inductor application using Bi-2212 tape conductors in a stacked pancake configuration. This coil would be constructed from four 56-turn pancake layers, each requiring 6.97 meters of conductor. Conductor and coil parameters for our design are listed in Table 1.

Magnetostatic and mechanical stress analyses were performed assuming a conductor current of 20.0 amps. Total calculated stored energy was 0.314 J. The peak field of 0.171 Tesla occurs at the inner surface of the coil. Maximum hoop stress at this point from Lorentz force is less than 60 kPa, compared to 30 MPa stress capability of the conductor at 300K.

CONCLUSIONS

We found that for powder-in-tube tape fabrication, addition of 38 μm silver particles to the Bi-2212 powder produced long, thin ribbons during rolling. This ceramic/metal composite has excellent properties during deformation. Tapes were rolled as thin as 51 μm with a 23 μm core with no necking, sausaging, or cracking in the superconductor core. The Ag ribbons also provide additional Ag/superconductor interfaces to help Bi-2212 phase formation and grain alignment. Improved critical current density on the order of 10^5 A/cm^2 was obtained at 4 K with a core thickness of 23 μm. Linear thermal expansion measurements indicate that the Ag is a dominant factor in the CTE behavior of these composite conductors at cryogenic temperature. Magnetic and mechanical stress analyses indicate good mechanical integrity of the 4-layer pancake coil design at rated operating conditions. The additional testing, coil assembly, and test are expected to be completed in the coming year.

ACKNOWLEDGMENTS

Work at Los Alamos National Laboratory was performed under the auspices of the U.S. Department of Energy, Office of Energy Management. We thank (at Los Alamos National Laboratory) J.F. Bingert, E.J. Peterson, and J.L. Smith for fruitful discussions and W.L. Hults and J.Y.Coulter for technical assistance.

REFERENCES

1. D.R. Smith and F.R. Fickett, "Cryogenic Properties of Silver", National Institute of Standards and Technology, Boulder.
2. M. Suenaga, Presentation at High Temperature Superconductivity Wire Development Workshop, St. Petersburg, Florida, (February 16, 1994).

CONSIDERATION OF USING HTS COILS AS A MAGNETIC CORE BIAS WINDING

J. X. Jin,[1] S. X. Dou,[1] R. Bhasale,[1] H. K. Liu,[1]
C. Grantham,[2] N. Savvides,[3] and M. Ionescu[1]

[1]Center for Superconducting and Electronic Materials
Dept. of Materials Engineering
University of Wollongong, NSW 2522, Australia
[2]Dept. of Electric Power
University of NSW, Sydney 2052, Australia
[3]CSIRO Division of Applied Physics
P.O.Box 218, Lindfield, NSW 2070, Australia

ABSTRACT

$(Bi,Pb)_2Sr_2Ca_2Cu_3O_{10+x}$ Ag clad wires, both single core and multifilament, were used to produce high T_c superconducting coils under a reaction and winding procedure. The magnetic fields generated by the produced coils are measured at 77 K and 4.2 K with and without a magnetic core. The superconducting coil made by the multifilament superconducting wire, used as a bias winding of a magnetic core, has critical current ampere-turns about 1591 A-t at 77 K and 13954 A-t at 4.2 K. The magnetic core influence on the superconducting coils is investigated by measuring I-V characteristic changes of the coils with different magnetic cores. The difficulty of using an existing $(Bi,Pb)_2Sr_2Ca_2Cu_3O_{10+x}$ superconducting wire at 77 K to generate high magnetic field, is identified as its poor magnetic field dependent critical current at 77 K. However, the achieved critical current ampere-turns of the coil at 77 K, is suitable for satisfying some low field applications.

INTRODUCTION

High T_c superconducting (HTS) $(Bi,Pb)_2Sr_2Ca_2Cu_3O_{10+x}$ Ag clad wires have been investigated for use in very high field applications. At 4.2 K, this HTS wire can maintain about 10^5 A/cm^2 critical current density under 23 T magnetic field.[1] As a hybrid magnet coil, a HTS coil made by this superconducting wire has been made to generate 1 T in a 20 T background field operated at 4.2 K, and as a magnet, the HTS coil has generated about 23 T at 4.2 K.[2,3] However, related to the flux pinning problems of this HTS wire,[4] currently this wire is not suitable for applications to generate very high field or to be worked under high field at 77 K.

However for some electrical applications, this HTS wire can be suitable to make a superconducting coil operated at 77 K. For example, devices such as a magnetic core saturable reactor or a HTS inductor,[5,6] in which the HTS wire is not necessary to be in a very high field, but a HTS coil with fairly high critical current ampere-turns is needed to replace normal copper windings on a magnetic core. In this case the generated magnetic field is related to its ampere-turns, but is normally much less than that of a magnet. The ampere-turns required for some applications can be achieved with presently produced $(Bi,Pb)_2Sr_2Ca_2Cu_3O_{10+x}$ Ag clad wires at 77 K.

A HTS $(Bi,Pb)_2Sr_2Ca_2Cu_3O_{10+x}$ Ag clad single core wire and a 27-core multifilament wire were used to produce HTS coils under the reaction and winding procedure. The $(Bi,Pb)_2Sr_2Ca_2Cu_3O_{10+x}$ Ag clad wires are tested for their practical properties, and their coils are studied with different applied magnetic cores. A pancake coil which consists of five units was made with the single core wire and connected by a double-pancake method, and a solenoid coil was made with the 27-filament wire. The critical currents and magnetic fields generated by both HTS coils are measured at 77 K, and the critical current of the 27-filament solenoid coil is also measured at 4.2 K. The critical current ampere-turns of the HTS coil, made by the 27-filament wire, were measured to be 1591 A-t at 77 K and 13954 A-t at 4.2 K. The influences on the HTS coils by the applied magnetic cores are investigated by testing the I-V curves of the HTS coils at 77 K, with variable applied magnetic core gaps.

HIGH T_c SUPERCONDUCTORS

A HTS $(Bi,Pb)_2Sr_2Ca_2Cu_3O_{10+x}$ Ag clad single core wire and a 27-core multifilament wire were used to produce HTS coils. The HTS wires were produced by normal powder-in-tube techniques. The powder used for the 27-filament wire has cation ratio Bi:Pb:Sr:Ca:Cu = 1.85:0.35:1.90:2.03:3.05. During the rolling-sintering procedure, the wire was sintered on a diameter ϕ = 75 mm Al_2O_3 tube at 830 °C to 834 °C in four cycles for a total of about 300 h. The final critical current of the total 46.3 m long wire is about 10 A at 77 K. A better critical current was achieved at a previous step i.e. with total 220 h sintering, for which sectional critical currents vary between 13.7 A to 16 A at 77 K over the entire 43 m length of wire. The single core wire is about 50 m long produced through three rolling-sintering procedures with total 240 h at 830 °C, and this wire was used to produce 5 pancake units.

The characteristics and testing results both of the single core wire and the 27-filament wire are listed on Table 1.

Table 1. High T_c superconductors

HTS 27-Filament Tape	HTS Single Core Tape
HTS: $(Bi,Pb)_2Sr_2Ca_2Cu_3O_{10+x}$ Ag clad tape CS_{tape}* = 4.15 mm × 0.22 mm CS_{HTS}**/CS_{tape} = 0.21	HTS: $(Bi,Pb)_2Sr_2Ca_2Cu_3O_{10+x}$ Ag clad tape CS_{tape} = 3.8 mm × 0.4 mm CS_{HTS}/CS_{tape} = 0.17
Normal short tape: I_c = 21 A at 77 K in 5 mT applied field I_c = 33 A at 66 K in 5 mT applied field	Normal short tape: I_c = 5.25 A at 77 K in 5 mT applied field I_c = 9.4 A at 66.2 K in 5 mT applied field
Winding tolerance: > J_c90% at 1% bend strain***	Winding tolerance: > J_c90% within 0.24 cm^{-1} in curvature

*CS_{tape} is the cross-section of the HTS tape.
** CS_{HTS} is the cross-section of the HTS tape core.
***The critical current density is an average value of five HTS tapes at 77 K. The bend strain is defined by t/2R, where t is the HTS tape thickness and 2R is the diameter of the bend former.

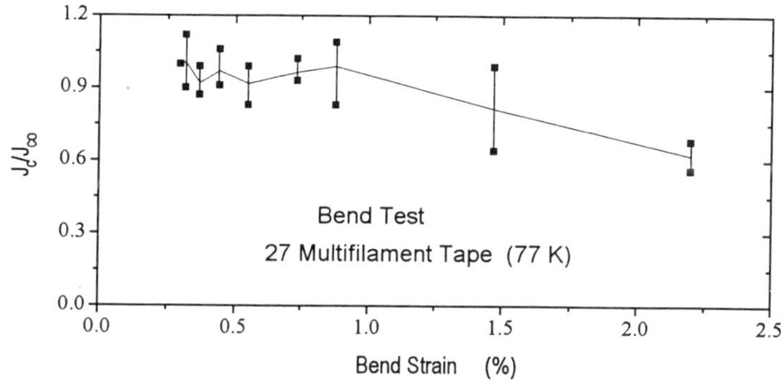

Figure 1. Mechanical bending property of the 27-filament HTS wire.

Mechanical properties of the 27-filament wire for winding the HTS coil were tested as shown in Figure 1, in which the critical current J_c and original critical current J_{co} were measured at 77 K. In Figure 1, the maximum and minimum values of five 40 mm length samples at each bending strain value are marked on the figure, and the curve represents the average of these five values. The samples have total cross-section 4.1 mm × 0.23 mm, and the samples were selected after 220 h sintering treatment. The bend strain is defined as t/2R, where t is the HTS tape thickness, and 2R is the diameter of the former on which the HTS tapes were bent.

The critical current of the 27-filament wire and the single core wire were tested at 77 K and 66 K with applied magnetic field, as shown in Figure 2. The applied magnetic field is parallel with the broad face of the HTS wires. The 77 K critical current of the 27-filament wire at 0.5 T maintains about 1/3 of the zero field value.

HIGH T_c SUPERCONDUCTING COILS

The HTS coils were wound both from the final sintered long wires, under the reaction and winding procedure, and wrapped with a thin Teflon tape as insulation. The solenoid coil is wound from the 46.3 m long 27-filament HTS wire. Five pancake unit coils were made from the Ag clad single core HTS wire and formed to be a 5 × 30 turns coil by a double winding method.

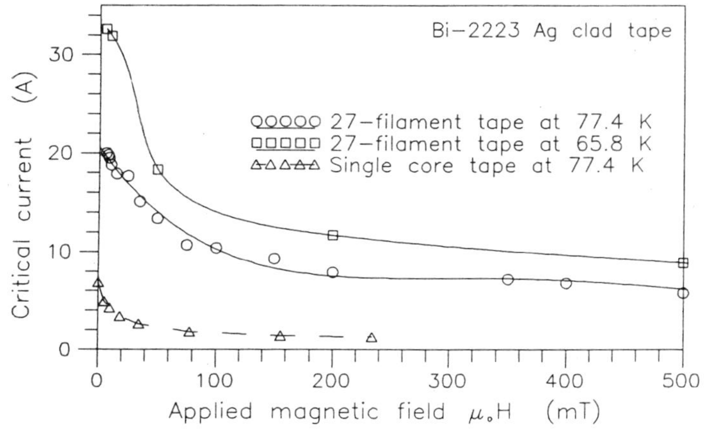

Figure 2. Critical currents of short pieces of the HTS wires with different applied magnetic fields, and at different temperatures.

Table 2. High T_c superconducting coils

	Solenoid coil	Pancake coil
Dimension	ϕ_{in} = 30 mm, ϕ_{out} = 55 mm, L = 65 mm, with total 306 turns	A 30-turn unit, ϕ_{in} = 57 mm, ϕ_{out} = 85 mm
Critical current	I_c = 5.2 A at 77K, and I_c = 45.6 A at 4.2 K in self field	I_c = 2 A at 77 K in self field
Critical current-turns	1591 A-t at 77 K 13954 A-t at 4.2 K	300 A-t at 77 K (5 units)
Calculated critical self field	0.031 T at 77 K (5.2 A - 306 t) 0.275 T at 4.2 K (45.6 A - 306 t)	0.015 T at 77 K (2 A - 150 t)

The details of the HTS coils are summarised in Table 2 with some testing results.

The maximum self fields of the HTS coils are decided by their critical currents. The five units pancake coil is about 300 critical current ampere-turns at 77 K generating 0.015 T at the coil centre. The critical currents of the 27-filament coil are measured at 77 K and also at 4.2 K as shown in Table 2. The calculated magnetic fields of the solenoid coil are about 0.031 T with 5.2 A (1 μV/cm) at 77 K, and 0.275 T (1 μV/cm) with 45.6 A at 4.2 K.

EXPERIMENTAL RESULTS

The HTS coil made with the single core wire was used as a bias coil of a magnetic core, and its critical current was measured at 77 K with varying magnetic core gaps. The calculated resistive losses ($I \times V$) of the HTS coil and generated magnetic flux densities with the different core gaps are shown in Figure 3. In Figure 3 a magnetic ferrite core with cross-section (CS) = 25 × 25 mm^2, and magnetic path length (MPL) = 135 × 2 + 75 × 2 mm, and a saturable flux density about 1.3 T was used. A search coil on the magnetic core was linked with a flux meter for the test. Figure 3 shows that for different core gaps there is a HTS coil low loss region to generate the magnetic fields prior to the losses increasing dramatically. For the closed magnetic core, it can be saturated by the HTS coil under its superconducting state, and there is a dramatic loss increase in the HTS coil above 1.35 T.

Figure 4 is the magnetic core flux densities generated by the 27-filament HTS solenoid coil with the required ampere-turns at different gap values of the magnetic core. The HTS coil loss with an air core is also represented in the figure by the HTS coil voltage drop with different current ampere-turns. The magnetic core used is CS = 13 × 11 mm^2, MPL = 140 × 2 + 70 × 2 mm. From this figure, the critical current ampere-turns of this HTS coil is capable of saturating the magnetic core with up to 1.3 mm double magnetic C-core gaps.

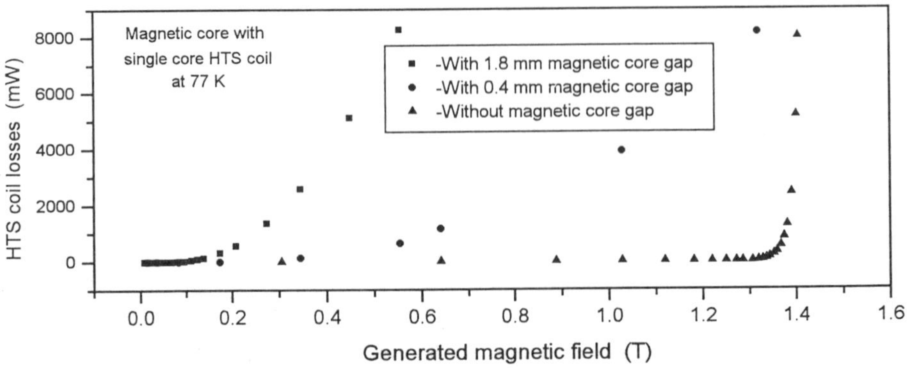

Figure 3. Losses of the 5-unit pancake HTS coil at 77 K with different magnetic core gaps.

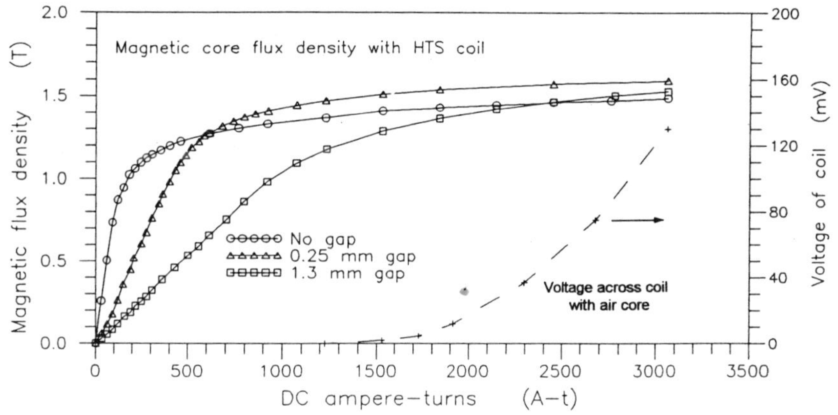

Figure 4. Magnetic core flux densities generated by the 27-filament HTS wire.

When the HTS coil is added with a magnetic core, the magnetic flux in the HTS coil will be re-distributed. To investigate the magnetic core influence on the HTS coil, the HTS coil I-V curves were tested with different magnetic core gaps and without the magnetic core. Figure 5 shows I-V curves for the HTS coil with an air core, two magnetic C-cores with different core gaps, and an iron bar core to present the magnetic core effect on the HTS coil made by the 27-filament wire. The two C-cores are the same as in Figure 4, and

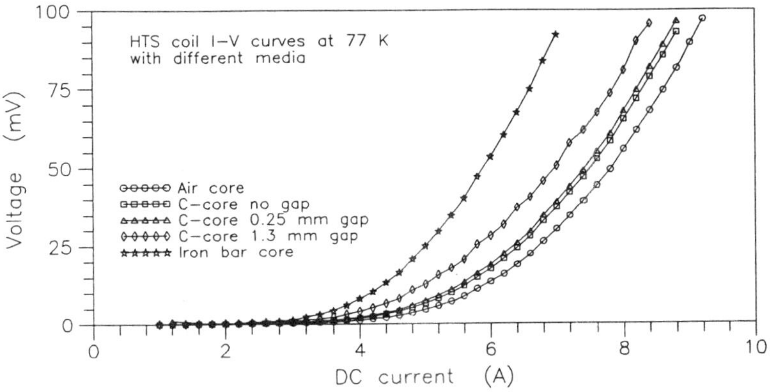

Figure 5. 77 K I-V curves of the HTS coil made by the 27-filament wire with different media.

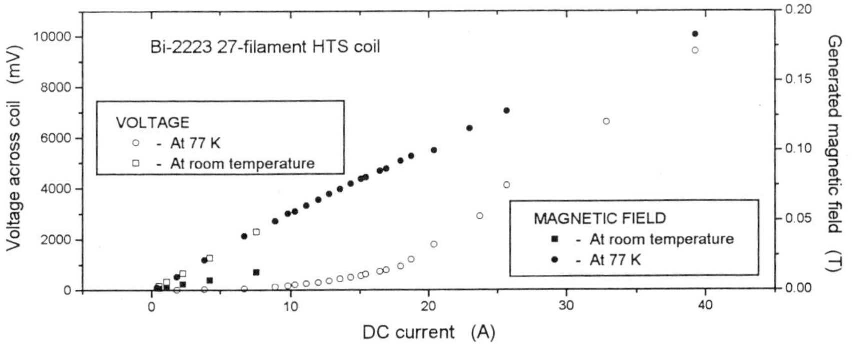

Figure 6. Generated magnetic field for the 27-filament HTS solenoid coil.

the iron bar is 16 mm in diameter and 150 mm in length. From Figure 5, it is indicated that there is only a slight critical current change of the HTS coil with a closed magnetic core and an air core.

The DC magnetic field of the 27-filament HTS coil tested at 77 K in air is shown in Figure 6. Both magnetic flux density measured at the HTS coil centre, and the voltage drop across the HTS coil are presented according to DC transport current. The magnetic field generated by the HTS coil is about 0.03 T at 5.2 A critical current. The voltage drop across the HTS coil at 77 K becomes dramatically increased after 17 A.

The magnetic field generated by the HTS coil is lower in room temperature as shown in the Figure 6; this may be caused by some shorted turns of the HTS coil, and in liquid nitrogen the DC current may be redirected to the superconducting path, therefore the field is increased. So for DC currents the short turns problem can be improved when the HTS coil is immersed in liquid nitrogen by redirecting the transport current to the HTS core from its Ag sheath.

CONCLUSIONS

The critical currents of the HTS coils made by $(Bi,Pb)_2Sr_2Ca_2Cu_3O_{10+x}$ Ag clad wires from both a single core wire and a multifilament wire, through a reaction and winding procedure, maintain a fairly high percentage of their short tape values when operated in a low field at 77 K. This enables the $(Bi,Pb)_2Sr_2Ca_2Cu_3O_{10+x}$ Ag clad wire to be used for some low field applications at liquid nitrogen temperature. Hitherto, this type of HTS wire has been considered for very high field magnet applications at 4.2 K, where a much higher critical current value is maintained by strengthen flux pinning of the HTS.

For some low field applications, the HTS wires are considered to work satisfactorily with a magnetic core as a bias coil. In a low field application, the self field of the HTS coil is much less than that in a high field magnet, therefore fairly high critical current ampere-turns of the HTS can be maintained. The applied magnetic core through a HTS coil, will cause the original magnetic flux of the HTS coil to re-distribute. Consequently the critical current of the HTS coil will be changed. However, this change is shown to be minor when compared to the I-V curve change of the HTS coil with a closed magnetic core loop and with an air core.

In addition the use of the HTS coils to produce a higher magnetic field has been achieved by reducing the temperature, such as at 64 K, 27 K and 4.2 K. However, to produce a HTS coil with a large number of critical current ampere-turns for high field applications at 77 K, further improvements in the critical current are necessary. The production of long length HTS wires with uniform properties is also required.

REFERENCES

1. K. Sato, T. Hikata, H. Mukai, M. Ueyama, N. Shibuta, T. Kato, T. Masuda, M. Nagata, K. Iwata, and T. Mitsui, High-J_c silver-sheathed Bi-based superconducting wires, *IEEE Trans. Mag.*, (1991) **27** 1231.
2. U. Balachandran, A.N. Iyer, J.Y. Huang, R. Jammy, P. Haidar, J.G. Hoehn, Jr., G. Galinski, and L.R. Motowidlo, Recent advances in bismuth-based superconductors, *JOM*, (1994) **46** 23.
3. K. Sato, N. Shibuta, H. Mukai, T. Hikata, M. Ueyama, T. Kato, and J. Fujikami, Bismuth superconducting wires and their application, *Cryogenics*, (1993) **33** 243.
4. S.X. Dou, H.K. Liu, Y.C. Guo, X.J. Jin, Q.Y. Hu, D.L. Shi, S. Salem-Sugui, and Z. Wang, Critical current density and irreversibility behaviour in Ag-sheathed Bi-based superconducting wires fabricated using a controlled melt procedure, *Appl. Supercond.*, (1993) **1** 1515.
5. J.X. Jin, C. Grantham, H.K. Liu, and S.X. Dou, Prototype fault current limiter with a high T_c superconducting coil, *J. Electr. & Electronics Eng., Australia*, (1995) **15** 117.
6. C. Grantham, J.X. Jin, S.X. Dou, H.K. Liu, and K.J.S. Holder, Novel electronic high voltage generator with high temperature superconducting coils, *Proc. IPEC*, Singapore, (1995) **2** 234.

QUENCH DEVELOPMENT IN A HIGH TEMPERATURE SUPERCONDUCTING TAPE

J. W. Lue[1], M. S. Lubell[1], D. Aized[2], J. M. Campbell[2], and R. E. Schwall[2]

[1] Oak Ridge National Laboratory
Oak Ridge, TN 37831-8040, USA
[2] American Superconductor Corporation
Westborough, MA 01581, USA

ABSTRACT

Normal zone propagation experiments have been performed on a long length of Bi-2223/Ag high temperature superconducting (HTS) tape. Tests were conducted with liquid nitrogen and gaseous helium cooling in temperatures from 5 to 77 K. No sustained expansion of a "normal" zone was observed with a short resistive heater. Non-uniform critical currents were, however, observed over the length of the conductor. When the conductor was charged and held at a current above the critical currents of weaker sections, a quench was being developed without distinctive "normal" zone propagation. Because of the high temperature margin and broad resistive transition of the superconductor, and the good thermal conductivity of the Ag-matrix, the quench process was very slow, and no large temperature gradient along the conductor was observed.

INTRODUCTION

Because of the higher operating temperatures of the high temperature superconductor (HTS), it is generally believed that a HTS would have a higher stability margin than a low temperature superconductor (LTS)[1]. It has also been shown analytically[2] and numerically[3] that a local normal zone in a HTS would propagate very slowly, on the order of millimeters per second. The analyses were based on a sharp resistive transition at the critical current as in a LTS. The resistive transition in a HTS is, however, very broad; that is, the n-value in the V-I power law of a HTS is very low[4] (often ≤ 10). Thus it is not clear how a local normal zone in a HTS coil would propagate, whether it would cause a local overheating or a runaway resistive transition — a quench of the coil.

Two series of normal zone propagation experiments were performed on a long length of HTS tape. During the first series of experiments no normal zone propagation of the kind similar to that in a LTS[5] was observed. It was, however, noticed that the critical current

Advances in Cryogenic Engineering, Vol. 41
Edited by P. Kittel, Plenum Press, New York, 1996

of the tape was lower over a section of about 1/3 of its length. The voltage taps were rearranged, and a second series of experiments were performed to locate the weak spot and to find out its consequence on the normal zone propagation of the conductor.

EXPERIMENTAL SETUP

Normal zone propagation experiments were performed on a 60-cm long Bi-2223/Ag HTS tape. The tape was spirally wrapped around a 6.4-cm diameter G-10 cylinder. A heater, two thermometers, and several voltage taps were installed for the experiment. Figure 1 shows the experimental arrangement of the conductor and the instrumentation. The multifilament Bi-2223/Ag conductor was made by a metallic precursor process[6]. It has a cross section of 0.533-cm × 0.016-cm and a 20% filling factor for the HTS filaments. The heater was made of a 120-Ω strain gauge that was 0.4-cm long by 0.3-cm wide. It was glued on the HTS with GE-7031 varnish. A Ruthenium Oxide (RO_2) thermometer, T_0 was laid in a groove on the G-10 cylinder, touching the back side of the conductor at the same position as the heater to monitor the hot-spot temperature of the conductor. A second RO_2-thermometer, T_1 was positioned 5.5 cm away. Several voltage taps were applied along the conductor.

The experiment was conducted inside a variable temperature cryostat. Controlled helium-gas flow through a capillary tube and a heater on the mounting block were used to control the sample temperature. For tests at liquid nitrogen temperatures, liquid nitrogen was transferred directly to the sample space.

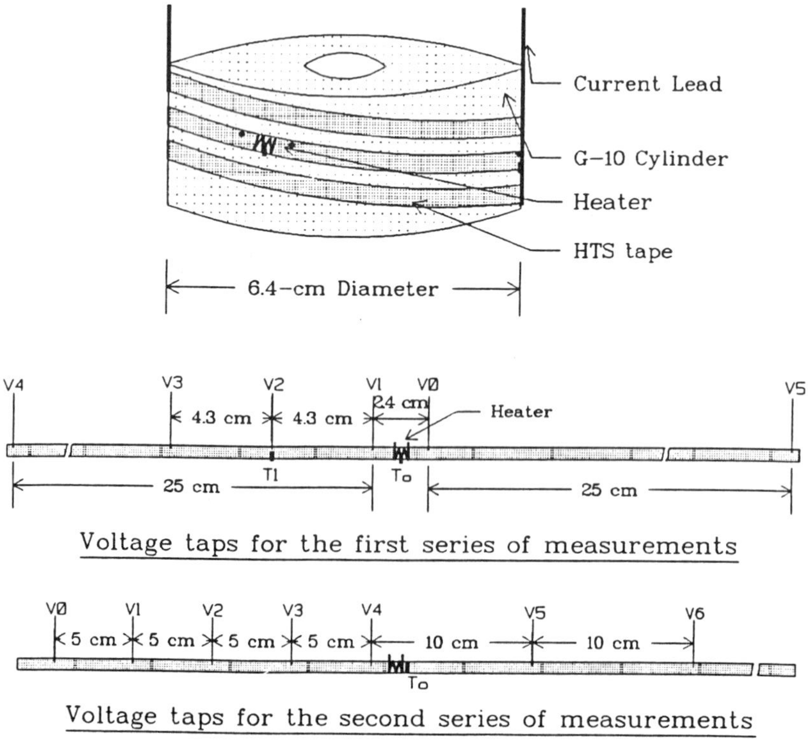

Figure 1. Experimental arrangement of the HTS tape for the normal zone propagation measurements. In test 2, the renamed voltage taps V_0 to V_4 cover about the same section V_4 to V_3 as in test 1.

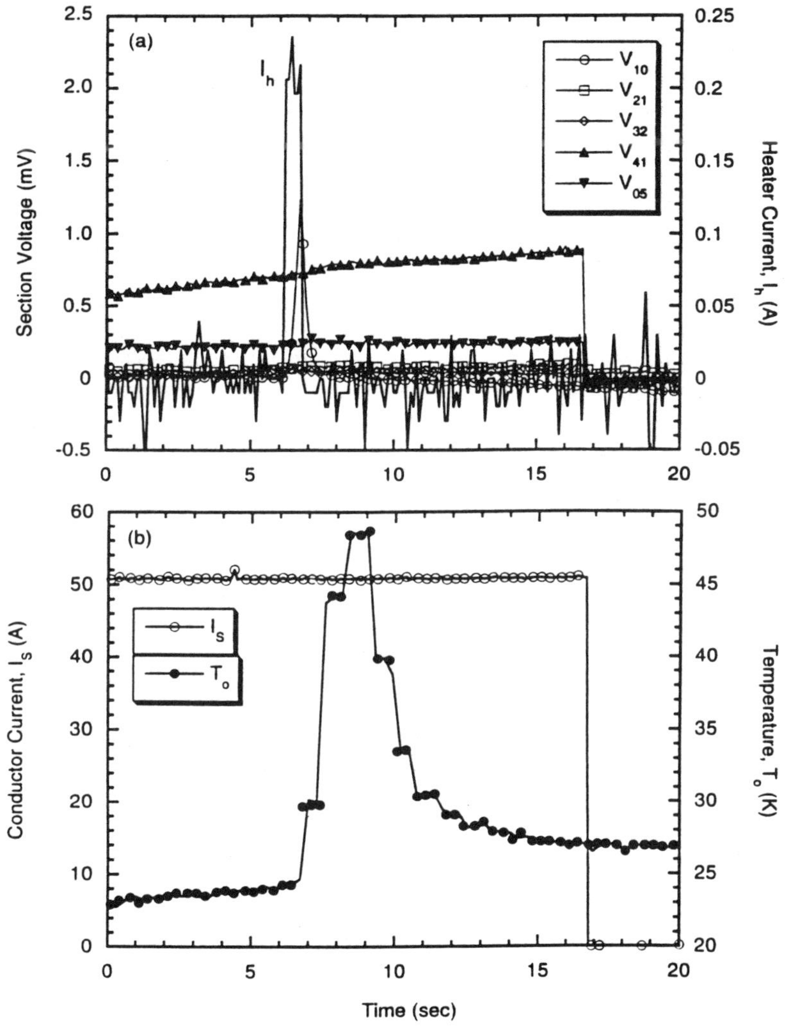

Figure 2. Conductor section voltages and the temperature T_o in response to a heating pulse.

FIRST SERIES OF EXPERIMENTS

Conventional ways of normal zone propagation experiments were attempted during the first series of experiments. The heater was located at the middle of the conductor length, and voltage taps were placed on both sides of the heater; more were on one side than the other as is shown in the test 1 instrumentation layout in Fig. 1. Measurements were made in liquid and solid nitrogen at 77 K to 63.5 K, and with gaseous helium from 40 K to 12 K. Based on the 1 μV/cm criterion over the whole conductor length critical currents, I_c, of 6.5 A at 77 K and 32 A at 20 K were measured.

The conductor was charged to a given current (sometimes higher than the overall I_c of the conductor), and a heater pulse was applied to create a local normal zone. The conductor voltage and temperature were monitored to measure the propagation of the normal zone. Figure 2 shows the voltages of the different sections of the conductor in response to a 0.21-A, 0.7-s heater pulse. The conductor temperature was about 23 K at the beginning of the test, and it was charged and held at 51 A. The heated zone voltage V_{10}

went up to 1.2 mV (as compared to 2.4 μV @ I_c defined by the 1 μV/cm criterion) during the heating. The thermometer, T_o at the middle of this zone indicated a temperature rise of about 26 K to 49 K. Thus a "normal' zone was created. But, it disappeared within 1 s. No other sections showed any appreciable voltage changes (and T_1 only raised by 4 K) as the result of the heating pulse or the initial "normal" zone. Similar results were observed at temperatures of 12 to 77 K and with heater pulse energy of up to 4 J. No sustained expansion of a normal zone, that is, sequential and continuous increases in voltages in successive sections of the conductor, was observed.

Note also in Fig. 2(a) that when the current was dropped at time 17 s, the voltages of the different conductor sections all went to about zero. Thus the 0.6 mV of voltage in V_{41} (and the 0.2 mV in V_{05}) at time zero that was increasing with time was real resistive voltage for that section at the current of 51 A. Furthermore, no appreciable voltages were seen in V_{10}, V_{21}, and V_{32}. Thus the resistive voltage in V_{41} was due mostly to section 4-3. Apparently, section 4-3 has lower current carrying capability and a lower critical current than the rest of the conductor. It would be interesting to locate the weak spot in this section and to investigate further its effect on the normal zone propagation of the conductor.

SECOND SERIES OF EXPERIMENTS

To locate the weak spot of the conductor, voltage taps were placed 5-cm apart in the previously identified weak section 4-3 of the conductor and renamed V_0 through V_4 and then placed 10-cm apart for the rest of the conductor for a second series of experiments. As is shown in the test 2 instrumentation layout in Fig.1, the heater and the thermometer T_o were also relocated right outside this weak section.

Voltages Beyond the Critical Currents

Voltages of the different sections of the conductor were recorded as the conductor was charged at different temperatures. Figure 3 shows the section voltages as the conductor current was charged toward 50 A at 24 K. One sees that section 3-4 started to show

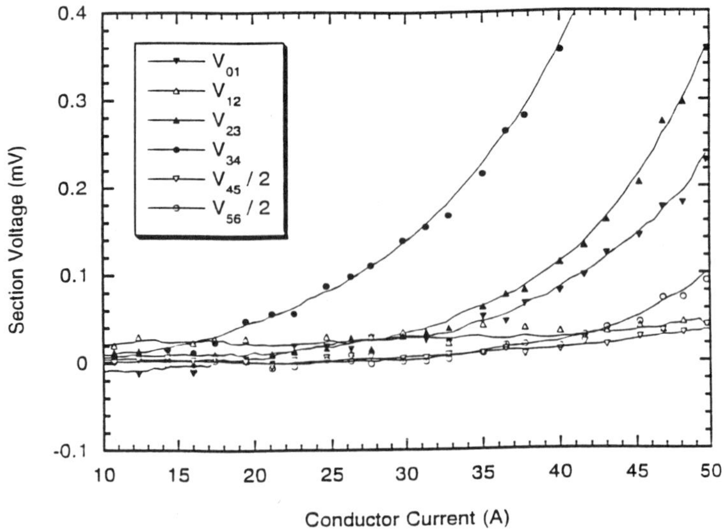

Figure 3. V-I characteristic curves of the different sections of the conductor at 24 K.

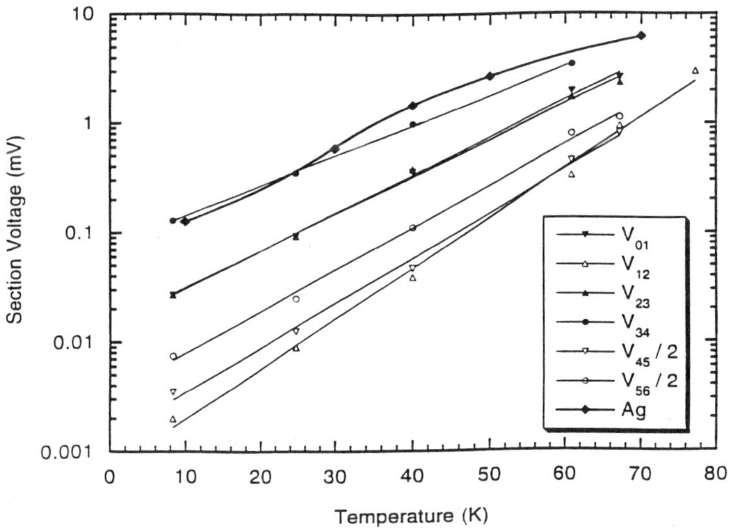

Figure 4. Section voltages of the conductor as a function of temperature at a current of 40 A.

resistive voltage at 18 A, followed by sections 2-3 and 0-1; while sections 1-2, 4-5, and 5-6 do not show any measurable voltage until about 40 A. Thus the current-carrying capability of the conductor varies over centimeter lengths. Critical current of the conductor based on a certain E-field (*e.g.* 1 μV/cm) would vary by more than a factor of 2 from section to section. It is also noted that the resistive transition of the superconductor is very broad for all the sections.

By measuring similar V-I curves at different temperatures, the temperature dependence of the section voltages beyond the critical current in the partially resistive state can be determined. Figure 4 shows the measured section voltages as a function of temperature at a conductor current of 40 A. Straight lines in this semi-log plot indicates that the resistivity

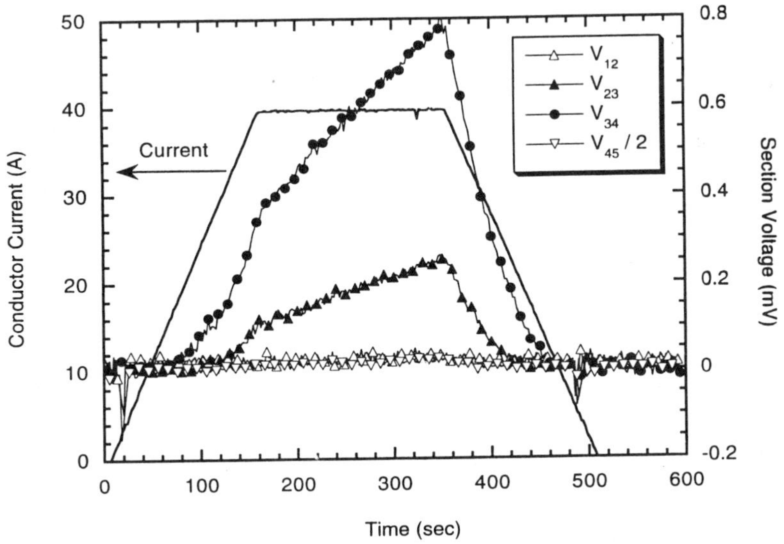

Figure 5. Section voltage excursions as the conductor was charged and held at 40 A.

of the superconductor in the partially resistive state appears to increase exponentially as a function of temperature. Also plotted in this figure for comparison is the calculated voltage of a 5-cm- long Ag tape of the same cross section as the present conductor at 40 A. It is noted that even the worst section of the present HTS conductor had equal or lower resistivity than Ag at this current (about 4.7 kA/cm^2 of current density over the whole conductor cross section) at temperatures up to 70 K. Apparently, the Bi-2223 core was not fully normal up to this temperature at the given current despite the existence of weak spots. The high critical temperature of the HTS (about 105 K) made this possible.

Quench Development from a Weak Spot

If a superconductor is charged and held at a current above the critical current of a weak spot, the joule heating would heat up that spot and spread the heat throughout the conductor. Thus a self induced (in contrast to a heater induced) normal zone propagation could be expected in a non-uniform superconductor. This thought was investigated with the present conductor. Figure 5 shows the current and the section voltages of a run with the conductor temperature of 23 K at the start of the current charge. The conductor was charged to 40 A (as compared to I_c of 32 A determined by the 1 μV/cm over the whole conductor length) in about 160 s, and held at that current for another 190 s. During this time the voltages in section 3-4 increased rapidly to about 0.8 mV, followed by section 2-3 to about 0.25 mV. Yet, the adjacent sections 4-5 and 1-2 barely show any voltage changes in this plot.

Measuring the section voltages at different times of the current charge and using the curves in Figure 4, one can determine the average temperature of each section of the tape during the test. Figure 6 shows the temperature profiles of the conductor at three different times of the above run. When the conductor was first charged to 40 A, the average temperature of the conductor was about 24 K. After holding the current at 40 A for 95 s, a hot spot centered around V_3 was seen to heat up to about 32 K. It was further raised to 37 K in another 95 s. This is evidently due to the joule heating associated with the voltages shown in sections 3-4 and 2-3 in Fig. 5.

Although voltages in sections 1-2 and 4-5 were barely measurable on the scale shown in Fig. 5, expanded scale showed small voltages and the temperature in these sections went

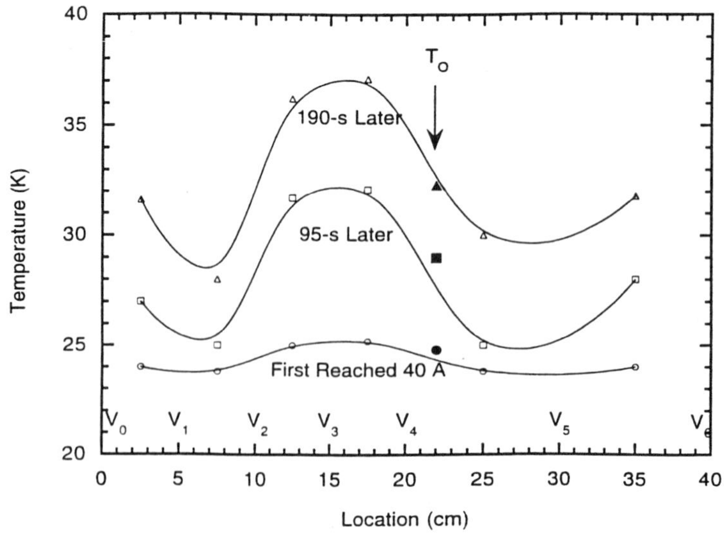

Figure 6. Temperature profiles of the conductor at three different times of the test in the run of Fig.5.

up to about 25 K and 29.5 K in the above two times. Apparently the high thermal conductivity of the Ag-matrix permitted conduction of the heat quite well along the conductor. It was sufficient (about 2.5 W/cm-K at 25 K) to keep the temperature differences between the weak spot and the good sections of the conductor below 10 K at all time during this test. Also shown in Figure 6 are the temperatures measured by the RO_2-thermometer, T_o placed near V_4. The consistency of these readings to the temperature profiles proves the validity of the present technique of calculating the average section temperatures from the section voltage readings.

It appears that the net effect of the weak spot was to heat up the whole conductor with a moderate temperature gradient. In the time duration of the test of about 3 minutes, no distinctive normal zone propagation similar to that in a LTS was observed. However, since the temperature and the section voltages were rising steadily, a quench is inevitable if the current were held longer.

DISCUSSION

Normal zone propagation process in a superconductor is governed by a heat balance equation which in one-dimension can be written as

$$C\frac{\partial T}{\partial t} = \frac{\partial}{\partial x}(\kappa\frac{\partial T}{\partial x}) + G_j(T) + G_d(t) - G_h(T). \tag{1}$$

The left-hand side represents the energy absorption through the volumetric heat capacity, C. The third term on the right hand side, G_d represents external disturbance such as the initial heater pulse. The last term, G_h represents the lateral cooling. With helium-gas cooling, the heat transfer is negligible (although the conduction to the G-10 support cylinder may be significant). Thus, after the transient the governing terms on the right hand side of Eq. (1) are the competing factors of joule heat generation, G_j and the conduction along the conductor.

A travelling-wave solution to Eq. (1) gives a speed with which a travelling temperature wave carries the heat along the conductor. When the neighboring conductor reaches the current sharing temperature the normal / superconducting boundary has advanced to that spot. Thus a normal zone propagation can also be deduced. Two factors complicate the propagation process in a HTS as compared to a LTS conductor. The first is related mostly with the Ag-matrix. Because of the higher HTS operating temperatures ($\geq$ 20 K), the heat capacity C, the thermal conductivity κ, and the resistivity of the conductors are not constant over the temperature range of interest. This factor has been included in previous analyses[2,3].

The second factor comes from the peculiar behavior of the HTS core. As is shown in Fig.3, the resistive transition of the HTS is very broad. When the superconductor is charged to and beyond its I_c the voltage increases roughly as a power law of the current. The n-value of the power is often very low (averaging about 3.2 for the present conductor). Thus I_c and the current sharing temperature are ill defined. The initial "normal" zone created by an external heat pulse or a current beyond local I_c is only partially resistive. The joule heating of this "normal" zone is much reduced as compared to that in a LTS conductor.

Furthermore, Fig. 4 indicates that the resistivity of the HTS in the partially resistive state has a exponential temperature dependence that extends to more than 70 K. Thus the arrival of a temperature wave does not translate into the advance of a distinctive normal / superconducting boundary. Therefore, we see a quench being developed through the spreading and heating of the whole conductor length without a distinctive normal zone propagation.

From the relatively smooth temperature profile and the non-distinctive "normal" front propagation one can conclude that thermal conductivity of the Ag-matrix dominates the joule heating in the quench development process. Furthermore, this quench process may take more than a few minutes to develop. As a matter of fact delayed spontaneous quenches were observed in a pancake coil made of HTS conductors similar to the present experiment[7].

CONCLUSION

Normal zone propagation experiments have been performed on a long length of HTS tape. No "normal" zone was sustained or propagated with up to 4 J of local pulse heating. The current carrying capability of the tape was found to be quite non-uniform along its length and the resistive transition was very broad. In the partially resistive regime, the conductor resistivity appears to increase exponentially with temperature from 5 to 77 K.

A new approach of studying the quench propagation of charging and holding the conductor at currents higher than the critical current of a weak spot showed that a quench was developed through the spreading and heating of the whole conductor length without a distinctive normal front propagation. This is thought as the result of thermal conduction dominating the joule heating in the quench process. A proper analytical or numerical quench model for HTS conductors should incorporate broad resistive transition and exponential temperature dependence as observed in the present experiment.

ACKNOWLEDGEMENT

This work was sponsored by the Office of Energy Efficiency and Renewable Energy, U.S. Department of Energy under contract DE-AC05-84OR21400 with Lockheed Martin Energy Systems, Inc. A portion of the work at American Superconductor Corporation was funded by superconductivity cooperative agreement number 86X-SK700V between Lockheed Martin Energy Systems, Inc., and American Superconductor Corporation.

REFERENCES

1. Y. Iwasa, "Design and operational issues for 77-K superconducting magnets", *IEEE Trans. Magnetics* MAG-24, 1211-1214 (1988).

2. L. Dresner, "Stability and protection of Ag/BSCCO magnets operated in the 20-40 K range," *Cryogenics*, Vol. 33, 900-909 (1993).

3. R.H. Bellis and Y. Iwasa, "Quench propagation in high temperature superconductors," *Cryogenics*, Vol. 34, 129-144 (1994).

4. J. W. Lue *et al.*, "Test results of two small high temperature superconducting coils," *Advances in Cryogenic Engineering*, Vol. 40, 327-334 (1994).

5. L. Dresner, J. R. Miller, and G.W. Donaldson, "Propagation of normal zones in composite superconductors," *IEEE Proc 6th Symp Eng Prob Fusion Res*, 266-270 (1975).

6. A. Otto *et al.*," Multifilament Bi-2223 tapes made by a metallic precursor route," *IEEE Trans. Applied Superconductivity*, Vol. 3, 915-922 (1993).

7. J. W. Lue *et al.*, "Spontaneous quenches of a high temperature superconducting pancake coil," paper presented at the 14th International conference on Magnet Technology, Tampere, Finland, June 1995.

COIL BOBBIN FOR STABLE SUPERCONDUCTING COILS

T.Kashima[1], A.Yamanaka[1], E.S.Yoneda[2], S.Nishijima[3], and T.Okada[3]

[1]Research Institute, Toyobo Co.,Ltd.
Ohtsu,Shiga,520-02Japan
[2]Research and Development Center, Toshiba Co.,Ltd.
Kawasaki,Kanagawa,210Japan
[3]ISIR Osaka University
Ibaraki,Osaka,567Japan

ABSTRACT

The coil bobbin for a.c. coils have been prepared with the high strength polyethylene fiber (DF) reinforced plastics (DFRP) or with hybrid composites reinforced by DF and glass fiber (GF). The coils with the bobbin were found to be markedly stable. The DF has a large negative thermal expansion coefficient and hence the circumferential thermal strain of bobbin can be designed by changing the ratio of DF to GF layer thickness (DF/GF). It was found that the thermal expansion coefficient in the circumferential direction of the outer surface changed from negative to positive with increasing DF/GF and became nearly zero at a DF/GF of approximately 5. 1 kArms class a.c. coils having a bobbin with a negative thermal expansion coefficient or small thermal contraction in the circumferential direction were fabricated and were confirmed to show higher quench current than that with a GFRP bobbin.

INTRODUCTION

Superconducting magnets have been developed for a.c. application such as electric power use. To reduce a.c. losses, it is necessary to decrease the diameter of the superconducting strand. Consequently, to obtain a large current capacity, an a.c. superconducting cable has to consist of a large number of strands twisted together. The coils using the a.c. superconducting cables, however, have been found to show a serious problem : the a.c. superconducting coils can not attain the short sample d.c. critical current at 50Hz[1].

The causes of the problem have been studied and clarified to be longitudinal-field losses[2], self-field losses[3], and mechanical factor such as frictional heat by wire motions. Among them the wire motion has been paid particular attention in this work.

When the bobbin showing expansion with decreasing temperature is used, the wire motion is expected to be reduced. High strength polyethylene fiber (DF)[4~8] has been developed and found to show a negative thermal expansion coefficient. Pipes made of DFRP (DF reinforced plastics) were confirmed to show expansion in the radial direction with decreasing temperature. The DFRP is also a nonmetallic material which is suitable for a.c. coil bobbin to avoid eddy current losses.

An a.c. coil with DFRP bobbin has been developed and its performances was compared with that of conventional coils. The coil has demonstrated higher performance. The thermal properties of the coil bobbin are presented together with the coil performance.

Table 1 Coil Parameters

Bobbin						
Inner Diameter (mm)	Outer Diameter (mm)	Height (mm)	Layers	Turns	Groove Pitch (mm)	N. B.
80	110	136	1	17	8	having grooves without impregenating

Conducter				
Strand Diameter (mm)	Filament Diameter(NbTi) (μm)	Matrix	Number of Strands	Winding Tension (kgf)
0.2	0.14	Cu30%Ni	36	30

EXPERIMENTAL

Sample Preparation

Pipe and Bobbin. Three kind of pipes were fabricated by means of the filament winding method. They are made of DFRP, GFRP and hybrid of DFRP and GFRP. The E-glass fiber (EG) was used as the reinforcement and hence the GFRP was named G_EFRP. In the hybrid material T-glass (TG) was employed and was called DG_TFRP. The thickness of pipes used in this work was 15 mm though the diameter was different. The thickness of the glass layer was indicated in the notation such as DG_T(4.5)FRP. It means the DG_TFRP was made with a TG layer of 4.5 mm thickness and DF layer 10.5 mm, respectively. The TG layer was arranged outside of the hybrid pipe. Grooves were made on the pipes for winding the superconducting cable. The grooved pipe is called the bobbin in this work.

Spiral grooves of 3 mm depth were made in the circumferential direction in which cables were embedded. Straight grooves were also made in the axial direction on the outer surface as a cooling channel having square shape of 4mm depth. The outer diameter (Do), inner diameter (Di) and Height (L) of the bobbin were 110 mm, 80 mm and 180 mm, respectively. Hereafter the size is expressed as 110/80x180.

Coil. Three kinds of coils using DFRP, DG_T(4.5)FRP and G_EFRP bobbins were wound with the 1 kArms-class superconducting cable. The superconducting cable was made of 36(6×6) strands containing ultra-fine NbTi filaments in a Cu30%Ni matrix and was wound under 30 kg tension without impregnating resin. The coil parameters are shown in Table 1. Photographs of coils using DFRP and DG_T(4.5)FRP are shown in Fig.1.

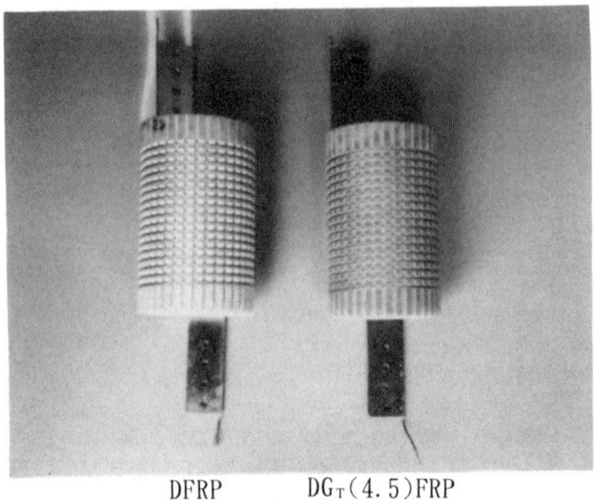

DFRP DG_T(4.5)FRP

Fig. 1. Developed a.c. coils using DFRP and DG_T(4.5)FRP bobbin.

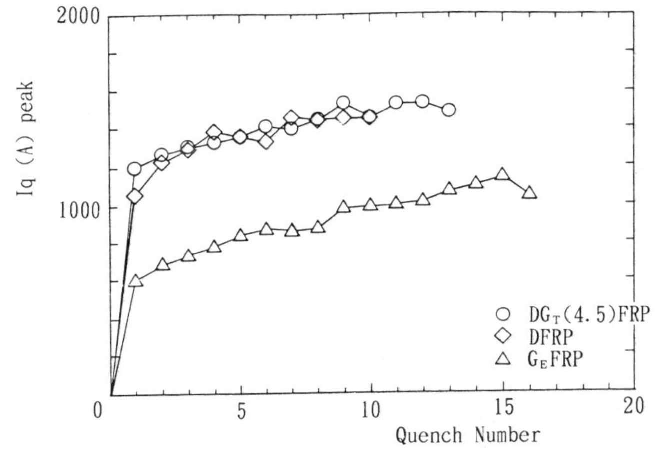

Fig. 2. Training behavior of each coils.

Thermal Strain Measurement

Thermal strains induced by cooling in each pipe and bobbin were measured on inner and outer surfaces. Strain gauges (Kyowadengyo) were bonded at the center of pipe or bobbin. The thermal strains were measured in circumferential and axial direction. Strain (ε) vs. time (t) curves were measured by dipping pipes into liquid nitrogen. Thermal strains was defined as following:

ε_{xo} axial strain on outer surface
ε_{xi} axial strain on inner surface
ε_{yo} circumferential strain on outer surface
ε_{yi} circumferential strain on inner surface.

Heat cycle between room (RT) and liquid nitrogen temperature (LNT) was also applied to the pipes in order to investigate the degradation of the pipes.

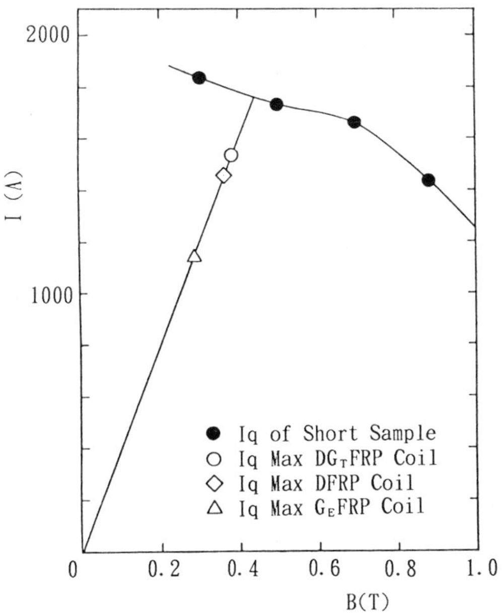

Fig. 3. Load line, critical current and Iq Max of each coil.

Coil Characteristics under a.c. 50Hz

The characteristics of the a.c. coils were measured. The coils were wound on the DFRP, DG_T(4.5)FRP and G_EFRP bobbins. Each coil was immersed in liquid helium and current was gradually increased from 0 A to the quenching current (Iq), which was defined as the current at which the resistive voltage starts to increase. The current was shut off within 3 cycles after quenching. Training curves were obtained by plotting the Iq against the quench number.

RESULTS and DISCUSSION

Coil Performance under A.C. 50Hz

Training characteristics of each coil are shown in Fig.2. The initial Iq of the coil with the G_EFRP bobbin was 610 A whereas those with DFRP and DG_T(4.5)FRP bobbin were 1034 A and 1208 A, respectively. The critical current Ic of the cable is 1736 A which was defined as the 36 times of that of a single strand measured with a.c. current. Consequently the Iq/Ic at the first quench of G_EFRP, DFRP and DG_T(4.5)FRP coil are 35%, 59% and 69%, respectively. Even after 15 quenchs the Iq of the G_EFRP coil was 1150 A, 66% of Ic which was still lower than the first Iq of the DG_T(4.5)FRP coil. The DFRP and DG_T(4.5)FRP coils reached 1452 and 1538 A of Iq at around 10 quenchs.

The load line for coils and critical current calculated as 36 times a.c. quenching current of short strand are shown in Fig.3. The attained Iq maximum in each coil is also presented. The superiority of the DFRP and DG_T(4.5)FRP bobbins has been confirmed experimentally. This strongly suggests that the wire motion was supressed by the bobbins with negative thermal expansion coefficient.

Thermal Strain of Bobbin and Winding Tension

The decrease of winding tension of a coil is closely related to the difference of thermal contraction between the superconducting cable and the coil bobbin. Winding tension of superconducting coils at LHeT before operating is roughly estimated as

$$F_1 = F_0 + \Delta F \qquad (1)$$

$$\Delta F = E_w \int_{RT}^{LHeT} \alpha_w dT - E_{by} \int_{RT}^{LHeT} \alpha_{by} dT \qquad (2)$$

where F_0 is winding tension of superconducting cables at RT and E_w, E_{by}, α_w and α_{by} are Young's modulus of the cable, circumferential Young's modulus of bobbin, thermal expansion coefficient (TEC) of the cable and that of bobbin, respectively. Here the Young's moduli are thought to be constant with temperature.

When the coils were energized, winding tension of coils decreased to F_2, due to the Lorentz's force in the outer radial direction, as

$$F_2 = F_1 - \Delta F_L \qquad (3)$$

The ΔF_L is the contribution of the Lorentz's force. An increase of F_2 would bring an increase of stability. An increase of F_0 and increase of ΔF contribute to the increase of F_1.

By choosing the bobbin material ΔF could be controlled. When the material has negative α_{by}, ΔF becomes larger and it results in the increase of F_2. This is the basic idea in selecting a bobbin for stable coil. This was confirmed experimentally, as mentioned above. The design method of the bobbin is discussed below.

Thermal Strain Behavior of Pipes and Bobbins

Fig.4 shows the thermal contraction of several kind of composite materials. The slope of the tangent at RT is defined as the thermal expansion coefficient. The materials are reinforced unidirectionally with approximately 50 volume percent of reinforcement. DFRP presents the negative expansion coefficient which shows expansion with decreasing temperature. Consequently the pipes made of DFRP could also show negative expansion coefficient and it results in the increase of the coil stability.

Setting the winding angle at 60 degree, the thermal strain ε_{y0} of bobbins and pipes were measured down to LNT and shown in Fig.5. The slitted pipe for winding is called a bobbin here. The measurement was made on 110/80x180 pipe or bobbin.

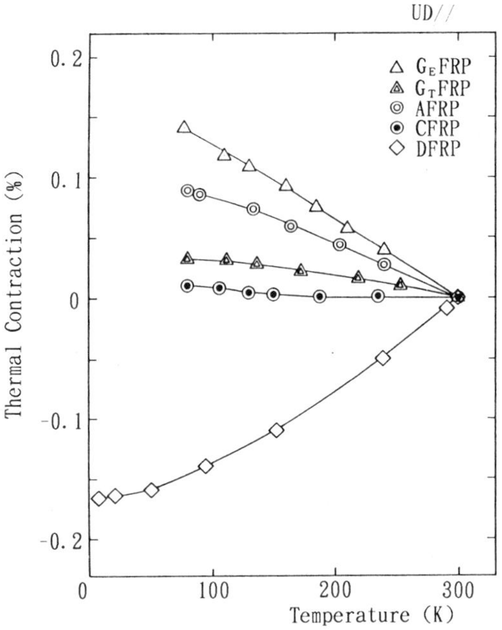

Fig. 4. Thermal contraction(strain) of various UD-FRP.

The ε_{yo} of GEFRP pipe and bobbin agree with each other and are negative. The outer diameter of the GEFRP pipe and bobbin decreased with decreasing temperature. Although the ε_{yo} of DGT(4.5)FRP are negative, the magnitude is smaller than that of GEFRP. This difference increased the coil stability. The ε_{yo} of DFRP pipe and bobbin are positive, the diameter increased with decreasing temperature. It was found that the ε_{yo} of bobbin is larger than that of pipe. This means that the constrain condition was changed by grooving the outer surface especially in DFRP. This suggests that the DF on the inner surface affects the positive thermal strain and hence the GTFRP put outside of the DFRP pipe makes the thermal strain negative as shown in DGT(4.5)FRP.

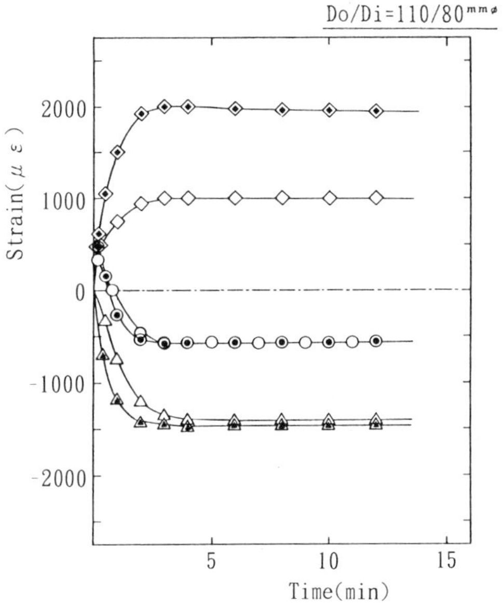

◈ Bobbin of DFRP ◇ Pipe of DFRP
◉ Bobbin of DGT(4.5)FRP ○ Pipe of DGT(4.5)FRP
▲ Bobbin of GEFRP △ Pipe of GEFRP

Fig. 5. Induced thermal strain in FW pipe by cooling down to LNT(ε_{yo}).

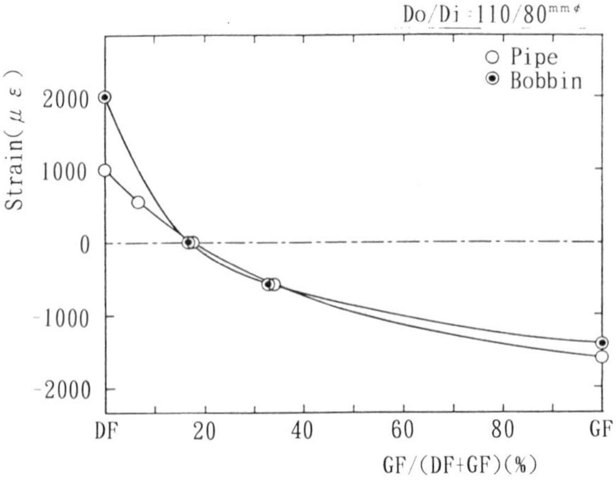

Fig. 6. GF thickness dependence of ε_{yo}.

Effect of GTFRP on Thermal Strain

GTFRP put outside of the DFRP was found to control the ε_{yo} as mentioned above. The effect of GTFRP was studied closely because GTFRP would be set outside of the bobbin to make the grooving easy. The thickness dependence of GTFRP on ε_{yo} is presented in Fig.6. The thickness is expressed as the fraction of GTFRP layer to the total thickness. The measurement was performed on the 110/80x180 pipes or bobbins. The ε_{yo} varied with the GTFRP thickness. The increase of the thickness brings the decrease of ε_{yo} and for thickness more than 20% ε_{yo} became negative. The difference between pipe and bobbin was marked especially at low GTFRP thicknesses. It was concluded that the thickness of GTFRP should be small to make the ε_{yo} positive or near zero.

Effect of Pipe Shape

It has been shown that the thermal strain of DFRP pipes changed with its shape, being different from that of GEFRP. The pipes were made with winding angle of 60 degrees, changing the ratio of inner diameter to thickness (Di/t). The ε_{yo} and ε_{yi} were

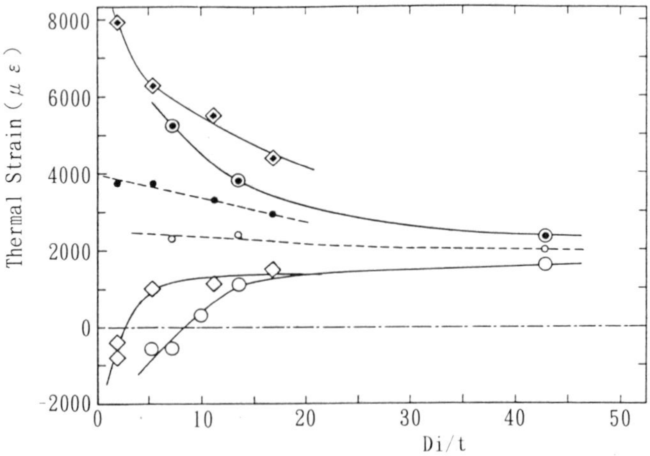

◇ ε_{yo} of DFRP; ◈ ε_{yi} of DFRP; ○ ε_{yo} of DG(4.5)FRP; ⊙ ε_{yi} of DG(4.5)FRP;
• $1/2(\varepsilon_{yo}+\varepsilon_{yi})$ of DFRP; ∘ $1/2(\varepsilon_{yo}+\varepsilon_{yi})$ of $DG_T(4.5)FRP$

Fig. 7. Dependence of thermal circumferential strain on the shape parameter.

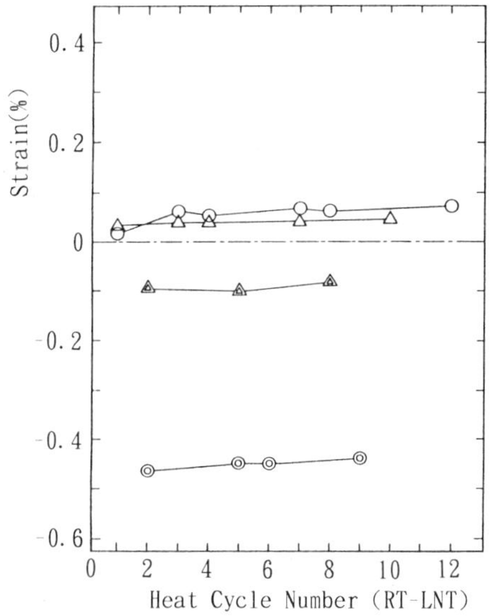

◎ 110/80$^{mm\phi}$ DF/GF(1.0) ε_{xo} ; ○ 110/80$^{mm\phi}$ DF/GF(1.0) ε_{yo}
⚠ 180/150$^{mm\phi}$ DF/GF(4.5) ε_{xo} ; △ 180/150$^{mm\phi}$ DF/GF(4.5) ε_{yo}

Fig. 8. Change of thermal strain of DG_TFRP pipes with heat cycle.

measured on DG_T(4.5)FRP and DFRP pipe. The relationship between Di/t and ε_{yo} or ε_{yi} is shown in Fig.7. It was found that the differences between ε_{yi} and ε_{yo} becomes large at the small Di/t region. At the region of large Di/t ε_{yi} and ε_{yo} begin to converge. The difference between pipe and bobbin is also noticeable at low Di/t.

The $(\varepsilon_{yo}+\varepsilon_{yi})/2$ is also presented in the figure. The Di/t dependency of $(\varepsilon_{yo}+\varepsilon_{yi})/2$ was found to be small and hence the design should be made based on $(\varepsilon_{yo}+\varepsilon_{yi})/2$, the average thermal strain. It means that at the given winding angle, the average thermal strain was fixed. Then the effect of Di/t should be considered. This is the basic method to design the DFRP or DG_TFRP pipes or bobbins.

Durability against Heat Cycle

The durability against thermal cycling was examined because it is very important for real applications. The thermal strain of UD/FRP was reproduced several times during thermal cycling between RT and LNT and suggested the durability against thermal cycling.

The thermal strain of pipes induced by cooling was measured with the number of heat cycles. The obtained strain is shown versus number of thermal cycles in Fig.8, for different sizes and shapes of pipes. The winding angle was set at 60 degree. Although a small change of thermal strain was observed during initial several cycles, the durability of the pipes could be confirmed. It was concluded that the developed coil bobbin can be applicable for practical use.

CONCLUSION

The coil bobbins for stable superconducting coils, especially a.c. applications, have been fabricated. The first a.c. quench current and the maximum a.c. quench current of coils using the bobbin composed of DFRP and DG_T(4.5)FRP were confirmed to be higher than that of GFRP. The ε_{yo} in the bobbins of DFRP or DG_TFRP were positive or, with small contraction with cooling down to cryogenic temperatures and hence the wire motions could be constrained. It was confirmed that the developed coil bobbins are effective to reduce training and are applicable for practical use. The design philosophy of the bobbin was also established.

ACKNOWLEDGMENT

The authors would like to thank Mr.T.Ohkuma, Mr.M.Nakade and Dr.T.Hara, Tokyo Electric Power Company for discussions on a.c. coil experiments.

REFERENCE

1. K.Funaki et al., IEEE Trans. Magn. MAG-28, 394(1992)
2. K.Funaki et al., Cryogenic Eng.,27,398(1992)
3. J.R.Duchateau et al., Cryogenics, 14, 545(1974)
4. G.T.Davis, P.K.Eby and J.P.Colson, J.Appl.Phys., 41, 4316(1970)
5. G.Dadobaev and A.I.Slutsker, Sov.Phys.Solid State, 23, 1131(1981)
6. G.K.White and C.L.Choy, J.Polym.Sci.Polym.Phys.Ed., 22, 835(1984)
7. L.Govaert, B.Brown and P.Smith, Macromolecules, 25, 3480(1992)
8. C.L.Choy, S.P.Wong and K.Young, J.Polym.Sci. Polym.Phys. Ed.,22, 979(1984)
9. A.Fevrier, A.Gueraud and J.P.Tavergnier, IEEE Trans. on Magnetics, 25, 1496(1989)
10. T.Okada, S.Nishijima, K.Takahata and J.Yamamoto, Cryogenics, 31, 307(1991)
11. T.Kashima, S.Nishijima, T.Okada, 46th Meeting on Cryogenic and Superconductivity, 245(1991)

EFFECT OF INCORPORATING COOLING CHANNELS INTO THE COIL SUPPORT STRUCTURE OF THE TPX TOROIDAL FIELD MAGNET

Yury M. Lvovsky, Gary W. Neeley, Wei Tong,
Kenneth E. Grut, and Timothy A. Antaya

Babcock and Wilcox, Accelerator and Magnet Systems
P.O. Box 785, Lynchburg, VA 24505, USA

ABSTRACT

Toroidal field (TF) magnets for the Tokamak Physics eXperiment (TPX) include superconducting cable-in conduit conductor (CICC) coils encased in a stainless steel support structure, which receives most of the heat generated in the magnet during operation. An efficient cooling of the structure is necessary to intercept the heat which otherwise would reach the winding pack and reduce temperature margin of the superconductor. Optimal cooling arrangement requires joint (coil+structure) analysis which considers thermal coupling between the winding pack and the structure.

A joint steady-state 3-D thermal-hydraulic model for toroidal field magnets is presented. It consists of the two submodels, coupled through iterative runs. First submodel analyzes one-dimensional helium flow in CICC double pancake accounting for thermal coupling netween turns. Second is a finite element model for the support structure, represented by 12 sections along the circumference of the coil. Model has demonstrated fast convergence and capabilities to analyze heat flux distribution in the winding pack and structure. Temperature profile in CICC and heat absorbed by case cooling channels are presented.

INTRODUCTION

Babcock and Wilcox Co. (B&W) is currently involved in the preliminary design and analysis of the Toroidal Field (TF) magnets for the TPX, based on the conceptual design developed at PPPL, MIT and LLNL. Sixteen D-shaped superconducting TF coils are assembled into a closed loop to produce a toroidal magnetic field confining the plasma. TF magnets will use cable-in-conduit conductor (CICC) which contains copper-stabilized Nb_3Sn strands enclosed by a rectangular conduit made of Incoloy-908, cooled by the forced flow of supercritical helium. Each coil (winding pack - WP) - is wound into 12 pancakes 7 turns each, insulated by polyimide wrap and glass-epoxy insulation. Hydraulically each WP is divided in 6 double pancakes (DP) with 175.3m long hydraulic paths for helium[1,2].

Advances in Cryogenic Engineering, Vol. 41
Edited by P. Kittel, Plenum Press, New York, 1996

The winding pack is placed inside the stainless steel coil case which provides structural support during magnet operation; adjacent coils with their coil cases are connected by the elements of intercoil structure (ICS) made of stainless steel. (Fig.1 shows a view of the octant of TF assembly which includes two adjacent coils.)

Helium cooling has to accommodate all the heat generated in the magnet by nuclear (neutronic) heating, eddy current heating due to pulsed operation of poloidal field (PF) coils, and radiation. The steady-state nuclear heating represents the biggest load, and is deposited in the WP (3.95 kW), ground wrap insulation (0.4 kW), coil case (3.44kW) and ICS (1.53 kW)[3].

Effective TF cooling has to be arranged to satisfy the temperature margin of CICC inside the winding pack (maximum difference between local current sharing temperature T_{cs} of Nb_3Sn and CICC temperature), and ensure stability against possible disturbances in the magnet. Helium flow inside CICC alone is not sufficient to absorb all the heat deposited in the magnet. Massive coil case and ICS elements are responsible for the large part of nuclear heating and almost all eddy current heating, which would eventually come to the winding pack; therefore coil case cooling channels (see Fig.1) are necessary to intercept most of this heat load.

Detailed analysis of the case cooling is needed to determine the fraction of the heat intercepted by the channels (previous estimates[1,4] yielded numbers from 91% to 55%). Not only will the intercepted portion depend on the design and location of the channels, but will also be different at different cross-sections, since both the geometry and the heat loads in the coil case and ICS vary widely along the circumference of the coil (see Fig.2). The analysis should determine heat flux and temperature distribution in the coil case / ICS, as well as in the CICC DP.

To estimate the effectiveness of case cooling, a preliminary 2-D finite element analysis in ANSYS has been performed for section 5 in the outer leg region of the TF coil, and also at the equatorial plane of the inboard leg. The nuclear and radiant heat loads acting on the TF structure were modeled with and without cooling channels to provide a comparison. The resulting temperature distribution in the TF structure and heat flux into the winding pack were the evaluation criteria to establish the effectiveness of the channels. Fig. 3 illustrates well how channels inside the coil case intercept heat coming to the WP.

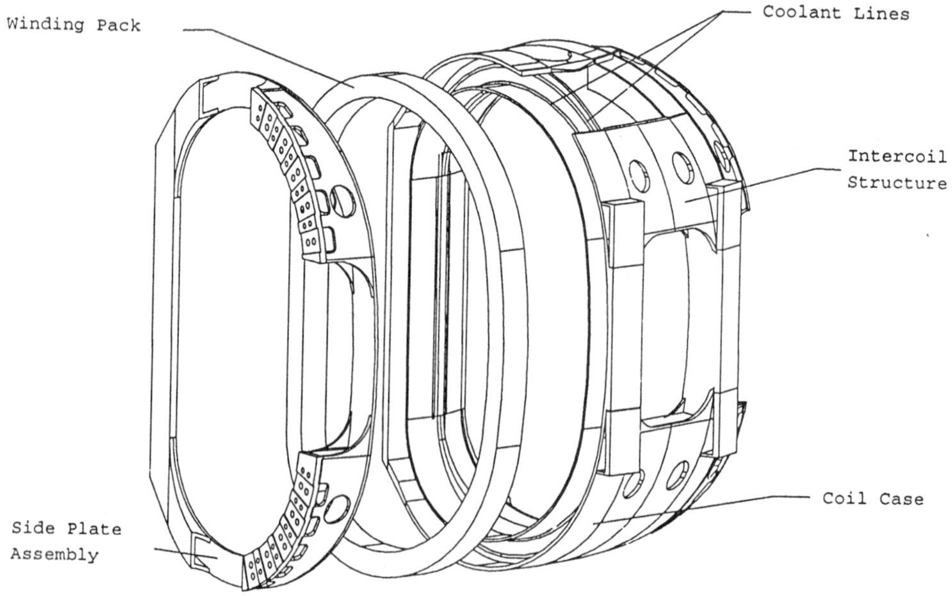

Figure 1. Schematic view of the TF assembly octant (two coils). Shown are: CICC winding pack, intercoil structure and coil case with channels for supercritical helium.

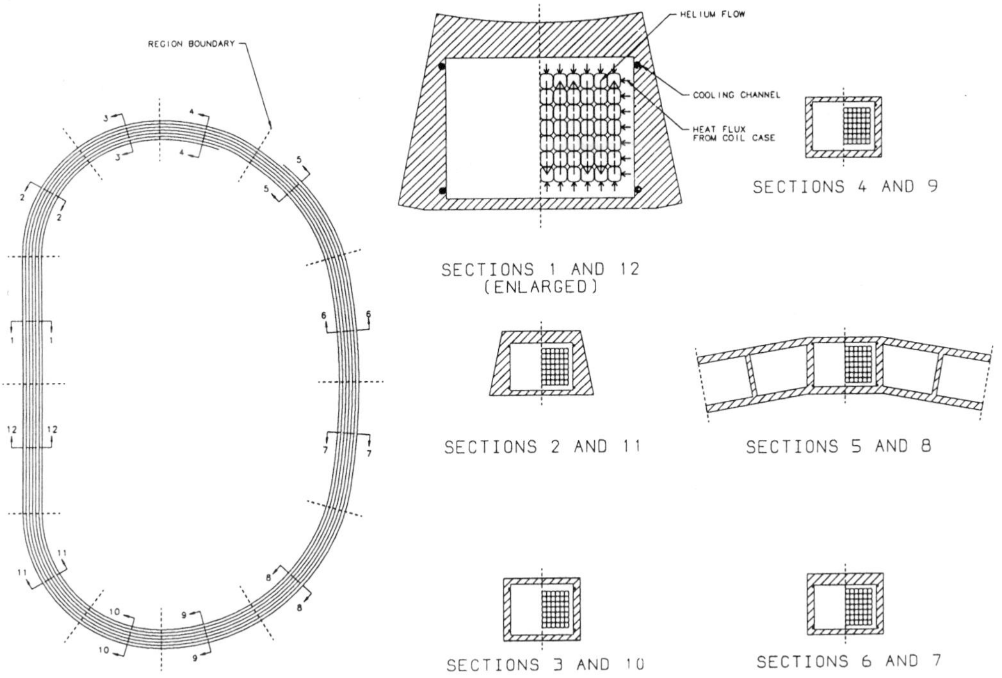

Figure 2. Twelve regions used in support structure modeling and their cross-sections (for each cross-section, plasma side is at the bottom).

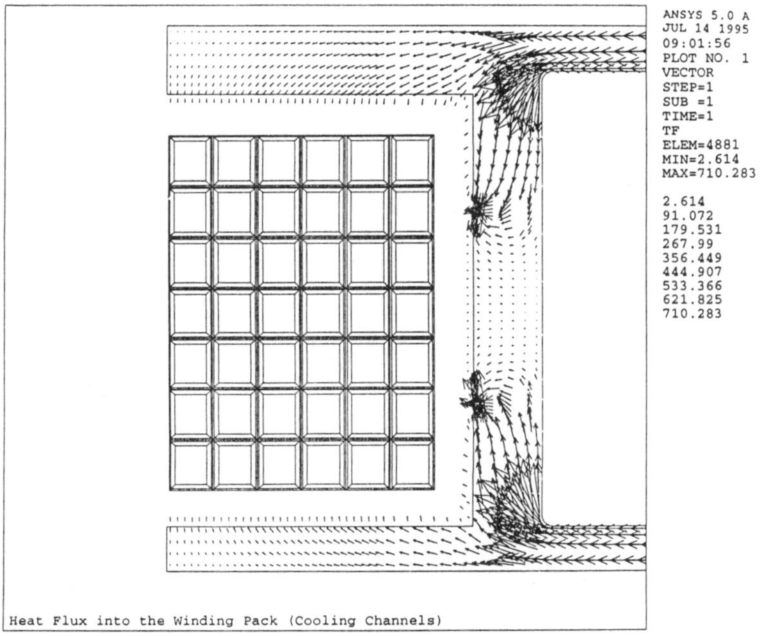

Figure 3. Heat flux distribution in the coil case with cooling channels inside the side walls (ANSYS analysis).

For further analysis, a combined 3-D steady-state thermal-hydraulic model for the TF case cooling and CICC double pancakes has been developed. Numerical implementation is described, example of the analysis is presented below.

APPROACH TO THE ANALYSIS

Full 3-D analysis of the TF WP, together with coil case and ICS is complicated by the fact that helium flow pattern in double pancakes requires modeling along the CICC hydraulic path (spiraling inward and outward in the DP). On the other hand, coil case / ICS modeling demands consideration of the heat conduction normal to the winding pack. The heat flux not intercepted by cooling channels in the coil case becomes an additional heat load, which is only applied to the inner pancakes and inner turns in outer pancakes (see Fig.2, section1). Our approach to the joint analysis uses two different submodels - for the winding pack and for the coil case / ICS, with iterative exchange of boundary conditions between them. Because of the symmetry, half of the winding pack and coil case /ICS is considered.

A Steady-State Thermal Hydraulic code (SSTH) has been created for the TF DP. It solves momentum, continuity and energy equations for one-dimensional helium flow in CICC. The interturn / interpancake heat exchange through the insulation increases maximum temperature in the double pancake because of counterflow in adjacent halves of DP[1,2]. It is incorporated in SSTH, thus bringing a 3-D capability to WP modeling.

To model the coil case with the ICS, a finite element code TEMPEST (Transient Energy, Momentum and Pressure Equation Solution in Three Dimensions) was used. Twelve sections around the TF coil circumference shown in Fig.2 were chosen to represent the TF structure. It includes six geometries only, since upper and lower parts of the coil are symmetric (sections 1 and 12 in the outer leg, 6 and 7 in the inner leg, etc.). However, all 12 sections have to be considered, since helium temperature in the case cooling channels and in CICC is different for symmetric sections. At each section, TEMPEST determines temperature distribution in the structure as well as heat fluxes coming to each of the cooling channels, and to the conductors in the WP. Temperature on the outside of the winding pack serves as a boundary condition on the inside surface of the coil case. Helium exiting the WP enters coil case channels at the beginning of region 5 and exits at the end of region 4. Therefore outlet temperature of the WP is taken to represent the inner temperature for the coil case channels.

The computational process commences with a run of SSTH, with nuclear heating inside the WP and zero heat flux coming from the case. After a solution is obtained, temperatures are determined at 12 CICC cross-sections for the turns facing the coil case, together with exiting helium temperature. These values are used as a boundary condition at the inside boundary for structure modeling, which is now being done by TEMPEST, section by section; heat coming to the cooling channels determines helium inlet/outlet temperatures for each section. Heat flux coming to CICC at 12 cross-sections, determined by TEMPEST, is now used in the next run of SSTH as an additional heat load distribution along the DP hydraulic path. The iteration is repeated until a converged solution is reached.

MODELING OF THE DOUBLE PANCAKE

Thermal-hydraulic modeling of the double pancake considered one-dimensional flow of supercritical helium, with thermal coupling between neighboring turns[1,2]. For the steady-state analysis, temperature of strands inside CICC is asumed to equal that of the helium. Longitudinal heat flux in strands due to thermal conduction is neglected since its characteristic length (few millimeters) is much smaller than the scale of changes in the CICC temperature profile. Helium flow is assumed to be highly subsonic. The walls of Incoloy conduit, together with the insulation, are treated as a transverse thermal resistance between adjacent turns. Under these assumptions, the continuity, momentum and energy equations for the steady-state helium flow take the form:

$$\frac{d}{dx}(\rho v) = 0 \tag{1}$$

$$\rho v \frac{dv}{dx} = -\frac{dp}{dx} - \xi \frac{\rho v^2}{2d_h} \tag{2}$$

$$\rho v \left[C_p \frac{dT}{dx} + \frac{T}{\rho} \rho_T \frac{dp}{dx} \right] = \frac{q(x)}{A_{he}} + \xi \frac{\rho v^3}{2d_h} \tag{3}$$

where x is coordinate along CICC; T is temperature; p is pressure; $\rho\,(T,p)$ and v are helium density and velocity; ξ is friction factor; $d_h = 4A_{he}/\Pi_w$ is hydraulic diameter of CICC, defined through the helium flow area A_{he} and total wetted perimeter of the strands Π_w ; $\rho_T \equiv (\partial\rho / \partial T)_{p=const}$, $\rho_p \equiv (\partial\rho / \partial p)_{T=const}$; $q(x)$ is heat generation per unit length of CICC.

We eliminate the velocity from eqs.(2), (3) by using $\rho v = \dot{m} / A_{he}$, where $\dot{m}$ is mass flow rate per channel. We also neglect the compressibility of helium for our subsonic flow ($\rho_p \dot{m}^2 / \rho^2 A_{he}^{\ 2} = (v^2/a^2)(C_p/C_v) << 1$; where a is velocity of sound), and the term, $T \rho_T^{\ 2} \dot{m}^2 / A_{he}^{\ 2} \rho^4 C_p$, which is of order of 10^{-2}-10^{-4}. With these simplifications, system (1)-(3) can be reduced to two ordinary differential equations

$$\frac{dT}{dx} = \frac{q(x)}{\dot{m}\, C_p} + \frac{\xi(1 + T\rho_T/\rho)}{2\, d_h C_p \rho^2 A_{he}^{\ 2} / \dot{m}^2} \tag{4}$$

$$\frac{dp}{dx} = \left(\frac{\rho_T\, \dot{m}^2}{\rho^2 A_{he}^{\ 2}} \right) \frac{dT}{dx} - \xi \frac{\dot{m}^2}{2\, d_h\, \rho\, A_{he}^{\ 2}} \tag{5}$$

with boundary conditions

$$T(x=0) = T_{in}; \qquad p(x=L) = p_{out} \tag{6}$$

where L is hydraulic length of the DP. The heat source

$$q(x) = q_{nucl}(x) + q_{case}(x) + \sum_n q_{tt}(T(x), T(x_n)) \tag{7}$$

consists of neutronic heat, heat coming from the coil case, and heat transferred between turns (summation in (7) is done over all neighboring turns). The neutronic heat load in the winding pack has been mapped along the DP hydraulic length and interpolated. The resulting dependence $q_{nucl}(x)$ is shown in Fig.4.

In order to cover the heat transport between adjacent flow channels (layer-to-layer and pancake-to-pancake), a 3-D node-mapping model was developed for the thermal analysis and case cooling design. As the result, heat conduction from adjacent pancakes, layers and side walls can be taken into account as the thermal boundary conditions in the analysis.

For each node on the winding pack, there are generally four neighboring nodes around it. These nodes may belong to the same or adjacent pancake, or to the wall of the coil case. The mapping of such nodes is rather straightforward. However, at the transition regions, the mapping becomes more complicated, requiring more detailed mapping at these regions.

Each DP consists of one pancake wound from the outside to the inside (out-to-in pancake), and one wound from the inside to the outside (in-to-out pancake).The layer-to-layer transitions for the out-to-in and in-to-out pancakes occur at the section 4-5 and 2-3, respectively. The pancake-to-pancake transition (within the same DP) and the DP-to-DP transition occur at the ID and OD, respectively[5]. Due to the existence of these transitions, only sections 1 and 12, 6 and 7 are strictly symmetric.

For the layer-to-layer transitions, the radius of the winding gradually varies from R_n to R_{n+1}, resulting in the change in contact areas, so that one layer partially contacts two adjacent layers. Similarly, the pancake-to-pancake transition travels from one pancake to another with a constant radius. Hence, it partially contacts two pancakes at its bottom. In the present analysis, regions with partial contact were not considered.

For each couple of neighboring turns, the interturn heat flux is given by $q_n = (T(x) - T(x_n)) / r_n$, where interturn thermal resistance r_n includes that of Incoloy wall and polyimide wrap of each conductor, and that of glass-epoxy insulation between the conductors.

Iterations were employed to solve system (4)-(7) numerically. Temperature values at neighboring turns participating in the interturn heat flux, are taken from the previous iteration (fast convergence with regard to temperature and pressure has been observed). In addition, inlet pressure was adjusted by bisections to satisfy given outlet pressure, p_{out}, and mass flow rate, $\dot{m}$.

COIL CASE / ICS MODELING

The 3-D thermal analysis of the TF coil case sections and the intercoil structure was performed with the time-dependent computational dynamics code TEMPEST. TEMPEST is a transient, three-dimensional, hydrothermal computer program designed to analyze a rather broad range of coupled fluid dynamic and heat transfer systems.

As discussed previously, the TF magnet coil case and associated intercoil structure were represented by twelve sections around the circumference of the TF coil case. A thermal-hydraulic analysis was performed on each section in which a temperature boundary condition was imposed on the thermal interface between the winding pack and the coil case at the Incoloy inner surface of the conduit walls. The temperature boundary condition was supplied by the SSTH code during the iteration process. Likewise, the heat fluxes computed by TEMPEST at the thermal boundary interface nodal volumes were used as input into the SSTH code.

Each section of the coil case was modelled in its exact 2-D cross-sectional geometric configuration with up to 6000 nodal volumes of unit thickness. The material configuration included all applicable ground wrap and epoxy layers along with the structural steel. The TF case cooling channels were modeled as square channels. The change in the helium temperature is computed over the length of the section and used as an input for the next section. The computational process commences with section 5 (helium entrance point) and progressed section by section along the length of the TF case with section 4 (helium exit point) being the last section calculated in the loop. Coolant pressure used in the analysis was 0.25 MPa.

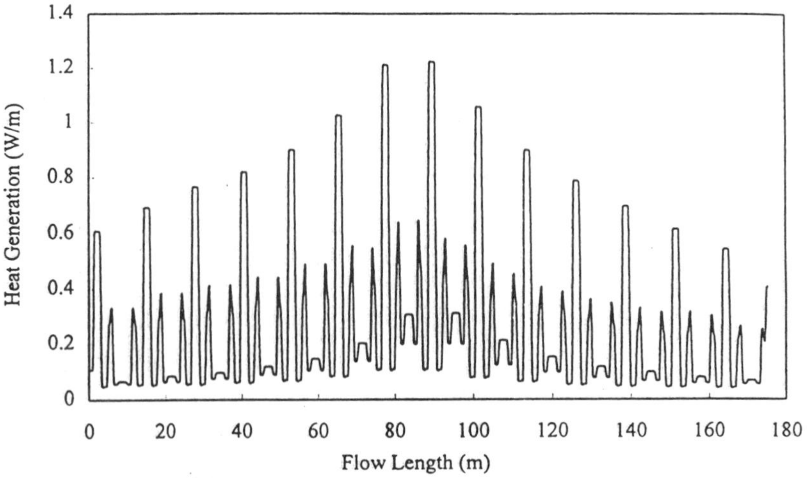

Figure 4. Nuclear heat load profile for TF coil (side double pancake).

Table 1. Parameters of the TF coil thermal analysis

Winding Pack	
Number of single pancakes / coil	12
Number of turns / pancake	7
Number of hydraulic channels / coil	6
Hydraulic length of double pancake (one channel)	175.3 m
Helium mass flow rate per channel	3.91 g/s
Inlet temperature	5 K
Outlet pressure	2.5 MPa
CICC outer dimensions	22.61mm×27.18mm
Number of strands in CICC	486
Strand diameter	0.78 mm
Helium flow area	146 mm^2
CICC hydraulic diameter	0.48 mm
Ground wrap insulation thickness	5.74 mm
Support Structure	
Material	SS 316LN
Number of regions in the model	12
Number of geometries	6
Number of channels per coil in preliminary modeling	4
Mass flow rate per channel	3.91 g/s

RESULTS

The preliminary modeling has been performed using only neutronic heat generation in the WP and coil case / ISC. Main parameters are listed in the Table 1. In this modeling, temperature of the helium exiting the outer DP was taken as an inlet temperature for cooling channels. Iterative runs of SSTH and TEMPEST has demonstrated rather fast convergence. As can be seen from the Table 2, values of the heat absorbed by cooling channels in different regions, as well as the maximum temperature inside CICC, show small changes after 3rd iteration. For the case considered, channels absorb about 65 % of the total heat deposited in the structure. A difference can be noticed in the heat absorbed at symmetric cros-sections, e.g. , those values at section 4 are smaller than at section 7. This is explained by the fact that helium in the channels has higher temperature at the end of the cooling loop, and absorbs less heat. The helium exit temperature is 8.7 K for the plasma side channel, and 7.5 K for the outer side channel.

Final temperature distribution along CICC in the outer double pancake is presented in Fig. 5 together with the current sharing temperature, which has been calculated according to the correlation by Summers et al.[6], using magnetic field distribution obtained by 3-D analysis by OPERA/TOSCA [7]. Most of the heat coming from the structure is received by the second half of the DP, which creates monotonous temperature profile. The temperature margin in the case modeled is about 2.0 K.

Table 2. Heat absorbed by cooling channels at different regions (Watts). Channel 1 is on the plasma side, 2 is on the outer side

Iteration	Channel	1	2+3+4	5	6	7	8	9+10+11	12	Total
1	1	6.98	5.03	13.93	7.25	7.25	13.76	5.54	7.09	66.83
	2	2.95	1.71	11.93	4.36	4.35	11.83	1.88	2.97	41.98
2	1	6.46	3.92	13.74	7.09	7.03	13.29	4.60	6.61	62.74
	2	2.64	0.96	11.83	4.24	4.20	11.54	1.17	2.67	39.25
3	1	6.39	3.78	13.70	7.06	6.99	13.22	4.47	6.55	62.16
	2	2.61	0.87	11.81	4.22	4.19	11.50	1.09	2.63	38.92

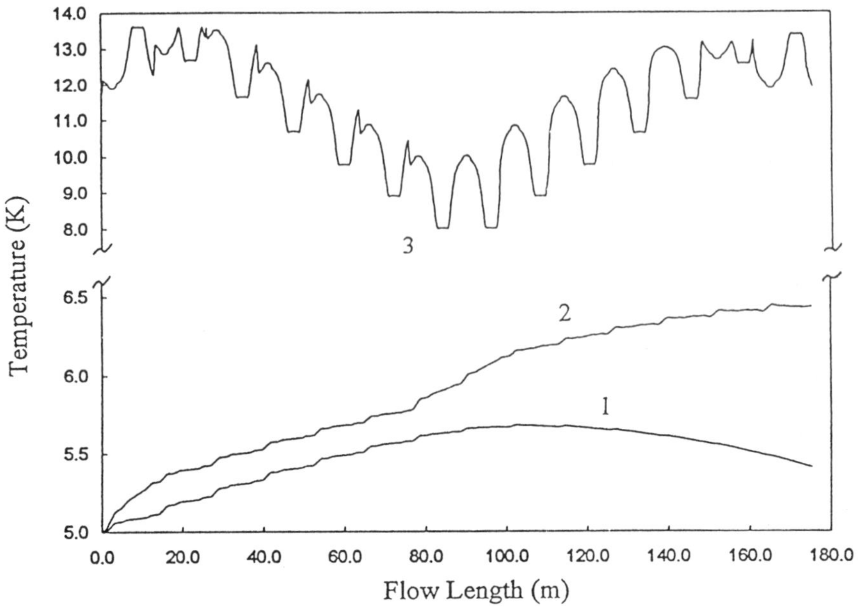

Fig. 5. Temperature profile in the side double pancake. *1* - first iteration (no heat from the coil case and ICS) ; *2*- final distribution; *3* - current sharing temperature

CONCLUSIONS

A joint 3-D steady-state thermal-hydraulic model for the winding pack and support structure of the TPX TF magnets has been built at B&W to evaluate the effectiveness of case cooling channels and its impact on CICC temperature margin. Preliminary runs have demonstrated the capability of the model to analyze temperature and heat flux distribution both in CICC and support structure, accounting for effects of their thermal coupling. Exchange of temperature / heat flux data between submodels results in fast convergence.

The model will be applied to assess main design parameters of the cooling arrangement of the TF magnets.

ACKNOWLEDGMENT

The authors wish to express their appreciation to G. T. Markham and D.T Stutzki for their valuable contribution in preparation of this paper. This work has been performed under DOE contract B235308.

REFERENCES

1. R.L. Wong, J.P. Zbasnik, and W.V. Hassenzahl. Thermo-hydraulic analysis of superconducting toroidal field magnets for the tokamak physics experiment, in: "Advances in Cryogenic Engineering", v. 39, Plenum Press, NY (1994), p.1715.
2. P.-W. Wang. SHE3D- a code for the three dimension steady-state helium-cooled CICC conductors , *IEEE Trans. on Appl. Supercond.*, 5-2: 600 (1995)
3. S.L. Liew. 3-D calculations of nuclear heating rates in TPX TF coils. -TPX Memorandum 94-940329 - PPPL/ SLiew-01, March 1994.
4. R.L. Myatt. Heat flow from the TPX TF coil support tsructure into the winding pack from 3-D nuclear and eddy current heat loads. - TPX Memorandum 13-940601-MIT/LMyatt-01, June 1994.
5. D. T. Stutzki, Winding Pack Modeling, Babcock & Wilcox report T3B-00001 (1995).
6. L.T.Summers, M.W.Guinan, J.R.Miller and P.A.Hahn. A model for the prediction of Nb_3Sn critical current as a function of field, temperature, strain and radiation damage, *IEEE Trans.on Magn.*, 27-2:2041(1991).
7. M. Xu. Private communication.

SUPPORT SYSTEM DESIGN INCORPORATING CARBON/EPOXY TENSION STRAPS FOR A FOUR TESLA, ONE METER BORE, MRI MAGNET

L.C. Watts,[1] K. Boulios,[2] K.T. Hartwig,[3] and F.R. Huson[2]

[1]RPM Engineering
Baton Rouge, Louisiana, 70809, USA

[2]Texas Accelerator Center
The Woodlands, Texas, 77381, USA

[3]Texas A&M University
College Station, Texas, 77843, USA

ABSTRACT

The Texas Accelerator Center (TAC) constructed a four tesla, one meter bore, MRI magnet. Graphite/epoxy tension straps with thermal intercepts at 77 K and 20 K were used for a safe, efficient, and cost effective support system. Sixteen straps were used to support the 10,400 kg cold mass. TAC used a superferric concept to restrict the fringe field. With proper alignment the attractive forces between the magnet and iron are designed to cancel. Therefore, easy adjustment for proper positioning of the magnet, and strength to withstand the magnetic forces due to any misalignment are requirements of the support system. Since the MRI magnet was shipped by truck, the support system had to withstand dynamic loads. The design conduction heat load to the cold mass is 0.05 W. Details of the support system are presented.

INTRODUCTION

The Texas Accelerator Center (TAC) has developed a superferric concept for superconducting magnets[1]. This concept restricts the fringe field, uses less superconducting material, and isolates the region of desired field from external disturbances. TAC has now designed and built three superferric NMR/MRI magnets for research. The second magnet, and the subject of this paper, has a field strength of four tesla and a one meter warm bore.

In TAC's design, much attention is given to reducing the heat transfer to the liquid helium (LHe) cooled portions of the magnet (cold mass). The power leads are

Advances in Cryogenic Engineering, Vol. 41
Edited by P. Kittel, Plenum Press, New York, 1996

disconnectable. There is a high vacuum and two radiation shields between the cold mass and room temperature components. The magnet support system suspends the cold mass inside the room temperature vacuum vessel while leaving room for the radiation shields. Improper design of the support system could allow heat flow that would negate the insulation achieved with the high vacuum and radiation shields.

The objectives of this paper are to define the specifications for the TAC MRI support system and describe the support system that was used.

SPECIFICATIONS

Cold Mass Positioning Accuracy

Figure 1 shows a cross sectional view of the MRI magnet and includes the cold mass, vacuum vessel, and surrounding iron. The magnet has a cylindrical shape and is symmetric about the axis.

In order for the large attractive forces between the iron and superconducting coils to cancel, the cold mass must be exactly centered within the iron. This cold mass must also be centered to achieve a homogeneous magnetic field. It is difficult to maintain this accurate position with the effect of thermal contraction. The bobbin, on which the superconducting wires are wound, is made of 316L stainless steel. This material contracts 0.3% when cooled from room temperature to LHe temperatures. The support system must position the cold mass to within ± 3 mm in order to keep off-center forces to within tolerable levels.

Space Limitations

The support system must fit in the radial space between the cold mass and vacuum vessel allowing room for the two radiation shields. In critical areas the radial space is as small as 10.8 cm and there should be a minimum of 4 cm for the two radiation shields.

Structural Conditions

The first structural condition is to suspend the cold mass inside the room temperature vacuum vessel for assembly. The second structural condition is for shipment. The magnet was shipped to its final destination on an eighteen wheel truck and trailer. Both the truck

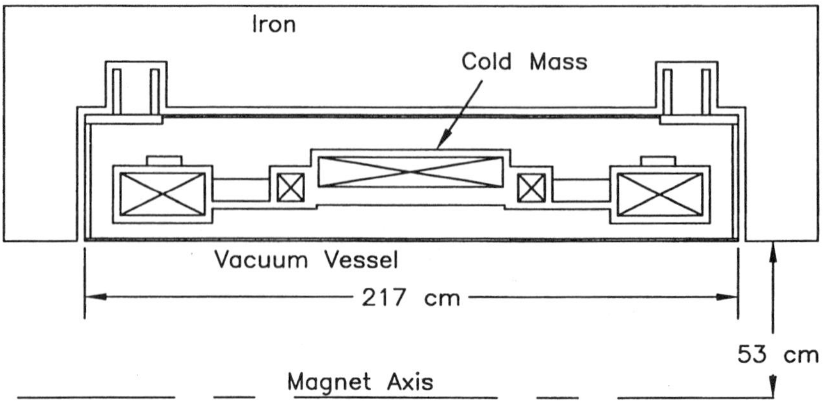

Figure 1. Cross section of TAC MRI magnet.

and trailer had air ride suspensions. The magnet axis was along the length of the trailer. The design static loads are based on studies[2,3] and industrial experience[4] with fragile cargo on this type vehicle. The loads take into account the accelerations normally seen on this type vehicle and the duration of these accelerations. These design loads are in terms of the earth's gravity, g. The static load requirements are 3.0 g vertical, 2.5 g axial, and 2.0 g lateral. In addition, the magnet's vibrational modes must be above those normally seen on this type vehicle. The truck studies indicate peak excitation energy at 2 and 10 Hz and much lower energy above 12 Hz. Therefore, an additional condition was established for the magnet system (vacuum vessel, support system, and cold mass). The magnet system must withstand a 0.5 g acceleration at 10 Hz in any direction.

The final structural condition is for normal operation of the magnet at four tesla. If the magnet is misaligned axially by 3 mm there will be 166,000 N of attractive force between the superconducting winding and surrounding iron. A radial misalignment of 3 mm will generate 33,000 N of attractive force. Any subsequent deflection of the support system will produce larger attractive forces. The support system must withstand these forces and have a large enough spring constant to minimize deflection.

Low Conduction

The magnet's superconducting windings are cooled with LHe to 4.2 K. A heat transfer rate of 1.0 W will boil off 1.41 liters per hour of LHe. This would cost thousands of dollars per month in LHe purchases and thousands more in trained personnel to perform the LHe unloading. The heat transfer to the cold mass directly effects the magnet's operating expenses.

General Electric Medical Systems[5] achieved a minimal 0.05 liters per hour (0.035 W) of LHe boiloff with a 0.85 meter bore, 0.5 tesla, MRI magnet. Although a much lighter magnet because of its lower field, it is a good example of today's skill in realizing low heat transfer.

TAC has incorporated a 20 K radiation shield surrounding the cold mass. This should reduce the radiation heat load to the cold mass to below 0.01 W. The power leads are disconnectable to eliminate their associated conduction heat load. The magnet support system is another heat load to the cold mass by conduction. This heat load should come as close as possible to the 0.01 W radiation heat load.

Low Cost

The six areas that contribute most to the cost of a magnet support system are: engineering, testing, material, tooling, machining, and assembly. One way to reduce the cost is to find a design that has been tested and is available with little or no additional tooling. Minimal machining time and ease of assembly will also lower the cost.

Designs should be compatible with production requirements. Costs for engineering, testing, and tooling are more significant for the production of a support system for one magnet. Although it is hoped that many magnets will be built using the TAC design, the current status is to produce only one. Therefore, the use of a tried-and-true method to support the magnet is preferred.

SUPPORT SYSTEM DESIGN

Support Concept

A review[6] was made of support systems for both superconducting magnets and cryogen storage vessels. Tension straps were chosen as the best support system for TAC's magnet.

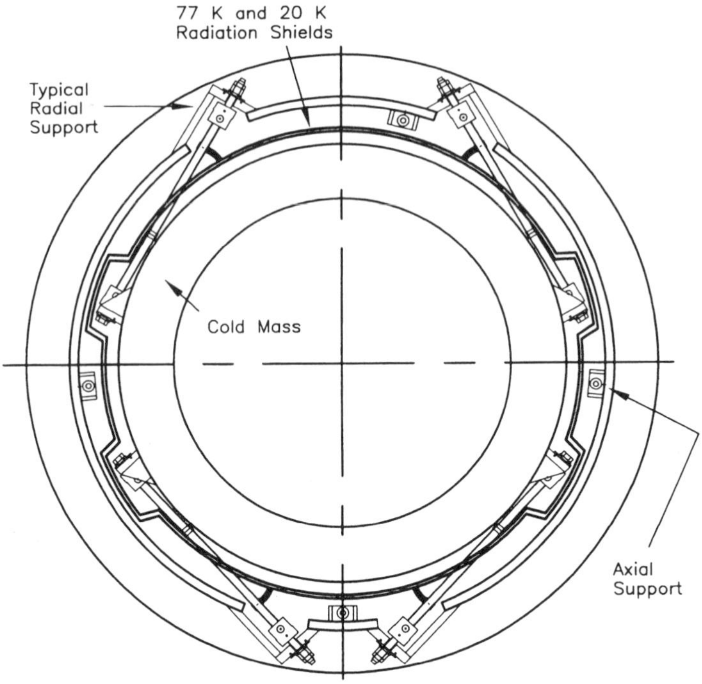

Figure 2. Arrangement for the radial supports.

Tension straps are continuous loops of fiber reinforced plastic that take advantage of the high unidirectional strength of composites. There are several reasons for choosing tension straps:

1) This method is tested. It has been used for both MRI magnet suspension[4] and in the suspension of LHe dewars in space[7].

2) There were companies available with experience in producing tension straps.

3) There was tooling available to produce the size tension straps needed for the magnet.

4) There was test data on a number of straps from the above mentioned tools.

The support system design includes two strap sizes. Each strap has a center to center distance of 73.4 cm, a radius of 1.27 cm, and a leg thickness of 0.28 cm. The only difference in the two straps is the width. One has a 2.54 cm width and the other 1.27 cm.

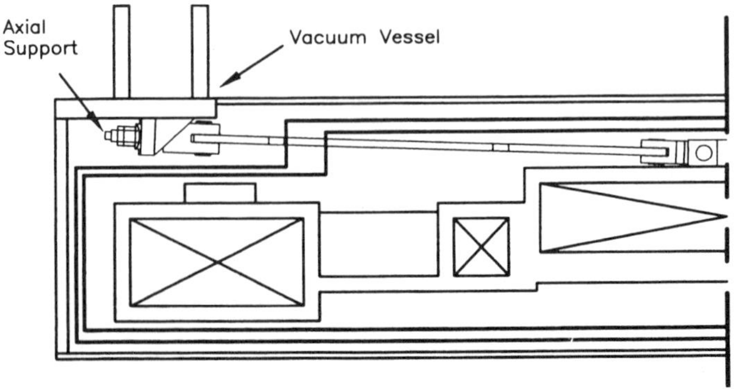

Figure 3. Arrangement for the axial supports.

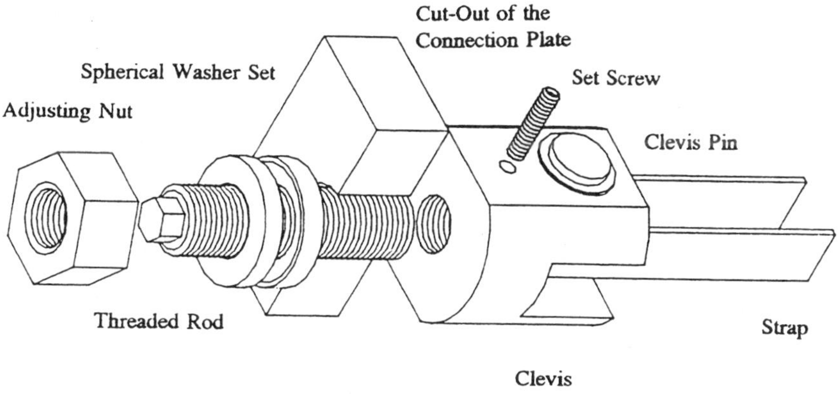

Figure 4. Strap tensioning device.

The material used in the straps is IM6 carbon/epoxy which has excellent properties for a magnet suspension:

1) With a 66% fiber volume, the IM6 carbon/epoxy strap has a 182 GPa elastic modulus[8]. This will provide a high enough stiffness for the dynamic load and attractive magnetic force requirements.

2) The thermal conductivity of carbon/epoxy has been shown to vary with the modulus[9]. Based on this relationship, the thermal conductivity of IM6 carbon/epoxy was estimated. The estimate gives a low thermal conductivity below 40 K that compares well to any of the currently used materials for magnet support systems, including G10.

3) Tests by the manufacturer show the tensile strength to be between 1600 and 1900 MPa. This will provide enough strength for the static load requirements.

4) The thermal contraction, when cooled from 300 K to 4.2 K , is very low. The cold mass must be properly aligned inside the surrounding iron by the support system. This alignment is performed at room temperature. Less movement of the support system reduces the uncertainty of the cold mass position after cooldown.

The support system uses a total of 16 straps. Figure 2 shows the radial support system looking through the bore of TAC's magnet. There are four straps at each end of the cold mass for radial support. Figure 3 is a view perpendicular to the magnet axis showing the axial straps. There are four straps at each end of the magnet for axial support. Figure 4 shows the warm end of one of the straps with the connection plate and tensioning device. The connection plate will be welded to the vacuum vessel at the vacuum vessel stiffener. Each strap has a similarly designed tension device at the warm end. At the cold end the threaded rod and adjusting nuts are replaced with a hex cap screw.

The rest of the materials used in the support system are annealed AISI 316 stainless steel, and age hardened Inconel 718. Both of these materials have a low magnetic permeability and retain their ductility at low temperatures. A yield strength of 207 MPa was used for the AISI 316. A yield strength of 1030 MPa was used for the Inconel 718.

Thermal Intercepts

The temperature of the strap at the vacuum vessel is 300 K and the temperature at the cold mass is 4.2 K. Because the thermal conductivity of IM6 carbon/epoxy is lower at low temperatures there will be a high temperature gradient at the cold end. High temperature gradients mean high heat transfer rates. With no adjustment to the strap temperature profile the conduction heat load to the cold mass would exceed the specification requirements.

Thermal intercepts are used to lower the temperature at strategic locations along the strap leg. These intercepts are a one inch wide, copper braid bonded to the strap leg by the strap manufacturer. Both 77 K and 20 K cooling are available from the same refrigeration system that cools the radiation shields. By employing two thermal intercepts, a 77 K intercept 12 cm from the warm end and a 20 K intercept 25.4 cm from the cold end, the temperature profile in the strap is modified to greatly reduce the conduction heat load to the cold mass.

If the intercepts were 100% efficient, the strap temperature at the intercepts would be that of the supplied cooling. However, experiments conducted at TAC[6] showed a 9 K temperature difference between the copper braid and the strap leg, with the copper braid at 105 K.

Considering the intercept inefficiencies, the maximum expected heat load to the cold mass from conduction through 16 straps is 0.05 W.

Stress Analysis

In order to increase the stiffness of the magnet system, opposing tension members are used. The straps are designed to remain in tension during each stage of the magnet life. Proper pretension of the straps during assembly was maintained with strain gauges on the strap legs.

The first three vibration modes for the magnet system have been calculated assuming the cold mass and vacuum vessels are infinitely stiff. Bracing will increase the stiffness of the vacuum vessel during shipment. The modes are 15.2 Hz lateral, 18.2 axial, and 20.4 vertical.

The strap manufacturer guarantees a minimum tensile strength in the straps of 1500 MPa. For the designed loading conditions, the period during maximum stress for the upper radial straps is during the 3.0 g vertical load. The straps have a stress of 880 MPa. The period of maximum stress for the lower radial straps is during the 2.0 g lateral load. The straps have a stress of 670 MPa. The period of maximum stress for the axial straps is during operation of the magnet at four tesla. The maximum stress is 920 MPa.

The remaining components of the magnet support system have been designed not to exceed their yield strength before the strap reaches 1500 MPa.

CONCLUSIONS

A number of methods for magnet suspension were reviewed and tension straps were chosen for the ability to meet all of TAC's requirements. A support system has been designed for TAC's four tesla, one meter bore, MRI magnet. The specifications include:

1. adjustments for cold mass positioning accuracy finer than $\pm$ 3 mm
2. the radial straps used 95 mm of radial space and the axial straps used 63 mm of radial space, leaving over 40 mm remaining for radiation shields
3. static loads on the 10,400 kg cold mass of 3.0 g vertical, 2.5 g axial, and 2.0 g lateral
4. dynamic loads of 0.5 g in any direction at 10 Hz
5. maximum expected conduction heat load through sixteen straps of 0.05 W
6. low cost

Although the conduction heat load is low, it is still the dominant heat load to the cold mass. Improvements in the thermal intercept performance could easily reduce the conduction load by half. Applying pressure at the thermal intercept and using aluminum filled epoxy is one technique that could be studied. The best overall concept from the support system review[6] was found to be tension straps connected in series with pins. This technique allowed the use of a high thermal conductivity material for a positive break at the thermal intercept. TAC used the straps in series concept for a subsequent magnet.

ACKNOWLEDGMENT

This research was sponsored by the National Science Foundation.

REFERENCES

1. F.R. Huson et al. High-field superferric NMR magnet, *Magnetic Resonance in Medicine*, (1992).
2. R.E. Glass and K.W. Gwinn, Design basis for resistance to shock and vibration, 9th International Symposium on the Packaging and Transportation of Radioactive Materials, 953 (1989).
3. R.E. Glass and K.W. Gwinn, Shock and vibration environments for truck transported nuclear waste, *Nuclear Materials Management*, 15:319 (1986).
4. Personal correspondence with B.C. Breneman of Applied Superconetics, San Diego, California.
5. K.M. Parker and Y.P. Tsavalas, Compact superconducting magnet with integral magnetic shield and low-loss cryostat for whole body MRI, in: "Advances in Cryogenic Engineering", 37A:355 (1992).
6. L.C. Watts, "Support System Design for a Superconducting, MRI Magnet", Texas A&M University, College Station, Texas, M.S. thesis, (1993).
7. A. Seidel and Th. PaBvogel, The iso-cryostat, in: "Advances in Cryogenic Engineering", 37B:1327 (1991)
8. V.L. Morris, Advanced composite structures for SSC, in: "IISSC", 1:525 (1986).
9. G. Hartwig and S. Knaak, Fibre-epoxy composites at low temperatures, *Cryogenics*, 24:639 (1984).

STUDY OF IMPREGNATING EPOXY RESINS FOR HIGH FIELD NMR SUPERCONDUCTING MAGNETS

G. Liang, [1,2] G. Luo,[2] L. Crowe,[1] J. Zeigler, [1] G. Shotzman,[1]
Y. Wu,[1] R. Teodorescu[1]

[1]Houston Advanced Research Center
The Woodlands, Texas 77381, USA
[2]Department of Physics, Sam Houston State University
Huntsville, Texas, 77341, USA

ABSTRACT

NMR magnet coils are usually quite long, thick, and tightly wound with thin superconducting wires. The successful vacuum/pressure impregnation of such kind of coils demands the use of epoxy resins with superior properties such as low viscosity, long pot life, and high cracking resistance etc.. In order to find the most appropriate impregnating epoxy resin for the fabrication of high-field NMR magnet coils, we have studied several promising epoxy resins by viscosity, thermal shock, bonding/de-bonding measurements. The results of these measurements are presented. Model coils have been vacuum/pressure impregnated with selected epoxy resins and analyzed with scanning electronic microscope (SEM). It was found that among all of the studies epoxy resins the CTD-101K epoxy resin is most suitable for impregnation of coils. The test results of the model NbTi superconducting coil show that coils potted with CTD-101K do not quench until critical current of the superconductor is reached. This epoxy and the impregnation technique have been successfully applied to the first 400 Mhz/89 mm actively shielded high resolution NMR magnet developed at Houston Advanced Research Center.

INTRODUCTION

Vacuum-pressure impregnation (VPI) of magnet coils with epoxy has been a widely used technique in building high field superconducting magnets. The purpose of epoxy impregnation is to prevent the conductor motion caused by the Lorentz force1 Such motion is one of the main sources for the premature quench of the superconducting magnets[1,2]. A high field NMR superconducting magnet usually involves the fabrication of long and thick coils which are tightly wound with very thin wires. Thus, the use of the most appropriate epoxy formulation and reliable VPI technique plays the key role in magnet construction.

In 1992, the Houston Advanced Research Center (HARC) started its project of developing the first actively shielded 400 MHz/89 mm bore NMR spectrometer magnet and other higher field magnets. Special coil design/winding features and conductor wire size of the 400 MHz magnet demanded the use of suitable epoxy formulation and impregnation technique. At that time, there had been only very few epoxy formulations reported in the literature[3] for vacuum impregnation of NMR magnet coils and they appeared not appropriate for our project. The goal of our research was to find the most appropriate epoxy resins for the successful impregnation of the 400 MHz NMR magnet coils.

SELECTION OF EPOXY FORMULATIONS

Criteria for Epoxy Selection

The selection of the epoxy resins was based on the following six criteria: 1. low viscosity (≤ 100 centistokes at 60 °C); 2. long pot life (≥ 20 hours at 60°C); 3. high crack resistance; 4. appropriate cure schedule and temperature (5 - 24 hours and maximum 125°C); 5. low toxicity (epoxy containing carcinogens should be avoided); and 6. Good vacuum and mechanical properties such as high fracture toughness and tensile strength etc..

The pot life here is defined as the time required for the viscosity of the epoxy to rise to 1000 centistokes (cSt). One of the main reasons for the criterion 4 above is that longer time high temperature curing could induce significant degradation of the critical current J_c, especially in high field. Our field dependent J_c measurement on the NbTi wires (for the 400 MHz magnet) had shown that there was an about 7% J_c degradation at 9 Tesla after the wires were annealed at 110-125 °C for about 20 hours.

Viscosity Measurements

Viscosity measurement is necessary for determining the pot life of the epoxy. The viscosity as a function of time and temperature were measured with a set of calibrated BOEKEL Zahn cup-type viscometers for more than ten epoxy formulations. These epoxide resins were pre-screened (based on the available information related to above criteria) from commercial products of Ciba-Geigy, the E. V. Robert&Associates, and Shell Chemical. In Fig.1, we show the viscosity data measured at 60 °C for seven epoxy formulations listed in Table 1. It is seen that the four samples A1 and A2 (Resin CY179 with accelerator XU183 ≤ 2.5 parts by weight), B (GY6010), and C (CTD-101K) have pot life longer than 20 hours at 60 °C and thus they were chosen for further model coil potting study and thermal shock test. All of the epoxy samples shown in Table 1 and Fig. 1 were from Ciba-Geigy except the sample C which was from the E. V. Robert & Associates. The CTD-101K is an anhydride cured epoxy originally developed for cryogenic applications down to liquid helium temperature[4].

VPI EQUIPMENT AND PROCEDURE

Equipment for VPI of Coils

Two VPI systems were built for this study. The first system was a smaller one for potting smaller model coils with size up to 8 inches in diameter and 14 inches in length. It consisted of an epoxy deaeration chamber (EDC) made of stainless steel and a vacuum-pressure impregnation chamber (VPIC) modified from a Galvanized steel paint tank which

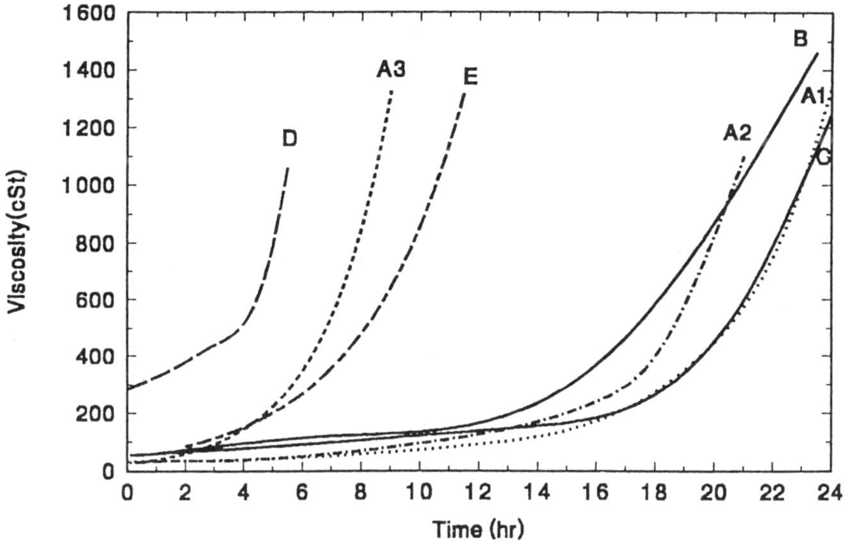

Figure 1. Viscosity as a function of time for selected epoxy samples at 60 °C.

can withstand 115 psi pressure. Both of them were equipped with vacuum/pressure, temperature controlling, heating, and window/lighting systems. An epoxy tank for mixing the epoxy and a mold tank for sitting the coil were placed inside the EDC and VPIC, respectively, both of these two tanks were wrapped with heating tapes. The second system is an enlargement of the first one with similar construction features. The larger system was used for potting the 400 MHz NMR magnet coils.

Table 1. Several selected epoxy formulations

Experiment Sample Code	Composition		Parts By Weight
A1	Araldite	CY179	100.0
	Hardener	HT907	105.0
	Accelerator	XU183	1.6
A2		CY179	100.0
		HT907	105.0
		XU183	2.5
A3		CY179	100.0
		HT907	105.0
		XU183	6.0
B		GY6010	100.0
		HT907	85.0
		XU183	1.4
C	Resin	PART A	100.0
	Hardener	PART B	90.0
	Accelerator	PART C	1.5
D		CY225	100.0
		HY227	100.0
E		GY6005	100.0
		HY906	80.0
		DY062	1.0

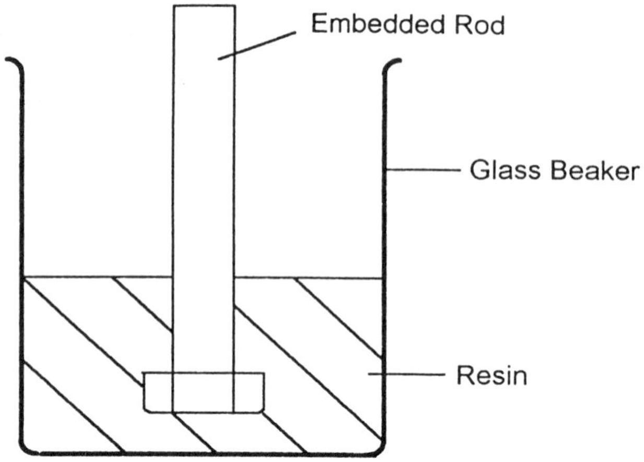

Figure 2. Schematic drawing of a epoxy sample.

Epoxy VPI Procedure

A typical procedure for VPI of model coils can be described as the following, taking epoxy CTD-101K as an example:

(1) Place the coil in the mold tank and bring the temperature up to about 70°C, degas the VPIC for about 12 hours at a vacuum of about 20 microns.

(2) Mix the resin, hardener, and accelerator together in the epoxy tank and warm the mixture up to about 60°C. Deaerate the mixture by a vacuum pump while stirring the epoxy with a motorized stirrer. Continues this process for about 45 minutes until the small bubbles completely cease to produce and pressure is about 20 microns.

(3) Transfer the epoxy from the epoxy tank (in EDC) into the mold where the coil is sitting. Stop the transferring as soon as the epoxy reaches a desired level. Set temperature of the coil at 60 °C.

(4) Deaerate the epoxy in the mold until vacuum goes down to about 20 microns. Pressurize the epoxy with nitrogen gas slowly for about 20 minutes until the pressure reaches 100 psi. Maintain this pressure and 60 °C for about 8 hours.

(5) Depressurize the VPIC to atmospheric pressure. Take out the coil and rotate it in a oven (or furnace) with temperature already set at 60 °C.

(6) Increase the temperature of the oven to 110 °C at a rate of about 2 °C/min. Cure the epoxy at this temperature for about 4-6 hours and then at 125 °C for about 15 hours.

THERMAL SHOCK EXPERIMENTS

Sample Preparation and Experimental

The purpose of our thermal shock experiment was to find the appropriate cure schedule which would yield high cracking resistance for the epoxy resins. Thermal shock specimens of the four long pot-life epoxy resins A1, A2, B, and C (see Table 1) were prepared in four different geometric configurations as following.

(1) The specimens were made approximately cylindrical shape using glass beakers as the epoxy mold. The diameters of the specimens have two sizes: 62 mm and 51 mm. The thickness ranges from 10 mm to 50 mm. Table 2 listed some selected specimens, where A1 and A2 specimens are the same epoxy resins as listed in Table 1; A3 is the same as A2 but

Table 2. Thermal shock result for selected samples in configuration (1)

Epoxy Sample	Dia, Length (mm), (mm)	Crack at First Cycle	Cycles prior to Failure	Curing Schedule
A1	51,15	N	>25	120°C × 8 h
A2	62,12	Y	0	100°C × 0.5 h + 120 °C ×6 h
A3	62,20	Y	0	80°C × 4.5 h +120°C × 15 h
B1	62,16	Y	0	120°C × 3 h
B2	62,30	Y	0	120°C × 3 h
B3	62,16	N	>26	120°C × 12 h
B4	62,30	N	>26	120°C × 12 h
B5	62,16	Y	0	120°C × 19 h
B6	62,30	Y	0	120°C × 19 h
C1	62,16	Y	0	135°C × 1.5 h
C2	62,16	N	>14	110°C × 6 h + 125°C × 2.5 h
C3	62,16	N	>14	110°C × 6 h + 125°C × 10 h
C4	62,30	Y	0	135°C × 1.5 h
C5	62,30	N	8	110°C × 6 h + 125°C × 2.5 h
C6	62,30	Y	0	110°C × 6 h + 125°C × 10 h
C7	62,14	N		120°C × 11 h
C8	62,32	Y	0	120°C × 11 h
C9	51,14	Y	0	110°C × 4 h + 125°C × 23 h

with a different cure schedule; B1-B6 and C1-C9 are resins B and C in Table 1, respectively, with different cure conditions.

(2) The specimens were the same as in configuration (1) except that each specimen contains a cylindrical steel insert of about 12 mm diameter, as shown in Fig. 2.

(3) The specimens were prepared using 1/4 inch diameter U-shaped copper tubing as mold. Only CTD-101K specimens were tested for this configuration.

(4) The specimens were the vacuum-pressure impregnated model coils which were wound with copper wires of various diameter (0.5-1.0 mm). They were cut and examined by scanning electronic microscope (SEM) after thermal shock test. Only CTD-101K and CY179 were tested for this configuration.

Mold release were applied on the interior surfaces of the mold so that the specimen can be removed from the mold after curing. The deaeration of the resins was as described in previous section. The thermal shock experiments were made by plunging the epoxy samples into liquid nitrogen and then warming them up. Sign of cracking was checked after each worming up.

Thermal Shock Results

Almost all of the specimens prepared in configuration (2) were cracked in the first shock cycle, and all the specimens prepared in configuration (3) were non-cracked (examined for 10 cycles). This result shows that the insert induce large stree in the epoxy and the specimens with smaller volume have much higher cracking resistance. The thermal shock results of the potted coils, i.e., the configuration (4), will be discussed later. Table 2 presents the thermal shock results for some of the selected specimens in configuration (1).

The data in Table 2. supports the following conclusion. First, moderate cure time and temperature could maximize the cracking resistance. For example, the B3 and B4 cured at 120 °C for 12 hours did not crack but those cured at shorter time (3 hours) and longer time (19 hours) cracked. Similar results were observed for the CDT-101K (C1-C9) specimens.

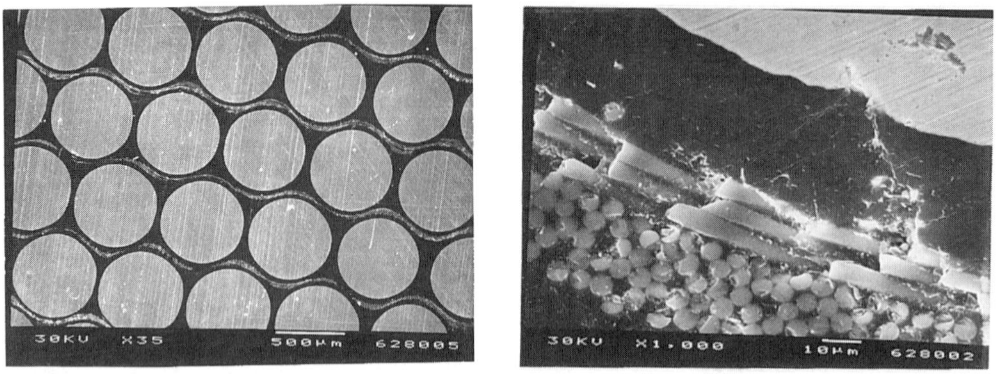

Figure 3. The SEM photographs of a section of the potted coil at two magnifications: 35 for the left and 1,000 for the right. Left photo: the circles are copper wire cross-section, the wavy bands are the E-glass cloth, the rest is the epoxy and formvar. Right photo: the up-right part is the copper wire, the left-bottom part is the mixture of E-glass and epoxy, and the band-like part between copper and e-glass is Formvar.

Second, larger specimens were easier to crack. Third, based on our overall cracked to non-cracked specimen ratio, it appeared that cracking resistance decreases in the order of CTD-101K , GY6010, and CY179. Thus, CTD-101K was chosen to be the primary candidate for other tests.

OTHER TESTS AND ANALYSIS

Bonding and De-bonding

Four kinds of mold-release have been tested. The surfaces of glass beaker, G-10 sheets, mylar, and steel rods were sprayed or coated with the mold release, put in contact with deaerated CDT-101K epoxy, and then annealed with the epoxy cure schedule. It was found that one of the FREKOTE mold-releases is very good for de-bonding the epoxy from steel, glass and mylar.

The bonding strength between CTD-101K epoxy and the formvar insulation layer were tested for some formvar insulated copper wires of about 1 mm diameter. The wires were inserted into the epoxy about 1 cm in a similar way as the embedded rod shown in Fig. 2. Tension was increasingly applied to a wire to pull it out of the epoxy. No slipping between the wire and the epoxy was observed until the wire was broken by the tension. This showed that the bonding between CTD-101K and the wires is extremely strong.

Coil Potting Performance Analysis

Eight coils wound with formvar insulated copper wires were potted according to the procedure described previously. Seven of them were potted with CTD-101K and one with CY179 formulation. The coils were fabricated exactly the same way as the designed 400 MHz NMR magnet coils: the same wire diameter range (0.64 and 0.89 mm) and thickness (about one inch). The inner diameter of the coil ranges from one inch to six inches and the length varies from 1.5 inch to 14 inches. E-glass cloth of two mil thickness was placed between neighboring conductor layers to facilitate the longitudinal percolation of the epoxy. Mold release was sprayed on the surface of the coil bobbin; mylar film was wrapped on the coil forms, and G-10 disks were placed against the flanges of the bobbin. Three potted copper coils were sliced and analyzed by SEM technique (with magnification up to 2000) after thermal shock cycles were made. Fig. 3 shows two of the SEM photographs.

Table 3. Specification of the NbTi test coil

Bore Material	Stainless Steel
Material Between Bore and Coil	Mylar
Material Between Coil Layers	E-glass
Material Between Coil and flange	G-10
Inner Radius	7.635 cm
Radius	10.235cm
Length	6.375 cm
Turns/Layer	160
Winding Prestrain	22 N
Material	NbTi / Cu
Bare Diameter	0.75 mm
Final Diameter	0.795 mm
Insulation	Formvar
Cu / Sc Ratio	1.35 : 1
No. Filaments	54
I_C (4.2K, 8T)	200A
I_C (4.2K, 9T)	103A

The following was found: Both CTD-101K and CY179 epoxide resins worked well for their penetration inside the coils. No voids was found inside the coils. All of the SEM photographs showed that there were no any micro cracks inside the epoxide resins after thermal shock. With the use of a specific mold release that we found through our research, the potted coil can be completely detached from the coil form.

TRAINING CHARACTERISTICS OF NbTi COILS

A 36 layer test coil was wound with Oxford formvar insulated NbTi wire and then potted with CTD-101K resin. The specifications of the coil are presented in Table 3. The coil was tested at 4.2 K with four excitations. The four quenching currents measured were 195.33 A, 195.12 A, 195.07 A, and 194.77 A. This 195 A current corresponds to a 8.0 Tesla peak field in the coil and a 6.5 Tesla field at the center. This shows that the quench current is 97.5% of the short-wire critical current (i.e., 200 A at 8 Tesla from Table 3) for the conductor. This test thus successfully demonstrated the excellent potting properties and performance of CTD-101K resin and our VPI technique.

The eight main coils of the 9.4 Tesla 400 MHz NMR magnet were also potted with CTD-101K resin. The specifications of the coils are listed in Table 4. The training characteristics of the magnet were measured and is shown in Fig. 4. The magnet quenched twelve times before it reached the full field current. The first several quenches are believed to be caused by unbalanced forces due to the axial misalignments during magnet assembly. Whether the rest of the quenches is related to the epoxy or not is currently unclear.

Table 4. Specifications of the main coils of the 400 MHz magnet

Parameter	Sec. 1	Sec. 2	Sec. 3	Sec. 4,5	Sec. 6	Sec. 7,8
Turns / Layer	460	732	118	373	98	114
Layers	24	38	38	38	30	30
Wire diameter (mm)	1.073	0.706	0.552	0.552	0.644	0.644
Cu / Sc ratio	1.35	2.0	3.0	3.0	7.0	7.0
Peak field (T)	9.49	7.75	3.40	4.71	2.41	2.33
J_c (A / mm^2)	122.0	261.0	539.6	405.0	327.6	331.4

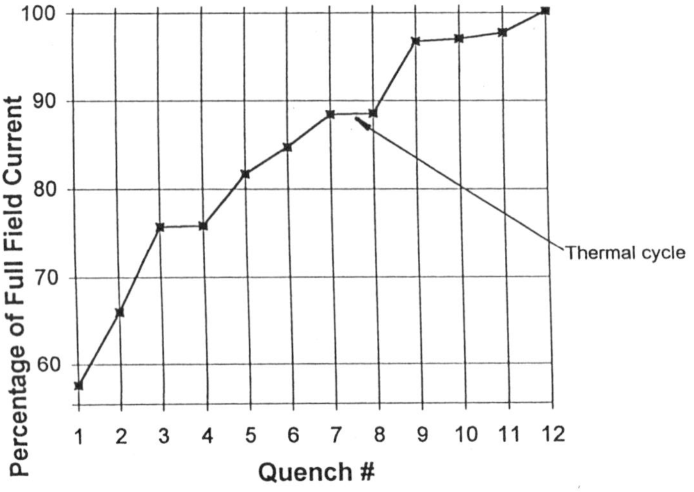

Figure 4. The training curve of the 400 MHz/900 mm NMR magnet.

CONCLUSIONS

Several epoxy formulations have been studied by viscosity, thermal shock, bonding/de-bonding, SEM measurements and model coil tests. Our study shows that among all of the epoxy formulations studied in this project, the CTD-101 K is the most appropriate resin for vacuum-pressure impregnation of high-field NMR magnet coils. The CTD-101K resin has been successfully applied to the fabrication of the 400 MHz/89 mm actively shielded high resolution NMR magnet. This research also developed a reliable vacuum-pressure impregnation technique which involves the combination of appropriate coil winding method, potting procedure, and potting equipment.

ACKNOWLEDGMENT

This research was supported by the Sealy Center for Molecular Science, University of Texas Medical Branch (UTMB) at Galveston in collaboration with Varian NMR Instruments, and in part by Sam Houston State University. We would like to thank the other team members from HARC for their support: J. Hunter, P. McIntyre, T. Mann, J. Colvin, D. Latypov, and C. Abshire.

REFERENCES

1. Y. Iwasa, Experimental and Theoretical Investigation of Mechnical Disturbances in Epoxy-impregnated Superconducting Coils, *Cryogenics* 25:304 (1985).
2. H. Fujita, T. Takaghi, E. S. Bobrov, O. Tsukamoto, and Y. Iwasa, The Training in Epoxy-Impregnated Superconducting Coils, *IEEE Trans. Magn. MAG*-21:380 (1985).
3. For example, MIT group used LBL formulation for their NMR magnets. For LBL formulation , See M. A. Green, D. E. Coyle, P. B. Miller, and W. F. Wenzel, Vacuum Impregnation with Epoxy of Large Superconducting Magnet Structure, *Nonmetallic Materials and Composites at Low Temperatures* 409 (1979).
4. N. A. Munshi, A Radiation-resistant Epoxy Resin System for Toroidal Field and Other Superconducting Coil Fabrication, *Adv. Cryog. Eng.* 38:255 (1992).

MECHANICAL AND THERMAL PROPERTIES OF SELF-BONDED SUPERCONDUCTING WINDING

H. Moriyama[1], H. Mitsui[1], J. Ohmori[1], S. Murai[1], S. Nishijima[2], and T. Okada[2]

[1]Toshiba corporation
Tsurumi, Yokohama 230, Japan
[2]ISIR Osaka University
Ibaraki, Osaka 567, Japan

ABSTRACT

The stability of self-bonded superconducting solenoids has been studied by investigating the mechanical and thermal properties of self-bonded superconducting windings. An adhesive-bonded superconducting solenoid can be fabricated by winding superconducting wire coated with self-bonding resin and by heating the winding. Although the interlaminar shear strength of the self-bonded winding was lower than that of the impregnated one, the self-bonded winding also exhibited high compressive modulus and slightly lower thermal contraction. It could be, therefore, assumed that the instability caused by friction between the winding and the former could be repressed. This was confirmed experimentally, that is, the high stability self-bonded test coils which did not show training were successfully constructed.

INTRODUCTION

In order to increase the stability of close-packed superconducting solenoids and/or to simplify the construction of the solenoids, the method by which the coil is wound with self-bonding resin coated superconducting wire and is bonded in a body by heating the coil (i.e. a self-bonding method) has been studied. Although this method has been often used for normal copper coils, the application of the method to the superconducting coil is not yet common. Though vacuum impregnation with adhesive resin has been used often for superconducting coils, the self-bonding method must be simpler than the impregnation method. The self-bonding method may become one of the promising methods if the magnet stability is high.

It was found that the stability was reduced by local or relative movement of the winding against the former (i.e. at the ground insulation area) during energizing [1]. This movement is caused by the thermal stress and the Lorentz force. The stability, therefore, should be increased by restraining the movement, especially the relative longitudinal movement at the ground insulation area. Although the winding displacement in the outer radius direction is restricted by the high tensile modulus of the superconducting wire, that in the longitudinal direction is controlled by the insulator with low modulus and/or the longitudinal frictional force at the ground insulation area. Complete prevention of the movement would not be possible and hence reduction of the heat generation induced by the movement was found to be effective, that is, reduction of the frictional force at ground insulation area [2]. This method was principally employed in this work.

Two self-bonded test coils were fabricated and were applied to the training test. Self-bonded specimens were cut from the test coil windings. The specimen cut from impregnated winding was also prepared for the comparison [3]. The compressive modulus, the interlaminar shear strength (ILSS) and the thermal contraction in the longitudinal direction of the winding were measured. On the basis of the properties, the stability of self-bonded close-packed superconducting solenoids is discussed in this work.

EXPERIMENT

Test Coils

Two test coils named 'aPr' and 'daPr' were fabricated by the self-bonding method. The coil named 'aBr' impregnated with epoxy [3] was also examined. Figure 1 shows the coil construction, where axes r, θ and z are the cylindrical coordinates. The z direction indicates the longitudinal one. The shape and size of the tested coils are identical. The specification of the test coils are presented in Table 1. The 'a' of the coil name stands for the aluminum alloy former machined in one body and 'da' that assembled with cylinder and flanges. The 'P' stands for the phenoxy resin and the 'B' the epoxy resin named B [4]. The 'r' means the release film [3] glued on the ground insulators.

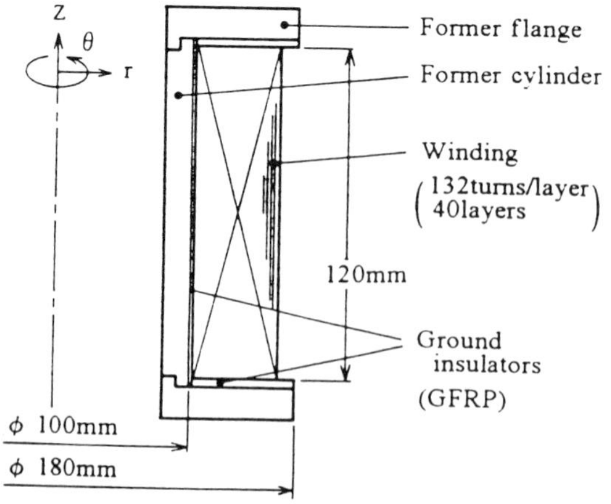

Figure 1. Constitution of the coil daPr.

Table 1 Specifications of the test coils

Coil	Aluminum former	Adhesive resin	No. of turns / layer
aBr	one body	epoxy	132
aPr	one body	phenoxy	130
daPr	divided	phenoxy	132

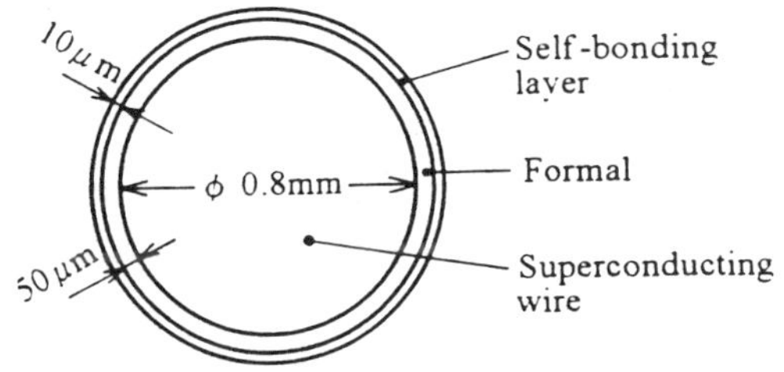

Figure 2. Constitution of the self-bonding superconducting wire.

Table 2 Winding tension of the test coils

Coil	Layers	Tension (N)
aBr	1~40	88
aPr	1~10	59
	11~20	69
	21~30	78
	31~40	88
daPr	1~9	59
	10~40	88

Figure 2 shows the constitution of the self-bonding superconducting wire. A formal insulator was baked on the bare superconducting wire [1] and the phenoxy resin coating was made on the formal layer.

When the test coil is cooled down, the aluminum alloy former contracts in the inner radius direction more than the winding and hence the normal reaction force at the ground insulation area should be decreased, resulting in the decrease of the frictional force [2]. The polytetrafuluoroethelene films having low frictional coefficient [5] were also used to decrease the frictional force.

If the longitudinal compressive modulus of the winding was low, a large frictional motion at the ground insulation area might be generated by the Lorentz force [1]. The winding of the coil daPr was, compressed in order to increase the modulus at 2MPa in the longitudinal direction by pressing the flange and was bonded in a body by heating the winding to 393 K. If the winding tension of the superconducting wire was increased, the normal reaction force at the ground insulation area should become higher, resulting in a higher friction force. On the contrary, if the winding tension was decreased, significant cracks among the bonded winding layers might occur at low temperature in the resin rich regions. The winding tensions of the coils are listed in Table 2. The tension of the inner layers of the coil aPr and daPr was set at 59N and that of the outer layer 88N. The tension of all layers of the aBr was 88N.

Training Test

The test coils aPr and daPr were applied to the training test. The training of each coil was repeated ten times.

Winding Specimens

After the training test, the specimens aPr, daPr and aBr were cut from the test coil windings.

Compression Test

The specimens machined to 18mm in width and 59mm in height were used. Figure 3 shows the compression test set-up. Two displacement transducers consisted of clip and strain gauges were used and the specimen was compressed up to 30MPa for measuring the compressive modulus. The compression test was made three times at LNT.

Interlaminar Shear Test

The specimen with two slits was applied to the interlaminar shear test [6]. Figure 4 shows the shape and size of the specimen. The specimen was compressed in the z

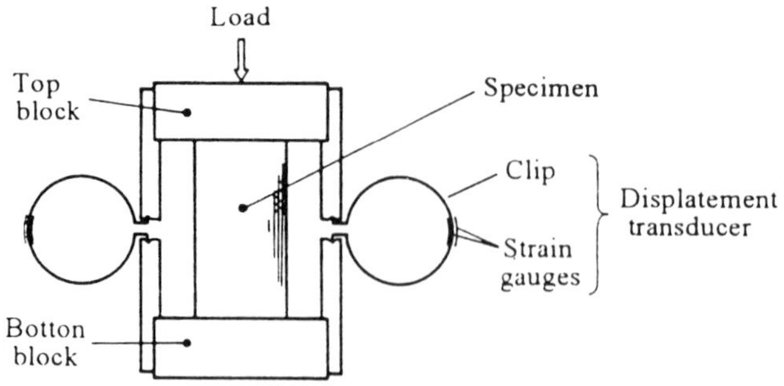

Figure 3. Compression test set-up.

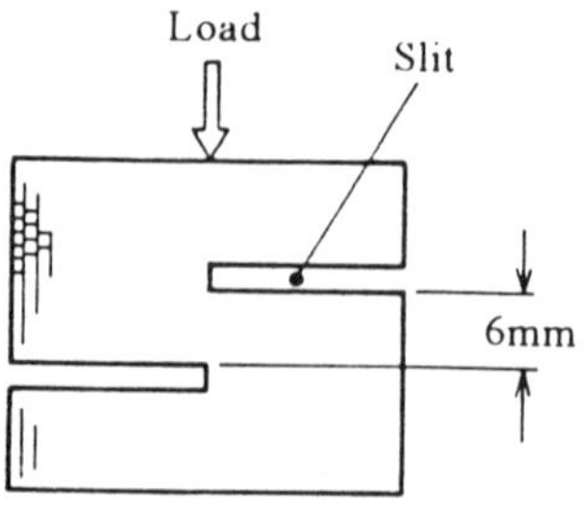

Figure 4. Shape of the interlaminar shear test specimen.

direction at LNT and ILSS τ_B was obtained by dividing the peak load by the fractured area (6mm × 18mm).

Thermal Contraction Measurement

The displacement of the specimen induced by cooling down from RT to LNT was measured. The thermal contraction of the quartz tube was also measured for the reference. The longitudinal thermal contraction α_z was obtained as the difference of the two. In order to confirm the accuracy of the method, the thermal contraction of stainless steel and aluminum alloy were also measured and was found to be accurate.

RESULTS AND DISCUSSION

Observation

Figure 5 shows the cross sectional view of the windings aPr, daPr and aBr. Although the microscopic cracks in the resin were found in aPr (see Figure 5d), any clear cracks were not observed in daPr and aBr. The epoxy resin B filled up between the superconducting wires as shown in Figure 5a. The holes were observed in aPr and daPr. The size of the hole in daPr was smaller than that in aPr. The volume fraction of hole V_h, that of bare superconducting wire V_{sc}, that of formal V_{fm} and that of resin V_r were calculated. The volume fraction of insulators V_{il} was defined as the sum of V_r and V_{fm}. The volume fractions are given in Table 3. The V_h of daPr was 1/2 as small as that of aPr. It was, therefore, confirmed that the longitudinal compression during the

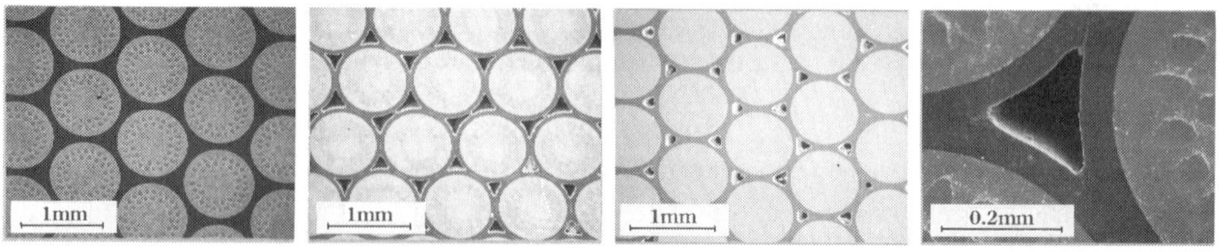

(a) (b) (c) (d)

Figure 5. Cross sectional view of the windings, (a) winding aBr; (b) winding aPr; (c) winding daPr; (d) expansion of (b), microscopic cracks occurred at the corners of the hole.

heat treatment of the coil was effective to reduce V_h. Using these volume fractions, elastic modulus and thermal contraction were calculated in following chapters.

Compressive Modulus

To calculate the compressive moduli E_z of the windings aBr, aPr and daPr the rule of mixture was modified by introducing V_h as equation (1). Elastic modulus of superconducting wire in radial direction E_{sc} and that of insulator E_{il} were assumed to be equal to those of copper and epoxy resin at LNT (i.e. 138Gpa [7] and 6.2Gpa), respectively.

$$1 / E_z = V_{sc} / E_{sc} + V_{il} / E_{il} \cdot V_{il} / (V_{il} - V_h) \qquad (1)$$

The measured and calculated elastic moduli E_z are shown in Table 4. The calculated elastic moduli were thought to coincide with the measured ones. Although the E_z of aPr was smaller than that of aBr, E_z of daPr was nearly equal to that of aBr. The longitudinal displacement of the self-bonded coil at the ground insulation area would be reduced by decreasing V_h or increasing V_{sc} and would lead to the increase of stability.

Interlaminar Shear Strength

The average values of the measured ILSS τ_B are given in Table 5. The ILSS of aPr and daPr were smaller than that of aBr. The effect of the τ_B on the coil stability will be described later. The ILSS of aPr and daPr were nearly equal to each other and hence the τ_B of self-bonded winding might not be changed markedly by decreasing Vh.

Thermal Contraction

The measured thermal contractions α of stainless steel and aluminum alloy together with α are listed in Table 6. The measured α s agreed well with those in the references [7]. The measurement with the displacement transducer was found to be valid. The dry coil's thermal contraction α_z calculated by equation (2) agreed well

Table 3 Volume fractions V of the windings

Winding	V_{sc}	V_{il}	V_h	V_r
aBr	0.72	0.28	0	0.052
aPr	0.69	0.22	0.09	0.039
daPr	0.72	0.23	0.05	0.041

Table 5 Interlaminar shear strength τ_B of the windings

Winding	τ_B (MPa)
aBr	96
aPr	58
daPr	60

Table 4 Compressive moduli E_z of the windings

Winding	Measured E_z (GPa)			Calculated E_z (GPa)
	1st load	2nd	3rd	
aBr	18	19	18	20
aPr	10	12	12	14
daPr	17	18	19	22

Table 6 Thermal contraction α of stainless steel and aluminum alloy

Material	α(%) Measured		Referenced[7]
stainless steel	0.29	0.31	0.29
aluminum alloy	0.37	0.39	0.39

with the measured one. The L_{sc} and L_{il} in equation (2) are the ratio of the diameter of the bare wire to the distance between the bare wires.

$$\alpha_z = \alpha_{sc} L_{sc} + \alpha_{il} L_{il} \qquad (2)$$
$$L_{sc} + L_{il} = 1$$

For the adhesive-bonded windings aBr, aPr and daPr, equation (2) was modified as equation (3), introducing the parameter k and the resin content Cr. The α_z increases with increasing C_r because k is positive. The α_{sc} and the α_{il} are assumed to be equal to those of copper and epoxy that is 0.31 and 1.1%, respectably. The k was decided to be 0.05.

$$\alpha_z = \alpha_{sc} L_{sc} + \alpha_{il} (L_{il} + k\, C_r) \qquad (3)$$
$$C_r = V_r / (1 - V_{sc} - V_{fm})$$

The measured and calculated α_z are given in Table 7. The measured values agreed with the calculated ones. The α_z of aBr, aPr and daPr were larger than those of stainless steel and aluminum alloy which are generally used as the former metal. The α_z of aPr and daPr were slightly smaller than that of aBr. These results indicate that the difference between α_z of the winding and that of the former might decrease with decreasing V_r. The holes in the self-bonded winding might contribute the reduction of the α_z.

Training Behavior

As previously reported[3], the quench current Iq of the first quench in the coil aBr was as high as 95% of critical current Ic and kept 100% from 2nd to 10th quench (i.e. Ic). The Iq of the coils aPr and daPr attained 100% of Ic at the first quench and did not degrade up to 10th quench. These results proved that the stability of the coils aPr and daPr was as high as that of the coil aBr. The stability of the coil aPr was found not to be affected by the microscopic cracking in the bonding material. The stability of the coils aPr and daPr was high in spite of low τ_B.

CONCLUSIONS

Two test coils of the self-bonded close-packed superconducting solenoid were fabricated. The longitudinal compressive moduli, ILSS and thermal contractions of the specimens cut from the self-bonded coil windings were measured. The training behaviors of the test coils were also measured. Although ILSS of the self-bonded winding was smaller than that of the impregnated winding, the self-bonded winding

Table 7 Thermal contraction α_z of the windings

Windings	α_z (%) Measured		Calculated
aBr	0.45	0.46	0.45
aPr	0.42	0.44	0.43
daPr	0.42	0.44	0.42

exhibited high compressive modulus and slightly lower thermal contraction compared with that of epoxy impregnated winding. It is, therefore, assumed that the instability originating from the longitudinal friction between the winding and the former is minor. This assumption was confirmed by the stable operation of the self-bonded test coils. The quench currents of the two self-bonded test coils were confirmed to be as high as those of the impregnated coil. The possibility of manufacturing highly stabilized close-packed superconducting coils with the self-bonding method was proved.

REFERENCES

1. H.Moriyama, H.Sekiya, H.Mitsui, S.Nishijima and T.Okada, Effect of Ground Insulation Contact on Stability of Epoxy Impregnated Superconducting Solenoids, ICMC Topical Meeting Hawaii, (1994)
2. H.Moriyama, H.Mitsui, J.Ohmori, S.Mine, S.Nishijima and T.Okada, Design and Fabrication of Highly Stabilized Close-packed Superconducting Solenoid, 14th Intl. Conf. Magnet Tech. F38, (1995)
3. H.Sekiya, H.Moriyama, H.Mitsui, S.Nishijima and T.Okada, Effect of Epoxy Cracking Resistance on Stability of Impregnated Superconducting Solenoids, ICMC Topical Meeting Hawaii, (1994)
4. H.Moriyama, Y.Inoue, H.Mitsui, Y.Sanada and Y.Kobayashi, "Advances in Cryogenic Engineering (Materials) 38", Plenum Press, New York(1992), p339
5. E.L.Stone, and W.C.Young, "Advances in Cryogenic Engineering (Materials) 26", Plenum Press, New York(1980), p315
6. T.Okada, S.Nishijima, T.Nishiura, K.Miyata and H.Yamaoka, "Advances in Cryogenic Engineering (Materials) 38", Plenum Press, New York(1992) , p241
7. H.M.Ledbetter and A.F.Clark, "Materials at low temperature", American Society for Metals, (1983),p1 and p75

CONSTRUCTION AND TESTING OF A GENERAL PURPOSE 50 LITER LIQUID HELIUM CRYOVESSEL

J. Schubert,[1] H. Blosser,[1] J. Kim,[2] L. Lee,[3] G. Stork,[1] and A. Zeller[1]

[1]National Superconducting Cyclotron Laboratory
Michigan State University
East Lansing, MI, 48824-1321, USA

[2]Argonne National Laboratory
Argonne, IL, 60439, USA

[3]University of Connecticut, Storrs, CT, 06269, USA

ABSTRACT

We have built and are operating a 50 liter liquid helium vessel with an access port for easy insertion of experiments. The vessel is primarily used to test components of superconducting magnet systems, such as cryogenic current leads. The access port is set at a 45° angle and the entire vessel may be tilted 45° in either direction so that current leads may be tested in any orientation from horizontal to vertical. The vessel was designed to be easily disassembled so that experimental apparatus too large for the access port may be mounted directly in the helium can. This way devices 40 cm in diameter may be tested. The heat load of the vessel has been measured under a variety of experimental conditions. A few devices tested so far in the cryovessel are briefly described.

INTRODUCTION

The heavy-ion beam program at the National Superconducting Cyclotron Laboratory (NSCL) uses forty-one superconducting magnets in separate liquid helium (LHe) cryostats. Until recently, the only general purpose cryostat available for testing magnet components was a commercially built 6000 L "dunking dewar."[1] Although useful for testing large magnet coils, the dunking dewar is too cumbersome and costly to operate for tests of small parts, such as superconducting wire joints, and it is not well suited to test vapor cooled current leads. We built the 50 L cryovessel to satisfy this need, and also to gain basic data for the construction of future magnet cryostats.

DESIGN

The mechanical design of the cryovessel is shown in Figure 1. With a current lead mounted in the test-port, helium vapor cools both the lead and the two radiation shields. The cold gas line that cools the radiation shields is also coiled around the neck

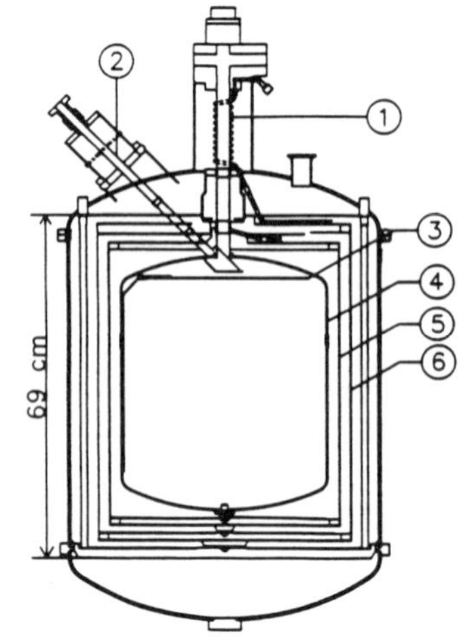

Fig. 1. The 50 L liquid helium cryovessel. 1: neck tube, 2: current lead tube mounted in test-port, 3: copper liner, 4: helium can, 5: inner radiation shield, 6: outer radiation shield.

tube in order to reduce the effect of heat conduction through the gas in the tube. No liquid nitrogen is used. Not shown in the figure are two liquid helium level sensors, one set parallel to the current lead and the other perpendicular, so that the helium boiloff may be studied in whichever direction the vessel is tipped. There is an array of 10 carbon resistor thermometers on the radiation shields, originally used to verify that the temperature of the shields varied by no more than 3 K from the bottom to the top. (Only the top is in direct contact with the exiting gas.) The thermometers are held in place with aluminum tape and may be easily reconfigured whenever the vessel is opened.

The inside of the helium can is lined with sheet copper on its top and sides. The primary purpose for the liner is to prevent stratification of the helium vapor so that the

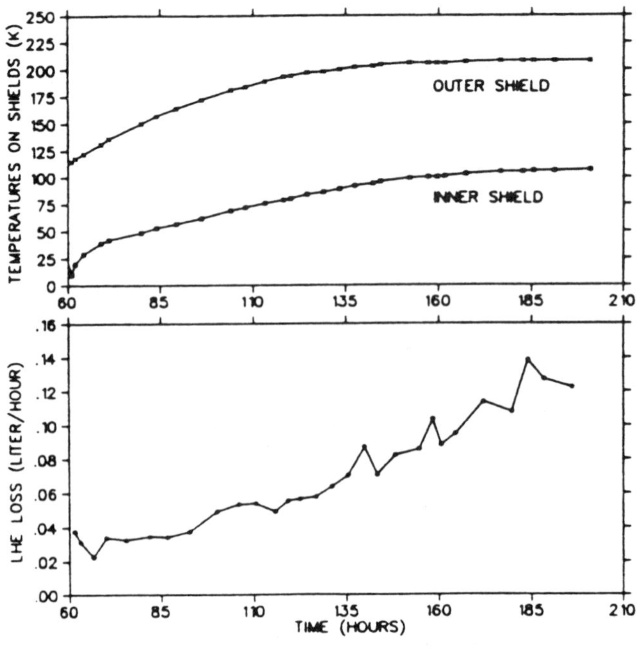

Fig. 2. Radiation shield temperatures and heatload with the test-port blanked off. Exhaust gas is vented to the atmosphere so that the helium vapor in the vessel is at a pressure of 101 kPa (1.0 atm.)

Table 1. Heatload of vessel (in L/hr of boiloff) for various experimental conditions

	test port horizontal	test port at 45°	test port vertical
Vessel with test port sealed by flange	—	0.12	—
Vessel with empty tube in test port	4.9	—	0.14
vessel with tube filled with smaller teflon tubes	0.47	—	0.27

exiting gas will have a constant temperature regardless of the liquid level. Also, since some experiments require only the cooling power of 4 K He gas, the liner provides a useful 4 K environment as long as there is any liquid in the vessel.[2]

The test port is set at a 45° angle so that current leads may tested in a horizontal or vertical orientation by rotating the entire vessel 45° in either direction. The vessel is mounted on a stand (not shown) that is balanced for all three positions. After filling the upright vessel and removing the transfer line, two people are required to tip the vessel to the desired orientation.

Although the test port is intended for current lead tests, any experiment smaller than 1.6 cm in diameter may be inserted through it. To test larger devices we must disassemble the cryovessel. The sides and bottom of the vacuum jacket are a single part, attached to the top by screws and sealed with an o-ring. Once the vessel is lifted out of its vacuum jacket, the bottoms and sides of the radiation shields may be similarly removed. All of the cooling pipes are soldered to the tops of the shields, so the plumbing does not interfere with access to the inner helium can. Only the helium can itself needs to be cut open and welded shut when a new experiment is installed.

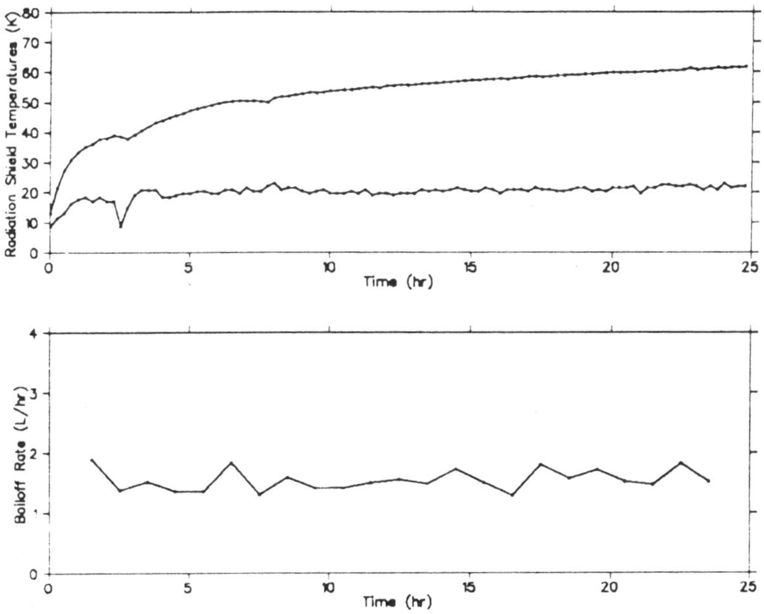

Fig. 3. Radiation shield temperatures and heatload with a pair of 400 A current leads mounted in the test port, and no current in the leads. Pressure is controlled at 122 kPa (3 psig) to force roughly half of the boiled off gas through the leads. The rest cools the radiation shields.

HEAT LOAD MEASUREMENTS OF CRYOSTAT AND EMPTY TUBE

The heat load on the cryostat was measured first with the test-port sealed off by a flange. The boiloff rate and radiation shield temperatures are plotted in figure 2. The radiation shields are cooled to nearly 4 K during the fill process and require several days to warm to equilibrium. The boiloff rate rises in proportion to the shield temperature, suggesting that the heatload is dominated by radiation.

The heatload to the cryostat was also measured with an empty tube mounted on the test port. The 1.5 cm diameter, 66 cm long tube simulates the quench vent tube in a typical superconducting magnet. Equilibrium boiloff rates for the vessel alone and with the empty tube are listed in table 1 where all of the boiloff tests were conducted at 101 kPa (1.0 atm) of pressure (where 1.0 W yields a boiloff of 1.39 L/hr.)

In the preferred vertical position the helium gas in the tube stratifies, and conduction down the tube is minimized. Recently, however, some magnets have been built which require horizontal vent tubes. The magnet in the Harper medical cyclotron, for example, must be rotated in a 360° vertical arc about the patient. The cryostat uses a quench vent which is always horizontal and a unique boiloff vent system in which multiple pipes wrap around the rotating cryostat.[3]

A horizontal pipe connecting a warm and cold region, and closed at the warm end, will experience heat convection due to counterflow. Gas from the cold end moves toward the warm end along the bottom of the pipe, and after being warmed, returns to the cold end along the top of the pipe. In our test a thermometer on the bottom of the horizontal tube read 20 K while an identical thermometer on the top read 41 K. In a similar test where the tube was filled with 1.6 mm diameter teflon tubes most of the counterflow convection was eliminated and no temperature difference was observed.

HEATLOAD MEASUREMENTS OF CURRENT LEAD

The first practical use of the cryovessel was to test the prototype current lead for an 8 T superconducting magnet.[4] This current lead is actually a pair of 400 A gas cooled leads and a gas cooled set of eighteen signal wires all in a 1.59 cm diameter package. Although this lead design is only used in the vertical orientation, it is very similar to the horizontal current lead in the Harper medical cyclotron.

With the 400 A current leads installed, the total heatload is increased by an order of magnitude. The additional boiloff gas lowers the equilibrium temperature of the radiation shields, and radiation heatload of the vessel may become negligible compared to the conduction heatload of the lead. Figure 3 shows the boiloff rate and shield temperatures as a function of time for the upright vessel with the lead tilted at a 45° angle, and no current flowing through the lead (the ohmic heating of the lead when energized is plotted in figure 4). Although current leads are rarely used in this

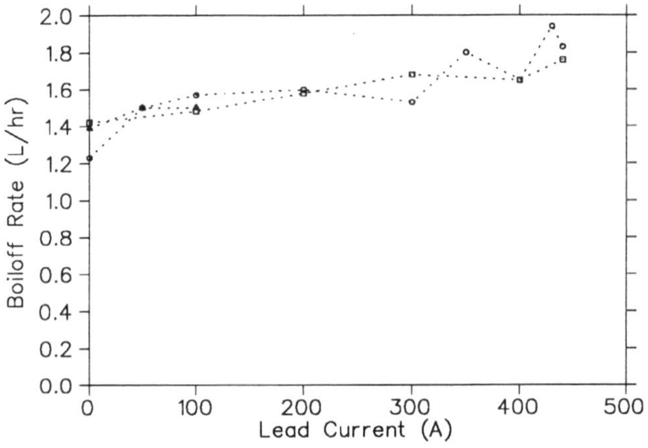

Fig. 4. Heatload of vessel and lead with various currents in the leads. Each boiloff rate is the average over approximately one hour.

Table 2. Heatload due to a pair of 400 A current leads (in L/hr of boiloff)

	test port horizontal	test port at 45°	test port vertical
Optimally cooled lead	—	1.4	1.6
Overcooled lead	—	1.4	—
Suboptimally cooled lead	2.6	1.7	1.7-2.4

orientation, the test provides a basis for comparison which is useful when separating phenomena related to the vessel from phenomena occurring in the lead.

The heatload of the current lead was tested in all three orientations. It was also useful to study the effect of coolant flow on the heat conduction through the lead. The equilibrium boiloff rates for all of the above conditions are listed in table 2. Where greater than one atmosphere of pressure was needed to force gas flow through the leads, the boiloff rates have been adjusted to an equivalent rate for the same heatload at one atmosphere. The high heatload for the horizontal lead is primarily due to the same counterflow convection that plagues horizontal vent tubes. In addition, the gas flow through the lead dropped below the optimum flow after the vessel was tipped. The effect of the gas flow rate on the horizontal lead will be investigated in future experiments. The difference in heatload between the vertical and 45° orientation of the lead is believed to be an effect of the cryovessel itself and is currently being studied.

ADDITION OF HELMHOLTZ RESONATOR

Thermal acoustic oscillations (TAOs) are a well known nuisance in the construction of dewars and cryostats. The low frequency (on the order of 10-60 Hz) oscillations which occur in pipes that connect warm and cold sections of a cryostat have been blamed for transferring energy to the cold end of the pipes and for rendering precise pressure measurements impossible.[5,6]

Fig. 5. Schematic of the Helmholtz resonator for damping thermoacoustic oscillations.

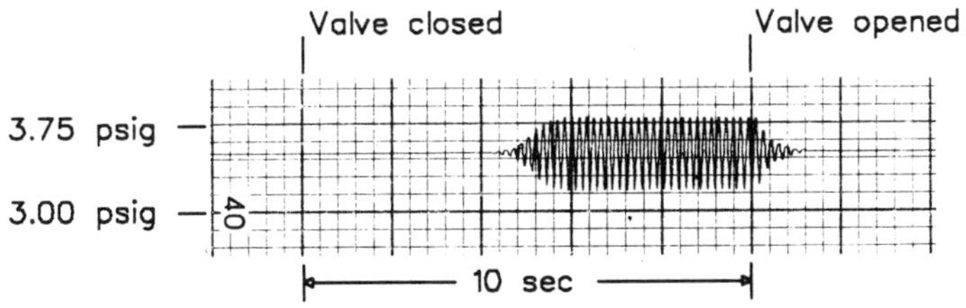

Fig. 6. The pressure of the exhaust helium gas, measured outside the vessel, as the valve to the Helmholtz resonator is closed and then opened. The frequency of the oscillation is 6.25 Hz.

Although most of the literature only discusses the existence of TAOs in static columns of gas, we believe that we have observed the same phenomenon in pipes with over 2 L/min of gas flow. At first the pressure of the gas exiting from the radiation shields oscillated with a frequency between 6 and 7 Hz, depending on the temperature of the shields. Although the oscillation did not noticeably effect the boiloff rate, it did interfere with both pressure and flow measurements.

We set out to damp the oscillations by installing a Helmholtz resonator as a side branch to the pipe that carries the helium away from the cryovessel. The Helmholtz resonator is simply a surge tank connected by a capillary to the side of the pipe that carries the TAO. This device destroys the resonant quality of the pipe so that acoustic standing waves in a given frequency range will be minimized. Since the resonator can work at the warm end of the vent pipe, it may be located outside the cryovessel, wherever space is available.

A schematic of the resonator is shown in figure 5. The volume of the surge tank and the dimensions of the capillary are based on the recipe reported by Tward and Mason,[5] with the exception of the radius of the capillary. Because there is a unit error in the equation reported in Tward and Mason's paper, we used the relationship $a^2 < \mu/\rho\omega$ where a is the radius of the capillary, μ and ρ are the viscosity and density of helium at the resonator, and $2\pi\omega$ is the frequency of the TAO.

As soon as the Helmholtz resonator was installed the TAO disappeared. In order to study the effect of the damper, we installed a valve between it and the rest of the system. Figure 6 is a strip-chart recording of the signal from a pressure transducer located near the Helmholz resonator. After the radiation shields reached equilibrium the valve was closed, removing the damper from the system. After a period of time that ranged from less than one second to seven or more seconds, the oscillations reappeared. When the valve was opened, the oscillations began to decay immediately and were undetectable after one second.

TESTS OF OTHER DEVICES

The cryovessel has proved convenient for testing a number of other small superconducting magnet components. A prototype sc-wire splice for the dipoles of the S800 spectograph[7] was mounted on the end of the 400 amp current lead in order to test its resistance. A simple ohm's law test yielded an upper limit on the resistance of $2X10^{-9}$ Ω.

We have also built and tested a 400 A persistent current switch (for a 7 mH superconducting test coil), copying the method of the Texas Accelerator Center.[8] By observing the rate of current decay in the coil, it is possible to measure the resistance of a pair of sc-joints (connecting the coil to the switch) with a resolution that is only limited by the amount of time that helium remains in the vessel. Since the vessel may be continuously resupplied from the lab's distribution dewar, there is, in principle, no lower limit to the joint resistance that may be tested.

In the near future we will use the vessel to test a new current lead which has been designed to operate with a minimal heatload in the horizontal orientation.

ACKNOWLEDGMENT

This work supported by the National Science Foundation.

REFERENCES

1. Janis Research Co., Stonham, MA.
2. J. Kim *et al.*, Cryovessel for the current leads test, *MSU-NSCL Ann. Report* 195 (1990).
3. H. Blosser *et al.*, Superconducting cyclotron for medical application, *IEEE Trans. ASC* 25,2:1746 (1989).
4. J. Kim, Eight tesla superconducting magnet for cyclotron studies, Ph.D. Thesis, MSU (1994).

5. E. Tward and P.V. Mason, Damping of thermoacoustic oscillations, *Adv. Cryo. Eng.* 27:807 (1982).
6. D.A. Ditmars and G.T. Furukawa, Detection and damping of thermal-acoustic oscillations in low-temperature measurements, *Jour. Res. NBS-C* 69C:35 (1964).
7. A.F. Zeller *et al.*, Construction of a large superconducting spectrometer dipole magnet with negative curvature, *IEEE Trans. ASC* 3,1:114 (1993).
8. J. Schubert, Construction and testing of a 400 A persistent current switch, *1994 MSU-NSCL Ann. Report*, in press.

EXPERIENCE ON SHEATHING 10 KM OF CABLE-IN-CONDUIT CONDUCTOR FOR THE NHMFL 45-T HYBRID

J.R Miller,[1] T.A. Painter,[1] A.L. Devernoe,[2] R.M. Schaedler,[3] G. Grabinsky,[4] J.E.C. Williams,[5] and E. Bobrov[5]

[1]National High Magnetic Field Laboratory, Tallahassee, FL 32316
[2]Intermagnetics General Corporation, Latham, NY 12110
[3]IGC Advanced Superconductors Inc., Waterbury, CT 06704
[4]Gibson Tube Inc., Bridgewater, NJ 08807
[5]Franics Bitter National Magnet Laboratory, Cambridge, MA 02139

ABSTRACT

The superconducting outsert magnet for the 45-T Hybrid Magnet System being constructed for the National High Magnetic Field Laboratory (NHMFL) in Tallahassee, Florida will contain almost 6.5 km of cable-in-conduit conductor (CICC) in three different grades. The project is a collaborative effort with the Francis Bitter National Magnet Laboratory (FBNML) and industry. Counting dummy CICCs manufactured during the development phase of the project and excesses for contingency, well over 10 km (over 11 MA·m) will have been produced. To fabricate these conductors, a special production line was assembled by the industrial team of Intermagnetics General Corporation (IGC) and Gibson Tube. Components in the line were adapted from modern cable-handling and welded-tube-manufacturing machinery, with the addition of a computer data-acquisition system for process monitoring. The versatility of the IGC/Gibson line has been demonstrated by the production of three different grades of CICC in which the external dimensions, wall thicknesses, corner radii, and even conduit alloy were variables. This paper describes special features and capabilities of the manufacturing line as well as production rates and tolerances achieved in the final product.

INTRODUCTION

The superconducting outsert of the 45-T Hybrid requires rectangular Cable-in-Conduit Conductor (CICC) with continuous lengths in excess of 700 meters. There are three subcoils in the outsert, two using Nb_3Sn-based CICCs and one using NbTi. The pertinent nominal dimensions of these conductors are listed in Table 1. The continuous seam-weld tube-mill method of manufacture was selected because of its proven record of reliability for commercial production of stainless-steel and special-alloy tubing. Technologies constantly in use at Gibson Tube to produce long lengths of empty tubing and small-scale jacketed cables for other applications have been successfully adapted in a cooperative arrangement between Gibson and IGC with funding from the NHMFL for encapsulation of superconductor cable. The process has now been proven in full-scale practice runs using copper cable, practice runs using prototype Nb_3Sn cables, and production runs using NbTi cables. The jacket materials used in these runs have included 304L and 316LN stainless steel and Incoloy 908. As of this writing, over 1000 meters of dummy and over 6000 meters of actual superconductor CICC have been successfully produced for the 45-T Hybrid program.

Table 1. Nominal dimensions of the CICCs for the three subcoils of the superconducting outsert for the 45-T Hybrid

Parameter	Coil A	Coil B	Coil C
Cable pattern	5 x 5 x 3 x (6 Nb_3Sn comp. + 1 Cu wires, 0.43-mm dia.)	5 x 3 x 3 x (3 Nb_3Sn comp. + 4 Cu wires, 0.51-mm dia.)	5 x 3 x 3 x (3 NbTi comp. wires, 0.81-mm dia.)
Cable cross-section	80.05 mm2	67.57 mm2	73.04 mm2
Void cross-section	50.96 mm2	44.03 mm2	36.07 mm2
Jacket cross-section	79.88 mm2 (16.24 mm x 13.69 mm, 1.65-mm wall, 3.65-mm outer corner)	74.10 mm2 (15.27 mm x 12.91 mm, 1.65-mm wall, 3.65-mm outer corner)	89.52 mm2 (15.88 mm x 13.72 mm, 2.00-mm wall, 4.75-mm outer corner)

MACHINE OPERATIONS

The components of the IGC/Gibson tube mill are shown in the schematic of Fig. 1. The basic procedure is to roll-form steel strip around the cable as both pass longitudinally through the first section of the tube mill, thereby forming around the cable a loosely fitting round tube, which is then continuously TIG welded on the seam. Following the welding and quenching stations, the tube is roll reduced, first almost round, then progressively more rectangular. The major considerations are protection of the superconductor while producing an essentially flawless and leaktight jacket with the most precise and consistent profile possible.

The strip payoff is a standard equipment item used to feed tube mills at Gibson. A pancake-wound package of strip, slit to a precise width, is mounted on a support with an adjustable center. Back tension is variable with an air-pressure brake. The required length of strip can be made up in advance by end-to-end splicing. The typical splice is a scarf weld crossing the strip at an angle of 30° from transverse, which causes the welded joint to curl around the circumference of the tube in helical fashion as the tube is formed. This improves the reliability of the joint. Making the splices off-line with a separate TIG welder allows a more thorough inspection, but splices can also be made "on the fly." As a final security measure, the strip is visually inspected during payoff for any potential contaminants or defects.

The cable payoff was designed for maximum flexibility in handling a reel of rope-style cable. The reel is mounted on a carrier which can be rotated in either direction to allow compensation for any tendency of the cable or the foil wrap to unwind. The cable is paid out through the rotating center and guided into the strip-forming section of the mill. Back tension is provided by a disc brake with variable air pressure. It was observed that when the cable is consistently well laid and well wrapped with stainless foil, continuous rotation during jacketing

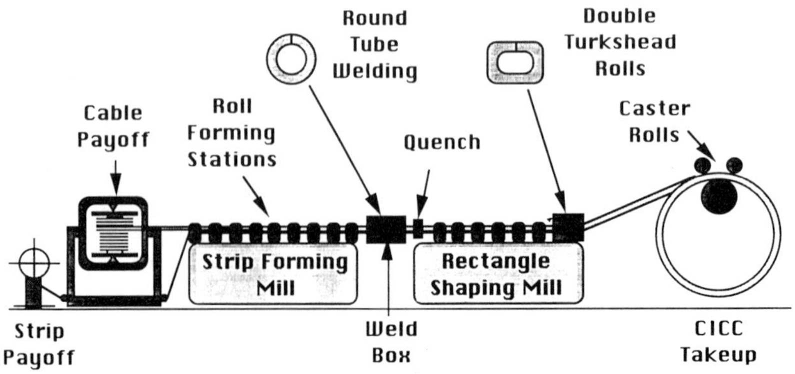

Figure 1. Schematic of the tube mill established for encapsulating conductor for the 45-T Hybrid.

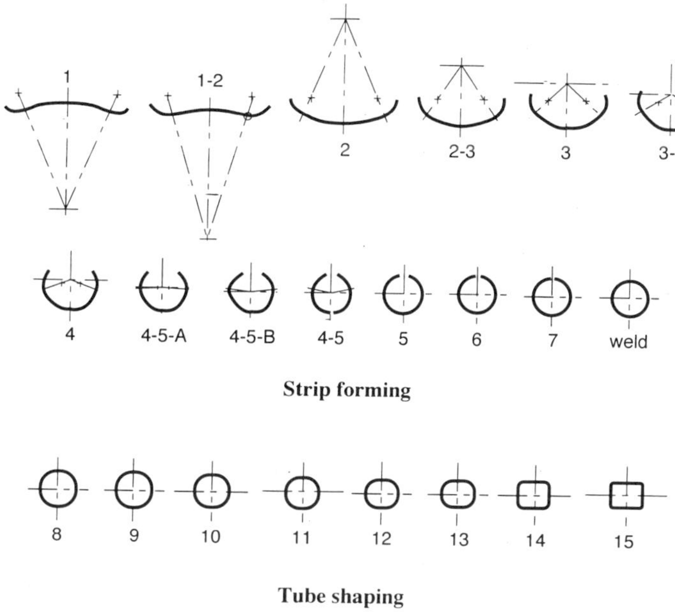

Figure 2. Profiles of the strip-forming and tube-shaping rolls in the IGC/Gibson tube mill.

is not necessary. However, this feature has been retained as a convenient contingency in case such compensation is required for any reason.

The cable enters the strip-forming section of the mill above the strip. The strip-forming section follows conventional tube-mill practice except for special guides, fixtures, and slotted rolls that are necessary to protect and direct passage of the cable. The shaping sequence in the strip-forming section is displayed in the upper part of Fig. 2. Note that the strip is first radius upward near the edges then gradually closed around the cable. By the time the strip enters the weld box it is a perfectly round tube. The cable is visually inspected at the entry to the mill to detect any inconsistencies in the cable lay or the stainless steel wrap that could disturb the welding process.

At the welding station, the round tube is continuously TIG welded to form a leaktight seam. The weld box contains a reducing atmosphere to protect the torch tip and the product. Cooling of the weld seam is accomplished by a water quench and air wipe. The weld arc is directed by a magnetic device that controls the shape of the weld bead. Welding in proximity to the foil-wrapped cable requires painstaking optimization of welder parameters to assure a consistent bead shape and prevent damage to the conductor.

Compaction of the tube around the cable and the formation of the rectangular profile is done in the tube-shaping section of the mill. The rolls are designed to take small increments of reduction from round to rectangle as illustrated in the lower portion of Fig. 2,. This distributes the strains of forming gently and avoids buckling or irregular shaping of the tube wall. The final rectangle size is controlled by a massive double "turkshead" roll set, which has all necessary adjustments to compensate for twist or cast of the output CICC.

To assure consistent quality, a PC-based data-acquisition system was designed and installed to monitor weld parameters, product speeds, and output dimensions of the conduit. The presence of the cable interferes with the eddy-current inspection techniques commonly used in standard tube-mill operations. However, tracking of the weld voltage was found to be valuable in that a disturbance in the arc, potentially causing or associated with a defect, generally gives rise to a detectable voltage spike. Measurements of the speed differential through the forming and shaping sections provides continuous data on material elongation, which is directly related to reduction in area and thickening of the material during its conversion from strip to conduit.

From the shaping section of the mill, the CICC is fed onto a takeup system that casts the conduit into a large-diameter helix and hangs it on a mandrel. The cast is imparted by a set of adjustable rolls, one of which is driven by DC motor. The weld seam is positioned on the outside of the helix to allow inspection along 100% of the length.

At the end of a run producing a single length of jacketed cable, terminations for valves and pressure gages are attached to the ends of the finished product by manual TIG welding. Then the

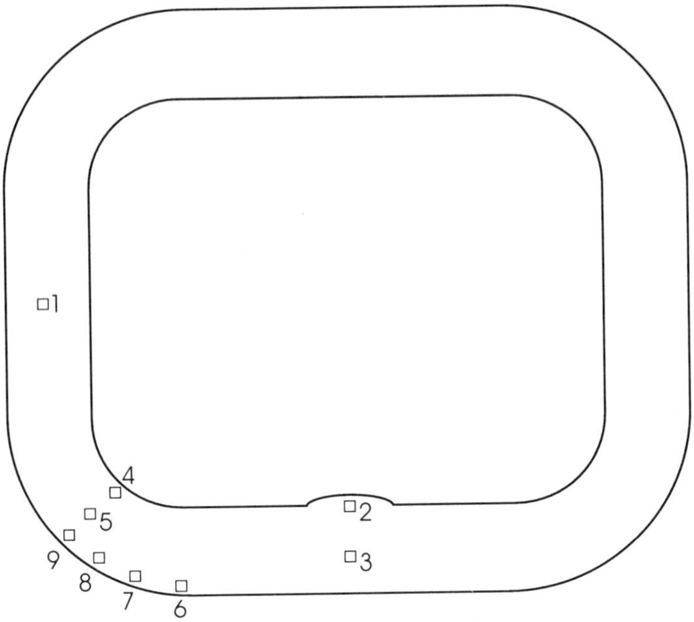

Figure 3. Cross-section of a sample of coil-C conduit showing the locations of points selected for test of hardness.

CICC is pressurized with helium to approximately 3.5 MPa. The leak tightness of conduit is checked by both pressure decay and by "sniffing" the entire seam length using a hand-held helium detector. In the event that a leak is found, a repair procedure has been developed using a hand TIG welder and heat sinks to minimize exposure of the conductor to heat.

PRODUCTION EXPERIENCE

As mentioned above, more than 7000 m of CICC have been fabricated to date on the IGC/Gibson tube mill. Of that quantity, well more than 6000 m contained real superconducting

Table 2. Measurements of hardness by the Vickers Pyramid method, at points indicated in Fig. 3 on the cross-section of a sample of coil-C conduit (pyramid load was 0.2 kg throughout)

Point Number	Indentation size (μm)	Vickers Pyramid Number
1	35	303
2	35.4	296
3	35	303
4	35.2	300
5	35.2	300
6	33	341
7	35	303
8	35.4	296
9	35.3	298

Table 3. Measured densities of jacket materials used in this program

Material	Density (kg/m3)
Incoloy 908	8096
316LN	7971
304L	7911

cable. Approximately 6020 m contained NbTi superconductor and will be used in the outermost (coil C) of the three coils making up the superconducting outsert of the 45-T Hybrid. Approximately 50 m of prototypical Nb_3Sn CICC was jacketed for developmental testing. The NbTi CICC was produced in 9 separate lengths – 8 lengths of just over 730 m each and 1 length of just over 180 m, which is enough for 33 double pancakes. Coil C requires 29, leaving 4 for spares. Five of the 8 long pieces used 316LN and the other 3 used 304L. The short length used 316LN.

Incoloy 908 was used to jacket 705 m of copper cable as part of the development for the conductor to be used in the innermost coil of the outsert (coil A). Much of the coil-A dummy conductor has since been used to wind a full-diameter, half-length model as a focus for developing the manufacturing processes for the two Nb_3Sn coils. Incoloy 908 was also used to jacket the Nb_3Sn coil-A prototype cables. However, because of concerns for SAGBO cracking during heat treatment, Incoloy 908 has since been replaced by an ultra-low-carbon version of 316LN as the choice for jacket material in the Nb_3Sn conductors slated for coils A and B of the outsert. In the process of gaining experience on the IGC/Gibson tube mill, several hundred meters more jacket material were processed – some containing copper cable, some empty. Most of this material was also absorbed in the development programs for practice on the coil fabrication processes. Final production of real superconducting CICC for the two Nb_3Sn coils is scheduled for late summer and early fall of this year.

From our experience to date, it is clear that, when strip stock is of good quality, very close tolerances can be maintained on the external dimensions of CICC produced on the IGC/Gibson mill. For example, on a typical production run, the standard deviation from the mean flat-to-flat dimensions of the conduit was less than 0.014 mm, while the maximum observed deviation was about 0.045 mm. On the same run, the corners were all 90° (within measurement error of approximately 0.3^o) and the corner radii were 3.0 ± 0.25 mm.

Of course, the wall thickness of the conductor jacket is directly dependent on the thickness and width of the strip stock. In general, there is a reduction of the mean perimeter as the stock is continuously deformed radially inward on its pass through the forming and shaping rolls. There is also a small reduction in cross-sectional area (and a corresponding elongation) as the material is extruded slightly by the squeeze of the rolls. We found the relationships between beginning strip and finished jacket dimensions to be very precise and repeatable for a particular combination of tooling and strip material, and they can be specified uniquely in terms of area and perimeter reduction factors. For the coil-A tooling and Incoloy 908 strip, the area reduction factor was 0.935 ± 0.002 and the perimeter reduction factor was 0.866 ± 0.004. For the coil-C tooling and either 304L or 316LN, the area reduction factor was 0.908 ± 0.003 and the perimeter reduction factor was 0.845 ± 0.003.

Of course, forming the strip into conduit introduces some coldwork, but the amount appears to be relatively small. This impression is partially substantiated by the relatively small elongation (in the range of 6 to 9% from the area reduction factors quoted above). To try to quantify the amount and to investigate the possibility of significant variations of the cold work, Vickers-Pyramid hardness measurements were made for a selected distribution of points on the cross-section of the coil-C conduit as shown in Fig. 3. The results are listed in Table 2. Except for one outlying point, the variation is not significant, indicating little variation in the yield strength and the cold work as well.

Exact measurements of wall thicknesses and cross-sections of finished conduit are very difficult to make, especially near the corners. We found it convenient to derive these values from measurements of external dimensions of a sample (flat-to-flat dimensions, external corner radii, and length), its weight, and the material density, where more accuracy is possible.

However, we found handbook values of our material densities too imprecise and made a separate measurement of the values, listed in Table 3, for the three different materials used to date.

At present, because of the developmental nature of this product and the comparatively small production quantities required, the unit costs are relatively high. The expense of the actual jacketing process is roughly $36/m in the quantities required for this project. Materials costs vary widely and their unit costs can be very expensive when special requirements and small quantities are specified. For example, the unit cost of Incoloy 908 was approximately $37/kg. The special 316LN that will be used for the Nb_3Sn coils has been priced at approximately $26/kg. In comparison, the 304L used for a portion of the coil-C conductor was under $6/kg. In addition to these costs, there are overhead and one-time tooling, setup, and handling charges. To get a feel for the impact that product maturity might eventually have on the cost of CICC produced by this process, it can be noted that Gibson's standard product in large quantities is presently priced at less than $1/m with materials costs added on at approximately $6/kg.

Table 4. Summary of Tube-Mill Capabilities

Cable payoff:	Reel o.d.:	1800 mm, maximum
	Reel width:	1100 mm, maximum outside flanges
	Weight capacity:	2.3 t
Strip payoff:	Capacity limited only by the available length of strip. Strip can be spliced, "on the fly" if necessary, or the mill can be stopped for re-load of strip and restarted without detriment.	
Tube mill capacities:	Length:	Limited only by the available cable length. The cable payoff can hold nearly 3 km of cable the size of coil-A conductor. A typical one-shift run can produce 600 - 800 m. Modifications to support equipment will allow greater lengths.
	Conduit outer dimensions:	Up to 38-mm square
	Conduit Wall:	Up to 2.5-mm thickness
	Cable o.d.:	Determined by conduit dimensions (approximately 1 - 2 mm less than tube i.d. at the weld head)
Tube-mill dimensional tolerances:	On flat-to-flat dimensions:	± 0.05 mm
	On corner radii:	± 0.25 mm on actual nominal
	On corner squareness:	90° (± 0.5^{o}, and within tolerance on the flat-to-flat dimension)
Takeup Looper:	Weight capacity:	2.3 t
	Coil diameter:	2 - 3 m typical
	In the future, traversing takeup on a reel will be preferred. The present system allows ready access to the product for inspection and leak checking. However, other means of measurement under development will allow reel takeup, which is much easier to manage during handling, shipment, and payout to a coil winder.	

The IGC/Gibson tube mill has demonstrated the capability to produce long lengths of CICC with minimal risk to the superconductor. Its features as presently configured are summarized in Table 4. With some minor modifications, an even broader range of parameters might be achieveable. Given consistent quality of materials, the continuous length of CICC that can be produced is limited only by the available length of cable.

During this entire experience of both development and production of active conductor for the 45-T Hybrid, there have been no observations of faults or other incidences that would appear to have put the conductor at risk. It has been demonstrated that the machine can be stopped any time and restarted without producing defects or damage to conductor. The on-line monitoring system has demonstrated potential for assuring consistency of the seam weld and tolerances on the conduit dimensions.

PROBLEMS/RESOLUTIONS

Bead Shape: In early trials with dummy materials, the presence of the cable wrapped with foil changed the interaction of the weld arc and solidification of the bead. Through long-length practice runs, optimization of welder controls, and use of certain Gibson proprietary techniques, a process was developed to give reliably a smooth, wide inside bead surface. With the present process, the superconductor was not disturbed, and in the worst case, the stainless steel foil was only lightly tacked to the bead.

Conduit defects: In the course of production jacketing for Coil C, several seam sinks and occasional holes were experienced. Assignable causes were the nature of the cable (low compaction) and some inconsistencies in the lay, which resulted in jerking of the payout. It is important to interleave the cable layers with cardboard to prevent pinching. Other problem areas should be noted prior to the tube mill run, such that compensation can be made or recovery actions planned. A number of options, including shutdown and restart, are available at the mill to prevent defeats, and an off-line repair procedure has been established and documented.

Seam wander: Occasional seam wander was caused by measurable inconsistencies in the properties of some of the strip stock. The strip must be uniformly annealed, and the surface finish must be consistent from end to end and across the width to assure that a top center seam position can be maintained. The slit edge should also be sheared square and free of burrs.

General: The need for thorough development with practice runs on empty tube and dummy cable cannot be understated. The actual jacket material to be used should be procured in quantity sufficient to provide for proofing of weld ability and strip dimensions. Strip finishing and slitting should be performed by suppliers well familiar with tube mill requirements.

CONCLUSIONS

In conclusion, the IGC/Gibson tube mill has fully demonstrated its capabilities for producing CICCs appropriate for large superconducting magnet applications, in long lengths, of superior quality, and free of defects. By length, the quantity of CICC production for the 45-T Hybrid Project on the IGC/Gibson tube mill already exceeds that for any other single application. By weight, the production is only slightly less than that for the largest single application, the Westinhouse LCP Coil. When production for the 45-T Hybrid Project is finished, more than 10 km and nearly 10 t of CICC will have been produced.

ACKNOWLEDGEMENTS

This work was supported by the National Science Foundation, through NSF Cooperative Grant No. DMR 9016241, and the State of Florida. The authors would also like to thank S. Pourrahimi for supplying measurements of hardness on samples of finished conduit.

DEVELOPMENT OF BENDING CHARACTERISTICS FOR THE TPX TF MAGNET COIL CABLE-IN-CONDUIT CONDUCTOR

K.E. Grut, R. L. Holbrook, E. Hook, and T. A. Antaya

Babcock and Wilcox - Accelerator and Magnet Systems
PO Box 785, Lynchburg, VA, USA

ABSTRACT

The conductor for the toroidal field (TF) magnet coils for the Tokamak Physics eXperiment (TPX) is an assembly of stranded Nb_3Sn superconductor sheathed by an Incoloy 908 conduit. The coil shape, when coupled with stiffness of the cable-in-conduit conductor (CICC) is such that conventional magnet winding techniques cannot be utilized. Therefore a bending and forming method will be employed in the TF coils. The cable will be reacted after bending because the reaction process hardens the conduit and also lowers the strain the cable can withstand without performance degradation. The Incoloy 908 conduit also work hardens quickly, necessitating the production of the required coil shape in one step without correction. This paper discusses the limiting processes for forming the TPX TF magnet geometry, the methods utilized in establishing the CICC bending characteristics and the methods employed to account for material springback so that a coil can be manufactured accurately and efficiently.

INTRODUCTION

The Tokamak Physics eXperiment is an experimental reactor in the development of fusion technology which is intended to produce energy by initiating a fusion reaction in a deuterium plasma for a sustained period of time. To accomplish this, superconducting magnets are used to create a toriodally shaped magnetic field which confines the plasma. The coils necessary to produce the peak magnetic field of 8T are "D" shaped and require accurate placement of the conductors so that the quality of the magnetic field is not compromised. The coils, composed of 12 pancake coils of 7 turns each, have Nb_3Sn superconductor as the cable and are cooled with single phase helium in a cable-in-conduit arrangement (often called CICC). Figures 1 and 2 show the TPX TF coil and a cross section of the CICC. The conduit material is Incoloy 908; a material specially designed to provide strength and toughness at low temperatures and have a coefficient of thermal contraction that would result in an overall compressive strain of the Nb_3Sn superconductor

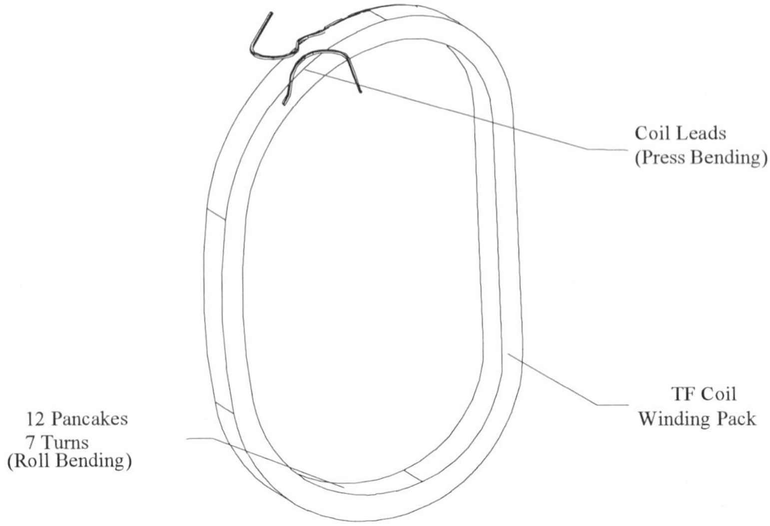

Figure 1 TPX Toroidal Field coil winding geometry. The bend radii to be formed vary for each turn. The forming radius must be accurately known so that the radii begin and end in the proper place and the total conductor payout is optimized.

at operating temperatures. Since the Incoloy 908 is a nickel alloy and quite stiff the coil must be wound by plastically deforming the CICC to the proper shape.

During the forming of the coils, the cable is still a conglomerate of niobium, tin and copper and must undergo a heat treatment which produces the Nb_3Sn superconducting alloy and its bronze matrix. This reaction cycle occurs at the precipitation hardening temperature of the Incoloy 908 which significantly increases the yield point of the conduit. Also, reacted cable has a limit on its axial strain so that the performance of the superconducting strands are not impaired.

It is for all these reasons that B&W, in its bending and forming development of the CICC attempted to accurately characterize the CICC bending behavior so that the coil can be successfully wound without correction after reaction. Mechanical springback, work hardening and cross sectional distortions must be well known so that the conductor can be placed in the proper location in its relaxed state. Any residual elastic strain energy remaining will result in a coil which will be difficult to handle and process further.

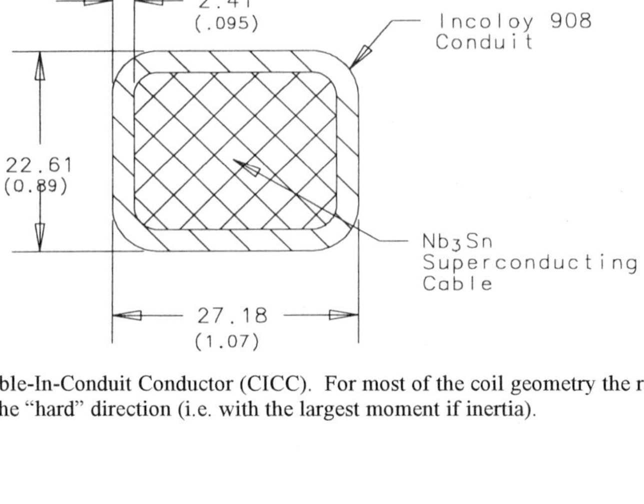

Figure 2 TPX Cable-In-Conduit Conductor (CICC). For most of the coil geometry the radius of curvature is to be formed in the "hard" direction (i.e. with the largest moment if inertia).

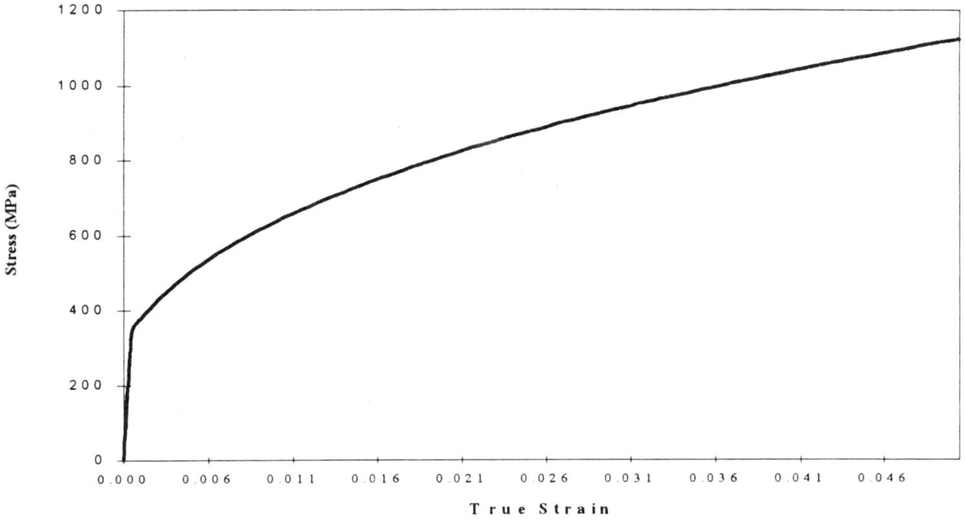

Figure 3 Stress Strain plot for Incoloy 908.

ANALYTICAL METHOD

The forming of the coils and the coil leads will be accomplished by either a roll bending or press bending operation. The main coil turns will be formed with the roll bender while the coil leads will be formed with a press and die. The method of estimating the springback in these two bend types differs slightly due to the differences in geometry. In order to better understand the behavior of the forming operations the following assumptions and clarifications are made:

1) The cable is much less stiff and allowed to move relative to the conduit, therefore it does not contribute significantly to the moment of inertia resisting the forming operation, however it will have an impact on the springback.

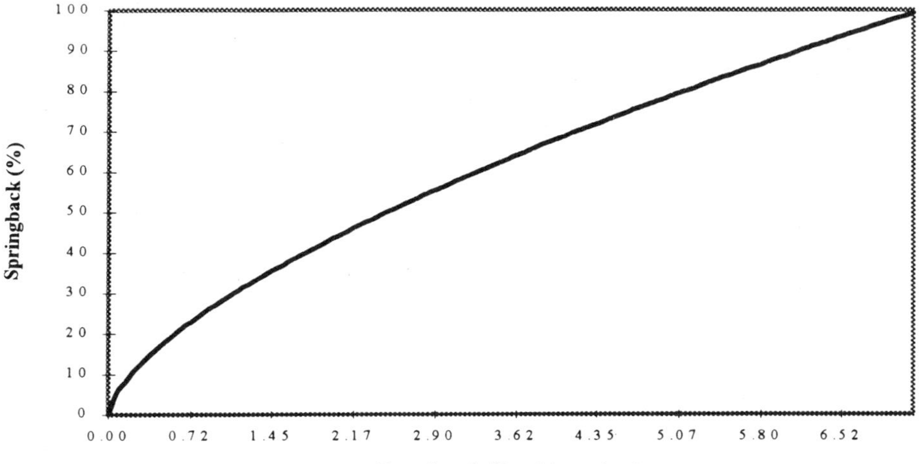

Figure 4 Springback of the TPX TF coil conductor as a function of the radius to be formed.

2) For roll bending, the springback is defined to be the percent difference between the radius of curvature in the loaded and relaxed states with respect to the radius in the relaxed state.

3) For press bending, the springback is a combination of the aforementioned mode of radius springback plus the elastic-plastic behavior of the free ends of the material not formed to the die radius of curvature.

4) For simplicity the free ends of the leads are assumed to be deflecting elastically. In other words the plastic deformation occurs in the conductor contact region with the die, and elastic behavior is assumed over the free span.

5) The strain, defined for a curved beam as the difference in circumferential length with respect to the undeformed length can be simplified to the following:

$$\varepsilon = Y/R \qquad (1)$$

Where Y is the distance to the outer fibers from the neutral axis of a beam and R is the radius at the neutral axis.

Roll Bending

The roll bending operation utilizes a series of rolls which deform the conduit to a radius of curvature which is tangent to all rolls. Once the conduit exits the roll bender the material relaxes elastically to a state of residual strain which is proportional the resultant radius of curvature. Simple beam equations can be used to solve for the bending moment necessary to produce a given curvature in the stressed state provided the elastic-plastic properties are known. Figure 3 is a plot of the stress strain curve for Incoloy 908 before it has been heat treated.

The yield strength is approximately 345 MPa as can be seen in Figure 3. Assuming the conduit acts like a straight beam at the point just prior to yielding, the maximum forming radius can be determined for the conductor. This limit is 7.25m (285 inches). Bending a TPX TF conductor to a radius larger than this limit will not result in any residual curvature after the forming load is released.

Since strain is proportional to the radius of curvature, the springback can be calculated as a function of bend radius and can be used to predict the radius in the stressed state that will produce a desired radius in the relaxed state. Springback is the elastic response of a material after its yield point has been exceeded. After a material is loaded past its yield point in tension and released, the material will unload elastically to a residual state of strain corresponding to the elastic modulus of the material (an offset parallel to the slope of the elastic region).

$$\varepsilon_r = \varepsilon_p - \frac{\sigma_p}{E} \qquad (2)$$

Where σ_p and ε_p are the stress and strain in the stressed state. The additional strain necessary to form a desired radius in the relaxed state is approximately the difference in strain between the stressed and relaxed states. Using the desired radius as a reference point the forming radius is just the radius corresponding to the additional strain.

$$\Delta\varepsilon = \varepsilon_p - \varepsilon_r \quad \text{and} \quad \varepsilon_f = \varepsilon_p + \Delta\varepsilon \qquad (3)$$

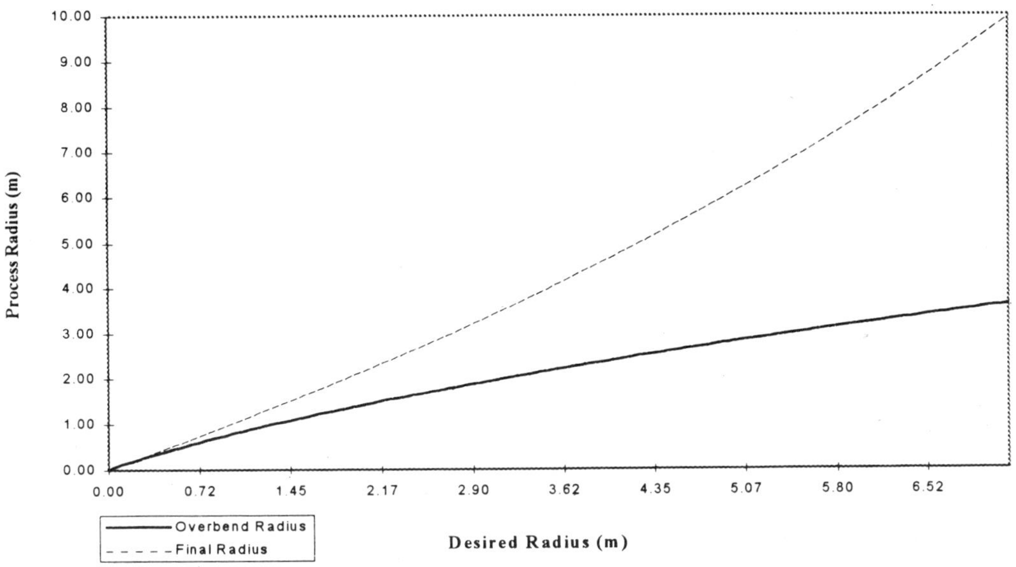

Figure 5 Necessary forming radius as a function of the desired coil radius (at the neutral axis). There is a slight diversion at the upper range of the data where the radius is near maximum limit.

B&W performed calculations to predict the amount of springback that the TPX TF conductor will exhibit and also estimate the forming radius necessary to produce a given bend radius. Figure 4 shows the springback of the CICC as a function of the desired bend radius. The plot is nearly linear in the areas in which the majority of the bends will be formed.

Figure 5 shows the forming radius as a function of the desired coil radius. This plot shows the springback of the conductor from the overbent condition and the actual radius after the conductor relaxes as a function of the desired radius to be formed. This demonstrates that forming the conduit using a tool radius indicated by the additional strain is a good method in the range of radii in which the TF coil will be formed. It also shows that for larger radii, the deformation is near the elastic limit of Incoloy 908 and that a slightly tighter forming radius may be required to produce these bends.

Press Bending of the Coil Leads

During press bending the CICC is bent around a die on a hydraulic ram. Rollers on either side of the die produce the necessary bending moment which deforms the conduit.

At the die, the deformation is similar to that of the roll bending process and if the conduit were bent to conform to the die without free ends, the springback would be identical. However, the ends are significantly distant from the die to allow elastic deformation to occur between the die and rollers. As the ram advances the angle between the ram and rollers decreases which increases the region of plastic deformation. The springback is therefore related to this angle and the radius of curvature.

The region near the "tangent" points of the free ends experiences a mixture of elastic and plastic deformation. For simplicity the regions near the tangent points are considered to be plastic on the die side and elastic on the side of the free end. The springback can then be established by estimating the free end length, modeling the free end as a cantilevered beam, determining the slope at the end and combining this with the angle from the radius springback.

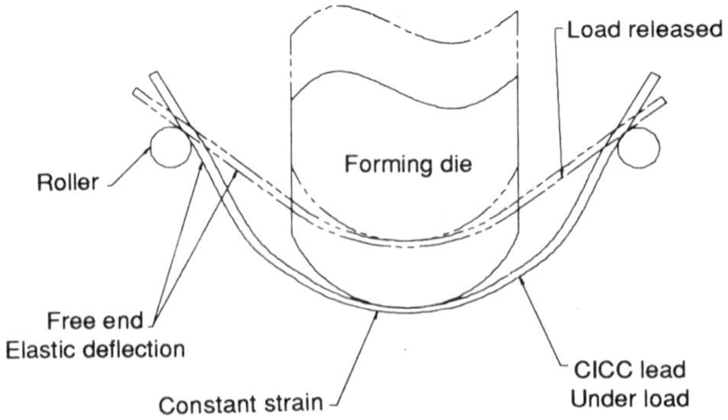

Figure 6 Schematic of a press bending operation demonstrating the loaded and unloaded conditions.

The springback was quantified for small bend radii (at 0.19 m and 0.229 m) and various angles of deformation to predict the results of B&W's forming development program. Figure 7 shows the relationship between the calculated bend angle in the relaxed state as a function of the bend angle of the conduit in the loaded condition. This relationship is linear because the radial springback in the curved region remains constant while the springback angle varies linearly with the bend angle.

EXPERIMENTAL RESULTS

The roll bending springback results support the trends calculated analytically but are slightly lower, probably due to the fact that the conductor adds a resistance to springback. There is also a dependency on the strain hardening of the material, that is, as the conduit is deformed, straightened and bent again, the resulting bend radius is larger for the same bend roll settings. This indicates that higher load and deformations are necessary to form a given radius in successive cycles.

The press bending results indicate a similar trend as to what is predicted. The mock-ups used for this set of trials were bent to successively tighter bend angles. Here work, hardening of the material is not a concern because as tighter angles are reached, the plastic strain remains constant while the region of plastic deformation is expanding during the sequence of the test and not being cycled. Figure 8 shows the springback of the TF lead in terms of the angle between the free ends of the mock-up for a .229 m (9 inch) radius bend.

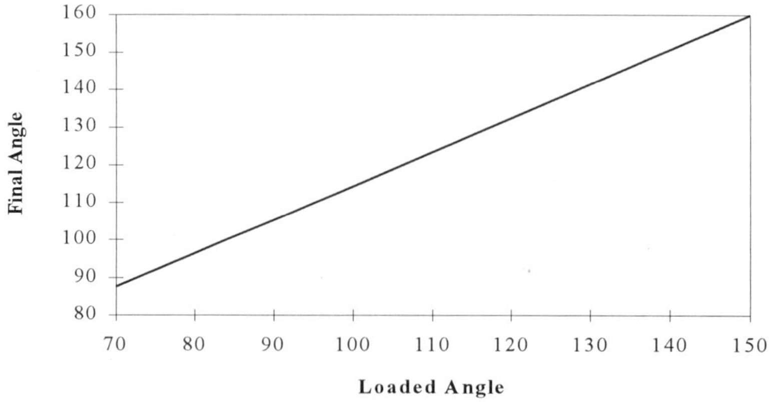

Figure 7 - Predicted angle between the free ends of a TPX TF coil lead after forming. The loaded angle refers to the angle between the free ends as the conductor is in the forming die and the hydraulic ram is pressurized.

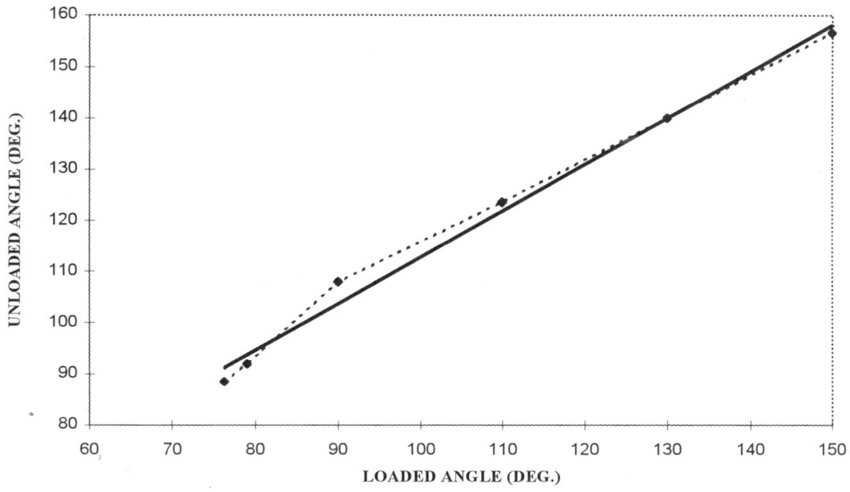

Figure 8 Angle between the free ends of TPX TF coil lead mock-up after forming. The experimental data confirms the analytical prediction of springback.

The press bending results indicate that there is little variability at these tight bend radii; probably because the material is experiencing nearly pure plastic deformation in the curved section. The press bending results also confirm that the method used to predict the springback for the roll bending process is valid at small bend radii.

It was also observed during the bending trials that material and configuration variability have an impact on the repeatability of the bending process. The variation of the seam weld location, local differences in material properties and tool run-out in the bending machine all have an impact on the ability to form a regular repeatable bend radius. Tool run-out seems to have the most impact in that it tends to produce a cyclic flattening of the cross section of the CICC which would result in a locally variation in radius of curvature. This emphasizes the need for rigidity and accuracy in the manufacture of the roll bending machine and forming tools. The machine used for these trials is not the one to be used in production and exhibited a considerable amount of variation during early bending trials. After investigation of the variations, B&W observed that the majority of the variation was cyclic and at a period consistent with a 0.006 inch run-out in the one of the rolls in the machine.

CONCLUSIONS

The theoretical calculations presented in this paper are a valid means of predicting the necessary tooling settings to be applied to the forming of the TPX TF coil geometry. The springback calculations are an accurate method of producing the additional deformation necessary to form a coil geometry accurately the first time. Variations in material properties and the forming configuration have an impact on the repeatability of the forming process and need to be mitigated if at all possible.

ACKNOWLEDGMENT

The author wishes to express appreciation to Dr. X.R. Huang, D.T. Stutzki, and W. Bensiek for their contributions in the preparation of this paper. This paper is being prepared under DOE contract B235308 for preliminary design of the TPX TF Magnet System at the Princeton Plasma Physics Laboratory.

REFERENCES

1. L.S. Toma , M.M. Steeves, MIT Plasma Fusion Center and R.P. Reed, Cryogenic Materials Inc "Incoloy Alloy 908 Data Handbook", March 1994.
2. Warren C. Young, "Roark's Formulas for Stress and Strain", Sixth Edition, McGraw-Hill, 1989.
3. James Sabine, Generating a springback model and adjusting the part geometry, *Tube International*, January, 1993.
4. W.Hassenzahl and T. O'Connor, LLNL, "TPX Toroidal Field System Design Description", March 16,1993.

RESULTS OF THERMAL HYDRAULIC MEASUREMENTS ON TWO 15 kA ITER STYLE CABLE IN CONDUIT CONDUCTORS

B. Rousset,[1] J-L. Duchateau,[2] A. Gauthier,[1] A. Martinez,[2] P. Seyfert,[1] and J. Weisend II,[3]
[1]DRFMC/SBT/CENG
Grenoble, 38054, France
[2]DRFC/STID/CEC
St Paul lez Durance, 13108, France
[3]DESY Laboratory
Hamburg, 22603, Germany

ABSTRACT

This paper describes the results of pressure drop and heat transfer measurements conducted on two CICC conductors similar to those proposed for the ITER project.

Both conductors consist of a conduit containing 6 petals of twisted subcables cabled around a central helium flow channel. In one conductor, a second tube separates the flow channel from the petals. This tube is absent in the second conductor. Pressure drop measurements were carried out at a mass flow rate between 0.4 & 4.3g/s. Heat transfer measurements using both single & multiple heat pulses of up to 2.5W were conducted at a flow of ≅1g/s. All experiments were conducted using supercritical helium (T≅5K, P≅6bar). The pressure drop results are compared with those of other CICC conductors. The heat transfer results are used to validate a numerical model developed by CE - Cadarache. In addition, the data can be used with the model to estimate the overall heat transfer coefficient from the subcables to the helium. The relative thermal hydraulic performance of the two conductors is also discussed.

INTRODUCTION

The magnetic system of the International Tokamak Experimental Reactor (ITER) currently being designed is made of superconducting magnets. Cable-in-conduit conductors (CICC) are the leading option for these magnets[1]. One of the proposed CICC designs has the superconducting cables arranged around a central cooling channel. This paper reports on efforts to understand the thermal hydraulic performance of two versions of this type of

Advances in Cryogenic Engineering, Vol. 41
Edited by P. Kittel, Plenum Press, New York, 1996

CICC. Specifically, measurements were made on the pressure drop across the conductors and on the thermal response of the conductors to a varying heat load similar to that caused by AC heating in the ITER magnets. The experimental results are combined with a numerical model developed at CE - Cadarache to estimate the overall heat transfer coefficient from the cables into the forced flow supercritical helium.

DESCRIPTION OF THE EXPERIMENT

Conductor Description

Two types of CICC studied are shown in fig. 1. Conductor A is encased by a stainless steel sleeve with an inner diameter of 17 mm. In the center of this tube is another smaller tube (I.D.=4.4mm, O.D.=6.4mm). The superconducting wires are bundled into 6 twisted subcables that are arranged in the annular space between the two tubes. Note that in the conductor A there are two separate helium flow spaces: 1) through the annular space containing the superconducting wires and 2) through the inside of the inner tube.

Conductor B is identical to A except that the inner tube has been removed. In this case the helium flowing in the annular space is directly connected to the helium flowing through the central channel. In our experiments, the subcables are made of copper rather than superconducting wires to allow the use of ohmic heating for simulating the periodic heating in the conductors.

Test Loop

The experiment is installed at CENG/SBT in a cryogenic test stand previously described[2]. The test loop, shown in figure 2, was assembled inside the vacuum vessel of the test stand. It consists, following the path **ABCDEF**, of 4 m of stainless steel tube between **A** and **B,** followed by 15 m of conductor B between **B** and **C**, followed by 15 m of conductor A between **C** and **D,** and finally by 3 m of stainless steel tube between **D** and **E**.

At Point **E,** the circuit enters a vessel containing saturated liquid helium which acts as a cold source for the installation. Between **E** and **F**, a copper tube is used to transfer to the cold bath the heat accumulated by the helium during its circulation through the test loop. Located between **F** and **A** is a centrifugal pump that drives the forced flow helium.

The test loop includes thermometers and pressure transducers. Six calibrated Allen Bradley carbon resistors are mounted on the outside of the two CICC sections. These sensors provide a good measurement of the temperature in the annular space of the conductor. Two additional carbon resistors are placed at the inlet and outlet of the first conductor section. Cold differential pressure transducers (from Validyne) allow the measurement of the pressure drop across each conductor and the total pressure drop across both conductors. The flow rate in the test loop is measured by a Venturi flowmeter. As a check of the Venturi measurements, a thermal flow metering technique is performed using heater W1 and the thermometers in the test loop.

For the thermal experiments, heat pulses are generated via ohmic heating of the copper subcables in the test sections. The current for this heating is delivered through current leads located at positions **B** and **D**. As a result of this arrangement, heating is carried out simultaneously in both conductors.

All experiments were conducted using supercritical helium with T~5K, P~$6x10^5$Pa (6bar). Pressure drop measurements were carried out at a mass flowrate between 0.4 & 4.3g/s. Heat transfer measurements using both single & multiple heat pulses up to 2.5W were conducted at a mass flowrate of ~1g/s.

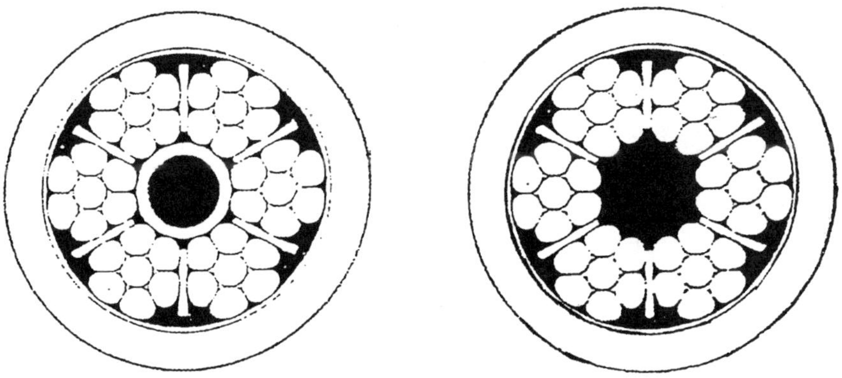

Figure 1. The respective cross section of the two types of cable in conduit conductor.

RESULTS

Pressure Drop Measurements

Rearranging the piping connections between experiments allowed us to measure three different pressure drops: the pressure drop across the conductor B, the pressure drop across the total cross section of the conductor A (i.e. with flow in both the annular space and the central cooling channel) and, by plugging the central cooling channel of the conductor A, the pressure drop in the annular space alone.

Figure 3 compares the measured pressure drop through both types of conductors. The pressure drop in the type A conductor is between 2.5 and 2.8 times larger than that of the type B conductor. This is not surprising as the internal tube in the type A conductor acts to reduce the flow cross section in the conductor as compared to that of the type B conductor.

The pressure drop through the internal tube in conductor A is given by:

$$\Delta P = \lambda \frac{L}{D} \rho \frac{v^2}{2} \tag{1}$$

where :

ρ = the helium density,
L = the length of the tube,
D = the inner diameter of the tube,
v = the flow velocity in the internal tube, and
λ = the friction factor which is given by the following correlations as:

$$\lambda = \frac{64}{\mathrm{Re}} \; \textit{for} \; \mathrm{Re} \leq 2000 \quad \textit{and} \; \lambda = \frac{0.3164}{\mathrm{Re}^{0.25}} \; \textit{for} \; \mathrm{Re} > 2000 \tag{2}$$

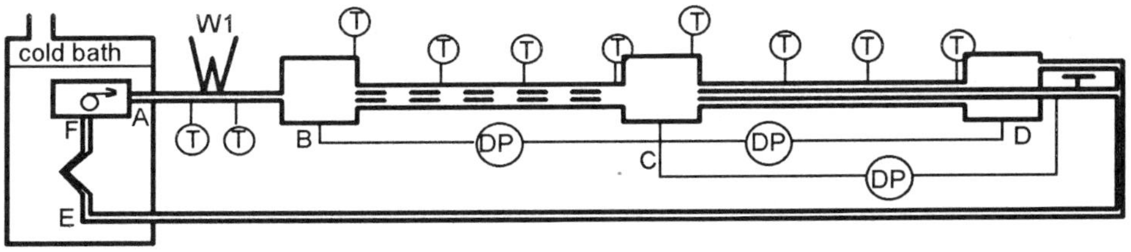

Figure 2. Forced flow cooling scheme.

where Re is the Reynolds number for the flow.

Using Eq. (1), the distribution of the mass flows between the internal tube and the annular space in conductor A can be calculated. It is found that there is approximately twice as much mass flow through the internal tube as in the annular space which seems reasonable.

With the internal tube of conductor A plugged, the pressure drop through just the annular space may be obtained. These results may be used with Eq. (1) to obtain a friction factor (λ) for the annular space alone. Note that for this analysis the diameter D is defined as the hydraulic diameter given by:

$$D_H = \frac{4A_a}{P_w} \tag{3}$$

where A_a is the cross sectional area and P_w the wetted perimeter of annular space. The annular space friction factor is shown on figure 4 as a function of the Reynolds number. Also plotted on this figure are results for a number of different CICC conductors which do not contain a central cooling channel[3]. The results of the estimation of the friction factor for the annular space of conductor A are in good agreement with the results from these other conductors. If we consider just the flow through the annular space then the geometry of the conductor is very similar to typical CICC conductors. Thus it is reasonable to think that the annular space measurements should be consistent with previous CICC studies.

We did not attempt to define an overall hydraulic diameter or friction factor for both of the two conductors tested. In fact, an earlier work[4] has indicated that length and hydraulic diameter do not fully characterize the geometry of this type of CICC. In order to predict fully the pressure drop behavior of the conductors used in ITER similar measurements should be made on the final conductor design.

Thermal Measurements

One of the main problems of a CICC with a central cooling channel is to understand the thermal behavior inside the CICC in order to predict the recooling time after a heat pulse. For a CICC without an inner hole, the time needed, for recooling to the initial temperature after a heat pulse, is simply given by the ratio of the length to the fluid velocity.

For a CICC with a central cooling channel, the time dependent temperature profile depends on two parameters: 1) the heat transfer coefficient (H) between the two channels and 2) the fluid velocity in each channel.

If H is very small, each channel is adiabatically isolated and the recooling time can be calculated using the flow velocity of the channel where heat has been deposited i.e. the annular channel.

If H is very high, which corresponds to an ideal case, the CICC behavior is similar to a one channel conductor with an equivalent velocity **V** given by the ratio of the total volumetric flow divided by the sum of each channel cross section. Considering a point **P** located far enough from the inlet of the conductor, the expected temperature evolution of helium during and after a pulse might be a trapezoidal curve. The temperature rise-time represents the heating of the conductor corresponding to the pulse duration **tc**, the plateau represents the time needed for a particle leaving the point **x=V*tc** to reach **P** and then the descent occurs until the last particle heated reaches **P**.

The recooling time includes the plateau and the descent, the slope of the latter being affected by **H**.

Experimental Method. To understand the CICC thermal behavior, we have performed a series of heat pulses in the loop described previously. Power was applied on the whole length of the conductor for a few tens of seconds in order to simulate the heating due to the variation of magnetic field in ITER. Unfortunately, it was difficult to exactly simulate

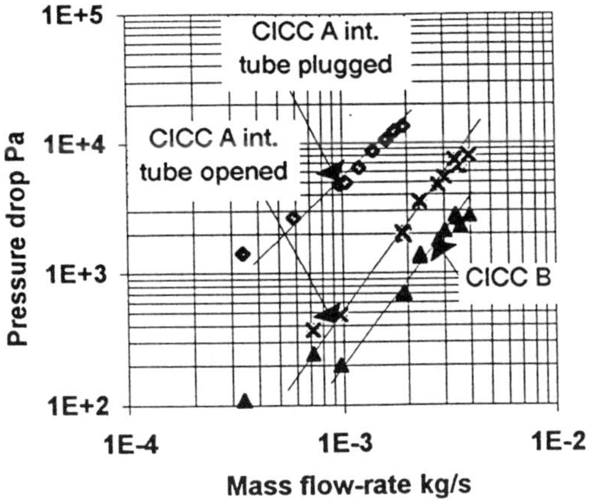

Figure 3. Pressure drop in the various sectors.

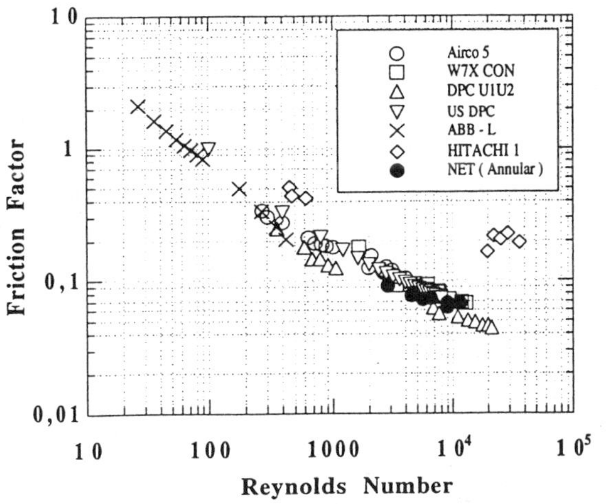

Figure 4. λ vs. Reynolds number for different CICC.

ITER conditions. The length of each dummy cable was 15m instead of 800m, the cross section was six times lower and to have maintained the ratio of the transit time over the heating time constant the mass flowrate (≅ 25g/s for ITER project) would have to be divided by a factor 320 which was not compatible with the experimental constraints. The total available power in each sector was 2.5W and we performed heat pulses at 0.9 and 0.66g/s, the latter corresponding to the minimum flow able to extract the intrinsic heat leak without significantly changing the temperature along the conductor.

As the current leads are located upstream of the first CICC sector and downstream of the second sector the power is simultaneously applied in both sectors. During the experiments, the pressure remained close to 6.5 10^5Pa. We observed a drift of temperature at the inlet of each sector during the heating phase. For the first sector, it may be caused by the poor performance of the current lead and perhaps also by a small decrease of the flowrate during the pulse. The second sector is directly affected by the heating coming from the first sector as the power is input simultaneously on both.

Numerical Model. In order to predict the temperature profile along the CICC, a numerical code[6] was developed with the assumption that the heat transport contribution of the stainless steel (jacket and tube) was negligible. Another approach has been made recently[5], but we maintain that the thermal behavior is governed by the two helium flows and their thermal coupling.

When the Joule-Thomson effect is neglected (which is reasonable for such low pressure drops), the equation of energy conservation applied to a one-dimensional incompressible flow may be written:

$$\rho Cp\left(\frac{\partial T}{\partial t}+U\frac{\partial T}{\partial x}\right)=\nabla(k\nabla T)+\phi \tag{4}$$

Neglecting the thermal conduction into the fluid, and assuming there is no mass exchange between channels, the equation (4) applied to each CICC channel becomes:

$$\frac{\partial T_1}{\partial t}+U_1\frac{\partial T_1}{\partial x}+\frac{HPm}{\rho_1 Cp_1 Sp_1}(T_1-T_2)=Q(t)\ \text{where}\ \frac{1}{H}=Pm\left(\frac{1}{P1h1}+\frac{1}{2\pi K}\ln\left(\frac{P1}{P2}\right)+\frac{1}{P2h2}\right) \tag{5}$$

$$\frac{\partial T_2}{\partial t} + U_2 \frac{\partial T_2}{\partial x} - \frac{HPm}{\rho_2 Cp_2 Sp_2}(T_1 - T_2) = 0 \tag{6}$$

where (with index 1 for the annular channel and index 2 for the inner channel):
Ti is the temperature, Ui is the average velocity in a section of channel i,
ρ_i is the density of fluid i, Cp_i the specific heat of fluid i,
Q(t) is the heat deposited in the annular region,
P1 is the central tube external perimeter, P2 the central tube internal perimeter,
Pm is the average perimeter of the central tube,
h1 is the transfer coefficient from the annular region to the central tube,
h2 is the transfer coefficient from the central tube to the central region, and
K is the thermal conductivity of the central tube

The code[6] solving equations (5) & (6) does not take account of the dependence on thermodynamical properties of helium on temperature. The properties are calculated for the most relevant temperature for both fluids implying a constant value for U1 and U2 in time and space during the calculation. Values for U1 and U2 are determined from the total mass flowrate and an assumption on the flow partition between the channels resulting from a preliminary calculation in accordance with pressure drop measurements.

In the special case of a step in inlet temperature at t=0 (with Q(t)=0), this system of equations has an analytical solution. A comparison between analytical and numerical solutions (figure 5) permits the determination of the mesh size required to prevent numerical diffusion.

Comparison of Model with Experimental Results. Before comparing experimental and computed results, it is useful to recall that for the calculation the adjustable parameters are the heat transfer coefficient H and the ratio of fluid velocity between the two channels and the initial conditions are the mass flowrate, temperature and pressure. The initial temperature and pressure are also used to define an average mass density which remains constant during the computation like mass flowrate and velocities in each channel. Consequently, the temperature must not vary too much along the line and temperature increases induced by heating must not be too high to avoid important changes in helium properties so that a good agreement between experiment and calculation is achieved. Moreover, it has been verified numerically that a large variation of the ratio of velocities in the two channels (eg a factor of 3) did not produce an important change in the response to a thermal pulse as the highest velocity remains almost constant.

The mass flowrate influences the transit time and, therefore, the length of the temperature plateau. Nevertheless, a variation of 10% of the mass flow-rate value results in insignificant modification of the curves.

To take account of the temperature drift during and after the heating phase, temperature at the inlet of the sectors (boundary condition for the code) has been taken with a value as close as possible of the experimental data. Furthermore, a constant power which corresponds to permanent thermal loads on the loop has been inserted into the computation and allows the fit of measured temperatures with computed ones before the beginning of heating. This correction is a contrivance since the model does not take into account the dependence of helium properties with temperature. It is impossible to expect a precise reproduction of experimental curves since thermal loads are not known and thermal sensor accuracy in the considered range of temperature is not better than 0.1K.

Nevertheless, it is possible to obtain a good agreement between calculation and experiment using H as a free parameter. Figure 6 (left side) shows comparison for temperature distribution on the CCIC with internal tube subjected to a monopulse of 2.25W held on for 50s. The mass flowrate was 0.84g/s. The best fit is achieved for a value of H between 50 and 200W/m²K, 100 giving the best agreement. For the other type of conductor,

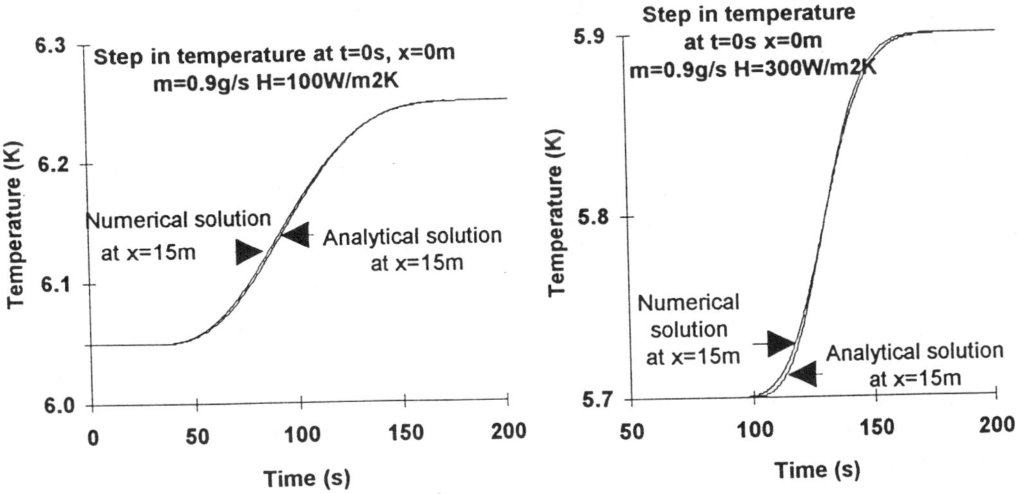

Figure 5. Comparison between analytical and numerical solution applied to conductor A and conductor B.

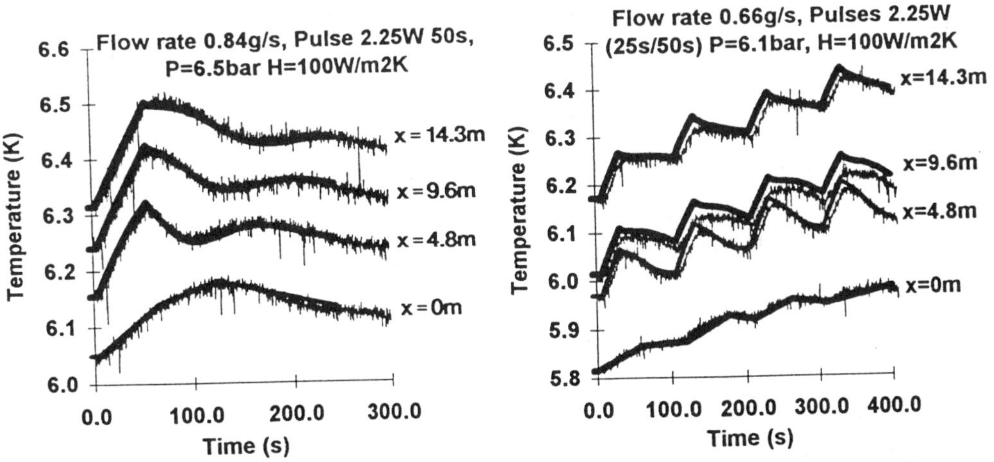

Figure 6. Conductor A, comparison of computed (thick line) and experimental (fine line) temperature.

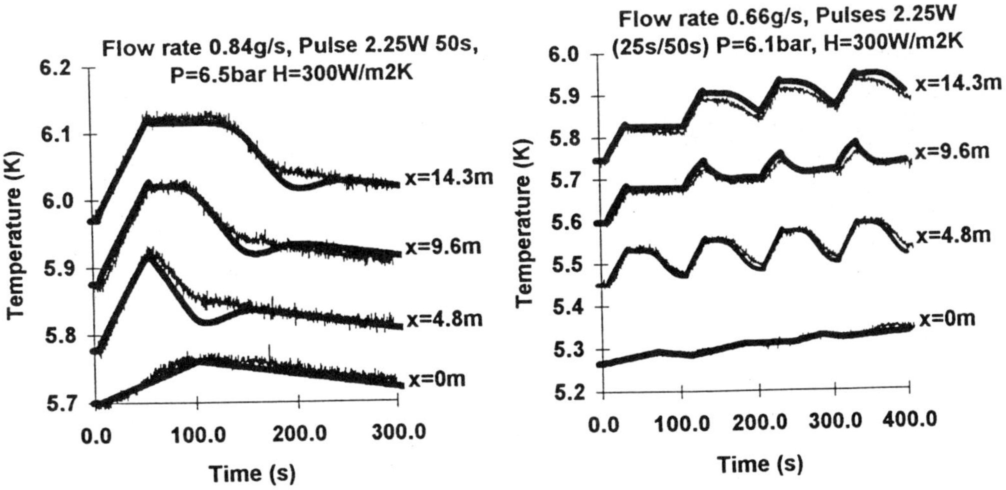

Figure 7. Conductor B, comparison of computed (thick line) and experimental (fine line) temperature.

under the same conditions, the best agreement is obtained for H comprised between 200 and 400 W/m^2K (see figures 7), 300 representing a good compromise.

The same type of tests are done with repetitive pulses with a constant frequency. Here again it is possible to obtain a good coincidence between computations and measurements and for the same values of transfer coefficient as in the monopulse experiments. Figures 6 & 7 (right side) show the results in case of pulses having an amplitude of 2.25W, a width of 25 seconds and a period of 100 seconds. The single phase mass flowrate was 0.66g/s.

Figure 7 displays the behaviour of conductor without internal tube. For the thermometer located at x=4.8 m, as the transit time is shorter than the time between two heat pulses, the temperature returns to its inital value before a new pulse is applied (in fact, in our experiment taking account of the drift of the upstream temperature, a value that is higher than the initial temperature). In contrast, for the thermometers located at x=9.6 m and x=14.3 m, a new pulse is heating when the temperature is still at a plateau resulting from the heating induced by the previous pulse. A periodic response is obtained after the second pulse only, a particle receiving two heat pulses during its transit along the CICC. The conductor with the inner tube has a smaller cross section, as a result the average helium velocity is larger but the same phenomena may be observed (see figure 7 right side).

CONCLUSION

The computer model developed at Cadarache does predict the general shape of the thermal response of these conductors to pulse heating and may be used in conjunction with experimental data to estimate the overall heat transfer coefficient. Concerning pressure drop, in case of conductor A, correlation can be applied separately to each channel, annular channel being described by reference 3. For CICC type B, the pressure drop results are not easily extrapolated to other size and tests on the ITER conductor design will be necessary.

REFERENCES

1. N. Michell "Conductor Designs For The ITER Toroidal And Poloidal Magnet Systems" IEEE Trans on Mag. 30:1602 (1994)
2. B. Rousset et al. "Pressure Drop And Transient Heat Transport In Forced Flow Single Phase Helium II At High Reynolds Numbers" ICEC 34:317 (1994)
3. H.Katheder "A General Formula For Calculation Of The Friction Factor For Cable In Conduit Conductors" The NET team, Internal note N/R/0821/26/A (1993)
4. R.Maekawa et al "Pressure Drop Measurements Of Prototype NET And CEA Cable In Conduit Conductors (CICCs)" Presented at Applied Superconductivity Conference (Boston 1994)
5. S.Depalo and L.Bottura "Computer Analysis Of The Thermo-hydraulic Measurements On CEA Dummy Cables Performed At CEN-Grenoble" The NET team, Internal note N/R/3500/47/A (1995)
6. A.Martinez et al. "Supercritical Helium Cooling Of A Cable In Conduit Conduit Conductor With An Inner Tube" ICEC 34:591 (1994) and CEA/DRFC, Internal Note P/EM/93.18

PREDICTED THERMAL-HYDRAULIC CHARACTERISTICS OF CABLE-IN-CONDUIT CONDUCTOR WINDINGS DURING STEADY-STATE OPERATION

Thomas Kupiszewski, Owen R. Christianson, Douglas Natelson

Westinghouse Electric Corporation
Science and Technology Center
1310 Beulah Road
Pittsburgh, PA 15235

ABSTRACT

A finite difference numerical model is developed to simulate steady-state supercritical helium flow and heat transfer within superconducting magnets using double-pancake coils of cable-in-conduit conductor (CICC). The model is implemented in computer programs which calculate global temperature and pressure distributions in winding packs subjected to time-averaged thermal loads. These programs are used to predict nuclear heating effects upon pressure drop and temperature rise along the forced-flow cooled superconductors of magnets in the Toroidal Physics eXperiment (TPX) tokamak. We present results suggesting superconductor temperature margin depends upon how effectively a coil design "surfs" the helium heat capacity "wave" and case-to-coil heat transfer interactions.

INTRODUCTION

Neutron fluence associated with D-D or D-T fueled plasmas is the principal source of internal heat generation for a subset of the superconducting coils used in magnetic confinement fusion devices. In the largest poloidal field (PF) coil of the Toroidal Plasma eXperiment (TPX), for example, analysis [1] indicates neutron heating is 24 times greater than the superconductor pulsed field loss [2] for a design fluence of 7.5×10^{16} neutrons per second. Total neutron energy dissipation in this PF coil amounts to 340 kJ during a thermonuclear burn lasting 1000 seconds, which is equivalent to a time-averaged heat input rate of 0.34 kW. Because cable-in-conduit conductors (CICCs) exhibit excellent thermal stability and an ability to remove steady-state heat input on the order of 0.1 Watt/m, they are attractive for fusion applications of superconducting magnet systems. Understanding the flow and heat transfer characteristics of CICC coils is important for optimizing the cooling design of fusion magnets from the standpoint of superconductor temperature margin. A thermal-hydraulic model which accurately predicts global temperature distributions in large scale CICC coils is of value as a design aid in this regard, especially for NbTi CICC coils which have relatively limited temperature margins compared to their Nb_3Sn counterparts.

Several previous analyses of CICC thermal-hydraulic response to tokamak-relevant heat inputs have been performed by various investigators. Kerns et al.[3] analyzed the cooling scheme for the toroidal field (TF) coils of the Tokamak Ignition/Burn Experimental Reactor

(TIBER) by solving the coupled ordinary differential equations [4] governing one-dimensional steady-state flow of supercritical (single phase) helium in a heated channel. Their calculations included effects of finite compressibility, CICC friction factors, and spatial variations in the steady-state heating distribution, but did not account for conduction heat transfer between adjacent turns. Wong [5] included inter-turn heat transport in his analysis of the transient thermal-hydraulic response of a single TPX TF double pancake coil to combined eddy current heating, pulsed field losses, and steady-state nuclear heating. His results revealed a "counter flow heat exchanger" effect whereby the peak helium temperature shifts towards the midpoint of the flow path. Recently, Wang [6] developed the computer code SHE3D which applies Kerns' computational technique to multi-pancake CICC winding packs with parallel-connected flow channels. SHE3D incorporates a simplified thermal-equivalent circuit representation of interturn heat conduction to update heat source terms appearing in the flow equations, thereby simulating the heat exchanger effect.

In this work, we have undertaken additional analytical studies providing further insight into cooling design performance of CICC fusion magnets. We have simulated how hot spot temperature in double pancake-wound CICC coils is effected by heat conduction between effluent-cooled case structures and multi-coil winding packs with parallel-connected flow channels. Furthermore, we have also investigated the impact of simultaneously extrapolating coolant channel length and steady-state thermal load into the 300 meter and 0.3 W/m regimes, respectively. During the course of analyzing these flow path and heat input scaling effects, a novel technique was discovered for post-processing computer model output. We believe it offers promise as a means of differentiating coil designs in ways not possible by comparison of temperature distribution plots. Examples of this technique, applied to various TPX coil designs, are presented and potential implications are discussed. The following sections describe our analysis methods, simulation input data, and results.

CASE-TO-COIL HEAT CONDUCTION EFFECTS

As part of Westinghouse's preproposal effort for the TPX Magnet Program, a preliminary scoping analysis of the TF coil configuration was performed to determine the effectiveness of case cooling in intercepting heat conduction across the ground wall insulation wrap. The TPX TF coolant flow path begins and ends in the outermost turns, with CICC coolant first spiraling inward towards the plasma, then indexing to an adjacent "layer" of the particular double pancake, and spiraling back outward. A total of six such flow paths are routed in a parallel arrangement through the entire winding pack, see Figure 1. Our analysis included a stainless steel case structure and its cooling by coil effluent. The assumed partitioning of steady-state heat load for a 7kW nuclear and eddy current heating scenario [7] is also presented in Figure 1, together with a plot of the conductor heat input profile based on an assumed radial attenuation distance of 3 conductor turns (roughly 7 cm).

Methodology

Helium pressure and temperature profiles are obtained from integrating the following nonlinear differential equations, derived from an enthalpy balance on a control volume:

$$\frac{dP}{dx} = \frac{-2fG^2/(\rho D) + 4qG\beta/(\rho D(Cp - u^2\beta))}{1 - (G^2/\rho)(K + \beta\phi)} \qquad (1)$$

and

$$\frac{dT}{dx} = \frac{-2fG^2\phi/(\rho D) + 4q/(GD(Cp - u^2\beta))}{1 - (G^2/\rho)(K + \beta\phi)} \qquad (2)$$

where the parameter ϕ is defined as

$$\phi = (\mu_j Cp + u^2 K)/(Cp - u^2\beta) \qquad (3)$$

Here, P is temperature, T is pressure, x is position along the flow path with x = 0 at the inlet,

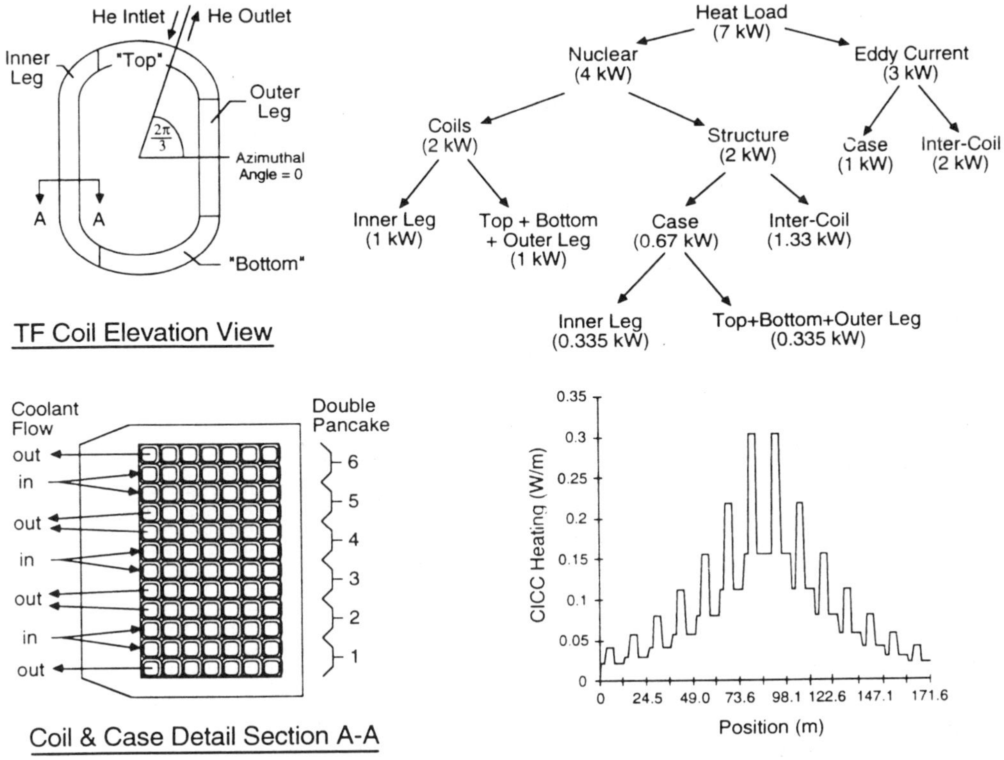

Figure 1. Notional steady-state heat distribution assumed for thermal-hydraulic simulation of full scale TF coil winding pack and case structure. Winding pack thermal loads of 2 kW total for all 16 TPX TF coils corresponds to an average conductor heat input of 0.12 Watt per meter.

f is Fanning friction factor, G is specific mass flow rate, ρ is density, D is hydraulic diameter, q is the heat input flux, Cp is constant pressure specific heat, u is the bulk flow velocity, β is bulk expansivity, κ is bulk compressibility, and μ_j is Joule-Thomson coefficient. Although CICC-relevant friction factors are given by Hooper's correlation [8] to Lue's data [9], our initial scoping studies [10] used smooth pipe correlations premultiplied by a correction factor between two and three for scaling pressure drops. Eq. (1) and Eq. (2) include the effects of viscous friction heating and are identical to those used in the numerical models of Kerns and Wang.

The convective heat input flux, q, in equations (1) and (2) is given by a control volume energy balance:

$$q = q_s - \Sigma q_c \tag{4}$$

where q_s is a spatially-distributed source term consisting of the net internal heat generation per unit convective surface area, q_c is the net conduction heat transfer flux between adjacent nodes, and the summation is taken over all neighboring nodes. For all winding pack and structure elements in thermal contact with the helium, Eq. (4) and the Giarratano correlation for fully-developed turbulent flow are solved iteratively to find node temperature and local heat transfer coefficient. For the conduction problem within the steel case, a similar procedure yields the equilibrium temperature of case elements, except Eq. (4) was modified by setting q equal to zero.

A FORTRAN computer program implementation of this model was developed to solve the relevant equations by integration using the Euler forward difference technique and relaxation for the conduction problem. Each element was represented by a single node located at its geometric center. Lumped parameters, like uniform temperature, were

Table 1. Geometry Parameter Values in TPX TF Case Cooling Simulation

average coil circumference	12.27 m
number of turns per double pancake coil	14
square CICC conduit dimension	22.89 mm
CICC cable space flow area	106 mm^2
wetted perimeter	827 mm
turn-to-turn & layer-to-layer conduit separation	1.63 mm
conduit wall thickness	2.41 mm
ground wrap thickness	7.31 mm

assumed. Flow boundary conditions consisted of helium temperature and pressure at the flow inlet. Symmetry considerations permitted modeling only half of the winding pack, so the program was based on an adiabatic boundary condition between double pancakes 3 and 4 such that the temperature profiles in double pancakes 1, 2, and 3 are identical to those in double pancakes 6, 5, and 4 respectively. The external case boundaries were also adiabatic. Other simplifying assumptions were the cable and conduit temperatures are identical at any given node and conduits have rectangular cross-sections with square corners.

Case Cooling Results

The effect of conductor effluent flowing through a single Mueller panel located on the inner leg case surface nearest the plasma was simulated. This location was of interest since it corresponded to the zone of maximum heat generation density, according to the heat partitioning scheme presented previously in Figure 1. For our study, inlet pressure and helium flow rate were set to 0.65 MPa and 4.17 g/s per double pancake, respectively. Epoxy-glass composite similar to G10/11-CR was assumed for the conductor and ground wall insulation wraps. Other relevant simulation parameter values are presented in Table 1.

Figure 2 presents four plots of the predicted CICC temperature versus position along the flow path from helium inlet to outlet. Plots for pancake 3 correspond to the flow channel closest to the vertical coil mid plane while plots for pancake 1 correspond to the flow channel closest to the case side wall. Inspection of these plots reveals roughly a 10% reduction in hot spot temperature rise which can be attributed to case cooling for the chosen simulation parameters. This modest reduction suggests a significant fraction of case heat is not being intercepted by coil effluent, a condition which is not surprising considering the Mueller panel on the inner leg covers only about 1/3 of the winding perimeter. The model-predicted value of 30 Watts per TF magnet for the product of mass flow rate times effluent enthalpy rise

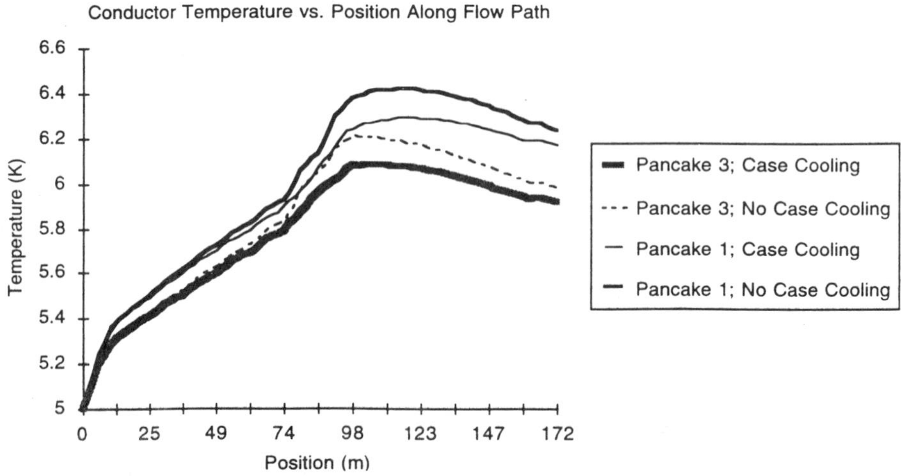

Figure 2. Predicted TF coil temperature profiles with and without case cooling.

Conductor Temperature, K AZIMUTHAL ANGLE 180.0

8.3723	5.9212	5.7662	5.6671	5.5683	5.4667	5.3632	in 5.1360	7.0809	7.4430	7.6044
8.4079	6.1171	6.2056	6.1886	6.1557	6.1105	6.0564	out 6.0066	7.1742	7.4738	7.6203
8.4366	6.1458	6.2279	6.2075	6.1719	6.1243	6.0682	out 6.0164	7.1937	7.4946	7.6412
8.4611	5.9529	5.7960	5.6944	5.5921	5.4863	5.3790	in 5.1411	7.1421	7.5074	7.6676
8.5455	6.0052	5.8566	5.7579	5.6540	5.5420	5.4229	in 5.1578	7.2014	7.5591	7.7131
8.7053	6.2368	6.3952	6.4179	6.4078	6.3733	6.3212	out 6.2657	7.3911	7.6498	7.7724
8.9321	8.4157	8.0293	7.7298	7.5047	7.3510	7.2796	7.3220	7.5377	7.7193	7.8158

inside radius — outside radius

Figure 3. Temperature distributions of TF inner leg at the horizontal mid plane without case cooling.

supports this notion. Compared to the 1.67 kW of combined nuclear and eddy current heating allocated to the case, the 0.48 kW dissipated by all 16 TF case coolant flow streams is only 29% of the total case heat generation for the simulated design scenario. An explanation can be gleaned from inspection of Figures 3 and 4.

Conductor Temperature, K AZIMUTHAL ANGLE 180.0

6.0719	5.8419	5.7248	5.6310	5.5379	5.4429	5.3468	in 5.1303	7.0489	7.4097	7.5708
6.0719	5.9955	6.0802	6.0756	6.0535	6.0179	5.9726	out 5.9330	7.1350	7.4381	7.5857
6.0719	6.0230	6.1014	6.0936	6.0689	6.0311	5.9840	out 5.9427	7.1534	7.4577	7.6053
6.0719	5.8723	5.7541	5.6578	5.5612	5.4621	5.3621	in 5.1353	7.1067	7.4702	7.6303
6.0719	5.9219	5.8118	5.7188	5.6213	5.5166	5.4054	in 5.1516	7.1628	7.5190	7.6731
6.0719	6.1019	6.2528	6.2881	6.2909	6.2702	6.2324	out 6.1926	7.3404	7.6039	7.7287
7.2612	7.4490	7.4475	7.3643	7.2639	7.1849	7.1610	7.2361	7.4763	7.6687	7.7692

inside radius — outside radius

Figure 4. Temperature distributions of TF inner leg at horizontal mid plane if case is ventilated.

Figure 3 presents predicted conductor and case temperature distributions along the horizontal mid plane of the magnet if the case is not ventilated. Heat flows along two paths from the case hot spot located at the inside radius corner edge. One path is toward the plane of vertical symmetry along the plasma-facing plate next to the innermost turns. The other path is radially outward along the case side wall plate, across the thick nose wedge region, and back in towards the cool outer turns. The temperature drop across the ground wall insulation (designated by the the thick darkened boundary) is noteworthy since it exceeds the typical turn-to-turn temperature drop by an order of magnitude, but is comparable to the inter-pancake temperature drop, implying significant heat flow across the thick ground wall barrier. Evidence of this heat flow is provided by the fact the predicted pancake 1 hot spot temperature rise, relative to the 5K inlet temperature, is about 18% greater than the corresponding pancake 3 value. Furthermore, the pancake 1 hot spot temperature location is shifted farther away from the plasma by an additional turn.

Figure 4 shows how these temperature distributions change when the case is ventilated with conductor effluent. Here, the case hot spot location has shifted to the outside radius corner edge in the thick nose wedge region. Note the plasma-facing plate, where the Mueller panel is located, is isothermal with a uniform temperature of 6.0719 K. However, most of the case nose wedge region exhibits negligible temperature depression due to the Mueller panel, indicating the lowest thermal impedance path for case heat dissipation is perpendicular to the ground wrap, not parallel to the ground wrap.

FLOW PATH AND HEAT INPUT SCALING EFFECTS

As part of Westinghouse's effort during the preliminary design phase of the TPX Magnet Program, a cooling design feasibility analysis [11] of the outer ring PF coils was performed. Unlike the encased TF coils, which use Nb_3Sn wire, the outer ring coils have no case and use NbTi wire since the peak field requirements are relatively low. Double-pancake flow path lengths and steady-state heat load are increased relative to the TF coils by roughly a factor of 2. The thermal-hydraulic engineering challenge is to limit wire temperature rise to $\leq$ 2K by sustaining sufficient coolant mass flow rates at increased inlet pressures. However, difficulties arise because specific heat of supercritical helium at the 5 K inlet temperature decreases as pressure increases. Table 2 presents a comparison of the coil cooling designs.

The same thermal-hydraulic model used in our earlier scoping analysis of TF case cooling was also applied to evaluating cooling design feasibility of the TPX PF coil baseline configurations. For this more recent study, a commercial spreadsheet program was used to run the model for a single double-pancake coil to facilitate design trade studies. Two refinements were implemented to improve modeling realism. A staggered-grid nodal mesh including both conduit and cable nodes was used to account for cable-to-conduit temperature differences arising from inter-turn heat flow and the relatively small wetted perimeter of the conduit interior. Since rounded corners comprise 40% of the conduit heat flow perimeter, the model's equivalent circuit representation of interturn heat transport was refined to account for changes in effective length and area associated with paths for heat conduction between

Table 2. TPX TF and outer ring PF coil cooling design parameter values

Parameter	TF Coil	PF Coil (PF7)
ave. coil circumference (m)	12.27	27.
no. of turns per double-pancake	14	12
steady-state heat (W) per channel	37.5	48.36 (peak:ave.=1:1) 100.2 (peak:ave.=2.1:1)
flow path length (m)	171.8	324
ave. heat (W/m) per unit length	0.22	0.15 to 0.31
cable space flow area (mm^2)	106	111.4
cable wetted perimeter (mm)	827	735
tot. no. of 0.78 mm dia. wires	405	360
conduit internal perimeter (mm)	65	65
CICC conduit material	Incoloy 908	stainless steel

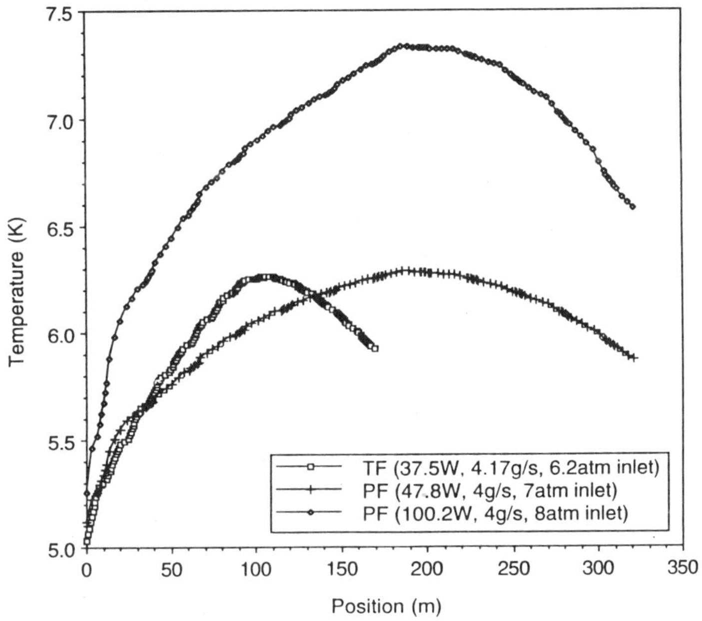

Figure 5. Predicted wire temperature profiles showing design scaling effects for the largest PF coil in TPX.

adjacent conduit corners and side walls. Figure 5 presents two plots of predicted PF wire temperature versus position along the flow path, with a third plot of the TF wire included for comparison purposes. Although PF coils are pulsed, the predicted helium residence time within the channels is less than the 1000 second duration of the plasma burn, thereby supporting applicability of a steady-state model. The hot spot temperature in the 7.3 to 7.4 K range for the 100.2 W heating case is noteworthy, considering the wire material is NbTi.

Investigation of high heating rates was motivated by the possibility of significant nonuniformity [12] in the coil nuclear heat distributions. All plots shown in Figure 5 are qualitatively similar, exhibiting the same convex profile characteristic of the counter flow heat

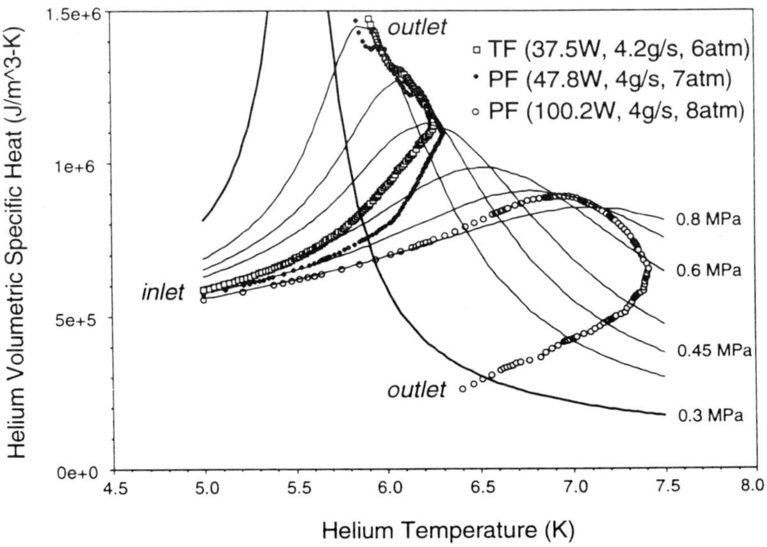

Figure 6. Plots of coolant state showing how different coil designs “surf” the helium heat capacity “wave”.

exchanger effect. These plots provide no indication that two distinctly different types of coil performance are predicted as heat input is increased from 47.8 W to 110.2 W. However, if one plots the product of helium density and constant pressure specific heat as a function of helium temperature predicted along the flow path, these different performance characteristics are revealed. As shown in Figure 6, the aforementioned product plots can be interpreted as design "trajectories" in helium heat capacity space, where the curling of the trajectory depends on how the design transits (or "surfs") a helium heat capacity "wave front" represented by a family of isobars. For one value of heat input, the PF coil response is virtually identical to that of the TF coil since their trajectories curl upward, intercepting and tracking the shifting crest of the heat capacity wave front. At the higher heat input value, the PF coil trajectory curls downward, passing up, over and far to the right of the wave front crest. We believe these results suggest the possibility of two different CICC performance regimes whose demarcation boundary may serve as a basis for a new CICC design criterion. However, more study is needed to evaluate this possibility and its potential significance.

CONCLUSIONS

Steady-state analysis based on a finite difference thermal-hydraulic model shows case structures can significantly influence thermal performance of pancake-wound CICC coils in large scale, high heating applications like fusion research. Case temperature distributions can produce pancake-to-pancake variations in hot spot temperature on the order of 10%. Also, a significant fraction of the case perimeter must be ventilated to effectively isolate the winding pack from undesirable, parasitic heat exchange with the case. Similar analysis indicates design scaling of double-pancake coils may be constrained by how channel flow impedance sets the inlet pressure required for a given mass flow rate, which determines whether the helium thermodynamic state resides on the left or right side of the heat capacity wave front.

REFERENCES

1. S.L. Liew, TPX memo 94-941128-PPPL/SLiew-01, Princeton Plasma Physics Lab, Princeton (1994).
2. R.D. Pillsbury, TPX memo 14-940802-MIT/RPillsbury-01, Plasma Fusion Center, Cambridge (1994).
3. J.A. Kerns, D.S. Slack, and J.R. Miller, *Adv. in Cryo Eng.*, 33:175-182 (1988)
4. V. Arp, *Adv. in Cryo Eng.*, 17:342-351 (1972)
5. R. L. Wong, *Adv. in Cryo Eng.*, 39:1715-1722 (1994)
6. P.W. Wang, *IEEE Transactions on Applied Superconductivity* , 5:600 (1995).
7. W. Hassenzahl and T. O'Connor, "TPX Toroidal Field System Design Description Document", TPX Conceptual Design Review, March 29 - April 1, 1993, LLNL, 58-71
8. R.J. Hooper, "Friction Factor Correlations for Cable-in-Conduit Conductors", Fusion Engineering Design Center memorandum FEDC-M-84-E/M-016, Oak Ridge National Laboratory, Oak Ridge (1984).
9. J.W. Lue, J.R. Miller, and J.C. Lottin, *IEEE Transactions on Magnetics*, 15:53-55 (1979).
10. D. Natelson, T. Kupiszewski, and O. R. Christianson, "A Thermal-Hydraulic Steady-state Model of the TPX Toroidal Field Coils," Westinghouse STC Report 93-1TM4-ADEMS-R1, Westinghouse Science & Technology Center, Pittsburgh (1993).
11. T. Kupiszewski, TPX report 14-950721-WEC-TKupiszewski-01, Westinghouse STC, Pittsburgh (1995)
12. D.R. Potter, TPX memo 14-950403-WEC-DPotter-01, Westinghouse Marine Div., Sunnyvale (1995)

TEST FACILITY FOR JOINTS OF SUBSIZE CABLE-IN-CONDUIT CONDUCTORS FOR NET/ITER WINDING STUDIES

B. Jager, P. Chaussonnet, D. Ciazynski, J.P. Serries, and R. Simon

Association EURATOM-CEA
DRFC, CE Cadarache
F 13108 Saint Paul Lez Durance, France

ABSTRACT

The joints between superconducting cables of the ITER coils present a major problem to be solved. A low resistance is required for the cooling, whereas small losses in variable magnetic field are needed for the stability. Different types of joints have already been suggested[1,2]. Some preliminary measurements[2,3] have already been made on prototypes of joints cooled in a helium bath at 4.2 K and for low currents.

The test facility presented herein allows testing joint prototypes at the 1/8th scale (current up to 10 kA). These joints are cooled under the same conditions as those of ITER, by circulation of supercritical helium. A dipole allows submitting the joint to magnetic fields up to 4 teslas and to field variation rates up to dB/dt = 42 T/s.

INTRODUCTION

Initiated for the European project NET, the study of joints for subsize cables by Europe has continued for the ITER tokamak. The European participation includes the proposal of a joint concept for the central solenoid and the toroidal field coils of ITER. The concept must be validated by tests on joints at the 1/8th scale and must lead to the design and manufacture of a full-size sample.

The reduced-scale samples are first tested in a helium bath at 4.2 K, to measure their essential characteristics, such as resistance and losses under variable magnetic fields. Once the concept is optimized, it is necessary to test samples cooled by circulation of supercritical helium. These tests enable, on the one hand, measuring the resistance with currents closer to the critical current (by increasing the temperature), and on the other hand, carrying out stability measurements under a variable magnetic field in cooling conditions identical to those found in a real coil.

GENERAL PRESENTATION

The general diagram of a facility for these tests is shown in Fig.1. The test facility mainly includes a supercritical helium source linked to a test cryostat through a vacuum insulated transfer line.

The supercritical helium source is made up of a tank of 206 liters of helium pressurized by helium at 80 K injected into the upper part. The supercritical helium is extracted from the lower part; the pressure can be regulated (0.3 MPa < P < 1 MPa), the temperature (4.5 K < T <12 K) and the flow rate (Q < 5 g/s).

The test cryostat is made of two parts :

- A part cooled by boiling liquid helium in which a superconducting dipole submits the joint to rapid variations of the magnetic field ($dB/dt_{max} = 42$ T/s).
- An insert cryostat positioned in the dipole bore in which the sample to be tested and the associated current leads are put. These current leads, sized for 10 kA, are also cooled by supercritical helium. They are made of a copper braid inserted into a stainless steel tube. The insert cryostat is kept under vacuum.

SUPERCRITICAL HELIUM SOURCE

The supercritical helium source (Fig.1) is essentially made up of a vertical helium tank (volume=206 liters, diameter=0.290 m, height=3.12 m, P_{max}=1 MPa). This tank is initially

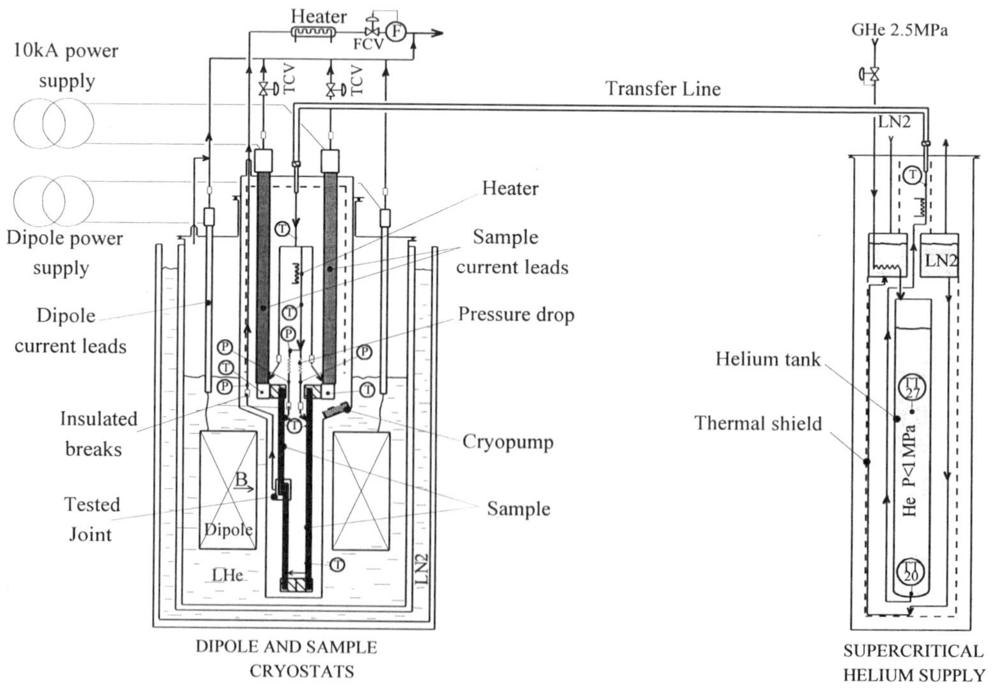

Figure 1. General diagram of the test facility.

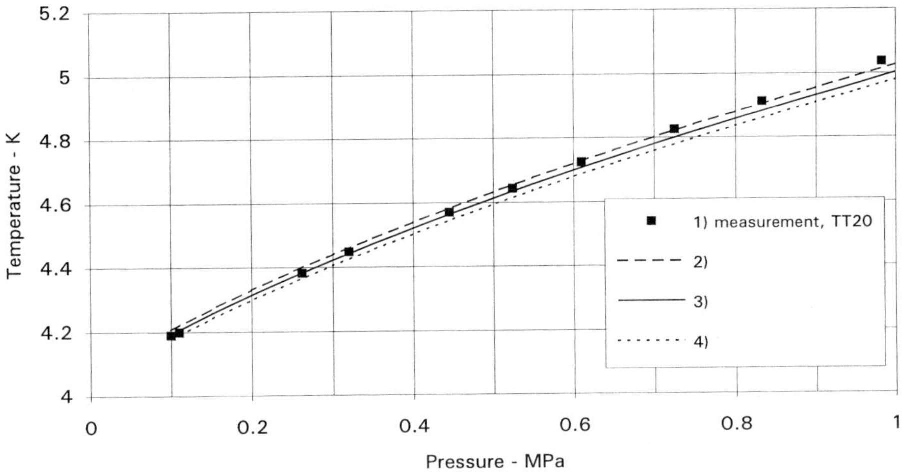

Figure 2. Variation of the temperature of supercritical helium depending on the pressure. 1) Measurements at the bottom of the tank (TT20). Evolutions calculated by assuming an isentropic compression for liquid helium saturated at: 2) initial P_i =100 kPa, 3) P_i = 98.5 kPa, 4) P_i = 97 kPa.

filled with liquid helium. The supercritical helium is obtained by pressurization of the tank by injecting helium at 80 K into the upper part.

A pressure control valve regulates the pressure at the desired value. The supercritical helium is extracted from the lower part. An electric heater included in the extraction line allows adjustment of the temperature of the helium. An 80 K thermal shield around the tank is cooled by a thermal siphon connected to a liquid nitrogen tank positioned in the upper part of the cryostat. This liquid nitrogen tank also contains the small precooling heat exchanger for the pressurization helium.

After the source has been filled, the pressurization of liquid helium to obtain supercritical helium ($P > 0.227$ MPa) leads to an increase in temperature. The 80 K helium injected in the upper part acts as a piston. The evolution of the temperature must, therefore, follow that of an isentropic compression. In Fig.2, we show the temperature (TT20, cf. Fig.1) measured at the bottom of the tank with varying pressure. The values are as expected.

The maximum helium mass in the tank is about 25 kg. For an average helium flow rate of 2 g/s, this corresponds to an autonomy higher than three hours. During this time, the temperature of the supercritical helium remains essentially constant (P=const.) as shown in Fig. 3. The thermal losses measured in the tank are about 1W , of which one half is due to the supports in the upper part, the other half is due to radiation. These losses and their location do not induce convection movement, thus the helium remains stratified.

TEST CRYOSTAT

The test cryostat is made up of two parts (Fig.1):

- An outer cryostat containing the dipole cooled in a helium bath at 4.2 K
- An insert cryostat containing the sample and its current leads cooled by the helium from the supercritical helium source.

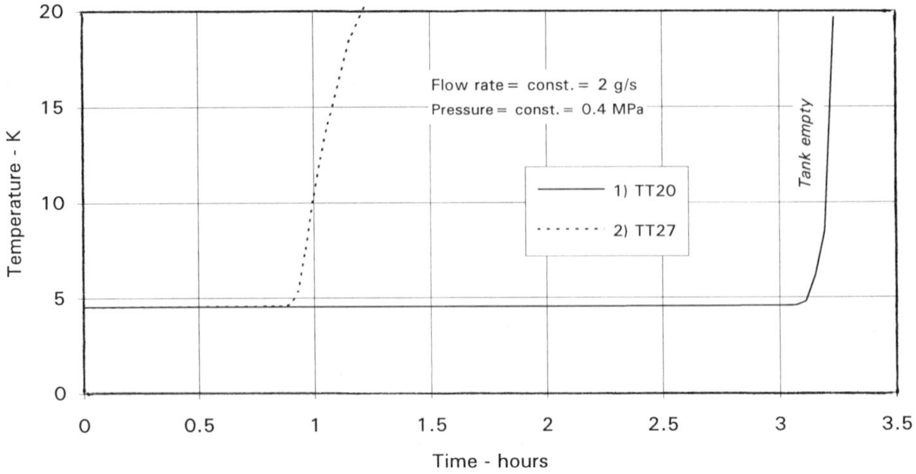

Figure 3. Variation of temperatures in the supercritical helium source depending on the time for a constant flow rate of 2 g/s, 1) at the bottom of the tank (TT20), 2) at 2/3 of the height (TT27).

Dipole Cryostat and Dipole Magnet

The magnetic field is produced by a superconducting dipole maintained at 4.2 K in a liquid helium bath. The main characteristics of the dipole are shown in Table 1. The dipole can be supplied by different sources of current :

- 2 kA / 15 V power supply, which provides 2 teslas and field variation rate of dB/dt = 3.2 T/s. It is also possible to make fast exponential discharges on resistance up to $(dB/dt)_{max}$ = 42 T/s corresponding to the maximum permissive voltage of 200 V across the magnet.
- The G0 power supply of the poloidal-field system for the Tore Supra tokamak (maximum current = 4 kA, maximum voltage of 1870 V). To limit the voltage to acceptable values for the dipole, it was necessary to connect copper coils in series[4]. This power supply allows increasing or decreasing the current in a pre-programmed way. During the tests, with dB/dt = 40 T/s, we had to limit the variation of the field to approximately $\Delta B = 1$ T so as not to induce a quench of the dipole.

The current leads of the dipole are optimized for 3 kA. They are made of a copper tube (30×20 mm) cooled by meshed washers[5]. In view of the quality of the copper (RRR ~ 5), these current leads have a great thermal inertia, enabling them to withstand a current up to 4 kA for several minutes without damage.

Table 1. Main characteristics of the dipole

field/current ratio	0.995 T/kA
maximum field	4 T
inductance	4.7 mH
aperture	∅ = 74 mm
useful length	l = 440 mm
mass	~ 200 kg
maximum voltage	200 V
$(dB/dt)_{max}$	42 T/s

Test Cryostat

The insert cryostat is positioned in the dipole bore. It is maintained under vacuum by a cryopump made by sticking coconut charcoal on the wall in contact with the helium bath (Fig.1). The supercritical helium arrives in the upper part and supplies the two current leads and the sample. The sample temperature can be adjusted by a small heater. All the entry and exit tubes are equipped with insulating electrical breaks[6]. The helium exiting the sample is used to cool a thermal shield in the upper part of the cryostat

The direction of the circulation in the prototype joint is identical to that for the joints planned for ITER. To minimize the thermal flux between the current leads and the sample, the helium supply is achieved through two tubes put at 0.1 m beneath the sample/current lead connection. The exit is made through a box surrounding the joint. So that the two flow rates supplying the sample are perfectly identical, we provided pressure drop much higher than those for the two branches upstream of the sample. They are made up of tubes 1 mm in diameter and 50 mm in length. At the exit of the cryostat, the sample flow rate is regulated by a FCV valve (cf. Fig. 1) placed downstream from an electric heater. This valve uses a mass flow control valve (Ref. Brooks Instrument - 5864E).

10 kA Current Leads for the Sample

The space available in the test cryostat is very small, which requires the most compact current leads possible. We used the same technology as that developed by C.Lesmond et al. in Saclay[7]. In Table 2, the main characteristics of these current leads are listed. They consist of a copper braid introduced into a stainless steel tube (external diameter = 44.5 mm, thickness = 2 mm). The braid is obtained by gathering tubular sheaths of increasing diameter, made directly one on the other. The tubular sheaths consist of wires that do not overlap one another. The ends of the braid are soldered in OFHC copper end pieces thanks to a tin soldering (ref. Castollin 1827 - Sn/Pb 60/40). These end pieces are soldered onto the stainless steel tube (soldering ref. Castollin 1802). The temperature at the cold end of these current leads is regulated by TCV valves (cf. Fig. 1), acting on the flow rate of the supercritical helium. The measured and calculated consumptions required to maintain specific cold-end temperatures are plotted in Fig. 4 as functions of the current. The calculations show that for 12 K, a hot spot appears for current values higher than 7.5 kA. If the temperature of the cold end drops, the helium flow rate increases and the hot spot moves towards higher currents. At 10 K, the hot spot appears for a current higher than 8.5 kA. At 6 K there is no hot spot as the ohmic power always remains lower than the sensible heat of the helium between 5 and 300 K. The current can then exceed 10 kA, without difficulty.

Table 2. Main characteristics of the 10 kA current leads for the sample

copper	Cual
measured RRR	$85 < RRR < 100$
wire diameter	0.3 mm
number of wires	7584
total section of copper	536 mm^2
diameter of the braid	40.4 mm
current density at 10 kA	18.66 A/mm^2
length of the cooled braid	1.405 m
total length including ends	2.115 m
LI/A	$2.62.10^7$ A/m
cooling	supercritical He, $0.3 < P < 1$ MPa, $T \geq 4.5$ K
regulation mode	temperature at cold end

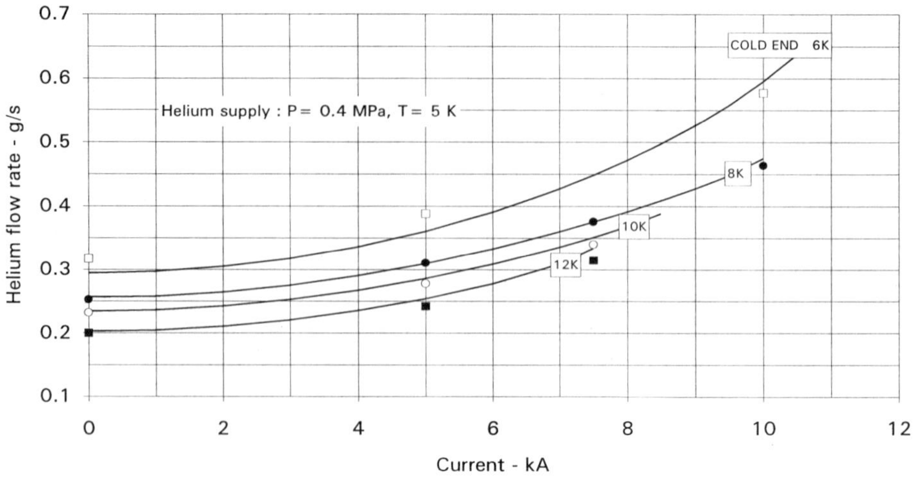

Figure 4. Consumption of the current leads depending on the current and the temperature at the cold end. In full lines, the calculated consumptions. □●○■ measured consumptions for a regulated temperature at the cold end respectively of : 6 K, 8 K, 10 K and 12 K.

These current leads present a low thermal inertia. In order to avoid thermal overheating, which could lead to their destruction, they have been equipped with safety devices. If the voltage between the cold end and the warm end exceeds 0.12 V, the current is automatically shut off.

The electrical connection between the current leads and the sample is ensured by bolted connections on flat copper surfaces of 180×60 mm. The surfaces in contact are perfectly machined and gilded. At cold temperature, to maintain a sufficient contact pressure, the bolts are equipped with titanium washers. Their thickness was calculated to compensate for contraction differences between the room temperature and 5 K. For a 10 kA current, the voltage measured at the terminals of the connection remains lower than 30 μV. This corresponds to a contact resistance lower than 3 nΩ. The power developed at the connection is then of 0.3 W, and is extracted mostly by the current leads.

SAMPLE

The sample is made starting with a cable-in-conduit manufactured by DOUR METAL with a square, stainless steel jacket, 13.8 mm each side. The manufacturing method of the sample has already been presented[1,2,3].

Table 3 Main Characteristics of the DOUR METAL cable

total number of strands	144
strand diameter	0.78 mm
non-copper area	30.6 mm^2
copper area in Nb_3Sn strands	38.25 mm^2
void fraction in cable space	40 %
steel jacket area	69.4 mm^2

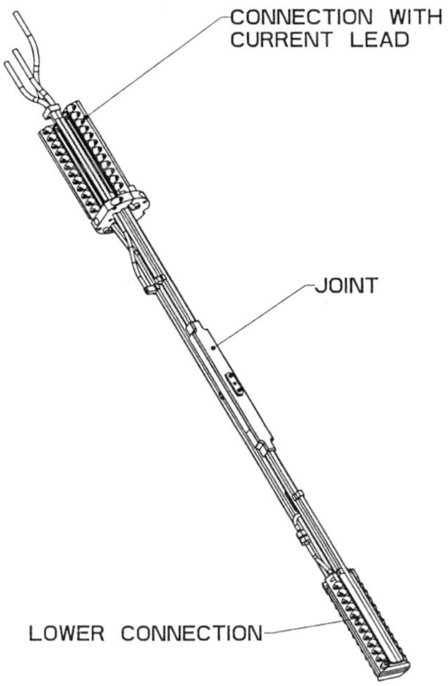

Figure 5. View of the *overlap* joint sample.

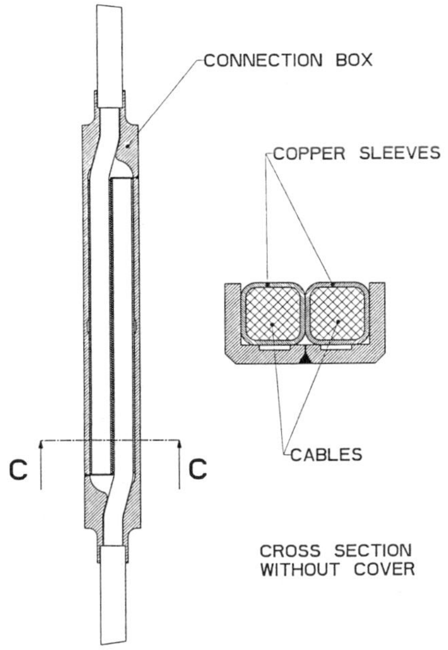

Figure 6. Cross section of the *overlap* joint.

The cable consists of 144 Nb_3Sn strands of 0.78 mm in diameter, made by VACUUMSCHMELZE using to the bronze route. These strands have been chrome plated (thickness=2 μm). In **Table 3**, the main characteristics of the cable are shown. A drawing of the joint sample is shown in Fig. 5 and 6. The joint is positioned in the middle of one of the branches of the sample. The orthogonal position of the high and low connection plates (cf. Fig.5), allows testing the same joint in two configurations of transverse fields : the field parallel or perpendicular to the joint plane.

FIRST TESTS OF A JOINT SAMPLE

The first test was done on an *overlap* type joint. A sample of this type had already been tested in a helium bath at 4.2 K[2]. The supercritical helium tests allowed confirmation of the independence of the resistance vs current and temperature (Table 4). For a 2 T field, we were thus able to measure up to 26% of the critical current (for I = 10 kA and T = 9.5 K), whereas the nominal value of this ratio in the ITER joints does not exceed 10% (for I = 46.6 kA, T = 5.5 K and B = 4.5 T).

Table 4 . Measurements of the resistance of the *overlap* joint depending on the temperature T(K) and on the magnetic field B(T). Resistance in nΩ

B(T) / T(K)	0	0.5	1	1.5	2
5.5	11.8	11.9	12.0	12.2	12.2
7.5	11.3	11.9	12.0	12.3	12.1
9.5	12.0	11.9	12.0	11.7	12.4

The loss measurements confirmed the values obtained in helium bath at 4.2 K as well as the existence of flux jumps beyond a certain rate of variation of the magnetic field $(dB/dt)_{fj}$. These tests showed the decrease of $(dB/dt)_{fj}$ with temperature, as predicted by the theory.

Stability measurements done at I = 6 kA and I = 10 kA, allowed us to obtain the quench of the joint by a variation of the magnetic field. These measurements showed the influence of temperature and of helium flow rate on the stability of the joint.

CONCLUSION

The test facility for joints has been in operation for 6 months. It allows joint prototypes to be tested under very severe conditions. Cooling by supercritical helium can vary widely (0.3 MPa < Pressure < 1 MPa, 5 K < Temperature < 12 K, 0 < Flow rate < 3 g/s). The magnetic field can reach 4 T and the field variation rate, $(dB/dt)_{max} = 42$ T/s.

Different basic characteristics are measurable on the joint such as the resistance and the losses. ITER relevant stability tests are also possible.

Different joints will be tested under conditions identical to those encountered in a real coil to lead to an optimized joint concept for the ITER coils.

REFERENCES

1. B.Turck,D.Ciazynski,P.Decool,B.Jager,P.Bruzzone, Conceptual and experimental studies of electrical joints for NET-ITER coils, Proceedings of the 17th Symp. Fusion Tech., Rome (1992),967.
2. D.Ciazynski,P.Decool,P.Hertout,B.Jager, Test results of subsize joints between Nb_3Sn cables as a first step for a design of joints for NET/ITER coils, Proceedings of 18th Symp. Fusion Tech., Karlsruhe (1994), to be published.
3. D.Ciazynski, "Task MWIN 1: Final Report on Phase II. Fabrication, Test Results and Analysis", CEA report DRFC P/EM94 55, oct. 94.
4. E.Rouanet, "Etude et Réalisation du Circuit d'Alimentation d'un Dipôle Supraconducteur Générateur de Champ Magnétique", thesis CNAM - Aix en Provence France, (nov. 1993).
5. G.Claudet,F.Viargues,G.Friesinger,R.Heller, Vapor cooled current leads for pulsed operation, Proceedings of the 13th Int. Cryog. Eng. Conf., Beijing China (1990) ,684.
6. G.Schenk et al, High voltage insulation and tests of cryogenic components for the superconducting model polo coil, Proceedings of the 15th Symp. Fusion Tech., Utrecht (1988), 1552.
7. C.Lesmond,J.C.Lottin,J.P.Lottin, Currents leads for ALEPH solenoid, Proceedings of the 12th Int. Cryog. Eng. Conf., Southampton (1988),247.

THERMAL HYDRAULIC CHARACTERISTICS OF A PROTOTYPE CEA CABLE-IN-CONDUIT CONDUCTOR

R. Maekawa, M.R. Smith, and S.W. Van Sciver

National High Magnetic Field Laboratory/FSU
Tallahassee, FL, 32306, USA

ABSTRACT

The thermal hydraulic characteristics of a prototype CEA Cable-in-Conduit Conductor (CICC) have been studied in steady state and transient conditions. The supercritical helium velocity in the central channel was measured with a Pitot tube located at the down stream end of the conductor. An inductive heater, located at the center of the conductor, initiated thermally induced transient flow of the helium within the conductor. The induced flow velocity was measured as a function of Reynolds number and heat input. A calorimetric calibration technique was used to estimate the total heat input to the conductor. In a separate part of the experiment, a thermometer array was installed in the central channel to record the helium temperature. The associated reduction of central channel flow area significantly affects the thermal hydraulic characteristics of the conductor.

INTRODUCTION

Cable-in-Conduit Conductors have been primarily used for the PF, TF and CS coils in Tokamak type experimental fusion reactors. New CICC designs have a hybrid cooling scheme, characterized by a central cooling hole surrounded by six bundles. The object of the central channel is to lower the flow impedance, and thereby improve cooling and recovery. The prototype CEA conductor is of this design class. Each bundle contains 144 copper strands wrapped by a stainless steel foil. Due to the complex geometry and flow in these conductors, computer simulations and analytic solutions of flow and heat transport problems in large superconducting magnet systems have remained in question without experimental data to back them up. This paper reports experimental results on the thermal hydraulic behavior of the conductor, using a non-intrusive inductive heating technique.

EXPERIMENTAL APPARATUS

The experiments were conducted in the Liquid Helium Flow Facility (LHFF).[1] A 3m long test section was installed in a closed helium flow loop and instrumented for temperature and absolute pressure at intervals along its length. The inductive heater was located at the center of the conductor, and a Pitot tube was installed at the discharge end to record the velocity in the central channel. Differential pressure sensors were used to record the overall pressure drop with varying flow rate, as well as the total mass flow at a venturi flow meter and the Pitot tube. A more detailed description of the experimental setup and LHFF has been reported by the authors elsewhere.[1]

Advances in Cryogenic Engineering, Vol. 41
Edited by P. Kittel, Plenum Press, New York, 1996

EXPERIMENTAL PROCEDURE

For the steady state and transient flow measurements, the speed of the centrifugal pump in our flow loop was changed from 0 Hz to 65 Hz in 5 Hz intervals. Data were taken once the flow through the venturi flow meter, located at the pump outlet, reached steady state.

Thermally induced transient flow experiments were performed as a function of steady state Reynolds number and total heat deposited by the inductive heater, which was energized by a signal from the computer. Data were taken from about 10 ms before the heat input until the system recovered or reached steady state conditions several seconds later. In all measurements, the supercritical helium is at ~4.3 K and ~0.3 MPa.

Calibration of the Inductive Heater

The inductive heating method is the most realistic and non-intrusive way to deposit heat into the CICC strands. A capacitor bank system was used as the pulsed power supply because of its rapid response and overall simplicity. The time varying field produced by the inductive heater induces eddy currents in the copper strands, resulting in Joule heating. The heat thus generated is subsequently absorbed by the HeII. In a closed system, the total energy deposited may be derived from the HeII enthalpy change. This type of calorimetric loss measurement was first employed for AC loss measurements on the SMES/CICC.[2]

For calibration, a 0.32m long conductor sample was filled with HeII at 1.8K and isolated by a press fit plug. The capacitor bank was subsequently charged to a particular voltage, then discharged through the inductive heater by a signal from the computer. An array of four Allen-Bradley thermometers in the central channel monitor the He II temperature change. Figure 1a shows a typical temperature profile after a heat pulse. Isothermal conditions were typically reached after approximately 0.4s.

The enthalpy may now be integrated between the initial and final temperatures, and using the volume of the conductor derived from image analysis of actual cable cross sections, one arrives at the total energy deposited as a function of initial energy stored in the capacitor bank. This is shown in Figure 1b, where the He II enthalpy change is a linear function of the stored energy in capacitor bank, effectively characterizing the coupling between the magnet (inductive heater) and the conductor strands. Although the heater efficiency is very low, it is sufficient to deposit more than 2 J/cm^3 to the conductor. The thermometer uncertainty is approximately ±2mK, so that the principal source of error lies in the measurement of the helium volume within the conductor. This figure is probably accurate to better than ±10%, making this a fairly accurate estimation of the energy input to the conductor.

Once the total energy input to the system is known, it is possible to calculate the power dissipation function Q(t), from simple eddy current considerations. Neglecting the response time of the conductor, we have from Faraday's law that $I(t)_{eddy}$ is proportional tothe rate of change of the magnetic field, dB/dt. Thus, the power dissipated may be written

$$Q(t) = I(t)^2_{eddy} R_{strands} \propto \left(\frac{dI(t)_{magnet}}{dt} \right)^2 \qquad (1)$$

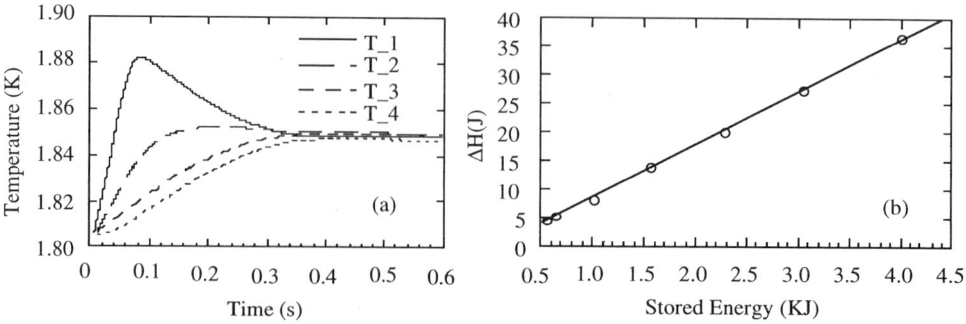

Figure 1. Short sample calibration of the inductive heater. A typical temperature profile following a heat pulse is shown in (a). In this case the initial energy stored in the capacitor bank was approximately 562J. Results of the calibration are shown in (b), where the stored energy refers to the initial energy in the capacitor band and the energy deposited by the inductive heater is plotted along the vertical axis

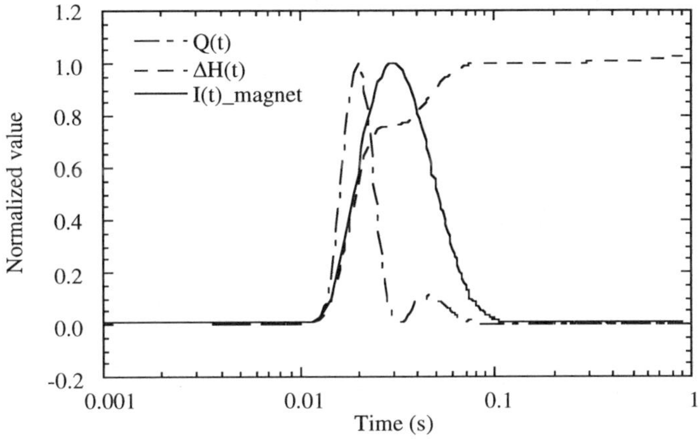

Figure 2. Estimated Q(t) and H(t) derived from the current through the inductive heater, which is monitored via a shunt resistor. A typical pulse lasts 50-60ms.

where the current through the inductive heater $I(t)_{magnet,}$ is easily monitored via a shunt resistor. The proportionality constant is determined by integrating this expression and setting it equal to the total enthalpy change of the HeII measured in the course of the calibration. Figure 2 shows $Q(t)$ calculated from $I(t)_{magnet}$ and the total He II enthalpy change as function of time.

RESULTS and DISCUSSION

Supercritical Helium Velocity in the Central Channel

The supercritical helium velocity in the central channel was measured by a Pitot tube located at the discharge end of the conductor. The mass flow rate in the central channel may be calculated by,

$$\dot{m} \cong A_{central}\sqrt{2\rho\Delta P} \tag{2}$$

where $A_{central}$ is the central channel area from image analysis, ρ is average density and ΔP is the measured pressure drop at the Pitot tube. Figure 3a shows the mass flow rate in the central channel as a fraction of the total mass flow rate. At larger Reynolds numbers, approximately 60% of the total mass flow rate is in the central channel , with the fluid velocity being approximately three times higher than that in the bundles, even though the area of the central channel is only 17% of the total flow area. This flow distribution makes better sense if one compares the hydraulic diameters of the central channel and the overall conductor; 3.9mm, and 0.6mm, respectively.

A model was developed which treats the conductor as a set of parallel, independent flow passages: a central channel; six bundles and six external channels. Areas of the constituent channels were estimated in two ways; from image analysis of actual CICC cross sections, and from simple geometric considerations. The results of the model(s) are plotted together with the experimental data in Figure 3b, indicating good agreement.

Thermally Induced Flow in the Conductor

Heat pulses of varying strengths were first applied to the conductor filled with non-flowing supercritical helium. The resultant thermally induced flow velocity in the central channel is presented in Figure 4 as a function of applied heat. The induced velocity shows a pronounced asymptotic behavior not reflected in Dresner's solution. The source of this disagreement may be the uneven surface of the central channel with its associated frictional retarding force, which was neglected by Dresner.[3]

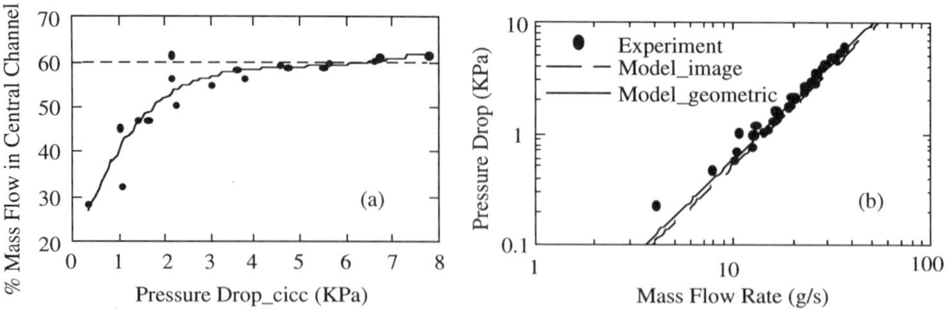

Figure 3. Steady state flow in the central channel. In (a), the mass flow in the central channel, as a percentage of the total mass flow, is plotted against the overall pressure drop along the length of the 3m test section. At higher flow rates, approximately 60% of the mass flow is in the central channel. The overall pressure drop is plotted against the total mass flow rate in (b), together with some simple flow modeling results. These flow models typically predict over 50% of the mass flow residing in the central channel.

Figures 5a and 5b illustrate examples of thermally induced flow in the central channel for two i mposed mass flow r ates. The corresponding Reynolds numbers are 5100 and 7050, respectively. Increasing the heat input for a particular imposed (steady state) mass flow rate clearly enhances the induced, or secondary flow within the central channel. Increasing the imposed mass flow however tends to suppress this effect, though the increased mass flow rate improves the overall heat transfer between the strands and the supercritical helium. The improved heat transfer at higher flow rates may be responsible for carrying heat away before it can induce large secondary flows. Thermally induced flow is primarily generated in the central channel, as seen in Figure 6. Although the figure suggests an initial flow reduction in bundle region, this may be caused by a slight over estimation of mass flow rate in the central channel. The induced flow dependence upon heat input and imposed mass flow rate is summarized in Figure 7a.

Figure 7b shows the total mass flow rate recorded at the venturi flow meter with the thermometer array installed in the central channel. A flow reduction is observed after the induced flow, which was not seen in previous measurements. Possibly, when the heat is input to the conductor, the helium absorbs it and expands in both the upstream and down stream directions. With the central channel blocked by the probe however, the helium cannot expand smoothly to the down stream end. At the same time, the expanded helium in the upstream direction creates a bulk flow resistance in the loop. These two effects may combine to produce the observed flow reduction.[4] This underlines the importance of an unobstructed central channel in the recovery process, particularly in the case of large systems.

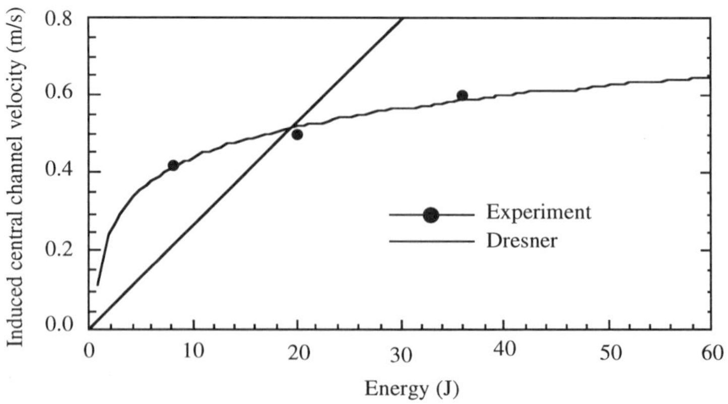

Figure 4. Thermally induced flow in static supercritical helium. Data are compared with Dresner's solution[3]. The line through the points represents a best fit to the data. The disagreement with Dresner may be explained by the irregular surface of the central channel and the added damping effect which this has upon the flow. The energy refers to the amount actually deposited in the conductor.

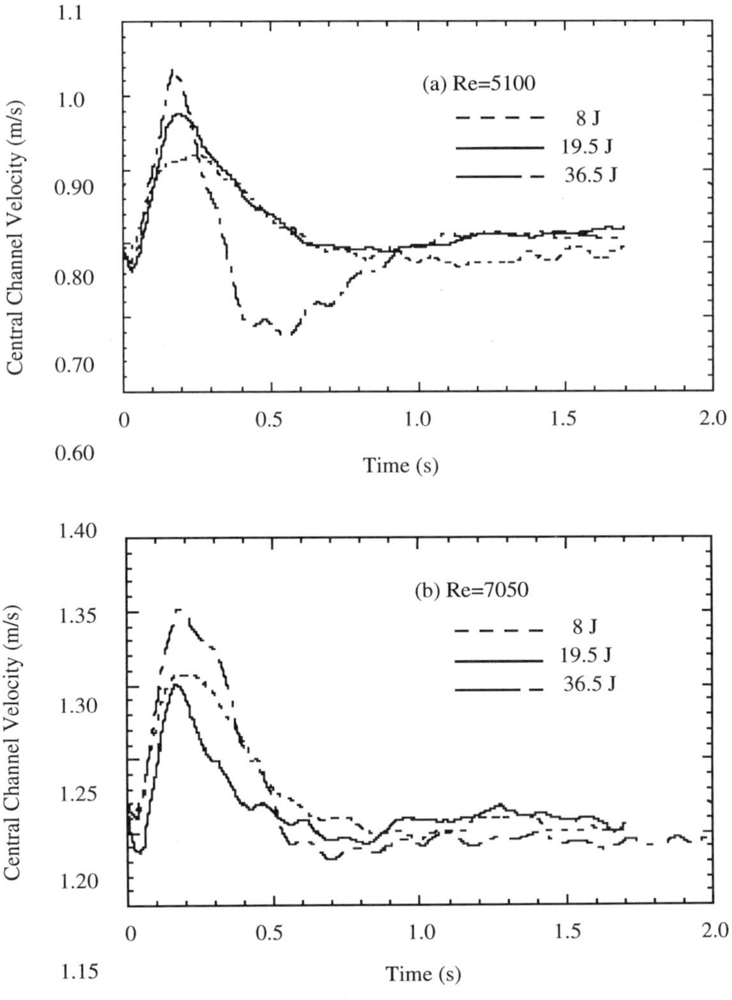

Figure 5. Thermally induced flow within the central channel in the presence of an imposed mass flow. Results are shown for two Reynolds numbers, with various amounts of heat deposited in the conductor as indicated. Larger heat pulses result in greater induced flows within the central channel, but increasing the imposed mass flow tends to suppress this effect.

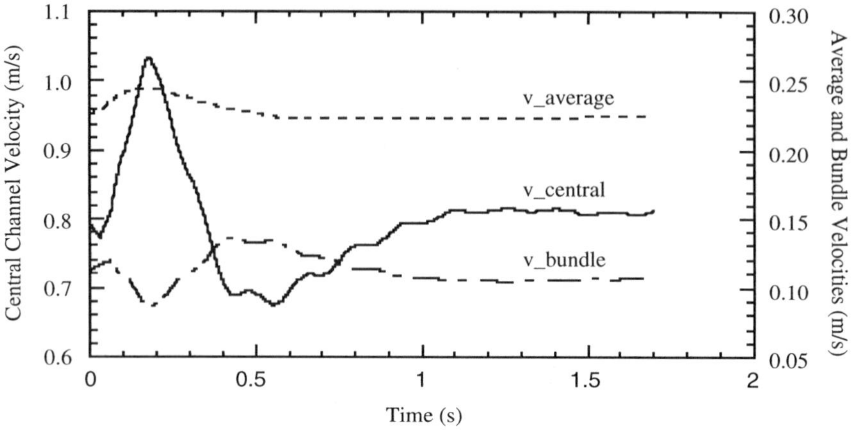

Figure 6. Induced velocity distribution in the cross-section. Values are calculated from the mass flows measured by the Pitot tube, and the venturi flow meter located at the pump outlet, using the areas of the various subassemblies derived from image analysis. Note the two vertical axes. The central channel velocity is plotted on the left axis. Clearly, induced flows are largest within the central channel. The slight temporary flow reduction apparent in the velocity trace for the bundle region may be due to a slight overestimation of the mass flow in the central channel.

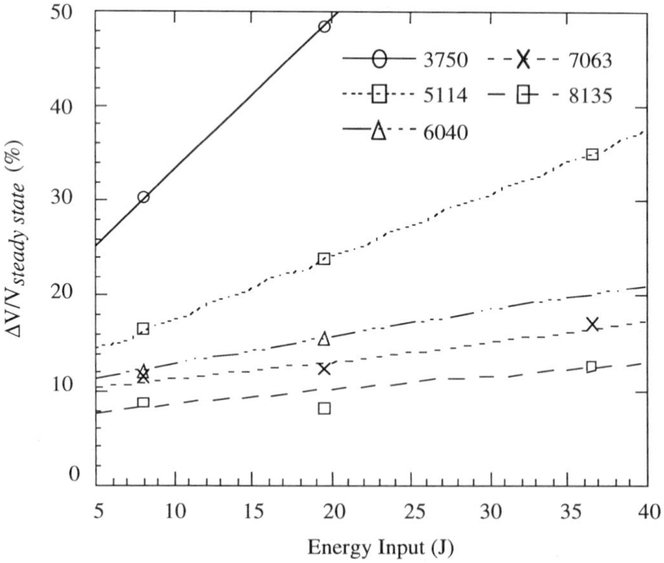

Figure 7. Summary of induced flow data and the effect of blocking the central channel. The induced flow dependence upon heat input and imposed mass flow rate is summarized. Different lines correspond to different imposed mass flow rates as indicated. The associated Reynolds numbers are, in order of increasing slope, 8135, 7063, 6040, 5114 and 3750.

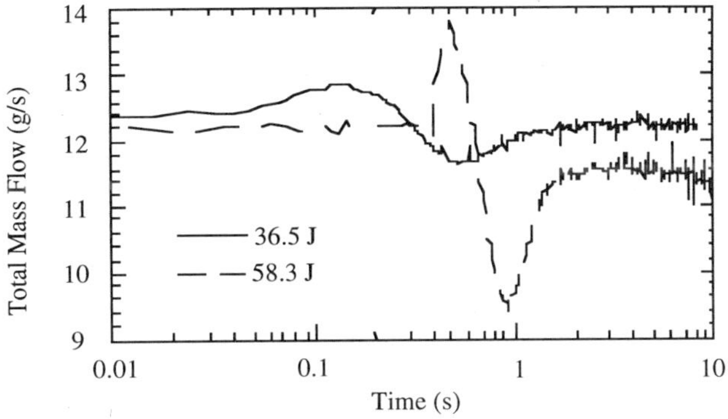

Figure 8. The effect of blocking the central channel. Thermally induced fluctuations in the total mass flow were recorded with the central channel partially blocked by a thermometer array (not present in previous investigations). The flow reduction is not seen with the central channel unobstructed.

CONCLUSION

Hydraulic characteristics of a prototype CEA Cable-in-Conduit Conductor have been studied in steady state and transient conditions. The central channel clearly dominates the characteristics of the conductor. The steady state velocity distribution in the conductor is entirely different between the central channel and bundle regions, with the flow velocity being approximately three times higher in the central channel at high Reynolds numbers. Flow models developed for the conductor agree well with experimental data. Thermally induced flow is primarily generated in the central channel. With the central channel effectively blocked by a thermometer array, thermally induced flows are significantly impacted. An unobstructed central channel clearly has an important role in cooling large superconducting magnet systems

ACKNOWLEDGMENT

Work supported in part by the Department of Energy under grant DE-FG05-93ER54192.

REFERENCES

1. R. Maekawa, M.R. Smith and S.W. Van Sciver, "Pressure drop measurements of prototype NET and CEA Cable-in-Conduit Conductors (CICCs)", IEEE Transactions on Applied Superconductivity, 5:741 (1995).
2. S.W. Van Sciver et al. " AC Loss Measurement of the SMES/CIC Conductor", Cryogenics, 34:615 (1994).
3. L. Dresner, " Heating induced flows in Cable-in-Conduit-Conductors", Cryogenics, 653 (1979).
4. R. Maekawa "Thermal Hydraulic Characteristics Study of Prototype NET and CEA Cable-in-Conduit Conductors (CICCs)", Ph.D Thesis, University of Wisconsin-Madison, 1995.

1995 CEC/ICMC Conference: The Chairs of the 1995 CEC/ICMC Conference: (from left to right) K. Theodore (Ted) Hartwig, ICMC Program Chair; James B. Peeples, CEC Conference Chair; E.W. (Ted) Collings, ICMC Conference Chair; and J. Patrick Kelly, CEC Program Chair.

COUPLING LOSSES OF FUSION CONDUCTORS : INFLUENCE OF SOME DESIGN PARAMETERS

J.L Duchateau[1], D. Ciazynski[1], A. Martinez[1], T. Schild[1], R. Navatel[1], and H. ten Kate[2]

[1]Association Euratom-CEA
Département de Recherches sur la Fusion contrôlée
CE Cadarache, 13108 St Paul Lez Durance Cedex, France

[2]Applied Superconductivity Centre, University of Twente
P.O Box 217, 7500 AE Enschede, The Netherlands

ABSTRACT

Superconductors for fusion are subjected to fast variations of magnetic fields. The effective diameter of the strands is of great importance to limit the level of losses appearing during theses variations. Another less controlled characteristic also plays a major role : the conductor time constant. A model is given to calculate this time constant. The void fraction, the twist pitches associated to the different stages, the chromium coating and the internal resistive barriers inside the strand influence greatly this time constant. The influence of the different parameters is discussed through practical considerations depending on the kind of cable considered. This model is applied to measurements performed at the University of Twente and at CEA on Nb3Sn subcables made of twisted strands relevant for ITER. The application of this model is also discussed for a typical NbTi strand for fusion application developed by GEC AISA. As a matter of fact, the NbTi is again present in the ITER design for most of the poloidal field coils, which represents hundreds tons of material.

INTRODUCTION

The time constant of the ITER conductor and the effective diameter of the filaments govern the losses appearing during the transient phases of ITER such as plasma initiation, plama control and plama disruption. Theses losses lead to an increase of temperature of the conductor which influences the design of the conductor. The effective diameter of the filaments is a well-known and predictable parameter depending on the kind of superconductor chosen. Internal tin process for Nb3Sn allows to obtain effective diameters of about 32 μm corresponding to the ITER specifications for HPI strands (600 mJ/cm^3 non Cu for a ± 3T cycle)[1]. The bronze route process for Nb3Sn leads to effective diameters far lower, typically 12μm corresponding to the ITER specification for HP II (200 mJ/cm^3 non Cu for a ± 3T cycle)[1]. On the contrary, the time constant of the conductor is a far less controlled parameter.

NECESSITY OF A MODEL FOR TIME CONSTANT CONTROL

Experimentally low time constants have been pointed out in the past on full size conductors, giving confidence in their design. The present ITER requirement for the conductor is to reach a value of $n\tau$ of 100 ms for the conductor corresponding to θ=50 ms to make a relation with the parameter used all in this paper and quoted as "the time constant" ($\theta=n\tau/2$). The CEA 40 kA conductors developed in the frame of the NET program, one in NbTi and one in Nb_3Sn (internal tin strand from TWCA) have given time constants in the range 10 to 30 ms[2]. Three Japanese full size conductors tested at CE Saclay have shown time constants in the range 3.5 ms to 65 ms, the higher value being related to a void fraction of about 30 %[3], whereas the ITER specification is 36 %. More recently an ITER benchmark[4] performed on a subsize Nb3Sn (bronze route from VACUUMSCHMELZE) conductor representative of the last but one stage of a typical ITER conductor has again shown low time constants of about 20 ms. On the contrary the extensive and very precise study performed by the University of Twente on 25 twisted samples has pointed out unacceptably high time constants. As an example a twisted Nb3Sn (internal tin) conductor made of 243 strands (void fraction 38 %) has shown a time constant of about 250 ms which is about 5 times the specified value for ITER. The physical origin of these high values values has to be understood very quickly.

THE HOMOGENEOUS MODEL

This model developped by CEA has already been presented[2]. Classically the potential method has been used to calculate the time constant of the elementary strands. This method provides an analytical formula which is in good agreement with the experimental losses, both for NbTi strands and for Nb3Sn strands. The same kind of approach is extended to the cable within an homogenous model. The main assumption of this model is that all the strands and successive stages running side to side inside the conductor experience a uniform contact resistance along the length of the cable. This situation is achieved in conductors where the void fraction is not too high. Under these conditions the strands are pressed against each other very firmly. This assumption can be made for the reference void fraction of ITER : 36 %. For higher void fraction this model cannot be applied and for typical 40 % void fraction conductors the strands are very loose and their movement can easily been observed. The main advantage of this model is that it leads, as for strands, to analytical formula where the influence of the different parameters can be easily identified.
This homogeneous model is particularly applicable for twisted cabled strands, while for braids, the situation is completely different. In Rutherford cables for accelerator magnets the contact resistance has been proven to establish essentially at the crossing of two strands between the two conductor layers. In this case a net model is more appropriate and related to numerical methods.

In the present model, the time constant is introduced by calculating the power per unit of composite volume created by these coupling currents when the regime is established :

$$P = \frac{2\theta\, B^2}{\mu_0}$$ B magnetic field outside the conductor

θ is the time constant of the conductor in relation with coupling current losses.

Under these conditions, the cable ac losses time constant θ can be written as a summation of N time constants θ_n , each of them being associated to the contribution of each cabling stage to the total AC losses (n = 1 for the basic strand) :

$$\theta = \sum_{n=1}^{N} \theta_n$$

In a general way we can write (for n > 1) :

$$\theta_n = \frac{\mu_0}{2\rho_n}\left(\frac{p_n^*}{2\pi}\right)^2 \frac{1}{1-\nu_{n-1}}$$

where p_n^* , ρ_n , ν_n are respectively the effective twist pitch length, the effective resistivity, and the average void fraction of cabling stage n.

We have :

$$p_n^* = p_n - \frac{r_{n-1}}{R_{n-1}} p_{n-1} \text{ , and : } \qquad \rho_n = \frac{\rho_b\, e_b}{\varepsilon_n\, R_{n-1}}$$

where p_n , R_n , r_n , ε_n are respectively the apparent twist pitch length, the outer radius, the twist radius, the contact area ratio of cabling stage n, and $\rho_b.e_b$ is the product (resistivity x thickness) of the contact resistive barrier. When $n = 2$, r_1 is the strand filamentary area radius and R_1 is the strand radius.

Comments on the application of the formula :

Twist pitch p_n

If the cable is cabled with back-twist (no torsion) which is the specification for ITER cables the apparent twist pitch p_n is equal to the cabling twist pitch of stage n, l_n.

contact area ratio (ε_n)

It is very difficult to estimate this value. This has to be done on a very clear micrography of the considered cable. For an accurate evaluation, the section had to be cut very cleanly by electro-erosion. The compaction has a great influence on this parameter which may be influenced by the coating of the strand. Chromium coating for instance is supposed to harden the strands and limit the contact area

the internal barrier ($\rho_b e_b$)

For the last stage of the ITER cable, the barrier is constituted by a 0.1 mm thick stainless steel wrapping. In this case the resistivity and the thickness of the steel strip have to be considered.

For the other stages the only barrier is the resistive barrier surrounding the outermost filament layer of the stage, which means the resistive barrier the coupling currents have to cross to circulate from one strand to another. This resistive barrier is in the case of a classical Nb_3Sn strand the bronze layer which separates the external filaments from the

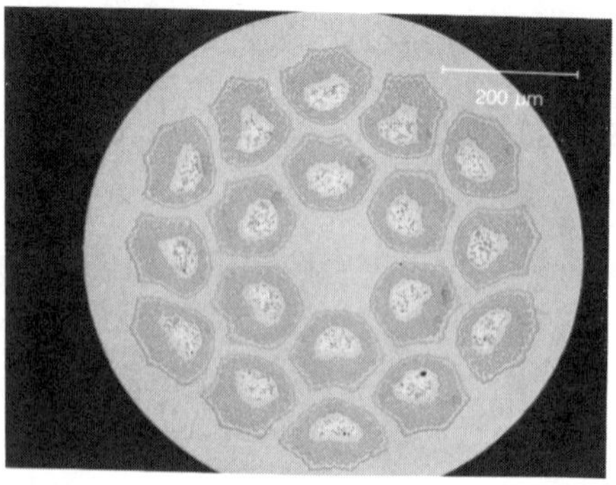

Fig.1 TWCA internal tin strand Φ=0.73mm.

anti diffusion barrier. This resistive barrier is in the case of a NbTi strand a CuNi barrier specially installedand designed to cut the coupling currents.

The importance of this internal barrier becomes crucial because it is the only limiting resistance, when the compaction ratio falls under 36 % in the case of metallic contacts between strands as it is the case for ITER conductors. On the contrary, for the US DPC conductor[6], the cable had a high void fraction and the strands were lubrificated , in this case the internal barrier was not playing any limiting role and the time constant of the cable was practically the time constant of the strand.

When the bronze barrier thickness is very small and if the RRR is very poor the copper shell itself has to be considered to calculate the internal barrier.

The bronze resistivity

The bronze resistivity is directly related to the remaining amount of tin after reaction. In fact this amount is quite large : about 6 % in case of the internal tin route[7]. For this value the resistivity of bronze is certainly around 10^{-7} Ωm[8].

This model has already been applied to the CEA full size NbTi and Nb_3Sn conductors[2] and found to be in good agreement with the experimental results.This exercise is repeated again here on two very different experimental results.

APPLICATION OF THE MODEL TO SUBSIZE CABLES TESTED AT CEA

The experiments have been carried out on 3 subsize cables of different compactions. The model is applied to the cable # 2.

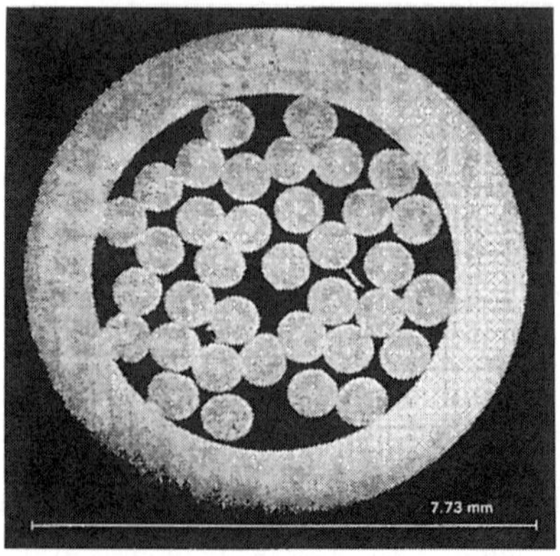

Fig. 2 36 % void fraction 1x3x3x4 twisted cable.

Table 1. Twisted subcables tested at CEA
TWCA strand

Cable sample	type	void fraction	time constant (measured)
#1	1x3x3x4	40%	4.1ms
#2	1x3x3x4	36%	3.7ms
#3	1x3x3x4	30%	12.4ms

Table 2. modeling of cable #2 and contribution of the different stages

stage	ν	p	p^*	ε	time constant (model)
1	-	10 mm	-	-	0.4ms
2	0.	20 mm	10.7 mm	0.17	0.23 ms
3	0.39	40 mm	29.5 mm	0.08	3.02 ms
4	0.37	80 mm	59.5 mm	0.02	6.1 ms

A section of the cable and of the strand is presented in Fig.1. and Fig. 2.

The strand is an internal tin Nb3Sn strand produced by LMI. The thickness of the bronze barrier can be evaluated to 5 μm and the RRR is 40.

Both experimental results and modelzation are presented in Table 1 and 2.

The time constant is evaluated around 9.75 ms which is bigger than two times the experimental value.

APPLICATION OF THE MODEL TO SUBSIZE CABLES TESTED AT UNIVERSITY OF TWENTE

The experiments have been carried out on 25 subsize twisted subcables of different compactions.

The model is applied to the cable # 8.

Table 3. Twisted subcables tested at University of Twente
LMI strand

Cable reference	type	void fraction	time constant (experiment)
#8	1x3x3x3x3	0.28	145 ms
#9	1x3x3x3x3	0.301	150 ms
#10	1x3x3x3x3	0.394	70 ms
#23	1x3x3x3x3	0.431	50 ms
#11	1x3x3x3x3	0.449	34 ms
#24	1x3x3x3x3	0.484	29.5 ms

Table 4. modeling of cable #8 and contribution of the different stages

stage	ν	p	p^*	ε	time constant (model)
1	-	10mm	-	-	4 ms
2	0.	25 mm	18 mm	0.17	1.7 ms
3	0.36	60 mm	47 mm	0.07	17.1 ms
4	0.17	105 mm	74 mm	0.09	65.4 ms
5	0.29	160 mm	110 mm	0.06	202.3 ms

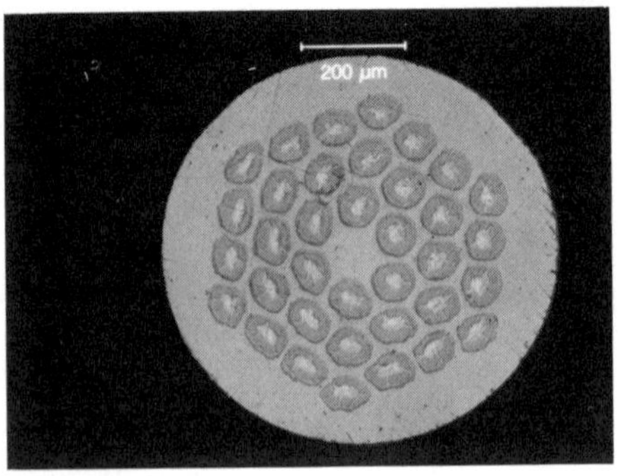

Fig. 3 EM-LMI internal tin strand Φ=0.73 mm.

A section of the cable and of the strand is presented in Fig.1. and Fig. 2.

The strand is an internal tin Nb3Sn strand produced by LMI. The thickness of the bronze barrier can be evaluated to be 1 µm on a very precise micrography centered on a bundle and the RRR is supposed to be 40.

Both experimental results and modeling are presented in Tables 3 and 4.

The time constant is evaluated around 290.5 ms, which is two times bigger than the experimental value.

DISCUSSION ON THE APPLICATION OF THE MODEL TO NBTI STRANDS FOR FUSION

NbTi has now an important place in the ITER project. Most of the poloidal field coils of ITER, in the regions where the field is lower than 5.5 T will be made of NbTi conductors instead of Nb3Sn. The major arguments for this [9], are the cheapness of this material which is 5 times less expensive than Nb3Sn, the impact on the costs of the magnets due to the absence of heat treatment and the possibility of coil replacement in case of failure.
Even if the industrial availability of NbTi has been demonstrated for a long time, the NbTi strand for fusion has to be carefully designed and optimized. For this purpose, CEA has decided to order 10 kg of such a composite to GEC ALSTHOM INTERMAGNETICS.

Again for a NbTi cable conductor the question of the internal barrier is determining.

As the Nb3Sn conductor time constant has an acceptable time constant it has been decided in a first approach to design the internal barrier in this case with a reinforced "$\rho_b e_b$" as compared to Nb3Sn. to limit the losses as the critical current density of NbTi is very sensitive to temperature.

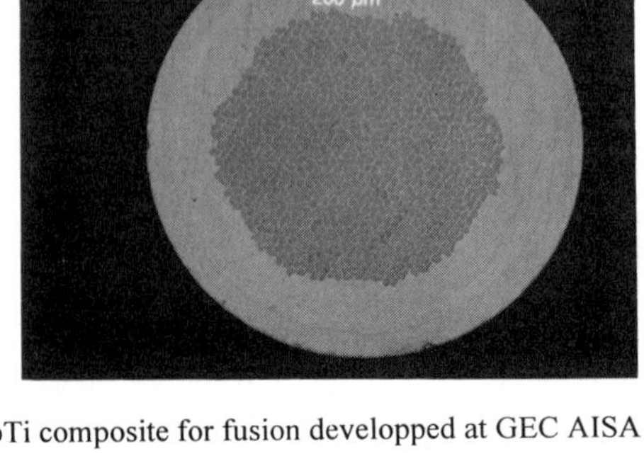

Fig. 4 NbTi composite for fusion developped at GEC AISA ϕ=0.8 mm.

Table 5 Characteristics of a NbTi composite developped by GEC AISA for fusion application.

composite diameter	0.8 mm
NbTi content	34%
CuNi content	6%
Cu content	60%
RRR	110
thickness of the CuNi barrier	15μm

Nb3Sn $\rho_b = 10^{-7}\Omega m$.(bronze) $e_b = 10\mu m$ (depending on the composite)
NbTi $\rho_b = 0.34\ 10^{-6}\Omega m$(CuNi) $e_b = 10\mu m$

The characteristics of the composite are presented in Table 5.
A section of the developped composite is presented in Fig 4.
As concerns the CuNi barrier a particular situation exists for the NbTi in comparison with the Nb3Sn. The position of the CuNi barrier is a free parameter at variance with the case of the bronze barrier. From a discussion with the industry the composite is cheaper if the CuNi barrier is located just in contact with the filamentary zone. On the contrary imposing an external CuNi barrier is certainly more expensive : a larger thickness will be needed for the fabrication and a peeling of the billett will be necessary to come back to the desired thickness. In this case more fundamental studies are needed to assess the influence of the position of the barrier on the value of the time constant. Certainly an external barrier is connected to lower time constants but the problem has to be examined in relation with the current repartition. Equal current repartition is certainly more ensured with an internal CuNi barrier.

Possible application to conductor for accelerators

This kind of internal barrier for NbTi strands is also of great interest for accelerators. In Rutherford cable the question of strand interresistance has always been in discussion in relation with the losses and field errors during the ramp phases. Introducing such a CuNi barrier gives a great opportunity to control and adjust the resistance between the strands.
In general the flexibility of this solution is quite large because the thickness and the percentage of nickel can be adjusted in relation with the design criteria.
The impact of this kind of solution on the conductor costs is very low, the integrity of this barrier along the time is ensured by its position inside the strand (contrary to coating solutions) and moreover a possibility of soldering the conductor appears offering an important mechanical stability which is determining for accelerator magnets.

CONCLUSION

The homogeneous model proposed in this paper gives the good order of magnitude for the time constants observed experimentally on subsize Nb3Sn conductors. It points out the importance of the design of the strand itself on the value of the conductor time constant through the effect of the bronze barrier. Additional investigations and in particular contact resistance measurements are planned on ITER strands to confirm that effect and explain the difference between the model and the experiment. Likewise, in NbTi conductors, the correct design of the CuNi shell offers a way to control both the losses and the current distribution between the strands.

ACKNOWLEDGMENT

These investigations are carried out as part of the NET contract 95/M29 between the European Union and CEA.

REFERENCES

1. ITER report Appendix A Conductor design April 1995.
2. D. Ciazynski, Approximate calculation of losses in a twisted cable made of superconducting composites. Internal CEA report DRFC/TS 41.85.06
3 C Meuris AC loss tests on full size superconducting cables Intermediate report : test of J-1, J-2 and J-3 samples. December 15, 1994 Internal report STEP 304 TC8.
4 P. Bruzzone and al Coupling current losses bench mark test of ITER subsize conductor. Conference on Magnet technology MT 24 Tempere Finland June 1995 to be published.
5 A. Nijhuis and al Parametric study on coupling loss in Subsize NET/ITER Nb_3Sn cabled specimen. Conference on magnet technology MT 24 Tempere Finland June 1995 To be published. Boston 1994.
6 T. A Painter and al. Test Data From the US Demonstration poloidal coil experiment. PFC/RR-92-1
7. Bruzzek Private Communication.
8. S.F Cogan, R. M. Rose Properties of CuSn Bronze at 4.2 K Cryogenics, June 1980, p. 313.
9 J.L Duchateau and P. Hertout Final report on NET Contract CNET 93-308. Assessment of the use of NbTi in the ITER coils.

TWO-DIMENSIONAL FINITE ELEMENT ANALYSIS OF THE STABILITY OF Ag/BSCCO TAPES

E. E. Burkhardt[1] and J. Schwartz[2]

National High Magnetic Field Laboratory
Florida State University
Tallahassee, FL 32306-4005
[1]Department of Nuclear Engineering
University of Illinois at Urbana-Champaign
[2]Department of Mechanical Engineering
FAMU/FSU College of Engineering

ABSTRACT

A two-dimensional finite element model has been developed to explore the stability of high-T_c Ag/BSCCO superconducting tapes operating at 4.2 K and 20 K. This approach allows the broad range of temperature during a transition and the strong temperature dependence of the material properties of both the superconductor and the Ag stabilizer to be considered explicitly. The magnitude of a 5 ms pulse required to quench the conductor is determined for several cases and it is shown that a three-dimensional model is needed.

INTRODUCTION

The stability of superconducting magnets subjected to external thermal disturbances or flux jumps depends on the interplay between the electromagnetic and thermal characteristics of the system. As stability can limit the performance of a superconducting magnet, the stabilities of NbTi and Nb_3Sn conductors have been well studied[1,2]. A recent study determining normal zone propagation in Ag/Bi-2223 superconducting tapes has shown the importance of heat dissipation along the length of the conductor[3], and a one-dimensional study in the longitudinal direction has also been done[4]. Here, we use the finite element method to analyze the stability of a Ag/Bi-2223 superconducting tape. The equations governing the entire system are Maxwell's equations and

$$\nabla \cdot (k(T)\nabla T) + \vec{E} \cdot \vec{J} + Q_{ext} = C(T)\frac{\partial T}{\partial t} \tag{1}$$

where $k(T)$ is the temperature dependent thermal conductivity, $\vec{E}$ is the induced electric field, $\vec{J}$ is the current density, $C(T)$ is the temperature dependent specific heat and Q_{ext} is any external heat source (e.g. a crack in epoxy). The most general case of a high-T_c tape conductor is shown in Figure 1. For this study, the epoxy layer is modeled as a perfect thermal insulator.

Eq. (1) must be solved in each region of the conductor. The boundary conditions are determined by the cooling mode: edge cooled, face cooled, or thermally insulated tapes. Although Eq. (1) describes the thermal behavior of magnetothermal systems in general, the $\vec{E} \cdot \vec{J}$ term explicitly describes the superconductor and its dissipation. For example, for $Bi_2Sr_2Ca_2Cu_3O_x$ tapes operating at elevated temperature and in a significant background magnetic field, flux creep may be of most interest. The voltage in this region will be exponentially increasing with respect to the difference between the operating current and the critical current. In that situation the source term is integrated over the volume into $V \cdot I$ and

$$V \propto \left(\frac{I}{I_c}\right)^n \tag{2}$$

is used, where V is the voltage, I is the operating current, and I_c is the critical current, and n is the index of transition (a higher n-value corresponds to a sharper transition from the superconducting state to the normal state). The anisotropic nature of high-T_c superconductors must also be taken into account when solving Eq. (1).

THERMAL PROPERTIES OF HTS CONDUCTOR CONSTITUENTS

The material thermal properties are temperature dependent over a significant temperature range for HTS stability analysis. As the temperature range of importance is expanded by HTS materials, it is necessary to include the following temperature and field dependent physical properties in the stability analysis:

- specific heat
- thermal conductivity
- magnetoresistivity.

Some of this required data is already available in literature, and other data have been or are currently being measured at the National High Magnetic Field Laboratory (NHMFL). Here we review these properties briefly.

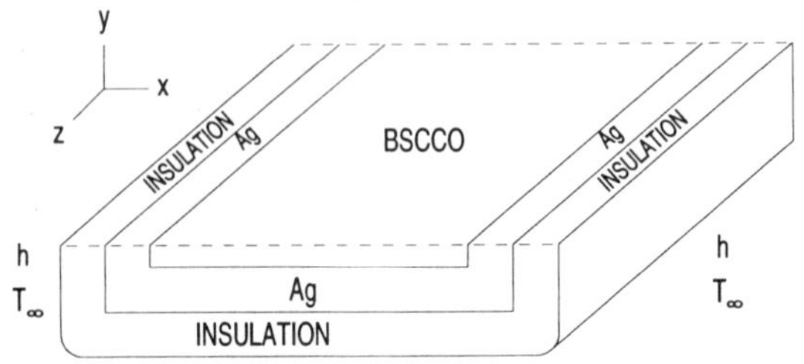

Figure 1. One-half cross-section of a Ag/Bi-2223 superconducting composite tape.

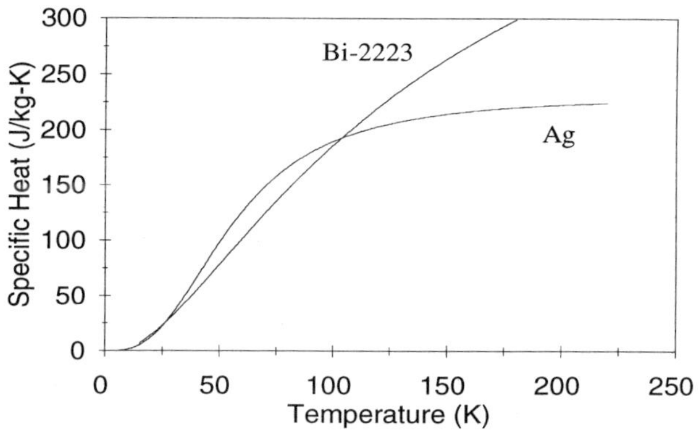

Figure 2. Specific heats of Bi-2223 and Ag.

Specific Heat

The specific heats of $Bi_2Sr_2Ca_2Cu_3O_x$ (Bi-2223)[5,6] and Ag[7] are shown in Figure 2. At 4.2 K and below (where conventional superconductors operate) the specific heat is quite low. As T increases, the specific heat increases rapidly. Thus the temperature dependence becomes more important for HTS systems.

Thermal Conductivity

The thermal conductivity of Bi-2223 (in pellet form) parallel to the a-b-plane is shown in Figure 3[8]. The thermal conductivity of pellets and tapes are expected to be quite different owing to the anisotropy of the crystal structure and varying degrees of grain alignment. This analysis uses the thermal conductivity shown in Figure (3) as the a-b-plane conductivity. The thermal conductivity in the c-direction is estimated as one-fourth of these values, which is approximately the fraction of J_c in the c-direction compared to the a-b-plane. Although the thermal conductivity of Bi-2223 is much lower than that of Ag, it cannot be ignored because heat generated in the Bi-2223 requires a mode of transport to the Ag.

The thermal conductivity of Ag for $0 \leq T \leq 130$ K is shown in Figure 4[9].

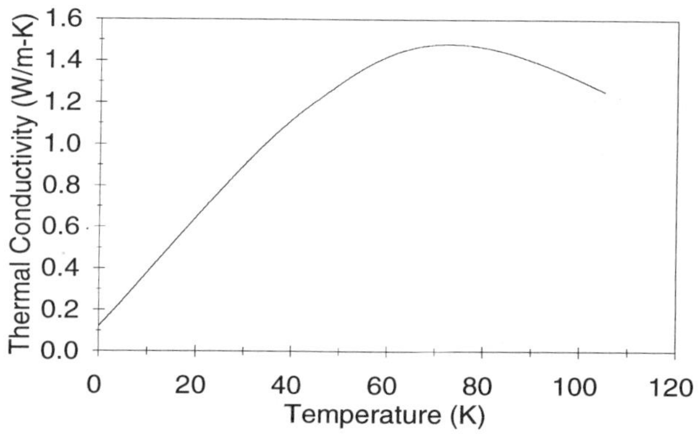

Figure 3. Thermal conductivity of Bi-2223.

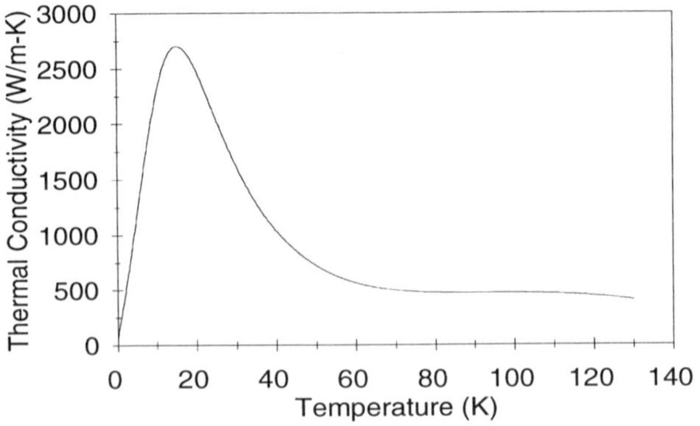

Figure 4. Thermal Conductivity of Ag.

Magnetoresistivity of Ag

Small changes in purity due to the contamination by BSCCO can cause large changes in the magnetoresistivity of Ag. The magnetoresistivity of a silver tape that was processed with $Bi_{1.4}Pb_{0.6}Sr_2Ca_2Cu_3O_x$ has been measured at the NHMFL[10]. The silver tape was peeled off the BSCCO and cleaned with nitric acid to remove excess BSCCO. Figure 5 presents ρ vs. T at 0, 1, 9, 17 and 20 T. The values above 150 K were obtained by extrapolation. The resistivity at 4.2 K was reported to be 2.82×10^{-10} Ω·m at 0 T. The RRR values at various fields were determined to be 105, 97, 33, 21, and 18 at 0, 1, 9, 17 and 20 T, respectively, which are considerably lower than the values reported by Iwasa *et al.*[11]. This is a result of processing in direct contact with BSCCO, whereas the sample used by Iwasa did not have such contact.

The thermal properties shown in Figures 2-5 were fit to analytic expressions using the TableCurve software package. This facilitated their direct inclusion into the stability model.

FINITE ELEMENT METHOD

The finite element method is used to solve (1) in each region of the conductor at each time step. The source is determined using the current, the critical current density,

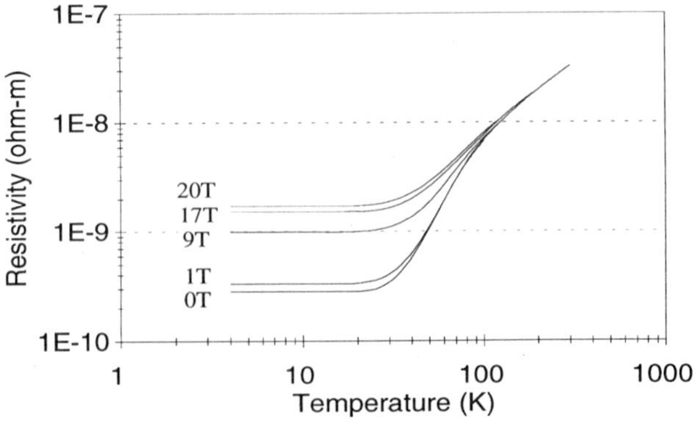

Figure 5. Magnetoresistivity of Ag for H parallel to the tape and perpendicular to the current.

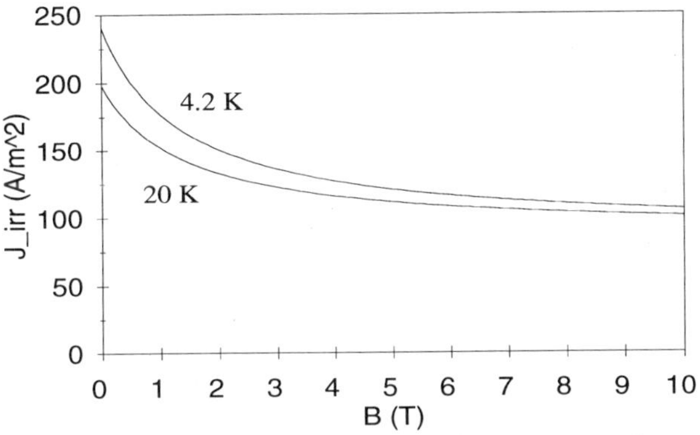

Figure 6. Critical current of Ag/Bi-2223 tape at 4.2 K.

and the resistivity of each element. Due to the strong temperature dependence of the thermal conductivity, the first term on the right-hand-side of (1) does not simplify to $k(T)\nabla^2 T$. This introduces non-linearities and motivates the use of the finite element method instead of finite differencing. Note that the temperature dependence of C_p and J_c are maintained as well. The Crank-Nicolson method (central difference scheme) is then used to analyze the models temporally. Triangular elements are used for this two-dimensional solution. Due to the temperature dependence of the material properties, average values are used for each element.

To illustrate the FEM approach, the stability of the tape shown in Figure 1 is determined. The current propagates in the z-direction and anisotropy of the Bi-2223 occurs in the y-direction relative to the x- and z-directions. The tape considered has a 2.5:1 Ag/BSCCO ratio and is 3.56 mm wide and 0.23 mm thick. At present, the insulation layer shown in Figure 1 is not used (an adiabatic boundary is used), and the resistivity of $YBa_2Cu_3O_7$ (YBCO) is used for Bi-2223.

For this two-dimensional solution, cross-sections in the x-y-plane and the x-z- and y-z-planes through the center of the tape are considered. Starting with the tape at steady-state with a current I operating in a magnetic field B, a heat pulse is introduced to raise the local temperature above T_c. Either a point source or a line source in the stabilizer (Ag) and/or the superconductor can be evaluated. For this study, a line source through the center of the tape is considered. The external source must be large enough to raise the temperature such that the critical current, I_c, becomes less than I locally. I_c for the tape is determined by summing I_c over all the Bi-2223 elements. The current that exceeds I_c contributes to the Joule heating resistively. After determining the current contributing to the heating, $I_s = I - I_c$, the fraction of I_s in each element, δI, is found via Ohm's law. Using the resistance of each element, δR, the elemental source due to the current is $\delta Q_s = (\delta I)^2 \delta R$. The tape will be considered unstable when the temperature rises sufficiently such that the tape remains normal and does not return to the initial state. If a pulse source is used, δQ_s is the only remaining source term after the pulse is finished; however, if a constant heat source is applied, that and the δQ_s coexist.

The critical current density is found for each field at the operating temperature (4.2 K or 20 K, see Figure 6) and the operating current is set to a fraction of I_c. A line source is then applied at the center of the tape, causing the temperature of the Bi-2223 to increase. If the pulse is sufficiently large, I_c will decrease to a value below I, thus initiating the Joule heating and eventually driving the conductor normal. A 5 ms heat

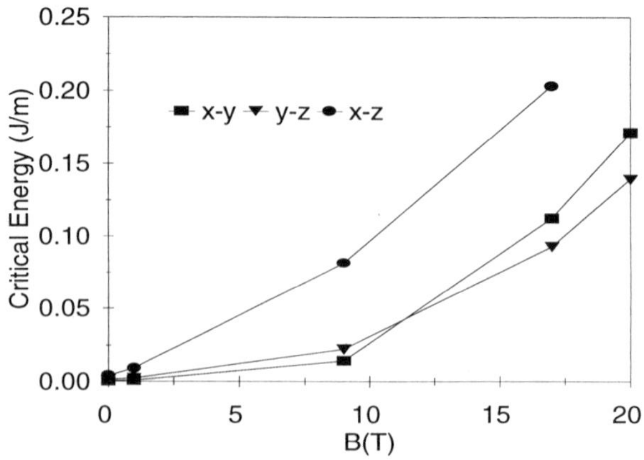

Figure 7. Critical heat pulse values for a conductor operating at 4.2 K, I/I_c=0.98. The heat pulse duration was 5 ms.

pulse is applied, and the magnitude of the pulse required to cause the tape to become critically unstable is determined.

Several cases have been evaluated with I/I_c=0.98 at 4.2 and 20 K (see Figures 7 and 8). The most stable case is the x-z-case, which is expected. Because the aspect ratio of the tape is approximately 15.6:1 and k_{xx}/k_{yy} is 4, the heat propagates through the Bi-2223 in the y-direction much faster than the x-direction. As a result, for a line source at the center of the tape, the temperature will become greater at the center for the x-z-case, which corresponds to an increase in the specific heat which will allow more heat to be deposited with a lower average temperature than in either the x-y- or y-z-cases. Because heat propagation in the y-direction is more significant than the x-direction, the x-y- and y-z-cases should be comparable, and Figures 7 and 8 show this except at high fields, 20 K. At 20 K, the average specific heat is much greater than at 4.2 K, and as a result, the critical energy required to quench is also greater. Figures 9 and 10 show the progression of the temperature of the tape during the pulse for the x-z- and y-z-cases. The slopes of the lines change at the center of the tape for the x-z-case because the line source does not have zero thickness, and the temperature profile for

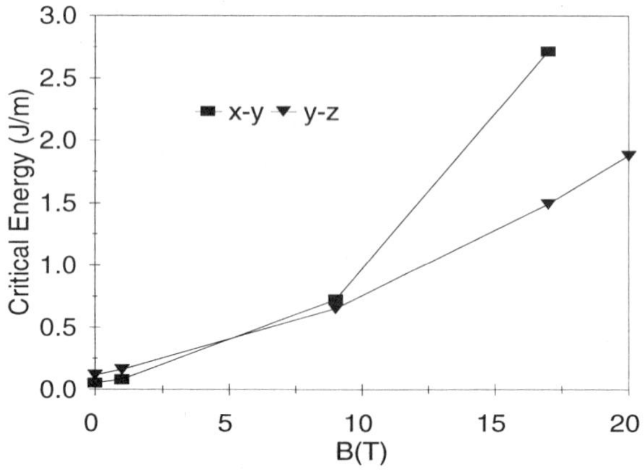

Figure 8. Critical heat pulse values for a conductor operating at 20 K, I/I_c=0.98. The heat pulse duration was 5 ms.

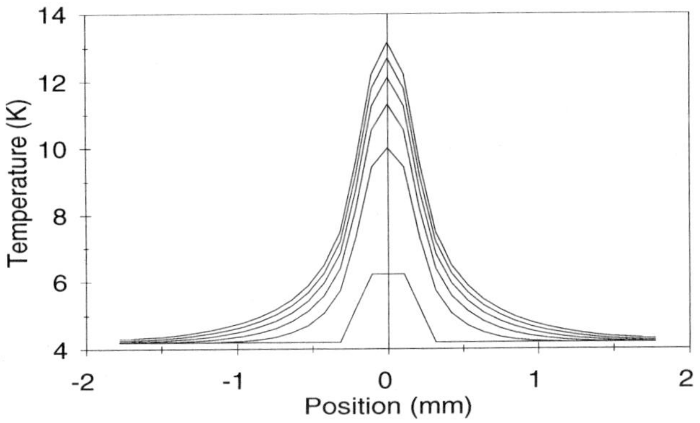

Figure 9. Time progression of the temperature of the tape (T vs. x) for the x-z-case with $B = 0$ T for 4.2 K, I/I_c=0.98 (T is increasing with time) for t:0-5 ms.

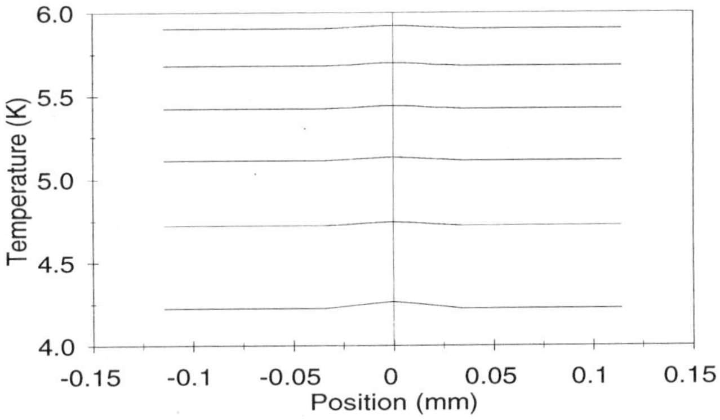

Figure 10. Time progression of the temperature of the tape (T vs. y) for the y-z-case with $B = 0$ T for 4.2 K, I/I_c=0.98 (T is increasing with time) for t:0-5 ms.

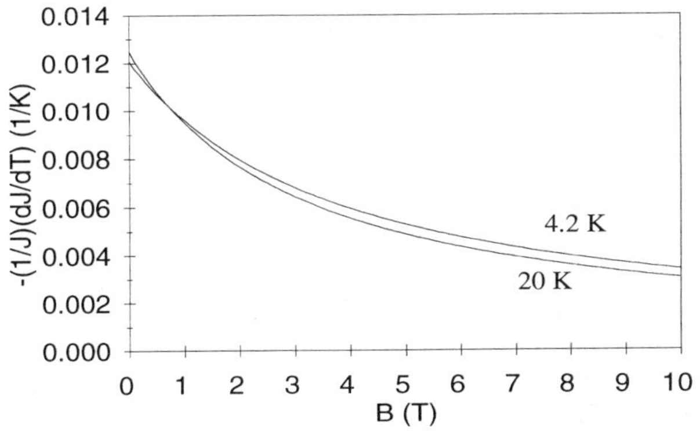

Figure 11. Inverse field-dependence of J_c at 4.2 K.

the y-z-case is almost flat because the heat travels to the Ag and spreads out quickly, whereas in the x-z-case, the heat remains almost entirely in the Bi-2223. For these cases, the limiting heat-pulse magnitude increases with field which is a result of the field-dependence of the critical current density as shown in Figure 11. At higher field the incremental reduction in J_c with increasing temperature is proportionally smaller and less current transfers into the Ag.

SUMMARY

A two-dimensional FEM has been developed to allow temperature dependent thermal properties and the analyses of a variety of geometries readily. Further results with predictive capabilities for developing magnets requires additional measurements of the electrical and thermal properties of the constituents. Further development using three-dimensional models and including different cooling modes will provide a FEM capable of more comprehensive analysis.

ACKNOWLEDGMENT

This work was supported by The Naval Research Laboratory and The Advanced Research Projects Agency. Earle Burkhardt's research was performed under appointment to the MFET Fellowship Program administered by Oak Ridge Institute for Science and Education for the U.S. DOE.

REFERENCES

1. S.Y. Seol and M-C. Chyu, Stability analysis for a composite superconductor subjected to a linear thermal disturbance, *Superconducting Science Technology*, 7:841-848 (1994).
2. A. Ünal and M-C. Chyu, Three-dimensional stability analysis for quenching and recovery of superconductor tapes or films, *International Symposium on Safety of Superconductors and Related Heat Transfer at Low Temperatures*, August 28-31 (1994).
3. S.B. Kim, Y. Ueno and A. Ishiyama, Experiment and numerical analysis of normal zone propagation properties in Ag sheathed Bi-2223 superconducting tapes, submitted to *IEEE Transactions on Magnetics* and presented at the MT-14 conference.
4. R.H. Bellis and Y. Iwasa, Quench propagation in high-T_c superconductors, *Cryogenics*, 34:129-144 (1994).
5. S.J. Collocott, R. Driver and C. Andrikidis, Specific heat of the ceramic superconductor $Bi_2Sr_2CuO_8$ from 0.4 to 20 K, *Physica C*, 173:117-124 (1991).
6. J.E. Gordon, S. Prigge, S.J. Collocott and R. Driver, Specific heat of $(BiPb)_2Sr_2Ca_2Cu_3O_{10-y}$ from 78 to 300 K, *Physica C*, 185-189:1351-1352 (1991).
7. L. Dresner, Stability and protection of Ag/BSCCO magnets operated in the 20-40 K range, *Cryogenics*, 33(9):900-909, (1993).
8. A. Jezowski, Thermal conductivity of Bi-based high-T_c superconductors in the vicinity of superconducting transition," *Solid State Communications*, 75(10):779-783 (1990).
9. R.W. Powell, C.Y. Ho and P.E. Liley,*Journal of Physics and Chemistry*, Ref. Data 3, Suppl. 1, Vol. 606 (1974).
10. S. Nakamae, J. Schwartz, Magnetoresistivity of Ag-tape co-processed with $Bi_{1.4}Pb_{0.6}Sr_2Ca_2Cu_3O_x$ superconductor, submitted to *Cryogenics*.
11. Y. Iwasa, E.J. McNiff, R.H. Bellis and K. Sato, Magnetoresistivity of silver over the temperature range 4.1-159 K, *Cryogenics*, 33(8):836-837 (1993).

A WAY TO INCREASE THE CRITICAL CURRENT IN THE STAINLESS STEEL JACKETED Nb_3Sn CONDUCTORS FOR FUSION

P.Decool[1], J.L.Duchateau[1], W.Specking[2]

[1]Association Euratom-CEA
Département de Recherches sur la Fusion controlée
CE Cadarache, 13108 St Paul Lez Durance Cedex, France

[2]Forschungszentrum Karlsruhe, Institut fur Technische
Physik, Postfach 3640, D-76021 Karlsruhe, Germany

ABSTRACT

The critical current density of Nb3Sn under magnetic field is highly dependent on the intrinsic strain ε, having a maximum value for ε=0 The performance of stainless-steel jacketed conductors is usually lowered by the differential thermal expansion between steel and the superconducting filaments (J is divided by 2 for B=12T, T=5.5K, and ε=-0.7%). This is due to the longitudinal compression appearing on the superconducting filaments during the cool down from the reaction temperature to 4K. To overcome this difficulty, in the reference design of ITER (International Thermonuclear Experimental Reactor), the conductors are jacketed with Incoloy. A progressive test program is foreseen to demonstrate the possibility to reduce the filament compression in conductors jacketed with A316 LN. This process is based on the use of a low thermal expansion material for the supporting tools during the reaction phase. The performance level of such a reacted conductor could, in principle, be similar to a conductor jacketed with Incoloy.

INTRODUCTION

At high fields,NbSn critical current density J is highly dependent on the strain ε of the superconducting filaments. The critical currents of the stainless-steel jacketed conductors are usually very degraded in comparison with the optimum value corresponding to a zero strain of the Nb3Sn (a factor of 2 for B=12T, T=5.5K, ε=-0.7%). The well known reason of this degradation is the longitudinal compression of the superconducting filaments during the cooldown from the heat treatment temperature to 4K. The high price of Incoloy and the uncertainty bound to its industrialization have lead us to see a process capable of limiting the

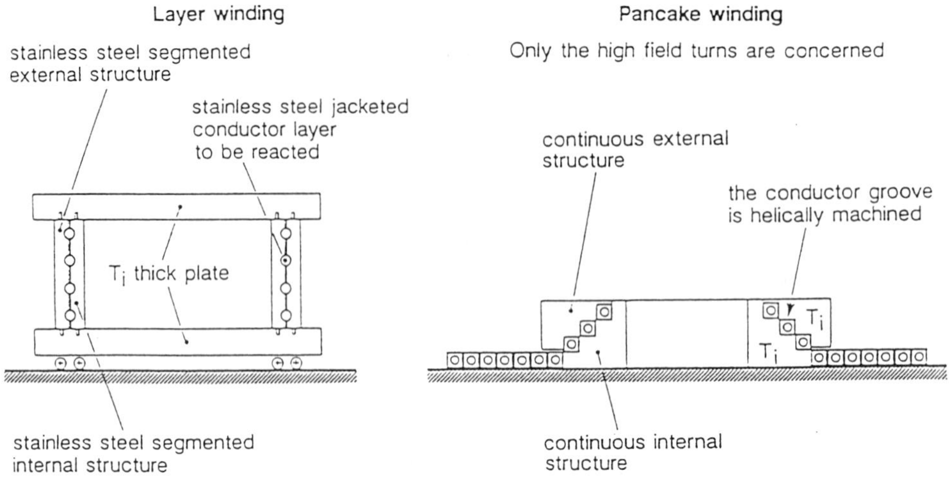

Figure 1

effects of the differential thermal contractions between the Nb3Sn and the more common alloys.

This paper describes a control process to control the longitudinal deformation of Nb3Sn during the temperature variations. This process, applicable to the full size ITER superconducting coils, could limit the current degradation to the same level as the Incoloy jacketed conductors. An associated experimental progressive program is presented.

PROCESS

The main idea is to limit the conductor deformations during all the heat treatment cycle 300K-923K-300K by using tooling of a low expansion material, such as Invar or Titanium, in order to minimize the Nb3Sn strain from 923K to 300K. Fig.1 shows such toolings applied to pancakes or layer windings.

In these two cases, the tooling cross section must be bigger than that of the jacket to make sure that:

a- the tooling deformation is neglectible comparing to that of the jacket

b- the tooling remains elastic even at 923K.

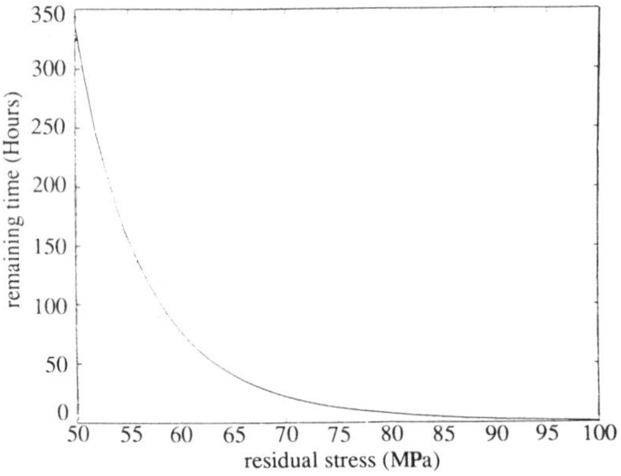

Figure 2. Creep effect at 600°C on a 316L steel for an imposed $\varepsilon = 0.6\%$.

The tooling imposes displacements during the temperature increase and at the 923K plateau during which the NbSn phase appears. The jacket is plastified under the compressive stress imposed by the tooling and this stress is partially reduced during the 923K plateau due to creep. Fig 2 shows the creep effect at 600°C for 316L steel, corresponding to a relative deformation ε representative of the differential thermal expansion for steel/titanium i.e 0.6%. After 200 hours, the last plateau duration, the residual stress in the jacket is about 55MPa.

During the cooling down to the room temperature, the jacket first recovers this compressive stress then is put into tension by the same low-thermal-expansion tooling. At room temperature, the tooling is removed and the conductor contracts to a value corresponding to the relaxation of the tensile stress in the jacket.

Such a process is quite passive and can be applied whatever the shape of the winding, and in particular for D shape windings as in the ITER Toroidal Field coils.

THEORETICAL MODEL

The estimation of the filaments strain ε_f, between 923K and 4K is composed of three parts $\varepsilon_f = \varepsilon_{1f}+\varepsilon_{2f}+\varepsilon_{3f}$ with :

ε_{1f} : strain during the cool down from 923K to 300K
ε_{2f} : strain due to the tooling removal at 300K
ε_{3f} : strain during the cooling down from 300K to 4K.

Our hypothesis is that the jacket is totally plastified during the last heat treatment stage leaving no residual stress at 923K.

The calculation of ε_1 and ε_3 is based on the model presented by Steeves[1].

From 923K to 300K, the NbSn filaments and the tooling are fully bonded and can be considered to have the same strain. By assuming the superconducting filaments and the tooling work in their elastic range,

$$\alpha_{tool}.\Delta T + \frac{\sigma_{tool}}{E_{tool}} = \alpha_f.\Delta T + \frac{\sigma_f}{E_f} \tag{1}$$

The force balance at 300K can be written:

$$A_{ss}.\sigma_{yss} + A_f.\sigma_f + A_{tool}.\sigma_{tool} + A_{cu}.\sigma_{ycu} + A_{bz}.\sigma_{ybz} = 0 \tag{2}$$

with the notations:

A_{cu} , copper area
A_{bz} , bronze area
A_f , filaments area
A_{ss} , stainless steel area
A_{tool} , tooling area
E_{tool} , tooling young's modulus at 300K
E_f , filament young's modulus at 300K
σ_{ycu} , copper yield stress at 300K
σ_{ybz} , bronze yield stress at 300K
σ_{yss} , steel yield stress at 300K $\rightarrow$ depending on the material work hardening

and by using Eq (1) and Eq(2):

Table 1

Conductor	TF	Sample tested at FZK
Strand diameter	0.78 mm	0.73 mm
non copper area (twisted)	199.7 mm^2	7.79 mm^2
total copper area (twisted)	580.6 mm^2	7.8 mm^2
jacket area	274.5 mm^2	21.0 mm^2

$$* \quad \varepsilon_{1f} = \frac{A_{tool} \cdot E_{tool} \cdot \left(\frac{\Delta L}{L_{tool}} - \frac{\Delta L}{L_f} \right)_{923 \rightarrow 300K} - A_{ss} \cdot \sigma_{yss} - A_{cu} \cdot \sigma_{ycu} - A_{bz} \cdot \sigma_{ybz}}{A_f \cdot E_f + A_{tool} \cdot E_{tool}}$$

* $\varepsilon_{2f} = -\sigma_{yss} / Ess$ (stress relaxation due to the tool removing)

* $\varepsilon_{3f} = -0.15\%$ $\rightarrow$ almost constant value for any considered solution (plates, casing, or jacketed conductors). This value represents the differential thermal contraction between steel and NbSn filaments from 300K to 4K.

APPLICATION

ITER Toroidal Field conductor

An application of this process has been considered in a first approach for the TF EDA conductor. The very reduced section of jacket of this conductor allows to use a limited section for the tooling.

The basic conductor used for the ITER TF coils is a circular cable-in-conduit made of NbSn strands. We take for example the TF conductor using High Performance 1 reference strands[2]. The main parameters are presented in Table 1.

We consider a low expansion material such as Titanium for the tooling. With a pessimistic value σ_{yss} of 550MPa at 300K, corresponding to a highly work hardened 316LN steel , the compressive stress on the filaments at 4K should be ε_f = **-0.49%** (-0.34 + -0.15). By supposing that ε_{3f} is compensated by the TF coil hoop strain, the strain to be used for the conductor dimensional calculations is $\varepsilon_{1f}+\varepsilon_{2f}$ then -0.34%. The calculation for Incoloy gives the same order of magnitude.

Typical subsize conductor used for testing

The sample is a NbSn subsize cable-in-conduit conductor used for experimental parameter study[1]. The main parameters are presented in Table 1.

We consider the use of a low expansion material such as Titanium for the tooling., the same yield stress : 550MPa σ_{yss} as in the previous paragraph induces at 4K a filament strain ε_f = **-0.49%**. This value is to be compared to ε = -0.7% actually measured on stainless steel jacketed samples manufactured with a normal heat treatment process. The critical current gain should be about 40% and could contribute to demonstration of the process.

EXPERIMENTAL RESULTS

The suggested process is dependent on the steel characteristics during the last 923K stage. In a first step, it appears necessary to characterize the jacket behaviour before starting applications with real superconducting cables.

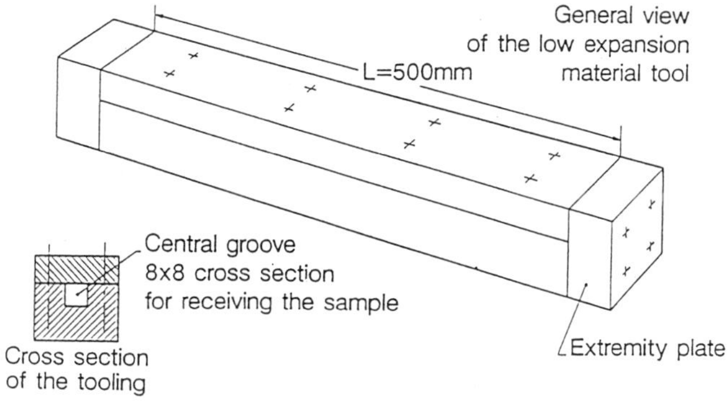

Figure 3. Low-expansion-material tooling .

We have started preliminary tests on straight samples in order to determine the residual stress and strain in the steel at 923K and 300K.

The tooling used is made of a low expansion material with a large square section and cut with a central groove to receive the sample. Steel plates are screwed on the two extremities to clamp the sample expansion (Fig.3). Two low-expansion materials have been tested for the tools: titanium (TA6V) and Invar.

A sample is made of a 316L 8mm inner diameter and 1mm thick tube inserted into the tool groove and adjusted in length. There are two kinds of samples (Fig.4):

Sample without shoulders where the thermal expansion during the heating from 300K to 923K is blocked by the tooling and the thermal contraction from 923K to 300K is free.

Sample with shoulders where both thermal expansion and contraction are blocked by the tooling during the heating and the cooling down.

For each test, the sample and the tool are measured before and after a heat treatment of 176 hours at 923K under pure Argon atmosphere.

From the tests on Sample without shoulders we obtain:

-The plastic strain of the steel at 923K

$$\varepsilon_{pl.923} = (L_1 - L_0) / L_0$$

L_0 : initial length
L_1 : final length

-The yield stress of the steel at 923K (reduced by the creeping at high temperature)

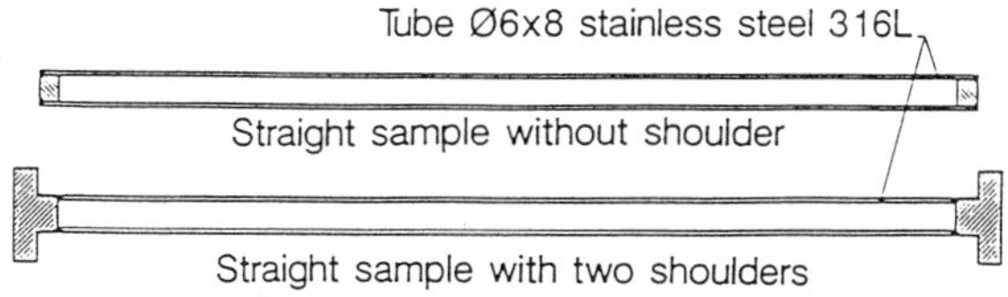

Figure 4. Basic straight samples .

Table 2

	SS jacket freely reacted	SS jacket reacted with Ti or Invar tooling	SS jacket reacted with TZM tooling	Incoloy jacket
ε_f (estimated working value)	-0.54%	-0.34%	-0.08%	-0.34%

$$\sigma_{ss.923} = \frac{\left(\frac{\Delta L}{L_{tool}} - \frac{\Delta L}{L_{ss}}\right)_{923\rightarrow 300\,K} - \frac{(L1 - L0)}{L0}}{\frac{1}{E_{ss\,923\,K}} + \frac{A_{ss}}{A_{tool} \cdot E_{tool\,923\,K}}}$$

From the tests on Sample with shoulders we obtain:

-The plastic strain of the steel at 300K

$$\varepsilon_{pl.300} = \frac{(L_1 - L_0) - (L_2 - L_0)}{L_0} = \frac{L_1 - L_2}{L_0}$$

L_2 : final length after dismantling of the tool

-The yield stress of the steel at 300K (reduced by the creeping)

$$\sigma_{ss.300} = \frac{\frac{L_2 - L_0}{L_0}}{\frac{1}{E_{ss\,300\,K}} + \frac{A_{ss}}{A_{tool} \cdot E_{tool\,300\,K}}}$$

And we obtain with an Invar tool : $\varepsilon_{pl.923} = -0.41\%$

$\sigma_{ss.923} = 28MPa$

$\varepsilon_{pl.300} = -0.17\%$

$\sigma_{ss.300} = 316MPa$

The theoretical differential thermal expansion ss/tool $\left(\frac{\Delta L}{Ltool} - \frac{\Delta L}{Lss}\right)_{923\rightarrow 300\,K} = -0.43\%$ is nearly equal to the value of $\varepsilon_{pl.923}$ and the residual yield stress at 923K is very small. These results confirm the hypothesis of total steel plasticity during the 923K stage.

The 300K yield stress of 316MPa is lower than the parametric value 550MPa which should give a 0.2% strain gain on the superconducting filaments. So, we have an important plasticity at 300K with this 316L steel.

The results with the titanium tooling have been perturbed by the plasticity of titanium during the 923 K plateau. The titanium alloy TA6V seems not usable for this process.

TOOLING MATERIAL

We have seen that the results of this heat-treatment process depend on the quality of the tooling:

Table 3. TF coil reduced gain brought by an Incoloy jacket (in case of a 2 mm thick jacket) (B=12.5T,T=4.75K)

	stainless steel jacket	Incoloy jacket
ε923K→300K	-0.535%	-0.34%
ε300K→4K	-0.15%	-0.15%
εhoop	+0.12%	+0.12%
ε(working assumption)	-0.5%	-0.3%
ΔT margin	2K	2K
safety factor(stability)	0.95	0.95
hot spot temperature	155K	155K
current density in the cable	43.8A/mm^2	46.2A/mm^2

-high yield stress at 923K
-high Young's modulus
-thermal expansion as low as possible between 300K and 923K

Molybden alloys, such as TZM, fullfill these conditions and give a supplementary increase of the critical current by decreasing the strain value. Table 2 give a comparison between the estimated working values of the filaments strain on the TF conductor for different cases:

FUTURE EXPERIMENTS

The partial results on straight steel samples confirms the model. This leads us to proceed farther with a test program to qualify this heat treatment process. This program will include:

-additional tests on a straight sample made with 316LN steel and use of better low-expansion-material tooling such as Molybden alloy.

-characterization of actual 316LN jacket by carrying out tensile tests on the jacket at 300 K and 923 K.

-comparative tests on superconducting subsize straight sample like experimental using superconducting Nb3Sn subsize cable steel jacketed. One will be freely reacted and the other will be blocked in a low expansion material tool during its heat treatment.

-final comparative tests a one layer coil using (1+2)×3×4 subsize cable steel jacketed (316L or 316LN). In order to have the capability to test such a coil, the number of superconducting strands will be limited to 12. The other strands will be made of copper. This coils could be tested in the ALINE facility (CE Cadarache) which offers a solenoid with a 300mm bore and a 9T background field. The tests will be done by increasing the helium temperature to reach the critical current and emphasize the strain influence.

TF COIL SYSTEM

As concerns the advantage of Incoloy vs conventional stainless steel, even with no special heat treatment, the particular case of the TF coil has to be considered.
As a matter of fact this advantage is not so determining for two reasons :
- the action of the differential thermal contraction has to be evaluated and compared for the two materials only between 923 K and room temperature and the full differential thermal contraction does not apply due to the limited section of the jacket (2 mm thick).
- the differential thermal contraction between room temperature and 4 K is the same because it is then governed in any case by the stainless steel of the coil plates.

CONCLUSION

The process proposed for the reaction of stainless steel jacketed conductors theoretically allows reduction of the final compressive stress of the superconducting filaments inherent to the use of steel. The strain increase should be greater than 0.2 % corresponding to a critical current gain of about 40 %. Bettre results can even be envisaged by use of low expansion material such as molybden alloys. the first experimental results confim the validity of the model. The next steps of this demonstration program will be to confirm the great interest of classical stainless steel for ITER conductors.

REFERENCES

1. Improvement of I in NbSn conductors by reduction of axial prestrain. ASC 1994- Oct. 16-21 Boston.
2. ITER report Appendix A Conductor design April 1995.
3. Consideration on the NbSn in relation with the different components of ITER conductors- Internal CEA report PEM 94.23

ON THE THERMAL STABILITY LENGTH DEPENDENCE OF HIGH T_C SUPERCONDUCTORS

B-Z. Maytal[1], S. Yang, J. Waldrop, and J.M. Pfotenhauer

[1]on leave from Rafael and Technion-Israeli Institute of Technology
Haifa, 32000 Israel

Applied Superconductivity Center
University of Wisconsin - Madison
Madison, WI 53706, U.S.A.

ABSTRACT

The margin of stability for high T_C superconductors may be characterized by the lowest thermal pulse that results in a quench. A numerical code, developed to investigate this stability margin, accounts for the temperature dependence of the thermal conductivity, electrical resistivity and heat capacity of the materials involved. The conductor is cooled solely by thermal conduction along its length, and its stability is studied as function of the length. It is found that the stability margin for a pure BSCCO 2212 conductor is independent of the length. However, the stability margin for a composite conductor obtained by adding 10 % (by volume) of silver to the BSCCO is found to be strongly length dependent. A transition length is identified, for which shorter lengths exhibit a dramatically higher stability margin. This feature results when the length dependent thermal diffusion time is shorter than the time required to determine the conductor's stability. This study confirms and explains the growing awareness that if normal zones appear in high T_C coils, they will remain fairly localized.

INTRODUCTION

The stability of high T_C superconductors has received significantly less attention than that of conventional superconductors. For the high T_C superconductors the primary focus has been on determining the normal zone propagation velocity, v, and quench protection. Dresner[1] established and solved an idealized model which includes temperature dependent thermal conductivity, k, heat capacity (per unit volume), C, and electrical resistivity, ρ. He found that v is remarkably smaller than that of low T_C conductors. Additionally, Dresner evaluated the energy pulse, E, required to initiate a quench. However, this analytic and closed form similarity solution refers to an infinitely long conductor and lacks a characteristic length dependence. Iwasa[2] has experimentally investigated normal zone propagation in high T_C superconductors and measured its velocity. In addition, Bellis and Isawa[3] have developed a one dimensional numerical model (averaging the properties of the superconductor and silver) which validates their results. With respect to the present study, we note regrettably that an investigation into the influence of the length between the heat source and sink was not included in their work. Finally we note that Soel and Hull[4] have analytically determined burnout times for superconducting elements in current leads and demonstrate a length dependence. The present work studies the issue of stability for high T_C superconductors from an alternate approach; that is, by numerically determining the maximum non-quenching thermal pulse which can be applied at the warm end of an HTS element as function of its length, while its other end is clamped at a cold temperature. The results for a length of pure BSCCO 2212 have relevance for current lead applications. However, the more significant results may be observed for the case of the silver stabilized conductor, where the length between hot spots and cold sinks significantly impact the stability of the conductor.

NUMERICAL MODEL

A numerical model has been developed using the ANSYS code[5] and is configured to produce time dependent temperature profiles along the length of the superconductor. In the initial state the conductor has a stable temperature profile with the cold and warm ends clamped at T_L (usually set to 10 K) and T_H (usually set to 77 K) respectively, while it conducts an electrical current density of $J = 700$ A/cm^2. This current represents a value just below a typical critical current density for BSCCO 2212 at 77 K and in a low magnetic field of B = 0.1 tesla[6, 7, 8]. Such a situation can occur for example when an HTS current lead is conductively cooled by a mechanical (GM) cooler[9 - 15].

At time t = 0 seconds, a pulse of 0.5 s duration is applied at the warm end, and the boundary condition for the warm end is changed from the clamped temperature of T_H to an adiabatic condition such as would exist at the center of a normal zone. Cooling is provided solely by thermal conduction to the lower temperature end at T_L. The size of the pulse is continuously increased until a critical value, E_{max}, is reached, which quenches the conductor. Equivalently, this is the lowest non-recoverable pulse, or minimum quench energy. The round BSCCO 2212 rod has a cross sectional area of 0.25 cm^2. In the silver stabilized version, the added cross sectional area due to the silver matrix which surrounds the BSCCO rod is only 10 % of the original area.

The model is ruled by the transient Fourier equation

$$C(T)\cdot\frac{\partial T}{\partial t}=\frac{1}{r}\cdot\frac{\partial}{\partial r}\left(k\cdot r\cdot\frac{\partial T}{\partial r}\right)+\frac{\partial}{\partial z}\left(k\cdot\frac{\partial T}{\partial z}\right)+J^2\cdot\rho(T) \qquad (1)$$

on a two-dimensional, cylindrically axisymmetric domain. The last term on the r.h.s. represents Joule heating in the normal state, with the current density and resistivity based on the silver area for the stabilized case. In the model, the resistivity of the silver matrix has been modified within a lookup table to reflect the temperature dependence of the BSCCO resistivity[17]. Thus, below T_c, the silver resistivity is artificially set to zero, and heat generation occurs for both the bare and the stabilized versions only when the local temperature exceeds T_c. The equations defining the initial conditions, and the boundary conditions after applying the pulse are respectively

$$T(x, t = 0) = f(x) \text{ and } T(x = 0, t = 0) = T_H. \qquad (2a)$$

$$T(x = L, t) = T_L \text{ and } \partial T/\partial z(x=0, t) = 0. \qquad (2b)$$

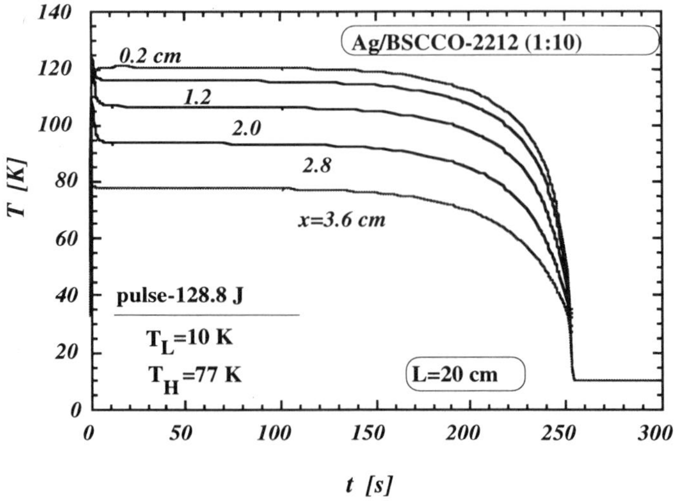

Figure 1. Evolution of temperature profiles of silver clad BSCCO-2212 in recovery situation.

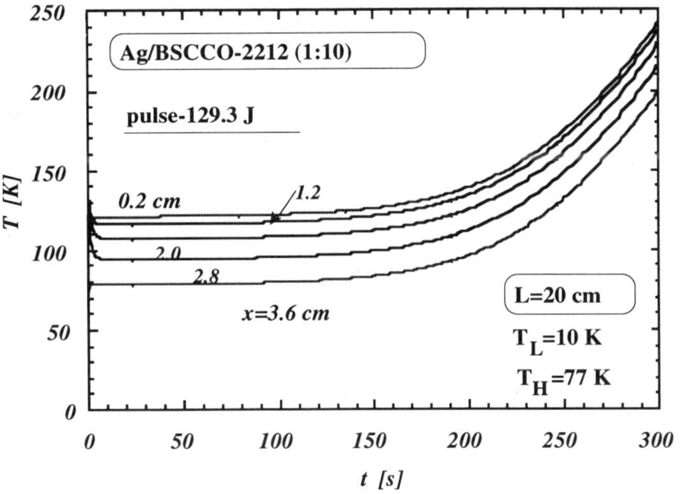

Figure 2. Evolution of temperature profiles of silver clad BSCCO in non-recovery situations.

Eq. (1) is solved separately in the BSSCO 2212 and silver domains and coupled by the identical temperature at their interface. In each domain the temperature dependent properties are included. To date the model has been used primarily to determine the stability boundary values, E_{max}, as a function of conductor length, L.

RESULTS

Maximum energy disturbance

The distinction between recovery conditions and non-recovery conditions allow a very sensitive determination of E_{max} as is demonstrated in figures 1 and 2. A pulse of 128.8 J is results in a recovery, while 129.3 J quenches a 20 cm long Ag/BSCCO sample. The distance along the conductor from the hot end where the pulse is applied is represented by the parameter x. The pulse was applied over two different lengths of 6 mm and 28 mm at the hot end with no difference to the results.

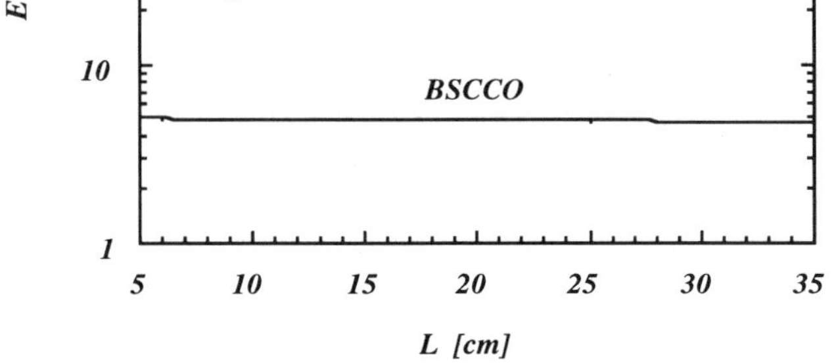

Figure 3. Dependence of minimum quenching energy on length between heat source and cold sink.

Figure 3 displays the length dependence of the lowest quenching pulses, E_{max}. Three conclusions can be drawn from the results:
(a) The stability of silver stabilized BSCCO-2212 composite conductors is strongly dependent on length, while for pure BSCCO conductors, such a dependence is not noticeable at all.
(b) Pure BSCCO-2212 conductors are less stable by more than one order of magnitude then the composite conductors.
(c) A thin stabilizing silver layer, even with as small of an area ratio as 1:10 adequately improves the stability. Conductors with larger silver content, such as are typically fabricated for magnet conductors, are expected to allow much larger pulses of energy without quenching.

Time to determine stability

The time at which the stability of the conductor is clearly evident also is significantly different for the two types of conductors investigated in this study. As is shown in figure 4 for the pure BSCCO-2212 conductor, stability is determined at approximately 1 second after initiation of the thermal disturbance. It is noteworthy that this is orders of magnitude larger than typical values for low T_c conductors. However, the results with the composite conductor are significantly larger than this, being closer to 50 s. Thus, the addition of the silver provides a much longer time for the conductor to respond to thermal disturbances.

DISCUSSION

The length independent value of E_{max} for the pure BSCCO-2212 conductor may be understood with reference to the process of thermal diffusion. Pure BSCCO-2212 exhibits a low thermal diffusivity, α, in comparison to silver. For example, at 80 K the values for the respective materials are 1.92 x 10^{-2} and 2.7 cm^2/s. During an elapsed time t , the thermal diffusion length, L_α, grows to a value on the order of magnitude of $\sqrt{\alpha \cdot t}$. Within the half second of the pulse duration, thermal energy is spread over 12 mm in the silver matrix, and by the end of the 50 seconds (at which time the stability is determined) the energy is spread over a distance of 12 cm . However, for the pure BSCCO-2212 sample the thermal energy is spread a distance of only 1.4 mm at the 1 second time at which the stability is determined.
One recognizes from these considerations that in pure BSCCO-2212, the released pulse creates a concentrated and local disturbance. Therefore, the transition to the normal state temperature, T_c, is quickly obtained. Since the thermal diffusion length, L_α, associated with the 1 second stability time is much shorter than any of the investigated lengths between the hot and cold ends, the sample experiences no benefit from cooling. In other words the hot zone does not "see" the cooling before the stability issue is determined. In this situation, the energy required to initiate the quench, E_{max},

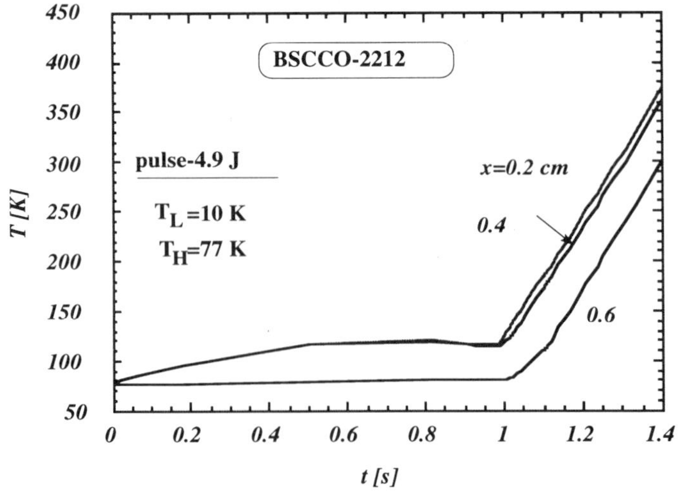

Figure 4. Evolution of temperature profile in pure BSCCO 2212 sample subsequent to minimum quench energy disturbance.

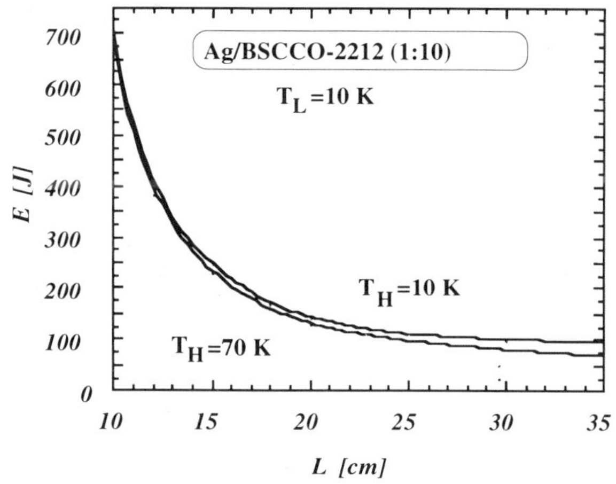

Figure 5. Demonstration that dependence of minimum quench energy on length between hot source and cold sink is insensitive to initial hot end temperature.

is only that required to raise the amount of BSCCO-2212 within the volume $\pi r^2 L_\alpha$ from the initial temperature T_H up to T_C. Indeed, our calculations of enthalpy increase based on the equation:

$$E = \delta \pi r^2 L_\alpha \int_{T_H}^{T_C} C_p dT \qquad (3)$$

result in an energy on the same order as that shown in Fig. 3. Here δ represents the material density. It can be further noted that all thermal disturbances larger than that defined by Eq. (3) in the pure BSCCO-2212 sample, and which are farther away from the heat sink than a distance L_α will initiate a local and rapid thermal run-away. The rapidity of the thermal runaway for the pure BSSCO sample is amplified by the fact that the resistivity of BSCCO-2212 is higher by a few orders of magnitude than that of silver.

Since the stability margin of pure BSCCO-2212 is set by the enthalpy gap between the high temperature and T_C, it is quite sensitive to the initial value of T_H. In contrast, the composite sample of Ag/BSCCO-2212 is not sensitive to T_H at all. The case of low and uniform initial distribution, $T_H = T_L = 10$ K is compared to a case of T_H =70 K and the same T_L = 10 K for a sample of 20 cm long. As shown in Figure 5, stability boundary for both cases are very similar. However, stability is still dependent on the cold source temperature, T_L, as shown in Figure 6.

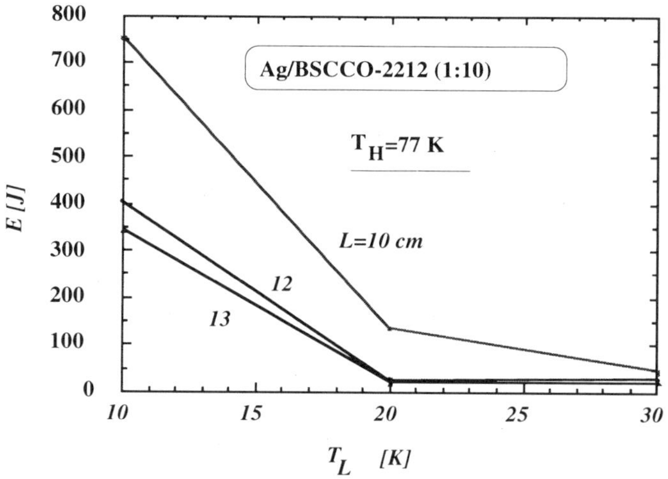

Figure 6. Sensitivity of minimum quench energy of the composite superconductor to the cold sink temperature.

SUMMARY

A numerical model developed with ANSYS, has been used to investigate the stability dynamics of high T_c superconductors. The model results demonstrate that stability is significantly improved by adding a small amount of silver to pure BSCCO conductors. Furthermore, the results show that the stability dynamics remain localized over long time periods, and that this is especially true of pure BSCCO conductors. The length independent quench energy of the pure BSCCO conductors, as well as other stability features may be well explained by the slow thermal diffusion rates within these conductors. For application purposes, this study demonstrates the futility of attempting to provide stability for high T_c magnets by conductive cooling. Because of the localized hot spots, the uncertainty of where they might occur, and the rapid thermal runaway if they do occur, cooling systems must provide a method to sweep the heat away from all regions of high T_c coil. To the degree to which stability is a concern for current leads (as for example in the situation of a temporary failure of the cooling system) this study demonstrates the benefit of adding a few percent of stabilizing silver to pure BSCCO elements in order to enjoy the sharp increase in stability at the cost of a somewhat elevated thermal leak.

AKNOWLEDGMENT

This study was funded by the U.S. Defense Nuclear Agency under DNA MIPR 92-719, work unit CD:00014, RCC: 7010, and by the University of Wisconsin-Madison.

REFERENCES

1 DRESNER, L., *Cryogenics*, ,Vol. 33, No. 9, pp. 900-909, (1993).
2. IWASA, Y., *Advances in Superconductivity*, Vol. 5, pp. 1205-1210, (1992).
3. BELLIS, R.H. and IWASA, Y.,*Cryogenics*, ,Vol. 34, No. 3, pp. 129-144, (1994).
4. SEOL, S.Y. and HULL, J.R., *Cryogenics*, ,Vol. 33, No. 10, pp. 966-975, (1993).
5. ANSYS (Engineering Analysis System), SWANSON ANALYSIS SYSTEMS, INC., P.O.Box 65, Houston, PA, 15342, U.S.A.
6. SATO, K., HIKATA, T. and IWASA, Y., *Applied Physics Letters,* Vol. 57, pp. 1928-1933, (1990).
7. SASAOKA, T., NOMURA, K., SATO, J. and KUMA, S., *Applied Physics Letters,* Vol. 64, pp. 1304-1305, (1994).
8. OBERLY, C.E., KOZLOWSKI, G. and FINGERS, R.T., *Advances in Cryogenic Engineering,* Vol. 38B, pp. 479, (1992).
9. VAN DER LAAN, M.T.G., TAX, R.B., ET AL., *Cryogenics*, Vol. 30, (September Supplement), pp. 163, (1990).
10. VAN DER LAAN, M.T.G., TAX, R.B., ET AL., "Advances in Cryogenic Engineering", Plenum Press, New York, (1994), Vol. 37B, pp. 1517,
11. OHKURA, K., MUKAI, H., et al., *Japanese Journal of Applied Physics*, Vol. 32 Part 2, No. 11A, pp. 1606, (1993)
12. SUKURABA, J., YAMADA, Y., et al., *Japanese Journal of Cryogenic Engineering*, Vol. 28, No. 9, pp. 519, (1993).
13 TAKAHASHI, M., HAKAMADA, R., et al., "Advances in Cryogenic Engineering", Plenum , New York, (1994), Vol. 39, pp. 343.
14 HAYASHI, S., EGI, T., ET AL., "Advances in Cryogenic Engineering", Plenum , New York, (1994), Vol. 40, pp. 319.
15 HASE, T., SHIBUTANI, K., HAYASHI, S., OGAWA, R. and KAWATE, Y., *Cryogenics*, Vol. 35, No. 2, pp. 127-133, (1995).
16. MATSUKAWA, M., TATEZAKI, F., NOTO, K., FUJISHIRO, H., et al. ., *Cryogenics*, Vol. 34, pp. 89-90, (1994).
17 MORI, K., SASAKAWA, M., IGARASHI, T. and ISIKAWA, Y. et al., *Physica C*, Vol. 162/164, pp. 512, (1989).
18. COLLINGS, E.W., 2nd International Symposium on Superconductivity, Tsukuba, Japan, (November 1989).
19. HERMAN, P.F. et al., *IEEE Transactions on Applied Superconductivity*, Vol. 3, pp. 876-880, (1993).
20. FLETCHER, L.S., PETERSON, G.P. and SCHAUP, R., *Transactions of the ASME-Heat Transfer*, Vol. 113, pp. 274-276, (1991)
21. REED, R.P. and CLARK, A.F., (edt.), *Materials at Low Temperature* , American Society of Metals, Metals Park, Ohio, (1983).

OPTIMIZATION OF THE INTERCEPT TEMPERATURE FOR HIGH TEMPERATURE SUPERCONDUCTING CURRENT LEAD

Suntao Yang and John M. Pfotenhauer

Applied Superconducting Center, University of Wisconsin - Madison
Madison, WI 53706

ABSTRACT

Current leads containing high temperature superconducting (HTS) components have the potential to greatly reduce the heat leak to liquid helium environment. However, significant heat leak reduction can only be achieved by intercepting heat at the joint between the upper (copper) and the lower (high temperature superconducting) sections of current leads. An optimization analysis has been performed to minimize the net room temperature refrigeration power required for operating HTS current leads. The analysis shows that competing factors of cryocooler efficiencies and the influence of HTS material properties on the required helium liquefaction power determine an optimum intercept temperature in the range between 70 and 80 K.

INTRODUCTION

Standard conventional current leads are self cooled, that is, cooled by the helium vapor generated from heat conducted to the liquid helium bath through the current leads. The cooling methods for hybrid superconducting-metal current leads can be quite different and will depend on the low temperature device to which the current is supplied. Typically a hybrid current lead will incorporate a copper section spanning the temperature range from room temperature to an intermediate temperature (T_{int}), where it will be joined to the superconducting section. The superconducting section will span the temperature range from T_{int} to the liquid helium temperature, 4.2 K. The intermediate temperature transaction provides the opportunity for a variety of cooling options. For example, in systems which avoid the use of liquid cryogens, the hybrid current leads are usually conduction cooled by a connection to a refrigeration system. However, if both liquid helium and liquid nitrogen are readily available, the current leads will most likely to be vapor cooled. Figure 1 shows schematically some of the most commonly chosen cooling methods for current leads, as well as the theoretical 4.2 K heat leak and room temperature power consumption for the corresponding cooling methods [1] - [4].

It is clear from figure 1 that in order to minimize the room temperature power requirement, it is always better to intercept heat at the copper to HTS junction for a hybrid current lead operating between 4.2 K and room temperature. In designs without intermediate heat interception (figure 1B), the heat flux to the liquid helium bath and the corresponding room temperature refrigeration power can only be reduced to 80% that of the best copper lead. This is because there is a relatively large amount of heat entering the high temperature superconductor from the copper part, which limits the performance of the superconductor. On the other hand, for practical designs with heat interception (figure 1C-1E), the heat flux to the liquid helium bath can be reduced to as little as 12.5%, and the corresponding room temperature refrigeration power can be reduced to as little as 26%, of that associated with the best copper lead. In general, the optimum intercepting temperature T_{int} will depend on the HTS material properties and the chosen cooling method for the whole lead.

The analysis which follows addresses the cooling choices represented by option D in figure 1. "Optimal" conditions are defined by a minimum total room temperature refrigeration power.

The total room temperature refrigeration power (W_{total}) required to cool a HTS current lead consists of two parts: the power required to intercept heat at the intermediate temperature, (W_1), and the power required to re-liquefy the liquid helium boiled by the heat leak at 4.2 K, (W_2). Thus,

$$W_{total} = W_1 + W_2 \tag{1}$$

Both W_1 and W_2 are functions of the intercept temperature T_{int}. With regard to the first term (W_1), one finds that the room temperature electrical power consumption of a cryogenic refrigerator is determined through a combination of the required power at the low temperature end, the Carnot ratio and the efficiency of the machine, all of which are strongly dependent on the temperature. As the intercept temperature T_{int} becomes lower, the electrical power required to intercept a given quantity of heat at this temperature (W_1) becomes larger. On the other hand, the electrical power required to re-liquefy the liquid helium boiled by the heat leak at 4.2 K (W_2) becomes smaller, due to the larger current density and hence smaller cross sectional area of the superconductors at lower temperatures. It is these two competing effects on the room temperature electrical power with the change of intercept temperature that allow an optimization of the intercept temperature.

CRYOCOOLER POWER

The room temperature power required to intercept heat Q_1 at temperature T_{int} is:

$$W_1 = \frac{1}{f_1}\frac{1}{\eta}Q_1 \tag{2}$$

where f_l is the efficiency, or figure of merit for a real cryocooler, and $\eta = \dfrac{T_{int}}{300K - T_{int}}$ is the Carnot coefficient of performance for cryocoolers.

For a copper current lead operating between 300K and T_{int} with no helium gas cooling (i.e., cooled by conduction only), the lead temperature profile can be described by the equation:

$$\frac{d}{dx}\left(k(T)A\frac{dT}{dx}\right) + \frac{I^2\rho(T)}{A} = 0 \tag{3}$$

	A	B	C	D	E
RT Power	100%	80%	28%	37%	26%
4.2 K Heat Leak	100%	80%	33%	12.5%	12.5%

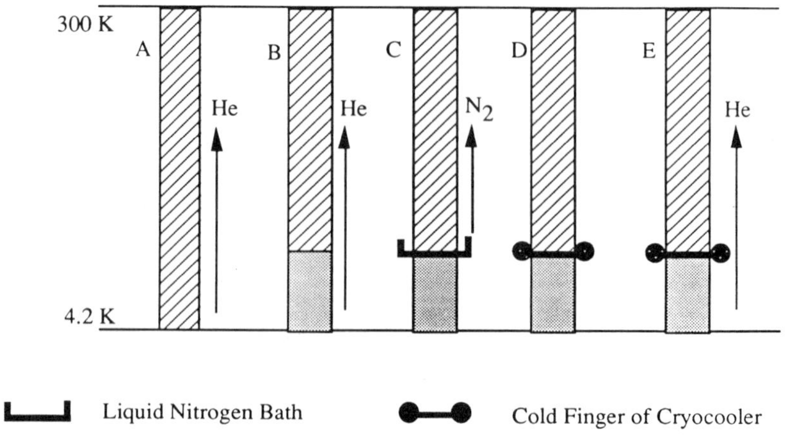

Figure 1. Some possible cooling modes for superconducting current leads. The theoretical estimation for room temperature power reduction and 4.2 K heat leak reduction are shown as a percentage to the best possible conventional leads.

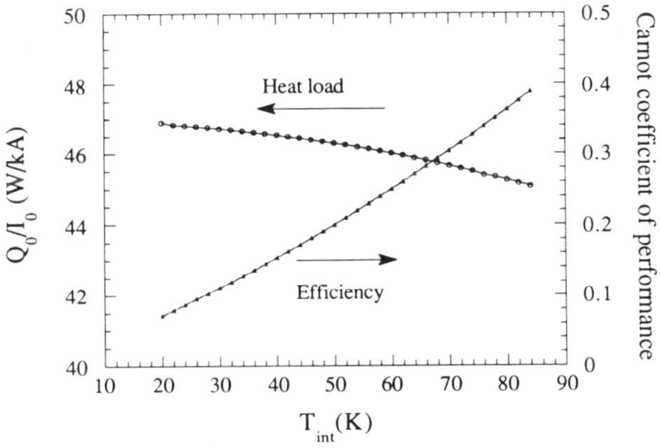

Figure 2. Cryocooler heat load at T_{int} for a conventional current lead cooled by conduction only and the Carnot coefficient of performance for cryocoolers.

where T(x) is temperature along the lead, A is the cross section area, I is the current passing through the lead, and k(T), ρ(T) are the thermal conductivity and electrical resistivity of the lead material respectively. This equation can be solved directly. For current leads comprised of materials which obeying the Wiedmann-Franz law ($k(T)\rho(T)=L_oT$), equation (3) can be solved directly [1]. The minimum heat load at the low temperature end for a given current is then given by

$$\frac{Q_0}{I_0} = \sqrt{L_0\left(300^2 - T_{int}^2\right)} \qquad \text{(W/A)} \qquad (4)$$

where L_0 is the Lorentz number, equal to 2.45 x 10^{-8} $W\Omega/K^2$, and this value of $\frac{Q_0}{I_0}$ corresponds to the minimum heat load at unit current. Figure 2 displays values of $\frac{Q_0}{I_0}$ as a function of T_{int}. Also shown in figure 2 is the Carnot coefficient of performance as a function of the cold end temperature T_{int}. For T_{int} in the range of 20 K to 80 K, $\frac{Q_0}{I_0}$ is between 45 and 47 W/kA.

Although the Carnot coefficient of performance depends only on the temperature, the figure of merit for cryocoolers also depends on the size of the cryogenic machine. The figure of merit here is defined in the standard sense, as the work ratio between a real cryocooler and an ideal cryocooler. For typical single stage G-M cryocoolers which deliver a cold head power in the 100 W range, the figure of merit f_1 ranges from 0.02 for a cold head temperature around 20 K to about 0.08 for cold

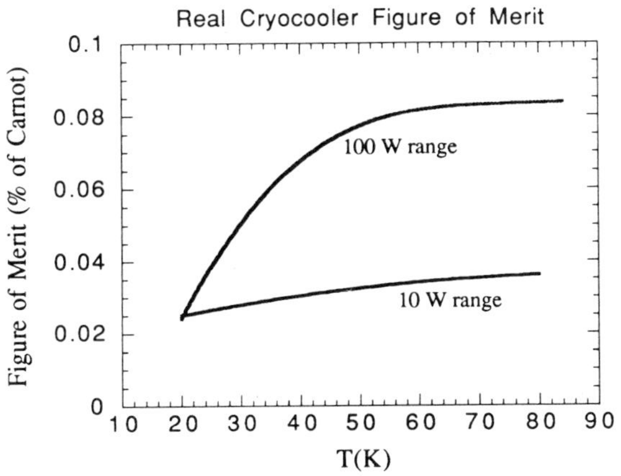

Figure 3. Cold head temperature dependence of cryocooler figure of merit as percentage of Carnot machine, for single stage G-M cryocooler with cold head power in the 100 W and 10 W range.

head temperature around 80 K [5]. For comparison, cryocoolers which deliver cold head power in the ten watts range have a figure of merit ranges from 0.02 at 20 K to approximately 0.04 for a cold head temperature at 80 K [5]. Figure 3 shows the figure of merit f_1 as a function of cold head temperatures for the real cryocoolers which deliver cold head power in the 100 W or 10 W range. By combining the results from equation 4, the cryocooler figure of merit and the Carnot coefficient of performance into equation 2, the room temperature power (W_1) required to intercept heat from an optimized, conductively cooled current lead at T_{int} can be obtained as follows:

$$\frac{W_1}{I} = \frac{1}{f_1} \frac{300 - T_{int}}{T_{int}} \sqrt{L_0 (300^2 - T_{int}^2)} \qquad (5)$$

Figure 5 shows the results obtained from these calculations.

LIQUEFIER POWER

Calculations of the refrigeration load due to operation of HTS elements spanning the temperature range between T_{int} and 4.2 K have been made for conditions below and above the 1 μV/cm criterion. In the simplest case, the superconductor operates below the 1 μV/cm criterion and the heat dissipation in the superconductor is negligible. The 4.2 K heat load is then determined exclusively by the heat conduction and losses due to contact resistance. For a carefully designed and fabricated contact, the contact resistance at 4.2 K is usually on the order of tens of nΩ [6],[7], resulting a heat generation of tens of mW for a current of 1 kA, this value being much smaller than the heat load at 4.2 K from conduction (hundreds of mW). Therefore the heat load due to contact resistance is also neglected in this analysis. The heat load from T_{int} to the 4.2 K end of a conductor with constant cross section A and length L is:

$$Q_2 = \frac{A}{L} \int_{4.2K}^{T_{int}} k(T)dT \qquad (6)$$

Assuming a constant current lead length L, Q_2 is a function of A and T_{int}. As T_{int} decreases, the critical current density of the superconductor increases. Therefore, a smaller cross section area A can be used to carrying a given current. In addition, the thermal conductivity integral will be smaller for a lower intercept temperature T_{int}. Thus as T_{int} becomes lower, Q_2 will be smaller. Figure 4 shows a typical temperature dependence for J_c (with J_c @ 77 K = 1000 A/cm^2), and the thermal conductivity integral as a function of temperature T for a typical BSCCO-2212 material [8]. The room temperature refrigeration power W_2 required to re-liquefy the helium gas vaporized by heat Q_2 at 4.2 K is [1]:

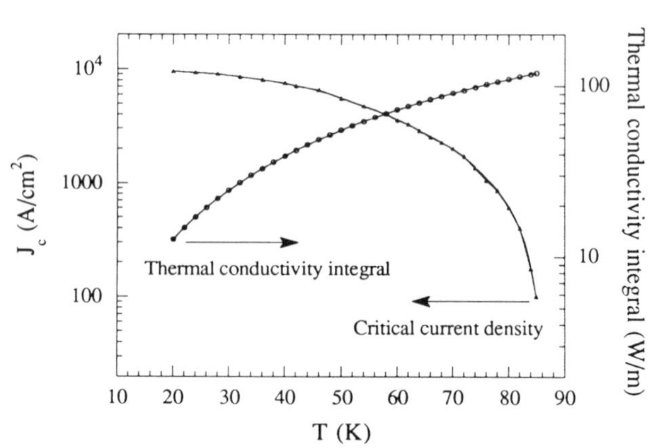

Figure 4. A typical critical current density at zero external magnetic field and the thermal conductivity integral of BSCCO-2212 as a function of temperature [3] - [7].

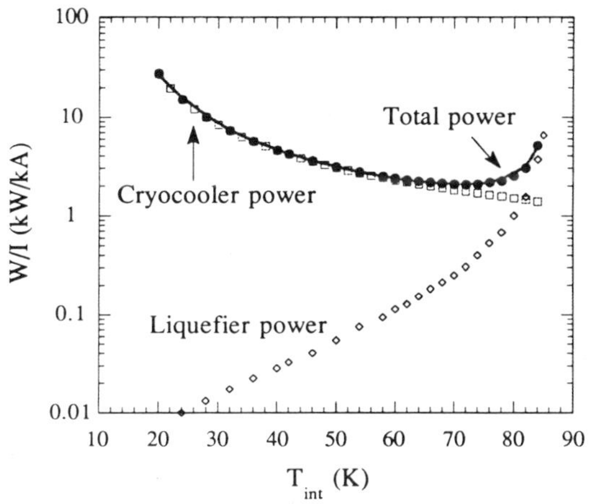

Figure 5. The total room temperature power required to operate a HTS current lead as a function of the intercepting temperature T_{int}. The total power is the sum of the cryocooler power required to intercepting heat in the transition and the power required to re-liquefy the boiled off helium gas at 4.2 K.

$$W_2 = \frac{1}{f_2}\frac{Q_2}{C_L}\left[T_{RT}\left(S_{RT} - S_{4.2}\right) - \left(H_{RT} - H_{4.2}\right)\right] = \frac{326 Q_2}{f_2} \tag{7}$$

where C_L is the latent heat of liquid helium, S and H are the entropy and enthalpy of helium. The subscripts RT and 4.2 refer to the room temperature and 4.2 K values respectively. This expression assumes that the helium vapor is warmed to room temperature before being re-liquefied. It represents the maximum liquefier power required to re-liquefy the helium vapor vaporized by the heat load at 4.2 K. If cooling method of 1D is chosen, then the liquefier power will be smaller because the helium vapor is returned at much colder temperature than room temperature.

The figure of merit for the refrigerator, f_2, is also strongly dependent on the size (capacity) of the machine as well as on the operating temperature. Typically machines designed to lift power at 4.2 K in the range of tens of watts will require 1 kW of electrical power for every watt of 4.2 K refrigeration power. Smaller machines, such as for lifting less than 5 watts at 4.2 K require as much as 10 kW of electrical power for every watt of 4.2 K refrigeration power. Since the heat load at 4.2 K for a hybrid current lead operating in the kA range is usually less than several watts, we have used the factor for the smaller machines, which results in the relation: $W_2 = 10000 Q_2$ [2],[3].

DISCUSSIONS AND CONCLUSIONS

Figure 5 shows the overall results of the combined calculations. Shown in figure 5 are the room temperature power required of the cryocooler to intercept heat at T_{int}, as determined by equation 5, and the room temperature liquefier power required to re-liquefy the helium vaporized by the heat load at 4.2 K. It is clear from this figure that the room temperature power required to operate the cryocooler is much larger than that required for the liquefier in the entire range between 20 and 80 K. It is only at temperatures higher than 80 K that the liquefier power becomes larger than the cryocooler power, because of the rapidly decreasing critical current density for the BSCCO-2212 material at these temperatures. The minimum total room temperature power occurs at about 72 K.

There are several very interesting points that grow out of this analysis. First, if the upper section of the lead is cooled by the generated helium vapor, then the amount of heat has to be intercepted at the intercept temperature will be smaller. This will lead to a lower optimal intercept temperature. Typically, the expected helium flow rate for HTS current lead is 1/3 to 1/8 of that of conventional current lead. With this flow rate, the amount of heat to be intercepted at the intercept temperature could be ten times smaller than that for the pure conduction cooled upper section (see figure 2). In this case, the required cryocooler power will be much smaller. The optimal intercept temperature will be at 64 K.

Second, the optimum intercept temperature T_{int} lies between 70 and 80 K because the critical current density J_c decreases rapidly above 70 K for the BSCCO-2212 material. If the overall critical current density J_c is increased (for example, from 1000 A/cm^2 to 2000 A/cm^2 at 77 K), but the temperature dependence of J_c remains the same, then the required liquefier power will decrease, and the optimum intercept temperature T_{int} will increase.

Third, if other HTS materials (YBCO, BSCCO-2223) in the bare forms are used for the lower section of the lead, the optimum T_{int} will remain between 70 and 80 K. This is due to the fact that even if the critical current density varies by a factor of 2 as compared to the value used here, the thermal conductivity integral for these materials only varies up to a factor of 2 also [3],[4]. Therefore, the optimum temperature T_{int} will remains roughly the same. On the other hand, if HTS materials with substrate were used for the lower section of current lead, then thermal conductivity of the substrate and overall critical current density of the material have to be evaluated to determine the optimal intercept temperature. For example, the thermal conductivity integral for the Ag-Au alloy (10 wt.% Au) is roughly 20 times that of the HTS materials [9]. If BSCCO-2223 tapes with Ag-Au sheath were used for current lead application, then the overall critical current density of the tape must be above 10^4 A/cm^2 at 77 K, to make the heat leak at 4.2 K to be roughly the same as that for current lead with bare BSCCO-2212 material. Nevertheless, the optimal intercepting temperature will still be around 72 K.

Fourth, the analysis above is for an HTS current lead designed to carry current in the kA range. If the HTS current lead is designed to carry current in the tens of kA range, the heat intercepted at the transition will be in the kW range, and the heat load at 4.2 K will be larger than several watts. In this case, the liquefier figure of merit will be larger due to the necessary larger machine and the resultant contribution to the total power due to the liquefier will be less significant. This will result in a shift of the optimal intercepting temperature to the high temperature side. For example, HTS current leads designed to carrying 20 kA will have an optimal intercept temperature around 80 K. This is in spite of the fact that the heat leak at 4.2 K with T_{int} at 80 K is about 4 times larger than that with T_{int} at 72 K.

To conclude, we have performed an optimization analysis that minimizes the net room temperature refrigeration power required for operating high temperature superconducting current leads. The analysis is based on the assumption that heat will be intercepted at the joint between the superconducting lower section and the conventional upper section. The room temperature cryocooler power required to lift all the heat at the intercept temperature (W_1) is determined by the cryocooler efficiency, the amount of heat to be intercepted and the intercept temperature. This power decreases as the intercept temperature increases. On the other hand, the room temperature power required to re-liquefy the vaporized helium (W_2) is determined by the liquefier efficiency and the amount of vapor to be liquefied (which in turn is a function of the critical current density and thermal conductivity integral of the material used). This power increases as the intercept temperature increases. For the HTS current leads applications considered in this report, the cryocooler power is the dominant factor in the two competing power requirements. The analysis shows that competing factors of cryocooler efficiencies and the influence of the HTS material properties on the required room temperature helium liquefaction power determine an optimum intercept temperature in the range between 70 and 80 K.

REFERENCES

1. M. N. Wilson, Superconducting Magnets, Oxford Science Publication, New York, 1993.
2. R. Wesche and A. M. Fuchs, "Design of superconducting current leads", Cryogenics, **33** (1993), 714-718.
3. P. F. Herrmann, E. Beghin, G. Bottini, C. Cottevieille, A. Leriche, T. Verhaege, and J. Bock, "Test results of a 1 kA (2kA) - 20 kV HTSC current lead model", Cryogenics **34** (1994), 543-548.
4. J. R. Hull, IEEE Trans. on Applied Superconductivity, **3** (1993), 869-875.
5. Products catalog of Cryomech, Syracuse, New York, 1995.
6. J. L. Wu, "High current capacity joints of HTSC and normal metals", Advances in Cryogenic Engineering, P. Kittel, Ed., Plenum Press: New York, **39** (1994), 1975-1981.
7. S. Elschner and J. Bock, "Electrical contact with low resistance to melt-processed high temperature superconductors", Advanced Materials, **4** (1992), 242.
8. S. Yang, B. Chen, E. Stires, E. Hellstrom,and J. Pfotenhauer, "Thermal conductivity and thermal contact conductance of BSCCO-2212", IEEE Transactions on Applied Superconductivity, Vol. 5, (1995), 1471.
9. T. Sasaoka, K. Nomura, J. Sato, S. Kuma, H. Fujishiro, M. Ikebe, and K. Noto, "Characteristics of Ag-Au alloy sheathed Bi-Pb-Sr-Ca-Cu-O superconducting tapes for current leads", Applied Physics Letters, 64 (10), 1304, March 1994.

A DESIGN METHOD FOR MULTIPLE TUBE GAS-COOLED ELECTRICAL LEADS FOR THE g-2 SUPERCONDUCTING MAGNETS

M. A. Green

E. O. Lawrence Berkeley National Laboratory
University of California
Berkeley, CA 94720

L. X. Jia, L. J. Addessi, J. R. Cullen Jr., A. J. Esper, R. E. Meier

Brookhaven National Laboratory
Upton, NY, 11973

ABSTRACT

This report presents a method for designing 5300 A gas cooled electrical leads for the g-2 solenoids and 2850 A leads for the g-2 self shielded inflector dipole magnet. Empirical design equations for annular tube gas cooled leads are presented. The leads are bundled tube leads which are cooled by helium flowing in annular cooling passages between tubes. Each tube in the bundle consists of nested circular copper tubes that can be cooled on both sides. Multiple current carrying tubes will increase the lead current capacity and cooling the tubes on both sides will increase lead efficiency for a given helium flow pressure drop. The design method presented here can applied to leads made from a variety of materials.

INTRODUCTION

The g-2 storage ring magnet system consists of four large superconducting solenoids that are up to 15.1 m in diameter[1]. The four solenoids will be hooked in series and they will be powered through a single pair of 5300 A gas cooled electrical leads. In addition there is an actively shielded inflector dipole that is separately powered through a pair of 2850 A gas cooled leads. Both sets of leads will be cooled using cold helium drawn directly from the forced two-phase helium cooling system used to cool the four solenoids and the inflector.

Tubular gas-cooled electrical leads have been used at various laboratories since the mid 1960s to power superconducting magnets. An insert was used to improve the heat transfer in tube leads in the late 1960s, allowing one to increase the lead current by as much as a factor of three. The use of annular flow leads was described in the literature by Smits in 1981[2]. The concept of bundling tube leads to increase their current was applied to curved gas cooled leads for the PEP-4 solenoid.

THEORY

Gas cooled leads can be described by a set of two coupled non-linear differential equations. These one-dimensional differential equations take the following form:

$$\frac{d}{dx}[k(T)\,A\,\frac{dT}{dx}] + \frac{\rho(T)\,I^2}{A} = h\,P\,(T - \theta) \tag{1a}$$

$$M\,c_p(\theta)\,\frac{d\theta}{dx} = h\,P\,(T - \theta) \tag{1b}$$

where x is the distance along the lead length; T is the temperature of the lead metal; θ is the temperature of the gas stream in the lead; A is the lead cross-sectional area; I is the current carried by the lead; h is the heat transfer coefficient between the metal in the lead and the cooling fluid; P is the wetted perimeter per unit length of the lead; k(T) is the temperature dependent thermal conductivity of the metal in the lead; $\rho(T)$ is the temperature dependent electrical resistivity of the metal in the lead; and $c_p(\theta)$ is the temperature dependent specific heat at constant pressure for the gas flowing up the lead. Different boundary conditions can be applied to T, dT/dx, at the two ends at x = 0 and x = L.

The helium flow through optimum tubular gas cooled electrical leads is always laminar (Re < 1500)[3]. The Nusselt number for well developed laminar flow in tubes is around four[4]. (Some authors claim the Nusselt number is eight, but this comes from their definition of hydraulic diameter. Regardless of the definition that is used, the heat transfer coefficient is the same.) Since the Nusselt number is around four regardless of the tube geometry, a tube with an insert that has an annular flow passage has a higher heat transfer coefficient at the tube wall (by as much as a factor of 60) than does the same size round tube without the insert to form an annular flow passage.

When one tries to solve the lead equations, one finds that an optimum lead (with minimum heat leak to 4.5 K and a minimum gas flow needed to cool the lead) will have a particular value of the current density length function IL/A. Once one knows the IL/A function for the lead and one can estimate the heat transfer coefficient between the tube wall and the helium, one can calculate an approximate value for the average temperature difference between the lead wall and the helium in the flow passage T*.

The Optimum IL/A function for an Efficient Lead

The IL/A function for gas cooled leads comes directly from the lead equations[5]. For a gas cooled lead IL/A is a function of a number of factors that include material properties and the lead heat transfer efficiency[6]. For perfect leads, the optimum IL/A function G_o can be estimated using the following approximate empirical expression:

$$G_o = \frac{IL}{A} = 286\,[\rho_{4.2}]^{-0.5} \tag{2}$$

where $\rho_{4.2}$ is the resistivity of the lead material at 4.2 K. Equation 2 can be applied to a wide variety of materials from OFHC copper to 304 Stainless Steel. Contrary to popular opinion, leads do not have to be made from good copper. A range of materials can be used for gas cooled leads as long the IL/A function is correctly selected and the product of the heat transfer coefficient times the heat transfer area is large enough to insure good heat transfer and a small temperature difference between the current carrying wall and the helium flowing through the lead. The advantages of using high low-temperature resistivity materials, such as phosphorous deoxidized copper[7] or 6061 aluminum[8], are 1) mistakes in selecting IL/A are not costly in terms of lead overall performance (unlike OFHC copper) and 2) the leads can be designed to operate at their root mean squared current rather than at their peak current. Table 1 presents the optimum IL/A for different materials.

The Average Temperature Difference between the Tube Wall and the Gas

The average temperature difference between the tube wall and the helium gas in a lead T* (for a lead without superconductor) for an annular tube lead can be estimated using the following expression:

$$T^* = \frac{J_o^2\,A\,E}{G_o}\;\frac{t}{2\pi\,N_s.N_t\,D_A\,k(T)} \qquad (3)$$

where J_o is the current density in the tube; G_o is the design IL/A for the lead; A is the cross-section area of the current carrying material in the lead; t is average thickness of the annuli between the tubes in the lead; k(T) is the average thermal conductivity of the helium in the tube (T=120 K for low RRR copper leads; for high RRR copper leads T<70 K); N_t is the number of current carrying tubes in the lead N_s is ratio the number of tube surfaces cooled by the gas to the number of tubes (N_s is between 1 and 2); and D_A is the average diameter of the tubes in the lead. E is the optimum voltage drop along the lead when it carries its design current. (Use 80 mV as design value for E[7].) The value of T* calculated by Equation 3 is approximate and should be regarded as a performance function for comparison purposes. A good rule of thumb is that the value of T* should be less than 5 K for a well designed annular tube leads. Lower values of T* are better.

Design Current Density for the Lead

From the IL/A constant, one finds that the length of an optimum gas cooled lead is proportional to its current carrying cross-section. The current carrying cross-sectional area is some function of the desired time for the lead to operate safely without gas flow. The most conservative approach is to assume the lead is adiabatic at its upper end. Using this approach and assuming a temperature change of 200 K at the upper end of the leads, the design current density for copper leads can be estimated using the following expression;

$$J_o = 1.78 \times 10^8\,[t_o]^{-0.5} \qquad (4a)$$

where t_o is the design time for the leads to operate without gas flow. The expression in Equation 4a is suitable for times t_o less than 100 seconds. Using this experimental data for phosphorus deoxidized copper leads with the upper end held at 300 K, the design current density for copper leads can be estimated using the following expression when t_o is greater than 100 seconds[2,3];

$$J_o = 4.48 \times 10^7\,[t_o]^{-0.2} \qquad (4b)$$

When leads are made of 1100 or 6061 aluminum, the values of J_o calculated using Equations 4a and 4b should be multiplied by 0.62. For 5465 aluminum multiply the copper J_o by a factor of 0.54; for leads made from brass multiply the copper J_o by a factor of 0.63; and for 304 Stainless steel leads multiply the copper J_o by a factor of 0.22.

The current density calculated using Equations 4a or 4b is the maximum allowable value of J_o. A second criterion is the allowable value of T* calculated using Equation 3. If having a value of T* less than a certain value (say < 5 K) is a design criteria, one should insert the value of J_o calculated using Equations 4a or 4b into Equation 3. If necessary, J_o can be adjusted downward so that the desired value of T* is reached.

Helium Mass Flow through the Gas Cooled Leads

One can estimate the design helium flow through the leads by using the assumption that a well designed lead, operating at its design current, will have no heat flow into or out of the room temperature ends of the leads. An estimate of the design mass flow M (in kilograms per second) through a lead can be estimated using following expression:

$$M = 6.5 \times 10^{-7}\,E\,i_o \qquad (5)$$

Table 1 Low Temperature Resistivity and the Optimum Lead IL/A Function for Various Materials

Material	4.2 K resistivity (ohm m)	IL/A Value (A per m)
OFHC Copper (RRR = 180)	8.7×10^{-11}	2.7×10^{7}
Copper (RRR = 30)	5.3×10^{-10}	1.1×10^{7}
1100-O Aluminum (RRR = 25)	1.0×10^{-9}	9.0×10^{6}
Type M Copper Pipe (RRR = 6.5)	3.1×10^{-9}	5.6×10^{6}
Phos. Deox. Copper (RRR = 3)	6.7×10^{-9}	3.5×10^{6}
6061-T6 Aluminum (RRR = 2)	1.4×10^{-8}	2.4×10^{6}
70 Cu 30 Zn Brass (RRR = 2)	2.8×10^{-8}	1.7×10^{6}
5456 Aluminum (RRR = 1.4)	3.3×10^{-8}	1.6×10^{6}
304 Stainless Steel (RRR 1.1)	3.7×10^{-7}	4.7×10^{5}

where i_o is the design current for the lead, and E is the optimum lead voltage drop. (Wilson suggests E = 80 mV[7].) When one applies the value of M from Equation 5 into the equation for the Reynolds number, one finds that the flow Reynolds number is always below 1500[3]. Leads that operate off their design current (in either direction) will require more gas flow per unit current than is suggested by Equation 5.

SELECTION CRITERIA FOR THE g-2 LEADS

Tube Material Selection

Standard phosphorous deoxidized copper pipe was selected as the current carrying material for the g-2 magnet gas cooled leads. The reasons for selecting standard copper pipe are as follows: 1) Copper pipe is generally available in the marketplace in a series of sizes that are compatible with nested tube lead construction. 2) The copper pipe has a low residual resistance ratio (the measured RRR for samples of the pipe to be used in the g-2 leads is about 6.5), which means that the design IL/A is a factor of five lower than for good copper. (The IL/A of the selected pipe material is about 5.6×10^6 A m^{-1}.) For a given lead current density, the leads will be more compact. Short leads are an advantage provided the heat transfer coefficient to the helium is high enough so that T* is small. 3) Leads made from low RRR material can be designed based on the root mean square current rather than the maximum current because they are stable when operated at currents well above the design current of the lead. If one has a stable lead, one can operate that lead safely above its design current by increasing the helium mass flow per unit current above the minimum design value. Leads made with phosphorous deoxidized copper will operate safely at a current 2.5 times the design current for the lead. (The gas flow per unit current is increased about forty percent.) 4) The gas usage at zero current is less for a phosphorous deoxidized copper lead than it is for a good (high RRR) copper lead.

Table 2 Lead Parameters for Copper Tube Annular Leads (Inside Cooling Only)
(IL/A = 5.6×10^{6} A m^{-1}, k(T) = 0.086 W m^{-1} K^{-1} at 120K, $J_o = 10^7$ A m^{-2}, $N_s = 1$, $N_t = 1$)

Copper Tube Type	OD (mm)	Wall thick (mm)	D_A (mm)	t (mm)	i_o (A)	T* (K)
0.375 inch Type L	12.70	0.89	10.92	0.70	330	5.6
0.500 inch Type L	15.88	1.02	13.84	0.57	476	5.2
0.625 inch Type L	19.05	1.07	16.91	0.52	604	4.9
0.750 inch Type L	22.23	1.14	19.95	0.44	755	4.4
0.500 inch Type M	15.88	0.71	14.46	0.88	339	5.5
0.625 inch Type M	19.05	0.76	17.53	0.83	438	5.5
0.750 inch Type M	22.23	0.81	20.61	0.77	545	5.4
1.000 inch Type M	28.58	0.89	26.80	0.70	774	5.3
1.250 inch Type M	34.93	1.07	32.79	0.52	1138	4.8

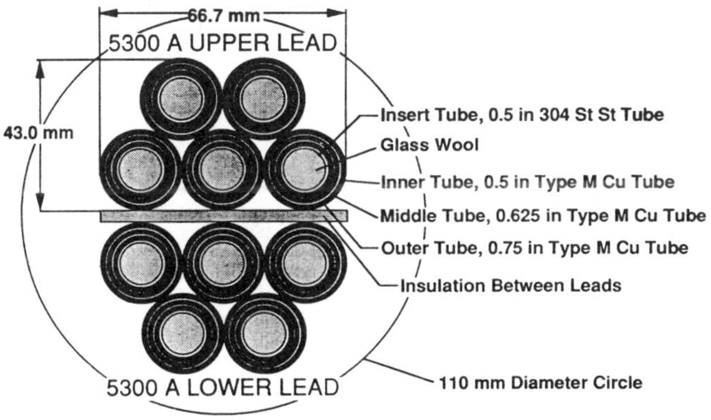

Figure 1. A Cross-section of the 5300 Ampere Gass Cooled Leads for the g-2 Solenoids .

Number of Tube Bundles and the Number of Nested Tubes per Outer Tube

The number of tube bundles was set at five for each 5300 A g-2 solenoid gas cooled lead. A design current of 1060 A per tube in the bundle was used to determine the design of the nested tube set. For each 2850 A g-2 inflector lead, the number of tube bundles can be set as low as two. Since the inflector has a low stored energy (about 9 kJ), the inflector lead current density can be set above 10 A per square millimeter. The design current per tube in the bundle is 1425 A when two tube bundles are used. Table 1 shows the parameters for single standard Type L and Type M copper tubes that can be used to fabricate leads[3,9].

Three nested tube lead options are apparent from Table 2: 1) A single 1.25 inch Type M tube with a stainless steel insert (N_s = 1.00 while N_t= 1) can be used for each tube in the bundle. For the g- 2 solenoid leads, L = 0.60 m; J_o = 9.31 A per square mm; and T* = 4.1 K. 2) A single 0.625 inch Type L tube inside of a single 0.75 inch Type L tube can be used for each tube in the bundle. The inner tube can have a stainless steel insert such that N_s = 1.5 while N_t = 2. For the g- 2 solenoid leads, L = 0.72 m; J_o = 7.80 A per square mm; and T* = 1.9 K. 3) Three nested Type M tubes with a 0.5 inch tube inside of a 0.625 inch tube, which is in turn inside of a 0.75 inch tube. A stainless steel insert can be installed inside the inner tube so that N_s = 1.67 while N_t = 3. For the g-2 solenoid leads, L = 0.70 m; J_o = 8.02 A per square mm and T* = 2.1 K.

Option 1 was eliminated on the basis of physical size (each lead would be 105 mm wide and 68 mm thick not including electrical insulation) and lower heat transfer efficiency. Options 2 and 3 are smaller (67 mm wide by 43 mm thick not including electrical insulation.) and they are more efficient. Option 3 was selected because there is enough space between the tubes to attach superconductor to all three of the current carrying tubes. Using option 3 on two tube bundle inflector magnet leads, L = 0.52 m; J_o = 10.8 A per square mm; and T* = 3.8 K.

CONCLUSION

Short gas cooled electrical leads for the g-2 solenoids and the inflector can be built using bundled tube leads where each of the tubes in the bundle consists of nested standard Type L or Type M copper tubes. Nested leads allow more current to be carried in a smaller physical space. Cooling of both sides of the copper tube increases the average efficiency of the gas cooled electrical leads. Both of the 5300 A solenoid leads will fit in a cylinder that is 110 mm in diameter by 750 mm long (see the cross-section of the solenoid leads in Figure 1). Both of the two tube bundle 2850 A inflector leads will fit within a cylinder that is 75 mm in diameter by 550 mm long. Superconductor will be attached to the tubes of both the solenoid and inflector leads in order to reduce the heat leak into the 4.5 K two-phase helium cooling circuit. The superconductor will reduce the amount helium needed to cool the leads by a few percent. It is estimated that 0.55 grams per second of helium is needed to cool the pair of 5300 A leads for the g-2 solenoid. The 2850 A inflector leads will require about 0.30 grams per seconds of helium to cool them.

ACKNOWLEDGMENTS

This work was performed at the Lawrence Berkeley Laboratory and at Brookhaven National Laboratory with the support of the Office of Energy Research, Office of High Energy and Nuclear Physics, High Energy Physics Division, United States Department of Energy under contract numbers DE-AC02-76CH00016 and DE-AC03-76SF00098.

REFERENCES

1. "Muon g-2 Design Report, BNL AGS E821, A New Precision Measurement of the Muon (g-2) Value at the Level of 0.35 ppm," Third Edition, Brookhaven National Laboratory Report, April 1994 Draft
2. R. G. Smits et al. "Gas Cooled Electrical Leads for Use on Forced Cooled Superconducting Magnets," Advances in Cryogenic Engineering **27**, p 169, Plenum Press, New York, (1981)
3. M. A. Green, "Some Annular Gas Cooled Electrical Lead Designs for the g-2 Superconducting Solenoid," Brookhaven National Laboratory g-2 Note 173, Oct. 1993
4. Frank Kreith, Principles of Heat Transfer, International Textbook Company, Scranton, 1958
5. L. X. Jia et al. "Design Parameters for Gas-Cooled Electrical Leads for the g-2 Magnets," to be published in the proceedings of the ICEC-15 Conference, Genova, Italy, 7 to 10 June 1994
6. D. P. Sekulic, Z. Uzeluc and F. J. Edeskuty. "Entropy Generation in a High Temperature Superconducting Current Lead," Cryogenics **32**, No 12, (1992)
7. Martin N. Wilson, Superconducting Magnets, Clarendon Press, Oxford, 1983
8. M. A. Green et al. "Measurements of Retractable Gas-Cooled 6061 Aluminum Electrical Leads Operating in a Vacuum," Advances In Cryogenic Engineering **37**, p 567, Plenum Press, New York (1991)
9. "Copper and Brass Sales Inc. Stock Catalogue and Metals Hand Book," 1986 Edition

ASSEMBLY AND TESTING OF A COMPOSITE HEAT PIPE THERMAL INTERCEPT FOR HTS CURRENT LEADS

M.A. Daugherty, D.E. Daney, F.C. Prenger, D.D. Hill, P.M. Williams, and H.J. Boenig

Los Alamos National Laboratory
Los Alamos, New Mexico, 87545, USA

ABSTRACT

We are building high temperature superconducting (HTS) current leads for a demonstration HTS high gradient magnetic separation (HGMS) system cooled by a cryocooler. The current leads are entirely conductively cooled. A composite nitrogen heat pipe provides efficient thermal communication, and simultaneously electrical isolation, between the lead and an intermediate temperature heat sink. Data on the thermal and electrical performance of the heat pipe thermal intercept are presented. The electrical isolation of the heat pipe was measured as a function of applied voltage with and without a thermal load across the heat pipe. The results show the electrical isolation with evaporation, condensation and internal circulation taking place in the heat pipe.

INTRODUCTION

In this paper we report on the assembly and testing of a heat pipe thermal intercept for HTS current leads. The thermal intercept is a key component of high performance (low heat leak) HTS current leads because it enables significant reductions in overall refrigeration requirements.[1,2] High performance current leads are an important component of the demonstration high gradient magnetic separation system we are building at Los Alamos.[3] The system is cooled by a cryocooler, so no liquid cryogens are used. The current leads for this project are therefore completely conductively cooled and must operate in a vacuum. The leads must be high performance to minimize the size of the refrigeration system and must also be rugged to withstand the shock and vibration experienced during transportation of the system.

The principal challenge in designing and building thermal intercepts for current leads is determining how to combine good electrical insulation with high thermal conductivity. Recent work at Argonne National Laboratory and Fermi National Accelerator Laboratory on low thermal resistance, high electrical isolation heat intercept connections has addressed this problem.[4] Niemann, Gonczy, and Nicol have developed a heat intercept design which involves clamping, by thermal interference fit, of an electrically insulating cylinder between an outer metallic ring and an inner metallic disc. The connections use the high clamping pressures made possible by the thermal interference fit to reduce the thermal resistance of joints between conducting and insulating materials.

We began examining composite heat pipes constructed of conducting and non-conducting materials in an effort to develop an alternative approach to high thermal conductivity intercepts with good electrical isolation. We believe the heat pipe thermal intercept offers significant performance advantages in many cases.

CURRENT LEAD DESIGN

Figure 1 shows our HTS current lead which incorporates a composite heat pipe thermal intercept. Because we have reported earlier on the detailed design,[5] we give only a brief description here. The lead has four major components: a flexible copper upper stage, the heat pipe thermal intercept, a flexible link, and the HTS lower stage. The flexible copper upper lead minimizes alignment problems and simplifies assembly. The heat pipe intercepts heat conducted down the upper copper lead and insures that the top end of the HTS conductors are adequately cooled. Heat is rejected from the heat pipe through its lower flange to a thermal shield, which is connected to the upper stage of the cryocooler.

A flexible link connects the heat pipe with the HTS lower lead assembly. The flexible link accommodates differential thermal expansion and reduces the strain on the HTS conductors. Most of the electrical resistive heating generated in the flexible link is removed by the heat pipe thermal intercept. A high thermal conductance is required to ensure that the resistance heating in the flexible link does not raise the temperature of the top of the HTS conductors above their maximum operating temperature. This lead design, with the flexible link located below the thermal intercept, would be difficult to implement without the enhanced thermal performance that can be obtained with the heat pipe thermal intercept.

The lower lead assembly is composed of two HTS conductors with copper end caps enclosed in a G-10 support tube. The top end cap is free to move vertically within the G-10 support tube to accommodate differential thermal contraction while providing lateral support. The lower copper end cap is connected to the HTS magnet by means of a clamped indium joint.

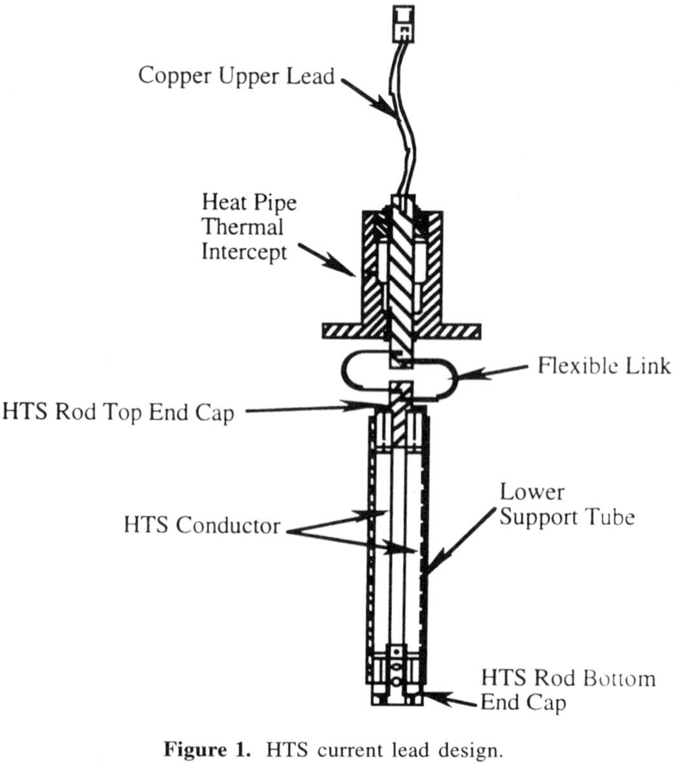

Figure 1. HTS current lead design.

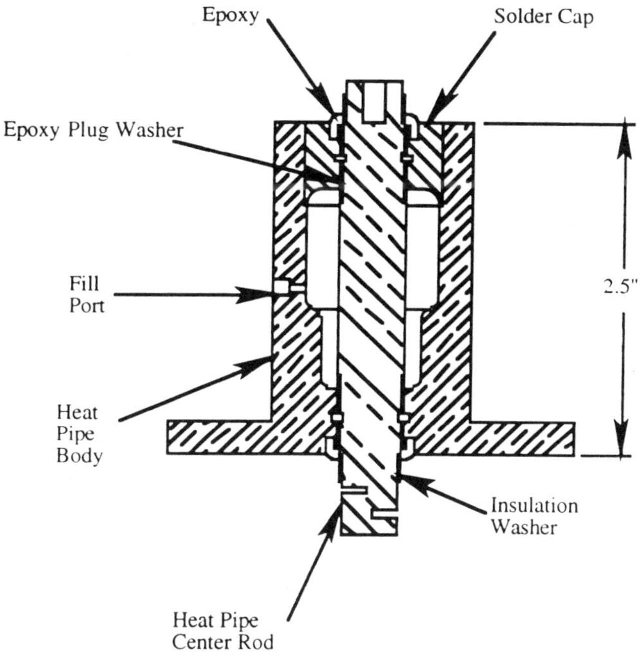

Figure 2. Heat pipe thermal intercept.

HEAT PIPE DESIGN AND ASSEMBLY

The most critical aspect of the design of the heat pipe is to ensure the mechanical and electrical integrity of the interface between the conducting and non-conducting portions of the assembly. This joint must be able to contain the 138 bar (2,000 psi) nitrogen gas with which the heat pipe is filled at room temperature. In addition, it must also remain leak tight during the cool down from room temperature to the 75-80 K operating temperature.

Initial designs used a disc of machinable ceramic material as the non-conducting portion of the heat pipe end caps. These discs were soldered into each end of the heat pipe body to insulate it from the center rod. In spite of careful attention to surface preparation and soldering techniques the solder joint could not withstand the 138 bar (2,000 psi) nitrogen fill pressure. These joints also failed to remain leak tight when cooled from room temperature to the 75 K operating temperature.

A revised design was developed using epoxy joints as shown in figure 2. The high pressure nitrogen charge is effectively contained by the solid heat pipe body and the solder cap. These two ETP copper components are assembled with a silver solder joint. The thickness of the epoxy joint between the heat pipe body and the center rod is minimized to reduce the forces exerted on it by the high pressure nitrogen gas. Further limitations on the joint are that it must be thick enough to maintain electrical isolation yet thin enough to insure that the joint does not fail due to differential thermal contraction on cool down.

The joint is made using epoxy which has been degassed in a vacuum chamber to remove any bubbles which could compromise electrical isolation. Any soldering done on the heat pipe after the epoxy joints are made must be with a low melting point solder to ensure that the nylon washers and epoxy are not damaged. The flexible upper copper lead is soldered into the heat pipe center rod using a soft silver solder before the epoxy joint is made. The flexible links are soldered to the bottom of the heat pipe center rod using a low melting point (58 C) solder after both epoxy joints are made. No soldering can be done after the heat pipe is filled with nitrogen due to the danger of exceeding design stresses by heating the nitrogen gas.

Filling the heat pipe with 138 bar (2,000 psi) nitrogen gas at room temperature is the final step in the assembly of the current leads. The shape of the cavity between the center rod and the heat pipe body was determined by thermal design requirements. When the heat pipe is cooled to its operating temperature the nitrogen will exist in the cavity as a two-phase fluid. The liquid nitrogen will just fill the well in the bottom of the cavity. This well was designed to balance the area required for boiling heat transfer on the center rod with the upper area required for condensation on the heat pipe body. By making the heat pipe body in two pieces the center cavity with the required geometry, as discussed above, can be easily machined. The rounded corners of the nitrogen cavity are intended to reduce electrical arcing which can occur at sharp projections.

HEAT PIPE PERFORMANCE

The outer copper cylinder serves as the heat pipe condenser and the center rod serves as the evaporator. During operation, heat is transferred from the center rod to the liquid nitrogen in the bottom of the heat pipe cavity. Heat is removed from the nitrogen vapor in the upper cavity by condensation. These mechanisms combine to allow high heat transfer rates with very low temperature differences.

Thermal Testing

The thermal performance of the heat pipe was determined by measuring the heat input to the center conductor and the resulting temperature difference between the center conductor and the outer cylinder. An electrical heater with voltage taps was attached to one end of the heat pipe center rod. The heater power was determined by measuring the voltage and current. Silicon diode thermometers, mounted on the opposite end of the heat pipe center rod and on the heat pipe outer cylinder, provided the temperature difference and absolute temperature. The heat pipe outer cylinder was clamped in a copper fixture and heat sunk to a liquid nitrogen bath .

The steady state temperature difference across the heat pipe was measured as a function of the heater power. The results are shown in figure 3. At low heater power heat transfer is by evaporation at the pool surface. Above approximately 3 watts boiling becomes the primary mode of heat transfer. Beyond this point additional heat transfer is achieved with relatively small increases in temperature difference. The temperature difference levels off to about 2.7 K for heater powers from 27 to 35 watts. Testing was stopped at 35 watts when the heater reached its maximum output. Additional testing is planned to determine the maximum heat transfer rate possible across the intercept.

The data from figure 3 was replotted as thermal conductance versus heat transfer rate. The results are shown in figure 4. The thermal conductance increases from of 2 W/K at low

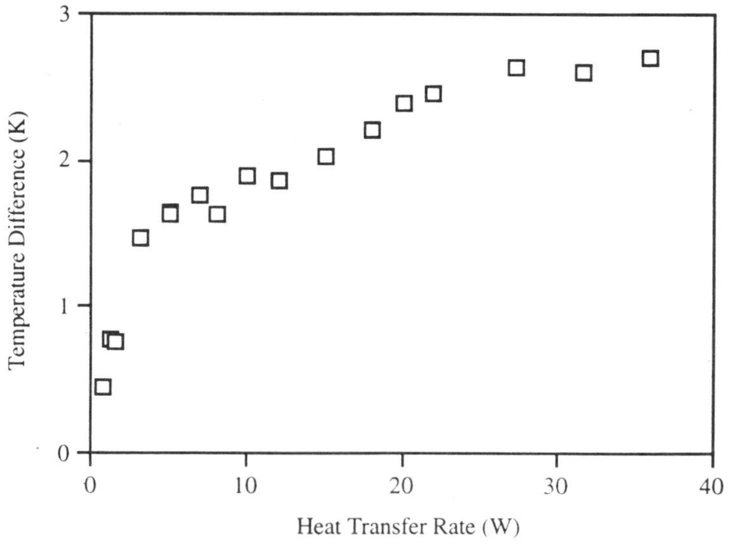

Figure 3. Heat pipe thermal performance at 80 K.

heat transfer rates to 13 W/K at 35 W. These values can be compared to a thermal conductance of 3.75 W/K (12 K temperature difference at 45 W heater power) reported by Argonne National Laboratory and Superconductivity, Inc. for a thermal interference fit current lead heat intercept.[6]

Electrical Testing

Electrical performance measurements were obtained by measuring the leakage current across the heat pipe as a function of the applied voltage. A positive voltage was applied to the heat pipe center rod with the heat pipe body connected to electrical ground. A large resistance (10 M Ω) was placed in the circuit to limit current flow in the event of electrical breakdown. The leakage current across the heat pipe was then measured.

Initial measurements were made with no heat flowing across the intercept, first at room temperature and then at 87 K. Once the integrity of the heat pipe electrical isolation was verified, measurements were made with heat loads on the heat pipe ranging from 5 to 15 W and with the temperature at 87 K. As figure 5 shows, there is essentially no difference between the room temperature and 87 K measurements with no heat load. Furthermore, a comparison of the data with an applied thermal load with the data with no thermal load indicate that the evaporation and condensation occurring in the heat pipe do not affect the electrical performance. No electrical breakdown was observed at voltages up to 8,000 volts. This performance is in excess of that required for current leads in large electric utility applications.[7]

The heat pipe greatly exceeded its electrical performance target of withstanding 1,000 volts with a 5 W thermal load. Initial concerns that the bubbles produced by boiling at the heat pipe center rod would significantly degrade electrical performance were demonstrated to be unfounded.

CONCLUSIONS

We have demonstrated the ability of a composite heat pipe, made of conducting and non-conducting materials, to function as an HTS current lead thermal intercept. The heat pipe transferred 35 W with a 2.7 K temperature difference, which gives a thermal conductance significantly higher than designs which rely on thermal conduction through solids. The heat pipe thermal intercept successfully withstood 8,000 volts without breakdown while carrying a thermal load of 15 watts. The heat pipe greatly exceeded its performance targets of 1,000 volts electrical isolation with a 5 watt thermal load.

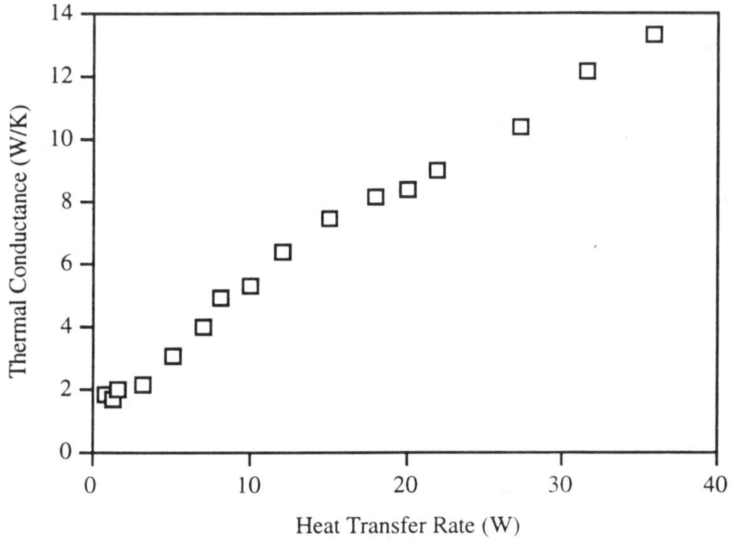

Figure 4. Thermal conductance of the heat pipe.

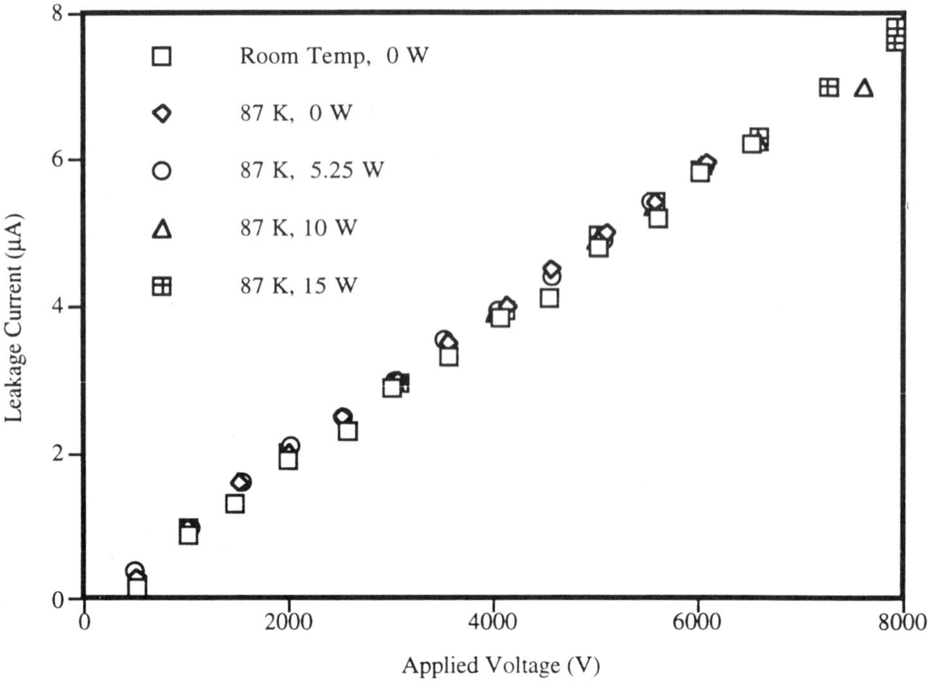

Figure 5. Heat pipe electrical performance.

The rugged and compact heat pipe design can be applied to a variety of current lead configurations. The ability to vary the shape and size of the evaporator and condenser heat transfer surfaces allows the heat pipe to be customized to meet a wide range of design requirements. It should be possible to cool very large current leads using this heat pipe design and maintain a very compact geometry. Finally, the heat pipe is built of inexpensive materials and does not require precision tolerances or expensive machining for its components.

ACKNOWLEDGMENTS

The authors would like to acknowledge the technical assistance of Claude Connor and Tim Pierce. This work was performed under the auspices of the United States Department of Energy, Office of Energy Management.

REFERENCES

1. K. Ueda, K. Takita, T. Ueda, M. Minura, N. Uno, and T. Tanaka, Thermal Performance of a Pair of Current Leads Incorporating Bismuth Compound Superconductors, *Fifth International Symposium on Superconductivity*, Kobe, Japan, November 16-19, 1992.
2. R. Wesche and A.M. Fuchs, Design of Superconducting Current Leads," *Cryogenics*, Volume 34, Number 2, p. 145, 1992.
3. M. Daugherty, F. Prenger, D. Hill, D. Daney, L. Worl, A. Schake, and D. Padilla, Application of High-Temperature Superconductors to High-Gradient Magnetic Separation, *Proceedings of the 29th Intersociety Energy Conversion Engineering Conference*, Monterey, CA, August 7-11, 1994, p. 1502.
4. R.C. Niemann, J.D. Gonczy and T.H. Nicol, Low-Thermal -Resistance, High-Electrical-Isolation Heat Intercept Connect, *Advances in Cryogenic Engineering*, Volume 39, p. 1665, 1994.

5. M.A. Daugherty, F.C. Prenger, D.D. Hill, D.E. Daney, and K.A. Woloshun, HTS Current Lead Using a Composite Heat Pipe, Paper LNC-3, presented at the 1994 Applied Superconductivity Conference, Boston, MA, October 16-21, 1994. Submitted to *IEEE Transactions on Applied Superconductivity.*
6. R.C. Niemann, Y.S. Cha and J.R. Hull, W.E. Buckles, B.R. Weber, and S.T. Yang, Performance Evaluation of High-Temperature Superconducting Current Leads for Micro-SMES Systems, Paper LNC-8, presented at the 1994 Applied Superconductivity Conference, Boston, MA, October 16-21, 1994. Submitted to *IEEE Transactions on Applied Superconductivity.*
7. R.C. Niemann, Y.S. Cha and J.R. Hull, C.M Rey, and K.D. Dixon, Design of High-Temperature Superconductor Current Lead for Electric Utility SMES, Paper LNC-7, presented at the 1994 Applied Superconductivity Conference, Boston, MA, October 16-21, 1994. Submitted to *IEEE Transactions on Applied Superconductivity.*

AC LOSS CALCULATIONS FOR HTSC CURRENT LEAD MODULES

H.P. Kraemer

Siemens AG, Corporate R&D
91050 Erlangen, Germany

ABSTRACT

In applications of low temperature superconductors losses are often dominated by the heat load due to current leads. Compared to conventional metal current leads the use of HTSC modules for the low temperature part can considerably reduce the heat released at the cold end of the current lead. In contrast to DC applications, the losses in the HTSC modules for AC application strongly depend on the cross sectional shape. In this work the temperature distributions of bulk HTSC current lead modules are calculated by numerical solution of the one dimensional heat conduction equation. Limitations of the critical state model, which is used to estimate the self field losses, are discussed. The impact of the module's geometry on the low temperature heat load is shown for circular rod and tube type modules. Guidelines for the geometrical layout of such modules are given.

INTRODUCTION

Nearly all superconducting devices need current leads (CL) as connections between the cold device and the power sources at ambient temperature. The ideal current lead material should be a good electrical but a poor thermal conductor. Materials used for conventional current leads are normal conductors and therefore essentially obey the Wiedemann-Franz law, which relates electrical resistivity, temperature and thermal conductivity of metals. This relation leads to a theoretical limit for the performance in terms of low temperature heat load, which is independent of the metal used. Especially for DC devices with conventional current leads the heat leak due to current leads often dominates the low temperature losses of the whole system.

The use of high T_c superconductor (HTSC) material for the low temperature part (below 80 K) of a current lead offers the combination of electrical superconductivity and low thermal conductivity typical for ceramics thus promising a distinct improvement of the CL performance. Many applications of superconductivity already utilize such hybrid current leads and especially for cryogen free DC magnets made of low T_c superconductors their use is essential.

The work presented here is part of a BRITE/EURAM cooperation between Alcatel Alsthom[1], Hoechst[2] and Siemens[3], which aims to design and build hybrid AC current leads for 5000 A_{rms}. Compared to DC-applications alternating currents give rise to additional losses due to the variation of the magnetic self field of the conductor leading to skin effects in normal conductors and hysteretic losses in superconductors. Both types of AC losses can be reduced by subdividing the conductor into filaments. As it is not practicable to produce unsheathed filaments from ceramic superconductors, the AC losses of rod and tube shaped HTSC modules have been estimated in order to find the optimum geometry.

DETAILS OF CALCULATION

Differential Equation

The temperature distribution $T(x)$ of a purely conduction cooled (no heat exchange with surrounding He atmosphere) HTSC module is calculated by solving the one dimensional stationary heat conduction equation

$$\frac{d}{dx}\left(\lambda(T)\cdot\frac{dT}{dx}\right)+S=0 \tag{1}$$

where λ is the thermal conductivity and S is the source term, which is the dissipated heat per unit volume. The temperatures at both ends (4.2 K and 77 K resp.) are taken as boundary conditions and the differential equation is numerically solved using the Tridiagonal Matrix Algorithm. Besides the temperature distribution the computer program yields the loss distribution $q(x)$ and the low temperature heat load Q_L, which is the heat released at the cold end and a parameter characterizing the performance of current lead.

AC Losses

The critical state model or Bean model[4] which is used to estimate AC-losses is based on the assumption that the current density j anywhere inside the conductor is either zero or equal to the critical current density j_c. The expression for AC losses per volume Q/V in circular rods given by Wilson[5] is generalized for concentric tubes with inner and outer diameters D_i, D_o as

$$\frac{Q}{V}=f\cdot\frac{\mu_o j_c^2}{2}\cdot R_o^2\cdot\beta\cdot\gamma\cdot\Gamma(\gamma) \tag{2}$$

with f: frequency (50 Hz)
$\mu_o = 1.256\cdot10^{-6}$ Vs/Am,
$\beta = I_{peak} / I_c$,
$\gamma = \beta(1 - (D_i / D_o)^2)$ and

$$\Gamma(\gamma)=\frac{2}{\gamma}-1+\frac{2}{\gamma^2}\cdot(1-\gamma)\cdot\ln(1-\gamma)\ . \tag{3}$$

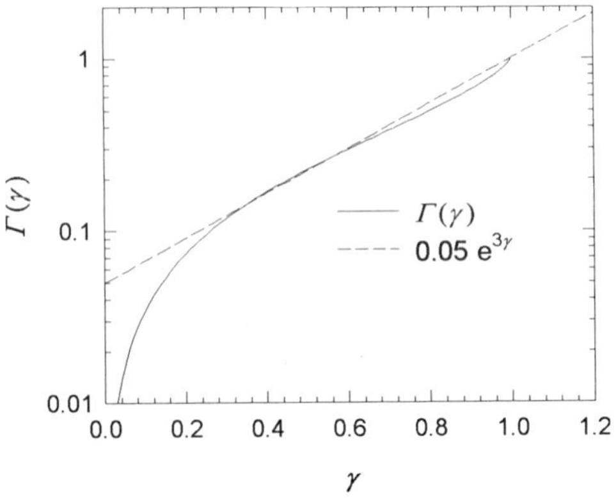

Figure 1. AC-loss function $\Gamma(\gamma)$ and its extrapolation for $\gamma > 1$

The major disadvantage of the critical state model is that it does not account for the flux flow resistance of a superconductor and therefore gives no solution for currents beyond the critical one I_c. To avoid numerical instability problems at currents close to I_c an extrapolation of the function $\Gamma(\gamma)$ is used for $\gamma>1$ as shown in Fig. 1.

Superconductor Properties

An analytical function $j_c(B,T)$ has been fitted to data measured on BSCCO 2212 samples small enough to neglect their self field[1]. To account for variations of j_c in different samples $j_c(B,T)$ has been scaled linearly to match the intrinsic current density at 77 K, $j_{c,o}$. In the calculations presented here a value of $j_{c,o}$ = 2000 A/cm², which is confirmed by experiments, has been used. The function for the thermal conductivity $\lambda(T)$ also has been fitted to data measured on BSCCO 2212 rods. The thermal conductivity integral between 4.2 K and 77 K is 65 W/m. Figs. 2 a), b) show experimental data and the corresponding fits for $j_c(B,T)/j_{c,o}$ and $\lambda(T)$ respectively.

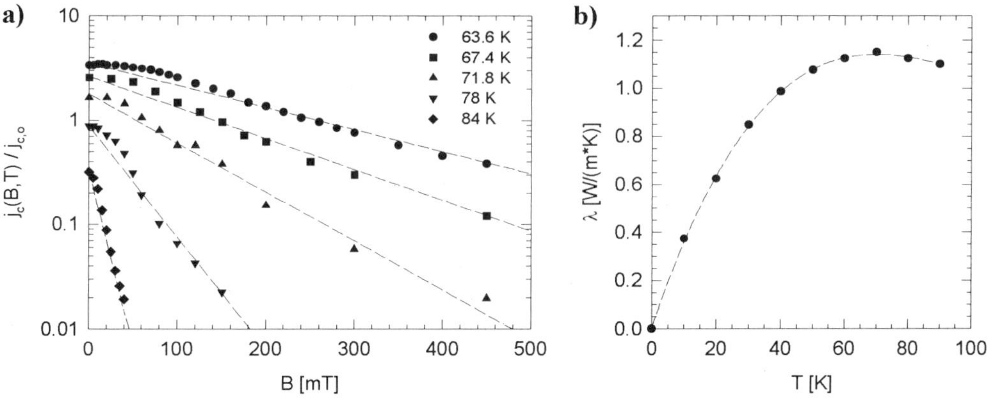

Figure 2. BSCCO 2212 experimental data and corresponding fits used in the calculations **a):** normalized critical current density $j_c(B,T)/j_c(0\ \mathrm{T},77\ \mathrm{K})$, **b):** thermal conductivity $\lambda\ (T)$

Self Field Effects

In current carrying rods or tubes the self field is maximum at the surface and zero at the center. Together with the field dependence of j_c this leads to a radial variation of j_c. As the one-dimensional model does not account for radial variations, an average j_c has to be used. Because of the nearly exponential decay of $j_c(B)$, in most cases the average value is close to the value at the surface[2] $j_c(B_{max})$ with $B_{max} = \mu_o I_{peak}/\pi D_o$. For this reason $j_c(B_{max})$ has been used in the calculations. External magnetic fields which may exist in some applications have not been considered.

RESULTS AND DISCUSSION

Temperature and Loss Distributions

Typical temperature and loss distributions of a HTSC module with a nominal current of 1 kA_{rms} are shown in Fig. 3 and compared to the values of a 1 kA_{DC} module. Both modules are rods of 200 mm length with diameters of 20.6 mm (AC) and 9.03 mm (DC). For the DC loss calculations experimental data for the flux flow resistance of the

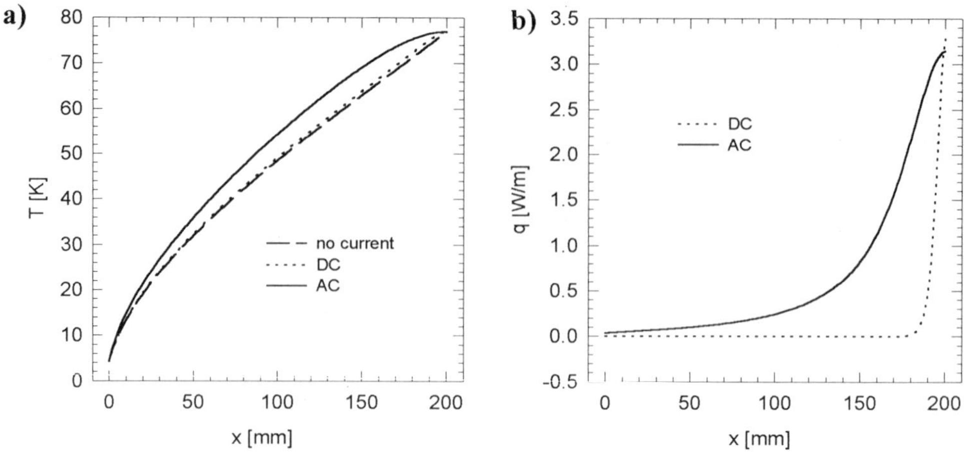

Figure 3. a) Temperature distributions in HTSC current lead modules for AC, DC and zero current. **b)** Loss distribution in HTSC current lead modules for AC and DC.

superconductor have been used. The dashed, dotted and solid lines in Fig. 3a) represent the temperature distribution $T(x)$ of the modules without, with DC and with AC operating current respectively. It can be seen that for the DC lead the temperature gradient at the cold end and thus the low temperature losses of about 21 mW are mainly independent of the load current. This is different for the AC lead which has standby losses of about 107 mW and operating losses of 132 mW. At nominal current the temperature gradient at the warm end is near zero for both versions which means that there is neither a release nor an absorption of heat from/to ambient. The comparison of typical AC and DC loss distributions is shown in Fig. 3b). In contrast to the strong concentration of the DC losses (dotted line) at the warm end of the lead, the AC losses exhibit a very smooth distribution with notable losses even at the colder part of the CL.

Multiple Rod Modules

The first current lead configuration studied consists of a number of circular rods arranged in parallel with spacings large enough to neglect the magnetic field generated by neighboring rods. Fig. 4 shows the low temperature heat load Q_L as a function of the total cross sectional area A for such arrangements with rod diameters between 10 and 50 mm. The two sets of curves correspond to rods 100 mm and 200 mm long. Both Q_L and A are normalized to the effective transport current I. Dots and rod numbers refer to the design current of 5 kA_{rms}. All curves reach the stability limit which is the lowest cross sectional area A for which the total length of the current lead is superconducting. The dashed lines stand for the low temperature heat load in absence of a current. Obviously for given length and rod diameter there is a minimum Q_L reached at an optimum cross sectional area A_o slightly above the stability limit. It can be seen that at a constant cross sectional area the reduction of the rod diameter leads to a smaller Q_L (e.g. from 4 rods of 20 mm∅ to 16 rods of 10 mm∅). Additionally the smaller diameter allows for a reduction of A (from 16 rods of 10 mm∅ to 10 rods) leading to an even greater decrease of Q_L. The comparison of arrangements with different lengths shows that the necessary cross sectional area increases less than the length and therefore further reduces the losses. This is in contrast to metal current leads where the losses are constant for equal L/A ratios. Thus the design rule drawn from the above is to make the rods as thin and long as possible in order to obtain the

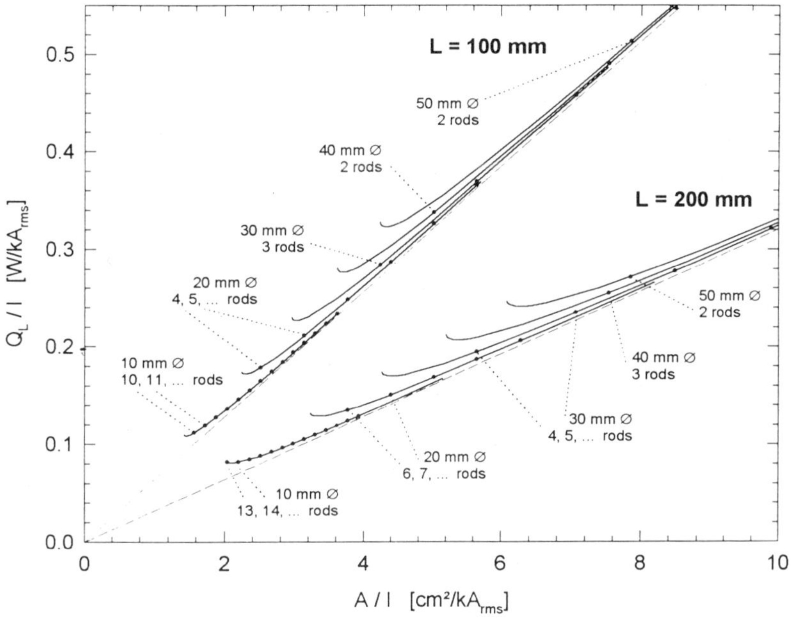

Figure 4. Specific low temperature heat load per pole of multiple rod configurations for various rod diameter and length values. Dots and rod numbers refer to $I = 5$ kA$_{rms}$.

minimum low temperature heat load. Since long and thin ceramic rods are expected to be very fragile, a compromise has to be found between the performance, the mechanical stability and the space requirements of the current lead.

A multiple rod configuration has the advantage that single elements can be tested before final assembly and that the CL can easily be adapted to other nominal currents by changing the number of rods. However there are some disadvantages. First, if the realistic magnetic self field distribution is regarded, the current carrying capacity of a cage of rods is clearly below the one of a tube of the same overall cross section. Second, due to different contact resistances and differences in the stray fields to which individual rods are exposed, uneven current distribution is likely under practical AC conditions which might cause an unstable behavior of the total module.

Single Tube Modules

The problems mentioned above and the availability of BSCCO tubes manufactured by a centrifugal melt casting process[2] were the motive to examine an alternative CL configuration consisting of a single tube.

The geometry of a tube is determined by three independent parameters which can be chosen as length L, outer diameter D_o and wall thickness W. If two of them are given within reasonable limits, the third one can be optimized for a minimum Q_L. In Fig. 5 a typical graph for Q_L as a function of L and D_o is given, the wall thickness being fixed at 5 mm in this example. It indicates that in contrast to metal current leads there is no lower limit for Q_L. Instead the losses steadily decrease with increasing length. However for a fixed diameter there is a relative minimum Q_L at an optimum length L_o and vice versa for a fixed length there is an optimum diameter which minimizes Q_L. The influence of the wall thickness becomes obvious from Fig. 6a). For W values of 3, 5 and 7 mm Q_L is shown as a function of length assuming that for each combination of L and W the outer diameter is optimized. As

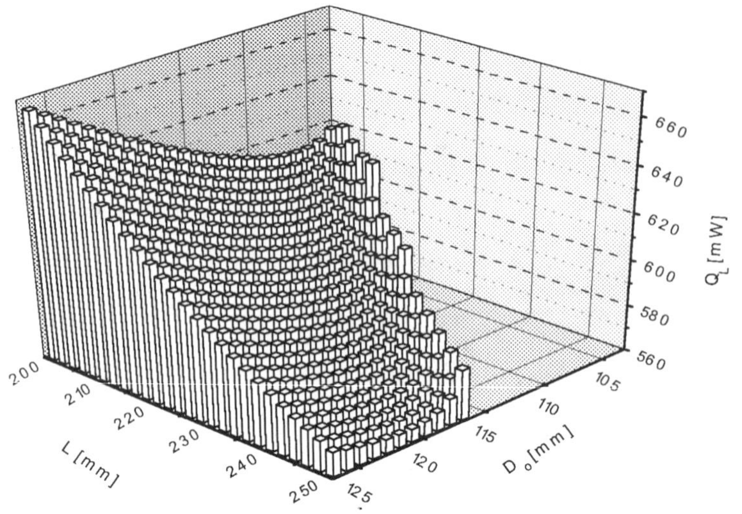

Figure 5. Low temperature heat load per pole as a function of length and outer diameter for tubes with a wall thickness of 5 mm at design current of 5 kA_{rms}.

can be seen losses can be reduced by increasing the length and by decreasing the wall thickness. The example in the diagram makes clear that the wall thickness is the more "effective" parameter: a reduction of the 790 mW losses of a 100 mm long and 5 mm thick tube to about 565 mW can be achieved either by increasing the length to 250 mm or by reducing the wall thickness to 3 mm.

The effect of variations of length and wall thickness on the optimized outer diameter is shown in Fig. 6b) as plots of L vs. D_o corresponding to the curves in Fig. 6a). As expected both ways to reduce the losses lead to an increase of the required diameter. To complete the above example: the outer diameter rises from 70 mm to 80 mm for the tube with reduced wall thickness and to 120 for the tube with increased length.

Although there is no general optimum geometry from theoretical reasons, in practice it is difficult to produce and handle very long and thin walled tubes with large diameters.

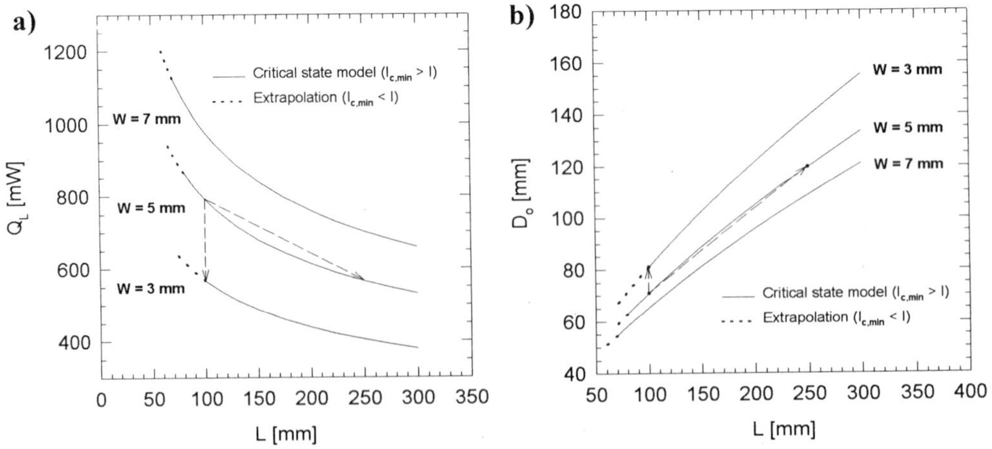

Figures 6. a) Low temperature heat load per pole as a function of length for tubes with different wall thickness values and optimized diameters. **b)** Optimum outer diameter as a function of length for tubes with different wall thickness. $I = 5$ kA_{rms}. See text for details.

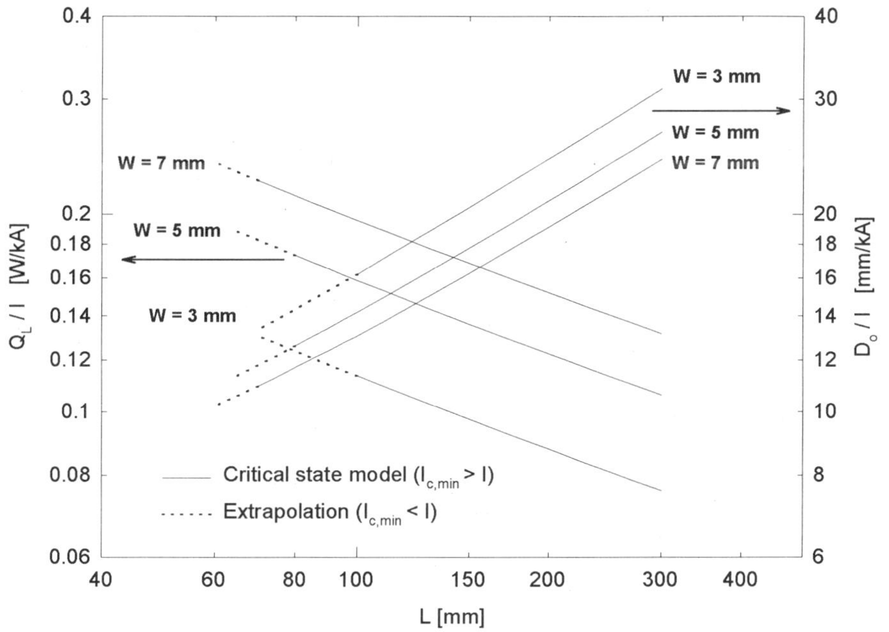

Figure 7. Specific low temperature heat load per pole and optimum outer diameter as a function of length for tubes with wall thickness 3, 5 and 7 mm.

Furthermore the space available for current leads is generally limited. Consequently the current lead geometry has to be optimized within the limits imposed by the design of a specific application and by the technology to produce HTSC elements.

The last diagram in this chapter (Fig. 7) gives rough guidelines to estimate the required optimum outer diameter as well as the low temperature losses for HTSC tubes of given length and tube wall thickness. The nearly straight lines in the log/log diagram can be approximated by $Q_L \propto L^{-0.375}$ and $D_o \propto L^{0.56}$. The values in the diagram are calculated for 5000 A_{rms} but should be scaleable to higher currents because the specific cross sectional area A/I and the self field at the surface essentially remain constant due to the high D_o/W ratio.

Limitations of the Critical State Model

The critical state model uses the most simple approximation for the $E(j)$ characteristic of a superconductor, that is $E(j<j_c) = 0$ and $E(j>j_c) = \infty$. Due to this the AC or hysteretic losses per volume can be described by an analytical function (Eq. 2). The thermal analysis of CL modules can thus be performed in a rather short time on a standard PC and variations of parameters are easily accomplished. A more realistic approximation for experimental $E(j)$ curves is $E(j) = E_c \cdot j_c(B,T)^{\,n(B,T)}$, which requires the application of a two-dimensional axisymmetric model[1] and leads to a considerable increase in the calculational effort.

The problem of the critical state model is, that it is only valid for currents below the critical one. As mentioned before, an extrapolation of $\Gamma(\gamma)$ is used when the critical current is exceeded. In Figs. 6 to 7 the use of this extrapolation is indicated by the dotted lines. The calculational results should be reliable as long as the system is "close" to the critical state model, that is the ratio of the peak operational and the minimum critical current reached at the warmest part of the module $\beta_{max} = I_{peak} / I_{c,min}$ is close to 1, and $\beta > 1$ only holds for a small part of the total length of the HTSC module. This is supported by the log/log-plot of Q_L/I and D_o/I as function of length in Fig. 6, where the results for $\beta_{max} > 1$ correspond well to a graphical extrapolation of the $\beta_{max} < 1$ results. Another justification for the use of the extended critical state model is the comparison with values calculated using experimental $E(j)$ curves[1]. The deviation of the optimized length and the corresponding Q_L is found to be less than 10 % for a module with $\beta_{max} = 1.2$ and about 30 % for $\beta_{max} = 3.5$. This suggests that the extended critical state model can be used reasonably well up to $\beta \approx 1.2$, which means that it is applicable for most cases of practical interest, but it should not be used for extremely thin and short CL modules.

CONCLUSIONS

For the thermal analysis of conduction cooled HTSC current lead modules a numerical tool has been developed. Due to the use of the critical state model including an extrapolation to currents above the critical one a fast variation of geometrical parameters is possible. It is found that in order to minimize the low temperature losses the HTSC modules should be long and thin rods or a long and thin walled tube. It is essential to use the lowest rod diameter or wall thickness that can safely be manufactured and handled. A compromise has to be found between low losses, geometrical constraints and mechanical stability. The use of the extended critical state model is justified by the comparison with a model using experimental $E(j)$ curves and is estimated to give reasonable results for most geometries of practical importance. So, for the 5 kA_{rms} current lead , a HTSC module is suggested with 70 mm outer diameter, 5 mm wall thickness and 100 mm length contributing about 0.16 W/kA per pole to the 4 K heat load. For very short and thin walled tubes and thin rods the extended critical state model becomes unreliable and the more sophisticated $E(j)$ model should be used.

ACKNOWLEDGMENTS

This work was supported by the European Community in the frame of BRITE / EURAM under contract number BE 7856. The author would like to thank C. Albrecht for valuable and stimulating discussions.

REFERENCES

1. P.F. Herrmann, C. Cottevieille, A. Leriche, S. Elschner, "Refrigeration load calculation of a HTSC current lead under AC conditions", *MT 14, Tampere, Finland*, June 11-16, 1995
2. S. Elschner, J. Bock, G. Brommer, P. Herrmann, "High currents in MCP BSCCO 2212 bulk material", *MT 14, Tampere, Finland,* June 11-16, 1995
3. C. Albrecht, J. Bock, P.F. Herrmann, "Stromzuführungen mit Hochtemperatursupraleitern", *VDI/GET, München*, 10 -11. Mai 1995
4. C.P. Bean, *Phys. Rev. Lett. 8*, 259 (1962) and *Rev. Mod. Phys. 36*, 31 (1964)
5. M.N. Wilson, "Superconducting Magnets" , *Clarendon Press, Oxford, UK*, 1983, pp. 194-197

DEVELOPMENT OF A 1 kA CLASS ALTERNATING CURRENT LEAD SYSTEM FOR A FAULT CURRENT LIMITER

M. Takahashi,[1] T. Kuriyama,[1] H. Nakagome,[1] K. Yamamoto,[1] Y. Yamada,[1] M. Nakade,[2] T. Okuma,[2] and T. Hara,[2]

[1]Toshiba, R & D Center
Kawasaki-ku, Kawasaki, 210 JAPAN
[2]Tokyo Electric Power Company, Power Engineering R & D Center
Tsurumi, Yokohama, 230 JAPAN

ABSTRACT

A thermal design of a 1 kA / 6.6 kV class superconducting fault current limiter was discussed, and specifications for a current lead system were obtained. According to these specifications, a 1 kA rms current lead system has been designed and tested. The current lead system was composed of $(Bi,Pb)_2Sr_2Ca_2Cu_3O_{10+x}$ tube leads (used in 77 - 4 K range), Cu leads (used in 300 - 77 K), a liquid helium vessel and a 77 K refrigerator, which cools the thermal anchor of the current leads and a radiation shield. The current lead system was placed in a vacuum vessel and cooled by thermal conduction. This lead system successfully transported 1 kA rms alternating current for 3 hours under steady-state temperature distribution. The total heat load to 4 K for 1 kA rms current was 3.5 W. We have a prospect of developing a small cooling system for a 1 kA class fault current limiter.

INTRODUCTION

A superconducting fault current limiter [1] to limit short-circuit currents in power systems is one of the promising alternating current (AC) applications of superconductivity. Recently, a 1 kA rms and 6.6 kV rms class current limiting device, to be used in transmission or distribution systems, has been developed and succeeded in limiting current [2]. But in this experiment, an open type liquid helium cryostat was used. To apply the fault current limiter for industrial use, a more sophisticated cooling system and current lead systems are required. The purpose of this work is to design a current lead system for a 1 kA rms / 6.6 kV rms class fault current limiter which is cooled by a small refrigerator requiring no replenishment of liquid helium, and to study the prospect of developing a fault current limiter system.

THERMAL DESIGN OF A FAULT CURRENT LIMITER

System Concept

A characteristic of this system is the use of large alternating currents, which cause a high heat load to liquid helium vessel where the current limiter is housed. Therefore, reducing the heat load is one of the most important problems in the design.

In addition, the 1 kA class fault current limiter is intended for use in a small underground substation, and as a result easy operation and compactness of the system are required. So the system could be thought of as a closed system with small refrigerators. This system needs no replenishment of liquid helium and exhausts no helium vapor, which can cause anoxia.

Figure 1 shows the schematic diagram of the 1 kA class fault current limiter. The technical objectives of this system are: reducing the heat load caused by large currents, and providing refrigeration by a small, high efficiency and high reliability refrigerator.

Current lead system

There are two type current lead systems. One is a conduction cooled system and the other is a gas cooled system. We selected a conduction cooled current lead system because gas evaporation can not be used in a closed system. Also, this type of system has a stable temperature distribution which improves the stability of high-Tc superconductor part of the current leads in the system.

The thermal conduction and ohmic heating of current lead is one of the major portion contributions to the heat load to 4 K. The heat load and the sizes of usual copper current leads can be optimized. The optimized heat load at 1 kA rms is 24 W for a pair of conduction-cooled copper current leads, in which the higher end temperature is 77 K and the

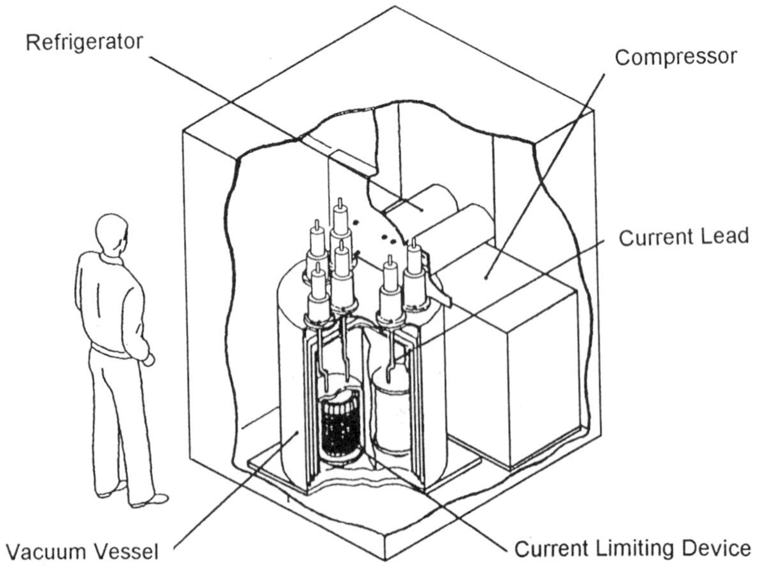

Figure 1. Scheme of a 1 kA class fault current limiter.

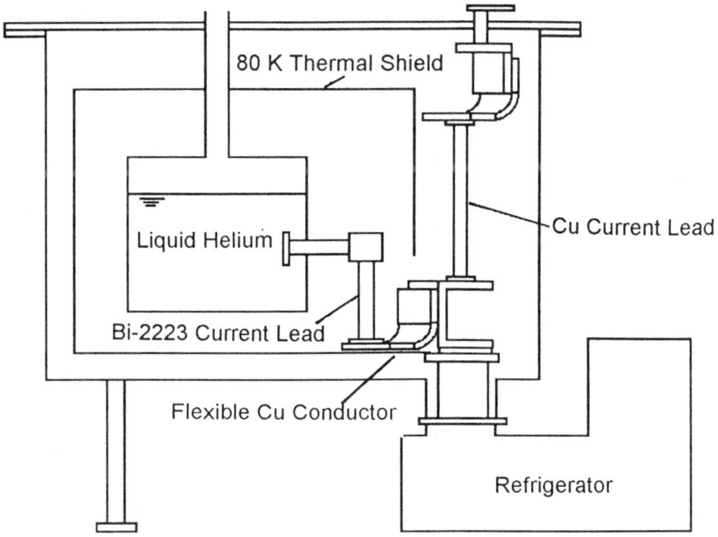

Figure 2. Schematic figure of the current lead system test apparatus.

lower end temperature is 4.2 K. The 24 W heat load is too large to handle by a small refrigerator.

Using high-Tc superconductor current leads is one of the best solutions to this problem. There are many studies related to high-Tc superconductor current leads, especially the gas-cooled type [3]. The conduction cooled type has been studied only recently [4], and very little research has been reported on AC applications. Tasaki et al. [5] reported Bi-2223 current leads cooled by a GM refrigerator that transported 400 A rms current, and the heat load of this lead was lower than that of a conduction-cooled copper current lead system.

The current lead system was manufactured and tested as described next section.

4 K Heat Load and 4 K Refrigerator

The major portions of heat load to 4 K were AC loss of the current limiting device and thermal conduction in the high-Tc current leads. The AC loss of the device was estimated to be about 1 W. The thermal conduction of high-Tc current leads was calculated from the thermal conductivity and was about 1.5 W for a pair of current leads. High-Tc superconductor leads also have a magnetization hysteresis loss or an AC loss generated by self-field. The AC loss was considerably reduced for a tubular bulk compared to the case of the cylindrical bulk having the same cross section and Jc value [6]. All other heat load such as resistive heating at connections and thermal conduction by supports are less than 1 W. The fact that there are three limiting devices and current lead systems in a three phase current limiter, the total heat load on the helium vessel was estimated to be about 10 W.

A 4 K refrigerator for recondensation of evaporated helium gas is necessary for a closed current limiter system. It could be a small refrigerator such as a GM / JT refrigerator or a Staring / JT refrigerator. The small refrigerator will be used not only to save the input power but also to reduce the refrigerator volume. We selected a GM / JT refrigerator for its high reliability. Also, Sumitomo Heavy Industries Ltd. and Daikin Corporation have developed a large refrigeration power and high efficiency GM / JT refrigerator for magnetic levitation train, with a refrigeration power at 4.3 K is 8 W, and the COP is smaller than 1000.

With a little improvement to reduce heat load, the current limiter system will be able to be cooled by only one GM / JT refrigerator. Therefore, we aimed at decreasing the heat load to under 8 W for the three-phase system.

77 K Heat Load and 77 K Refrigerator

A 77 K refrigerator is needed to cool the thermal anchor for the current leads and a radiation shield. Most of the heat load to 77 K was from the copper current leads. The calculated minimum heat load of a pair of copper leads was 90 W at 1000 A rms current. The total heat load to 77 K in a three phase system was calculated to be about 280 W.

There are many types of 77 K refrigerators and selected a Staring cycle refrigerator was selected for its high efficiency.

CURRENT LEAD SYSTEM TEST APPARATUS

Construction

Figure 2 is a schematic diagram of a current lead system test apparatus, which was designed to demonstrate the actual lead system of the current limiter. It consisted of a helium vessel, a pair of current leads and a 77 K refrigerator, which were placed in a vacuum vessel for thermal isolation. A radiation shield cooled by the 77 K refrigerator was placed to reducethermal radiation. The current leads were shorted in the liquid helium vessel and this experimental apparatus had no current limiter device. Also, there is no 4 K refrigerator, and evaporated helium gas was allowed to flow out.

Current Lead System

The current lead system consisted of a pair of copper leads , a 77 K thermal anchor and a pair of Bi-2223 current leads.

The Bi-2223 current leads were placed between the 77 K thermal anchor and the liquid helium vessel. They were fabricated by Sumitomo Heavy Industries, Ltd.[7] using the CIP (cold isostatic press) and sintering technique. To enhance the critical current density Jc, a

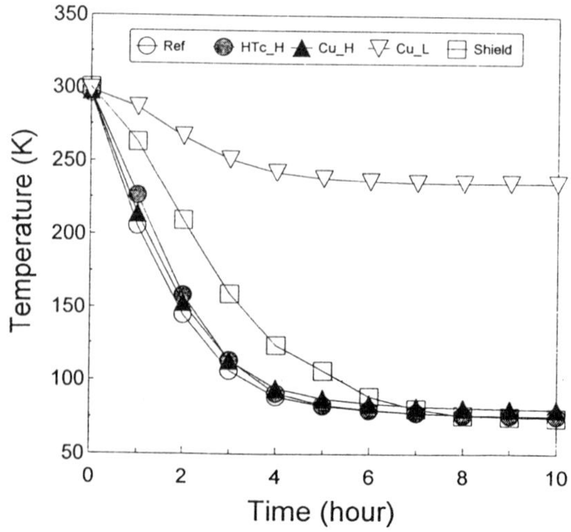

Figure 3. Temperature at selected positions as a function of time.

tubular shape was used; 34-36 mm in outer diameter, 30 mm in inner diameter and 190 mm in length. Both ends of the bulk were coated with silver and soldered to copper terminals. The typical value of DC critical current at 77 K and 0 T was 1800 A, and contact resistance at a connecting terminal was 2×10^{-8} (Ω). The Bi-2223 current leads were supported by stainless pipe to protect the leads from vibration.

Copper leads were used between the room temperature region and the 77 K thermal anchor; these leads were pipe-shaped. Alternating current causes surface effects and the current is contained only to the skin depth on the surface of the leads. Calculated skin depth of copper was 4 mm at room temperature and 2 mm at 77 K. Therefore, we selected 3 mm thick pipes. Inner and outer diameter of the copper pipes were 19 mm and 25 mm respectively, and length was 340 mm. They were optimized at 1700 A, which was larger than the testing current for safety.

Some flexible copper conductors were used to reduce the vibration from the refrigerator. They consisted of 200 stacked copper films with 0.1 mm thickness and binding blocks, and had enough flexibility.

77 K Refrigerator and Thermal Anchor

The 77 K refrigerator was a Staring cycle refrigerator of Carrier Corporation make. Its refrigeration power at 77 K was 250 W with 3.6 kW input power. Its operating speed was variable from 900 rpm to 1800 rpm, and refrigeration power varied according to the operating speed. The refrigerator temperature was held at 77 K by controlling the operating speed. The operating state of an industrial current limiter was simulated. Also, the refrigerator was operated avoiding resonance frequencies to damp the vibration of the Bi-2223 leads.

The 77 K thermal anchor consisted of a copper plate, two copper blocks and 12 AlN tips. The copper plate was cooled by the 77 K refrigerator. The copper blocks were soldered to the current leads, and screwed on the copper plate with electrical insulation. Between each block and plate, 6 AlN tips were clamped. Here, indium sheets were inserted between each contact surface for good thermal contact. The size of each AlN tip was 20 mm $\times$ 20 mm $\times$ 4 mm. The measured thermal conductivity of this AlN tip at 77 K was 5 W / cm K, and the contact thermal resistance was estimated as 0.1 K cm^2 / W. The calculated thermal resistance of the thermal anchor was 0.0058 K / W.

EXPERIMENTAL RESULTS

Cooling Characteristics

After turning on the refrigerator and supplying liquid helium into the vessel, the test apparatus was cooled as shown in figure 3. The current lead system was cooled down to steady state in 5 hours. But the radiation shield temperature remained over 100 K. The entire system was cooled down to steady state in 8 hours. The temperature of the refrigerator was 72 K and both thermal anchor and radiation shield were cooled to about 77 K. The main reason for the temperature difference was caused by thermal resistance of the copper plate which connect the refrigerator and the thermal anchor. The temperature difference in the thermal anchor was 1.0 K, and the thermal resistance was 0.008 K / W. Here the heat load to 77 K was 120 W which was estimated from the temperature difference in the copper plate. It is the same order as the design value.

The transporting current successfully reached 1 kA rms, following a holding test of more than 3 hours duration. Figure 4 shows elapsed time profiles of the temperature at

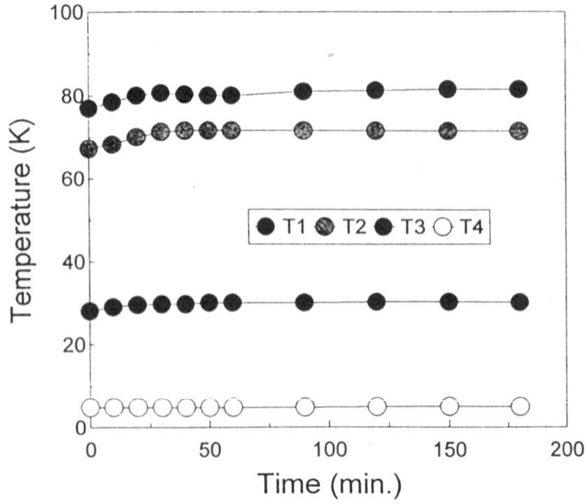

Figure 4. Temperature profile of the Bi-2223 lead during a 1000 A rms transporting test.

several points during the 1 kA rms holding. Each measured point is shown in figure 2. T_1 rose to 82K in the first 40 min, and became stable. T_2-T_4 show the same patterns as T_1. This result will fully satisfy the required level for the current lead of the 1 kA rms current limiter.

Heat Load

Heat load on the liquid helium vessel was measured by a calorific method. The flow rate of evaporated helium gas was measured by a mass flow meter, and the heat load was calculated by the latent heat. The total heat load at 0 A was 1.1 W, and the total heat load at 1000 A rms was 3.4 W.

Voltage of several points were measured and resistive heat loss was obtained.

AC loss was obtained as a product of current and voltage measured at the lead surface. Approximating the voltage as the following simple function, AC loss of the lead was obtained as 25 mW. On the other hand, AC loss could be estimated from the change of temperature gradient at the lower temperature end of the lead. Figure 5 shows the temperature

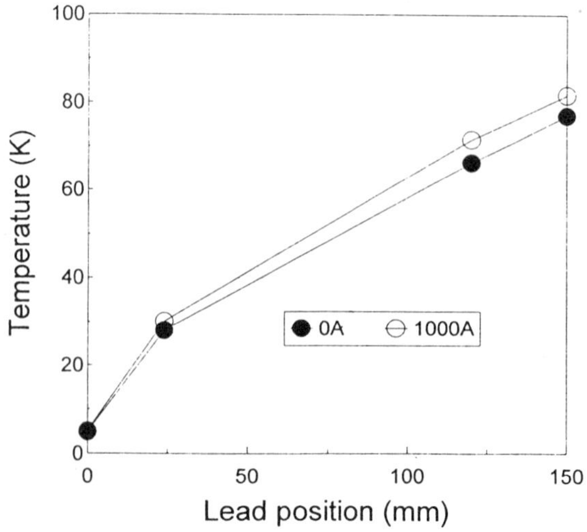

Figure 5. Temperature distribution in the Bi-2223 lead for 0 A and 1000 A rms.

Table 1. Heat load to 4 K

Current	0 (A)	1000 (A)
Radiation	0.09 (W)	0.09 (W)
Conduction of		
current leads	0.48 (W)	0.52 (W)
lead supports	0.40 (W)	0.44 (W)
supports	0.11 (W)	0.12 (W)
Heating of		
lead connection		0.06 (W)
feed through		0.40 (W)
short bar		0.23 (W)
AC loss		0.06 (W)
Total	1.08 (W)	1.92 (W)
Measured	1.1 (W)	3.4 (W)

distribution in the Bi-2223 lead for 0 A and 1000 A rms. The obtained AC loss of the pair of leads was 60 mW. This value is in good agreement with that estimated from the voltage measurement.

There was 1.5 W of unknown heat load in these experiments. Later, it was realized that 1.0 W was induced inductive current heating of Kovar used to feed through to the liquid helium vessel. Changing Kovar to CuNi, the induced current heating was decreased to 0.1 W. So, we have the prospect to decrease the heat load of the current lead system to 1.5 W.

Vibration

The vibration of the refrigerator was reduced by operating the refrigerator to avoid the resonance frequency of the cryostat. The acceleration at the end of the Bi-2223 lead was measured. It was less than 0.3 g, and the Bi-2223 leads had no damage from the vibration.

CONCLUSION

We have designed and tested a 1 kA rms current lead system for a fault current limiter within the framework of the current limiter program being jointly carried out by Tokyo Electric Power Co. and Toshiba Corp.

The current lead system successfully transported 1 kA rms for 3 hours under steady-state temperature distribution.

Heat load to liquid helium vessel was 1.1 W at 0 A, and 3.4 W at 1000 A rms. Even though these exceed the design value, since 1.0 W can be attributed to the induced current heating of Kovar which could be decreased to 0.1 W by changing the material. The prospect that the heat load can be reduced to 1.5 W has been established.

The vibration of the Bi-2223 leads caused by the refrigerator was reduced 0.3 g, and the Bi-2223 leads had no damage from the vibration.

ACKNOWLEDGMENT

The authors would like to thank Sumitomo Heavy Industries Ltd. for supplying of the Bi-2223 bulk.

REFERENCES

1. T. Hara et al., *Advances in Superconductivity V* (1992) 1255
2. T. Yazawa et al., *Pro. 14th International Conference on Magnet Technology*. (1995)
3. J. L. Wu et al., *IEEE Trans. Mag.* 27, 2 (1991) 1866-1869
4. K. Yamamoto et al., *Pro. 6th Workshop on HTSC Supercond.* (1993)
5. K. Tasaki et al., *Pro. Workshop on AC Supercond.* (1992) 85-88
6. M. N. Wilson, "Superconducting Magnets" chapter 8 (1983) UK
7. Y. Yamada et al., *IEEE Trans. Appl. Supercond.* 3 (1993) 923

LIQUID HELIUM BOIL-OFF MEASUREMENTS OF HEAT LEAKAGE FROM SINTER-FORGED BSCCO CURRENT LEADS UNDER DC AND AC CONDITIONS

Y. S. Cha,[1] R. C. Niemann,[1] J. R. Hull,[1] C. A. Youngdahl,[1]
M. T. Lanagan,[1] M. Nakade,[2] and T. Hara[2]

[1]Argonne National Laboratory
Argonne, Illinois 60439, USA
[2]Tokyo Electric Power Company
4-1 Egasaki-cho Turumi-ku
Yokohama 230 Japan

ABSTRACT

Liquid helium boil-off experiments were conducted to determine the heat leakage rate of two sinter-forged BSCCO 2223 high-temperature superconductor current leads. The experiments were carried out under both DC and AC conditions and with and without an intermediate heat intercept. Current ranges were 0-500 A for DC tests and 0-1,000 A_{rms} for AC tests. The leads were self-cooled. The results show that magnetic hysteresis (AC) losses for both the BSCCO leads and for a low-temperature superconductor current jumper were small for the current range covered in the present experiments. It has been demonstrated that significant reduction in heat leakage rate (liquid helium boil-off rate) is realized by using BSCCO superconductor leads. At 100 A, the measured heat leakage rate of the BSCCO/copper binary lead was ≈29% of that of a conventional copper lead. Further reduction in liquid helium boil-off rate was achieved by using an intermediate heat intercept. For example, At 500 A, the measured heat leakage rate of the BSCCO/copper binary lead was only 7% of that of the conventional copper lead when an intermediate heat intercept was employed.

INTRODUCTION

Recent experiments have demonstrated that, compared with conventional copper leads,[1-3] high-temperature superconductor current leads significantly reduce liquid helium boil-off rates. The reduction in liquid helium boil-off rate is the result of both the relatively low thermal conductivity and the extremely low resistive heating of the ceramic superconductors at currents below the critical current. The liquid helium boil-off rate can be reduced further by employing an intermediate heat intercept at the junction of the upper stage conventional copper lead and the lower stage high-temperature superconductor lead at temperatures below the critical temperature. High-temperature superconductor current leads can be used in both DC (superconducting magnets) and AC (fault-current limiter) applications. Under AC conditions, additional dissipation due to magnetic hysteresis is also present.

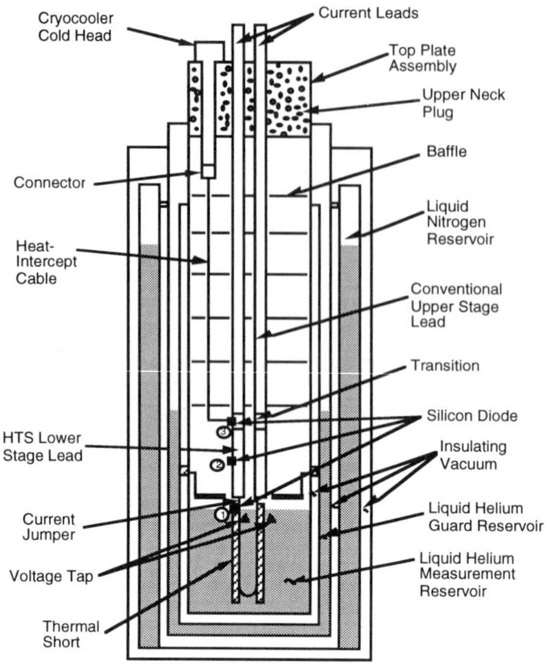

Fig. 1. Schematic diagram of liquid helium boil-off experimental apparatus at ANL.

In this paper, we describe the results of two separate experiments in which liquid helium boil-off rates were measured when sinter-forged BSCCO 2223 current leads were used under DC and AC conditions. The test sample was made of BSCCO 2223 bars. A detailed description of powder synthesis and the processes for forming the superconductor bars can be found in reference 4. Recent experiments show that the DC critical current densities of these leads are between 950 and 1300 A/cm^2 at 77 K and are >5000 A/cm^2 at 4 K.[4] The experiments reported here were carried out at Argonne National Laboratory (ANL) and

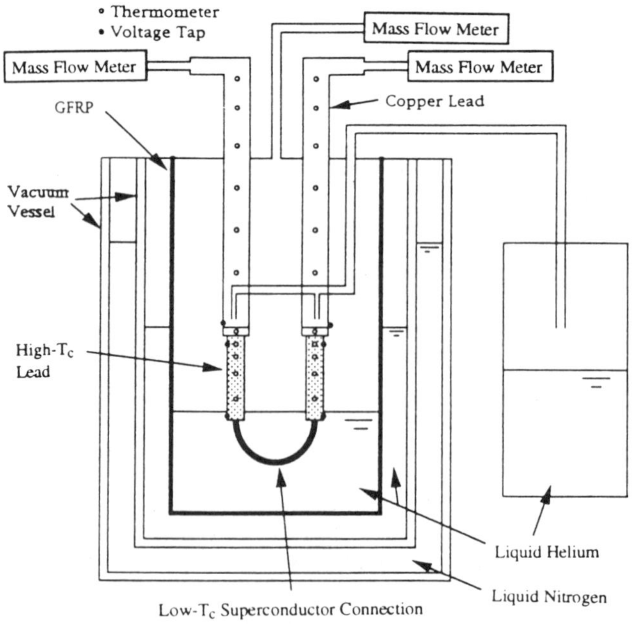

Fig. 2. Schematic diagram of liquid helium boil-off experimental apparatus at TEPCO.

Table 1. Parameters and conditions for liquid helium boil-off experiments conducted at ANL and TEPCO

Parameter	ANL	TEPCO
DC current	0-100 A	0-500 A
AC current	0-260 A_{rms} f = 60 Hz	0-978 A_{rms} f = 50 Hz
Cooling	Self-cooled	Upper stage copper leads cooled by external vapor source
Heat intercept	No	Yes
No. of BSCCO bars per lead	1	4
Sinter-forged BSCCO bars		
Cross-sectional area	0.48 cm^2	0.50 cm^2
Length	18 cm	25 cm

Tokyo Electric Power Company (TEPCO). Table 1 shows the parameters and the conditions under which these experiments were conducted. Experiments at ANL were conducted at relatively low currents (≤260 A) and those at TEPCO were conducted at higher currents (up to 1,000 A). The major difference between the tests conducted at the two facilities is that an intermediate heat intercept was used at the TEPCO facility whereas an intermediate heat intercept was not employed in the ANL tests, even though such capability was available.

EXPERIMENTAL APPARATUS

Figure 1 shows a schematic diagram of the liquid helium dewar at ANL. The dewar consists of an inner liquid helium reservoir within an outer liquid helium reservoir that is thermally shielded from the room-temperature environment by multiple insulation layers in a vacuum environment and a liquid nitrogen reservoir. Although a cryocooler is shown in Fig. 1 for intercepting the heat conducting down the copper leads, it is not used in the present experiments. A detailed description of the characteristics of the dewar and its performance can be found elsewhere.[2,5] The instrumentation system includes devices to measure temperature, pressure, voltage drop, vapor mass flow, and liquid level. Three silicon diodes are employed to measure temperature distribution along one of the high-temperature superconductor leads, as shown in Fig. 1. Two relatively large copper plates were employed as thermal shorts between the cold end of the high-temperature superconductor leads and the low-temperature superconductor jumper. The thermal shorts provide a large heat transfer area and serve to reduce contact resistances and maintain temperature at the cold end of the high-temperature superconductor leads at ≈4.2 K. Two voltage taps were soldered to the copper thermal shorts, as shown in Fig. 1. All data are fed to a computer for on-line monitoring and recording.

Figure 2 is a schematic diagram of the liquid helium dewar at TEPCO, which also consists of two liquid helium reservoirs within a liquid nitrogen reservoir. In addition, the TEPCO facility has an external liquid helium dewar that provides helium vapor to cool the upper stage copper leads and serves as a heat intercept at the junction of the copper and high-temperature superconductor leads. The temperatures at the junctions can be controlled by adjusting the vapor mass flow from the external liquid helium dewar. Mass flow from the helium boil-off in the test dewar and that from the external dewar are measured separately, as shown in Fig. 2. Locations of temperature sensors and voltage taps are also shown in Fig. 2. Four BSCCO bars were employed per lead in the TEPCO experiments.

EXPERIMENTAL RESULTS AT ANL

Figure 3 shows the variation of the total heat leakage rate with current (DC and AC). At low currents (<50 A), the total heat leakage rate with the two currents is about the same (within experimental error), because hysteresis (AC) losses in the superconductors are small at these low currents. At 100 A or A_{rms}, the difference in boil-off rate between DC and AC tests is small but measurable, as shown in Fig. 3. The difference in total heat leakage rate between AC and DC tests at a current of 100 A or A_{rms} is ≈10 mW. Total heat leakage rate increases with AC current; however, the relationship is not linear, as shown in Fig. 3. The total heat leakage rate includes all of the heat that crosses the liquid helium boundary in the inner helium reservoir plus any heat sources that are submerged in the liquid helium. The former includes conductive (through current leads and vessel walls), convective (between vapor space above the liquid and the liquid helium), and radiative heat transfer between the liquid helium and the surrounding environment. The latter includes dissipation caused by contact resistances (soldered joints) between a short copper transition piece and the low-temperature superconductor current jumper, Joule heating in the copper piece, and AC (magnetic hysteresis) loss in the low-temperature superconductor current jumper.

In the present experiments, two relatively short copper plates with fairly large surface areas were used as transitions between the high-temperature superconductor current leads and the low-temperature superconductor current jumper. Because of the large surface areas of the copper plates and the relatively low current, dissipations due to contact resistance and Joule heating are probably negligible compared to other forms of heat leakage into the liquid helium. Losses due to magnetic hysteresis of the low-temperature superconductor current jumper are also small at these low currents. For example, the voltage drop across the thermal shorts is ≈30 μV at a current of 100 A_{rms}. At this current, the combined heating rate due to contact resistances, Joule heating, and magnetic hysteresis of the low-temperature superconductor is <3 mW. In any case, at these low currents, the total heat leakage is small (<0.20 W at a current of 260 A_{rms}).

Experiments were conducted to determine the heat leakage rate when current leads were absent. The heat leakage rate obtained in this manner is referred to as the background heat leakage rate. By subtracting the background heat leakage rate from the total heat leakage rate we obtain the heat leakage rate through the high-temperature superconductor current leads only. The results are shown in Fig. 4. At low currents, the background heat leakage rate is a significant portion of the total heat leakage rate. As the current increases, the heat conducted down the leads increases correspondingly and the background heat leakage rate becomes only a small fraction of the total heat leakage rate. The temperature distribution along one of the superconductor lead is shown in Fig. 5.

The heat leakage rate shown in Figs. 3 and 4 is calculated from the measured vapor mass flow rate

$$Q = h_{fg}(dm/dt) / (1 - \rho_g/\rho_f), \tag{1}$$

where Q is the heat leakage rate, h_{fg} is the latent heat of vaporization, dm/dt is the measured vapor mass flow rate, and ρ_g and ρ_f are the vapor and liquid density of helium, respectively.

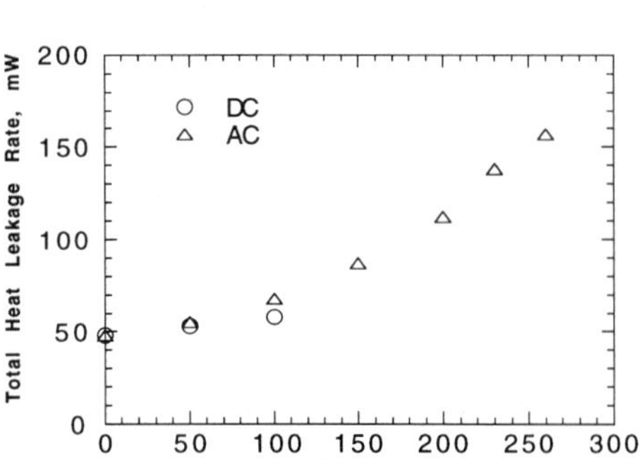

Fig. 3. Variation of total heat leakage rate with current in tests conducted at ANL.

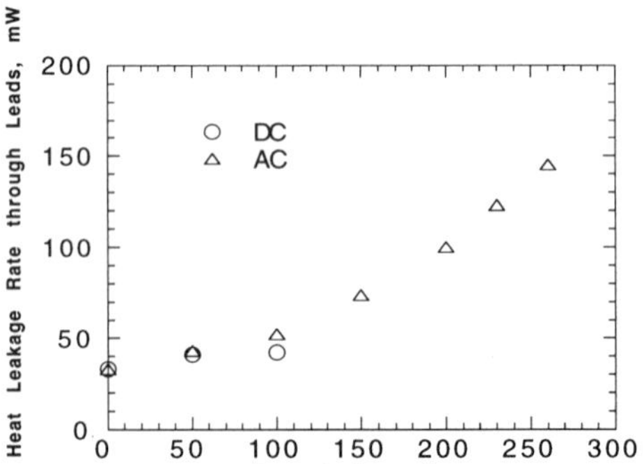

Fig. 4. Variation with current of heat leakage rate through BSCCO superconductor leads in tests conducted at ANL.

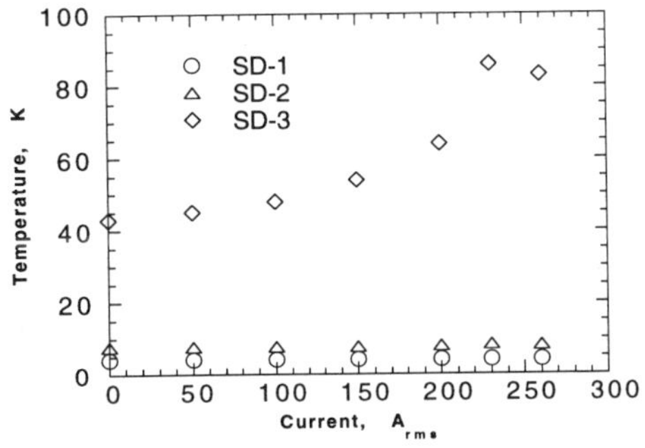

Fig. 5. Variation with current (AC) of temperature along BSCCO superconductor lead (measured by silicon diodes SDs) in tests conducted at ANL.

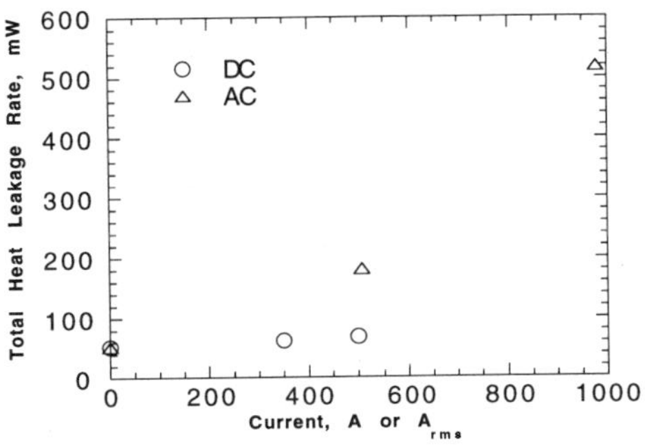

Fig. 6. Variation of total heat leakage rate with current in tests conducted at TEPCO.

Table 2. Measured total heat leakage rate and heating rate due to contact resistance for DC tests conducted at TEPCO

Current (A)	Total heat leakage rate (mW)	Heating rate due to contact resistance* (mW)
0	52	0
350	62	11
499	68	22

*The contact resistance determined from voltage drop measurements is ≈0.09 μΩ.

Table 3. Measured total heat leakage rate and heating rate due to contact resistance for AC tests conducted at TEPCO

Current (A)	Total heat leakage rate (mW)	Heating rate due to contact resistance (mW)	Heating rate due to contact resistance** (mW)
0	52	0	0
507	181	146	23
978	517	616	86

**Calculated from the contact resistance (0.09 μΩ) determined in the DC tests.

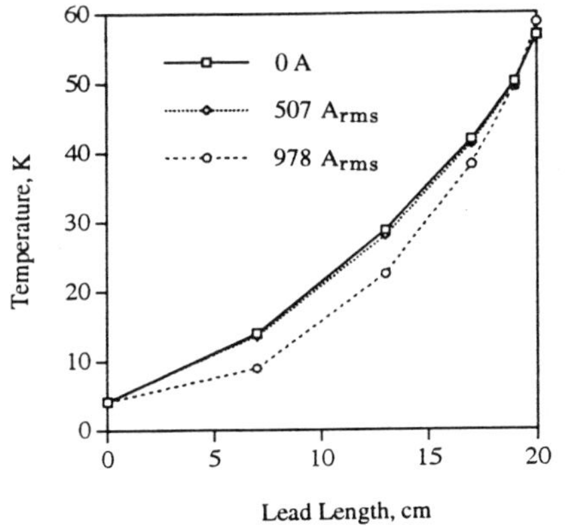

Fig. 7. Variation of temperature along length of BSCCO superconductor lead in tests conducted at TEPCO.

Derivation of equation (1) and discussions of its significance for liquid helium boil-off experiments are given in references 5 and 6.

EXPERIMENTAL RESULTS AT TEPCO

Figure 6 shows the variation of total heat leakage (calculated from the measured helium boil-off rate) with DC and AC currents. The total heat leakage data is also shown in Tables 2 and 3, together with the heating rate generated by contact resistances (soldered joints) calculated from the measured voltage drop between the lower ends of the two BSCCO leads (the locations of the voltage taps are shown in Fig. 2). The total heat leakage rate consists mainly of the heat conducted down the leads and the heat generated by the heat sources in the liquid helium. Table 3 indicates that, at a current of 978 A_{rms}, the heating rate calculated from the measured voltage drop is ≥ the total heat leakage rate, which means that the entire boil-off comes from the heat sources in the liquid helium. In other words, very little heat conducted down the current leads reaches the liquid helium. This observation is consistent with the temperature distribution along one of the BSCCO leads shown in Fig. 7, which shows that, at a current of 978 A_{rms}, the temperature gradient at the lower end of the BSCCO bar is very small, indicating that very little heat reaches the liquid helium by conduction (estimated to be <10 mW, based on the measured temperature gradient at the lower end of the BSCCO bar).

Figure 7 also shows that, at the same vertical location, the temperature is lower when the current is higher. This can only occur if there is more vapor cooling at higher currents. The additional vapor flow obviously comes from the heat sources in the liquid helium. These heat sources include contact losses and AC losses of the low-temperature superconductors. The AC losses of the low-temperature superconductors are relatively small at these currents. For example, at 1,000 A_{rms}, the AC losses are estimated to be ≈20 mW, which is only a small fraction of the total heat leakage rate of 517 mW.

The DC critical currents (determined from separate measurements of individual short samples) of the low-temperature superconductors and the BSCCO bars are not exceeded in the present tests. Therefore, a majority of the heat must come from the contact resistances of the soldered joints. If both the low-temperature superconductors and the BSCCO bars are in their superconducting states over their entire length, the contact resistances of the soldered joints should not change with current. This is clearly shown in Table 2, where the contact resistance is approximately constant (0.09 $\mu\Omega$) for all of the DC tests. Also, the contact resistances should be the same for the DC and AC tests, provided that the superconductors are in their superconducting states. Based on the contact resistance measured from the DC tests, we calculated the heating rate generated by the contact resistance for the AC tests. The results are shown in the last column of Table 3.

If we compare the values in the last two columns of Table 3, we observe a large difference between the heating rate due to contact resistance determined directly from the measured voltage drop and that inferred from the DC contact resistance. This large difference indicates that the contact resistance of the AC tests has drastically changed from its DC value. The reason for this change is not clearly understood. One possibility is that, even though the bulk of the BSCCO bars and the low-temperature superconductors are in their superconducting states, current densities near the soldered joints between the BSCCO bars and the low-temperature superconductors might have been increased sharply as a result of local current redistribution.

Closer examination of the test setup indicates that there is a transition between the lower ends of the BSCCO leads and the low-temperature superconductor current jumper. The transition consists of 32 small-diameter (0.66 mm) low-temperature superconductor wires (four for each BSCCO bar). Two types of soldered joints are present, one type at the junction of a BSCCO bar and the low-temperature superconductor wires, the other type at the junction between the low-temperature superconductor transition wires and the low-temperature superconductor current jumper. At the junction of the BSCCO bar and the transition wires, there are four small-diameter (0.66 mm) low-temperature superconductor wires attached to one BSCCO bar. The cross-sectional area of a BSCCO bar is ≈50 mm^2, whereas

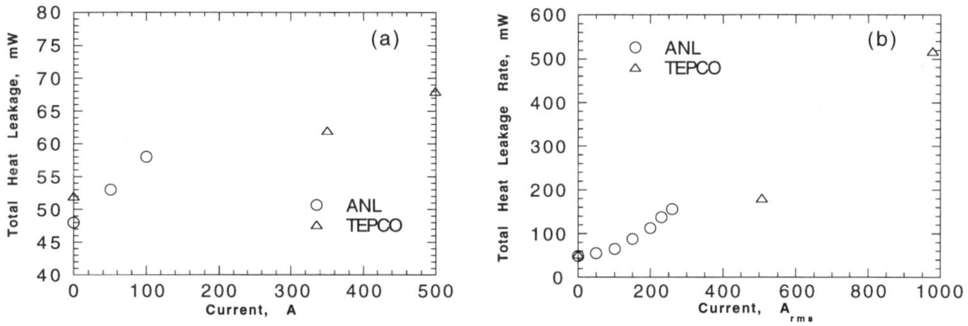

Fig. 8. Total heat leakage rate at different currents in tests conducted at ANL and TEPCO; (a) DC tests; (b) AC tests.

that of the four wires is only 1.4 mm^2. Because the cross-sectional area is reduced by a factor of ≈36, current redistribution must occur in the BSCCO bar before it is transferred to the low-temperature superconductor wires. This current redistribution could cause the current density to increase locally. The result would be a large increase in contact resistance if the local current density exceeds the critical current density.

At the junction between the transition wires and the low-temperature superconductor current jumper, eight groups of wires (each group has four transition wires) are soldered to the low-temperature superconductor current jumper. The current distribution in the low-temperature superconductor current jumper immediately adjacent to the soldered joint might not be uniform, and local current density could increase drastically and could approach or exceed the DC critical current density of the superconductor. For example, the DC critical current of the low-temperature superconductor current jumper is 1600 A, which is equivalent to 1131 A_{rms}. For the AC test with a current of 978 A_{rms}, a 20% increase in local current could cause the superconductor to become dissipative. This conjectured increase in contact resistance can be avoided by redesigning the soldered joints.

COMPARISON OF TEST RESULTS

Figure 8 shows the total heat leakage rate observed during tests conducted at ANL and TEPCO. Even though the two experiments were carried out within different current ranges, the results clearly demonstrate the advantage of using an intermediate heat intercept, because the slope for the ANL data is steeper than that of the TEPCO data, which means that boil-off is less if a heat intercept is employed. This comparison may not be fair because either an additional liquid helium (or nitrogen) source or a cryocooler is required to intercept the heat conducted down the upper stage copper leads. However, in most applications, liquid helium or nitrogen are readily available. In certain applications, a cold shield, maintained by a cryocooler or cryo-refrigerator, is present in the device (such as a micro superconducting magnetic energy storage system with a 60 K cold boundary). Under these circumstances, it is worthwhile to tap into the existing heat sink and provide a heat intercept at the copper/ superconductor junctions.

Finally, we can compare the total heat leakage rate obtained in the present experiments with that of conventional copper lead. Only comparison under DC condition will be made here. The data in Fig. 3 (for two leads) show that, at a current of 100 A, the total heat leakage rate for the BSCCO lead is 29 mW per lead. The heat leakage rate for a copper lead at 100 A is ≈100 mW. Thus, without using an intermediate heat intercept, the heat leakage rate of the BSCCO lead is 29% of that of the conventional copper lead. The data in Fig. 6 (for two leads) show that, at a current of 500 A, the total heat leakage rate for the BSCCO lead is 34 mW per lead. The corresponding heat leakage rate for a copper lead is 500 mW. Thus, by using an intermediate heat intercept, the heat leakage rate for the BSCCO lead is only ≈7% of that of the copper lead.

SUMMARY AND CONCLUSIONS

Liquid helium boil-off experiments were conducted to determine the heat leakage rate of BSCCO 2223 high-temperature superconductor current leads made by sinter forging. The experiments were carried out under both DC (0-500 A) and AC (0-1,000 A_{rms}) conditions and at two experimental facilities, ANL and TEPCO. At low currents (<100 A), the differences in heat leakage rate between DC and AC conditions are small. The difference increases as the current is increased well above 100 A. The experimental results demonstrate that significant reduction in heat leakage rate (liquid helium boil-off rate) can be achieved by using an intermediate heat intercept at the junctions between the upper stage copper leads and the lower stage superconductor leads. Without the intermediate heat intercept, the heat leakage rate of the BSCCO/copper binary current lead is ≈29% of that of the conventional copper lead. With the intermediate heat intercept, the heat leakage rate of the BSCCO/copper binary lead is only ≈7% of that of the conventional copper lead. In the TEPCO experiments, it was determined that most of the helium boil-off is due to the heat generated by the contact resistance at the soldered joints between the BSCCO bars and the low-temperature superconductors (used as current jumpers in the experiments) for the AC tests, although it is not clear what causes the contact resistance to increase. The magnetic hysteresis (AC) losses of both the BSCCO leads and the low-temperature superconductor current jumper are small compared to the heat generated by contact resistance in the AC tests of TEPCO experiments.

ACKNOWLEDGMENTS

The work performed at ANL has been supported by the U.S. Department of Energy, Energy Efficiency and Renewable Energy, as part of a program to develop electric power technology, under Contract W-31-109-Eng-38. The effort of A. Wantroba in building the insert for the ANL test is appreciated. The efforts of T. Ohkuma, S. Hirano, and T. Fukushima in carrying out the experiments at TEPCO are gratefully acknowledged.

REFERENCES

1. J.L. Wu, J.T. Dederer, P.W. Eckels, S.K. Singh, J.R. Hull, R.B. Poeppel, C.A. Youngdahl, J.P. Singh, M.T. Lanagan, and U. Balachandran, Design and testing of a high temperature superconducting current lead, *IEEE Trans. Mag.* 27(2):1861 (1991).
2. R.C. Niemann, Y.S. Cha, and J.R. Hull, Performance measurements of superconducting current leads with low helium boil-off rates, *IEEE Trans. Appl. Superconductivity* 3(1):392 (1993).
3. K. Ueda, T. Bohno, K. Takita, T. Uede, I. Itoh, M. Mimura, N. Uno, and T. Tanaka, Design and testing of a pair of current leads using bismuth compound superconductor, *IEEE Trans. Appl. Superconductivity* 3(1):400 (1993).
4. U. Balachandran, C.A. Youngdahl, M.T. Lanagan, S.E. Dorris, J.J. Picciolo, J. Cluff, T. Brent, M. Marinelli, B. Fisher, P. Winandy, J. Wenzlaff, T. Askew, R. Niemann, R.B. Poeppel, M. Nakada, and T. Hara, Design and production of efficient current leads for 1500-A, 50 Hz service in a 77-4 K temperature gradient, paper presented at the 7th Int. Symp. on Superconductivity, Fukuoka, Japan (1994).
5. Y.S. Cha, R.C. Niemann, and J.R. Hull, Thermal characteristics of a low-loss liquid helium dewar, in "Proceedings of the ASME/JSME Thermal Engineering Conference," Vol. 1, p. 115-212 (1995).
6. Y.S. Cha, R.C. Niemann, and J.R. Hull, Thermodynamic analysis of helium boil-off experiments with pressure variations, *Cryogenics* 33(7):675 (1993).

PERFORMANCE EVALUATION OF HIGH-TEMPERATURE SUPERCONDUCTING CURRENT LEADS FOR MICRO-SMES SYSTEMS*

R. C. Niemann,[1] Y. S. Cha,[1] J. R. Hull,[1] W. E. Buckles,[2]
B. R. Weber,[2] and S. T. Yang[3]

[1]Argonne National Laboratory
Argonne, IL 60439, USA

[2]Superconductivity, Inc.
Middleton, WI 53705, USA

[3]University of Wisconsin
Madison, WI 53706, USA

ABSTRACT

As part of the U.S. Department of Energy's Superconductivity Technology Program, Argonne National Laboratory and Superconductivity, Inc., are developing high-temperature superconductor (HTS) current leads for application to micro-superconducting magnetic energy storage systems. Two 1500-A HTS leads have been designed and constructed. The performance of the current lead assemblies is being evaluated in a zero-magnetic-field test program that includes assembly procedures, tooling, and quality assurance; thermal and electrical performance; and flow and mechanical characteristics. Results of evaluations performed to date are presented.

INTRODUCTION

A superconducting magnetic energy storage (SMES) system stores electrical energy to provide power during voltage sags and momentary power losses. SMES devices store electric energy as a circulating current in a magnetic coil; because superconducting coils offer virtually no resistance to the current, they are far more efficient than normal conducting coils, achieving overall energy conversion efficiencies of >95%. In addition, these devices have very high response rates and contain no internal moving parts, so they are highly reliable and promise long service life.

HTS current leads are particularly advantageous for SMES devices because they have the potential to reduce refrigeration requirements to values significantly below the best values achievable with conventional leads.

*Work partially supported by the U.S. Department of Energy, Energy Efficiency and Renewable Energy, as part of a program to develop electric power technology, under Contract W-31-109-Eng-38, and by Superconductivity, Inc., Middleton, WI, USA.

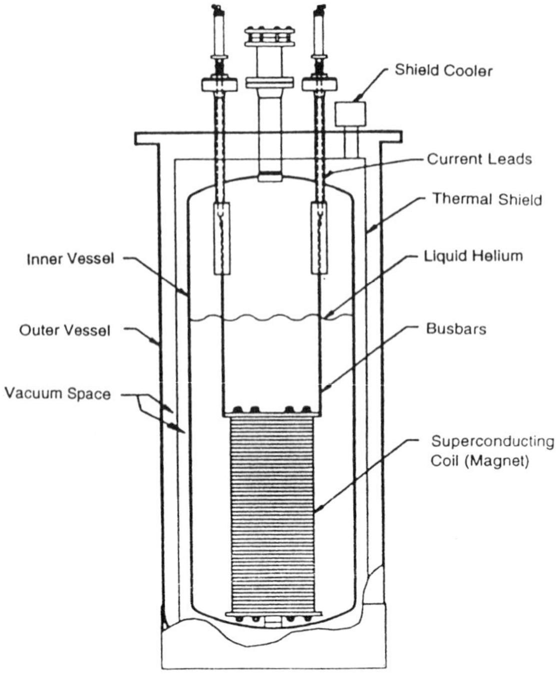

Figure 1. Current lead installation geometry.

As part of the U.S. Department of Energy's Superconductivity Technology Program, Argonne National Laboratory is developing, with Superconductivity, Inc. (SI), HTS current leads suitable for micro-SMES applications.

Superconductivity, Inc., is engaged in the design, construction, operation, and commercial sales of micro-SMES systems for application in the power-quality market.[1] This

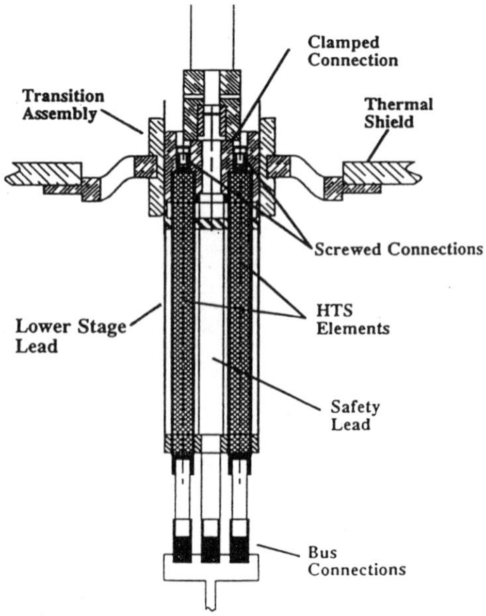

Figure 2. General arrangement of current lead.

Table 1. Major performance requirements for HTS current leads

Parameter	Design value
I_{op}, V_{max}, B_{max}	1.5 kA, 7.5 kV, SMES + self
Dimensions	0.1 m diam. x 1 m length
HTS Temperature	$T_{warm} \leq 60$ K, $T_{cold} \cong 10$ K
Heat leakage @ 10 K	< 0.7 W/lead
300-60 K cooling	Helium gas + refrigerated intercept
60-10 K cooling	Helium gas + conduction

application requires relatively low energy storage and comparatively high power delivery. Units store ≈1 MJ and are able to discharge at a rate of 1 MW. The integrated system is factory-built for low cost and consistent quality and is transportable overland.

The economic justification for SI's system is not based on the cost of the energy stored in the unit, but on the value of avoiding an interruption of the power supply. The cost of shutdowns due to sags and outages has been estimated to be in the billions of dollars per year in the U.S. alone. While long-term outages have been effectively reduced by electric utilities, momentary voltage disturbances have actually increased in many power systems across the country. At the same time, many industries have dramatically expanded their use of electronically controlled equipment, which is particularly sensitive to momentary voltage disturbances. These two factors have resulted in an increased need for quality power in many segments of the economy.

PERFORMANCE SPECIFICATIONS FOR CURRENT LEAD

The current lead installation geometry is shown in Fig. 1, and major performance requirements are given in Table 1.[2]

CURRENT LEAD DESIGN

General Arrangement

The general arrangement of the current lead is shown in Fig. 2. Major subassemblies are the conventional upper stage, the transition assembly, and the HTS lower stage.

Conventional Upper Stage

The design parameters of the upper stage include a 1500-A operating current; warm- and cold-end temperatures of 300 and 60 K, respectively; and helium-vapor flow equivalent to 1.2 L/hr per lead. Estimated heat leakage to 60 K is 47 W per lead. The upper stage was designed and constructed by a commercial vendor.

Transition Assembly

The transition assembly provides electrical and mechanical connections between the lower and upper stages, flow paths for the helium-vapor cooling stream, a heat intercept connection to a cryocooler, electrical isolation between the lead and the adjacent cryostat components, and pressure vessel continuity within the cryostat. The geometry, materials, and methods of fabricating the transition assembly are selected to minimize the temperature difference between the warm end of the HTS element and the cryocooler intercept. The transition assembly must also provide effective and reliable electrical isolation, i.e., very low leakage current and high voltage standoff.[3,4]

HTS Lower Stage

The lower stage consists of six parallel current-carrying YBCO HTS rods containing 15 vol.% silver. The rods, 1.3 cm in diameter and 20 cm long, are connected to the transition assembly at their warm ends and to the bus collector assembly at their cold ends. Each rod is enclosed in an epoxy/fiberglass tube that channels the helium-vapor coolant and provides structural support for the rod. A safety lead is in the center of the HTS rod array. The estimated heat leakage to 4 K is 0.9 W per lead.

A safety lead is incorporated in the lower-stage assembly as an alternate current path in the event of HTS malfunction. The safety lead consists of a stainless steel tube connected in parallel with the conductor elements.

The lower ends of the six HTS elements are connected to a bus collector by six low-temperature superconductor (LTS) composite elements. The collector is connected to the terminal of the magnet coil by LTS buswork. Voltage isolation along the length of the lead is provided by epoxy/fiberglass composite tubing that surrounds the lead elements.

CONSTRUCTION OF CURRENT LEADS

Conductor Elements

The YBCO conductor elements incorporate silver-painted and sintered end caps, which were joined to the gold-plated copper warm and cold end caps with 60/40 Sn/Pb solder.

The critical currents of the HTS rods were measured at 77 K. The critical current for the 12 rods ranged from 215 A to 250 A. The rods were divided into two balanced groups, one

Figure 3. Conductor elements-transition assembly.

Figure 4. Current lead assembly.

for each lead, based on their 77 K critical currents. The rods in a group were then azimuthally oriented to minimize the effects of the SMES magnet's fringe field on the current-carrying capacity of the rods.

To reduce the effects of atmospheric moisture on the HTS material, shrinkable Mylar tubing was applied over the HTS rods. The small gap between the Mylar tubing and each end cap was sealed with a coating of Shell 820/Versamid 140 epoxy. An epoxy/fiberglass tube was then installed over the HTS rod. The tube was bonded to the warm end cap and allowed the cold end cap to slide in order to accommodate differential thermal movement.

Because final assembly of the current leads was delayed for ≈1 year due to human resource availability, and because during that period the rods were exposed to a non-humidity-controlled environment, it was decided to remeasure the critical currents. The 77 K critical currents ranged from 214 A to 282 A. This demonstrated the effectiveness of the humidity-proofing measures employed.

The conductor element assemblies, i.e., HTS rod, end caps, and reinforcing, surrounding epoxy fiberglass tube, proved to be structurally robust, having survived repeated 300-to-77-to-300 K thermal cycling, common carrier shipments, and the handling of two assemblies and one disassembly.

Upper Stage - Transition Assembly

The upper stage and the transition assembly were assembled by a clamped connection. The clamped members were gold-plated copper and incorporated an indium gasket at their interface.[5] Clamping was achieved with a torqued bolt and conical spring washers that were used to maintain the bolt load.

Table 2. Steady-state and upset thermal and electrical tests

Measured quantity	Current	Lead gas flow	Cryocooler status
Zero current heat leakage	0	Unrestricted	On
Nonzero current heat leakage	25-100% of design	Unrestricted	On
Zero flow current	50-100% of design	Off	On
Zero current, zero cryocooler heat leakage	0	Unrestricted	Off
Nonzero current, zero cryocooler heat leakage	75-100% of design	Unrestricted	Off
Loss of flow upset	25-100% of design	Unrestricted to off	On
Loss of cryocooler upset	25-100% of design	Unrestricted	On to off

Transition Assembly - Lower Stage

The conductor elements were connected to the transition assembly by gold-plated-copper screwed connections.[5] The conductor elements-transition assembly is shown in Fig. 3. The safety lead and the low-temperature superconductor cold end transitions were installed. The current lead assembly is shown in Fig. 4.

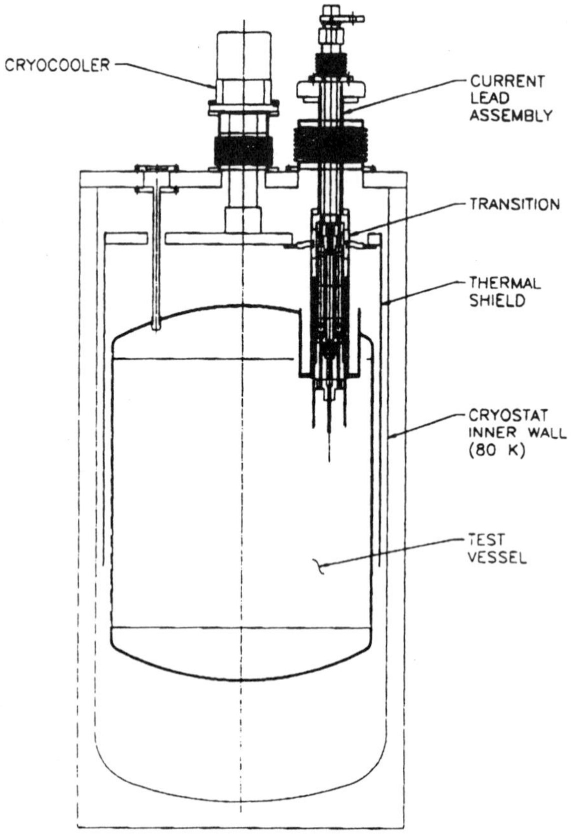

Figure 5. Liquid-helium test cryostat.

High-Voltage Isolation

After installation of voltage and temperature sensors, high-voltage isolation members in the form of epoxy/fiberglass tubing were installed.

OPERATION IN ZERO MAGNETIC FIELD

Objectives

A program of system performance evaluation has been developed to evaluate the performance of the current lead assemblies. The test program objectives include evaluation of thermal performance, including heat leakage, temperature distribution, steady-state and upset response, and shield cooler performance effects; electrical performance, including voltage drops, voltage isolation, upset conditions, safety lead, and lead-to-lead differences; flow characteristics, including pressure drop vs. mass flow rate, voltage drop vs. mass flow rate, voltage drop vs. inlet temperature, and lead-to-lead differences; and mechanical characteristics, including thermal contraction, HTS rod strain, upset conditions, and stiffness of upper- and lower-end connections.

Test Program

Evaluations are performed in a liquid-helium test cryostat as shown in Fig. 5. Upper-stage (heat meter) and lower-stage (boil-off) heat leakage values are calculated from test data. Measured electrical values include upper-stage, transition assembly, rod, and collector assembly voltage drops. Tests are planned for several steady-state and upset conditions, as shown in Table 2. The test program includes runs at reduced cryocooler capacity and elevated cooling-gas inlet temperature.

Initial Operation

The integrated performance measurement system has been initially operated to evaluate operational readiness. The liquid helium and liquid nitrogen vessels and piping were leak-tight during cryogenic operations. The cryocooler is functional and provides effective shield and intercept cooling. Cryostat external insulating vacuum feedthroughs were found to be temperature-sensitive and require corrective measures. Instrumentation and data acquisition system are functional, as is the power supply system.

The integrated system has been operated at lead currents to 1150 A during checkout of the measurement system.

OPERATION WITH CURRENT

Initial operation with current was at the 600 A level. The intermediate temperature intercept, i.e., HTS conductor element warm end, was operated at 45 K and 55 K. Lead flow was varied from 100% to 30% of the total flow, i.e., lead flow plus dewar flow.

The cryostat inner vacuum wall failed by buckling during a pumpdown and purge operation preceding the 900 A test run. The balance of the testing will be performed in another cryostat. Preparations for the conversion are underway.

CONCLUSIONS

- Two of 1500-A HTS current leads for micro-SMES application have been constructed.
- Each lead employs a parallel cylindrical array of six 1.3-cm-OD x 20-cm-long YBCO w/15 vol.% Ag sintered HTS conductor elements. The average measured critical current was 234 A @ 77 K @ 0.005 T applied field.

• Demountable connections join the conventional upper stage lead to the HTS lower stage lead. Both clamped and screwed connections are employed and result in an apparent joint resistivity of ≤1.8 that of the base material (ETP copper) at modest assembly torques.

• Heat intercepting is via a transition assembly that provides an electrically isolating and thermally effective connection. The measured ΔT for a prototype assembly across the insulator is ≈12 K @ 45 W heat flow @ 70 K operating temperature.

• The transition assembly cables have been calibrated as heat meters to allow measurement of the heat load to the transition intercept.

• The lead pair is undergoing electrical, thermal, and structural evaluations under simulated operating conditions. The integrated measurement system has been operated to evaluate functional readiness. The leads have been initially operated at currents to 1150 A. Performance has been evaluated at 600 A. Further testing at higher currents has been delayed by a failure in the test cryostat.

ACKNOWLEDGMENTS

The authors gratefully acknowledge the engineering contributions of M. A. Daugherty; the technical expertise of M. Bordson, D. J. Evans, A. S. Wantroba, and K. Ziemer; the manuscript preparation skills of J. A. Stephens; and the editorial contributions of C. A. Malefyt. Work at ANL was supported by the U.S. Department of Energy, Energy Efficiency and Renewable Energy, as part of a program to develop electric power technology, under Contract W-31-109-Eng-38. Work at Superconductivity, Inc., was conducted as a part of the corporation's internal-systems development program.

REFERENCES

1. W.E. Buckles, M.A. Daugherty, B.R. Weber, and E.L. Kostecki. The SSD: a commercial application of magnetic energy storage, *IEEE Trans. Appl. Supercond.* 3:328 (1993).
2. R.C. Niemann, Y.S. Cha, J.R. Hull, W.E. Buckles, and M.A. Daugherty. Engineering design of a high-temperature superconductor current lead, in *Supercollider 5*, P. Hale, ed., Plenum Press, New York (1994), pp. 629-634.
3. R.C. Niemann, J.D. Gonczy, T.H. and Nicol. 1994c, Low-thermal-resistance, high-electrical-isolation heat intercept connection, in *Advances in Cryogenic Engineering*, Plenum Press, New York, Vol. 39 (1994), pp. 1665-1673.
4. R.C. Niemann, J.D. Gonczy, and T.H. Nicol, Thermally effective, electrically isolating heat intercept connections, in *Thermal Engineering 1995*, Vol. 1, L. S. Fletcher and T. Aihara, eds., ASME, New York, (1995), pp. 43-50.
5. R.C. Niemann, Y.S. Cha, J.R. Hull, W.E. Buckles, M.A. and Daugherty. Performance evaluations of demountable electrical connections, in *Advances in Cryogenic Engineering*, Plenum Press, New York, Vol. 39 (1994), pp. 1153-1160.

PERFORMANCE EVALUATION OF HIGH-TEMPERATURE SUPERCONDUCTING CURRENT LEADS FOR ELECTRIC UTILITY SMES SYSTEMS

R. C. Niemann,[1] Y. S. Cha,[1] J. R. Hull,[1] C. M. Rey,[2] and K. D. Dixon[2]

[1]Argonne National Laboratory
Argonne, IL 60439, USA

[2]Babcock & Wilcox
Lynchburg, VA 24506, USA

ABSTRACT

As part of the U.S. Department of Energy's Superconductivity Technology Program, Argonne National Laboratory and Babcock & Wilcox are developing high-temperature superconductor (HTS) current leads for application to electric utility superconducting magnetic energy storage systems. A 16,000-A HTS lead has been designed and is being constructed. An evaluation program for component performance was conducted to confirm performance predictions and/or to qualify the design features for construction. Performance of the current lead assemblies will be evaluated in a test program that includes assembly procedures, tooling, and quality assurance; thermal and electrical performance; and flow and mechanical characteristics. Results of the evaluations to date are presented.

INTRODUCTION

A superconducting magnetic energy storage (SMES) system stores electrical energy to provide power during voltage sags and momentary power losses. SMES devices store electric energy as a circulating current in a magnetic coil.

High-temperature superconductor (HTS) current leads are particularly advantageous for SMES devices because they have the potential to reduce coolant refrigeration requirements to values significantly below the best values achievable with conventional leads.

As part of the U.S. Department of Energy's Superconductivity Technology Program, Argonne National Laboratory is developing, with industry, HTS current leads suitable for application to SMES systems.

MID-SIZED SMES APPLICATION

Babcock & Wilcox is engaged in a project to design, build, and demonstrate a mid-sized SMES system, which has been sized in response to the requirements of Anchorage (Alaska) Municipal Light and Power. The 0.5-MWh energy-capacity system, with a discharge power of 30 MW, is a basic unit that could be grouped with other modules at a utility site to meet much larger energy storage needs.

Table 1. Performance requirements of HTS current leads

Parameter	Design value
I_{op}, V_{max}, B_{max}	16 kA, 5 kV, SMES + self
Dimensions	0.1 m diam. x 1 m length
HTS temperature	T_{warm} < 60 K, T_{cold} = 5 K
Heat leakage @ 5 K	< 15 W/lead
300-60 K cooling	Helium gas
60-5 K cooling	Conduction

Performance Specifications

Major elements of the system's performance are given in Table 1.[1]

General Arrangement

The general arrangement of the current lead is shown in Fig. 1.

Upper-Stage Current Lead

Gas-cooled leads (GCLs) are designed to efficiently transmit current from 300 K to the upper-stage-to-middle-stage transition that operates at ≈60 K. Helium gas at 42 K, supplied by a refrigerator, removes the resistive and conductive heat from the GCL itself. The GCL is

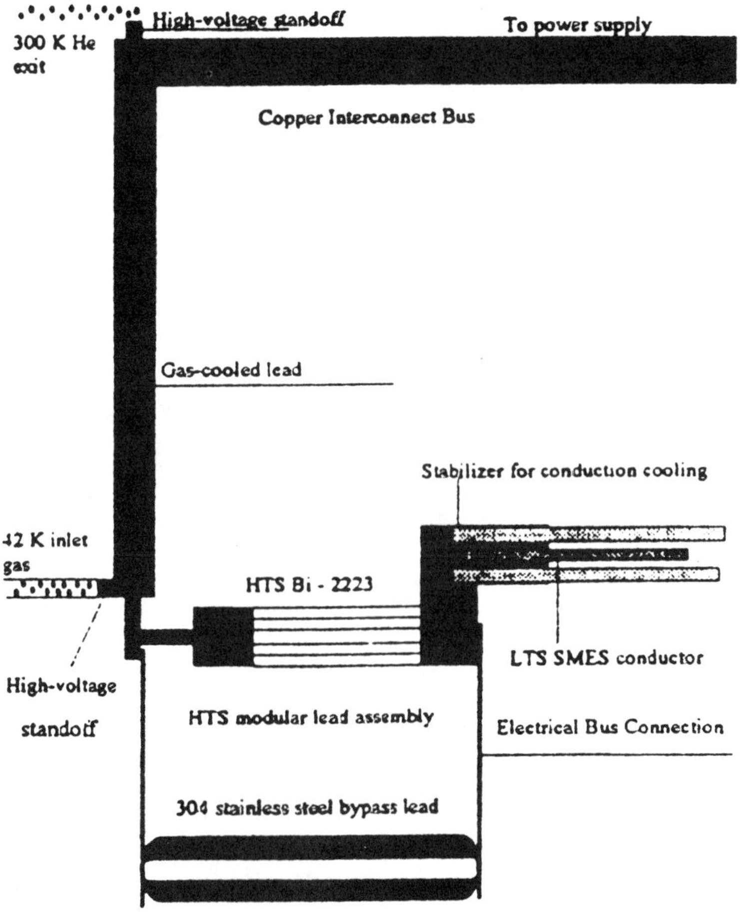

Fig. 1. General arrangement of current lead.

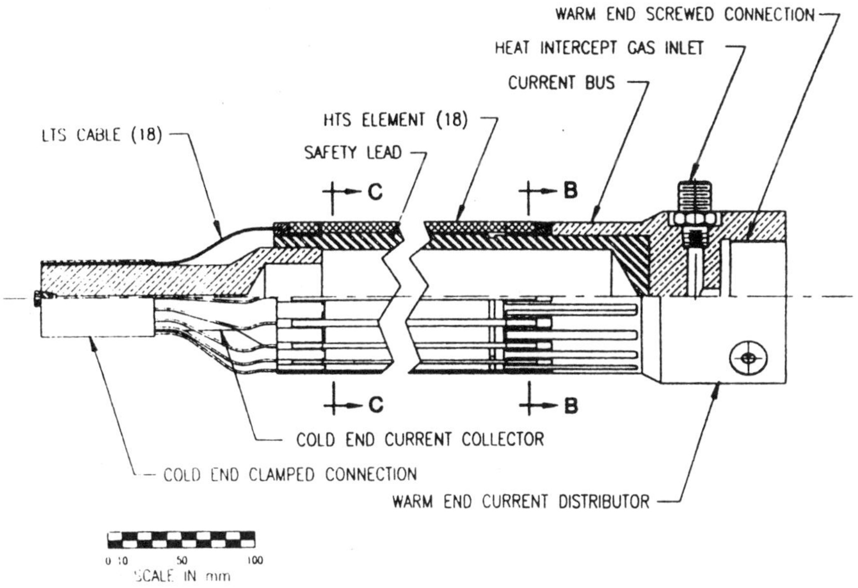

Fig. 2. Axial cross section of middle-stage current lead.

an adaptation of a vendor-manufactured vapor-cooled lead with a cross section and wetted perimeter based on a standard vapor-cooled design optimized for minimum lower-end heat flow.

Upper-Stage-to-Middle-Stage Transition

Electrical and structural connection to the upper stage is by a large, fine-pitch screw thread. The mating threaded sections are copper and are plated with nickel and gold to improve electrical and thermal conductance across the joint. The male thread is tinned with indium.

Current flow through the transition can follow two paths. During superconducting operation, all current flows through the female portion of the threaded connection to a circular array of current buses. To accommodate nonuniform contact resistance to the HTS elements, the buses are made of copper and their geometry, area, and length are selected to control the current-sharing among the HTS elements. Bus resistance is designed to be five or more times that of the sum of the other resistances in the current path of the conductor element. During an upset condition corresponding to the HTS elements being in the normal state, a fraction of the current flows through the female portion of the threaded connection to the middle-stage safety lead, while the remaining fraction flows through the conductor element's Ag-Au sheath. The amount of current flowing through the safety lead will depend on the magnitude of the disturbance, which controls the amount of HTS that has become a normal conductor.

Helium gas at 42 K with a mass flow rate of 2.5 g/s is injected at the lower end of the transition. The gas absorbs the ohmic heating of the normal metal components and electrical connections and then enters the upper stage to be cooled.

Middle-Stage Current Lead

The axial cross section of the middle stage is as shown in Fig. 2. The middle stage incorporates 18 parallel HTS elements configured in a cylindrical array. The conductor elements are supported by an internal safety lead and are connected at their ends to current collectors. Most of the middle stage is cooled by conduction from its 5-K end.

The major features of the selected conductor element are PIT tape with a Bi-2223 multifilamentary core in an Ag + 3 at.% Au sheath, composite conductor of stacked PIT tapes sintered together, critical current of >1100 A @ 60 K in an applied magnetic field, 4 K-

Table 2. Performance evaluation matrix for preproduction conductor element

	Sequence of events, performance evaluation[a]	Sample
1	I_c @ 77 K @ 1 μV/cm • B = 0 T • B = 0.2 T ∥ a • B = 0.1 T ∥ b • B = 0.02 T ∥ c	L94-345 & L94-346
2	I_c @ 60 K @ 1 μV/cm (in variable-temperature cryostat) • B = 0 T • B = 0.1 T ∥ b • B = 0.02 T ∥ c	L94-345 & L94-346
3	I_c vs. 300 K → 77 K → 300 K for 50 thermal cycles	L94-345
4	I_c vs. 0-1000 A for 600 cycles	L94-346
5	End connection joint resistivities @ 77 K	L94-345
6	Irreversible strain limit @ 77 K	L94-345 & L94-346

[a]I_c = critical current

heat load of ≤0.4 W/element, irreversible strain limit of ≥0.1%, thermal cycling results in <4% predicted degradation in critical current, size = 0.48 x 1.07 x 54 cm, and end caps of unalloyed silver.

The safety lead is the core of the middle-stage assembly, which provides an alternate current path and, more importantly, additional thermal mass in the event of HTS element quench. The current path is through a 304 stainless steel tube that is slotted in order to guide and mechanically support the conductor elements. The external surface is nickel- and gold-plated. The conductor elements are fixed axially at their warm ends, strain-relieved at their cold ends, and soldered into the slots of the stainless steel cylinder.

The safety lead is designed to limit the temperature of the HTS to <400 K during a quench and has a conductive heat leak <2 W to 5 K. In the event of catastrophic failure of

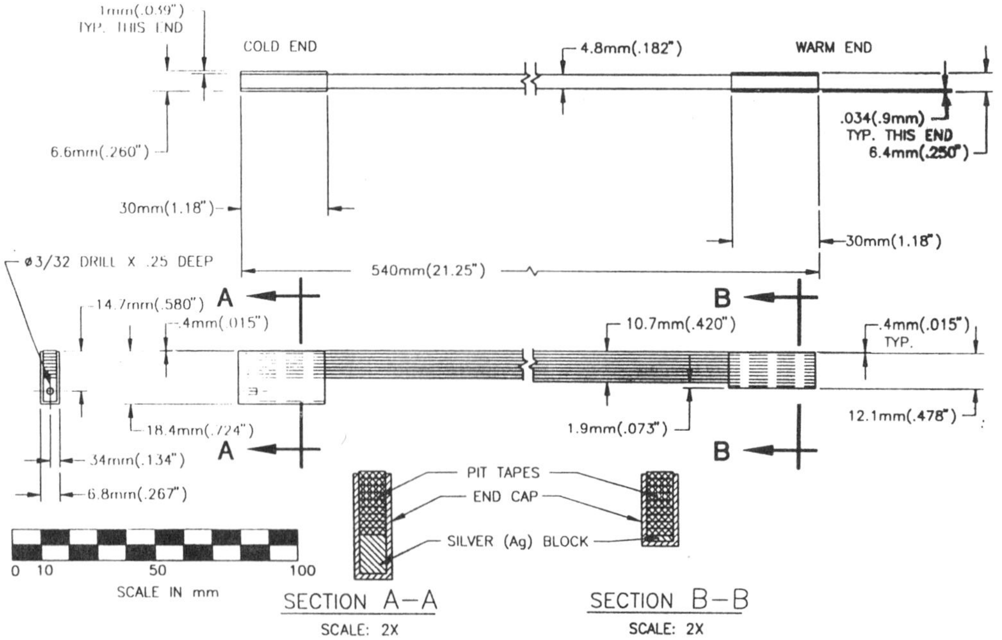

Fig. 3. HTS conductor element geometry.

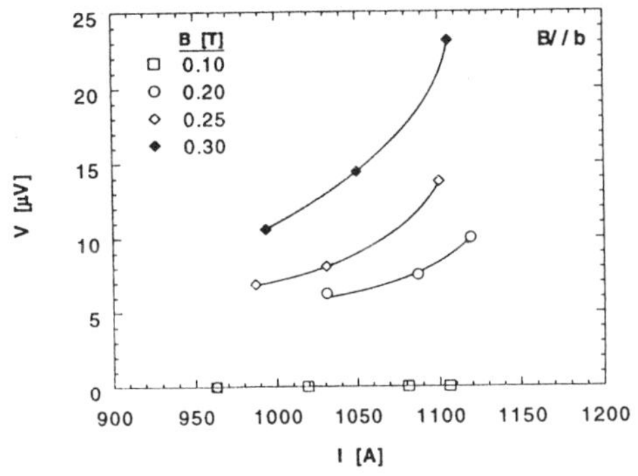

Fig. 4. Sample L94-345 voltage drop vs. current @ 60 K @ B//b ·

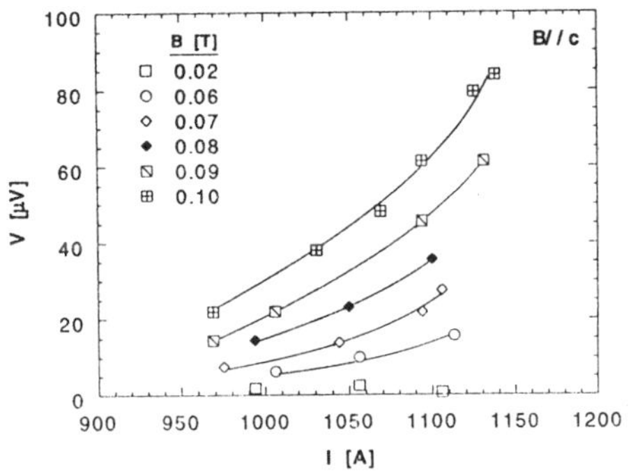

Fig. 5. Sample L94-345 voltage drop vs. current @ 60 K @ B//c.

the HTS leads, the burnout time of the safety lead is >100 s so that current in the SMES magnet can be brought down to safe levels without causing damage to the magnet and its subsystems.

Middle-Stage-to-Lower-Stage Transition

The electrical connection to the lower-stage current junction consists of seven twisted parallel low-temperature superconductor (LTS) strands for each conductor element. The strands are shaped to provide axial strain relief. The 5-K heat load due to connection ohmic heating is 1 W.

Lower Stage

Electrical and structural connection to the middle/lower-stage transition is by a clamp connection. The cross-sectional area and length of the lower-stage copper bus are selected to provide adequate thermal conduction so as to limit the temperature rise of the middle/lower-stage transition to <1 K. Associated overall ohmic heating is 1.23 W.

Thermal Performance

Numerical analysis of the HTS lead with a 60-K intercept shows a reduction in a 5-K equivalent heat load by nearly 50% that of a conventional vapor-cooled lead. This reduction translates to <100 W equivalent heat load at 5 K.

EVALUATIONS OF COMPONENT PERFORMANCE

Upper-Stage-to-Middle Stage Transition

A threaded connection identical to that of the upper-stage-to-middle-stage transition was constructed and employed to measure the electrical resistance of the joint. The joint resistance vs. ambient assembly torque was measured in LN_2. The measured joint resistance was 0.0023 μΩ, as compared to the predicted value of 0.0015 μΩ. The resistance is flat with respect to assembly torque for torques greater than 400 N·m (≈300 ft-lb).

HTS Conductor Elements

The conductor elements were constructed from sintered stacks of multifilamentary composite conductor; geometry is as shown in Fig. 3. The conductor is made by the PIT method, and is in the form of an aspected tape[5] composed of fine BSCCO-2223 filaments in a silver/gold alloy matrix.

A test matrix was developed to evaluate the performance of the preproduction conductor elements relative to their performance specifications. The test matrix is given in Table 2.

The current used to evaluate the critical current (I_c) of the preproduction conductor elements was pulsed at a low duty cycle to avoid errors caused by effects of current-contact heating. The I_c measurements were made in a facility consisting of a variable-temperature helium cryostat, a superconducting magnet, and a current pulser. The variable-temperature cryostat was configured to contain two samples and was rotatable in the magnet to provide B//b and B//c applied field orientations.

Results of the 60 K I_c tests of Sample L94-345 in transversely applied magnetic fields are shown in Figs. 4 and 5. As expected, the superconductive performance of both bars was most sensitive to magnetic fields applied in the B//c direction.

A thermal cycling test was performed on Sample L94-345 after the testing in magnetic fields had been completed. Critical current in a 45-cm-long segment of the bar at 77 K was measured by the pulsed current technique. Fifty thermal cycles were applied. No significant changes attributable to thermal cycling were found.

An electrical cycling test was performed on Sample L94-346 at 77 K after the testing in magnetic fields had been completed. The initial I_c was found to be 990 A. A total of 600 square-wave current pulses, each of 1000-A amplitude and >300 ms length, was then applied at a low duty cycle. Afterward the I_c was found to be 978 A. The values before and after electrical cycling differed by only 0.6% from their averages.

Specific resistivities of metallic contacts to the superconductive cores of American Superconductor Corporation (ASC) tape samples comparable to those used in the composite bars were measured by the lap-joint method.[6,7] Half-joint resistivities at 77 K ranged from 0.031 to 0.037 $\mu\Omega\cdot cm^2$. The values consisted of contributions from the silver-gold/superconductor interface, one thickness of the silver-gold sheathing, and 1/2 thickness of a minimal quantity of tin-lead solder used to form the lap joints tested.

Lap-joint resistivity tests were done also with the specially formed silver metal connectors sintered to the tapes in Sample L94-345. The average resistivity of the half-joints at 77 K, including the contribution from the silver connector and its sintered interface to the tapes, was $\approx 1\ \mu\Omega\cdot cm^2$.

Strain tolerance evaluation of the composite conductor elements was carried out by subjecting the conductor elements to room-temperature tensile strains and then testing for I_c. The specimen was loaded at 300 K to predetermined strains, unloaded, and then measured for I_c in liquid nitrogen at 77 K and 0 T. This process was repeated with increasing strains until sample failure. Sample L94-346 fractured completely at a strain of ≈5%. Figure 6 shows the variation of retained I_c (%) with axial strain. Figure 7 shows a typical variation of retained I_c (%) with axial strain for Sample L94-345. Greater than 90% I_c was retained at 1.2% strain.

PROTOTYPE CURRENT LEAD ASSEMBLY

A prototype middle-stage assembly was designed and is being constructed. The prototype assembly is being employed to evaluate safety lead brazing, middle-stage machining, conductor element soldering, LTS cable fabrication, and quality control methods.

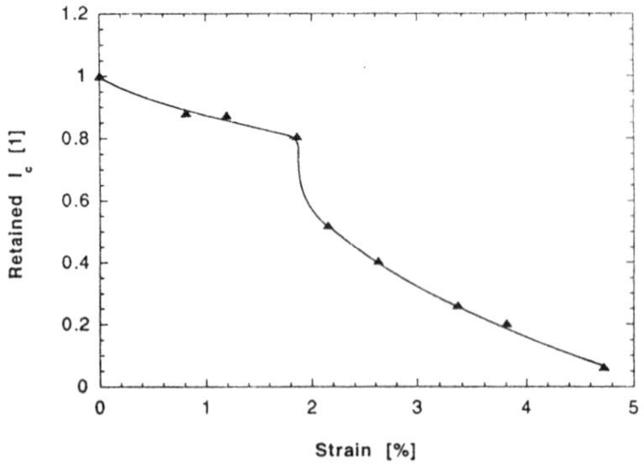

Fig. 6. Sample L94-346 variation of retained I_c with axial strain.

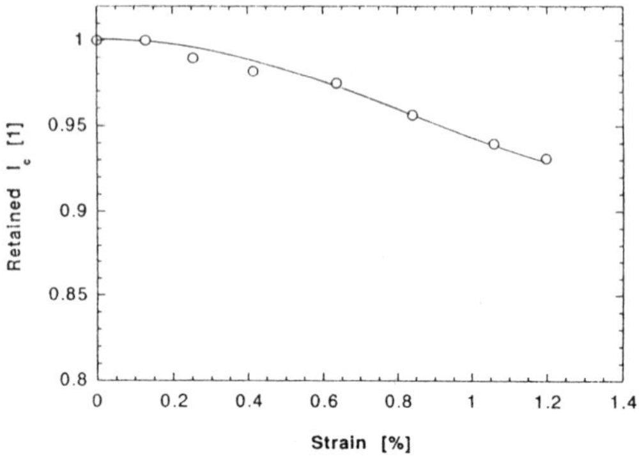

Fig. 7. Sample L94-345 variation of retained I_c with axial strain.

CURRENT LEAD ASSEMBLY PERFORMANCE EVALUATION

The HTS lead assembly, consisting of the conventional gas-cooled upper stage and the HTS middle stage, will be tested in two phases. The first phase will consist of performance evaluations of the conventional upper stage beginning with initial evaluations of the background heat load of the test cryostat vessel without the upper stage present. Once the background heat load of the test cryostat has been established, the upper stage will be installed and heat load measurements will be performed at both zero current and operating current (16,000 A) plus an additional 10%. Initial helium mass flow rates in the gas-cooled lead will be based on design calculations for the modified lead. A preliminary parametric study of the optimal mass flow and inlet temperature will be performed. These values will be used initially during the second phase of testing.

During the second phase of testing, the HTS portion of the current lead will be attached via a screw connection to the conventional upper stage. Initially, mass flow rates and inlet temperatures to the gas-cooled upper stage will be based on results obtained during the first phase of testing, with adjustments based on the additional calculated heat loads of the HTS portion of the current lead. Testing will begin by ramping the current lead assembly to operating current plus 10% (17,600 A) in zero applied field. The next test will consist of applying an external magnetic field of 0.2 T perpendicular to the axis of the lead assembly and ramping the current lead to 17,600 A. The external magnetic field will be supplied by a split solenoid magnet. Requirements for the external field are that it provide 0.2 T field with an active region of 10 cm in the axial direction, homogeneous to within 20%, in a clear bore > 15 cm. The active region of the field will be positioned at the warm end of the middle stage. A final parametric study to determine the optimal helium mass flow rate and inlet temperature will be performed at both zero current and operating current in an 0.2 T applied field. Once the optimal flow rate and inlet temperature have been established, heat load measurements will be performed at zero current and operating current in an 0.2 T field.

The next tests will be for reliability. The current lead assembly will undergo three separate sets of five thermal cycles and 70 electrical cycles, for a total of 15 thermal cycles and 210 electrical cycles. A thermal cycle is defined as room temperature → 100 K → room temperature. An electrical cycle is defined as zero current → operating current → zero current at a ramp rate of 50 A/s increasing and 200 A/s decreasing.

The final tests of this phase will be the safety lead test, which will use two different methods. The first will be at operating current, magnetic field, and optimal inlet gas temperature. Mass flow will initially start at the optimal flow rate and slowly be decreased to a predetermined minimum value. The second method will be at operating current, magnetic field, and optimal flow rate. Inlet gas temperature will initially start at the optimal value and slowly increase toward the critical transition temperature (B,I) of the HTS material.

CONCLUSIONS

- An HTS current lead with a 16-kA operating current has been designed for incorporation into a near-term, 0.5-MWh-capacity SMES system. The helium-gas-cooled upper-stage lead operates between 300 and 60 K. Features include multiple gas flow passages and two-stage heat exchange. The 60-K heat load for both leads is 196 W. The conduction-cooled middle-stage lead operates between 60 and 5 K. Current flow is shared by a cylindrical array of 18 parallel composite conductor elements fabricated from Bi-2223 PIT tapes and having a silver-with-gold alloy sheath. Protection is provided by a stainless steel safety lead installed in parallel to the conductor element array. The 5-K heat load for both leads is 11.2 W.
- Two preproduction conductor elements have been evaluated relative to their performance specifications. Critical current @ 60 K @ B//b and B//c @ 1 μV/cm was measured in a pulsed-current mode. One sample conductor element was subjected to 50 thermal cycles of 300 K → 77 K → 300 K, with no significant change in I_c. One sample conductor element was subjected to 600 electrical ramp cycles corresponding to those predicted for the SMES current lead application, with no significant change in I_c. Specific resistivities of metallic

contacts to the superconductive cores of tape samples comparable to those used in the composite bars were measured. Half-joint resistivities at 77 K ranged from 0.031 to 0.37 $\mu\Omega\text{-cm}^2$. Lap-joint resistivity tests were done with the specially formed silver metal connectors sintered to the tapes. The average resistivity of the half-joints at 77 K, including the contribution from the silver connector and its sintered interface to the tapes, was about 1 $\mu\Omega\text{-cm}^2$. Both sample conductor elements were evaluated for their irreversible strain limit. Greater than 90% I_c was retained at 1.2% strain. The sample conductor elements displayed a high degree of structural and electrical integrity, especially because they each were handled many times by various individuals but displayed no discernible performance degradation.

- A prototype middle stage assembly is being constructed for the purpose of evaluating manufacturing methods.
- Current lead assembly performance evaluation program will be carried out. Included will be performance of the upper stage, the integrated lead assembly, and safety and reliability studies.

ACKNOWLEDGMENTS

The authors gratefully acknowledge the technical contributions of Y. S. Cha, K. Duncan, D. J. Evans, E. Hudak, T. E. Kasprzyk, and J. Ulandy, as well as the manuscript preparation skills of J. A. Stephens and the editorial contributions of C. A. Malefyt. Work at ANL was supported by the U.S. Department of Energy, Energy Efficiency and Renewable Energy, as part of a program to develop electric power technology, under Contract W-31-109-Eng-38. Work at Babcock & Wilcox was conducted as a part of the company's SMES development program.

REFERENCES

1. R.C. Niemann, Y.S. Cha, J.R. Hull, C.M. Rey, and K.D. Dixon, Design of a high-temperature superconductor current lead for electric utility SMES, *IEEE Trans. on Appl. Superconductivity* 5:789 (1995).
2. H. Fujishiro, M. Ikebe, K. Noto, M. Matsukawa, T. Sasaoka, K. Nomura, J. Sato, and S. Kuma, Low thermal conductive Bi-2223 tapes sheathed with Ag-Au alloys, *IEEE Trans. Magnetics* 301:645 (1994).
3. T. Sasaoka, T. Umezawa, A. Nomoto, N. Tatsumi, and M. Seido, Power current leads using Ag-sheathed high-T_c superconducting tapes, in *Advances in Superconductivity IV, Proc. 4th Intl. Symp. on Superconductivity (ISS '91),* H. Hayakawa and N. Koshizuka, eds., Springer-Verlag, New York (1992), pp. 621-624.
4. S.W. Van Sciver, Y.S. Hascicek, W.D. Markiewicz, L.R. Motowidlo, D.R. Hazelton, and P. Haldar, Strain dependence of Bi-2223 tape conductors for application in high field magnets, in *Advances in Superconductivity V, Proc. 5th Intl. Symp. on Superconductivity (ISS '92),* Y. Bando, ed., Springer-Verlag, New York (1993), pp. 819-822.
5. K.H. Sandhage, G.N. Riley, and W.L. Carter, *Journal of Metals* 43:21 (1991).
6. H. Chi-Shiung, G.H. Haertling, and M.D. Sherrill, *Rev. Sci. Instrum.,* :1317 (1991).
7. R. Birkhahn, N. Browning, A. Youngdahl, M. Lanagan, and R. Poeppel, *Applied Superconductivity* 2:67 (1994).

TEST OF COPPER-BRAID-STABILIZED BUS LINES FOR SUPERCONDUCTING DIPOLE MAGNETS

M. Doi,[1] A. Kabe[2], Y. Kojima[2], H. Nakai[2], K. Hara[2], Y. Morita[2],
K. Hosoyama[2]
[1]Engineering Research Center, NKK Corporation, Kawasaki, JAPAN
[2]National Laboratory for High Energy Physics (KEK), Tukuba, JAPAN

ABSTRACT

A high cryogenic stability suprconducting bus-line has been developed to connect a superconducting dipole magnet with a full length of 13 m to a current lead approximately 2 meters from the magnet. The superconducting bus-line is made of NbTi strand cables for magnet use soldered to copper braid. The copper braid has a large surface area to improve cooling efficiency and increase cryogenic stability. Three kinds of bus-line are prepared on experimental basis: a bare superconducting cable, a superconducting cable joined copper braid with a thin layer of solder, and one made by filling the inside of copper braid with solder. Cryogenic stability tests confirmed that a bus-line equipped with a copper braid provides twice the cryogenic stability as a bare superconducting cable.

INTRODUCTION

At the National Laboratory for High Energy Physics (KEK), a superconducting dipole magnet for use on the Superconducting Super Collider (SSC) has been developed by a Joint U.S. Japan research project underway since 1990. In preparation for excitation testing of a prototype superconducting magnet with a full length of 13m, a superconducting bus-line has been made to connect the magnet to a current lead approximately 2 meters from the magnet. The superconducting bus-line could possibly receive the heat from the current lead, and it is not firmly fixed in place with GFRP. To prevent quench caused by these problem, it is necessary to stabilize the superconducting bus-line. The one method to stabilize is that a copper plate is attached to a superconducting cable to reduce the joule heating and increase cryogenic stability [1)], but a report on the results of a simulation [2)] indicate that it is possible to achieve effective cryogenic stability by attaching a copper braid to increase the cooling surface area. Therefore, three kinds of superconducting bus-lines were prepared: a bare

Advances in Cryogenic Engineering, Vol. 41
Edited by P. Kittel, Plenum Press, New York, 1996

superconducting cable, a bus-line made by soldering a copper braid to a superconducting cable with a thin layer of solder, and another made by filling the copper braid interior with solder. Cryogenic stability tests of all three were conducted, confirming that attaching a copper braid is an effective way to improve cryogenic stability.

FABRICATION OF THE SUPERCONDUCTING BUS-LINE

A superconducting bus-line consisting of a superconducting cable attached with a copper braid has been developed to connect a superconducting dipole magnet to a current lead. The superconducting cable is an NbTi strand cable (wire diameter: 0.809 mm, number of wire: 30, Cu/Su ratio: 1.4) used to form the inner layer coil of the magnet. Almost the same size as the superconducting cable, the copper braid was made of 490 copper strands with a diameter of 0.12 mm. The copper braid was connected to the superconducting cable with foil-shaped solder made by adding a small amount of silver to tin-lead solder. Only the contact surfaces of the copper braid and the superconducting cable had to be connected so that the addition of the copper braid would increase the cryogenic stability by expanding the cooling surface area. The optimum thickness of the solder foil were 40 μm. Figure 1 shows the cross-section of the superconducting bus-line.

CRYOGENIC STABILITY TESTING EQUIPMENT

Figure 2 shows the schematic of the testing equipment. The superconducting bus-line prepared for the test had a full-length of 3 m, and was firmly fixed with GFRP. 30 voltage taps were attached at 10 cm intervals, and two foil heaters with differing dimensions were installed. Heater 2 (dimensions: 9 mm × 4 mm) was installed between the number 9 and 10 taps, while heater 1 (dimensions: 11.5 mm × 22 mm) was installed between the number 21 and 22 taps. The cover was installed for cryogenic stability tests in boiling helium. The boiling helium filled into the cover with the heater which is under the cover .It adjusted the amount of the heater input to keep a liquid surface constant. During the testing, a rectangular pulse current was supplied to the foil heaters as the pulse width was varied, and measurements were made of the minimum heater input and the propagation velocity of the normal zone when quench occurred. The propagation velocity was found from the start-up time interval of the voltage signal from the voltage taps. The test was conducted for three bus-lines: the bare superconducting cable , one made by soldering a copper braid to a superconducting cable with a thin layer of solder, and one made by filling the interior of a copper braid with solder.

MINIMUM HEATER INPUT OF QUENCH

Figure 3 shows the relationship between the minimum heater input and the current when quench occurred. During the test of the bus-line with copper braid attached ("Braid" in the figure), the minimum heater input was about half as much as during the test of the bare superconducting cable ("Bare" in the figure). But there was a discrepancy in the amount of heater input between the two heaters during the test of the bare superconducting cable, with the minimum heat input by heater 1 smaller than that of heater 2. And after the current exceeded 5000A, the minimum heat input of the heater 2 were almost identical at approximately 0.5J. The larger heating surface area of heater 1 is considered to be the cause

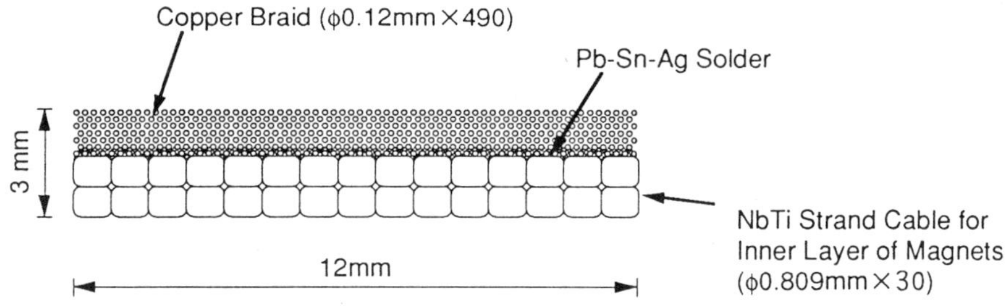

Figure 1. Cross-Section of the superconducting bus-line.

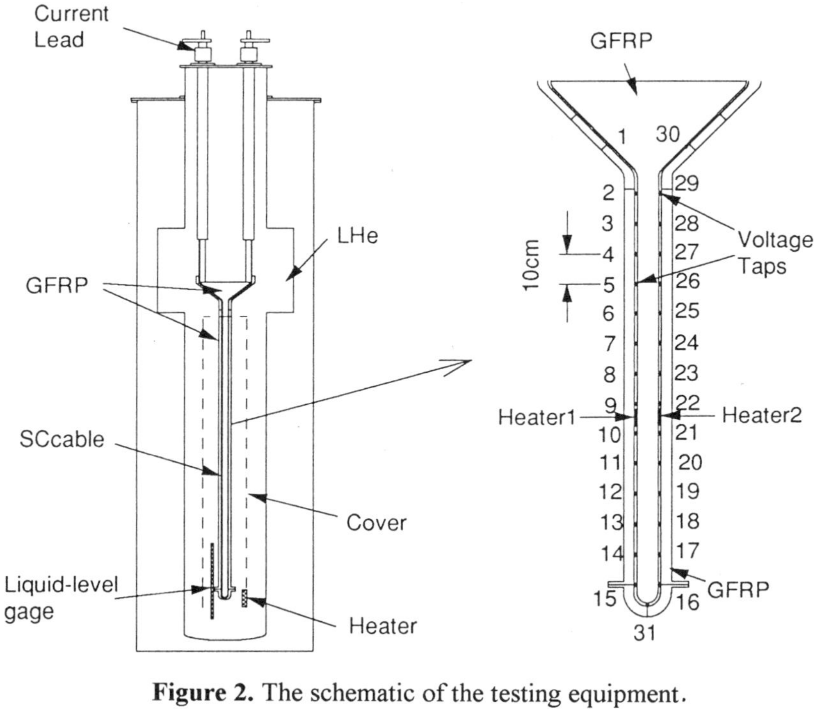

Figure 2. The schematic of the testing equipment.

of this result. The amount of the heat leakage increases when the heating surface area becomes big and the small heater input can not be transferred to the bus-line.

The results of the test in boiling helium ("in GHe" in the figure) is that the minimum heater input of the bus-line with copper braid in boiling helium was as much as the one of the bare cable in liquid helium. The minimum heater input of the bare cable in boiling helium were almost identical at approximately 0.5J. It is considered that this result was caused by being described above, too.

PROPAGATION VELOCITY OF THE NORMAL ZONE

Figure 4 shows the results of measurements of the propagation velocity of the normal zone. The propagation velocity of the normal zone increases as the current rises, but like the results for the minimum heat input of the heaters, the propagation velocity was about half as high for the test of the bus-line with the copper braid as it was for the test of the bare superconducting cable. These results indicate that the attachment of a copper braid increases the margin of the quench energy to about twice that of the bare superconducting cable. But no differences were revealed between the test using a bus-line made by attaching copper braid to the superconducting cable with thin solder (○ △ symbols in the figure) and that using a bus-line made by filling the copper braid with solder (∗ symbols in the figure). These results were considered using the propagation velocity of the normal zone calculation formula.[3)] The perimeter is influenced by the size of the flow channel of the liquid helium. Therefore the results of fitting the calculation formula with the perimeter P as the parameter, are curves 1 and 2 in Figure 4. Table 1 shows the input data at this time.

These results show that the attachment of the copper braid increased the cooling perimeter 2.5 times. This increase in the cooling perimeter is about 1/7 of the full length of the copper braid. This fact and the results of an inspection of the cross-section of the bus-line made by filling the copper braid with solder, indicate that because it was not actually filled with solder, there was no discrepancy in the results of the tests performed with the bus-line made by filling the copper braid with solder and tests done with the bus-line made by attaching the copper braid to the superconducting cable with a thin layer of solder. Curve 3 represents a case in which the cross sectional area of the conductor cable was doubled while

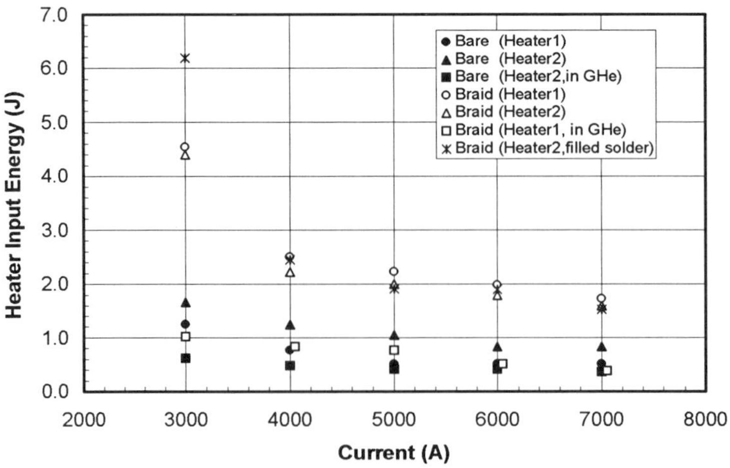

Figure 3. Heater input energy required to initiate a quench versus the current.

Table 1. Parameters used for calculations

	Curve1	Curve2	Curve3	Curve4
NbTi Cross Section(mm^2)	6.42	6.42	6.42×2	6.42×2
Cu Cross Section(mm^2)	8.98	8.98	8.98×2	8.98×2
Braid-Cu Cross	—	5.54	—	—
Cooled Perimeter (mm)	18	45	18	18×2

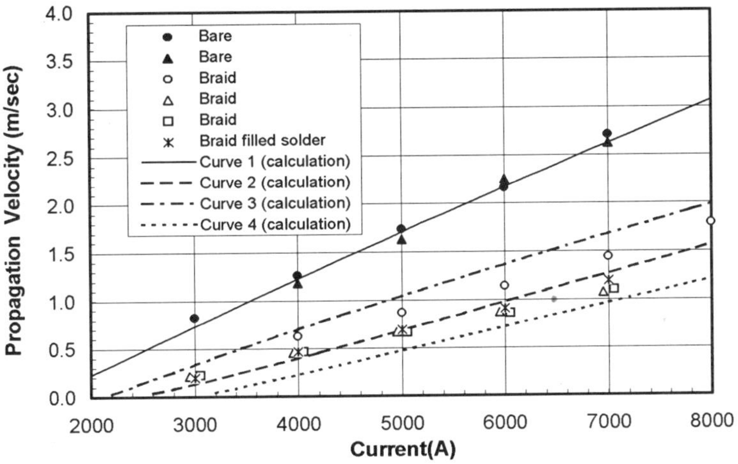

Figure 4. Measured propagation velocity compared with calculations.

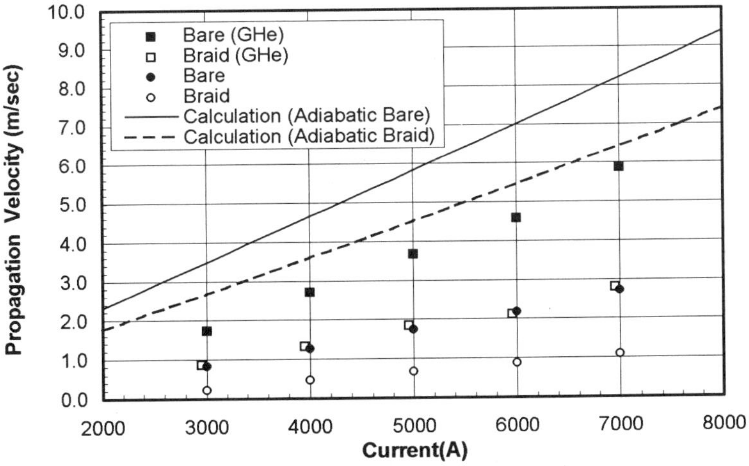

Figure 5. Measured propagation velocity in boiling helium compared with calculations under adiabatic conditions.

the cooling perimeter remained identical to that of curve 1, and is a result calculated with the cross sectional area of the conductor increased to a size much larger than the cross section area of the copper braid. Curve 4, a case in which the cooling perimeter and area of the cross section of the superconducting cable are double those for curve 1, represents the results of calculations when two layers of superconducting strand cables used in the test are overlapped. Because curve 2 stands between these, the attachment of the copper braid does not increase the surface area of the conductor, and the increase in the cooling surface area doubles the cryogenic stability.

Figure 5 shows the results of measurements of the propagation velocity in boiling helium. Both of the propagation velocity of the bare cable and the one of the copper-stabilized cable in boiling helium ("GHe" in the figure) was about twice as high as the propagation velocity in liquid helium. The propagation velocity of the copper-stabilized cable in boiling helium was as high as the one of the bare cable in liquid helium. The propagation velocities in boiling helium is lower than the results of calculations under adiabatic conditions.

CONCLUSION

A study was made of the cryogenic stability of a copper-braid-stabilized bus-line. In liquid helium and boiling helium the minimum heater input and the propagation velocity was measured in the following cases: a bare superconducting cable, a superconducting cable soldered copper braid with a thin layer of solder, a superconducting cable soldered copper braid filling the inside with solder. As a result, this bus-line provides twice the cryogenic stability of a bare superconducting cable. The merits of this bus-line can be fabricated at low cost. It was actually installed to connect a 13m dipole magnet to a current lead.

REFERENCES

1. A. A. Olyunin et al. " Study of Quench Processes in Copper-Stabilized Superconducting Cables ", Proceedings of ICFA Workshop on AC Superconductivity, 1992, p93
2. W. Clay et al. " Design of A Thermal Expansion Joint for The SSC Magnet Leads ", Supercollider 3, 1991, p1045
3. M. Wilson, " Superconducting Magnets ", Clarendon Press Oxford, 1983, p205

COMMISSIONING AND OPERATION OF THE CEBAF END STATION REFRIGERATION SYSTEM

D. Arenius, B. Bevins, W.C. Chronis, V. Ganni
D. Kashy, M. Keesee, J. Wilson Jr.

Cryogenic Systems Group
Continuous Electron Beam Accelerator Facility
Newport News, Virginia, 23606, USA

ABSTRACT

The CEBAF End Station Helium Refrigerator (ESR) System [1] provides refrigeration at 80 K, 20 K and 4.5 K to three End Station experimental halls. The facility consists of a two stage helium screw compressor system, 4.5 K refrigerator, cryogen distribution valve box, and transfer lines to the individual experimental halls. The 4.5 K cold box and compressors were originally part of the ESCAR 1500 W, 4 K refrigeration system at Lawrence Berkeley Laboratory which was first commissioned in 1977. The compressors, 4.5 K cold box, and control system design were modified to adapt the plant for the requirements of the CEBAF experimental halls. Additional subsystems of cryogen distribution, transfer lines [2], warm gas management, and computer control interface were added. This paper describes the major plant subsystems, modifications, operational experiences and performance.

INTRODUCTION

The End Station Refrigerator is one of three CEBAF cryogenic facilities. These facilities are the Central Helium Liquefier (CHL), located in the center of the accelerator; the End Station Refrigerator (ESR), located next to the End Stations Halls A, B and C; and Cryogenic Test Facility Liquefier (CTF) which is located adjacent to the Cryomodule Test Laboratory. While the purpose of the CHL and CTF is to provide cryogens to superconducting cavities in the accelerator ring and testing areas, the ESR is designed to provide cryogens to the CEBAF experimental halls. These halls have superconducting magnets, cryogenic targets and various equipment which require refrigeration at 4.5 K, 20 K, and 80 K. The ESR can provide refrigeration to one or all three halls with at any given time.

Table 1. End Station Cryogenic Heat Load Budget [3]

	Device	Lead Flow	4 K	20 K	80K
Hall A					
	Magnet	64 L/HR	130 W		UNK*
	Transfer Line	-NA-	76 W		50 W
	Cryotarget	-NA-		250 W	
Hall B					
	Torus Magnet	40 L/HR	220 W		UNK*
	Transfer Line	-NA-	20 W		30 W
	Cryotarget	-NA-	0 W		0 W
Hall C					
	Magnets	60 L/HR	135 W		400 W
	Transfer Line	-NA-	46 W		50 W
	Cryotarget	-NA-		250 W	

* Unknown -NA- not appliable

ESR Cryogenic Heat Loads. The three Experimental End Stations (Halls) contain superconducting quadrupole magnets, a large six coil toroid, and miscellaneous user's equipment. The CEBAF End Station cryogenic refrigeration load allowance[3] for the experimental halls is presented in Table 1. These requirements include refrigeration at 4.5 K, 20 K, and 80 K. The 80 K refrigeration is provided by LN_2 from the CHL. The CHL LN_2 storage capacity is 75,700 L. The CHL facility also provides gaseous helium storage and purification for the facility. The ESR facility can accept 10 G/S at 4.5 K and 2.8 bar from the CHL as a secondary helium cryogen source for the End Stations.

Refrigerator History. The refrigerator (cold box) was originally designed for Lawrence Berkeley Laboratory (LBL) in 1977 for two principle modes of operations. As a liquefier, it provided 11.7 G/S at 4.42 K. In the refrigerator mode, it was designed to provide 1450 watts of refrigeration and 3 G/S of liquid at 4.42 K. The refrigerator was one of two identical refrigerators which were built at approximately the same time for separate laboratories. At the second laboratory the refrigerator had a performance design mode requirement which could provide 950 watts of refrigeration plus 1.7 G/S of liquid at 4.42 K and 1250 W at 21.5 K. The second plant, at FERMI, is still in operation today. In June of 1990 the refrigerator system was shipped to CEBAF from LBL. The shipment included the compressor skids, motor starters, a local analog control console, 10,000 L helium dewar, and the refrigerator. The plant was placed into storage until the building for the plant could be finished. An outline of the plant's history is presented in Table 2.

The number of operating hours at LBL were less than 1000 hours and the refrigerator had been inactive from 1981 until 1994. The technical concerns were therefore centered on the condition and refurbishing of the major equipment. This included the compressor motors, motor starters, compressors, control system and the turbine expanders. Consideration was given to the anticipated maintainablity and the experiences of the LBL operations. Many of the component manufacturers, including the turbine expanders and compressors, were still providing spare parts. Original LBL spare parts were transfered to CEBAF with the refrigerator and compressors. Other spares, such as turbine cartridges,were purchased later by CEBAF. Much of the initial operational maintenance

Table 2. ESR System Timeline

SYSTEM	DATE	REMARKS
LBL		
Compressors Received	Aug-76	
Coldbox Recieved	Jun-77	
Acceptance Test	Nov-77	
Last Used At LBL	Sept-8	~960 hours
CEBAF		
LBL System Received	Jun-90	
Installation Start	Feb-92	
Compressor Startup	July-94	~3300 hours to date
Cold Box Startup	Dec-94	~2000 hours to date
CEBAF Modifications		
Gas Management, Utilities	Feb-92	New To System
Refrigerator Controls	Feb-92	PLC and Computer
Compressor Controls/Tubing	Jun-92	PLC/Welded Lines
Cold Box LN_2 Boiler Installation	Jun-94	New To System
External 80 K Purifier Beds	Feb-95	New To System
Cold Box 4.5 K Separater/Subcooler	Jun-95	New To System
Dewar Internal Subcooler	Jul-95	New To System
Compressor Oil Coolers	Jul-95	Replacement
Spare Compressor	Aug-95	New to System

insight for the plant came from the LBL logbooks, documentation and operator's manual. New subsystems were also needed for the CEBAF facility. This included a cooling water system, electrical distribution, cryogen distribution valve box, transfer lines[2], compressor oil removal, gas management, instrument air, gas contamination monitoring, and computer interface cabinets. Design, fabrication, and installation of the system modifications and new subsystems were performed by CEBAF personnel.

CEBAF PROCESS DESCRIPTION

Overall Design

The CEBAF End Station Refrigerator System is presented in Figure 1 for the CEBAF full 4.5 K refrigeration design capacity mode. Two stages of compressors compress helium from 1 bar to 17 bar with a discharge flow rate of 214 G/S. During cooldown, system makeup gas is brought into the compressor system interstage or first stage suction at 2.5 bar or 1 bar respectively. A 2nd stage discharge massout control valve provides the means to remove excess gas from the system during a warmup process. Compressor bypass valves are provided around each stage to regulate the compressor suction pressures. Flow from the refrigerator returns 113 G/S to the interstage and 101 G/S to the first stage compressor suction. The compressor discharge flow enters the refrigerator where 5 G/S is diverted to a GN_2 vent warmup heat exchanger. The diverted flow is then recombined with the main high pressure flow. The flow stream is then precooled in a LN_2 pool boiling heat exchanger to 80 K, passing through one of two new 80 K adsorber beds. As the discharge flow continues

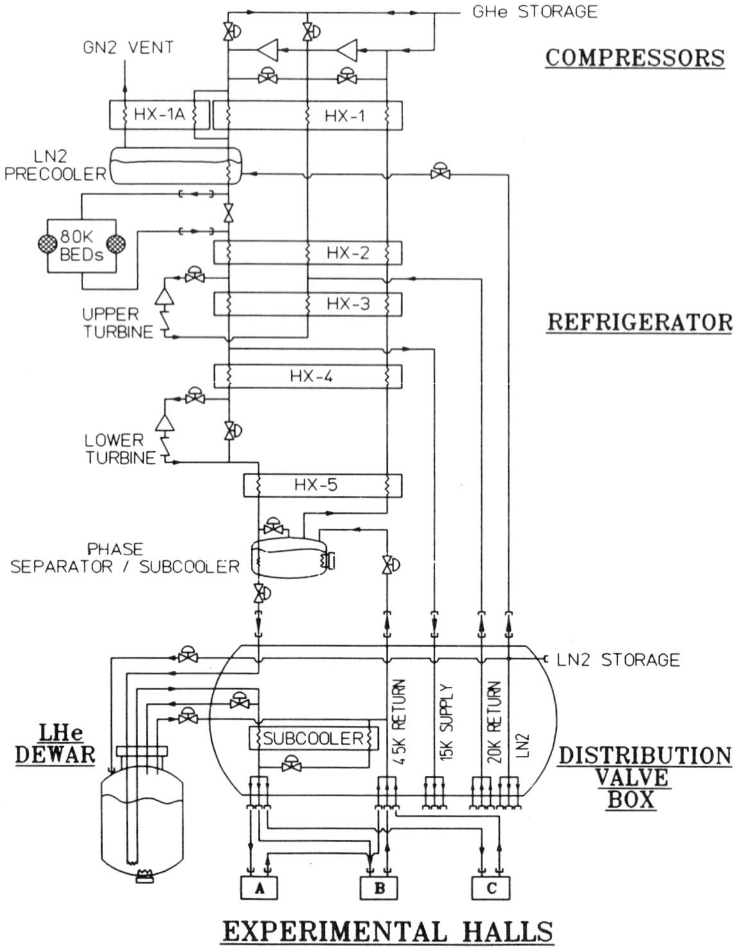

FIGURE 1. ESR Facility Simplified Flow Diagram indicating the major subsystems of compressors, refrigerator, liquid helium dewar and distribution valve box.

through the refrigerator 113 G/S is diverted through the upper turbine at an inlet temperature of 18.5 K. The turbine flow exhausts at 2.5 bar and 12 K and is returned to the interstage of the compressor system.

The remaining 101 G/S flow is expanded through a parallel combination of the lower turbine (75 G/S) and a bypass valve (26 G/S) to 2.9 bar and 5.7 K. The process flow is then subcooled to 4.5 K by utilizing 10 g/s of the process stream in a new subcooler/phase separator which is internal to the refrigerator. Additional subcooling capacity is provided by a 10,000 L liquid helium dewar which was modified with internal cooling coils.

A refrigerator process connection is provided for supplying 15 K He to future hall cryogenic target loads at reduced 4.5 K refrigeration loads. The target load flow returns via a 20 K refrigeration process return connection to the refrigerator interstage at 2.5 bar.

Compressor System

The compressor system is comprised of two stages of oil flooded screw compressors. Each stage has a single electrically driven compressor and was a part of the original LBL facility. The first stage is a 3600 rpm, 150 KW Sullair C-25-LB compressor. The second stage is a 3600 rpm, 750 KW Sullair C-25-LA compressor. Each compressor is mounted on an individual skid with an oil cooler, bulk oil separator, 480 volt drive motor, local control panel and a gaseous helium aftercooler. The original compressor skid mounted relay logic control panels, instrumentation, instrument tubing, and oil injection lines were replaced. This was to eliminate threaded and compression fittings as possible sources of system contamination. A small programmable control system was installed on each compressor skid and integated with the CEBAF computer control system for unattended operation. The remodeled compressor controls have worked very well with the exception of an initial loose control fuse holder which resulted in a few sparatic compressor trips.

The first stage compressor and the oil coolers of both stages were disassembled and inspected for possible corrosion which may have occurred during the system inactivity and storage period. There was very little corrosion evident. Compressor oil seals, oil, and motor bearings were routinely replaced. Initial static motor winding testing indicated no adnormalities. After approximately 1000 hours of compressor operation, however, both the first and second stage motors required rewinding or replacement because of the development of electrical winding failures. The second stage compressor motor has been increased from a 600 KW to a 750 KW drive to allow for a larger operational load margin. The compressor gas management system is a new design. The system utilizes relatively small equal percentage gas makeup valves (<25 G/S). This provides a finer degree of control during normal cooldown or steady state refrigeration operations. The compressor bypass valves were sized for the compressor startup and system pressure building phase of operation. During this phase of operation the valves must bypass approximately 40 percent of the design flow with a fully unloaded compressor and low initial compressor differential pressure. This conservative design approach allows compressor starting when the available helium makeup gas storage pressures are relatively low. There has not been any difficulties associated with the gas management control system.

Flow orifice plates were installed on each main process flow stream to the refrigerator. Additional flow instrumentation provisions were made to test the performance of the compressors independently of the operation of the refrigerator. This was accomplished by placing flow orifice plates upsteam of each of the compressor bypass valves. A performance check of the compressor operation indicated nominal design flow rates with a flow balance indication error of 3 percent.

A new compressor oil removal system consisting of three stages of Monsanto oil coalescers, a single stage of carbon adsorber and a final particulate filter were added. Although the design allows the oil to be returned to either compressor stage, the oil is normally returned to the first stage since there is a normal oil carry-over from the first to the second stage compressor. Similiar oil removal systems used at CEBAF have been in continuous operation for eight years with great success. The oil removal system is functioning well with the majority of the oil removal occurring in the compressor bulk oil separator and the first coalescer.

During the initial running of the compressors at CEBAF the water contamination level in the helium gas process stream increased whenever the compressors were stopped. An investigation showed the compressor oil coolers were of the swaged tube design. Some tubes were leaking cooling water contamination into the helium gas system whenever the helium gas stream pressure was below that of the water system. Successful reswaging was completed but replacement oil coolers with a welded tube design were procured as a precautionary measure.

Compressor Summary. Currently the compression system, oil removal and gas management are working reliably with 3000 hours of operation. Much of those operational hours were unattended. Improvements, such as the replacement of the oil coolers and the addition of a third compressor (which may used as a first or second stage backup) are being made.

Refrigerator

Control System. The refrigerator or cold box was modified in a number of ways. The original control system was based on an obsolete programmable controller. Valve control was accomplished with separate electronic analog controllers. A determination was made that the system would be difficult to maintain due to the age of the devices and the availability of spare parts. The programmable controller system was replaced and the analog controllers were eliminated completely. The new programmable controller requires only 1/16th the space of the earlier model. It is identical to other controllers which have proven to be reliable at CEBAF. The duplication of the type of controller eliminated additional spare part inventories and personnel training. The former function of the analog loop controllers was replaced with direct CAMAC computer interface control.

The control logic of the refrigerator was modified to include the LBL operational experiences and concerns. Pressure switches were added to the outlet of the lower turbine to stop the turbine after starting if the outlet pressure drops below 2.3 bar or rises above 5 bar. This safety interlock prevents turbine damage should the downstream turbine flow be restricted or there is a danger of forming liquid on the turbine blades.

A pressure equalization valve was formerly designed to automatically depressurize the refrigerator high pressure supply stream whenever a compressor shutdown, power outage, or loss of instrument air occurred. The depressurization was changed to a manual operation to minimize the rate of turbine inlet pressure change as the turbine was being stopped.

The former control logic included a vapor pressure thermometer with a pneumatic amplifying relay for an automatic shutdown of the upper turbine in the event its inlet temperature became too cold. The accuracy of the turbine trip point was dependent on the turbine temperature ratio and the hystersis of the pneumatic devices. The design is being modified to monitor the turbine outlet temperature electronically as a more accurate monitoring means. All refrigerator silcon diode temperature sensors were replaced and all critical control sensors were spared. The control logic was also modified to isolate the cold box from the hall loads in the event of a refrigerator, utility, or compressor loss.

80 K Adsorber Beds. Originally the refrigerator did not have 80 K adsorber beds. Two redundant adsorber beds have been added to the system since the refrigerator is now being used for extended periods of time. The beds have been installed external to the existing cold box and are connected to the refrigerator by bayonets. Each bed is equipped with isolating valving, regenerative heaters, and instrumentation. An external manual gas management control panel provides the operator with the means to cooldown, warmup, and regenerate the beds. The adsorber beds will be used for the first time in August, 1995.

80 K Precooler. During the LBL plant operations, there was concern about accidently damaging the upper heat exchanger by freezing the LN_2 passage of the heat exchanger. Such an event could occur during a helium flow imbalance where excess cold helium gas or liquid was returned through the low pressure return piping of the refrigerator. To prevent this, the LN_2 precooling in the upper heat exchanger was eliminated. An existing auxillary internal rapid cooldown heat exchanger is now used in combination with a new LN_2 pool boiling heat exchanger to provide helium gas precooling. Flow diverter valving was added to warm the nitrogen venting from the refrigerator. Under operation, the new design has performed well and is very stable.

Phase Separator. During initial CEBAF operation of the refrigerator, the internal phase separator with its liquid helium level instrumentation would not function properly. Liquid was carried over into the refrigerator low pressure return piping, resulting in a rapid oscillating lower turbine outlet pressure. This caused the lower turbine expander to shutdown. A replacement phase separator with a new subcooler has been designed and is presently being installed in the refrigerator. The new separator and subcooler will prevent liquid carryover and double the existing liquid volume capacity .

Refrigerator Summary. The refrigerator has been modified with a number of improvements. These have included the addition of a new LN_2 precooler, controls, turbine safety interlocks, 80 K adsorber beds, and the design of a new 4.5 K phase separator/subcooler. The refrigerator has been operated for 2000 hours, utilizing single upper turbine and dual turbine operation. The operations have included cooling the Hall B Torus Magnet and a Hall A dipole magnet to 4 K. During most of these operations, the refrigerator was unattended and it operated reliably.

CONCLUSIONS

The ESR system has been successfully commissioned. Although some of the major subsystems are approximately 18 years old, there were a number of factors which have helped the commissioning. These include low former operating hours, a design and capacity which had been previously proven suitable for our needs, the equipment was well cared for, and the existance of very accurate documentation of both the equipment and previous operating experiences.

ACKNOWLEDGEMENTS

The authors would like to thank all the members of the CEBAF Cryogenic Systems Group for their work in the design, fabrication, installation, and operation of the End Station Refrigerator Facility. They would like to also extend a special "thank you" to the staff of the CEBAF experimental halls and Lawrence Berkeley Laboratory for their support. This work is supported by DOE Contract #DE-AC05-84ER40150.

REFERENCES

1. C. H. Rode et al. "2.0 K CEBAF Cryogenics", in: "Advances in Cryogenic Engineering," Vol 35A, Plenum Press, New York (1990), p. 275
2. D. Kashy et al. "CEBAF Transfer Line Systems", in: "Advances in Cryogenic Engineering," Vol 37A, Plenum Press, New York (1992), p. 577
3. "End Station Parameters", in: "CEBAF Design Handbook," Chapter 6.2.2, Table 6.23, Rev. January 14, 1991, p. 6.2-4

1995 CEC/ICMC Conference: The 1995 CEC Board of Directors: (from left to right) back row: William A. Little, Peter Kittel, Jeffrey R. Bergen, John A. Barclay, Klaus D. Timmerhaus, Peter E. Gifford, Jay C. Theilacker, and A. Robert Urbach; front row: James B. Peeples, Al Zeller, John R. Hull, D. Michael Coffey, J. Patrick Kelley, Geoffrey F. Green, and Robert Witt. Not present: Peter V. Mason, Gerald R. Karr, Walter F. Stewart, and Bascom W. Birmingham.

PROCUREMENT AND COMMISSIONING OF THE CHL REFRIGERATOR AT CEBAF*

W. C. Chronis, D. M. Arenius, B. S. Bevins,
V. Ganni, D. H. Kashy, M. M. Keesee, T. R. Reid,
and J. D. Wilson

Continuous Electron Beam Accelerator Facility
12000 Jefferson Avenue
Newport News, VA 23606-1909 USA

ABSTRACT

The CEBAF Central Helium Liquefier (CHL) provides 2K refrigeration to the 338 superconducting niobium cavities in two 400 MeV linacs and one 45 MeV injector. The CHL consists of three first stage and three second stage compressors, a 4.5K cold box, a 2K cold box, liquid and gaseous helium storage, liquid nitrogen storage, and transfer lines. Figure 1 presents a block diagram of the CHL refrigerator. The system was designed to provide 4.8 kW of primary refrigeration at 2K, 12 kW of shield refrigeration at 45K for the linac cryomodules, and 10 g/s of liquid flow for the end stations. In April 1994, stable 2K operation of the previously uncommissioned cold compressors was achieved. The cold compressors are a cold vacuum pump with an inlet temperature of circa 3.0K. These compressors operate on magnetic bearings, and therefore eliminate the possibility of contamination due to any air leaks into the system. Operational data and commissioning experience as they relate to the warm gaseous helium compressors, turbines, instrumentation and control, and the cold compressors are presented.

INTRODUCTION

The cryogenic effort started with converting the 1986 CEBAF Conceptual Design Report into the detailed specification for the CHL. This was the highest priority due to the long lead time associated with the cold compressors and the need for a projected two-year operating period prior to full accelerator operation to allow a 98% availability to be achieved.

The process of awarding the contract took 11 months starting with the draft of the specification being sent to the vendors, with the awarding of the contract in January 1988. This was CEBAF's first major procurement and remains the second largest technical contract (SRF cavity production is the largest). Problems surfaced during the installation and commissioning. The 10 months scheduled for installation and commissioning turned into four and one-half years. Table 1 presents the Cryogenic Timeline.

* Supported by U.S. DOE contract DE-AC05-84ER40150

Table 1. Cryogenic Timeline

DATE		DATE	
2/86	CEBAF CDR	12/91	T4 turbine failure
2/87	CHL Specification to vendor		Start LN2 cooldown
1/88	CHL Contract awarded	1/92	Valve repair
10/89	Start installation	2/92	Valve repair
2/90	Cold Compressor #3 arc over in France	3/92	Various problems, valves controls, and T4 testing
7/90	Scheduled acceptance test	4/92	CC#4 position sensor failure
10/90	Main compressors ready to operate	6/92	Leak identified in 2K box
11/90	Rebuild oil coolers		T4 testing
12/90	Main compressors ready to operate		CC testing
	Start compressor test	7/92	Leak identified in 4K box
	Deliver first 2K system		CHL start-up
1/91	Main compressors ready to test		CC testing
2/91	Cooldown of 4K cold box	8/92	CC testing, surges and rotor displacement
	T4 turbine failure, radial bearing		CC#4 field coil arc over
	Produce liquid, cool N LINAC		T4 turbine operational
3/91	CHL auto transformer shorts, C4 and C5	3/93	S LINAC supply and return TL cooldown
	Jumper all second stage auto-transformers		Rebuilt CC on site
4/91	T4 turbine failure, thrust bearing	7/93	Unstable 2.9K operation
6/91	Cooldown of 2K box	9/93	CEBAF assumes CC commissioning
7/91	4K box looses vacuum with T4 on line		New control concept incorporated
	T4 has leaks inside 4K box		30 minute operation at 2.2K
	T4 testing	12/93	CHL contract closed
8/91	Start cooldown of 2K box	1/94	Additional 4.5K heat exchanger installed
	Testing 2K valves, leaks present		3750 W @ 2.1K realized
9/91	CEBAF changes valves on dewar, leaking	2/94	Last second stage compressor replaced
10/91	Start cooldown of N LINAC, 4K and 2K boxes		Cooldown first ESR magnet
		4/94	Stable 2.3K operation
	Extreme heat leak 2K box, iceball	5/94	Start final beam commissioning
	Problems with valves on 2K box		Stable 2.1K operation
	T4 testing, 18th to 31st	7/94	First beam on target
11/91	Trouble shut 2K box valves & controls	8/94	32 day continuous CC operation

4.5K SYSTEM

The 4.5K system was delivered approximately two months behind schedule; however, the commissioning had not been initiated by the scheduled 2K acceptance test date. In December 1990, the warm screw compressors were operational. The "normal" 168 hour performance test was attempted. The flowmeters were not installed with the proper L/D ratios, and the readings were not correct. Therefore, it was questionable at the time of the test what the capacity was of the warm compressor system Also, it was established prior to the test that the oil coolers were assembled incorrectly. CEBAF diagnosed the problem as leaking seals between the first and second passes of the oil coolers. Therefore, hot oil was leaking directly from the oil inlet of the exchanger to the outlet. This resulted in undesirable oil injection temperatures.

The next major problem was the cold turbine, T4. The bearings failed three times. This coupled with the "turn around" time of the required modifications resulted in schedule slippage. Both the T4 and T3 turbines experienced a high wheel inlet pressure during operation. At first contamination was suspected, however, when the turbine was warmed and operated again, the wheel inlet tripped almost immediately. This is not the scenario associated with contamination. The problem was identified as a leaking seal. The specified seal and the installed seal were not the same. After the proper seal was incorporated, the problem was solved on both the turbines.

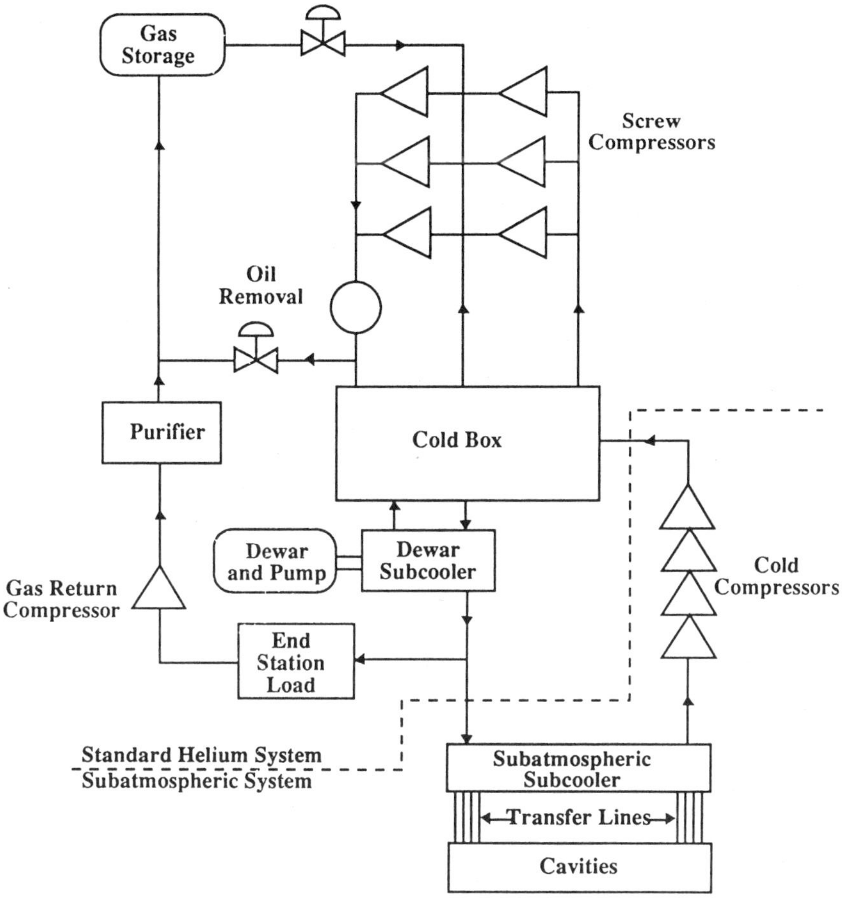

Figure 1. Block diagram of refrigerator.

Other problems caused difficulty in the commissioning of the cold compressors:

- The contract required that the system could be operated at full capacity with two first stage and three second stage warm compressors. The proposal had the first stage loaded to 97% and the second stage loaded to 77%. At reduced capacity, only two second stage compressors were to be required. We were unable to operate the cold compressors with all six warm compressors loaded to 100% prior to a modification of the control parameters. In fact, the second stage compressors would not support the operation of the 4K system; only three of the four turbines could be operated. The solution was to install larger compressors. This required removing the Howden 321/132 machines and replacing them with 321/165 machines. This also required the oil injection pumps and motors to be larger. Since the oil piping change to accommodate the larger pump would be very difficult to incorporate due to operating schedule, the speed of each of the existing pumps was increased to 2250 from 1750 rpm. This was accomplished by installing a vari-drive on each of the second stage oil pumps. This of course reduces the MTBF of the pumps. The failure rate of the oil pumps definitely has been increased due to their increase in speed. The pumps on the second stage warm compressors operate at a higher temperature in the area of the bearings. The first stage oil pumps have not experienced a single failure. They have been operating upwards of 30,000 hours. The second stage pumps have each been overhauled at least once per machine. The bearings have been the failed part each time.
- The heat exchangers between the 30 and 4K level were sized for steady state operation only. During cold compressor pumpdown, the return side of the exchangers had such high pressure drop that the cold compressors were not able to remain

operational. Replacement of the heat exchangers would require a three month shutdown, and this was not possible due to schedule slippage and accelerator commissioning. This "fix" might be completed in the future.

- The 4.5K heat exchanger has two additional problems. A) Two phase flow was attempted in a platefin exchanger; in this case it resulted in oscillations in the 4.5K system. B) The exchanger is undersized by approximately a factor of five in heat transfer. Therefore, CEBAF added a tube-in-shell boiler to correct the undersized heat capacity of the exchanger and the two phase flow problem. This was installed in the outlet U-tube of the 4.5K box leading to the 2K box.
- In an attempt to fix the previously discussed T4 turbine, the flow nozzles were reduced in size by about 8%. This results in an undersized unit, and the cold compressors will suffer when operated at full load during pumpdown. In order to "protect" the 4.5K box under this situation, a code for the throttling of the J-T valves has been developed. This prevents the turbines from overloading and warming up the cold box. A spare turbine of what is believed to be the correct size has been delivered. However, it has remained untested so far due to operational requirements of the accelerator.

2.0K SYSTEM

The 2K cold box consists of four stages of cold compressors and a two pass heat exchanger. The exchanger cools the GHe supply to 2.3K and the return flow at 2K increases to 3.2K at the inlet of the first stage cold compressor. Each of the cold compressor stages has its own variable speed drive and its own magnetic bearing set. The motors are LN_2 cooled, and the bearings have five degrees of freedom backed by mechanical bearings in case of power and battery back-up failure.

The 2K cold box suffered a series of failures. The cold compressor design was based upon the Tore Supra installation and scaled by a factor of three in size, and by a factor of ten in power. The Tore Supra installation had operated for circa 50,000 hours without a major failure, while the CEBAF cold compressor train had an MTBF of <<100 hours and an MTTR of >>1000 hours.

There were eight major electrical failures, and they were caused by two problems:

1) High voltage in the low pressure He (2 failures)
2) Differential contraction (6 failures)

The problem of voltage breakdown in He is well known to superconducting magnet fabricators but not to industry in general. The Tore Supra cold compressors were scaled by a factor of three in both voltage and current, which results in voltages of 380 V in the third and fourth stages. In 1989 the third stage arced over during component testing in France. Isolation transformers and spike filters were incorporated in the third and fourth stage compressors. This was the primary reason that the 2K cold box was delivered 14 months late.

The second problem was in the fine wire that was potted for the position and speed sensing coils. These coils were reported to be identical to the Tore Supra design, except for a slight increase in diameter. The wire would open circuit during cooldown and in at least one case corrected itself at warm-up. There were three failures of the position sensing coils which required an average of 1000 hours to repair. After the second failure all of the upper sensing coils were replaced with unpotted coils. Upon cooldown, a lower coil failed, resulting in the replacement of these also.

Two failures of the speed sensors did not halt the testing. The speed request was wired to supply the actual speed signal. These were rebuilt during the motor refabrication.

The last failure was experienced after the new motors were installed and cooled down. The upward axial thrust coil was a dual coil that was unknown to CEBAF. It incorporated another fine wire coil to provide the dc force to compensate for gravity. This coil had not been replaced and resulted in an intermittent ground fault. The electronics were modified to eliminate this coil and use the main coil to provide the dc biases as well.

The second arc occurred in 1992 at CEBAF in the fourth stage. The fourth stage had passed all electrical integrity tests after high voltage testing within a He environment. An inspection of the motor revealed it had been subjected to over compression of the motor windings during the coil forming process, the motor rotor had a crack in it, and there was damage to the upper backup radial bearings.

To allow for adequate space for coil forming the stator core needed to be shortened which resulted in a motor redesign for the third and fourth stage. During the redesign the motor voltage was also lowered to 170 volts as a secondary precaution. The redesign and refabrication took eight months.

By May 1993, serious testing of the cold compressors was initiated. However, the cold compressors reached 3.35K and were unstable. The next unstable run was in July and we reached 2.9K. CEBAF then decided to incorporate two major changes:

1) CEBAF assumed the responsibility for the commissioning of the cold compressors in order to accelerate the commissioning process.
2) A philosophic error was identified in the cold compressor control. This was corrected and a 30 minute run was completed at 2.2K on September 13, 1993. The contract was closed in December 1993.

Throughout the testing of the cold compressors the accelerator commissioning was being accomplished by the use of warm vacuum pumps. This system was composed of a 2000 cfm ring pump with a 10000 cfm blower. CEBAF started with one of these systems to commission the injector. As time progressed, another system was purchased for the north linac commissioning. However, as the operation of the accelerator approached, four additional bare ring pumps and blowers were purchased as a backup for the cold compressors that were not yet operational. CEBAF purchased the remainder of the required components, namely the heat exchangers, motors, and the oil pumps. CEBAF designed skids for the vacuum pumps and fabricated them at the tech shop. During this time the building was being erected for the additional vacuum pumping system. During the construction of the building to house this additional system the testing of the cold compressors continued in earnest.

The remainder of 1993 was devoted to the investigation of the problems in the 4K cold box. Meanwhile, the 4.5K portion of the system was operated in liquefier mode to support the accelerator. As the accelerator "turn-on" time approached, an intense effort was made to develop a reliable procedure for pumpdowns and stable operation of the 2K system. Cryogenics personnel operated the system for pumpdown development around the clock. A procedure was completed, and the system was able to operate at 2.3K by the start of operations as planned.

However, after about three weeks of operation, there was concern that since operations above lambda resulted in nucleations and this could result in cavity vibrations, causing problems with the RF control, beam was stopped for three days and we spent this time developing a procedure for operation at 2.1K. This task was successful and the machine has been operational at 2.08 to 2.1K.

Since July 1994, effort has been placed on improving the availability of the system, reliable restarts, regulation, and a fully automatic pumpdown controlled by the computer. We have, as will be discussed later in this paper, increased the availability to approach 98%, and the auto pumpdown code has been in operation. Details are in reference 3. Regulation has been improved with a code called auto heaters. This code maintains a constant total heat for the cold compressors. When the RF load is increased, the ohmic heating in all the cryomodules is decreased; therefore a total load is maintained. The cold compressors are operated at 85% of the 236 g/s design flow. The compressors have been operated as low as 180 g/s; however, during a perturbation such as an RF shutoff, the flow fluctuates and a trip can occur. Therefore, operation at 200 g/s has been incorporated. Further testing is scheduled to investigate the possible operation at 180 g/s.

The initial pumpdowns of the linacs, after being at 4.5K for long periods of time, have required about eight hours to complete. The normal pumpdowns experienced, where a restart is possible within three to four hours, require about 2.5 hours. The pumpdown procedure has been modified several times and the latest procedure is to delay adding heat until a pressure of 40 mb has been reached. Additional information on the pumpdown is presented in reference 4. A typical pumpdown curve is presented in Figure 2.

SYSTEM RELIABILITY

The 35,000 hours of operation have resulted in typical reliability problems encountered with the commissioning of a new system. Loss of the utilities is one of the most catastrophic events. This, of course, results in the entire system being shut down. The CEBAF CHL has a high reliability power feed. Over the past years power outages

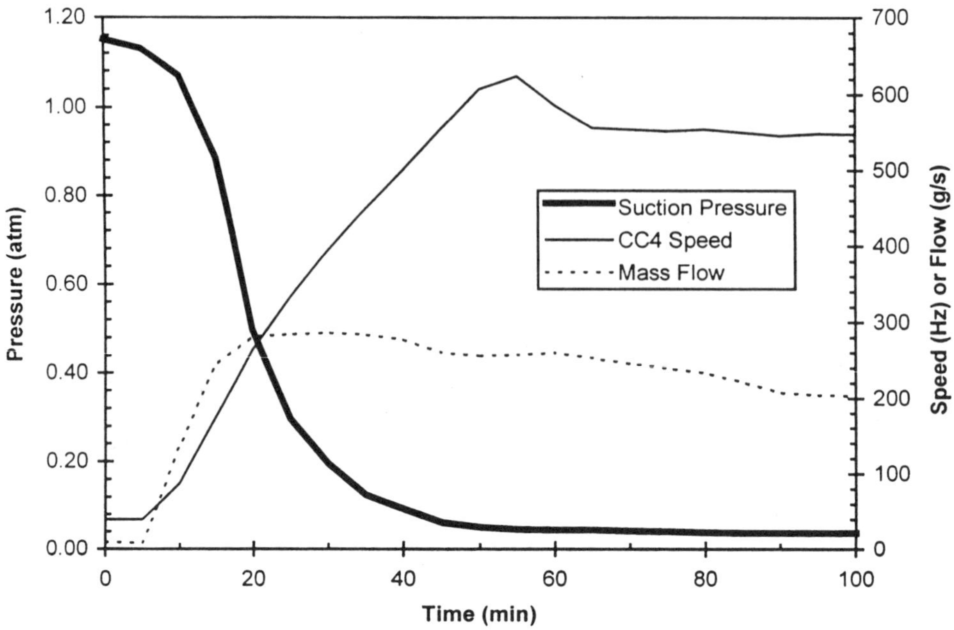

Figure 2. Repair and pumpdown cycle.

have been minimal. When a planned power outage is occurring, a pumpdown of every system except the linac's is completed with the linac's isolated. This allows the system to be down for up to four hours without the loss of helium.

The CHL water system also has an outage on the average of once per year. The cooling towers are open and after a period of time soil and other foreign material collects in the water sump. This results in clogging of certain heat exchangers and cooling water filters.

The system that has not resulted in down time directly is the computer power supply system. This supply has an UPS backup, two power feeds from the site grid, and an auto starting engine generator capable of 20 hours of operation from the propane supply tank.

The 4.5K system has been quite reliable. Problems were experienced with the turbine seals during the first operational period. However, after this was corrected the cold box operated well. The main problem is the warm compressors. Due to the undersizing of the machines by the contractor, the required redundancy of the second stage is not available.

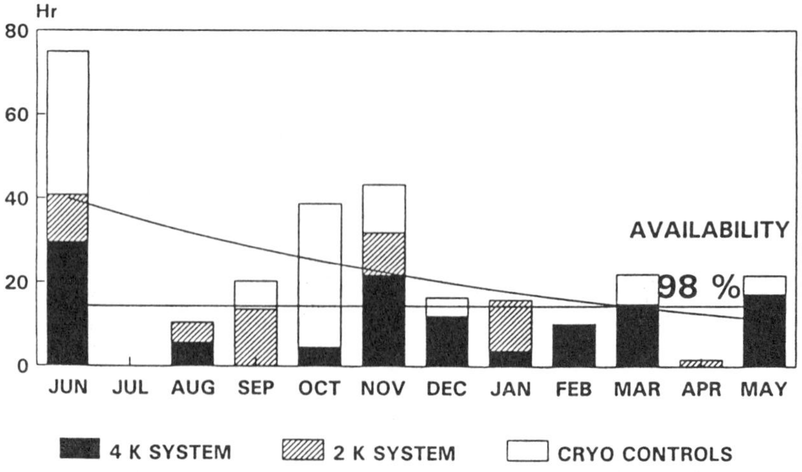

Figure 3. Cryogenic downtime (June 1994 - May 1995).

Three second stage compressors must be operated presently to maintain the cold compressors. Work is in progress to improve this situation. CEBAF is in the process of planning for the installation of additional second stage compressors. With 30,000 hours on all six machines, time is starting to run short. At 10,000 hours, there were two premature failures that were a result of improper alignment of the motors to the compressors and improper coupling installation. These potential problems were pointed out to the subcontractor at the time but went unheeded.

The system has also experienced the normal valve failures, electrical faults, and startup contamination problems.

AVAILABILITY

The cryosystem availability as defined at CEBAF, is the amount of time the accelerator is at 2K and proper operating liquid level is present in the cryomodules. This is whether beam is being produced or if other testing is in progress.

Cryogenic availability for the previous year has averaged 96.9% and 98.0% for the last six months; the downtime and its causes are shown in figure 3. The causes are split between the 4K system (1.4%), and the 2K system (0.6%), and the cryogenic controls (1.2%). The cryogenic controls category includes cryogenic software and hardware, as well as linac cryogenic instrumentation for the cavities. Not included in the downtime is another 0.8% of nonavailability charged to other sub-systems such as utilities; these included site power, city water, end station errors, and MCC problems.

During this period there were 44 unscheduled cold compressor trips plus three additional downtimes which did not trip the cold compressors. The cold compressors have a 186 hour MTBF and a 5.9 hour MTTR. The longest cold compressor run was 766 hours, while the shortest was 5 hours. About half of the 5.9 hour MTTR was in the response and repair time, while the other half was the accelerator pumpdown time.

The primary 4K system downtime was caused by contamination tripping the 25K and 15K turbines; the turbine trip in turn causes a temperature transient, which would trip the cold compressors. Other causes included the warm screw compressor trips and some control valves.

Only one of the 2K system downtimes was associated with the cold compressor hardware; with a valiant 14 hour all-night effort, it was possible to get the magnetic bearing electronics operational again. Six downtimes were due to excursions of the cold compressors out of their stable operating regions, and were not traceable to any equipment failures.

The unreasonably large cryo control downtime was due to three root causes: A) a failure of a supervisory LAN connection and/or board, B) intermittent failures of the linac serial highways which transmit load liquid level information, and C) overloading of memory allocations due to adding the third refrigeration system, ESR, to the network. The first was fixed by replacing several boards and reworking all the terminations, the second problem still remains, and the third has been partially fixed.

The effort on cold compressor restarting procedures had major effects on the availability. In June 1994 a very good cold compressor restart took 5 hours, while bad ones took three to four times longer. During the fall, procedures improved and increased the probability of successfully pumping down. During the last six months, the average downtime was 4.6 hours, with the pumpdown time being 2.5 hours for cold compressor trips lasting 3 hours or less. During the last four months, this has been fully automated, including the jumping of turbines through their critical speed range.

REMAINING TASKS

The primary need is to be able to shut down one of the warm compressors for maintenance. Presently, one of the first stage compressors can be shut down for repairs; however, the three second stage compressors all must be operational or the cold compressors will not operate. If additional second stage compressors can be added, then the system availability will be greatly improved. We cannot operate the cold compressors for more than two hours without all three second stage machines.

The primary weakness of the cold compressor system is the repair time. Even with 50,000 hours compressor and 40,000 hours controller MTBF's, we cannot achieve the required 98% availability. To achieve this, we need one week maximum repair time for the compressors, and eight hours on the controllers. We found that not all of the printed circuit boards were calibrated correctly, and the supplier was required to spend a week at CEBAF to determine the proper values and set up the boards. Also, some of the boards were not fabricated correctly, and several were required to be returned to the factory to be reworked.

CEBAF is in the process of securing a complete set of redundant cold compressors and controllers. During the following year these will be incorporated in a redundant cold box system.

The remaining major problem is the 30K to 4K heat exchangers. The present exchangers are costing efficiency and cold compressor restart delays. Since there is not a planned three month shutdown for the next few years, work-arounds will be continued. We are planning to order the exchangers to have them on hand; in the event of a shutdown they can be installed.

LESSONS LEARNED

This contract contained one high tech element, the cold compressors. CEBAF originally was planning to make two separate contracts; however, all the potential suppliers wanted the two procurements combined. CEBAF's agreement to combine the two contracts appears to have been a mistake. The contract required that the system be designed with separate 4K and 2K cold boxes, which would have permitted the commissioning of the 4K box prior to the 2K box. This would have saved time and also cost to CEBAF due to the replacement of the warm compressors.

Two independent contracts, each with its own acceptance requirements, would have saved a minimum of two years of the nine year effort. The gains would have been primarily from the 4K system:

1) The 4K contract would have specified the interface parameters eliminating at least some of the design errors.
2) The 4K acceptance test would have identified the problems in 1991 and forced the vendor to correct these prior to the delivery of the 2K box.
3) The commissioning would have been independent and would have eliminated the delays of the 4K box due to problems with the 2K box.

The second mistake was not to have an inspector/engineer at the site of the vendor. This would have allowed the problems to be more readily visible.

ACKNOWLEDGMENTS

This paper represents the work of the entire cryogenic group presently at CEBAF and all the term and contract persons no longer employed which were involved in the construction of the CHL system and the transfer lines. Thanks to the personnel at L'Air Liquide for the design and fabrication of the rotating machinery. Also, thanks to James Howden of America for the timely delivery of the larger second stage compressors.

REFERENCES

1. W. C. Chronis, D. Arenius, D. Kashy, M. Keesee, and C. H. Rode, "The CEBAF Cryogenic System," *Proceedings of the 1989 IEEE Particle Accelerator Conference,* pp. 592-594, March 1989.
2. CEBAF *Design Handbook,* ch 11.
3. C. H. Rode, "CEBAF Cryogenic System," to be published in the proceedings of the 1995 Particle Accelerator Conference.
4. B. S. Bevins, W. C. Chronis, and M. M. Keesee, "Automatic Pumpdown of the 2K Cold Compressors for the CEBAF Central Helium Liquefier," to be published in the proceedings of the July 1995 Cryogenic Engineering Conference.

THE CEBAF CONTROL SYSTEM FOR THE CHL*

Marie S. Keesee and Brian S. Bevins

Continuous Electron Beam Accelerator Facility
Newport News, VA 23606-1909 USA

ABSTRACT

The CEBAF Central Helium Liquefier (CHL) control system consists of independent safety controls located at each subsystem, CAMAC computer interface hardware, and a CEBAF designed control software called Thaumaturgic Automated Control Logic (TACL). The paper describes how control software was interfaced with the subsystems of the Central Helium Liquefier. Topics of configuration, editing, operator interface, datalogging, and internal logic functions are presented as they relate to the operational needs of the helium plant. The paper will also describe the effort underway to convert from TACL to the Experimental Physics and Industrial Control System (EPICS), the new control system for the CEBAF accelerator. This software change will require customizing EPICS software to cryogenic process control.

SYSTEM DESCRIPTION

The CEBAF cryogenic system consists of three separate facilities - the Central Helium Liquefier (CHL), the End Station Refrigerator (ESR), and the Cryogenic Test Facility (CTF). The CTF provides refrigeration for use in CEBAF's test laboratory, the ESR provides refrigeration for the various cryogenic equipment in the experimental halls, and the CHL provides shield cooling and 2K refrigeration to components in CEBAF's injector and two superconducting linear accelerators. CEBAF's Cryogenic Scope is shown in Figure 1.

The CHL is a 4.8 kW, 2.0 K refrigerator and transfer line system which supplies the linacs with 2.0 K and 12 kW of 50 K shield refrigeration. The CHL also supplies 10 g/s liquid for the experimental physics halls. The system block diagram is shown in Figure 2. The main components of the system include two stages of warm helium screw compressors, a 4.5 K Cold Box, a 2.0 K Cold Box (Cold Compressors), and various support equipment.

---

* Supported by U.S. DOE contract DE-AC05-84ER40150

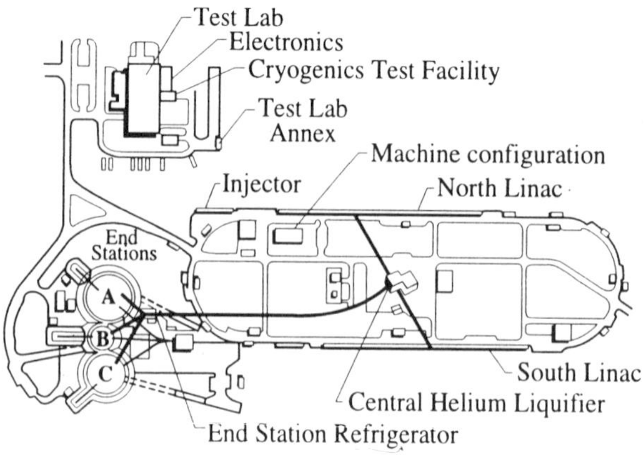

Figure 1. CEBAF Cryogenic Scope.

INSTRUMENTATION AND INDEPENDENT SAFETY CONTROLS

The CEBAF cryogenic system has two levels of control. The first level of control is equipment protection and safety, and the second level is process control. The cryogenic group's philosophy is to separate the first level of control from the accelerator control system. Therefore, the CHL contract required that no piece of equipment shall depend on CEBAF's control system for equipment protection and safety. Each piece of equipment was supplied with all the instrumentation and controls required to protect the equipment in the event of equipment failure or off-design conditions that could damage the equipment. Most

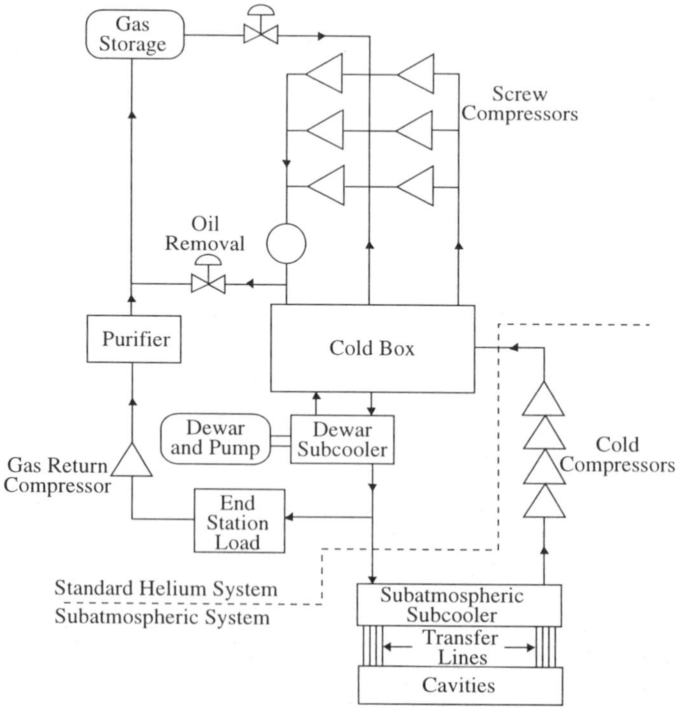

Figure 2. Block Diagram of CHL Refrigerator.

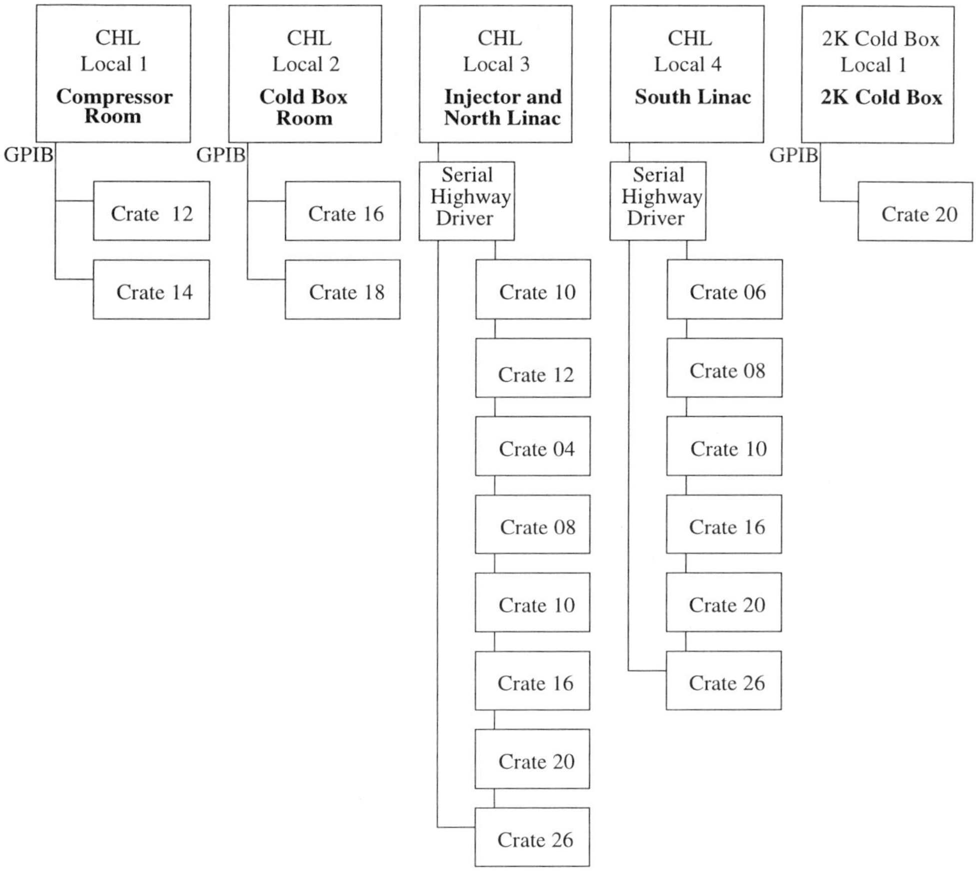

Figure 3. The CHL CAMAC Crate Layout .

of this is accomplished with local instrumentation and relays, or programmable logic controllers.

Responsibility for the second level of control was split between CEBAF and the equipment supplier. All equipment was supplied with the vendor's recommended instrumentation. The vendor then wired the instrumentation to a designated location in the CEBAF supplied interface racks. CEBAF supplied the wiring from that interface location in the racks to the I/O crates and then to the control system computers.

CAMAC INTERFACE

The I/O crates that the cryogenic system uses are CAMAC crates. Off the shelf CAMAC modules are used whenever possible to encourage standardization. The standard CAMAC modules in use include Analog to Digital Converters, Digital to Analog Converters, Digital Input modules, and Digital Output modules.

Three custom-built CAMAC modules were necessary to support the needs of the cryogenic system. The first two custom CAMAC modules are electric and pneumatic valve drivers that were designed and built at CEBAF. The electric valve driver module sends DC signals to valve actuators driven by 24 volt DC motors. The pneumatic valve driver module sends 4-20mA signals for pneumatically actuated valves. These two modules also support an LVDT or a potentiometer (or voltage) valve position readback. The third custom module is a silicon diode and platinum resistor temperature scanner that was contracted to be designed and built for CEBAF. This module sends the appropriate current to the temperature resistor or diode, reads back the resulting voltage, and converts the voltage to a temperature using the proper calibration curve data.

There are over 1150 I/O channels currently interfaced through the CAMAC crates for the CHL system. The CHL building accounts for 770 I/O channels in five CAMAC crates. The CAMAC crates in the CHL building are connected to the control system in parallel by a GPIB interface. There are two crates for the compressor room, two crates for the cold box room, and one crate to support the 2.0 K cold box (cold compressors).

The cryogenic equipment in the two superconducting linacs accounts for 384 I/O channels distributed through fourteen CAMAC crates. These CAMAC crates connect to the control computers via two serial highways. There are 6 crates for each linac, and two crates for the injector area. The CAMAC crate layout is shown in Figure 3.

THE TACL SYSTEM

Overview

TACL is a logic-based software control system that was developed at CEBAF and has been in use since November, 1987[1]. TACL supports a CAMAC I/O interface, and provides for easy configuration of control logic databases and generation of custom screens for operator interface. The software runs on a distributed network of Hewlett-Packard UNIX workstations.

Configuration

The TACL control system for the CHL uses an architecture of 4 supervisor computers and 7 local computers. The control system layout is shown in Figure 4.

The supervisor computers handle the network communications between other supervisors and different parts of the control system, and run logic databases for necessary supervisory tasks.

The local computers run the logic databases that include most of the engineering conversions, control logic, PID calculations, and input/output operations with the CAMAC crates. The local computers only communicate with their own supervisor, and therefore are not overloaded by network traffic. The disadvantage of this layout is that the local computers cannot communicate with each other. They must communicate by way of their supervisors.

Editing and Operator Interface

TACL consists of tools that allow users to build the control logic databases and operator interface screens for their particular application. This reduces the load on the controls group staff, and allows the control system to be configured by users knowledgable in each application.

The TACL Logic Editor is an icon-based database editor. Menus are used to select and describe all the hardware interfaces in the system. This editor also allows the user to symbolically define the logical flow of control. The editor generates a database from the user defined information that is executed at run-time.

The TACL Display Editor allows the user to create control screens for operator interface using a wide variety of drawing tools. The user may create custom symbols to supplement the available drawing functions. A custom symbol was created and added to the drawing functions list for monitoring and operating the many PID loops in the cryogenic

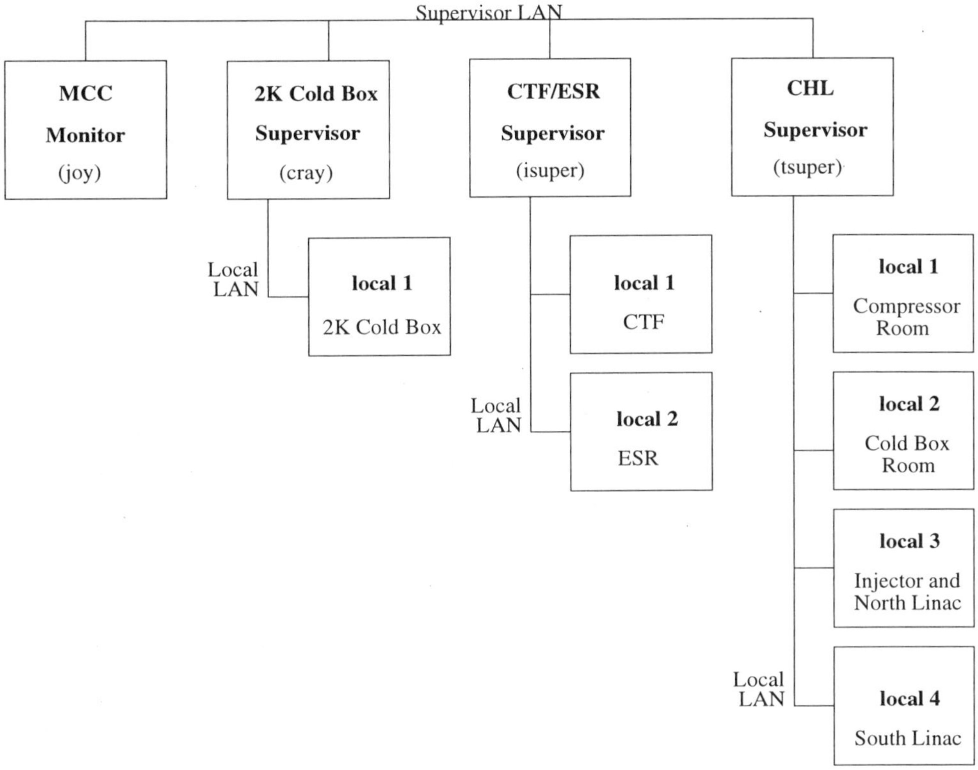

Figure 4. Cryogenic Control System Diagram .

system. Most of the control screens for the CHL were created to look like the system process diagrams. This makes it easier for the engineers and technicians familiar with the hardware to monitor and control the system.

Special Logic Functions

Since the CHL is a process control application, PID loops are widely used in the system. PID loops are software proportional, integral, and differential feedback loops. The CHL control system uses over 125 PID loops to control the cryogenic equipment in the CHL building and in the two linacs. A special PID logic function was added to TACL for the CEBAF cryogenic systems. This PID loop allows for all of its parameters, including gains and limits, to be changed during runtime, allowing for easy runtime tuning of the loops.

Utility Programs

Utility programs have been written to perform datalogging, strip chart graphing, and saving and restoring settings.

The datalogger allows periodic logging of process signals in samples from one second and up in one second increments. A menu driven editor is supplied to configure the datalogging file.

The graphs program plots process signals in strip chart format. Up to ten signals may be plotted on one graph. Each signal may have its own range, and the time between plotted samples may be changed. This program can plot signals in real time or can plot datalogged signals.

There is also a program that can save current operational settings so that they can be restored after a computer reboot or failure, or to support a different operational mode. Restoration of PID settings is very important to a process control system like the CHL.

THE EPICS CONVERSION EFFORT

EPICS is a control system toolkit being developed by a collaboration of several DOE laboratories[2]. EPICS will support existing CAMAC hardware through serial highway to VME based I/O Controllers (IOC) that run the EPICS software. UNIX workstations are used to execute high level applications, and XWindows format is used for the operator interface. The differences between TACL and EPICS and reasons for CEBAF's conversion of control systems from TACL to EPICS are discussed in references 2 and 3.

The CHL system will be converted to EPICS late in 1996, after the Cryogenic Test Facility (CTF) is successfully converted. CTF will be used as a test bed for the CHL conversion to work out conversion details relating to cryogenics.

A new EPICS PID is currently being written to support CEBAF's needs. The new PID will act like the TACL PID, so that all of the CEBAF cryogenic PIDs will not have to be retuned.

A graphs program is also being developed to support the cryogenic conversion to EPICS. Since most cryogenic processes are relatively slow, a good real-time graphs program assists the operators in monitoring the system.

CTF is scheduled to be converted to the EPICS control system in August of this year. The programming and configuration for the control system is 75% complete and is waiting on completion of the new PID. All of the existing instrumentation and CAMAC hardware will be used under the EPICS system with the exception of the CAMAC crate controllers. The crate controllers will be changed to serial controllers to support the serial highway mentioned above. The operator interface screens are in the process of being redone since EPICS uses an XWindows operator interface.

Since most of the control system runs on VME based IOCs, the computers used for the operator interface will be X-terminals. These X-terminals will run off of one workstation that will also run the higher level applications such as graphs, archiving, and save/restore programs.

REFERENCES

1. E. Navarro, M. Keesee, R. Bork, C. Grubb, G. Lahti, J. Sage, Use of the TACL System at CEBAF for Control of the Cryogenic Test Facility, *Proc. 1989 Conference on Real-Time Computer Applications in Nuclear, Particle, and Plasma Physics*, June (1989).
2. K. White et al., *The Migration of the CEBAF Accelerator Control System from TACL to EPICS*, CEBAF Control System Review, May (1994).
3. S. Schaffner et al., *Device Control at CEBAF*, Particle Accelerator Conference, May (1995).

QUALIFICATION AND CRYOGENIC PERFORMANCE OF CRYOMODULE COMPONENTS AT CEBAF*

J. Heckman, K. Macha, J. Fischer, E. Folts,
J. Preble, M. Wiseman

Continuous Electron Beam Accelerator Facility
Newport News, VA, 23606 US

ABSTRACT

At CEBAF an electron beam is accelerated by superconducting resonant niobium cavities which are operated submerged in superfluid helium. The accelerator has 42 1/4 cryomodules, each containing eight cavities. The qualification and design of components for the cryomodules under went stringent testing and evaluation for acceptance. Indium wire seals are used between the cavity and helium vessel interface to make a superfluid helium leak tight seal. Each cavity is equipped with a mechanical tuner assembly designed to stretch and compress the cavities. Two rotary feedthroughs are used to operate each mechanical tuner assembly. Ceramic feedthroughs not designed for superfluid were qualified for tuner and cryogenic instrumentation. To ensure long term integrity of the machine special attention is required for material specifications and machine processes. The following is to share the qualification methods, design and performance of the cryogenic cryomodule components.

INTRODUCTION

The CEBAF accelerator uses 338 Superconducting Radio Frequency niobium cavity assemblies at 2 K to provide a continuous electron beam of 200 μA and 4 GeV after 5 recirculations through 2 anti-parallel linear accelerators. The accelerator has 42 1/4 cryomodules, each containing 8 cavities which are submerged in superfluid helium. A cryomodule is made up of 4 cryounits and 2 end cans. Each cryounit contains two SRF cavities with mechanical tuners housed in a helium vessel and surrounded by a 50 K thermal shield and vacuum vessel (Fig. 1). The end cans supply and return the 1400 liters of superfluid helium to each cryomodule. The physical length of a cryomodule is approximately 9 meters.

Three separate vacuum systems are contained in each cryomodule. The RF waveguide vacuum is defined by a warm polyethylene window at room temperature and pressure, and by a cold ceramic window between the waveguide and the cavity fundamental coupler, which is located inside the helium vessel. The insulating vacuum encompasses all the space between the helium vessel and the vacuum vessel, excluding the cryomodules RF waveguides. This vacuum is maintained at 13 to 1330 mPa (10E-07 to 10E-05 Torr.). The third vacuum system is defined by the SRF cavities and the beam pipe, which traverses the entire accelerator. The beam pipe connects the cavities between helium vessels and between cryomodules. The cavity and beam pipe vacuum is maintained between 0.013 to 0.13 mPa (10^{-10} to 10^{-09} Torr.). All three vacuum systems may be subjected to a loss of vacuum due to an air or helium leak.

The cryomodule uses several different kinds of seals for the connection of components. The cavities are assembled with approximately 96 indium wire seals, which equates to roughly 4,000 indium wire seals installed in the accelerator. The SRF cavities and RF waveguide vacuum use indium wire seals at the ceramic window assembly which must be leak tight to 0.013 mPa m^3/sec. (1.0^{-10} Torr liters/sec). The cold seals must survive numerous thermal cycles at low temperatures. The beam line is made up of conflat-type flanges utilizing OFHC copper gaskets (oxygen free high conductivity) and has to be leak tight in a superfluid environment. The insulating vacuum utilizes several sealing methods, such as, o-rings, VCR fittings, and welding. These different sealing configurations and seals were subject to stringent leak check specifications, had to be qualified for superfluid service, and operate reliably.

Two types of electrical feedthroughs are mounted on the helium vessel via conflat-type flanges. These feedthroughs are an SMA-type for RF field measurement and an 8 pin connector used for temperature diode signals, heater, and tuner position instrumentation. The SMA-type is rated for 77 K service and the 8 pin connector are rated for 4 K. Both feedthroughs are stainless steel and ceramic and must be able to withstand superfluid temperatures. These feedthroughs were purchased as off the shelf items and were qualified for use in a superfluid environment.

Also installed on the helium vessel and vacuum jacket are the rotary feedthroughs which in conjunction with the associated tuner assemblies, are used to compress or stretch the cavities to change their frequency. The rotary feedthroughs installed on the helium vessel operate in superfluid helium and must be able to perform without binding.

Material failures were a problem with 3 cryomodule components. In the end cans, stainless steel tubing is used as conduit for process temperature diode signals. This tubing suffered stress corrosion failure in several locations. The second material failure was in the bellows used between the cavity pair and helium vessel dished head to compensate for pressure changes. These developed chemical stress related cracks in the stainless steel. The third component failure was in the cold RF ceramic window assembly. The ceramic in the windows were subject to cracks resulting from improper grinding. These cracks result in vacuum leaks between the cavity and RF waveguide vacuum systems.

Each cryomodule was designed to have a reliable service life of 10 years. To achieve this goal, commercial components had to be qualified for superfluid service and stringent quality control measures had to be implemented. This paper describes that effort and our experiences with these components.

Vacuum Seals

To ensure the long term reliability of the vacuum systems of the cryomodule in the 2 K environment, as many connections as possible were welded. When welding was not

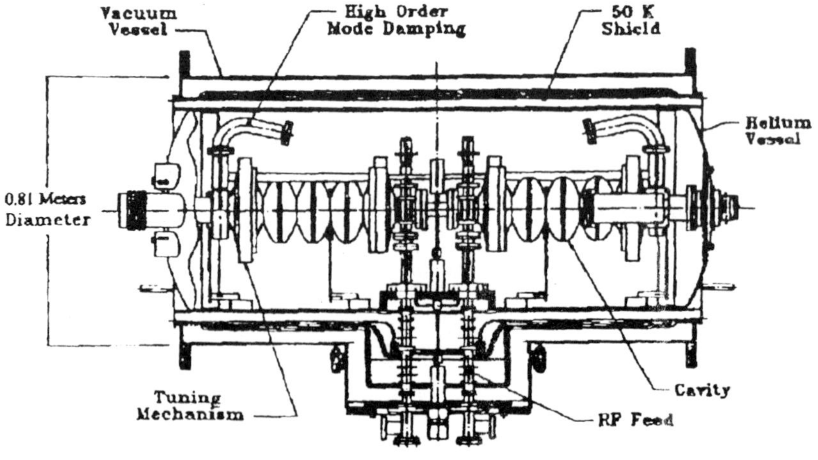

Figure 1. Cryounit Cross-section.

possible, o-rings, conflat-type seals, and indium wire seals were used. Several o-ring seals were used on the vacuum vessel shell with minimal amounts of vacuum grease. For the stainless steel connections on the waveguide, 2 K systems, beamlines, and feedthrough assemblies on the cryomodule, conflat-type seals were used. Conventional assembly techniques were utilized on the conflat-type flanges. Silver plated bolts were used to prevent gauling and to minimize personal exposure to wet lubricants that could be activated by electron beam interactions. The niobium RF waveguide vacuum window assembly and the SRF cavities used indium wire seals. The indium wire sealed flanges were of special concern because of their prolific use throughout the cryomodule and in the cavities. Indium seals were used because they are easily formed, provide a good RF joint, and require a low sealing load. Early problems with He leaks in the indium seals resulted in several changes to the assembly procedure. These changes included, changing the indium wire sizes, incorporating special flange preparation and cleaning procedures, and the incorporation of high resolution leak checking methods. Originally, only 60 mil indium wire seals were used for all the RF vacuum seals. The seals were torqued until the flanges were metal to metal. It was discovered that many of the Nb flanges were too thin to reliably support this bolt load. It was deemed too expensive to change the flange design so 40 mil indium wire seals were used for the thinner flanges. New torquing procedures were also implemented to insure that the seal loading was uniform throughout the rectangular flanges. The bolts were torqued so that same seal loading pressure was achieved for both the 40 and 60 mil indium wire and the flanges were not yielded. Both size seals were compressed to approximately 11 mils thick. To insure that the seal loading was constant from 300 K to 2 K, the appropriate number of belleville washers were used. All of the high purity indium wire seals used were ultasonically cleaned and alcohol rinsed prior to using.

The surface preparation for the indium sealed flanges consisted of a lapped surface to ensure that the flange surfaces were planer and scratch-free with a finish of 1.3 mm (32 μin.).[1] The lapping is done in several steps with wet sanding discs on a rotary table with 120, 240 grit, 14 μm (360 μin.), and 1.2 μm (30 μin.). The cleaning procedure consisted of a rinse in hot water, immersed in an ultrasonic cleaner for 5-10 minutes, rinse with hot water again and let dry. The flanges were then dipped in a solution of BCP (H_3, PO_4, HNO_3, HF in 1:1:1) for less than 1 minute and then rinsed with ultra pure water. We experienced a few vacuum failures of RF window flanges when hand lapping was used to remove visible scratches on the window flanges and the BCP was not applied. Several tests were made which compared the leak integrity of the lapped vs. the BCP etched Nb flanges. To summarize the test results: the lapped flanges leaked at rates of 1.33 μPa m^3/sec. to 13.3 μPa m^3/sec. (1.0^{-08} to 1.0^{-07} Torr liters/sec.) at ambient temperatures, while the etched flanges were leak tight to < 0.013 μPa m^3/sec. (1.0^{-10} Torr liters/sec.) both at ambient and LHe temperatures.

After assembly, each individual indium joint was bagged in He for five minutes and had to be leak tight to 0.013 μPa m^3/sec. (1.0^{-10} Torr liters/sec.). The typical leak detector used in the leak checking of cryomodules and components is a 240 liter/sec Balzers turbo molecular pumping station with an Alcatel EA10 leak checking module mounted to the inlet of the turbo. The cavity pairs underwent an additional leak check by a helium desorption method and will be discussed later in more detail.

The long term reliability of these cryogenic indium seals was proven by subjecting them to repeated thermal cycles (10-20) between superfluid helium temperature (2 K) and room temperature.

For these high sensitivity helium desorption leak checks, the leak detector consists of an RGA (residual gas analyzer) and a manifold with an ion pump as seen in Fig. 2. As the assemblies are cooled to liquid helium temperatures the RGA is turned on to monitor the pressure. In the absence of a large leak, no trace of helium is seen because the helium is being adsorbed onto the cold metal surfaces. The liquid helium is then boiled off slowly and the partial pressures of H_2 and He are monitored as a function of the temperature. The He increases with increasing temperature due to desorption and levels off in the temperature range of 10-11 K, while the partial pressure of H_2 remains constant. At a temperature greater than 11 K the partial pressure of H_2 and He suddenly starts increasing. During the cavity pair cool down, the ion pump on the test set-up is isolated from the test set-up at 100 K to avoid adsorbing the degassing He from the ion pump onto the Nb cavity walls. The leak rate of the system is calculated from the partial pressure of He, the volume of the system, and the time the cavity pair was kept at helium temperature after cool down. Each production cavity pair was leaked checked using the desorption method. The acceptable integrated leak rate for all of the indium seals on the pair was 1 X 10^{-8} Torr-liters/sec.

Electrical Feedthroughs

The electrical feedthroughs purchased by CEBAF were off the shelf items from Ceramaseal (Part No. 808B9340-3, for the SMA and 808B11881-04-cf for the 8 pin). Our requirements are that a feedthrough at superfluid temperatures to be leak tight to 0.013 μPa m^3/sec (1.0^{-10} Torr liters/sec). To qualify these two feedthroughs for production, it was decided to cryogenically cycle and leak check 10% of the lot, which was 34 SMAs and 67 eight pin feedthroughs. The thermal cycling leak checking process submerged each feedthrough into a liquid nitrogen bath for 1 minute, then warmed to ambient, which took approximately 10 minutes. The parts were then leak checked with our typical leak detector described earlier. We initially cycled these components to 4K using a small helium dewar. This was stopped when we found the results were the same as for liquid nitrogen cycling which is a more cost effective approach.

Mechanical Tuner Assemblies

The mechanical tuner assembly and rotary feedthroughs are used to compress and stretch the cavities. The fundamental frequency of the cavity is 1.497 GHz ± 100 Hz. The

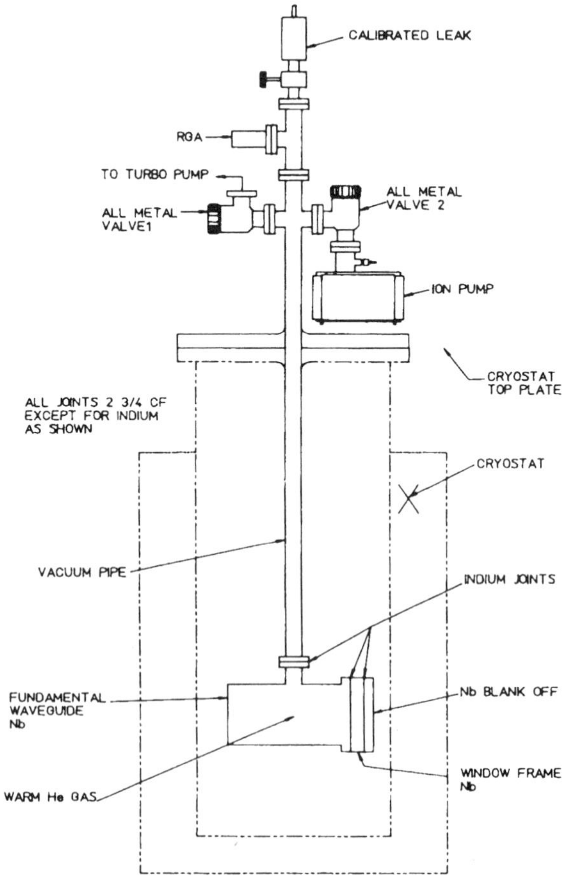

Figure 2. RF window flange thermal cycle test setup .

Figure 3. SEM scan of a 1/8" stainless steel tube corrosion .

tuner provides 200 kHz of adjustment. It is paramount that the rotary feedthroughs on the tuner assembly operate in superfluid helium in a non-binding configuration. The mechanical tuner is driven by a stepper motor mounted in air on the cryounit top hat (Fig. 1). The torque from the motor is delivered to the tuner using two rotary feedthroughs, one mounted on the cryounit vacuum jacket, and the other mounted on the helium vessel. At the start of production for CEBAF, no rotary feedthrough had been qualified to operate at 2 K. A modified Varian rotary feedthrough was selected for prototyping. The mechanical tuner and rotary feedthrough mechanisms were stress tested to twice their anticipated life at 4 K.[3] Early failures of the rotary feedthroughs during these tests prompted an improved feedthrough design from Varian Vacuum Products. Thus far, during operation at CEBAF there has only been one known failure of a cold rotary feedthrough, which was a failure of the welded bellows.

During testing of the first production tuner assemblies in the cryomodules, it was found that some of the tuner drives were binding when cooled to low temperatures. To troubleshoot the binding problem, each individual part of the tuner assembly was cryogenically tested. Each part was submerged in liquid nitrogen and operated manually using a handle to control tuner movement.[4] As the tuner assembly got colder it became difficult to operate the control handle in either direction because of binding. When the assembly was removed from the cryogen and warmed to room temperature, the assembly began to work freely again. This imitated the problems we experienced in the cryomodules. Through this empirical testing of each component, the rotary feedthroughs and gear box were found to be OK, and the problem was finally determined to be in the gimbals bearing on the shaft. The bearing had a tight pressed fit, which when cooled, caused the bearing to bind internally. We eliminated the press fit and the binding at cryogenic temperatures was corrected. The remaining 308 production tuner assemblies were then retrofitted with this new configuration for cryomodule production. We currently have two tuners that bind when cooled to operating temperature. It is not known at this time what the failure mode is.

Material Failures

Three material failures occurred during production that were the direct result of improper machining and cleaning methods. The first material failure of note was in the end cans, which supply the cryogens to the cryomodule helium vessels. The failures occurred in 1/8" S.S. capillary tube that routes diode wiring from the cold process pipes to the warm electrical feedthrough. The conduits were leak tight when shipped from the manufacturer. After storage at CEBAF for approximately one year, several leaks were discovered in the conduits during the final cryomodule assembly. After checking all of the end cans in storage a total of 16 leaks were found, some of them large enough to be seen by visual inspection. The capillary tubes were sent to NASA for SEM (scanning electron microscope) analysis and found to be corroded, see Fig. 3. The corrosion was caused by fluoride and chloride contamination leading to interangular corrosion. The tubes had corroded from the inside out due to a manufacturers cleaning and assembly process.[5] This problem with corrosion was rectified at the manufacturer and by CEBAF personnel.

The pressure compensation bellows is between the cavity and the helium vessel and provides a flexible connection separating the liquid helium in the helium vessel from the insulating vacuum space. The bellows is made from 3 stainless steel components which are tungsten inert gas (TIG) welded. These vacuum leaks were also discovered during the final leak check of the cryomodules. During this leak check, the interior of the helium vessels are pressurized to 2 atm with helium while the insulating vacuum is evacuated and monitored for helium. To achieve reasonable leak check sensitivity it was necessary to use a nitrogen cooled sorption pump before the helium mass spectrometer. This reduces the water partial pressure at the mass spectrometer while not affecting the helium partial pressure. Earlier leak checks of these components do not allow for as high a sensitivity and the leaks were missed. Additionally, it was concluded that moisture freezing in the cracks during pump down prevented their detection.[6] The leaks were found near the joint of the weld cuff and bellows. The causes of the failures were determined to be that extruded piping material was used for the weld cuff and that the material had been cleaned with a chlorinated solvent, eventually leading to a chemical stress failure. A high degree of weld zone contamination in the form of elongated stringers can be expected by the nature of the pipe extrusion process. When the fusion line of the weld is aligned to be closely parallel with the extrusion direction of the cuff pipe, the stringers become activated as crack initiation sites. Under the influence of residual stresses, the cracks propagate between the stringers forming a through-thickness crack that serves as a vacuum leakage path.[7] Other mitigating factors involved in the failure of these parts were that subsequent machining involved a chlorinated solvent and the fabrication-to-installation period was almost a year.

The RF window ceramics that separate the fundamental power coupler and cavities developed vacuum leakage through the ceramic. Of the first 70 window assemblies, 4 developed small leaks (1.0^{-07} to 1.0^{-08} Torr liters/sec) directly through the ceramic. In each of the 4 cases, the leak was through the thin portion of the ceramic at a radius which provide the transition to the thick edge of the ceramic (an hour glass design). The vacuum integrity of these assemblies may have been compromised by high stresses during machining. Subsequent dye penetrate tests by the Non-Destructive Testing group at NASA revealed numerous grinding-induced defects.[8] The ceramic used for these windows consists of 99.5% aluminum oxide with 0.5% glassy phase and is commonly used in RF window applications. To minimize ceramic window vacuum leakage in future window assemblies for production, improved grinding and machining procedures were implemented. A dye penetrate test was also required.

CONCLUSIONS

The unique properties and requirements for the CEBAF cryomodules resulted in several commercial products being qualified for use in environments that they were not designed for. Procedures that were developed during the early production stages of CEBAF cryomodules resulted in the continuous improvement of the cryomodule performance and reliability. The data is still being taken on these components and the procedures implemented during the operation of the CEBAF accelerator.

ACKNOWLEDGMENT

*This work supported by U.S. DOE Contract No. DE-AC05-84ER40150. The authors acknowledge the able assistance of the following colleagues with the preparation of this paper: Larry Phillips, Charles Reece, William Schneider, Gayle Sundeen and Gloria Fleming.

REFERENCES

1. J. Brawley, F. Folsom, S. Morgan, B. Jones, R. Bennett, and C. Reece, CEBAF experience with in-house fabrication of niobium parts, Proceedings of the 6th Workshop on RF Superconductivity, Oct., 1993, Vol. II.
2. M. G. Rao, High sensitivity He desorption leak detection method, J. Vac. Sci. Technical, A (4), Jul/Aug 1993, 1993 American Vacuum Society.
3. M. Wiseman, et al., Cryobench - Apparatus for Testing CEBAF Cryogenic Subcomponents, Applications of Cryogenic Technology, Vol. 10., 1991
4. E. Folts, A history of production cryomodule tuners, March 22, 1995.
5. D. Henkel, Report on the investigation of vacuum leakage in the CEBAF cavity pair bellows assembly, a report prepared for CEBAF by the NDE Lab. of NASA Langley Research Center, 11/27/92.
6. W. Schneider, private communication.
7. D. Henkel, Ref. 5.
8. W. Schneider, Review recommendations re: charge to working group I: flange and seal modification, memo to F. Dylla, July 3, 1991.

AUTOMATIC PUMPDOWN OF THE 2K COLD COMPRESSORS FOR THE CEBAF CENTRAL HELIUM LIQUEFIER*

B. S. Bevins, W. C. Chronis, and M. S. Keesee

Continuous Electron Beam Accelerator Facility
12000 Jefferson Avenue
Newport News, VA 23606-1909 USA

ABSTRACT

The 2K system at CEBAF includes four stages of centrifugal cold compressors that operate between a suction pressure of 0.031 atm and a discharge of ~ 1 atm. Starting the cold compressors and pumping the system down to operating pressure is a complex and time consuming operation. Many months of rigorous testing were required to develop a reliable startup procedure. This procedure has been automated through the development of an Autopumpdown program that intelligently coordinates the necessary steps across the boundaries of the existing cryogenic and accelerator control systems. This program has greatly reduced both the time and effort required to recover from failures and restart the system. The paper describes the requirements addressed by the program, the steps it takes while operating, the historical development of the program, the spin-off benefits, and the course of anticipated future developments.

INTRODUCTION

The Central Helium Liquefier (CHL) provides refrigeration at three temperature levels for the operation of CEBAF's electron beam linear accelerator (linac) and experimental physics halls. 12 kW of refrigeration at 38K is provided for shield cooling in the 42¼ cryomodules that make up the accelerator, 10 g/sec of 4.5K liquid is supplied to the experimental halls to cool superconducting magnets and cryogenic beam targets, and 4.8 kW of refrigeration at 2K is supplied to cool the superconducting radio frequency (RF) cavities inside the cryomodules. The cryomodules are arranged as parallel loads on the refrigerator, each with its own independent Joule-Thompson (JT) inlet valve. When operating, each cryomodule holds approximately 1440 L of liquid helium.

The CHL is divided into two main sections that were fabricated separately, as shown in Figure 1. The shield cooling and liquefaction loads are supplied by a conventional 4K cold box which also feeds the more specialized 2K cold box. Operation at 2K requires the cryomodule liquid baths to be maintained at an absolute pressure of 0.031 atm. This pressure is achieved by four stages of centrifugal cold compressors (CCs) in the 2K cold box. Each compressor stage has its own magnetic bearings and LN_2 cooled variable speed electric motor.

* Supported by U.S. DOE contract DE-AC05-84ER40150

The pumpdown procedure is the transition from 4K operation at 1 atm to 2K operation at 0.031 atm. This requires coordinating the speeds of the four cold compressors with a variety of other system adjustments. The main goal of the pumpdown is to minimize the time required to prepare the cryomodules for accelerator operations. This involves both reaching 2K operating pressure and satisfying the minimum liquid levels required in the cryomodules, since the liquid levels drop during pumpdown. This must be accomplished within the constraints of the 4K system capacity and the cold compressor surge limits. The centrifugal cold compressors require a minimum inlet volume flow to develop the required pressure ratios without surging. Early in the pumpdown, at high suction pressures, higher mass flow rates are required to satisfy this volume flow requirement. This unloads the 4K system, because the 4K cold box is running in an unbalanced mode, recovering more gas from the cryomodules than it is supplying as liquid. As the pumpdown progresses, the load on the 4K system increases to a maximum (when the return flow through the cold compressors decreases and the liquefaction load to refill the cryomodules is high) and then settles into balanced operation near the completion of the pumpdown.

The CHL Autopumpdown program began as an effort to make the existing 2K pumpdown procedure less labor intensive. The manual procedure was the result of months of intensive development by the CEBAF cryogenics staff, often working around the clock to accomplish as much testing as possible during the limited time windows allowed by accelerator commissioning efforts. This initial manual procedure required at least three operators in two different locations coordinating by telephone. One operator was responsible for the operation of the 4K cold box and warm compressors, one for the 2K cold box, and one for the heaters and JT valves associated with each of the cryomodules. Progress was often slowed by the difficulty in communicating every new setting between shifts and the resulting non-uniformity of different pumpdown attempts. The

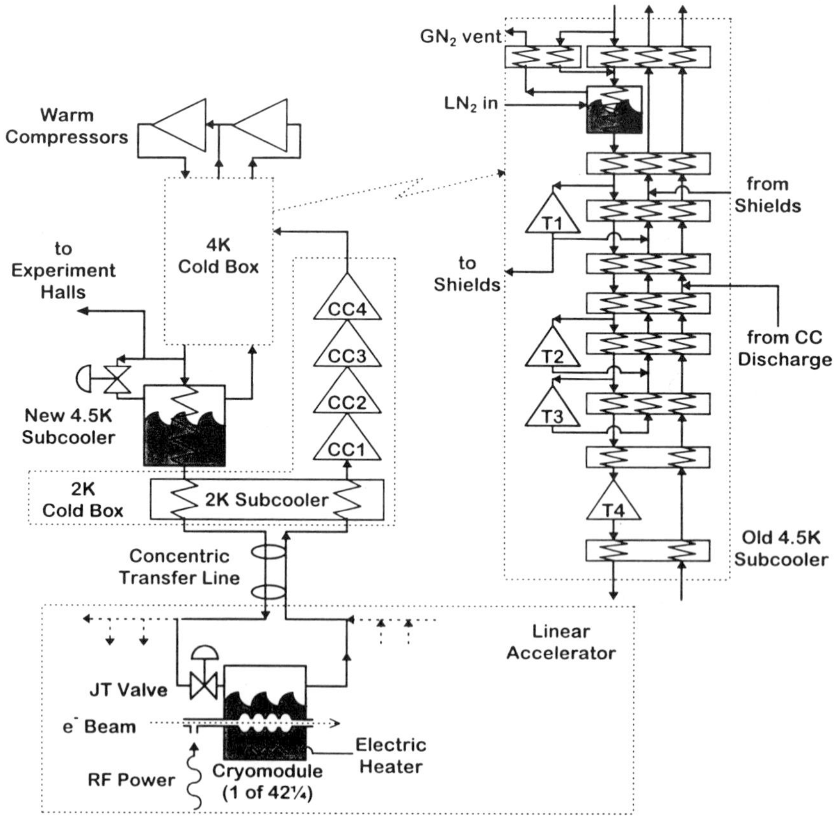

Figure 1. Partial flow diagram for the CEBAF Central Helium Liquefier with an expanded view of the 4K cold box. The new 4.5K subcooler was installed in the piping between the 4K cold box and the 2K cold box. For clarity, not all system components are shown.

Autopumpdown program helped to resolve this problem and free the operators from tedious repetitive tasks to concentrate on refining the procedure. Once it was first implemented, the Autopumpdown software continued to evolve hand in hand with the manual procedure. New procedures were tested manually and those that proved useful were incorporated into the software as quickly as possible, usually the same day.

During the development of the pumpdown procedure, CEBAF was in the midst of changing control systems from TACL (Thaumaturgic Automated Control Logic) to EPICS (Experimental Physics and Industrial Control System). Today the cryogenic systems are still entirely in TACL while the accelerator systems are almost completely in EPICS. Bridging the gap between these systems as they evolved over time was another challenge for the Autopumpdown software. Continuing efforts to convert the cryogenic controls to EPICS are described in Reference 1.

HISTORY

The first phase of what became the Autopumpdown program was a set of scripts to control the JT valves or heaters in all of the cryomodules simultaneously. These scripts reduced many frantic minutes of typing to a few mouse clicks and confirmations. As the pumpdown procedure firmed up, it became obvious that these operations would be required at certain repeatable pressures at the suction of the cold compressors. As a result, CC suction pressure was chosen as the sequencing variable, and these operations were automated with a preliminary set of independent programs that adjusted the heaters and JT valves at preset pressures as the linacs were pumped down. This automation proceeded in phases. First the heaters were controlled, then the JT valves, and then the cold compressor mass flow setpoint as well.

As the pumpdown procedure continued to become better defined, these simple programs were combined into a master Autopumpdown program that coordinated all of the changes and allowed for other operator actions to be synchronized with them. Cold compressor suction pressure has remained the primary sequencing variable, but individual actions can have their own predefined triggers. For example, as described in the next section, the 30K turbine (T2) is jumped through its critical speed range only when the 4K cold box has warmed sufficiently due to the increasing load to make it necessary, but is prohibited from doing so until the CC suction pressure is below a given value.

Because the Autopumpdown program evolved along with a continuously changing procedure, very few of its control parameters are hard coded. Most of them are accessible through TACL display pages that are available in both the CHL control room and the accelerator Machine Control Center. This allows for integrated operation of the Autopumpdown with the existing cryogenic controls software and allows the Autopumpdown parameters to be saved and restored along with the rest of the machine control parameters.

THE CURRENT AUTOPUMPDOWN

Currently the Autopumpdown program performs the following actions, most of which can be controlled from interactive display pages:

- **Cold Compressor #4 Speed:** The speed of CC4 is controlled by a standard PID (Proportional-Integral-Derivative) control loop to maintain a specified mass flow through the CC train. It is this mass flow (minus the heat load applied in the cryomodules) that determines the rate of pumpdown. However, the flow must be capped due to the limits imposed by the 4K plant. The effects of undersized heat exchangers in the 4K cold box and undersized warm gas compressors are discussed in Reference 2. The Autopumpdown controls this flow setpoint (and hence CC4 speed) as a function of CC suction pressure.

- **Cold Compressor Speed Ratios:** The speeds of CC's 1-3 are controlled as ratios to the speed of CC4. These ratios are determined by theoretical performance curves supplied by the manufacturer and verified to some extent in bench testing, though the scale of the machines limited the bench test capability. Due to stringent accelerator commissioning schedules, there has not yet been sufficient time to re-verify performance data over the full operating ranges of the machines. Operating experience, however, has indicated that there are substantial deviations from the theoretical predictions. The Autopumpdown program accounts for this deviation by allowing the ratios to be controlled on the theoretical curves only for specific "safe" intervals. In the earliest phase of the pumpdown the ratios are held fixed, during the intermediate phases they are adjusted with offsets from the theoretical curves, and in the last phase they are limited to experimentally determined values. See also **Final Adjustments** below.

- **JT Valve Positions:** The JT valve positions for each of the cryomodules are controlled by separate PID loops to maintain liquid level in each module. The Autopumpdown adjusts cryomodule liquid level setpoints (and hence JT valve positions) as a function of CC suction pressure. This is necessary both to provide enough flow to start the pumpdown and to slightly overfill the cryomodules so that the final liquid level after mass has been removed by the pumpdown is within range for accelerator RF operations to begin as soon as possible.

- **Cryomodule Heaters:** Each cryomodule contains a string of four heater elements that are used for load-leveling purposes. Software within the RF control system monitors the calculated heat load deposited in each cryomodule by the RF power and adjusts these heaters to keep the total load per module at a requested value. The Autopumpdown program communicates with the RF control software to adjust the requested heat in each cryomodule as a function of CC suction pressure. This feature was used extensively in early pumpdowns to slow the rate of pumpdown in order to analyze the problems encountered at particular pressure ranges. Currently it is used exclusively to perform a "soft landing" at the end of the pumpdown, when heat is applied gradually to halt the pumpdown at a specific target pressure.

- **4.5K Subcooler:** One of the crucial breakthroughs of the development phase of the pumpdown procedure was the determination that the cold compressors were much more susceptible to fluctuations in inlet temperature during the early stages of a pumpdown than during the later stages. The inlet temperature to the cold compressor train is closely coupled through a 2K subcooler heat exchanger to the temperature of the helium being supplied to the load. This temperature in turn is controlled by conditions within a special 4.5K boiler/subcooler installed by CEBAF when it was found that the originally supplied 4.5K subcooler was inadequate. (See Reference 2.) Keeping this subcooler at low liquid levels and high pressures during the early part of a pumpdown (in order to elevate the CC suction temperature) greatly enhances the stability of the cold compressors. The boiler must eventually be filled, however, or the CC exhaust becomes too hot for the 4K cold box and the supply temperature to the load becomes too high. The Autopumpdown program makes this transition gradually as a function of CC suction pressure when the cold compressors are operating in a range less sensitive to inlet temperature.

- **Turbine Speeds:** The 4K capacity required to support a pumpdown varies significantly over the course of the pumpdown. Capacity is normally controlled by PID loops that regulate the inlet valves to the turbines in the 4K cold box. However, due to the fixed brake pressures and critical resonant frequencies of the turbines, large portions of their speed ranges are forbidden. This results in each turbine (except the coldest) having two operating modes, above and below its critical speed range. Critical speeds are avoided by limits placed on inlet valve motion. When it is necessary to cross the critical speed ranges, rapid changes are made to inlet valve position to quickly "jump" the rotor speed up or down. If the rotor spends too long in the prohibited range, it is shut down by the local turbine controller. Because the critical speed ranges are so wide, a turbine that is "jumped" up in speed because its outlet temperature is too high may immediately trip itself off on low outlet temperature. The Autopumpdown manages the speeds of the second and third turbines to meet the capacity requirements of the pumpdown within the operating constraints of the turbines themselves. This includes controlling the "jumps" through critical speeds.

- **RF Power Interlocks:** During a pumpdown, operation of RF power in the cryomodules is prohibited in order to stabilize the heat load. The Autopumpdown notifies the RF control system through a set of software interlocks when RF is prohibited. Once the target CC suction pressure is reached, the Autopumpdown notifies the RF control system that RF power is again allowed.

- **Controlling Pressure Device:** A chronic problem during cold compressor testing was the failure of various pressure transducers in the 2K cold box. As a result, the Autopumpdown allows the selection of controlling pressure transducer from a list of devices tied to various points on the CC suction path. Optionally, it will automatically switch from the selected high range device to the default low range device at a specified pressure.

- **Final Adjustments:** It has been found that the cold compressors operate more stably and robustly at steady state when they have more nearly equal pressure ratios across the stages than those determined by the theoretical performance curves. Accordingly, once the target pressure has been reached, the Autopumpdown gradually adjusts the CC speed ratios to specified values to equalize the pressure ratios across the four stages. Once the speed ratio adjustments are complete, the cold compressor flow is reduced to a lower value made possible by the more stable operation.

BEYOND AUTOPUMPDOWN

Besides the Autopumpdown itself, a small suite of additional control programs has grown out of the techniques developed. The net result is a dramatic decrease in both the human intervention required to operate the CHL and the time to recovery in the event of a cold compressor trip.

Auto Heater Control

The linac pressure is regulated during 2K operation by a 990 W "swing" heater made up of the heaters contained in six of the cryomodules. This swing heater is under PID regulation from the main cryogenic control software. The Automatic Heater Control program continuously monitors the "swing" heater performance and augments it by adjusting the heaters in the remaining 36 cryomodules that are available for control. These remaining heaters are controlled by the accelerator EPICS control system and are not directly accessible to the cryogenics TACL control system, except through the Auto Heater Control program. These are the same heaters that are adjusted by the Autopumpdown program and they are accessed in the same way. This effectively extends the useful range of the swing heater from 990 W to over 4 kW to compensate for changes in heat load, such as a sudden loss of RF power in the accelerator. This capability has prevented many shutdowns of the 2K system, especially during the early phases of the EPICS conversion when RF control software was unreliable.

The Auto Heater Control program also has the responsibility of shutting off all heat and RF power in the accelerator in the event of a cold compressor shutdown. When the cold compressors trip, the linac begins to pressurize and any heat load on the cryomodules tends to warm and expand the 2K liquid. Since accelerator operation requires the cryomodules to be 90% full at all times, expanding liquid will overflow into the return transfer line. Overflowing liquid pools in the return line where it reduces the flow area and boils off, causing flow instabilities. When there has been substantial overflowing into the return line, the time required for a pumpdown increases dramatically and the probability of success for an immediate pumpdown is greatly reduced. Both of these effects are mitigated by automatic shutoff of all heater and RF loads immediately after a trip.

Autobackfill

The Autobackfill program was designed to automate the tasks required to prepare the system to be pumped down again after a trip. It is not initiated automatically to avoid losing any information that might be useful in diagnosing the cause of the trip. An operator must launch the program after investigating (and correcting if necessary) the cause of the cold compressor trip. When the cold compressors shut down for any reason, isolation valves on the 2K cold box close automatically. In order to protect the cold compressors, the programmable logic controller (PLC) on the 2K cold box will not allow these valves to be opened or the CC motors to be started until the cryomodules have been re-pressurized through bypass valves.

The Autobackfill program will reset all latched faults on the cold compressor controllers, open bypass valves to pressurize the cryomodules and return transfer line, open cooldown valves to bring the cold compressor suction temperature into the acceptable range for a pumpdown, open the isolation valves around the 2K cold box, and start the cold compressor motors. While backfilling and cooling down, the program monitors for and avoids undesirable conditions of bringing the main compressor suction header subatmospheric and producing liquid in the cold compressor suction header. When all parameters are in range, the program closes bypass and cooldown valves and exits. The machine is then ready to pump down. In the event of a cold compressor trip due to some transient condition that doesn't damage any hardware, the system can be restored to full 2K operation by simply activating the Autobackfill followed immediately by the Autopumpdown.

Autoramp

This simple utility program allows any operator input parameter in the cryogenic system to be smoothly and automatically ramped to a target value at a fixed rate. The operator invokes the program and then specifies the name of the signal to be controlled, the target value, and the ramp rate. Once the input is confirmed, the ramp is carried out and the program exits.

FUTURE DIRECTIONS

The long evolutionary process has resulted in an Autopumpdown program carefully tailored to perform all the complex operations of a cold compressor pumpdown. More than 90% of the time no operator intervention is required once the Autopumpdown has been initiated, and very little intervention is usually required in the remaining runs. Effort is still underway to improve this hands-off success rate. After each pumpdown where operators are required to intercede, the reasons for the intervention are analyzed. If the causes are sufficiently likely to occur again, a standard response is developed and integrated into the Autopumpdown code.

As the list of control parameters that have been determined to have a direct impact on the cold compressor pumpdown process has grown, more and more signals have come under the control of the Autopumpdown. Many of these signals, however, do not require repeated updates during the pumpdown and need only be set once on startup. Currently this process is still largely manual; an operator compares values from the computer against a standard set in hardcopy, looking for discrepancies. A future extension of the Autopumpdown program will keep these settings in human-readable files, separate from the bulk of control system saved settings. The program will then be able to ensure that all necessary startup values are correct.

More radical modifications to the pumpdown procedure are under consideration for testing when the accelerator operation schedule permits. The most interesting of these options concern manipulations in the speed ratios of the various cold compressor stages.

ACKNOWLEDGMENTS

Development of the cold compressor pumpdown procedures was a group effort involving the entire CEBAF Cryogenics staff (D. Arenius, V. Ganni, D. Kashy, T. Reid, J. Wilson, and the authors) with the assistance and direction of C. Rode. D. Wetherholt of the CEBAF Instrumentation and Controls group developed a real-time plotting program to monitor the cold compressors in dimensionless phase space that was invaluable during development of the pumpdown procedures.

REFERENCES

1. M. S. Keesee et al. "The CEBAF Control System for the CHL," to be published in the 1995 *Advances in Cryogenic Engineering*.
2. W. C. Chronis et al. "Commissioning of the CHL Refrigerator at CEBAF," to be published in the 1995 *Advances in Cryogenic Engineering*.

THERMAL PERFORMANCE OF THE CEBAF HALL A CRYOGENIC SYSTEM

W.J. Schneider[1], E. Folts[1], J. Miller[1], A. Langhorn[2], M. Morgan[3], S. Stoy[3]

[1]CEBAF, Newport News, VA 23606
[2]STARTECH Inc., Solana Beach, CA 92075
[3]Ability Engineering Technology, S. Holland, IL 60473

ABSTRACT

CEBAF, The Continuous Electron Beam Accelerator Facility, is centered on a 4 GeV recirculating linac which has recently started routinely producing beam to the first of three experimental halls. Hall A, the second of three experimental halls, houses two high resolution spectrometers, HRS, and is due to become operational by the end of 1995. Each HRS consists of two nuclear physics detectors, with optics determined by a superconducting dipole and three superconducting quadrupoles cooled by the End Station Refrigerator, ESR. Minimization of the total heat load from Hall A to the refrigerator is essential since the capacity is fixed. The design, construction, and initial test of the cryogenic distribution system and how the overall heat load is controlled are discussed in this paper.

INTRODUCTION

CEBAF, The Continuous Electron Beam Accelerator Facility, is centered on a 4 GeV recirculating linac [1], which has recently started routinely producing beam to the first of three experimental halls. Hall A, the second of three experimental halls, houses two high resolution spectrometers, HRS, and is due to become operational by the end 1995. The two HRS are designed to study fundamental particle physics with a 200 μA CW electron beam. Each consists of two nuclear physics detectors, with optics determined by superconducting magnets [2] cooled by the End Station Refrigerator, ESR [3].

Figure 1 shows the major components that make up the spectrometers in the hall. Each spectrometer consists of a combined-function dipole magnet, mounted on a trio of 400 ton jacks which sit atop a cradle that revolves around a scattering chamber with target. Three quadrupoles, two in front and one after the dipole, make up the remaining optical elements, and are mounted on a box beam which functions as an optical bench and moves in unison with

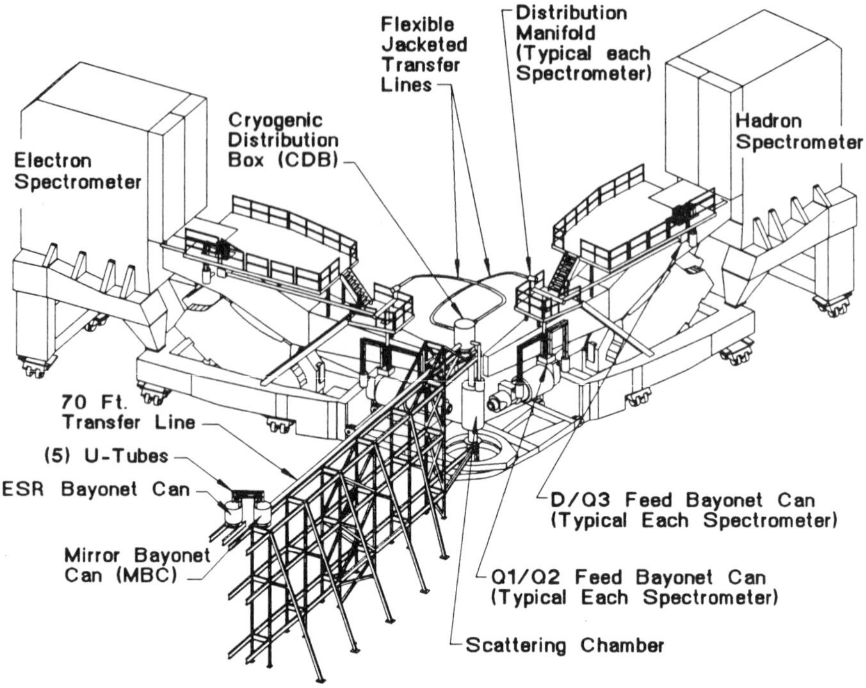

Figure 1. Hall A cryogenic distribution system.

the dipole. All eight of the superconducting magnets—Q1, Q2, D1, and Q3 on each arm—are of the pool boiling type and are supported by the cryogenic distribution system.

The cryogenic distribution system described in this paper is designed to cool these magnets from ambient to liquid helium temperatures and then maintain them efficiently at liquid helium temperatures for long periods of time while particle physics experiments are conducted using the spectrometers. All of the magnets, as well as the majority of the transfer lines, have a liquid nitrogen heat shield to minimize the 4 K heat load. This is essential, since the ESR capacity is fixed and must eventually handle all three experimental halls. The

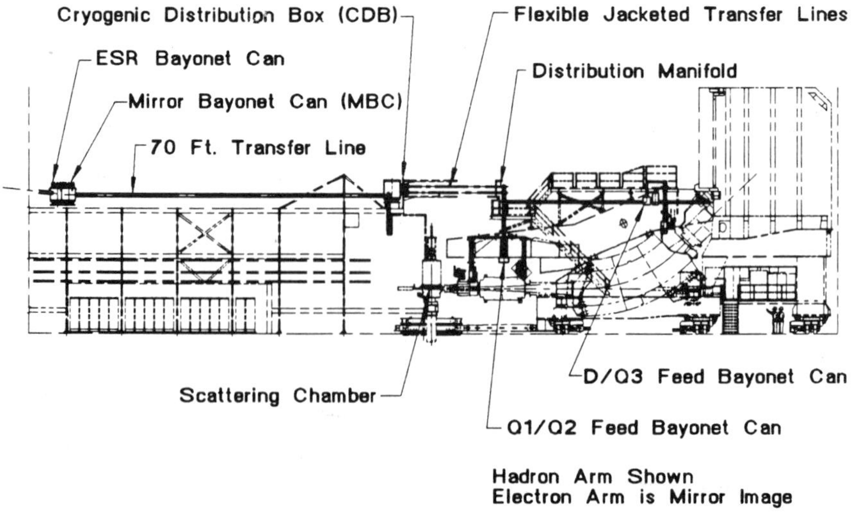

Figure 2. Hall A elevation view.

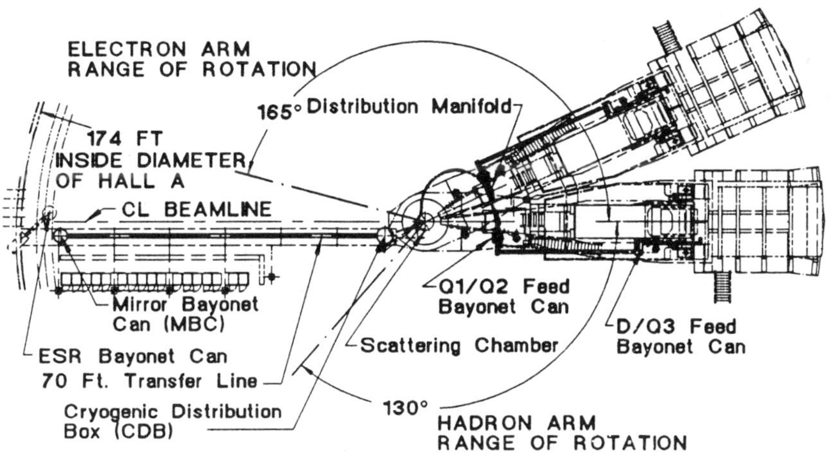

Figure 3. Hall A plan view.

scattering chamber can be equipped with a cryo target capable of handling a beam power of up to 1 kW at 15 K which is also serviced through the ESR. The cryogenic system has been optimized to reduce capital expenditure at the expense of some inconvenience during operations such as cooldown and warmup. This is not expected to be much of a factor, since these evolutions are anticipated infrequently, perhaps annually.

Instrumentation and control of the cryogenic system including the magnet liquid levels is centralized with the ESR and controlled by the cryo group through their central computer at the Central Helium Liquefier, CHL. The magnet power supplies, magnet lead flow, and spectrometer position are independently controlled by the Hall A computer networked with both the cryo and the accelerator computers. Safety interlocks such as quench protection are hard wired and provided locally to the power supplies.

DESIGN

The reference design of the transfer lines for Hall A is based upon the experience gained with the design, manufacture and experience of transfer lines and components installed in the accelerator [4,5]. These lines and components are updated versions developed at other labs such as BNL, Fermi and the SSC. Ability Engineering modified the reference design, in some instances, to optimize performance and cost. One of our main concerns was the integration of the distribution system into the spectrometer structure. A considerable effort was spent on designing the piping runs to clear the myriad of details associated with the spectrometer and to integrate the connections to all of the magnets.

The cryogenic distribution system comprises five major sub-systems. These include: 1) Cryogenic feed and cryogenic distribution box CDB; 2) Flexible transfer lines; 3) Spectrometer distribution; 4) Target cooling; 5) Gas management system.

The cryogenic services from the ESR enter the hall at the balcony level, elevation 31 ft above the floor, Figure 2. The ESR bayonet can, comprised of five U-tubes, is matched by a mirror bayonet can, MBC, and a 70 ft long transfer line extension to the CDB located at the same elevation near the center of the hall. This is the feed to both spectrometer distribution systems via the flexible transfer lines and the feed to the beam target cooling system.

The CDB houses a liquid nitrogen to gaseous helium cooldown heat exchanger with associated control valves and instrumentation. The box and valves are shielded with liquid

Table 1. Cryogenic system heat load budget

Parameter	Refr. W@4.4 K	Liq'f. g/s	Targ. W@20 K	LN_2 L/hr
Electron Arm				
Magnets	75	1.5		5
Piping	10			3
Hadron Arm				
Magnets	75	1.5		5
Piping	10			3
Pivot				
Flex Lines	10			4
Balcony Feed				
Distribution	35		100	100*
Beam Target			1000	
Piping Subtotal	65		100	10
TOTAL	215	3.0	1100	20

* During Cooldown only

nitrogen to minimize the heat leak to the 4 K circuit. As indicated, a separate header from the CDB carries cryogens to the beam target using standard unshielded transfer line techniques.

Flexible vacuum-jacketed transfer lines, manufactured by Kabelmetal, supply the cryogens to each spectrometer. The flexing duty of these lines is quite severe, since the spectrometers are required to rotate nearly 150 degrees about the pivot. Figure 3 depicts the plan view and shows the range of movement of the two spectrometers. Careful consideration has been given to their design because of their arduous duty. A design life of 5000 cycles has been set. By combining the LHe supply and the cooldown line the number of lines was minimized. End positions were selected to prevent torsion to the corrugated tubes and to minimize sag and bending. The LHe supply (and cooldown), LHe return and LN_2 supply are

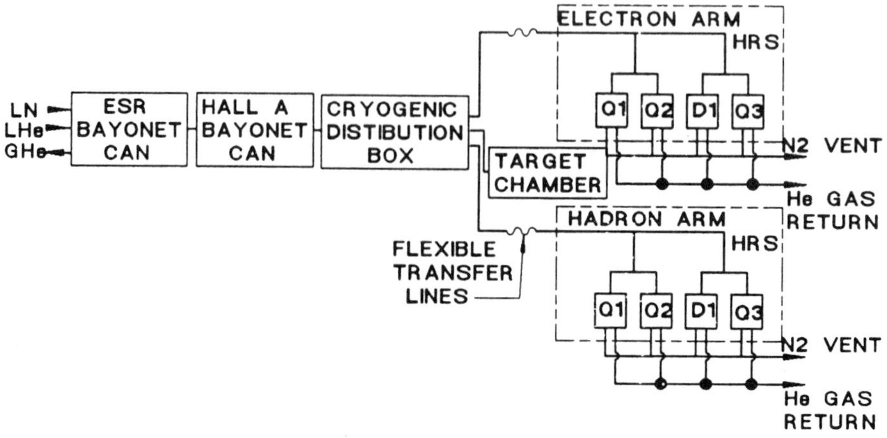

Figure 4. Simplified cryogenic system schematic.

Table 2. Cryogenic supply conditions

Parameter	Supply Cold	Supply Warm	Return Cold	Return Warm
Helium steady state				
Temperature	4.5 K	None	Saturated	300 K
Pressure	4—2.5 atm		1.2 atm	1.1 atm
Flow (max)	7.5 g/s			
Helium Cooldown				
Temperature	4.5—300 K	300 K	None	300 K
Pressure	3 atm	4 atm	> 5 K	1.1 atm
Flow (max)	7.5 g/s	10 g/s		
Helium Target				
Temperature	15 K	21.3 K		
Pressure	20 atm	19 atm		
Flow (max)	22.5 g/s	22.5 g/s		
Nitrogen Liquid				
Temperature	80 K			300 K
Pressure	3 atm			1 atm
Flow (max)	22 g/s			Vent

combined coaxially into one line using the proven Kabelmetal design with specially designed end fittings. This design reduces the LHe heat leak to near zero.

The distribution manifold located atop each spectrometer carries cryogens to the four magnets. The manifold acts as the flexible transfer line terminator and feeds the magnets from two extension arms. One arm feeds quadrupoles Q1 and Q2. These quads provide focussing and are required to move axially along the beam path by ±68 cm to change the spectrometer tune. This pair is connected to the cryogenic distribution system using separate flexible transfer lines. The extension line that feeds D1 and Q3 is designed using standard shielded transfer line construction techniques. Figure 4 shows schematically the cryo system.

The gas management system consists of 4 atm helium gas to the cooldown heat exchanger, low pressure helium gas from the cooldown manifolds, nitrogen gaseous vent, current lead gas return and quench gas recovery. The majority of this piping is welded stainless steel.

Meeting the 4 K heat load and providing enhanced operational flexibility along with reduction of overall cost were CEBAF's goals for this system. Overall heat leak and pressure drop budget goals including anchors, expansion joints, and spacers for the transfer lines are indicated in Table 1. This budget was based upon the design conditions indicated in Table 2. CEBAF budgeted a total of 65 W to the 4.4 K circuit and 450 W (10 L/hr) to the 80 K shield for the entire distribution system piping. The budget for the target circuit piping is limited to 100 W. As a goal, we attempted to minimize these loads to 2/3 of the budget.

CONSTRUCTION

All vacuum vessels and vacuum jackets are stainless steel, designed and constructed in accordance with the ASME Code for external pressure. No vessels were required to be code stamped. The base line insulation system consists of a blanket made up of 1/4 mil doubly aluminized Mylar separated by a nonwoven 3 mil spunbonded polyester spacer such as Reemay. On all primary circuits with a thermal shield, 3 six layer blankets were provided. For lines without shields, or for the secondary circuit, 4 fifteen layer blankets were provided.

Piping was sized in accordance with the design conditions listed in Table 2. Allowable pressure drops did not exceed 0.5% of the operating pressure at the maximum flow. All piping subject to cryogenic temperatures was made of stainless steel type 304L using all welded construction. At connections to the relief valves and warm piping, epoxy joints were made to ensure a leak tight system. Strict attention was given to stresses due to thermal contraction during cooldown and upset conditions. The use of expansion loops was preferred over flexible hose. The cryogenic piping and penetrations were designed to avoid thermal acoustic oscillations. Cryogenic check valves were installed at the cold end of susceptible lines.

All sensors were separated from control wires and both were terminated in local junction boxes attached to the vacuum vessel. Wires are marked with tags and identified on a wiring diagram generated by Ability Engineering. The internal piping and cold components were supported to withstand without damage an omnidirectional 2 G shock and vibration due to shipping and handling.

The piping was tested pneumatically to 110% of design pressures at Ability's shop. A final acceptance test will be performed by CEBAF after installation in Hall A. The maximum allowable leak rate in all cases was less than 10^{-9} atm cc/s per ASTM E498-90. Tests were performed with all internal pipework and exterior joints in a helium environment. Tests were conducted in internal and external piping and the vacuum jackets.

The transfer lines were shown to meet the heat leak requirements by calculations and by a sample boil off test (LN_2) for one typical component such as a bayonet can. Before production on all components started, a first article was submitted and approved by CEBAF. Upon receipt at CEBAF, the transfer lines and components supplied were assembled into the configuration as shown on the assembly drawings. Initial tests included dimensional checks and subsequently a pressure and a mass spectrometer leak test was performed.

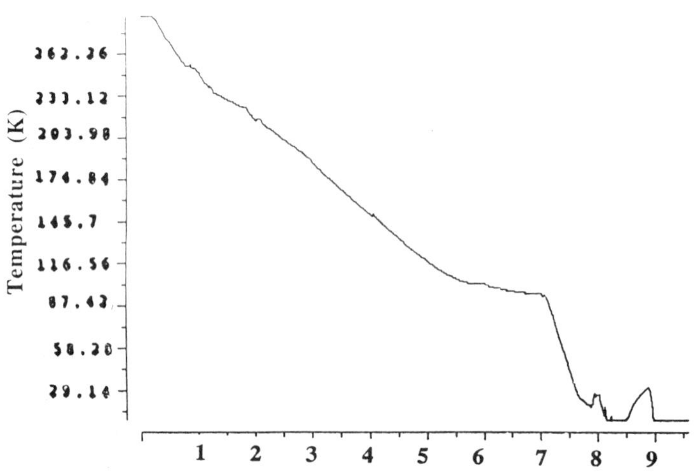

Figure 5. Magnet cooldown.

OPERATING MODES

The major feature of the magnet cooldown/warmup system is the helium/LN_2 heat exchanger built into the cryogenic distribution box (E) and the common cooldown/liquid supply manifolds for each spectrometer. During cooldown of a single magnet, the other magnets on the affected spectrometer are de-energized and isolated. Boil off of the isolated magnets can be directed through the current leads or cooldown return.

The stand alone cooldown system uses liquid nitrogen as the refrigerant. Since there is no opportunity to recycle the sensible heat from the cooldown gas, the gas is returned to compressor suction through the cooldown return. This reduces the refrigeration load on the ESR. Thermal stress on the magnet between 300 and 100 K is controlled by minimizing the gradient across the magnet to less than 75 K and limiting the incoming gas temperature to some 50 K below the average. At 5 g/s the dipole magnet was cooled to 100 K in about 180 hr (Figure 5). Cooldown to liquid helium temperatures is accomplished by isolating the heat exchanger and cooling the rest of the way with the ESR at 2.8 atm and 4.5 K. At 5 g/s flow the cooldown from 100 K was 12 hr. When the gas temperature dropped below 5 K, the return flow was routed from the cooldown line through the transfer lines back to the ESR so as to take advantage of the sensible heat in the return stream. Warmup can be achieved by reversing the process.

During steady-state conditions the magnet cryostats are supplied with supercritical gas at 2.8 atm from the ESR. The gas is expanded through a JT valve yielding nearly 100% liquid. Boil off is returned through the transfer lines to the ESR. A portion of the flow is returned to compressor suction at 300 K and 1.1 atm through the gas cooled leads.

Liquid nitrogen, supplied from bulk storage, is used to cool the radiation shields on the lines and in the magnets and the heat exchanger in the CDB.

Upset conditions can be categorized as isolated (such as quench) or global (such as refrigerator failure). Typically during an isolated incident the supply shuts off and the gas flows back to the refrigerator. If the return temperature increases excessively, return flow can be diverted to compressor suction.

INSTRUMENTATION AND CONTROL

As indicated in the introduction, the cryogenic system control including the magnet liquid levels is centralized with the ESR and controlled via CAMAC by the cryo group through their central computer at the Central Helium Liquefier, CHL. All of this information is available on the network and is read out in the Hall A control room via EPICS by a Hewlett Packard series 700 computer. The magnet power supplies, magnet lead flow, spectrometer position are independently controlled by the Hall A computer networked, via EPICS, with both the cryo and the accelerator computers. Safety interlocks such as quench protection and lead flow are hard wired and provided locally to the power supplies. The system operates in the steady-state mode via remote control as the hall is not manned during beam operations. Cooldown of the individual magnets is permitted to be manual operation.

The cryogenic distribution box, located near the pivot, includes valving and a LN_2/helium heat exchanger. A motorized valve throttles a flow of warm helium gas through the exchanger producing a variable flow stream of 100 K gas. This is automatically balanced against a parallel flow of warm gas bypassing the exchanger. The bypass flow is adjusted to limit the supply temperature to less than 50 to 75 K of the average magnet coil temperature. This limits the thermal stresses on the magnets during cooldown and warmup and controls the cooldown rate, which is limited by mass flow and the cooling configuration within the magnets. The liquid nitrogen flow is controlled by a motorized valve which maintains a constant outlet temperature.

ACKNOWLEDGMENT

Many people have contributed to the successful development, design, procurement, fabrication, assembly, testing, commissioning and operational support of the Hall A cryogenic system. Particular credit goes to J. Alcorn and S. Nanda for their work designing and refining the magnets, the core components of the spectrometer. Assembly and commissioning of the system were accomplished by G. Dezern, G. Parker, E. Smythe and R. Zarobinski. Controls were coordinated by J. Proffitt, C. Rowe, M. Stevens and H. Whitehead. The cryo group and Hall A engineering have been very helpful during initial operations of the magnet acceptance tests. Many thanks to M. Bennett for graciously typing the manuscript. This work was carried out under DOE Contract #DE-AC05-84ER40150.

REFERENCES

1. H.A. Grunder et al., "The Continuous Electron Beam Accelerator Facility", in Proceedings of the 1987 IEEE Particle Accelerator Conference (Washington, DC 1987) Vol. l, pp. 13—18.

2. J. Alcorn, "Large, Precision, Superconducting Magnet Systems for the GeV/*c* High Resolution Spectrometers of CEBAF's Hall A", Advances in Cryogenic Engineering, Vol. 41, Plenum Press NY 1996.

3. D. Arenius et al., "Commissioning and Operation of the CEBAF End Station Refrigerator System", Advances in Cryogenic Engineering, Vol. 41, Plenum Press NY 1996.

4. W.J. Schneider, et al., "Thermal Performance of the CEBAF Superconducting Linac Cryomodules", in Advances in Cryogenic Engineering, Vol. 39, pp. 589—596, Plenum Press NY 1994.

5. D. Kashy et al., " CEBAF Cryogenic Distribution System", Advances in Cryogenic Engineering, Vol. 37, pp. 537—543, Plenum Press NY 1992.

CRYOGENIC TESTING AND ANALYSIS ASSOCIATED WITH TEVATRON LOWER TEMPERATURE OPERATION

J. C. Theilacker

Fermi National Accelerator Laboratory*
Batavia, IL, 60510, USA

ABSTRACT

An upgrade of the Tevatron cryogenic system was installed and commissioned in 1993 to allow lower temperature operation.[1] As a result, higher energy operation of the Fermilab superconducting Tevatron accelerator is possible. Following the installation and initial commissioning, it was decided to continue the current colliding beam physics run at the previous energy of 900 GeV. This has allowed us to perform parasitic lower temperature tests in the Tevatron over the last year and a half.

This paper presents the results of operational experiences and thermal and hydraulic testing which have taken place. The primary goal of the testing is to better understand the operation of the cold compressor system, associated instrumentation, and the performance of the existing magnet system during lower temperature operation. This will lead to a tentatively scheduled higher energy test run in the fall of 1995.

The test results have shown that more elaborate controlling methods are necessary in order to achieve reliable system operation. Fortunately, our new satellite refrigerator controls system is capable of the expansion necessary to reach our goal.[2] New features are being added to the controls systems which will allow for more intelligent control and better diagnostics for component monitoring and trending.

INTRODUCTION

It is desirable to increase the energy of the superconducting Tevatron accelerator from its current level of 900 GeV to 1000 GeV. The current carrying capability of the superconductor used in the Tevatron improves by 15% per degree Kelvin in the range of interest. A cryogenic system upgrade was installed in 1993 which has the capability of reducing the magnet temperature by as much as 1K. Centrifugal cold helium vapor compressors were installed in each of the twenty-four satellite refrigerators to accomplish this.[3]

Each of the satellite refrigerators cools two 125m long magnet strings. The superconducting coils are bathed in subcooled liquid at 0.22 MPa. In each magnet, half of the single-phase (1ø) stream bathes the superconducting coils and half is cooled by an annular two-phase (2ø) helium passage. The two 1ø streams mix at the end of each magnet. Static (heat leak) and dynamic (AC loss) heat loads are removed by the latent heat of the

* Work supported by the U.S. Department of Energy under contract No. DE-AC02-76CH03000.

liquid in the 2ø passage. The remaining 2ø helium from both magnet strings returns to a subcooling dewar at the refrigerator.

Figure 1 shows a schematic layout of a satellite refrigerator with one of it's two magnet strings. Components new to the lower temperature upgrade include the cold compressor, 130 liter subcooling dewar and provisions for a void fraction sensor on the 2ø return from the magnet strings.

The subcooling dewar serves two purposes. First, it separates the liquid and vapor in order to ensure that the centrifugal cold compressor only pumps on saturated vapor. Second, it provides a thermal buffer between the refrigerator and the magnet strings. The returning liquid is vaporized by a heat exchanger which subcools the outgoing stream of the refrigerator. The level of liquid in the subcooling dewar is maintained by adjusting the amount of liquid helium drawn from the Central Helium Liquefier (CHL).

Tests were performed on the cold compressors and the behavior of the magnet strings during the current 900 GeV Collider Run 1B. Considerable testing was performed parasitic to collider operations. However, all of the tests presented in this paper required dedicated use of the Tevatron. In most cases, this time was made available while another system was down, making physics or beam related studies in the Tevatron impossible.

Figure 1 also shows the instrumentation available for characterizing the behavior of the Tevatron magnet systems during our tests. Thermometry used is shown as a "T" and is generally a carbon resistor thermometer. The redundant thermometer shown on the 1ø inlet and outlet is a vapor pressure thermometer as is the thermometer on the cold compressor inlet. Pressure measurements are made by standard capacitance type transducers and are shown as a "P" or "DP" for differential. Flow is measured with a liquid helium venturi and utilizes a warm differential pressure transducer, shown as "F" in Figure 1.

The layout and instrumentation available for a typical 125m long magnet string is also shown in Figure 1. Temperature of the 1ø is measured every half cell. A half cell consists of four dipole magnets ("D" in Figure 1), a quadrupole magnet (Q), and a spool piece (S).

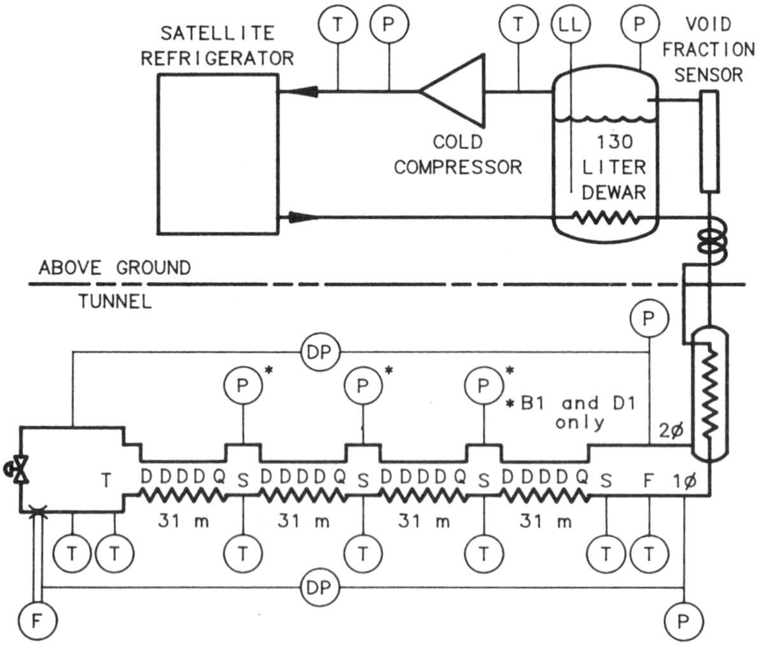

Figure 1. Schematic layout of a satellite refrigerator and one of its two 125m long magnet strings. The temperature, pressure and flow instrumentation available for the testing is shown.

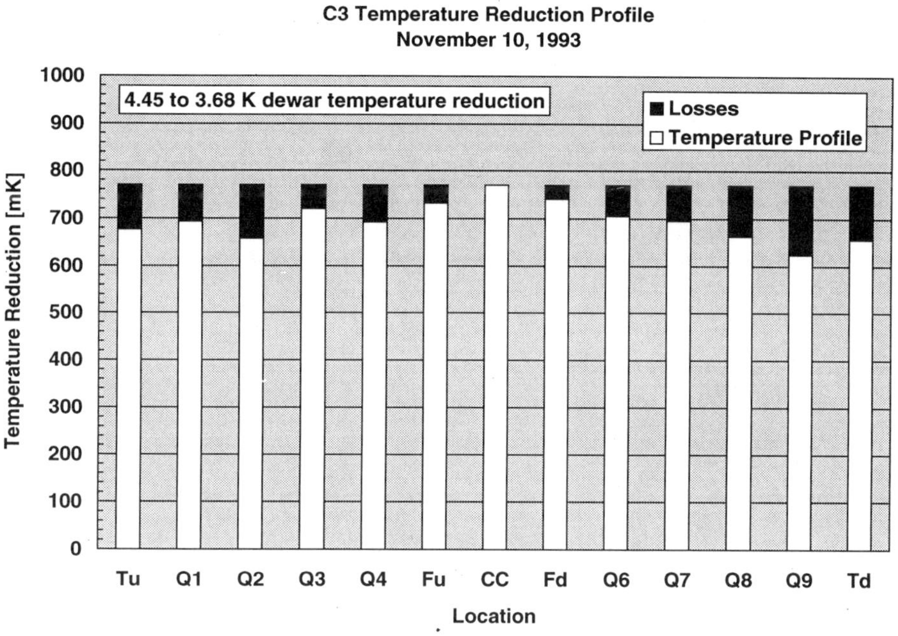

Figure 2. Temperature reduction profile for the C3 magnet system.

There are four half cells in a typical magnet string. The refrigerator interfaces with the magnet strings at the feedcan (F). At the end of the 125m magnet string, the 1ø flow is throttled into 2ø in the turnaround box (T).

TEMPERATURE REDUCTION PROFILES

Cold compressors regulate the temperature in a subcooling dewar located in each refrigerator. The temperature along the two 125m long magnet strings is then somewhat above this temperature due to 2ø pressure drop and imperfect heat exchange. Temperature measurements are made in the 1ø approximately every 31m using carbon resistor thermometers. Calibration of the thermometers were made over twelve years ago, and then only down to a temperature of 4.2K.

Some attempts have been made to recalibrate the thermometers in place by depressurizing the 1ø, allowing it to go two-phase, while measuring the pressure. This becomes a very tedious task, requiring considerable downtime and large swings in helium inventory. The calibration results have been mixed, depending on the piping geometry in the components which house the thermometer.

Experience has shown that long-term changes in the thermometer calibration do not effect the slope of the response. As a result, temperature changes measured by the same thermometer remain accurate even though absolute measurement is off. We used this fact to our advantage by considering temperature reductions from a known baseline instead of absolute temperatures. This baseline is "known" only through magnet quench performance and not absolute temperature. Since the ultimate goal of the project is to increase the quench energy of the Tevatron, this is an acceptable characterization for a baseline. The baseline for the Tevatron is a quench energy of 935 GeV at a dewar temperature of 4.50K.

In order to estimate the increase in quench energy of the Tevatron for a given reduction in dewar operating temperature, we need to know how effectively we reduced the temperature throughout the magnet strings. Figure 2 shows a temperature reduction plot for the C3 satellite magnet system. The cold compressor was used to reduce the temperature of the subcooling dewar from the nominal 4.45K to 3.68K. This 770 mK reduction is shown in the center of the figure at location "CC" (cold compressor). Left and right of this point represents the 125m long upstream and downstream magnet strings, respectively.

Columns labeled with a "F" and "T" represent the feedcan and turnaround boxes, as in Figure 1. Columns labeled Q# represent the thermometer in the spool piece, located every half cell (approximately every 31m). The lighter colored column represents the actual temperature reduction realized at each location. The dark section represents pressure drop and heat transfer losses from the maximum temperature reduction, in this case 770 mK.

Figure 2 shows that over 80% of the temperature reduction was realized at all locations throughout the two magnet strings. This is consistent with predictions for 2ø pressure drop and the change in helium specific heat. This is not the case, however, in Figure 3 for the B1 satellite magnet system. In this case, only 50% of the temperature reduction was realized over much of the upstream string. The only significant difference between C3 and B1 is that B1 has additional quadrupole magnets installed. The most significant of these quadrupole magnets are located at Q3, Q2, Q1, and LB, corresponding to the highest loss in temperature reduction.

In order to better understand the losses, additional pressure transducers were added to the 2ø circuit. This allowed us to separate the losses into pressure drop and heat transfer components. Included in the heat transfer component is the effect of reduced helium heat capacity. From Figure 3, it is shown that a local pressure drop between Q8 and Q9 accounts for the discrepancy on the downstream string. Over the history of the Tevatron, several obstructions (rags) have been removed from the cryogenic circuits of the magnet system. Recently, one was removed from between Q2 and Q3 at B1. During that time, all 2ø interfaces in B1 were inspected and no further obstructions were found. It is suspected that an obstruction may be deep inside a component, probably a spool piece, and not visible from the interface.

A 2ø pressure drop is also shown on the upstream string between Q4 and Q5. Although this loss is quite significant, representing 11% of the temperature reduction, there is even larger heat transfer loss term. This loss is not currently understood, but has also been observed in another satellite magnet system which also has additional quadrupole magnets installed. Further temperature reduction profile testing in other satellite magnet systems is required in order to understand if this large loss is only associated with the additional quadrupole magnets.

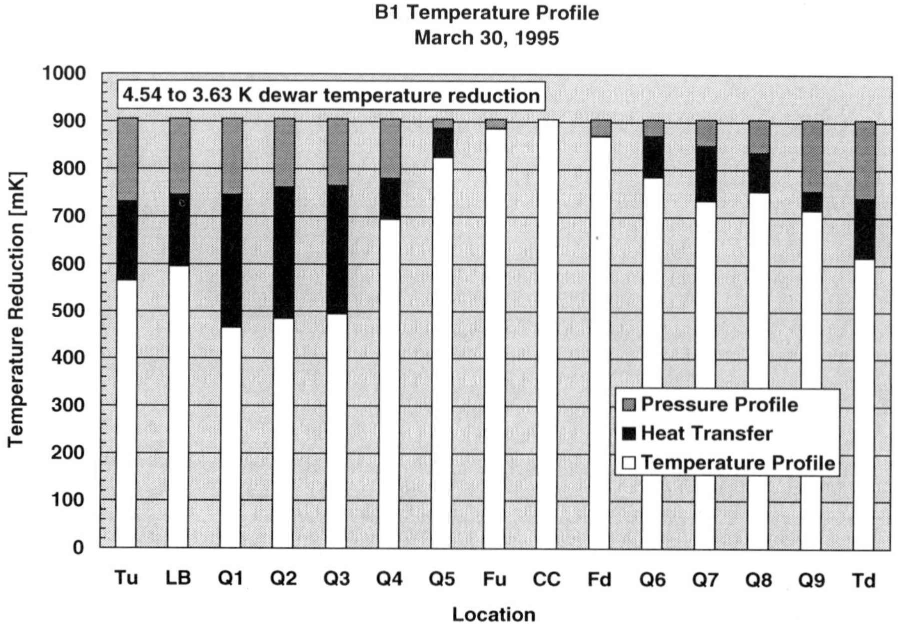

Figure 3. Temperature reduction profile for the B1 magnet system including special quadrupole magnets.

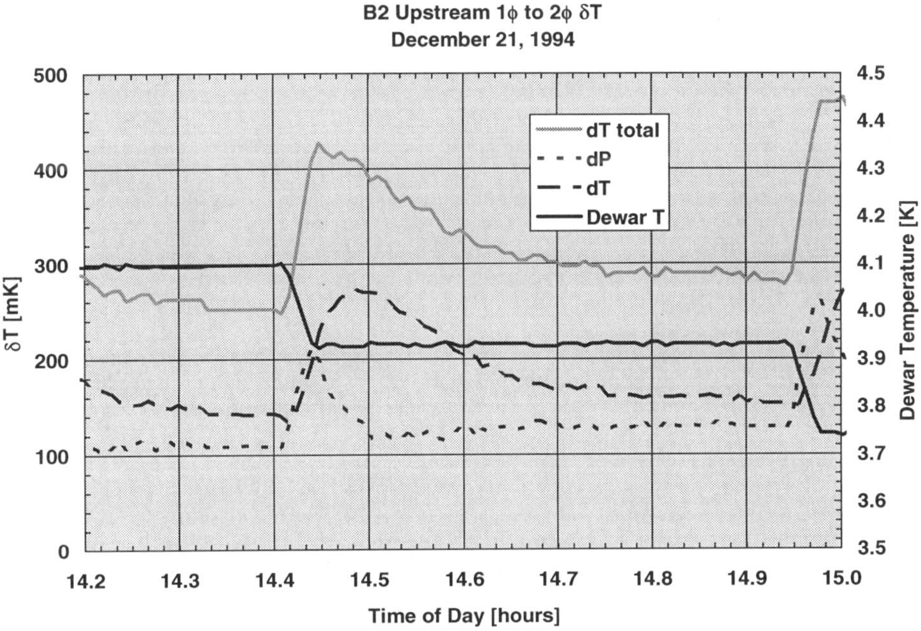

Figure 4. System Response to a step change in dewar pressure at constant mass flow rate.

MAGNET SYSTEM TIME RESPONSE

The time response of the 2ø circuit was measured and calculated when the Tevatron was commissioned to be about 6 minutes. The time response has to do with adding or removing helium inventory caused by a change in flow rate or heat load (AC losses due to cyclical ramping of the accelerator). It was desirable to investigate whether the time response changes significantly at lower temperature in order to develop an appropriate controlling scheme.

Figure 4 shows the time response of the 2ø pressure and 1ø temperature at the end of a magnet string following a 170 mK reduction in the cold compressor regulation point. The dark solid line represents the dewar temperature, as measured by the vapor pressure. It took approximately 2.5 minutes for the cold compressor to pump down the dewar to the new set point. The figure also shows that the regulation is quite stable.

The line labeled dP represents the temperature difference due to pressure drop along the 2ø circuit of the magnet string. The time response of the 2ø pressure drop is about 6 minutes, consistent with previous measurements at 4.5K.

The line labeled dT represents the temperature difference between the 1ø at the end of the magnet string and the 2ø at the end of the string. The time response is considerably longer at 21 minutes. This time is curiously similar to the travel time of helium through the 1ø of a magnet string. It is believed that this is only a coincidence. The relatively high heat leak and continuous heat exchange design of the magnets make it unlikely that the helium travel time would effect this.

The sum of the dP and dT loss is plotted in the solid gray line. The long time response of the heat transfer to the 1ø will require a waiting period prior to restarting the accelerator to high energy after a cool down following a quench or maintenance period.

SYSTEM ΔT

Tests in the previous section were performed with a constant mass flow rate and varying the 2ø pressure. This section addresses tests performed with a constant 2ø pressure and varying mass flow rate. The primary purpose of the test was to measure the 2ø pressure drop versus flow rate. This was being investigated as a possible explanation for differences between quench levels attained in 1991 and 1993 testing.

Varying the mass flow rate in the Tevatron 2ø changes the void fraction throughout the length of the magnet string. Vapor generation is due to heat load, which is fixed when the Tevatron is not powered. Decreasing the mass flow rate reduces the amount of liquid which travels the length of the 2ø circuit, but the vapor mass flow rate remains constant.

Figure 5 is a time plot of the same three differential temperature parameters from Figure 4 plotted with the magnet string mass flow rate at a constant dewar temperature of 3.84K. The mass flow rate was decreased by approximately 2 g/s increments from 23 g/s to 14 g/s. Each step was allowed to stabilize for a half an hour based on the experience gained from the time response tests.

For the first three mass flow steps, the 1ø/2ø mass flow rate was reduced by 26%. Over the same range, the ΔT corresponding to pressure drop decreased by 36%. What hadn't been anticipated was that the ΔT between the 1ø and 2ø increased by 56%. This heat transfer effect dominated, resulting in a net increase in ΔT between the 1ø and the dewar as the mass flow rate was reduced, as shown by the dT total line in Figure 5.

The last mass flow rate reduction shown in Figure 5 resulted in a considerable increase in 1ø to 2ø ΔT. It is believed that this mass flow rate is very close to the flow necessary to maintain the static heat leak of the magnet string. Operation in this range becomes very unstable when changes, positive or negative, are made in the system.

The results of this test suggests that higher flow rates will allow for lower coil temperatures and thus higher quench energies. This result could explain the difference in quench energies observed between the 1991 and 1993 testing. Between that time, the injection point for the liquid helium flow from the CHL was moved from the 1ø to the subcooling dewar. This results in a net reduction of 1ø/2ø mass flow rate of about 6 g/s per satellite, or 3 g/s per magnet string.

CONCLUSIONS AND RECOMMENDATIONS

Further testing is required in order to understand the operation and control of the Tevatron lower temperature upgrade. Enhancements to the controls are necessary for both operations and further testing. Automatic cool down schemes need to be refined to consider the results of mass flow rate and time response effects.

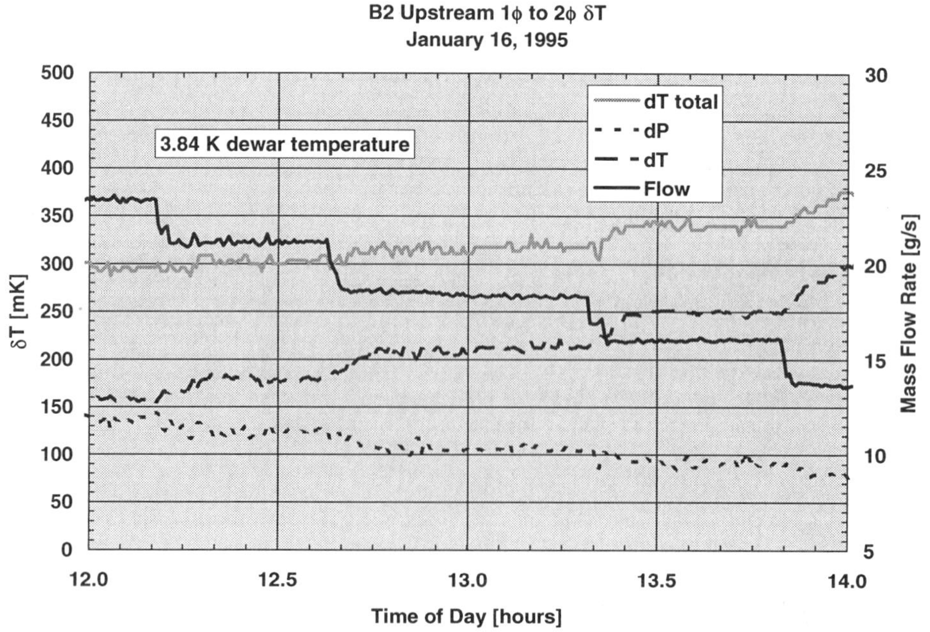

Figure 5. System response to a step changes in mass flow rate at constant dewar operating pressure.

Enhancements also need to be made to automate further testing in other satellite magnet systems. Automatic control schemes need to be developed for carbon resistor calibration and temperature reduction profiles. This is the only way to ensure uniformity in technique while eventually testing twenty-four satellite magnet systems.

Additional tests are planned to understand the heat transfer in a Tevatron half cell. A stand-alone half cell string test is planned with highly instrumented magnets. The goals of the test are to understand the heat leak to the 1ø and 2ø circuits, the 1ø to 2ø heat transfer, the 1ø inner and outer flow split, 1ø and 2ø stratification, and AC losses.

ACKNOWLEDGMENT

The author acknowledges the dedication shown by the Accelerator Division Cryogenics Department in supporting the operation the Tevatron cryogenic system and the tests presented in this paper. The data logging and plotting tools supplied by the Accelerator Division Controls Department were invaluable to this work.

REFERENCES

1. J. C. Theilacker, Upgrade of the Tevatron cryogenic system, in: "Advances in Cryogenic Engineering, " Vol. 39, Plenum Press, New York (1994), p. 517.
2. B. L. Norris et al. New cryogenic controls for the Tevatron low temperature upgrade, in: "Advances in Cryogenic Engineering, " Vol. 39, Plenum Press, New York (1994), p. 1185.
3. J. C. Theilacker, Tevatron cold compressor operating experience, in: "Cryogenics," Vol. 34, ICEC Supplement, Butterworth-Heinemann, Oxford (1994), p. 107.

REFRIGERATED HYDROGEN GAS JET FOR THE FERMILAB ANTIPROTON ACCUMULATOR

D. H. Allspach,[1] C. L. Kendziora,[1] M. Marinelli,[2]
M. Macri,[2] E. Robutti,[2] and G. Boero[3]

[1]Fermi National Accelerator Laboratory
Batavia, Illinois, 60510, USA
[2]University of Genoa
Department of Physics
16146 Genoa, Italy
[3]CERN
Geneva 23, Switzerland

ABSTRACT

A hydrogen gas jet has been built for use at Fermilab for the study of charmonium spectroscopy in proton-antiproton annihilations (Experiment 835). The hydrogen gas jet is part of an upgrade to a previous experiment (Experiment 760) which ran in the Fermilab 1990-1991 fixed target program utilizing a jet cooled to 80 K with liquid nitrogen. The jet delivers a defined stream of hydrogen gas which travels through a series of vacuum chambers and then intersects the circulating antiproton beam. The goal of the upgrade is to provide a hydrogen gas stream at least twice as dense as used for the earlier experiment to increase the interaction rate and allow an improved study of rare processes. This is achieved by cooling the stream to below 30 K using a Gifford-McMahon refrigerator. The jet apparatus is designed to allow motion in the plane perpendicular to the gas stream as well as angular positioning at the jet nozzle to provide a means of optimizing the interaction rate. Two skimmers located in the vacuum chambers are used to define the gas stream dimensions. The jet target vacuum chambers require constant pumping with turbomolecular pumps. The vacuum space around the jet is designed to have a large system pumping speed so that the chamber pressure can be maintained below an absolute pressure of 1 Pa. The jet will operate in the next fixed target run at Fermilab. Details of the design and test results are discussed.

INTRODUCTION

The hydrogen jet target, located on the Fermilab Antiproton Accumulator, will provide Experiment 835 with a jet stream of hydrogen to be utilized as the target for the antiproton beam. The jet stream is produced by flowing hydrogen gas through a converging-diverging nozzle with a small throat diameter (37 μm). During the expansion through the nozzle the hydrogen molecules undergo a process of condensation causing them to collect as clusters. Each cluster contains about 10^4 to 10^5 atoms.[1] A jet of clusters is thus formed which constitutes the core of the hydrogen jet stream.

The antiproton beam is produced in the Fermilab Antiproton Source. A 120 GeV proton beam focused on a fixed target yields approximately one antiproton for every 10^6 protons.

The initial energies of the antiprotons cover a wide range but they are "cooled" such that they are made equal. Roughly 10^{12} antiprotons can be collected over a period of 20 hours. With this quantity of antiprotons circulating in the Antiproton Accumulator, the beam has a current of 100 mA. In the absence of the hydrogen gas jet stream, the antiproton beam lifetime is around 400 hours.

Hydrogen is chosen as the target material rather than a heavier gas because the experiment depends on achieving a very well defined collision energy between target and beam. Even the spread in energies due to nucleon motion in a helium atom nucleus would be too large for the experiment to work. During experimentation an interaction rate of 5 MHz between the jet stream and the antiproton beam is chosen resulting in a beam lifetime of about 30 hours. Experiment 835 is designed to have an interaction rate at least five times greater than Experiment 760 in order that the study of the charmonium spectrum may be completed during the upcoming Fermilab Fixed Target Program. The upgraded hydrogen gas jet, producing a denser jet stream of hydrogen atoms, will help in meeting this goal.

OVERALL JET TARGET SYSTEM DESCRIPTION

The jet target (see Fig. 1) consists of several vacuum chambers which surround the Antiproton Accumulator beam-pipe (AA). The jet stream, originating in chamber J1, passes through each chamber and across the beam-pipe which are connected by small holes or "skimmers". The communication between chambers is limited to prevent uncondensed hydrogen gas from diffusing into the beam-pipe. This gas would act to reduce the beam lifetime without contributing to the experiment. Several vacuum pumps are utilized to maintain a very high vacuum in the beam-pipe. Ten turbomolecular pumps (TMP's) are installed directly onto the chambers. Eight of these have a capacity of 1000 L/s and two are rated for 3500 L/s. The pump names indicate their capacity, i.e., TMPAA1 1000 has a capacity of 1000 L/s. Two additional turbomolecular pumps are used as booster pumps downstream of the pumps on the J2, J3, AA1, AA2, R1, R2 and R3 chambers. Also included are two positive displacement blowers and two roughing pumps.

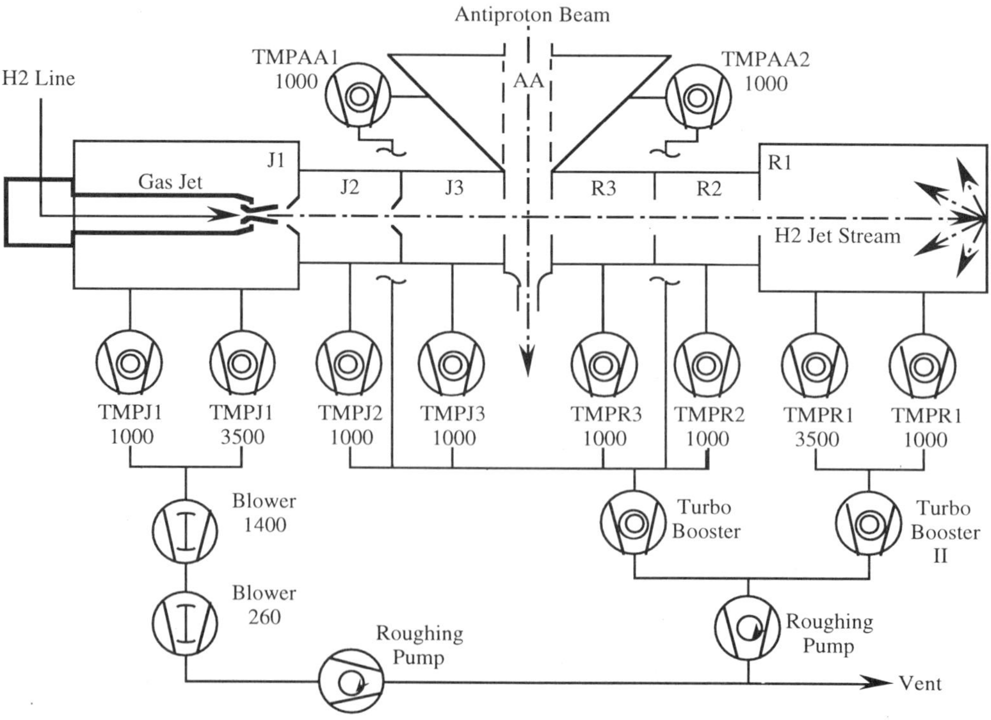

Figure 1. General Jet Target system components.

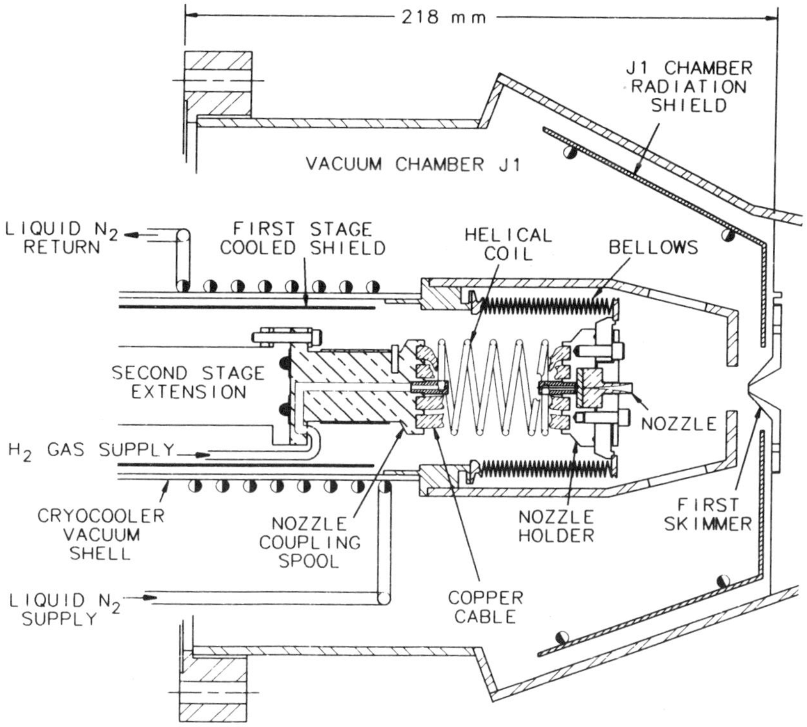

Figure 2. Hydrogen Gas Jet components located in the J1 Vacuum Chamber .

The nozzle, located in chamber J1, has a hydrogen gas flow which can reach a maximum rate of 13 cm^3/s STP. The refrigeration system used to cool the hydrogen gas and the nozzle is located in chamber J1 as well. Roughly 10% of the hydrogen flow make up the jet stream of clusters that passes through the first skimmer. This skimmer is located on the wall dividing chambers J1 and J2 and has an aperture of 2.3 mm. The remaining 90% of the hydrogen gas exiting the nozzle is pumped from chamber J1 via TMPJ1 1000 and TMPJ1 3500. This "background gas" consists of the nozzle throughput which remains uncondensed and that which is "skimmed off" by the first skimmer. A second skimmer is located on the wall dividing chambers J2 and J3. With an aperture of 4.3 mm, this skimmer defines the diameter of the jet stream as it intersects the antiproton beam inside the Antiproton Accumulator beam-pipe. The jet stream measures roughly 7 mm in diameter at this point. After passing through the skimmers, much of the jet stream terminates in the R1 chamber. It follows that the chamber required to evacuate the second greatest amount of hydrogen gas is the R1 chamber. Immediately downstream of the interaction area the Antiproton Accumulator beam-pipe narrows in diameter and its wall thickness is significantly reduced. It is here that the detectors for the experiment are located.

DETAILS OF THE GAS JET UPGRADE

There are primarily three steps which have been taken to improve the performance of the hydrogen gas jet. This upgrade program includes:

(1) Lowering the stagnation temperature of the hydrogen gas supplied at the nozzle such that the density of the hydrogen jet stream is increased.
(2) Improving the system pumping speed of the vacuum chamber in which the gas jet is located to reduce the interaction of the background gas with the jet stream.
(3) Controlling the nozzle position in the plane normal to the jet stream (x-y plane) and its angular position to correctly locate the jet stream with respect to the antiproton beam.

The following paragraphs summarize the details of the upgrade program implementation.

Gas Jet Description

Lowered Gas Jet Operating Temperature. Through testing of the upgraded jet target system it has been determined that a successful hydrogen gas operating temperature is near 27 K. This distinguishes Experiment 835 from Experiment 760 for which the hydrogen was cooled to about 80 K. The operating pressure of the gas for the upgraded system is slightly less than the saturation pressure of hydrogen at 27 K to avoid condensation in the hydrogen circuit. The new gas jet is refrigerated by means of a cryocooler employing the Gifford-McMahon refrigeration cycle. The thermodynamic state of the hydrogen gas is precisely controlled to avoid oscillations in the density of the jet stream.

An APD Cryogenics Inc. closed cycle cryogenic system provides the refrigeration required to cool the hydrogen gas and the nozzle. This system includes an APD model DE-204SL cryocooler (commercially rated for 9 W at 20 K) and an APD helium compressor model HC-4. This system was chosen for its capacity, physical design and dimensions. The cryocooler includes two stages of cooling. In testing our system, we have found the cryocooler capable of cooling the nozzle to a temperature very near 10 K.

Gas Jet Design. The first stage of the cryocooler is used to precool the hydrogen gas on its path to the nozzle. The hydrogen flows through a 3.2 mm diameter copper tube which is wrapped and soldered around the first stage to achieve thermal coupling. Here, the hydrogen is cooled from room temperature to almost 50 K. The gas then continues through a stainless steel tube which attaches to the nozzle coupling (see Fig. 2). The nozzle coupling consists of a copper spool, a helically coiled stainless steel tube, copper cable and the "nozzle holder". The copper spool is bolted to the cryocooler second stage extension. An indium seal between the spool and the extension decreases the contact resistance between the two parts. The spool is bored to provide a flow path for the gas. The hydrogen circuit continues from the spool to the nozzle holder through the helically coiled tube. Twelve lengths of copper cable thermally connect the nozzle holder to the spool. The helically coiled tube and the copper cable are used in the nozzle coupling design to allow for angular and x-y positioning.

Although the cryocooler is mounted into the J1 vacuum chamber, it is contained within its own insulating vacuum shell. This vacuum shell is attached to the cryocooler stainless steel mounting flange and extends to the nozzle. A welded metal bellows connects the cylindrical portion of the cryocooler vacuum shell to the nozzle holder. This bellows allows the angular motion required of the vacuum shell for nozzle angular positioning. A 50 L/s turbomolecular pump is used to evacuate this insulating vacuum space. Without a vacuum shell, the cryocooler would have a heat load imposed on it due to conduction through the background gas existing in the J1 chamber. Liquid nitrogen cools the insulating vacuum shell near the bellows. This cooling acts to intercept heat in the form of conduction between the room temperature portion of the vacuum shell and the 27 K nozzle.

The nozzle is a replaceable component secured with a retainer. An indium seal is used between the nozzle and the nozzle holder in order to prevent hydrogen gas from leaking into the cryocooler insulating vacuum from either the hydrogen gas supply circuit or from the J1 chamber.

The second stage, its extension and a portion of the nozzle coupling are protected from thermal radiation with a nickel plated copper shield. This shield is attached to and thus cooled by the first stage of the cryocooler.

Cryocooler Heat Load. As stated above, the first stage of the cryocooler precools the hydrogen gas flowing to the nozzle and provides refrigeration to the radiation shield surrounding the second stage, its extension and much of the nozzle coupling. Approximately 1.5 W of refrigeration is required for these heat loads.

The second stage of the cryocooler receives heat from several sources. The radiation heat load is minimal as the components attached to the second stage are well shielded by the first stage cooled shield, the liquid nitrogen cooled cryocooler insulating vacuum shell and a third shield cooled by liquid nitrogen in chamber J1 located in the nozzle area (see Fig. 2). The conductive heat load is of more significance. The nozzle holder thermally communicates with the liquid nitrogen shielded portion of the cryocooler vacuum shell via the welded metal bellows. The 0.127 mm thick diaphragms of the stainless steel bellows provide a good resistance to conduction. Both of the nozzle angle positioner arms are attached to the nozzle holder. Each of these arms is intercepted with a liquid nitrogen temperature copper braid to minimize the heat added to the nozzle holder. The face of the nozzle holder exposed to the J1 chamber vacuum space has a heat load due to the presence of the J1 background gas. The pressure of the gas in this location is such that free molecular conduction occurs. Barron[2] discusses the energy transfer rate due to molecular conduction. This heat load is determined with Eq. (1):

$$\dot{Q} = GpA_1(T_2 - T_1) \tag{1}$$

It is seen that the energy rate is dependent on properties of the hydrogen gas (G), the pressure of the gas (p), surface area of the nozzle holder (A_1) and the temperature difference between the nozzle holder and the J1 chamber radiation shield in the nozzle area (T_2 -T_1). Lastly, the second stage of the cryocooler provides the cooling for the hydrogen gas supply to the nozzle. This occurs primarily inside the nozzle coupling spool. The summation of these heat loads equals approximately 2 W. Note that the available capacity of the APD model DE-204SL cryocooler provides the gas jet with a great deal of contingency.

Temperature and Pressure Control. It is critical to Experiment 835 that the hydrogen jet stream be a target of consistent density. For this reason, the state of the hydrogen gas supplied to the nozzle must be stable. Testing of the gas jet has shown that temperature control to within ± 0.1 K and pressure control to within ± 3500 Pa (0.5 psi) is required.

The nozzle temperature is sensed with a calibrated germanium resistance thermometer located on the nozzle holder. The sensor is positioned in a cavity whose dimensions create a close sliding fit with the sensor. A vacuum grease with good thermal conductance is used to minimize the contact resistance. A 16 bit temperature controller reads the nozzle temperature and provides an analog voltage output to a heater foil wrapped around the nozzle coupling spool. The temperature of the hydrogen gas entering the nozzle coupling spool is also controlled. The temperature here is measured with a platinum resistance thermometer. A temperature transmitter sends a conditioned signal to a multiple loop controller from the platinum resistor. The loop controller sends an output signal via a remote sensing power supply to heater foil wrapped on the tube leading from the first stage of the cryocooler to the coupling spool. Control of the gas temperature entering the coupling spool is necessary due to the temperature difference between the nozzle holder and the spool (about 2 K lower due to its 2 W heat load). The operating hydrogen pressure is very near the saturation pressure of the operating temperature. Thus, providing temperature control at the spool inlet avoids condensing the hydrogen as it passes through.

The multiple loop controller is used to control the hydrogen gas pressure as well. A high performance pressure transmitter provides the pressure reading to the controller. An electromagnetically operated flow control valve is then positioned by the controller in order to maintain the gas set pressure.

Hydrogen Gas Purification and Filtering. Before entering the refrigeration system, the hydrogen gas is purified. This is to avoid plating contaminants in the refrigerated hydrogen circuit and to prevent the small aperture of the nozzle from becoming partially or completely plugged.

Hydrogen is provided to the experiment from gas cylinders at a purity of 99.995%. Water vapor in the hydrogen is removed using a commercial water vapor purifier containing a 4A molecular sieve cartridge. The hydrogen next passes through a liquid nitrogen cooled activated carbon adsorber. The vacuum insulated adsorber contains about 2 kg of activated carbon (coconut) type PCB 4 x 8 inside a 10.2 cm tube. At both the inlet and outlet of the adsorber container, a combination of spun fiberglass, felt pad and screen are utilized to avoid carrying carbon dust throughout the hydrogen circuit. Given the small hydrogen flow rate and the relatively large amount of activated carbon, a purity better than 1 ppm is predicted. Pressure drop through the adsorber bed is insignificant. Additional filtering completely prevents particles from reaching the nozzle.

J1 Vacuum Chamber System Pumping Speed

The Experiment 760 J1 vacuum chamber was replaced with a larger chamber for Experiment 835 in order to maximize the evacuation of the J1 chamber background gas. As previously mentioned, two turbomolecular pumps are mounted on this chamber (see Fig. 1).

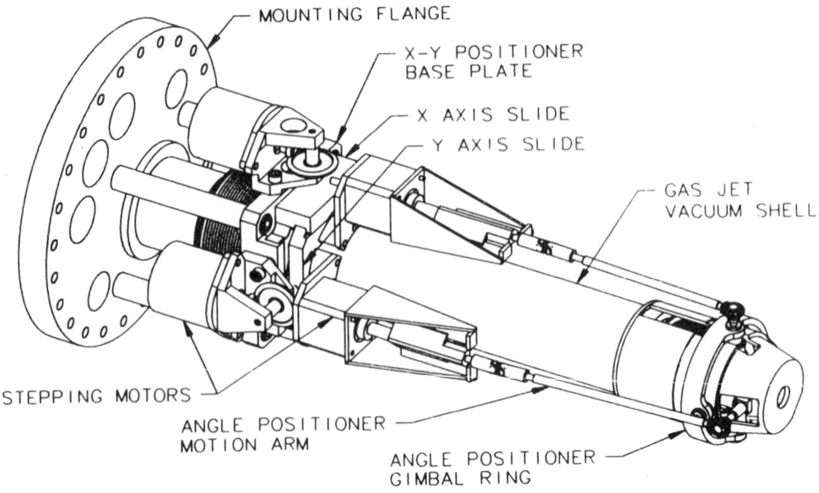

Figure 3. Components of the Nozzle X-Y and Angular Positioners .

Effective removal of the these hydrogen molecules from the space between the nozzle and the first skimmer minimizes their interaction with the jet stream. The new J1 chamber was designed to maintain a pressure below 1 Pa in the nozzle area. J1 system pumping speed measurements indicate an actual rate of about 1450 L/s.

Nozzle X-Y and Angular Positioning

Alignment of the jet stream through the skimmers to the interaction area inside the Antiproton Accumulator beam-pipe is accomplished with x-y and angular positioning of the nozzle (see Fig. 3). The positioners assure that the jet stream (along the z axis) interaction with the antiproton beam is maximized. The x-y positioner consists of a base supported by the cryocooler mounting flange, an x-axis slide and a y-axis slide. The x-axis and y-axis slides are each driven by a stepping motor. The x-y positioner range of motion is 3 mm in any direction from the gas jet centerline. Motion sensitivity is better than 10 μm. The angle positioner is supported from the y-axis slide of the x-y positioner. With its gimbal-like design, it allows nozzle positioning up to a 15 degree angle from the gas jet centerline in any direction. Two stepping motors are used to drive the apparatus. Arms extend from the motors to the gimbal rings which connect to the nozzle holder. Angle sensitivity is better than 0.05 degrees.

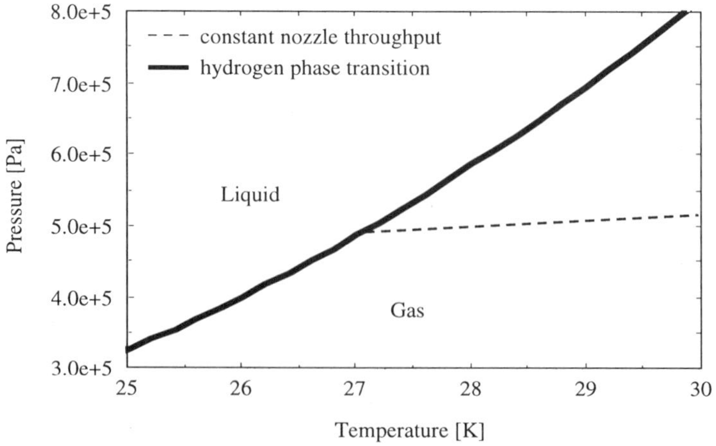

Figure 4. Hydrogen phase transition curve and constant nozzle throughput at 8×10^{20} atoms/s.

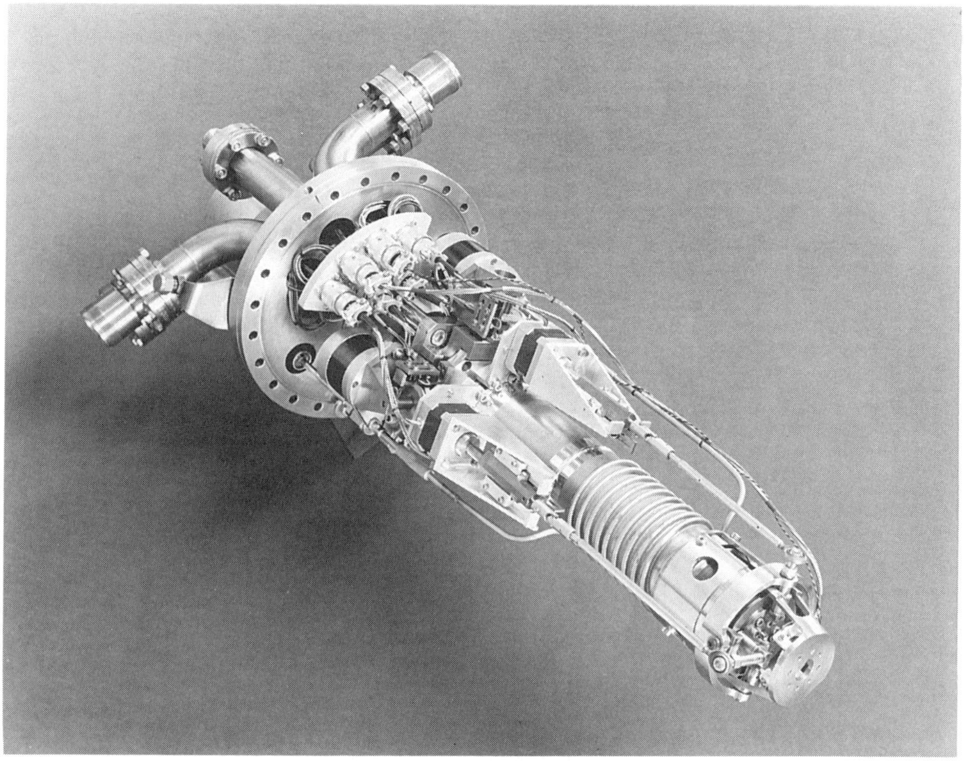

Figure 5. Experiment 835 Gas Jet.

TEST RESULTS

Testing has been performed to evaluate the effectiveness of the upgraded gas jet on its ability to produce the specified increase in jet stream density.[1] To determine this increase the hydrogen supply operating temperature and pressure are optimized to avoid a nozzle throughput which is too high. Optimizing the nozzle throughput maintains the J1 pressure below 1 Pa and thus limits the amount of gas which diffuses through the skimmers to the Antiproton Accumulator beam-pipe to an acceptable level. Given the J1 chamber system pumping speed and nozzle characteristics, the nozzle throughput is limited to 8×10^{20} atoms/s for any hydrogen gas supply temperature to the nozzle. This value of throughput is plotted for a nozzle with a 37 μm aperture against supply temperature and pressure in Fig. 4. Also shown is the liquid-gas phase transition curve for hydrogen. Our operating point must be chosen below the constant nozzle throughput curve and to the right of the hydrogen phase transition curve. Condensation inside the hydrogen supply circuit will cause operational instability.

The average density (atoms/cm^3) of the jet stream is given in Eq. 2:

$$\bar{\rho} = \frac{\phi}{A\nu} \tag{2}$$

where ϕ is the jet stream throughput (atoms/s), A is the jet cross sectional area in the interaction area and ν is the average speed of the jet stream over its cross sectional area. To increase the jet density, one must maximize the jet stream throughput and minimize the jet stream speed. The jet stream speed is dependent on the gas temperature supplied to the nozzle. This speed has been determined in our tests from 90 K to 16 K by performing "time of flight" measurements of the jet stream using a variable speed beam chopper. It was found that the jet stream speed was reduced from about 1300 m/s to roughly 300 m/s due to the lowered operating temperature. At 27 K the jet stream speed was measured at 600 m/s.

Coupled with the increased throughput of the upgraded gas jet, the available increase in the average jet stream density exceeds five times that obtainable in the previous gas jet system. The upgraded gas jet is pictured in Figure 5.

CONCLUSIONS

The performance of the gas jet to be used in Experiment 835 in its ability to produce an increased jet stream density surpasses the experiment specifications. The refrigeration system and its control, a J1 vacuum chamber with improved pumping speed and nozzle positioners have all been successfully implemented into the jet target system. In addition, the upgraded system offers the possibility of maintaining a constant interaction rate between the antiproton beam and the jet stream. As the intensity of the antiproton beam decreases over its 30 hour lifetime during which data is taken, the density of the jet stream may be gradually increased by properly raising the hydrogen supply pressure in order to achieve this constant rate. A constant rate of data production creates efficient use of the data detection system.

ACKNOWLEDGMENTS

This work is sponsored by the University Research Association, under contract with the U.S. Department of Energy and the Italian National Institute of Nuclear Physics (INFN) in Genoa, Italy. The authors gratefully acknowledge the technical support received from J. Barilla, R. Davis, D. Friend, M. McKenna, D. Miller, J. Peifer, S. Pordes, J. Rauch, J. Seeman and E. Villegas of Fermilab, and E. Bozzo, E. Cavanna, F. Conforti, G. Franzone, A. Manco, P. Pollovio and P. Pozzo of the Genoa INFN.

REFERENCES

1. G. Boero et al., "The E760 Jet Target: Measurements of Performance at 77 K," Fermilab-TM-1915 (1994).
2. R.F. Barron, Cryogenic-fluid storage and transfer systems, in: "Cryogenic Systems," Oxford University Press, New York (1985), p. 356.

INITIAL PERFORMANCE OF UPGRADED TEVATRON CRYOGENIC SYSTEMS

B. L. Norris

Fermi National Accelerator Laboratory*
Batavia, Illinois 60510

ABSTRACT

Fermilab began operating a re-designed satellite refrigerator system in November 1993. Upgrades were installed to operate the Tevatron at a magnet temperature of 3.5K, approximately 1K lower than the original design. Refrigerator upgrades included new valve boxes, larger reciprocating expanders, the installation of cold vapor compressors, new sub atmospheric instrumentation and an entirely new distributed controls system.

Cryogenic system reliability data for Colliding Physics Run 1B is presented emphasizing a failure analysis for each aspect of the upgrade. Comparison to data for Colliding Physics Run 1A (previous to the upgrade) is presented to show the impact of a major system overhaul. New operational problems and their solutions are presented in detail.

INTRODUCTION

After 10 years of successful operation, from 1983 to 1993, the Tevatron Cryogenic Satellite Refrigeration System was re-designed to lower the operating temperature of the Tevatron magnet system down to a two-phase temperature of 3.56K from the original 4.45K temperature. The Low Temperature Upgrade, as this work is referred too, is required to increase the operating colliding beam physics energy of the Tevatron to 1000 GeV from the nominal 900 GeV energy. Modifications to the satellite refrigeration system included the installation of new valve boxes, installation of cold compressors, hardening of the two-phase circuit to make it leak tight for subatmospheric operation, conversion of wet expanders to 7.5 cm piston machines, installation of 5.6 kW AC variable frequency drive for wet expander continuous braking, and installation of a new distributed control system. These modifications have been discussed in detail elsewhere[1] and will not be discussed here. A flow diagram of a satellite refrigerator is shown in Figure 1.

To maintain reliable accelerator operations, it is imperative that the satellite refrigeration system downtime be minimized. This paper attempts to compare the statistics of the last two colliding beam physics runs: a before and after look at reliability. The statistics shown are a comment on how well the Low Temperature Upgrade was designed, installed and operated.

* Work supported by the U. S. Department of Energy under contract No. DE-AC02-76CH03000.

LOW TEMPERATURE/HIGHER ENERGY STATUS

After the installation of the Low Temperature Upgrade equipment in the Summer and Fall of 1993, a series of power tests were performed to get to 1000 GeV operation.[2] The conclusion of these tests proved we could achieve 975 GeV operation with a cold compressor inlet temperature of 3.84K. This energy was attained in a DC condition (a "STORE") for over an hour. Attempts to go higher in energy resulted in a quench.

Before achieving this one hour "STORE" condition, we had attempted to ramp the machine to 1000 GeV and found that we experienced a significant number of training quenches. The highest energy we achieved at any time during these studies was 997 GeV (this was on ramp up; flattop was never achieved).

Because of the uncertainty in being able to easily get to 1000 GeV and because of a desire to proceed with Physics Run 1B, laboratory management decided to continue with accelerator physics at 900 GeV. Thus cold compressor and low temperature operation have not been of a "critical" nature to this Run. We have, however, operated cold compressors as we could. All of this means that the statistics for cold compressor reliability and impact to operations are not yet well established. Although we intend to perform more studies to understand how to attain 1000 GeV operation, Low Temperature operation will not become an everyday reality until Collider Run 2 presently scheduled for 1998-9.

RELIABILITY DATA BEFORE AND AFTER UPGRADES

A real concern that we recognized throughout the design process for all aspects of the Low Temperature Upgrade was continued high reliability. Our goal was to better our downtime numbers where possible.

The Accelerator Division Operation Department keeps statistics on all aspects of accelerator downtime. This data has been used as the basis for the statistics shown in this

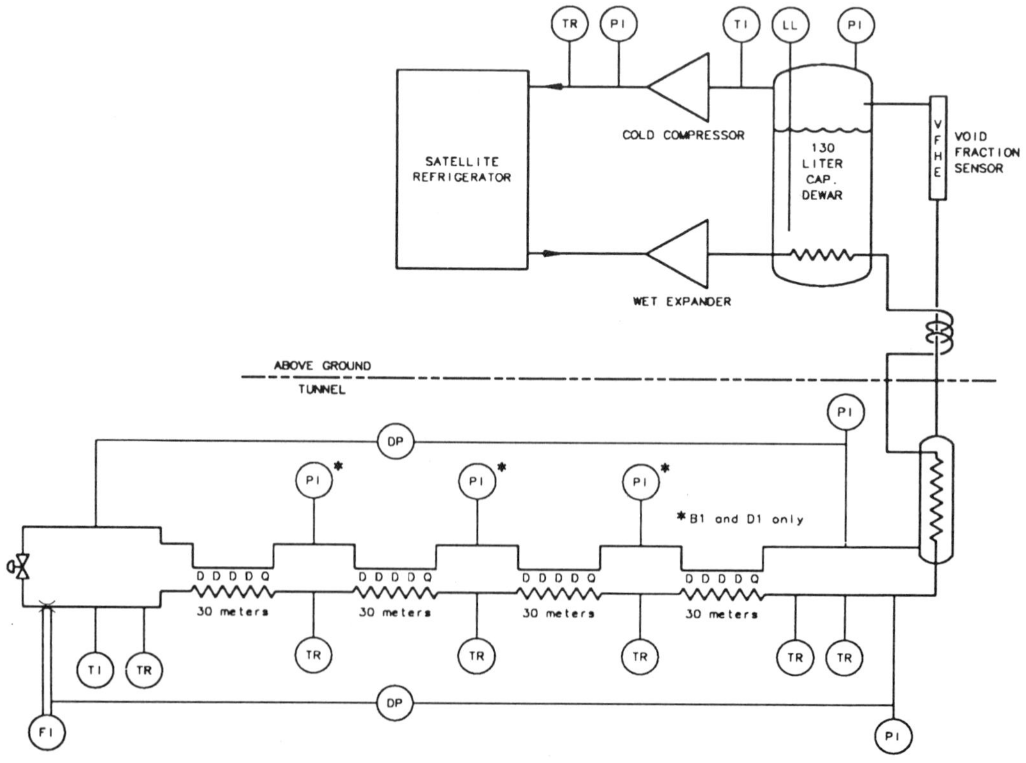

Figure 1. Flow diagram showing satellite refrigerator and one half of magnets (D = dipole magnet, Q = quadrupole magnet).

Table 1. Comparison of Reliability Data for Run 1A and Run 1B

	Run 1A June 15, 1992-May 31, 1993		Run 1B Dec. 15, 1993 - June 11, 1995	
Category of Failure	% Of Cryogenic Downtime	Total # Of Hours Down	% Of Cryogenic Downtime	Total # Of Hours Down
Expanders (Mechanical)	18.80%	15.1	50.6%	61.1
Expanders (Electrical)	10.70%	8.6	3.1%	3.7
Kautzky Reliefs	35.70%	28.9	11.1%	13.4
Distributed Controls	6.90%	5.6	4.5%	5.4
Cryo Control/Electronic	1%	0.8	5.5%	6.7
Cold Compressors	NA	NA	0.4%	0.6
Instabilities	4.3%	3.4	18.8%	22.7
Power Lead	22.6%	18.3	6.0%	7.3
Total Cryo Downtime	NA	80.7	NA	120.9
Total Calendar Time	NA	6720.0	NA	13,032.0
TeV Unavailable		336.0		1632.0
TeV Available		6384.0		11,400.0
Cryo Downtime Total %		1.2%		1.0%

paper. To better understand the impact of the Low Temperature Upgrade on cryogenic reliability, data in Table 1 shows the present run, Run 1B, and the last pre-Low Temperature Upgrade run,Run 1A. Table 1 is broken down into major categories of cryogenic failures. These categories have been chosen based on experience of typical Tevatron failures over the years. Each failure category will be discussed individually in the paper.

Table 1 includes the amount of calendar time and Tevatron availability time. The Tevatron availability time during Run 1B is about twice that of Run 1A. (Note that problems such as power outages or failed Tevatron magnets are accounted for in the Tevatron Unavailable Category.) However the total cryogenic downtime for Run 1B has been the same amount as the previous pre-Upgrade Run, about 1.0-1.2%. This translates into averaging approximately two hours of satellite downtime per week. As shown in Figure 2, both Runs began with significantly higher Average Weekly Downtime but converged to the two hour level over a number of weeks. After the installation of the Low Temperature Upgrade equipment, it has taken longer than the previous Run to converge to the two hour point. This is explained by two issues: a learning curve needed for operators on the new equipment and a much higher than normal failure rate with our wet expanders (see section on expansion engine failures).

The fact that the last two Runs have the same general weekly downtime average is quite encouraging and illustrates that a major Tevatron system upgrade can be successfully implemented without adding to the overall downtime of the accelerator. Of course, only after Low Temperature Operation becomes an everyday reality will we truly be able to determine how well we have done in terms of reliability.

WET EXPANDERS

The refrigerator cycle for the Upgrade was studied in detail and a special Fermi designed Macintosh simulator was used to understand the best cycle we should create.[3] This work led to the wet expander exhausting into a subcooled dewar of liquid rather than exhausting directly into the single phase of the magnets. With the cold compressor in use, the expander now operates with an inlet temperature of 7-8K where in the past the expander operated with an inlet <5K. The supply-side density change forced an increase from 5 cm to 7.5 cm diameter pistons. This increase translated into a need to have a motorized brake sized for 5.6 kW (up from the old 1.5 kW). To accomplish this we purchased AC variable frequency drives from Mitsubishi that included a regenerative unit for converting the energy of the expander into 60 Hz electrical energy and placed back onto the power grid.

Mechanical Statistics for Expanders

At the start of Run 1B, we experienced an abnormally high number of mechanical failures with the expanders.[4] There were five failures that caused downtime within the first two months of the run. These failures were caused by cyclic metal fatigue of the main crankshaft. An independent study showed we were experiencing fretting of these crankshafts. The failures were brought on by upgrading to the larger piston machine. The same outside study suggested using a higher strength steel and surface treating the shafts, either ion-nitriding or a special chroming. Since doing so we have experienced only one other broken crankshaft and this shaft had been treated with an inferior surface treatment soon realized to be inadequate.

Besides this rash of crankshaft failures, our expander downtime has increased in general. This increase seems to be a result of the larger piston diameter which places more force and stress on bearings, shafts, and frames overall. All of these items are warm-end failures. Cold-end failures such as o-ring blowby, piston, connecting rod, or internal leak problems have remained about the same. Figure 3 shows the weekly downtime average for this category.

Electrical Statistics for Expanders

The motor and motor drive system used on the wet expander are used as a continuous brake. A PID closed loop tries to maintain the proper speed for pressurizing the magnet's single phase pressure to 0.22 MPa. This loop outputs a speed setting to the Mitsubishi drive which then "brakes" the expander to the correct speed. This speed is adjusted, if necessary, every two seconds. The motor is allowed to make maximum speed changes each sample time of 100 rpm.

Since we began using the new drives in November, 1993, we have accumulated approximately 329,000 hours of time on the motors and drives. As shown in Table 1, this category of downtime amounts to 3.1% of our total cryo downtime and is a significant improvement from our previous motor drive systems in Run 1A. We have had one catastrophic motor failure and had to replace one Mitsubishi Drive system after a lightening storm appeared to have damaged some open-collector transistors used for status. There have been no failures of an in-house built interface box used to connect the Mitsubishi drive to the new distributed control package.

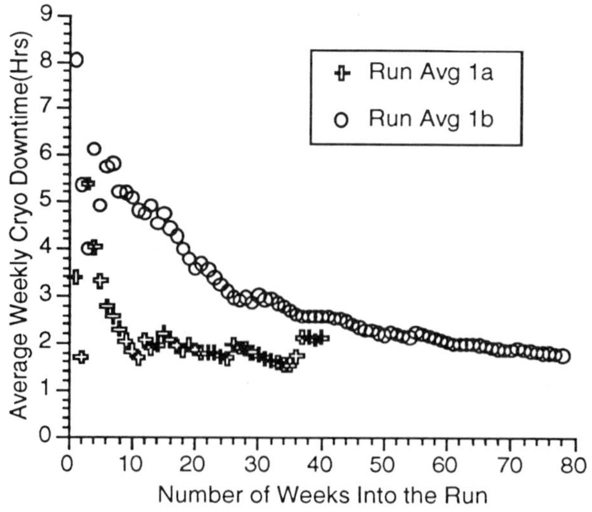

Figure 2. Comparison of average weekly downtime for Run 1A and Run 1B.

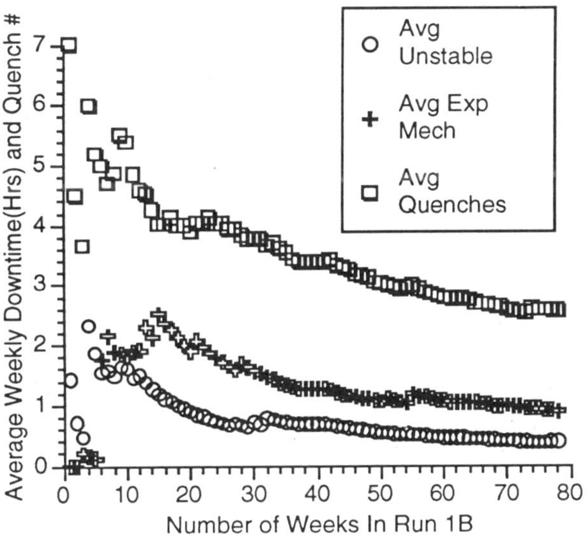

Figure 3. Average weekly statistics for expanders (mech.), instabilities, and quenches.

KAUTZKY RELIEFS

The Tevatron uses a bellows operated relief valve, known as a Kautzky valve, on the single-phase, two-phase, and nitrogen circuits of the Tevatron magnet system. All circuits are relieved at 0.3 MPa. During the low temperature upgrade, the two-phase circuit of the system had to be made leak-tight, including the Kautzky valves. These valves now use a low pressure helium guard and a double o-ring configuration. A total of 250 valves were modified. We did not alter reliefs on the single-phase or nitrogen circuits.

These valves have a history of high failure rate. These failures generally occur in two ways. The first way is that something (a physical object like a loose screw) gets lodged in the valve and forces it to stick open. If the object cannot be dislodged the valve must be changed out at 20K or greater temperature conditions. The second common failure is that the bellows tears typically at its first convolution, and control pressure gas used to set the relieving pressure, cannot seal the valve. This problem is caused by repeated cycles of the bellows.[5]

Nothing that we did in the upgrade would explain the statistical differences seen between the two Runs. We did not change the bellows nor enhance the performance of these reliefs in anyway. Even more, the failures are normally single-phase Kautzkys as these experience the very high cyclical stress placed on them during quench episodes (see Figure 3 for average number of quenches). It is believed that we simply do "accelerator maintenance" on reliefs in a better way. That is, when the accelerator is down for scheduled maintenance we actively look for reliefs that may have a damaged bellows and replace as necessary. Doing this means we remove damaged reliefs before they catastrophically fail and cause accelerator downtime. It also means we have the same number of failures as in Run 1A, we just aren't causing unscheduled downtime because of them.

DISTRIBUTED CONTROLS

To reliably control all the old and new cryogenic equipment used in the Tevatron, it was necessary to upgrade our distributed control system. Fermilab engineers designed and installed a new system based on a Multibus II platform using Intel 80386 and 186 processors. Token ring is used as the link between 6 primary crates while Arcnet is used as a LAN between the primary crate and individual I/O crates located at each refrigerator or compressor building (there are 32 such buildings in the Tevatron). This control package has been detailed by Norris.[6,7]

In terms of actual downtime, the new distributed control system has been a success. In fact, it wasn't until the week of March 13, 1995 (the 66th week of the Run) that this control system accounted for any downtime.

The downtime listed in Table 1 is three separate events. Two of these events were related to a loss of communication from Token Ring to the 386 processors of the refrigerators. In both situations, the cryogenic system continued to operate properly because the communication from 386 to the local I/O crate was not disturbed. Only when rectifying the problem by rebooting the master crate and thus the Token Ring board did we encounter problems. Both cases were situations where software bugs were discovered on boot-up. These bugs have since been fixed. The third downtime event has been charged to our control system but may, in reality, not be a control system failure. During the week of April 17, 1995 we had a magnet JT valve suddenly step closed. This valve controls the flow through the magnet string of half of the house. This resulted in a quench. All other possibilities were pursued but an exact conclusion was never reached. This has been charged to the controls category simply because it is the most probable mechanism for such a valve action. Software has been upgraded to prevent such an occurrence in the future. This event accounts for over 71% of the distributed controls downtime. If you assume there was not a controls failure in this case the distributed control system downtime would be reduced to a total of 1.5 hours.

CRYO CONTROL/ELECTRONIC

The category of Cryo control/electronic was created to encompass all the electronics that is not a part of the distributed control system but is required to operate these systems. These devices include all types of cryogenic transducers, electric motor actuators with LVDT readbacks for power lead solenoids used to manipulate flow, plus many other monitoring and control devices.

As can be seen in Table 1 the total number of downtime hours is significantly higher in Run 1B. However this is only two downtime events. In the first case a tunnel valve had a damaged position readback and led to inaccurate valve positioning and finally a quench. This problem arose after work had been done in the same area as the valve so it is believed this work led to the problem. The second problem led to more serious work by the engineering staff.

A key element in the operation of a satellite is to maintain proper dewar level. This dewar is located between the refrigerator and the magnet single-phase and two-phase circuits. It acts to buffer the load from refrigerator upsets and the level of this dewar is an indicator of how well the refrigerator is performing overall. Liquid level is measured in 2 different ways, with a differential pressure measurement (DP) across the dewar and with a superconducting liquid level probe (LL). In the case of the downtime event, we were using the DP as a level measurement. Based on its value a control valve opens and closes adding liquid helium from the Tevatron's transfer line system. On this occasion the DP cell began to oscillate wildly causing growing instabilities in the dewar and finally a quench of the magnet strings. It was recognized early that this it is imperative to have adequate level sensing for refrigerator stability and protection of the IHI cold compressors. Ingestion of liquid droplets will at a minimum cause the cold compressor to trip on overload and potentially damage the turbo machine.

A lot of time was spent understanding our superconducting LL probe behavior. In the case above, we had been using the DP cells because we did not trust the signal the probes were giving out. After weeks of studies it was determined that the probes we were using were becoming noisy whenever a fill valve would spay liquid into the dewar. It was realized that the LL probe, which has holes along it's G-10 structure, could be rotated in a direction away from the spray to help it's performance. All of this work led to us putting a sheath around the probe and establishing small pin holes in the sheath. This simple idea has proven very reliable and has led us to sheath all 24 probes that we use. There has been no more downtime related to this control issue.

COLD COMPRESSORS

In January of 1994 the lab decided to continue Physics at 900 GeV and not pursue higher energies until later. That decision, plus a decision to limit the amount of compressors that were to be operated at the Central Helium Liquefier for budgetary reasons, limited how many cold compressors we could operate and at what speed. Therefore, the downtime category for the cold compressors is really not a true representation of what downtime rate we might have had we run at lower temperatures.

However it must be noted that we have put many hours on these machines. Beginning in January, 1994 we operated 3 to 5 cold compressors at all times to understand their behavior in quench conditions. During earlier power testing we discovered the cold compressors would sometimes trip when their exhaust side pressure was quickly elevated, as happens in a Tevatron quench. We found this could be controlled by changing the rate of which the IHI was allowed to change speeds. This testing lasted until August 1994.

In August 1994, the lab experienced a delivery problem with liquid nitrogen and was forced to go into an extended maintenance period. At this time we decided to install new journal bearings from IHI that could be operated at lower speeds (20,000 rpm instead of 40,000) that would cause less of a perturbation to the refrigerator on startup. From early September until April of 1995, we accrued >93,000 hours on these new bearing machines (the original bearings had accrued >27,000 hours). However during that period we experienced 3 machine failures, all experiencing bearing problems on startup. At this point these failures are not completely understood and Fermilab and IHI are working to resolve this issue.

INSTABILITIES

The instabilities' category is used to account for any downtime that isn't necessarily an equipment failure. This includes items such as unstable refrigerators, low warm compressor discharge, plugged cryogenic instrumentation used to measure magnet pressures, or human error situations.

In reviewing the statistics for this category it is apparent we suffer mostly from two failure conditions, both cold instrumentation. The first failure is a plugging of the capillary tubing used for sensing the single-phase pressure at the feedcan in the tunnel. This is the process variable the wet expander tries to maintain. Any partial plugging of this line and the refrigerator becomes unstable. The second instrumentation failure is similar to the first in that we have plugged instrumentation. In the Tevatron we use a differential pressure gauge to compare two-phase pressure and two-phase temperature (using a vapor pressure bulb). This reading is an indicator of the amount of subcooling in the magnets two-phase. If the pressure measurement tubing plugs we get inaccurate readings that cause magnet JT valves to be positioned inaccurately.

These two failures dominate this category of downtime and reinforce our need to have a clean helium system. During each period of scheduled long term maintenance we warm-up to 80K and purify the system to <5 ppm Nitrogen in Helium. See Figure 3 for the average weekly downtime of this category.

POWER LEADS

This category of failure has before this run been a constant problem. Our 5000 amp power leads require cooling taken from the single phase of our magnet strings. This flow can be selected to be three different flow rates depending on the energy of the machine. Solenoids are used to control pre-defined flow orifices and give the 3 ranges of flow. In Run 1A we spent a lot of time adjusting flows for various houses. This led to a significant amount of downtime. During Run 1B, we have had minimal downtime due to these lead flows. In fact, all of this downtime is related to special low beta quadrupole leads at the B0 and D0 interaction areas. There have been six total downtime entries.

FUTURE WORK

In late July 1995 we plan to continue low temperature testing with five days dedicated to understanding higher energy operation. This second test period should give us enough data to understand if and how we can achieve 1000 GeV energy levels. Until we accomplish this goal, engineering time will continue to be spent on perfecting the equipment we installed for this upgrade. Note, Run 1B is scheduled to continue until February 15, 1996.

CONCLUSION

We have shown during the last year and a half that a major cryogenic upgrade could be undertaken on the Tevatron without significantly altering the average downtime of the machine. Although lower temperature operation on a continuous basis is not yet a reality, this work and the reliability it has accomplished is encouraging to us and should offer others in the field encouragement about their success chances when the time comes for them to upgrade major cryogenic facilities.

ACKNOWLEDGEMENTS

The author wishes to thank the personnel of the Cryogenic Systems Group for their dedicated efforts during the design, installation, and operating stages of this project. Specifically, J. Theilacker, A. Martinez, J. Fuerst, W. Soyars, J. Brubaker, and J. Panek (now a student at Florida State) need to be commended for their dedication to this upgrade.

REFERENCES

1. J.C. Theilacker, Upgrade of the tevatron cryogenic system, in: "Advances in Cryogenic Engineering," Vol. 39, Plenum Press, New York (1994), p. 517.
2. J.D. Fuerst and J.C. Theilacker, increasing the energy of the fermilab tevatron accelerator: presented at the EPAC94, London, England.
3. W.M. Soyars, Simulating the tevatron liquid helium satellite refrigerators, in: "Advances in Cryogenic Engineering," Vol. 39, Plenum Press, New York (1994), p. 1231.
4. W.M. Soyars, Operation and maintenance of fermilab's satellite refrigerator expansion engines, to be presented this conference.
5. J.C. Theilacker, Tevatron quench pressure measurement, in: "Advances in Cryogenic Engineering," Vol. 39, Plenum Press, New York (1994), p. 469.
6. B.L. Norris, New cryogenic controls for the tevatron low temperature upgrade, in: "Advances in Cryogenic Engineering," Vol. 39, Plenum Press, New York (1994), p. 1185.
7. B.L. Norris, Control system for fermilab's low temperature upgrade, presented at the Particle Accelerator Conference, May 1995, Dallax TX.

REQUIREMENTS AND INTERFACES TO CRYOGENIC AND POWER SUPPLY PLANTS FOR THE ITER MAGNET SYSTEM

K. Yoshida[1], V. Kalinin[1], S. Stoner[1], T. Kato[2]

[1]ITER Joint Central Team, Naka Joint Work Site
Naka-machi, Naka-gun, Ibaraki-ken, 311-01 Japan
[2]Japan Atomic Energy Research Institute
Naka-machi, Naka-gun, Ibaraki-ken, 311-01 Japan

ABSTRACT

This paper describes the requirements and interfaces of cryogenic and power supply plants for the ITER magnet system. The heat loads and mass flow rates of the magnet system is evaluated to keep coil temperature within 6.5 K during the plasma operation of 2200 s. The helium consumption in the current leads is estimated for a current feed of 4.5 MA from 4 K to 300 K. The location of the electrical insulation breaks is compared near winding and outside of cryostat. The terminal box is designed for interfacing the cryogenic and power supply systems. Separate vacuums for the cryostat and the cryolines facilitate maintenance of all cryogenic components without warm-up of the whole system. A superconducting busbar is used to supply current from a current connector at the coil to the vapor-cooled lead in the coil terminal box. The layout of cryogenic lines and equipment is designed to consider space limitations and routing restrictions in the tokamak hall and the electrical termination building. The main parameters of the cryogenic plant (cryoplant) are discussed.

INTRODUCTION

The European Atomic Energy Community (Euratom), and the Governments of Japan, the Russian Federation and the United States of America have agreed to cooperate on the Engineering Design Activities (EDA) of the International Thermonuclear Experimental Reactor (ITER). The superconducting magnet system[1] for ITER consists of 20 toroidal field (TF) coils, a central solenoid (CS) coil and 7 poloidal field (PF) coils[2]. The magnet system requires a cryoplant[3] with a heat load removal capacity of about 75 kW at 4.5 K and liquid helium supply of about 6000 l/h.

A closed loop forced flow cooling system with liquid helium mechanical pumps is shown in Figure 1 (a). Cooling with the cryo-pump and heat exchanger to a reservoir is more practical than using direct flow from the helium refrigerator / liquefier (R/L) because the required cold mass flow rate is much higher than the mass flow rate supplied from room temperature compressor. The separation of fluids between the cryoplant and the magnet system cooling circuits is a safety feature to protect the cryoplant from tritium contamination in case of accident.

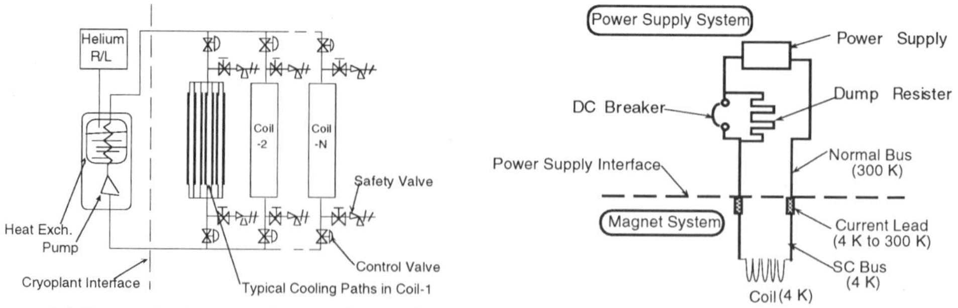

(a) Connection between the cryoplant and superconducting coils

(b) Connection between the power supply system and superconducting coils

Figure 1 Interfaces to cryoplant and power supply systems

Long superconducting busbar connections from the coils to the power supply buses extend from the tokamak hall to the electrical termination building. The current leads are located in the electrical terminal building next to the tokamak hall. Conventional vapor-cooled current leads with liquid helium baths are used to feed current from room temperature (300 K) to 4 K. The top of the vapor-cooled current leads is the interface between the power supply system and the magnet system as shown in Figure 1(b).

REQUIREMENTS OF THE MAGNETS

The helium outlet (return) temperature is a function of the variation in heat load for a given plasma cycle. The helium outlet temperatures and the heat loads generated in the coils have been evaluated for a 2200 s cycle with a 1000 s plasma burn phase. The helium mass flow rates are set such that the maximum temperature increment in the helium forced flows does not exceed 2.0 K. The helium mass flow rates and the number and length of cooling channels for the TF and PF coils are summarized in Table 1. The pumping loss calculation assumes a 0.02 MPa pressure drop in the pipes and a pump efficiency of 0.7.

The heat loads associated with the coils are summarized in Table 2. The heat loads are calculated assuming the following conditions:

1) AC loss and nuclear loads: AC losses in the conductors and structure are included. Nuclear heating of all TF coils is 17 kW during burn (1,000 s).
2) Radiation: The thermal radiation from the 80 K thermal shield is estimated using an emissivity of 0.2 at cold and warm surface condition.
3) Gravity support to room temperature: The thermal conduction from the 80 K thermal anchor is estimated from 1 m long stainless steel supports with an area of 10 m^2
4) Friction load: The frictional heat generation in the coil support system is estimated at 3.5 MJ per cycle.
5) Refrigerator efficiency: As indicated by preliminary estimates, it is assumed that the periodic pressure oscillation in the helium baths of the heat exchanger will not be high enough to significantly influence the efficiency of the refrigerator. This will be verified by detailed thermo-hydraulic analysis.

Table 1 Mass flow rates and pumping loads

Components of magnet system	Number of cooling paths	Mass flow rate (kg/s)	Pressure drop (MPa)	Pumping loss (kW)	Length of cooling paths (m)
TF winding	240	3.1	0.150	6.0	566 - 418
TF Case	20	13.1	0.020	6.0	45
Central Solenoid	56	1.4	0.250	4.4	690-1020
Inner PF coils (Nb_3Sn)	48	0.7	0.150	1.4	980
Outer PF coils (Nb-Ti)	234	4.8	0.075	5.2	470-510
Inner Structure	20	10.0	0.020	4.6	45-90
Outer Structure	20	5.0	0.020	2.3	45-90
Total	638	38.1		29.8	

Table 2 Average heat loads (kW) removed by the refrigerator during repeated 2200 s cycles

Heat Loads (kW)	TF Coils	CS Coil	Outer PF Coils	TF Structure	CS Structure
Scenario A.C. loss	2.4	4.2	4.4	4.5	1.8
Plasma control A.C. loss	1.5	0.2	2.8	2.5	-
Joints*	1.5	0.3	0.7	-	-
Mechanical friction			1.6 (load to TF)		
Nuclear heating	7.7	0.0	0.2	0.5	-
Liquid He pumps*	6.0	4.4	6.5	8.3	4.6
Thermal shields*			6.5 (load to TF)		
Gravity supports*			2.0 (load to TF)		
Static loads*	16.0	4.7	7.2	8.3	4.6
Pulse loads (averaged)	13.1	4.4	7.4	7.5	1.8
Sum of loads	29.1	9.1	14.6	15.8	6.4
Total			75.0		

* Static loads

The vapor-cooled current leads provide the electrical connection between the superconducting busbar at 4K and the power supply busbar at room temperature. There are 45 pairs of current leads. The number of terminals in each coil is set by the terminal voltage limitation and the number of modules. The capacity of the current leads are 60 and 43 kA for the TF and PF coils, respectively. The load factor of a PF coil current leads are assumed to be 0.6 - 0.01 depending on the current profile required for the operating scenario. A current lead needs liquid helium of 1.69 l/h at 1 kA and of 1.01 l/h at zero current. The liquid helium consumption associated with operation in normal phase with currents and in stand-by phase without current are 6600 l/h and 4600 l/h, respectively.

INTERFACES

All coil components located inside the cryostat must be maintained remotely. Hands-on maintenance is possible outside of the cryostat. The optimum location for several components of the coil has not been determined, but will be based on assessment of the risk of voltage break down and helium leakage versus maintainability. A schematic drawing showing all interfacing components between a coil, the power supply system and the cryoplant system is shown in Figure 2.

Connection to the Cryoplant

Location of the insulation breaks. The inlet and outlet pipes connected to the coil electrical joints will be at the electrical potential of the joint. Since the pipes must ultimately be connected to the cryoplant at ground potential, it is necessary for each coolant line to have an electrical break or insulating section. The lines on the high potential side of the break must be insulated for ground voltage and the lines on the other side of the break will be at ground potential. The electrical insulation break in each cooling line is a possible point for helium leakage. Approximately 600 insulation breaks will be used in the magnet system.

A study is being done to determine the optimum location for the insulation breaks. As they may fail during a lifetime, it must be possible to replace the breaks. The insulation breaks may be located outside of the cryostat (far from the coils) or near the coils. A location outside of the cryostat requires many long, small, insulated pipes penetrating the cryostat. Replacement of the breaks in this case can be done 'hands-on'. A location near the coil means the high potential lines are short with a manifold on the coil side of the breaks. Two large-diameter grounded lines per coil module exit the manifold to penetrate the cryostat. In the latter case, the breaks must be replaced remotely.

Terminal box. Each coil terminal box containers the cryogenic components which connect a coil to the cryoplant and the power supply system, as shown in Figure 2. The connections to the cryoplant, power supply and utilities are:

1) Inlet and outlet of supercritical helium (4.5 - 6.5 K, 0.6 MPa)

2) Liquid helium supply and room temperature return for current leads
3) Inlet and outlet of thermal shield gas (80 K, 1 MPa)
4) Vent line for coil quench
5) Vacuum line for terminal box vacuum
6) Two current lead terminal connections to power supply
7) Instrumentation

Each coil terminal box includes two current leads, cold and warm control valves, relief valves for coil quench, insulation breaks for current lead, and instrumentation sensors. A total of 45 coil terminal boxes are installed in the electrical terminal building next to the tokamak hall. The detailed layout of the cryolines and the normal busbar will be performed with the layout of the power supply and the cryoplant.

Cryostat Feedthough. The terminal box components (valves, current leads and many instrument sensors) may require maintenance. It is possible to repair them without warming up the entire magnet system because they are located in a terminal box with a separate vacuum. Separation between the coil terminal box and the cryostat is accomplished using a feedthrough as shown in Figure 3. The material of the feedthrough is fiber reinforced plastic (FRP). The current bus with ground insulation and the pipes can pass through the feedthrough.

The cooling pipes and bus bars shrink during cool-down by the thermal contraction of 0.3 % of length. The bellows in supercondiucting (SC) busbar is difficult to design to prevent thermal expansion because SC busbar of cable-in-conduit conductor has no expansion flexibility in length The U-shape bending at the outside of the cryostat is designed to release thermal stress inside of the cryostat. The U-shape bending or the spring loop are applied for a long pipe line (about 100 m) from the cryostat to the terminal box at the electrical terminal building.

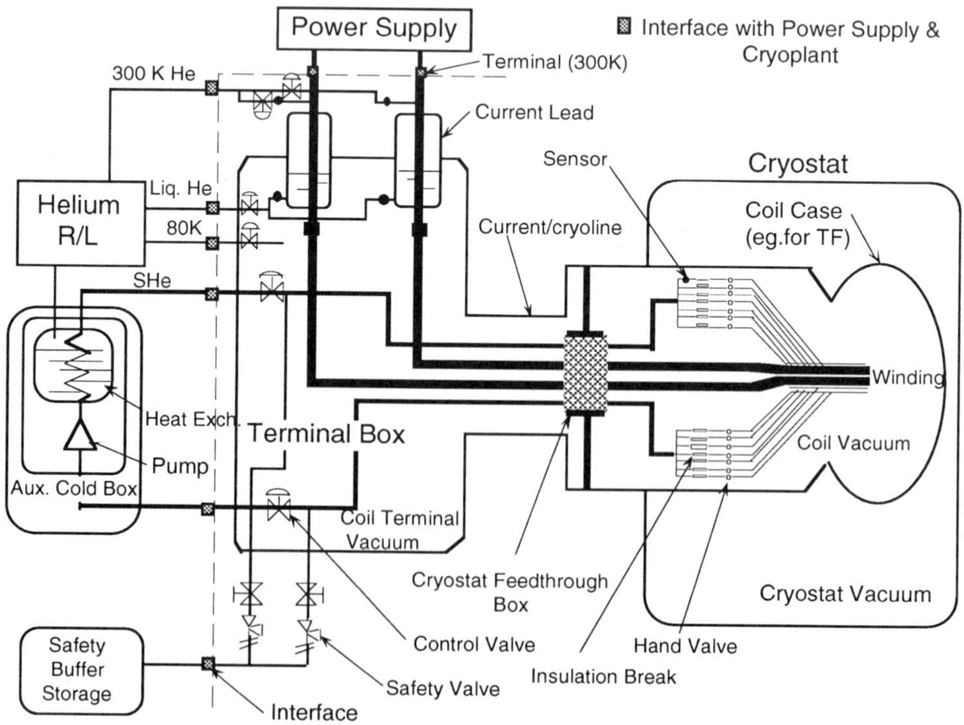

Figure 2 Schematic drawing of all components interfacing with the power supply, cryoplant and a superconducting coil.

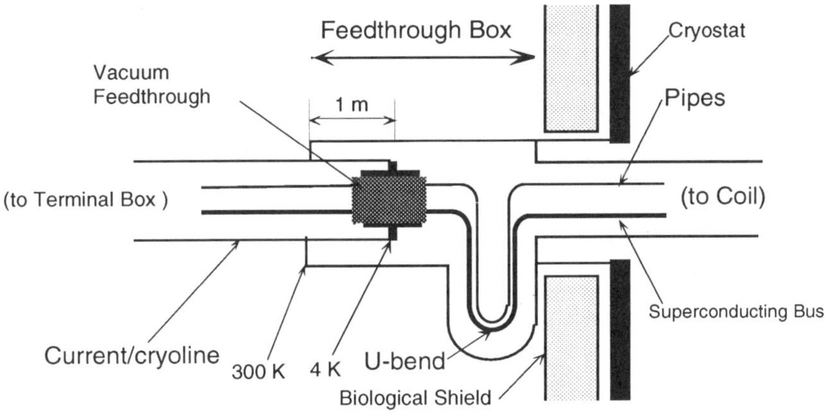

Figure 3 Separation of vacuum with feedthrough between the cryostat and the current/cryoline to the terminal box.

Connection to the Power Supply

Superconducting Busbar. A Nb-Ti cable-in-conduit conductor (CICC) will be used to supply current from a current connector at the coil to the vapor-cooled lead in the coil terminal box. The maximum length of SC busbar is 100 m. This conductor will use a cable similar to that of the outer PF coils. The superconducting busbar design assumes the following conditions:

1) The hotspot temperature of a conductor shall be less than 300 K with a dump time constant of 200 s. The busbar can be operated for 90 s after it quenches.
2) Strands consist of Nb-Ti with CuNi barrier (10 mm) and copper wires with CuNi barrier. Strand diameter is 0.75 mm.
3) The length of cooling paths is approximately 100 m.
4) The TF / PF operating currents of are 61/ 43 kA at 4.5 K and 3T
5) The temperature margin of the conductor is more than 2 K (Tcs = 6.5 K)

A typical cross section of a TF coil busbar is shown in Figure 4. For assembly, a SC busbar will need several joints within its length because the unit length is limited to approximately 30 m.

Current Leads. Each vapor-cooled lead (VCL) has a helium reservoir which can provide a few hours of operation without refilling from the liquefier. The top of the vapor-cooled lead is the interface to the normal bus from the power supply. Local supports for the normal bus and a system to prevent condensation of moisture near the VCL will be required.

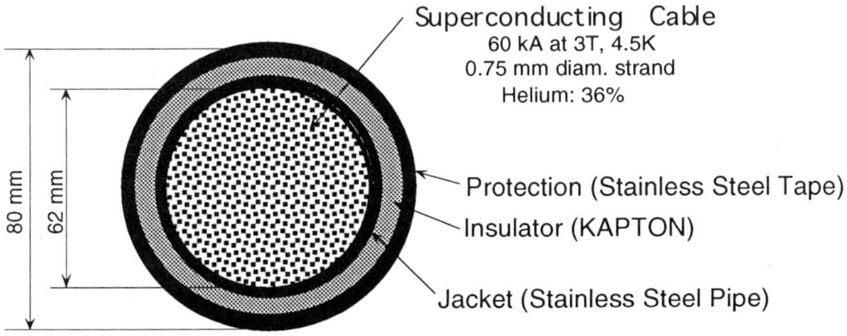

Figure 4 Cross section of a TF coil superconducting busbar.

LAYOUT OF THE CURRENT/CRYOLINES

From each coil, the current/cryolines are routed through the cryostat and the biological shield via the cryostat feedthrough box as shown in Figure 5. These boxes are attached to the outside of the biological shield, which completely surrounds the machine. The function of the biological shield is to limit the amount of radiation from the tokamak. Where possible, from the connection points on the coils, the current/cryolines are routed radially outward to penetrate the cryostat and biological shield so that installation of the cryostat feedthrough box and the current/cryoline as a single unit is possible as shown for the TF current/cryolines. From the cryostat feedthrough boxes, the current/cryolines are routed by the most direct route feasible to the building west of the tokamak hall, which houses the coil terminal boxes.

The TF current/cryolines are routed to exit radically through the cylindrical portion of the cryostat and the biological shield at 20 symmetric locations. Next, the current/cryolines run vertically along the biological shield, then those on the east side of the machine are routed around the biological shield to the west, where all current/cryolines exit the tokamak hall.

All CS and PF current/cryolines exit through the biological shield on the west side at various elevations consistent with the coil positions. In most cases, routing inside the cryostat is non-radial in order to bypass other components and to minimize length. Intermediate joints will be required for installation of these cryostat feedthrough boxes and current/cryolines. The current/cryolines for the poloidal field coils[2] PF-5, PF-6, and PF-7 exit the biological shield at the same elevation, near the bottom of the machine. They are routed vertically as a group to the elevation where they exit the tokamak hall. PF-4 and PF-8 exit at intermediate elevations, and PF-2, PF-3, and CS coils exit the biological shield between the TF current/cryolines on the west side.

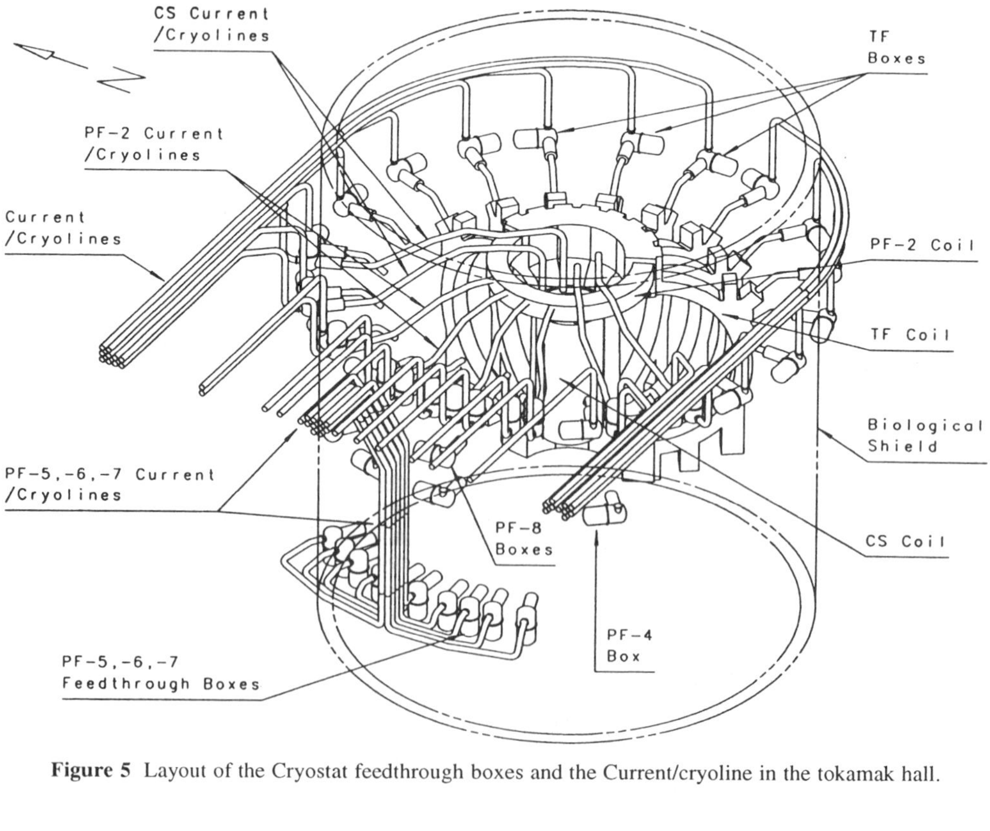

Figure 5 Layout of the Cryostat feedthrough boxes and the Current/cryoline in the tokamak hall.

CRYOPLANT

The main parameters of the ITER cryogenic system are shown in Table 3. The flow/ component diagram of the cryogenic system is illustrated in Figure 6. The ITER cryogenic system is subdivided into three types of equipment: cryoplant, cryo-distribution system and individual cooling loops of the ITER cryogenic users.

Cryoplant

The functions performed by the cryoplant are; to produce liquid and 60-100 K helium for ITER users; to supply the helium plant itself with liquid nitrogen for its normal operation; and to cool-down the ITER components. The cryoplant is comprised of the following equipment:

1) helium and nitrogen compressor stations and warm medium pressure tanks for storing the gaseous helium;
2) refrigerator modules, helium purification units and liquid helium storage dewar;
3) 60-100 K helium installation for cooling of the thermal shields;
4) liquid nitrogen delivery and storage subsystem .

Liquid Helium Plant Modules

The helium cryoplant shall operate concurrently in the combined liquefaction and refrigeration modes according to ITER plant demands. The maximum cooling capacity of the cryoplant is used for sustained operation at 100 % availability with 2200 s plasma cycles during 24 hour periods.

The helium cryoplant shall consist of 5 modules of the same cooling capacity. One module is a redundant unit that can supply the ITER machine when another module is being maintained or repaired, without interruption of the ITER plant normal operation. One or two modules will be used for on-site, pre-assembly coil testing, before incorporation into the complete cryoplant. The final choice of the required number of modules for the liquid helium plant will be made after a trade study to meet manufacturing and shipping constraints for large capacity cryo-equipment and cryogenic system reliability.

Operation with Reduced Cooling Capacity

The cryoplant capacity may be provided in the a staged way, with a minimum capacity, when plasma cycle requirements allow the ‘silent hours’ between plasma operation shifts to be used for storing liquid helium. During ‘silent hours’, two modules would be operated in the liquefaction mode, to produce liquid helium for later current lead cooling, and a third module would be in stand-by mode to remove the heat loads due to thermal radiation and operation of the liquid helium pumps. Some of the liquid helium pumps would be stopped or adjusted to reduce the flow rate to about half of the in-pulse value, and the total heat load to the cryoplant could be reduced during ‘silent hours’ from 125 kW to 41 kW.

Table 3 Main parameters of ITER cryogenic system parameters

Maximum cooling capacity at 4.5 K	125	kW
Magnet system (including liquid helium)	100	kW
Other cryogenic users	20	kW
Cryo-distribution system	5	kW
Maximum cooling capacity at 60-100 K	300	kW
Mass flow rate of the He compressor station	16.4	kg/s
Flow capacity of the He gas purification units	1.4	kg/s
Total inventory of the warm He storage	50	ton
Liquid nitrogen consumption :		
Normal operation	11	ton / h
Cool down from 300 K to 4.5 K (27 days)	11 to 20	ton / h
Total electric power consumption	60	MW

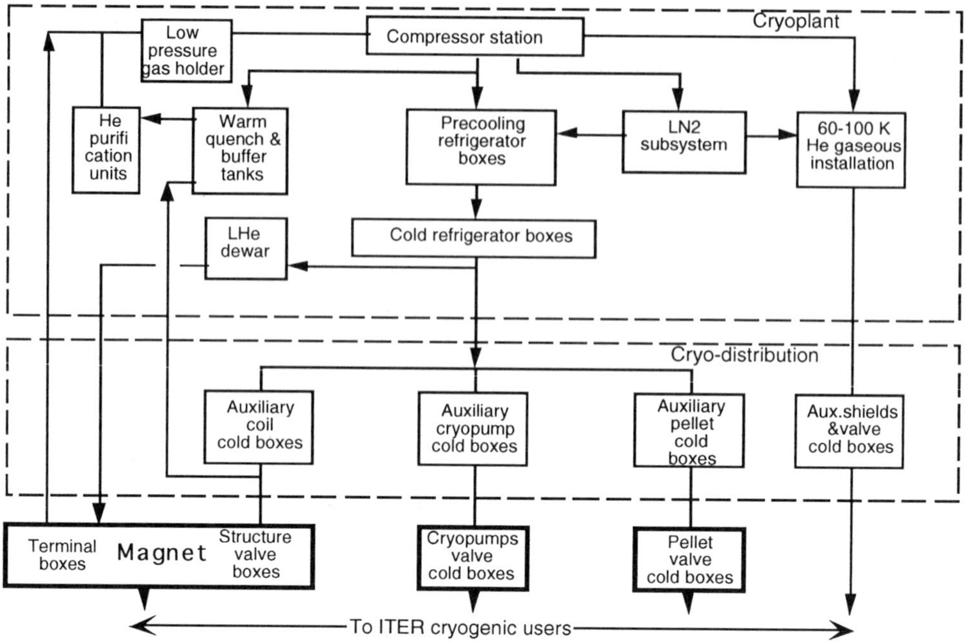

Figure 6 The flow and components diagram of the cryogenic equipment.

SUMMARY

An average heat load removal capacity of 75 kW is required for the magnet system refrigeration and the liquid helium of about 6000 l/h is consumed in the current leads. The mass flow rate of 38 kg/s is required to keep coil temperature within 6.5 K. The forced-flow cooling system with helium mechanical pumps is selected for the magnet system because of the huge mass flow required by the magnet system. Accessibility of cryogenic lines in the cryostat is considerably restricted in the ITER. More study of remote handling is required to determine the location of electrical insulation breaks. Separate vacuums for the cryostat and the cryolines facilitate maintenance of all cryogenic components without warm-up of the whole system. The vapor-cooled leads have a helium reservoir which can provide a few hours of operation without refilling from the liquefier. The layout of cryogenic lines and equipment in the tokamak hall and the electrical termination building is designed to accommodate space limitations and routing restrictions. The total capacity of the cryoplant is 125 kW for the magnets, the cryo-pumps and the distribution system.

ACKNOWLEDGMENT

This report is an account of work undertaken within the frame work of the ITER EDA Agreement. The views of authors do not necessarily reflect those of the ITER Director, the Parties to the ITER EDA Agreement, or the International Atomic Energy Agency.

REFERENCES

1. R. J. Thome, "Design & Development of the ITER Magnet System", presented at ICEC-15, Genova, June (1994)
2. K. Yoshida M. Shimada, M. Matsukawa, et al., "Design of the Outer Poloidal Field Coil System for ITER", presented at 14th Int. Con. on Magnet Technology, Tampere, June (1995)
3. L. Bottura, M. Hasegawa, J. Heim, et al, "ITER MAGNET", ITER Documentation Series, No. 26, IAEA, Vienna (1991), p161

CRYOGENIC SYSTEM FOR ITER CS MODEL COIL

T. Kato[1], K. Hamada[1], K. Kawano[1], T. Hiyama[1], K. Matsui[1], K. Nishida[1], T. Honda[1], S. Sekiguchi[1], K. Ootsu[1], H. Tsuji[1], S. Shimamoto[1], K. Okuno[2], Y. Watanabe[3], E. Tada[3], Y. Nakayama[3], and M. Shimoda[3]

[1]JAERI, Naka Fusion Research Establishment
801, Mukaiyama, Naka-machi, Naka-gun, Ibaraki-ken, 311-01, Japan
[2]ITER Joint Central Team, ITER Naka Joint Work Site
801, Mukaiyama, Naka-machi, Naka-gun, Ibaraki-ken, 311-01, Japan
Kobe Steel , LTD.
[3]3-1, Araicho-Shinhara, 2-chome, Takasago, Hyogo-ken, 676, Japan

ABSTRACT

A 5-kW/4.5-K helium refrigerator has been developed, which will be used for the test of ITER Central Solenoid Model Coil that is under fabrication by an international collaboration under the framework of the ITER Engineering Design Activity. Its acceptance test was recently finished and its specified refrigeration power of 5 kW with a surplus liquefaction rate of 114 l/h was demonstrated. A major feature of the refrigerator is that newly developed components, such as helium compressor and turbo-expander, are adopted in its simple refrigeration process in order that the results will be transferred to the ITER helium cryogenic system.

INTRODUCTION

This cryogenic system will be used for the test of the ITER Central Solenoid (CS) Model Coil[1] that will ramp up to 13 T at more than 0.5 T/s , composed of Nb_3Sn cable-in-conduit conductors with a total weight of around 200 tons including coil supporting structure, cooled by a forced flow of supercritical helium (SHe) at 4.5 K, 0.6 MPa. The cryogenic system is mainly composed of a helium refrigerator and a cryogenic pump unit.[2] The maximum refrigeration power of 5 kW or the maximum liquefaction capacity of 800 l/h is specified for the refrigerator in terms of coil heat load design, test requirements, and contingency. The cryogenic pump unit, which can generate SHe of 4.0-4.5 K, 0.5-1.0 MPa and circulate it to the coil system at up to 500 g/s, was already developed as the part of the Demo Poloidal Coil test facility[3] and is now re-installed in this cryogenic system. The cryogenic system is identified to be one of the largest systems in the field of the fusion magnet R&D test facility. This paper is focused on the newly developed refrigerator, where the design features and the developments of a helium compressor system and turbo-expanders, including the encountered troubles, are introduced. Finally, it is reported that the specified refrigeration power has been confirmed through its acceptance test.

DESIGN FEATURES

Recent large scale helium refrigerator developments have been enhanced through the construction of large superconducting accelerators. Several large, high performance refrigerators have been successfully developed, which are apt to use many stages of turbo-expander in order to achieve higher Figures Of Merit (FOM). Indeed, liquid nitrogen precooling is replaced with a few stages of turbo-expander and a SHe or a wet turbo-expander is now developed instead of a Joule-Thomson (JT) valve. Accordingly, the high performance refrigerator has been realized with FOM of less than 300 W/W.[4]

Since the fusion reactor is an energy production machine, the better FOM refrigerator is one of the most important subjects for the ITER refrigerator. However, we didn't treat this subject as a key issue of this refrigerator design but focus on the development of highly reliable, highly efficient, large size refrigerator components which can be further scaled up for the ITER refrigerator. We, consequently, challenged to adopt an applicable large scale component, such as turbo-expander and helium compressor, in the refrigerator aiming to realize scaling up to a unit refrigerator for the ITER, where five units of the 30-kW scale refrigerator will compose the total of 150-kW refrigeration power. A conventional and simple Claude type refrigeration cycle has been chosen with liquid nitrogen precooling, only two stages of turbo-expander connected in a series arrangement, and double JT valves as shown in Fig. 1. The key component size is increased to its maximum possibly. Indeed, the capacity of the turbo-expander is the largest ever developed with a dynamic gas bearing type. The helium compressor is also one of the largest oil lubricated screw compressor. The first stage (low pressure stage) compressor is designed to have a flow capacity of 250 g/s. The compressor unit is composed of three first-stage compressors (0.103 to 0.91 MPa) and one second-stage (high pressure stage) compressor (0.91 to 1.9 MPa), generating a total mass flow rate of 750 g/s, sufficient to produce more than 8-kW refrigeration capacity. In fact, the designed refrigerator will be able to be modified to have 8-kW of refrigerator if the first JT valve is replaced with a SHe turbo-expander.[5] The general layout of the cryogenic system is shown in Fig. 2.

COMPONENT DEVELOPMENTS

Helium Turbo-expander Development[6]

Design criteria and features. The development of the turbo-expander is the key issue to determine the specified refrigerator capacity. The following design criteria are applied.

1) To ensure a large flow: a maximum mass flow rate of more than 170 g/s.

2) To achieve high isentropic efficiency at the design point: more than 75% for both the first stage and the second stage

3) To apply self acting dynamic gas bearing for both a journal and a thrust bearing.

To achieve such design criteria, the following designs were adopted.

(a) A three-dimensional radial inward flow impeller. The impeller is equipped with a three-dimensional shroud cover by using diffusion bonding, and should aid in attaining high isentropic efficiency by providing low loss expansion and by reducing bypass flow between the impeller and the impeller casing.

(b) Titanium alloy is used for the impeller and the rotor material due to low weight, high tensile stiffness, and low thermal conductivity. The total weight of the rotor including the impeller is around 600 g.

(c) A pump-in type spiral grooved bearing is adopted for the thrust bearing where pressured gas is fed through the holes in the thrust bearing to float the rotor, avoiding the bearing surface damage when rotation begins. The gas is supplied until 1,000 rps is attained and then stopped, at which time the dynamic bearing start to function. Accordingly, the static bearing part and the dynamic bearing part are arranged together in the concentric circle, a plane land for the static part is located inside circle and a spiral groove is outside, respectively. Electromagnetic bearing can't be used due to non-magnetic material. On the other hand, the journal bearing adopts a tilting pad type self-acting gas bearing where the equal sized pads with the diameter of 25 mm are used.

The designed parameters are listed in Table 1 and the view of the second stage turbo-expander is shown in Fig. 3, respectively.

Table 1. Design parameters for the developed turbo-expanders

	Liquefaction (800 l/h)		Refrigeration (5 kW)	
	First Stage	Second Stage	First Stage	Second Stage
Inlet Temperature (K)	48	24.8	36	18.6
Inlet Pressure (MPa)	1.6	6.5	1.6	6.5
Outlet Pressure (MPa)	0.678	0.12	0.678	0.12
Revolution (rps)	1815	1674	1571	1432
Mass flow rate (g/s)	154.4	154.4	178	178
U/Co	0.62	0.64	0.62	0.64
Isentropic Efficiency	0.75	0.75	0.75	0.75
Power (kW)	8.06	6.81	6.97	5.79

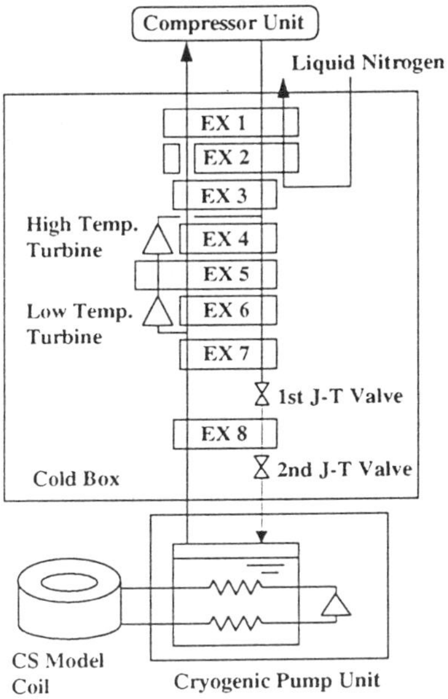

Figure 1. Adopted refrigeration cycle.

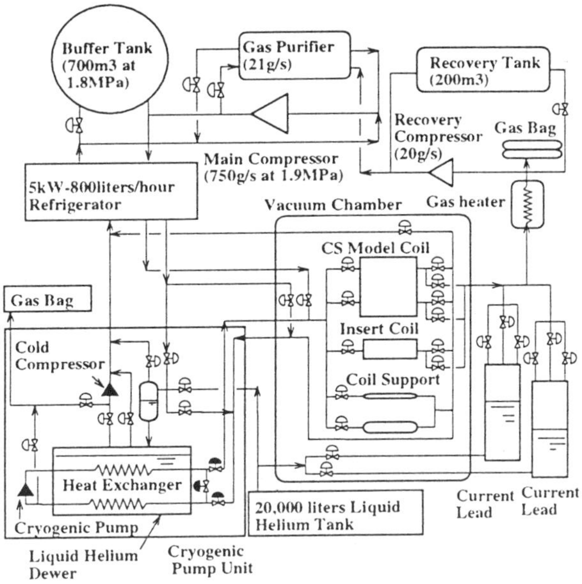

Figure 2. General layout of the cryogenic system.

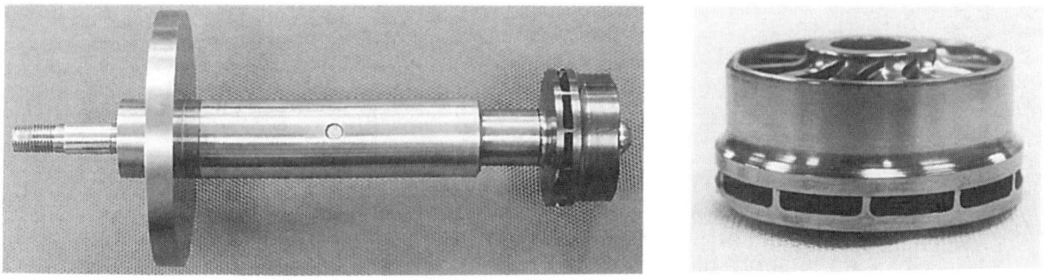

Figure 3. The view of second stage turbo-expander, the rotor (left) and the impeller (right) are shown.

Table 2. Investigation of the first stage turbo-expander thrust bearing data

	Rated condition (Expected)	Damaged condition (Measured)	Achieved condition (Measured)
Thrust force (N)	287	244	220
Floating gap (μm)	15	15*	18
Rotation speed (rps)	1600	around 1600*	around 1550
ΔP** (MPa)	0.28	0.25	0.23

* The value was measured just before the turbo-expander stopped.

** ΔP indicates the differential pressure between the inner bearing case and the outlet turbo-expander.

Encountered trouble and its measures in the acceptance test. When the turbo-expander was operated close to the rated conditions in the acceptance test, the first stage turbo-expander was suddenly stopped and damaged, which was the most significant trouble experienced through the development. The thrust collar was damaged at both the inner most part of bottom surface and the outer most part of the upper surface, the corresponding parts of the thrust bearing were also damaged. This would suggest that the thrust collar was bent downward like a parabola, and the expected bearing force was insufficient. The thrust bearing data were checked as listed in Table 2. The damage occurred at 224 N which is 85% of the rated thrust force of 287 N (the maximum design is 304 N). The floating gap of 15 μm was observed just before damage. It would be expected that this value would be practically zero. Specially, there would be a 3-μm gap at the achieved condition. Such phenomena may be complexly related with the pressure and temperature conditions of the bearing system. The detail investigation has started to clear the problem. Some verification tests are now being planned.

Since it takes a long time to change and correct the thrust bearing, we decided to provisionally use continuous pumped-in gas together with the dynamic bearing in the high speed region. The pumped-in gas flow was determined using the following consideration; First, the required floating gap is determined to be 20 μm at the rated condition, expecting a practical gap of 5 μm. The dynamic thrust bearing stiffness at the 20-μm gap becomes 157 N based on the design data. Consequently, a stiffness of more than 130 N should be required for the pumped-in static bearing. A pumped-in gas flow of 20 Nm^3/h was selected, where a stiffness of 167 N is assumed at a gap of 20-μm. This method was tested and verified to be possible. Indeed, the measured thrust floating gap was more than 30 μm at a thrust force of up to 230 N, which is an increase of 10 μm compared to the previous values. Since the additional pumped-in gas would flow to the cold process line and the turbine isentropic efficiency would be worse, such mass flow was provisionally removed by the gas removing line located at the turbine bearing casing which was used at the initial turbine rotation up to 1,000 rps as mentioned in the design features.

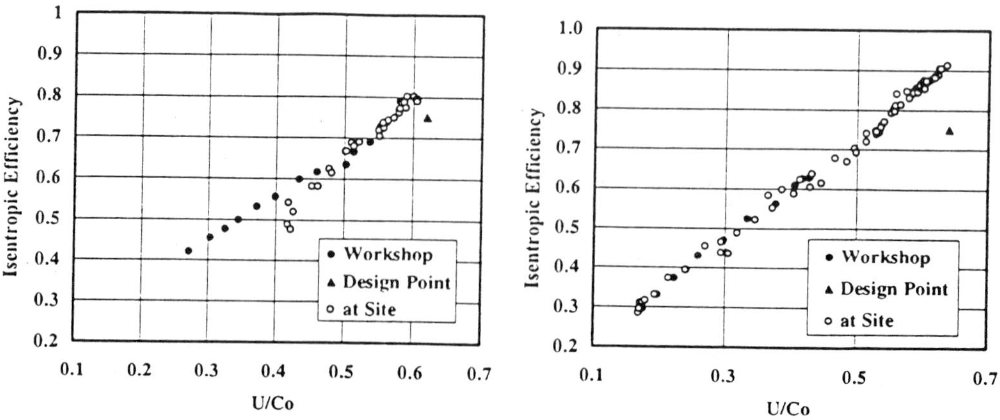

Figure 4. Measured isentropic efficiency as a function of speed ratio, the first stage turbo-expander is left figure and the second stage turbo-expander is right figure.

Table 3. Measured turbine performance at the specified operation mode

	Liquefaction (800L/Hr)		Refrigeration (5kW)	
	First Stage	Second Stage	First Stage	Second Stage
Inlet Temperature (K)	48.21	24.28	36.51	18.26
Outlet Temperature (K)	39.17	15.1	29.18	10.8
Inlet Pressure (MPa)	1.426	0.516	1.426	0.496
Outlet Pressure (MPa)	0.672	0.114	0.695	0.115
Revolution (rps)	1533	1423	1420	1318
Mass Flow Rate (g/s)	132.5	132.5	148.5	148.5
U/C0	0.553	0.572	0.601	0.624
Isentropic Efficiency(%)	72.5	82.6	80.1	91.1
Power (kW)	6.417	6.233	5.785	5.55

However, a journal bearing instability was then observed when the pressure difference between the inner pressure of the bearing case and the turbo-expander outlet pressure increased over 0.23 MPa, which may be caused by a self-acting oscillation generated by hot gas flow from the bearing case to the cold process line through the impeller part where the gas would suddenly contact, even if the gas removing line was used. This phenomenon was related to the pressure difference, the inlet pressure condition, and the gas removing line control. Through out the test, it was found that stable operation was restricted an the inlet pressure of less than 1.45 MPa, where the pumped-in gas would not flow to the cold process line. On the other hand, the second stage turbo-expander has not shown any problems. Indeed, the thrust bearing force was not over 200 N during any operation. However, the same treatment was performed against unknown failures.

Development results. The verification tests, required for the development, were performed in the workshop by using the existing test stand which could supply only one-third of the helium mass flow rate against the rated condition, providing limitation to verify the design performance. However, it could successfully simulated more than 70% of the design value for the design parameters such as efficiency, speed ratio, pressure ratio and modified flow, which could provide sufficient data to estimate the full performance.

The isentropic efficiency as a function of the speed ratio (U/Co), measured in both the verification test and the acceptance test, is shown in Fig. 4. Note that the second turbine data at the workshop could not be covered at the high U/Co region, because the test stand at the workshop could not reduce the turbine inlet temperature at the sufficient degree due to the capacity. The results near the design point show so high, excellent efficiencies, verifying the success of the development program except for the problem of the thrust bearing. The data measured at 800-l/h of liquefaction and 5-kW of refrigeration are listed in Table 3. Such high efficiency significantly helped to generate the specified refrigeration capacity even though the inlet pressure was limited by around 10% less than the design. Note that the accuracy of the measurements will be within $\pm 2\%$ according to the temperature measurement accuracy of less than ± 0.2 K.

Helium Compressor System Development

Design criteria and features. An oil lubricated screw compressor is selected due to its wide spread acceptance for helium service. The following design criteria are applied, aiming to obtain information to be applied for the ITER refrigerator system;

(1) To aim at the isothermal compression efficiency of more than 60%.
(2) To have the largest mass flow for a unit compressor with down size casing.
(3) Minimize or eliminate the generation of impurities.

To attain the above criteria, the following design features are incorporated;

(a) The compressor system is separated into two stages; the first stage and the second stage. Then, the first stage compressor has two stages of compression produced by one driving moter in one casing, so called a tandem or a compound compressor. The compressor system had practically three compression stages, which will improve the isothermal compression efficiency. On the impurities, three-stage arrangement is expected to reduce the risk of impurities production because the temperature rise at each stage should be smaller than that of the conventional two stages compressor system. Indeed, the outlet temperature at each stage, calculated by the isentropic compression process model, can be kept lower by 100℃ than that of the initiating molecular decomposition. A mineral oil is used as a lubricating oil and its molecular decomposition temperature is measured to be around 350℃. Further, three staged oil removers are adopted, with a charcoal absorber as the final stage. (b) The compressors are modified from those used for a chemical gas plant. The clearance between the rotors and the casing were reduced as small as possible. The oil injection ports were also improved to provide efficient cooling for helium gas and rotors. The first stage rotor size is 408-mm in diameter and 673.2-mm in length, which is the largest rotor ever applied to a helium gas screw compressor. The design parameters of the compressor system are listed in Table 4 and the view of the compressor system layout at the site is shown in Fig. 5. The installation area is 12 m x 14 m.

Development results. The verification test was carried out in the workshop. In the test, a single low stage and a single high stage compressor are connected in series, instead of the full arrangement. A bypass line with a flow control valve was prepared for the high stage compressor to supplement the lack of the suction flow. No impurities were detected at a sensitivity of less than 0.1 ppm after 28 hours running. The test could simulate the full performance for each stage compressor and the obtained results are listed in Table 5. The total isothermal efficiency was barely 60%. However, the efficiency should improve by a few percent when the interstage pressure is slightly reduced. A mechanical rotation test with air was carried out for the remaining two low pressure stage compressors. As a result of the development, the specified design criteria was successfully verified.

The measured flow performance in the acceptance test was similar to the results of the verification test, a mass flow rate of around 800 g/s was measured. On the other hand, the isothermal compression efficiency was determined to be around 64% by measuring the total consumption of the practical electric power (2.19 MW) at 791 g/s. Note that the accuracy of the measurement is assumed to be around ± 2%. As was aimed, the world largest screw compressor with a high isothermal compression efficiency of more than 60% has been successfully developed.

Encountered trouble and its measures in the acceptance test. In the acceptance test, every compressor rotor was turned a few times by hand to check for rotor damage before operating. After a few weeks running with many stops and starts, the manual turning was getting so heavy for one of the low pressure stage compressors, indicating a damage of the rotor. This compressor was shipped to the workshop and overhauled. There was very light scratch at the outlet part of the screw rotor. The basic reason for this problem is shown in Fig. 6. To correct the problem, the cooling oil control valve and piping lines were changed to larger ones. The cooling oil flow was increased to cover the lack of cooling for the rotors. After taking such measures, the outlet temperature of the compressor became lower by around 10℃ than before and the heavy turning disappeared.

Table 4. Design parameters of the compressor

	Low stage	High stage
Inlet pressure	0.103 MPa	0.91 MPa
Outlet pressure	0.910 MPa	1.86 MPa
Inlet temperature	≤ 308 K	←
Outlet temperature	≤ 363 K	←
Flow rate	≥ 250 g/s	≥ 750 g/s
Revolution	2030 rpm	2980 rpm
Rotor diameter	408/321 mm	279 mm
Rotor length	673/250 mm	446 mm
Isothermal efficiency	≥ 60%	

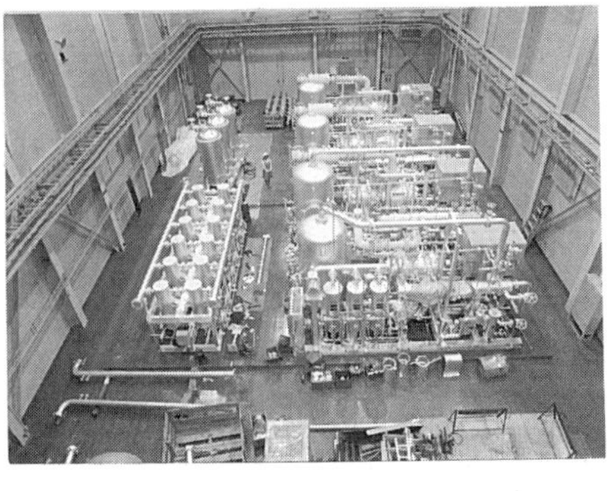

Figure 5. The compressor system layout at the site.

Table 5. The Measured performance of the verification test at the workshop

	Low Pressure Stage		High Pressure Stage	
	Specification	Measured	Specification	Measured
Flow Rate	255 g/s	273 g/s	761 g/s	854 g/s
Volumetric Efficiency	84%	90%	80%	90%
Electric Power	626 kW	620 kW	582 kW	593 kW
Isothermal Efficiency	Specification		Measured	
	56.0 %		60.1 %	

ACCEPTANCE TEST RESULTS

The refrigerator acceptance test was concentrated to achieve three specified operation modes: the 800-l/h mode, 3-kW/4.5-K and 380-l/h mode, and 5-kW/4.5- K mode. The practical test procedure is quite simple; First the refrigerator operates to liquefy and the liquefaction rate is observed at the steady state. Second the heaters, equipped in both the cold end of the cold box and the liquid helium dewar, are turned on to keep 3 kW totally and the liquefaction rate is measured. Finally the heater power is increased to be 5 kW. The test results are plotted at the refrigerator power as a function of the liquefaction rate as shown in Fig. 7 where the 2-kW and the 4-kW operation results are superimposed. Through the tests, the inlet pressure of the first stage turbo-expander was kept to be less than 1.45 MPa due to the thrust bearing limitation. The refrigerator capacities for the specified operation modes were 801 l/h at the full liquefaction mode, 383 l/h with the total heater power of 3,106 W, and 114 l/h with heater power of 5,025 W, respectively. Consequently, the design specification has been successfully confirmed. Note that the liquid helium dewar loss of 15 l/h, which was measured before the tests, was considered in determining the liquefaction data.

Fully automatic control is provided to the liquefaction operation by the computer-aided distributed control system. In the acceptance test, automatic control was performed and confirmed. Indeed, the only manual operations was the control of the input electric power to the heater after touching the automatic operation select button on the computer CRT.

DISCUSSION

The designed FOM of the refrigerator is 346 W/W at a refrigeration power of 5 kW.[2] According to the acceptance test data, FOM was calculated to be around 450 W/W at 5 kW. In the calculation, the compressor power at 800 g/s and the liquid nitrogen consumption (where 1 g/s of liquid nitrogen consumption is corresponding to 2.87 kW). However, the actual cold box flow rate at 5-kW operations is smaller than 800 g/s. Some of the

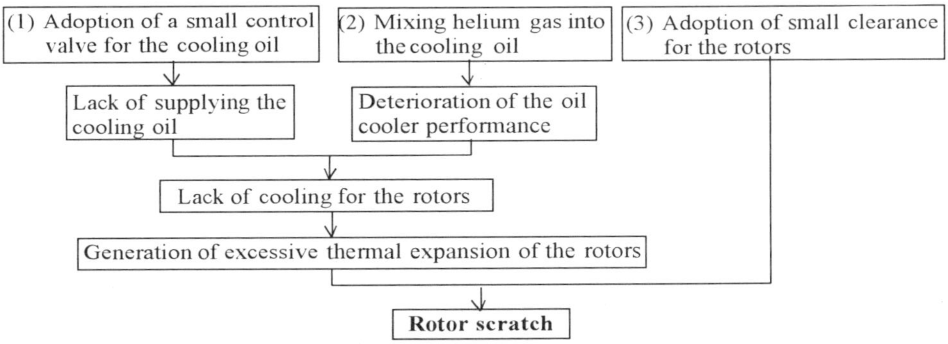

Figure 6. Fault tree for the low pressure stage compressor trouble.

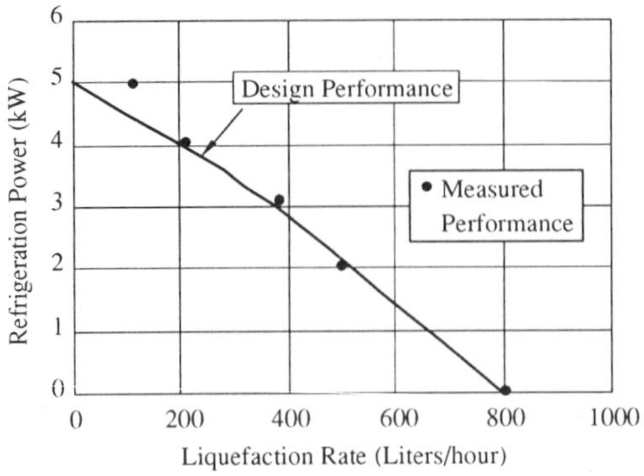

Figure 7. The obtained refrigeration vs. liquefaction.

Figure 8. The view of the cold box.

compressor flow bypassed the cold box in this acceptance test. There isn't a flow meter on the bypass line. The cold box flow rate couldn't be estimated. If the cold box flow was estimated by the percent that the bypass line valve is open, it would be roughly 380 to 450 g/s. If the FOM is calculated by using cold box flow rates these numbers imply (350-420g/s) and the compressor efficiency of 64%, it would be from 220 to 258 W/W. The liquid nitrogen consumption, in fact, was less than 22 g/s which is smaller by at least 40% than that of the process design calculation. Thus the cycle performance of the cold box is expected to be better than the design performance. The view of the cold box is shown in Fig. 8. In the acceptance test, very high turbine isentropic efficiency, high compressor isothermal efficiency were measured and the FOM of the cold box could not be measured accurately. We will try to measure them again more detail after the turbo-expanders are replaced by new one that will modify the thrust bearing system. Consequently, more detail data acquisition and analysis will be performed to determine the accurate refrigerator cycle performance.

CONCLUSION

The refrigerator acceptance test was successfully carried out to confirm the specified cryogenic capacity, even though the turbo-expander provisionally used this time provided operational limitations. The component developments also succeeded in achieving most specified design criteria. Such results could lead to more efficient large helium refrigerator. For example, an 85% turbine isentropic efficiency and a 64% isothermal compressor efficiency are applied to the ITER refrigerator design, the expected FOM will be less than 250 W/W. If another turbo-expander such as a SHe or a wet type is applied, the cold box efficiency will be more improved. The improvement of the compressor efficiency at the large scale will also provide an impact for scaling up the refrigerator.

ACKNOWLEDGMENT

Authors would like to thank Drs. N. Shikazono and Y. Tanaka for their continuous encouragement on this work. The authors are also grateful to the staff of the Superconducting Magnet Lab. for their supporting of this work.

REFERENCES

1. F. Iida et al., "Progress in the ITER Model Coil Program," Proc. of 14th Inter. Conf. Magnet Tech., Tempere (1995).
2. K. Hamada et al., "Final Design of a Cryogenic System for ITER CS Model Coil," ICEC supplement, Cryogenics, Vol. 34 (1994), p. 65.

3. T. Kato et al., "Operation Performance of DPCF in The Test of The Nb-Ti Demo Poloidal Coil," Proc. of 11th Inter. Conf. Magnet Tech., Vol. 2 (1989), p. 1350.
4. S. Claudet et al., "Four 12kW/4.5K Cryoplants at CERN," ICEC supplement, Vol. 34 (1994), p. 99.
5. T. Kato et al., "Design of a 8-kW Refrigeration System for ITER Common Test," ICEC supplement, Cryogenics, Vol. 32 (1992), p. 60.
6. T. Kato et al., "A Large Scale Turbo-expander Development and Its Performance Test Results," Advances in Cryogenic Engineering, Vol. 35 (1990), p. 1005.

OPERATIONAL EXPERIENCE AND RELIABILITY OF THE CRYOGENIC SYSTEMS FOR THE TRISTAN INSERTION QUADRUPOLE MAGNETS

K. Tsuchiya, N. Ohuchi, Y. Morita, A. Kabe, R. Sugahara, and T. Ogitsu

KEK, National Laboratory for High Energy Physics
Tsukuba, Ibaraki, 305 Japan

ABSTRACT

Four sets of helium cryogenic systems for the mini-beta insertion quadrupole magnets were installed near the interaction points of the TRISTAN main ring in 1990. Each system consists of a helium compressor, a cold box, a subcooler, transfer lines, two magnet-cryostats, two helium gas tanks and a liquid nitrogen storage tank, and its nominal cooling capacity is 140 W at 4.2 K + 25 L/h. The four systems are controlled automatically by a process control computer system. The first operation started in 1991 and by the middle of 1995, the total operating time for each system reached about 28,000 hours. In this paper we report the experience gathered during 28,000 x 4 operating hours in running four cryogenic systems together with the control system. Maintenance experience and statistics of failures of different components are also described.

OUTLINE OF THE CRYOGENIC SYSTEMS

The TRISTAN main ring, whose circumference is 3018 m, has four experimental halls, Oho, Tsukuba, Nikko, Fuji, and three of them are equipped with large experimental detectors housing superconducting solenoids, AMY, TOPAZ, and VENUS. In 1990, four cryogenic systems for insertion quadrupole magnets were installed in the experimental halls[1]. A typical layout of the cryogenic equipments at an experimental site of the TRISTAN main ring is illustrated in Fig. 1 and a simplified flow-diagram of the cryogenic system is shown in Fig. 2.

The system consists of a helium compressor, a cold box, a subcooler, transfer lines, two magnet-cryostats, two helium gas tanks and a liquid nitrogen storage tank with a capacity of 9800 L. The compressor, a helium gas tank of 20 Nm^3 volume, and the liquid nitrogen tank are placed on the ground-level while the others are located near the collision point down to about 15 m below the ground-level. In order to reduce the requirement of refrigeration capacity of the system, the two magnets on either side of the experimental detector, subcooler and coldbox are set as close as possible and they are connected via transfer lines. A recovery helium tank with a capacity of 5 Nm^3 is placed near the coldbox to recover evaporated gas from the magnet-cryostats in case of quench.

These cryogenic systems are controlled by a process-control computer comprising central computer units and four substations located near each cryogenic system[2]. These computers are connected to each other via a duplex optical-fiber data link along the

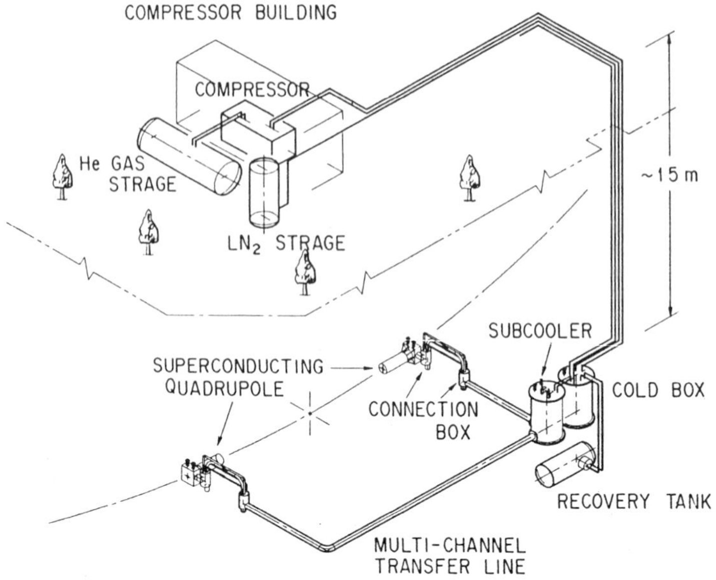

Figure 1. Schematic layout of cryogenic equipments at an experimental hall.

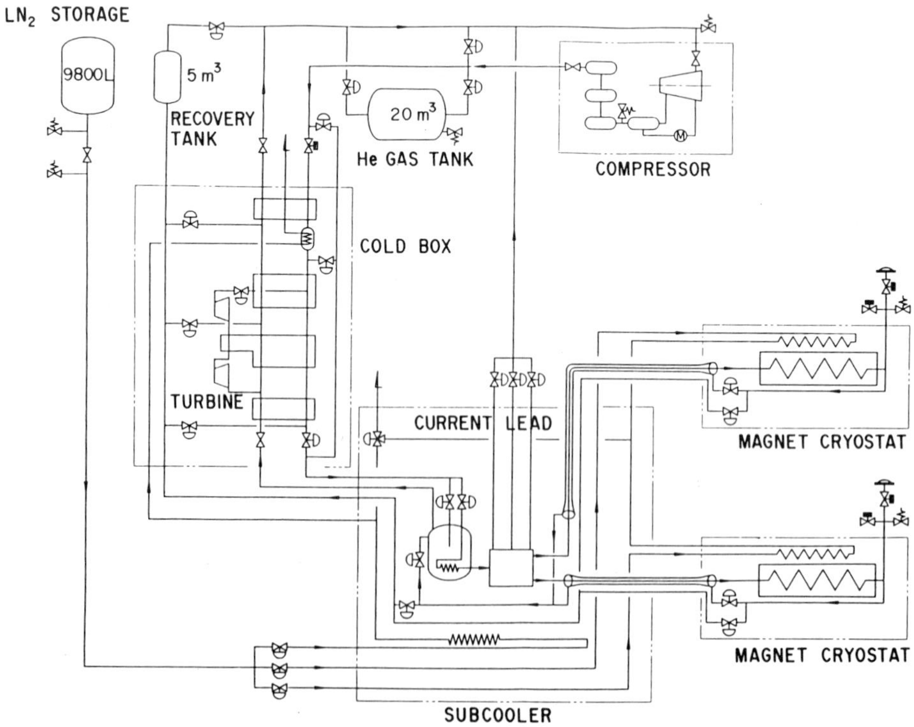

Figure 2. Flow-diagram of the cryogenic system.

TRISTAN ring. By using this control system we can do both cool down and warm up of the four cryogenic systems automatically. The control system also responds to emergencies and returns the cryogenic system to normal operating conditions.

HISTORY OF OPERATION

Following the successful operation of a prototype[3] at KEK, the design and construction of four sets of superconducting quadrupole systems were started in 1989. The manufacturing of the components and the performance test of the superconducting magnets were completed by the middle of September 1990. The installation of four systems started in August, and proceeded in parallel until the end of 1990. The performance test of each of the cryogenic system which included the automatic operation test[1,2,4,5], was finished by the end of January 1991, completing the installation.

In February 1991, the accelerator operation with these insertion quadrupole magnets was started. At the beginning of beam tuning, it was found that the rotational alignment errors of the quadrupoles caused a serious beam loss, therefore, re-alignment of the magnets was necessary. For that purpose we had to modify the lengths of a few transfer lines. However, after the re-alignment, we succeeded to double the luminosity of the TRISTAN. Since then, the four cryogenic systems have been routinely operating without serious troubles. The systems are operated annually during two periods: first, from February to June and then from October to December, and the maintenance work have been carried out during the long shutdown periods in between.

In Fig. 3 (a) and (b) we show the integrated operating hours of each cryogenic systems during the past five years.

MAINTENANCE AND OPERATION EXPERIENCE

In order to assure reliable long-term operation and to minimize the downtime of the systems, we performed routine maintenance once or twice a year depending upon the components. Naturally, moving machines in the cryogenic systems have the highest potential for attrition. In our case, these were the helium compressors, air compressors, turbines and valves. However, we did not pay any attention to the turbines since the turbines were well consummate components. Others, which needed maintenance are the components of the process control computer system and the battery back up systems against power failure.

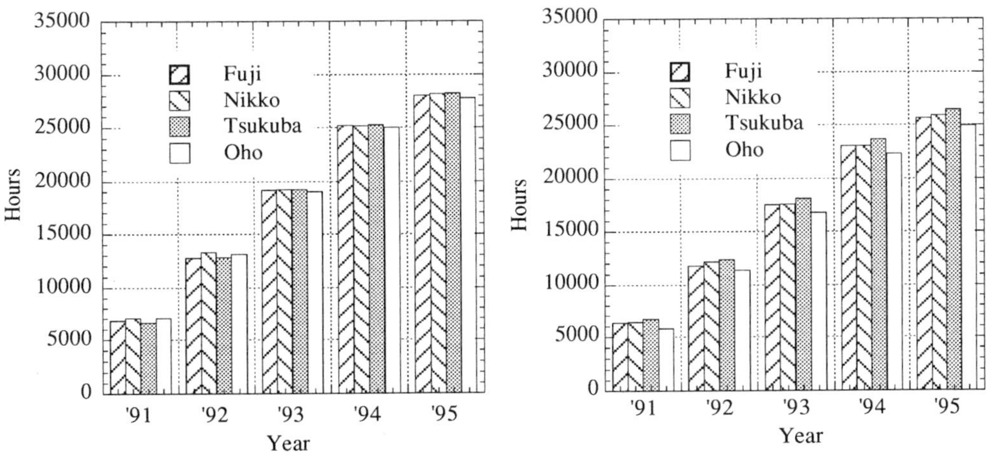

(a) compressors (b) coldboxes

Figure 3. Integrated operating hours of each cryogenic system during the past five years.

Compressors

The helium compressor is an oil injected two-stage screw type with two slide valves. It has a maximum flow rate of 1250 Nm^3/h and is coupled to a 250 kW electric motor. The compressor, an electric motor, an oil removal system and an oil circulation system, all are on the same skid. The oil removal system consists of a bulk oil separator with a demister, two coalescing filters mounted in series and a final oil separator. The final separator contains two different types of adsorbents. Those are charcoal adsorbent (150 kg) and molecular sieve 13X (40 kg). The separator is designed to raise the temperature up to 150 °C when the adsorbents are regenerated.

The main items which we had to pay attention to during the maintenance were the shaft seal, the monitor of the slide valve position and the oil adsorbents.

The specified leak tightness of shaft seals was $< 3 \times 10^{-3}$ L/h. The tightness is checked during every 8-hour shift. A seal is replaced if the leak rate becomes larger than the specified value. During the past five years, fifteen seals of four compressors were replaced with new ones: out of these, eight were changed before we adopted fine alignments of the coupling between motors and compressors in August 1992. In Fig. 4 we show the lifetime of shaft seals which started leak after the replacements. As can be seen in this figure, the contribution of the fine adjustment of the coupling is great. However, still the lifetime of the changed shaft seal distributed a lot; some could live up to a few or many thousands of hours but some only several hundred hours.

The monitor system of the slide valve position was a weak point of the compressors because the sensors always suffered from severe vibrations due to high speed rotation of the screws. Almost half of the number of compressor-troubles were due to this monitor system. Recently we installed a new monitor system and have used it for about 2,800 hours. The results were very satisfactory.

The oil separation is a very important activity so far as the long-term operation of the cryogenic systems is concerned, so the oil concentration was checked routinely at the end of each operation period. Until now, we have no experience to detect an oil concentration which exceeded 0.1 ppm (vol). First replacement of the charcoal and the molecular sieve was done after about 7,000 hours of operation. This is because the manufacturer recommended to change those every 8,000 hours. However, the second replacement of those adsorbents was performed when the integrated operation time of the compressor reached about 19,000 hours. After we replaced the adsorbents, we regenerated the charcoal with nitrogen gas at 120 °C. The flow rate of nitrogen gas was about 30 Nm^3/h and the regeneration time was about 150 hours.

Cold Boxes and Other Cryogenic Equipments

Each of the cold boxes houses heat exchangers, six control valves, and two turbo-expanders. In addition, a charcoal adsorbent of 22 kg is placed at the 80 K temperature stage

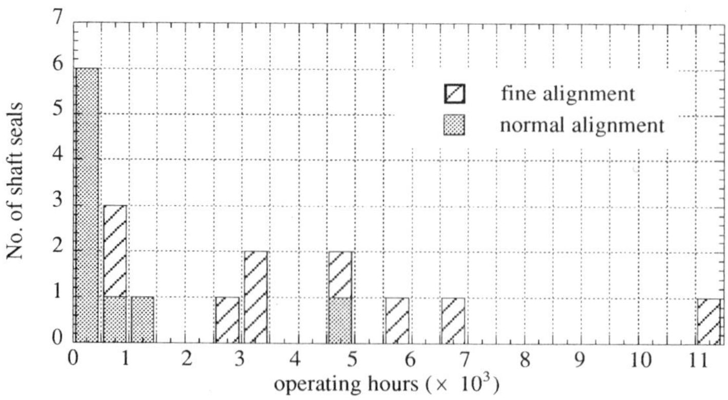

Figure 4. Lifetime distribution of the shaft seals of the helium compressors.

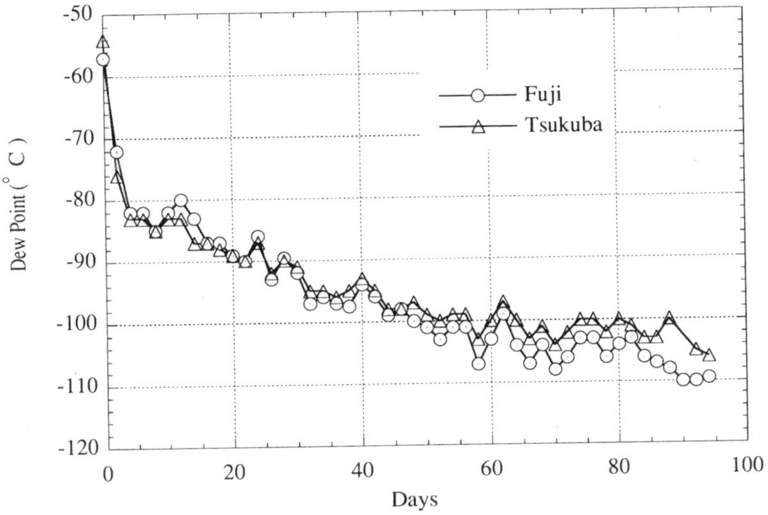

Figure 5. Trends of the dew point after starting the purification with a cryogenic purifier. After 10 days purification, the systems were cooled down and reached steady state after 2 days more.

to remove impurities in helium gas that might be introduced into the circuit. The turbines are static gas-bearing type and arranged in series. There is a 10 μm filter in front of the warm turbine. The turbines needed no maintenance.

Before starting the operation of the system, we always did purification of the helium gas in the system with a cryogenic purifier until the impurity level was less than 1 ppm and the dew point was below –85 °C. After completing the purification, the cryogenic purifier was detached from the system and the system was cooled down to liquid helium temperature. Usually, the dew point reached less than –90 °C while the system was kept in a steady state condition. During past five years of operation, there was no trouble due to the impurities in the process gas. In Fig. 5 we show typical trends of the dew point during 100 days after starting the purification.

Another item, we paid attention to, was gas leak. Since we designed the system, so as to minimize the helium inventory in one system to about 250 Nm^3, we did severe leak checking frequently. Owing to these efforts, we could reduce the helium loss to the level of <0.3 Nm^3/month.

Process Control

A process control computer system was installed to automatize the operation of the four cryogenic systems. The system consists of multi-controllers, which are processors for doing loop control, sequence control etc., operator consoles with 20 inch color CRT, color hard copy devices and printers. Those were installed in each local control rooms as well as in a central control room. All transducer signals are read into the computer which stores the data and is capable of displaying process measurements up to the operator's consoles. The computer system performs all interlock functions. The total number of process I/O signals of one system was 245 (980 in the four systems). About 120 digitized analog signals for one cryogenic system were gathered at intervals of both 2 minutes and 10 seconds. The data collected are stored in the hard disk and kept for 2 days. In order to save these data for a long time, we had to copy them onto floppy disks every two days. For a reliable operation, the easy and extensive access for the operators to the process information in the form of process flow diagrams with the actual process status is essential. Displays of historical trends were also very helpful for a safe and reliable operation. Protection against power failures was achieved by a battery back up (20-30 min.).

The maintenance of the computer system and the battery back up system is routinely done every year, especially the maintenance of the former takes about two weeks and goes into particulars which cover from general clean up to calibration of all I/O channels.

Air Compressors

Since the compressor was a reciprocal type, it required a lot of maintenance. The compressors were overhauled thoroughly twice a year and all parts of them were cleaned. During the continuous operation, these compressors required cleanup of the inlet filter every month.

FAILURES OF THE CRYOGENIC SYSTEMS

There were failures of components which did not interrupt the cryogenic system operation and failures which switched off the "cryo-ready" signal. When this signal was missing, the current of the quadrupole magnets and the beam in the accelerator were dumped. The down-time caused by failures of this kind is shown in Fig. 6. We classified the failures in five categories indicating if the failure occurred in the cryogenic plant (compressors, cryogenic equipments which include valves and monitors, etc.), in main electric power system (by thunderstorms, transformers, etc.), in cooling water system, other sources (quadrupole magnet power supply, accelerator safety interlock, etc.) and from quenches. In the category of the cooling water, the time which was required to repair the cooling water system was also included. As can be seen in Fig. 6, the main failures which took longer down-time are the electric power system failure and the cooling water trouble.

Table 1 lists the summary of all failures which occurred during the past five years. Though total number of failures is not a small number, however, most of the failures didn't interrupt the continuous operation of the systems. The numbers of failures which interrupted the normal operation of the systems or required an operator's assistance to return to the normal operating condition are listed in parentheses. The ratio of this kind of trouble to all kinds of failures is about one fourth.

A prominent category in Table 1 is the category (D), i.e. quench and magnet power supply failures. Actually some quenches did happen, however, those quenches were all

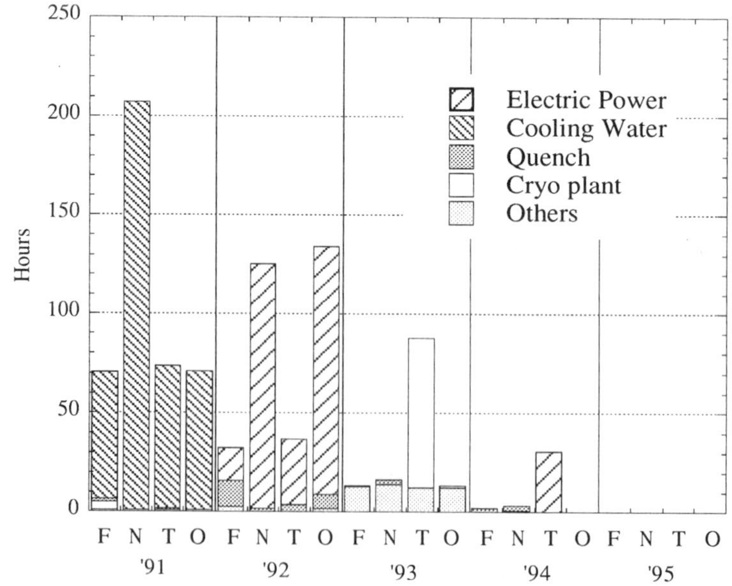

Figure 6. Downtime of the cryogenic systems during past five years.
F=Fuji, N=Nikko, T=Tsukuba, O=Oho

Table 1. Summary of the failures

failures	number of failures	
A) Utility		
Cooling water	8	(2)
Electric power	3	(3)
Air compressor	8	
B) Cryogenic plant		
Compressor	24	(1)
Turbine	7	(3)
Valve	14	(3)
Monitor	18	(4)
(flow meter, pressure transducer, thermometer etc.)		
Gas leak	8	
Vacuum	5	
C) Control system		
Computer system	9	
Battery back up system	2	
D) Quench and Magnet power supply	17	(17)
E) Operation miss	2	
Total	125	(33)

beam induced quenches which were triggered when the accelerator beam hit the superconducting coil. In the insertion quadrupole magnet, the beam size is considerably larger than in other accelerator elements, therefore, the probability of beam hitting is very large if other accelerator elements were broken. We have no occurrence of a quench which is triggered by either cooling problem or by magnet problem up to now. All failures in this category were due to either beam induced quenches or due to power supply troubles.

SUMMARY

The four cryogenic systems have been operated during the past five years and their total operating time reached about 28,000 x 4 hours. The availability of the systems as compared to the scheduled time in which steady state cooling conditions were required is about 99.1 %. The experience in the operation shows that the present cryogenic systems have enough potential for reliability to perform a continuous operation of more than 5,000 hours at a stretch, if we do proper maintenance. In order to increase the duration of this continuous operation time, we would like to have better reliability of the control valves, the monitor equipments as well as the compressors.

ACKNOWLEDGMENT

We are very grateful to our colleagues of the TRISTAN magnet group for their contributions to this operation. The authors would like to express gratitude to Professors Y. Kimura, K. Takata, and S. Kurokawa for their continuous support. We also would like to thank the operators of Hitachi Ltd. for their skillful work and Dr. N. C. Bhattacharya for critical reading of the manuscript.

REFERENCES

1. K. Tsuchiya, N. Ohuchi, Y. Morita, et al., "Advances in Cryogenic Engineering," Vol. 37, Part A, Plenum Press, New York (1992), p.667.
2. N. Ohuchi, A. Kabe, K. Tsuchiya, et al., "Advances in Cryogenic Engineering," Vol. 37, Part A, Plenum Press, New York (1992), p.675.

3. K. Tsuchiya, N. Ohuchi, A. Terashima, et al., "Advances in Cryogenic Engineering," Vol. 35, Plenum Press, New York (1990), p.941.
4. Y. Morita, K. Tsuchiya, N. Ohuchi, et al., "Advances in Cryogenic Engineering," Vol. 37, Part A, Plenum Press, New York (1992), p.551.
5. K. Endo, K. Tsuchiya, N. Ohuchi, et al., IEEE Trans. Magn. Vol.28, (1992), p.542.

CONSTRUCTION AND COMMISSIONING TESTS OF A 10 kW CLASS HELIUM REFRIGERATOR FOR THE LARGE HELICAL DEVICE

S. Satoh,[1] T. Mito,[1] S. Yamada,[1] J. Yamamoto,[1] O. Motojima,[1] S. Moriuchi,[1] I. Ohtake,[1] M. Katada,[2] and T. Fukano [2]

[1]National Institute for Fusion Science
Furo-cho,Tikusa-ku, Nagoya, Aichi, 464-01, Japan
[2]Nippon Sanso Corporation
6-2 Kojima-cho, Kawasaki-ku, Kawasaki, Kanagawa, 210, Japan

ABSTRACT

The Large Helical Device is a heliotron fusion experimental device in which magnetic fields are controlled exclusively by superconducting coils: a pair of helical coils and three pairs of poloidal coils[1]. The helium refrigerator in the LHD cryogenic system has a design capacity of 5700 W refrigeration at 4.4 K, plus 650 L/h liquefaction and 20600 W refrigeration between 40 and 80 K. The contract was signed in July 1992 and the construction at the site (Toki-city, Japan.) was completed at the end of 1994. Subsequently, a series of commissioning runs were made and various design performances are being confirmed. These include the independent and combined runs of 8 compressors at full power under the helium operating conditions, a demonstration of the helium purifier, and the manual or automatic liquefaction operations that use VME controllers with a newly developed distributed control system software. These results demonstrate that the refrigerator has a satisfactory thermodynamic performance in general.

INTRODUCTION

The Large Helical Device (LHD) is a heliotron fusion experimental device in which magnetic fields are controlled exclusively by superconducting coils. As shown in Figure 1, around the plasma vessel that has a dumbbell-shaped cross section, there run a pair of helical coils and three pairs of poloidal coils which are supported by the thick-walled support structure that has a toroidal shape. The weight of the entire cold mass is 950 tons. Liquid helium for the pool-boiling cooled helical coil, supercritical helium for the forced flow cooled cable-in-conduit conductor of the poloidal coil, and two-phase helium for the support structure and the superconducting bus line are supplied by the external refrigerator through the lunch-box shaped cryostat (called Belljar).

The various estimated heat loads to the coldbox are listed in Table 1. The refrigerator described in this paper covers the load for PHASE I (3 Tesla plasma operation); that is, the refrigerator should have a capacity of 5700 W refrigeration at 4.4 K, plus 650 L/h liquefaction and 20600 W refrigeration between 40 and 80 K. A thermodynamically equivalent value is 9100 W @ 4.4 K or 2465 L/h. The heat load for PHASE II (4 Tesla) may require another similar size machine, however, the LHD upgrade for PHASE II will be made after PHASE I experiments which will be initiated at the beginning of 1998.

Table 1. Heat loads to the coldbox

	Phase I (Watt)			Phase II (Watt)	
	40 to 80 K	4.4 K	4.4 K	4.4 K	1.8 K
Conduction	5500	300	-	300	120
Radiation	5500	130	-	130	-
Conduit connection	-	400	-	300	100
AC loss	-	-	-	2100	280
SC-Bus Line, Power Leads	1000	1070	650 L/h	1070	-
SHE Circulation Pump Power	-	-	-	2500	-
Cryostats, Transfer tubes, Others	8600	3800	-	4300	200
Total	20600	5700	650 L/h	10700	700

DESIGN AND CONSTRUCTION

Design Considerations

A flow diagram of the LHD cryogenic system is shown in Figure 2. In this Claude cycle refrigerator, a pair of coldboxes, A and B (diagrammed as two rectangles, using dashed lines) have seven turboexpanders (T1...T7), 14 staged aluminum plate-fin heatexchangers (rectangles), and 1 liquid nitrogen reboiller/heat exchanger (left upper corner). The reboiller heat exchanger is used only for the initial cool-down of the superconducting coil system. The thermal radiation shields of the LHD are cooled by helium at temperatures between 40 and 80 K instead of 80-K liquid nitrogen. The upper and lower coils of the three poloidal coil pairs (hidden in the right middle dotted square) are cooled sequentially by 1 MPa circulating supercritical helium (SHE) which is driven by the room temperature compressors through the coldbox. On the other hand, two-phase helium is supplied to both the helical coils *via* the coil supporting structure and 9 sets of superconducting bus lines (also hidden) through individual valve boxes or cryostats (lower 4 dotted boxes). Eight compressors (C1...C8) are divided into two groups: a middle suction pressure train (C7 and C8) and an atmospheric suction pressure train (C1...C6).

Coldbox. Figure 3 shows a cycle scheme drawn on a temperature-entropy diagram of helium (not scaled). Three turboexpanders (T1,T2,T3) are driven by the compressors (C7 and

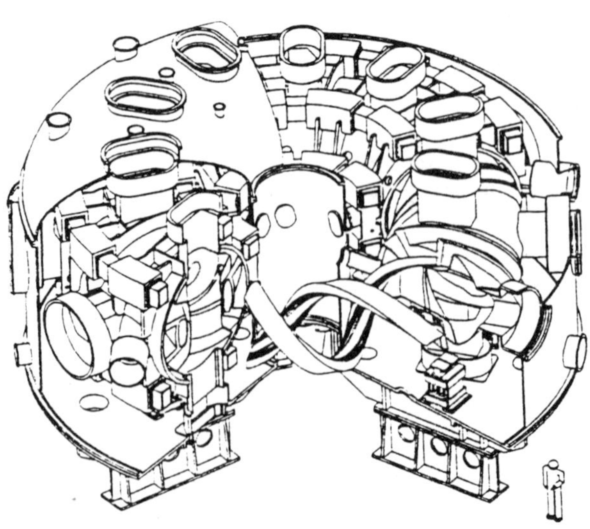

Figure 1. The large helical device.

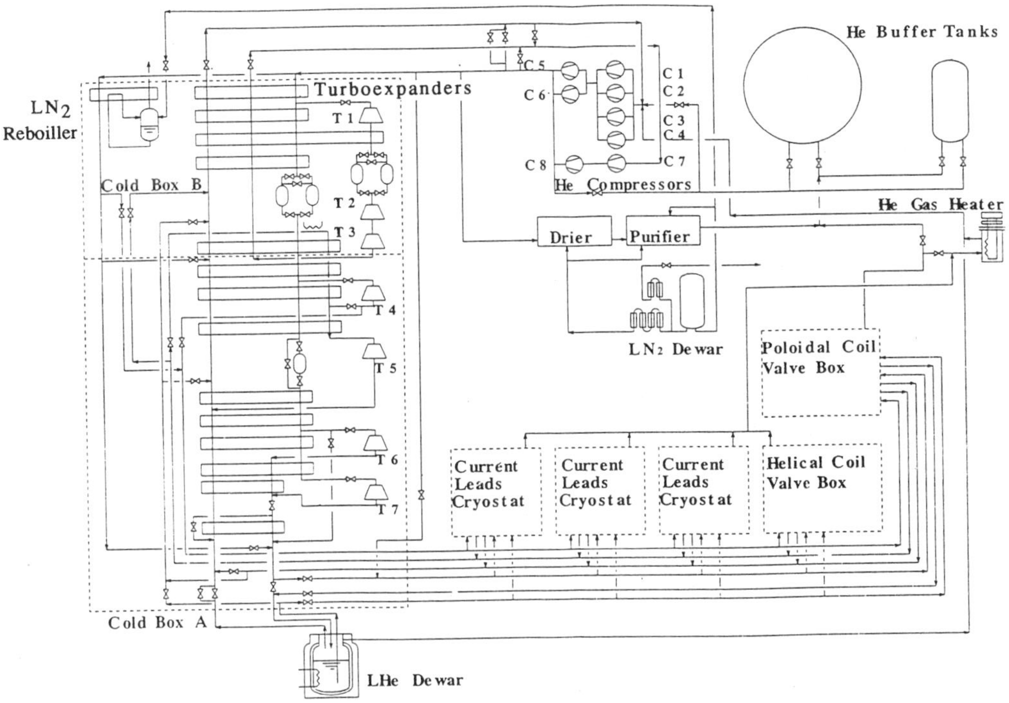

Figure 2. Flow-diagram of the LHD cryogenic system.

C8) in the middle suction pressure loop (218 kPa or 2.15 atm) and provide cooling which otherwise must be supplied by liquid nitrogen during normal refrigeration duties. The 21 kW heat load between 40 and 80 K is compensated mainly by the cold generated at T4 and T5. Two supercritical expanders (T6 and T7) exhaust SHE at a pressure of 1 MPa (9.96 atm), which is the supply pressure to the first half of the three poloidal coil pairs. In the high pressure lines at temperatures near 84 K (between T1 and T2), 74 K (between 4th and 5th heat

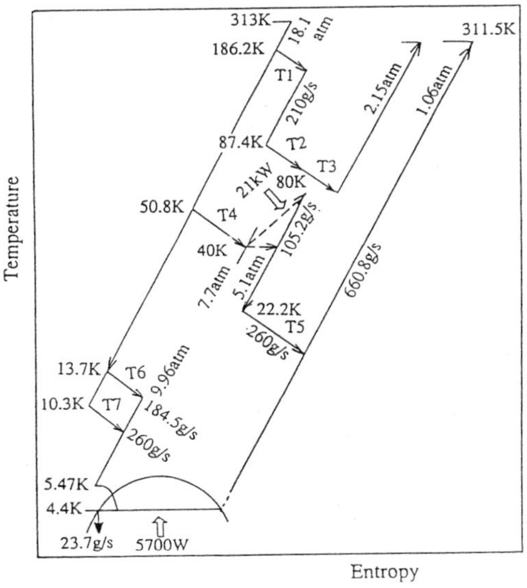

Figure 3. Cycle scheme.

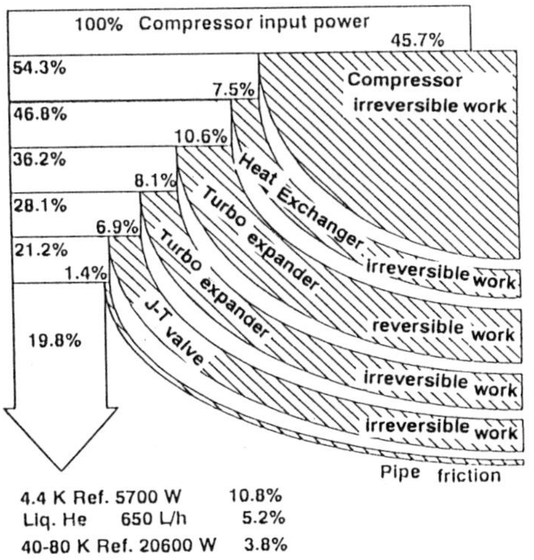

Figure 4. Exergy analysis.

Table 2. Compressor specification

Speed 3550 rpm

	Low-pressure loop			Middle-pressure loop	
	1st stage C1 to C3	1st stage C4	2nd stage C5, C6	1st stage C7	2nd stage C8
Number	3	1	2	1	1
Diameter, Male rotor (mm)	320	250	250	250	250
Discharge rate (g/s)	215	105	375	210	210
Inlet/Outlet Pressure (kPa)	101/412	101/412	397/1935	203/535	520/1935
Internal Pressure Ratio at He	3.87	3.87	4.73	3.01	3.35
Rated power (kW)	400	210	740	270	330

exchangers) and 20 K (between 8th and 9th heat exchangers), there are charcoal absorbers through which all helium is decontaminated continuously while the coldbox is operated during normal refrigeration duties.

Figure 4 shows a summary of the exergy analysis on the coldbox at its design specification. As is well known, exergy is a required power input to a refrigerator working at ambient temperature to obtain a given sequence of (or single) unit operations between an environment at ambient temperature and an environment at a significantly lower temperature; it is defined using only the property value of the state as:

$$e = h - h_u - T_u (s - s_u) \qquad (1)$$

where T, s, h, and e are temperarute, entropy, enthalpy, and exergy and safix u denotes ambient state. Regarding Figure 4, some information is worth noting here:

1) Only 54.3% of compressor net input power(3204 kW=100%) is effective at the warm end of the coldbox. In other words, the average isothermal efficiency of 8 compressors is assumed here to be 54.3%.
2) The sum of exergic work for refrigeration (5700W) at 4.4 K, liquefaction (650 L/h) and refrigeration (20600W) between 40 and 80 K corresponds to 19.8% of the compressor net input power, which means that the percent Carnot efficiency of the present coldbox is 19.8%. This figure is marginal at best if one considers the size effect mentioned below.
3) From the exergic sum, an equivalent capacity during single duty is calculated to be 9.1 kW refrigeration at 4.4 K or 2465 L/h liquefaction. Those values support the claim that this coldbox currently ranks as one of the largest refrigerators in the world.
4) The total exergy for gas expansion through 7 expanders amounts to 18.7%, of which 10.6%. is reversible (hence recoverable) and 8.1%. is irreversible work.
5) The irreversible work for J-T valve expansion is 6.9%. This is favorably improved value mainly due to SHE expanders T6 and T7; however, the value may be further reduced if the SHE supply pressure to the poloidal coils (or exhaust pressure of the SHE expanders) can be reduced from the current 1 MPa to 0.7 MPa, which is a value that corresponds to the use of a SHE circulation pump on the occasion of a later LHD upgrade.

Table 3. Turboexpander specification

	Middle-press. loop			Low-press. loop		Supercritical	
	T1	T2	T3	T4	T5	T6	T7
Expander/Brake Wheel dia(mm)	45/63	45/80	45/80	32/56	40/80	18/50	18/50
Journal bearing dia. (mm)	45	45	45	32	45	32	32
Expander flow rate (g/s)	210	210	210	260	260	184.5	260
Inlet/Exhaust press. (MPa)	1.78/1.25	1.25/0.62	0.62/0.24	1.78/0.78	0.52/0.12	1.76/1.02	1.76/1.01
Speed (rps)	1700	1700	1700	2500	1700	1800	1500
Design expansion efficiency(%)	69	69	70	76	72	65	65

Table 4. Other key components

1.	Liquid Helium Storage Tank	Capacity 4.4 K Heat leak(measured)	20000 L 0.71 %/day
2.	Liquid Nitrogen Storage Tank	Capacity 80 K Heat leak(measured)	50000 L 0.24 %/day
3.	External Gas Purifier (Twin Molecular Sheeve Bed Dryers and Twin Charcoal Bed Cold Absorbers)	Capacity Switch over cycle	1000 Nm^3 /h 12 h
4.	Middle Pressure Gas Holders	Capacity A 1-Spherical B 1-Cylindrical Design pressure	700 m^3 100 m^3 2.0 MPa
5.	80K Radiation shielded, Vacuum insulated Multi-line assembling Transfer tubes	4.4 K Heat-inleak(estimated) TRT3: 6 lines-650 mm OD-68 m TRT2: 5 lines-550 mm OD-50 m	 5.0 W/m 6.4 W/m
	Vacuum insulated Multi-line assembling Transfer tube	TRT1: 3 lines-500 mm OD-21m	16.0 W/m

Manufacturer: 1. Cryenco (USA), 2. 3. 5. Nippon Sanso, 4A. Tohoku Elc. Mach. Works; 4B. Tsurumi Ind. & Const.

Compressors and Turboexpanders. The eight helium circulation compressors are oil injected variable-capacity Leishorm (screw) type, individually driven by electric motors and manufactured by Mayekawa Mfg Co. As shown in Table 2, in the low-pressure loop, the 4 first-stage compressors take 750 g/s helium at 101 kPa and successively the 2 second-stage compressors discharge it at 1935 kPa. In the middle-pressure loop, a couple of 1st and 2nd compressors compress 210 g/s helium from 203 to 1935 kPa. The three compressors C1..C3 have the maximum rotor size the manufacturer can supply. The sum of rated power of the electric motors is 3700 kW to circulate 960 g/s or 19365 Nm^3/h helium. The oil removal system consists of 8 individual primary separators after each compressor, each one 2nd & 3rd oil-filter and 4th charcoal absorber on a discharge collecting line. The system removes oil and water from helium leaving measured impurities at less than 1 ppm wt. The oil contamination (as well as N_2 and H_2O) is continuously monitored as C_XH_Y using an optical emission spectroscopy gas analyzer with an oil pyrolyser. The manufacturer of the analyzer is Linde Kryotecnik AG. Cooling water for injected oil is always assured by a redundant water pump and an emergency power generator running in the water cooling-tower system.

Each of the seven turboexpanders is composed of a rotor with a mixed-flow type expansion wheel and a similar brake compressor impeller supported by gas-lubricated bearings. The manufacturer is Linde Kryoteknik AG. Some of the design specifications are listed in Table 3. The four expanders T1..T3,T5 have the maximum machine size the manufacturer can supply. Flow rates to the expanders are controlled by flow controlling valves; one for the train of T1,T2,T3 and one for each of the rest. The 54 cryogenic valves for the coldbox are supplied by Weka AG. Cooling water for removing heat from a brake compressor loop is always assured by the same redundant water pump mentioned in the previous paragraph.

Miscellaneous. The specifications on other key components are listed in Table 4. The external gas purifier removes contaminants from the surfaces of coil-conduits, pipes, etc, at the initial start-up, or after maintenance or repair works. The measured residual values of contaminants are less than 1.2 ppm. The flow rate of 1000 Nm^3/h is about 5% of compressor discharge flow.

Two middle pressure gas holders are used for different reasons. A 700 m^3 holder is used for balancing the helium inventory in the LHD, and it may also be used for buffering a sudden pressure increase due to a coil quench or an urgent current dumping. Since the density ratio of saturated liquid to normal state gas for helium is roughly 700, the volume of the holder gives an operator a convenient tip that 1 atm pressure decrease in the holder corresponds to a 1000 L storage increase in LH_e dewar. A 100 m^3 holder is used to resume vaporized/superheated gas within the LHD during a complete compressor shut-down or maintenance period, although a complete balance of gas between the holders and the inventory at coils may require several hundred cubic meters additionally. In the mean time, most of the helium inventory must be liquefied and stored in the LH_e dewar before a long shut-down period of the refrigerator is scheduled. A multi-line transfer tube TRT-3 connects the coldbox in a machine room and the valve boxes for coils which are in the LHD experimental room. Although the two rooms are located adjacently on the ground level, they are separated by a 2-m thick concrete wall, and all pipe and wiring must be connected together *en route* underground to reduce the neutron irradiation to the exterior. TRT-2 connects the coldbox to the current lead cryostats which are underground beneath the coldbox. TRT-1 connects the coldbox and the LH_e dewar.

The estimated 4.4 K heat leaks shown in Table 4 are rather moderate values considering that TRT-3 and TRT-2 contain many vacuum partitions with detachable tube-connectors which easily introduce a lot of heat-inleak by thermal conduction. The listed values were attained by repeated R&D, and compared to the initial design they give savings of 440 W at 80K-level and a 570 W at 4.4 K.

Lastly, the control system incorporates recent novelties both in hardware and software. It utilizes the excellent performance of a modern VME-bus controller so the programs for a distributed control can be written easily by any user with EWS, LUN and the widely prevailing OS such as UNIX, Vx-Woks, etc. A new multipurpose digital-instrumentation software is developed for process operation. In addition to easy accessibility, it offers a lower price, plus system flexibility and an expandability which enables on-line/off-line maintenance and remote monitoring.

Construction at the Site

Construction started in March 1994 and was completed in December of the same year. The coldbox is divided into two sectors due to a transportation size limit. Each compressor mounted on a common bed with its motor, oil-cooler, and oil-scrubber took only a week to be installed. The compressor room is 50 m from the coldbox so connecting pipes ran in a tunnel underground. Other than a flooding accident in the tunnel on the occasion of a typhoon in September, the construction progressed smoothly. In Figure 5 a bird's-eye view of the system is shown. For vessels having wider projection-area to the LHD such as the coldboxes, and LN_2 or LH_e tanks, the exterior shell materials are limited to austenitic or non-magnetic steels that will not magnetize and radiate an error in magnetic field during the plasma confinement. Also, the foundations of those tall vessels are designed earthquake-proof assuming a 0.3 G horizontal acceleration. An oxygen deficiency monitoring system is introduced to avoid hazards to humans in the machine room and the basement where LN_2 and LH_e tanks are installed, even though the blowdown gases from safety valves (including the ones for superconducting coils) are exclusively guided outside through a duct of 950 mm diameter and 150 m length.

Figure 5. Coldboxes, External Gas Purifier, Storage Tanks and Transfer tubes.

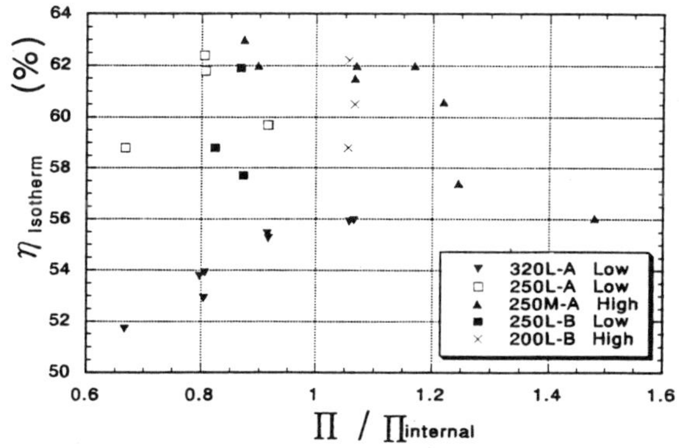

Figure 6. Compressor efficiencies.

PERFORMANCE

Component Tests by the Fabricators

In addition to routine inspections of materials and structures, several thermodynamic performance tests were conducted on the machines while they were at the fabricator's shops. Typically, several reasons such as power or size limitations, very low temperature, and the expensive of helium may limit a performance test at the shop, especially when it is a test of the largest machine they ever made. The compressors were run individually using Freon and air, turbines were run using room temperature helium, and the multi-line transfer tubes were cooled using liquid nitrogen. Those tests showed some basic machine characteristics. Results from those tests are published elsewhere[2].

Commissioning Run after On-site Assembly

Compressors. Some combined runs in helium were conducted using three groups of compressors:

Group A-1: 1 high-press. stage (C5) and 2 identical low-press. stages (C1,C2).
Group A-2: 1 high-press. stage (C6) and 2 different low-press. stages (C3,C4).
Group B : 1 high-press. stage (C7) and 1 middle-press. stage (C8) on the middle pressure suction loop.

Here, C1, C2, and C3 are identical machines, and so are C5 and C6. Considering the difference of the intermittent pressure, which is the pressure between the 1st and 2nd stages

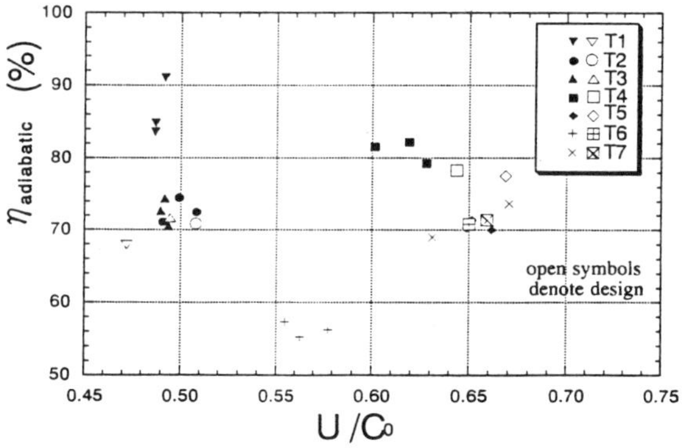

Figure 7. Turboexpander efficiencies.

for each of the above three groups, the independent run as single group, as well as the combined runs with two/three groups, give an efficiency curve for the compression pressure ratio for a given machine. Normalizing the compression pressure ratio with the internal rotor pressure ratio, which is produced by the stroke volumes of a given machine with a given gas, and assuming that identical machines have the same efficiency, the isothermal efficiencies for the 5 different machines can be obtained as shown in Figure 6. Considering the rather narrow range of ordinate, we see a general tendency, that the maximum isothermal efficiency is highest near the normalized compression ratio of 1, and all efficiencies at full operation (slightly below the ratio of 1) fall between 55 and 62%, which is higher than the assumed average of 54.2% adopted in Figure 2. Other tests on mechanical soundness, noise levels, and functions of the oil-redistribution system were also conducted satisfactorily.

Turboexpanders and Coldbox. After the instrumentation tuning and gas preparations, the first refrigeration operation was conducted in March, 1995. Without liquid nitrogen precooling, it took about one day to cool the coldbox and LH_e dewar down to the liquefaction temperature. In this operation, built-in electric heaters in the coldbox and LH_e dewar simulate the heat loads in the 40-80 K range and at 4.4 K respectively. After setting heater currents at their design heat loads, we measured the liquid level increase rate for a continual four-hour steady state operation while pressures, temperatures, and flow rates were surveyed every 1 minute and their histories were recorded. The isentropic (adiabatic) efficiencies of the seven turboexpanders are calculated using those property values and are shown in conjunction with their speed ratios in Figure 7. A speed ratio,U/C_0 indicates how close each expander is running to its optimum speed which gives the highest efficiency and is defined as the ratio of a peripheral velocity, U of the expander wheel to a spouting velocity C_0, which is obtained by the ideal expansion of the potential energy across the expander. Since at U/C_0 around 0.65 the highest efficiency can be obtained for a mixed-flow expander in general, the measured efficiencies of T2..T5 are reasonable values as obtained by the present measurement accuracy. On the other hand, the efficiencies of T1, T6, and T7 are suspect. This may be claimed due to a possible transient excess thermal flow into T1 from a large heat capacity heat exchanger at a higher temperature level, whereas a residue between two small temperature values involved a large error for T6 and T7. Nevertheless, it may be pointed out that the efficiency measured here are generally 2 to 4 points higher than the values expected in the exergic analysis in Figure 3.

Finally, an actual "Percent Carnot Efficiency" and the real capacity of the refrigerator are estimated using the measured property values, heater load applied, and the increase rate of liquid level. The results are summarized in Figure 8 which indicates a larger capacity and a higher efficiency than the design values. The cause of the difference may be attributed to the contingency made in the design process and a coservative assumption on the efficiencies of the component machines.

SUMMARY

The construction of the cryogenic system for the LHD took 35 months, 19 for fabrication and inspection at the manufacturer's shop and the other 16 for installation, piping, instrumentation, and testing. Since the helical coil set and the largest poloidal coil set are still in the coil-winding process in the LHD experimental hall, the refrigerator will be operated

Summary of the Helium Refrigerator for LHD

1.Capacity;	Measured Heat Input	Exergy
Ref. @ 4.4 K	6.0 kW...	403.1 kW
Liq.	643 L/h...	146.9 kW
Ref. @ 40-80 K	22.1 kW ..	103.9 kW
Total Exergy		653.8 kW

Equivalent Capacity <u>9.7 kW Ref. @ 4.4 K</u>

or <u>2860 L/h LHe</u>

2.Plant Efficiency;	Measured Power Input
Compressor @ 295K	2699 kW
Total Exergy	653.8 kW

Percent Carnot Efficiency=653.8/2699

=<u>0.242</u>

Figure 8. Summaries of measured cycle efficiency and capacities.

intermittently for machine tuning, developing coil cooling mode control programs, and training operators in preparation for plasma experiments to be initiated at the beginning of 1998. In summarizing the paper, we note the following;

(1) A 10-kW-class helium refrigerator for the LHD is introduced here by showing design considerations, thermodynamic analysis, and operating features.

(2) The thermodynamic performances of the coldbox are measured and the results are discussed with separately measured compressor performances and simultaneously obtained turboexpander efficiencies, which showed a high "Percent Carnot Efficiency" of 24.2% as well as a large equivalent capacity of 9.7 kW at 4.4 K compared to estimates made in the design stage.

ACKNOWLEDGMENT

The authors wish to acknowledge the staff of the NIFS LHD construction group for their continuous support in every aspect of the construction, Mayekawa Mfg Co., for their cooperation in fabricating and testing the helium compressors, and Linde Kryoteknik AG., for their cooperation in fabricating and testing the turboexpanders.

REFERENCES

1. Motojima, O. et al., Physics and engineering design studies on LHD, Fusion Engineering and Design, 20, 3-14 (1993).
2. Satoh, S. et al., Construction of a 10-kW-class helium cryogenic system for the Large Helical Device, Cryogenics, 1994, vol. 34, ICEC supplement 95-98, (1994).

AN EXPERIENCE IN THE MAINTENANCE OF A LIQUEFIER FROM THE T-15 CRYOGENIC SYSTEM, RESULTS OF ITS RELIABILITY AND CAPACITY ENHANCEMENT

A.Baranov, V.Duzhev, G.Kashirskikh, A.Mikhailov, A.Ugrovatov, and V.Zhulkin

RRC Kurchatov Institute , Institute of Nuclear Fusion
Moscow, 123182, Russia

ABSTRACT

For the production of the liquid helium and realization of technological operation modes of cooling down, steady state and warming up cryogenic objects at Tokamak T-15 facility liquefiers with the present flow rate, 0.222 L/s (800 L/h) each, are used. In 1988 the liquefier was put into operation under conditions without implementation of the third turbine stage, at the liquid helium flow rate 0.0694 - 0.0972 L/s (250 - 350 L/h), for the time of continuous operation of 200 hours long. Since 1988, work has been done to enhance its reliability and capacity. This includes:

Simplification pneumatic-hydraulic circuit-diagram of the liquefiers for increasing reliability.

Start-up of the third turbine stages of the liquefiers. Piston expander and two variant of turbines were tested, only the last variant has demonstrated reliable and effective operation.

Work out and realization of technological and circuit-diagram solution (usage of preliminary helium purification from neon, development and assembly of a neon adsorber) against freezing neon and hydrogen impurities - changing the mass flow ratio - in the valves and heat exchangers.

Test of various prototype main turbines. Many prototype main turbines with various work pressure and various mass flow were tested. Turbines with optimal parameters have been selected.

At present, the liquefier has the flow rate equal 0.222 L/s (800 L/h), at the compressed helium flow rate - 0.222 kg/s (800 kg/h), and the inlet liquefier pressure - 2.2 MPa.

INTRODUCTION

The liquefier, OГ-800, was designed and manufactured at the Scientific Industrial Corporation "Cryogenmash" specially for the cryogenic system at "Tokamak-15" facility. In the OГ-800 diagram a two-loops-included cryogenic system (CS) idea was realized, i.e. the idea of separating the liquid helium production from its utilization for refrigerating purposes in various apparatus. Thus, the main task of the liquefier is to

Advances in Cryogenic Engineering, Vol. 41
Edited by P. Kittel, Plenum Press, New York, 1996

produce liquid helium. Utilization of the liquefier as a refrigerator is limited to the regimes of cooling down in the temperature range 110 - 20 K.

The CS-diagram provides the simultaneous helium purification and its preliminary cooling to 80 K in an independent low temperature helium purification unit. The liquefier diagram includes heat exchangers operating in the temperature range 80 - 4.5 K, set of the main turbines, final stage expander, locking and controlling valves, as well as temperature, pressure, level, and expenditure sensing elements. The initial liquefier diagram is given in Fig.1.

A great number of valves was provided by the designers for maximal-effective use of the turbo expanders at all stages of cooling. The design allowed the expanders to be switched "ON" optimally for each temperature level. It was also expected that the replacement of the main turbines by the spare ones would be realized by a simple resetting of the valves. The use of a piston expander with an opportunity to regulate its capacity over a wide range as a expander at the final stage. The liquefier diagram did not provide the use of any adsorbers for additional helium purification to protect the turbines and valves against any damage by solid impurities in helium. Mounting for filters with different surface areas of filtering and with different mesh cell dimensions was provided.

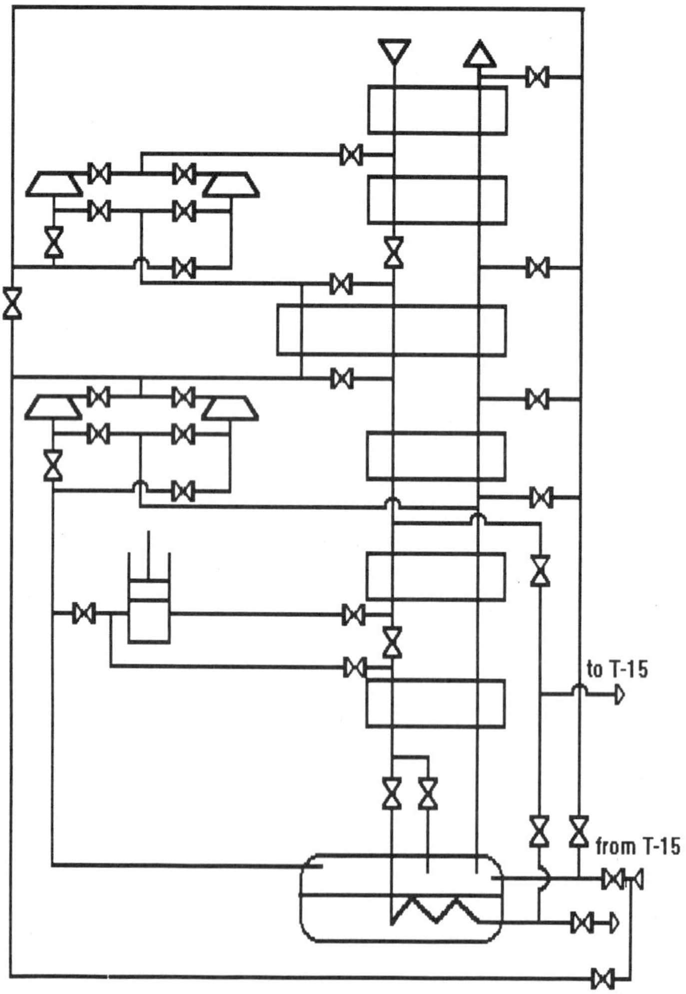

Figure 1. The initial diagram of liquefier.

Table 1. Design parameters of turbines

Parameters	1 stage	2 stage	3 stage
P_0 (MPa) the pressure in front of a nozzle	2.30	0,71	2.30
P_2 (MPa) the pressure beyond the impeller	0.73	0,12	0.25
T_0 (K) the temperature at the entry to the turbine	58.3	27.5	10.2
G (kg/sec) the mass flow	0.111	0.111	0.111
Isoentropic efficiency	0.67	0.7	0.85

The structural realization of liquefiers was oriented to the maximal performance of assembly and test at the plant site. The liquefier is a single unit including the frame upon which the unit of heat exchangers, unit of the main turbines, final stage expander unit were mounted and connected via pipelines. The heat protective and vacuum-sealed cases of the units during mounting/dismounting, repair and tests are elevated with a crane. The heat exchangers within the unit are placed at the same level. Storage for liquid helium is located at the top (diagram is opposite to the traditional one).

MODIFICATION OF LIQUEFIERS

Simplification of a diagram

As mentioned, a great number of valves was installed in the liquefier for the most effective realization of the cool down regimes. The second side of this was a noticeable reduction in the liquefier reliability, as a whole. Even an insignificant passage of helium through the locking valve results in negative consequences (considerable reduction in the liquefier efficiency). As a result of the changes done together with the "Cryogenmash"-specialists the valves of spare turbines were excluded. The replacement of turbines in case of their failure is realized by disassembly of broken one and by the assembly of a new machine. This excluded the variety of regimes for turbines, and it simplified their control.

A great number of filters with underestimated surface and with overestimated requirements of the size of particles passing through them resulted in an increase in the hydraulic losses within the liquefying path and, respectively, in reduction in their capacity. The doubling filters with small surfaces were removed. This allowed us normalize the hydraulic losses.

Selection and modification of turbines

Turbines for the liquefiers were developed and manufactured by "Cryogenmash".

The mowing part consists of bearings, brakes, and rotor. The bearings are hydrodynamic, three-wedged ones, they have a separate oil feed to the radial and end-face surface. A low viscosity oil is used.

Table2. The calculated parameters of turbines

Parameters	1 stage	2 stage	3 stage
P_0 (MPa)	2.30	0.90	2.28
P_2 (MPa)	0.95	0.122	0.25
T_0 (K)	54.3	21.6	8.6
G (kg/sec)	0.130	0.130	0.0927
Isoentropic efficiency	0.7	0.7	0.66

The power in a turbine is removed by friction in bearings and a cylindrical oil brake, the oil from turbines is drained to an oil tank. In the turbines of the second stage the brake is not used (the whole energy is removed in bearings).

The channeling part of turbine includes a directing apparatus and an impeller. The directing apparatus has wing like blades. The impeller has a covering disc. The impeller, a radial-axial type, is fastened to the upper cantilever of the rotor.

The end-face labyrinth seals were done on the covering disc to reduce the leakages of helium from its flow. The labyrinth seals are produced by cutting ridges in a graphite bushing. At the place of the shaft entry to the cool zone (from a warm one) there is also a step-like labyrinth seal. An oil rejecting bushing is locate at the bearing side. The feed of warm sealing helium into the middle of the labyrinth seal is organized to reduce the leakage of cool helium.

The necessary pressure of sealing helium is provided by a pressure drop regulator which controls the pressure beyond the directing apparatus and feeds the warm helium, at a pressure 0.005 - 0.01 MPa lower than that beyond the directing apparatus.

A piston expander with electromagnetic inlet and outlet valves was used as the third stage. Using a piston expander in the end stage of a liquefier turns out to be rather problematic. Some attempts of its triggering have shown that the expander design is far from perfect, and the machine does not provide the desired performance.

The start-up of liquefier maintenance manifested a number serious deficiencies in the main turbines. They are:

- low efficiency (for the first stage and for the second was 0.6 -0.65):
- turbines (at the inlet pressure 2.0 MPa) only allowed the passage of helium to the facility at 0.153 kg/s (550 kg/h) (insufficient flow rate characteristics of turbines);
- initiating strong noise under turbines operation that is characteristic of a given design of dynamic bearings;
- reduction in the cooling capacity in connection with wear in graphite labyrinth seal on the covering disc and an increase in leakage past the impeller;
- possibility of the entry the oil entry into the channeling part from the oil bearings.

An average capacity in 1988 - 1989 was equal 0.0778 L/sec (280L/h); in the run during 1990, after a small modification in the turbines, ОГ-800, at the flow rate of 0.167 kg/s (600 kg/h) the capacity was increased to 0.1 L/s (360 L/h).

The turboexpander third stage, was designed and manufactured in 1991, for utilization of ОГ-800 in the gas-expansion machine regime, by "Cryogenmash". In its structure, the turbine of the third stage differs from the turbines of the first/second stages by implementation of gasostatic type bearings, (without oil usage), with separate helium feed to the radial, radial thrust and thrust bearings. The third stage module of turboexpander is mounted on the aggregate of piston expander (impeller down). The channeling part consists of a directing apparatus, channel type, including six channels, and the impeller with covering disc. The produced power is spent on compression of helium in turbocompressor placed upon the rotor cantilever, opposite to the impeller. The labyrinth seal are similar to those in the turboexpanders of first/second stages. The ОГ-800 with the third stage was tested in 1991. On average, the capacity was 0.119L/s (430 L/h) at the helium consumption 0.161 kg/s (580 kg/h) (input pressure at the facility was 2.15MPa). In that case the third stage efficiency was about 0.4.

The detected deficiencies in operation with the third stage of "Cryogenmash":

- rapid wear in graphite end-face labyrinth seal and, as a result, large leakage from the working flow past the impeller, through the end-face labyrinth seal;
- complexity in the triggering procedure;
- unstable operation of the moving part in the turboexpander;
- passage cross-sections in the directing apparatus and in the impeller did not provide a helium flow rate to the liquefier greater than 0.167 kg/s (600 kg/h).

Table 3. The parameters of turbines under gas-expansion regime

Parameters	1 stage	2 stage	3 stage
P_0 (MPa)	2.10	0.95	2.1
P_2 (MPa)	1.07	0.145	0.223
T_0 (K)	58	26	9.3
G (kg/sec)	0.122	0.122	0.0944
Isoentropic efficiency	0.56	0.7	0.55

Thus, the implementation of a turboexpander at the third stage did not result in a noticeable increase in capacity.

Such maintenance parameters were evidently insufficient for the tasks of T-15 maintenance. In order to increase the OΓ-800 capacity to 0.222 L/s a mathematical model of liquefier heat exchangers, taking account of real geometry for liquefiers, has been developed and the calculations of new turbines (1,2 3- stages) have been done. For the capacity of 0.222 L/s , at the helium flow rate 0.222 kg/s through the liquefier, the optimal parameters for the turboexpander calculations have been produced and given to a few Firms as RFP on the development and manufacture of highly-effective and reliable turboexpander sets. Then, the comparative tests of prototypes were done. On the basis of comparison, the turboexpander design produced by "Heliummash" was chosen. The calculated parameters are given in Table 2.

Description of the turboexpander design, produced by "Heliummash". Each stage of the turboexpander has the similar design, differing from each other by dimensions only. Helium is fed to the directing apparatus along the annular channel. The directing apparatus has nozzles of the channel type. The turboexpander is a centripetal turbine with the impeller cantilever-located upon the shaft. The impeller is of radial-axial type without covering disc.

The shaft rotates in the combined gas-oil supports. The upper bearing (located closer to the impeller) is a gasostatic one, the lower radial bearing and two thrust bearings are hydrostatic ones. The radial gaseous and oil bearings have the same diameter and the same radial gaps that allows one to manufacture the moving part of the turboexpander with high accuracy. To reduce the leakage of cool helium some labyrinth seal are done. Into their middles warm helium is fed, preventing the cool leakage.

The design of bearings and that of a shaft allow one to essentially reduce the gap in the labyrinth seal and to reduce the cool leakage to a minimum.

The results of the first OΓ-800 - tests with the turbines made by "Heliummash" before optimization of the parameters.

Capacity under throttle regime: 0.125 L/s (450 L/h), at the OΓ-800 flow rate equal 0.225 kg/s, under inlet pressure 2.2MPa;

Capacity under gas-expansion regime: 0.181 L/s (650 L/h), at the OΓ-800 flow rate 0.217 kg/s, under inlet pressure 2.2 Mpa.

The characteristic parameters of turbines under gas-expansion regime are given in Table 3. The calculation results were corrected by the results of testing the turbines within the OΓ-800 helium liquefier structure in order to optimize the parameters of a

Table 4. The final parameters of turbines under gas-expansion regime

Parameters	1 stage	2 stage	3 stage
P_0 (MPa)	2.15	0.95	2.18
P_2 (MPa)	1.10	0.145	0.242
T_0 (K)	58	27	9.6
G (kg/sec)	0.132	0.132	0.0944
Isoentropic efficiency	0.6	0.71	0.68

turboexpander channeling part. The optimized calculated data were given to "Heliummash".

After modification of turbines the following results were produced: capacity under gas-expansion regime 0.222 L/s (800 L/h)at the compressed helium flow rate through ОГ-800 0.226 kg/s under pressure of 2.25 MPa at the liquefier entry.

Characteristic parameters of turboexpanders under gas-expansion regime after modernization are given in Table 4.

The ОГ-800 - liquefier capacity was determined by calculations, using reference points, and by sinking the gasholder bell under regimes of filling the liquefier storage.

To exclude the possibility of damaging the turbines because of an oil pressure drop in the bearings and because of an excess over the maximal rotation speed an instrumentation protection system closing the working helium flow entry to turbines was developed for the case when the following parameters emerge:

- Oil pressure in the oil feed collector is below 1.4 MPa;
- Helium pressure in radial bearing (1 stage) is lower than 1.9 MPa;
- Rotor rotation frequency (1 stage) is higher than 18850 rad/s (180000 turns/min);

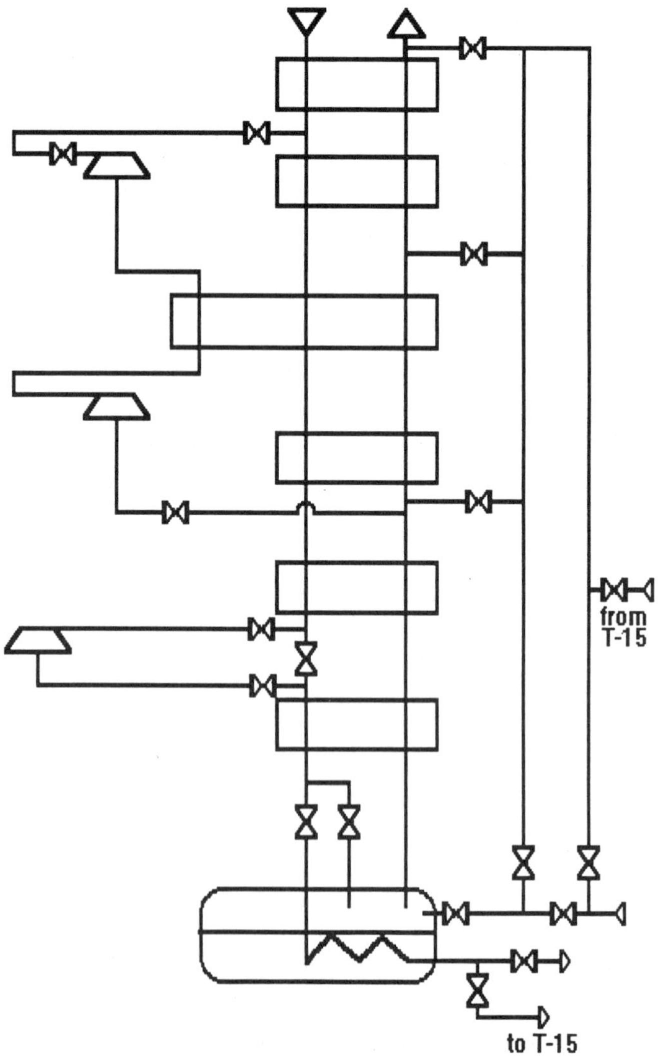

Figure 2. The final diagram of liquefier.

- Rotor rotation frequency (2 stage) is higher than 17300 rad/s (165000 turns/min).

Further maintenance has shown a high reliability of apparatus protections that allows is to provide a reliable operation of turbines and avoids the emergency of faults on them.

From the results of tests the rules for preparing turbines for operation, their start-up and shut down were developed.

New elements in the diagram

An experience in the work with liquefiers during long experimental turns has shown that the neon included in helium as an impurity is accumulated on the walls of heat exchangers pipes and in the cross-section of the valves, changing the characteristics. The designed system of preliminary helium purification turned out to be inconvenient for maintenance and did not perform its task completely. Neon adsorbers directly in the liquefier (ahead of two last heat exchangers) were developed and mounted as an alternative. The realized tests showed the reliable helium purification of neon and hydrogen.

In the process of testing and maintenance the turbines made by "Heliummash" some damages to gaseous bearings in them took place. The analysis done has shown that those damages were caused by an insufficient purity of the helium given to the bearing under transitional operating conditions (at a low pressure beyond the compressor the helium feed is performed through a reductor from the stored gas vessels). The helium purification system was additionally developed and mounted at this line.

CONCLUSION ABOUT THE WORK DONE

A considerable volume of work has been done for the period since the first start-up of liquefier. It allowed us to produce the liquefier capacity equal 0.222 L/s, to considerably-simplify its maintenance and to increase its reliability. At present, the liquefier operates reliably and stable, providing technological regimes in the T-15 cryogenic system.

TOKAMAK-15 MODERNIZATION AND AN ANALYSIS OF CRYOGENIC SYSTEM OPERATION FOR THE PERIOD FROM 1988 TO 1994

V.E. Duzhev , V.F. Zhulkin , G.M. Kashirskikh, and
A.E. Ugrovatov

RRC Kurchatov Institute , Institute of Nuclear Fusion
Moscow, 123182, Russia

ABSTRACT

The T-15 cryogenics system has been designed for cooling down, cryostatting, warming up of superconducting, cryoresistive and cryogenics T-15 objects. Maintenance of the cryogenics system has been on going since 1988.

For the mentioned period, in the cryogenics T-15 system

- The capacity of screw compressor was increased from 0.181 kg/s to 0.236 kg/s (third stage compressors with increased capacity were developed and manufactured), their reliability was also enhanced.
- The capacity of liquefiers was increased from 0.0833 - 0.0972 L/s (300-350 L/h) to 0.222 L/s (800 L/h) due to replacement of turboexpanders by more effective ones and due to introduction of an end-stage turboexpander into maintenance.
- The heat influxes to the cryogenics pipelines were reduced by 50%.

For the same period some technological regimes of cryogenics system have been developed to produce the maximal output of cold.

The cooling down from 110 K to 15 K is done, when one or two liquefiers are in operation under refrigerating conditions with the reverse flow splitting.

The further cooling is performed under joint operation of two liquefiers; one them operates in the liquefying mode, another, in the refrigerating one with excess reverse flow. A change in the operating conditions was necessary because of the impossibility of regulating the distribution of the reverse helium flow between two liquefiers at the temperature below 15K.

The main regime at the level of 4,5 K is a two-loop operating diagram, when one liquefier and a passive refrigerator with excessive reverse flow are in operation, the refrigerating capacity is about 3 kW.

INTRODUCTION

T-15 cryogenics system was developed by Scientific Industrial Corporation "Cryogenmash" to provide the technological regimes for T-15 facility. The first tests of the system elements and the first experimental run at TOKAMAK-15 took place in 1988.

Advances in Cryogenic Engineering, Vol. 41
Edited by P. Kittel, Plenum Press, New York, 1996

The cryogenics system includes:

-The helium compression plant based on screw compressors. The plant includes three compressor aggregates with the discharge pressure 2.4 MPa. The pumping-in compressors at the discharge pressure 32 MPa and with the capacity 0.05 m^3/s are used for filling the storage vessels.

- System of helium storage and thermostatting. The system has two lines of purification and liquefying, each of them includes an oil/water purification unit, low temperature purification of gaseous impurities, helium liquefier. Each liquefier has the rated capacity 0.222 L/s, at the gaseous helium flow rate 0.222 kg/s. Moreover, there is a passive refrigerator with the capacity 2000 W and a reservoir for liquid helium with the capacity of 40000 L.

- Nitrogen system. The nitrogen system has liquid nitrogen storage with the capacity of about 100000 kg, nitrogen liquefier for condensing 1 l/s of liquid nitrogen, centrifugal pumps for providing the liquid nitrogen circulation through the channels of electromagnetic coils and heat-protective screens.

- Two systems for providing the auxiliary plasma heating: microwave heating and injectors of neutral atoms.

- Control system. During the whole period of maintenance for provision of rated parameters and for the reliability enhancement the equipment underwent further development and modernization.

MODERNIZATION

Reliability and capacity of screw compressors

The screw compressor aggregates “КАСКАД-80/25” - used in the T-15 cryogenics - were developed specially for large cryogenics system. This aggregate consists of two compressor aggregates “КАСКАД-40/16” used as the first two stages, and of one aggregate “КАСКАД-6/16-25” for the final compression. In total, there are five compressors in the aggregate “КАСКАД-80/25”: two compressors at each first and second stages, one compressor of the third stage. All the compressors have an oil spraying in the compression cavity.

The first start-ups of the compressors have shown their insufficient reliability and capacity. Their reliability have been improved by cleaning of all the communications of contaminants, development of electrical motors (replacement of bearing by those of longer service life), change in the design of bearing units in the compressors. Compressor oil coolers using finned pipes were replaced by similar ones with smooth pipes that allowed one to exclude the entry of water into the oil (finned pipes used to be broken rather often because of vibrations). All the realized works allowed one to prolong the service life under continuous operation of the compressor aggregates up to 1000 hours before the first failure. Unfortunately, for the initial period of maintenance, as a result of strong wear, the serviceability and fitness to repair of compressors was exhausted up to a considerable degree.

The compressor aggregates “КАСКАД-6/16-25”, used as the third stage, were made as experimental ones. The tests have shown that their design was far from being perfect. It did not allow one to use, in full measure, the capacities of the first two stages (three stage aggregate capacity did not exceed 0.167- 0.180 kg/s). After realized compressor modernization the total aggregate capacity is determined by the capacities of the first two stages, being equal 0.236 - 0.250 kg/s.

Helium purification system

As known, the utilization of screw compressors for helium compression with oil spray into the compression cavity requires a perfect system to purity helium of oil. In

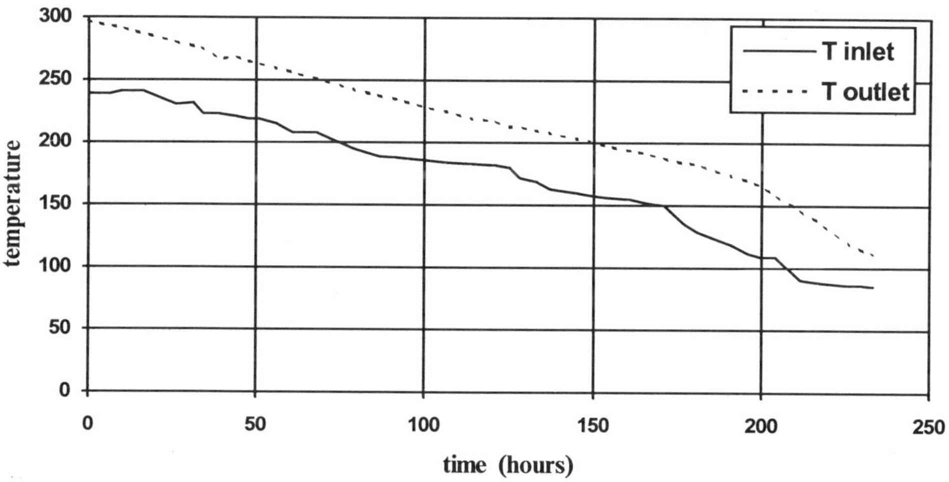

Figure 1. Cooling of the toroidal field coils from 300 K to 110 K.

T-15 cryogenics system such a purification is performed at a few stages, two of them are at the compressor plant; others, in a special unit of oil purification and drying of helium. The gas, after its passage through the compressor, passes through a centrifugal separator, where it is purified of oil drops, then, through the rough purification filter, and, after that, gaseous helium enters the unit MO-800 through a long connecting pipeline. Three filters providing the gas purification of oil in drops and aerosols are placed in series, one after another, then two carbon adsorbers, located in parallel, follow. They provide the purification of oil vapors. Beyond them, two zeolite adsorbers, are placed providing the helium purification of humidity. Then, the gas enters the unit of low temperature purification, HO-800. There, passing through, adsorbers with activated carbon at 80 K, it is purified of gaseous impurities (nitrogen, oxygen).

Analyses on the content of oil in drops are realized by the weighing technique, fixing the weight of filters before and after the passage of a definite amount of gas through the filter, a luminescent technique, using the instrument "KBAHT-5", is applied for measuring the amount of oil vapors.

An analysis of helium on the content of nitrogen, oxygen, neon, hydrogen impurities is performed with gaseous chromatographs.

As result of realized modernization - inclusion of additional filters at the entry to the purification units - the design requirements on the purity of helium were attained. The initial period of maintenance manifested also significant hydraulic losses along the path of purification. The places, where such losses could be reduced due to insignificant structural changes, were found and the changes were done. As a result, the pressure in front of the turboexpanders was increased to 2,5 MPa (instead of 1,9 MPa present previously) that allowed one to increase the capacity of liquefiers.

Helium liquefiers

For liquid helium production and for realization of technological regimes for cooling, cryostatting and for warming cryogenics objects at the T-15 facility the liquefiers with the rated capacity 0.222L/s (800 L/h) each are used.

In 1988 the liquefier was put into operation under conditions without usage of end gas-expansion machine, having the capacity 0.0833 - 0.0972 L/s (300-350 L/h) of liquid helium, at the time of continuous operation up to 200 hours.

Since 1988 the works on enhancement of its reliability and capacity have been on going. They are:

- simplification of pneumatic-hydraulic liquefier diagram directed to an increase in the reliability;
- introduction of the end gas-expansion machine into operation. Piston expander and two modifications of the turboexpander design were tested. Only the last modification of the turboexpander design has shown reliable and effective operation;
- development and realization of technological and diagram solutions (usage of preliminary helium purification of neon, development and assembly of a neon adsorber) directed against the frost injury by neon and hydrogen impurities in the armature and in the heat exchangers. That injury changed the mass flow ratio;
- tests of various turboexpander modifications (designs with various operating pressures and with various mass flow rates, the turboexpanders with optimal parameters have been chosen).

At present, the liquefier has the capacity equal 0.222 L/s (800 L/h), at the compressed helium expenditure of 0.222 kg/s, under the inlet pressure of 2,2 MPa to the liquefying unit.

Cryogenics pipelines

The cool helium and nitrogen transportation is realized through the pipelines with super insulation. The design of such pipelines assumes the fabrication manufacture of the units at the plants with subsequent mounting and tests at the site. In the process of mounting separate pieces of the inner pipeline are welded, then the welded joints are tested for strength and for hermetic sealing, the joint is covered with insulation, and the external case is closed. After welding of the external casing, the heat-isolated volume is pumped out and its sealing is tested.

The usage of such technology assumes high skill in the production of units, their keeping and mounting. Unfortunately, some pipelines, after completion of mounting, did not attain the necessary values of heat influxes. For the replacement of defect pieces new units were manufactured; some of them, used nitrogen screens.

ANALYSIS OF CRYOGENIC SYSTEM OPERATION

Regime of cooling from 300 K to 110 K

The regime of cooling from 300 K to 110 K is realized by helium circulation through both the superconducting magnetic system channels and the channels through screens and through cryoresistive control coils. Such a solution allows one to more precisely sustain a temperature gradient respective to various objects. The helium feed is performed from the unit of low temperature purification and preliminary helium cooling. The temperature regulation occurs by to the mixing of helium fluxes with different temperatures, the range of the temperature regulation is from 300 K to 80 K.

Table 1. Content of oil in helium

Place of measurement	Content of oil according to the design	Actual content of oil before modernization	Actual content of oil after modernization
After the first stage of purification at the compressor plant	1,5 mg/m^3	4,2 mg/m^3	1,5 mg/m^3
Before the carbon adsorbers in the purification unit	0,004 mg/m^3	0,2 mg/m^3	0,0035 mg/m^3

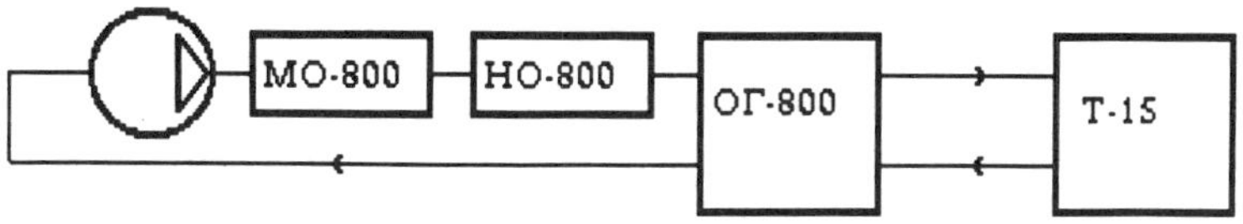

Figure 2. The diagram of cooling down from 90 K to 20 K.

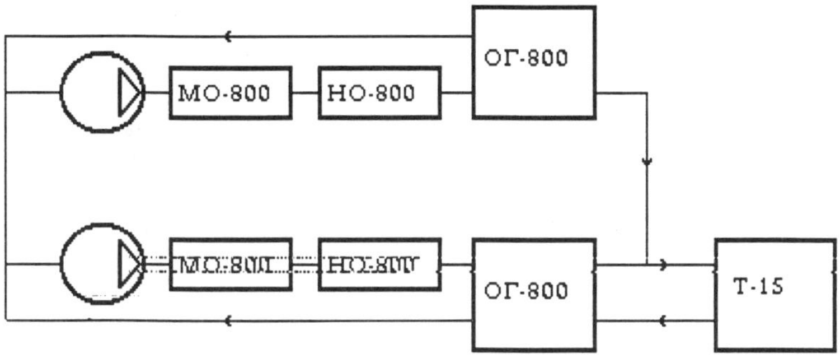

Figure 3. The diagram of cooling down from 20 K to 5K (first variant).

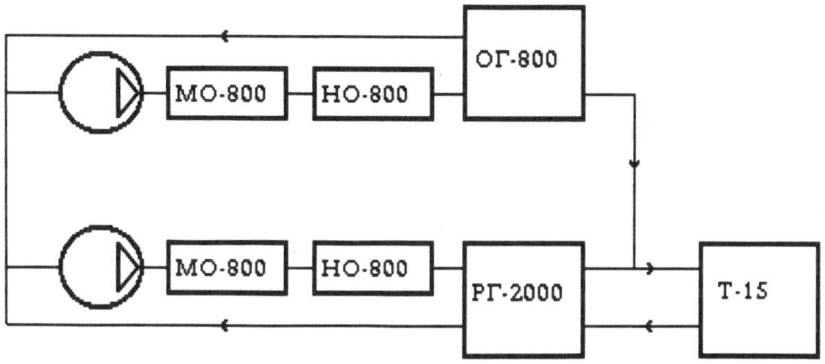

Figure 4. The diagram of cooling down from 20 K to 5K (second variant).

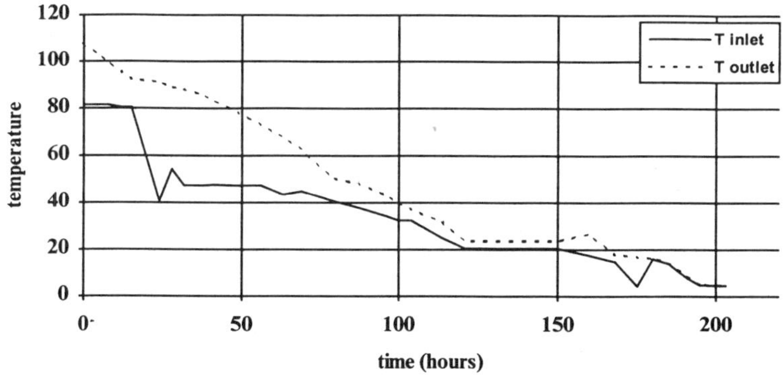

Figure 5. Cooling of the toroidal field coils from 110 K to 5 K.

The rate of cooling is determined by the passage rate through T-15 - objects, by the magnitude of a flow through the T-15 - objects. The graph of changes in the inlet and outlet helium temperatures is depicted in Fig. 1. A change in the cooling rate at the last stage is explained by an increase in the consumption by the objects.

Regime of cooling below 110 K

The cooling below 110 K is realized after replacing the coolant (helium) in the channels through screens and through cryoresistive coils by liquid nitrogen. The liquid nitrogen circulation is performed with centrifugal pumps.

As a rule, the toroidal field coil cooling to 90 K is realized according to the above-mentioned diagram, i.e. using the units of purification and preliminary helium cooling only. For the further cooling the liquefier, ОГ-800, operates in the refrigerating mode. Attaining the coil temperature of 20-25 K, the diagram of cooling is modified once more. A passive refrigerator, РГ-2000, is introduced into the diagram, and the liquefier, ОГ-800, is transferred to the liquefying mode. The passive refrigerator uses the liquid helium fed from the liquefier. In that case, the reverse flow rate through the refrigerator exceeds the direct flow rate by a value of the liquid helium addition.

The diagrams of cooling down from 90 K to 20 K and from 20 K to 5 K are depicted, respective by, in Figs. 2, 3 ,4.

Cryostatting mode

In the cryostatting mode the diagram depicted in Fig. 4 is used. Heat evading are:

- to the nitrogen system (from screens and from cryoresistive coils) from 60 kW to 150 kW that depends on the temperature of T-15 discharge chamber heating;

- to the helium system of about 2,7 kW.

In the inductive heating mode of the discharge chamber an increase in the loading on the helium system takes place that speaks about the presence of closed electric discharge in the helium header.

Warming up

Warming up of T-15 cryogenics objects is realized at two stages; at the first stage the toroidal field coil warming up to 120 K occurs. The warming is done by the helium feed, at 80-120 K, from the unit, HO-800. When the temperature is raised higher than that, nitrogen and oxygen gas is released accumulated on the cool surfaces of toroidal plasma-facing field coil frames. Vacuum deterioration occurs in the cryostat. At that instant, the rate warming is moderated by recovering the required pressure in the cryostat.

At the temperature of superconducting toroidal field coils equal 120 K, the liquid nitrogen drains from the channels through the screens and through the cryoresistive coils. The nitrogen is replaced by helium. Later, the warming of all the cryogenics objects is done by circulation of helium. The permissible temperature difference at various T-15 - objects, equal 50 K, is set under both cooling and warming (for the temperatures higher than 80 K). The warming is stopped after attaining the temperature of toroidal field coils equal 250 K.

CONCLUSION

During the maintenance of T-15 cryogenics system the equipment was developed and modernized, technological regimes of cooling, cryostatting and warming were developed that allowed us to realize the program of physical experiments at T-15.

PERFORMANCE OF THE SUPERCRITICAL HELIUM COOLING LOOP FOR THE JET DIVERTOR CRYOPUMP

W. Obert[1], C. Mayaux[1], K. Barth[1], L. Herblin[2]

[1]JET Joint Undertaking, Abingdon, OXON, OX14 3EA, UK
[2]CERN, CH Geneva

ABSTRACT

A supercritical helium cooling loop for the two JET divertor cryopumps has been tested, commissioned and is operational practically uninterrupted for over one year. Operation experience under a number of different boundary and transient conditions have been obtained.

The flow of the supercritical helium (6 g/s, 2.7 bar) is driven by the main compressor of the JET helium refrigerator passing a heat exchanger where it is subcooled to 4.1 K before entering the two cryopumps which are an assembly of two 60 m long and 20 mm diameter corrugated stainless steel tubes. By using a dedicated cold ejector which is driven by the main flow and where the expansion from 12 bar to 2.7 bar takes place increases the flow of supercritical helium up to ~17 g/s. The steady state thermal load to the cooling loop of the cryopump is < 80 W but during transient conditions in particular due to nuclear heating in the active phase of JET considerably higher transient heat loads can be accepted by the loop.

Details about the steady state and transient thermal conditions as well as the cooldown and warm up behavior of the loop and the interaction of the supercritical loop with the operation of other plant equipment will be discussed in the paper.

INTRODUCTION

For a new program stage of JET[1] (Joint European Torus, the world's largest fusion experiment) a cryopump with 500 000 l/s pumping speed for Deuterium has been installed inside the torus in conjunction with a magnetic 'divertor' to control plasma impurities to give a so called *pumped divertor*. The operation of the in-vessel Pumped Divertor (PD) cryopumps proved to be very beneficial for many aspects of the plasma experiments. Their availability has become a major requirement for routine operation of JET in order

Table 1 Calculated maximum thermal load of the JET in-vessel cryopumps

Steady state heat load include. cryolines	< 80 W
Transients during 10 sec Plasma pulse	
Particle condensation	< 60 W
Plasma radiation	< 500 W
Energetic particles	< 1 kW
Nuclear heating	< 4.5 kW

to achieve optimum vacuum conditions of the torus together with other methods such as Glow Discharge Cleaning and Beryllium evaporation. However, any change in the operation conditions of the cryopumps that causes an increase of the cryopump temperature are not compatible with the plasma operation as a uncontrolled temperature rise can cause an uncontrolled gas release into the torus. This can trigger a potentially harmful plasma disruption (similar to a quench of a superconducting device) and therefore the stable and reliable operation of the helium loop is of paramount importance. In contrast to many supercritical devices the thermal mass of the cryopumps structure is small which had been a consequence of the design requirement to make the system compatible with nuclear heating from the neutron and gamma pulse in the Tritium phase of JET. This also led to the decision to use supercritical (sc) helium for the cooling of the loop.

THE JET PUMPED DIVERTOR CRYOPUMPS

The design of the JET pumped divertor cryopumpsystem is described elsewhere[2,3,4], the pumping performance of the in-vessel PD cryopumps are discussed in references 5 and 6. The system consists of two independent 180 ^{0} circular shaped systems which are installed directly inside the JET vacuum vessel which is operated at 250 -300 ^{0}C and in close vicinity of the 200 Million ^{0}C plasma. In order to protect the cryopump against this hostile environment the cryopumps are installed behind a water cooled structure (see Figure 1).

Helium panel: In order to minimize the amount of structure material (for minimum nuclear heating) and to provide mechanical stability the cryocondensation panels of the PD

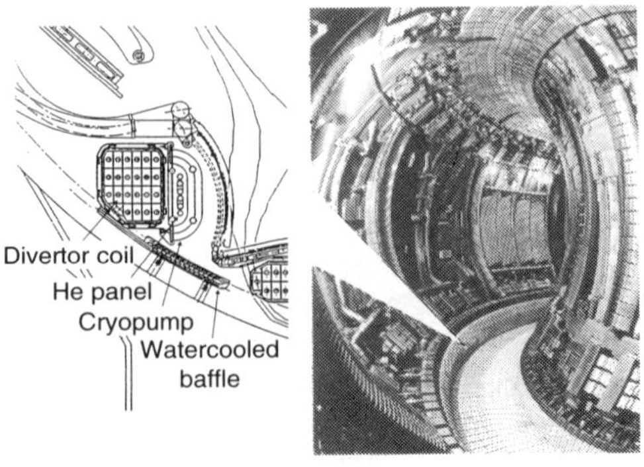

Figure 1. The JET vacuum vessel with the PD cryopump.

Figure 2. The helium panel of the JET PD cryopump.

cryopumps are a zig-zag arrangement of 6 thin wall (0.3mm) slightly (1mm) corrugated stainless steel tubes (see Figure 2) with 20mm diameter, a total length of 60 m and a total weight of ~5kg. This provides a total cryocondensation area of ~2m^2 per pump.

Thermal load: The thermal loads to the cryopump under various operation conditions are listed in Table 1[1]. The steady state heat load is < 40W per loop. Thermal peak loads of up to 4 kW for 10 sec will be absorbed by the thermal capacity of the helium content.

Operation cycle: The operation duty cycle for the PD cryopumps is a continuous run of both cryopump systems for at least two shifts (15 hours) per day with occasional regenerations overnight if required. The cryopumps can run continuously for over a week and their operation is limited only by the condensed gas inventory which must not exceed 2000 bar·l for safety reasons (ignition in case of an air accident). In addition the cryopumps have to be regenerated before the torus wall conditioning by Glow Discharge Cleaning.

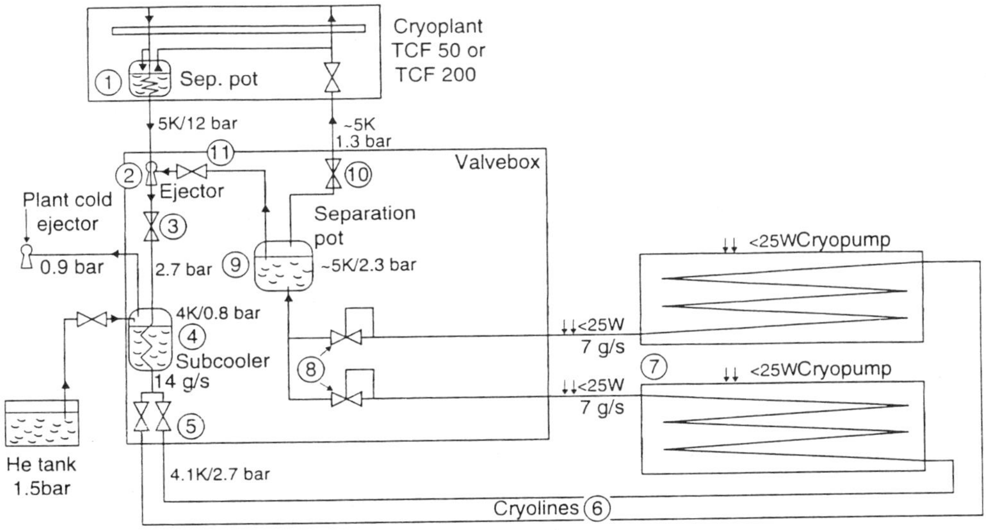

Figure 3. Flow diagram of sc helium cooling loop of JET Pumped Divertor cryopump.

Argon frost/helium pumping: A special feature of the cryopump is the facility to spray Argon directly onto the cryocondensation panel and to provide this way an active surface for the cryopumping by cryosorption of helium gas which cannot be pumped by cryocondensation. A ratio of 20:1 in number of atoms of Argon to Helium has been found to provide optimum pumping characteristic. Before each helium experiment the sc helium cooled cryopanel is loaded with Argon and typical load rates are 100 mbar l/s of Argon which correspond to a heat load from both the gas enthalpy (c_p= 0.5 J/gK) and heat of condensation (163 J/g) of <50 W (this corresponds for a total Argon load of 50 bar l when loaded over a period of 500 sec to a total deposited energy of <25 kJ).

THE SUPERCRITICAL HELIUM CRYOSUPPLY

The design of the cryopumps with long horizontal pipework and no space for a liquid helium manifold for phase separation called for a supply with supercritical (sc) helium. The process diagram for the sc helium supply for the JET pumped divertor cryopump is shown in Figure 3 and the main elements of the loop are:

(1) **Cryoplant:** High pressure (12 bar) helium is taken from the high pressure loop of either one of the two helium refrigerators at JET (LINDE, TCF200 with 850 W/4.5K or TCF 50 with 350 W/4.5 K) after passing the heat-exchanger in the separation pot of the refrigerator where it becomes subcooled to ~ 5K.

(2) **Cold ejector:** This subcooled helium enters the distribution box and there a cold ejector where it is expanded to ~ 2.7 bar. The cold ejector acts as the Joule-Thompson expansion nozzle and in addition as driving element for a booster flow which enhances the flow by approximately a factor of ~3. This flow is taken from the separation pot in the return circuit (9). The main dimensions of the cold ejector are: Joule-Thompson nozzle diameter 0.9 mm, mixing chamber diameter 2.8 mm - length 65 mm, diffuser cone length 100 mm. The ejector was carefully designed in order to provide a) the high suction capacity required b) to optimize the maximum pressure drop for maximum efficiency and c) to restrict the flow taken from the cryoplant to 6 g/s which is the maximum liquefaction flow which can be provided by the TCF50 when in pure liquefaction mode. This is to avoid that the sc helium flow for the PD cryopumps would take excessive flow from the main flow.

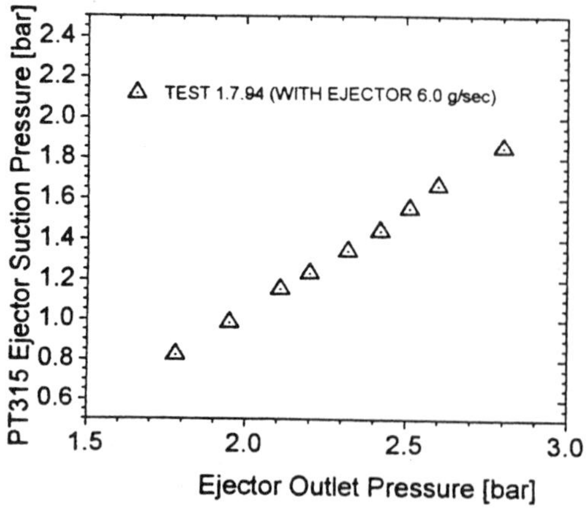

Figure 4. Cold ejector performance: suction pressure vs. outlet pressure at 12 bar inlet pressure.

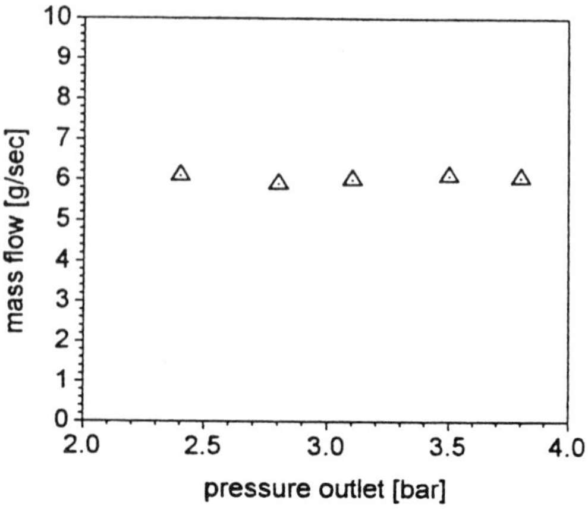

Figure 5. Cold ejector performance: mass flow vs. outlet pressure.

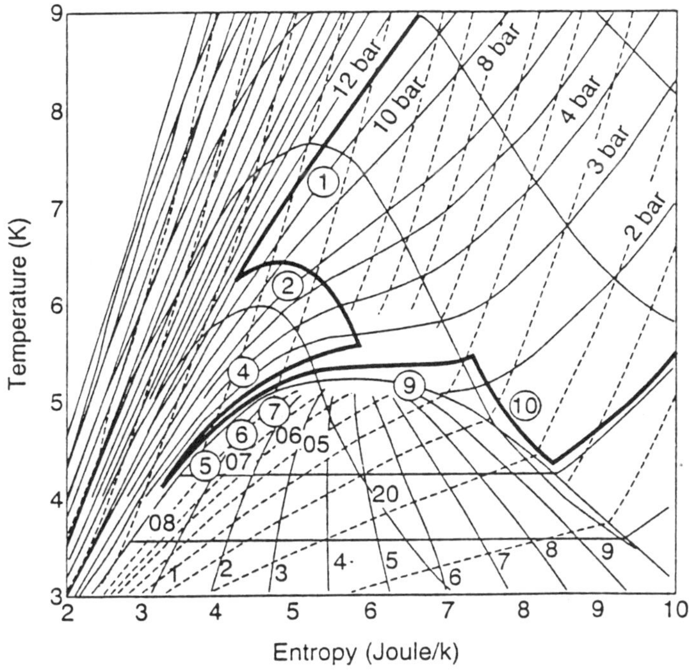

Figure 6 . Temperature entropy diagram with process variables for the JET PD cryopump sc helium cooling loop.

To use the ejector as main flow restriction allows to minimize the control efforts for the sc helium loop and avoids the use of control valve (3) in Figure 3 which stays fully open during operation. It would otherwise produce additional pressure drop and reduce the efficiency of the ejector. The setting up of the ejector required careful balancing in particular for the distance between the expansion nozzle and the drift channel.

Performance data of the cold ejector can be seen in Figure 4 where the suction pressure p_s is shown as a function of the outlet pressure for 12 bar inlet pressure. p_s is typically 1 bar below the outlet pressure. Figure 5 shows the ejector mass flow throughput as a function of the outlet pressure at 12 bar inlet pressure and demonstrates that the ejector acts in a wide range as flow stabilizer.

The flow rate through the ejector with closed suction valve has been measured directly with a flow meter after the helium gas has been warmed up to ambient in a water

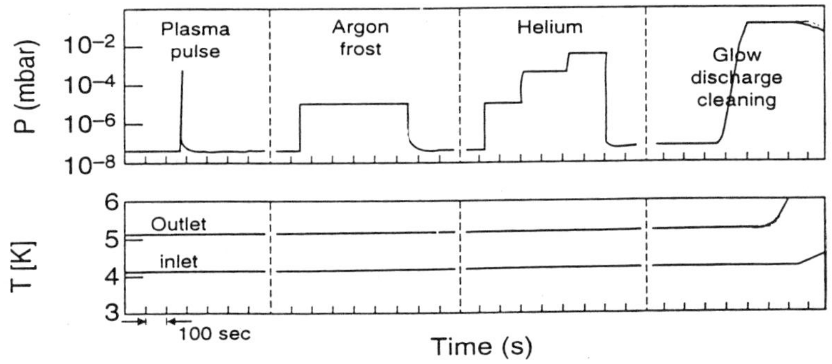

Figure 7. Inlet and return temperature of the JET in-vessel cryopumps as function of time together with the torus pressure and the main thermal loads.

bath and found to be ~ 6 g/s in a wide range of conditions. The flow rate through the ejector with open suction valve (11) and under fully established sc helium flow conditions has been indirectly derived from the thermal balance of the helium subcooler (fall in level with given in and outlet flow conditions) and by calorimetric measurements in a separate test facility. It was found that the flow with open suction increased to ~17 g/s which corresponds to a flow boosting factor of ~ 3.

(4) **Liquid helium subcooler:** After the cold ejector the sc helium flow enters a subcooler with liquid helium at ~0.8 bar (~ 4 K) and becomes subcooled to ~ 4,1 K.

(6) **Cryolines:** The sc helium is distributed to the two in-vessel cryopumps via two 55 m long flexible cryolines[4] which are integral part of the sc cryoloop. They are of coaxial, boil off and liquid nitrogen shielded design[5]. The corrugated supply line has a diameter of 14/18 mm the annular return has 21/25 mm and 39/44 mm. The heat load is on the inner supply and return line is ~ 10 mW/m and ~150 mW/m respectively

(7) **Cryopump and return flow:** The sc helium loop enters the cryopumps where it is exposed to the heat load as listed in Table 1. The return valves (8) after the cryoline are controlling the pressure of each pump in order to guarantee stable conditions and flow balancing. Then the helium enters the separation pot (9) where a part of the return flow is returned to the loop by the pumping of the cold ejector - the other part returns to the refrigerator after expansion to the suction pressure of the cryoplant compressor (10).

PROCESS

The thermodynamic process variables during the standard steady state conditions are shown in Figure 6 where the bold line follows the design of the thermodynamic process and the numbers refer to the numbers shown in the flow diagram of Figure 3 with

(1) Isobaric subcooling of 12 bar supply in refrigerator separation pot
(2) Isenthalpic Joule-Thompson expansion in cold ejector to 2.7 bar
(4) Isobaric subcooling in He-subcooler to 4.1K
(7) Heat load from cryopump and transferline
(10) Expansion to 1.1 bar of cryoplant (TCF50) inlet pressure

This process variables have been chosen in order to achieve the lowest reasonable temperature together with stable flow conditions and is a compromise between the need to achieve a low D_2 vapor pressure[2] ($<10^{-6}$ mbar which requires a temperature <6K) and to stay sufficiently away from the TCL (Transposed Critical Line).

OPERATION EXPERIENCE

Stability: The two parallel cryopump systems run absolutely stable once they are cooled down as can be seen in Figure 7 where the in- and outlet temperatures of the PD cryopump are shown as function of time together with the pressure readings of the torus pressure. No signs of flow instabilities can be observed under the chosen operation conditions and various transient heat loads such as:

- During the *normal operation* period a plasma pulse of typically 10 sec. length and a repetition rate of ~20 minutes creates a short heat load to the cryopump (see Table1). The deposited energy from the normal plasma pulses without the nuclear heating from the DT reaction is comparatively small and not detectable at the cryopump outlet.
- The heat load during the *Argon seeding* is higher (< 40 W) but again not detectable at the cryopump outlet temperature.

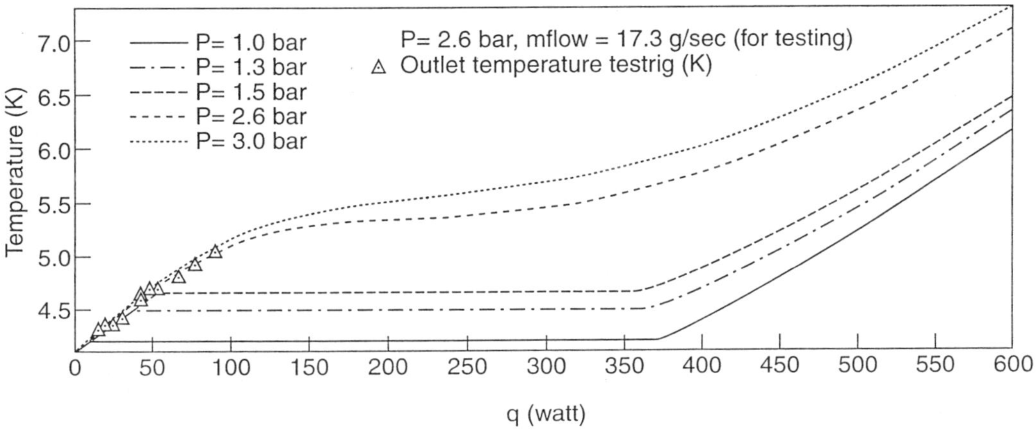

Figure 8. Calculated return temperature of JET PD cryopump as function of applied heat load for a cooling flow of 17 g/s sc helium and various operation pressures together with measured temperatures.

- Test runs with elevated torus pressure (up to 10^{-3} mbar for few hundred seconds) corresponding to heat loads relation of up to ~100W calculated from Knudsen showed no change in return temperature.
- conditions under Glow Discharge Cleaning with a base pressure of ~10^{-1}mbar triggered a uncontrolled regeneration after ~ 100 seconds.

All these different conditions are summarized in Figure 7.

Transients: The cooldown of the cryopumps takes usually place from liquid nitrogen temperature in particular after regeneration when the helium panel temperature equilibrates fully with the surrounding LN cooled radiation shield. The cryopumps can be cooled down within 30 minutes, the complete loop is cold and stable after ~ 90 minutes.

The regeneration of the cryopumps is a fully automatic process and can be achieved by stopping the helium supply which results in a slow warm up of the cryopump and regeneration (gas release) takes place after approximately 30 minutes. In order to accelerate the regeneration a forced flow of ambient temperature helium is normally blown through the pipework of the helium loop. This way the pump can be fully regenerated within less than 10 minutes.

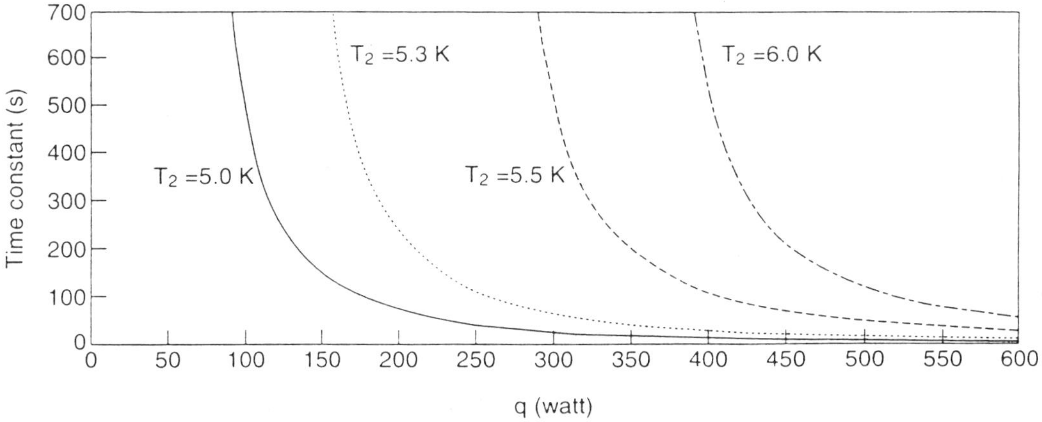

Figure 9. Time constant for sc helium loop of JET PD cryopump to reach temperature T_2 from T_1= 4.1 K for a mass flow of 17 g/s, and an inventory of 5 kg sc helium at 2.7 bar.

Interaction with other systems: At the same time the sc helium loop for the PD cryopumps is kept running from the cryoplant series of other users[7] such as the three large JET Injector cryopumps and the LHCD cryopump are supplied with liquid helium in a closed refrigeration loop from the same cryoplant together with series of liquid helium bulk users. A large (15000 l) liquid helium storage capacity allows to decouple the liquid helium users from the sc loop.

DISCUSSION

Steady state thermal load: The acceptance of a steady state thermal load is governed by the throughput of sc helium. The specific heat of the sc helium at the operation conditions is extremely non linear and the calculated temperature rise (using "HEPAK") of the cryopump return is shown in Figure 8.

The same figure shows measured return-temperatures for various heat loads. The measured values fit well with the calculation. However the measured return temperatures for Argon seeding and elevated torus pressure are consistently below the calculated temperatures for the steady state conditions and indicate that the thermal behavior of the cryopump is dominated by the heat capacity of the helium within the cryopump.

Dynamic thermal load : The helium content of the cryopump was chosen to comply with the expected 10 seconds of up to 4 kW for the DT phase of JET. This thermal capacity allows to "digest" other smaller heat loads by the system. Figure 9 shows the calculated transient time t to reach a certain temperature as a function of the applied heat load according to the relation:

$$t = m_{He} \int_{T_1}^{T_2} c_p \, dT \; / \; (q - \dot{m} \int_{T_1}^{T_2} c_p \, dT)$$

with m_{He} = sc helium inventory, c_p= specific heat, q = dynamic heat load, $\dot{m}$ = sc helium cooling flow rate (g/s).

It can be seen that the time constants to achieve equilibrium temperature are in excess of the time spans for the thermal loads of the PD cryopump from Plasma pulses, Argon seeding and transient pressure excursion of <400 sec. However the conditions of Glow Discharge Cleaning are not compatible with the given flow conditions.

CONCLUSION

- For a new program of JET a cryopump with 500 000 l/s pumping speed has been installed.
- For the 4K cryocondensation loop cooling by supercritical helium is used.
- The helium loop operates under various operation conditions absolutely stable even under considerable thermal peak loads.
- A cold ejector is used to boost the helium flow by more than a factor of ~3 proved to be very effective.

REFERENCES

1 The JET team (presented by D.Stork) 'The New Experimental Phase of JET and Prospects for Future Operation', Procceedings 15th Int. Conf. on Plasma Physics and Contr. Fus. Res., Seville (1994)
2 W. Obert et al.'JET Pumped Divertor Cryopump' Proceedings of Symposium on Fusion Technology, SOFT 16, London 1990, pp 488-492

3 W. Obert 'The Liquid Nitrogen and Supercritical Helium Cooling Loop for the JET Pumped Divertor Cryopump' Adv.in Cryogenic Engineering , Vol. 39, pp1569-1576, Edit. P. Kittel, Plenum Press, NY 1994
4 S. Papastergiou, W. Obert and E. Thompson 'Material Selection, Qualification and Manufacturing of the In-Vessel Divertor Cryopump for JET', Adv. in Cryog. Eng. Vol 40B, pp1429-1436, Plen. Press, NY 1994
5 G. Saibene et al.'Effects of Active Pumping and Fuelling on Divertor Plasma Discharges at JET' Proceedings of 22nd Europ. Phys. Soc. Conf., Bournemouth (1995) to be published
6 W. Obert et al. 'Performance of the JET Pumped Diveror Cryopumpsystem' Proceedings of 16th Symposium on Fusion Engineering (SOFE), Champaign, USA (1995), to be published
7 W. Obert et al.'The JET Cryosystem, Overview and Experience', Adv. in Cryog. Eng., Vol 39,pp 493-500, Edit. P. Kittel, Plenum Press, NY

CONCLUSIONS FROM PROCURING, INSTALLING AND COMMISSIONING SIX LARGE-SCALE HELIUM REFRIGERATORS AT CERN

M. Barranco-Luque, S. Claudet, W. Erdt, P. Frandsen, Ph. Gayet, D. Güsewell, Ph. Lebrun, J. Schmid, N. Solheim, Ch. Titcomb, U. Wagner, G. Winkler

Accelerator Technology Division, CERN
CH-1211 Geneva 23, Switzerland

ABSTRACT

Between 1990 and 1994 CERN procured, installed and commissioned six large-scale helium cryoplants for its programme of superconducting acceleration cavities in the electron-positron collider LEP. Two European suppliers were selected to each provide one plant of 6 kW and two plants of 12 kW equivalent cooling power at 4.5 K. All installations are now commissioned and operational, some have been running continuously for several years. The concepts applied to specification, tendering, sharing of responsibilities for infrastructure and controls, installation, and commissioning are presented. Conclusions are drawn from the experiences during the different phases of this project and applied to acquisition of plant upgrades and additional plants required for LHC, CERN's new project for a proton-proton collider in the LEP tunnel using superconducting magnets.

INTRODUCTION

LEP is currently the main particle accelerator machine at CERN. Installed in a 27 km-long underground ring tunnel, it has been producing electron-positron collisions at four experimental areas (called below also "LEP points") since 1989. LEP uses normal magnets with modest fields for beam guidance and water-cooled copper cavities for particle acceleration. LEP2 is an upgrade project aimed at doubling the particle energy of LEP to study the existence and properties of new particles. As synchrotron radiation losses increase with the fourth power of the particle energy, LEP2 requires at each of the four experimental areas superconducting (SC) accelerating cavities with a total active length of about 100 m and the associate helium cryogenics of some 12 kW cooling power at 4.5 K. Apart from further upgrades, all LEP2 cryogenics have been ordered, installed and brought into operation during the past 5 years. With LEP2 the helium refrigeration capacity of CERN was boosted by one order of magnitude. At a time when the next major project at CERN, known as LHC, has been approved and will require a further tripling of the helium refrigeration at CERN, it is appropriate to draw some conclusions from the LEP2 exercise.

THE LEP2 PROJECT AND ITS CRYOGENICS

The LEP2 programme[1] was initiated in 1989. The same year the basic LEP collider machine was brought into operation with beam particle energies of 45 GeV and started producing the

Advances in Cryogenic Engineering, Vol. 41
Edited by P. Kittel, Plenum Press, New York, 1996

neutral Z^0 particle by millions, thus allowing verification of the predictions of the standard theory of elementary particles with unprecedented precision. The electrically charged sister particles $W^{\pm}$ can only be produced in pairs and for production in significant numbers require beam energies above 90 GeV. More recently energies close to 95 GeV or even above were requested in an attempt at finding some of the elusive 'Higgs' particles. For this reason there was considerable pressure from CERN researchers to increase the final number of accelerating cavities to be installed for LEP2; it went up from the 192 planned in 1989 to the current target of 224, and may even reach 272 if another addition is approved in the coming months.

LEP2 cavities are 4-cell cavities of 75 cm diameter and 1.7 m accelerating length operating at a radiofrequency (RF) frequency of 352 MHz; four of them are assembled, tested and installed as one 11-m long module. In the beginning they were made from niobium sheet and later from copper with a 1 μm thick sputtered niobium layer. Transfer of the special manufacturing technology to industry was a long learning process[2], but now most of the cavities are available, module assembly is progressing and also technological problems with the RF couplers and their suitability for higher beam currents could be solved. 48 SC cavities in 12 modules are currently operational with LEP. With 160 expected to run from autumn 1996, the $W^{\pm}$ threshold should be reached for the first time, and all 272 cavities or 68 modules operational from spring 1998. LEP is expected to stop and make room for LHC from the year 2000.

The LEP2 cavities in each module are cooled[3] by a bath of boiling LHe at a pressure of 1.2 - 1.3 bar A fraction of the evaporated GHe is used at a rate of 0.8 g/s for cooling heat intercepts and tuning the cavity frequency. The design refrigeration load per module was 750 W (incl. 100 W for reliquefaction flow, 150 W for insulation and distribution losses, 300 W of RF losses at nominal RF field and cavity quality factor, and 200 W of spare capacity for higher fields or lower quality factors), which led to a requirement, for each of the four LEP2 acceleration points, of helium refrigeration of 12 kW equivalent power at 4.5 K.

A study launched as early as 1984 by CERN, in collaboration with the three potential European suppliers of helium cryoplants in the kilowatt range, AIR LIQUIDE (F), LINDE (D) and SULZER (CH), had shown that 6 kW at 4.5 K was the upper power limit for which coldboxes could still be lowered as one piece through the LEP shafts and installed in existing

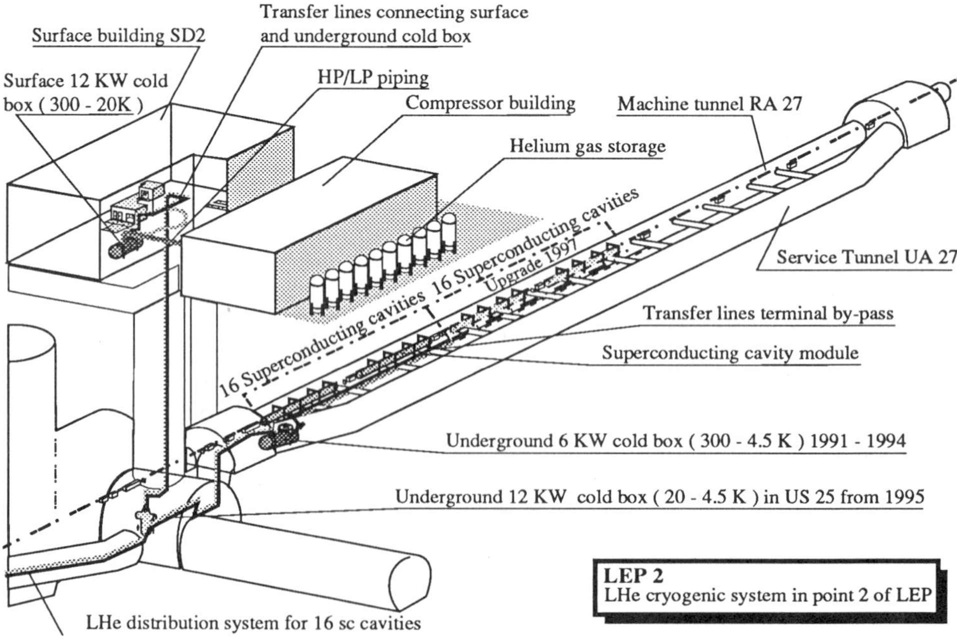

Figure 1. Typical layout of LEP2 cryogenics (here for LEP point 2) with symmetric implementation of first 2 x 4 and later 2 x 8 (or 9) superconducting modules in the LEP tunnel and integration of the 12 kW refrigerator into the existing access shaft with split into an upper coldbox at surface level and a lower coldbox at tunnel level.The prototype solution using a compact 6 kW coldbox in a bypass tunnel near the cavities is also shown.

Table 1. Milestones of procurement for LEP2 cryogenics

Plants	Action	History: Dates	History: Timescale (months)
6 kW plants	Discussion with suppliers, in particular on space requirements, and writing of specification	March-Aug. 89	- 6
	Call for Tenders	Sept. 1st, 89	0
1 x AL	Offers in	Nov. 20, 89	+ 2.6
	Letter of Intent out	May 90	+ 8.5
+	Contract signed	Aug. 90	+ 11.5
	First compressors running	Oct. 91	+ 25
1 x SU	First cooldown	March 92	+ 30.5
	Provisional Acceptance Test	May 92	+ 32
	Last point settled on punch list of open problems	End 94	≈ 60
12 kW plants	Discussion with suppliers, in particular on coldbox split, and writing of specification	March-May 90	-3
	Call for Tenders	June 6, 1990	0
2 x AL	Offers in	Sept.19, 90	+ 2.5
+	Letter of Intent out	Dec. 19, 90	+ 5.5
2 x LI	Contract signed	May 91	+11
	Provisional Acceptance Test	Pt.2, AL: Dec.94 Pt.4, LI: Nov.94 Pt.6, LI:Sept. 93 Pt.8, AL: Nov.93	+ 39 to + 54

caverns without expensive civil engineering. When in 1989 the requirements of LHC became clearer, the concept of one single 12 kW cryoplant per LEP point with upgrade potential to 18 kW was born[4], and the space problem underground was solved without additional civil engineering, by splitting the coldbox into a bigger box with temperatures down to 20 K at surface level, and a smaller coldbox at tunnel level for all components in the 20 - 4.5 K range (Figure 1).

PROCURING, INSTALLING AND COMMISSIONING THE REFRIGERATORS FOR LEP2

History of decisions. When in 1989 the funding for the LEP2 upgrade programme became available, the development of the SC niobium cavities chosen for use in LEP was already in an advanced stage. After years of prototype development at CERN, first orders could be placed in industry and, since early 1988, one of the cavities had been successfully operated with the SPS accelerator. The assembly of the first three 4-cavity modules was under way and it was anticipated that eight of them could be ready early 1992 and installed at LEP point 2 for cooling by a 6 kW cryoplant; it seemed further likely that finally all 8 or 16 modules would be ready for cooling at each of the 4 LEP points from early 1994, and this set the very tight schedule for the definition and the acquisition of the LEP2 cryogenics. In addition to the cryoplants, transfer line systems with about 400 m length had to be provided at each of the four LEP points, for distribution of the cold helium to the strings of cavity modules[3].

Although in 1989 the refrigeration requirements of LHC were already known to be close to 8 x 18 kW equivalent power at 4.5 K and the concept of 12 kW plants for LEP2 with preparation for a later upgrade to 18 kW seemed the best way of using CERN funds, it was decided to order as quickly as possible two of the 6 kW plants, for which a concept was available, one for the use with a prototype string of cavities in LEP and the other one for a new test area in which first LEP2 cavities and later LHC magnets could be tested.

6 kW plants. The history of actions for the 6 kW plants is shown in the upper part of Table 1. Three offers were received and orders placed in May 1990 for one plant each with the two lowest bidders AIR LIQUIDE and SULZER. Two years later both plants started operation, and as there was an urgent need for both tunnel operation with first cavity modules and series testing of cavities in the new test area, CERN took over both plants after the first satisfactory performance tests and accepted settlement of open points at a later stage. Both plants gave full satisfaction, the LINDE plant with 7 turbines[5,6] running in the LEP tunnel until end of 1994 (low failure rate also visible in Table 3), and the AIR LIQUIDE plant with 4 turbines[7] in the test area supplying continuously liquid helium to cavities and, since 1994, also to prototype SC magnets for LHC. As a consequence of the early takeover by CERN, however, there was little pressure for a fast settlement of open problems and often no access to the plants The acceptance procedure could only be concluded at the end of 1994.

12 kW plants. Under the pressure of the cavity programme the tendering and ordering of four 12 kW plants had to start immediately (lower part of Table 1) once the letters of intent had been sent out for the two 6 kW plants. For this reason the same concepts had to be applied and only slight corrections resulted from the short experience with the first order. Again three offers were received and the order for the 4 plants split 2 and 2 between AIR LIQUIDE and SULZER as the lowest bidders. In June 1992, however, LINDE took over the helium cryogenics department of SULZER and with it the contract for two LEP2 plants.

Both suppliers, AIR LIQUIDE and LINDE, handled their contracts quite well; one plant from each firm was accepted in late 1993, but modules to be cooled by a 12 kW plant only existed at that time at LEP point 6 (Table 3). In fact, the programme of module production and installation suffered a considerable delay in the years 1991-1994, mainly due to the introduction of new cavity technology, niobium sheet cavities being replaced by sputter-coated copper cavities and 3 suppliers having to learn the new technique; in addition, the RF couplers on the cavities had to be matched to considerably increased LEP requirements, and this led to developments and corrections which blocked series production of accessories until recently. As a consequence, the cryogenics for LEP2 disappeared from the critical path of the programme, and the completion of the 12 kW contracts could be modulated in the most economic way.

We now give a closer look at a few particularities of the tendering and main conclusions from the different phases from ordering to operation:

Special Features Adopted for Tendering the LEP2 Cryogenics

Performance specification - no process details. As the CERN team in charge of LEP2 cryogenics was small and had full confidence in the competence of the selected suppliers, only performances and interfaces with CERN services and CERN loads were specified. However, for ancillary equipment with heavier maintenance requirements, like screw compressors, electric motors and instrumentation, clear preferences were expressed to achieve, wherever possible, economies by standardization between the two suppliers and with other cryoplants running at CERN.

Incentive for aiming at high power efficiency and good part-load behaviour. The tenderers were asked to declare the electric power necessary to achieve the nominal performances. For adjudication a formula was given which combined the price quotation with the estimated electricity cost from running over 10 years at CERN. The acceptance tests had to be done at the specified electricity input, and on the results a bonus or malus could be applied. Unfortunately, a tolerance band down to -5% below nominal value was allowed to cover errors in measurements, and this had only the effect, under conditions of competitive tendering, that the tenderers designed their plants not for the nominal performance, but for 5% less.

Upgrade potential of 50%. The tenderers were asked to dimension piping, valves and heat exchangers in such a way, that the performance could be upgraded from 12 to 18 kW, without changes inside the coldboxes, by just adding compressor flow and upgrading accessible turbine components.

Controls separately tendered by CERN. The tenderers were informed that CERN intended to provide a control system, including hardware and software, of its own choice to achieve uniformity of the operator's interface for all new plants and compatibility with the existing systems at CERN and to best utilise existing experience; they were asked to provide all instrumentation with cabling to a standard interface and all hard-wired securities to avoid damage to their equipment from software errors of the CERN supplied control system.

Inclusion of LHC requirements. Wherever possible, known LHC requirements were taken into account, in particular for additional entry points for helium at intermediate temperature, higher flow rates and higher design pressure values. This added constraints to the design, increased the cost and in many cases it will not correspond to the exact needs of the final LHC.

Experience from Installation & Commissioning Phase

General results. Reports on the design and first commissioning results of their plants were published by both suppliers[8,9]. CERN described its experience at ICEC15[10]. The main performance results are summarized in Table 2, where the total performance for 3 of the 4 plants

Table 2. Main performance results for the four 12 kW plants

	Pt.2	Pt.4	Pt.6	Pt.8
Supplier	AIR LIQUIDE	LINDE	LINDE	AIR LIQUIDE
Total equivalent refrigeration power [kW] @4.5 K, incl 6.7 kW screen load @50-70 K and 13 g/s liquefaction load, as measured at the nominal grid power input given in the offer	(9.5)*	11.7†	11.7†	11.5†
Power factor [W/W] at nominal grid power input ("100% load")	NM	215	218	232
Power factor [W/W] at "75% load"	NM	249	263	274
Power factor [W/W] at "50% load"	NM	337	354	~450
Residual 4.5 K refrigeration power [kW] if at "50% load" full nominal screen and reliquefaction loads are applied	NM	3.8	3.7	2.8

*Problem identified by vendor in top heat exchanger block; replacement proposed for late 1995. †Within the -5% tolerance band of the Technical Specification. NM = Not yet measured.

is within the specified -5% tolerance band. The only remaining problem is the lower performance of the plant at LEP point 2; AIR LIQUIDE and their heat exchanger manufacturer think that this is related to abnormal temperature profiles observed on the heat exchanger blocks at the room temperature end of the plant; they propose to replace the block by an improved unit with all possible attention given to an equal distribution of flow over all parallel passages in the plate-fin exchanger.

Table 2 also displays the excellent power ratio between electricity input and refrigeration performance as well as the remarkable performance at reduced load.

Controls and software production. For all LEP2 cryoplants control hardware was ordered from ABB (CH/S) after competitive tendering. ABB also did most of the software programming, with process and logics specifications from the plant manufacturers and specification of the control logics for the cavity modules from CERN. Lack of experience with object-oriented programming and absence of suitable standards for defining process information led to a considerable waste of effort in the beginning. Now the principles for clear programme structures are established, a library of about 40 objects to cover all common types of equipment is available, and standard sheets are in use as input for programming. Whereas for LEP2 cryogenics about 7% of the investment went into controls hardware and some 6% into software programming by ABB, a substantial saving of software cost can be expected from the acquired experience for any future project.

First Experience from Running with the LEP2 Cavity System

Currently two of the 12 kW cryoplants are running with modules in LEP; a third will become active in October this year, and the fourth will have modules connected from May next year. Niobium cavities cannot be left on the beam line without being superconducting. This is why a high degree of reliability of the LEP2 cryoplants is of prime importance. Table 3 and other results presented at PAC95[12] show that cryogenics related failures are occurring only very exceptionally and that the overall availability of cryogenics is mainly determined by the reliability of utilities like grid power and water cooling; further improvements can be expected from systematic preventive maintenance of components and full use of the potential of the control system.. The duration of utility-forced interruptions are being progressively shortened by analyzing safety conditions established by the vendors, reducing risk situations, finding more specific signals and developing efficient, automatic restart procedures.

THE LHC PROJECT AT CERN AND ITS CRYOGENICS

In December 1994, the CERN Council approved the construction of the Large Hadron Collider (LHC) project[13] with installation in the LEP tunnel starting at the end of LEP2 operation, probably in the year 2000. This machine, the largest and most powerful ever built,

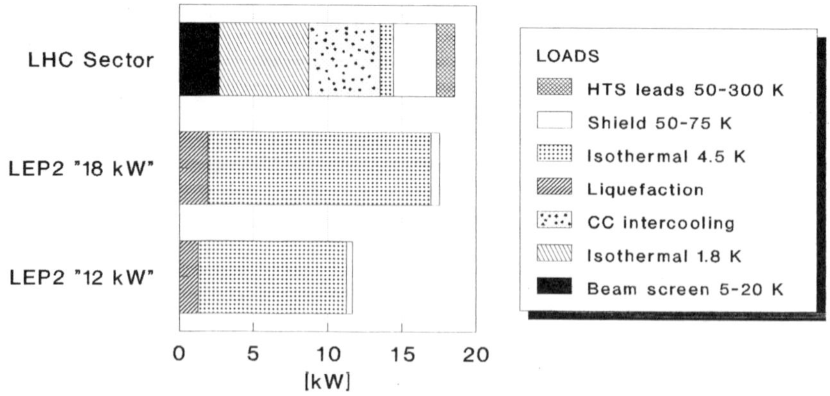

Figure 2. Equivalent 4.5 K refrigeration requirements of a LHC octant sector as compared to the capacity of the LEP2 plants in their current "12 kW" version and of the LEP2 "18 kW" upgrade as offered in 1990.

will bring into collision intense counter-rotating beams of protons or ions, at the never reached before center-of-mass energy of 14 TeV. To achieve these ambitious goals at an acceptable cost, the latest developments in applied superconductivity will have to be used: most of the 26.7 km circumference of the machine will consist of twin-aperture, high-field (8.4 T) superconducting magnets operating in superfluid helium below 2 K.

Main features of LHC cryogenics

The cryogenic system[14] of the LHC will thus require large-capacity refrigeration at several temperature levels down to 1.8 K, to be provided by eight octant helium cryogenic plants. LEP's four existing 12 kW cryogenic plants will constitute the first building blocks of the LHC cryogenic system, a configuration which allows staging of investments and optimum use of financial and manpower resources. Although the overall entropic load requirements for LHC are not far from the 18 kW capacity defined in 1990 as possible upgrade of the LEP2 refrigerators, the mix of refrigeration duties is quite different (Figure 2). While conserving the main architecture, cryogenic valves, pipework and heat exchangers of the existing coldboxes, substantial modifications of the overall process and of various components will be required. To produce multi-kilowatt cooling power at 1.8 K, the lower coldboxes will be complemented, at their lower temperature end, by subatmospheric cold compressor boxes based on the novel technology of cryogenic centrifugal compressors, which will maintain the required 16 mbar saturation pressure.

For cooling down the 31×10^6 kg cold mass of the LHC ring, each octant cryoplant will also need to supply up to 300 kW refrigeration at gradually decreasing temperature; this power,

Table 3. Operational experience with SC modules in the LEP tunnel

Year	Plant (LEP point)	Modules	Compressor hours	Number of failures Total	Cryo. origin	Loss of time for LEP experiments due to cryogenics related failures
1992	6 kW (P2)	2	6200 h	7	1	0
1993	6 kW (P2)	2	3800 h	8	2	5 h
	12 kW (P6)	1	1400 h	2	0	0
1994	6 kW (P2)	2	4600 h	14	3	4 h
	12 kW (P6)	3	1500 h	3	1	1.5 h
1995	12 kW (P2)	2 -> 4	2400 h	2	0	0
	12 kW (P6)	6 -> 8	2500 h	2	0	0

incommensurable with the steady-state capacity requirements quoted above, will be provided by vaporization of liquid nitrogen and transported to the string of magnets in the tunnel by a forced flow of gaseous helium under pressure. The helium inventory in the LHC is expected to be around 10^5 kg of which about 30% can be kept stored in gaseous form at 20 bar and ambient temperature during shutdowns of the cryogenic system, using the buffer tanks needed to recuperate the helium evaporated during quenches of whole magnet octants.

Application of the Lessons Learned from the LEP2 Experience

Basic features adopted for LEP2 will be maintained for LHC. Also for LHC cryogenics, new civil engineering work will have to be kept to the minimum. For this reason, the successful vertical split of main coldboxes with interconnecting transfer lines at 20 K level will also be applied to the four additional cryoplants. In view of the cost, no real redundancy will be possible, except for selected vulnerable and vital components, and full operation of LHC possible only with all plants providing full performance; CERN will thus continue to count on reliability rather than redundancy.

Close collaboration with manufacturers. LHC will be a much more complex load than the SC cavities with only thin-walled structures and small inertia. More system studies have already been done for LHC cryogenics than ever done for LEP2. Nevertheless the LEP2 line of specifying performance requirements and leaving process details for definition by the suppliers will be maintained.

Acceptance tests. Also for LHC, acceptance tests will have to be decoupled from the availability of the final magnet strings and will have to be done with a simulation of final loads. In order to avoid the problems of the dead-band bonus-malus system adopted for LEP2, it is envisaged to make the supplier responsible for all acceptance test equipment, including the definition, in his offer, of the overall accuracy. He then will have to guarantee full nominal performance at the lower edge of his accuracy band.

Control system and automatic operation. With plant stations dispersed in the French countryside near CERN, distances of up to 10 km between stations and routine operation being sub-contracted more and more to external firms, a powerful, standardized control system for automatic operation and remote access from various control rooms, office or home consoles will also be vital for LHC. CERN will thus continue to keep control of it and make full use of experience gained on the LEP2 programme.

CONCLUSIONS

In general the experience with LEP2 cryogenics is seen as a success at CERN. It became reality within a short time-scale, cost has been kept under control, and now highly reliable plants are running with automatic operation under only one operator on call at a time.

For this reason we think that the main elements of this exercise can also be kept for the procurement of production and distribution of helium refrigeration for the LHC, in particular the matching of coldbox geometry to existing tunnel space, incitements to power saving and good part-load capability, reliance on reliability instead of redundancy, and use of industrial control equipment with separate procurement by CERN. Preparatory studies in close collaboration with the potential suppliers have already started on the basis of the existing equipment for LEP2 and with the idea of keeping the principles adopted for LEP2, not only for the LHC upgrade of existing equipment to the level of 18 kW refrigeration power, but also for the four additional cryoplants of similar power which will have to be installed at the beginning of the next century.

ACKNOWLEDGEMENTS

We acknowledge the excellent collaboration with our two supplier firms and their experts in charge of helium cryogenics. They have successfully adopted quite a number of CERN's unconventional suggestions despite difficulties in making their work for CERN profitable business. The fact that, finally, most difficulties could be overcome in a satisfactory way for both parties and that the six LEP plants could start production on time for the cavity progress must also be credited to the continuous support from project management under C.Wyss and the dedicated effort of our team of technicians and operators.

REFERENCES

1. S. Myers, "LEP Status and Plans", *Proc. of the 1995 Particle Accelerator Conference*, held in Dallas 1-5 May, 1995.
2. G. Cavallari et al., "Status of RF Superconductivity at CERN", *Proc. of the 6th Workshop on RF Superconductivity* held at CEBAF, Newport News, 4-8 Oct., 1993.
3. D. Güsewell et al., "Cryogenics for the LEP200 Superconducting Cavities at CERN", *Proc.IEEE of the 1993 Particle Accelerator Conference*, 4, p.2956-2958.
4. M. Barranco Luque et al., "The Refrigeration System of the LEP2 Energy Upgrade", Proc.ICEC13, *Cryogenics* 30, September Supplement, pp.136-142 (1990).
5. K. Kurtcuoglu et al., "Plant Performance of the Highly-Efficient 6 kW Helium Refrigerator for the LEP200 Project at CERN", paper presented at CEC93, Albuquerque (1993).
6. M. Barranco Luque et al., "First Operational Experience in Running a New 6 kW Cryoplant for the Superconducting Cavities in LEP at 4.5 K", *Proc. of the 6th Workshop on RF Superconductivity* , held at CEBAF, Newport News, 4-8 Oct., 1993.
7. G.M. Gistau and M.Bonneton, "A 6 kW at 4.5 K Helium Refrigerator for CERN's Cryogenic Test Station", paper presented at CEC93, Albuquerque (1993).
8. B.Chromec et al., "A Highly-Efficient 12 kW Helium Refrigerator for the LEP200 Project at CERN", *Supercollider 5*, Plenum Press, New York (1993).
9. G.M. Gistau and J.Veaux, "A 12/18 kW at 4.5 K Helium Refrigerator for CERN's LEP Superconducting Accelerating Cavities", *Cryogenics* 34, ICEC Supplement, p. 103-106 (1994).
10. S. Claudet et al., "Four 12 kW / 4.5 K Cryoplants at CERN", *Cryogenics* 34, ICEC Supplement, p. 99-102 (1994).
11. Ph. Gayet et al., "Architecture of the LEP2 Cryogenics Control System: Conception, Status and Evaluation, *Cryogenics* 34, ICEC Supplement, p. 83-86 (1994).
12. G. Winkler et al., "Cryogenic Operation and On-Line Measurement of RF Losses in the SC Cavities of LEP2",*Proc. of the 1995 Particle Accelerator Conference*, held in Dallas 1-5 May, 1995.
13. L.R. Evans, "The Large Hadron Collider",*Proc. of the 1995 Particle Accelerator Conference*, held in Dallas 1-5 May, 1995.
14. Ph. Lebrun, "Superfluid Helium Cryogenics for the Large Hadron Collider Project at CERN", *Cryogenics* 34, ICEC Supplement, p.1-8 (1994).

R&D OF HIGH RELIABLE REFRIGERATION SYSTEM FOR SUPERCONDUCTING GENERATORS

T. Hosoya,[1] S. Shindo,[1] H. Yaguchi,[1] T. Ichikawa,[1]
K. Tanimoto,[2] M. Kazumori,[3] H. Yanagi,[4] H. Asakura[5]

[1] Super–GM(Engineering Research Association for Superconductive Generation Equipment and Materials), Osaka, Japan
[2] AIST(Agency of Industrial Science and Technology of MITI), Tokyo, Japan
[3] NEDO(New Energy and Industrial Technology Development Organization) Tokyo, Japan
[4] Mayekawa Mfg. Co., Ltd., Ibaraki, Japan
[5] Ishikawajima–Harima Hevey Industries Co., Ltd., Tokyo, Japan

ABSTRACT

Super–GM carries out R&D of 70 MW class superconducting generators (model machines), refrigeration system and superconducting wires to apply superconducting technology to electric power apparatuses.

The helium refrigeration system for keeping field windings of superconducting generator (SCG) in cryogenic environment must meet the requirement of high reliability for uninterrupted long term operation of the SCG.

In FY 1992, a high reliable conventional refrigeration system for the model machines was integrated by combining components such as compressor unit, higher temperature cold box and lower temperature cold box which were manufactured utilizing various fundamental technologies developed in early stage of the project since 1988. Since FY 1993, its performance tests have been carried out. It has been confirmed that its performance was fulfilled the development target of liquefaction capacity of 100 L/h and impurity removal in the helium gas to < 0.1 ppm. Furthermore, its operation method and performance were clarified to all different modes as how to control liquefaction rate and how to supply liquid helium from a dewar to the model machine.

In addition, we have made performance tests and system performance analysis of oil free screw type of and turbo type compressors which greatly improve reliability of conventional refrigeration systems. The operation performance and operational control method of the compressors has been clarified through the tests and analysis.

INTRODUCTION

Super–GM was established in September, 1987 to develop technology for superconducting generation equipment and materials as a national project.

As a part of the project, we have been developing two type of highly reliable refrigeration systems for SCG; a conventional refrigeration system using oil–injection–type screw compressors and an advanced refrigeration system using oil free compressors[1].

Table 1. R&D Schedule of Refrigeration Systems

	'88	'89 ~ '90	'91 ~ '92	'93~ '95	'96~ '98
R&D Items, Conventional Type	Element techniques	Partially trial manufacture & test	Manufacture of trial machines	Factory test	Verification test with model machine
Advanced Type	Element techniques (Compressor)		Partially trial model manufacture & test (Compressor)		Research of Comp. unit and System (proposed)

The major subjects in developing the conventional refrigeration system for verification tests of the model machines are to develop technology for removing impurities in helium gas which largely affects reliability of the refrigeration system and to further improve reliability of each components of the system.

The major subject in developing the advanced refrigeration system is to develop both screw type of and turbo type of oil free compressors which can fundamentally solve failures caused by impurities, in order to greatly improve reliability of the refrigeration system.

R&D SITUATION OF REFRIGERATION SYSTEM

1. R&D schedule

R&D schedule of the refrigeration system is shown in Table 1. For the conventional refrigeration system, performance tests of the system has been carried out at the factory since 1993 fiscal year. The system will be transferred to Verification Test Center of Super-GM in 1995 fiscal year and verification test wil be performed in 1996 fiscal year by combining the system with the model machine. For the advanced refrigeration system, research works for developing technologies to design and manufacturing oil free compressors will be completed in 1995 fiscal year.

2. Conventional refrigeration system

We integrated the refrigeration system, combining a prototype of the compressor unit, a higher temperature cold box and a lower temperature cold box with equipments such as dewar. We have carried out its performance tests at the factory and confirmed liquefaction capacity and impurities removable capacity. Furthermore, we confirmed operation method of the refrigeration system such as the control methods of liquefaction rate and transferring liquid helium from the dewar to the model machine responding to its heat load.

2.1 Configuration of the conventional refrigeration system

The system flow diagram of the conventional refrigeration system is shown in Fig. 1. The compressor unit is composed of a canned compressor with a main compressor and a motor integrated, and an oil separator for removing impurities at room temperature [2, 8]. The higher temperature cold box houses a low temperature adsorber for removing air, which might penetrate into the cryogenic portion, with a capacity of purifying all the amount of circulated helium gas. For pre cooling, freon refrigerators are employed instead of liquid nitrogen in steady operation mode as a feature of the refrigeration system [3, 8]. The lower temperature cold box is equipped with a developed expansion turbine with high efficiency and high expansion ratio [4, 8]. Effective storage volume of LHe dewar is a 1600 L. The J-T valve is installed inside at the top of the dewar to decrease evaporating loss from liquefied helium as much us possible and three-fold vacuum-superinsulated transfer lines were employed for connecting between the lower temperature

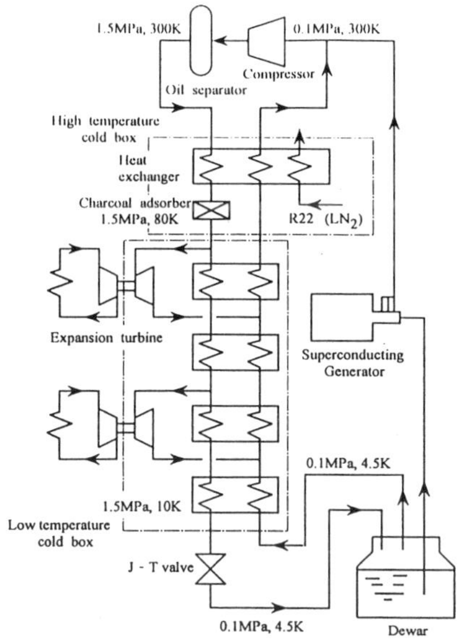

Fig. 1 System flow diagram
(Conventional refrigeration system).

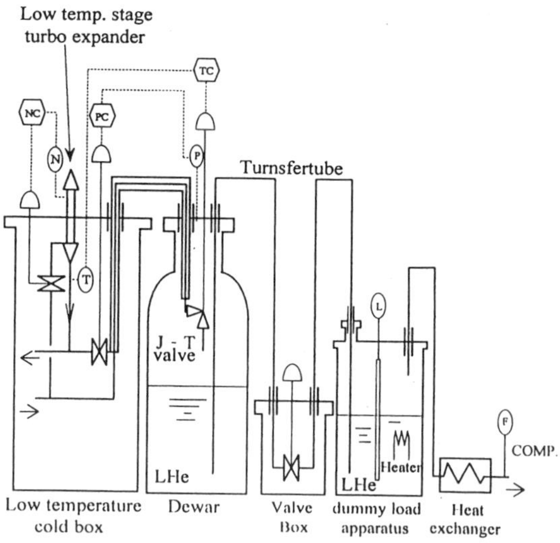

Fig. 2 Testing flow diagram at the factory
(Liquefaction and supplying LHe).

cold box and the dewar, and between the dewar and the model machine, aiming for the reduction of heat leak and pressure loss.

2.2 Performance and operation control of the conventional refrigeration system

As shown in Fig. 2, we have carried out liquefaction and liquid helium supplying tests of the refrigeration system, using a simplified dummy load apparatus which simulates heat load during operation of the model machine.

Table 2 summarises performance test results. Through the tests, we have confirmed that each of performance of the developed components satisfies the designed values.

For liquefaction capacity, the efficiency of the expansion turbine increased by approximately 8 % over the designed value, therefore, we could obtain the specified liquefaction rate of 100 L/h with

Table 2. Performance test results of conventional refrigeration system

ITEM		Design	Result
Comp - ressor	Total Isothermal efficiency (%)	54.1	55.1
	Canned Motor efficiency (%)	≧80	80.3
Impurities remove capacity	Dew point (K)	183	≦183
	Impure Gas (ppm)	0.1	≦0.1
Adiabatic efficiency of turbo expander (%)		≧65	73.5
Liquefaction capacity (L/h)		100	100
Mass flow rate (g/s)		69.17	68.0
Cooling Capacity (W)	High temp. stage turbo expander	4043	3950
	Low temp. stage turbo expander	1168	1300
Precooling Capacity (W)		2724	2630

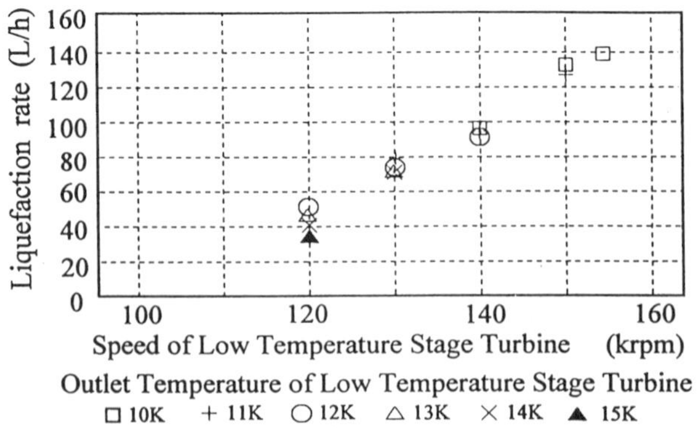

Fig. 3 Liquefaction test results.

less circulating flow than the designed amount. As shown in Fig. 3, it was confirmed that the liquefaction rate can be stably controlled within a range of 40 to 140 L/h by operating the expansion turbine with low–temperature turbine speed as a main parameter and with low–temperature turbine outlet temperature as supplementary parameter for stabilizing operation of the refrigeration system.

A LHe supply from the dewar to the model machine was adjusted by controlling openness of the liquid supply valve in the valve box and dewar pressure. It was found that the liquid supply valve can control the refrigeration system more stably than the dewar pressure control, with the smaller fluctuation of the operation conditions of the refrigeration system.

Liquid supplying test was carried out by providing heat load to a simple dummy load with an electric heater inside and by operating the refrigeration system so as to maintains the liquid helium levels in the dewar and in the simplified dummy load apparatus constant simultaneously. As a result, we have confirmed, as shown in Fig. 4, liquid supplying capacity at the normal operation corresponding to 48 W of assumed heat load for each of the model machines. Further, we have confirmed the maximum capacity of 1.4 times the normal capacity and found the operation method to make supplied liquid volume and liquefaction rate equal. We have also confirmed that the liquid level can be controlled constant by automatically adjusting supplied liquid volume based on the liquid level signals of the simplified dummy load apparatus when heat load given by the heater is changed.

2.3 Verification test of the conventional refrigeration system

The refrigeration system for the model machines will be installed in a test facility for the model machines and the refrigeration system in FY 1995. The model machines will be tested in the next year [5]. The model machines will be tested under actual loading conditions using the Motor–Generator method as shown in Fig. 5. The test facility is located at the Osaka Power Station of The Kansai Electric Power Co., Inc. in Japan, and the construction work was started in June 1994 [6]. Fig. 6 shows the test facility under construction. A control building, a refrigeration system building and a service building were already constructed.

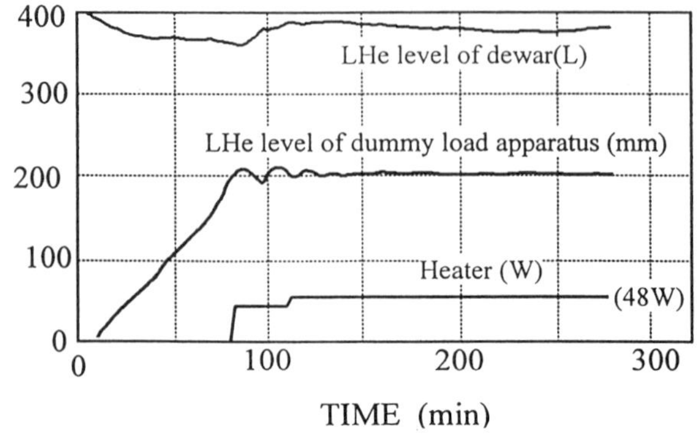

Fig. 4 LHe supplying test results.

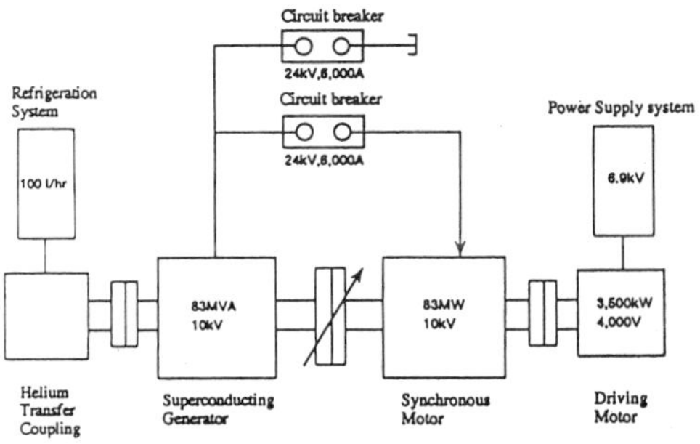

Fig. 5 Test system for model machine.

Fig. 6 Test facility under construction.

3. Advanced refrigeration system

For R&D of the advanced refrigeration system, operation performance and operation method of the compressor have been studied based on performance tests of the oil free compressors and system performance analysis .

3.1 Oil free screw compressor

The oil free screw compressor compresses helium gas, retaining clearance between screw rotors. We have made trial rotors, considering heat deformation caused by compression heat. As shown in Fig. 7, seal performance was confirmed by carrying out rotation tests with the seal system incorporated in the trial rotors [7].

Fig. 7 Rotor and seal system.

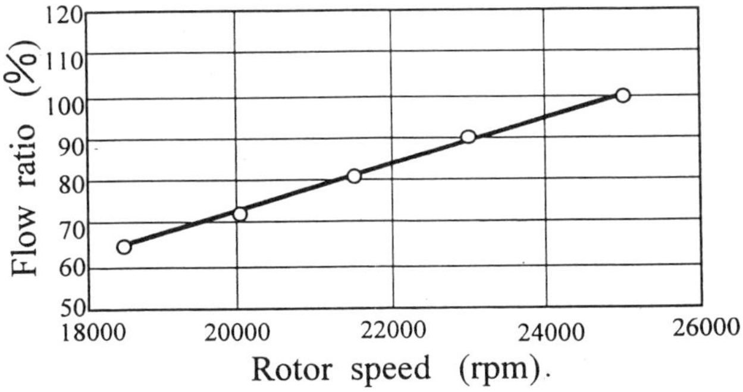

Fig. 8 Variance of flow ratio versus rotor speed.

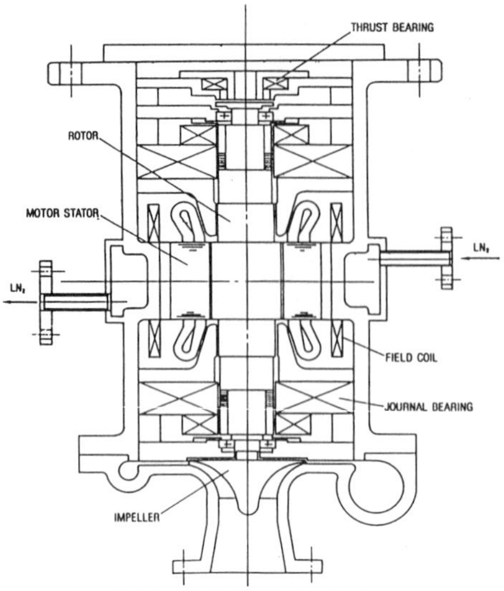

STAGE	INLET PRESS. MPa	OUTLET PRESS. MPa	ROTATION SPEED RPM	FLOW RATE g/s	IMPELLER DIA. mm
1	0.101	0.196	99500	64	109
2	0.194	0.342	99500	113	101
3	0.340	0.549	99500	113	92
4	0.547	0.811	99500	113	92

Fig. 9 Cross section of the turbo compressor.

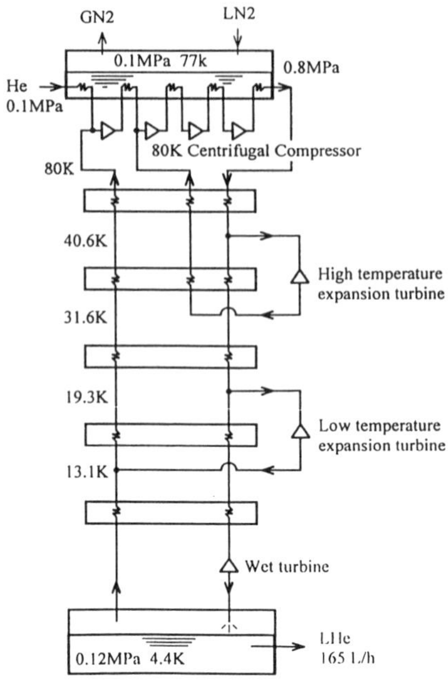

Fig. 10 Refrigeration system flow.

To increase compression efficiency, clearance between the rotors should be minimized and the rotors should be rotated at high speeds. It was found that two rotors rotating can be operated stably up to the targetted 25,000 rpm on non-contact condition by adjusting pre-loaded pressure of the thrust bearings supporting the rotors.

Further, performance tests of the trial rotors was made, employing an element test apparatus with air used as a working fluid. Compression efficiency of the trial rotors attained 0.72 at the converted volumetric efficiency on the condition of a rotor speed of 25,000 rpm and a designed compression ratio of 4, exceeding tha target value. After checking capacity characteristics caused by rotor speed control, we have found as shown in Fig. 8 that discharged flow rate is proportional to rotor speed. Further, we have found operation method of the compressor by operating the compressor on the assumption of actual operation pattern ranging from start to stop.

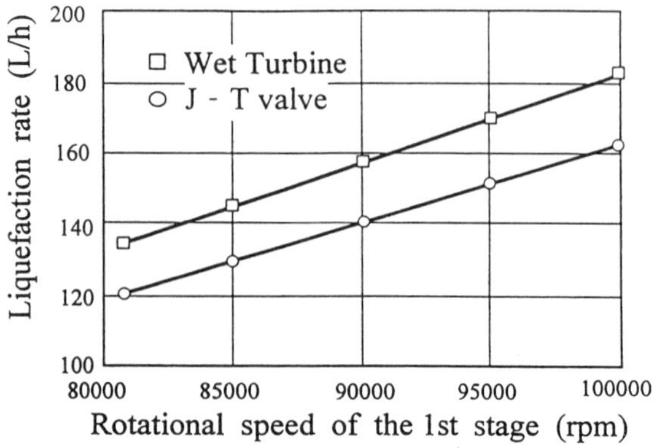

Fig. 11 Result of liquefaction rate analysis.

3.2 Oil free turbo compressor

An oil free turbo compressor with 4 stages have been under development, which compresses helium gas of inlet temperature of 80 K from 0.1 to 0.8 MPa, with reliability increased through decrease of the number of compression stages [8].

Low temperature performance tests were carried out with use of the prototype compressor (single stage) as shown in Fig. 9 and resulted in satisfying flow rate at each stage and pressure ratio of design specifications at lower speed than the designed speed of 99,500 rpm. We have also obtained 62.3% of adiabatic efficiency with total 4 stages.

We have analyzed performance of a refrigeration system including the turbo compressor and checked the relation between compressor speed and liquefaction rate. The refrigeration system flow is shown in Fig. 10. In this system a wet turbine is employed to improve the liquefaction rate and system efficiency. Performance analysis was made on the condition that inlet and outlet gas temperature of the compressor was constant at 80 K and that only the first stage compressor speed was variable and the 2nd to the 4th stage speed was fixed at 90,600, 98,000 and 88,000 rpm respectively. An efficiency of wet turbine was here assumed to be constant at 54% and comparative study of two cases by using a wet turbine and by a J−T valve was carried out.

Fig. 11 shows analysis result. Liquefaction rate varies proportionally to speed of the first jp3 stage compressor, which shows that the compressor speed control is the effective way to control liquefaction rate. It was also found that the case of using a wet turbine increases liquefaction rate approximately 10% compared with that of using the J−T valve.

CONCLUSION

1. Conventional refrigeration systems

Specified liquefaction capacity value of 100 L/h of the refrigeration system for the 70 MW class superconducting generators is confirmed.

We have confirmed that impurities removable capacity satisfies the target capacity removing impurities of 0.1 ppm or less.

We have confirmed the operation methods of the refrigeration system were established, concerning control method of liquefaction rate made by rotational speed control of the low−temperature turbine, and liquid helium supplying method made by the liquid supply valve control and dewar pressure control.

And we are planning to verify operability and reliability of the system incorporating the verification test facility.

2. Oil free screw compressor for advanced refrigeration system

Compression efficiency of the trial rotors in the air tests attained 0.72 at the converted volumetric efficiency on the condition of a rotor speed of 25,000 rpm and a designed compression ratio of 4, exceeding the target value.

Discharge flow rate is in the relationship proportional to rotor speed, showing better characteristics of volumetric control.

3. Oil free turbo compressor for advanced refrigeration system

The following were obtained from the simulation analysis of the refrigeration system employing a multi− stage compressor.

The liquefaction rate can be controlled effectively by varying the first stage compressor speed.

The liquefaction rate varies proportionally to the first stage compressor speed.

The wet turbine contributes to increase the liquefaction rate approximately 10% compared with the J−T valve.

ACKNOWLEDGEMENTS

This research has been carried out as a part of R&D on superconducting technology for electric power apparatuses under the New Sunshine Project of AIST, MITI, being consigned by NEDO.

REFERENCES

1. H. Ohya, K. Fujima, and D. Kato, R&D of High Reliability Refrigeration Systems for Superconducting Generators, in: " Advances in Cryogenic Engineering, " Vol. 37, Plenum Press, New York(1992), P. 747
2. T. Kishi, M. Kudo, H. Iisaka, and N. Ino, High Efficiency, High Reliability Oil Injection Type Helium Screw Compressor, in: " Advances in Cryogenic Engineering, " Vol. 37, Plenum Press, New York(1992), P. 771
3. H. Yanagi, H. Yaguchi, T. Yabana, K. Fujima, N. Ino, and A. Yasuda, The Planning of a Low Temperature Purifier for Super- GM helium Refrigeration, in: " Advances in Cryogenic Engineering, " Vol. 37, Plenum Press, New York(1992), P. 739
4. N. Ino, A. Machida, K. Tsugawa, Y, Arai, H. Hashimoto, and A. Yasuda, Development of Externally Pressurized Thrust Bearing for High Expansion Ratio Expander, in: " Advances in Cryogenic Engineering, " Vol. 37, Plenum Press, New York(1992), P. 817
5. T. Ageta, T. Ichikawa, Y. Tsunoda, M. Kazumori, H. Taniguchi, Y. Yagi, A. Ueda, and T. Kitajima, Recent development progress of 70MW class superconducting generators (part1), in: No. 11-102, 1994 CIGRE Session, 28th Aug. - 3rd Sept., 1994, Paris, France.
6. A. Katagiri, S. Ohshima, B. Ikeda, K. Kimura, T. Ichikawa, K. Tanimoto, M. kazumori, T. Sonobe, K. Suzuki, and T. Kitajima, Development of 70MW class Superconducting Generators, 14th International Conference on Magnet Technology, June 12, 1995
7. K. Arai, T. Ogawa, A. Yasuda, R. Toyama, D. Kato, and K. Fujima, R&D of High Reliability Refrigeration System for Superconducting Generators, in: " Advances in Cryogenic Engineering, " Vol. 39, Plenum Press, New York(1994), P. 581
8. H. Asakura, D. Kato, N. Saji, and H. Ohya, 80 K Centrifugal Compressor for Helium Refrigeration system, in: " Advances in Cryogenic Engineering, " Vol. 37, Plenum Press, New York(1992), P. 787
9. H. Asakura, D. Kato, N. Saji, H. Ohya, M. Kubota, Y. Kaneko, S. nagai, and R. Toyama, Performance Test Results of 80 K Centrifugal Compressor for Helium Refrigerator, in: " Advances in Cryogenic Engineering, " Vol. 39, Plenum Press, New York (1994), P. 893

THE SUPERFLUID HELIUM CRYOGENIC SYSTEM FOR THE LHC TEST STRING: DESIGN, CONSTRUCTION AND FIRST OPERATION

A. Bézaguet, J. Casas-Cubillos, B. Flemsaeter, B. Gaillard-Grenadier,
Th. Goiffon, H. Guinaudeau, Ph. Lebrun, M. Marquet, L. Serio,
A. Suraci, L. Tavian and R. van Weelderen

Accelerator Technology Division, CERN
CH-1211 Geneva 23, Switzerland

ABSTRACT

A major milestone in the preparation of the Large Hadron Collider (LHC) project is the testing and operation of a 50-m long superconducting magnet string, representing a half-cell of the machine lattice. This also corresponds to the length of the elementary cooling loops providing refrigeration at the 1.9 K, 4.5-to-20 K, and 50-to-75 K levels to the LHC cryomagnets. Based on existing large-capacity cryogenic infrastructure, we have designed, built and are operating a dedicated cryogenic system feeding the LHC Test String, with installed capacities of 120 W @ 1.8 K and 10 g/s supercritical helium at 4.5 K. The system also includes 15 kA, 1.6 kA, 500 A, 250 A and 50 A current lead pairs for powering of main and auxiliary magnet circuits, as well as a 120 kW liquid nitrogen vaporizer for controlled cooldown of the 10^5 kg cold mass. The system is fully instrumented, controlled by dedicated industrial PLCs connected to an industrial supervision system. We report on performance in operation, including response of the system to transients such as current ramp and discharge, as well as magnet resistive transitions.

INTRODUCTION

The Large Hadron Collider (LHC) project, approved by the CERN Council in December 1994, will be the next major research facility in high-energy physics.[1] Based on a ring of high-field, twin-aperture superconducting magnets[2] operating in superfluid helium below 2 K,[3] to be installed in the 26.7 km circumference tunnel of the present LEP collider, it will provide proton-proton collisions with a center-of-mass energy of 14 TeV and at an unprecedented luminosity of 10^{34} $cm^{-2}.s^{-1}$. Therefore, the LHC also represents a major engineering challenge in applied superconductivity and cryogenics,[4] and has thus required a specific R&D programme in these domains.

An important milestone in the validation of the basic technical choices, as far as main accelerator systems - magnets, cryogenics and vacuum - are concerned, is the testing and operation of a full-scale magnet string, representing a half-cell of the machine lattice. After recalling basic operating principles of test string cryogenics, we describe system layout, report on the spring 1995 campaign and present main results concerning cryogenics.

PRINCIPLES OF OPERATION AND SYSTEM LAYOUT

The first version of the LHC test string was assembled, commissioned and operated for the first time at the end of 1994.[5] Due to delays in the delivery of the 10 m-long dipole prototypes, one quadrupole (Q) and only two dipoles (D1 and D2) were installed in the string, on a concrete beam with a slope of 1.4 % (see figure 1), to exactly reproduce the extreme conditions that will be encountered in the LHC tunnel. Even though the results obtained with this configuration are sufficient to validate the basic technical choices, the other two dipoles will be installed as soon as they become available.

The principal components of the cryogenic system are the **S**tring **F**eed **B**ox (SFB), the **S**tring **R**eturn **B**ox (SRB), and the saturated superfluid helium circuit coupled to the pressurized (1 bar) superfluid helium bath through a corrugated heat exchanger tube.[6,7] This acts as a linear cold source of large developed heat exchange area and at quasi-constant temperature, irrespective of the magnet position in the string. Furthermore, the high apparent thermal conductivity of superfluid helium, pressurized at 1 bar and permeating the superconducting windings thanks to its low viscosity, is exploited to transport the heat to the two-phase saturated helium II, flowing inside the heat exchanger tube.

Figure 1. General view of LHC Test String .

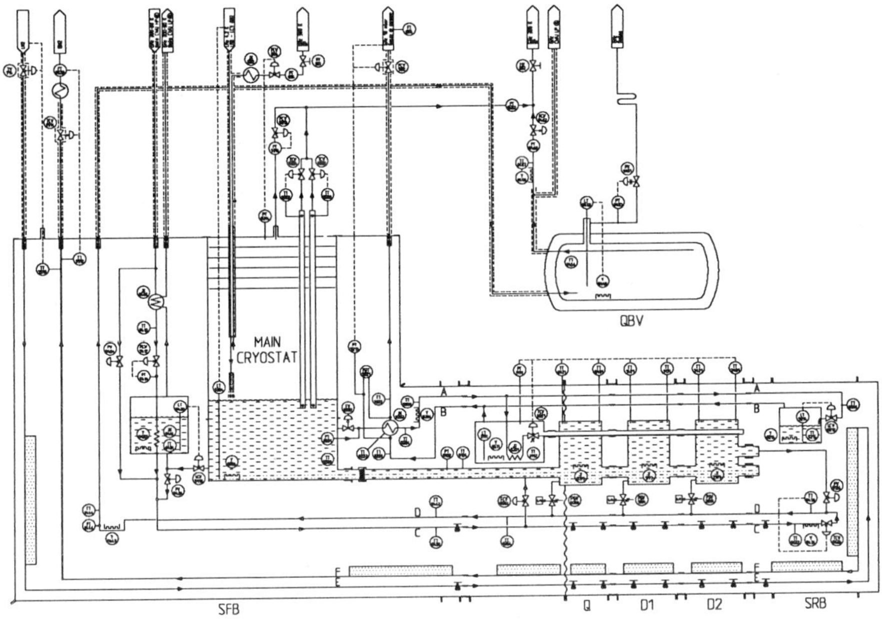

Figure 2. Cryogenic flow-scheme of LHC Test String (Phase 1).

The SFB external mechanical assembly, capable to withstand the pulling force of 8 t during evacuation of the insulation vacuum, contains the main cryostat, used to distribute the liquid helium to different circuits in order to perform refrigeration and transient phases (i.e. cooldown and warm-up). This cryostat is equipped with several pairs of current leads for powering of dipole and quadrupole (15 kA) and later correction elements (50, 250, 500 and 1600 A). Furthermore, the SFB contains a gas counterflow heat exchanger, a liquid/gas heat exchanger, a very low pressure liquid subcooler, as well as all the valves and instrumentation necessary for the cryogenic operation of the string. The SRB is the end module of the string and is used to close the cryogenic loops as well as the insulation vacuum. The design of completely autonomous modules allowed the prior testing of the full cryogenic loop without magnet.

The cooling scheme is implemented in independent cooling loops, each extending over the total length of the string (figure 2). Liquid He I taken from the main cryostat is subcooled in a very low pressure heat exchanger and distributed through line A, where it is expanded to saturation through a Joule-Thomson valve and fed to the far end of the heat exchanger tube. The low saturation pressure is maintained on the flowing two-phase helium by pumping the vapor through line B and the very low pressure heat exchanger. Line C supplies supercritical helium through the magnet supports to intercept residual conduction at 4.5 K and through the magnet cold mass to perform the operation of cooldown, filling and warming-up (in the latter case high pressure gaseous helium is used).

Line E and F are used to supply liquid nitrogen to the thermal shield and magnet support to intercept heat at the 90 K level.

Main features of the dipole cryostat[8] are the integration of all pipes distributing refrigeration, the support and positioning of the large cold mass by cylindrical post made of non-metallic composite materials and the use of multilayer reflective insulation.

The short straight section cryostat[9] is based on the same design principle and houses the local cryogenic components of the superfluid cooling loop and an insulation vacuum barrier.

After a magnet resistive transition,[10] the resulting pressure rise is contained below the 2 MPa maximum design pressure of the cold mass by opening the quench relief valve[11] on the quench trigger and discharging helium in line D.

The system also makes full use of the pre-existing cryogenic infrastructure[12] that comprises a 6 kW cryogenic plant capable of producing 18 g/s of liquid helium, a **W**arm **P**umping **U**nit (WPU) to pump on the saturated superfluid helium up to 6 g/s, a **Q**uench **B**uffer **V**essel (QBV)(2000 L) capable of recovering the total volume of liquid discharged from the magnets after a quench, the balloon system (300 m^3) to recover the gaseous helium expelled after a quench and the **C**ooling and **W**arming-up **U**nit (CWU), which provides temperature-controlled gaseous helium in the range 300 to 80 K, using a liquid nitrogen/gaseous helium heat exchanger and a 25 kW heater.

INSTRUMENTATION AND PROCESS CONTROL

The LHC test string is equipped with the following cryogenic instruments: temperature and pressure sensors, liquid level gauges and mass flowmeters. Of these only the liquid level gauges and platinum thermometers have a standard response curve and do not need to be calibrated individually.

To measure liquid helium mass flow in the superfluid helium cooling circuit a Venturi flowmeter[13] at 4.5 K and a thermal flowmeter at 2.2 K are used. The thermal mass flowmeter confirmed the correct operation of the Venturi in spite of possible electromagnetic interference between cold pressure transducer and the main 12 kA current bus feeding the superconducting magnets. Tests were also done to compare the two cryogenic flowmeters with ambient temperature differential pressure and thermal flowmeters at the outlet of the WPU.

Magnet temperature measurements are done in the 300 K to 1.6 K range by using a platinum and Allen-Bradley resistance thermometer pairs. The Allen-Bradley thermometers had to be recalibrated in situ because they presented temperature offsets that could be as high as 0.2 K while the magnets temperature needs to be regulated to within better than 0.05 K at an absolute temperature that could range between 1.8 K and 1.9 K. This calibration was done for temperatures below 2 K to within better than 0.01 K thanks to the extremely good thermal conductance of pressurized superfluid liquid helium. The temperature reference was given by the pressure of the saturated liquid helium flowing inside the distributed heat exchanger. No effort was done to make in-situ temperature calibrations in normal liquid helium because of the difficulty to have an isothermal liquid helium I bath; furthermore the magnets are never operated in this temperature range.

The magnet pressure is measured by using piezoresistive cold sensors[14] with a typical range from 0 to 3 MPa and a response that strongly depends on the temperature. After the occurrence of a quench they all yielded a reading 60 % below the one expected and this for a few tens of minutes. This effect, not completely understood, might be due to a sensor temperature higher than the one given by the thermometers installed in the magnets.

The cryogenic process control is performed using an industrial Programmable Logic Controller (PLC) in which the number of I/O channels is of the order of 250 and there are about 30 closed control loops. The automatic operation modes include the string cooldown, normal operation, quench recovery and warm-up. This equipment is interfaced with a supervision system provided by CERN's industrial control group.[12]

In addition, around 140 cryogenic transducers are recorded for analysis purposes by the data acquisition system with 1 kHz capability.

COOLDOWN AND WARMUP

The string cooldown is started once the insulation vacuum is around 10^{-2} Pa. The first phase from ambient to 80 K is achieved passing pressurized GHe from the CWU at progressively decreasing temperature, regulated on the overall string gradient in order to

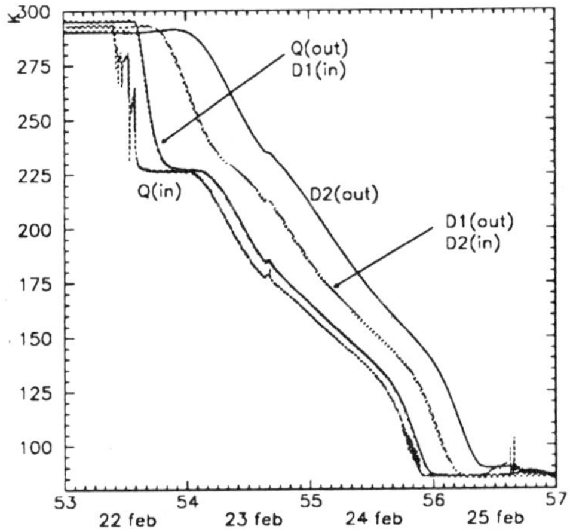

Figure 3. Forced-flow cooldown, 300-80 K.

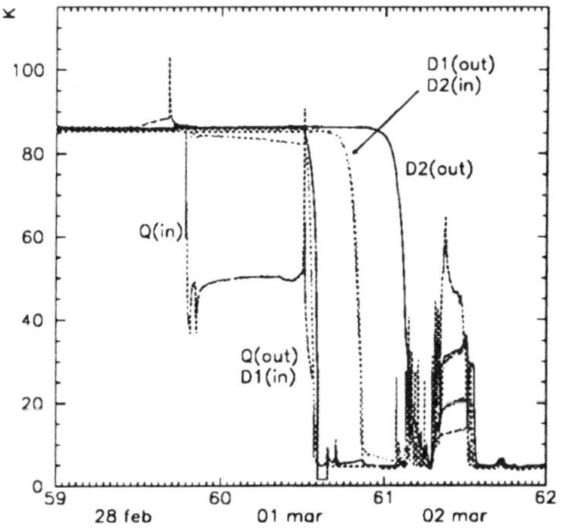

Figure 4. Supercritical helium cooldown, 80-4.5K.

avoid thermal stresses on the magnets. In the two complete cooldown phases, the flow was set to 50 g/s with 50 K and 60 K overall gradient respectively, with almost no difference in cooling speed (figure 3 [60 K gradient]). At the same time liquid nitrogen is sent and vaporized in the thermal shield. When the cold mass reaches 80 K, the cooldown to 4.5 K can be done in two ways: one using liquid helium directly from the cryostat, the other using supercritical helium (figure 4). In the latter case, a 80 K GHe flow is provided by the CWU and subcooled through two heat exchangers down to [4.5 K, 2.5 bar] and then expanded through the feeding valve in the cold mass. At the end of this phase, the 700 L volume of the cold mass is filled up and the WPU pumping rate limits the cooling speed down to 1.9 K (figure 5). The gained operational experience has reduced the time of cooldown from 70 K to 4.5 K down to 9 hours using the LHe I version.

Warming-up is achieved by pressurized GHe at progressively increasing temperature flowing in the system from the CWU and regulated on the overall string gradient. Figure 6 shows a warm-up at 80 g/s and with 60 K maximum overall gradient.

The results obtained, summarized in table 1, can be extrapolated to estimate the time required for cooling, warming and intervention on a full octant of the final LHC machine. The different cold/warm fronts show the good heat exchange capacity of the magnets and

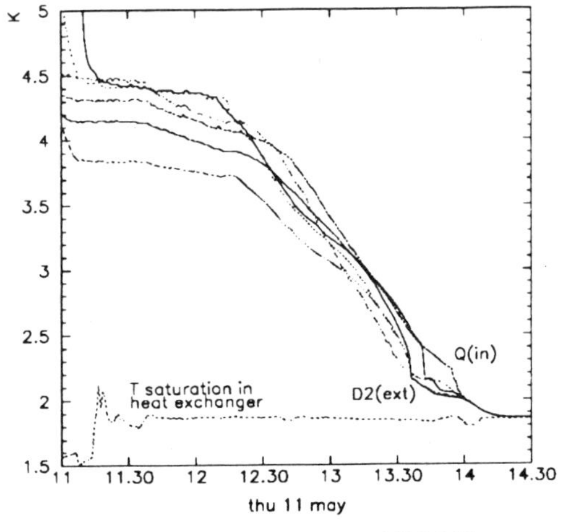

Figure 5. Final cooldown, 4.5-1.8 K.

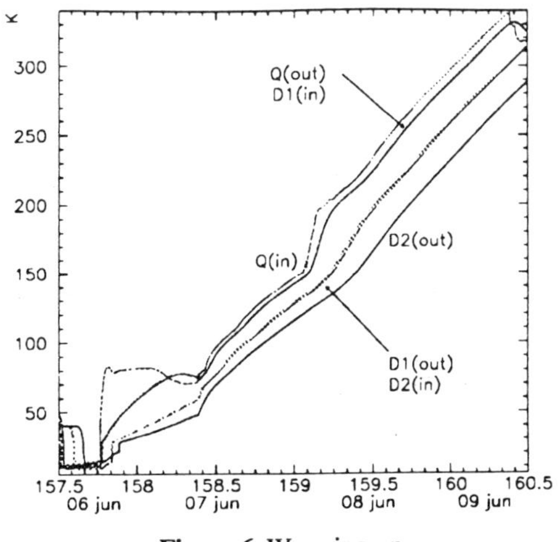

Figure 6. Warming-up.

Table 1. Cryogenic system performance summary

Phase		Time	Consumption
300 - 80 K	Cold mass	3.5 days	GHe 50 g/s
	Thermal shield	2 days	N2 145 L/min
80 - 4.5 K			
	using LHe	<3 days	LHe 5 g/s + 700 L
	using SHe	<3 days	GHe 20 g/s + LHe 2 g/s
4.5 - 1.8 K		2h30m	LHe 10 g/s
Quench Recovery (**35-1.8 K**)		5-6 hours	LHe 10 g/s + 700 L
4.5 -300 K		3 days	GHe 80 g/s

underline the limiting factors, i.e. maximum gradient on the magnets and cryogenic plant characteristics.

STEADY OPERATION AND CURRENT RAMPS

The superfluid helium cooling loop controls the temperature of the static pressurized helium II bath of the superconducting magnets by gradual vaporization of the flow of saturated helium II along the wetted length of the corrugated heat exchanger tube, fed by the Joule-Thomson valve. This process involves separated two-phase flow of helium II, transverse Kapitza and solid conductance across the tube wall as well as longitudinal conduction in pressurized helium II, and thus results in strongly non-linear behaviour.

Particular care and several simulation tests with applied heat loads were devoted to the optimization of the control law for the Joule-Thomson valve, in order to maintain the magnets at their operating temperature along the string length and throughout steady and transient operating modes.[15] Heat loads up to 1 W/m were applied on the cold mass to simulate a larger heating power than the one that will be deposited by the proton beams. The maximum temperature excursions registered were below 30 mK and during nominal ramping rates (i.e. 10 A/s) the temperature increase was of the order of 6 mK (figure 7).

Even with such a short response time, the system stability was not compromised and the temperature oscillations over a period of 48 hours were smaller than a few millikelvin.

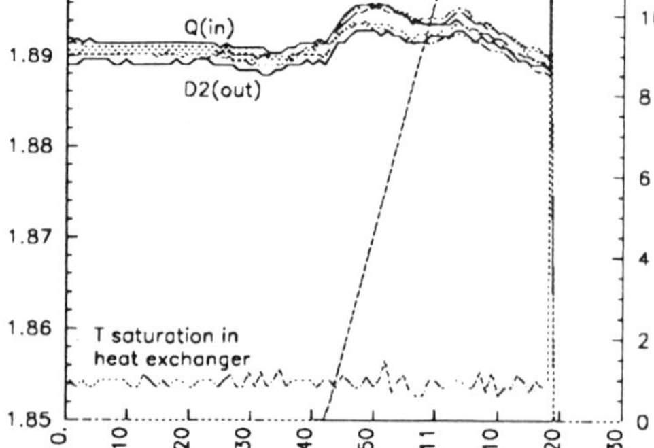

Figure 7. Stability of magnets temperatures along the string and response to current ramping .

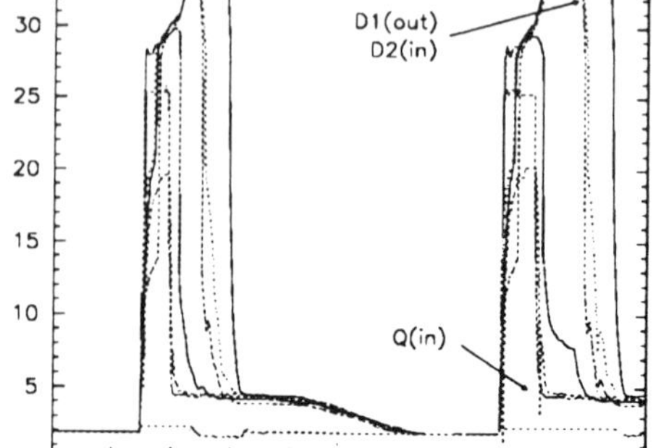

Figure 8. Recoveries after resistive transition .

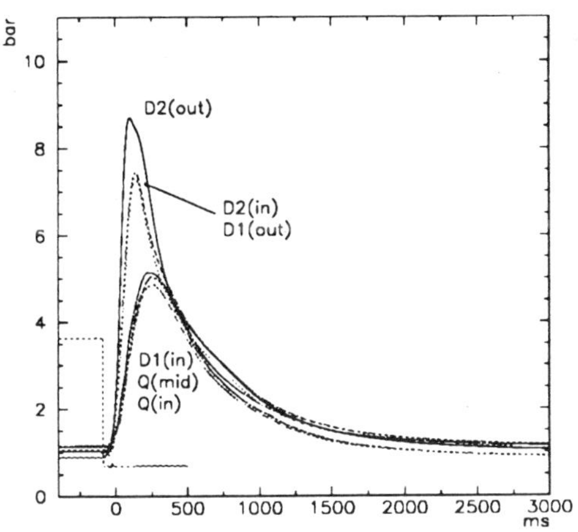

Figure 9. Thermohydraulic effects of a resistive transition with all discharge valves opened-12.8 kA .

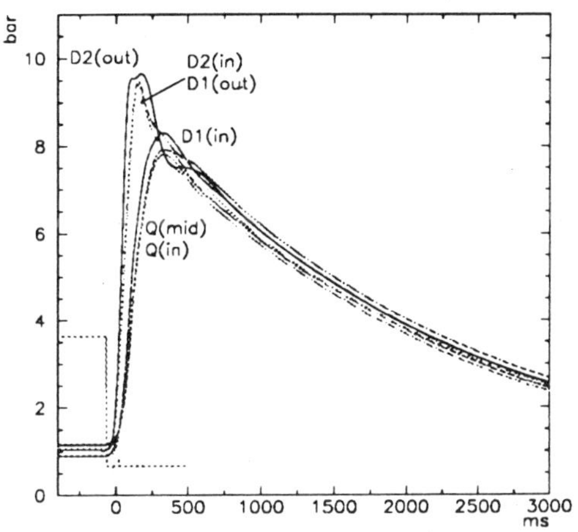

Figure 10. Thermohydraulic effects of a resistive transition with 2 discharge valves blocked-13.1 kA .

Current discharge at nominal rate (around 130 A/s) resulted in a magnets temperatures increase of 50 mK, well below the transition temperature of 2.2 K.

MAGNET RESISTIVE TRANSITIONS AND RECOVERY

One of the main goals of this campaign was the analysis of thermohydraulic effects of a resistive transition of a magnet (figure 8), when the energy stored in the coils (several MJ) is suddenly discharged in the liquid helium bath. Tests were performed mostly at a current of 13 kA (corresponding to a field of 9 T) varying mainly two parameters: the number of discharge valves used and the sequence of firing the magnet heaters. The ratio of liquid to gaseous helium recovered after being expelled from the cold mass through the discharge valves and the recovery line, respectively, in the QBV and in the balloons, was always of the order of 4 to 1, while practically no liquid was left in the cold mass. Figures 9 and 10 show two typical pressure transients varying the number of discharge valves opened. The contained increase in peak pressure in the case of a single valve opening has justified increasing the spacing between these valves, thus gaining in cost and reliability.

CONCLUSIONS AND FUTURE OUTLOOK

The first operation of the string was very successful and has validated the basic design choices of LHC cryogenics.

After a short summer shut-down, the string will be upgraded with a new dipole and the cryogenic system slightly modified to allow the verification of the performance of the superfluid helium cooling loop in counter-current two-phase flow. At the same time the control system strategies will be updated profiting from the operational experience gained. In line with the results obtained, only two discharge valves will be installed, one at each end of the string.

ACKNOWLEDGEMENTS

The authors wish to thank the LHC String Team for the valuable work performed in preparing and run the first experimental campaign of the string.

REFERENCES

1. L.R. Evans, "The Large Hadron Collider", invited paper at the Particle Accelerator Conference, Dallas (1995).
2. R. Perin for the LHC Magnet Team, "Status of LHC Programme and Magnet Development", invited paper at the Applied Superconductivity Conference, Boston (1994).
3. Ph. Lebrun, "Superfluid Helium Cryogenics for the Large Hadron Collider Project at CERN", *Cryogenics* 34 ICEC Supplement, pp.1-8 (1994).
4. L. Evans, "Advanced Technology Issues in the LHC Project", Proc. EPAC'94 Vol.1, World Scientific, pp. 429-433 (1994).
5. P. Faugeras, for the LHC String Team, "Assembly and Commissioning of the LHC Test String", paper presented at the Particle Accelerator Conference, Dallas (1995).
6. J. Casas-Cubillos, A. Cyvoct, Ph. Lebrun, M. Marquet, L. Tavian and R. van Weelderen, "Design Concept and First Experimental Validation of the Superfluid Helium System for the Large Hadron Collider Project at CERN", *Cryogenics* 32 ICEC Supplement, pp.118-121 (1992).
7. A. Bézaguet, J. Casas-Cubillos, Ph. Lebrun, M. Marquet, L. Tavian and R. van Weelderen, "The Superfluid Helium Model Cryoloop for the CERN Large Hadron Collider", *Adv. Cryo. Eng.* 39A, pp. 649-656 (1994).
8. J.C. Brunet, J. Kerby, Ph. Lebrun, P. Rohmig, B. Szeless and L.R. Williams, "Design of LHC Prototype Dipole Cryostats", *Cryogenics* 32 ICEC Supplement, pp. 191-194 (1992).
9. W. Cameron, B. Jenny, G. Riddone, P. Rohmig and R. van Weelderen, "Design and Construction of a Prototype Superfluid Helium Cryostat for the Short Straight Sections of the CERN Large Hadron Collider", *Adv. Cryo. Eng.* 39A, pp. 663-670 (1994).
10. G. Gerin, B. Vullierme and R. van Weelderen, "Measurement of the Thermo-Hydraulic Behaviour of LHC Dipole Prototypes after Quench", paper presented at this conference.
11. H. Danielsson, G. Ferlin, B. Jenninger, C. Luguet, S.E. Milner and J.M. Rieubland, "Cryogenic Performance of a Superfluid Helium Relief Valve for the LHC Superconducting Magnets", paper presented at this conference.
12. V. Benda, G. Duraffour, A. Guiard-Marigny, Ph. Lebrun, F. Momal, R. Saban, V. Sergo, L. Tavian and B. Vullierme, "Cryogenic Infrastructure for the Superfluid Helium Testing of LHC Prototype Superconducting Magnets", *Adv. Cryo. Eng.*, 39A pp. 641-648 (1994)
13. A. Rivetti, G. Martini and G. Birello, "Metrological Performances of Ventury Flowmeters in Normal, Supercritical and Superfluid Helium", *Adv. Cryo. Eng.*, 39B p 1051-1058 (1994)
14. Siemens KPY
15. B. Flemsaeter, "Contribution to the Dynamic Analysis and Optimal Control of the Superfluid Helium Cooling Loop for the LHC Magnet String", Diploma Thesis, Norwegian Institute of Technology, Trondheim (1995).

MEASUREMENT AND ANALYSIS OF THERMAL PERFORMANCE OF LHC PROTOTYPE CRYOSTATS

V. Benda, L. Dufay, G. Ferlin, Ph. Lebrun, J-M. Rieubland, G. Riddone, B. Szeless, L. Tavian, and L. Williams

Accelerator Technology Division, CERN
CH-1211 Geneva 23, Switzerland

ABSTRACT

The heat inleaks into the cryostats represent a significant fraction of the total heat loads to the Large Hadron Collider (LHC). In view of the large size of this machine and the high thermodynamic cost of low-temperature refrigeration, it is essential to maintain the heat inleaks within budgeted values. Following experimental validation of thermal design of critical components (e.g. multilayer reflective insulation, support posts, cold valves), thermal performance assessment of complete cryomagnets has been undertaken. Thermo-mechanical and cryogenic behaviour was measured on full-scale prototype units.

The heat loads in nominal steady-state conditions at the three temperature levels: 1.9 K, 4.5-20 K, 50-75 K have been analysed. The variation of temperature of cryostat components has allowed a thorough investigation of the total heat inleak sensitivity to temperature.

The effect of vacuum degradation on heat inleaks (from 10^{-3} Pa to 10^{-1} Pa) has also been measured.

Measured values have been compared to calculated values and to budgeted heat loads.

INTRODUCTION

The Large Hadron Collider (LHC) project, now approved, will represent a unique facility in high-energy physics research[1]. To be installed in an existing 26.7 km circumference tunnel, it is based on the large-scale use of high-field twin-aperture superconducting magnets[2] and will operate below 1.9 K in pressurized helium II[3]. The high cost of low-temperature refrigeration imposes that heat loads are maintained within budgeted values and the search for a cryostat of economic thermo-mechanical design renders essential the evaluation of the heat loads into complete cryomagnets. This has been carried out on full-scale prototype units, in the course of single-magnet tests[4], on a dedicated Cryostat Thermal Model (CTM)[5] and during the recent string tests[6].

Advances in Cryogenic Engineering, Vol. 41
Edited by P. Kittel, Plenum Press, New York, 1996

In collaboration with European industry, CERN is running an accelerator magnet development program which includes the testing of prototype 10 m-long twin-aperture high-field superconducting dipole magnets. Since February 1994, the thermal, mechanical and magnetic behaviour of four of these dipole magnets has been evaluated.

The string test phase 1, comprising one quadrupole and two dipoles, connected to a refrigeration unit and powered to nominal LHC conditions, has been operating since March 1995 and has allowed basic choices concerning LHC cryogenics to be validated.

The sources of heat loads are the heat inleaks, the resistive dissipation in the non-superconducting cable splices and the beam-induced loads produced by synchrotron radiation, beam image currents and the loss of particles by nuclear inelastic scattering with residual gas. The heat inleaks constitute 19 % of the heat loads at 4.5-20 K and 53 % at 1.9 K. Precise calorimetric measurements are difficult to perform during single-magnet tests as additional heat loads due to the cryogenic plant, high-current feedthroughs and equipment for "warm" magnetic measurements in the apertures dominate. The CTM has been finely instrumented and exploited to provide precise measurements of heat inleaks at 1.9 K, 4.5-20 K and 50-75 K without the contribution of the interconnection regions and other equipment.

CRYOSTAT THERMAL DESIGN

The currently tested prototype dipole cryostats[7] (Fig. 1a) house the cold mass of a twin-aperture dipole magnet, of 10.25 m length and 0.56 m outside diameter. Two cylindrical composite support posts maintain the cold mass in position. Top and bottom stainless steel flanges glued to each tubular column interface to the cold mass and the vacuum vessel. Two heat intercept plates in aluminium alloy are shrink-fitted and glued to the column to allow thermalisation at the two intermediate temperature levels. The cryogenic piping for helium distribution is fixed on both sides of the cold mass via low heat conduction supports. This cold mass and piping assembly is wrapped with 10 layers of reflective aluminized mylar superinsulation (MLI), thermally insulated from the cold mass at 1.9 K and in contact with the cryogenic pipes operating between 5 and 20 K.

The thermal shield, in commercially pure aluminium alloy consists of a 5 mm thick lower tray clipped and tack-welded to 3 mm-thick upper shell parts. This structure is wrapped with 30 layers of MLI.

The vacuum vessel, a 10 mm thick mild steel cylinder with stainless steel end flanges, several auxiliary ports and two reinforced load points for the cold mass and screens, contains an insulation vacuum in the 10^{-4} Pa range during normal operation. A bellows element attached to the vacuum vessel at one end houses the interconnect region to the adjacent magnet. The instrumentation feedthrough comprising two stainless steel tubes 8 mm and 12 mm-diameter by 3.5 m long, filled with instrumentation cables and rendered leak tight by epoxy impregnation, is welded to the cold mass and the vacuum vessel.

The cold mass of the CTM (Fig. 1b), which simulates that of a real magnet, is formed from a low-mass double-walled stainless steel cylindrical structure, spot-welded and pre-formed plastically to give adequate hydraulic resistance for the helium flow. A collector in stainless steel housing the copper heat exchanger and a dummy cold bore is placed on the bottom of the cylinder and is connected hydraulically in series with it. The two 1.9 K tubes are simulated by copper sheets approximating their shape and surface area and are fixed mechanically and thermally to the dummy cold mass. All other elements are identical to those in a real prototype dipole cryostat.

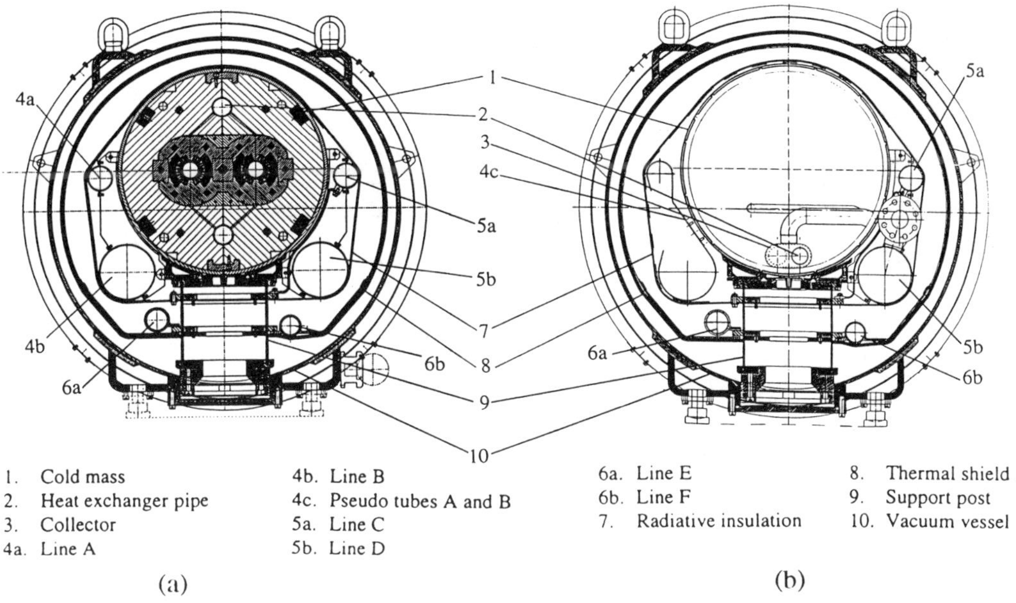

Figure 1. Cross sections of the prototype LHC dipole cryostat (a) and the Cryostat Thermal Model (b).

CRYOGENIC CIRCUITS

The simplified hydraulic circuit of the LHC half cell, the smallest repetitive LHC machine module, and of the string test, designed to simulate the behaviour of the half cell is shown in Fig. 2.

Magnets are cooled from ambient temperature to 4.2 K by forced circulation of gaseous helium tapped via the cooldown-and-fill valve (CFV) from line C and returned through the valve FV via line D.

Helium at 2.2 K and 1 bar from line A first passes through a heat exchanger and after expansion to saturation in a Joule-Thomson valve (TCV1) is expelled through the inner tube of the heat exchanger at 1.9 K. During its return through the outer heat exchanger tube, the helium vaporizes and cools the cold mass from 4.2 K to below 1.9 K.

The 4.5-20 K circuit, lines C (supply) and D (return), is cooled by forced flow of supercritical helium. It intercepts heat from the support posts and from the thermal shield through the radiative insulation. In the case of a superconducting magnet quench, helium is discharged from the cold mass into the recovery line D through the safety relief valve (SRV).

The circuit at 50-75 K, lines E (supply) and F (return), will be cooled in the LHC by forced flow of gaseous helium. Line E cools the support posts and line F cools the thermal shield which represents a first heat intercept of the heat inleak from room temperature.

The CTM maintains the temperature levels envisaged for the LHC cryogenic system. It is not equipped with the safety relief valve (SRV) and its flow scheme can be found in reference 5.

For single-magnet and string tests, the 50-75 K circuits are cooled by saturated nitrogen around 80 K.

The budgeted heat loads[8,9] for the prototype cryostats under steady-state nominal conditions at the three temperature levels are listed in Table 1.

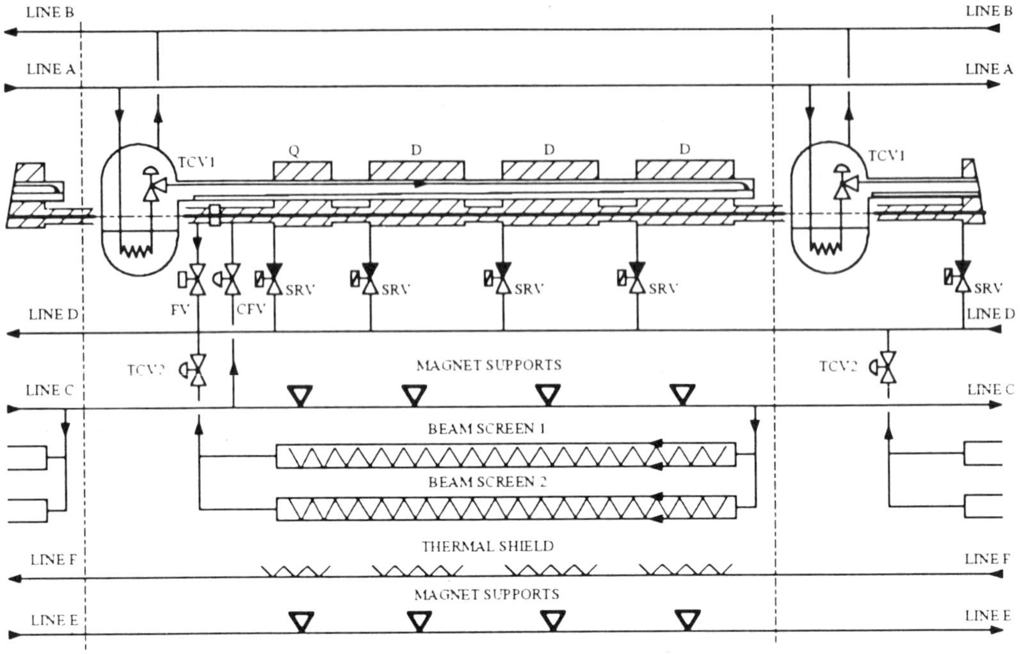

Figure 2. Flow-scheme of the LHC prototype half-cell.

TEST CONDITIONS AND HEAT LOAD MEASUREMENT METHODS

Cryostat Thermal Model

The heat inleaks under nominal LHC operational conditions have been measured and a parametric sensitivity study aimed at identifying the most critical cryostat components has been carried out. The parameters varied were temperature, mass flow and residual gas pressure in the insulation space.

The first evaluation of steady-state heat inleak to the CTM cryostat was made without the contributions from the end boxes and additional components, such as safety valves and instrumentation feedthroughs. The residual gas pressure in the cryostat was 10^{-3} Pa and nominal temperatures on the cryogenic lines have been obtained with the following mass flows: 0.5 g/s at 50-75 K, 0.2 g/s at 4.5-20 and the 0.08 g/s at 1.9 K. Under steady-state conditions at 50-75 K the longitudinal temperature gradient in the thermal shield was less than 0.3 K/m. At 1.9 K the thermal gradient between the pressurized and the saturated helium was 100 mK and the longitudinal thermal gradient on the cold mass was below 50 mK under nominal operating conditions.

The mass-flow at each temperature level 4.5-20 K and 50-75 K and the inlet and outlet temperatures have been measured and the heat inleak on each of these cryogenic lines

Table 1. Distributed heat loads under steady-state conditions at nominal beam energy and luminosity [W.m^{-1}]

Heat load	50-75 K	4.5-20 K	1.9 K
Heat inleak	6.27	0.17	0.19
Resistive and beam-induced heating		0.71	0.17
Total	6.27	0.88	0.36

calculated from enthalpy balance. Maintaining the nominal residual gas pressure value the temperature of the circuit E-F has been varied from 50 K up to 150 K in steps of 10-15 K and the temperature of the circuit C-D from 4.5 K to 35 K in steps of 5-10 K. Off-design measurements were particularly difficult to control since the mass flows were low and as a consequence longer time was required to reach steady-state conditions.

Heat inleak to 1.9 K was determined firstly by measuring the steady-state mass flow-rate of vapour boil-off at 1.9 K, and secondly by evaluating the temperature drift with time of the quasi-isothermal helium II bath over the interval between 1.9 K and the lambda point, and calibrating it with an electrically applied heat source of known value.

Under nominal temperature conditions the vacuum has been degraded in steps from 10^{-3} Pa to 10^{-1} Pa by introducing dry helium gas into the cryostat insulation volume.

Single-magnet tests and string test

The single-magnet test heat load measurements have been carried out at zero magnet excitation under nominal steady-state conditions with the equipment for “warm” magnetic measurements in the apertures maintained at 1.9 K. The residual gas pressure was 10^{-4} Pa and the lines E and F were cooled by saturated nitrogen at 80 K.

String test heat loads have been measured at all the three temperature levels without current in the superconducting magnets. The temperature of the nitrogen circulating in the lines E and F varied between 85 K at inlet and 100 K at outlet.

CRYOSTAT THERMAL PERFORMANCE

Cryostat thermal model

Heat inleak measurements from the CTM are in agreement with those calculated. Table 2 lists the measured and the calculated heat inleaks of the cryostat in steady-state conditions without the contribution of the end boxes (which amounts to 40 W at 50-75 K, 4.2 W at 4.5-20 K, 0.1 W at 1.9 K).

The cryostat instrumentation permits the heat inleak contribution at 4.5-20 K and at 50-75 K from the end boxes to be separated from those of the cryostat.

Table 2. Measured and calculated heat inleaks under steady-state conditions

Heat inleak	**50-75 K**		**4.5-20 K**		**1.9 K**	
	Measured	Calculated	Measured	Calculated	Measured	Calculated
[W]	47.8±3	45.8	2.3±0.15	2.4	1.82±0.1	1.92
[W/m]	4.66±0.29	4.47	0.23±0.01	0.23	0.18±0.01	0.19

Heat flux, from 300 K to 50-75 K through the superinsulation only, was estimated to 32 W, corresponding to an average of 1.2 W/m^2. Radiation heat inleak from 50-75 K to 4.5-20 K confirmed the value of 90 mW/m^2 already found during tests of multilayer reflective insulation[10]. The heat inleak to the cold mass from the support posts[11] has been maintained low by two heat intercepts at 10 K and at 85 K.

The cold-mass heat inleak is given by radiation from the thermal shield (line F) and conduction through the support posts (line C). Fig. 3 and fig. 4 show the heat inleak at 1.9 K as a function of the line C and thermal shield temperatures.

Table 3. Measured heat loads on single-magnet and string tests

Heat load	80 K [W]	4.5-20 K [W]	1.9 K [W]
Single-magnet tests (1 dipole+auxiliaries)	150	40	10
String test (2 dipoles+1quadrupole+auxiliaries)	250	15	20

Heat inleak to 4.5-20 K depends on line E and line C temperatures by conduction through the support posts and line F temperature by radiation from the thermal shield. Fig. 5 shows the heat inleak at 4.5-20 K as a function of the thermal shield temperature.

Radiation from the vacuum vessel (line F) and conduction through the support posts (line E) constitute the main contributions to the heat inleak to 50-75 K. Fig. 6 shows the heat inleak to 50-75 K as a function of the thermal shield temperature.

Following a vacuum degradation the 1.9 K level is the most affected since in nominal steady-state conditions the heat transfer through the radiative insulation represents 80 % of the total heat inleak. Fig. 7 shows the heat inleak to 1.9 K as a function of the residual gas pressure.

Single-magnet tests and string test

Table 3 lists the heat inleaks at the three temperature levels measured in the single-magnet tests and in the string test under steady-state conditions. Heat inleak contribution from auxiliary components (end boxes, quench valves, warm bore anticryostats and current leads) largely exceed those from the cryostat and renders their measurements difficult. This justifies a posteriori the interest of the CTM for proper validation of cryostat design. The heat loads below were measured with a precision of around ± 5 %.

CONCLUSION AND FUTURE PROGRAM

Good agreement has been obtained between measured heat loads and those calculated over a large range of variation of parameters and all results lie within the limits imposed by the refrigeration capacity foreseen for the LHC project.

Future work will include evaluation of new types of superinsulation with a hard screen at 4.5-20 K and the performance of different thermal contacts will be investigated.

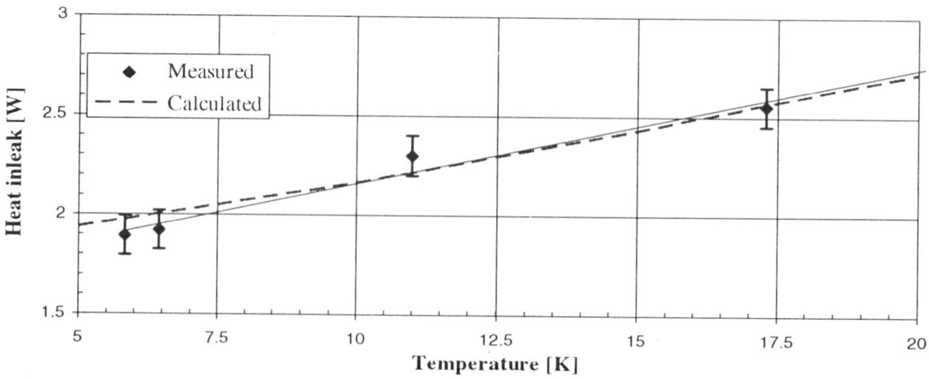

Figure 3. Heat inleak at 1.9 K versus line C temperature (residual gas pressure=10^{-3} Pa, thermal shield temperature=75 K).

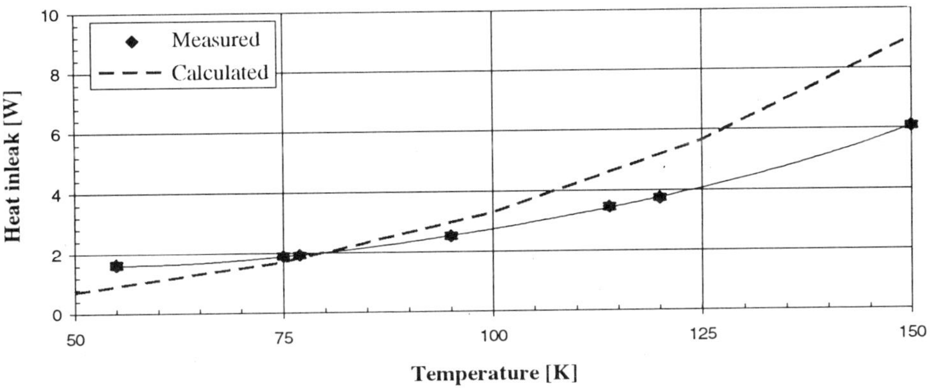

Figure 4. Heat inleak at 1.9 K versus thermal shield temperature (residual gas pressure=10^{-3} Pa, line C temperature=5 K).

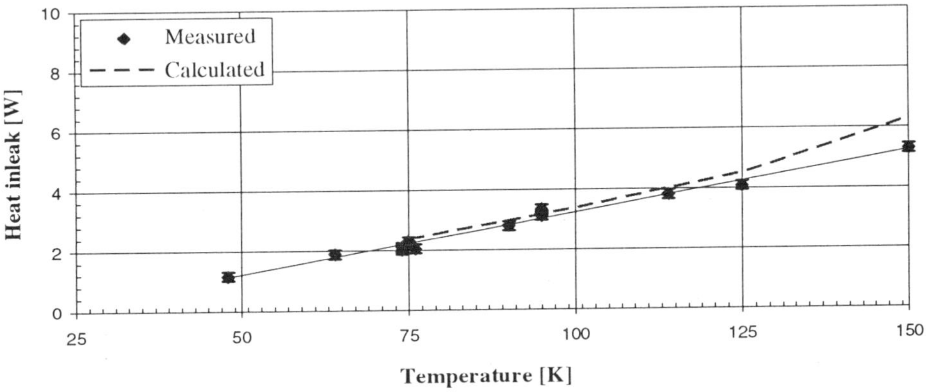

Figure 5. Heat inleak at 4.5-20 K versus thermal shield temperature (residual gas pressure=10^{-3} Pa).

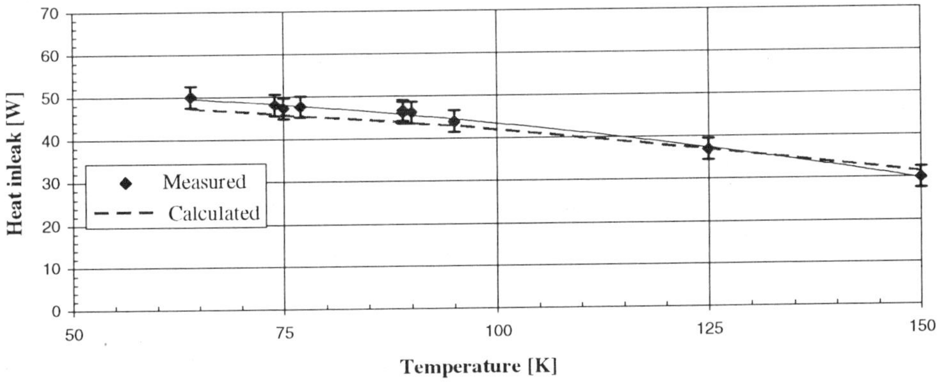

Figure 6. Heat inleak at 50-75 K versus thermal shield temperature (residual gas pressure=10^{-3} Pa).

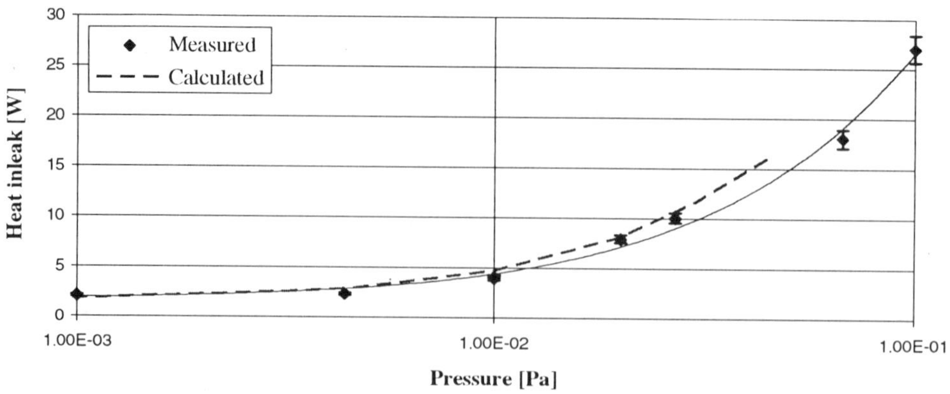

Figure 7. Heat inleak at 1.9 K versus residual gas pressure (thermal shield temperature=75 K, line C temperature=5 K).

ACKNOWLEDGMENTS

The authors would like to thank the String Team, in particular B. Gaillard-Grenadier, Th. Goiffon and L. Serio for their support during the string heat load measurements.

REFERENCES

1. L. R. Evans, "The Large Hadron Collider" invited paper at the Particle Accelerator Conference, Dallas (1995).
2. R. Perin for the LHC Magnet Team, " Status of LHC programme and Magnet Development", invited paper at the Applied Superconductivity Conference, Boston (1994).
3. Ph. Lebrun, "Superfluid Helium Cryogenics for the Large Hadron Collider project at CERN", Cryogenics 34 ICEC Supplement, 1-8 (1994).
4. L. Rossi, V. Sergo, B. Szeless, L. Tavian, B. Vullierme, R. van Weelderen, L. Williams, "Thermal Behaviour and Cryogenic Performance of the First CERN/INFN Prototype Dipole Cryomagnet for the CERN LHC project", Cryogenics 34 ICEC Supplement, 697-700 (1994).
5. L. Dufay, G. Ferlin, Ph. Lebrun, G. Riddone, J.-M. Rieubland, A. Rijllart, B. Szeless, L. Williams, "A full-scale thermal model of a prototype dipole cryomagnet for the CERN LHC project", Cryogenics 34 ICEC Supplement, 693-696 (1994).
6. A. Bézaguet, J. Casas-Cubillos, B. Flemsaeter, B. Gaillard-Grenadier, Th. Goiffon, H. Guinaudeau, Ph. Lebrun, M. Marquet, L. Serio, A. Suraci, L. Tavian, R. van Weelderen, "The superfluid helium cryogenic system for the LHC test string: design, construction and first operation", paper presented at this conference.
7. J.-C. Brunet, J.Kerby, Ph. Lebrun, P. Rohmig, B. Szeless, L.R. Williams, "Design of LHC Prototype Dipole Cryostats", Cryogenics 32 ICEC Supplement, 191-194 (1992).
8. G. Riddone, "Bilancio termico della semicella prototipo dell' acceleratore di particelle L.H.C.", Tesi di Laurea in Ingegneria Nucleare, Politecnico di Torino (1993).
9. The LHC Study Group, "The Large Hadron Collider Project", CERN Report AC/93-03 (LHC), (1993)
10. Ph. Lebrun, L. Mazzone, V. Sergo, B. Vullierme, "Investigation and qualification of thermal insulation systems between 80 K and 4.2 K", Cryogenics 32 ICMC Supplement, 44-47 (1992)
11. M. Blin, H. Danielsson, B. Evans, M. Mathieu "Design, construction and performance of superconducting magnet support posts for the Large Hadron Collider", Adv. Cryo. Eng. 39A, 671-678 (1994)

A COMPOSITE VACUUM BARRIER FOR THE LHC SHORT STRAIGHT SECTION

B. Jenny[1], P. Rohmig[1], J.-M. Uriarte[2]

[1]CERN, 1211 Geneva 23, Switzerland
[2]FIBERTECNIC, 01013 Vitoria-Gasteiz, Spain

ABSTRACT

The lattice of the CERN Large Hadron Collider (LHC) will contain 384 Short Straight Section (SSS) units, one in every 53 m half-cell. The SSS is composed of a twin aperture high-field superconducting quadrupole and of two combined-function corrector magnets operating in pressurized helium at 1.9 K. The SSS cryostat contains also a barrier for sectorisation of the insulation vacuum. The vacuum barrier is mounted between the helium vessel and the vacuum enclosure. Its functions are to limit the extent of eventual helium leaks and to facilitate the leak detection and the pumping-down from atmospheric pressure. During installation of the LHC, the vacuum barrier permits independent testing of the half-cells, thus enabling higher installation rates. In parallel to a conventional barrier made out of austenitic stainless steel, a barrier of composite material was developed, taking advantage of the lower thermal conductivity of glass fibre reinforced epoxy resin, and with the aim of reducing costs for LHC. The thermo-mechanical design together with the conception and the moulding techniques used for the manufacture of the prototype are described. Bonding techniques for the leak tight stainless steel composite interfaces are presented and test results shown. Results on the mechanical performance and on the helium tests carried out on the prototype are given.

INTRODUCTION

CERN is currently undertaking design studies and prototype development work leading towards the construction of the Large Hadron Collider (LHC)[1]. This multi-TeV range collider will consist of a ring of twin aperture superconducting magnets, operating in a pressurized superfluid helium bath at temperatures below 2 K and housed in vacuum enclosures.

The LHC of a total circumference of about 27 km, is subdivided in eight standard arcs, each about 3.2 km in length. The insulation vacuum of the helium enclosures of each standard arc is sectorized by vacuum barriers. The functions of these barriers are to limit the propagation of eventual helium leaks and to facilitate leak detection. During installation of the LHC, they enable independent testing of the half cells, and promote high installation rates.

Placed within the Short Straight Sections (SSS)[2] they sectorize the insulation vacuum every 53 m (Fig. 1).

These vacuum barriers must present a high thermal impedance to limit the thermal load to the 1.9 K helium bath, while retaining the high mechanical strength necessary to sustain atmospheric pressure loads and thermal stresses.

Advances in Cryogenic Engineering, Vol. 41
Edited by P. Kittel, Plenum Press, New York, 1996

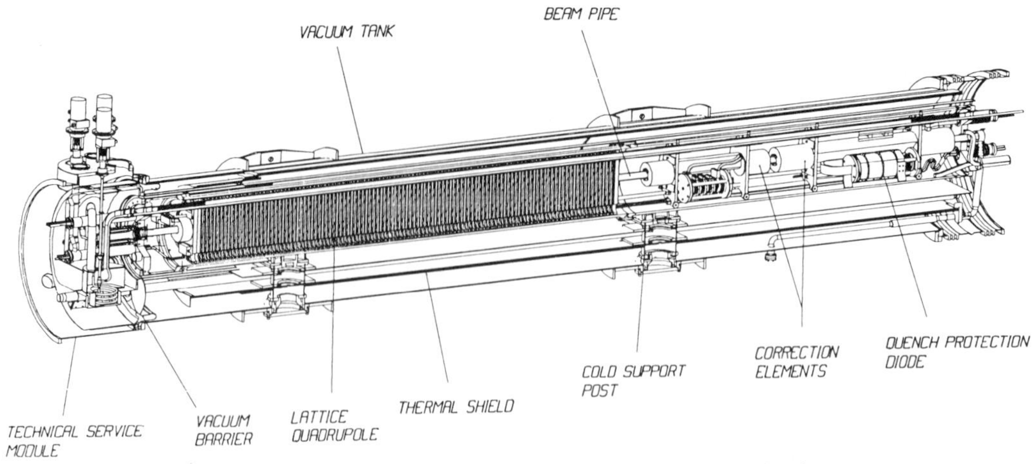

Figure 1. Longitudinal section of the LHC Short Straight Section.

Figure 2. Vacuum barrier made out of stainless steel.

Figure 3. Vacuum barrier made out of composite.

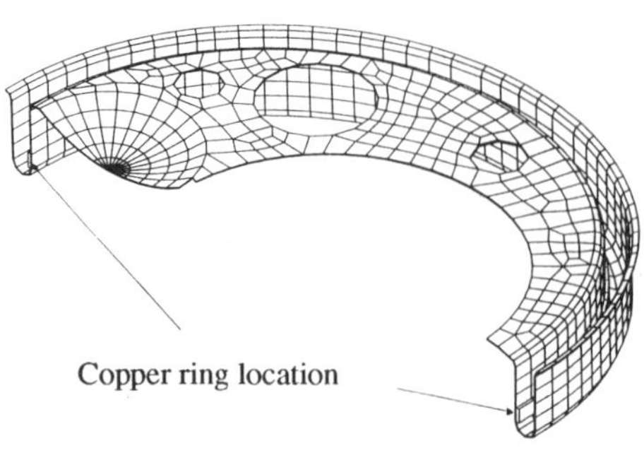

Figure 4. FEM mesh of composite vacuum barrier.

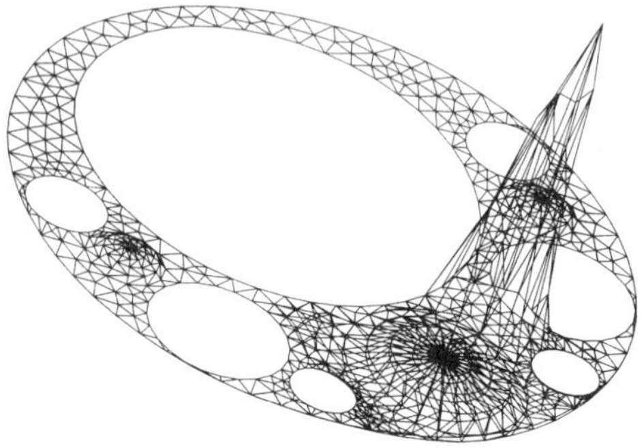

Figure 5. Result of FEM buckling calculations.

Table 1. Von Mises maximum mechanical stresses and TSAI-WU criteria

	Lower Ply	Intermediate Ply	Upper Ply
Von Mises (MPa)	254	54	303
TSAI-WU (< 1)	0.55	0.04	0.84

Table 2. Von Mises maximum thermal stresses and TSAI-WU criteria

	Lower Ply	Intermediate Ply	Upper Ply
Von Mises (MPa)	81	69	127
TSAI-WU (< 1)	0.2	0.08	0.2

Table 3. Thermal losses [Watts]

			Calculated	Specified
Thermalization		(50 K)	4.61	10.8
He - enclosure		(1.9 K)	0.44	0.33
Cryogenic tube	4	(5 K)	0.07	0.13
	5	(5 K)	0.19	0.13
	6	(50 K)	- 0.09	----
	7	(50 K)	- 0.08	----
	8	(1.9 K)	0.15	0.09
	9	(2.2 K)	0.06	0.04

1) Extension of the Von Mises criterion to a quadratic one, based on scalar products of stress components

The barriers must also be leak tight to helium, have a low helium permeation rate, and they must present a low outgassing rate under vacuum.

A prototype vacuum barrier made out of corrugated stainless steel tubes separated by a thermalized eccentric disk (Fig. 2) is installed in a so-called test string[3]. This string, composed of a series of prototype superconducting magnets, is dedicated to testing and validating choices concerning essential functions (cryogenics, powering, quench protection etc.) of the future LHC.

In parallel, a vacuum barrier made out of composite material has been developed with industry, taking advantage of the high resistance and low heat conductivity of this material. Another aim is to achieve low cost series production for the LHC.

Pre-impregnated fibre glass epoxy resin segments, forming the membrane, are placed in a mould and polymerised in a furnace together with stainless steel tube interfaces. Designed to fit in the prototype Short Straight Section, a prototype was built and mechanically tested (Fig. 3).

CERN is currently mounting this composite vacuum barrier into the test string to simulate cryogenic cycling and to verify its thermo-mechanical behaviour.

THERMO-MECHANICAL DESIGN

Several load cases influence the design of the vacuum barrier. During installation atmospheric pressure may act in either direction, while vacuum is maintained for leak detection on the opposing side. During normal operation there is no pressure difference, but thermal contraction during cooldown results in a displacement of the inner flange fixation by 0.5 mm vertically and about 3 mm longitudinally. Various tubes at working temperatures between 4.5 and 75 K pass through and are connected to the vacuum barrier.

The geometry consists mainly of a flat membrane with spherical reinforcements to improve buckling resistance and a large undulation which increases the thermal conduction path and provides the necessary flexibility for the thermal contraction displacements. To intercept heat from the 300 K flange, a copper ring is integrated in the inner part of the contraction wave and is thermally connected to the 50 K line.

The structure was designed and optimized on the basis of acceptable mechanical and thermal stresses calculated with Finite Element Method (FEM)[4]. The maximum Von Mises equivalent stresses and failure criteria "TSAI-WU"1) are shown in Table 1 and 2.

The highest stresses occur in the vicinity of the junctions with the 50 K tubes, and a maximum displacement of 6 mm is located next to the spherical reinforcement. The buckling resistance was calculated and after optimization of the membrane a safety factor of 5 was achieved. Thermal losses were also calculated with FEM (Table 3) and results conform to the values stipulated in the specifications. To meet the stringent thermal requirements a pre-impregnated E-glass fibre/epoxy resin was chosen for the barrier, taking a tensile strength of 420 MPa and thermal properties of G-10 CR laminate as reference.

MANUFACTURING TECHNIQUES

Multi-Layer Design

For manufacture a two dimensional anisotropic (monolithic) structure formed by the superposition of composite layers was selected. These layers are formed from fabric, consisting of two perpendicular thread systems, the warp and weft, previously impregnated by epoxy resin "prepreg". A satin weave was chosen for its better drapability. This fulfils best the mechanical strength requirements through addition of the strengths of the different single layers and the multi-directionality of the layer fibres.

The "prepreg" concept favours an homogeneous and constant resin impregnation of the fibres. At the same time it permits an accurate orientation and alignment of the fibre tissue.

Moulding Technique

The moulding process was performed in a fully air-conditioned clean room where the relative humidity was kept at 50% ± 3 and the temperature at 21 °C ± 1 to avoid water contamination of the "prepreg" material. The precise directional alignment of the fibres, required in particular in the complicated parts of the mould (corners), imposed the adoption of a manual process, called "hand lay up". This process was also necessary to permit introduction of the metallic parts between the plies.

In order to obtain a correct compactness between the plies, and avoid porosity (air bubbles), the barrier was vacuum pumped twice for periods of 30 minutes during layer assembly.

After the build up of all plies in the mould a vacuum bag (Fig. 6) was made, composed of :

- a release agent film; to separate the plies from the mould,
- a breather; to facilitate air pumping during polymerization,
- a thermocouple; to measure continuously the temperature and to control the process,
- a vacuum film; to close the bag with a sealant zinc chromate tape,
- a vacuum connection; to allow a controlled pump down to be carried out.

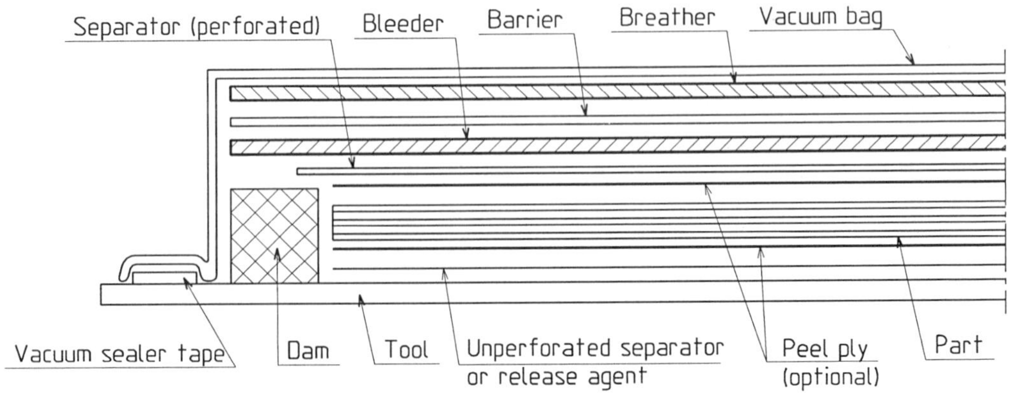

Figure 6. Schematical layout of the moulding technique.

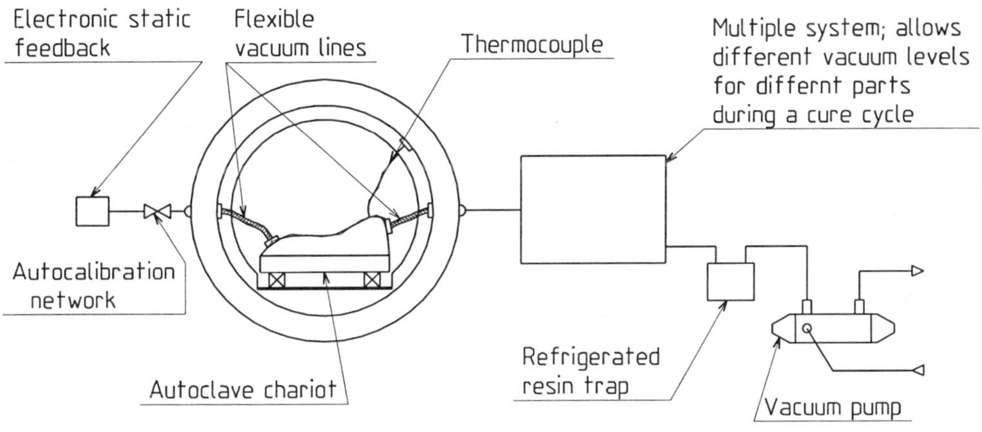

Figure 7. Curing set-up scheme.

Curing Process

The curing of the vacuum barrier was carried out in an autoclave. An external pressure was applied while vacuum pumping of the mould with the barrier and holding the polymerisation temperature (Fig. 7). Constant process provides a maximum compactness between the fibres thereby yielding optimum mechanical properties.

During the thermal process the viscosity was controlled. The most critical part of the cycle is reached when the viscosity is minimum. At this point of the curing cycle, the pressure applied defines the final location of each fibre and thus fixes the final compactness and consequently the mechanical properties of the vacuum barrier.

Prototype manufacturing tests

The manufacturing of the prototype was preceded and followed by test programs ensuring the optimum choice of materials and technologies.

Helium permeation tests[5] on 2.5 mm thick samples confirmed that outgassing values of 8×10^{-14} m^2 s^{-1} at 300 K can be obtained without special surface coatings.

The leak tight connections of the composite membrane with the stainless steel tubes represent critical areas of the vacuum barrier. Results of tension and shearing tests of flat samples with different adhesion are shown in Table 4.

Subsequent thermal cycling on model tube composite connections, from 300 K to 77 K (LN_2), showed good results concerning leak tightness, but also revealed the sensitivity to rapid thermal shocks. In this cases cracks in the composite bounding occurred.

A hydraulic pressure test of a test barrier showed good agreement with the calculated estimates of failure. Final reception tests at 300 K of the vacuum barrier confirmed the leak tightness under atmospheric pressure and a background signal of 1×10^{-8} mbar l s^{-1}.

Table 4. Results of sample tests

	T (K)	Shear (MPa)	Tensile (MPa)	Contraction (//) % (293 - 77 K)
Adhesive only (FM 73M Cytec)	293	23.78	48.10	
	77	18.52	97.36	1.1
"Prepreg" only (MXB ICI Fiberite)	293	33.34	63.69	
	77	36.99	114.44	0.24
Adhesive + "Prepreg" + Adhesive	293	28.26	51.25	
	77	28.10	87.08	

CONCLUSIONS

The vacuum barrier made out of composite material (GFRER) while presenting similar heat loads and mechanical properties as the stainless steel version, is considerably lighter (25 kg against 260 kg) and more compact (150 mm against 750 mm in length). The reduced weight enables easy handling during assembly. The moulding technique opens the way for semi-automatic series production. Further studies on the bonding technique for the composite to stainless steel interfaces are required to improve the reliability of these junctions. This in particular in view of a fully welded link to the vacuum and helium enclosures.

The future LHC string test program foresees integration of a composite vacuum barrier to gain information units on the cryogenic behaviour.

ACKNOWLEDGEMENTS

This project was carried out in the framework of a collaboration contract between CERN and FIBERTECNIC, initiated and financially supported by the Centre for the Development of Industrial Technology (CDTI), Madrid, Spain.

The authors wish to thank the teams involved in the design, the calculation and the manufacture of the prototype vacuum barrier at CERN and at FIBERTECNIC. Special thanks to P.-E. Faugeras for his support, A. Mathewson for the fruitful discussions concerning the vacuum requirements and A. Toukhtarov for his design work on the string test set up.

REFERENCES

1. The LHC Study Group, Design Study of the Large Hadron Collider (LHC), CERN 91-03, May (1991).
2. W. Cameron, B. Jenny, G. Riddone, P. Rohmig, R. van Weelderen, "Design and Construction of the Prototype Cryostat for the Short Straight Sections (SSS) of the CERN LHC", CERN MT/93-04 (ESH), Presented at the Cryogenic Engineering Conf., Albuquerque, July (1993).
3. P.-E. Faugeras for the LHC String Team, "Assembly and Commissioning of the LHC Test String" Geneva, May (1995).
4. Design Calculations FIBERTECNIC / INASMET, Spain, July (1993).
5. V. Baglin, A. Grillot, "Test de Perméation à l'Hélium d'Echantillons pour une Barrière à Vide en Composite", Note Technique 94-05, Février (1994).

EXPERIMENTAL INVESTIGATION OF ACCIDENTAL LOSS OF INSULATION VACUUM IN AN LHC PROTOTYPE DIPOLE CRYOSTAT

Ph. Lebrun, B. Szeless, L. Tavian, L.R. Williams

Accelerator Technology Division, CERN
CH-1211 Geneva 23, Switzerland

ABSTRACT

A sudden loss of insulation vacuum in an accelerator cryomagnet system be it due to a helium leak from a cryogenic circuit or leak from atmosphere has consequences for the cold mass as well as for the vacuum enclosure. The vacuum enclosure for the LHC dipole cryostat is made of carbon steel to reduce costs and to provide magnetic shielding. In the case of a loss of insulation vacuum the vacuum enclosure will be cooled rapidly but must not reach temperatures low enough to cause embrittlement of the wall material.

Helium gas at room temperature was admitted into the insulation space of a modified prototype cryomagnet, the cold mass of which had been previously cooled to 80 K. The evolution of temperature with time of the cold mass and of the vacuum enclosure has been measured and the corresponding heat influx estimated from a simple model.

INTRODUCTION

The Large Hadron Collider (LHC) project, currently under design at CERN [1] relies on extensive use of high field superconducting magnets operating below 1.9 K in a pressurized bath of helium II.

Superconducting magnet cold masses for accelerators are housed in horizontal cryostats [2] which provide mechanical support and limit heat inleaks [3] to a level compatible with the LHC refrigeration capacity.

To limit convective heat transport, under normal operating conditions the outermost cryostat envelope or vacuum enclosure contains an insulation vacuum in the 10^{-4} Pa range. To reduce costs and to shield stray field the current LHC dipole vacuum enclosure design is based on a cylindrical shell 10 mm thick of diameter 1 m by 10 m long in ISO 430 carbon steel.

A sudden loss of insulation vacuum in a cold cryomagnet will cause a rapid increase in convective heat transfer and cooling of the vacuum enclosure wall towards a safety limit defined by the low temperature embrittlement of carbon steels. Loss of insulation vacuum may be due to an air leak from the ambient surroundings or to an internal dry helium leak from a cryogenic circuit. In the first instance rapid formation of an insulating ice layer on the vacuum enclosure and on all internal components will progressively reduce heat transfer to the cold mass. In the second case, only an external insulating frost layer will form. The significantly higher thermal conductivity of helium gas compared to that of air increases the internal heat transfer from the vacuum enclosure to the cold mass and leads to proportionately lower vacuum enclosure wall temperatures.

Following sudden degradation to atmospheric pressure of the insulation vacuum with dry helium gas at room temperature, the evolution of cold mass, helium vessel and vacuum enclosure temperatures with time has been measured on a representative full-scale 10 m-long dipole cryostat.

EQUIPMENT AND TEST PROCEDURES

The T.A.P. cryomagnet [4] (Fig. 1a), in common with derivatives currently proposed for LHC, operates under normal conditions at 4 distinct temperature levels: the cold mass at 1.9 K, the inner radiative insulation at 4.5 to 20 K, the thermal screen at 50 to 75 K and the vacuum enclosure at ambient temperature.

Tests were conducted using a simplified version of the T.A.P. cryomagnet reassembled as shown in Fig. 1b. The helium vessel, supported to the vacuum enclosure via its 3 composite support posts, was left uninsulated at each end but was wrapped circumferentially along its entire length with a 30-layer superinsulation blanket. No thermal shield was mounted.

An initial insulation vacuum of around 10^{-2} Pa was pumped which on cooling the cold mass from room temperature to 80 K improved into the 10^{-4} Pa range. Throughout the tests, ambient conditions in still air with a constant relative humidity of 45 % and temperature of 294 K were maintained. About 6 m^3 of dry helium gas at room temperature were admitted into the insulation space at an average rate of 13 $L.s^{-1}$, up to a pressure of 0.115 MPa to compensate the effects of gas cooling. The average cold mass temperature was measured with one platinum resistance thermometer (PRT) placed internally. Six PRT attached to the vacuum enclosure outer wall measured its temperature at the locations indicated in Fig. 2. Additional temperature monitoring was performed with a contact thermometer. Following the admission of helium gas the evolution of cold mass and vacuum enclosure wall temperatures was recorded over a 24 hour period.

MATHEMATICAL MODEL

Basic Assumptions

The evolution with time of heat flux and temperature in the different cryostat components has been simulated with a one-dimensional radial model, shown schematically in Fig. 3. The vacuum enclosure at temperature T1 receives heat Q01 from the environment at temperature T0 by natural convection in air and by radiation. The helium vessel at temperature T2 receives heat Q12 from the vacuum enclosure by convection in the helium gas at 0.1 MPa which was admitted to break the insulation vacuum, as well as by conduction in the helium trapped between superinsulation layers. The cold mass at temperature T3 receives heat Q23 from the helium vessel by conduction in the static helium gas which fills the helium vessel. No convection can occur in the restricted spaces between

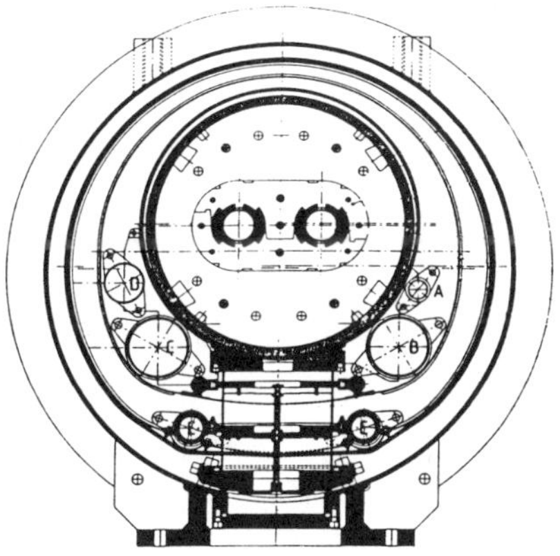

Figure 1a. Transverse cross-section of the T.A.P. cryomagnet.

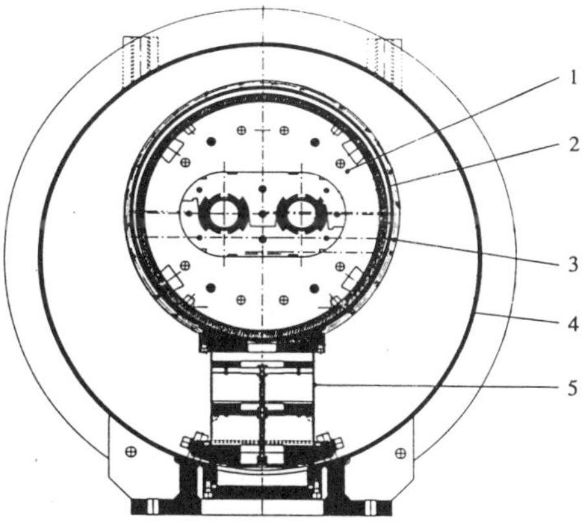

Figure 1b. Transverse cross-section of the test cryomagnet.

1. Cold mass
2. Helium vessel
3. 30 layer superinsulation blanket
4. Vacuum enclosure
5. Support post

superinsulation layers and between the helium vessel and cold mass. This model takes no account of water condensation or ice formation on the outside surface of the vacuum enclosure.

Equations

The heat Q01 coming from the environment is the sum of natural convection and radiation where h is the natural convection heat transfer coefficient [5] and σ is the Stefan-Boltzmann constant.

$$Q01 = h \cdot A1 \cdot (T0 - T1) + \sigma \cdot A1 \cdot \left(T0^4 - T1^4\right)$$

$$h = 1.3 \cdot \left(\frac{T0}{D1}\right)^{0.25}$$

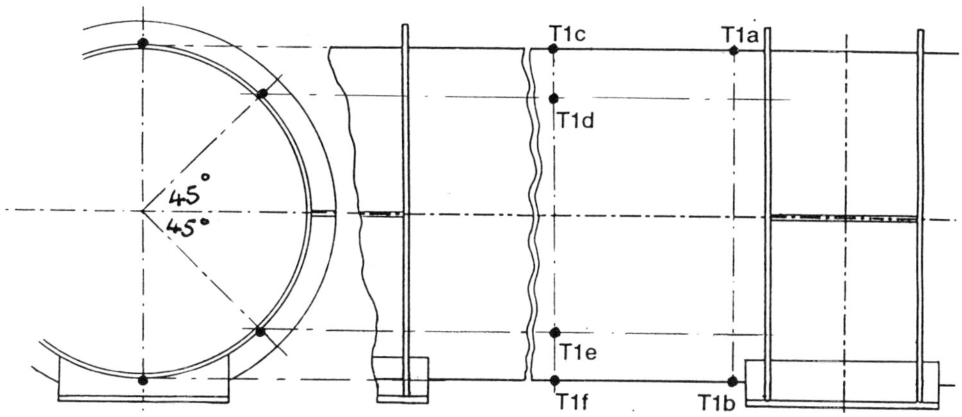

Figure 2. Location of vacuum enclosure temperature measurements.

The heat Q12 between the vacuum enclosure and the helium vessel is given by convection in closed volume [5] and by conduction in helium. The convection in closed volume is determined by the thermal conductivity k, and the convection factor f depending on the Prandtl Pr and Grashof Gr numbers which are functions of the isobaric thermal expansion α, the density ρ and the viscosity η.

$$Q12 = \frac{2 \cdot \pi \cdot L \cdot k\left(\frac{T1+T2}{2}\right)}{\left[\frac{1}{f} \cdot \ln\left(\frac{D1}{D2+2 \cdot e}\right) + \ln\left(\frac{D2+2 \cdot e}{D2}\right)\right]} \cdot (T1 - T2)$$

$$f = 0.18 \cdot [Pr \cdot Gr]^{0.25}$$

The heat Q23 between the helium vessel and the cold-mass is given by conduction in helium.

$$Q23 = \frac{2 \cdot \pi \cdot L \cdot k\left(\frac{T2+T3}{2}\right)}{\ln\left(\frac{D2}{D3}\right)} \cdot (T2 - T3)$$

The differential equation system which links the temperatures T1, T2 and T3 is:

$$\begin{cases} M1 \cdot C(T1) \cdot \frac{dT1}{dt} = Q01(T1) - Q12(T1, T2) \\ M2 \cdot C(T2) \cdot \frac{dT2}{dt} = Q12(T1, T2) - Q23(T2, T3) \\ M2 \cdot C(T3) \cdot \frac{dT3}{dt} = Q23(T2, T3) \end{cases}$$

Results

This system is numerically solved using the Runge-Kutta algorithm. Fig. 4 shows the computed evolution of the different temperatures. For comparison, the measured values are plotted on the same graph (No measurement of the helium vessel temperature T2 is

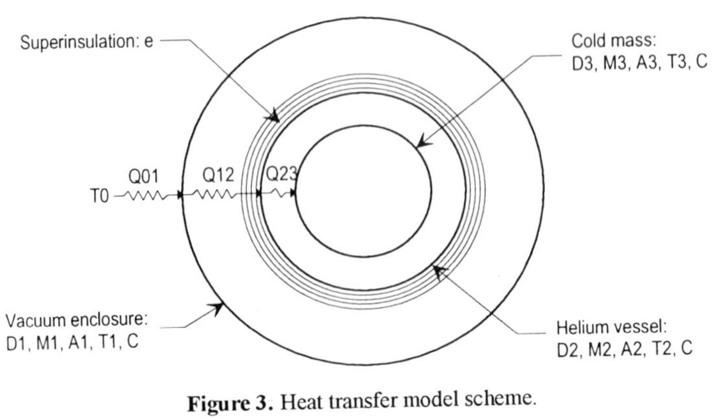

Figure 3. Heat transfer model scheme.

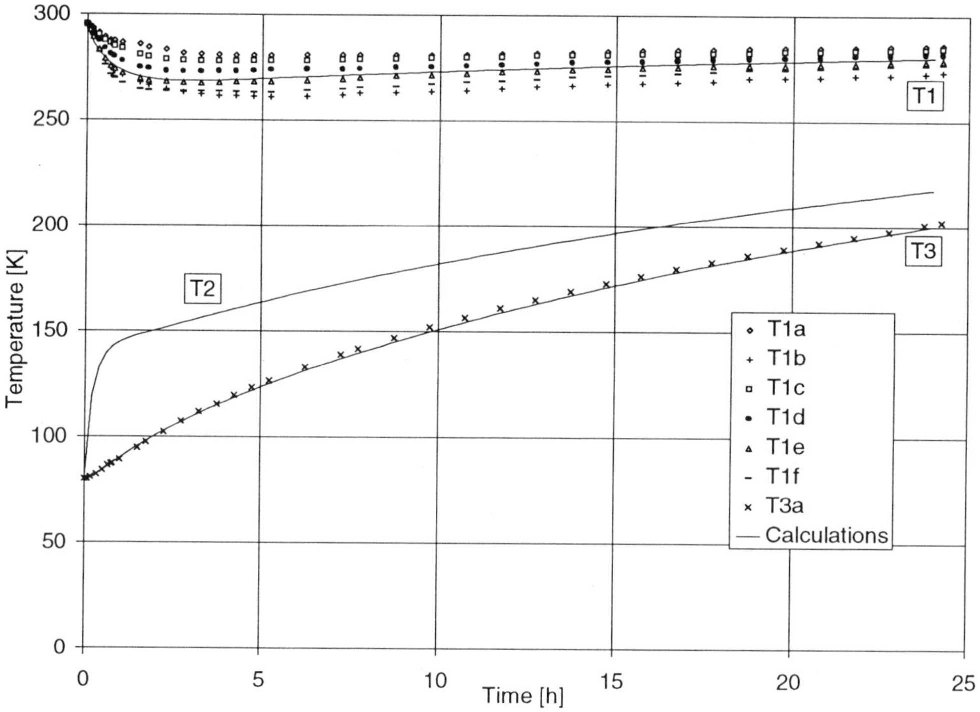

Figure 4. Calculated and measured temperature evolution.

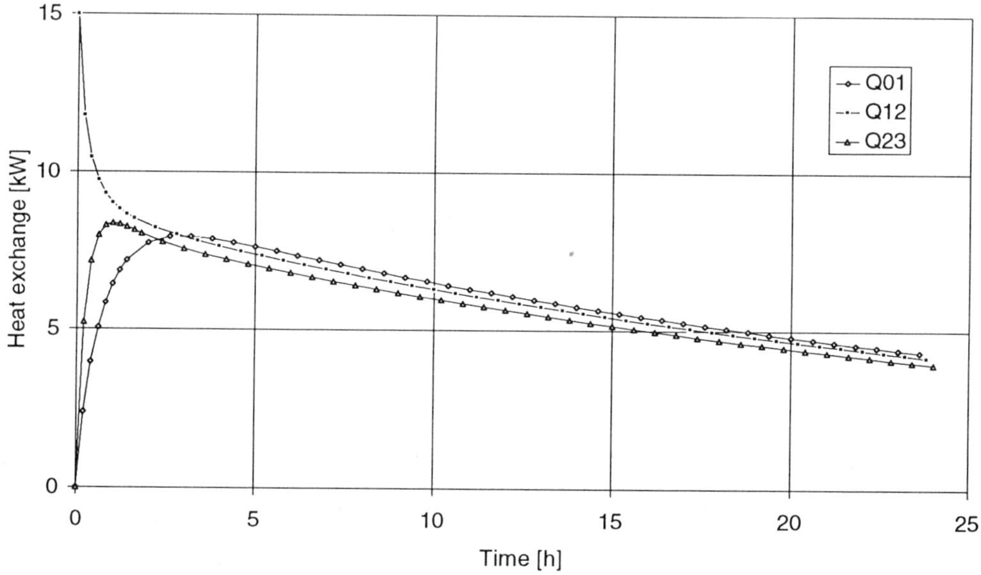

Figure 5. Calculated heat exchange between environment, vacuum enclosure, helium vessel and cold-mass.

available). Concerning the cold mass temperature T3 the calculated values are in accordance with experiment. Because the model is one-dimensional, it takes no account of the asymmetric heat exchange between the top and the bottom parts of the vacuum enclosure but the calculated values of T1 predict very closely an average of the six temperatures measured. Fig. 5 shows the corresponding calculated heat exchange between the different components. The maximum heat received by the vacuum enclosure is around 8 kW and occurs 3 hours after the insulation vacuum is broken. The maximum heat flux received by the cold mass is around 8.3 kW and occurs after 1 hour. The heat flux received by the helium vessel has its maximum of 15 kW at the beginning.

VISUAL OBSERVATIONS

From the time of helium admission, condensation first appeared on the underside of the vacuum enclosure after about 20 minutes and frost formation followed after about 35 minutes. Frost formation continued spreading upwards around the vacuum enclosure from the underside to cover a maximum 70 % of the circumference after 6 hours and from this point a gradual melting from the top down began. Frost growth up to a maximum thickness of about 3 mm after about 3 hours was measured at the location of the lowest vacuum enclosure temperatures and the thickness then remained approximately stable until the onset of melting.

CONCLUSIONS

The effects of an abrupt loss of vacuum on a full-scale accelerator cryomagnet initially at 80 K with only one 30-layer superinsulation blanket directly wrapped onto the helium vessel have been studied. A simple mathematical model adequately predicts an average of the temperature evolution with time of the vacuum enclosure. Unaccounted for in this model, the external formation of an ice layer in the quantities described does not significantly affect the rate of heat transfer across the vacuum enclosure wall.

A simple model of this kind can be used to predict the behaviour of the LHC machine cryostats after a sudden loss of insulation vacuum.

It can be concluded that the minimum temperature of 260 K attained by the vacuum enclosure wall is well above the limit generally accepted for carbon steel embrittlement. Its temperature evolution, initially of about 35 $K.h^{-1}$, is not considered catastrophic and leaves sufficient time to render equipment safe within the scope of normal accelerator operations. Moreover, the heating power received by the cold mass is small as compared to that dissipated after a magnet resistive transition, and its consequences can therefore be easily handled by quench discharge devices.

REFERENCES

1. The LHC Study Group, "The Large Hadron Collider Accelerator Project", CERN Report AC/95-05 (LHC), 1995.
2. J.C. Brunet, J. Kirby, Ph. Lebrun, P. Rohmig, B. Szeless, L.R. Williams, "Design of LHC Prototype Dipole Cryostats", *Cryogenics 1992*, Vol. 32, pp. 191-194, ICEC Supplement
3. V. Benda, L. Dufay, G. Ferlin, Ph. Lebrun, J.-M. Rieubland, G. Riddone, B. Szeless, L. Tavian, L.R. Williams, "Measurement and Analysis of Thermal Performance of LHC Prototype Dipole Cryostats", Presented at this Conference.
4. M. Granier, Ph. Lebrun, M. Mischiatti, "Design and Construction of a Superfluid Helium Cryostat for a 10 m Long High Field Superconducting Dipole Magnet", *Cryogenics 1990*, Vol. 30, pp. 98-102, September Supplement
5. L. Weil, "Elements des Echanges Thermiques," Gauthier-Villars, Paris (1965).

CRYOGENIC PERFORMANCE OF A SUPERFLUID HELIUM RELIEF VALVE FOR THE LHC SUPERCONDUCTING MAGNETS

H. Danielsson[1], G. Ferlin[2], B. Jenninger[2], C. Luguet[2], S-E. Milner[2], J-M. Rieubland[2]

[1]The University of Lund, S-22100 Lund, Sweden
[2]CERN, CH-1211 Geneva 23, Switzerland

ABSTRACT

The high-field superconducting magnets of the Large Hadron Collider (LHC) project at CERN will operate below 1.9 K in static baths of pressurized helium II. In case of resistive transition ("quench"), the resulting pressure rise in the cryostats must be limited to below their 2 MPa design pressure. This is achieved by discharging helium at high flow-rates into a cold recovery header, normally maintained at 20 K. For this purpose, we have designed, built and tested a cryogenic quench relief valve with a nominal diameter of 50 mm and an opening time of below 0.1 s. The valve, which can be opened on an external trigger, also acts as a relief device actuated by the upstream pressure when it exceeds 0.4 MPa. In normal operation, the closed poppet must be helium-tight, for hydraulic and thermal separation of the magnet baths from the recovery header. Following mechanical qualification tests under vacuum, we have mounted the relief valve in a dedicated cryogenic measuring bench, in order to perform precision thermal measurements with pressurized helium II.

INTRODUCTION

The Large Hadron Collider (LHC) at CERN is based on the large-scale use of superconducting high-field magnets [1]. They will be operated at a temperature below 1.9 K in pressurized helium II. The high cost of low-temperature refrigeration imposes strong requirements on cryogenic components to minimize heat loads. Independent studies of these components in dedicated heat leak measuring benches are carried out in order to optimize the system [2]. The quench relief valve developed at CERN [3], has been tested in a thermal measuring bench. The objective of these tests was to measure heat and mass leak of the valve under normal operating conditions.

VALVE DESIGN

In LHC, two quench relief valves are installed inside the cryostat vacuum every 53 m. In case of a quench, each valve has to discharge about 400 l of liquid helium into a recuperation line kept at 20 K. The valve is controlled by the quench detection signal via an electro-pneumatic actuator mounted outside the cryostat. The valve and actuator are connected through a metal rod sealed to the cryostat tank by a bellows feed-through. If the quench signal fails the valve opens once the pressure setting is reached. A cross-section of the relief valve is shown in figure 1.

EXPERIMENTAL SET-UP

The measuring bench

The quench relief valve was mounted in a specially built cryostat to simulate LHC conditions (see Figure 2). The cryostat contained three cryogenic reservoirs filled with nitrogen, normal liquid helium (4.2 K, 1 bar) and saturated superfluid helium (~1.6 K). Thermal screens at 80 K and 5 K shielded the valve from thermal radiation. The cell (magnet side of the valve) was cooled by the saturated He II - bath of the cryostat through a copper plate which served also as a calibrated thermal impedance (heatmeter) [4]. In order to maintain one bar in the cell, it was connected to the He I - reservoir of the cryostat via a capillary tube of 1.5 mm inner diameter and a length of about 1m.

In our experiment, the heat flow was measured as a function of the temperature difference of the two He II - baths (T1 and T2 in figure 2). For the calibration of the heatmeter, the valve was replaced by a steel cylinder, in order to avoid parasitic heat load. A heater inside the cylinder was used for the calibration to generate the temperature difference over the two temperature sensors. As the main thermal impedance of the heatmeter comes

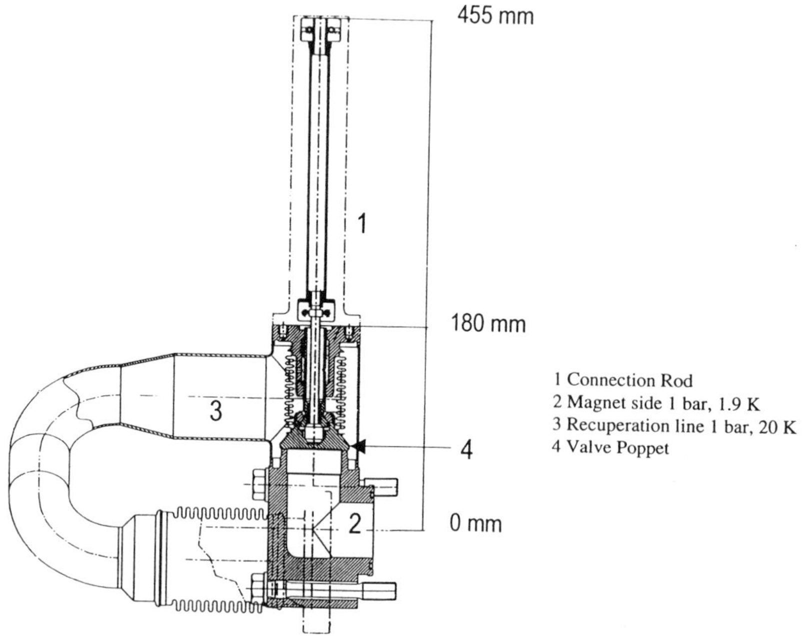

Figure 1. Cross-section of the quench relief valve.

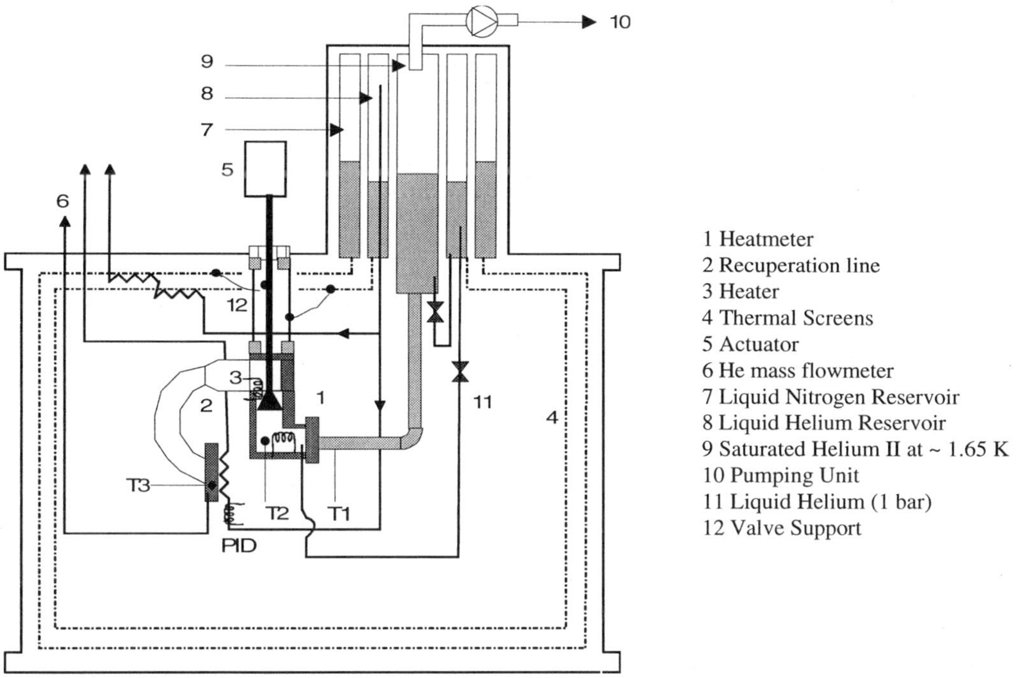

Figure 2. Schematic view of the quench valve mounted in the cryogenic measuring bench.

from the Kapitza resistance on the two faces of the copper plate, which varies strongly with the temperature, the temperature T1 was kept constant at 1.65 K. With T2 = 1.80 K, the temperature difference over the heatmeter was 0.15 K, which corresponds to a heat flow of 800 mW. The temperature in the recuperation line, T3, was controlled with a heater piloted through a PID regulator.

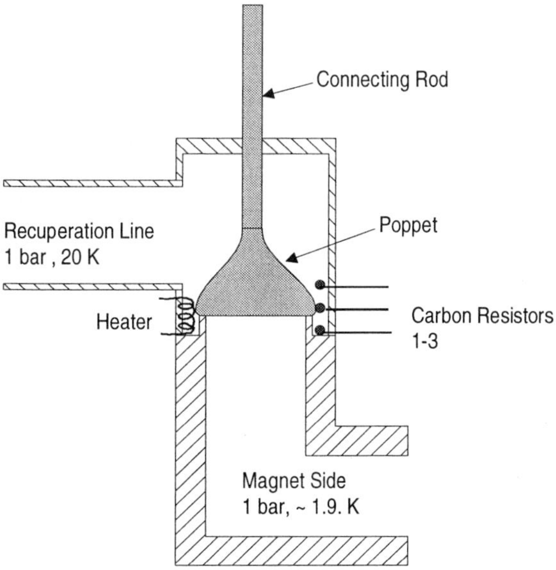

Figure 3. Schematic view of the positions of the heater and the three level indicators (carbon resistors) on the outside of the poppet.

Instrumentation and data acquisition

Temperatures were measured with PT100 platinum sensors above 30 K and Allen-Bradley carbon resistors below 30 K. Two precision current sources were used to send a current of 10 μA and 100 μA through the carbon and platinum sensors respectively. A personal computer equipped with a multifunction card and IEEE card performed the data acquisition under LabVIEW 2.2 [5] where the multifunction card was used for monitoring and regulating temperatures. For the heatmeter measurements, a digital multimeter/scanner (DMM) was used to ensure a high accuracy.

RESULTS AND DISCUSSION

Heat leak

Experiments were performed to measure the mass leak and the heat leak as function of the temperature at the end of the recuperation line, T3 in Figure 2. With the recuperation line closed, i.e. no flow of helium through the valve, the pressure became stabilized at 80 mbar. This overpressure originates from the fountain effect [6]. The liquid helium level behind the poppet was measured with three carbon resistors (Figure 3) where the lowest one was found to be at the λ-point. The heat load was measured as function of the temperature at the end of the recuperation line and the result is shown in figure 4. At 20 K, the heat load reached about 850 mW.

Later, the recuperation line was opened and the mass flow through the valve was measured outside the cryostat with a mass flowmeter. The mass flow through the valve was 8×10^{-6} kg/s (160 L/h STP). The overpressure in the recuperation line was now 40 mbar due to the flow resistance of the flowmeter, and the heat leak dropped to 600 mW (T3 = 20 K). This decrease of the heat load can be explained with a change of the convection conditions in the recuperation line when there was a mass flow. Helium leaking through the valve has to be replaced by helium from the 4.2 K reservoir. The supplementary heat load due to cooling that helium from 4.2 K to 1.8 K (~65 mW) was totally offset by forced convection.

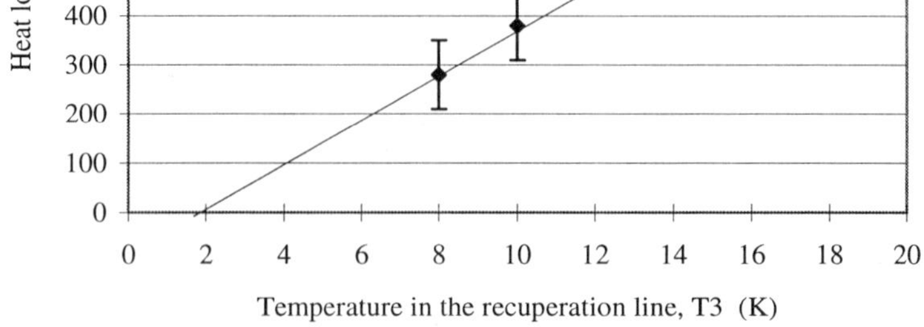

Figure 4. The heat load as a function of the temperature in the recuperation line, T3, with no mass flow across the valve.

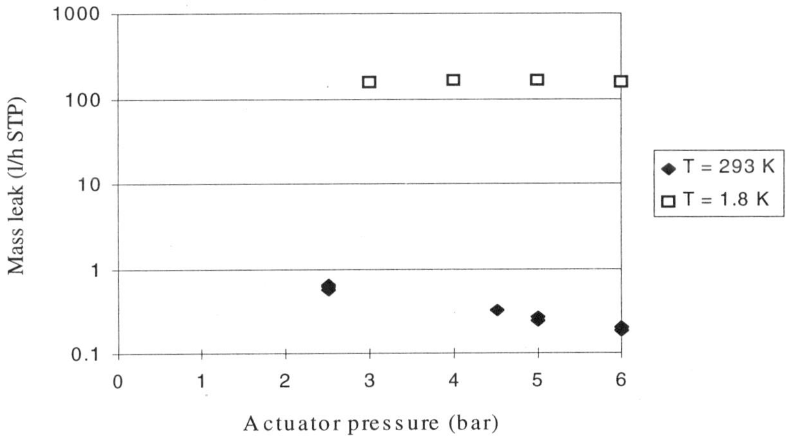

Figure 5. Mass flow through the valve as a function of poppet pressure at room temperature and under superfluid helium conditions. Note that during the measurement at 293 K the pressure in the recuperation line was 10^{-2} mbar while in the cell the pressure was 1 bar.

Leak rate as function of poppet pressure

The leak rate was first measured for different poppet pressures at room temperature and with a pressure drop over the valve of 1 bar. A pump with a capacity of 8 m^3/ h at 10^{-2} mbar was connected to the recuperation line (see Figure 3). The leak rate as a function of poppet pressure is showed in figure 5. There were no measurements below ~1.5 bar as the valve spring starts to open the valve at this value.

The mass leak was then measured at 1.9 K for poppet pressures between 3 and 6 bar with T3 = 20 K with a mass flowmeter connected outside the cryostat (see Figure 2). The mass leak was found to be 8 x 10^{-6} kg/s (160 L/h STP) and independent of the poppet pressure within 10 %, which is two orders of magnitude higher than at room temperature. The difference in pressure between the magnet side the recuperation side was 40 mbar caused by the pressure drop over the flowmeter. When the counter pressure over the valve poppet was eliminated, the mass flow increased to 16 x 10^{-6} kg/s (280 L/h STP).

CONCLUSIONS

The heat leak at nominal LHC conditions, i.e. T3 = 20 K and no mass leak through the valve, was measured to 850 mW. The pressure difference over the valve poppet was measured to be 80 mbar. As there was superfluid helium present behind the poppet, this pressure difference can be explained by the fountain effect. Therefore, a counter pressure was needed to prevent a flow of superfluid helium. When the counter pressure was eliminated, a significant mass leak of 16 x 10^{-6} kg/s (280 L/h STP), while the heat leak decreases to 600 mW.

The tests in the measuring bench simulated ideal operating conditions of the LHC dipoles. The tested valve is the first prototype and improvements will be necessary. Under real operating conditions, impurities will deposit on the poppet seat while the valve is open and change the tightness of the valve. Therefore, further investigation of the real conditions in the dipole magnets has also to be undertaken.

ACKNOWLEDGMENT

The authors would like to thank A. Cambon for his help during design and construction of the measuring bench and the operating team of the central cryogenic laboratory at CERN (H. Bouyaya, S. Chevassus, D. Urli and S. Guido) for their steady support.

REFERENCES

1. The LHC Study Group, The large Hadron Collider accelerator project, CERN Report AC/93-03(LHC) (1993).
2. H. Danielsson, Ph. Lebrun, J-M. Rieubland, Precision heat inleak measurements on cryogenic components, *Cryogenics 1992*, Vol 32, ICEC Supplement, 215-218.
3. S-E. Milner, LHC quench relief valve, CERN MT-ESM/93-01, 1993.
4. M. Kuchnir, J.D Gonczy , J.L Tague, Measuring heat leak with a heatmeter, Adv. Cryo. Eng., Vol 31 (1986) 1285-1290.
5. LabVIEW 2.2, National Instruments Corporation
6. S. W. Van Sciver, Helium Cryogenics, Plenum Press, New York, 1986

MEASUREMENT OF THE THERMO-HYDRAULIC BEHAVIOUR OF LHC DIPOLE PROTOTYPES AFTER QUENCH

G. Gerin, B. Vullierme, and R. van Weelderen

Accelerator Technology Division, CERN
CH-1211 Geneva 23, Switzerland

ABSTRACT

The superconducting dipole magnets under design for the Large Hadron Collider (LHC) at CERN operate in pressurized static baths of superfluid helium at a pressure of 1 bar and at a temperature of 1.9 K[1] . Several prototype magnets have been tested. We have measured for different quench conditions the development of pressure, temperature and helium content of the cryostat after a quench. Temperature data of the collar, yoke and the helium, both inside the cryostat and after discharge from the cryostat, were taken. Pressures were measured in the cryostat end-volumes and at three locations along the beam pipe. We present the results obtained as a function of the relevant quench parameters like energy released to the cryostat and discharge valve opening time.

INTRODUCTION

When a LHC dipole magnet quenches its full stored magnetic energy will be dissipated by ohmic heating in the coils as they become resistive[2] . This heat will eventually end up in the cold mass, and unless one expells the helium contained inside the cold mass sufficiently fast it will drive the helium pressure an order of magnitude beyond the 2 MPa (20 bar) cryostat design pressure. Therefore after quench detection a quench relief valve (QRV) will be be opened, fast enough to allow the helium to discharge and to limit the pressure rise to below this cryostat design pressure

Four prototype magnets (denoted A1,A2 and A3 for ANSALDO1 to ANSALDO3, and N1 for NOELL1)[3] from two different manufactures have been tested on a specially constructed magnet test bench[4] (MTB).

We have measured the development of pressure, temperature and helium content of the cryostat after a quench, with the aim of answering the questions relating to: the magnitude of the pressure rise in the bulk helium volume, the magnitude of the pressure

Advances in Cryogenic Engineering, Vol. 41
Edited by P. Kittel, Plenum Press, New York, 1996

rise around the beam tube, the QRV sizing and the influence of QRV-opening time on maximum quench pressures.

EXPERIMENT

Installation Layout

The magnets are tested on the so called Magnet Test Bench, MTB, which is described in detail in a separate article[4] . It consists of the Magnet Feed Box, MFB, the interconnect and the dipole cryostat. The superconducting dipole magnet operates in a pressurized static bath of superfluid helium at a pressure of 1 bar and at a temperature of 1.9 K. The heat load on this superfluid helium bath is extracted by conduction along the magnet length and along the interconnect to a lumped cold source inside the MFB. A 4.5 K saturated helium bath, which intercepts residual heat at the lower end of the 18 kA vapour cooled current leads, is thermally and hydraulically separated from the pressurized superfluid helium enclosure by two "λ-plates", which allow feedthrough of the superconducting current bus and instrumentation wiring. Emergency pressure relief and helium discharge after a magnet quench is performed through a DN50 electro-pneumatically-actuated, industrial-type, cold valve: the QRV. The helium is discharged into the 4.5 K vessel wherefrom the two-phase mixture will be decanted.

Instrumentation

Figures 1 and 2 show the placement of sensors inside the magnet cryostat. Temperature sensors (Allen-Bradley 100 Ohm carbon resistors) are placed in the end volumes and midlength in the collar and yoke. Silicon-based pressure sensors are placed in the end volumes (Siemens KPY16), and at three locations along the length in the collar noses (Siemens KPY47), see figures 1 and 2. The pressure sensors in the end volumes give information about the bulk cryostat pressure rise. The pressure sensors at the collar noses were put because we expect the highest pressures to occur near the coils, where the helium trapped inside absorbs the heat.

Data was recorded with a sampling rate of 1 ms for a time period of 3 s and with a sampling rate of 10 ms for a time period of 40 s, using a commercial based transient recorder (Visual Designer) running on a presonal computer.

RESULTS

Pressure Distribution

The rate of pressure increase is determined by a combination of the way the dissipated power in the coil changes when the current is being discharged, the transfer of this power to the helium trapped inside and in close contact with the coils, and of the hydraulics of the cold mass. The differences between the magnets are thus reflected in their respective pressure versus time curves.

Figures 3 and 4 show a typical measurement of pressure development after a quench as function of time and at different measurement locations. Figure 3 also shows the current and the QRV opening (normalized on a scale of 1 to 10). The quench is detected at $t = 0$ s. The magnet is then discharged within 600 ms. The QRV starts to open 120-130 ms after the quench detection signal and reaches 90 % opening within 210-250 ms.

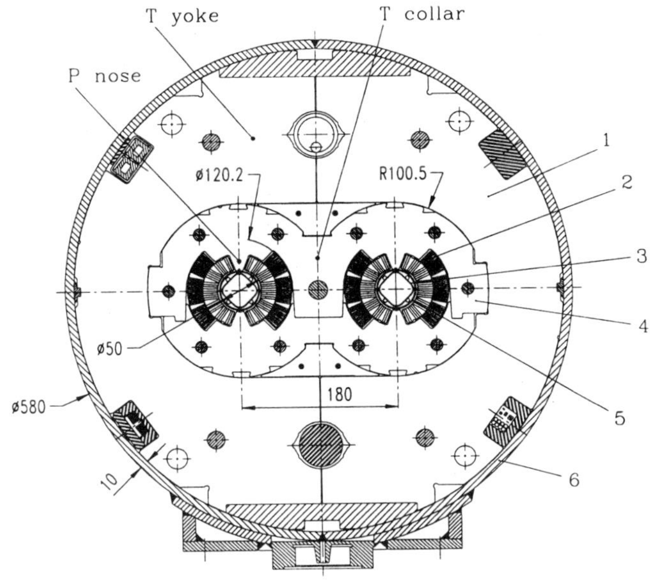

Fig. 1. Cross section of a 10 m long LHC dipole prototype magnet, showing the placement of sensors at midlength: $P_{nose,m}$ in the collar nose, T_{collar} and T_{yoke} in collar and yoke respectively. 1:yoke, 2:coils, 3:beam screen, 4:collars, 5:beam tube, 6:shrinking cylinder.

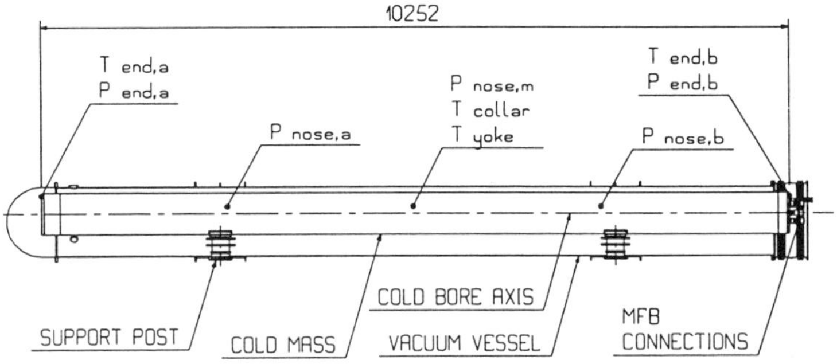

Fig. 2. Simplified longitudinal cut of a 10 m long LHC dipole prototype magnet inside its cryostat, showing the placement of sensors: $T_{end,a}$, $P_{end,a}$, $T_{end,b}$ and $P_{end,b}$ in the end volumes, $P_{nose,a}$, $P_{nose,m}$ and $P_{nose,b}$ in the collar noses, T_{collar} and T_{yoke} in collar and yoke respectively. The helium discharge is at the MFB connection side.

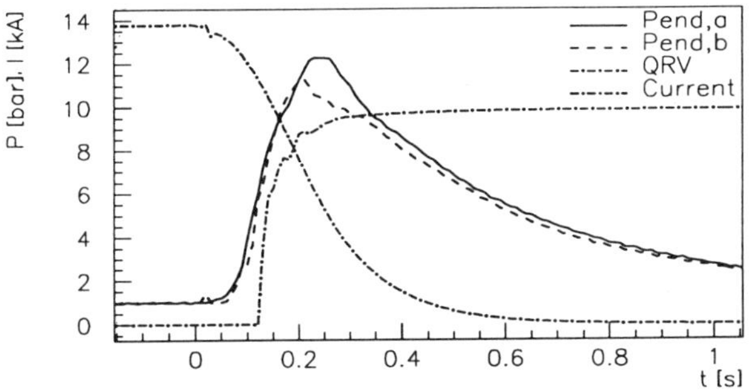

Fig. 3. Pressure, Current and QRV opening versus time for A3.

Table 1. Characteristics of quench pressure rise

magnet	t_{rise} [ms]	$(dP/dt)_{max}$ [bar/s]	$t_{changeover}$ [ms]	$(dP/dt)_{medium}$ [bar/s]	$t_{discharge}$ [ms]	t_{rise}^{collar} [ms]
A1	80 ± 10	39 ± 5			210 ± 10	210 ± 10
A2	60 ± 10	96 ± 5	150 ± 10	35 ± 5	210 ± 10	120 ± 10
A3	60 ± 10	99 ± 5	160 ± 10	45 ± 5	210 ± 10	130 ± 10
N1	80 ± 10	100 ± 5	150 ± 10	0	210 ± 10	150 ± 10

For the four dipoles tested we grouped together some characteristics of the pressure development curves in table 1. The meaning of the table becomes clear as one follows the pressure versus time curve of figure 3.

We observe that after quench detection and with the QRV still closed all pressures start to rise gently at t_{rise}. This time corresponds to the transition from superconducting to normal of all the conductors in the outer layer[2] . We measure for A2 and A3 a 20 ms faster t_{rise} time than for A1 and N1.

Afterwards, with the QRV still closed, all pressures start to increase sharply. This rate of increase of pressure with time, $(dP/dt)_{max}$, is significantly lower for A1 (39 bar/s), than for all other magnets (96-100 bar/s). There is a change in slope of pressure increase from $(dP/dt)_{max}$ to $(dP/dt)_{medium}$ at $t_{changeover} = 150$ and 160 ms for A2 and A3 respectively. There is no change in slope for A1, and the slope becomes zero at $t_{changeover} = 150$ ms for N1.

Next the pressure nearest to the QRV starts to go down, indicating that the helium begins to discharge from the cold mass. This is confirmed by the mass flow measurement using a cryognic Pitot tube located just after the QRV. It corresponds to the times indicated with $t_{discharge}$, and to a QRV-opening of 90 ± 5 %. All magnets show the same behaviour and start to discharge at $t_{discharge} = 210$ ms.

The pressures at locations further away from the QRV continue to increase until there is a maximum pressure difference over the magnet length, ($P_{end,a}$-$P_{end,b}$), of about 1 to 1.5 bar. After 2 seconds the pressure peak has passed and the pressures are down again to practically 1 bar.

We know from earlier measurements[5] and from numerical simulations[6] . that if the helium which is trapped inside the coils is forced to flow through the annular space between beam tube and coil in order to reach the end volumes, the collar nose pressures will be very high with respect to the end volume pressures. Our measurements (figure 4) show that all pressures, in the ends and in the collar noses, rise simultaneously and that no extreme pressure differences develop. Therefore we conclude that for these

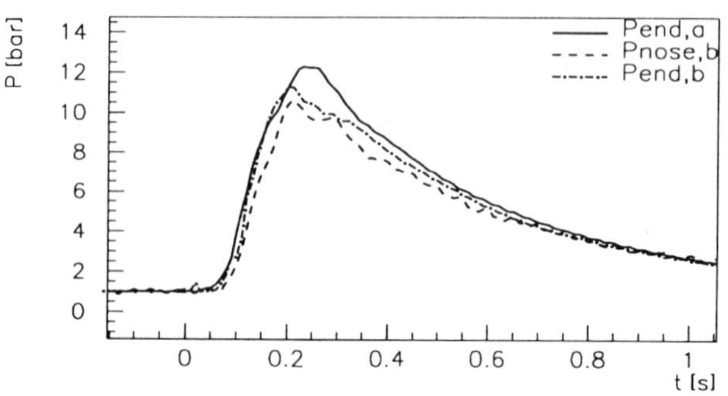

Fig. 4. Collar nose pressure compared with end pressures versus time for A3.

LHC prototype magnets the principal escape path for the helium trapped inside the coils is radially through the collar noses, the collaring and the yoke and then longitudinally towards the end volumes.

Maximum Pressure versus Stored Energy

Figure 5 shows for the A1 magnet the maximum pressures, P_{max}, after a quench as function of dissipated magnetic energy and at different measurement locations. The maximum pressures depend linearly on dissipated energy. Our measurements show that similar linear relationships hold for the other three magnets. Furthermore they show that for all magnets all pressures remain below the 2 MPa (20 bar) cryostat design limit. Conform the observations from the pressure versus time curves, table 1, the pressures are highest in A2 and A3 (12 and 12 bar at 4000 kJ) and lowest in A1 and N1, (8 and 7 bar at 4000 kJ).

Influence of Valve Opening Time on Maximum Pressure

On A3 and N1 we measured P_{max} when the quench relieve valve was already open before the quench, see figure 6. This was easily achieved during current ramps by opening the valve a few minutes before the current would reach quench values. Opening the valve caused only a very slight rise in bath temperature, of the order of 10's of mK. Since quenches in these magnets have a structural cause these measurements are representative for which maximum pressures one could get with a very fast valve. They give the lower bound for P_{max}. Furthermore these measurements provide an opportunity to observe the time at which the helium would start to discharge when there is no obstruction in the discharge line. It gives the time below which a faster opening QRV would no longer result in a decrease in P_{max}. For the A3 magnet we observed that all pressure peaks are reduced by approximately 2.5 - 3 bar. The maximum pressure is reached after 180 ($P_{end,a}$) ms. $P_{end,b}$ and $P_{end,a}$ join eachother after 250 ms. The mass starts to discharge after 130 ± 10 ms. At the same time of the mass discharge we see that the temperature in the collar starts to increase, see t_{rise}^{collar} in table 1. For the N1 magnet we observed no change in P_{max}, except near the QRV where $P_{end,b}$ went down by approximately 1.5 bar. The maximum pressure is reached after 160 ($P_{end,a}$) ms. Again $P_{end,b}$ and $P_{end,a}$ join eachother after 250 ms. The mass starts to discharge after 150 ± 10 ms, while the collar temperature starts to increase after $t_{rise}^{collar} = 150 \pm 10$ ms. Both measurements show that the increase in collar temperature is correlated with the onset of the mass discharge.

On A3 we have also done some measurements with longer QRV opening times. The opening time being defined as the total time from quench detection until the valve reaches a given aperture. Because of the QRV-characteristic we choose the time at which it has opened by 95 % as reference. Figure 7 shows the effect of opening time on maximum quench pressure for A3. Using the measured linear relationship between maximum pressure and dissipated energy for A3 (like figure 5), all pressures are rescaled to an energy dissipation of 4732 kJ corresponding to a 13 kA quench current. The rate of increase, 48 bar/s for $P_{end,a}$, and 42 bar/s for $P_{end,b}$ agree with the expected rate of $(dP/dt)_{medium}$ from table 1.

Based upon the mass discharge measurements with the QRV open we would choose an opening time for A3 of 130 ms. and for N1 150 ms. However extrapolation of the measurements shown in figure 7 indicate that for A3 we could probably use an QRV-opening time of about 180 ± 20 ms, without significantly increasing the P_{max}. This is close to the measured discharge time with QRV opening on the quench trigger, $t_{discharge} = 210 \pm 10$ of table 1.

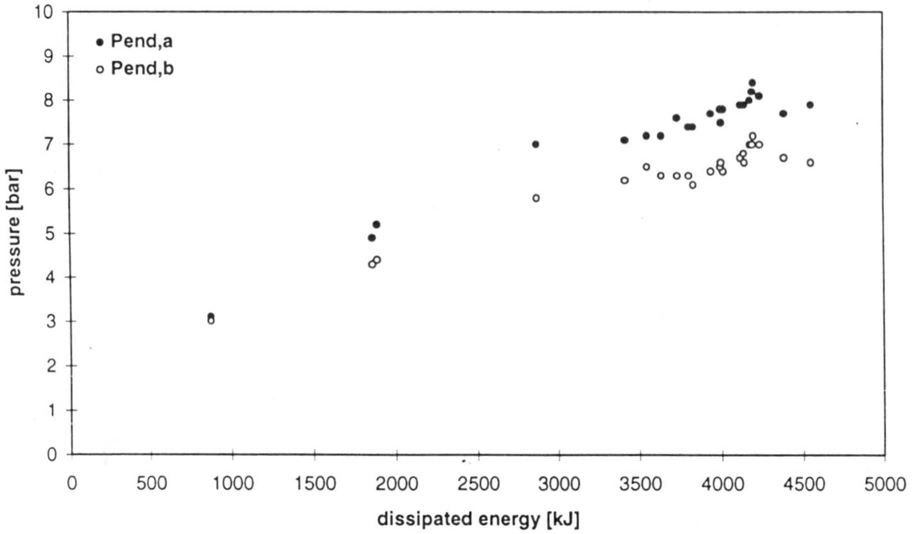

Fig. 5. P_{max} versus dissipated energy for A1.

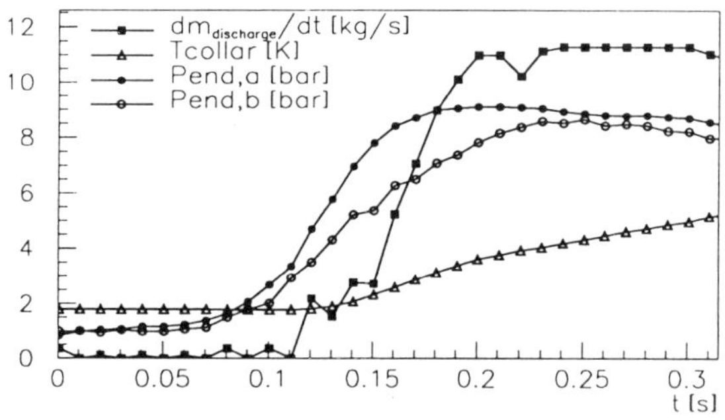

Fig. 6. Pressure versus time for A3, QRV open.

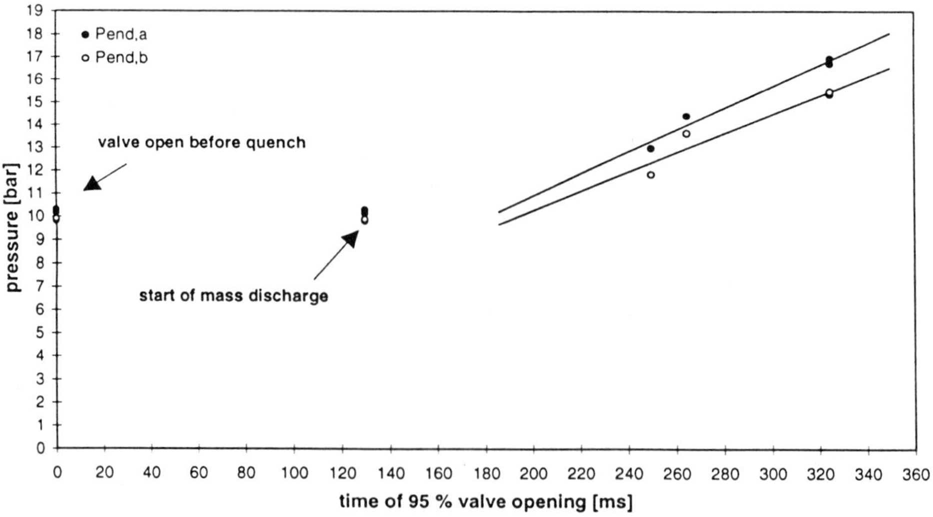

Fig. 7. Pmax versus valve opening time for A3. Dissipated energy 4732 kJ, corresponding to a 13 kA quench current.

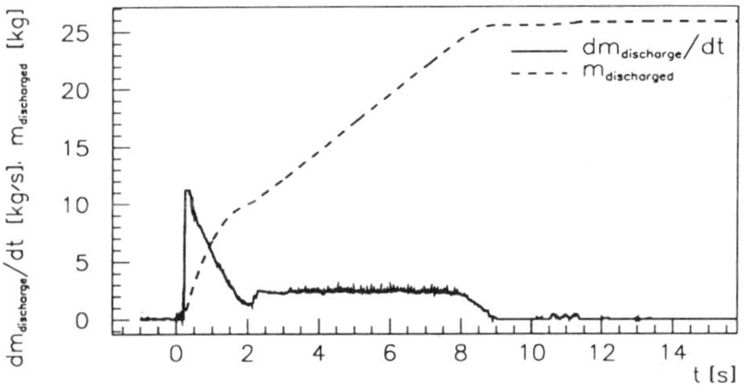

Fig. 8. Mass discharge. A3, dissipated energy 4222 kJ, 13.804 kA quench current.

Mass Discharge

The mass discharge was measured using a cryogenic Pitot tube. Figure 8 shows a typical mass discharge curve. The onset is at 210 ± 10 ms. It reaches a maximum of 11.5 kg/s 240 ms after the quench. At 350 ms this initial burst goes down and it reaches a steady 3 kg/s after 2 s, where it remains until practically the whole cryostat (84 %) is emptied.

Temperatures

Figure 9 shows typical temperature development curves. The end plate and helium temperatures rise to 3-4 K. The yoke follows until after 17.5 s (8.5 s after all the helium is discharged) it starts to rise, reaching 16 K after 40 s. The collar temperature increases 120 ± 10 ms after the quench and reaches 30 K after 40 s. This onset of collar temperature corresponds to the onset of mass discharge and collar temperature as measured with the QRV open before the quench. As a consequence we can take the onset of collar temperature t_{rise}^{collar} measured with the QRVclosed as the time below which a faster opening QRV would no longer result in a decrease in P_{max}, see table 1. t_{rise}^{collar} for A1 is significantly higher than for the other magnets.

CONCLUSIONS

Our measurements show that all quench pressures remain below the 2 MPa (20 bar) cryostat design limit.

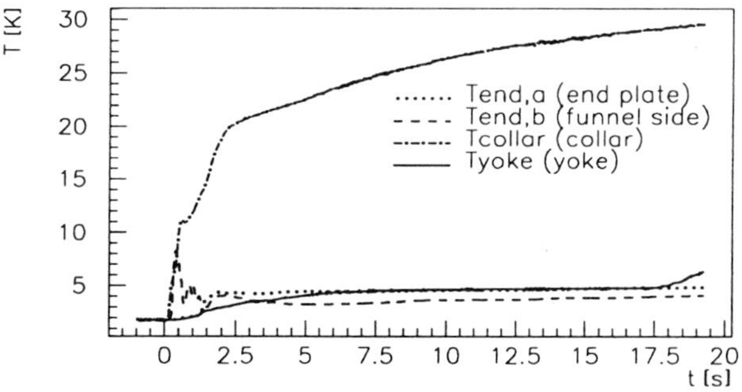

Fig. 9. Temperatures. A3, dissipated energy 4222 kJ, 13.804 kA quench current.

The maximum pressures reached are linearly dependent on dissipated energy.

The pressures in the ends and in the collar noses, rise simultaneously and no extreme pressure differences develop. This is in sharp contrast with a scenario in which the helium escapes longitudinally along the annular space between beam tube and coils. Therefore in these LHC prototype magnets the principal escape path for the helium trapped inside the coils must be radially through the collar noses, the collaring and the yoke and then longitudinally towards the end volumes.

The onset in collar temperature, t_{rise}^{collar}, is correlated with the onset of the mass discharge in case the QRV is open before the quench. The t_{rise}^{collar} time is therefore the lower limit, below which a faster QRV opening time will no longer be an effective means to reduce maximum quench pressures. For opening times exceeding $t_{discharge}$ the pressure increase can be estimated using the dP/dt values deduced from the pressure versus time measurements.

There are differences between the different magnets. The low rate of bulk pressure increase for A1 and its significantly higher t_{rise}^{collar} remain unexplained as well as the leveling off of P_{max} after 150 ms for N1.

ACKNOWLEDGMENT

The authors would like to thank the crew of the LHC test bench for all their effort in getting the measurements taken. Thank are also due to W. Cameron for providing the schematics.

REFERENCES

1. Ph. Lebrun, Superfluid Helium Cryogenics for the Large Hadron Collider Project at CERN, *Cryogenics* **34** ICEC Supplement (1994), pp. 1-8.
2. F. Rodriguez-Mateos, G. Gerin and A. Marquis, Quench Protection Test Results and Comparative Simulations on the First 10 Metre Prototype Dipoles for the Large Hadron Collider, *Fourteenth International Magnet Technology Conference (MT-14),* Tampere (1995).
3. R. Perin, Status of LHC Programm and Magnet development, *Applied Superconductivity Conference (ASC),* Boston (1994).
4. V. Benda, M. Granier, Ph. Lebrun, G. Novellini, V. Sergo, L. Tavian and B. Vullierme, Cryogenic Benches for Superfluid Helium Testing of Full-Scale Prototype Superconducting Magnets for the CERN LHC Project, *Cryogenics* **34** ICEC Supplement (1994), pp. 733-736.
5. M. Granier, D. Hagedorn, K.N. Henrichsen, Ph. Lebrun, D. Leroy, R. Perin, J. Vlogaert, J. Deregel, P.Genevey, J.P. Jacquemin, F. Kircher, J. Le Bars, Performance of the Twin Aperture Dipole for the CERN LHC, *European Particle Accelerator Conference (EPAC),* Berlin (1992) and CERN Internal Note AT-MA/9204 (1992).
6. Ph. Lebrun, Th. Wahlström, R. van Weelderen and L. Williams, Investigation of Quench Pressure Transients in the LHC Superconducting Magnets, *Cryogenics* **34** ICEC Supplement (1994), pp. 705-708. Genova (1994).

CONCEPTUAL DESIGN OF THE CMS 4 TESLA SOLENOID

J.C. Lottin,[1] H. Desportes,[1] C. Lesmond,[1] C. Lyraud,[1] C. Pes,[1]
A. Hervé[2]

[1]CEA-DSM-DAPNIA, Saclay, 91191, Gif sur Yvette, France
[2]CERN, CH-1211, Geneva 23, Switzerland

ABSTRACT

The detection of new physics signals at the highest luminosities available in proton-proton collisions at LHC requires identification and precise measurement of muons, photons and electrons. Toroidal and solenoidal fields were considered at the beginning of the design. For the CMS detector, the choice of a compact design led to the choice of a strong magnetic field. The most practical magnet that can generate a strong magnetic field is a solenoid. A long (about 13 m) superconducting solenoid of large radius generating a magnetic field of 4 T guarantees good momentum resolution. The magnetic flux is returned via a 1.8 m thick iron yoke of a weight of 12 000 tonnes. The magnetic stored energy is 2.52 GJ and the coil total weight is 500 tonnes. The coil main design features are indirect cooling, pure aluminium stabilisation and mechanically reinforced conductor. It is a four layer winding, composed of 4 axial sections bolted together.

INTRODUCTION

The choice of magnet system was the starting point for the CMS detector design.[1] A long superconducting solenoid of length 13 m and useful diameter 5.9 m, producing with the iron yoke a magnetic field of 4 T, has been chosen. The favourable aspect ratio of the solenoid allows efficient muon detection and measurement up to a rapidity of 2.4 without the need of forward toroids. Thus the muon spectrometer uses a single magnet, simplifying the detector design. Since the magnet is the main element of CMS in terms of size, weight, and structural rigidity, it is used as the principal structural element to support all other barrel detector components. The existing detector large solenoids [1,2] produce magnetic fields up to 1.5 T. The 4 T requirement for the CMS solenoid increases the overall parameters in terms of magnetic forces, ampere-turns, stored energy and necessitates some departures from the previous designs.

Table 1. CMS coil characteristics

Warm bore	m	5.9
Overall OD	m	7.3
Overall length	m	13
Magnetic induction at interaction point	T	4.0
Peak magnetic induction on the conductor	T	4.6
Magnetic length	m	12.48
Stored energy	GJ	2.52
Magnetomotive force	MAt	42.24
Magnetic radial pressure	MPa	6.47
Axial compressive force at mid plane	MN	122

DESIGN CRITERIA

Concept guidelines. The CMS solenoid parameters given in Table 1 have been determined as a realistic limit with regard to feasibility and cost, considering available and proven technology.

The conceptual design of the CMS coil builds on the experience gained over the last fifteen years with superconducting magnets for high energy physics, particularly in the areas of quench protection and enthalpy stabilisation. The main features which have demonstrated the high quality and reliability of such magnets are the following:

- enthalpy stabilisation using high purity aluminium,
- a monolithic external cylinder made from aluminium alloy,
- indirect cooling,
- full epoxy impregnation.

The 4 T field specified for CMS requires the introduction of two important new features to provide the number of ampere-turns needed and to resist the greater magnetic forces:

- a four layer winding, as opposed to one or two layers in most coils,
- a larger and more complex conductor incorporating a mechanical reinforcement.

In addition, for ease of fabrication and transportation the winding is divided in 4 longitudinal sections bolted together. The design parameters of the superconducting winding are summarised in Table 2.

Quench Protection Method. Coil protection in the case of a quench is considered to be the most critical issue because of the very large stored energy. The coil protection system is based on a winding/cylinder design which will initiate a 'quench-back' by rapidly discharging the coil to external resistors.

Table 2. CMS coil parameters

Operating current	kA	20
Inductance	H	12.6
Total number of turns		2112
Total length of conductor	km	43
Number of axial sections		4
Number of layers		4

Table 3. Basic conductor parameters

External dimensions (bare)	mm^2	62 × 22.3
Operating current	kA	20
Critical current at 5 T and 4.2 K	kA	61.5
Superconducting central cable		
Number of strands		40
Strand diameter	mm	1.3
Cu/Sc. (typical)		1.33
Aluminium stabiliser		
Cross-section	mm^2	640
Residual resistivity ratio (RRR) (0 field, 0 strain)		1000
Aluminium-Alloy reinforcing sheath		
Material		AA 6061
Cross-section	mm^2	693

The winding/cylinder configuration is essentially a current transformer, in which the winding acts as the primary and the cylinder as the secondary. During normal coil operation or slow dumping, the induced current in the coil cylinder is negligible due the very low rate of current variation. When a fast current dump occurs the induced loss is sufficient to heat the coil and drive it everywhere to a normal state, within a few seconds. To estimate the post-quench-back coil temperature distribution, computations have been made with varying values of cylinder resistivity and discharge resistance. The important effect of field strength and mechanical strain on the resistivity of pure aluminium was taken into account. These calculations were used to determine the operating current of 20 kA, the dump voltage of 500 V and the amount of pure aluminium in the conductor (see Table 3), leading to the following principal characteristics:

- quench-back delay: 10 s,
- average final coil temperature: 80 K,
- maximum hot spot temperature: 105 K,
- maximum ΔT within the coil during quench: 40 K.

Thermosiphon cooling Method. A consequence of the protection method is that any fast dump, or interruption of the liquid helium supply, driving the coil to its normal state will deposit a large fraction of the stored energy into the winding. With a refrigerator sized on the normal operation cryogenic load, this will necessitate about ten days for re-cooling. Consequently the coil cryogenic system must be extremely reliable and protected against sudden power failure, fast dumping being used only in an emergency.

Thermosiphon cooling is chosen for the CMS cryogenic system because it operates continuously without external supply and without moving components. Furthermore, in case of quench or fast dumping, it spontaneously switches off. This mechanism results in a limited pressure rise (typically 3 bars) without the need of helium recovery system. The thermosiphon cooling method has already been used successfully with the ALEPH solenoid. [4]

Enthalpy stabilisation. A superconducting magnet is stable if the winding thermal capacity is sufficiently large to prevent excessive temperature rise. In contrast to total cryo-stability, enthalpy stabilisation only allows limited thermal disturbances. For detector magnets, such as the CMS solenoid, the only perturbation mechanism is mechanical, either due to epoxy cracking or to conductor micro slippage, both of which are related to the stress

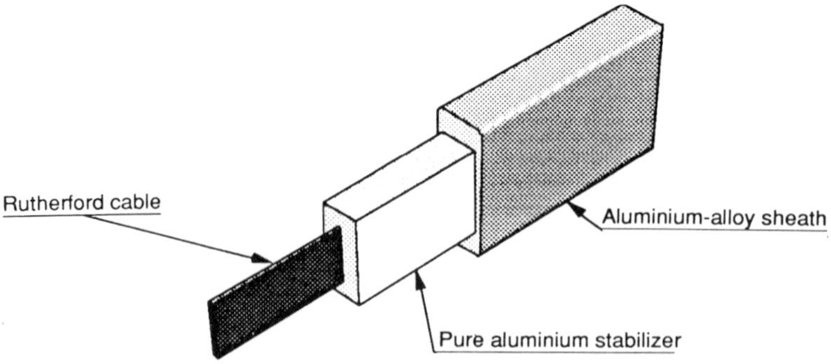

Figure 1. CMS reinforced conductor.

distribution in the winding. Consequently the stability will be determined by the mechanical stability and by the enthalpy margin.

Mechanical stability has been the principal issue in the design of the reinforced conductor and winding structure, in order to maintain shear stress at levels well below critical values. It also requires a high-quality winding with full impregnation. This quality can be achieved with the inner winding technique of a dry conductor [2] and an impregnation under vacuum. Resin properties at low temperature and impregnation process are now under experimental investigation at Saclay, in order to enable the proper selection of the impregnation method. [5]

The choice of a low current density (30% of the critical value at 4.6 T) results in a relatively high critical temperature of 6.5 K. At a working temperature of 4.5 K, an enthalpy margin of 3300 J/m^3 (4.6 J/m of conductor) is obtained. For comparison, the ALEPH enthalpy margin is 0.5 J/m. The energy density margin is even higher for localised thermal disturbances. Because of heat diffusion within the winding, local transient temperature excursions into the resistive domain are supported. This mechanism involves parameters related to heat generation and diffusion which are dependent on temperature, field and time. A working group is conducting theoretical and experimental studies on this complex issue.[6]

COMPOUND CONDUCTOR

The classical aluminium stabilised conductor is a soft component not able to sustain high working stresses. In the existing solenoids the magnetic stresses are small enough to be

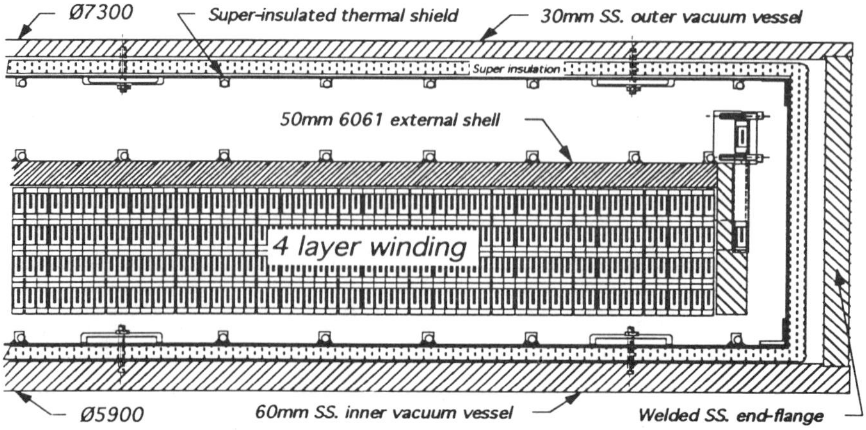

Figure 2. Cross-section through the end of the solenoid.

transferred to the aluminium alloy external cylinder. For CMS this force containment method would lead to a thick external cylinder and to excessive shear stresses especially at the winding/cylinder interface. By associating directly to the conductor most part of the required structural material, the winding becomes a force self supporting structure in which the shear stress level is low (5 MPa), with respect to the fibre glass epoxy resin limit larger than 20 MPa. This mechanical concept allows higher working strain under magnetic load and thus a reduction of the overall coil thickness and mass. In CMS the radial strain is 0.15 %, the cold mass is 260 tonnes and the stored energy per mass unit 11 J/g.

As shown in Figure 1, the conductor consists of three concentric parts: the central flat superconducting cable, the high purity aluminium stabiliser and the external aluminium-alloy reinforcing sheath. The main parameters are given in Table 3.

Two contracts for 100 m long samples of the CMS reinforced conductor have been placed with GI in Slovak Republic and LMI in Italy. Both manufacturers have proposed extruding the compound conductor. The first one will extrude the pure aluminium around the Sc cable and the sheath around the pure aluminium in two stages. The other manufacturer will combine the two operations in a single pass. A third method, in which the outer aluminium-alloy sheath will be attached using electron beam welding, is also being investigated. Conductor performance measurements on development samples supplied by the manufacturers are cross-checked at Saclay[7] and the critical current of the whole conductor (61.5 kA) will be measured using an inductive method at the University of Geneva.[8]

WINDING

As already mentioned, the winding is divided into four sections of equal length, assembled with bolted flanges. With no special requirements on the field homogeneity, the four sections can be identical and have uniform current density. The sections have been designed to enable complete fabrication in industry before transportation to CERN. The end of the coil inside the cryostat is shown in Figure 2.

Each coil section consists of four concentric layers of conductor supported by an external aluminium-alloy cylinder on which aluminium cooling tubes are welded. After winding, each section will be fully impregnated with epoxy resin. Each layer is made of a continuous conductor length of about 3 km and the conductor junctions lie outside the external cylinder with a helium cooled heat sink, in a zone of low field.

At the end of the section, the conductors pass through a groove machined in the end-flange to permit connection to the adjacent layer or section. The electrical insulation is fibreglass cloth which will be impregnated with epoxy resin after the winding operation. The turn-to-turn insulation consists of two layers of 0.135 mm thick fibreglass ribbon wrapped round the conductor and the layer-to-layer insulation includes an additional 0.2 mm thick fibreglass cloth. Insulation at the winding/cylinder interface is reinforced using three sheaths of identical fibreglass cloth.

HOLDING SYSTEM

In order to facilitate the overall assembly and access to the detector components, the coil vacuum tank is entirely supported from the yoke centre section. The cold mass is attached by a set of 16 tie rods made from titanium alloy connecting the coil external cylinder to the outer wall of the vacuum tank as shown in figure 3. In addition to the 260 tonnes weight of the cold mass the tie rod system has to contain any magnetic forces due to coil misalignment inside the yoke. A 3D FEA shows that any misalignment results in unstable forces. Each tie rod is designed to withstand a maximum load of 1.5 MN and consequently the maximum

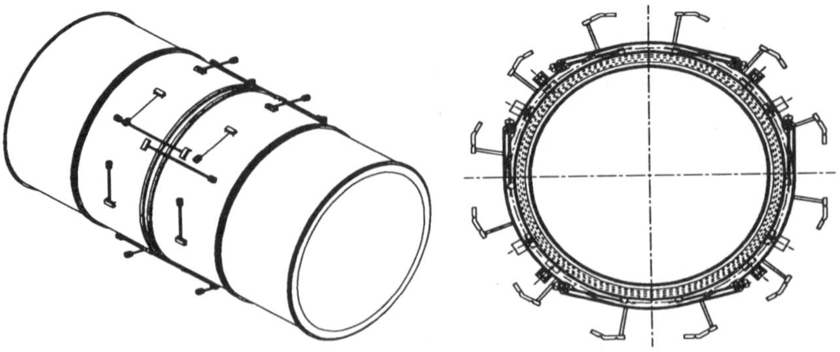

Figure 3 CMS cold mass holding system.

misalignment must be less than 10 mm. In order to achieve accurate centring it is planned to measure the unbalanced forces, with limited current in the coil, and then to correct the coil position. This fine centring adjustment necessitates strain gauge sensors on the rods and access to their warm ends from outside the vacuum tank.

CRYOGENICS

The calculated cryogenics loads are 540 W at 4.4 K plus a 2.5 g/s LHe flow for the two 20 kA currents leads and 3400 W for the radiation shield. This corresponds to a 1 kW refrigerator.

The external cryogenic system will insure all the functions of cooling down, normal operation, re-cooling after fast discharge and warming up. It includes a pre-cooler, fed with 500 l/h of liquid nitrogen which provides additional refrigeration power for the cool-down, which is expected to last approximately 40 days from room temperature. Re-cooling from the post-fast-discharge temperature of 80 K will last ten days.

The internal cryogenic system contains two circuits, one for the winding and the other for the thermal screen. A vessel located on top of the magnet serves both circuits and contains the connection to the external cryogenic system. This vessel includes a helium phase-separator for the coil cooling circuit.

The coil cooling circuit is ensured by 8 independent loops operated in a thermosiphon circulating mode. Each of them is a set of semi-circular aluminium tubes welded over the coil cylinder and connected in parallel to the phase separator through bottom and top manifolds. The common supply line of the 8 loops incorporates a single isolation valve, which closes automatically when the helium pressure exceeds 2 bars. Thus, during cool-down, this valve is closed, allowing forced-flow helium to circulate through the coil. In the case of a fast dump this valve prevents thermosiphon re-circulation and so acts as a thermal switch.

REFERENCES

1 CERN/LHCC 94-38 " The Compact Muon Solenoid Technical Proposal " 15 dec 1994

2 Desportes et al. " Design, Construction and test of the large Superconducting Solenoid Aleph ", IEEE TRANSACTIONS ON MAGNETICS, Vol. 24, N°2, March 1988

3 P.T.M. Klee and D.E. Baynham. " Toward the Realisation of Two 1.2 TESLA Superconducting Solenoids for Particle Physics Experiments " MT 11 Tsukuba 1989

4 J.C. Lottin, R. Duthil, " ALEPH Solenoid Cryogenic System " ICEC-12 Southampton 1988

5 J.M. Rey et al. « Epoxy Resin Developments for the ATLAS and CMS Superconducting Magnet Impregnation » ICMC, Columbus 1995

6 F.P. Juster et al. " Stability of Aluminium Stabilised Conductor for LHC Detector Magnets ", ASC4, 1994

7 F. Kircher et al. " Conductor developments for Atlas and CMS superconducting Magnets ", MT 14

8 P. Fabbricatore et al. " Critical Current Measurements on the Cables for LHC Detector Magnets " MT 14

1995 CEC/ICMC Conference: Two of the winners of the 1995 Russell B. Scott Memorial Award for Cryogenic Engineering Applications: (from left to right) G.J. Laughon and K.M. Schaubel.

TECHNICAL CHALLENGES OF SUPERCONDUCTIVITY AND CRYOGENICS IN PURSUING TESLA-TTF

Quan-Sheng Shu, for the TESLA Collaboration

Deutsches Elektronen-Synchrotron (DESY)
Notkestrasse 85, 22607 Hamburg, Germany

ABSTRACT

TESLA (TeV Energy Superconducting Linear Accelerator) Collaboration is an international R & D effort towards the development of an e^+e^- linear collider with 500 GeV center of mass by means of 20 km active superconducting accelerating structures at a frequency of 1.3 GHz. The ultimate challenges faced by the TESLA project are (1) to raise operational accelerating gradients to 25 MV/m from current world level of 5-10 MV/m, and (2) to reduce construction costs (cryomodules, klystrons, etc.) down to $2,000/MV from now about $40,000/MV.

The TESLA Collaboration is building a prototype TESLA test facility (TTF) of a 500 MeV superconducting linear accelerator to establish the technical basis. TTF is presently under construction and will be commissioned at DESY in 1997, through the joint efforts of 24 laboratories from 8 countries. Significant progress has been made in reaching the high accelerating gradient of 25 MV/m in superconducting cavities, developing cryomodules and constructing TTF infrastructure, etc. This paper will briefly discuss the challenges being faced and review the progress achieved in the technical area of superconductivity and cryogenics by the TESLA Collaboration.

INTRODUCTION

There is a widespread consensus within the high energy physics community that the next electron positron collider would be built with a center of mass energy of 500 GeV and luminosity of a few times 10^{33} $cm^{-2}s^{-1}$. Such a collider would provide for top analyses and discovery reach up to a Higgs mass of ≈ 350 GeV. Worldwide, there are a number of groups pursuing different linear collider designs. The TESLA collaboration is an international R & D effort to develop a linear collider using superconducting accelerating structures (25 MV/m, $Q_o=5x10^9$) at low frequency (1.3 GHz)[1,2,3]. The TESLA collaboration consists of 24 institutes from 8 countries.

Advantages of TESLA

The technical advantages of superconducting RF cavities is their high Q value and low RF wall losses (less than Cu cavities by a factor of 10^5). It allows us to use large aperture structures operating at low frequency (1.3 GHz, L-band) with long macro pulse length and low peak power requirements. The large aperture has a beneficial consequence of

Table 1. Parameters of TESLA-500

Linac energy	2x250 GeV	Luminosity	$6x10^{33}$ $cm^{-2}s^{-1}$
Beam pulse current	8 mA	Bunch separation	0.7 μs
Number e/bunch	$3.63x10^{10}$	Bunch length	0.7 mm
Nu. of bunches/pulse	1130	Rep rate	5 Hz
RF pulse length	1.3 ms	Total Length (incl. inter. reg.)	32 km
Beam pulse length	0.8 ms	RF Freq.	1.3 GHz

substantially reducing transverse and longitudinal wake field effects, leading to relaxed Linac alignment and tolerances.

Challenges of TESLA

Despite the attractive feature of the TESLA design, a major effort is needed to demonstrate that a linear collider can be built at a cost competitive with its normal conducting counterparts. The two technical challenges being faced and the key approaches to reach the ultimate goals can be summarised as follows [4,5]:

(1) Increase operational accelerating gradients of SRF cavities to 25 MV/m from current levels of 5-10 MV/m by eliminating field emission and thermal breakdown.

(2) Decrease structure cost by utilising multicell structure, long cryostat, high efficiency klystron and cost effective fabrication techniques.

We believe that the construction cost can be reduce to $2,000/MV compared to the current $40,000/MV. The TESLA Collaboration is building a test facility (TTF) of a 500 MeV superconducting linear accelerator to establish the technical base.

BASIC INFORMATION ABOUT TESLA AND TTF

TESLA Layout

The overall layout of the TESLA 500 is illustrated in Figure. 1 [1,6]. The electron beams (e^+ & e^-) are accelerated to 250 GeV by RF fields in each main linac. Each linac has an active accelerating legnth of 10,000m consisting of about 10,000 superconducting RF cavities. Total length of TESLA-500 is about 32 km. A discussion of the layout is introduced in the reference [6]. An overview of the main parameters of TESLA-500 is given in table 1.

As shown in Figure 1. TESLA is composed of many components. However, here we only introduce the technical issues relevant to cryigenic engineering. For convenience of discussion, we consider the TESLA-500 as a superconducting and cryogenic structure which is energized through thousands of high power RF couplers and cooled by a huge superfluid-He refrigeration system. The structure of the 20-kilometer active accelerating machines is summarised in Table 2 with a SRF cavity (E_{acc} = 25 MV/m, $Q_o = 5x10^9$) as the smallest unit. The basic cryogenic structure unit is a cryomodule consisting mainly of 8 cavities. 4 cryomodules share a klystron of 10 MW peakpower. The heat load to 2K from a cryomodule is 21.4 W including dynamic (RF) and static losses [7]. To keep the TESLA machine in superconducting state, the total estimated cooling power is: 51 kW at 2 K, 37 kW at 4.5 K and 314 kW at 40-80 K [8]. Sixteen 2 K-refrigerators are needed to provide the cooling power distributed over the machine. The total liquid He inventory estimated in the TESLA machine is about 87,200 - 102,800 kg (depending on the version of cooling loops to be selected).

TTF Layout

Figure 2 is a schematic layout of the TTF consisting of an injector with a capture cavity, 15 MeV beam analysis area, TTF Linac with four cryomodules (the design requirement of the TTF cavities is E_{acc} = 15 MV/m and $Q = 3x10^9$) and a high energy

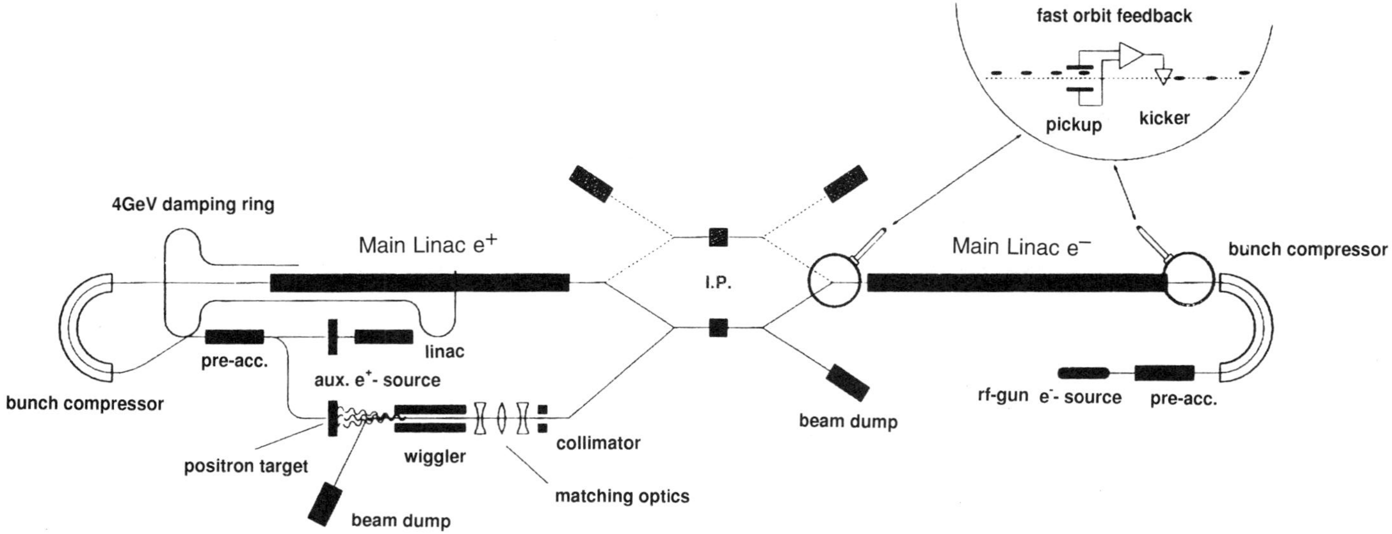

Figure 1. An overall layout for TESLA-500.

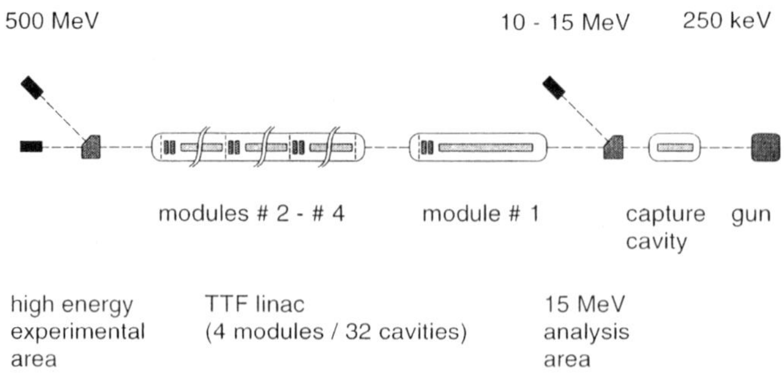

Figure 2. Schematic layout of the TESLA Test Facility (TTF).

experimental area. The TTF is refrigerated by a He II refrigerator: 100 W at 1.8 K, 400 W at 4.5 K and 2000 W at 70 K (the refrigeration power will be expanded to 200 W at 1.8 K).

With the TTF, we will demonstrate and verify many technical issues, such as, cavity processing technologies, cavities performance, design and performance of couplers and cryostat, cryogenic operation, energy and position feedback, alignment, etc.

The TTF is also being considered for use as a Free Electron Laser Facility.

TESLA SRF CAVITIES

The TESLA SRF cavity is shown in Figure 3 and an overview of the main parameters is given in table 3. The TESLA cavities are made of high purity Nb sheets (2.8 mm thick)

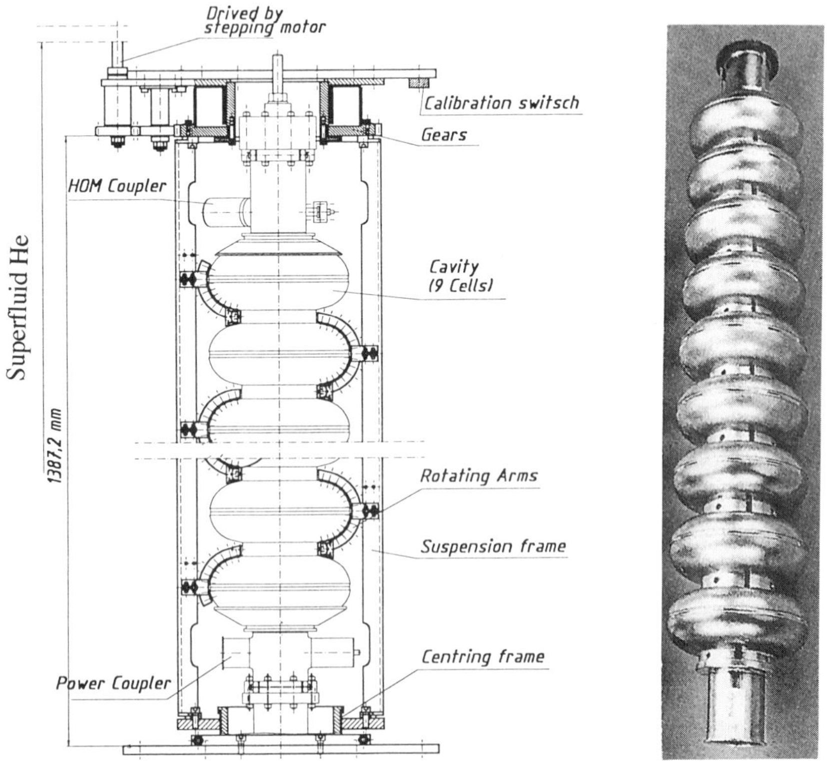

Figure 3. TESLA cavity with LHe vessel & 2K forward line, and a picture of the prototype cavity.

Table 2. TESLA-500 as a superconducting and cryogenic machine (consider SRF cavity as basic unit)

Unit	Group
9 SRF cavity cells	A cavity (including LHe vessel, RF couplers)
8 cavities (incl. quadr. packages)	A cryomodule of 12.2 m, and (4 cryomodule share one 10 MW klystron)
12 cryomodules	A string of 148 m, has an individual cryo-loop
12 strings	A TESLA subunit of 1830 m (one subunit has a 3.2 KW/2K cryo-plant)
16 subunits	The TESLA-500 (an active superconducting accelerating length of 20,000m)

Table 3. Parameters of the TESLA cavity

RF Freq.	1.3 GHz	Gradient	25 MV/m
Cavity cell	9	Cavity aperture	35 mm-radius
Effective length	1.035 m	Epeak/Eacc	2
Bpeak/Eacc	4.2 mT/MV/m	R/Q	1011 Ω/cavity
Peak RF power	206 kW/m	Stored energy	$0.127\ J/(MV/m)^2$
Cryo load/module	21.4 W	G (Rs=G/Qo)	271
Coupling cell-cell	1.87%	HOM K(ll)	9.24 V/pC

with a RRR of about 300. The cavity shape combining an elliptical iris and a spherical equator results in more satisfactory electromagnetic parameters than other combinations. Another consideration to TESLA cavity design is to minimize the ratio both of Epeak/Eacc (2.1) and Hpeak/Eacc (4.2 mT/MV/m) in order to reach the highest gradients [3,7].

The maximum cell number of 9 is determined by effective HOM damping requirement. There is one input power coupler and two HOM couplers for a cavity. Each cavity has a liquid He vessel with small LHe inventory which reduces the coldmass and allows a fast cooldown and warmup of the cavity.

QUEST FOR HIGHEST ACCELERATING GRADIENT -- AN ULTIMATE TASK

The TESLA collaboration aims for the highest operating gradients Eacc in SRF cavities using economically affordable approaches.

Theoretical Limit of Eacc

The limit magnetic field in RF is larger than Bc (type I superconductors) and Bc1 (type II) respectively, and is called superheating critical field Bsh. Table 4 gives the Bsh values of typical materials studied in SRF technology compared with the maximum surface field

Table 4. Critical field in DC and RF superconductivity

Materials	Tc [K]	Bcth [mT]	Bc1 [mT]	Bc2 [mT]	Bsh [mT]	Bexp [mT]
Sn	3.7	30.9	-	-	68	30.6
In	3.4	29.3	-	-	104	28.4
Pb	7.2	80.4	-	-	105	112
Nb	9.2	200	185	420	240	160
Nb_3Sn	18.2	535	20	2400	400	106

Bcth - thermodynamic critical field
Bsh - superheating critical field
Bexp - experimentally obtained maximum field
All data refer to values at T = 0 K

Bexp experimentally obtained so far[9]. Using a rule of thumb that 40 Gauss (4mT) is equivalent to 1 MV/m accelerating gradient in cavities with the TESLA shape, the theoritical limit for Nb by Bsh is then about 60 MV/m, and for Nb_3Sn, 100 MV/m[9].

Achievements and Limit in Operating Accelerators

Since the early 1970s, significant progress in the state of the art SRF accelerating cavities have been achieved. The operating accelerating gradients in more than 10 laboratories (such as Argonne[10], CEBAF[11], CERN[12], Cornell[13], Darmstadt[14], DESY[15], KEK[16], Saclay[17], etc.) and 400 structures reached 5-10 MV/m, compared to design goal of 5 MV/m. These achievements are attributed to anti-multipactor, round cell shape, high thermal conductivity Nb to avoid thermal breakdown (TB) and clean surface preparation to avoid field emission (FE). However, the excellent performance of the operating cavities is not adequate for the proposed TESLA machine. Tremendous efforts are still needed to comfortably reach the TESLA goals of Eacc $\geq$ 25 MV/m and Q $\geq 5\times10^9$. The main obstacles still to overcome are FE and TB. Investigation and elimination of FE and TB have become one of the highest priorities in TESLA and TTF projects.

Understanding FE and TB

Significant efforts have contributed to understading and defeating FE and TB by the TESLA Collaboration.

Electrons on the cavity surface can be pulled out and accelerated in the cavity vacuum by RF electrical fields. These field emitted electrons absorb energy from RF fields and deposit the energy in their landing area on the cavity, resulting in degradation of Q value and limiting the Eacc. Studies with DC FE scanning microscope and in SRF cavities [18,19,20,21] indicate that most FE electrons come from submicro-size foreign particles of a metallic nature with irregular shapes. Some studies also found that the condensed gases and heat treatment at 200-600° C will activate FE [21, 22]. Thermal breakdown is due to imperfections of the RF surface. For a given imperfection, the thermal breakdown field

Figure 4. Rotating T-R mapping system for diagnostics of TESLA 9-cell cavities.

Figure 5. The ultahigh vacuum oven for Ti-gettering purification of cavities.

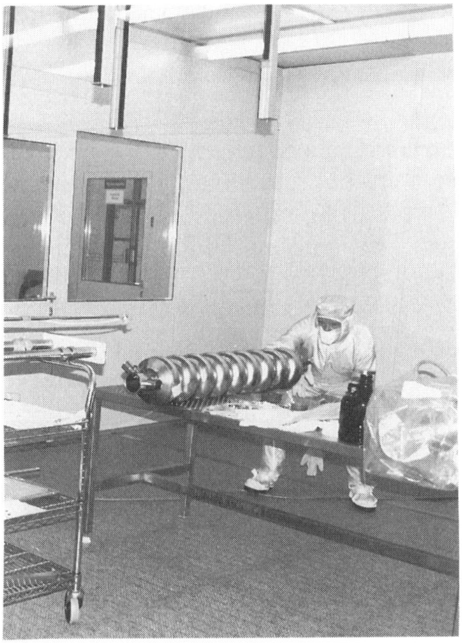

Figure 6. The clean room for cavity processing and assembling.

Figure 7. Chemistry and rinsing of cavity in the clean room.

scales roughly as the (thermal conductivity)$^{1/2}$. The residual resistance ratio (RRR) is used to characterise the thermal conductivity in a convenient way.

It is impossible to directly observe the FE and TB on the inner surface of cavities during RF operation. Therefore, the main approach to understanding the FE and TB of cavities is to study the hot spots and X-rays (induced by impacting FE electrons) generated on the cavity surfaces during RF operation. At DESY a rotating T-R mapping system for TESLA 9-cell cavities has been developed and more than 10,000 spots on the cavity can be investigated in one turn with 10° stepping. We have used it to successfully identify the TB location, trace the dynamic prograss of cavity quench, locate the heating areas by FE electrons and simulate the emitter locations [23].

Reducing FE and TB

For years in many laboratories around the world, there have emerged many comprehensive approaches for cavities processing in order to reach the highest possible gradient. Many of these approaches have been adopted, further developed and used by TESLA collaboration at DESY.

Reducing TB with UHV Oven

Over the last 10 years the RRR of sheet Nb delivered by industry for cavities has been improved to 300 from 30 by better melting practices removing most of the dissolved impurities of O, N, H, C, etc. Cavities using these sheets produce a range of TB at 13 - 19 MV/m. Higher RRR of > 500 is desired for reaching a Eacc > 25 MV/m. Currently the way to increase RRR is to employ solid state gettering [19, 24]. In this technique, both the inside and outside surface of a prepared cavity are exposed to Ti vapors at 1400° C for 2-4 hrs. The oxygen which diffuses to the cavity surface is gettered by the evaporated layer. After improvement of RRR, both surfaces are etched to remove the foreign metal layer. Figure 5 is the UHV oven used at DESY (1400°C at 10^{-7} mbar). The control samples Nb indicate that RRR of 250-300 is raised to > 500 after the Ti-purufication treatment. Several TESLA 9-cell cavities show Eacc increase to 20 - 26 MV/m.

Also the UHV oven has played an important role in reducing the FE and the hydrogen content of Nb, eliminating the "H-desease".

Cleaning Technology Defeating FE

Cleanning techniques similar to those utilized in semiconductor industry are used to remove all FE particles from the cavity's RF surfaces [25]. Cleanliness during chemical etching, water rinsing (high purity water of 15 MΩ-m) and assembling have played a importamt role in higher Eacc. The high pressure water rinsing (HPR) device made by CERN for the TESLA collaboration is very helpful in removing foreign particles which are difficult to remove by regular water rinsing. Several prototype TESLA cavities after HPR (without Ti-purification) produced Eacc > 15 MV/m.

We built a clean room of 300 m^2 with class 10 and 100 areas in 1993 for chemical etching, HPR, and cavity assembling (also we can load cavities into the UHV oven in a clean room area). The capability of HPR is as high as 200 bar. Figure 6 is a part of the clean room and Figure 7 shows the clean chemical etching of a TESLA cavity.

High Peak Power RF Processing - Last Chance to Reduce FE

Despite how good a job performed to eliminate FE, there is always a possibility that particles escape removal and stay on cavities surfaces. Therefore a technique that can eliminate the emmiters in situ is highly desirable. The technique, called HPP - high power processing developed at Cornell University, is to apply a high power RF pulse to the cavity in situ and eliminate the FE through an explosive process [26]. The idea is to raise the electric field at an emitter as high as possible in a short time (μs - ms) which generates a very high FE current. The transient high current melts, evaporates and activates a RF spark to destroy the FE emitters. The high RF pulse power processing of cavities provides a final, effective way to destroy the remaining FE emitters[22]. At DESY a HPP facility has been used successfuly in raising Eacc. The peak klystron power is 4.5 MW with pulse length of 2 ms.

The key technologies we have used to rduce FE and TB can be summarized in Table 5.

Encouraging Achievements with TTF Project

With the above techniques, we significantly increase thermal conductivity and educe FE. Two 9-cell TTF prototype cavities reached respectively Eacc = 16 MV/m and 21 MV/m both with Q > $6x10^9$ in CW mode. The first series TTF 9-cell cavity (RRR=350) reached Eacc=26 MV/m in 1 ms RF power length (TESLA operational condition) and 22 MV/m in CW [27]. Also, a TTF injector cavity provided by Saclay reached 21MV/m, Q >

Table 5. Key technologies of cavities processing for highest Eacc

Techniques	Short Description	Impact on FE	Impact on TB	Results
Clean cavity handling	Class 10-100 clean room for chemistry, rinsing & assembling	Eliminating foreign particle contamination	Deep chemistry may remove some impurities	Good for Q and Eacc
Heat treatment + Ti-purification in UHV oven	Heat cavity at 1400 C for 1-4 hrs, Ti vapor coated in vacuum	Greatly reduce emitters existing on cavities	Improve RRR by a factor of 2, (improve K)	Substantially increase Q and Eacc
High pressure water rinse[24]	Rinse cavity by high purity water (15 MΩm) at 200 bar	Reduce emitters density		Increase Q and Eacc
High RF power processing	Apply high RF power pulse (1-2 MW, μs-ms) to cavity surface	Eliminate emitters in situ		Substantially increase Q and Eacc

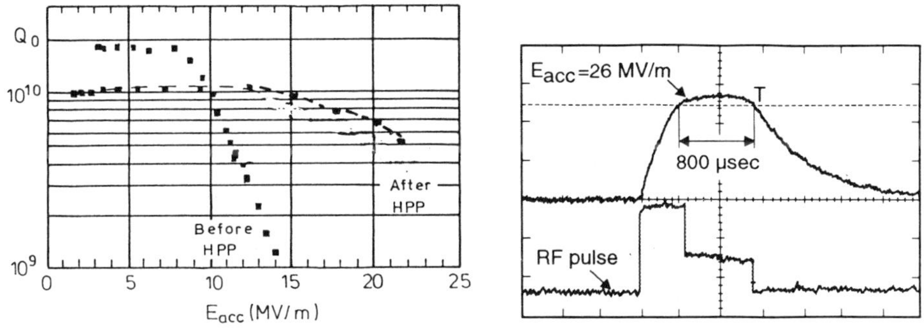

Figure 8. (A) Q vs. Eacc plots of TTF 9-cell cavity -1. (B) The performance of cavity D2 in pulse RF condition similar to TTF operation. 26 MV/m of Eacc is reached.

4×10^9. Compared to the TTF goal of Eacc=15 MV/m and Q=3×10^9, these initial result is very encouraging. Continuing progress is expected with improvements in processing technologies and diagnostic testing. Some representative results are plotted in figure 8.

TESLA PROTOTYPE CRYOMODULE

The cryomodule is 12 m long and consists of 8 SRF cavities with LHe vessels, one quadrupole magnet package, associate fixtures (RF couplers, tuner, alignments, etc.) and a cryostat in which all the above components are housed. The TTF cryomodule also has some special requirements due to SRF cavity technologies. For instance:

(1) To eliminate FE, the eight cavities and quadrupole package must be assembled together in a clean room and be inserted in the cryostat as a single UHV tight unit.

(2) To improve cavity Q, magnetic shields are needed to reduce residual earth magnetic fields to less than 20 mGauss.

(3) Due to beam dynamic consideration, the axes of the cavities to the ideal beam axis need to be within ≤± 0.5mm.

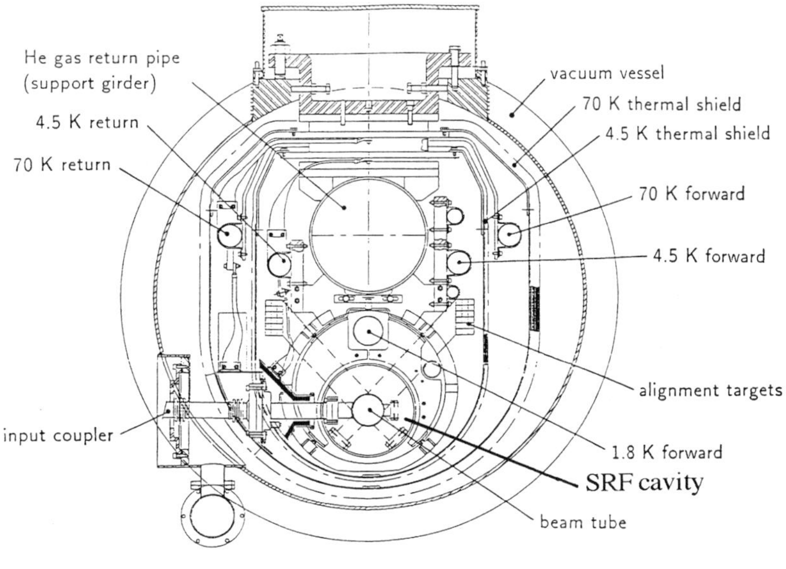

Figure 9. A cross section of TTF cryomodule.

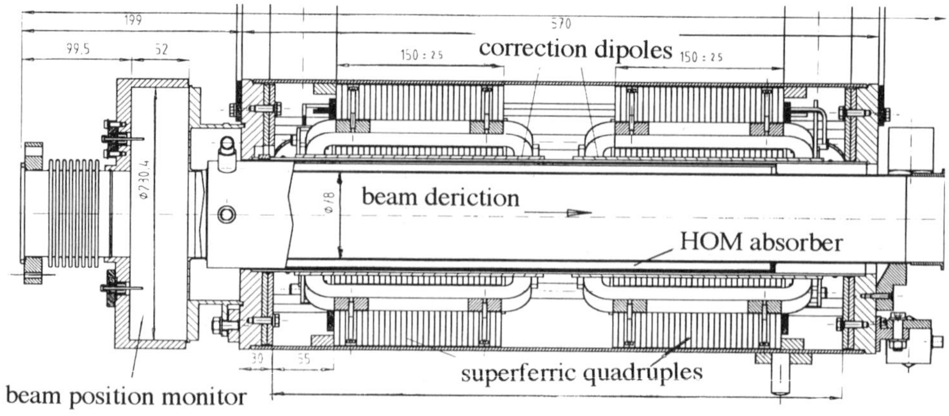

Figure 11. Overview of the TESLA Test Facility (TTF) at DESY.

Based upon cost effective design philosophy, most of the needs for keeping such special technical features must be addressed in the cryostat design. Figure 9 is a cross section of TTF cryomodule [3].

TTF Cryostat

The technical features and design details of the cryostat are discussed in the references [28, 29]. The total heat load to 2K from a cryomodule is estimated to be 21.4 W. The anticipated static heat load budget for one cryomodule is 4W at 2K, 14W at 4.5K and 120 W at 70K [28, 29 30]. Since dominate heat load in cryomodule is RF dynamic heating, nott the static heat leak, the primary interest of measuring the heat load is to verify the RF dynamic load. With the heat attributed to the RF loss, we can calculated the Q of each individual cavity or cavities. A comprehensive verification plan have been developed[3] (with about 135 thermometers, 2 accelerometers, etc.) that will allow us to study the heat leak through the power and HOM coupler and the thermal performance of the cryostat as well.

Special Magnetic Shield

The trapped magnetic flux from ambient fields (even less than 3 gauss) during cool down will seriously impact the cavity performance. The trapped flux will raise the RF

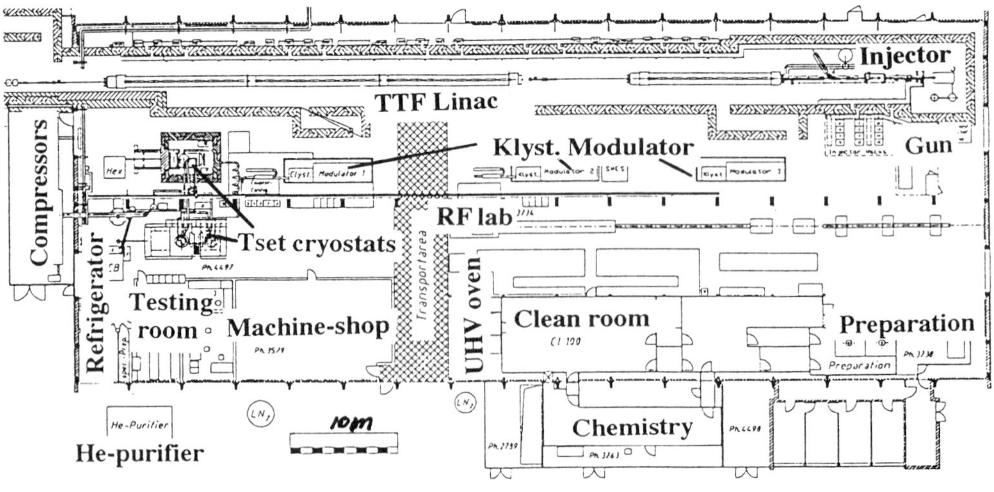

Figure 10. A quadrupole package.

power dissipation and gives an equivalent local residual resistance about Rs ≅ Rn(H/Hc2)sinα. For a TESLA cavity (RRR=300, f=1.3 GHz), a conversion factor of 0.35 nΩ/mGauss closely matched experimental data. In order to get Q = $5x10^9$ as specified for TESLA, surface resistance can not be larger than 25 nΩ, equivalent to 70 mGauss. Considering other contributions to the residual resistance and possibly increasing the Q, the remaining field around the TESLA-TTF cavities should be ≤ 20 mGauss. There are two shielding approaches in use: passive shields made of Cryoperm and active cancellation coil, both of which have been designed and examined[31].

Quadrupole Package

There is a quadrupole package [32] as shown in Figure 10, located at the end of each cryomodule, which includes: (1) a superferric quadrupole doublet (maximum gradient of 20 T/m and integrated gradient of 3T at 55.7 A) enclosed in a 4K LHe vessel, (2) two pairs of single layer dipole steering windings inside the quadrupole yoke bore, (3) a section of beam tube equipped with a beam position monitor (BPM) and HOM absorber (20 W), and (4) the He gas cooled power leads (0.1 g/s) for energising the magnets. At the end plate, there are two arms each holding reference targets for alignment.

Power and HOM Couplers

The TESLA input couplers [33, 34, 35]. are articulated with bellows to allow for the cavity moving during cooldown. The couplers also have an adjustable external Q of over a factor 10 range. The allowed RF heat load and heat leak are very low. All the above features make the coupler mechanically complicated. The input coupler is directly connected to the cavity cut off tube. It does not penetrate the LHe vessel, but is thermally anchored to the radiation shields with radiation cones. The thermal budget is 6W at 70 K, 0.5 W at 4.5 K and 0.06 W at 2 K. For TTF, two 5 MW peak power klystrons TH 2104 C from Thomson will power the four cryomodules and an extra 300 kW pulse klystron is needed for the injector .

To restrict the multi-bunch phenomena due to wakefields, the higher order modes (HOM) of the TESLA cavities must be damped (with Qext = 10^4-10^5) by two coaxial HOM couplers for each cavity [35]. The accelerating mode is not affected by the HOM couplers.

Capture Cavity

Besides the standard cryomodule there is an important cryo-component, called the capture cavity [36] in TTF. The capture cavity terminates the bunching upstream and provides the necessary energy for injection into the first cryomodule. The capture cavity is a standard TESLA 9-cell cavity and is installed in a separate cryostat at the end of the injector. It shares a common cryogenic feed box with the other four cryomodules. The capture cryostat only has an 80K insulation shield with 40 MLI layers. Epoxy-fiberglass rods are used instead of posts. The focusing superconducting solenoid magnet is conduction cooled from the 4K loop.

TTF STATUS

A 3000 m^2 of building, Hall III, has been assigned to host the TTF Linac and TTF infrastructure [5, 25]. We plan to deliver a beam through injector and first cryomodule with energy of about 140 MV/m by the end of 1995 (or early 1996). The complete 500 MeV TTF Linac will be commissioned in 1997. Figure 11 is an overview of the TESLA Test Facility.

For cavity processing, a clean room of $300m^2$ (including a class 10 area); a chemistry etching facility with purity standards equal to semiconductor industry; a high pressure ultra-clean water rinsing system (18 MΩ-m, 200 bar); an UHV furnace with Ti-gettering purification fixture at about 1400° C, 10^{-7}mbar; a high power RF pulse processing system with up to 2 MW peak power; and a modulator and klystron of 4.5MW, 2ms have been installed and commissioned.

For testing, a LHe II cryogenic system with a warm vacuum compressor assembly for handling 10 g/s He at a pressure of 10 mbar and cooling capability of 100 W at 1.8K (at maximum liquefaction of about 5 g/s), 400W at 4.5K and 2000W at 70K is in normal operation. In the future, the plant will be expanded to 200 W at 1.8 K with 10 g/s liquefaction [30]. Two vertical test cryostats (one in use) and a horizontal test cryostat with RF system are installed. A rotating temperature and radiation mapping diagnostic system is also used to analyse the cavity performance. Control system is under development.

ACKNOWLEDGEMENTS

The author sincerely thanks many colleagues for the fruitful discussions and valuable information from CEBAF, CERN, Cornell, Darmstadt, DESY, INFN, KEK, Orsay and Saclay in the TESLA collaboration. This assistance enabled the author to contribute this paper to CEC/ICMC-95, Columbus Ohio, USA.

REFERENCES

1. H. T. Edwards, TESLA Parameters Update - TESLA 94-22, DESY, August, 1994.
2. R, Sindelin and M. Tigner, TESLA Parameters & Comparisions, Proc. 5th workshop on RFS, Ed. by D. Proch, 1991, Hamburg, Germany
3 D. Edwards (editor), TESLA TEST Facility Linac - Design report, TESLA 95-01, DESY, March 1995.
4. D. Proch (editor), 5th workshop on RFS, ED. by D. Proch, 1991, Hamburg, Germany
5. M. Leenen, 4th European particle accelerator conf. 1994. to be published.
   H. Weise, 4th European particle accelerator conf. 1994. to be published.
6. R. Brinkmann, Proc. Particle Accelerator Conf. Dallas, USA, 1995, to be published.
7. D. Proch, DESY, private communication, 1995.
8. G. Horlitz, CEC/ICMC-95
9. W. Weingarten, CERN, private communication, 1995.
10. K. W. Shepard et al. Proc. 6 th workshop on RFS, ed. by R. Sundelin, CEBAF, USA, 1994.
11. C. Reece. Proc. PAC-95 conf. Dallas, TX. USA, to be published.
12 G. Gavallari et al, ibid ref. 10, 1994
13. H. Padamsee et al, ibid ref. 10, 1994
14. H. D. Graef et al, ibid ref. 10, 1994
15. B. Dwersteg et al, 4th European particle accelerator conf. 1994. to be published.
16. K. Akai, ibid ref. 4, 1992
17 B. Bonin et al, ibid ref. 5, 1994
18. H. Padamsee, Applied superconducting conf. Boston, 1994
19. Q. S. Shu et al, IEEE transaction, Vol. 27, No. 2, 1991.
20. B. Bonin et al, ibid ref. 10, 1994
21. G. Mueller , R. Roeth, ibid. ref. 4, 1992 and ref. 10, 1994.
22. Q. S. Shu et al, IEEE transaction on magnetics, Vol. 25, No 2, 1989
23. Q. S. Shu et al, Proc. PAC-95 conf. Dallas, TX. USA, to be published.and TU-A3 this conf.
24. H. Padamsee, et al, IEEE Tran. Mag. -21, 1007, 1985
   P. Kneisel, J. Less Common Metal, 139, 94, 1973.
   D. Trines, A. Matheisen, et al, TESLA - TTF Meetings, 1994 and 1995.
25. S. Wolff, Proc. PAC-95 conf. Dallas, TX. USA, to be published.
26. J. Graber, Ph. D. Dissertation, Cornell Univ. 1993.
27. W-D. Moeller et al, TESLA internal meeting and TTF meeting, 1995.
28. T. Nicol, this conf. 1995.
29. F. Alessandria, this conf., 1995.
30. B. Petersen et al, this conf. 1995.
31. M. Bolore et al, TESLA Report, 94-23, 1994.
   H. Haiser et al, TESLA internal meeting, 1995.
32 A. Koski, B. Bandelman, S. Wolff, MT-14, Finland, 1995.
33. K. Koepke, this conf., 1995.
34. M. Kuchnir et al, this conf., 1995.
35. B. Dwersteg, TESLA internal meeting, 1994 and 1995.
   A. Sekutowicz. TESLA internal meeting, 1994 and 1995.
36. S. Buller et al, this conf., 1995.

TESLA TEST FACILITY ALTERNATE CRYOSTAT DESIGN

T.H. Nicol

Fermi National Accelerator Laboratory
Batavia, IL, 60510, USA

ABSTRACT

Collaborators on the design of a Tevatron Superconducting Linear Accelerator (TESLA) are working toward construction of a test cell consisting of four full length cryostats, 12 meters long, each consisting of eight, 9-cell superconducting RF cavities. The purpose of this facility is to test all aspects of the accelerator system design; vacuum, cryogenics, RF, and electron source, prior to initiating construction of the full linac. The design for these cryostats pose many interesting challenges to cryostat designers. The systems must be capable of supporting all eight cavity structures within tight alignment tolerances, impose very low thermal heat loads on the 1.8K cryogenic system, provide strength and stiffness to resist structural loads during fabrication, shipping, and installation, and be manufactured at low cost. Several design options are being explored, each of which attempt to address requirements imposed by the reference design guidelines. This paper describes the design and analysis of one design alternative.

INTRODUCTION

During the course of early work by the TESLA collaboration, two working groups formed charged with the design of cryostats for use in the TESLA Test Facility (TTF). One group worked toward final development of ideas originating at DESY prior to the formation of the full collaboration. A second group worked on ideas based on the collective experiences gained during development of various cryostats at Fermilab, CEN-Saclay, DESY, INFN-Frascati, and E. Zanon SPA. The latter is an Italian industrial company which fabricated cryostats for HERA at DESY. This paper describes the baseline design, now fully developed and being fabricated for use in TTF as well as alternatives to various elements of that design in various stages of completion

Table 1. TTF Cryostat Miscellaneous Design Parameters	
Heat load to 1.8K	3 W
Heat load to 4.5K	14 W
Heat load to 70K	80 W
Vertical dynamic load	1 g
Lateral dynamic load	1 g
Axial dynamic load	1.8 g
Primary structural resonance	25 Hz
Cavity alignment tolerance	±0.5 mm

which are available for use if full testing of the baseline indicates any need for further development. Some of the more critical design specifications applicable to any cryostat design effort are outlined in Table 1.

OVERVIEW OF THE BASELINE DESIGN

The baseline design is described in the literature, but is briefly outlined here for reference.[1,2] Figure 1 illustrates a typical cross section through the baseline cryostat. Figure 2 illustrates a longitudinal view of the same design. These figures are similar to, but not necessarily identical to the final design and do not include a depiction of the quadrupole due to lack of detail by this author.

Each cryostat contains eight superconducting RF cavities, their respective helium containment vessels, RF tuning systems, and input couplers, a gas helium (GHe) return header, two thermal radiation shields, cryogenic service pipes and connections, a structural support system, a single superconducting

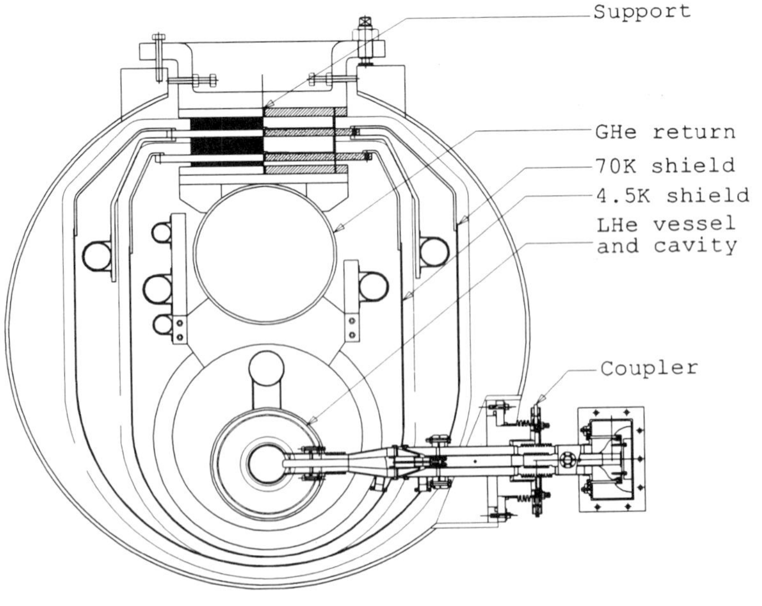

Figure 1. Baseline cryostat cross section.

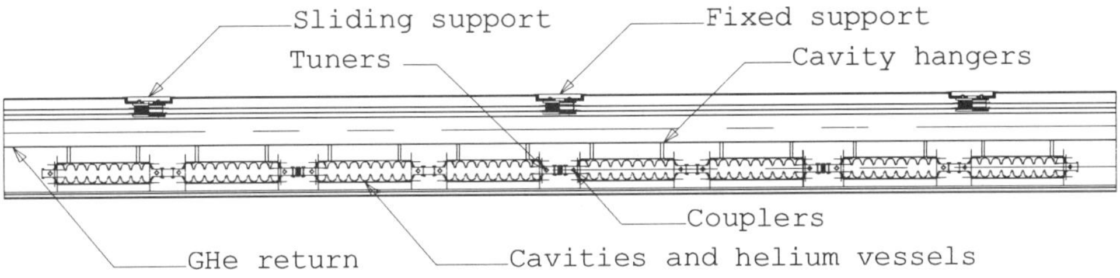

Figure 2. Baseline cryostat design longitudinal view.

quadrupole, and vacuum vessel. The GHe return tube is attached to the vacuum vessel by means of three support posts similar to those developed for use in the Superconducting Super Collider (SSC) dipole magnets. The posts also serve to support the 4.5K and 70K thermal radiation shields, multilayer insulation, and cryogenic pipes. The cavities, helium vessels, and quadrupole are all hung from the GHe return tube. Input couplers which serve as coaxial connections between the RF system and the cavities are attached to one end of each cavity at each of the eight internal cavity interconnections. By virtue of there being three external supports and the use of the GHe return as a structural girder, the design of the input couplers must accommodate up to 15 mm of axial contraction during cooldown of the cryostat assembly.

GOALS OF THE ALTERNATE DESIGN

Alternatives to the baseline design have been encouraged throughout the collaborative effort as a means of ensuring that options exist should they become necessary at some point in the project. The following represent the main goals of the alternate cryostat design effort as well as a brief rationale for each. The content of the goals do not represent areas of concern with the baseline design, but rather are areas in which different design philosophies can be exercised without making significant perturbations to the general cryostat cross section.

- The design should accommodate a rigid or nearly rigid input coupler. This requires a minimum of four and possibly five supports and most likely implies “back-to-back” mounting of the input couplers, that is, two couplers at every other internal interconnect as opposed to one coupler at each internal interconnect as in the baseline design. The hope is that by minimizing the axial contraction during cooldown at the input coupler locations one can simplify the design of the coupler itself. These devices are very sophisticated and are prone to sparking and multipacting, some of which may be due to bellows required to accommodate axial movement. No design may be truly rigid due to radial shrinkage and required allowances for assembly tolerances, however, flexibility requirements for these allowances are small.

- The design should attempt to structurally decouple the GHe return tube and the cavity vessels. Some concern exists that thermal distortions during cooldown will affect assembly alignment of the cavities.[3] It seems plausible to connect the cavity and helium vessel assembly directly to the supports as a means of minimizing changes in cavity alignment during cooldown.

- An attempt should be made to minimize the liquid helium inventory. Early helium vessel designs were substantially larger in design than the cavity to allow for two-phase flow over the top of the cavity. This resulted in a helium inventory significantly higher than that necessary to maintain the cavity temperature at 1.8K. Very early in the process of developing alternate designs, the helium vessel diameter was reduced significantly and a small connecting tube added above the vessel to allow for the same two-phase flow.[4] Since that time a very similar configuration was adopted for the baseline design. The revised vessel is shown in Figure 1. The large annular space around the helium vessel was necessary for the earlier design and has been retained in the TTF modules to minimize required design changes in other system components.

- The overall cryostat (vacuum vessel) diameter should be reduced as much as possible out of consideration for reducing overall cost and facilitating tunnel accessibility.

- Every effort should be made to minimize cost. It is safe to say this has also been a goal of the baseline design. Cost minimization in the alternate design is simply a continuation of that same philosophy.

- The axial and lateral suspension stiffnesses should be made as high as possible, while maintaining heat loads to the thermal intercepts at or below design limits. Although difficult to quantify specific goals, recent design studies and field tests on SSC dipoles at Fermilab indicate that high suspension system stiffness simplifies alignment and minimizes perturbations to system components during shipping and handling. Flexure of the GHe return defines the lowest frequency dynamic modes of the internal cryostat structure.[5] If the number of supports are increased and the GHe return is structurally decoupled from the cavity helium vessels it is possible to drive up the lowest modal frequencies.

CONSTRAINTS ON ALTERNATE DESIGNS

The goals in any design project must be consistent with a set of design constraints. The alternate cryostat design is no different. The constraints listed below represent a starting point for any responsible cryostat design effort which is aimed at providing alternatives to the TTF baseline.

- Any new design must be compatible with cavities being procured by DESY and INFN.

- The helium vessel should be modeled after that currently being designed at DESY, i.e. it should be of the same material, be nearly the same diameter, and interface to the cavity in the same way.

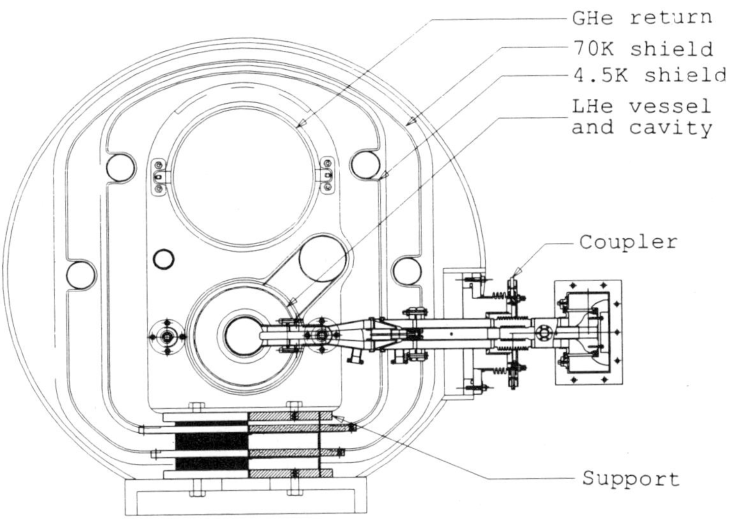

Figure 3. Alternate cryostat design end view.

- The helium vessel and cryostat layout should be compatible with the cold tuning system being developed at CEN-Saclay.

- The piping system in any new design must be compatible with the TTF cryogenic system. It is too severe to dictate that any new design have exactly the same interconnect. However, it should have the ability to connect to a baseline cryostat with an adapter assembly.

- The design must include consideration of the quadrupole being designed at DESY.

- Cryomodule-specific design requirements must be met. For example, shipping and handling requirements must be met, heat loads should be consistent with those outlined in the design criteria, pipe sizes should be consistent with required flow rates, etc.

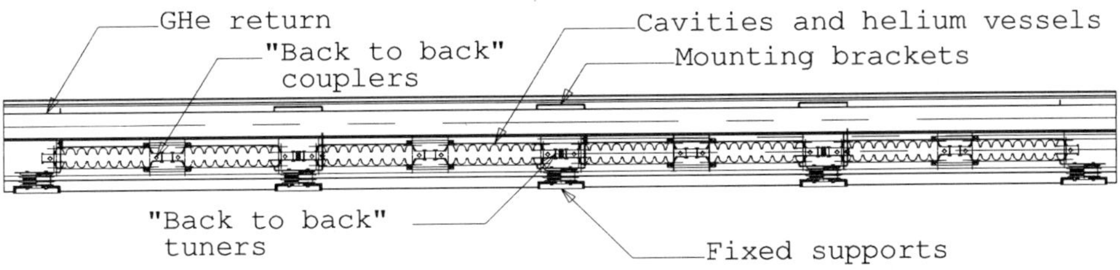

Figure 4. Alternate cryostat design longitudinal view.

OVERVIEW OF THE ALTERNATE DESIGN

Figures 3 and 4 illustrate a cross section and longitudinal view of the alternate cryostat design for TTF. Clearly they are similar to their baseline counterparts in Figures 1 and 2, but there are some important differences. A detailed comparison of the baseline and any alternate cryostat design is not possible due to the fact that few detailed alternate designs exist. However, it is possible to describe design differences in general terms. The primary departures in design originate in the orientation and location of suspension system components. This departure is key because it gives rise to the ability to control cooldown contractions, the ability to decouple the GHe return from the balance of the internal components, and the ability to control the dynamic response.

As described above and shown in Figures 1 and 2, the baseline uses three supports located on the top of the vacuum vessel and connected to the GHe return at three points. The center attachment serves as the axial anchor point. The outboard two attachments are fixed vertically and laterally, but are free to slide axially to accommodate cooldown contraction. The cavity helium vessels are suspended from the GHe return by hangers.

In contrast, the alternate design uses five supports fixed rigidly to the bottom of the vacuum vessel. As shown in Figure 4, this allows rigid connections at one end of each cavity helium vessel which, in turn, enables one to contemplate couplers which need be flexible at levels significantly less than the 15 mm required by the baseline design. The GHe return is attached to a mounting bracket attached to the top of each support and allowed to "float", decoupled structurally from the cavity helium vessels. As a direct result of the bottom mounted supports, the effective stiffness of the suspension is increased from the point of view of the cavity vessels due to the reduced lever arm between the base of the support and the centerline of the cavity. The individual supports for each design are identical with the exception of the composite tube thickness. The thickness is scaled by the number of supports so that the conduction heat loads for each case are identical.[6]

There are of course potential problems with the alternate design. Support for the free ends of the cavity helium vessels and allowance for cooldown contraction are not as straightforward as in the baseline due to the alternate design not relying on the GHe return for structural support. The ends of the cavity helium vessels between supports must be connected to each other by a bridge to effect a rigid interconnection. In addition, the increased number of supports implies more penetrations through the thermal radiation shields, more supports, and more penetrations in the vacuum vessel. The advantages in terms of simplified input couplers and greater suspension system stiffness may prove worth the added complications. A responsible determination can only be made after a detailed design, analysis, and cost estimate is done.

SUMMARY

Fabrication of baseline design cryostats for TTF is well under way. The first completed cryomodule is scheduled to be shipped to DESY from Zanon in Schio, Italy in the fall of 1995. Every expectation exists that the baseline design will perform its intended function well and will meet the design specifications. As with any design and production process, the possibility exists that changes will be indicated for either the balance of the TTF cryomodules or for production modules for the remainder of the linac. In the event that either of those things happen, it is hoped that the body of work done on alternate designs can serve as a basis for implementing any required changes or incorporating added functionality.

ACKNOWLEDGMENTS

The author would like to thank the TESLA Collaboration for the opportunity to work on this project and especially fellow collaborators at Fermilab, DESY, INFN, CEN-Saclay, and Zanon for their constructive criticism, guidance, invaluable feedback, and sharing of experiences throughout the course of this design effort.

REFERENCES

1. D.A. Edwards, editor, "TESLA Test Facility Linac - Design Report," TESLA Collaboration Report 95-01, March 1995.
2. F. Alessandria et al. "Design, Manufacture, and Test of the TESLA TTF Cavity Cryostat," presented at CEC/ICMC 1995.
3. T.H. Nicol, "TESLA Test Facility Cryostat Gas Helium Return Tube Thermal Gradient Analysis," TESLA Collaboration Report 94-12, May 1994.
4. H. Kaiser, Ge. Meyer, H.B. Peters, and G. Weichert, "Helium Vessel for the TTF Cavity," TESLA Collaboration Report 94-26, October 1994.
5. T.H. Nicol, "TTF Cryostat Modal Analysis," TESLA Collaboration Report 94-30, December 1994.
6. T.H. Nicol, "TESLA Test Cell Cryostat Support Post Thermal and Structural Analysis," TESLA Collaboration Report 94-01, January 1994.

STATUS OF THE TTF CRYOGENIC SYSTEM

G.Grygiel[1], U.Knopf[1], R.Lange[1], B.Petersen[1],T.Peterson[2], D.Sellmann[1], and J.Weisend II [1]

[1]Deutsches Elektronen-Synchrotron (DESY)
Hamburg, 22603, Germany
[2]Fermilab, Batavia, Illinois 60510, U.S.A.

ABSTRACT

As part of the TESLA-Collaboration, components of a 1.8 K cryogenic plant, designed for testing superconducting RF cavities at the DESY laboratory, have been designed and built at the CEN SACLAY, LAL ORSAY, FERMILAB, INFN FRASCATI and DESY. Some additional components are being fabricated at IHEP PROTVINO and industry. The general layout of a 1.8 K Test Facility for the test of superconducting RF cavities has been described earlier[1].

The first stage of the cryogenic facility has been brought into operation. During tests of the prototype cavities, the cryogenic performance of the main components of the cryogenic plant was monitored and is presented. This includes vacuum compressors, vertical and horizontal test dewars, a collection box and a helium heater.

For the second stage, required for the operation of the TTF-500 MeV-linac, the construction of the first module cryostat, the capture cavity cryostat, the 6-fold transfer line and the TTF-feedbox is almost finished. The construction of TTF-module transfer lines, TTF-module end caps and a large low pressure heat exchanger has begun. The design of these components and the present status of the installation at DESY is reported.

INTRODUCTION

The cryogenics of a 1.8 K TESLA Test Facility (TTF) for the test of superconducting RF cavities and the operation of a superconducting 500 MeV test linac has been described recently [1,2]. After the rebuilding of a 900 W 4.4 K helium plant and the installation of a vacuum compressor system and a helium distribution system, a vertical test dewar[3] and a horizontal test dewar (CHECHIA)[4] were brought into operation.

The vertical dewar is used to test the RF performance of single superconducting RF 9-cell cavities at a temperature of about 1.8 K in a liquid helium bath. In addition High Peak Power (HPP) processing and temperature mapping measurements take place in this dewar[2]. In CHECHIA single cavities are tested again, after they have been equipped with their individual helium tanks , the main RF couplers and the higher order mode (HOM) couplers. After the installation of the main couplers, the RF quality Q_{RF} can only be detected by cryogenic heat loss measurements.

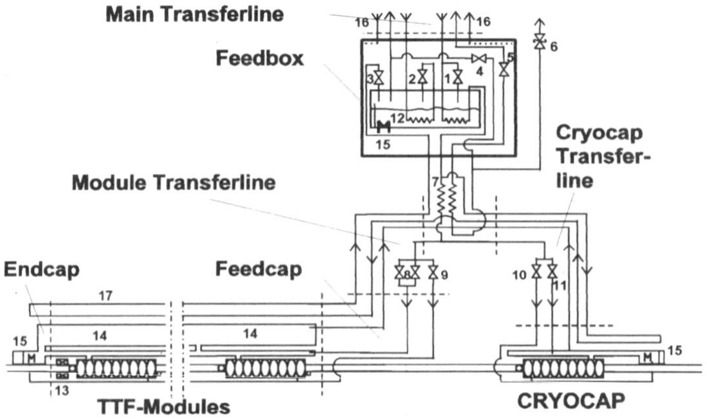

Figure 1. Simplified flowscheme of the cryogenic supply of the TESLA Test Facility.
(1) 5 K supply Joule Thomson (JT) valve, (2) 7.5 K supply JT-valve (from large heat exchanger), (3) 4.4 K return JT-valve, (4) cool down bypass valve, (5) 1.8 K return isolation valve, (6) relief valve system, (7) low temperature heat exchanger, (8) 1.8 K supply JT-valves, (9) cool down bypass valve, (10) cool down bypass valve, (11) 1.8 K supply JT-valve, (12) subcooler, (13) module quadrupole, (14) ∅ 300 mm module return tube, (15) liquid level sensor and heater, (16) 70/80 K thermal shield supply (details not shown), (17) 4.4 K thermal shield.

The continuous supply of 1.8 K liquid helium which is precooled by means of a low temperature heat exchanger and expanded via a Joule-Thomson valve into a helium bath at a constant level is common for all test cryostats of the facility as well as for the later TTF-test linac. During the first tests of cavities experience can be gained in the operation of the Helium plant, which is valuable for the later operation of the other components.
The complete TTF-linac will consist of a beam-injector including a superconducting capture cavity (CRYOCAP) and four module cryostats (see fig. 1). The detailed status of CRYOCAP and the module cryostats is reported separately in this conference[5,6,7]. Each module cryostat includes eight superconducting cavities and one superconducting quadrupole. The first module cryostat[6,7] is under construction at INFN. The module

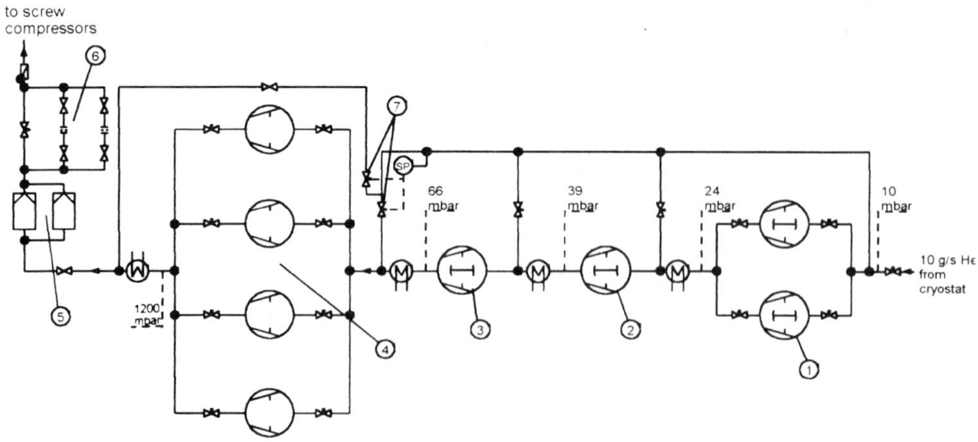

Figure 2. Vacuum compressor assembly. The vacuum compressor assembly consists of two parallel RA16 000 (1), one RA 13000 (2), one RA 9001 (3) rootsblowers and four parallel SV 1200 rotary vane pumps (4), a set of oil filters (5), warm mass flowmeters in the range 0 - 2 g/s and 2 - 20 g/s (6) and two bypass valves (7). Pressures are indicated for the nominal inlet pressure of 10 mbar and maximal massflow of 10 g/s. (Simplified flow scheme).

vacuum vessel, thermal shields , supports, tubing and a 300 mm diameter return tube will be delivered to DESY for final assembly, were also a chain of eight eqipped cavities and a quadrupole will be inserted into the vacuum vessel.

The operation of the linac will start in several steps. As a first step the injector and one module cryostat will be operated. With the installation of a 100 m long transferline, a TTF-feedbox and the transferlines to CRYOCAP and the first module the cryogenic supply will be similar to the supply of the test cryostats. For the supply of the complete test linac the liquefaction mode of the plant will be changed into refrigeration mode by means of an additional large heat exchanger . The heat load capacity at 1.8 K will thus be enhanced from about 100 W to 200 W .

VACUUM COMPRESSOR SYSTEM

The vacuum pumping assembly is designed to meet the demands of all phases of the TTF project. For helium mass flows in the range of 0.2 to 10 g/s the pressure can be regulated from 10 mbar to 30 mbar with a relative stability better than 1 %. For mass flows lower than 10 g/s the pressure can be lowered to about 1 mbar.

Although cold compressor systems for this pressure range and mass flows up to 10 g/s are available, warm vacuum compressors were chosen for the TTF due to their flexibility and simplicity of operation. After crossing low temperature heat exchangers and a gas heater in the first stage or a large low pressure heat exchanger in the second stage the helium gas is at a temperature of more than 253 K before entering the vacuum compressors.

As shown in fig. 2 the vacuum compressor system consists of four stages. At the low pressure end the gas is compressed by two roots blowers (LEYBOLD RA 16000) in parallel followed by two stages of single roots blowers (RA 13000/RA9001). All roots blowers are dry pumps, only the bearings are oil lubricated. In the last stage four rotary vane pumps (LEYBOLD SOGEVAC 1200) in parallel feed the gas back to the main gas circuit of the helium refrigerator at a pressure of 1.05 to 1.3 bar, which corresponds to the suction pressure of the screw compressors.

The helium mass flowrate is measured at room temperature by one of two parallel flowmeters at the outlet of the vacuum compressor assembly.

Under full stationary load the maximum electrical power consumption does not exceed 150 kW (with 340 kW power installed).

All pump stages are separated by gas coolers.The rotary vane pumps are cooled in addition by oil/water heat exchangers . The oil contamination of the gas is lower than 60 ppm at the outlet of the rotary vane pumps and is further reduced to <1 ppm by additional filter elements to avoid any risk of oil condensation in the cold parts of the refrigerator. Breox B 35 oil is used in all vacuum compressors as well as in the screw compressors of the refrigerator.

The suction pressure of the vacuum compressor assembly is regulated by two bypass valves in parallel to the rotary vane pumps and the roots blowers respectively.

The whole system is operated by an internal process control system (SIEMENS SIMATIC S5) which communicates with the process computer of the refrigerator. With the desired suction pressure set, the system reacts to changes of mass flow and pressure by operating the bypass valves and switching pumps on and off.

The vacuum compressors are housed in an extra building attached to the TTF-experimental hall but separated by its own foundation with a distance of about 20 m to the test linac. The beam of the linac will be extremely sensitive to any kind of external vibrations. To avoid vibrations as far as possible, the frame of the vacuum compressors is separated from the floor by special damping elements (BARRYMOUNT SLM) in addition, resulting in more than 90 % damping of the vibrations . (The RMS oscillation velocity of 2 mm/s at the frame is reduced to less than 0.1 mm/s at the floor).

The complete vacuum compressor assembly was designed and constructed by the

LEYBOLD AG in Germany and delivered to the TTF-collaboration in time and on schedule.

After the vacuum compressor assembly had passed initial acceptance test on the manufacturer's side and at DESY and an operational test of 1000 hours without severe failures the plant has now reached about 4000 hours of operation.

VERTICAL DEWAR

The vertical test dewar has a depth of 3.8 m and an inner diameter of 0.6 m. During the RF tests of 9-cell cavities the dewar is filled with 620 liters of liquid helium at a temperature of 1.8 K.

The vertical test dewar consists of the dewar itself, the outer vacuum vessel with a thermal shield cooled by liquid nitrogen and the top plates. The supply and return tubes, cryogenic valves and the low temperature heat exchanger are installed outside the helium dewar and inside the outer vacuum vessel, leaving free all space inside the helium dewar for the cavity insert, the RF connections and the vacuum equipment. The transferlines including the vacuum insulated helium pumping tube are welded to the outer top plate of the dewar. A more detailed description of the vertical dewar can be found elsewhere [3]. The dewar has been designed and built at FERMILAB.

Helium of about 4.4 K and a pressure of 1.2 to 3 bar is supplied by the 900 W plant, precooled by a low temperature heat exchanger to about 2.2 K and expanded via a Joule-Thomson valve into the helium dewar. The operation of the JT Valve is controlled so as to keep the liquid level constant during steady state operation. The vapor pressure of the liquid helium bath is lowered to a pressure of 16 mbar by means of the vacuum compressor system. The helium is pumped through the low temperature heat exchanger, the collecting box and the helium heater.

The helium dewar and the insert are equipped with several temperature sensors, a level sensor and pressure gauges. Additional heat load of up to 100 W can be supplied by

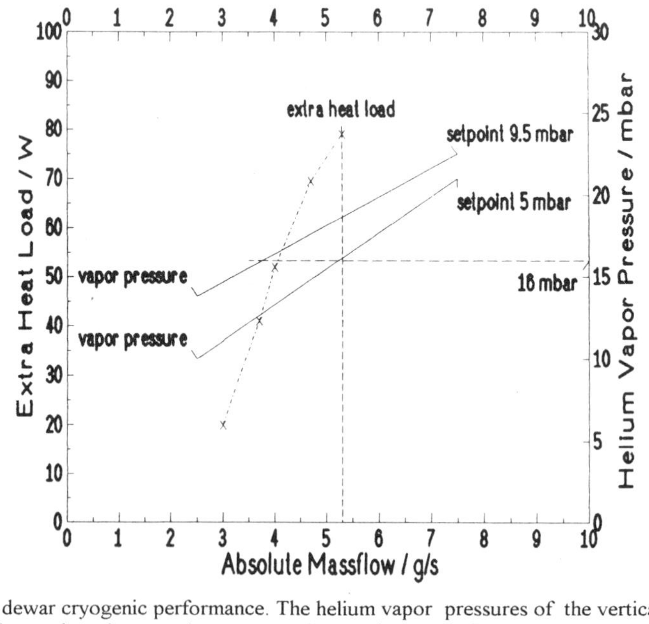

Figure 3. Vertical dewar cryogenic performance. The helium vapor pressures of the vertical dewar and the external heat loads are plotted versus the corresponding helium massflows at two setpoints of pressure at the vacuum compressors. For a given vapor pressure in the dewar the maximum external heat load can be obtained from the diagram. For the design vapor pressure of 16 mbar (1.8 K) 79 W external heat can be applied.

an electrical heater in the bath. This heater can be used complementary to the changing RF-Power to obtain constant heat loads and more stable conditions.

For the performance tests different helium massflows were adjusted by the heat load of the electrical heater in the helium bath. The liquid helium level was kept constant at a reading of 95 % of the level sensor (this means that a 9-cell cavity is covered with about 20 cm of liquid at its top end). Each setpoint was maintained for several hours to insure stationary thermal conditions.

In fig. 3 the maximum obtainable extra heat load as well as helium bath vapor pressures are plotted versus the pumped helium massflow at 9.5 mbar and 5 mbar pressure setpoint at the vacuum compressors. For a vapor pressure of 16 mbar (1.8 K) the maximum extra heat load turns out to be about 79 W at a setpoint of 5 mbar at the vacuum compressors (about 40 W at a setpoint of 9.5 mbar).

Heat loads in the vertical dewar as well as in CHECHIA can be obtained from the measured liquid level drop while the supplies are closed, or from the absolute pumped helium massflow measured at constant liquid level of the helium bath. Using the second method, it has to be taken into account that a fraction of the pumped massflow is caused by the quality of the fluid at the JT-expansion. This quality can be calculated from thermodynamic state functions if temperature and pressure of the helium in the supply of the JT-valve is known.

The total heat loss results from the static heat loss of the helium dewar including the cryostat insert (cavity, waveguide, couplers, vacuum tubes, supports and cables), thermal radiation and thermal conduction across the helium vessel and the tubing connected to the vessel, and the extra heat loads due to RF operation of the cavity (or the electrical heater): During continuous wave RF operation of a nine cell cavity in the vertical dewar the Q_{RF}-value was calculated from the corresponding heat loss measurements in the helium bath. The deviation of the Q_{RF} calculated from electrical measurements to the Q_{RF} calculated from the heat losses is smaller than 10 %. In CHECHIA the Q_{RF} can only be monitored by cryogenic measurements after the main RF coupler has been installed[4].

From the data in fig. 3 a static heat load of 20 - 25 W results for the vertical dewar with a 9-cell cavity if the liquid helium level is in the range of 95 % . This static heat loss can be lowered to about 10 W if the level is decreased further. With a maximum of about 80 W for continuous RF cavity heat load left at 1.8 K, there is enough cryogenic capacity for all RF measurements, HPP preparation and temperature mapping. In addition, was demonstrated during the first cw-tests of the 9-cell prototype cavities that the regulation system will return to stable operation even if heat loads up to 300 W are transferred into the helium bath for some seconds. With minor changes in the design, a second vertical test dewar is under construction at Fermilab.

The continuous supply of helium into the dewar resulting in a constant helium level during the subatmospheric operation has some important advantages in comparison to the more common method of lowering the level during pumping on a helium bath: the time to test superconducting RF cavities at 1.8 K is not limited by an emptied helium bath and about one meter of cryostat length corresponding to about 200 l of liquid helium could be saved. Detailed results of heat loss measurements of the CHECHIA cryostat are presented in a separate paper at this conference[4].

MAIN TTF TRANSFERLINE

For the cryogenic supply of the TTF-linac a 6-fold 100 m transferline is required to connect the 900 W refrigerator, the helium heater (stage I) or a heat exchanger (stage II) and a feedbox close to the first TTF module cryostat. This transferline has been designed and is under construction now at the DEMACO company in the Netherlands. Fig. 4 shows a cross section of this transferline. At the 70 K level 2 W/m and on the 4.4 K level 0.2 W/m are specificated for the overall heat losses of the transferline.

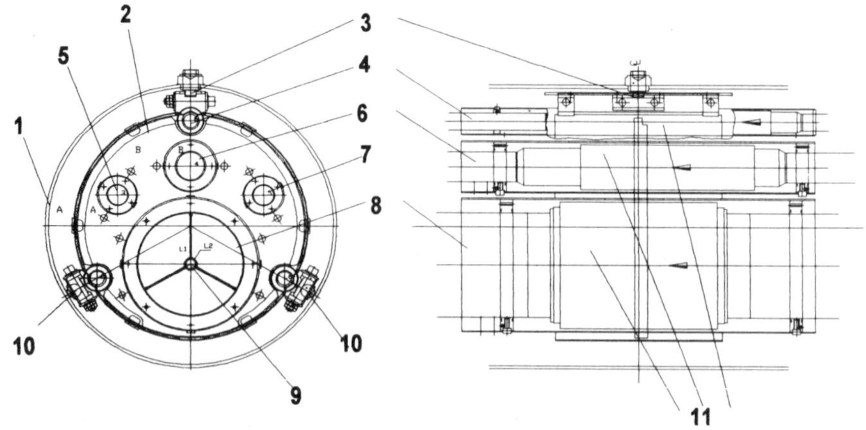

Figure 4. Cross section of main transferline. (1) Vacuum vessel, (2) 80 K thermal shield, (3) glass fiber reinforced plastic (GFRP) suspension, (4) 80 K return tube, (5) 70 K supply tube, (6) 4.4 K return tube, (7) 5 K supply tube, (8) 3.5 K / 16 mbar return tube, (9) 7.5 K supply tube (from large heat exchanger), (10) moving supports, (11) tube compensators.

TTF FEEDBOX

The feedbox is needed to branch the 70 K, 4.4 K and 1.8 K helium circuits to the first module cryostat and CRYOCAP. The simplified flow scheme of the feedbox is shown in fig. 1. In addition the 4.4 K has to be recooled in the feedbox to get stable conditions for both the 4.4 K shield of the modules as well as for the JT-supply of the 1.8 K bath. The

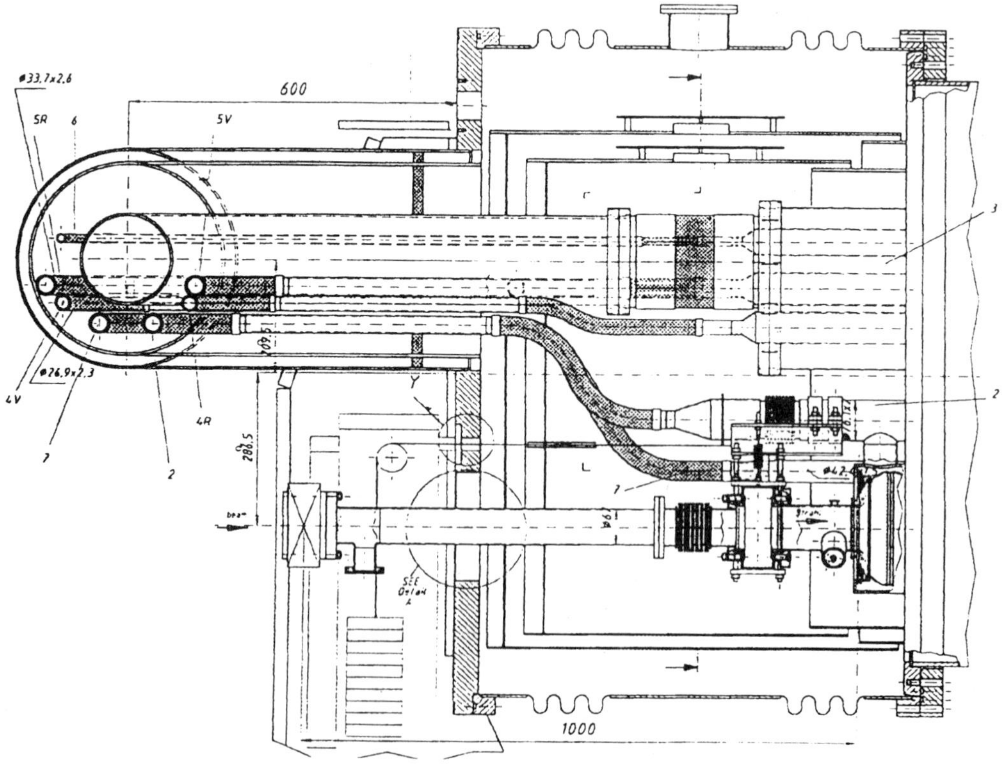

Figure 5. The connection of the module transferline to the first TTF-module (feedcap)

return gas from the 4.4 K shield of the modules is expanded via a JT-valve into the liquid bath of the subcooler. After the installation of a large low pressure heat exchanger in the second stage 7.5 K/14 bar helium coming from this heat exchanger will also be preecooled to about 5 K and than be expanded into the subcooler bath. The low temperature heat exchanger for the supply of the modules is attached to the feedbox. The feedbox has been designed and is under construction at FERMILAB.

MODULE TRANSFERLINES, FEED & END CAPS

Transferlines from the feedbox to the first module (8 m) and to cryocap (3 m) include JT-valves and bypass valves for cool down and warm up (see fig. 1) . By means of flexible hoses only small transversal forces will act on the suspensions of the module cryostat. A sketch of the feed cap connection to the first module is shown in fig. 5. The flanges of feed and end cap have to support 14 000 kg vacuum force. The end flanges are equipped with feedthroughs for optical alignment and a stretched wire monitor system. The module transferlines and the feedcap are designed and constructed at DESY and the endcap at INFN.

LARGE LOW PRESSURE HEAT EXCHANGER

To increase the cooling capacity from about 100 W to 200 W at 1.8 K for the later supply of the complete test linac, the enthalpy of the low pressure gas has to be transformed to a pressure and temperature level which can be used in the cryoplant by means of a Large Low Pressure Heat Exchanger.
In this heat exchanger incoming warm 14 bar helium gas is cooled to a temperature of less than 7.5 K by the counterflow of low pressure gas which enters the heat exchanger at a temperature of about 3.5 K.

At a maximum mass flow of 10 g/s of the low pressure gas the high pressure mass flow has to be tuned to the performance of the heat exchanger resulting in a mass flow of about 9 g/s. The high pressure gas is expanded via a JT valve into the subcooler of the TTF-feedbox (see fig. 1). The high pressure gas flow rate is controlled by the gas temperature in the supply of this JT valve.

At a low pressure mass flow of 10 g/s a pressure drop of only 3 mbar can be allowed, a large thermal stress due to the high temperature gradient is expected. The performance of the heat exchanger will be seriously affected by internal heat flow along the tubes. A small gas heater of 1.5 kW at the warm outlet of the low pressure flow is added to give more flexibility.

The Large Low Pressure Heat Exchanger is installed into a vacuum isolated box,separated from the 4.4 K Cold-Box and from the feedbox of the test linac. IHEP Protvino is in charge for the design of the heat exchanger which will be built in Russian industry.

CONCLUSION

The cryogenic system for the TESLA Test Facility is well underway. The first vertical dewar and CHECHIA have been successfully operated. A second vertical test dewar, the TTF feedbox, main transferline, module transferlines, feedcap and endcap are to be delivered by the end of 1995. Delivery of the large low pressure heat exchanger is planned for summer 1996. Cryogenic operation of the first part of the TTF Linac (first module and CRYOCAP) is expected in the first quarter of 1996.

REFERENCES

1. Horlitz, G. et al.,'A 1.8 K Test Facility for Superconducting Rf Cavities', Adv. in Cryogenic Engineering, Plenum Press, New York, 1993, Vol. 39 , (Proc. of the CEC/ICMC Conference 1993 in Albuquerque, N.M.)
2. Grygiel, G. et al., 'A 1.8 K Test Facility for Superconducting RF Cavities', TESLA-COLLABORATION, TESLA 94-08, April 1994
3. Nicol, T.H. et al, ' TESLA Vertical Test Dewar Cryogenic and Mechanical Design', IEEE Proceedings of the 1993 Particle Accelerator Conference, Vol. 2, pp. 989-991, Piscataway, N.J., 1993
4. Duthil, R. et al.,' Cryogenic and Electrical Test Cryostat for Instrumented Superconductive RF Cavities (CHECHIA)', Adv. in Cryogenic Engineering, this conference
5. Buhler,S.,'Status Report of the TTF Capture Cavity Cryostat', Adv. in Cryogenic Engineering, this conference
6. Nicol,T.H.,' TESLA Test Facility Cryostat Design', Adv. in Cryogenic Engineering, this conference
7. Alessandria,F. et al.,'Design,Manufacture and Test of the TESLA-TTF Cavity Cryostat', Adv. in Cryogenic Engineering, this conference

DESIGN , MANUFACTURE AND TEST OF THE TESLA-TTF CAVITY CRYOSTAT

F. Alessandria[1], G. Cavallari[2], M.Minestrini[3], T.H.Nicol[4], C. Pagani[1], R.Palmieri[5], S. Tazzari[6], G. Varisco[1]

[1]INFN sez. Milano, Milano,Italy
[2] INFN sez. Roma2, on leave of absence from CERN, Geneva
[3] INFN Laboratori Nazionali Frascati, Italy
[4] Fermi National Accelerator Laboratory, Batavia, IL. USA
[5] ENEA Centro di Frascati, Italy
[6] University of Roma Tor Vergata and INFN sez. Roma2, Roma Italy

ABSTRACT

The aim of the TESLA Test Facility (TTF) is to build and to operate a superconducting (SC) linear accelerator accelerating electrons to an energy of approximately 450MeV, to establish the technological basis for a SC accelerating structure of competitive cost, that could reliably achieve accelerating field of about 15MV/m. Eight individual nine-cell, 1300Mhz superconducting cavities, their RF equipment, a superferric quadrupole doublet and dipole steering coils and all diagnostic elements are combined in a 12m long accelerator module, housed in a single cryostat. The TTF linac consists of four such modules. Cavities are cooled by superfluid two-phase Helium at 1.8K and the quadrupole windings at 4.5K.

The liquid Helium distribution and cold gas recovery system are incorporated into the cryostat providing additional savings by the reduction of the number of transfer lines and other hardware on the outside of the cryostat enclosure.

The TESLA project

Factors determining the high cost of present SC cavity installations are their comparatively low accelerating field and that a single or few cavities have been housed in a separate cryostat with its own refrigeration circuit, thus requiring a relatively large number of cold to warm transitions and expensive distribution cold boxes; static losses, proportional to the number of transitions are a further cost factor. The TESLA accelerating structure and cryostat design philosophy has thus been to make individual accelerating modules as long as possible and combine then in strings fed by a single cold box, a technique similar to that used for SC magnets in large proton accelerators.

In the TESLA proposal eight individual nine-cell, 1300Mhz SC cavities, their RF equipment, a superferric quadrupole doublet and dipole steering coils and all diagnostic elements are combined in a 12m long accelerator module, housed in a single cryostat. Twelve cryomodules are combined into a 145 m subsection, individually fed by two phase He; finally a separate cryoplant supplies a string of 12 subsections. A cryogenic unit would have a length of 1880 meters in TESLA.

The TTF accelerating structure will consists of four such modules.

Advances in Cryogenic Engineering, Vol. 41
Edited by P. Kittel, Plenum Press, New York, 1996

Cryostat design

The scheme for cooling the TESLA prototype modules in the TTF test linac corresponds to the cooling process which is envisioned for a future TESLA linear accelerator. The cavities are in a 1.8K helium bath. Through the string of 4 modules between the feed box and the end box, an approximately uniform helium level is maintained. The regulation of the JT valve in the entrance of the modules is via the measurement of the liquid level in the end box. The He vapor at 1.8K returns back to the feed box via a 300mm diameter return line (HeGRP), which at the same time serves as the mechanical support of the cold mass. The design of this return line and the other process line is based on the dimensions of TESLA.

The cavity production technique has been carefully designed and specified to produce absolutely clean surfaces and to keep them such throughout the whole manufacturing and assembly process. This poses significant constraints on the design and assembly together in a clean room as they have to be inserted in the cryostat as a single UHV tight unit.

The design solutions have been chosen for simplicity and ease of fabrication; thus putting some constraints onto other RF components. To minimize the costs, machining on large items such as vacuum vessel and He gas return pipe has been limited to a few vacuum flanges requiring precise positioning. The long string of accelerator modules with a common insulation vacuum minimizes the number of warm to cold transitions. All refrigeration piping run inside the cryostat thus avoiding costly external transfer lines.

The cryostat vacuum vessel, Fig. 5, of carbon steel is evacuated to a pressure of 10^{-6} mbar. The design critical over pressure is 1.25 bar. It is the main mechanical structure and a number of constraints have been put on the shape: position of the power coupler flanges ± 1mm, cylindricity tolerance ≤ 3mm. The actual shape of the prototype vessel is shown in Fig. 1.

The 300 mm Helium gas return pipe, required to recover the 16mbar cold gas evaporated from the 1.8K He bath, is also used as the main support beam for the accelerator module. The overall weight of items hanging from the pipe is about 2300Kg. The HeGRP operating pressure is specified to be 2 Bar, the test pressure 3 Bar .

The HeGRP is supported from above by three support posts that provide the necessary thermal insulation; posts are fastened to large flanges on the upper part of the vacuum vessel, by means of adjustable suspension brackets. The scope of the adjusting mechanism is to allow for the accelerator module axis to be brought into its proper position, independent of the absolute position of the vacuum vessel flanges on which it rests.

In the longitudinal direction the center post is fixed to the vacuum vessel while the two end brackets are allowed to move to accommodate differential shrinkage during temperature cycling. The posts consist of fiberglass pipe terminated by two shrink fit stainless steel flanges. At intermediate positions, optimized to minimize heat leak, two additional shrink fit flanges are provided to allow intermediate heat shield connections to 4.2K and 70K.

Seven additional cooling pipes run through the cryomodule (Fig. 5):

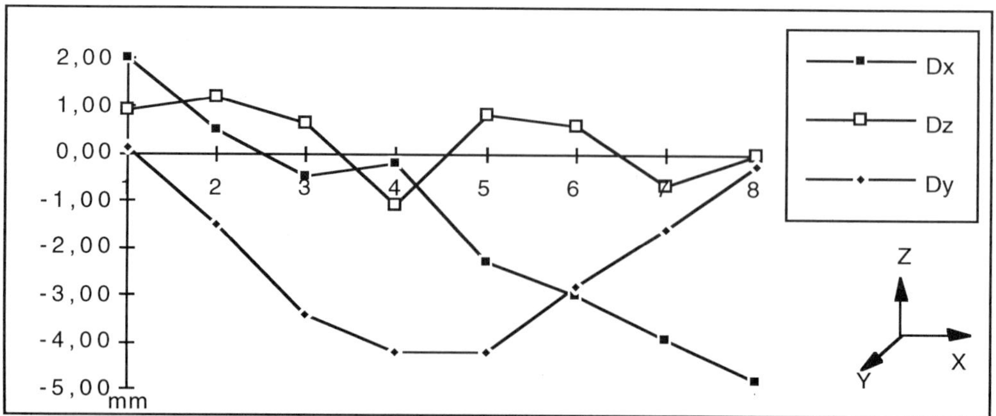

Figure 1. Vacuum vessel: Measured position errors of the Main Coupler flanges The vessel is aligned on the x axis, on the abscissa the cavity number.

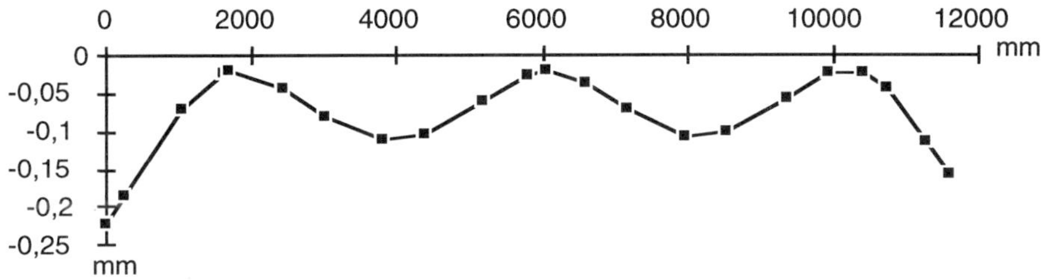

Figure 2 - HeGRP design: computed static deflection of the pipe loaded by the cold mass and resting on three posts.

1.8K two phase line forward supply
2.2 K pressurized He forward, to supply downstream modules
4.5 K pressurized He forward, to cool the quadrupole lens
4.5 K Pressurized He return, to cool the 4.5K shield
70 K He forward, also used to cool the quadrupole beam tube
70 K He return, also used to cool the 70K radiation shield
a warm-up/cool-down line

All pipe connections in between modules are welded; bellows are provided to mechanical decouple modules and to allow for thermal contraction.

Alignment

Another very significant requirement, deriving from beam dynamics considerations, is the overall alignment requirements. The axes of the 8 cavities must stay aligned to the ideal beam axis to within $\pm$ 0.5 mm and those of quadrupoles to within + 0.1 mm over the entire length of the accelerator; in addition the vertical mid plane of the lens must stay aligned to the vertical direction to $\leq \pm$ 0.1 mrad. Cavity positions will be monitored during cool down with optical marks and wire positions monitors (WPM) installed on each cavity of the first prototype. The cavities and the quadrupole are aligned at room temperature before insertion in the vacuum vessel. Reference Taylor-Obson spheres on the posts are then used to align the cold mass in the vacuum vessel.

Modal analysis

The accelerator repetition rate is 10Hz and the RF cycle lasts 1.4 msec. Lorenz forces at 10Hz, induced in the cavities by the RF radiation pressure, may excite mechanical resonance. The aim of the TTF cryostat modal analysis [1] is thus twofold. First to evaluate the static behavior, mainly deflection, due to the lumped masses of the cavities, the quadrupole and the weight of the GHe return tube itself. Fig 2 gives the results.

Secondly to extract all the translational modes. The accelerator repetition rate being 10Hz, all mode above 20 Hz are considered non dangerous. The analysis yielded 12 modes below 100Hz. The modes, frequencies and deflections planes are summarized in table 1. Nearly all the first 12 modes are associated with flexure of the GHe return pipe. The exception is mode 6 at 26.7Hz. This mode is almost exclusively associated with lateral deflections of the support posts. It is of course possible to determine deflections at resonance only with a specific input load spectrum.

Heat treatment

Concern has been expressed about thermally induced distortions of the HeGRP. Specifically, the issue involves thermal gradients which can occur in this tube from flow stratification during cool down. These thermal gradients induce mechanical distortions and thermal stresses in the tube and impose structural loads on the support system[1]. In a stress free tube the bending stresses resulting from thermal bow are small compared to the yield

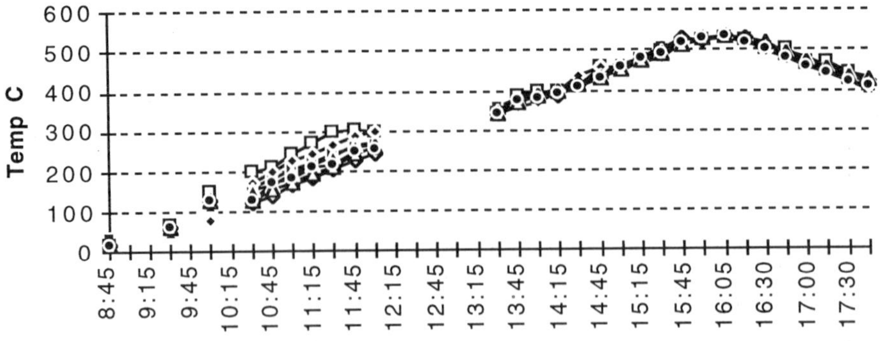

Figure 3. Heat treatement of the HeGRP. The time above 450C was limited to less than 3 hours.

Table 1 - HeGRP Modal Analysis

mode no.	freq. (Hz)	defl. plane
1	18,7	x-y
2	20,3	x-z
3	21,7	x-z
4	22,1	x-y
5	25,4	x-y
6	26,7	x-z
7	27,6	x-z
8	41,9	x-y
9	49,6	x-y
10	54,0	x-z
11	76,0	x-z
12	101,3	x-z

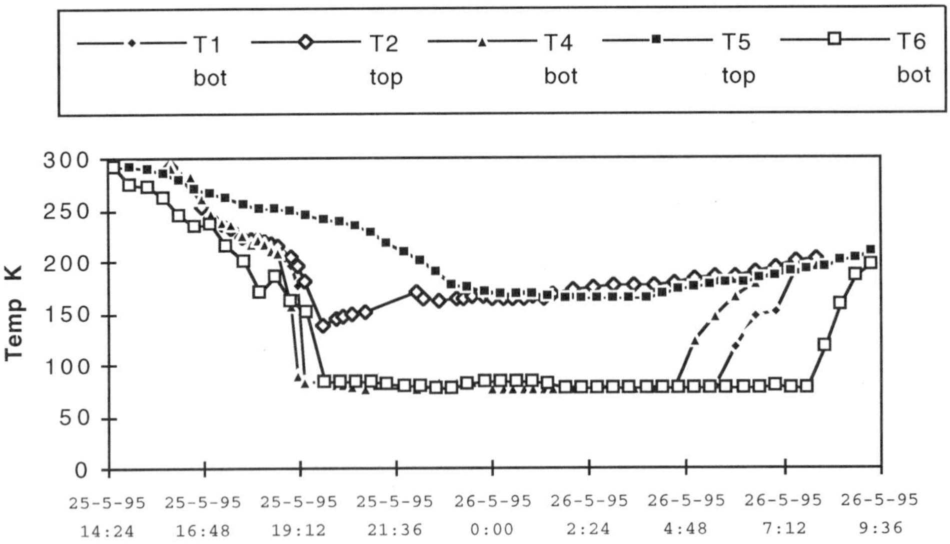

Figure 4. HeGRP cooldown at LN_2. The lower part was at 78K while the upper part was limited to 180K due to stratification of the GN_2.

Table 2.	Estimated cryogenic losses (W)		
	2 K	4.5 K	70 K
Static losses	2,80	13,90	76,80
Total losses at 15MV/m	15,40	15,20	125,00
Total losses at 25MV/m	21,40	16,00	136,00

strength. However, residual stresses from fabrication of the tube and welding of various cryostat components to the tube imply that some parts of the tube may in fact be on the verge of yield. Any additional stresses will result in permanent deformation of the tube. The finished HeGRP has been heat treated to 550 C to relieve stresses, Fig. 3. The thermal cycle was specified such as to avoid carbide precipitation.

After heat treatment, the shape of the HeGRP standing on two supports has been measured with optical means[3], cooled down with LN_2 and re-measured after warm up.

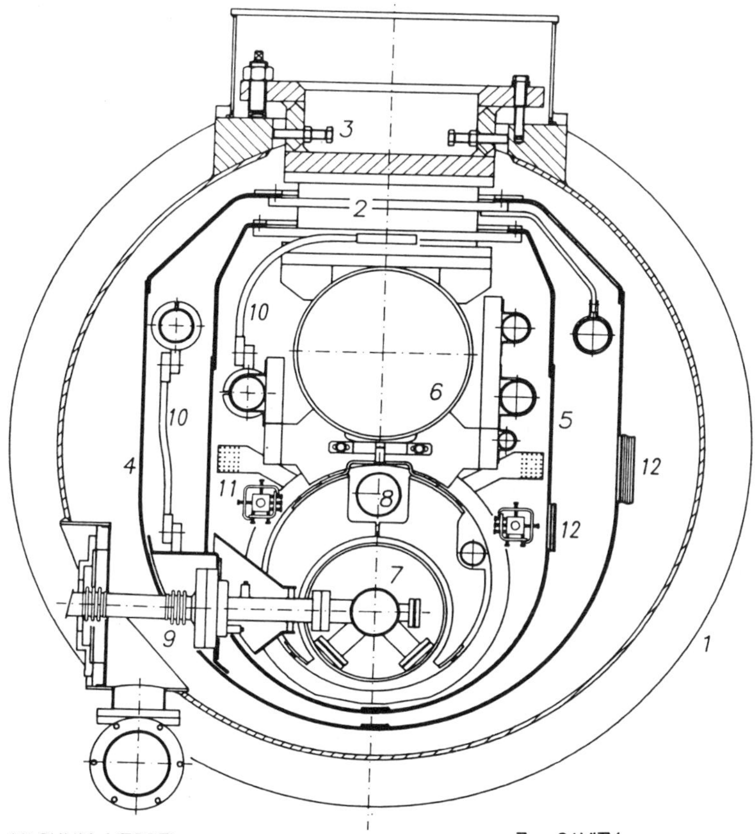

1	VACUUM VESSEL	7	CAVITY
2	COMPOSITE POST	8	1.8 K LHe SUPPLY
3	SUPPORT BRACKET	9	MAIN COUPLER
4	70 K SHIELD	10	Cu BRAIDS
5	4.5 K SHIELD	11	ALIGNEMENT TARGETS
6	1.8 K He GAS RETURN PIPE	12	ML INSULATION

Figure 5. Cryostat cross section at cavity.

Within the measurement precision of 0.15mm no permanent deformations induced by the cool down cycle have been spotted but a deviation from straight line of $\pm$ 1.5mm. The HeGRP was closed with welded flanges, connected to inlet and outlet lines and inserted in a horizontal vacuum vessel. The inlet pipe went all along the HeGRP at the cylinder axis with nozzles every 200mm. Filling with LN_2 was stopped when liquid poured out of the exit line, i.e. when the HeGRP was half full. Temperature gauges were installed at the outer side of the pipe, three on the top and three on the bottom. While the bottom part was at the LN_2 temperature for more then 8 hours, the top part cooled down to only 160K, Fig. 4. It can be explained by a full stratification of the GN_2 inside the HeGRP due to the very slow gas speed.

Radiation shields

Two aluminum radiation shields are provided at intermediate nominal temperatures of 4.2K and 70K. Each shield is made in two halves. The half-shield upper sections, 5mm thick, are supported by the post intermediate flanges: they are bolted onto the center post but can slide on the end posts, to avoid build up of excessive forces on the posts during thermal cycling, when temperature differences of 100 K can develop between different points of the shield sections. The shields are cooled by means of flexible copper braids at regular intervals to the centerline of the shield upper section; the braids are directly connected to the cooling pipes and their cross section is such as to keep the ΔT between shield connection and pipe to $\leq$ 10 K and $\leq$ 2K for the 70K and the 4.5 K levels respectively. The lower half-shields, 3mm thick, are bolted to the upper-half and are cooled by conduction.

The Al 6082 thermal conductivity been measured at 4.2K[3]. The 3mm and the 5mm thick Al sheets have 11 W/mK and 18W/mK respectively. Measurements on different samples cut across and along the rolling direction have given the same values: we conclude this difference to come from the bulk material itself. These values are above the figures assumed for the thermal budget design by at least a factor of 5.

Super insulation

All surfaces at 70K and 4K facing surfaces at ambient temperature or 70K respectively are covered with multi layer insulation (MLI). Ten and thirty layers of MLI are provided on the 4 K and 70 K shields respectively. The estimated heat loads are $0.1 W/m^2$ and $0.9 W/m^2$. The He vessel, the HeGRP and the quadrupole 4K feed lines have a 5 layer blanket, to reduce heat transmission in case of vacuum failure.

The anticipated static and dynamic heat load budget for a single cryomodule is shown in table 2.

Diagnostics

PT1000 and Cernox-1030 temperature sensors will be installed at various places of the cold mass. The Cernox sensors are thin film resistors, are very stable over repeated thermal cycling and can withstand the high radiation doses expected around the SC cavities. The sensors are calibrated from room temperature to 1.7K with a number of calibration points according to the final operating temperature. In the first cryomodule we will install 92 Cernox and 43 PT1000 sensors to qualify the cryogenic design.

Magnetic shield

In order to reach their nominal performance level [5], the superconducting cavities of TESLA must be cooled in a magnetic field $\leq 2\mu T$, this corresponds to a field 20 to 40 times smaller than the ambient magnetic field. To achieve this shielding level it is intended to use a mixed scheme combining a Cryoperm tube around the cavity Helium vessel and a string of Helmoltz coils around the cryomodule vacuum vessel. The former, and the carbon steel vacuum vessel will mainly shield the transverse component of the ambient field, the latter will shield the longitudinal component. The longitudinal demagnetization factor of a long cylinder, as it is the case for a string of TESLA cryomodules, approaches unity.

Acknowledgments

The TTF community at DESY, Hamburg, have participated to the cryostat design with many fruitful discussions and advises. W. Weingarten and C. Dalmas of CERN have kindly performed the thermal conductivity measurements. One of us (G.C.) thanks INFN for the hospitality.

References

1. T. Nicol: TTF Cryostat Modal Analysis; TESLA Rep. 94-30
2. T. Nicol: TTF cryostat: HeGRP thermal analysis; TESLA Rep. 94-12
3. F. Wei: Private communication; 1995
4. W. Weingarten: private communication, 1995
5. M. Bolore et al.: Magnetic Shielding of the TESLA TTF SC Cavities; TESLA Rep. 94-23
6. D.A. Edwards ed.: TESLA Test Facility Linac - Design Report; TESLA Rep. 95-01

STATUS REPORT OF THE TTF CAPTURE CAVITY CRYOSTAT

S. Bühler, P. Blache, R. Chevrollier, T. Junquera[1],
N. Colombel, R. Panvier[2]
J. Gastebois[3]

[1]Institut de Physique Nucléaire
Orsay, 91406 cedex, France
[2]Laboratoire de l'Accélérateur Linéaire
Orsay, 91405 cedex, France
[3]CEA /DAPNIA / SEA
Gif-sur-Yvette, 91191 cedex, France

ABSTRACT

We describe the capture cavity cryostat, for the Tesla Test Facility at DESY in Hamburg, which was designed and is presently under assembly in France. We also discuss the construction of an ancillary feed box which is required for a preliminary cryogenic test prior to delivery.

INTRODUCTION

The capture cavity is the first cryogenic device on the beam line of the Tesla Test Facility (TTF) at DESY in Hamburg and will connect the "warm" injector to the first cryomodule.

Design, manufacture and integration of the different parts of the cryostat are substantially done in France and will be completed locally in order to perform two cryogenic tests prior to the delivery to DESY. This decision, however, implies the construction of major ancillary equipment for supply and control of the cryogenics in this first phase.

REFRIGERATION

The capture cavity is mounted in a separate cryostat at the end of the injector. For its final installation at DESY (1) a common feed box provides the cooling for Cryocap as well as for the subsequent cryomodules, connected partly in parallel, partly in series at the following three temperature levels :

- a supply of two-phase He at 1.8 K, 16 mbar which maintains a constant level of superfluid LHe around the cavity,
- a second loop, fed with supercritical He at 3 bar, enters at 4.5 K and cools a heat sink on the RF input coupler,
- a third loop uses He gas at 14 bar and 60 K for cooling a single radiation shield.

For initial cool-down or final warm-up of the cavity a small feed line is connected to the bottom of the LHe tank.

CRYOSTAT

Figure 1 shows the composition of Cryocap.

LHe tank with cavity

The capture cavity consists of a single but standard 9 cell 1.3 GHz cavity and features the new concepts developed for the TESLA project, i.e. integration into a cylindrical helium vessel with a reduced capacity (23L), with all RF couplers located outside in the vacuum space and thus cooled by conduction only through the niobium walls of the cavity.

The LHe tank is made from titanium, easy to weld to niobium and of a similar thermal contraction. But as niobium turns brittle at very low temperature, the composed titanium-niobium reservoir is not accepted as a pressure vessel, therefore the outer vacuum tank has to assume the duty of a containment volume.

Fine RF tuning of the cold cavity is obtained with a mechanism which compresses the whole 9-cell assembly by approximately ± 1.5 mm.

Passive magnetic shielding of the cavity is provided with a sheet of Cryoperm placed on the LHe tank.

10 layers of superinsulation are wrapped around the fully equipped LHe vessel.

Suspension system

Two split suspension rings are clamped onto the LHe vessel. A radial suspension device with 2 x 4 rods of epoxy-fibreglass, in an antagonistic array which permanently blocks any thermal contraction, maintains the cavity in its adjusted initial position. Such a mounting is acceptable since the low Young's modulus of the composite induces only moderate thermal stresses which may occasionally be measured with a set of 4 strain gauges mounted in the upper fixture of the suspension rods.

In the axial direction, the LHe tank is blocked with a finger device which penetrates from the vacuum tank into one of the suspension rings.

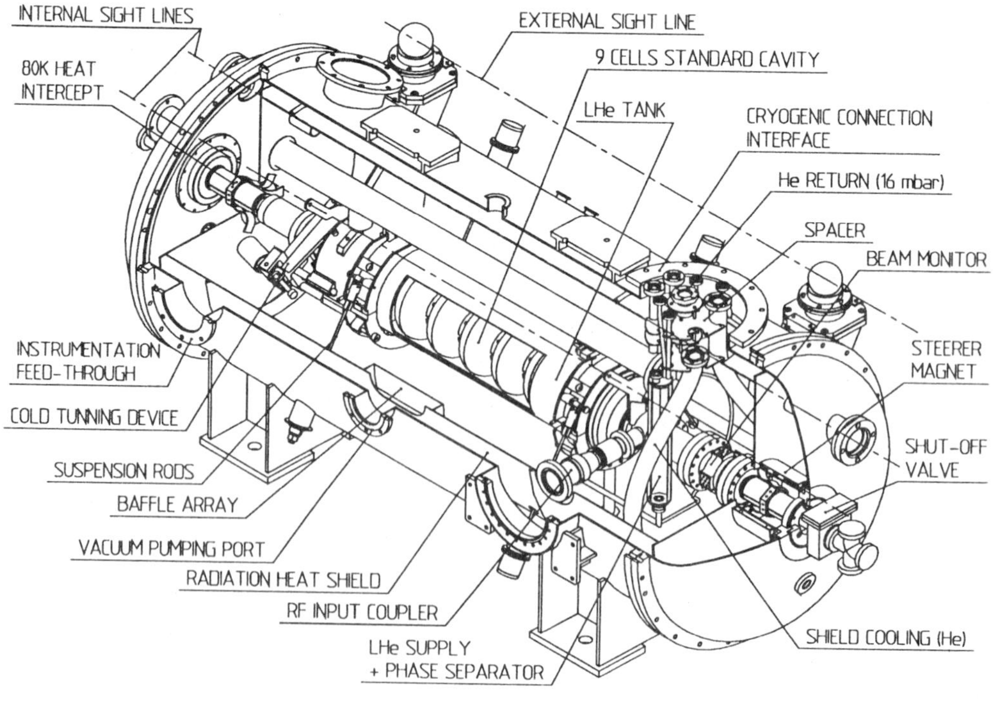

Figure 1. Capture cavity cryostat (CRYOCAP).

Table 1. Heat load estimate for cryocap

Temperature level (K)	Nature of heat flow	Static heat flow with no RF power (W)	Static heat flow with full RF power (W)
80	Conduction (supports, heat intercepts)	15	18
	Radiation	37	37
	Total	**52**	**55**
4.5	Intercept on RF input coupler	**0.3**	**0.5**
1.8	Conduction (supports, electric leads)	1.1	1.1
	Radiation	0.4	0.4
	RF power dissipation in cavity	-	1.3
	Total	**1.5**	**2.8**

Radiation heat shield

The thermal radiation from ambient temperature to the cavity is intercepted by a single copper radiation shield cooled to 80 K with forced flow He or LN circulation on its cylindrical part. It integrates into the vacuum tank where it stands on 4 nylon bases. Both removable end covers are conduction cooled.

Heat flow from the surroundings is reduced with a blanket of 40 layers of superinsulation.

The insulating vacuum is autonomous ; a specially devised barrier, located in the cryogenic connection line, separates it from the feed box vacuum space.

A blackened baffle array on the radiation shield allows an efficient evacuation of the innermost vacuum space.

Its basic function apart, the radiation shield also serves other purposes :

- it provides a heat sink for thermal intercepts,
- it supports a solenoidal coil for active magnetic shielding during cool down,
- it may be used to accelerate a warm-up of the cryostat with a thermocoax resistor soldered onto the copper cylinder.

Vacuum tank

The vacuum tank, made from stainless steel, is designed for a service pressure of 1.2 bar abs. As a containment vessel for the fragile cavity reservoir it is protected with a big safety valve against any excessive overpressure.

The removable end caps define the primary alignment axis for the cavity with two permanent sight lines through the entire cryostat, thus allowing an occasional check of the cavity position even in a cold state.

Heat load

A thermal budget estimate limited to the fully equipped CRYOCAP is given in Table 1.

ANCILLARY FEED BOX (AFB)

For the first cryogenic tests in France, a special ancillary feed box was built which substitutes for the DESY feed box and provides the cooling for Cryocap at 1.8 K and 4.5 K with liquid helium (LHe) and at 80 K with liquid nitrogen (LN) corresponding to the flow scheme of Figure 2.

LHe under 1.2 bar is siphoned continuously from a standard storage vessel, passes a recondensing coil immersed in a LHe pool at 4.2 K and then divides into two flows.

The first flow is only used occasionally for an initial cool-down of the cavity.

The second flow supplies the 4.5 K cooling loop in Cryocap where it partially vaporizes, then returns to AFB where the residual liquid maintains a constant LHe level of the 4.2 K bath, adjusted with the control valve LCV1.

LHe for the 1.8 K bath is taken from the 4.2 K reservoir. A passage through heat exchanger E1 cools it to 2.2 K and thus substantially reduces the flashing losses during expansion to 16 mbar in LCV 2. This valve remotely controls a constant LHe level of the 1.8 K cooling bath in Cryocap.

A second storage vessel provides LN which is essentially used for radiation shield cooling. Thus a continuous feed of LN passes, first in an annular gap through the coaxial LHe transfer line, then through the cooling loops soldered onto the radiation shields of both AFB and Cryocap and finally cools the charcoal adsorbers, alternately E30 or E31, where it entirely vaporizes.

The cold He vapour from Cryocap and those from the 4.2 K cooling bath in AFB may be handled in two different ways :

a) they are simply warmed up to ambient temperature in an electric heater,
b) their potential cooling capacity is recovered in an economizer.

With the construction of the feed box we wanted to combine our long term interests with our short term obligations, so we decided to equip the cold box from the very beginning with a line of heat exchangers which might be operated arbitrarily as an economizer or as a simple heater.

The economizer principle consists essentially in a counter-current heat exchange of the available enthalpy in the escaping low pressure stream to an inflowing high pressure stream whose final Joule-Thomson expansion produces a substantial amount of LHe which in turn reduces the supply from the external storage for the same useful cooling capacity.

Our arrangement consists of two separate Hampson type heat exchangers of high efficiency, operating between 4 to 80 K and a single heat exchanger with a comparatively poor

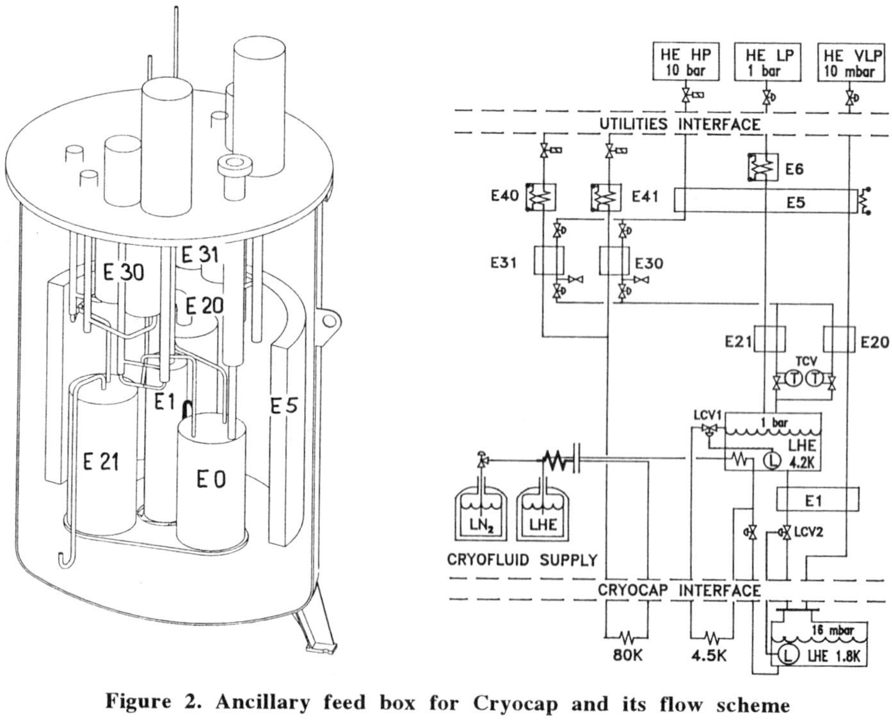

Figure 2. Ancillary feed box for Cryocap and its flow scheme

efficiency which operates between 80 to 300 K as a counter-current heat exchanger. The latter is made from two coaxial flexible hoses of 20 m length wound on a cylindrical mandrel. High (HP) and low pressure (LP) gas flows inside of the flexibles hoses, whereas the very low pressure (VLP) gas passes externally and perpendicular to the windings which are, moreover, equipped with an electric heater of 3 kW. Therefore, if no economizer is used, the same apparatus might only be used to warm up the cold vapours to ambient temperature.

Two sets of commutative charcoal traps are provided to purify the incoming high pressure He and to maintain its temperature down to 80 K.

For optimum performance of the economizer the HP mass flow has to be permanently adjusted to the fluctuating LP and VLP cold gas returns. Two identical autonomous control valves have been developed which maintain the temperature of the incoming HP flow at an optimum value (i.e. 6 K).

PRESENT STATUS (July 1995)

Some problems in developing the cavity and its RF coupler induced a substantial delay in our initial planning. Thus assembly of Cryocap has just started and will be completed in next autumn.

The ancillary feed box has been assembled and a first cryogenic test of this particular unit is in progress.

REFERENCES

(1) G. Grygiel, U. Knopf, R. Lange, B. Petersen, D. Sellmann, J. Weisend "Status of the TTF Cryogenic System", presented at the CEC Conference, July 17-21, 1995, Columbus OH.

CRYOSTAT FOR TESTING RF POWER COUPLERS

M. Kuchnir, M.S. Champion, K.P. Koepke and J.R. Misek

Fermi National Accelerator Laboratory*
Batavia, IL, 60510, USA

ABSTRACT

Similar to the power leads of accelerator superconducting magnets, the power couplers of accelerator superconducting cavities are components that link room temperature to superfluid helium temperature for the purpose of energy transfer. Instead of conducting kiloamperes of current they guide megawatts of RF power between those two temperatures. In this paper we describe a cryostat designed for testing the performance of these components and measuring their heat loads. A special feature of this cryostat is its minimum liquid inventory that considerably simplifies safety related requirements. This cryostat is part of a Fermilab facility contributing to the international collaboration working on TESLA (TeV Electron Superconducting Linear Accelerator). This facility is now operational and we will be presenting specifications as well as performance data on the cryostat as well as the first pair of power couplers tested with it.

INTRODUCTION

One of the TESLA (TeV Electron Superconducting Linear Accelerator)[1] essential components is the **Input RF Power Coupler** that transmit the 1.3 GHz Power from room temperature to the 1.8 K environment of the superconducting accelerating cavities. These articulated coaxial transmission lines have a wave-guide to coax transition at room temperature, a ceramic window at 78 K, and an outer conductor extremity connected to the superconducting beam-tube between the cavities.

In order to test the power handling capability of these power couplers and measure their heat loads on the refrigeration system we designed the cryostat described here. It measures their heat loads by thermal sinking them through heatmeters .[2] A resonant ring containing a single cell superconducting cavity between two power couplers is used with the available 100 kW power source in order to test them RFwise up to the 1 MW power level required for HPP (High Power Processing)[3] of the TESLA cavities in situ. This arrangement is schematically shown in figure 1. The cavity together with the colder segments of the couplers are attachable to the cryostat as an evacuated unit. A special container for shipment of this unit was also built. This allows for the chemical surface treatment of the cavity and clean room assembly of this three components unit to be performed at already existing facilities of participants in the TESLA collaboration.

In this paper we describe the cryostat, its operation and one of the first tests using it. This test verified the cryostat operation and measured the static heat load of the two

*Fermi National Accelerator Laboratory is operated by Universities Research Association, Inc. under contract with the U.S. Department of Energy.

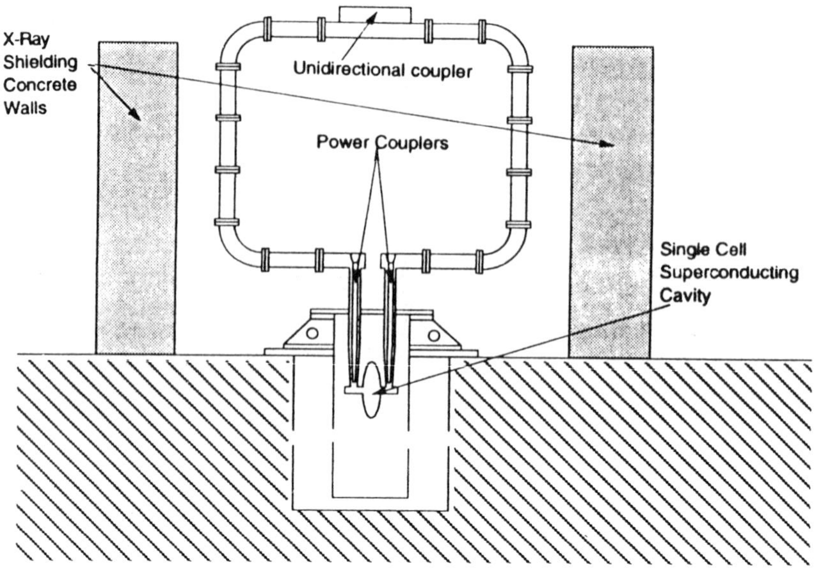

Figure 1. Resonant ring testing concept .

power couplers in use. Performance under RF power was precluded by the lack of the required chemical treatment of the superconducting cavity.

CRYOSTAT DESCRIPTION

The length of the cryostat is dictated by the power couplers length. As a consequence it is rather short (0.7 m), also the separation between the couplers is dictated by simulation to the TESLA cavity cell-coupler distance resulting in a axis to axis distance of (0.203 m). This cryostat fits in an existing sunken dewar of 0.44 m internal diameter that is used here just as a vacuum vessel.

In order to make the cryostat inherently safe and avoid need for boiler code certifications etc. the liquid He inventory is minimized to 1.6 liters and the operation is based on continuously flowing cryogens from a liquid N_2 trailer and a liquid He 500 liter storage dewar.

The cryostat is formed by a **stainless steel top plate**, a **copper LN_2 plate** with its **aluminum shield** that surrounds a **copper LHe plate** and its **aluminum shield** that in turn

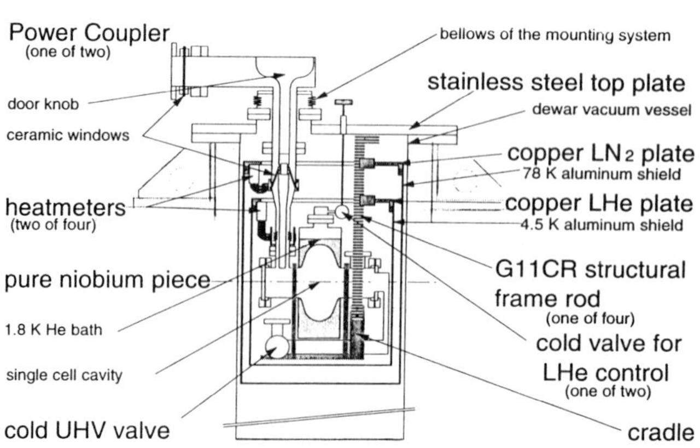

Figure 2. General mechanical structure of the cryostat .

surrounds the **pure niobium piece**. The surrounding is not complete since race track shaped holes in the copper plates are needed for the couplers, plumbing lines and instrumentation cables. The gaps remaining in these holes are closed for infrared radiation purposes with closer fitting multilayered insulation mats. Four G11CR fiberglass-epoxy rods form the **structural frame** that holds together the plates and the cradle holding the niobium piece. Figure 2 presents these components in a simplified schematic. The simplification required for figure 2 substitutes by a single stainless steel bellows the rather sophisticated mounting system of the **power coupler** to the top plate. This mounting system allows the motions required for RF tuning of the coupler to the TESLA cavities as well as the resonant ring in this test arrangement besides taking care of thermal expansion differentials.

The **stainless steel top plate** is 2.54 mm thick. Besides the circle of bolts that seals it to the dewar and the two flanges for the coupler adapters it has 10 penetrations and 4 blind screw holes for the set of G11CR fiberglass-epoxy rods. Pairs of these penetrations are for the LHe transfer lines, LN_2 lines, rotary feedthroughs, electrical wires and fiber optic feedthroughs, vacuum and pump out ports. The penetrations for LHe transfer lines are equipped with Cajon fittings for 12.7 mm OD transfer lines. These rigid U shaped transferlines have 6.35 mm ID and are used for both the incoming liquid and the exiting cold gas from the copper plate and cryopump (Exhaust).

The penetrations for the LN_2 lines and the LHe Vent line are double walled elbows using the insulating vacuum to prevent their cold inner tube (6.35 mm diameter) from cooling the plate. Since thick frosted coating on these lines was not a concern, "Hytron" thermoplastic hoses (manufactured by Imperial Eastman) have been used here with good results although these temperatures are well below their specifications . The rotary "Ferrofluidic" feedtroughs are used to permit the control of the two cold valves through 6.35 mm diameter G11CR rods that extend their stems inside the cryostat.

The **copper plates** are 9.53 mm thick discs of OFHC copper to which are brazed two copper-magnet-bus-hollow conductors and two small OFHC copper discs. Through the hollow conductors will flow the cryogens that determine the temperature of the plates. One of these is 2.16 m long and has 5.79 x 5.79 mm cross section with a 3.18 mm diameter hole, the other has a 14.3 x 25.4 mm cross section with a 6.35 mm diameter hole and form the rim of the plate. The outside of this rim is then machined at a 2.0 ^{o} angle to form a conical surface for thermal contact with its aluminum shield. The two small discs add thickness to the plates locally for the 1/2-14NPT pipe threaded hole used for the attachment of the heatmeters

The **78 K** and **4.5 K shields** are made from 3.18 mm thick aluminum plate rolled and welded to form flat based cylindrical cans. The rims are internally machined at a 2.0 ^{o} angle to closely contact the conic surfaces of the respective plate rims. Three race-track shaped holes equally spaced around are machined partially intersecting the conic surface to permit the introduction of a tool for the uncoupling of this binding conic contact. These holes are also used by special fasteners to prevent gravitational induced de coupling. The conic surfaces are lightly coated with Apiezon L grease to improve thermal contact and ease of disassembly. The thermal contact of these conic surfaces, helped by the copper/aluminum differential thermal expansion coefficient proved to be completely satisfactory.

The **pure niobium piece** was formed at the Laboratory for Nuclear Studies of Cornell University and further modified by e-beam welding to it the LHe reservoir, the pulled-out tubes and flanges . These latter operations were done at Fermilab. Noteworthy are the seals to the stainless steel mating flanges that use Helicoflex "O-rings". In particular, one of them seals the superfluid He containing reservoir from the insulating vacuum. This piece together with the Ultra High Vacuum (UHV) manifold and its all metal valve are held to a cradle that allow us to position it easily on a table or attach it and the lower parts of its two couplers to the cryostat frame or to a shipping frame. So the niobium piece consists of a beam tube with two radial pullouts (for the couplers) between which it expands in an elliptical single cell cavity. This cavity is coaxially surrounded by a cylindrical liquid He container also with a radial pull-out which is parallel and in between the other two. The extremities of the beam tube and pull-outs have shoulders that fit into special stainless flanges designed to seal with Helicoflex metal "O-rings". The entire piece is made exclusively of high purity niobium stock and electron-beam welded together. The niobium piece, its cradle and the ceramic-window-sealed colder halves of the two couplers, plus a manifold ending in an all metal valve, V1, are detachable as one UHV system that can be shipped for chemical polishing and assembly in a clean room.

The four supports that form the **structural frame** of the cryostat were machined out of a G11CR plate. They have the shape of an inverted long and narrow L. The foot of each L is attached to the bottom of the stainless top plate by means of a 1/4-20 screw into a threaded hole 12.7 mm deep. The long vertical legs have a 12.7 x 25.4 mm cross section. Horizontal threaded holes are used to hold the copper plates and heat-sink the supports at the copper plates temperatures. This is accomplished by means of copper coated screws and copper coated stainless adapters. The LN2 cooled copper plate is at 76.2 mm below the stainless top plate and the LHe cooled copper plate is 102. mm further down. The horizontal holes used for the attachment of the cradle and its pure niobium piece (at 1.8 K) are 241. mm further down.

The **four heatmeters** are very similar. The two, designed for the 80 K region, have a 2.1 mm long stainless steel body segment between copper segments containing 100 Ω Pt RTDs (Resistance vs. Temperature Devices). The other two, designed for the 5 K region, have a 0.5 mm long stainless steel body segment between copper segments containing 100 Ω, 1/8 w Allen Bradley carbon composition RTDs.

Each heatmeter has a 45 Ω calibration heater and an aluminum thermal shield. The contact to the heat sinking plates is by means of 1/2-14 NPT pipe threads (copper to copper) and to the couplers through copper braided cables and copper clamps. Copper powder loaded grease is used in these contacts and care is taken to prevent the braided cables from touching anything.

Not shown in figure 2 is the **cavity ultra high vacuum pumping line,** it is composed of three sections. The section from 300 K to 80 K is formed by a zigzag of three coaxial vertical tubes in order to reduce heat conduction. The middle section is a 31.75 mm diameter 0.11 m long horizontal stainless tube joining the other (vertical) sections through mitered welds. The lowest section has a brazed copper sleeve .32 m long around it. This sleeve has a cooling channel that makes it act as a cryopump. This line terminates with Conflat 2.75 flanges. The flange at room temperature is after an elbow adapted to the top plate and the flange at low temperature connects to the all metal valve of the detachable UHV system.

The **1.8 K bath pumping line**, also not shown in figure 2, starts as a 6.35 mm diameter stainless tube leaving radially from the cap of the bath and connecting to the shell side of the **heat exchanger**. The shell of this heat exchanger is made out a 15.9 mm diameter, .89 mm wall, 0.53 m long stainless tube curved in a semi-circle. This semi-circle lays below and along the inner edge of the copper LHe plate without touching it. Inside the heat exchanger, seven stainless capillary tubes each 0.79 mm OD x 0.15 mm wall x 1.70 m long are connected in parallel and coiled. They carry and cool the liquid helium to the JT valve. After the heat exchanger this line continues as a 19.0 mm

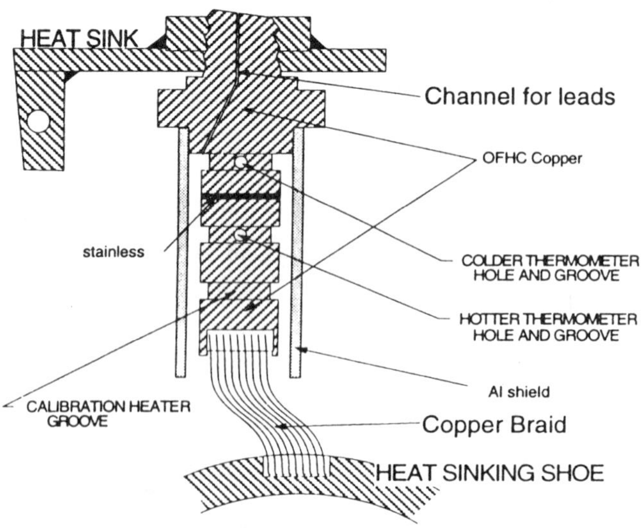

Figure 3. Heatmeter mechanical schematic.

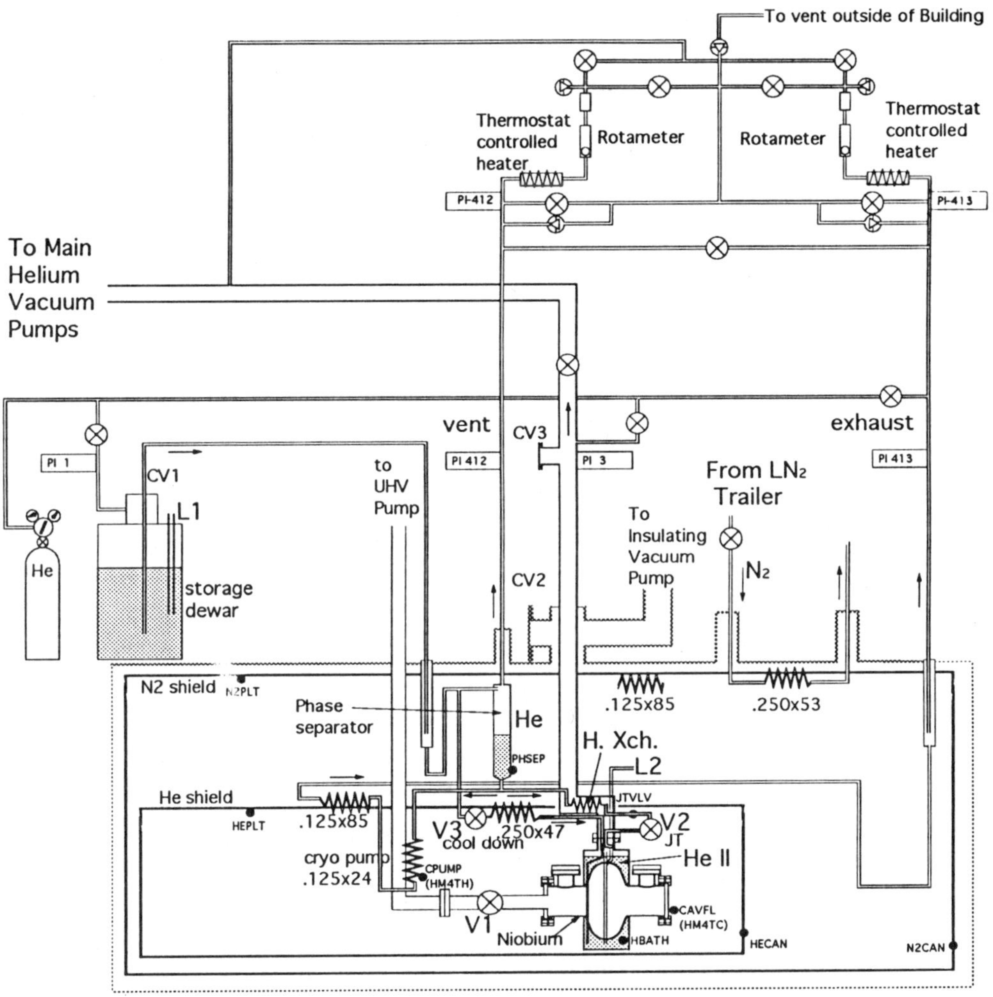

Figure 4. Flow diagram of the cryostat.

OD stainless tube to a bayonet type port on the top plate. Care was taken to insure that this line slopes up all the way to prevent formation of liquid pools above the bath.

The **cap of the 1.8 K** bath is a rather busy component that contains an electrical feedthrough with 9 wires, the pumpout tube, the refill tube, the cooldown tube and the supports for three electrical connectors one inside and two outside besides the shoulder for the sealing flanges. The feedthrough is made from Stycast 2850 in a well tested style[4]. One of the outside connectors is of SMA type and is used for the twisted pair leads of the capacitive liquid Helium level sensor. The other two are commercial 9 pin connectors based on diallyl phthalate. So, before sealing the cap to the bath, a cable with the 4 leads from a RTD, 2 leads from the level sensor and two leads from the immersed bottom heater is plugged in the inner connector. Also, before sealing the cap to the bath, a Teflon tube gets connected to the cooldown tube so that the cold helium is directed to the bottom of the bath. Therefore, the end of the cooldown tube inside the cap points directly down ready for connection to the Teflon tube that will extend it to the bottom of the bath. The refill tube however brings the expanding cold mixture from the JT valve pointing it horizontal and tangential to the inner wall for a phase separation action and minimum disturbance of the accumulated liquid. The pumpout tube leaves the cap from the cylindrical side wall radially.

The **capacitive liquid helium level sensor** is made out of 3 stainless steel strips (.13 mm thick, 12.7 mm wide shims 260. mm long) separated with 1.59 mm x 9.52 mm spacers, 25.4 mm apart, each made out of 7 layers of adhesive Teflon tape buildup to a thickness of 0.62 mm. The outer strips form the external grounded plate of the capacitor. This flexible probe is inserted in the bath space along the external equator of the single cell cavity. Therefore its response to the level is non-linear in a calculable way. The circuit used has been described by S.J. Collocott[5] . In this system we placed the front end frequency to voltage converter (an integrated circuit) on the connector inside the cryostat to reduce the parasitic capacitance and reduce noise pick up.

MODES OF OPERATION AND FLOW DIAGRAM

The cryostat is cooled and run by continuously flowing liquid He from a 500 liter storage dewar and liquid N_2 from an existing storage trailer. Figure 4 is the flow diagram of the system showing these components and the cryostat plumbing. Non conventional schematic symbols and lines are used to give a more pictorial representation of the actual system. Valve V1 is an all metal valve which is operated only when the system is at room temperature and is kept open when the system is in the dewar for cooling or running. It is not part of the helium flow system. The liquid from the storage dewar and the part of it that is vaporized by the transfer tube heat leak are handled differently according to the operational mode.

After establishing the plumbing connections, vacuum leak testing the different volumes, leaving them with the appropriate gas and valve settings the cool down starts by flowing nitrogen from the storage trailer through the LN_2.copper plate. In the first stage of the *cool down mode*, the vent plumbing, valves V3 and V2 are closed directing the cold helium to cool the lower section of UHV pumpline (cryopump) and the copper LHe plate exiting through the **exhaust exit plumbing**. In the second stage, once the UHV cryopump is cold, valve V3 is open directing the helium to further cool the copper LHe plate and the pure niobium piece leaving the system via the bath pumpout line. This stage should be paced by the conduction cooling of the heavy end-flanges of the cavity.

In the *operating mode*, valve V3 is closed and the incoming mixture reaches the **phase separator** where the gas is directed to the now open **vent exit plumbing** while the liquid continues to be used to cool the cryopump section of the UHV pumping line and the copper LHe plate leaving the cryostat via the exhaust exit plumbing. Part of this flow is diverted to the 1.8 K bath via the cold heat exchanger and the V2 valve which here acts as a the Joule Thompson expansion valve (JT). The aperture of this valve is adjusted to maintain the level in the bath stable above the cell wall. The time constant characteristic of this control is suitable for manual control, but future upgrade includes an automatic control for this function.

In figure 4, the heat exchanging parts of the flow lines are indicated with their diameter and length in inches. The connections are made with 6.35 mm OD x 0.89 mm wall stainless tubes welded or brazed. The heat exchanger before the V2 valve has been described above as part of the 1.8 K bath pumping line. The phase separator has a volume of less than 50 cm^3 and is probably not optimal. The vent tube coming out of it is being redesigned to go trough a serpentine cooling line in the LHe shield prior to exiting the cryostat.

The general control of the operation is done by the pressure in the LHe storage dewar and the flow limiting rotameter valves downstream in the exhaust and vent exit plumbing. These helium flows determine the 4.5 K shield temperature and enable the JT valve (V2) effective operation, which together with the pumping on the helium bath surrounding the cavity and the heat load determines the cavity temperature.

Besides the thermometers and heaters in the heatmeters that have been described above and the sensors in the helium bath there are RTDs installed in the shields and plates. Sixteen RTDs are multiplexed into four hardware channels. Other instruments installed are four pressure gauges, two flow meters, and two liquid level indicators.

The **data acquisition system** is based on hardware and software designed and built at Fermilab for control of its Linac[6]. The hardware consists of two modules both of which are nodes in the Internet. One of them is an RF wave form digitizer and the other is the Internet Rack Monitor (IRM) which has been used here for the cryogenic data acquisition. A special set of instrumentation amplifiers[7] and IRM software[8] was developed and implemented to handle RTD thermometry. The Parampage[9] software used is specifically written for MacIntosh computers and communicates with the rest of the hardware by local network or the Internet. So data acquisition (and control if properly authorized) can be monitored independently and simultaneously from anywhere. Most of the time, trouble shooting and resetting operations have been done by the specialists in the Linac Control group at their desks a few kilometers away.

RESULTS

Besides a commissioning run without power couplers, only one attempt has been made to run this cryostat in the proposed resonant ring mode but the low Q due to the yet unpolished cavity surface prevented resonance. So the data acquired so far are the heat loads of the first two power couplers (carrying no RF power) and the operational parameters for the cryostat itself.

These heat loads include contribution of the infrared radiation exchange between the braided copper connection and its surroundings. They are a function of the temperature distribution throughout the cryostat and are shown in Table 1. We estimate the order of the uncertainty here to be less than 30 % . More running experience and tests are needed to reduce this estimate uncertainty. At this time we are not in need of a more precise result.

Table 1. Power Coupler heat loads

	window load	at	base load	at
Coupler #1	6.7 W	102 K	1.01 W	6.7 K
Coupler #2	5.8 W	101 K	0.96 W	6.8 K

Measurements were also made of the temperature of the He bath surrounding the cavity as a function of the power deposited in its internal heater. The steady state temperature was 1.43 K and Table 2 presents its value as a function of power. The level of the superfluid in the bath went down under the 2 W and 4 W loads but it took of the order of 15 minutes to empty.

Table 2. Cavity He Bath temperature under several power loads

Watt	0.00	.129	.254	.380	.498	1.00	2.00	4.01
K	1.43	1.45	1.48	1.50	1.52	1.60	1.71	1.86

The operation of the cryostat is best when it is not left to warm up much overnight. This involves an overnight liquid helium consumption rate of 4 liters/hour but it prevents contamination problems and long cooldown times. Under typical measuring conditions, the consumption rate is 32 liters/hour. One feature of the system worth mentioning is how this rate is measured: The American Magnetics superconducting wire level sensor used here is part of an adapter to the commercial 500L liquid He container. Its analog voltage output seen as a function of time and the zooming capabilities in the Parampage graphical display provide great sensitivity to the level of the Liquid He supply (± 0.1 mm). This sensitivity allows for the observation of many factors affecting the level. The most obvious the consumption rate reflected in the slope.

CONCLUSION

A special cryostat has been built and tested at Fermilab capable of measuring the heat loads of TESLA input power couplers. The relevance of these measurements is related with the number of these couplers required. The test facilty now under construction in Hamburg requires 32 and the projected accelerator over 20,000.

ACKNOWLEDGMENTS

We thank the Cornell U. Laboratory for Nuclear Studies for providing the niobium cavity. A large number of people from Fermilab contributed to this project. In particular we want thank Helen Edwards and Frank Turkot for general support. The following contributions are acknowledged: design and drafting by Clare Miller; machining and welding by Fermilab village machine shop; coordination of assembly and operation by Al Rusy; level capacitive sensor, heatmeters wiring, cryostat wiring and superinsulation by Dan Assell; level probe electronics by Dyrrell Lewis; heat meter drafting by Clark Reid; general support by Fermilab Lab 2 personnel; data acquisition system setting up and maintenance by Mike Kucera and Bob Florian.

REFERENCES

1. H. T. Edwards, TESLA Parameters Update - A Progress Report on the TESLA Collider Design, in: "6th Workshop on RF Superconductivity," R.M. Sundelin, ed., CEBAF, Newport News, VA (1993), vol. 1, p. 361.
2. M. Kuchnir, J. D. Gonczy, J. L. Tague, Measuring heat leak with a heatmeter, in: "Advances in Cryogenic Engineering," vol. 31, Plenum Press, New York, (1986), p. 1285.
3. J. Graber et al., High peak power RF processing studies of 3 GHz niobium cavities, in: "5th Workshop on RF Superconductivity," D. Proch, ed., DESY, Hamburg (1991), vol. 2, p. 758.
4. M. Kuchnir, Fabrication of cryogenic electrical feedthroughs, Fermilab TM-596 (1975).
5. S. J. Collocott, *Cryogenics*, 327 (1983).
6. R. W. Goodwin, M. J. Kucera, M. F. Shea, *Nucl. Instr. and Meth. in Phys. Res. A* 352, 189 (1994).
7. R.D. Oberholtzer and B.M. Wisner - private communication
8. R.W. Goodwin - private communication
9. R. E. Peters - private communication

DESIGN OF POWER AND HOM COUPLERS FOR TESLA

Karl Koepke, for the TESLA Collaboration

Fermi National Accelerator Laboratory
P.O. Box 500, Batavia, Illinois 60510, USA

ABSTRACT

The basic TESLA accelerating structure is a 1 m long, 9-cell superconducting cavity operating at a frequency of 1.3 GHz and an accelerating gradient of 25 MV/m. Each of these cavities has one fundamental mode power input coupler and two higher order mode (HOM) couplers. The rf requirement of the input coupler is a peak power of 200 kW for 1.3 ms at a 10 Hz repetition rate. The input coupler must also accommodate a 15 mm cavity motion during cryostat cool down and must minimize the cryogenic heat load with and without rf power present. The HOM couplers must extract from the cavities the beam induced HOM energy that does not propagate to the rf absorbers located at the end of each cryomodule. Two distinct input couplers and two distinct HOM couplers designs have been produced. Their design criteria and performance will be reported.

INTRODUCTION

TESLA [1] (TeV Energy Superconducting Linear Accelerator) is one of a family of proposed e^+e^- linear colliders[2] with a CM collision energy of 500 GeV. TESLA is unique among these proposed accelerators in that it uses superconducting rf accelerating cavities. The basic TESLA accelerating cavity[3] is shown in Figure 1. It consists of 9 cells that are dimensioned to operate in the standing wave π mode of the TM_{010} passband at a frequency of 1.3 GHz. The cells, bore tube and flanges are fabricated out of high purity niobium sheets (RRR=300).

The basic rf power source[4] for TESLA is a 1.3 GHz klystron capable of producing a peak power up to 5 MW (100 kW average) at the TESLA duty cycle. The klystron rf output passes thru a circulator and is then split approximately equally by directional couplers into 16 WR650 waveguide channels to power 16 cavities. The input couplers transmit the rf power from the waveguide at room temperature to the cavity at 1.8 K. The circulator protects the klystron from the rf power that is reflected due to mismatches in the rf circuit, primarily from the cavity, but also due to imperfections in the waveguides and input couplers.

Each 9-cell cavity is energized by a single power input coupler to which 1.3 ms long rf pulses with a peak power of 200 kW are applied at a 10 Hz rate. The accelerating gradient of 25 MV/m is reached 0.5 ms after the rf pulse is applied. At this time, 800 beam bunches, spaced 1 μs apart, are injected into the cavities and accelerated. Each beam bunch contains 5 10^{10} particles. The input coupler must also tolerate a 1 MW peak power pulse at reduced pulse length and repetition rate for in situ peak power reprocessing of 9-cell cavities after installation in the cryostat or after a vacuum failure.

As the beam bunches traverse the cavities, they excite higher order mode (HOM) resonances in the cavities. The electromagnetic fields of this induced HOM energy interact with the beam, causing beam instabilities, emittance growth and energy spread. The HOM energy is also an unacceptable 1.8 K heat load and in extreme cases can cause thermal breakdown of the superconducting cavities (quench) unless the HOM energy is extracted from the cavities. Approximately 70 % of the HOM energy in the TESLA cavities is expected to be above the beam tube cutoff frequency and will be absorbed by broad band rf absorbers located at the end of each cryomodule. The remaining HOM energy and some HOM energy above cutoff but trapped between the cavities will be extracted by the HOM couplers. Each 9-cell cavity has 2 HOM couplers for this purpose.

The basic TESLA cryogenic unit is the 12 m long cryomodule[5]. Each cryomodule contains eight 9-cell cavities plus beam focusing and steering magnets. Each 9-cell cavity within the cryomodule is contained within its own 1.8 K vessel. These vessels are filled by gravity with 1.8 K liquid He from the 1.8 K liquid He supply pipe located above the cavities. The cryomodule also contains 4.5 K and 70 K radiation shields. The He supply lines for these shields also serve as heat intercepts for devices that penetrate to the 1.8 K region, e.g., supports, input couplers, etc. The input coupler and 2 HOM couplers of each cavity are attached to the bore tube of the cavity outside of its 1.8 K vessel. This results in a simplified mechanical structure and eliminates a source of He leaks as all the coupler joints within the cryomodule have vacuum on both sides. However, the large number of cells per cavity makes it more difficult to obtain a uniform accelerating field in time for the next accelerated beam bunch and to couple adequately to all the HOM excited within the cells.

An international collaboration has assembled a TESLA Test Facility (TTF) at DESY, Hamburg, Germany, to demonstrate the technical viability of TESLA. TTF already has the infrastructure to clean, assemble and test superconducting cavities. The present effort is to produce 4 complete cryomodules that will be part of a 500 MeV test linac. As part of this effort, DESY[6] and Fermilab[7] have each developed an input coupler, and DESY[8] and Saclay[9] have each developed a HOM coupler.

INPUT COUPLERS

The two TESLA input coupler versions are shown in Figure 2 and Figure 3. Although they differ in detail, they both satisfy the same electrical , mechanical and cryogenic requirements.

Waveguide-Coax Transition

This component transforms the TE_{10} mode transmitted by the WR650 waveguide into the TEM mode transmitted by the coaxial line of the coupler. This transformation needs to be achieved with minimum reflections (matched at 1.3 GHz) for rf efficiency and diagnostic reasons. The magnitude and phase of the reflected power are an important indicator of coupler and cavity performance. The Fermilab input coupler utilizes a "doorknob" transition. The Hewlett Packard code, HFSS (High Frequency Structure

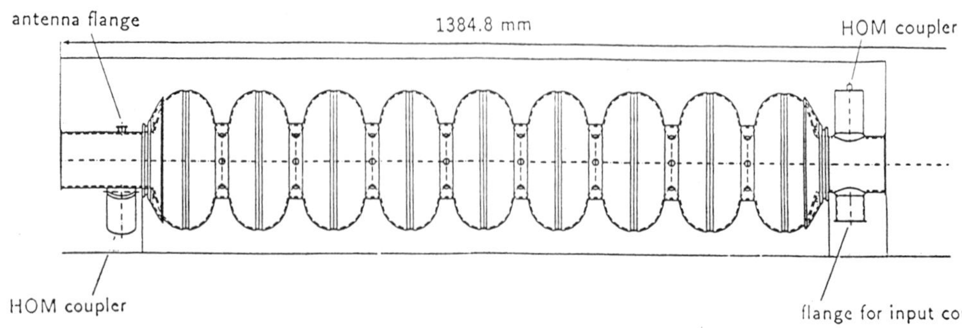

Figure 1. TESLA 9-cell superconducting cavity with coupler locations.

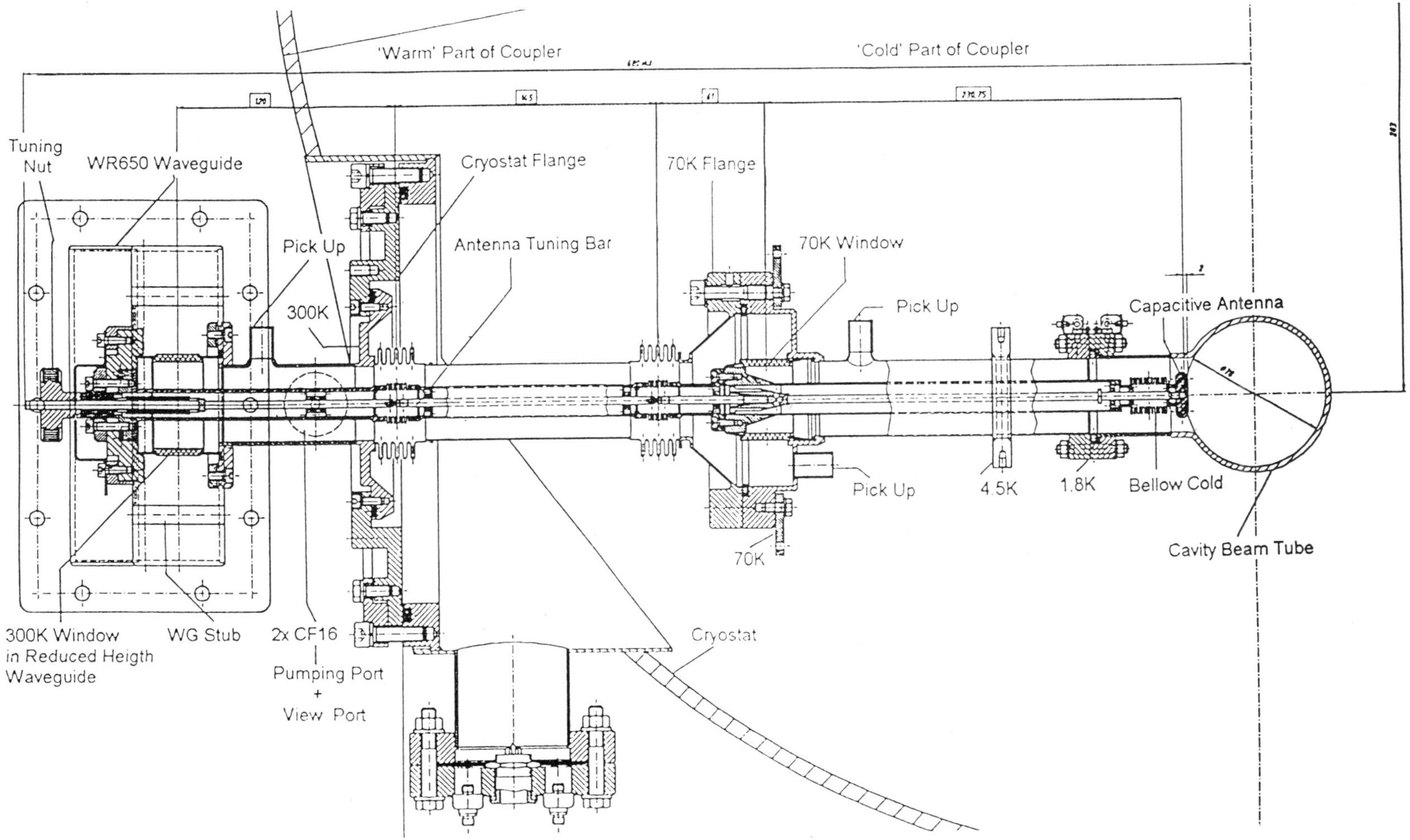

Figure 2. DESY designed TESLA power input coupler.

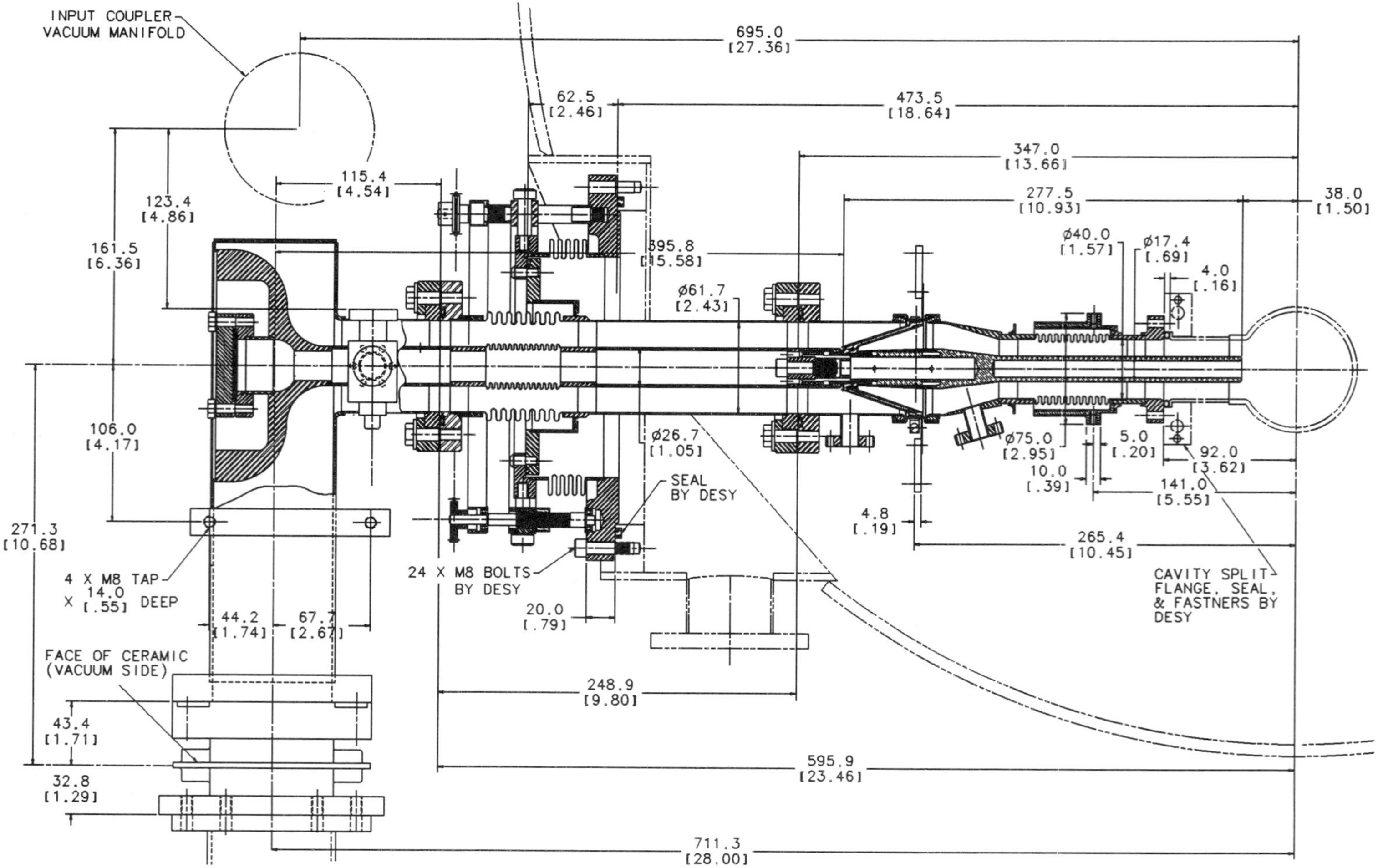

Figure 3. Fermilab designed TESLA power input coupler.

Simulation), has been used to calculate the shape of the "doorknob" to assure a match at 1.3 GHz. The DESY transition uses a "flattened doorknob"; the WR650 waveguide tapers to a reduced height WR650 section. It is matched by two posts located in the reduced height waveguide upstream of the warm rf window.

Warm Window

This component isolates the atmosphere in the waveguide from the upstream vacuum of the coupler. The Fermilab warm window is a commercial "pillbox" type, located immediately upstream of the "doorknob" transition and matched by two posts upstream of the window. The DESY warm window is cylindrical and mounted within the waveguide transition and concentric with the inner coax conductor. Both windows are fabricated out of high purity Al_2O_3 (>99%) and are "thin" to minimize rf losses.

Cold Window

Both coupler versions have a cold rf window located close to the cavity. This window isolates the coupler vacuum from the ultraclean vacuum of the cavity. The cold window is attached to the cavity in the clean room and prevents contamination of the cavity by micron sized particles during subsequent handling and installation into the cryostat. The windows are "thin" and constructed out of high purity Al_2O_3. Both the Fermilab and DESY cold windows are matched to the coaxial line at 1.3 GHz. The Fermilab window is conical in shape. The DESY window is cylindrical.

Coaxial Lines

Both coupler versions utilize a 50 Ω coaxial line to transmit the rf from the waveguide transition through the vacuum vessel of the cryostat. The coaxial line terminates on the bore tube at a point outside of the 1.8 K vessel that surrounds the cavity. The cold end of the outer coaxial conductor is cooled to 1.8 K by conduction through the niobium bore tube, connection tube, and flange. The static heat load and rf heat loads to the 1.8 K volume are minimized by 4.5 K and 70 K heat intercepts located slightly upstream of the 1.8 K flange and at the cold window respectively. The Fermilab coaxial line upstream of the cold window has an outer diameter of 6 cm and is fabricated out of thin walled stainless steel for mechanical rigidity and minimum heat conduction. The inner surfaces in contact with rf are plated with approximately 10 μm of copper. The coaxial line tapers to 4 cm after the cold window. The inner conductor of the 4 cm coaxial section is thick walled copper to keep the temperature at the inner conductor tip approximately equal to the 70 K temperature at the cold window. The DESY coaxial line diameter is 4 cm over its whole length. It is also constructed out of copper plated stainless steel with a copper inner conductor tip at its end.

Bellows

Depending on their position in the cryomodule relative to the centrally located fixed suppport, the 9-cell cavities move up to 15 mm during cool down. Both coupler versions have bellows on the inner and outer coaxial parts of the coupler to allow this motion. When warm, the coupler is kinked. When cold, the coupler straightens to its nominal operating position. The bellows are hydroformed stainless steel and copper plated where necessary to reduce the rf surface resistance.

Coupling Adjustment

Coupling to the 9-cell cavity needs to be adjustable to compensate for cavity performance variations. The nominally required external Q is $3\ 10^6$. The external Q of both coupler versions is adjustable approximately a factor of 10 around this value by adjusting the position of the coupler's inner conductor relative to the bore tube center. The Fermilab coupler achieves this by varying the length of the outer conductor bellow nearest to the bore tube. The DESY coupler adjusts the inner conductor bellow nearest to the bore tube. As the impedance of the 9-cell cavity changes as a function of accelerating field and accelerated beam current, the coupler is matched for the nominal accelerating voltage and beam current. During the rise time of the accelerating voltage in the 9-cell cavity and for beam currents other than the nominal beam current, a fraction of the rf power applied to the input coupler is reflected at the cavity-coupler interface and is absorbed by the rf load attached to the circulator. When the input coupler is matched to the cavity, the rf power going into the coupler is completely absorbed by the accelerated beam.

Heat Load

The heat loads into the 1.8 K, 4.5 K and 70 K heat sinks have been estimated by a one dimensional program that divides the coaxial conductors into segments and then performs an energy balance on each segment. The program uses temperature dependent tables for the thermal conductivity and electrical resistivity of stainless steel and copper to calculate heat conduction and rf heating. At cryogenic temperature, the 10 μm copper plating dominates not only the rf loss but also the heat conduction. The thickness of the copper plating, the RRR value of the copper, and the copper surface quality are therefore critical. The rf skin depth must also be adjusted for the anomalous skin effect. The calculated heat load[10] during full rf duty cycle operation is essentially identical for the two input coupler designs; approximately .06 W, .6W, and 6.0 W to the 1.8 K, 4.5 K, and 70 K heat intercept points.

Standing Wave

The input couplers are overcoupled until beam is injected into the cavity. At the beginning of the rf pulse, all of the rf is reflected back from the tip of the inner conductor. The resulting standing wave pattern has a voltage maximum equal to twice the voltage of the traveling wave. The maximum repeats every half wave length with the first occurring approximately at the tip of the inner conductor. The voltage amplitude modulation of the standing wave decreases as the cavity accelerating field rises, reaches zero, and then rises again with a quarter wave phase shift. Beam is injected at the time when the standing wave modulation amplitude is zero and the beam maintains the coupler matched for the remaining duration of the rf pulse. The rf windows of the Fermilab coupler have been placed at voltage minimum locations. The DESY coupler has its cold window at a voltage maximum, the warm window at a voltage minimum. All of the rf windows have been coated with a few nanometers of TiN to reduce multipactoring. The TiN reduces the secondary electron coefficient of the surface. The Fermilab couplers will also have the copper plated surfaces coated with TiN.

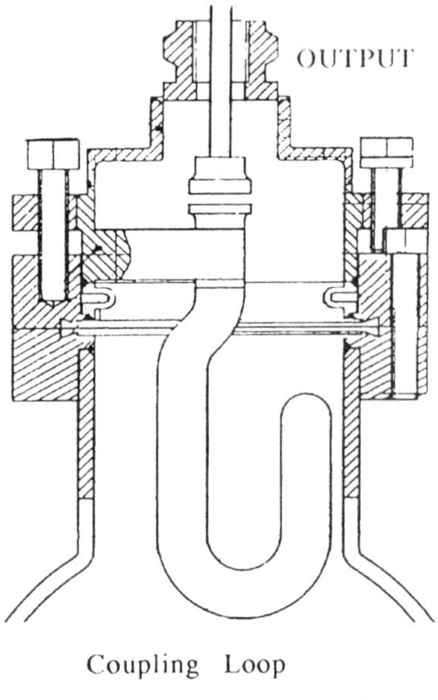

Figure 4. Demountable HOM coupler.

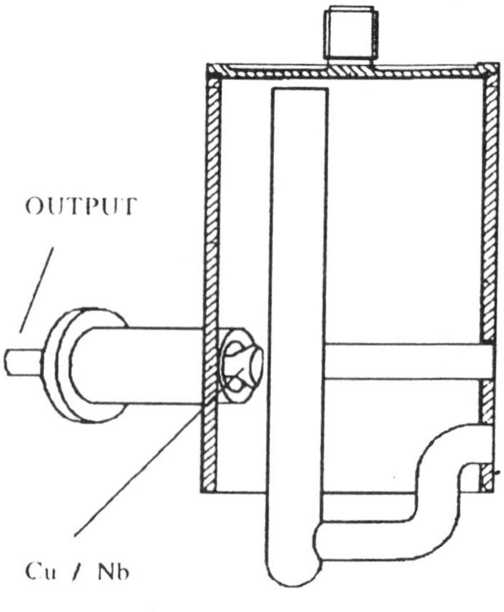

Figure 5. Welded HOM coupler.

HOM COUPLERS

Two types of HOM couplers have been developed for testing at TTF, one with a flange to allow demounting of the HOM coupler from the bore tube, the other welded to the bore tube. The demountable HOM coupler was developed by CE Saclay and is similar to the HOM coupler used on the LEP cavities but scaled to the 1.3 GHz TESLA fundamental frequency. A sketch of the demountable HOM coupler is shown in Figure 4. The welded HOM coupler (Figure 5) was developed at DESY. It is based on the HOM coupler used on the 4-cell HERA cavities.

Unlike the fundamental mode input coupler that needs to be matched at 1.3 GHz to maximize the power available for beam acceleration, the HOM couplers contain a rejection filter at this frequency to prevent loading of the cavity's accelerating gradient and to prevent an excessive heat load in the HOM cable that connects the HOM coupler to the load external to the cryostat. This filter must have sufficient bandwidth to permit pretuning at room temperature without impairing damping of adjacent HOM frequencies.

The mode families identified as possible sources of beam blowup are: TE_{111} (1622 MHz - 1789 MHz, TM_{110} (1800 MHz - 1889 MHz), TE_{121} (3076 MHz - 3114 MHz), TM_{011} (2380 MHz - 2454 MHz), and TM_{012} (3720 MHz - 3857 MHz). The HOM couplers can provide damping over the frequency range, 1.8 GHz to 10 GHz, as their transfer functions are essentially flat over this range. The external Q for a particular HOM passband frequency depends on the HOM field amplitude at the location of the HOM coupler and the orientation of the HOM coupler. Depending on the HOM passband, coupling can be predominantly to the electric or magnetic HOM fields. To assure proper coupling to the above HOM families, two HOM couplers per 9-cell cavity are placed on the bore tube outside of the cavity helium vessel. The demountable HOM couplers are separated by an azimuthal angle of 115°, the welded HOM coupler at an angle of 150°. These angles are the optimal position for these couplers relative to the summed impedances of the eight highest (R/Q) HOM dipole passbands.

The fundamental frequency rejection filter of the demountable HOM coupler is formed by the inductance of the loop and the capacitance of the loop end to the outer conductor. Final tuning of the rejection frequency is accomplished with a single convolution bellow located above the demounting flange joint. A capacitive probe extracts the HOM power at the top of the coupler. The fundamental frequency rejection filter for the welded HOM coupler is formed by the inductance of the inner conductor pin and the capacitance of the pin gap to the top surface of the coupler. This rejection filter is tuned by an inelastic deformation of the coupler's top. A capacitive probe extracts the HOM power at the side of the coupler. Both HOM couplers can adjust Qext at 1.3 GHz to larger than 10^{11}.

The HOM couplers are cooled by conduction to the bore tube. The demountable HOM coupler has an additional thermal shunt to 1.8 K at the coupling loop support post location as the thermal conduction through the demounting flange joint proved to be inadequate. To minimize rf heating, caused primarily by currents at the fundamental mode frequency, the inner and outer conductors of the HOM couplers are fabricated out of niobium. The coupler temperature must therefore be maintained below 9 K to avoid a quench and thermal runaway.

PERFORMANCE

Two prototypes of the Fermilab input coupler have been tested up to a peak power of 70 kW, the highest 1.3 GHz power currently available at Fermilab. The tests were performed without a cavity. The cold windows of the couplers were cooled by immersing the couplers in an open dewar containing LN up to the 70 K intercept point and the coupler ends were terminated with a 4 cm diameter closed pipe. The first coupler was tested with a 50 kW standing wave and two inner conductor tip lengths; the first placed the cold window at a voltage maximum of the standing wave pattern, the second at a minimum. This exposed every part of the coupler to the voltage equivalent of a 200 kW traveling wave. The second prototype input coupler was tested up to 70 kW with the cold window placed at a standing wave minimum. The vacuum trip level during conditioning was set at 2 10^{-6} Torr. Photomultipliers and electron probes detected light and electrons early in the conditioning period. This activity was conditioned away after 2 weeks of operation. These couplers are presently at DESY where they are scheduled to be tested at full power.

The cold rf window of the DESY input coupler has been tested in a three half wavelength long coax resonator. The window was placed at the standing wave voltage maximum at the resonator's center and tested with a 100 kW peak power standing wave. This is equivalent to the voltage of a 400 kW peak power traveling wave. The observed vacuum, light, and electron activity was removed by conditioning.

Both HOM couplers have been tested at DESY and CE Saclay with copper models of the 9-cell TESLA cavity. The copper cavity tests confirmed that both HOM couplers have sufficient damping of the HOM passbands that were considered as possible emittance growth sources. 1.5 GHz models of the HOM couplers were also tested on a single cell 1.5 GHz superconducting cavity at CE Saclay. Both couplers worked properly up to 12.5 MV/m in cw mode. This represents a significant safety margin for 25 MV/m operation at the 1.3 % TESLA duty cycle.

SUMMARY

Twelve Fermilab designed input couplers are presently being assembled at Fermilab. Three DESY designed input couplers have been ordered from industry. These input couplers are expected to start arriving at DESY within a few weeks. They will be conditioned in a fixture that tests two couplers in series, and after successful testing, will be assembled to 9-cell cavities that have been preconditioned in a vertical dewar. Testing of the HOM couplers is already in progress at full rf power during conditioning of 9-cell cavities in a vertical dewar.

ACKNOWLEDGMENTS

Fermilab is operated by the Universities Research Association under contract to the U.S. Department of Energy. The couplers were developed at CE Saclay, DESY and Fermilab primarily by the following individuals: M. Champion, S.Chel, B. Dwersteg, A. Mosnier and J. Sekutowicz.

REFERENCES

1. H. Edwards, Proc. 6th Workshop on RF Superconductivity, CEBAF, Vol.1, p.361 (1993).
2. J. Rossbach, Proc. 1992 Linear Acc. Conf., Ottowa, Vol.1, pg. 25.
3. D. Proch, Proc. 6th Workshop on RF Superconductivity, CEBAF, Vol.1, p.382 (1993).
4. A. Gamp, Proc. 6th Workshop on RF Superconductivity, CEBAF, Vol.1, p.492 (1993)
5. A. Tazzari, F. Alessandria, F. Castagna, L. Catani, M. Ferrario, M. Minestrini, T. Nicol, Proc. 6th Workshop on RF Superconductivity, CEBAF, Vol.1,p.440 (1993).
6. B. Dwersteg, G. Weichert, N. Adelt, Proc. 6th Workshop on Superconductivity, CEBAF, Vol.2, p.1144 (1993).
7. M. Champion, Proc. 6th Workshop on RF Superconductivity, CEBAF, Vol.1, p.406 (1993).
8. J. Sekutowicz, Proc. 6th Workshop on RF Superconductivity, CEBAF, Vol.1, p.426 (1993).
9. S. Chel, A. Mosnier, M. Fouaidy, T. Junquera, contributed to 4th European Particle Accelerator Conference, London, June 1994.
10. T. Peterson, TESLA Report 93-37, available from DESY.

VARIOUS METHODS OF MANUFACTURING SUPERCONDUCTING ACCELERATING CAVITIES

C. Benvenuti, Ph. Bernard, D. Bloess, E. Chiaveri, C. Hauviller and W. Weingarten

CERN, European Organization for Nuclear Research
1211 Geneva 23, Switzerland

ABSTRACT

We report on experience in superconducting cavity production methods gained in shaping, joining and thin film coating with various materials and techniques (Pb, Nb, Nb_3Sn, NbN, NbTiN) with emphasis on their potential to reduce mass production costs.

INTRODUCTION

Production technology of superconducting (sc) cavities has entered a new phase: the number of manufactured cavities has jumped from a low value (recyclotrons, heavy ion accelerator facilities, intermediate sc RF systems such as in TRISTAN, HERA) to a large series of hundreds (CEBAF, LEP2) and even thousands (proposed TESLA). This necessitates reconsideration of production and quality control methods with respect to performance requirements and cost effectiveness.

The maximum attainable gradient is not necessarily that for which the construction cost is a minimum. If the total RF voltage, V, and the current are fixed, the total investment costs are given by three terms. The first one is proportional to the total length L of the cavities and cryostats ("linear" costs), the second one is proportional to the total RF power P_{RF} to be transmitted to the beam (independent of L), and the third one is proportional to the cooling power ("cryogenic" costs), which is essentially the power P_d dissipated in the RF cavities. Since $V = E_a \cdot L$ is constant, E_a being the accelerating gradient, the first term is inversely proportional to E_a. If E_a is increased, the RF losses will go up quadratically with E_a. On the other hand the length L is inversely proportional with E_a, because V is constant. Hence P_d increases only linearly with E_a and consequently the cooling power to be installed. The cost minimum is located where the first and third term are equal. The following example is based on numbers, which are close to CERN's experience[1] for LEP. Fig. 1 displays these three terms and their dependence on E_a. It shows that a gradient between 8 and 10 MV/m is an optimum choice for this particular application.

However, having in mind even larger applications, such as the sc linear collider TESLA under design, at higher frequency (1.3 GHz) and higher energy (500 GeV), costs must be reduced substantially. The cooling power may be reduced (technical Q-values being in the 10^9 range) by slowly pulsing the accelerator, which reduces the cryogenic costs. One can see from Fig. 1 that for a duty cycle of 1% (close to that envisaged in TESLA) a gradient of 25 MV/m would yield the cost minimum provided that the linear costs go down by a factor of 16. To obtain such a factor would be a challenge. Under these assumptions the linear and cryogenic costs (for $V = 2$ GV) would go down by a factor of 40.

Advances in Cryogenic Engineering, Vol. 41
Edited by P. Kittel, Plenum Press, New York, 1996

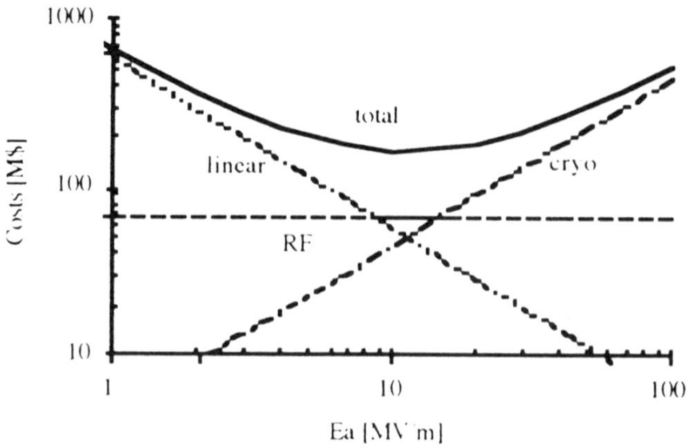

Figure 1. Estimate of investment costs for an sc RF system (f = 352 MHz, V = 2 GV, 12 MW RF power, 5 W/m standby heat load, $Q_0 = 4 \cdot 10^9$): The total costs are the sum of three terms, linear, RF and cryogenic costs.

The purpose of this paper is to review the state of the art of producing sc cavities. It will focus on the manufacturing of the sc accelerating structure itself. Methods with the potential to save costs will be looked at more closely.

THE MECHANICAL STRUCTURE

The sc cavity consists of two essentials: the "mechanical structure" and the layer which carries the RF current, the "sc surface".

The mechanical structure has to provide three features. At first it must be sufficiently stable and rigid at low temperatures to sustain with a safety margin the pressure in the helium tank (usually 1 bar) outside, the radiation pressure inside and to block mechanical vibrations. Secondly the thermal conductivity must be such that heat is effectively removed without generating too large temperature gradients (thermal stability). Thirdly it must be shaped economically with reasonable effort and joints between pieces must be vacuum tight and reliable.

Up till now two materials have been used for the mechanical structure; niobium and copper (Table 1). Niobium can simultaneously provide the mechanical structure and the sc surface. In addition it is the element with the largest transition temperature T_c. Copper is relatively cheap, can be manufactured and shaped without tremendous effort and has a large thermal conductivity.

Table 1. Comparison of niobium and copper [1)]

	Cavity grade Nb	OFE Cu
Thermal conductivity @ 4.2 K [W/(mK)]	10 (RRR = 40) - 100 (RRR = 400)	400 - 500
Electrical resistivity @ 295 K [10^{-8} Ωm]	14.4	1.68
Density [g/cm^3]	8.57	8.96
Linear expansion (L_{273}- $L_{4.2}$)/$L_{4.2}$ [‰]	1.43	3.3
Atomic weight	92.91	63.54
Atomic number	41	29
Melting point [°C]	2468 - 2497	1080
Recrystallization temperature [°C]	830 - 1230	400 - 600
Stress relieving temperature [°C]	780 (1h)	200 - 250
Tensile strength [N/mm^2]	207 - 274	196 - 245 [3)]
Elongation [%]	6 - 42	6 - 50
Yield point [N/mm^2]	20 [2)]- 196	15 - 300

[1)] Some data are taken from refs. 2, 3
[2)] annealed for several hours at 1800 [°C] and 10^{-8} Torr
[3)] annealed

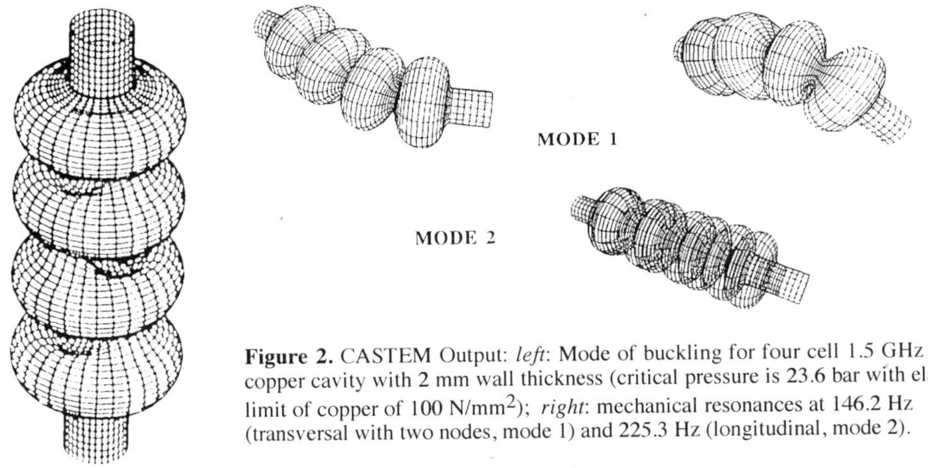

Figure 2. CASTEM Output: *left*: Mode of buckling for four cell 1.5 GHz copper cavity with 2 mm wall thickness (critical pressure is 23.6 bar with elastic limit of copper of 100 N/mm^2); *right*: mechanical resonances at 146.2 Hz (transversal with two nodes, mode 1) and 225.3 Hz (longitudinal, mode 2).

Mechanical Stability and Rigidity

The wall thickness of the mechanical structure assures its mechanical stability. It has to be designed according to contradicting constraints. Thin walls are less costly, easier to handle and effectively cooled, but on the other hand prone to buckling from pressure spikes inside the He tank and mechanical and ponderomotive oscillations (cf. below). Modern accelerating structures are made from several mm thick material (niobium or copper). A Finite Element Analysis software such as CASTEM[4] allows the mechanical behaviour of the structure, in particular the buckling modes, the maximum tolerable pressure, and the natural resonant frequencies to be understood (Fig. 2)[5]. Main modes of vibration can be suppressed by imposing forces of constraint on well-chosen points of maximum amplitude. A complete suppression of all the mechanical vibration modes is impossible. These modes add phase shifts on the tuner feedback loop and can make it unstable. A way out is to shift their resonant frequencies to larger values beyond the tuner bandwidth. This could be done by increasing the rigidity of the stiffening bars and hence the resonant frequency.

Ponderomotive forces[6] depend on the square of the accelerating gradient. By increasing the accelerating gradient, the cavity is driven out of resonance, for example by the radiation pressure acting on the cavity wall (Lorentz force). It takes a certain time before the wall moves (there is a delay). The RF amplitude will fall, and the resonant frequency will be shifted back to tune again, which closes the loop. The delays present can make the system unstable.

Thermal Conductivity

The temperature T near a localized heat source falls inversely proportional to the thermal conductivity λ and the distance r from that heat source. This may be a normal-conducting inclusion, a poorly-cooled welding bead, peeled-off flake etc., or a region of electron impact. If the source dissipates too much heat, the transition temperature of the sc surface can be surpassed, by which effect the stored energy (100 J for a LEP cavity at 6 MV/m accelerating gradient) is dumped near the defect (quench). The cavity can only be operated at a gradient well below the quench. Therefore the thermal conductivity near T_c of the material in the intimate vicinity of the heat source is the important parameter.

It becomes immediately clear that, to increase the maximum gradient, the number and size of the surface defects have to be reduced, and, on the other hand, sc metal of increased thermal conductivity must be made available (thermal stabilization of defects). It is known[7,8] that the thermal conductivity of niobium is most severely affected by interstitial impurities such as oxygen, nitrogen and carbon. These impurities can be considerably reduced by repeated furnace treatments on the manufacturer's premises[9-11] and/or solid-state gettering by wrapping a foil of metal with a large affinity to these impurities around the cavity in a furnace at 1200–1400 °C. Yttrium and titanium are used as getter material[12,13].

Shaping and Joining Techniques

There exists a wealth of knowledge on shaping and joining techniques[2,14-25], which will not be repeated here. Niobium can be machined by the usual high speed steel tools. It turns very much like lead or soft copper. It has nevertheless the tendency to gall; tool angles, cutting speed and lubrication have to be carefully chosen. Normal techniques of metal spinning are applied, and the metal takes up the contour of the former. Deep drawing is also used and gives satisfactory results (no "orange peel") for fine grain material (ASTM <4).

Hydroforming is a common manufacturing procedure (e.g. production of bellows). The principle is to push the part against a rigid die by applying a large pressure through a liquid. Monolithic pieces can be produced, and welding can be avoided. Reproducibility is superior compared to welded parts, but the tooling is expensive. This fact is of no consequence for a large series. However, repeated annealing is necessary and increases the cost. Cavities in the frequency range from 300 MHz to 3 GHz have been produced from OFE (Oxygen Free Electronic grade) copper tubes[22]. The frequency of the fundamental mode of 1.5 GHz five-cell cavities has a standard deviation of ±0.3‰[26]. The final deformation is more than 200%. The ultimate elongation of copper being only of the order of 50% (Table 1), this process has been achieved in several stages: swaging has been followed by three expansion and intermediate annealing steps at 600 °C for one hour.

Swaging means that an oil-pressurized membrane pushes the annealed copper tube on to the internal core creating a toroidal groove. The problem is to prevent plastic buckling due to high compressive stresses creating ripples. The process has been modelled by the BOSOR5[27] software and experimentally checked. Good correlation has been obtained for the critical buckling values in the case of uniform thickness. Thirty-five percent reduction of the iris diameter has been achieved in only one stage with a pressure up to 650 bars.

The expansion is obtained by a multi-part die which is initially open and closed progressively during expansion. The closed die has the exact external shape of the final cavity. The internal hydraulic pressure progressively increases up to 200 bar. The number of expansion steps depends on the radial deformation needed and on the behaviour of the annealed copper. CASTEM software[4] was used for modelling. A precise prediction of the behaviour during all stages has been possible.

OFE copper has high purity and high conductivity at cryogenic temperatures. The parameters of the heat treatment were determined using standard tensile tests. The heat treatment starts to influence material properties from 400 °C. The hydroforming phases were then simulated on test samples (initial annealing, 32% elongation and second annealing, 32% elongation and third annealing).

Hydroforming niobium was first used at HEPL (Stanford)[28]. This metal has very similar properties to copper (Table 1) except the characteristic temperatures such as, for example, the melting point. Heat treatment should be done at 1000°C for about one hour. The procurement of high quality niobium tubes is not easy. Seamless tubes and rotary swaged longitudinally welded ones are presently being investigated.

Spinning is another method for cheap mass production. A seamless nine-cell 1.5 GHz cavity starting from a planar disk of Al has been spun without annealing[29], and Cu and Nb nine-cell cavities are under preparation. The thickness uniformity was about a factor of two.

Galvanoplastic shaping[30] is another technique to build the whole structure without welds in one stride: an organic glass mandrel (shaped like the inner surface of the cavity) is sputter coated with a copper layer, on to which the bulk copper is deposited by the galvanoplastic technique. Subsequently the mandrel is dissolved. A 3 GHz nine-cell cavity has been produced and sputter-coated with a niobium film inside. RF results are not yet published.

THE SUPERCONDUCTING SURFACE

The superconducting surface has to comply with the following three features: it must tolerate high electric fields and currents at sufficiently low RF losses (large Q-values); it must be operated at a temperature which can be produced by the evaporation of a cryogenic liquid with reasonable effort; it must be operated at a temperature sufficiently below the critical temperature T_c of the sc material (safety margin). There are three different ways of complying simultaneously with these demands, which reflect the historical progress of the technology. The most straightforward way is to use the structure material as the sc surface as well. The

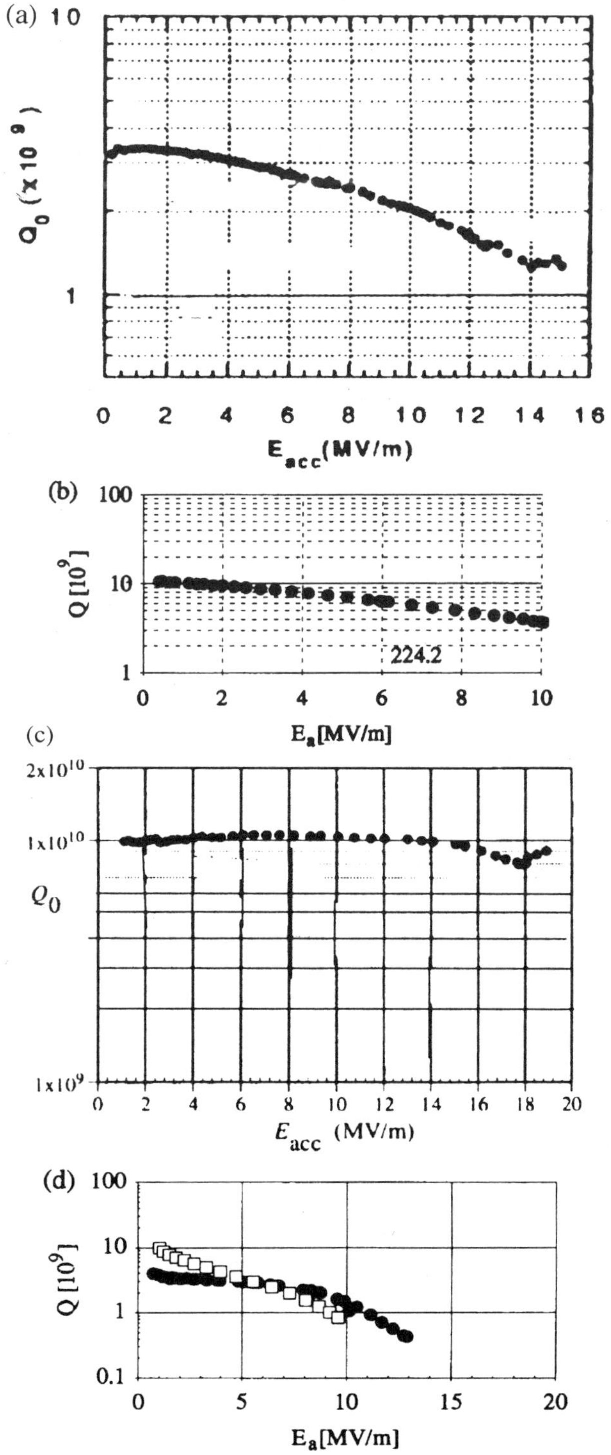

Figure 3. Q_0-E_a plots in vertical tests (best results) for cavities from serial production in industry [except (d)]: (a) a 508 MHz five-cell niobium sheet cavity for TRISTAN[31] (4.2 K); (b) a 352 four-cell niobium coated copper cavity for LEP[32] (4.2 K); (c) a 1500 MHz niobium sheet cavity for CEBAF[33] (1.8 K); (d) 1500 MHz five-cell hydroformed niobium coated copper cavity (from industrial production feasibility study)[21] (1.8 K).

second one is to modify the structure material (towards larger T_c and Q-value which means less effort for cooling). The third one separates the function of the structure and the sc surface completely and opens up an avenue towards new materials for the sc surface (thin films).

Identical Superconducting Surface and Structure Material

The metal (exclusively niobium) has been machined as, for example, for the CERN-Karlsruhe RF separator[25], or has been shaped by deep drawing and machined, as for the Stanford HEPL recyclotron[28], or is shaped by deep drawing or spun from sheet metal like, for example, the CEBAF, CERN-SPS-LEP (first phase), Darmstadt-Wuppertal, DESY-HERA, KEK-TRISTAN structures. Some effort with TIG welding has been made, but EB welding of the half cups[23] (preferably from the inside[24]) and irises gave more reliable results. The best results in Q-value and maximum gradient E_a, obtained in an industrial series production, are shown in Fig. 3 (a) and (c).

The processing of the surface. When the structure and the sc surface are identical, the properties of the sc surface pay tribute to those of the mechanical structure (inclusions from rolling and shaping, thermal conductivity). Nevertheless, the methods of surface processing have been refined in such a way that the maximum gradients obtained in multi-cell accelerating structures have been pushed up to 16 MV/m in a five-cell 508 MHz cavity at KEK[31] and about 22 MV/m in a nine-cell 1.5 GHz cavity in the TESLA collaboration[34]. The first result was obtained by electropolishing, heat treatment (1.5 h at 700°C) to remove the embedded hydrogen, and rinsing with ultra-pure water under ultrasonic agitation. The second one was obtained by chemical polishing, solid state getter furnace treatment (to improve the thermal conductivity) before welding, rinsing with ultra-pure water (under low and high pressure), and pulsed operation under high RF power (high power pulsed processing)[35].

In the presence of hydrogen (accumulated by slow chemical polishing), cavity grade Nb with high thermal conductivity forms compounds which during slow cool down precipitate as hydrides at low temperatures (~ 100 K) and degrade the Q-value[36]. These cavities have to be annealed to chase the hydrogen or the cool down has to be done fast.

Cavities with identical mechanical structure and sc surface could be taken out one by one, coated by compounds based on the structure metal with improved surface properties (as Nb_3Sn, see below), and retrofitted into the accelerator again.

Acceptance rates in initial vertical tests in an industrial series production for CEBAF are 98% (E_a = 5 MV/m and $Q = 2.4 \cdot 10^9$)[33].

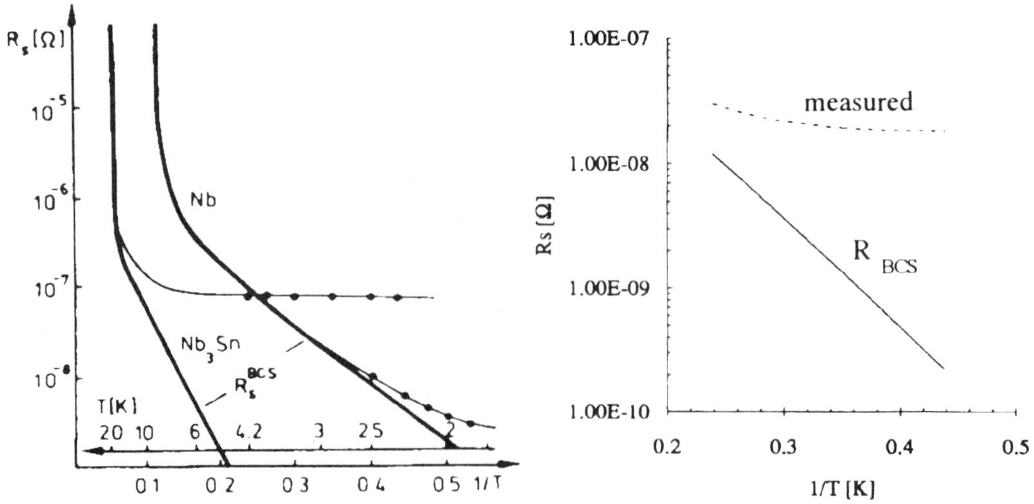

Figure 4. The surface resistance at 500 MHz of a re-entrant cavity with niobium and Nb_3Sn surface (left)[39], and that of a NbTiN layer on a copper accelerating mono-cell cavity (right)[40].

Table 2. BCS surface resistance of some metals

	T_c [K]	A	n	Δ'
Nb (RRR = 100)	9.25	105	2	18
Nb_3Sn	18.1	105	2	40
Pb (Ref. 37, Fig. 1)	7.2	74	1.75	14.6
PbSn (4 at% Sn)[38]	7.5	68.5	1.9	15.1

Superconducting Surface from Reprocessed Structure Material

A15 and B1 compounds. The surface resistance R_s is composed of the theoretically well understood BCS surface resistance R_{BCS} (which can be calculated by using the BCS theory) and the residual surface resistance R_{res}. At a given operating temperature R_{BCS} falls exponentially with increasing Cooper pair energy 2Δ (2Δ is the energy gap of the sc metal). The BCS theory tells us that Δ is proportional to T_c, $\Delta = 3.5\ kT_c$, hence superconductors with a larger pairing energy Δ should allow a smaller surface resistance at the same temperature, provided the pre-exponential constant is about the same. This is the reason why both the B1 and A15 compounds NbN and Nb_3Sn are also being investigated for RF applications.

The BCS surface resistance (in $\mu\Omega$) of sc metals for $T < T_c/2$ is given by $R_{BCS} = A \cdot f^n \cdot \exp(-\Delta'/T))/T$, with the temperature T measured in K, the frequency f in GHz, and the constants from Table 2. There is a large potential of attaining a low surface resistance in these compounds, provided that R_{res} can be made sufficiently low (Fig. 4).

Thin coatings of these materials on RF cavities are produced by thermal diffusion out of the vapour phase at a temperature of 930-1160 °C and 1400°C for Nb_3Sn[39,41,42,43,44] and NbN[45], respectively. Recently, CEBAF has obtained in a Nb_3Sn coated single-cell 1.5 GHz cavity (in collaboration with Wuppertal) a Q-value at 4.2 K, which exceeds the design Q-value for Nb at 2 K[46] at the design field of 5 MV/m.

Thin Superconducting Films on a Separate Structure

In an sc metal the London penetration depth λ of the RF field is independent of the frequency in a wide range up to tens of GHz, 39 nm (Nb) and 37 nm (Pb). Therefore, to keep the RF losses low, an sc layer of say ten times the penetration depth is largely sufficient. Such films can be produced by several methods, as follows (we do not consider composite materials, as for example niobium clad copper, the RF performance of which is very similar to bulk niobium).

Coating by electroplating. Lead can be deposited on a copper substrate by electrodeposition. Although niobium is superior to lead (larger T_c, B_c and lower BCS surface resistance, Table 2), nevertheless lead is particularly used for heavy ion structures[37] in small series at low frequencies (where BCS losses are low), if the geometry is such that high thermal gradients may occur, or when the full potential of RF superconductivity technology is not of paramount importance. Several pre-plating, plating and post-plating treatments have been developed, including plating with Pb-Sn[38] and Pb-Bi[47] alloys.

Coating by sputtering. At CERN[48], coating a cavity made of OFE copper by a niobium film has been chosen for several reasons. First, a thermal breakdown (quench) is improbable. Secondly, the material costs would have amounted to about 22 % of the total cavity and cryostat production costs if made from niobium sheet. They are negligible if made from copper sheet. Sputtering has been chosen because the sputtered atoms have a large kinetic energy, which favours the adhesion. DC Magnetron sputtering has been preferred to diode sputtering because of relatively large deposition rates, easy application to multi-cell cavities, and good thickness uniformity. In the final phase of the industrial series production for CERN, the acceptance rates in initial vertical tests is 60 - 70 % (E_a = 6 MV/m and $Q = 4 \cdot 10^9$)[32]. Q vs. E_a curves are shown in Fig. 3 (b) and (d). Compounds as NbN and NbTiN have also been investigated [Fig. 4 (b)][48,49].

CONCLUSION

Cost-effective mass production is a key issue for accelerating structures for linear colliders. A strong development effort is needed for the following:

- fabrication methods which allow a substantial reduction in production costs (such as, for example, hydroforming, spinning, galvanoplating) and
- more stringent quality control in a fully automated production chain.

REFERENCES

1. W. Weingarten, Superconducting cavities, *Int. J. Mod. Phys. A (Proc. Suppl.)* 2B:678 (1993), or Proc. XVth Int. Conf. High Energy Acc., J. Rossbach, ed., World Scientific, Singapore (1993), p. 678.
2. W. Bauer, Fabrication of niobium cavities, Proc. 1st Workshop RF Supercond., M. Kuntze, ed., Karlsruhe (1980), KfK 3019, p. 271.
3. M. G. Rao and P. Kneisel, Thermal and mechanical properties of electron beam welded and heat-treated niobium for TESLA, Proc. 6th Workshop RF Supercond., R. Sundelin, ed., CEBAF, Newport News VA, USA (1993), p. 643.
4. A. Combescure, Manuel d'utilisation CASTEM 2000, Rapport CEA DEMT 88/176.
5. J. Genest, private communication.
6. P. H. Ceperley, Ponderomotive oscillations in a superconducting helical resonator, *IEEE Trans. Nucl. Sci*. NS-19:217 (1972).
7. K. K. Schulze, Preparation and characterization of ultra-high-purity niobium, *Journ. of Metals* 33:33 (1981).
8. H. Padamsee, The technology of Nb production and purification, Proc. 2nd Workshop RF Supercond., H. Lengeler, ed., CERN Geneva (1984), p. 339.
9. M. Hörmann, The production of high thermal conductivity niobium on a technical scale for high frequency superconductors, available from W.C. Heraeus GmbH, Produktbereich Sondermetalle, P.O. Box 1553, D-6450 Hanau 1, Germany.
10. E. Drost and M. Hörmann, Mass Production of Superconducting Materials Nb and NbTi, available from W.C. Heraeus GmbH, Produktbereich Sondermetalle, P.O. Box 1553, D-6450 Hanau 1, Germany.
11. P. Kneisel, Improvements in materials and fabrication methods - large scale production and testing - Proc. 5th Workshop RF Supercond., DESY 1991, D. Proch, ed., Hamburg (1992), DESY Report M-92-01, p. 163.
12. H. Padamsee, A new purification technique for improving the thermal conductivity of superconducting Nb microwave cavities, *IEEE Trans. Magn. MAG-21*:1007 (1985).
13. P. Kneisel, Use of titanium solid state gettering process for the improvement of the performance of superconducting RF cavities, *J. Less Common Metals* 139:179 (1988).
14. Columbium (Niobium) Alloys and special products from Teledyne Wah Chang Albany, OR, USA.
15. Columbium (Niobium) from Kawecki Berylco Industries, Inc., Reading PA, USA.
16. Niobium / Columbium from Companhia Brasileira de Metalurgia e Mineração, São Paolo, Brazil.
17. J. P. Bacher et al., Brazing of niobium to stainless steel for UHV application in superconducting cavities, Internal Report CERN/EF-RF 87-7, Dec. 1987.
18. J. Susta, Development of Fabrication Methods, ibid. ref. 8, p. 597.
19. J. Kirchgessner, Forming and Welding of niobium for superconducting cavities, Proc. 3rd Workshop RF Supercond., K. W. Shepard, ed., Argonne Nat. Lab. (1988), ANL-PHY-88-1, p. 533.
20. T. Furuya, Preparation and handling of superconducting RF cavities, Proc. 4th Workshop RF Supercond. KEK, Tsukuba, Japan, 1989, KEK Report 89-21 (January 1990), p. 305.
21. D. Bloess et al., Superconducting, hydroformed, niobium sputter coated copper cavities at 1.5 GHz, Proc. 4th European Particle Acc. Conf. London 1994, V. Suller, Ch. Petit-Jean-Genaz, eds., Singapore (1994), p. 2057.
22. C. Hauviller, Fully hydroformed RF cavities, Proc. Particle Acc. Conf. Chicago 1989 (IEEE, New York, 1989), p. 213.
23. E. Chiaveri and H. Lengeler, Welding of Niobium Cavities at CERN, ibid. ref. 8, p. 611.
24. E. Chiaveri et al., Welding of radio frequency cavities with an internal EB gun, Proc. First Int. Conf. Power Beam Technology, Brighton, UK, 1986, The Welding Institute, London.
25. A. Citron et al., The Karlsruhe-CERN superconducting RF separator, *Nucl. Instr. Meth.* 164:31 (1979).
26. Ph. Bernard et al., Superconducting hydroformed niobium sputter coated copper cavities at 1.5 GHz, ibid. ref. 3, p. 739.
27. D. Bushnell, BOSOR5 Program for buckling of complex, branched shells of revolution including large deflections, plasticity and creep, Lockheed Applied Mechanics Lab.
28. J. Turneaure et al., Performance of 6-m 1300 MHz superconducting niobium accelerator structures, *Appl. Phys. Lett.* 25:247 (1974).
29. V. Palmieri et al., A new method for forming seamless 1.5 GHz multicell cavities starting from planar disks, ibid. ref. 21, p. 2212.
30. A.I. Ageev et al., Development of production technology for weldless copper shells of sc cavities, ibid. ref. 3, p. 802.
31. E. Kako et al., Long term performance of the TRISTAN superconducting RF cavities, Proc. 1991 IEEE Particle Acc. Conf. San Francisco, USA, 1991, (IEEE, New York, 1991), p. 2408.

32. E. Chiaveri et al., Analysis and results of the industrial production of the superconducting Nb/Cu cavities for the LEP2 project, Proc. XVIth Int. Conf. High Energy Acc., Dallas TX, USA, 1995, in press.
33. C. Reece et al., Performance of Production SRF Cavities for CEBAF, Proc. 1993 IEEE Part. Acc. Conf., Piscataway, NJ, USA (1993), p. 1016.
34. B. Aune, private communication.
35. M. Leenen, The infrastructure for the TESLA Test Facility (TTF) - a status report, ibid. ref. 21, p. 2060.
36. B. Bonin and R. W. Röth, Q-degradation of niobium cavities due to hydrogen contamination, ibid. ref. 11, p. 210.
37. J. R. Delayen, RF properties of superconducting Pb electroplated onto Cu, ibid. ref. 19, p. 469.
38. L. Dietl and U. Trinks, The surface resistance of a superconducting lead-tin alloy, *Nucl. Instr. Meth.* A284:293 (1989).
39. G. Arnolds-Mayer and N. Hilleret, Field emission and secondary electron emission from Nb_3Sn surfaces, *Advanc. in Cryog. Eng. Mat.* 28:611 (1982).
40. C. Benvenuti et al., (NbTi)N and NbTi coatings for superconducting accelerating cavities, ibid. ref. 11, p. 518.
41. G. Arnolds-Mayer, ibid. ref. 8, p. 643.
42. M. Peiniger et al., Nb_3Sn for superconducting accelerators at 4.2 K, Proc. First European Acc. Conf., Rome 1988, S. Tazzari, ed., Singapore (1989), p. 1295.
43. G. Arnolds-Mayer and E. Chiaveri, On a 500 MHz single cell cavity with Nb_3Sn surface, ibid. ref. 19, p. 491.
44. B. Hillenbrand et al., Superconducting Nb_3Sn cavities *IEEE Trans. Magn.* MAG-11:420 (1975).
45. G. Gemme et al., RF surface resistance measurements of binary and ternary niobium compounds, *J. Appl. Phys.* 77:257 (1995).
46. H. Piel, private communication.
47. J. Sikora et al., Lead/Tin Resonator Development at the Stony Brook Heavy-Ion Linac, ibid. ref. 19, p. 419.
48. C. Benvenuti et al., Niobium films for superconducting accelerating cavities, *Appl. Phys. Lett.* 45:583 (1984); C. Benvenuti et al., Superconducting 500 MHz accelerating copper cavities sputter-coated with niobium films, *IEEE Trans. on Magn.* MAG-21:153 (1985).
49. S. Cantacuzène et al., Elaboration et caractérisation de couches minces de $(Nb_xTi_{1-x})N$ pour des applications en hyperfréquences, thesis, Université de Paris Sud, centre d'Orsay (1995).

A NOVEL ROTATING TEMPERATURE AND RADIATION MAPPING SYSTEM IN SUPERFLUID He AND ITS SUCCESSFUL DIAGNOSTICS

Q. S. Shu[1], T. Junquera[2], A. Caruette[2], G. Deppe[1], M. Fouaidy[2]
W-D. Moeller[1], M. Pekeler[1], D. Proch[1], D. Renken[1], C. Stolzenburg[1]

[1]Deutsches Elektronen-Synchrotron (DESY)
Notkestrasse 85, 22607 Hamburg, Germany
[2]Institute of Nuclear Physics
(CNRS - IN2P3) 91406 ORSAY cedex, France

ABSTRACT

An novel rotating temperature and radiation mapping system in He II has been developed to investigate field emission (FE) & thermal breakdown (TB) in TESLA 9-cell SRF cavities. More than 10,000 spots on a cavity surface can be analyzed in one turn with 5° stepping. 116 special surface scanning thermometers have been developed to measure surface temperature in He II. 32 photodiodes are employed to study the X-rays induced by FE electrons. Each rotating arm holds 14 thermometers and 4 photodiodes. A unique driving and suspension system is designed to gently turn the 9 arms around the cavity and uniformly press the thermometers against cavity surfaces. A moving adapter device (pancakes) is designed for rotating a large number of electronic cables which become inflexible in superfluid He.

The T-R mapping system has successfully detected and diagnozed serious problems caused by FE and TB, and has played a significant role in cavity processing.

INTRODUCTION

Main Obstacles of High Gradient Cavities

Field emission (FE) and thermal breakdown (TB) are still the main obstacles preventing SRF cavities from confidently reaching Eacc = 25 MV/m (TESLA's goal) from existing operating levels of 5-10 MV/m[1,2,3]. Most of the FE sources and TB defects on the inner RF surfaces of cavities were found to be submicro-sizes [4,5] and activated only at high RF fields while cavities are in a superconducting state. It is impossible to directly observe the FE and TB on the inner surface of cavities during RF operation. Therefore, the main approach to understanding the FE and TB of cavities is to study the hot spots and X-rays (induced by impacting FE electrons) generated on the cavity surfaces during RF operation.

DESY's Rotating T-R Mapping System

Various temperature (T) mapping and X-ray (R) mapping systems have been developed at many laboratories around the world[3,4,6,7,8]. The systems can be classified into two categories: (1) Fixed Mapping - thermometers or photodiodes are fixed on the surfaces of the cavity. (2) Rotating Mapping - thermometers or photodiodes are rotating against the surfaces of the cavity.

The Rotating T-R mapping system developed at DESY for TESLA 9-cell cavities combines measurements of T & R and employs special rotating scanning thermometers which were developed at INP Orsay.

Advantages.

(1) Greatly reduce the number of sensors: The DESY T-R mapping analyzes 10,000 spots on the cavity surface using only 116 scanning thermometers. A stystem using fixed thermometers would require 10,000 thermometers to analyse the same spots on cavity.

(2) Once an area is suspected, the sensors in DESY mapping system can be relocated to the suspect location during cryogenic-RF operation for additional analysis.

(3) The mapping combines T-R diagnostic systems to give information on both heating and x-rays for understanding the dynamic progress of cavity processing.

Challenges.

(1) Fixed contacts and use of grease as bounding agent to enhance thermal contact between thermometer and surface being measured are essential to reach a high efficiency (particularly, in the case of He II). In a rotating system, neither fixed contact nor grease can be applied. A new type of thermometer was needed[9].

(2) Due to TESLA cavity structure, thermometers can not reach the high risk areas of FE at cavity irises. A combined measurement of T & R was required.

(3) The TESLA cavity has 9 cells (the largest cell number for low frequency cavities) with complex surface curves. The space in the cavity test cryostat is tightly constrained. Assuring satisfaction of 3-dimension tolerances at all moving contact points is a challenge.

(5) A fast data acquisition system was also needed to trace the dynamic progress.

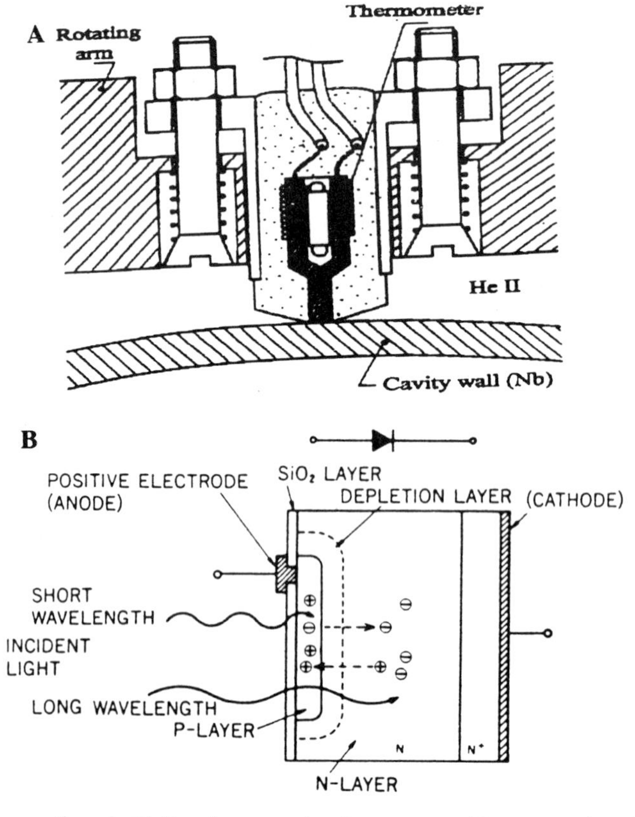

Figure 1 (A) Cross section of a HeII surface scanning thermometer, (B) cross section of a photodiode.

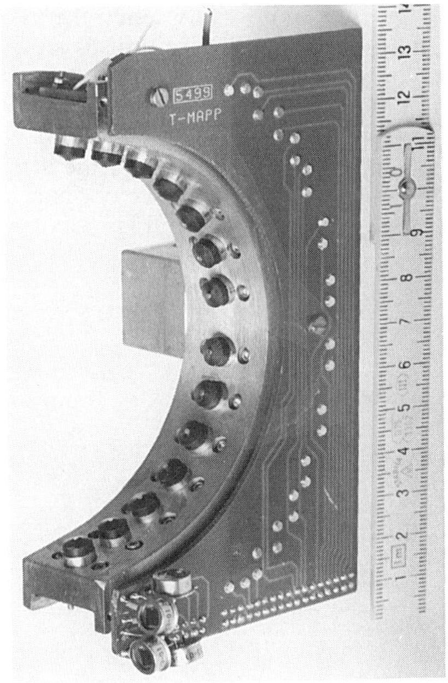

Figure 2. A picture of the rotating arm.

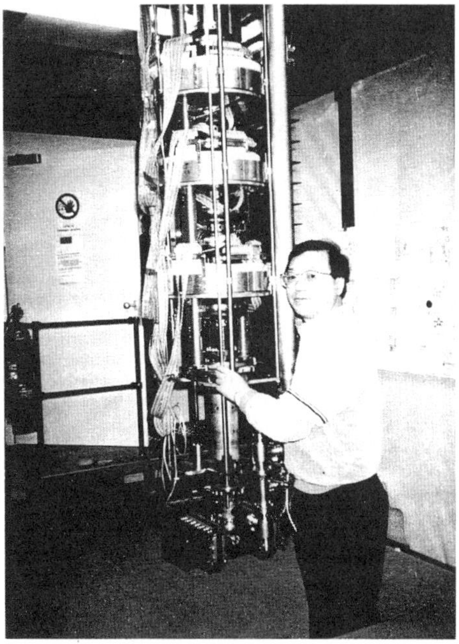

Figure 3. A picture of the T-R mapping system.

We have overcome the above challenges and developed the rotating T-R mapping system. Since December 1994, the system has been successfully employed in diagnostic tests and played a significant role in cavity processing[10].

TECHNOLOGIES DEVELOPMENT

Surface scanning Thermometer and Photodiode

The surface thermometer design as shown in Figure 1(A) is very close to the model developed earlier for the CERN's SRF cavity project by INP Orsay[10,11,12]. The sensitive part is an Allen-Bradley carbon resistor (100 Ohm, 1/8 W) housed in a silver block with a sensor tip of 1 mm diameter for the thermal contact to the external surface of the cavity. This housing is thermally insulated from the surrounding He II by an epoxy envelope (Stycast) moulded around the silver block and into a bronze piece which allows the sensor to be mounted in the rotating thermometric arm. The thermometer's tip must present a good contact with the cavity wall when scanning. Each thermometer has two independent manganin wires thermally anchored to the silver block with ~ 15 cm free length for connecting to each cell board (14 thermometers).

The complementary calibration test was performed by mounting the thermometers in the real operating conditions of the scanning device at different spring pressures and heat fluxes in He II and subcooled He (2.3K and one bar). The detailed results are presented in another paper[9]. In the case of He II, the efficiency is heater power dependant.

Commercial PIN silicon/S 1223-01 photodiodes are used as x-ray detectors in the mapping because of their small size (3mm x ϕ10 mm) and ultra-fast response. A simplified photodiode cross-section is shown in Figure 1(B).

Rotating T-R Arms

As shown in Fig. 2, 14 thermometers and 4 photodiodes are mounted in each arm which is precisely machined to have the same curved surface as the cavity cell. Due to the reinforced structure of TESLA cavity, the thermometers can not directly touch the surfaces of the cavity iris. Considering that the electrical fields reach maximum at the iris, 4 photodiodes are located in the end of each arm to monitor FE induced X-rays while 14 thermometers are used to monitor the temperatures in the entire region between the irises of each cell[13]. Two springs located inside two holes in the body of the rotating arm are used to adjust the contact pressure. A printed circuit board is mounted on the side of the arm. All cables for the sensors are fed through a device, called moving adapter device or "pancakes", and then connected to a feedthrough on the top flange of the cryostat. Fig. 3 is the T-R mapping system.

Driving and Suspension Frame (DSF)

A total of 116 surface scanning thermometers (14x7 + 9x2)[13] and 32 photodiodes are assembled into 9 rotating arms which are mounted in the DSF as shown in Figure 4 (A).

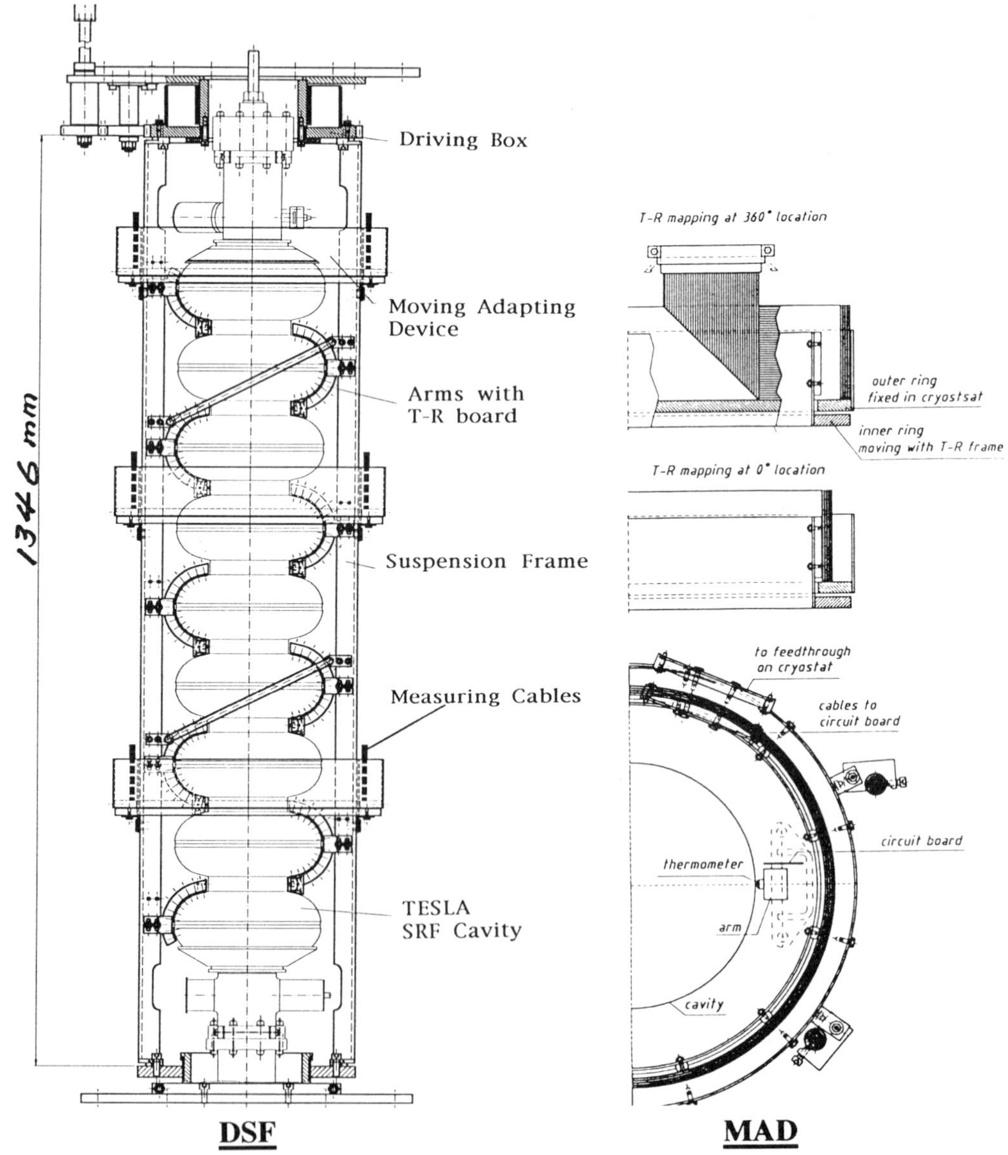

Figure 4. Schematic cross sections of the DSF and MAD.

The most important consideration in the mechanical design is to assure the three dimension tolerance between the cavity surface and the tips of the 116 thermometers over the entire cavity surface (i.e. more than 10,000 points) are within ±1mm. The DSF has two centering rings to allow the axis of DSF as close to the axis of the cavity cells as possible. The rotating frame is suspended on two disks made of low friction materials. These structures enable the DSF in superfluid He to gently turn the arms around the cavity and uniformly press the thermometers (through the spring-holder structure, force=100 g per thermometer) against the cavity surfaces. Driven by a computer-controlled stepping motor, the T-R arms can be automatically turned to any position on the cavity surface with a accuracy of ±1 degree.

Moving Adapter Device (MAD)

A large number of electronic measuring cables have to move with the rotating arms when the T-R mapping rotates. These cables become very rigid in LHe. A moving adapter device[13] was successfully designed to overcome the problem as shown in Figure 4 (B). Each pancake has two rings. The inner ring is mounted in the moving DSF and turns with the DSF around the cavity. Its outer ring is fixed with cavity supports. One end of each 64-wire-cable is connected to the inner moving ring and the cables make 9 turns around the inner ring of the MAD while the other end of the cables connects to the outer fixed ring. When the DSF turns 360° around the cavity, the cables only make relatively short movement inside the MAD. The space in the TTF vertical cryostat is very constrained which makes the MAD design even more difficult. The MAD has been tested and functions well in many cavity experiments.

Fast Data Acquisition and Test Procedure

Two Ge-thermometers and three additional scanning thermometers are used to monitor the change of bath temperature during measurement. To check the photodiodes two small lights are placed in the DSF. Maps can be taken with auto-scanning of entire cavity surface or scanning with time in a fixed position. The temperature change ΔT is made by comparison of measurements of RF power on and off (also, the bath temperature changes are subtracted from the total ΔT). The effective resolution of temperature measurement is less than 5 mK. One longitudinal measurement in a fixed angular position can be completed in less than 10 ms. All data taken, control and display are performed through a multiplexer by a Sun-station computer with a LabView™ language program[14].

RESULTS, ANALYSIS & DIAGNOSTICS

The T-R mapping has been successfully employed in the diagnostic testing of TESLA SRF cavities. We will briefly introduce some of the more interesting results here. A detailed test report will be presented in the "1995 SRF Superconducting Workshop".

Detection of Thermal Breakdown

We use the TESLA 9-cell cavity (D-6) as an example shown in Figure 5. This cavity has been heat treated with Ti-purification at 1400° C for 4 hrs and a test sample has a RRR of about 500. At first, the cavity was limited by severe FE at Eacc = 4 MV/m and Q dropped to 8×10^7, Fig. 5 (A). With RF power processing, the FE events were eliminated and the T-maps show the heating areas by FE gradually disappeared. Finally the cavity reached Eacc = 12.5 MV/m through a high RF pulse peak power processing (HPP) and then limited only by a quench.

While scanning the entire surfaces of the 9-cell cavity, the T-mapping located a strong heating area centred at the equator of the cell-5 over 10 thermometers between the 10° to 50° longitude as shown in Fig. 5 (C). To further study the TB event, we relocated the T-R arm to the heating area, moved it by 5° angular steps and found the hottest spot to be near 35°. Finally we moved the arms to 35°, turned on the RF power in CW mode and observed

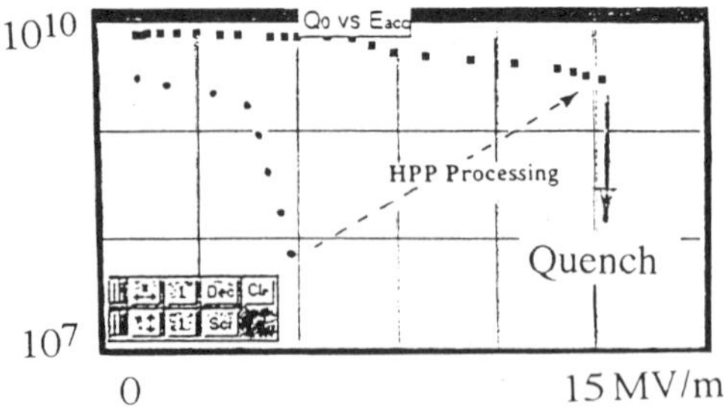

(A) Overall performance of the 9-cell cavity D6.

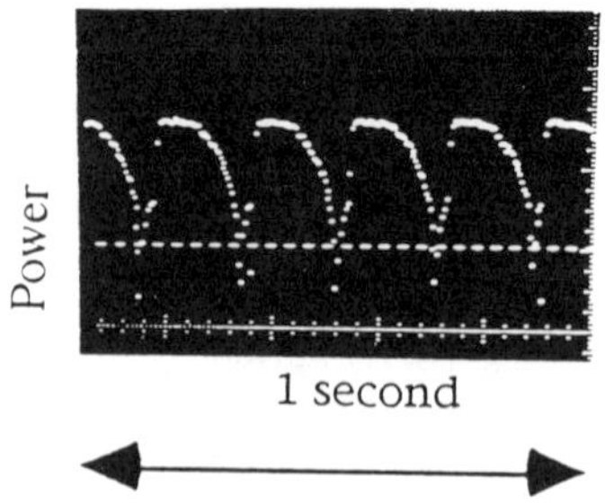

(B) Oscilloscope traces of transmitted power during continuing TB (quench) of the cavity.

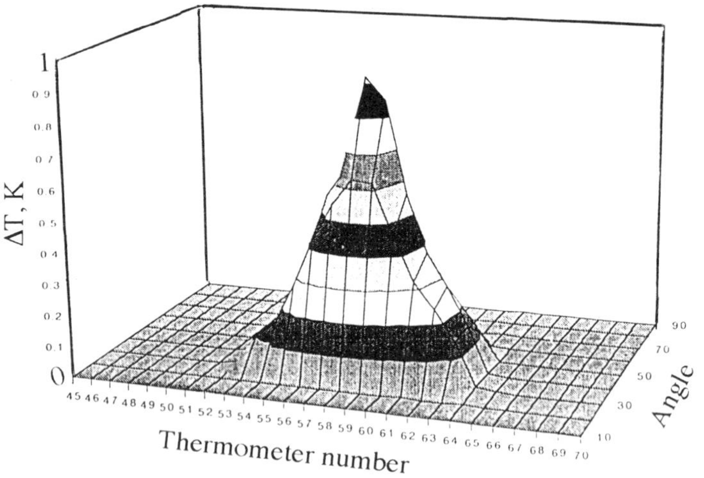

(C) The detected heating area due to TB (quench) at the equator of the cell-5.

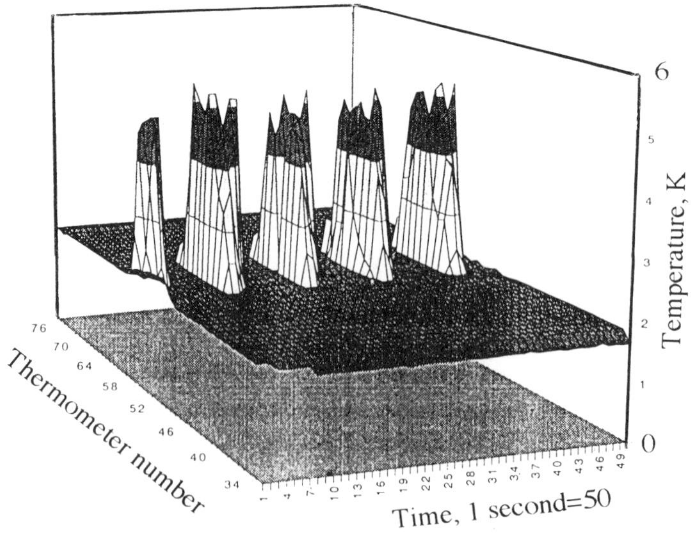

(D) Temperature changes as a function of time at longitudinal of 35° of the cell-5 during continuing quench & recovery.

Figure 5. The dynamic progress of a cavity thermal breakdown (quench) detected by the T-R Mapping.

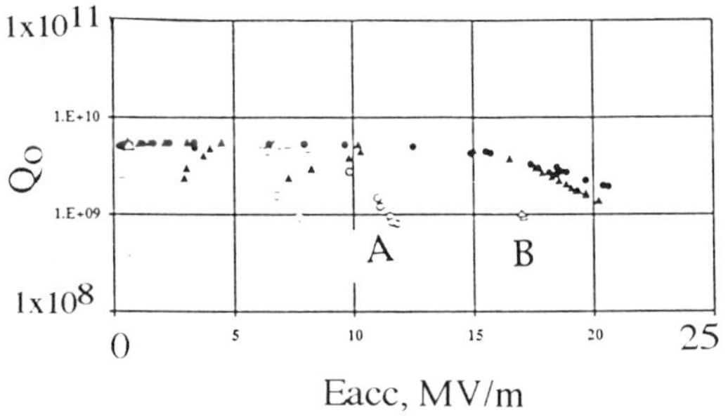

Figure 6. (A) Overall RF performance of the TESLA cavity -1.

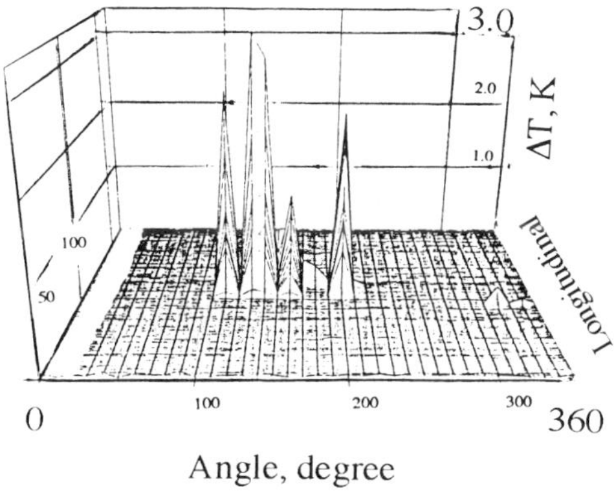

(B) FE heating area responsible for the Q dropping and limit of Eacc.

continuing quenches and recoveries of the cavity as shown in Fig 5 (B). Simultaneously, we continuously took temperature measurements at this fixed location. Fig. 5 (D) shows a dynamic progress of the temperature changes at 35° as a function of time. The highest temperature measured on the outer surface of the cavity is above 5 K and the quench was limited to one cell and did not propagate to adjacent cells.

These results indicate that HPP is very effective in eliminating FE[15], but not for TB. The tests tell us that there may be a combination of local defects at cell-5 and existence of a low thermal conducting thin layer (due to Ti-purification) on the cavity. Optical observation after test showed a deep scratch on the inner surface near the equator and close to 35° of cell-5 . We are planing to take an additional 50 μm of material from the inner surface and test it again.

Identification of FE Heating and Emitter Location

In investigating FE, we first use T-R mapping to identify the landing areas of FE electrons and then find out the locations of emitters with simulations of FE electron trajectories. we use a TESLA prototype 9-cell cavity (-1) as an example. It has reached 20 MV/m as shown in figure 6 (A). Previously, cavity -1 had been limited by thermal breakdown at about Eacc = 10 MV/m. Afterwards, the cavity -1 was heat treated at 1400° C with Ti-purification. We then removed 80 μm of material from the inner RF surface and 30 μm from outer side by chemistry, followed by high pressure rinsing.

Locating of Heated Areas and Intensity

In the test, cavity -1 was initially stopped by heavy field emission at point A (figure 6) at 11.2 MV/m with a Q of 8.5×10^{8}. The T-map, figure 6 (B) indicates an important heated region delimited by 12 thermometers (#53 to #64) centred close to the equator of the 5th cell, between the 110° to 200° angles. Outside of this region the heating is very low. The ΔT value in this region is 100mK - 3.3K. The y-axis of figure 6(B) is the thermometer number from 0, close to the top iris of cell-1, to 116, close to the bottom iris of cell-9. The x-axial represents the angular location on the cavity surface.

Analysis of Thermal Performance

The experimental data obtained with the T-R mapping is consistent with the thermal analysis if we make the following assumptions: high efficiency of thermometer at high heat flux, heat transfer governed by Kapitza regime, and electron trajectories impacts over a large area. The magnetic field heating at equators of the 5th cell, for Eacc=11.2 MV/m and Rs=30 nΩ, gives only ΔT=5 mK. The power related to the electron FE, Pelec=173 W, is focusing on local region.

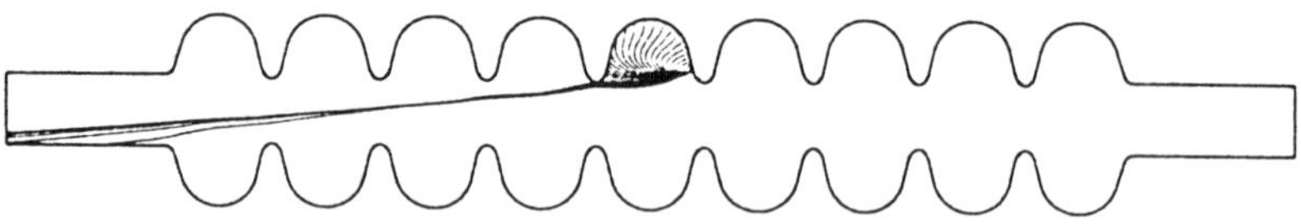

(A) FE electron trajectories of the emitter located at So = 8 cm and the emitter is responsible for the Q degradation.

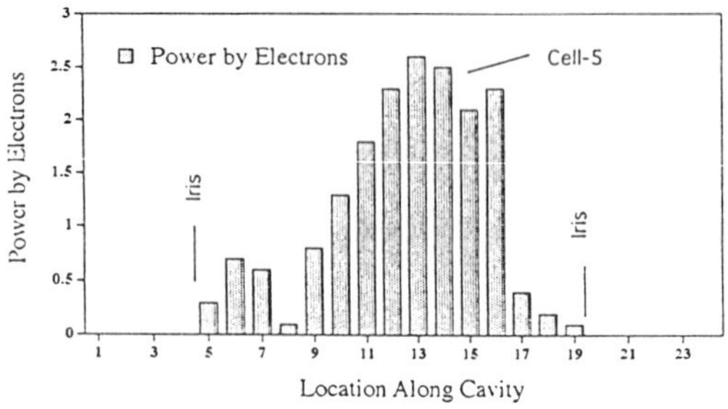

(B) Power distribution contributed by impacting FE electrons from So = 8 cm.

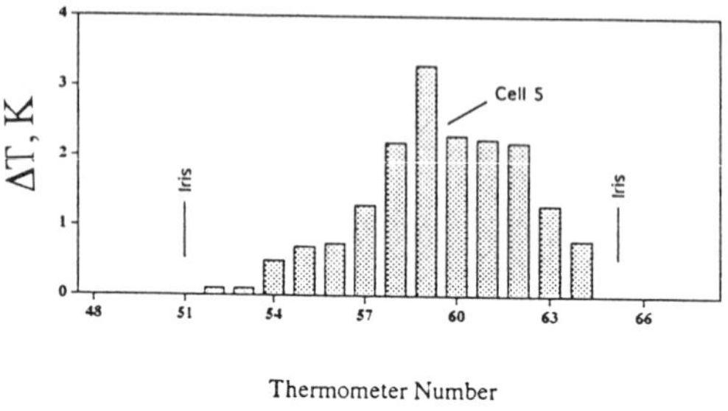

(C) Experimental longitudinal ΔT plot from T-map data of Figure 6 B fixed at 140°.

Figure 7. A comparison of computer simulation of the FE emitter with the experimentally heated areas.

The very high value ΔT measured in this region (100mK-3.3K) can only be explained by assuming that the efficiency of a scanning thermometer increases strongly with the heat flux density at the interface between the cavity wall and HeII. Such a high heat flux density is slightly less than the critical heat flux densities reported in experiments with metallic flat heaters in HeII[16]. So it is believed that the heat transfer is in the regime governed by Kapitza conductance. The integration of the product of Kapitza conductance and ΔT over the heated region leads to a total heat power going to He bath: Q~100 W. This value is consistent with the RF measurements of the experiment.

Identifying of FE Emitter Location

Locating origins of FE and TB is very important to understand the influence of various cavity processing and also to a guided reparation of defected cavities. However, the measured hot spots only indicate the landing of impacting FE electrons, but not the emitter.

The simulation of FE electron trajectories demonstrate the following interesting results: FE electrons from an emitter can impact over a very large area, the shape of trajectories are sensitive to emitter location (So), and emitters responsible for the heating areas (shown in figure 6 B) can be successfully identified. Since electron trajectories, impacting electron energy and power deposition distribution (dP/ds vs. s) are controlled by the Eacc and So. A series of simulations are performed by changing So at Eacc=11.2 MV/m, assuming emission enhancement $\beta = 200$, Se (emitter area) = $1x10^{-13}m^2$. It is found that an emitter located at So=8 cm (at the iris area) has electron trajectories shown in figure 7 (A). Its power distribution (dP/ds vs. s) in Figure 7 (B) seems to be very close to the shape of the measured temperature distribution, Figure 7 (C). It indicates that heated areas at the equator (usually by defects) can also be caused by FE. The β and Se of the candidate emitter were adjusted to fit with the thermal analysis and RF experimental data. For instance, at Eacc=11.2 MV/m, if Se=$1x10^{-13}$ m^2, $\beta = 400$, the total mean power landed over RF period is 10W.

Finally, the high pulse RF power processing (HPP) was introduced to the cavity (150KW) and successfully eliminated the field emitters. Another T-map also witnessed the FE elimination. After HPP, the cavity finally reached 20 MV/m in CW mode.

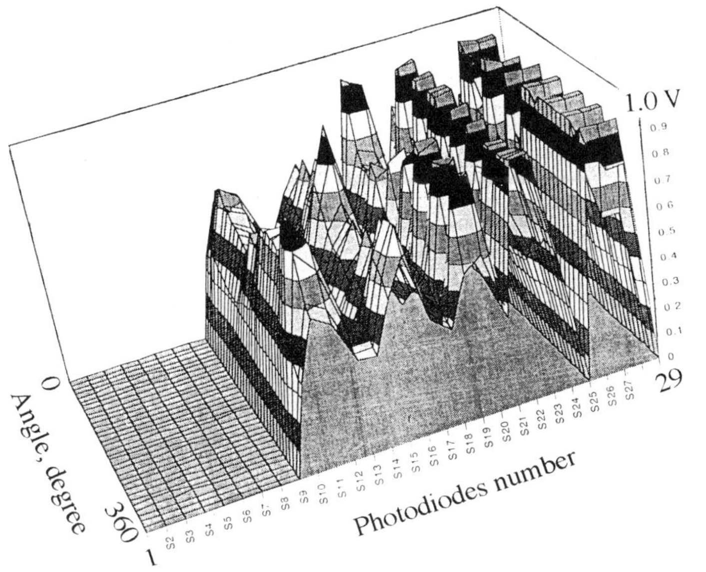

(A) An X-ray map in which the strong radiation closed to the cavity irises.

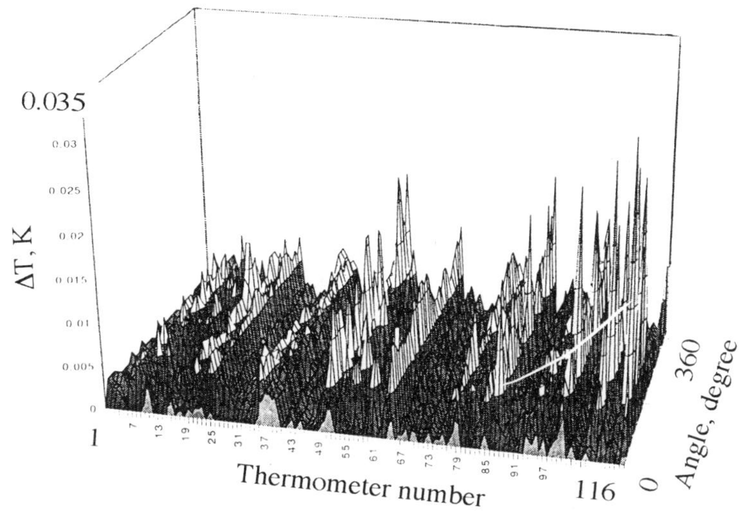

(B) A T-map taken in the same time as Fig 8(A) shows that the FE heated areas are also located mainly in cavity irises.

Figure 8. A comparison of the X-ray map with the T-map in the same 9-cell cavity.

Information from X-Ray Maps

A large number of radiation maps of X-rays induced by FE electrons were also observed. In general, information from X-ray maps are in consistent with that obtained from T-maps. As we mainly discuss the technical areas relevant to low temperature science here, the X-ray maps and analysis will be presented at the 1995 SRF superconducting workshop (Saclay, France). Figure 8 (A) shows an X-ray map recorded during a cavity processing, while figure 8 (B) presents the T-map taken on the same cavity and in the same time.

CONCLUSION

The T-R mapping system for TESLA 9-cell cavities was commissioned and successfully analysed and diagnosed the problems with the TESLA cavities caused by FE and TB. The information learned from T-R mapping results has played a significant roles in cavity processing and will be very valuable in further guided reparation of some cavities.

ACKNOWLEDGEMENT

We sincerely thank P. Kneisel (CEBAF), W. Weingarten (CERN), H. Padamsee, M. Champion (Fermilab), C. Pagani (INFN), B. Bonin (Saclay) and G. Wueller, R. Roeth (Wuppertal) for many fresh discussions and hints. Sincere thanks are also presented to our DESY colleagues in the cryogenic group, vacuum group, mechanical group, MHF group for their support.

REFERENCES

1. D. Proch (editor), Procceedings of the 5th workshop on RFS, Hamburg, Aug., 1991. (status report, p5, p23, p37, p44, p84, ans p245) published by DESY.
   R. Sundelin (editor), Procceedings of the 6th workshop on RFS, Newport News, Oct., 1993 (status report, p33, p49, p67, p77, p131, p173) published by CEBAF.
2. H. Pademsee, Applied Supercond. Conf., Boston, 1994.
3. Q.S. Shu et al., IEEE transaction, Vol. 27, No. 2, 1991
4. B. Bonin et al., Procceedings of the 6th workshop on RFS, 1993.
5. R. Roth and G. Muller et al., Procceedings of the 5th (and 6th) workshop on RFS, DESY 1991 (and CEBAF, 1993).
6. Ph. Bernard et al, Nucl. Inst. & Methods in Phys. vol 190.
   Ph. Bernard et al. & S. Buhler et. al, the 5th (and 6th) workshop on RFS, DESY 1991 (and CEBAF, 1993).
7. Q.S. Shu et al., Nucl Inst & Methods in Phys A278. 1989.
   J. Knobloch et al., SRF 94-0419-03, Cornell Univ., 1994
8. M. Fouaidy et al Proc 5th workshop on RFS, DESY,1991.
9. T. Junquera et al. TTP 14, Dallas, TX. PAC/95, May 1995.
10. Q.S. Shu et al., TTP 19, Dallas TX. PAC/95, May 1995.
11 R. Romijn, W. Weingarten, IEEE Trans. on Magnetics, Mag. p. 1318, 19 (1983).
12. S. Buehler et al., Procceedings of the 6th workshop on RFS, Newport News, published By CEBAF, Oct., 1993.
13. Q. S. Shu et al. TESLA Weekly Meeting, February, 1995, DESY, Hamburg.
14. M. Pekeler, internal tech. note to be in Ph. D. thesis, 1995.
15. J. Graber et al., Nucl. Inst. & Methods in Phy. A278. 1989.
16. A. Kashani, S. W. Van Sciver, Cryogenics 25, 1985.

CRYOGENIC AND ELECTRICAL TEST CRYOSTAT FOR INSTRUMENTED SUPERCONDUCTIVE RF CAVITIES (CHECHIA)

P. Clay,[1] J.P. Desvard,[1] R. Duthil,[1] J. Gastebois,[2] G. Grygiel,[3] U. Knopf,[3] R. Lange,[3] F. Lejars,[1] C. Mayri,[1] P. Pailler,[1] B. Petersen,[3] D. Sellmann[3]

[1]CEA - CE/Saclay - DSM/DAPNIA/STCM
91191 Gif-sur-Yvette Cedex (France)

[2]CEA - CE/Saclay - DSM/DAPNIA/SEA
91191 Gif-sur-Yvette Cedex (France)

[3] DESY - Notkestrasse 85
22603 Hambourg (Germany)

ABSTRACT

This facility has been designed to carry out cryogenic RF quality factor (Q_{RF}) measurements, cold tuner tests, high order modes coupler tests, main coupler tests and full RF tests of superconducting cavities under real accelerator conditions (pulse mode, feedback loops, etc). The cryostat is built to receive one horizontally positioned cavity equipped with its helium vessel, its tuner and various RF couplers. The maximum heat load is ≈ 58 W (2.5 g/s) at 2 K for high power peak processing and ≈ 6 W (0.25 g/s) under normal conditions. Q_{RF} will be obtained from heat load measurements by LHe level variation and gas flow measurement at known temperature and pressure. The calibration is given by a DC electrical heater. The paper describes the design and the preliminary results of cryogenic performance tests of the cryostat itself.

INTRODUCTION

The cryostat is built to receive a 9 cell cavity fully equipped with its welded helium vessel, main coupler, HOM couplers and cold tuner before final assembly of 8 cavities together in the TTF (Tesla Test Facilities) cryomodule.

It will allow measurement of cavity performances, tuning range, accuracy and speed of the cold tuner, depending on bath pressure variations and Lorentz forces action in the cavity.

It may also allow a more detailed study of RF electrons trajectories and X ray production, due to surface emitters.

In this simplified layout the quench behaviour of cavities at high fields can be studied and first tests on the handling of quenches and adequate protection systems can be performed, involving not only RF signals, but also the bath pressure and cavity vacuum.

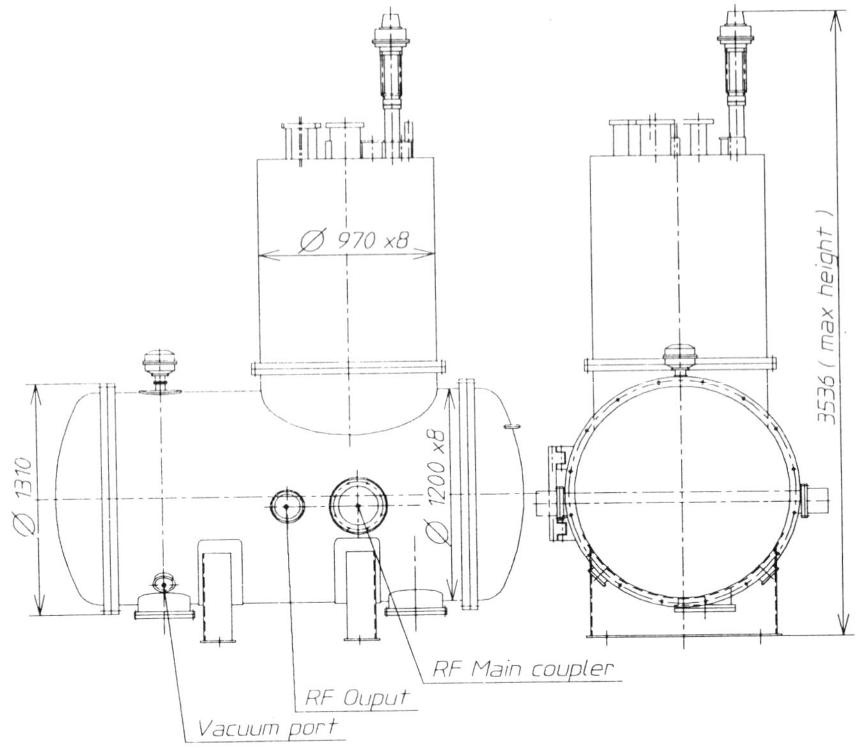

Figure 1. Outside view of the cryostat.

MAIN FEATURES OF THE CRYOSTAT

The vacuum vessel made of mild steel is equipped with a large hinged cover, giving full access for the cavity assembly. All connections are placed close to the cover (input coupler, helium supply vacuum pipe and cabling).

Two thick aluminum shields, one cooled at 77 K with liquid nitrogen, the second at 4.5 K with liquid helium insure the thermal insulation of the cavity (fig 1).

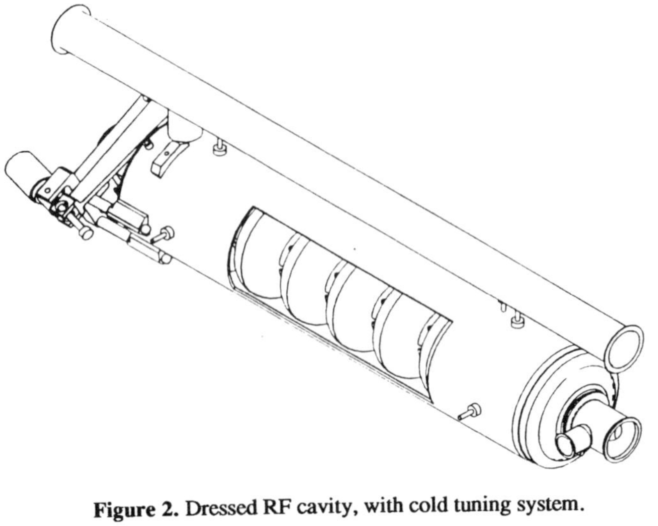

Figure 2. Dressed RF cavity, with cold tuning system.

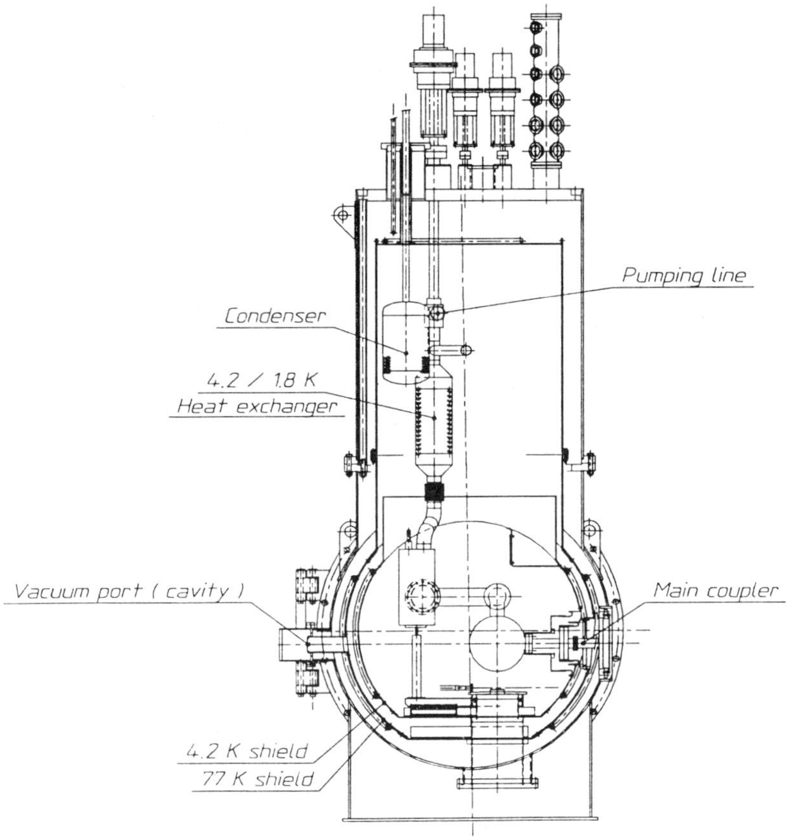

Figure 3. Cross section of the turret.

The cavity is fixed on a bench supported by two insulating posts. Each post has two intercepts one at 4.5 K and one at 77 K.

The cryogenic equipment is placed in a vertical turret . All components are easily accessible by lifting the vacuum turret sleeve (fig 2 & 3).

Table 1. Main characteristics of Chechia

Item	Unit	Value
Length	mm	2 612
Height	mm	3 536
4.5 K shield dimensions	mm	Ø 954 x 1850
Design pressure		
1.8 K section	bar abs	2
4.5 K section	bar abs	3
77 K section	bar abs	3
Heat consumptions		
1.8 K section	g/s	0.012
	W	0.28
	l/h	0.29
4.5 K section	g/s	0.49
(including transfert line)	W	9.9
	l/h	14.25
4.5 K shield temperatures		
Temperature inlet	K	≈ 4.5 K
Temperature outlet	K	≈ 6.5 K

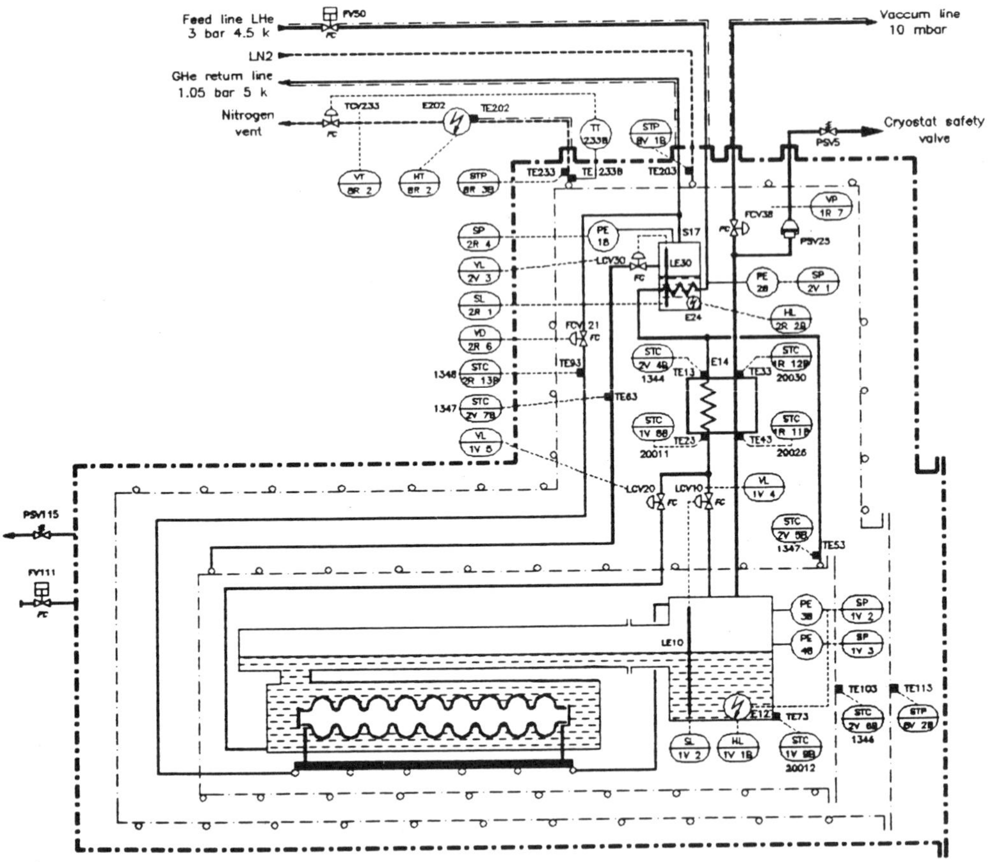

Figure 4. General flow chart of CHECHIA.

Table 2. Instrumentation

Sensors	Quantity	Type
Thermometers		
HOM	2x2	Carbon sensor RIVA®
Main coupler	2	»
Spare	5	»
He vessel		
Shields		
J.T. valves	1	Carbon sensor RIVA®
Pumping line	1	»
Cold tuner	2	»
Heater (2 K circuit)	1	»
Level sensors		
2 K sump	1	SC wire sensor
2 K sump	1	»
Pressure sensors		
1.8 K circuit	1 + 1	WIKA® MKS®
4.5 K circuit	2	WIKA®

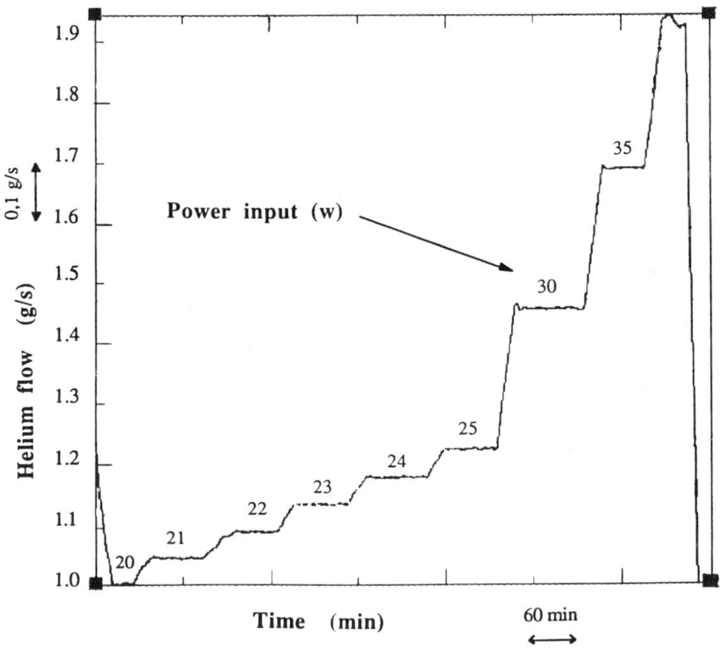

Figure 5. Flow measurements function of Q_H input (computed), with mass flow method.

CRYOGENIC FLOW CHART

The flow schematic for CHECHIA is shown in fig. 4.

The helium is supplied by a 4.4 K/900 W refrigerator at a pressure of about 3 bars and a temperature of 4.4 K.

Before being expanded into the two phase volume, the mass flow is passed through the 5 K heat shield. After having absorbed the heat load of the shield, the helium returns at about 5K to the condenser inlet and expands through the valve LCV 30. The vapor is returned to the refrigerator or to the helium recovery circuit (fig. 5).

From the liquid phase of the condenser a flow of about 2.5 g/s is forced into the heat exchanger where it is subcooled to about 2.2 K by means of the 2 K vapor return flow from the cavity helium vessel.

The final pressure in the cavity will be 16 mb at 1.8 K, after expansion through the Joule-Thomson valves LCV 10 and 20.

UTILIZATION OF THE CRYOSTAT

Cavity performance measurement

The aim is to measure a few points of the curve Q_{RF} versus accelerating field. As the input coupler will not have the right coupling coefficient, cryogenic measurements have to be done to determine the heat load at 2 K with and without RF.

The Q_{RF} input power is simulated by an electrical heater placed in the 2K sump.

The goal is to reach an accuracy of ≈ 0.4 W which means that the static heat load has to be at most about 4 W.

Two methods will be used:

1. For a steady liquid level, one measures the flow rate out of the cryostat. A good accuracy is expected in measuring the pumped helium gas flow, with mass flow meters, provided the measurement is made during sufficiently long time.

2. With no helium flow in, one measures the liquid helium level change. To increase the accuracy, the level probe is installed in the 2 K sump. This container is placed higher than the cavity vessel which can be kept full of helium during the measurement. The cross section of the sump is small enough to amplify the level variations.

Table 3. Measurement of Q by level difference at 1.8 K

Q input (w)	0	1	2	18	34
Total evaporated He flow (g/s)	0,012 (1)	0,053	0,095	0,82	1,48
Corrected flow (g/s)		0,042	0,084	0,81	1,47
Q measured (w)		0,97	1,95	18,2	34,2

High peak processing treatment

The HPP technique offers the possibility of cleaning up residual emission. Controlled exposures that test the survival of the benefits of HPP will be carried out to establish a protocol for the assembly of cavities. The helium consumption at 2 K will then be much higher than under normal RF processing. A compromise has been set at 2.5 g/s for the maximum helium flow at 2 K, corresponding to 58 watts.

RESULTS

Determination of Q_{RF} with the flow meters

Two mass flow meters are used, one covering the range of 0 - 1 g/s and the second the range of 0 - 20 g/s.

Equivalent Q_{RF} input is computed from the voltage on the heater. The flow was measured for various values of Q_H input power. The steady flow is obtained after 1 hour stabilization. For 1 watt after integration of the flow-meter signal during 1 hour, the precision is better than ≈ ±0,2 W in the range of 20 to 40 watts of input power.

If a calibration is made before the test of the cavity, an accuracy of ≈ ±0.1w can be reached.

Determination of Q_{RF} with level difference

The level is measured with a superconducting sensor in the sump, the internal diameter is 150 mm.

The precision of level measurement is ≈ 1 mm over a height of about 80 mm, which gives an accuracy better than ≈ ±0.2 w (over 4 mn). This measurement requires no additional calibration.

(1) Static losses

The precision of this method is altered by the decrease of heat losses coming from the piping when the input power increases.

CONCLUSION

Both mass flow and level difference measurements can reliably be used to determine the Q_{RF} values. The accuracy of mass flow measurements depends on the calibration before the tests. The level difference method gives a better accuracy in the low range values (0 - 10 watts).

REFERENCES

1. A proposal to Construct and Test Prototype Superconducting R.F. Structures for Linear Collider, DESY/Hamburg, April 1992.
2. Tesla Test Facility Linac - Design Report, March 1995, TESLA 95-01.

THE TESLA 500 CRYOGENIC SYSTEM LAYOUT

G. Horlitz,[1] T. Peterson[2], D. Trines[1]

[1]Deutsches Elektronen-Synchrotron DESY
Notkestraße 85, 22603 Hamburg, Germany
[2]Fermi National Accelerator Lab.
P.O. BOX 500, Batavia, IL 60510, USA

ABSTRACT

The superconducting 2 x 250 GeV e^+e^- linear accelerator TESLA is composed of a series of high frequency resonators (cavities), fabricated of high purity niobium. It has to be operated at a temperature of ca. 2 K. A nominal refrigeration power of about 51 KW at 2 K, 37 KW at 4.5 K and 314 KW at 40/80 K is required. The cooling power has to be distributed over a distance of more than 30 km. Different aspects of the cryogenic system are discussed.

INTRODUCTION

The TESLA project has been presented previously [1]. Two linacs of about 15 km length, producing 250 GeV electrons and positrons respectively, will be installed collinearly, separated by a ca. 2 km long interaction zone (fig. 1). The cryogenic system, maintaining the operating temperature of 2 K, will be described in the following chapters.

GENERAL LAYOUT

Cavities

Basic elements of the linac are cavities, consisting of nine cell high purity niobium RF-resonators, welded into cylindrical containers of titanium (fig. 2). Details of this design are presented in other sessions of this conference [2; 3]. The space between the cavity and the cylinder will be completely filled with superfluid liquid He at temperatures below 2 K, in order to maintain the desired operating temperature.

Modules

A module (12.2 m length) is an assembly of 8 cavities mounted into a common vacuum cylinder together with one quadrupole and one absorber for higher order mode (HOM) losses at an end. Two concentric heat shields, one at 4.5 K, and a second one at a

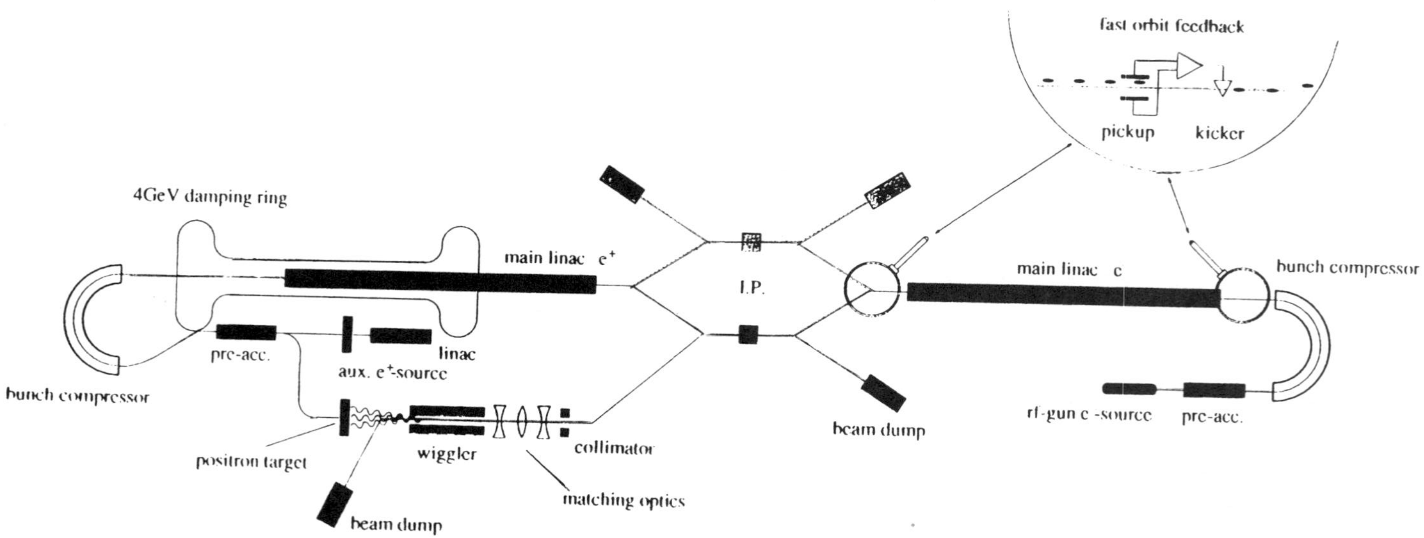

Fig. 1 TESLA General Layout

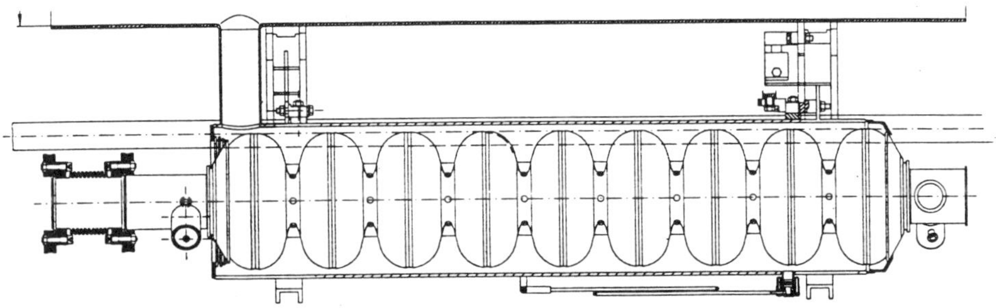

Fig. 2 TESLA Cavity.

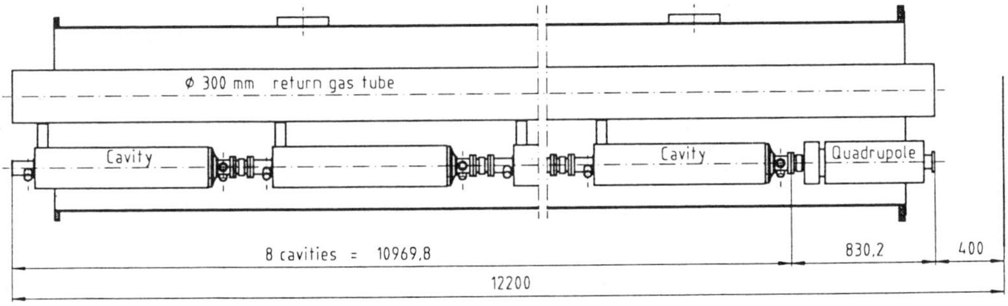

Fig. 3 TESLA Module.

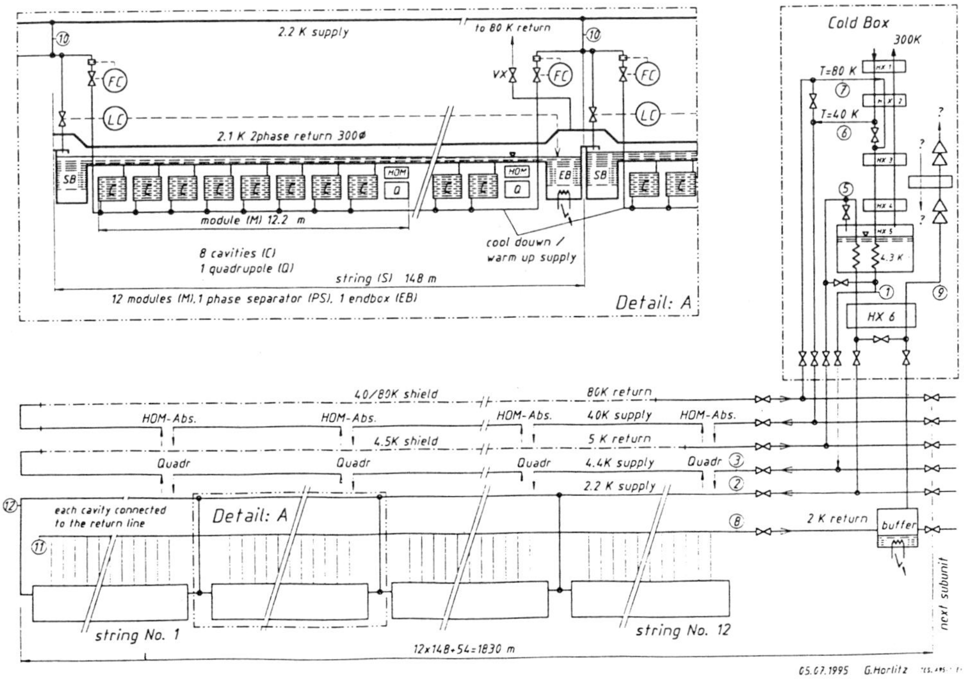

Fig. 4 TESLA Cryogenic Circuits.

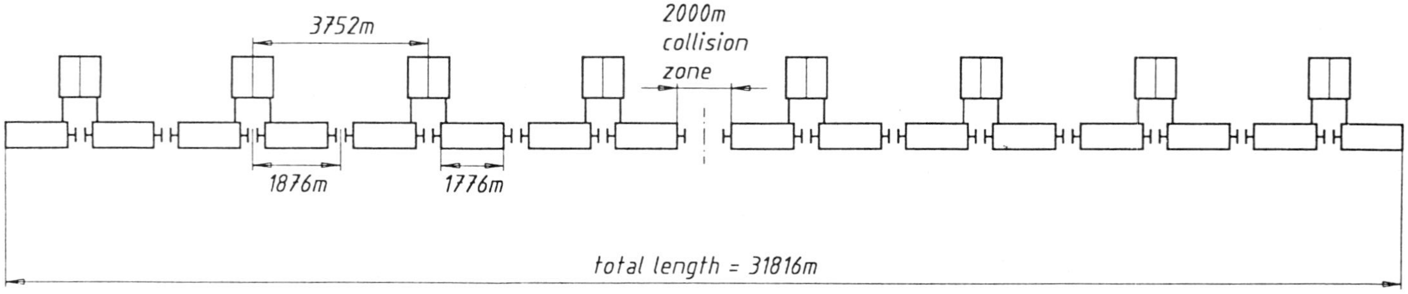

Fig. 5 TESLA Cryogenic Subdivsion .

Table 1 TESLA heat loads

(n = normally expected values; d = design values; stat = HF-power off; dyn = HF-power on)

Temperature	2 K		4.5 K		40/80 K		
	stat	dyn	stat	dyn	stat	dyn	W
Module n	2.8	21.4	13.9	16.0	76.8	136.0	W
d	4.2	32.1	20.9	24.0	115.2	204.0	W
String n	43.0	266.2	166.8	192.0	921.6	1632.0	W
d	64.5	399.3	250.2	288.0	1382.4	2448.0	W
Subunit n	521	3200	2007	2309	11070	19595	W
d	782	4800	3011	3500	16605	30000	W
TESLA n	8336	51200	32112	37000	177120	313500	W
d	12512	76800	48168	56000	265680	480000	W

temperature between 40 K and 80 K, protect the cavities. All cold supply and return lines go through the modules as well [fig. 3]. The TESLA 500 cavities will be connected directly to the 0.3 m return tube The 100 mm two phase tube of the TTF cryostats is eliminated. The exact number of quadrupoles in the machine is not yet fixed. Probably only one quadrupole every second or every third module will be installed.

Strings

12 modules are arranged in a string. At the end of the string there will be a supply box, the other end is terminated by an end box (fig. 4, Detail A). The total length of a string including the boxes is 148 m.

Subunits

An assembly of 12 strings is a subunit (fig. 4). It will be supplied by one common refrigerator . The resulting length of 1830 m for each subunit is tolerable from the point of view of pressure drops within reasonable tube sizes.
The "geographical" length of subunits in fig. 5 differs somewhat from the "cryogenic" length in fig. 4 due to the location of the refrigerators.

The machine

Based on an accelerating gradient of 25 MV/m, for an end energy of 250 GeV an active length of 10000 m per linac is required, which means a mechanical length 14908 m (8 subunits, space factor 0.671). The total length of the system, including 2 km interaction region, amounts to ca. 32 km (Fig 5).

HEAT LOADS

An estimate of heat loads for modules and cavities has been given recently [4]. Static and total (dynamic) loads are summarized in Tab. 1. Higher order mode (HOM) losses are partially transferred to the cavity helium, a large fraction escapes out of the cavities and travels through the beam pipe from where it has to be removed by means of absorbers in the quadrupoles, cooled by the 40/80 K shield helium.

The values marked "n" in Tab. 1 represent the calculated actual estimates ("nominal"). The whole system was designed with safety factors of about 1.5 for all three temperature levels. The "design" values are marked "d".

The equivalent room temperature power inputs have been evaluated, taking into account conversion factors of 800; 250; 25 W/W for T = 2.0; 4.5; 60 K respectively. They are listed in Tab. 2.

Table 2 TESLA ambient temperature electrical power input in mega watt

Temperature		2 K		4.5 K		40/80 K		Σ input	
		stat	dyn	stat	dyn	stat	dyn	stat	dyn
Subunit	n	0.42	2.56	0.50	0.58	0.28	0.49	1.2	3.6
	d	0.63	3.84	0.75	0.88	0.42	0.75	1.8	5.5
TESLA	n	6.7	41.0	8.0	9.3	4.4	7.8	19.1	58.1
	d	10.0	61.4	12.0	14.0	6.6	12.0	28.6	87.4

Table 3 Thermodynamic data at different points of the flow scheme

normal heat loads n				design heat loads d		
No.*	m [kg/s]	T [K]	p [bar]	m [kg/s]	T [K]	p [bar]
1	0.371	4.5	3.9	0.560	4.5	5.4
2	0.160	2.2	3.0	0.242	2.2	3.0
3	0.211	4.5	3.9	0.318	4.5	5.4
5	0.210	5.3	3.04	0.317	5.43	3.1
6	0.0945	40.0	18.0	0.142	40.0	18.0
7	0.0945	80.0	17.58	0.142	80.0	17.0
8	0.160	1.98	0.0297	0.242	1.958	0.0276
9	0.160	3.38	0.027	0.242	3.38	0.025
11	0.0133	2.00	0.0313	0.0202	2.00	0.0313

* position numbers of flow scheme Fig. 4

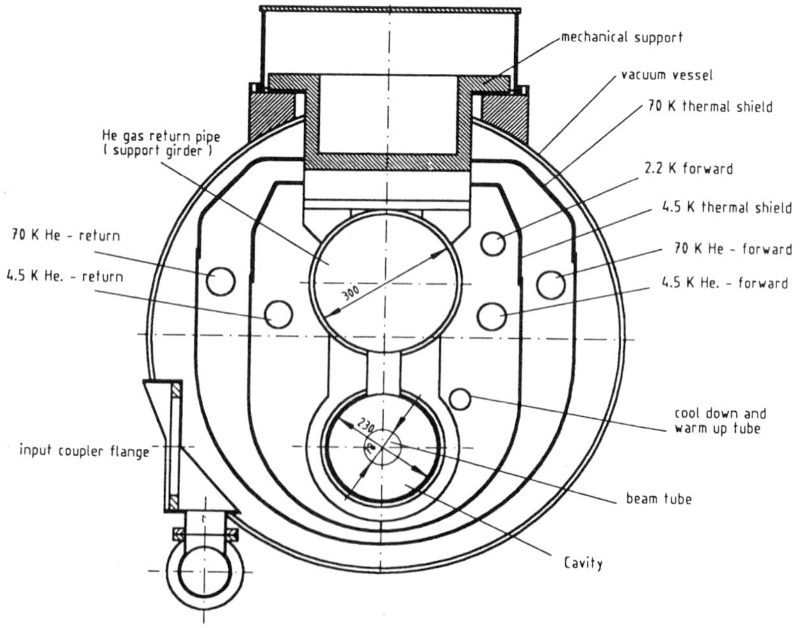

Fig. 6 TESLA Cryostat Cross Section.

THE CRYOGENIC SYSTEM

The generalized flow scheme is shown in fig. 4. A refrigerator (to be described elsewhere in this conference [5]) produces supercritical helium ($p \cong 5.4$ bar, $T \cong 4.5$ K) in point 1.
The whole mass flow enters first the 4.5 K quadrupole line at point 3, passes all quadrupoles in the subunit and returns from the end through the 4.5 K shields to point 5. A fraction is further cooled to 2.2 K in the heat exchanger HX6. It enters the 2.2 K supply line through point 2. The rest of the flow is expanded into the bath precooler HX5. ($T \cong 4.3$ K, $p \cong 1.1$ bar).

From point 6, a 40 K/18 bar flow is deviated for the 40/80 K shield cooling. After the passage through the shields and the HOM-absorbers in series it returns with $T \cong 80$ K, $p \cong 17$ bar at point 7.

At every string, about 1/12 of the 2.2 K supply flow is expanded through Joule Thomson valves into the supply boxes. From here, a two phase flow moves through the 0.3 m diameter return tube back to the end box of the string.
The heat from the cavities is transferred to the liquid in the return tube through vertical connection tubes between cavities and return tube (fig. 6) by means of the internal heat conduction process in He II [6]. In the tube, all heat flow is removed by evaporation of liquid at its surface.

For steady state operation the flow rates in the strings have to be controlled depending on the heat loads. At the end of each string, all liquid has been converted into vapor. The level is kept constant by means of a level controller LC acting on the JT-valve.

The end box of each string is mechanically combined with the supply box of the next one. A wall in the lower part seperates the two adjacend liquid levels, the vapor passes through (fig.4 Detail:A).

In point 8 the vapor returns to the heat exchanger HX 6 via a connection tube (0.3 m diameter, ca. 54 m length) and a buffer tank. The return gas temperature is about 1.9 K, its pressure ca. 30 mbar.

At point 9, the helium is warmed up to ca 3.4 K. Cold compressors (4 stages) are used to raise the pressure to the inlet pressure of the main compressor group (ca. 1.0 bar).

Driving forces for the liquid through the strings are the gradients of vapor pressure $\partial p/\partial x$ and liquid level $\partial h/\partial x$. Various flow conditions have been investigated under the assumption of independent flow of liquid and vapor in the return tube [7]. Fig. 7 displays pressure, temperature and liquid level in the return tube under different loads.

Table 3 gives massflow rates ($\dot{m}$), temperatures T and pressures p at different points in the system.

Cool down/warm up procedures

Cooling will be performed in several steps. Starting at room temperature, helium of about 80 K will be supplied through the 2.2 K line. At point 10 (fig. 4, Detail A) cool down lines parallel to the strings are branched. Every cavity is connected to these lines by means of narrow tubes, the flow resistances of which are large compared to all longitudinal resistances. This assures an approximately uniform flow distribution through the cavities, returning through the common 0.3 m return line. A parallel cooling of all cavities of one subunit is estimated to reach 80 K within some hours. Faster cooling, if desired, is possible, if the cooling capacity is concentrated to only one or two strings at the same time. In order to prevent cold exhausting gas penetrating to the next strings not yet to be cooled, the 2 K line must be closed at the end and the gas returns through valve VX and the 80 K return line. The cooling will be continued with running turbines and decreasing temperatures until the cavity temperatures are at about 4.5 K. At this level the filling of the cavities can be executed through the JT valves and the return line. Once having filled the whole system with 4.5 K liquid the cold compressors can start to establish final pressure and temperature.

Alternative circuit

An interesting alternative solution is presented in fig. 8. Here, the whole massflow for the string is injected through one single JT-valve at the end in point 12. Subdivision into strings is only necessary for cool down/warm up processes. The advantages of this system are: only one JT-valve with control loop hardware and software instead of 12, only 2 boxes at the ends of the subunit. The overall length of a subunit reduces to 1812 m. Flow conditions

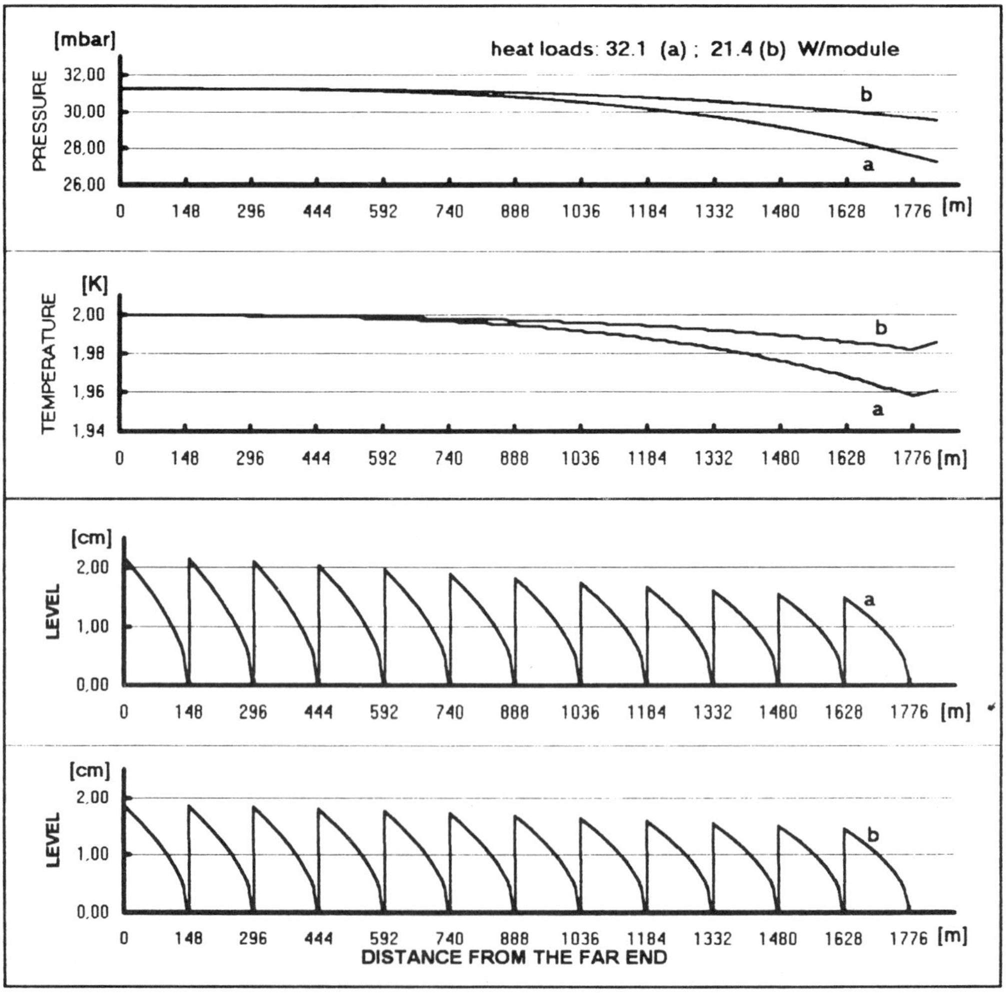

Fig. 7 TESLA Flow Parameters (1 JT-valve per 148 string).

Table 4 Helium stored in different sections of a TESLA unit

Circuit section	A [kg]	B [kg]
Liquid in the return tube	262.2	1630.7
Vapor in the return tube	95.7	85.8
Cavaties and other tubes	6662	6624
Total He-mass per subunit	7025	8346

A = 12 JT-valve version
B = 1 JT-valve version

have been calculated for this system as well [6]. Pressures and temperatures are very similar to fig. 7, the much smoother level behavior is shown in fig. 9.

A disadvantage may be that a higher liquid level at the JT-valve end of the string is required for driving the total string liquid rate through the full length of the unit. This causes a higher amount of liquid being stored in the return line.

The estimated amounts of helium in a subunit are listed for both solutions in Tab. 4 under design conditions.

SUMMARY

A possible cooling scheme for TESLA has been worked out and is proposed here. Also many requirements are unique, the limits of the present state of the art are certainly reached, but not exceeded. For future optimisation of the system, experiences from CEBAF, LEP 2, LHC and especially from TTF will be of essential value. Furthermore, it has to be investigated, whether the simplicity of the alternative scheme has drawbacks for operation or cooldown.

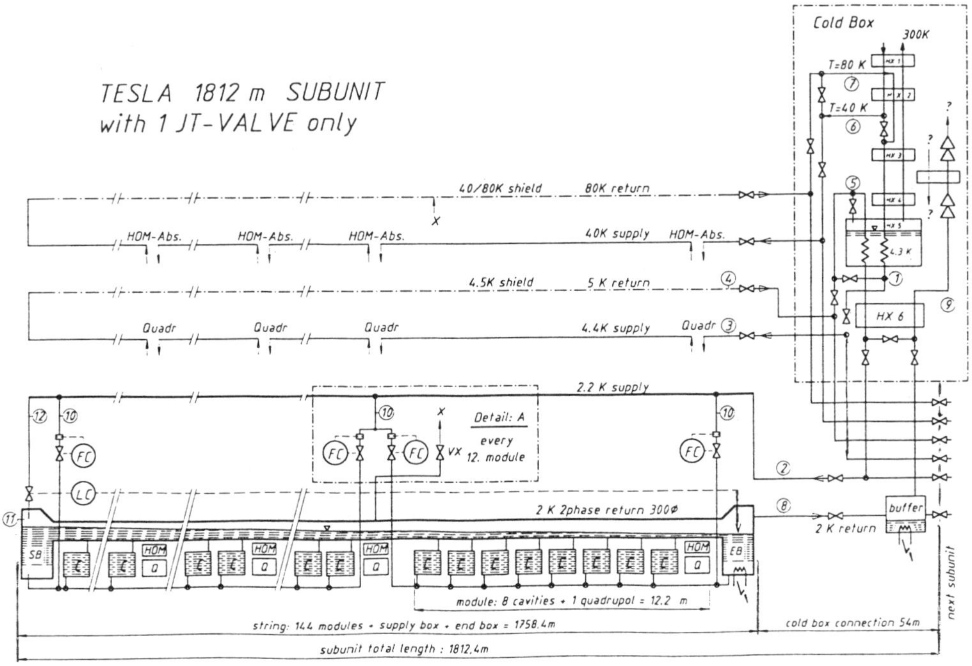

Fig. 8 TESLA Alternative Scheme.

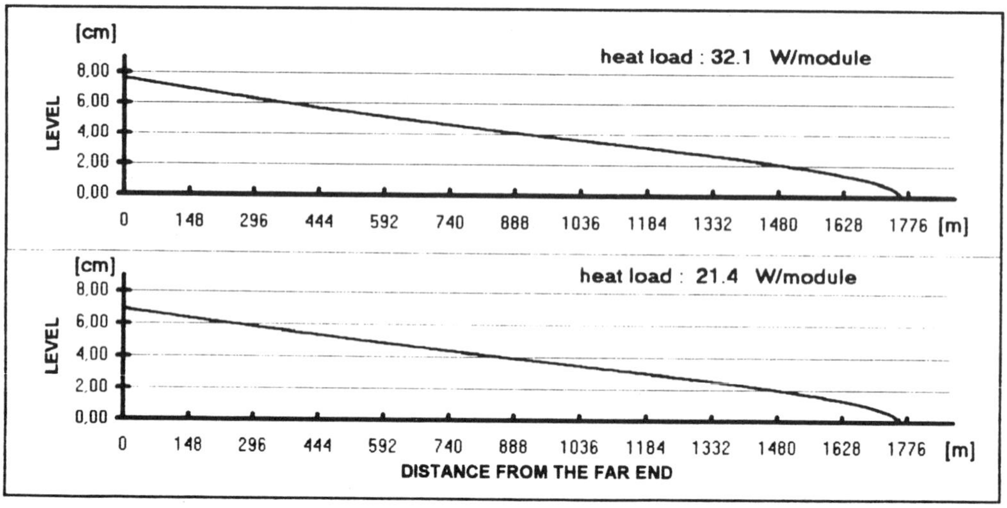

Fig. 9 TESLA Liquid Levels (1 JT-valve per 1812 m subunit).

REFERENCES

1. G. Horlitz, T. Peterson, D. Trines, A 2 Kelvin Helium II Distributed Cooling System for the 2 x 250 GeV e^+e^- Linear Collider TESLA, Proc. 15. Intern. Cryog. Eng. Conf. Genova, 1994, p. 131
2. Q. S. Shu, D. Trines, Technical Challenges of Superconductivity and Cryogenics in Persuing TESLA Test Facility (TTF), Proc. CEC/ICMC 1995, to be published
3. T. Nicol, TESLA Test Facility Cryostat Design, Proc. CEC/ICMC 1995, to be published
4. D. A. Edwards (Editor) TESLA Test Facility LINAC-Design Report, TESLA-Report No. 95-01, p. 228
5. M. Kauschke, C. Haberstroh, H. Quack, Safe and Efficient Operation of Multistage Cold Compressor Systems, Proc. CEC/ICMC 1995, to be published
6. T. Peterson, Notes about the Limits of Heat Transport from a TESLA Helium Vessel with a Nearly Closed Saturated Bath of Helium II, TESLA-Report No. 94-18
7. G. Horlitz, Computer Simulation of Pressures, Temperatures and Liquid Levels in a 1830 m TESLA Subunit under different Conditions and System Configurations, TESLA-Report No. 94-17, to be updated and published Sept. 1995

SAFE AND EFFICIENT OPERATION OF MULTISTAGE COLD COMPRESSOR SYSTEMS

M. Kauschke, C. Haberstroh, H. Quack

Technische Universität Dresden
Lehrstuhl für Kälte- und Kryotechnik
D-01062 Dresden, Germany

ABSTRACT

Large refrigeration rates in the temperature range of super fluid helium can only be obtained with the help of centrifugal cold compressors. For the large 2 K systems, four compression stages are necessary to reach atmospheric pressure. Centrifugal cold compressors are quite sensitive to mass flow and suction temperature variations; but these have to be expected in a real system. The first step in the systems design is to find safe and efficient quasi-stationary modes of operation.

The system which is being proposed for the TESLA refrigerators relies on two features. The first is to allow the room temperature screw compressor, downstream of the cold compressors to work occasionally with a subatmospheric suction pressure. The second is to stabilise the suction temperature of the third stage of compression at about 10 K. With these features it is possible, that in all modes of operation all four compressor stages operate exactly at their design point.

INTRODUCTION

In the linear collider project TESLA electrons will be accelerated up to a collision energy of 500 GeV. TESLA will work with superconducting cavities operating at 1.3 GHz. This e^+e^--collider will be built from nine-cell cavities with an acceleration gradient of about 25 MV/m. For a final energy of 500 GeV a total accelerator length of 32 km is required.
The liquid helium for the linac will be made available by 16 refrigerators in 8 refrigerator stations. This means that each refrigerator supplies a length of 1.8 km of the linac. The refrigerator specification is mainly given by the cooling demands of the cavities and the quadrupoles. The superconducting cavities have to be cooled below 2 K [1]. The quadrupole magnets for focusing the beam are designed for a working temperature of 4.4 K. An 80 K cooling shield is proposed in order to reduce radiation losses.

The provisional specifications for one refrigerator are given in Table 1.

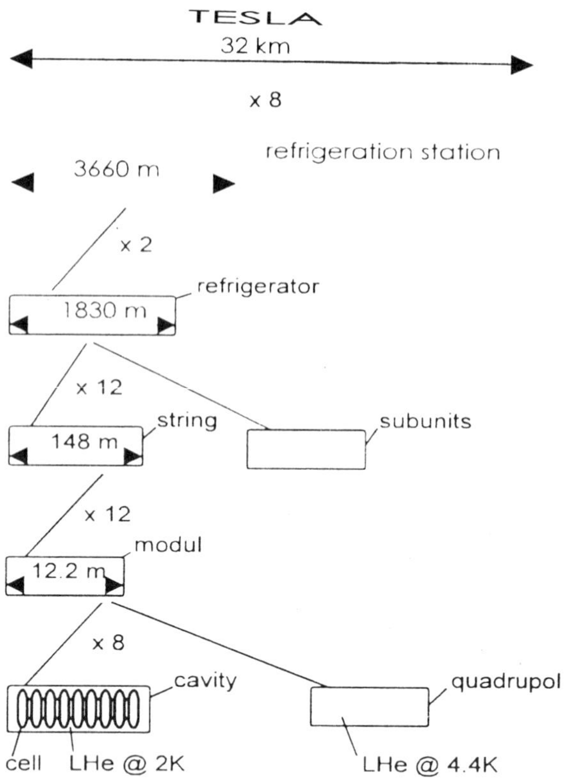

Figure 1. TESLA refrigeration scheme.

Table 1. Heat loads of TESLA at the three temperature levels per refrigerator

	static load	total load
2 K	720 W	4.8 kW
4.4 K	3 kW	3.5 kW
80 K	16.6 kW	30 kW

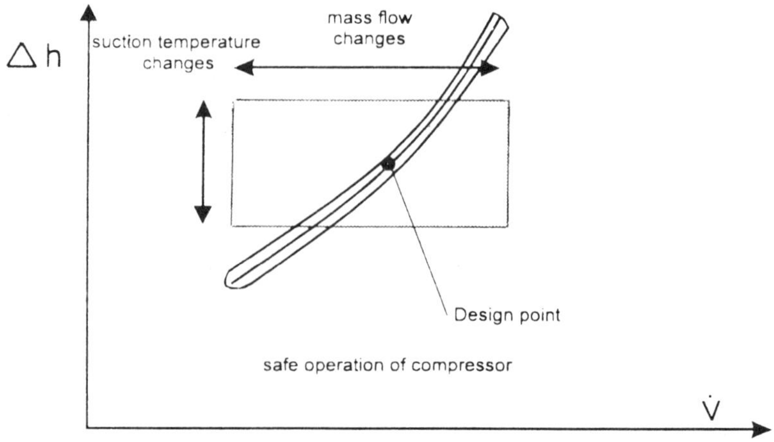

Figure 2. Required duties and safe operational regime of cold compressors.

It is important to note that additionally to the static load, dynamic loads are induced by the 1.3 GHz excitation of the cavities. So the refrigerators will have to cope with strongly and rapidly changing loads. Nevertheless the refrigerators should have a very good efficiency, because the running costs of the collider are a major cost factor and they contribute strongly to the viability of the total project.

So the challenge is to invent a refrigerator scheme, which is thermodynamically efficient, but also flexible to cope with rapidly changing loads.

COLD COMPRESSORS

The temperature of 2 K corresponds to a vapour pressure of helium of 31 mbar. The large refrigeration capacity required for TESLA can only be obtained with a multi-stage cold compression system [2]. The total pressure ratio up to atmospheric pressure is about 36, so using a four stage system an average pressure ratio of 2.5 per stage is required. Actually the colder stages will have a somewhat larger pressure ratio.

The main difficulty of the task comes from the fact, that such turbocompressors have a very narrow safe operational band, quite different from the load changes, which one has to expect. The cold compressors are very sensitive to changes in the suction temperature, because the suction temperature influences strongly the necessary enthalpy head which has to be covered by the single compressor.

Figure 2 demonstrates the dilemma caused by the ranges of mass flow and suction temperature which have to be covered. Figure 2 shows even only half of the problem, because it does not yet describe the dynamics, i.e. the speed at which the changes may occur.

QUASI-STATIC SOLUTIONS

The task has to be solved in two steps:
First a solution has to be found for the quasi- static operation, i.e. for the required operational range shown in figure 2, where the speed of change is not yet taken into account.

In figure 3, a number of previous approaches are shown [3]. Figure 3a shows the starting system with the characteristic of figure 2. In figure 3b the suction temperature of the first stage of compression is stabilised by a thermal anchoring to a 4.4 K bath. By this method the disturbing influence on the required enthalpy head is removed, but

- the changes in the mass flow rate are not addressed
- in the heat exchanger thermodynamic losses due to heat transfer from 4.4 K to 3.5 K are produced
- and this is a very expensive heat exchanger causing additional pressure drop for the low pressure gas.

Figure 3c shows the proposal for a bypass valve from the vapour space of the 4.4 K bath to the suction side of the first compression stage. This way the fluctuations of the mass flow rate are reduced, to a certain degree also the suction temperature is stabilised. But still this operation is thermodynamically not acceptable over longer periods of operation.
Figure 3d shows a bypass valve over all four stages of compression. This may stabilise the suction mass flow, but it has a detrimental effect on the suction temperature and efficiency.

Our approach was led by the search for a solution with a minimum of loss producing features at least for the quasi-static operation. To cope with the mass flow variation, we intend to work with the speed control of the compressors only: If less flow comes the speed is reduced, if more flow comes the speed is increased. And if the suction temperature rises or falls we live with that pressure ratio, which the compressor is able to handle under such flow and speed conditions.

With this control method, the total pressure ratio cannot anymore be reached under all mass flow conditions, the exhaust pressure of the fourth stage is being reduced at smaller than design flow rates. But this fits well with the volumetric suction characteristic of the main refrigerator screw compressor. At lower mass flow rates it may go to subatmospheric suction pressure. At higher than design flow rates, the cold turbo compressors provide a higher than design pressure ratio and the screw compressor may operate at an elevated suction pressure.

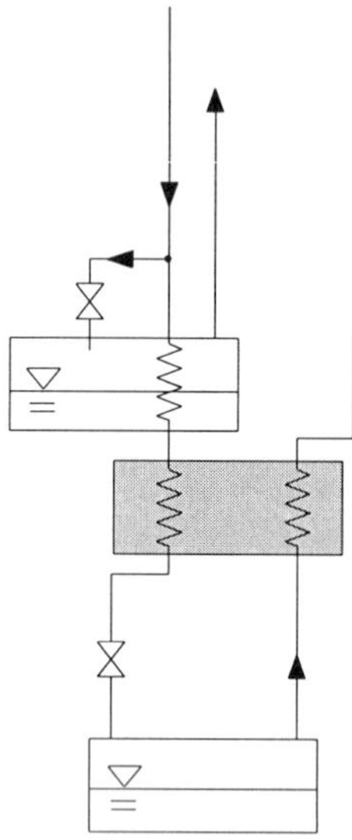

Figure 3a

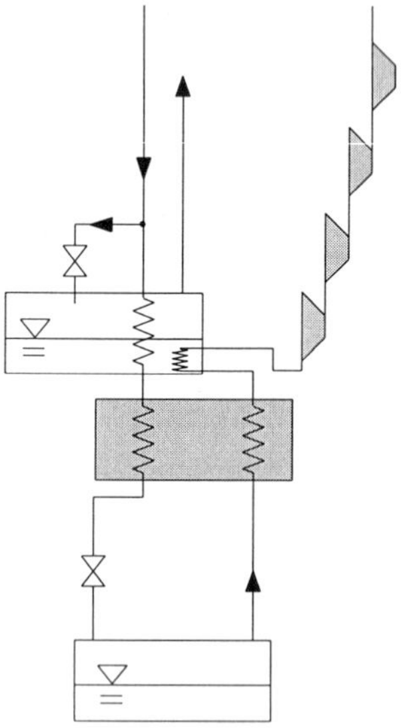

Figure 3b

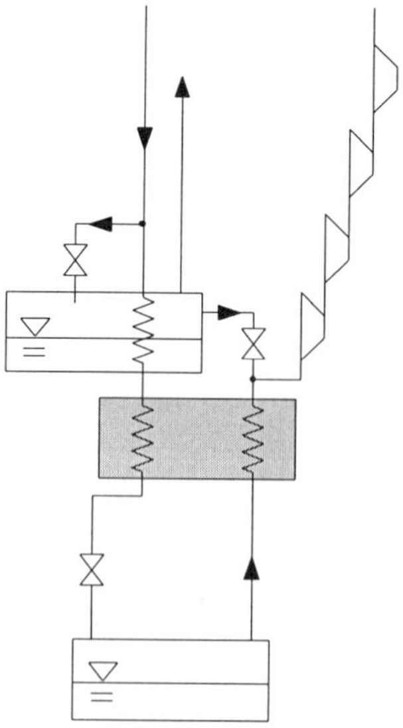

Figure 3c

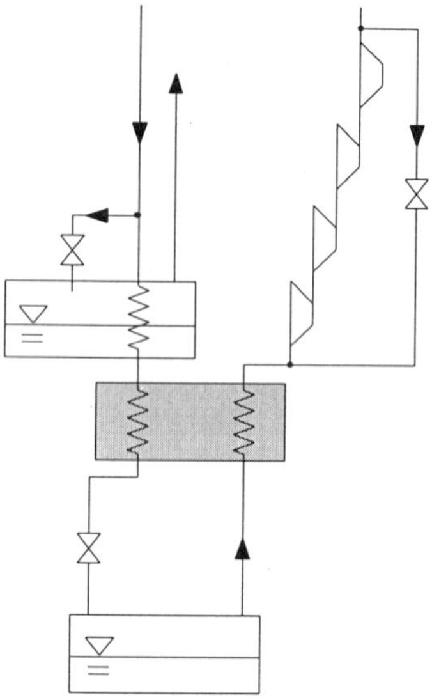

Figure 3d

Figure 3. Diverse approaches of controlling safe operation of cold compressors.

To make this scheme possible, the flow between the cold compressors and the associated screw compressor needs its own channels in the plate-fin heat exchangers.

DAMPENING HEAT EXCHANGER

The pressure ratio which can be reached with this scheme can be stabilised by anchoring the suction side of the third compression stage to the 10 K level of the main refrigeration cycle. This is the temperature range, where the high pressure stream has its peak specific heat. Therefore it is a quite stable range - similar to a condenser in a vapour compression system. At low flow rates, the first two stages of compression provide a relatively low pressure rise and therefore also a small temperature rise to the low pressure stream. Thus the low pressure stream will be warmed up in the 10 K anchor exchanger. So the third stage gets a larger suction volumetric flow rate and can produce a higher pressure ratio.

At high flow rates, the low pressure stream comes at a higher pressure and temperature to this heat exchanger. It is cooled down to 10 K and the duty of the third and fourth stage of compression is reduced.

For the design of the anchoring heat exchanger, we propose to use a plate fin exchanger in cross flow configuration. Since no internal distributors are needed in this design, pressure drop is minimised. Because this heat exchanger is situated at a relatively high pressure and temperature level, the thermodynamic losses are minimal.

FLOW DIAGRAM

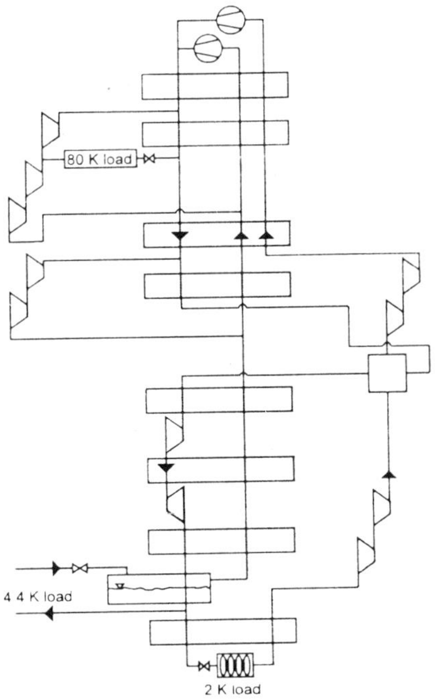

Figure 4. Proposed flow diagram.

In figure 4 the flow diagram of one of the refrigerators is shown. The screw compressor downstream of the cold compressors is allowed to work with a suction pressure below atmospheric pressure.

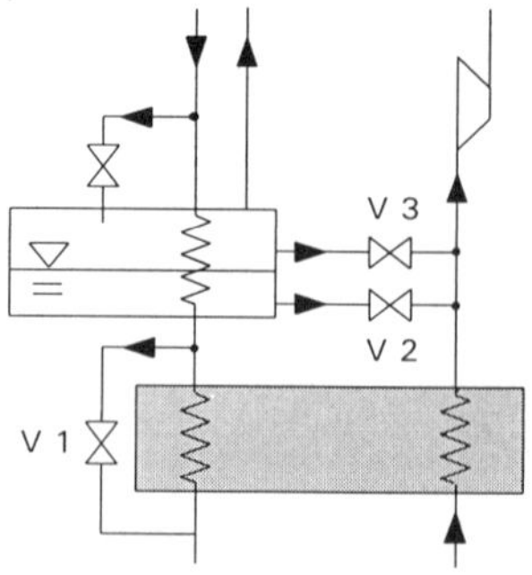

Figure 5. Means for handling of dynamic loads.

HANDLING OF DYNAMIC LOAD CHANGES

Having established a scheme, which allows efficient and safe operation at all foreseeable quasi- static modes of operation, further dampening means have to be provided which give the refrigerator enough time to adjust to quick load changes. The operation of 3 valves is provisionally foreseen for this purpose:

- V1- a HP bypass valve around the final heat exchanger
- V2- an injection of liquid helium into the cold compressor suction line
- V3- the throttling of 4.4 K vapour into the cold compressor suction line

The arrangment of the valves is shown in Figure 5. The questions, which of the these valves will be operated under what conditions will be studied in a dynamic numerical simulation investigation.

ACKNOWLEDGEMENTS

We would like to thank DESY (Deutsches Elektronen-Synchrotron) for supporting this research project.

REFERENCES

[1] D. Trines et al., Conceptual design of the cryostat and cryogenic supply for a superconducting linear collider, High Energy Accelerator Conference, 1992

[2] H. Quack, Cold compression of helium for refrigeration below 4 K, Advances in Cryogenic Engineering, Vol. 33 (1987), 647-653

[3] Jean- Philippe Guignard; Contribution a l'etude de la stabilite de fonctionnement et de l'adaptation de charge des compresseurs centrifuges cryogeniques multietages, CERN AT Group Report 93-05 (CR)

DEVELOPMENT OF A NEW GAS BEARING HELIUM EXPANDER

E. Thomas[1]

[1]CCI Cryogenics
1176 N. Irving Street
Allentown, PA 18103, USA

ABSTRACT

This paper will address the design, construction and limited testing of a small centripetal expansion turbine to be used for helium refrigeration.

The major design features are externally pressurized gas lubricated bearings, fixed adjustable nozzles and a compressor wheel load device.

The design is directed toward the simplest construction that will be consistent with the accuracy required for gas bearing operation.

INTRODUCTION

The turbine under discussion is the "warm" unit of a pair. It is designed for the following service:

Gas - Helium
Flow - 190 gm /sec (25 lb/min.)
P_{in} 15 Atm
P_{out} 4.5 Atm
T_{in} 38K
T_{out} 27.02K

The "warm" and "cold" turbines will use the same bearings, bearing housings, shafting and nozzles. The impellers will be unique to their service conditions.

Externally pressurized bearings were chosen because they generally have higher load capacity and greater stability than self-acting gas bearings. The lubricating gas is from the process stream and bearing discharge is returned to the process at an appropriate pressure level.

The external pressurization of the bearings eliminates shaft to bearing contact during start up which is necessary to maintain the very small bearing clearances encountered in gas lubricated bearings. The major elements of the turbine will be briefly discussed.

Advances in Cryogenic Engineering, Vol. 41
Edited by P. Kittel, Plenum Press, New York, 1996

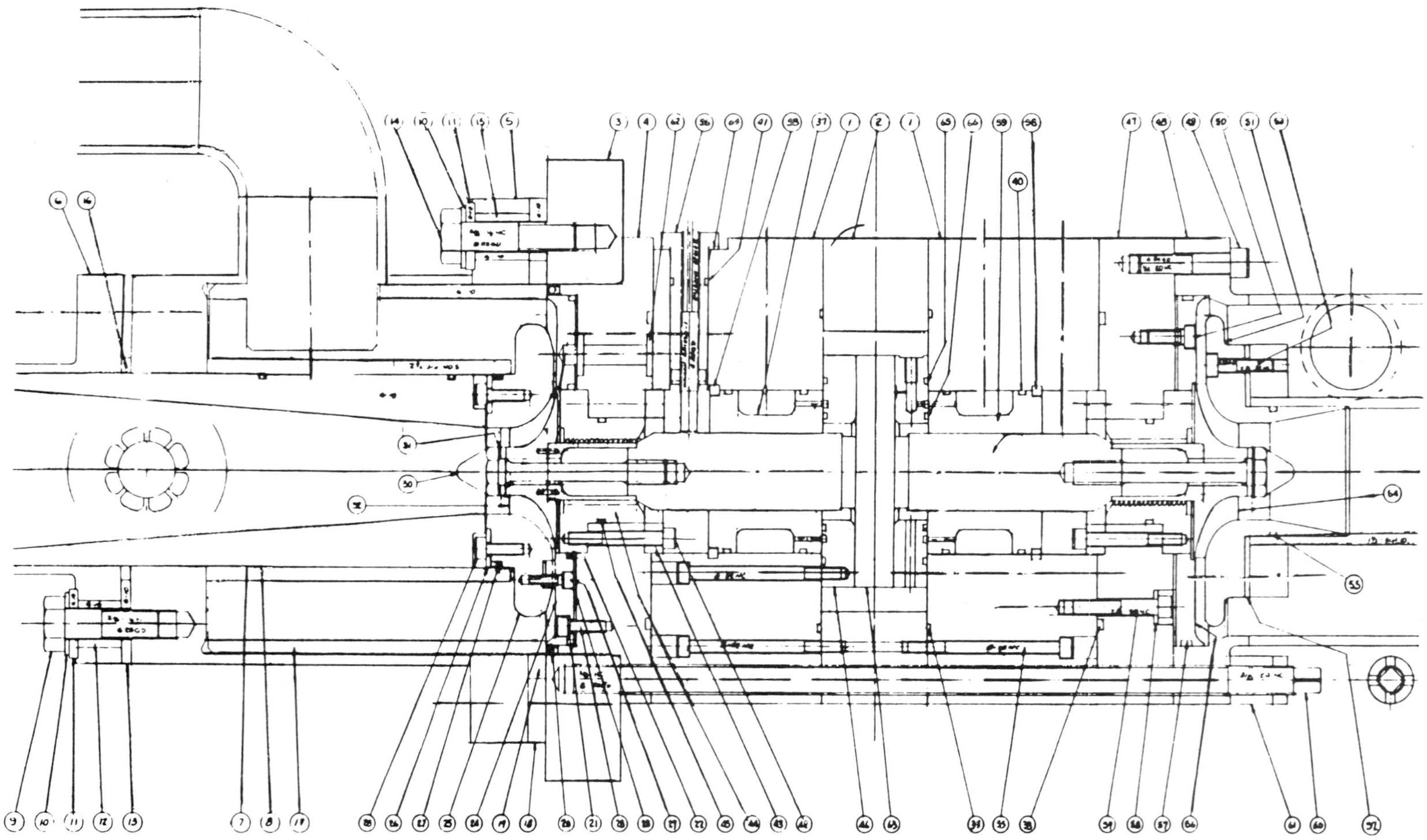

Fig. 1. Cross-section of expansion turbine.

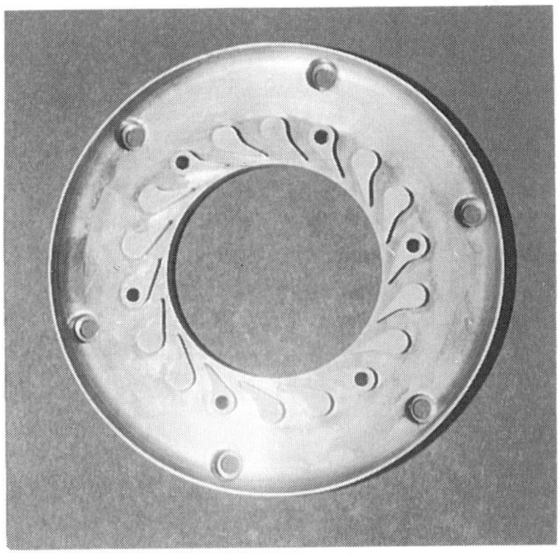

Fig. 2. Back plate showing vanes.

Fig. 3. Shroud showing cavities that match vanes on back plate.

A cross-section of the turbine is shown in Fig. 1, and Fig. 4 and Fig. 5 show the turbine and compressor ends of the bearing housing.

INLET HOUSING

The inlet housing is a weldment of 304 stainless steel. This housing will be permanently installed in a vacuum box and will not be disturbed if turbine removal is necessary. The housing permits the turbine to be removed as a "plug-in" unit and the process piping will not be disturbed. Referring to Fig. 1, the inlet housing is identified by 3.

IMPELLER

The impeller is a precision investment casting of C-355-T6. It has 13 radial blades each with a circular exducer. The rough casting has an outside diameter of 63.5 mm (2.5 ins.) and has a maximum tip width of 6.4 mm (.25 ins.). The maximum operating speed is 125,000 RPM with a outside diameter of 57.2 mm (2.25 ins.). The same casting is used for the turbine impeller and the compressor (brake) impeller. On Fig. 1 the turbine impeller is identified by 32, the compressor is identified by 54.

INLET NOZZLE

The inlet nozzle is the fixed angle, adjustable width type. The nozzle area is a adjusted by widening the flow passages.

The vanes are milled integrally with the back plate. The vane height is 6.4 mm (0.25 ins.) and the opposite plate is milled with cavities that match the vanes. By inserting or retracting the vanes from the cavities the area of the nozzle is changed. The nozzle is manufactured from type 310 stainless steel.

See Fig. 2 and Fig. 3 for detail photos.

TABLE 1. BEARING VIBRATIONS VS SHAFT RPM

	Turbine Bearing Vibration (x 1.e-03) Inches		Compressor Bearing Vibration (x 1.e-0) Inches	
RPM	"X" Probe	"Y" Probe	"X" Probe	"Y" Probe
88000	0.32	0.24	0.58	0.42
90270	0.32	0.24	0.58	0.44
92500	0.32	0.24	0.58	0.44
93500	0.32	0.24	0.58	0.44
95000	0.32	0.24	0.58	0.44

BEARINGS - JOURNAL AND THRUST

The journal bearings are designed for a 25.4 mm diameter shaft. The L/D ratio is 1.5 and the shaft to bore clearance is .0200 to .0254 mm. From test runs where the orbit diameter was measured to be .0102 mm the bearing stiffness was determined to be 175 Kg/mm. The journal bearings are 3 piece construction consisting of a center section with end caps. This construction permits making the feed jets easily since they are milled on the ends of the center section. Drilling 0.25 mm diameter holes is eliminated.

The thrust bearings are the flat plate type with a circular feed groove.

The bearings are made from SAE64 which is a leaded bronze. This material was selected for its anti-friction properties which would reduce scoring during initial testing.

SHAFT

The shaft is made from a nitriding steel and hardened to 600 Bn on the bearing surfaces. The thrust disc is integral with the shaft and carries a small projection on its periphery for speed measurement.

The impellers are fitted to the shaft with a slight shrink fit and the torque is transmitted by means of 4 pins at each end. The impellers are retained by a center bolt torqued to 8.5 Nm.

Fig. 4. Turbine end of bearing housing.

Fig. 5. Compressor end of bearing case with shroud removed.

The shaft and impellers and center bolts were dynamically balanced to 3 mg at a radius of 20 mm.

TESTING PROGRAM

Testing was conducted primarily to determine if there were bearing instabilities in the operating range. Since two turbines are required for the proposed applications and they will operate at 85,000 RPM and 60,000 RPM, the testing covered this range.

The test gas was nitrogen gas from a liquid tank. The gas was supplied to the bearing at 8 bar and controlled by needle valves to the desired bearing gas pressure.

The test consisted of four runs of 10 hours duration each run.

The bearing vibrations were measured with X-Y probes on outboard end of each bearing. During each run the speed was varied from 40,000 to 95,000 RPM. The turbine was dismantled three times during the testing to inspect for bearing condition.. The tear downs were for informational purposes since shaft vibrations were not in evidence during operation. The inspections indicated that shaft to bearing contact had not occurred.

The ability to pressurize the bearing before rotation eliminated contact and was confirmed by the X-Y probes, which showed the shaft was centered in the bearings before rotation began.

IMPROVEMENTS IN THE EFFICIENCY OF TURBOEXPANDERS IN CRYOGENIC APPLICATIONS

R.R. Agahi, M.C. Lin, and B. Ershaghi

Atlas Copco Rotoflow
Gardena, California 90248, USA

ABSTRACT

Process designers have utilized turboexpanders in cryogenic processes because of their higher thermal efficiencies when compared with conventional refrigeration cycles. Process design and equipment performance have improved substantially through the utilization of modern technologies.

Turboexpander manufacturers have also adopted Computational Fluid Dynamic Software, Computer Numerical Control Technology and Holography Techniques to further improve an already impressive turboexpander efficiency performance.

In this paper, the authors explain the design process of the turboexpander utilizing modern technology. Two cases of turboexpanders processing helium (4.35°K) and hydrogen (56°K) will be presented.

INTRODUCTION

Hydrogen and helium demand and production are on the rise [1]. Space exploration and alternate fuel programs were and are the major markets for hydrogen and helium. In industrial applications, the largest demand for hydrogen is in ammonia production and petroleum refineries. Demand for high purity hydrogen (99.5%+) has been on the rise in the past decade [1]. Since liquid helium is the coldest substance available, it has been used in the field of cryogenics, and super conductivity studies too. Helium is also used in gas refrigerators and experiments at very low temperatures (isotope 3He) which is produced in nuclear reactors, has reached 0.3k). Helium is mostly produced from natural gas containing more then 0.4% helium. The production process consists of lowering the temperature of natural gas low enough so that all the components, except helium and traces of hydrogen are condensed. Hydrogen is mainly produced from oil refinery gas and catalytic steam-hydrocarbon reforming processes.

Turboexpanders are the most economical method for cryogenic temperature production [2]. Turboexpanders are very suitable for the liquefaction and purification of gases, such as H_2 or

Advances in Cryogenic Engineering, Vol. 41
Edited by P. Kittel, Plenum Press, New York, 1996

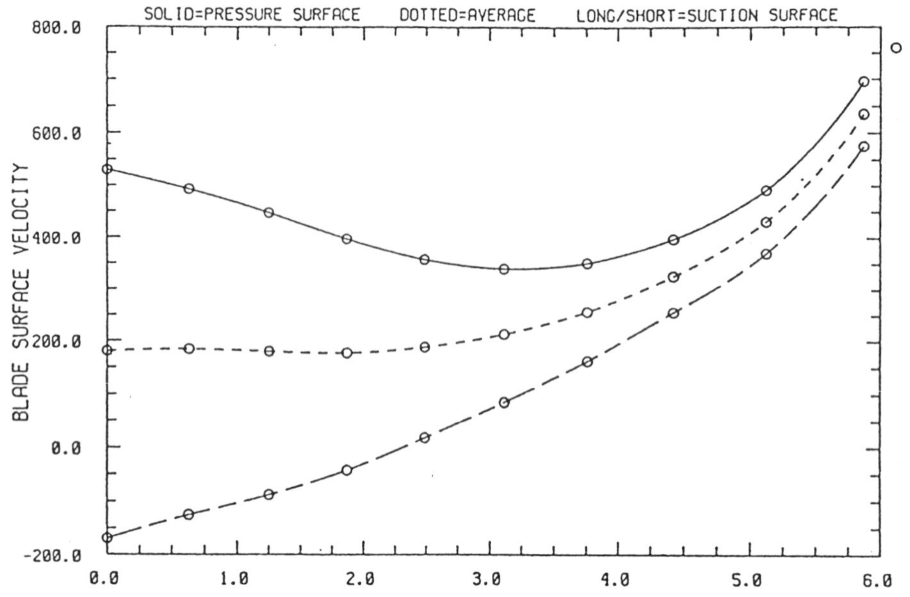

Figure 1. Camberline Distance Loading Diagram at Hub Streamtube.

He by producing deep cryogenic temperatures. These units are usually small in size, 5 to 50 HP, operating at speeds ranging from 45,000 to 100,000 rpm and mostly with oil brakes. Sixty years of development in turboexpander technology has resulted in highly efficient machines which are utilized in applications such as the above. In the past, due to mechanical limitations, such as rotational speed, turboexpander efficiency was limited and that rendered it unsuitable for hydrogen and helium applications. Utilizing Computation Fluid Dynamic software, Computer Numerical Control Technology, and Holography Techniques has resulted in turboexpander isentropic efficiency improving to about 90% within the last 10 years. Improvement in efficiency is mainly due to:

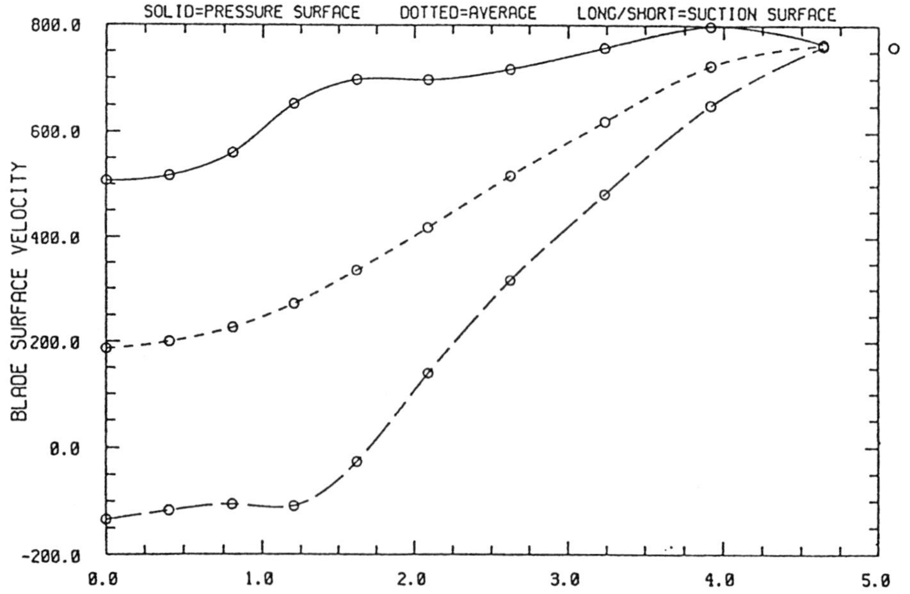

Figure 2. Camberline Distance Loading Diagram at Shroud Streamtube.

Table 1. Design Feature of a Hydrogen Expander

Gas	Hydrogen rich gas (M.W. = 2.04)
Discharge Temperature	56°K
Head	190 kj/kg
Outlet Flow	320 Nm^3/hr
Speed	100,000 RPM
Wheel Diameter	75mm
Efficiency (Calculated)	88%
Horsepower	16 kw
Wheel Tip Speed	390 m/sec.

(1) Better understanding of flow patterns through expander inlet guide vanes (nozzles) and wheels,
(2) More accurate numerical methods to define blade geometry and manufacturing methods of the turboexpander wheels, and
(3) Less mechanical constraints by utilizing new bearing technologies and advanced rotor dynamic computer codes.

Turboexpanders have replaced reciprocating expansion engines in most large systems (which have been used in the last 70 years) and have the following advantages over reciprocating equipment:

(1) Turboexpanders are capable of producing very high isentropic efficiencies, in some cases up to 90%,
(2) The sizes of these units may be reduced (by increasing operating speeds), which results in less investment by the end user,
(3) Wheels and shafts are the only moving parts. Maintenance of turboexpanders is easy and simple, and hence maintenance cost is low, and
(4) Turboexpanders can handle condensation of impurities without loss of efficiency.

FLUID DYNAMICS CONSIDERATION

Designing of an expander wheel is an involved task because two incompatible requirements must be satisfied. The first requirement is that the wheel should be designed for minimum loss when process gas is passing through it. The second requirement is that the wheel should provide acceptable distributions of relative velocity on both leading and trailing surfaces of the blades in order to minimize the possibility of flow separation and the associated loss due to turbulence [3].

The first requirement is usually satisfied by utilizing manufacturer's experience. Manufacturers have developed regression models based on several hundred wheels they have designed and produced. For the second requirement, it is understood that losses in a turboexpander are partially caused by boundary layer and flow separation in the expander and its subsequent effects in the discharge diffuser. Appropriate velocity distributions are particularly important in the expander wheel where the flow must closely follow the blade, hub,

Table 2. Distribution of the gas dynamic losses in hydrogen turboexpander wheel

Primary Guide Vanes (Nozzle)	4.2%
Secondary Nozzle	1.3%
Disk Loss	3.0%
Wheel Seal Loss	1.2%
Channel Friction	1.3%
Inducer & Turning	1.0%
Total Losses	12%

and shroud surfaces in order to extract the energy efficiently from the fluid and enter the discharge diffuser with little distortion. Modern computer technology allows a designer to modify parameters such as blade turning, shroud geometry and blade length in order to obtain a good velocity distribution.

Figure 1 and Figure 2 show the surface velocities along the hub and shroud Camberline distances. Figure 3 shows the blading of a hydrogen expander wheel produced from blade geometries [3]. In order to reduce blade stresses, as it is shown on Figure 3, this expander wheel is equipped with radially elemented blades.

As an example, the optimum design of a hydrogen expander is illustrated below. This hydrogen expander is one of a 4-stage expander for a hydrogen purification plant. Table 1 shows the design data of the expander. At the design point, the expander head is 190 kj/kg and discharge volume is 320 Nm^3/hr. The combination of low volume and high head determines

Figure 3. Picture of Hydrogen Expander.

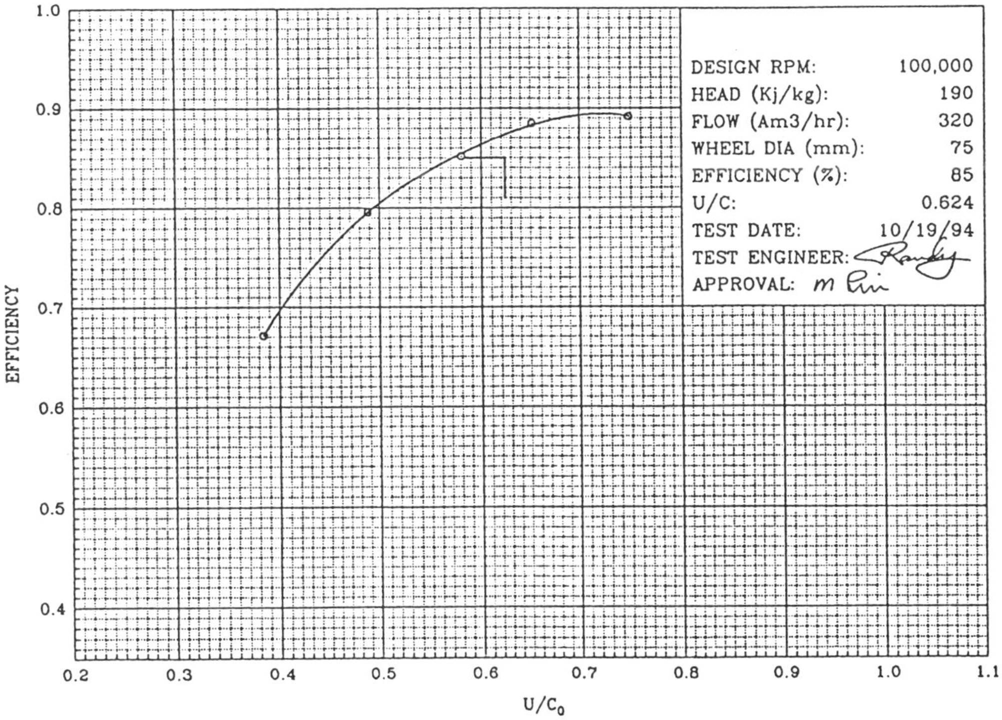

Figure 4. Hydrogen Expander Performance.

that the optimum thermal efficiency can be attained at the rotational speed range of 90,000 to 100,000 RPM. As it is important to have high efficiency for this application, the turboexpander is designed to operate at near its upper range of speed of 100,000 RPM.

Table 2 illustrates the losses of the optimized turboexpander which has a 75mm diameter wheel and operates at a tip speed of 390 m/s.

The expander efficiency with 12% loss will be 88%. Figure 4 shows the efficiency of the expander obtained during factory testing with open loop air. This figure shows that the efficiency can be further improved by increasing the speed.

Table 3. Design Data of the Helium Turboexpander

Gas	Pure Helium (M.W. = 4.003)
Discharge Temperature	4.35°K
Pressure Ratio	17
Design Outlet Flow	25 Nm^3/hr
Speed	60,000 RPM
Wheel Diameter	40mm
Efficiency (Calculated)	83%
Liq. Wt. % Condensed	7%

Another example is a turboexpander in the liquefaction of helium. In this example, the helium expander operates at the two-phase region (4.35°K) and condenses about 7% of the stream. Table 3 shows the operating data of the turboexpander.

The two-phase expansion is highly desirable thermodynamically, but the liquid must neither bombard (and erode the wheel blades, hub and shroud), nor accumulate in the wheel since that would cause efficiency loss.

Computer Fluid Dynamic Software provides the capability to shape the blade, hub and shroud parallel to the flow stream passing through the turboexpander, and to operate safely and efficiently with condensing stream. This helium expander was built in the mid-1980's and has operated satisfactorily since its installation. The test results are shown in figure 5.

MANUFACTURING AND GEOMETRICAL CONSIDERATION

Radial inflow type turboexpanders are designed and manufactured with a power range from 1 hp to 16,000 hp. The expander wheel sizes vary from 15 mm to 1250 mm. Generally speaking, helium and hydrogen expanders need high operating speeds to attain the required thermal efficiencies. As rotational speed is increased, wheel size will be reduced. In small turboexpanders, problems are related to miniaturization, Reynolds number effects, heat transfer, seals, bearings and critical speed. When the wheel size is small, it is questionable whether the wheel can be produced economically with the required accuracy to reduce losses.

For a high speed expander, brazing of the wheel is not acceptable because of high tip speeds. Wheel casting patterns are very expensive and are very difficult to cast, particularly for small wheels when the surface of the blade has an arbitrary shape that requires close tolerances.

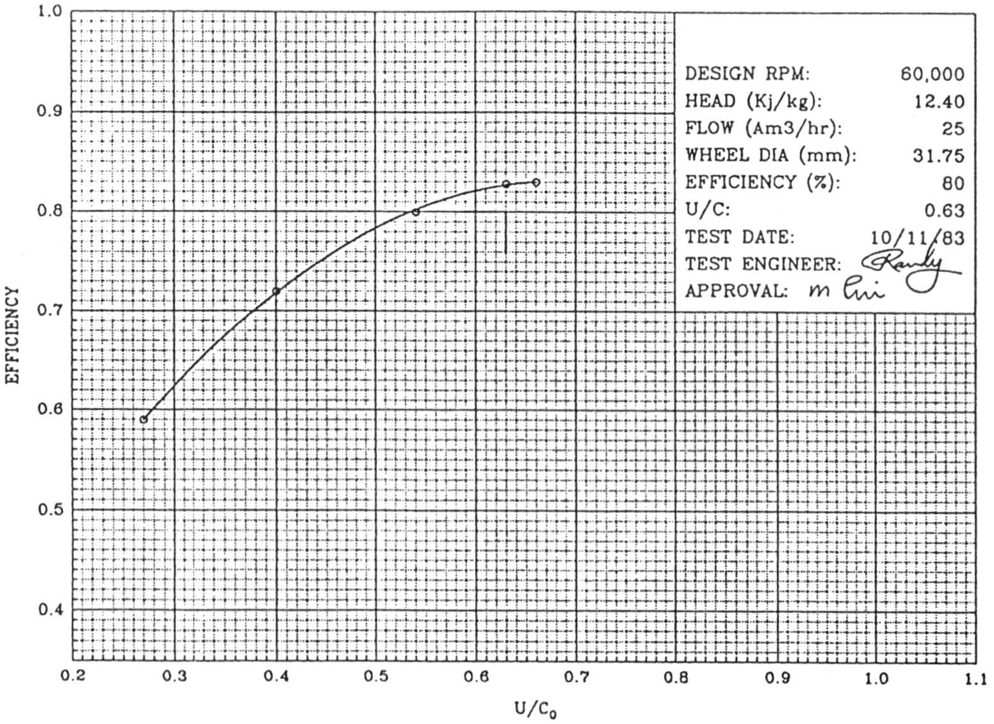

Figure 5. Helium Expander Performance.

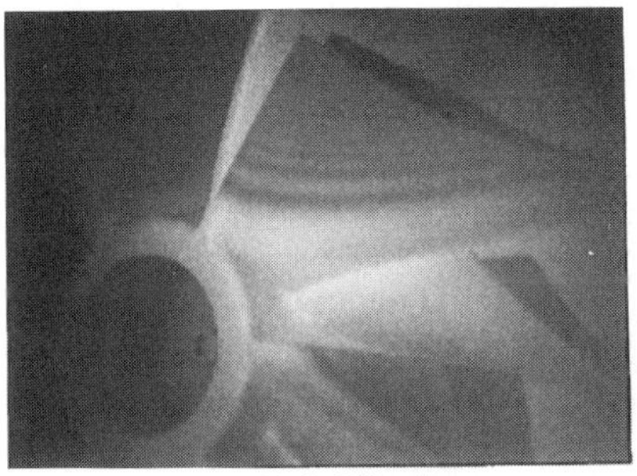

Blade Vibration

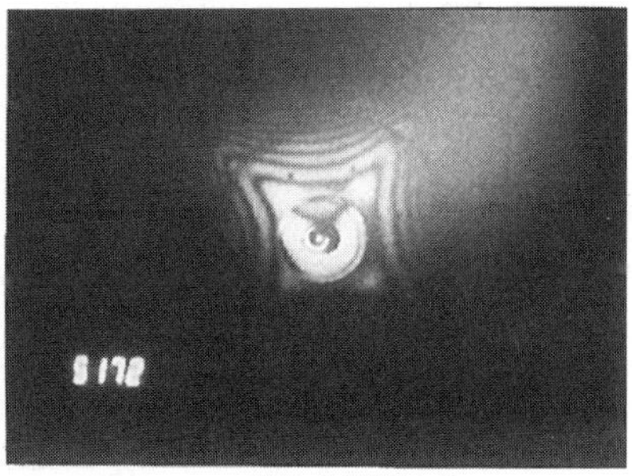

4 Nodal Vibration

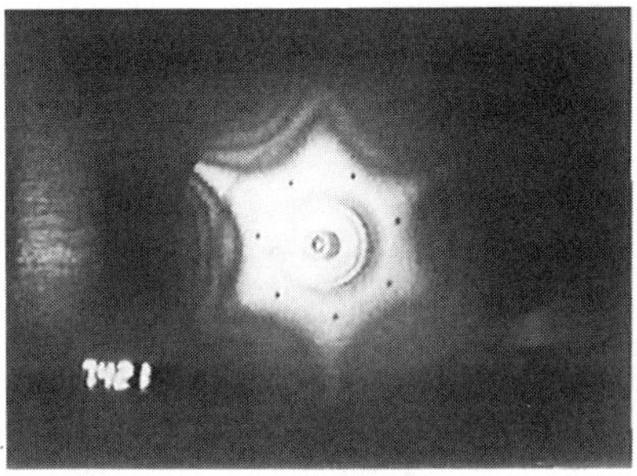

6 Nodal Vibration

Figure 6. Typical Picture of Holograph.

Interactive graphic assisted tool selection software [5] has produced satisfactorily the geometric shape per design specification. This computer program is intended for straight-line element blading and features optimal flank milling.

MECHANICAL FEATURES

The components of a turboexpander may fail due to a variety of causes including abrasives, stress rupture, creep, corrosion, partially condensing steams, weak bearing, imbalance of thrust force, wheel resonance and imbalance of the wheel.

The most serious problem is wheel resonance that results in fatigue failure. All the expander wheels have many modes of resonance vibration which occur at quite high frequencies (several thousand per second). These frequencies are multiple times of the running speed and the danger is that one may coincide with the passing frequencies of the number of nozzles or number of blades. If this happens, it will crack the wheel. Therefore, such resonances must be avoided.

With high tip speed wheels, which is the case in hydrogen and helium expanders, the situation becomes more complex. Practically, it becomes impossible to completely avoid all natural frequencies. The frequencies are driven not only by the first and second nozzle harmonics, but higher harmonics that will excite increasingly more complex modes in the wheel.

Traditional methods of analysis are somewhat blind to detect the presence of nodes and anti-nodes, and cannot easily find the interaction between complex blade modes and disk modes of vibration, or where to effectively remove material for tuning purposes. One of the most accurate detection methods is Holographic measurement.

The Holographic method determines all mode shapes with great sensitivity and accuracy and easily records images for reference. This method can also be used for analysis of torsional modes of wheels and shafts, natural frequencies of wheel shaft combinations or any other parts exposed to vibration. Figure 6 shows the typical holograph of a wheel.

CONCLUSION

Successful applications of turboexpanders in liquefaction and purification of hydrogen and helium have been increased in the last two decades. Turboexpander fluid dynamics design, manufacturing of components and detection and elimination of mechanical constraints have been improved substantially by utilizing modern technologies in recent years.

In summary, the state-of-the-art modern turboexpander has reached a very high level of reliability and thermal efficiency.

REFERENCES

1. U.S. Department of Commerce, Bureau of the Census Reports 1980-1990.
2. J.S. Swearingen, "Turboexpanders and Processes that Use Them," Chemical Engineering Progress, Vol. 68, No. 7 (1972).
3. W. Jansen and A. M. Kirschner, Impeller Blade Design Method for Centrifugal Compressors, 1970 Symposium on Fluid Mechanics, Acoustics, and Design Turbomachinery.
4. R. Agahi and B. Ershaghi, "Ethylene Plant Turboexpander Improvements," 1992 AIChE Spring National Meeting.
5. NREC, An Interactive Graphics System for the Design of Radial Turbomachinery, Northern Research and Engineering Corporation, Woburn, Massachusetts, USA.

HIGH PERFORMANCE CRYOGENIC TURBOEXPANDERS

R. R. Agahi, B. Ershaghi, and M. C. Lin

Atlas Copco Rotoflow
Gardena, California, 90248, USA

ABSTRACT

The use of turboexpanders for deep cryogenic temperatures has been constrained because of thermal efficiency limitations. This limited thermal efficiency was mostly due to mechanical constraints. Recent improvements in analytical techniques, bearing technology, and design features have made it possible to design and operate turboexpanders at more favorable conditions, such as of higher rotational speeds. Several turboexpander installations in helium and hydrogen processes have shown a significant improvement in plant performance over non-turboexpander options.

INTRODUCTION

Liquefaction of hydrogen at 20°K (-423°F) by James Dewar in 1895 and liquefaction of helium at 4°K (-452°F) by Kammerling Onnes in 1908 made it possible to study nature near absolute zero 0°K (-460 F) [1]. A major breakthrough in cryogenic science took place in the late 1940's with development of the helium liquefier [2]. Figure 1 shows the cryogenic temperature range of some pure and mixed gases at atmospheric pressure [2]. Applied research and the commercialization that followed paved the way for the design of equipment suitable for these ultra low temperatures.

Turboexpander technology started in air separation with a low temperature of 88°K (-300°F). This was subsequently improved in industry's quest for colder temperatures. Improvements were mostly made in materials, in the ability to manufacture parts with closer clearances, and in better understanding of rotordynamics so that turbomachines with much higher rotational speeds could be designed.

Turboexpanders are among the most economical methods of cryogenic temperature production in gas processing plants [1]. Expanders can also function as a control valve for process control, thus also simplifying the process design.

Advances in Cryogenic Engineering, Vol. 41
Edited by P. Kittel, Plenum Press, New York, 1996

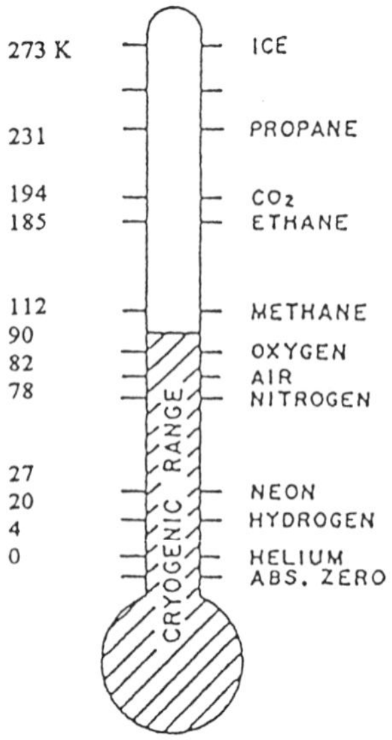

Figure 1. Cryogenic Thermometer Showing Normal Temperature At Atmospheric Pressure.

Liquefaction of Helium

Liquid helium is a remarkable substance and has made a new category of quantum liquid in physics [4]. The normal boiling point of helium is 4.3°K·(-452°F) and the critical temperature, the highest temperature at which the liquid phase can exist, is only 1°K (1.8 F) higher at -450°F. In the old liquefaction method of helium, compressed helium at about 30 bar was cooled to about 15K (-432°F) by contact with liquid hydrogen and passed through an expansion valve. The new process design replaces the hydrogen coolant with a turboexpander. Figure 2 shows a flow diagram of a helium liquefier.

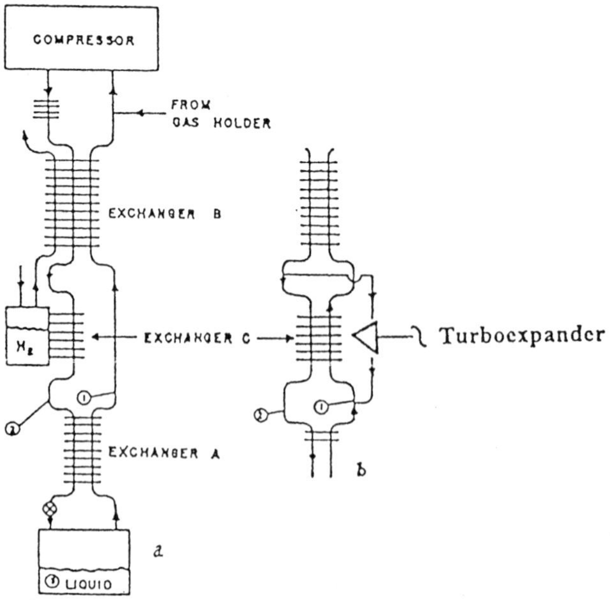

Figure 2. Flow Diagram of Helium Liquefier.

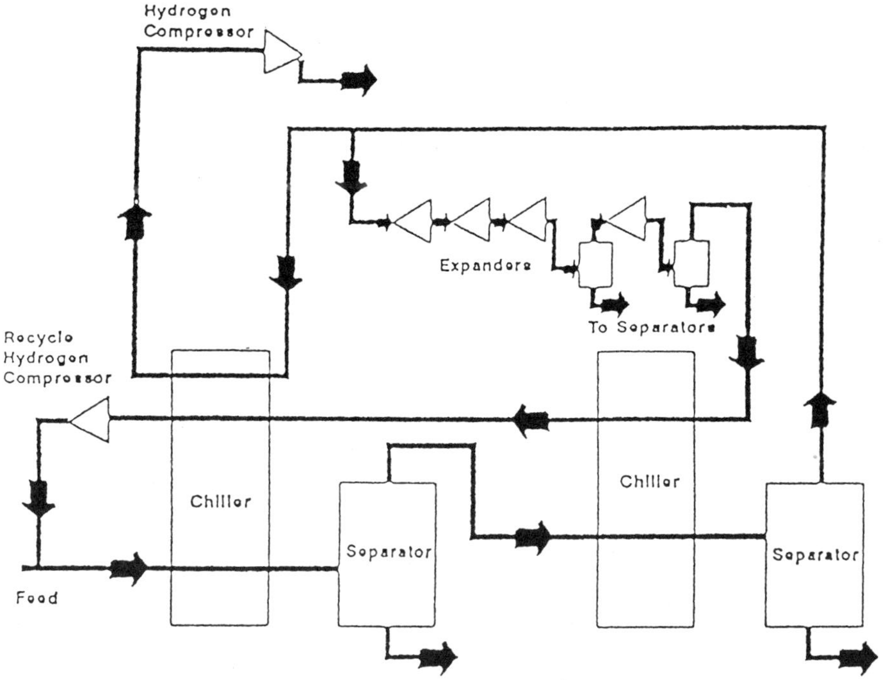

Figure 3. Typical Hydrogen Purification.

Liquefaction of Hydrogen

Hydrogen liquefaction is not feasible with a cascade process. One problem with the expansion valve is the inversion temperature for hydrogen at 204°K (-92 F). Above this temperature the Joule-Thompson effect changes the sign causing warming rather than cooling to take place. A process design with a prechilled turboexpander has proven to be the simplest and most economical method [1].

The prechilled turboexpander cycle requires three or four stages of expansion depending on the recycle compressor power. The advantages are a low power demand, low pressure operation, and no need for a precooling liquid. The turboexpanders utilized in such a process have to have a high thermal efficiency in order to make the process cycle competitive with the other methods. In large liquefaction plants, the difference in cycle efficiency with turboexpanders should be significant to justify the use of turboexpanders over other methods. Turboexpander technology has advanced to the level that makes this feasible. Figure 3 shows a hydrogen liquefaction cycle [5].

Turboexpander Design

The thermal efficiency of the turboexpander is an essential performance parameter in the process design of hydrogen purification and helium liquefaction plants. Since the process requires a low flow and a high head, the expander thermal efficiency can only be improved by increasing the rotational speed of the unit. High rotational speeds require smaller diameter bearings and this in turn is often constrained by rotordynamic characteristics. The cryogenic process is normally very sensitive to the expander efficiency. Therefore, turbomachinery manufacturer's have focused their efforts on enhancing the turboexpander performance.

Conventional tilt pad bearings provide low cross-coupled stiffness through tilting on a point, line, or spherical pivot. The contact stresses can result in significant fretting and brinelling under certain combinations of speed and load conditions. Eventually, this will result in an increase in the bearing set clearance and subsequently a decrease in the preload, which leads to further deterioration in vibrations and bearing performance.

Furthermore, conventional tilt pad bearings are comprised of different parts which are assembled together. Each part has manufacturing tolerances associated with its critical dimensions. As a result, the critical speed and stability can vary over a wide range making it difficult to meet certain specifications and requirements. Another limitation is due to pad flutter and instability which prevented the use of four and three pad bearings on many high speed turbomachinery in spite of their better rotordynamic characteristics. The new bearing designs are made possible through the use of state of the art wire Electric Discharge Machining (EDM) technology. The precision and flexibility of this novel approach to the manufacturing of fluid film bearings makes it possible to obtain high performance FLEXURE PIVOT bearings [7].

The geometric configuration of this bearing style and the manufacturing technique which uses state of the art wire Electric Discharge Machining eliminates the stack-up in manufacturing tolerances. The precision accessible with the latest wire EDM CNC manufacturing technology expands the practical use of tilting pad bearings to smaller bearing sizes. This is not possible with conventional style bearings and the machine tools typically used in their manufacturing. Furthermore, the set bore and the pad bore in FLEXURE PIVOT bearings are obtained during the same cutting process and not separately, as is the case with conventional tilt pad bearings. Therefore, the manufacturing tolerances are further reduced since the pad bore and set bore are not independently acquired. This was crucial in the case of the high speed turboexpander application which runs at 100,000 rpm with a shaft diameter of only 16mm [7].

The higher rotational speed, increases the wheel stress. Finite element analysis (FEA) permits stress prediction with precision never possible in the last decades. The capabilities of FEA are in static dynamic analysis and thermal analysis of the rotating elements. In addition natural frequencies, mode shapes and buckling loads are calculated.

Proper sealing of the process gas, especially in a small turboexpander, is a very important factor in improving machine performance. The most common sealing systems are labyrinth type, floating carbon rings, and dynamic dry face seals. Due to the extreme cold temperature, commercial dry face seal materials are not suitable for helium and hydrogen expanders and a special design is needed. Floating carbon rings, on the other hand, will increase the shaft overhang and therefore limits the rotational speed.

Considering the above, the expander shaft seal is normally a close non-contacting conical labyrinth seal . The operating clearances are in the order of .02-.05 micron (.001-.002 mils) . With a close clearance shaft seal, it is important that the shaft is maintained in its rotating position very accurately, and thus a flexible shaft design is not acceptable [6]. In other words, bearings which maintain the closest alignment of the shaft are the best candidate for such applications.

Process flow variation is a normal factor of operation in any plant. Such variations may be due to demand fluctuations and/or operational considerations of the plant. In such circumstances, variable inlet guidevanes (variable nozzles) are required to maintain an optimal flow angle into the turbine wheel.

The variable nozzles acting as a flow control device should provide a high efficiency over a wide range of flows. The plate covering the nozzle segments is equipped with special spring supports to provide the proper clamping force and prevent blow-by over the entire range of pressures and flows, from the fully closed to the fully open position [6].

When turboexpanders are used in condensing streams, the potential problems are solved with a very simple concept. The blades of the expander wheel are shaped so that their walls are parallel to flow path at every point [6]. This design shape minimizes any potential erosion due to the vector resultant of the forces acting upon suspended fog or liquid droplets. Operational

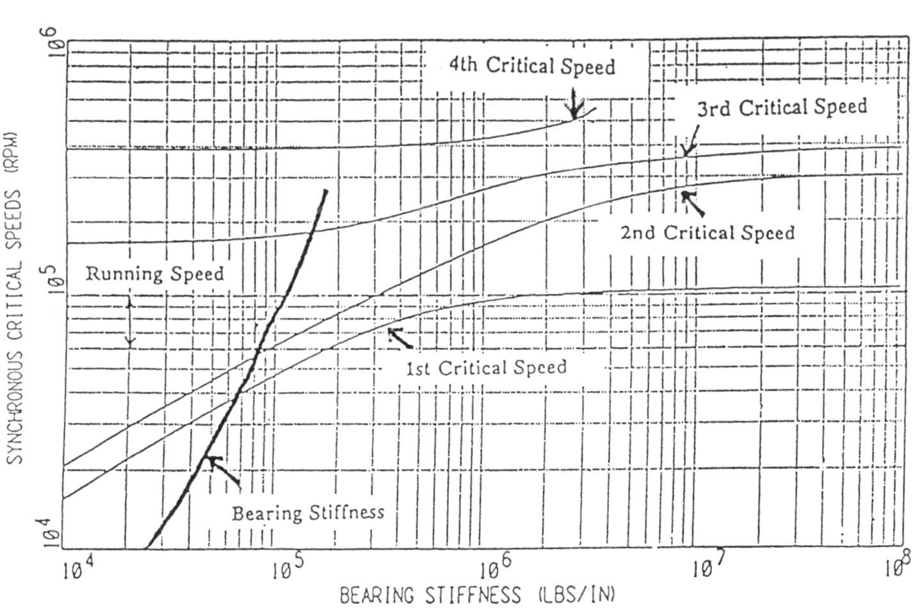

Figure 4. Undamped Critical Speed Plot for Hydrogen Turboexpander.

experience has shown that for this design, suspended fog particles do not drift toward the walls, do not form large drops, and neither interfere with performance nor erode the blades [6].

Turboexpander wheel complex shapes require the application of five-axis milling machines in order to meet high design tolerances. The geometry which is available from CAD system is converted into tool movements in CAM system.

In high performance turboexpanders, current CAM systems are too general and special methods are applied in the milling of the leading and trailing edges, and in the inaccessible locations on the blades.

Turboexpander utilized in hydrogen purification and helium liquefaction are generally of low power, and typically less than 50 kw. Recovery of such a low power is not economical at present, and so the power is normally absorbed by a small oil turbulence rotor which operates in a pressurized cone filled with oil. The oil is agitated, resulting in the oil being heated and self circulated through a heat exchanger in a closed loop. In very low power turboexpanders (less than 5 KW) the oil flow to the oil brake is controlled to maintain the turboexpander speed for achieving the optiminal thermal efficiency.

Thermal Insulation of the Cold Box

Turboexpanders and separators, (when there is liquid to separate), are located in a cold box to minimize heat leakage. The cold box may be filled with powders, fibers or foam for insulation. Depending on the application, the cold box may be vaccumed or be purged by nitrogen or helium. The turboexpander design should accommodate its installation in the cold box, yet allow easy access for maintenance.

Field Experiences

Two turboexpander designs with approximately the same gas dynamic design features but with different bearing systems are presented in this section. Four turboexpander stages are installed in a hydrogen purification plant, while the other four turboexpander stages are installed in a helium liquefaction plant.

Hydrogen Turboexpanders. The application of turboexpanders of in hydrogen plants dates back to the early 1970's. These early turboexpanders, were operating at about 72,000 rpm and had a maximum thermal efficiency of only 72%. The machine described in this paper, featuring the efficient design improvements was manufactured and installed in 1994.

The last stage expander discharge temperature is -215°C (-355°F). Hydrogen contaminants, such as, methane etc. are separated between each stage until the hydrogen purity reaches to 99.99%. Condensation take place at the expander discharge without any performance penalty due to a special and patented blade design. The overall thermal efficiency of these high head and low volume expanders operating at 100,000 RPM is higher than 80%. Heat leakage from the cold box is included in this calculation. During the initial start up, impurities in the gas were freezing and causing a blockage in the second stage expander discharge. This freezing problem (and subsequent blockage) was identified by a reduction in the speed of the next stage expander together with an increase in the second stage discharge pressure. This situation was mainly due to the fact that seal gas was methane rich and the expander efficiency were quite high. As a result the expander discharge temperature reached below the freezing temperature of methane, something which had not previously occurred. This problem was solved by switching to an alternate source of seal gas with a lower methane content.

These expanders were also designed with flexible pad bearings. Flexible pad bearings were selected to overcome the limitation of conventional tilting pad or fixed geometry bearings. The rotor dynamic map of one rotor is shown in figure 4.

Helium Turboexpanders. In this application, the turboexpanders are operating at 85,000 RPM in a helium liquefaction plant. These turboexpanders are equipped with fixed geometry, pressure dam bearings. The overall expander package thermal efficiency is more than 82%. A major challenge in the design of these units was the individual adjustment of load to provide the desirable speed and hence refrigeration. Extremely low shaft power levels of approximately 5 kW made the oil brake design challenging and its control a sensitive issue. A control loop with a multiple hole orifice was designed and proved its effectiveness during operation. Oil flow to each rotating brake disk was controlled by a speed signal from the expander. This control system provides the optimum speed for the turboexpander.

CONCLUSION

The most economical method of purification or liquefaction of most gases is by utilizing turboexpanders. The trend toward higher performance turboexpanders in helium and hydrogen applications have resulted in several innovative design features. Turboexpanders with high performance bearings and other unique features have proven to be efficient and successful in operation with rotational speeds above 100,000 RPM.

REFERENCES

1. Davis, Marcel; Physical Principals of Gas Liquefaction and Low Temperature Rectification, London (1949).
2. ASHRE Data Book (The American Society of Refrigeration Engineers), Vol.1, No.1 , Section 45 (1959).
3. R. L. Powell, W. A. Blanpied, and D. O. Coffin, Journal of Research, National Bureau of Standards, Vol. 59, (1957).
4. J. C. Slater, Quantum Theory of Matter, McGraw-Hill Book Co., New York (1968).

5. R. R Agahi, B. Ershaghi, M. C. Lin, and A. Mansoori, Thermodynamic Behavior of Hydrogen/Natural Gas Mixture, 74th Annual GPA Convention, March 13, 1995.
6. J. S. Swearingen and J. Holm, Application of Turboexpanders for Energy Conservation, ASME 77-Pet-17, June 8, 1977.
7. F. Y. Zeidan, The Use of Flexure Pivot Bearings in High Speed and High Performance Turboexpanders, Turboexpander user Council, Houston, Texas, June 1995.

EXPERIENCE WITH SMALL TURBOMACHINERY IN A 400 WATT REFRIGERATOR

J. D. Fuerst
Fermi National Accelerator Laboratory*
Batavia, Illinois 60510 USA

ABSTRACT

A refrigerator similar to one of the Fermilab Tevatron satellites was re-configured to use turbomachinery instead of the reciprocating equipment typical of the installations. A Sulzer dry turboexpander, Creare wet turboexpander, and IHI centrifugal cold compressor have been installed and operated for about 8000 hours. Experience was gained both with the rotating machinery and with the refrigerator itself as it interfaced with the load. Equipment was set up to regulate in the same manner as the reciprocating devices had. Heat loads and operating mode were adjusted and evaluations made regarding the behavior of the devices.

Individual equipment performance is described, as well as system behavior and overall integration of the machinery. In particular, attention is paid to the Creare wet turboexpander. This device is operated for the first time as part of a full scale refrigeration system, testing not only its performance at the design point but also its off design characteristics and behavior in transient situations.

INTRODUCTION

Fermilab's Accelerator Division/Cryogenics Department is responsible for the maintenance and operation of both the Central Helium Liquefier (CHL) and the system of 24 satellite refrigerators which provide 4.5K refrigeration to the magnets of the Tevatron synchrotron.[1] The satellite refrigeration group also operates several additional helium refrigerators which are generally tied to the Tevatron system but may not be directly involved with operation of the accelerator. One of these systems is located at the antiproton source. This refrigerator (known as PR) is essentially identical to the Tevatron satellites and originally supplied LHe to some electronics associated with antiproton production. Advances in electronics eliminated the need for 4.5K operation which allowed the Division to make the PR refrigerator available to the Cryogenics Department for use as a test bed. In the last few years PR has been used for a variety of tests including operator training and various small experiments. Most recently, PR has been home to several pieces of high speed turbomachinery. This equipment has operated for the past year providing data both on individual component performance and on system behavior. Figure 1 is a photograph of the PR refrigerator.

The rotating expansion engines used in the Tevatron satellites are reciprocating devices manufactured by Process Systems, Inc.[2] These expanders achieve mid - 70% efficiency and

* Work supported by the U.S. Department of Energy under contract No. DE-AC-76CH03000.

Figure 1. Photo of PR refrigerator building showing (l-r) Sulzer dry expander, Creare wet expander, 160L dewar, and IHI cold compressor. The valve box and horizontal heat exchanger are visible in the background.

are well integrated with our existing system. However, they possess drawbacks which make them undesirable in certain other applications. These drawbacks include physical size, maintenance requirements, and flow pulsation created by the positive displacement nature of the units. The Cryogenics Department is interested in gaining experience with high speed rotating machinery, in part to advance the state of the art but also to provide a working system where experiments may be conducted free of flow pulsation. There is considerable experience at the CHL with larger scale turbomachinery; the unique aspect of PR is the use of small devices in a 50 g/s, 400+ watt refrigerator. This paper describes the system layout, the details of the turbomachinery, and the performance both of individual components and of the overall refrigerator in various operational and upset modes.

SYSTEM DESCRIPTION

The PR refrigerator was built to supply LHe to three dewars housing electronics associated with the antiproton source. The plant is based on the Tevatron satellite refrigerators, using the same heat exchanger and (originally) reciprocating expanders. Following an electronics upgrade which obviated the need for LHe cooling, the refrigerator was idled or used for training or experimentation. In early 1994 the reciprocating expanders were replaced with high speed turbomachines which had formerly been in storage. Our prototype centrifugal cold compressor as well as a 160L dewar were added to create a complete refrigerator. A 1000W heater in the dewar provides the heat load. Figure 2 shows a simplified schematic of the system.

The Tevatron high pressure header supplies about 75 g/s of 2 MPa, ambient temperature helium to the tube side of the heat exchanger. Part of the supply is precooled in a LN_2/GHe exchanger before joining the main stream. The gas is then cooled in counterflow with the low pressure (0.13 MPa) return to about 6K at the cold end. The dry expander is

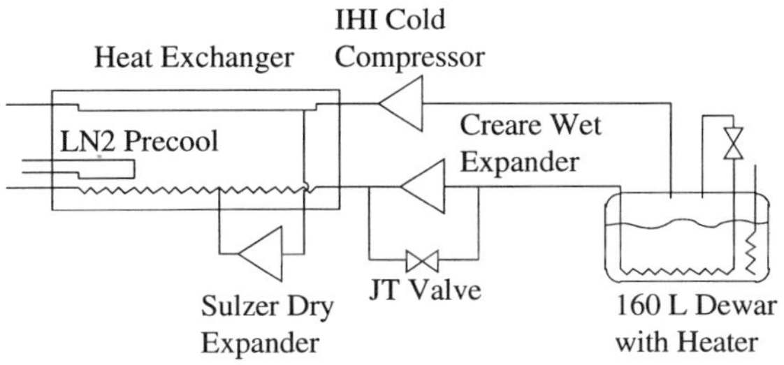

Figure 2. Simplified schematic of the PR refrigerator showing location of turbomachinery.

located at the 20K point in the exchanger train and takes a (design) flow of 30 g/s, discharging into the shell side. Tube side pressure drop with these flow rates is about 0.3 MPa, giving a 45 g/s, 6K, 1.7 MPa wet expander inlet. This pressure is less than design for the wet expander and may partially explain the reduced efficiency measured on that unit. The wet expander exhausts at pressures at or below 0.3 MPa depending on how the system is configured. Flow passes through a finned tube coil where it is subcooled by the liquid in the 160L dewar before being throttled into the dewar itself. A heater in the dewar maintains liquid level. The cold compressor regulates dewar pressure by drawing vapor from the dewar and discharging to the shell side of the heat exchanger. Cold compressor inlet pressure ranges from 140 kPa (compressor off) to about 50 kPa with the compressor operating at design.

ROTATING EQUIPMENT

Sulzer Dry Expander

The standard reciprocating dry (20K) expander is replaced with a Sulzer turboexpander consisting of two TGL 16 turbines in series. The units are housed in a common free standing cryostat with all required instrumentation and controls on board, including speed pickups for both shafts. Dynamic (self-acting) gas bearings are used which require no external gas supply. Flow rate is controlled by an electric actuated inlet valve which is controlled via the Fermilab distributed control system. Valves on the turbine brake circuits are used to set shaft speeds. They are manually positioned and require adjustment for optimum efficiency at given inlet conditions. Care must be taken to avoid overspeeding the turbines on startup, either by improper brake valve position or excessive gas flow. The turbines are protected by overtemperature and overspeed interlocks to limit the risk of damage. The brakes are water cooled although we have added supplemental fins and fans to reduce operating temperature.

This expander has been at Fermilab since the mid-1980's and has been installed at various locations within the Division, including (for awhile) a Tevatron satellite refrigerator. At that time the unit was not configured for remote control of the inlet valve so operations were cumbersome, requiring visits to the refrigerator to adjust flow rate and turbine speed. In the current installation at PR, the unit automatically regulates flow rate based on discharge temperature. Brake circuits are still tuned by hand but this has proven acceptable as most brake adjustment occurs during startup and cooldown - once the unit is cold, the turbine speeds are not strongly affected by small changes in inlet valve position as the expander regulates. Table 1 lists dry expander parameters and operating data.

IHI Cold Compressor

The cold compressor at PR is the same unit described in an earlier paper.[3] It is manufactured by IHI Co., Ltd. and served as our prototype for units eventually purchased in support of the Tevatron low temperature/higher energy upgrade.[4] It consists of a single stage impeller and shaft supported by dynamic foil gas bearings and powered by a water cooled variable frequency induction motor. Table 2 shows selected parameters for the compressor. Performance data is given in the above reference.

Table 1. Sulzer Dry Turboexpander

Turbine Type:	T1: TGL 16-11/B22	design speed 4.7 kHz	
	T2: TGL 16-14/A28	design speed 4.4 kHz	
	Fluid: pure helium		
	Flow: 30 g/s		
	Inlet: 1.82 MPa, 25K		
	Outlet Pressure: 0.122 MPa		
Measured Performance:	Inlet:		1.58 MPa, 24.32K
	Exhaust:		0.15 MPa, 14.77K
	Flow Rate:		31.6 g/s
	Speeds:		T1: 4.35 kHz, T2: 4.10 kHz
	Efficiency:		60.8% isentropic

Note: efficiency is measured from upstream of the inlet valve to downstream of turbine 2. Inlet valve ΔP is 0.16 MPa, giving a turbine inlet condition of 1.42 MPa, 24.24K and actual turbine isentropic efficiency of 62.6%.

Creare Wet Expander

Creare Inc. developed a miniature wet turboexpander in the late 1980's which was tested at Fermilab[5]. This research was funded by the U.S. Department of Energy, which took possession of the expander at the conclusion of the project. In turn, Fermilab requested and was granted ownership of the device on behalf of the DOE. Expander performance during the initial tests is reported in the above reference. The unit at that time was essentially a laboratory device which required continuous supervision during operation. This was acceptable for initial operation and proof of concept but further controls and interface work were required to allow "hands free" operation in an industrial setting.

Most important for unattended operation was a means of automating a balancing circuit which maintains equal pressure on either side of the shaft seal. The expander is designed with a labyrinth seal at the lower end of the shaft between the lower journal bearing and the turbine. This seal is exposed to system pressure on one side and is ported to a regulated supply of warm helium on the other side. Since pressure at the system side of the seal varies depending on inlet conditions, the pressure at the other side must be adjusted to maintain zero differential pressure across the seal. Leakage from the process side up into the expander risks freezing out the bearings while leakage in the other direction adds warm gas to the process stream and destroys efficiency. During initial expander testing the labyrinth seal was balanced by observing a differential pressure gauge and manually adjusting a regulator. This action was necessary every few minutes in order to maintain pressure differentials less than ± 2.5 kPa. Automatic operation of this labyrinth balancing circuit has been accomplished by motorizing the regulator and driving it with a signal from a differential pressure transducer. A Dwyer Capsu-Photohelic gauge with a center zero, ± 3.7 kPa range, and adjustable limit switches on both the positive and negative side is wired to a Hurst stepper motor attached through a telescoping linkage to the regulator. Control is

Table 2. IHI Cold Compressor

	Design	Off Design	Standby
Inlet Pressure:	51 kPa	41-81 kPa	142 kPa
Inlet Temperature:	sat. vap.	sat. vap.	sat. vap.
Outlet Pressure:	142 kPa	142 kPa	
Flow Rate:	60 g/s	40-70 g/s	55 g/s
Speed:	1.3 kHz	1.5 kHz max	0 Hz
Input Power:	1.25 kW	1.5 kW max	

Table 3. Creare Wet Turboexpander

	Design	Measured
Flow Rate:	50 g/s	43 g/s
Inlet:	1.9 MPa, 6K	1.8 MPa, 6K
Outlet Pressure:	0.22 MPa	0.25 MPa
Pressure Ratio:	8.6	7.2
Speed:	6.4 kHz	5.9 kHz
Isentropic Efficiency:	72.2%	60 ± 10 %*

*calculated across expander and inlet valve with valve at 100% (zero ΔP); does not include effect of 0.8 g/s warm helium flow for bearings, brake, and labyrinth circuit

a simple on/off circuit which drives the stepper motor in the desired direction when the differential pressure across the labyrinth seal exceeds the position of one of the limit switches. Motor step angle is 0.1 degree, with an adjustable frequency of between 5 and 480 Hz. This simple, self contained system has proven sufficiently versatile to provide accurate control of seal differential pressure to within ± 1.5 kPa without serious overshoot during normal operation. A more sophisticated control scheme may be implemented in the future which makes use of the Tevatron distributed control system, giving the labyrinth balancing circuit all the capabilities of a standard refrigerator control loop.

Wet expander speed is variable up to 7 kHz, with a design speed of 6.5 kHz. Shaft speed depends on inlet conditions and the pressure in the brake circuit. The brake consists of a pressurized, water cooled, closed loop helium circuit driven by an impeller attached to the top end of the expander shaft. Brake circuit pressure is controlled with a regulated supply of warm helium to optimize speed for maximum efficiency and to prevent overspeeding when the expander is warm. Shaft speed is sensed with a capacitance probe and displayed on an oscilloscope. Currently the brake pressure is manually adjusted

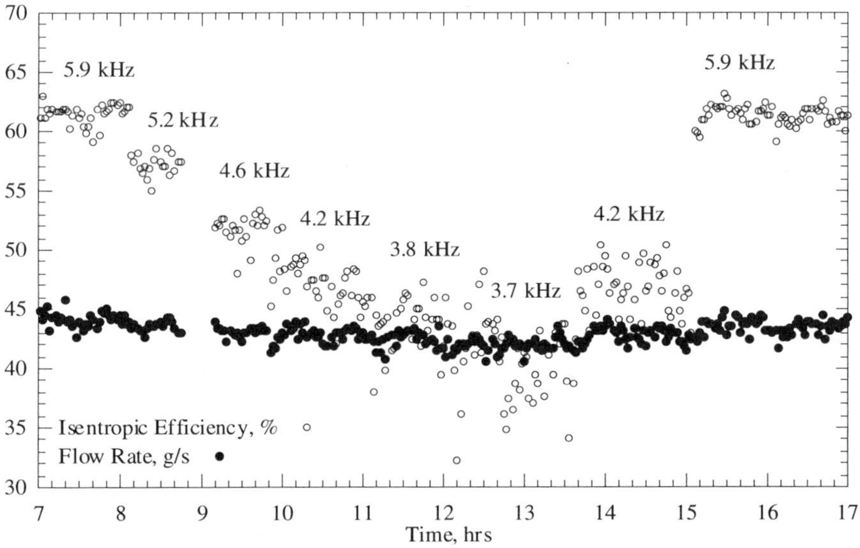

Figure 3. Creare wet turboexpander efficiency and flow rate over a ten hour period. The inlet throttle valve is wide open. Inlet and exhaust pressures are a constant 1.8 and 0.26 MPa. Inlet temperature varies from 5.5 to 6.5K. Frequency numbers correspond to turbine speed as brake pressure is varied between the 0.45 MPa minimum (producing a maximum speed of 5.9 kHz - less than the design of 6.4 kHz) and 1.3 MPa, the highest safe pressure given the thrust bearing instability. Note mass flow is under the 50 g/s design point, perhaps due to reduced inlet pressure caused by high pressure drop in the heat exchanger.

although future plans call for automatic adjustment based on expander speed. For the most part the brake circuit runs at its minimum pressure of 0.45 MPa so adjustments are infrequent. The exception is during cooldown when higher brake pressure must be maintained to prevent shaft overspeed. Maximum brake pressure is rated equal to the supply for the expander's pressurized (static) gas bearings (about 2 MPa) although we have been limited to 1.3 MPa due to a thrust bearing instability which causes severe, audible shaft vibration at certain ratios of bearing supply to brake supply pressure. Wet turboexpander parameters and performance are listed in Table 3 and Figure 3.

SYSTEM PERFORMANCE

Capacities

Liquefaction and refrigeration capacities were measured directly using the 160L dewar and built in 1 kW heater. Both quantities were measured with and without the wet turboexpander, substituting an adiabatic throttle valve. Table 4 lists maximum capacities achieved at PR after experimenting with various tunes. Refrigeration at LHe temperature is given directly by heater power required to maintain a stable liquid level while liquefaction rate is measured by the rate of liquid accumulation in the dewar with the heater off. Refrigeration results are in reasonable agreement with Tevatron satellite refrigerator performance in "stand-alone" mode (no LHe injection from the CHL). This is expected since PR is essentially identical to a Tevatron satellite except for the rotating machinery and the load. Liquefier capacity is about 70% of a satellite in stand-alone mode with wet expander and 40% of a satellite in stand-alone/JT operation.

Controllability

A key element of the PR turbomachinery test is to document the behavior of these small units as they interact with the cold box and each other over changes in flow, heat load, and tune. Efforts have been made to provide each unit with regulating capability, allowing it to adjust flow rates and speeds as conditions warrant. The IHI cold compressor adjusts speed based on its inlet pressure, effectively controlling the temperature of the liquid in the dewar.

Tevatron satellites use the speed of the reciprocating wet expander to control expander exhaust pressure, thereby fixing the pressure in the "single phase" circuit of the Tevatron magnet strings. The analogy at PR is to allow the Creare wet turboexpander inlet valve to throttle the expander flow based on exhaust pressure. This works quite well from a control standpoint given an appropriate valve/actuator combination. However, overall efficiency of the expander package suffers since some portion of the expansion is performed across a valve with 0% efficiency. With the inlet valve at 62% open, the expander inlet pressure is reduced from 1.8 MPa to 1.3 MPa. Although efficiency across the expander itself is maintained at 55%, the overall efficiency of the unit (valve plus expander) is measured to be about 30%.

The Sulzer dry expander operates out of a more loosely coupled control loop than the other two machines. This expander, like the Creare, adjusts flow rate with a throttle valve upstream of the turbines. The control point is the dry expander exhaust temperature as measured by a hydrogen vapor pressure thermometer (VPT). By careful selection of control loop parameters, this system was made to control to within ± 34 kPa on the VPT - a temperature resolution of a few degrees. This is a satisfying result given the limited useful range of the VPT and the steepness of the pressure vs. temperature curve for that device. Of course the Sulzer expander suffers the same efficiency penalty as the Creare imposed by inlet throttle valve positions less than 100%.

Table 4. PR Refrigerator Capacities

	Wet Expander Mode	JT Valve Mode
Liquefaction	84 L/hr	22 L/hr
Refrigeration	575W	310W

Transients & Upsets

There have been ten interruptions in PR operation in the last 16 months, three of which were caused by site-wide power outages at Fermilab. The other stoppages were intentional for equipment maintenance or modification. Each interruption involved a shutdown of all three turbomachines and subsequent restart. This has provided experience with warmup and cooldown of these units. In addition, the safety interlocks were tested during the power outages. In particular, the Creare wet expander is vulnerable in a power outage because it requires a pressurized helium supply for the bearings, brake, and labyrinth balancing circuit. The inlet valve is fitted with a fail closed actuator which shuts off flow to the expander. A bearing gas accumulator provides sufficient ballast to allow the shaft to spin down before the helium bearing supply pressure dwindles (warm gas for bearings, etc. is taken from the Tevatron high pressure header; a power outage shuts down our compressors causing this header to depressurize.) The Sulzer dry expander and IHI cold compressor use self-acting gas bearings which are pressurized by the process stream and as such are not at risk during an outage.

The wet expander survived the first two power outages with no perceived ill effects. In both cases, the unit was successfully restarted after resumption of power and performance was equal to that before the outage. However, the unit failed to spin up after the third outage which occurred on 22 June 1995. A loud vibration typical of the thrust bearing instability was heard as we re-pressurized the brake circuit in preparation for opening the inlet valve. During this vibration, the speed trace on the oscilloscope went flat, suggesting seizure of the shaft. There has been no shaft motion since. The wet turboexpander operated for about 7500 hours with no maintenance prior to this failure.

CONCLUSIONS

The PR refrigerator at Fermilab has operated for over a year using high speed turbomachinery instead of the standard reciprocating hardware found in a Tevatron satellite refrigerator. Although efficiencies are less than those measured on reciprocating devices, the units have met the modulation requirements of a satellite. In addition, they are low (or no) maintenance devices which are (or could be) significantly more compact than their reciprocating counterparts. This is important where space constraints, remote location, or accessibility problems outweigh the need for maximum efficiency. Finally, the continuous flow nature of these devices eliminates the flow pulsations and noise inherent in a positive displacement reciprocating machine.

REFERENCES

1. J.C. Theilacker, Current operating experience and upgrade plans of the Tevatron cryogenic system, in: "Advances in Cryogenic Engineering," Vol. 35b, Plenum Press, New York (1990) p. 917.
2. W.M. Soyars, Operation and maintenance of Fermilab's satellite refrigerator expansion engines, to be presented this conference.
3. J.D. Fuerst, Selection of cold compressors for the Fermilab Tevatron, in: "Advances in Cryogenic Engineering," Vol. 39a, Plenum Press, New York (1994), p. 863.
4. B.L. Norris, Status report of the Tevation lower temperature upgrade, "Cryogenics" Vol. 34, Butterworth-Heinemann Oxford UK (1994), p. 73.
5. H. Sixsmith et al, A miniature wet turboexpander, in: "Advances in Cryogenic Engineering," Vol. 35, Plenum Press, New York (1990), p. 989.

OPERATION AND MAINTENANCE OF FERMILAB'S SATELLITE REFRIGERATOR EXPANSION ENGINES

William M. Soyars

Fermi National Accelerator Laboratory*
Batavia, Illinois 60510

ABSTRACT

Fermilab's superconducting Tevatron accelerator is cooled to liquid helium temperatures by 24 satellite refrigerators, each of which uses for normal operations a reciprocating "wet" expansion engine. These expanders are basically Process System (formerly Koch) Model 1400 expanders installed in standalone cryostats designed by Fermilab. This paper will summarize recent experience with operations and maintenance of these expansion engines. Some of the statistics presented will include total engine hours, mean time between major and minor maintenance, and frequent causes of major maintenance.

INTRODUCTION

A Fermilab reciprocating expander is shown in Figure 1. The original design and the initial operational experience for Fermilab's expansion engines was described by Peterson.[1] Further operating experience has been described for the first six years of the Tevatron's life, which was commissioned in 1983.[2,3] This paper will update some of these statistics.

There are two different styles of engines used by Fermilab, "wet" engines used to produce liquid helium and "dry" engines used for intermediate cooling of the heat exchanger column. One major change in wet and dry engines has occurred. Prior to summer of 1993, the wet engine had twin reciprocating 5 cm diameter pistons while the dry engine had 7.5 cm pistons. In the summer of 1993, the piston sizes were reversed so that wets now have the larger diameter pistons. This was done to accommodate system changes related to the cryogenic system lower temperature upgrade, resulting in warmer expander inlet temperatures. Furthermore, changes have been made to the controllers. The original wets used a 1.5 kW DC controller, while the new wets now have a 5.6 kW AC regenerative controller.[4] These statistics presented here will help find any changes in maintenance trends since the conversion of 7.5 cm piston machines into wet engines.

Since normal Tevatron cryogenic system operation uses only the wet engine, run hours and maintenance statistics presented here will be based exclusively on wet engine observations. One would expect maintenance trends to be similar for the dry engine, due to the similar nature of their design.

---

* Work supported by the U.S. Department of Energy under contract No. DE-AC02-76CH03000

Figure 1. Fermilab wet expansion engine

Table 1. Fermilab modifications to engines

Realignment of frame (top end of the engine, which includes the flywheel assembly)	Alignment controls the main bearing location so that the shaft is appropriately located with respect to the cylinder centers. Sacrifice a slight bit of alignment to fit definition of an "average" alignment.	Makes frames interchangeable, any frame can be used with any top plate.
Crosshead guide inserts	Bore out the original steel crosshead guides to insert a bronze bushing (upper end) and a chrome sleeve (lower end, where O-ring slides).	Avoids surface wear and damage in the more expensive crosshead guide.
Crosshead replacement	Aluminum piston crossheads replaced with ones that are chrome plated steel.	Longer wear life Easier to reuse the crossheads by having them rechromed
Wrist pin bushing	Bore out the wrist pin hole in the connecting rod to accept a bronze bushing	Avoids wear in the connecting rod.

FERMILAB MODIFICATIONS TO PURCHASED ENGINES

To completely assess these maintenance statistics, one must be aware of the fact that significant changes are made by Fermilab before installation of an engine into the cryogenic system. Based on our experience, these changes enhance the life, maintainability, and safety of these machines. The modifications that are only for operational preference or safety will not be addressed here. Table 1 details those changes that increase the life and maintainability of our expansion engines.

GENERAL MAINTENANCE ROUTINE

The reliable operation of the wet expanders is essential to uninterrupted operation of the cryogenic system, and therefore the Tevatron accelerator. Because the consequences of an engine failure during physics operations are so severe in terms of lost experimental time, an experimental run typically begins with all the wet engines completely overhauled. A complete overhaul includes rebuilding the frame, pistons, crosshead guides, valves, and jackshaft. The new components used to rebuild each of these subsystems are shown in Table 2.

During a physics run, routine maintenance consists of weekly inspections of engines and monthly greasing of bearings and piston crosshead guides. Such routine maintenance is not included in these maintenance statistics. Repairs are made if possible during accelerator down periods, otherwise done on demand.

FAILURE EVENTS TABLE

Records of engine operation from 1988 to 1995 have been reviewed. During this period, there have been six accelerator cryogenic expander runs of varying duration. An expander run is defined as a period that concludes with an extended shutdown in which the liquid helium system is turned off and engine overhaul maintenance occurs. Do not confuse these expander runs with a physics experimental run; a physics run may have different beginning and ending dates. Statistics for this report were collected up to the February 1995 system maintenance shutdown. Information past this point is not included since the present expander run is scheduled to last until the end of July 1995.

Failures of wet engines have been counted and grouped by failure type. See Table 3. Failures are grouped into two different mechanical categories: major failures and minor failures. A major failure requires the engine to be warmed up from cryogenic temperatures, while a minor failure can be repaired while the engine remains cold (but it must be shut off).

The shaded portion of Table 3 summarizes the failure statistics. Total hours reflect those amassed by all 24 expanders, while average hours represent a typical machine. If one assumes a machine that is repaired upon failure and whose failure rate is constant over time, it is meaningful to define a "mean time between failure" (MTBF) as total operating time divided by number of failures.[5] However, it is not a simple matter in this case to define a MTBF since it assumes that a given machine runs until at least one failure. Quite often in our case an engine will last the entire run with no failures. Nevertheless, since engines are overhauled prior to beginning the next run, one can define a "mean time between maintenance" (MTBM) whether or not there was a failure. This time reflects how long an engine has been allowed to operate before performing maintenance due to either a failure or an accelerator shutdown. For a given run, for this system of 24 engines:

$$\overline{MTBM} = \frac{1}{24}\sum_{i=1}^{24}\frac{hours_i}{failures_i + 1} \tag{1}$$

Notice that a MTBM for minor repairs was not calculated for the 1993-94 run period. In this case, the majority of warm components, which are those involved in minor repairs, were not overhauled at the end of this run. Since this led to additional service hours, this run's minor failures are accounted for in the 1994-95 minor MTBM calculation. This calculation also includes four frame components that were overhauled as normal at the run's completion. Its interesting to note that ten frame assemblies survived the 10,800 hours during the combined last two runs with no failures.

Table 2. Parts replaced during engine overhaul

Frame	Pistons	Crosshead Guides	Valves	Jackshaft
main crankshaft	O-rings and felts	bronze bushing	valve arm bearings valve arm bushings	shaft
main bearings	wrist pins	chrome sleeve	cam bearings	bearings
connecting rod bearings connecting rod bushings	wrist pin bearings	O-ring seal	valve seats O-rings and felts	belts

Table 3. Failure Incidences for 24 wet engines during accelerator operations

Problem	repair time (hours)	94-95	93-94	92-93	90-92	90	88-89
O-ring Blow by	*4*	*2*	*4*	*5*	*4*	*0*	*8*
Genl Pist overhaul	*4*	*0*	*0*	*2*	*0*	*1*	*2*
Connect. Rod	*4*	*0*	*1*	*1*	*0*	*0*	*0*
Wrist Pin	*4*	*0*	*2*	*2*	*1*	*1*	*0*
Internal Leak	*6*	*0*	*0*	*0*	*2*	*0*	*0*
Frame-main brg	1	4	1	0	0	0	2
Frame-general	1	2	1	3	0	0	0
Crankshaft	1	0	5	0	0	0	0
Jackshaft and brg	1	0	1	4	2	1	0
Connect. rod brg	1	1	1	1	4	0	0
Valve rod/arm	0.5	2	2	1	4	3	2
Valve change	0.5	0	2	1	1	5	19
tot hrs=		88645	170934	246824	230819	140497	235301
avg hrs=		3694	7122	10248	9617	5854	9804
MTBM	(major)	3540	6083	8113	8215	5610	7762
MTBM	(minor)	5258	*	8113	7413	4756	5106

italics- indicate "major" failures, requires engine warmup for repair
plain - indicate "minor" failures, engine does not require warmup for repair
MTBM- average "Mean Time Between Maintenance" in hours for the 24 engines (see eqn 1)
*-Separate lifetime not applicable. Statistics included in 94-95 run value.

ANALYSIS OF FAILURE RATE ASSUMPTION

As noted above in defining a MTBM, a constant failure rate is assumed. Quite often a more realistic situation is to have a somewhat high initial failure rate that decreases over time (reflecting start-up "infant mortality" failures), followed by some period of a constant failure rate (random failures), followed by a final period of increasing failure rate (wear-out failures). Often for individual components of a system, bearings for instance, wear-out failures will dominate. But for entire systems, such as an expansion engine, often a constant failure rate can be assumed over the majority of the operational life. This assumption can be checked by looking at a time-series chart in Figure 2. To make one clarification in this figure, recall that many of the warm end components during the 1994-95 run had not been rebuilt prior to this run. Since cold ends were indeed rebuilt, this run has different effective end-of-run times for the cold end and warm end components, as shown in Figure 2.

This data can be used to perform a "centroid test" to see if any trends are present.[6] If there are n events, occurring at time x_i, for an observation period of x_0, then the test statistic for trend is:

$$U = \frac{\sqrt{12n}}{x_0}\left[\sum_{i=1}^{n}\frac{x_i}{n} - \frac{x_0}{2}\right] \tag{2}$$

This compares the centroid of the observed arrival times with the mid-point of the observation period. If U=0, there is no trend, failure rate tends to be constant. If U<0, the trend is for a failure rate that decreases with time, that is the time between failures tends to become larger. If U>0, the trend is for a failure rate that increases with time. Further analysis can indicate whether or not a constant failure rate is a reasonable assumption for that period. The value of U can be used to test the null hypothesis that there is no trend in the ordered data by use of the standard cumulative normal distribution function. For example, at U= ± 1.65, there is a 5% chance that such chronological ordering occurred with a constant failure rate; thus the null hypothesis can be rejected at the 5% statistical significance level. In engineering it is often typical to consider 5% or less statistical significance to be sufficient evidence upon which to reject a null hypothesis. But Table 4 shows that in many cases, one does not have this amount of confidence to reject the null hypothesis. Therefore, there is no highly statistically significant evidence to preclude the use of a constant failure rate assumption.

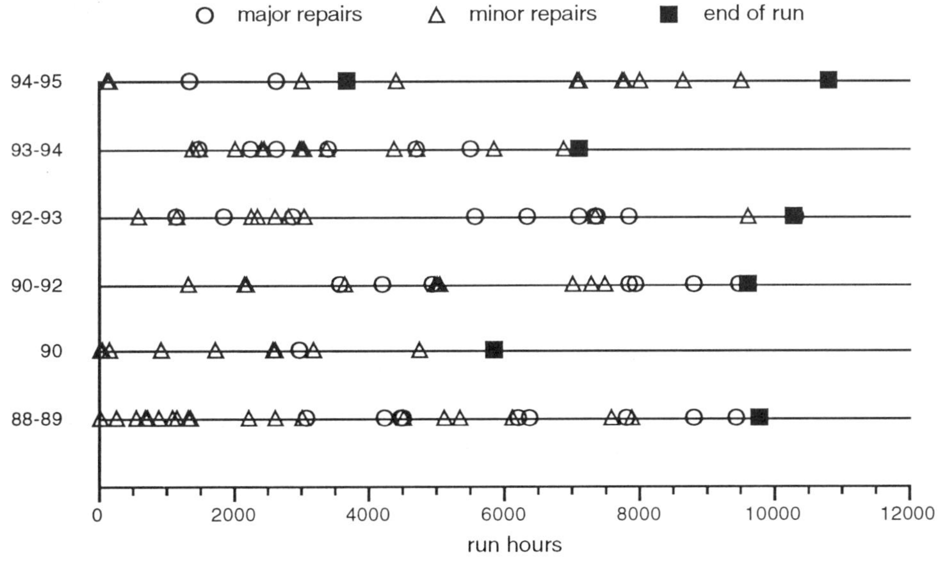

Figure 2. Time-series chart of failures for system of 24 wet expanders.

Table 4. Centroid U values and significance levels for trends

		94-95	93-94	92-93	90-92	90	88-89
U value	(major)	+0.19	-0.03	+0.69	+1.80	+0.05	+1.55
significance	for trend, %	42	49	24	3.6	48	6.1
U value	(minor)	-1.17	---	-1.96	-0.49	-0.95	-3.90
significance	for trend, %	12	---	2.5	31	17	0.005

However in general, major failures can be seen to most probably have a failure rate that slightly increases over time due to wear-out. Minor failures most probably have a slightly decreasing failure rate trend, dominated by start-up troubles. Nevertheless, neither failure rate consistently shows an overwhelming trend. Overall, assuming a constant failure rate is adequate.

FAILURE RATE ASSESSMENT

For each run, an overall failure rate can be found by looking at the total number of failures divided by the total run hours. This is shown in Figure 3. This failure rate represents the average failure rate for one engine, based on a sample of 24 engines. One can use this to make judgments on the reliability of reciprocating cryogenic engines. It has an advantage over the MTBM values in that MTBM is influenced by when shutdown maintenance occurs. This figure offers a good comparison of maintenance trends over the last several years. Also, one can compare the reliability of the new larger piston wets to the original smaller piston wets.

In general the failure rate for minor maintenance items, those warm end parts that do not require engine warmup to repair, has increased during the last two runs. Recall, these are the runs using larger diameter pistons. Table 3 reveals that the last two runs have experienced increased failures of main bearings, connecting rod bearings, shafts, and frames in general. This was a consequence of the conversion to larger piston wet engines, which result in more force applied to the rotating shaft and bearings.

A specific illustration of this can be seen in Table 3 by noting the rash of crankshaft failures during 1993-94. The ultimate problem was fatigue failure, accelerated by excessive fretting. The fretting corrosion mechanism will cause small pits and cracks at contacting surfaces, which serve as fatigue crack initiation sites. In response to this, the crankshaft now

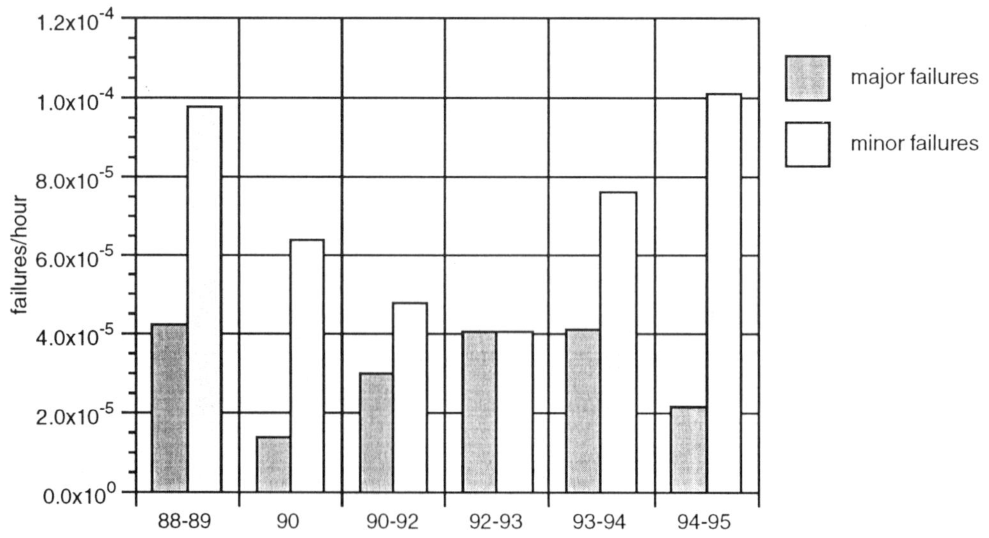

Figure 3. Wet engine failure rates

is made of a higher strength steel and, most importantly, receives a surface treatment for hardening. The surface treatment induces residual compressive stresses at the surface, inhibiting fatigue crack growth. There has not been a failure of a crankshaft with the superior varieties of surface treatments, ion-nitriding and a special chroming (a conventional chroming process could lead to hydrogen embrittlement and reduced fatigue strength). Crankshaft fretting certainly occurred in the original wet engines, but with the smaller forces there was more margin of safety.

Figure 3 shows that the failure rate for major repairs has remained about constant. Recall, these failures mostly consist of piston O-ring failures. This failure rate in fact is a function of run length. Notice that for the shorter runs in 1990 and 1994-95, failure rates are less. This indicates that wear-out failures dominate, as expected. There are virtually no changes in the major failure rate since the conversion to larger piston wets.

One line of Table 3 that bears explaining is the very high number of cold valve changes that took place during the 88-89 run. This was a consequence of a large number of contamination problems at the beginning of this run. This serves to elevate the minor failure rate for this run and also shift many of these failures to early in the run.

CONCLUSIONS

Through the years, reliable operation of Fermilab's reciprocating expansion engines has been demonstrated. To date about two million hours have been experienced at Fermilab. One now expects a mean time between major maintenance of about 7000 hours. This is an improvement on the mean times of 2000 to 3000 hours originally reported.[7]

In general, it is apparent that overall reliability with the 7.5 cm piston wets used since late 1993 is slightly less than that obtained with the 5 cm piston wets. This is attributable to an increase in minor maintenance, repairs that do not require engine warmup. Major failure rates are virtually unchanged. Since there has been a slight increase in minor failure rate, some improvements could be made in strengthening the frame components that now experience larger stress. Another alternative would be to perform warm component overhauls more frequently.

ACKNOWLEDGMENTS

The author wishes to thank the current staff of the Accelerator Cryogenics Engine Group: Ronald Norton, Michael Barbor, Daniel Freeman, Barry Skinner, and Derrick West. It is their skill and dedication, as well as the contributions made by earlier engineers and technicians in this group, that keep our expansions engines working so reliably.

REFERENCES

1. T. J. Peterson, Fermilab's satellite refrigerator expansion engines, "Advances in Cryogenic Engineering," Vol. 29, Plenum Press, New York (1984), p. 443-451.
2. J. C. Theilacker, Current operating experience and upgrade plans of the Tevatron cryogenic system, "Advances in Cryogenic Engineering," Vol. 35B, Plenum Press, New York (1990), p. 917-23.
3. B. L. Norris and J. C. Theilacker, Tevatron operational experiences, "Supercollider," Vol.1, Plenum Press, New York (1989), p. 321-28.
4. B. L. Norris, Status report on the Tevatron lower temperature upgrade, "Cryogenics," Vol. 34 ICEC supplement, Butterworth-Heinemann, London (1994), p. 73-76
5. P. D. T. O'Conner. "Practical Reliability Engineering," 3rd ed., John Wiley and Sons, Chichester England (1991), p. 4, 7, 38.
6. Ibid., p. 60-61.
7. Peterson, expansion engines, p. 449.

OPERATION AND MAINTENANCE OF FERMILAB'S 300 kW HELIUM SCREW COMPRESSORS AFTER TWO-MILLION HOURS OF OPERATION

A. Martinez and C.B. Pallaver

Fermi National Accelerator Laboratory*
Batavia, Illinois 60510

ABSTRACT

Fermi National Accelerator Laboratory uses 34 identical compressor systems connected to a common header to supply high pressure gas to the 24 satellite refrigerators of the Tevatron. These compressor packages consist of a two-stage oil injected Mycom helium screw compressor with associated oil removal system. This report summarizes (after 2,000,000 hours of operation) the problems, modifications and experiences associated with the operation of these compressor systems.

INTRODUCTION

The Tevatron refrigeration system uses thirty-four Mycom screw compressors installed in compressor buildings at nine locations around the Fermilab ring. Each of the thirty-four compressors consists of an oil-injected, two-stage, Mycom screw compressor driven by either a 260 or 300 kW (350 or 400 hp) motor and associated oil removal system. The system is designed to provide clean dry helium gas at 20 atm (~285 psig) to a 7.24 km (4-1/2 mile) header supplying 24 satellite helium refrigerators located around the Tevatron Accelerator.

Proper maintenance of such a large system is essential to reliable Tevatron operation. Typically, a minimum of 26 out of the 34 compressors are required during Tevatron operations. This redundancy in compressors allows maintenance to be carried on at all times regardless of the Tevatron status. This report will attempt to summarize after approximately 12 years and 2,000,000 total hours of operation, the problems, modifications and experiences associated with the operation of these compressor systems.

MAINTENANCE / EXPERIENCE

Compressor

The Mycom 2016C compressor consists of two sets of mating helically grooved rotors, a low stage and a high stage. The low stage male rotor is used as the driver and is coupled to the high stage rotor with a gear type coupling. Each male rotor consists of 4 lobes. The two female rotors consist of 6 lobes and are directly driven by the male rotors without the use of a

* Work supported by the U. S. Department of Energy under contract No. DE-AC02-76CH03000.

timing gear. Both sets are made of heat resistant cast iron, a material that has proven difficult to repair. The rotors turn at 3600 revolutions per minute and are positioned with preloaded precision thrust angular ball bearings. Babbitt journal bearings support the rotors at each end. Capacity control is accomplished through the use of a sliding valve in the rotor casing which moves parallel to the rotor axis and changes the area of the opening in the bottom of the rotor casing. This, in effect, lengthens or shortens the compression zone of the rotor and further acts to bypass gas to the suction side. This mechanism is operated by a hydraulic piston and cylinder assembly located within the compressor itself. The piston is actuated by oil and driven in either direction according to the operation of a four-way solenoid valve. Operating experience has shown that if the high stage is loaded and the low stage unloaded, a vacuum develops between the stages. By adding a 20 percent minimum position metal spacer block on the low stage slider, positive pressure is retained with minimal increase in power. The slider hydraulic valves have been a high maintenance item over the years because of the cycling of the pistons. The constant cycling causes fatigue failure of stop washers within the valve assembly. This problem has been reduced by operationally locking the compressors in their fully loaded condition and only allowing a select number of compressors to regulate. Other areas which have shown significant wear are the slider guide groove and the guide blocks.

Main shaft seals are one of our lowest maintenance items when compared to industry, with an average of 21,000 hours per seal. Oil leakage is measured once a month. Total losses for all machines are less than 7.2 liters per month (7.625 quarts).

Over the last several years, Fermilab has been involved in a joint venture with Mycom Corporation in testing of an improved compressor at Fermilab[2]. Improvements to the compressor include a new modified rotor profile which reduces internal gas leakages, a new oil relief modification within the final compressed space reducing the oil compression within the compressor and a new generation lubricant with lower viscosity properties. These enhancements translated to a 58 kW reduction in shaft power while increasing efficiency by 33%. The results of these tests have paved the way for a complete retrofit of our entire compressor system to the new modified compressors. These modifications are tentatively scheduled to be incorporated within the next two years.

The main objective with any moving piece of equipment is to maximize the operating hours without damaging the device. In the first few years after commissioning, the compressors were regularly overhauled at the 20,000 hour mark as recommended by the manufacturer. Our operating statistics indicate 50,000 hours is a safe limit for the bearings.

Indications are that the second stage drive rotor thrust bearings determine the life of the machine. Our experience shows that the machine is in a self destruct mode when a bearing wears more than 0.0508 mm (0.002 in.). The normal supply of oil masks the vibrations, noise, and cutting action of the rotors striking the walls when the bearings fail. Higher power is the final message from an unattended, unmonitored machine. Inspection of several compressors with over 50,000 hours of service shows signs of spalling on the outer races of the second stage drive rotor thrust bearing. Because of this we have chosen 50,000 hours as the appropriate time to overhaul a compressor. A new or overhauled compressor typically has clearances of 0.0508 mm (0.002 in.) between rotors and 0.1778 mm (0.007 in.) rotor to housing. Inspections have shown that the high stage rotor to rotor backlash and rotor to housing clearance show significant wear after 50,000 hours corroborating the choice of 50,000 hours as a limit for these machines.

We use two methods to detect deteriorating bearings. A simple shaft contact probe placed on the second stage male rotor detects any axial movement of the shaft. When movement of the thrust bearing occurs contact is made between the second stage rotor and the shaft probe completing a circuit and subsequently tripping off the compressor. This device is set at 0.1016 mm (0.004 in.) to 0.1524 mm (0.006 in.). This works well but only protects the main drive rotor. Should this probe circuit fail and a bearing failure occur, the rotors would eventually strike the interior casing causing an increase in motor power and eventually tripping off the compressor and possibly damaging the motor. This device is a simple, valuable and inexpensive instrument preventing costly damage to the compressor.

The second method for detecting damaged bearings are monthly vibration readings taken with a hand held meter. Normal readings average 0.127 to 0.381 cm/sec rms. (0.05 to 0.15 in/sec rms.). Experience has shown that a compressor is in a self destruct mode when vibration readings consistently reach over 0.508 cm/sec rms. (0.20 in/sec rms.). Experience gained with the use of vibration readings has encouraged us to pursue the installation of accelerometers on one machine for testing.

To a limited degree we have fitted several compressors with thermocouples on each thrust bearing outer race, a total of four on each compressor. These temperatures are read continuously and registered on a chart recorder. During a thrust bearing failure at one of these locations, a rise of 1 to 2 degrees Celsius (~2 to 3 degrees Fahrenheit) was registered[3]. The temperature differential encountered was insufficient for incorporation of an alarm or shut down protection. Testing continues at the other installations.

Table 1 shows the overhaul history of all 34 compressors. Overhauls involve complete dismantling of the compressor and typically includes replacement of all wearing items such as bearings, gaskets, seals and rotors as needed depending on condition. Repairs done in place, such as shaft seal replacement which does not require complete dismantling of the compressor, are not considered overhauls and are not reflected in the overhaul statistics. The table indicates that of the 57 overhauls, roughly half can be attributed to bearing related problems and the other half to general 50,000 hour preventative maintenance removal. The remaining overhauls are attributed to gasket failure and damage caused by electrical failure. Compressors with partially blown high stage gaskets are typically kept in service. These compressors will run with a higher interstage pressure and run less efficiently but are typically not overhauled until the problem becomes severe enough to warrant removal. Surface grinding of the casing walls to make them flatter during overhauls has alleviated this problem. Several compressors have been damaged due to electrical failure of the motor. Although typically this involves compressor bearing damage it has been separated in the statistics to distinguish between the two.

Figure 1 shows the distribution of overhauls over the last ten years. In the first six years of operation all compressors were overhauled at 20,000 hours for preventative maintenance. This rate has slowed since the decision to overhaul compressors at 50,000 hours. The trend shows a decline in failures over the life of the machines. We believe that this can be attributed to experience gained, improved preventative maintenance and the reliability of our shaft sensor probes.

Drive Motor

The compressor drive is a 260 or 300 kW (350 or 400 hp) 3600 rpm horizontal induction motor with an open drip-proof enclosure. It is connected to the compressor by a double-flexing disc coupling of the spacer type. The motors are lubricated and vibration analyzed every six months. Whenever the coupling is removed, the motor shaft is checked for vertical ball bearing play. By placing a dial indicator on the shaft and lifting it, we can determine the condition of the bearing or bearings outer race housing. A new motor bearing has a 0.0508 to 0.1016 mm (0.002 to 0.004 in.) play. The bearing at the drive shaft end takes most of the wear. The manufactured tolerances of the bearing are a maximum of 0.1016 mm (0.004 in.) We replace any motor above that number if it has high vibrations. A rebuilt motor exceeding maximum tolerances is sent back to the vendor. Once a month vibration is measured with a handheld meter. Readings average 0.127 to 0.381 cm/sec rms. (0.05 to 0.15 in/sec rms.). Bearing case noise varies from machine to machine and seems to have no effect on the life of a motor bearing.

Table 1. Compressor Overhaul Summary

Reason for Overhaul	# of Overhauls
High Hours/Prev. Maint.	26
Bearing Related Problems	26
Blown High Stage Gasket	3
Damage caused by Electrical Failure	2
Total	57

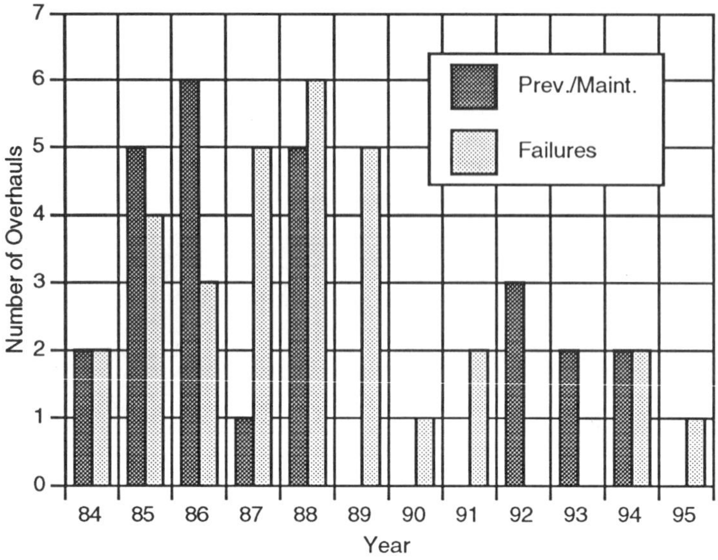

Figure 1. Compressor overhaul history

Since 1983, we have had 30 motors fail as figure 2 shows. Twenty-six have been attributed to winding failure with the remaining six failing due to bearing deterioration. Ten out of the thirty-four compressors are running with their original motors. Initially during the first few years of operation, many of the 260 kW (350 hp) motors were replaced with 300 kW (400 hp) high efficiency motors. Currently only seven of the 260 kW (350 hp) motors remain in service. Unlike compressors, the motors have been allowed to run continuously without periodic overhaul or reconditioning. The only maintenance performed are the monthly vibration readings and greasing every 6 months. Within the last few years we have begun to recondition motors which have over 80,000 hours. This limit was instituted after consulting with several bearing manufacturers, who define 100,000 hours as the limit to their bearings under normal operating conditions. Motor manufacturers are noncommittal in defining the life of a motor. As many of our motors were beginning to reach 70,000 hours of service we decided to begin reconditioning at 80,000 hours. Motor failures have been somewhat unpredictable as seen in figure 3. Figure 3 displays motor replacements with respect to location and indicates the number of hours each motor was in service with respect to the overall hours of the compressor skid. Certain locations have had as many as five motors fail while others are still running with their original motors. Some failures have occurred within as little as 1000 hours of initial operation while others have as many as 87,000 hours of operation without failure. This random failure rate may be attributed to differences in make and model of motors or perhaps in differences in starting conditions at time of failure such as high suction pressure causing the motor to run in its service factor, or several consecutive starts causing heat buildup and subsequent winding failure. In any case there seems to not be enough data to pinpoint weaknesses.

Lube Oil System

Oil plays an integral part in the performance of a screw compressor. It is required to seal, cool and lubricate the compressor as well as actuate the capacity control valves. The complete oil system consists of all the components necessary to store, circulate and cool the oil which is used as a lubricant and coolant for the compressor. Oil is stored in a pressurized vessel which serves as an oil reservoir and oil separator. This reservoir contains 340 liters (~90 gal.) of oil with an electric immersion heater keeping the oil above 15.6°C (60°F).

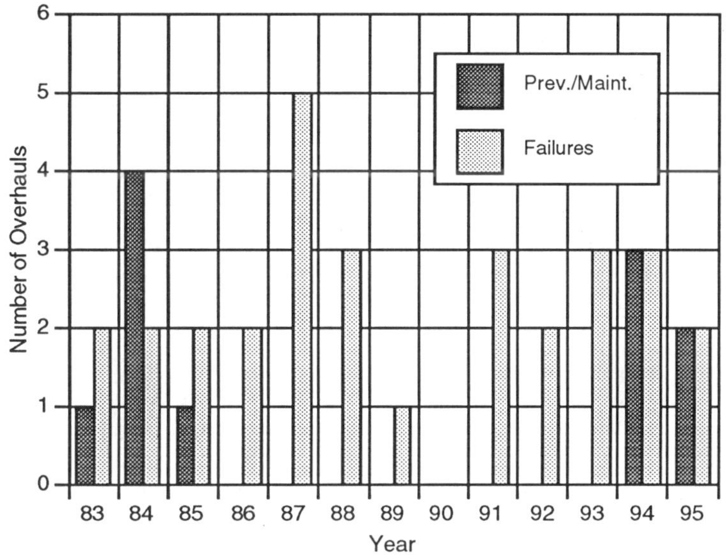

Figure 2. Motor overhaul history

An oil pump draws this oil through an oil cooler, a pump suction strainer, and drives it through a 10 micron filter and injects it into the compressor at several locations for bearing lubrication, heat dissipation and to provide a seal between the rotors and between rotors and the compressor housing. Oil is injected into the helium stream only during the first stage of compression and is discharged from the first stage entrained in the gas stream and then injected into the second stage. After the second stage of compression, the gas/oil mixture is delivered to the oil separator where the oil is allowed to settle out into the oil reservoir. Any

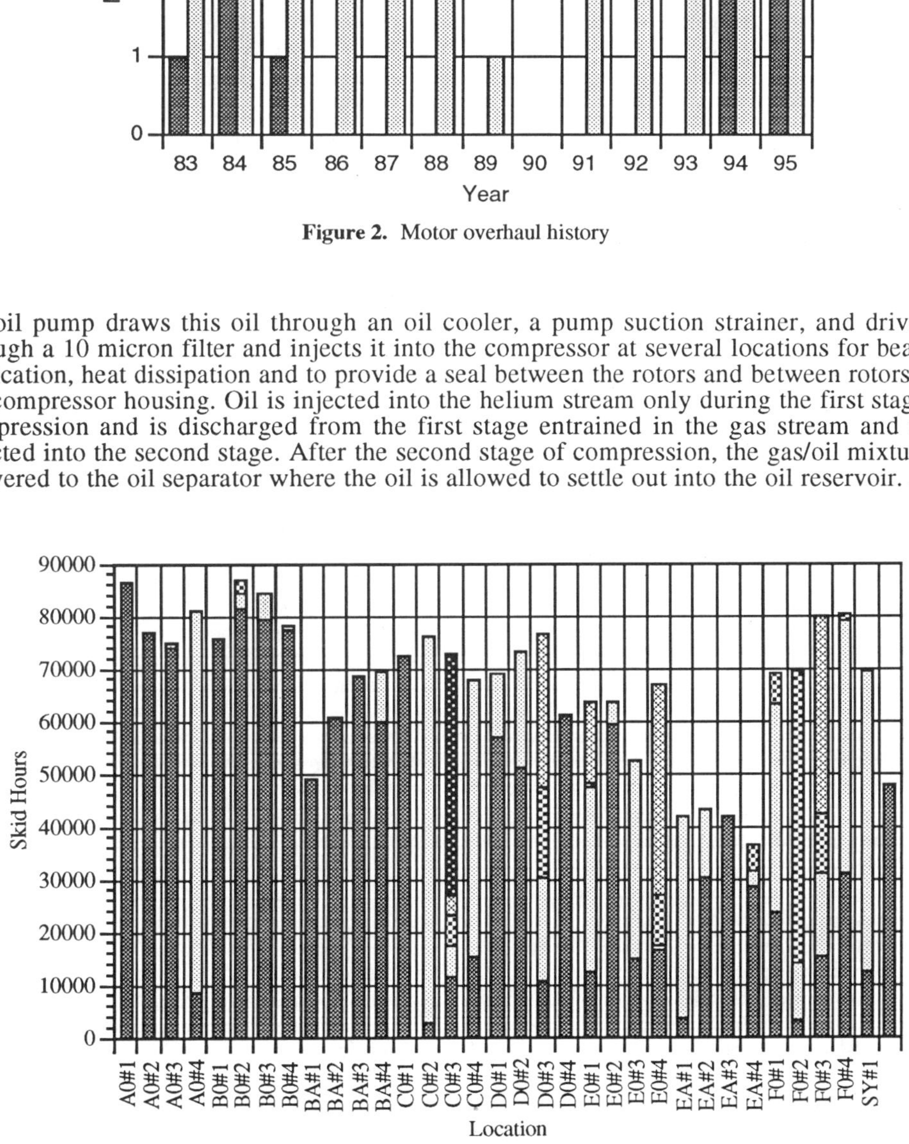

Figure 3. Motor replacement history

oil remaining in the gas is further removed by an oil and moisture removal system downstream of the compressor.

The oil inventory consists of Union Carbide LB-170X synthetic oil. With over 80,000 hours on many skids with original oil we have noticed no degradation in the oil except for some discoloration. Every two years selected compressors are sampled for viscosity and debris in the oil.

The vulnerability of the cryogenic system to contamination requires that the oil have a low water content in order not to cause plugging of heat exchangers and other equipment associated with the refrigeration system. The Union Carbide LB-170X oil as shipped from the manufacturer contains over 1500 parts per million of water by weight. Fermilab reprocesses the oil to reduce the water content to levels suitable for helium cryogenic service. Reprocessing consists of heating the oil to 121°C (250°F) and pulling a vacuum during the processing. When processing is complete the oil will contain less than one part per million of water by weight.

As mentioned earlier, Fermilab has been involved in a joint venture with Mycom Corporation in testing a new and improved compressor[2]. One of the improvements is in the use of a new generation lubricant with lower viscosity properties than the Union Carbide oil. Tests indicate that the use of this lower viscosity oil provides a reduction of approximately 14 kW in shaft power with a corresponding increase of 6.5% in isothermal efficiency and without any significant degradation in volumetric efficiency. The new oil (Gargoyle Arctic SHC 218HE) is a synthetic polyalphaolefin base manufactured by Mobil USA specifically developed for use in helium gas compressors.

Oil is circulated using an oil pump with a screw-configuration design and is rated at 300 LPM (80 GPM) of flow. The pumps have been a low maintenance item. Oil pump seals have averaged over 30,000 hours and require less maintenance than the main shaft seals. The oil pump is directly driven by a 5.6 kW (7.5 hp), 1750 rpm motor. These motors have an average of over 60,000 hours on them and are beginning to show their age. We are scheduling the replacement of these motors and plan on reconditioning all of the pumps within the next year.

Heat Exchangers / Water System

Oil cooling is achieved through an oil cooler heat exchanger where the cooling medium is water. The oil cooler is a four pass shell and tube heat exchanger with the water passing through the tubes and the oil through the shell. Water fouling of these heat exchangers is an on going problem. Every two years the heat exchangers are descaled using a biodegradable solution and the tubes are reamed as necessary.

The water system is a high maintenance area. Water erosion plays havoc with water valve seats and plungers. Recently the water valve plunger material has been changed from bronze to a plastic material with promising results. Water valve operation has been changed from gas ballast actuation to an electric actuated type. This modification allows more precise temperature control and also allows for periodic cycling of the valve to remove scale formation build up at the seat of the valve. Manual blowdown lines have also been added to assist in cleaning out strainers. In one location, at certain times of the year, the strainers on four compressors must be blown down at least once a week because of the influx of fresh water clams.

An aftercooler is installed at the discharge side of the compressor. It is used to cool the discharge gas before entering the oil removal system. Cooling of the helium is necessary to efficiently coalesce the oil out of the helium. Failure to properly coalesce the oil would result in oil carry-over to the refrigeration system. Such an event would be very time consuming to repair and unpredictable. The original aftercoolers were of a design which did not allow the water circuit to be cleaned out. Consequently, the water side of these heat exchangers had become blocked with clam shells and mud from the cooling water ponds. Most aftercoolers had lost their heat exchange properties. A new aftercooler has been purchased which reestablishes good heat transfer and allows for adequate scale removal. Installation of these new aftercoolers is currently underway.

Oil and Moisture Removal System

The purification system removes oil mist, oil vapor, water vapor, and particulates from the compressed helium. The units are designed with consideration of modularity, reliability and necessary redundancy. The system is designed for oil and moisture removal to below 0.1 part per million by weight and is discussed in detail by Satti[4].

The oil in the gas/oil mixture is initially allowed to settle out as it enters the oil separator/reservoir located downstream of the compressor discharge port. The reservoir contains a demister element which effectively removes the oil to 2500 ppm by weight. The next level of purification involves three stages of oil mist removal using Monsanto "Mist Eliminator" coalescers. The first two stages remove the oil mist to levels acceptable for cryogenic service while the third redundant stage is used as a guard coalescer. Solenoid valves installed at these units are used to recycle oil back to the compressor's interstage. A charcoal adsorbent vessel is installed downstream of the coalescers to remove any oil vapor entrained in the helium gas. The vessel contains 270 kg (~600 lbs.) of carbon pellets. A molecular sieve vessel is installed downstream of the charcoal bed and is used to adsorb water vapor left in the helium gas. Finally, a final filter is installed to retain any particle of 1 micron and larger which may have come from the adsorbent beds.

With over ten years of operation, the purification system has not shown any signs of oil carryover. Oil carryover at the third coalescer is monitored every six months

SUMMARY

The two-stage Mycom Corporation gas screw compressor has proven to be a reliable, low maintenance machine. The improvements initiated by Fermilab have contributed to this reliability. Upgrades such as the complete retrofit of existing compressors to new compressors with modified rotor profiles and new generation lubricant will greatly enhance the efficiency and performance of the compressor system. Experience gained over the years has drastically reduced the number of overhauls performed on the compressors. Much of this can also be attributed to the installation of the shaft movement sensor probe. The use of this probe has allowed technicians to take problem compressors off-line before catastrophic compressor failure can occur, thus saving time and money. Time invested in the design and testing of oil removal and gas purification resulted in the selection of components requiring virtually zero maintenance.

ACKNOWLEDGMENTS

The authors wish to thank the staff of the Accelerator Cryogenics Compressor Group; Bill Martin, Ted Roberts, Mike Benignetti, Terry Cross and Bruce Wilson for their skill and dedication in helping keep our compressors working so reliably.

REFERENCES

1 C.B. Pallaver, The Design, Fabrication, Operation and Maintenance of (41) 400 H.P. - 600 SCFM Helium Screw Compressor Systems (Five-Year Operation Report), Fermilab Technical Memo 1561, Fermi National Accelerator Laboratory, (1988).

2 A. Martinez and C.B. Pallaver, Test of an Improved Oil Injected Helium Screw Compressor at Fermilab, in: "Advances in Cryogenic Engineering," Vol. 39a, Plenum Press, New York, (1994), p. 887.

3 C.B. Pallaver and T. Cross, Screw Compressor Bearing Failure at C0#4, Fermilab Technical Memo 1931, Fermi National Accelerator Laboratory, (1995).

4 J.A. Satti and R.A. Andrews, Satellite Refrigerator Compressors with Oil and Moisture Removal Systems, in: "Advances in Cryogenic Engineering," Vol. 29, Plenum Press, New York, (1984), p. 453.

AN ACTIVE MAGNETIC BEARING WITH HIGH T_c SUPERCONDUCTING COILS AND FERROMAGNETIC CORES

G.V. Brown, E. DiRusso, and A.J. Provenza

NASA Lewis Research Center
Cleveland, OH 44135, USA

ABSTRACT

A proof-of-feasibility demonstration showed that high-T_c superconductor (HTS) coils can be used in a high-load, active magnetic bearing in LN_2. A homopolar radial bearing with commercially wound HTS (Bi 2223) bias and control coils produced over 890 N (200 lb) radial load capacity (measured nonrotating) and supported a shaft to 14000 rpm. The goal was to show that HTS coils can operate stably with ferromagnetic cores in a feedback controlled system at a current density similar to that for Cu in LN_2.

The bias coil, wound with nontwisted, multifilament HTS conductor, dissipated negligible power for its direct current. The control coils, wound with monofilament HTS sheathed in Ag, dissipated negligible power for direct current. AC losses increased rapidly with frequency and quadratically with AC amplitude. Above about 2 Hz, the effective resistance of the control coils exceeds that of the silver which is in electrical parallel with the oxide superconductor. These results show that twisted multifilament conductor is not needed for stable levitation but may be desired to reduce control power for sizable dynamic loads.

INTRODUCTION AND BACKGROUND

It is now possible to purchase HTS coils that carry 1000 A turns in LN_2, produce a few hundred gauss and are as small as copper coils operating in the same conditions. These are roughly the requirements for coils in active magnetic bearings. There are concerns about feasibility which center around the poor physical strength of the ceramic superconductors, their poor J_c at LN_2 temperature, and the prospect of high dynamic losses in HTS control coils that must carry currents with frequency components of several hundred hertz. These concerns have previously confined HTS magnetic bearing development to passive types.

Two distinct motivations for using superconducting coils in magnetic bearings instead of normal coils are to save power and to reduce coil cross section. At LN_2 temperature at this

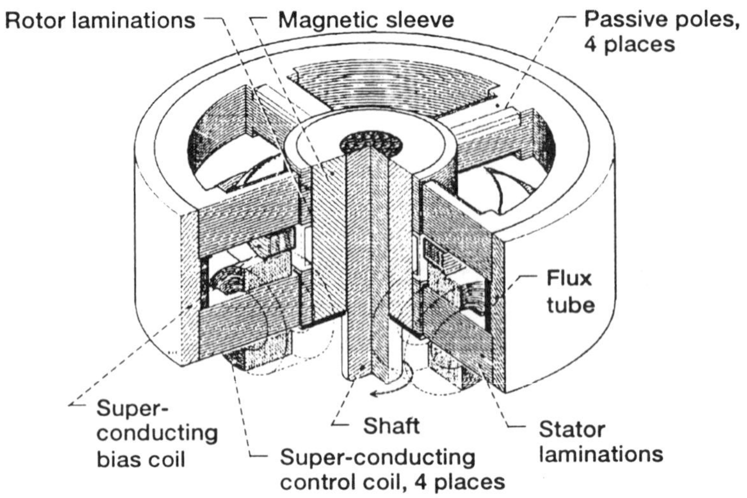

Figure 1.—Active magnetic bearing with high temperature superconductor bias and control coils.

time only the Bi 2223 material can carry enough current in a small coil (limited by the self-field of the coil) to compete with normal conductors.

Under a NASA Lewis Research Center grant to the University of Wisconsin,[1] a substantial increase of load capacity was predicted if HTS windings are used in magnetic bearings. Three approaches were studied analytically. In the first ferromagnetic cores were retained, but high current density was assumed, yielding high load capacity by driving the cores into saturation. The second approach was to abandon the stator cores and use only stator coils. The third approach used HTS coils on both stator and rotor. The second and third approaches will lead to higher loads when HTS conductors can tolerate higher stress and higher self field in the coils. At present it is feasible to use only the first approach, which does not require high stress or high field strength in the coils.

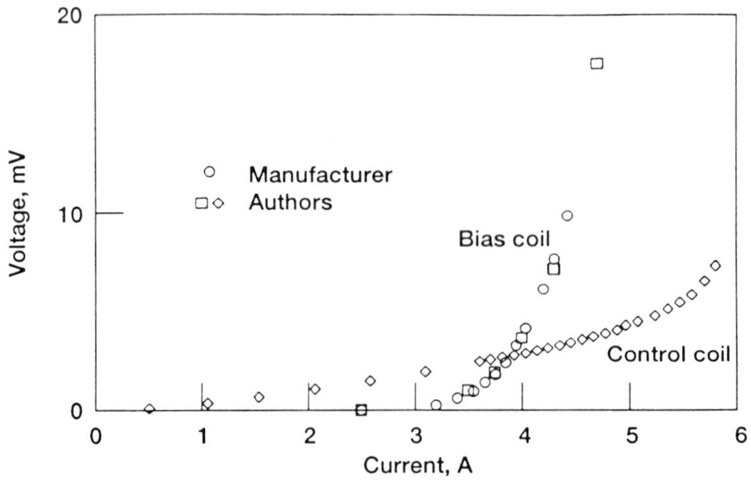

Figure 2.—Voltage across bias and control coils as functions of steady current in LN_2.

PHYSICAL DESCRIPTION OF MAGNETIC BEARING

The bearing is a homopolar radial bearing with a separate bias coil. Successful operation of the bias coil, which carries only direct current, was expected and would save the majority of the power consumed in a normal bearing. Successful operation of the control coils, which must carry a combination of DC and AC to produce steady and dynamic forces, was less assured. There was concern about the ability of the HTS composite to carry rapidly changing currents.

A drawing of the bearing is shown in Fig. 1. The overall bearing size is 18 cm O.D., 11.2 cm long with a 7.57 cm diameter journal. The bias coil encircles the ferromagnetic shaft. Flux passes through the shaft, through four air gaps to the poles of one stator, to the exterior flux tube to the other stator and through that stator's air gaps to the rotor. We used only four control coils to reduce expense, installing them on one stator and leaving the other stator poles empty. We compromised the design of the other bearing parts to accommodate externally-wound, circular coils. The rotor and stator laminations are 2V Permendur.

THE HIGH-T_c COILS

Bias Coil

The bias coil in a homopolar radial magnetic bearing produces a steady (DC) magnetic field in the air gaps. The gap between the journal and stator is 0.5 mm. Since there are two gaps in series in the magnetic circuit, about 1000 A-turns are required to produce 1 T in the gaps, allowing for leakage and fringing.

The commercially-produced circular bias coil (electrical size: 8.3 cm I.D., 11.8 cm O.D. and 1.36 cm long) was layer-wound on an aluminum coil form. A single length of multifilament conductor (19 filaments in a 0.2 mm by 2.5 mm Ag tape) was reacted before winding on the coil form. The silver-to-superconductor ratio is about 3. The 244 turn coil has a critical current (1 μV/cm criterion) of 3.9 A. The coil could be operated at 4.7 A with less than 20 mV potential drop. Without a ferromagnetic core the coil produces a calculated maximum axial magnetic induction of ~400 G at its inner bore. The voltage across the coil vs. steady current through the coil is shown in Fig. 2. The voltage was measured with voltage taps attached to the superconducting composite tape inside the current leads. The measured steady power dissipated in the coil vs. current is shown in Fig. 3. This measurement was made in the bearing (with ferromagnetic cores). Also shown is the power that would be required if only the Ag matrix were present, to give a rough idea of the power saved by using the HTS composite.

Control Coils

The control coils in a homopolar magnetic bearing with a separate bias coil must each provide about half the number of ampere-turns (500 A-t) as the bias coil (1000 A-t). This allows the two control coils on an axis to double the field of one pole (in the absence of core saturation) and to reduce the field of the opposite pole to zero. This gives the maximum net radial force toward the double-strength pole. The control coil must carry a combination of DC and AC currents to support steady loads (such as shaft weight) and dynamic loads (such as those due to shaft unbalance). We over-designed the control coils with 1000 A-t to allow heteropolar operation, if desired.

The four commercially-produced circular control coils were 3.6 cm I.D., 6.2 cm O.D. and 2.0 cm long. They were layer-wound from a single length of Ag tape (tape cross section

0.11 mm by 4.7 mm) with a monofilament HTS core and reacted after winding. The epoxy-impregnated coils fit over the laminated stator poles without any coil form to conserve space and to avoid eddy currents. Each coil has between 230 and 232 turns, and the critical currents (1 μV/cm criterion) varied from 5.1 A to 5.6 A.

The voltage across one control coil as a function of steady current is shown in Fig. 2. The voltage was measured with taps on the Ag tape leads. The voltage increase at low currents comes partly from those leads, but it increases faster than linearly. The nonlinear part may be due to segments of superconductor with reduced critical currents, which force current into the Ag. The coil reached 5.8 A with under 8 mV potential drop.

The measured power dissipated at LN_2 temperature in two of the control coils is shown in Fig. 3, as well as the power calculated for the pair if only the Ag were present. Note the ordinate log scale. The HTS power is seen to be two to three orders of magnitude lower for steady current, which would be required in control coils to support steady loads such as the weight of a horizontal shaft.

AC LOSS MEASUREMENTS ON CONTROL COILS

Control coil power dissipation was measured for single frequencies, with and without superimposed direct current, and with and without ferromagnetic cores. The average power P_{av} was obtained from the product of the instantaneous current I and the voltage V, sampled at up to 12.8 kHz and averaged over a time T much larger than the period of the AC frequency f

$$P_{av} = (1/T) \int_{0}^{T} V(t) \quad I(t) \quad dt \tag{1}$$

where $T >> 1/f$ and $I = I_o \sin(2\pi ft) + I_{offset}$. When there is no DC offset current, we calculate an effective resistance R_{eff} by

$$R_{eff} = 2P_{av} / I_o^2. \tag{2}$$

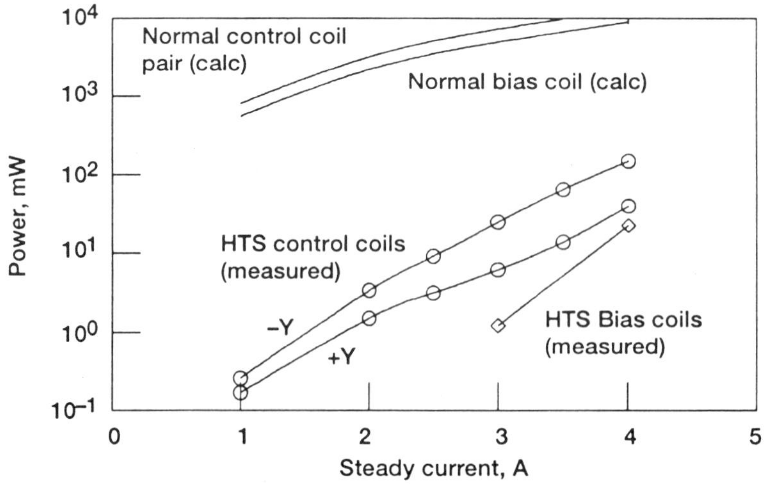

Figure 3.—Comparison of power consumption of HTS coils with normal coils (based on Ag cross sections in coils).

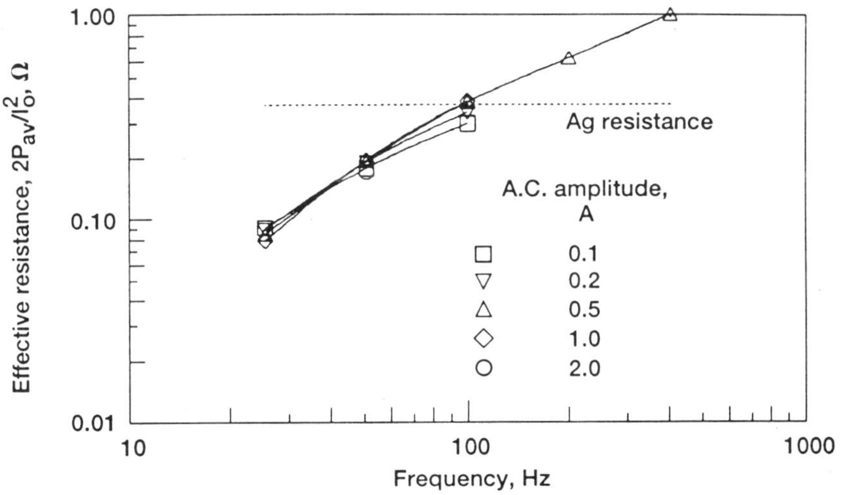

Figure 4.—Effective resistance of control coil as a function of frequency for several values of AC amplitude with no DC offset. No ferromagnetic core.

We compare this effective resistance below to the calculated resistance of the Ag matrix at LN_2 temperature. That resistance is based on the measured room temperature resistance and the resistivity ratio, 5.97, for Ag at LN_2 temperature.

This method of measurement may be subject to errors, especially for higher frequencies. A calorimetric method would be preferred (such as by measuring rate of LN_2 boil-off), but it is not feasible in our apparatus. Sensitive lock-in amplifiers are used to measure losses in short samples (of the order of 1 cm long) where the resistive voltages may be small compared to noise. But the length of conductor in our coils is about 3500 cm and the resistive component of the voltages is easily observed. Our method of averaging instantaneous power over many periods of the sinusoidal current should therefore be equivalent to the action of a lock-in amplifier.

However a more difficult problem arises from the fact that, in the frequency range of the data we present, the magnitude of the inductive voltage L dI/dt is comparable to or up to about five times as large as the magnitude of the resistive voltage (as determined by our measurements). If there are no relative phase shifts between voltage and current signals, caused in the measuring circuits, then the product of the sinusoidal current and the inductive component of the voltage would average to zero over time. If there is a relative phase shift, a spurious contribution to the resistive component results. Our loss measurements and our interpretations of them are presented with these cautionary comments.

An additional factor that may affect the AC measurements is the possiblity of normal (non-superconducting) turn-to-turn shorts within layers of the coils. Such shorts were a possiblity when these "wind then react" coils were made, but would be unlikely with insulation techniques now in use.

Air Core AC Losses

The average power dissipated in one of the control coils with no ferromagnetic core as a function of frequency with no DC offset current for several current amplitudes I_o was measured. The effective resistance, calculated from the data by using Eqs. (1) and (2), is shown in Fig. 4. In the range of this data the effective resistance is roughly independent of AC amplitude. The figure also shows the resistance of the Ag matrix of the conductor for comparison. The resistance of the HTS composite surpasses the Ag resistance above about 100 Hz. The

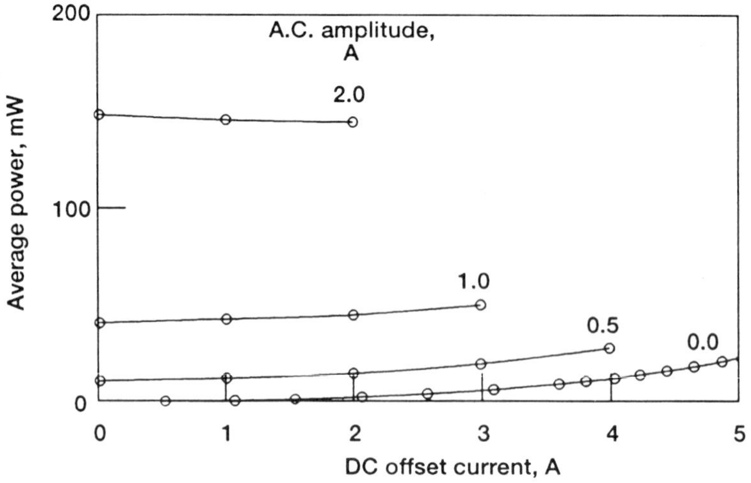

Figure 5.—Average control coil power as a function of DC offset at 25 Hz for four values of AC amplitude. No ferromagnetic core.

coils would have less resistance above that frequency if the oxide were not present. This result may be due to hindered flux penetration into the composite tape, predicted by a Bean-type model.[2] The currents and fields cannot freely penetrate the entire cross section of the tape, but must enter from surfaces as the current and field increase. Hence at high enough frequency, induced supercurrents prevent parts of the Ag from carrying transport current, increasing the apparent resistance. This problem may be reduced by twisted multifilamentary conductor, which is well developed for low temperature superconductors but was beyond the HTS art when these coils were made.

The effect of DC offset current added to AC is shown in Fig. 5. There the frequency is fixed at 25 hz and data for AC amplitude values of 0.0, 0.5, 1 and 2 A are shown. At this frequency the effect of DC offset is relatively minor and appears to be mainly due to the increase in resistance seen without any superimposed AC component. The measured power actually decreased with DC offset for an AC amplitude of 2 A. This might be due to the DC

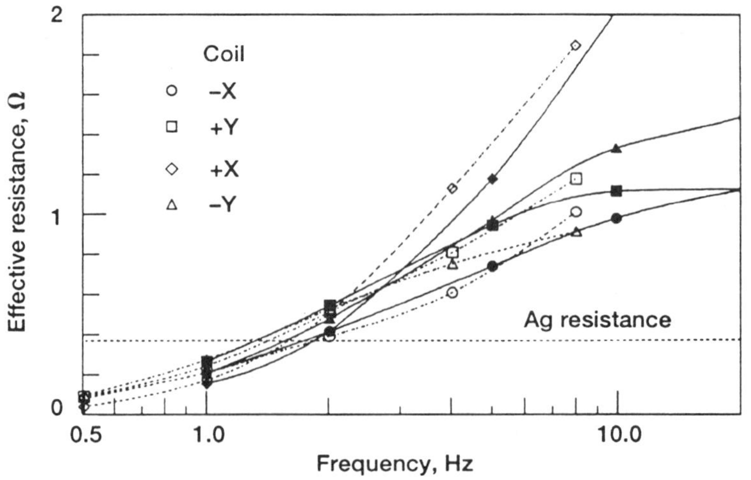

Figure 6.—Effective resistance of control coils vs. frequency with ferromagnetic cores (filled symbols) and vs. frequency divided by 50 for coils without cores (open symbols). AC amplitude is 0.5 A.

offset reducing the ability of the composite conductor to support shielding currents, allowing a greater portion of the Ag to carry transport current.

AC Losses with Ferromagnetic Core

The effective resistance of all of the control coils was measured in the magnetic bearing for pure frequencies. The results are shown in Fig. 6 with filled symbols and solid lines. There are substantial differences between coils, which is not surprising at this stage of HTS conductor and coil technology. All AC results before Fig. 6 were for the –X coil, one of the lowest loss coils. The +X coil stands out as significantly worse than the others at high frequencies. The increase in resistance occurs at much lower frequencies than it did without ferromagnetic cores. With cores the resistance surpasses the Ag resistance below 2 Hz rather than between 60 and 100 Hz. This is apparently related to the much greater change of flux through the coils.

The behavior of the resistance in the two cases with respect to frequency is remarkably similar with a simple change of frequency scale. Figure 6 shows (in open symbols and dotted lines) the effective resistance of the four coils from air-core tests with the frequencies divided arbitrarily by 50 (a number selected to best superimpose the two sets of data). With this rescaling there is considerable correspondence between the results for the same coil, with and without cores. Inductance measurements were made on the coils between 1 and 10 Hz in the superconducting state with and without the ferromagnetic cores. Those measurements yielded 75 and 2.2 mH, respectively. The ratio of these values is 34, a number of the same order of magnitude as the value of 50 used above to rescale the air-core frequency data. These results suggest that the losses may be largely linked to the rate of flux change through the coils. It is therefore hoped that twisted, multifilamentary conductor will substantially reduce the losses, making their onset occur at higher frequency and limiting their maximum values to approximately what would occur in the Ag.

The effect of DC offset current on AC power dissipation was measured at 1 Hz for AC amplitudes of 0.5, 1 and 2 A for two of the coils in the bearing. The results are shown in Fig. 7. At the lowest amplitude there is not much increase in power dissipation below 3 A offset current. For the higher amplitudes the dissipation first falls with offset current, perhaps because of a suppression of shielding supercurrents. As the sum of offset plus AC amplitude approaches within an ampere or so of the critical current, the dissipation begins to increase.

MAGNETIC BEARING OPERATING RESULTS

The LN_2 test rig has been described previously.[3,4] The shaft is vertical, supported near the top by a conventional ball bearing pair operating at room temperature. The magnetic bearing near the lower end of the shaft can be submerged in LN_2.

The bearing was operated to the maximum speed of the rig, 14,000 rpm, both at room temperature at 2A bias and at LN_2 temperature at various bias currents up to 4A. For stable operation it was necessary to use values of damping several times as high as required for other bearings we have run of similar size. At this time we do not know if this was due to the superconducting coils, the uncontrolled stator poles, or some other cause.

A maximum load capacity of 890 N (200 lb) was produced by the bearing, as shown in Fig. 8. An approximately linear load range extends (during load increase) to about 556 to 668 N (125 to 150 lb). The figure shows complete force hysteresis loops (each taken over a 5 sec period) for maximum control currents of ±4 and ±2 A. The higher current carries the force to a saturation level. The hysteresis is probably due to the Co-Fe core material rather than to superconducting effects. Loads were measured with the shaft nonrotating with a load fixture with small load cells.

With approximately 3 cm increase in bearing length, control coils could be added to the empty stator, increasing the maximum load capacity to 1780 N (400 lb).

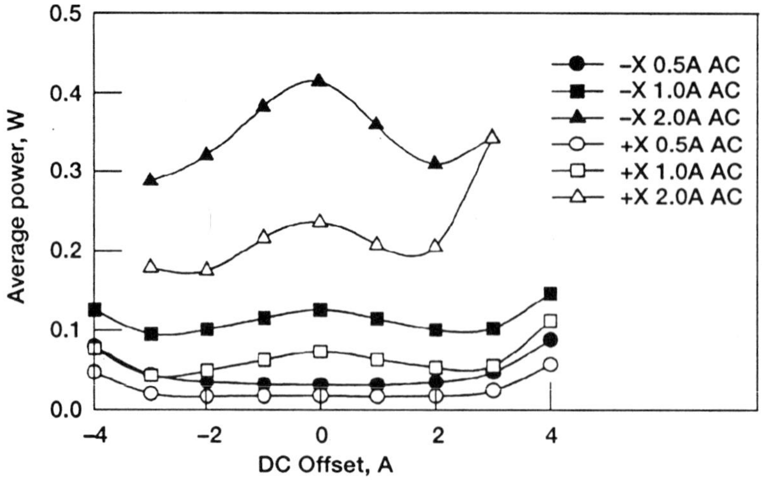

Figure 7.—X axis HTS control coil power dissipation vs. DC offset at 1 Hz with ferromagnetic cores.

CONCLUDING REMARKS

We have shown that an active high-load HTS magnetic bearing can be operated at LN_2 temperature with several performance measures approaching those possible with conventional windings. Comparable current density and hence coil size were obtained. For a lower temperature application (e.g., in LH_2) the coils could be substantially smaller, subject to stress limitations, of course. Power to supply the bias field was reduced essentially to zero, the same advantage provided by permanent magnet biasing. Power dissipated in the control coils to support steady loads was reduced essentially to zero. The control coils support dynamic currents sufficiently well to permit stable levitation up to the top speed (14,000 rpm) of our rig. A.C. losses appear to be related to coil inductance, suggesting that a twisted multi-filamentary conductor could reduce the losses substantially.

We found that the power dissipated in the control coils above about 2 Hz exceeded that which would be dissipated in normal coils.

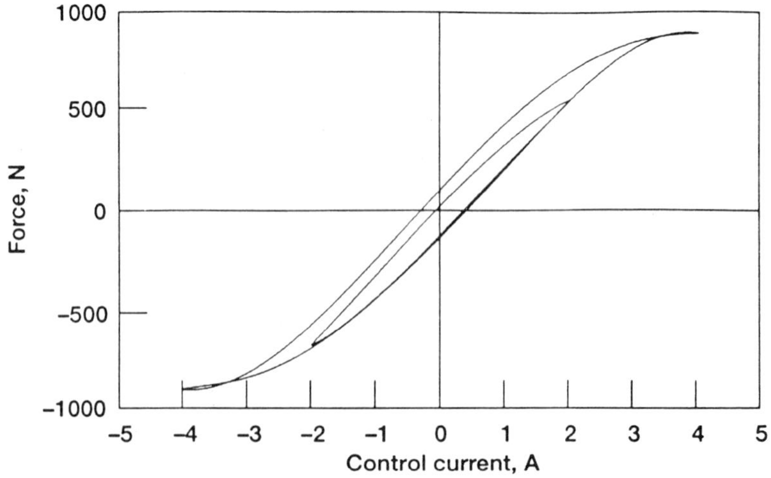

Figure 8.—X axis force as a function of control current at 4 A bias. Liquid nitrogen temperature.

Critical currents of the HTS conductors are improving and smaller coils are already possible. The primary technical improvements needed for active bearings that retain ferromagnetic cores are noncircular coils and twisted, multifilamentary conductor.

REFERENCES

1. Y.M. Eyssa and X. Huang, High pressure superconducting radial magnetic bearing, in "IECEC–90", Proceedings of the 25th Intersociety Energy Conversion Engineering Conference, Reno, Nev., Aug. 12–17, 1990. Vol. 3. American Inst. of Chem. Engin., New York, (1990). Work performed under NASA grant NAG3–1041.
2. C.P. Bean, Rev. Mod. Phys., 36:31 (1964).
3. E. DiRusso and G. V. Brown, Performance tests of a cryogenic hybrid magnetic bearing for turbopumps, in "Proceedings of the Third International Symposium on Magnetic Bearings," Paul Allaire, ed., Technomic Publishing Company, Inc., Lancaster, PA, (1992) p. 386.
4. E. Dirusso, G.V. Brown, and A.J. Provenza, Tests of a cryogenic electromagnet biased homopolar magnetic bearing, in "Advanced Earth-to-Orbit Propulsion Technology—1994," NASA CP–3282, Vol. II, (1994), p. 385.

HYBRID SUPERCONDUCTING MAGNETIC BEARING FOR KINETIC ENERGY STORAGE APPLICATIONS AND ITS FRICTIONAL ENERGY LOSS

Zule Xia, Ki Ma, Quark Chen, Rodger Cooley, Paul Fowler and Wei-Kan Chu

Texas Center for Superconductivity at the University of Houston
3201 Cullen Blvd.
Houston, Texas 77204

ABSTRACT

A hybrid superconducting magnetic bearing (HSMB) has been designed, constructed and implemented in a flywheel energy storage (FES) prototype. The HSMB design uses magnetic forces from permanent magnets for levitation and high temperature superconductor *YBCO* in between the magnets for stabilization. A *19* kg (*42* lb.) flywheel currently can rotate up to *6,000* RPM with kinetic energy of *8* Wh stored. To identify the factors of frictional energy loss in the bearings, we have conducted a series of spin-down experiments in different operational conditions. The result from the recent test under the air pressure of 10^{-5} torr indicates an average frictional energy loss <*2%* per hour, with the imperfect system alignment and an unbalanced rotor.

INTRODUCTION

Flywheels, as one of many energy storage devices, were intensively studied in the 70's and 80's[1,2]. An integrated flywheel system using advanced technology is mainly composed of magnetic bearings, a rotor made of composite materials and non-contact, electronically commuted motor/generator[1]. The feasibility of superconducting magnetic bearings (SMB) for industrial applications has been investigated since the discovery of high temperature superconducting materials in the late 80's[3,4,5]. The efforts have been concentrated on building superconducting bearings of industry-scale capability with totally passive levitation and extremely low frictional loss. A promising near-term application using SMB may be flywheel energy storage (FES) designed for utility industries, since any energy storage systems for utility industries must satisfy the requirements of very high diurnal storage efficiencies (>90%) for economic reasons[6].

This work represents an initial study on a research program "*Flywheel Energy Storage Using Hybrid Superconducting Magnetic Bearings.*" We have designed a HSMB and built a FES prototype as shown in Fig. 1 for characterizing the bearing performance. The final goal aims to construct an industrial module of a FES system which is able to maintain *1~2* kWh kinetic energy. The paper will first present the concept behind a HSMB design and then discuss the experimental result.

HYBRID SUPERCONDUCTING MAGNETIC BEARING

Superconducting bearings make use of a unique physical interaction between HTS and permanent magnets for levitation, known as 'flux pinning' effect[7]. A commonly used cooling procedure to activate a superconducting bearing, namely the "field-cooled" method, is to cool a HTS in the presence of a magnetic field. Because of the flux-pinning effect, as a HTS is being cooled below the critical temperature, the flux lines are trapped within the HTS, as in the configuration of a conventional SMB design shown in Fig. 1, then any attempt for subsequent field change in the vicinity of the superconductor will be resisted[8]. If a proper size of the gap between the magnet and the HTS is arranged, the magnet will be maintained in its position statically as well as dynamically in all directions, by the force and stiffness from the HTS and magnets. Furthermore, if the magnetic field has axial symmetry, a rotor consisting of the magnet can rotate freely about the symmetry axis with almost no energy dissipation. Such superconducting bearings are expected to have a friction loss of *~0.1%* of the stored energy per hour[6].

It has been commonly recognized that the insufficient levitating force from the HTS alone in SMB is one of the main obstacles for practical industry-scale designs. The typical lifting pressure observed is only *70~140* kPa as the gap varies from *2~0* mm. In this work, we use a hybrid design concept shown in Fig. 1 to achieve a high levitating force. In this design, the rotor weight is levitated by forces from the magnets, rather than from the HTSs.

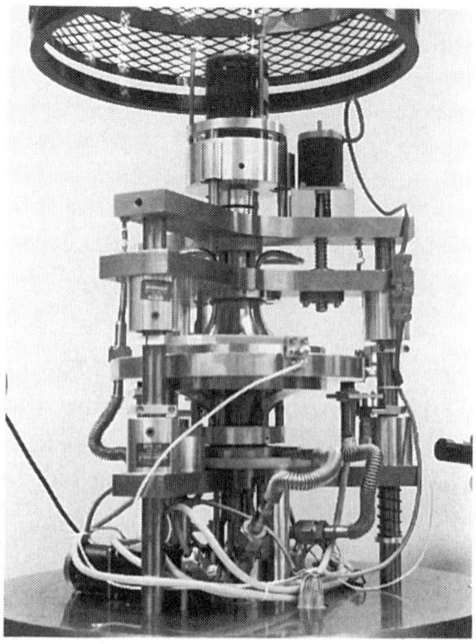

Figure 1. FES prototype using hybrid superconducting magnetic bearings.

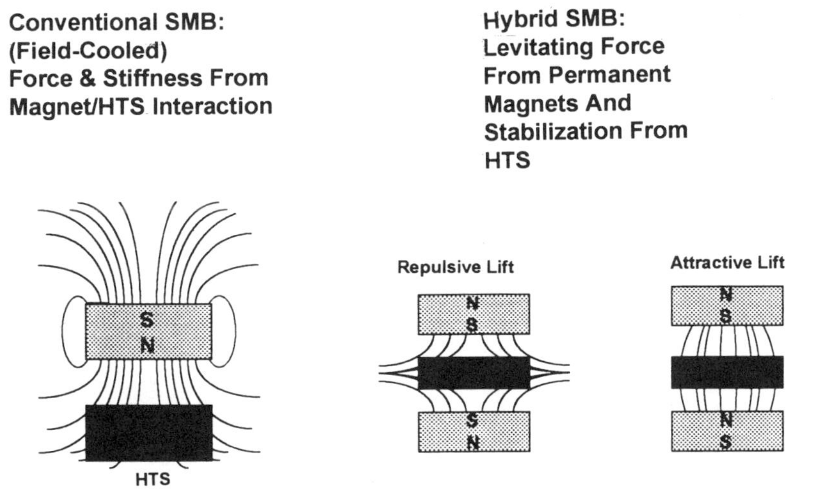

Figure 2. Configurations of conventional and hybrid SMB designs.

The HTSs are placed in between the magnets just for stabilization in both axial and radial directions. Such a configuration offers a much higher levitating force than that from the direct interaction between a magnet and a HTS. Besides, with the "field-cooled" procedure applied, the system is less susceptible to flux creep because the differences between the external and internal fields are small, due to flux pinning.

The forces and stiffnesses from the top and bottom bearings have been measured and listed in the reference[9]. To enhance the overall stiffness, an additional magnet of *10.16* cm (*4* inch) outer diameter is attached to the bottom of the rotor, which will interact with a superconducting ring consisting of *12* HTS pellets aligned in a circle.

To make the inherent instability from magnets to magnets weak enough for HTS to compensate, we shared the total weight of rotor equally between the top and bottom magnets, as shown in Fig. 3, so that the load can be carried in a marginally unstable position, each bearing providing about *94* Newtons of lifting force.

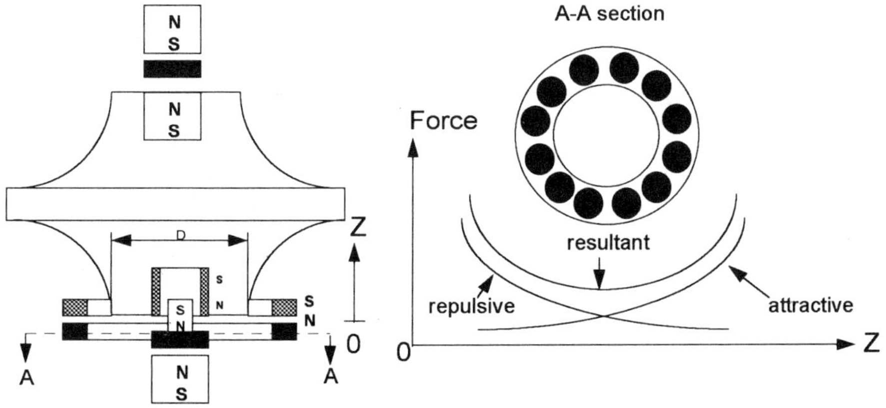

Figure 3. Rotor-bearing configuration.

ROTOR SPIN-DOWN TESTS AND ENERGY LOSS DISCUSSION

1. Coefficient of friction and energy loss per hour

To estimate the friction loss in the bearings, we can calculate the average coefficient of friction and the frictional energy loss per hour by the following formulas[11],

$$\bar{f}_c = \frac{3\pi I}{W_L R \Delta t}(f_1 - f_2) \tag{1}$$

$$\bar{E}_{loss} = \frac{1}{\Delta t}\left[1 - \left(\frac{f_2}{f_1}\right)^2\right] \tag{2}$$

where

I = Moment of Inertia (kg.m^2)
W_L = Load (N)
R = Radius of Flywheel Bottom Surface (m)
Δt = Time interval (hour in (2) and second in (1))
f_1 = Initial Rotating Frequency (Hz)
f_2 = final Rotating Frequency (Hz)

Fig. 4 shows several rotor spin-down experiments as the vacuum was gradually improved. The first test is conducted under a low vacuum of 10^{-2} torr and with poor magnet uniformity. The speed drops from *410* RPM to *340* RPM within about *40* minutes, indicating an average loss > *40%* per hour. The latest two experiments, one of which runs for about *40* hours and another for *15* hours, are conducted in a vacuum of 5×10^{-4} torr and 10^{-5} torr, respectively. Both are started from the same speed of about *6,000* RPM. The later experiment shows an average loss of *4* % per hour, compared to a loss of *5* % per hour during the first *15* hours from the earlier experiment. However the average loss during the whole time span of the earlier experiment comes down to only *2.5%* per hour. Then the average coefficient of friction under a pressure of 10^{-5} torr by using (1) is

$$\bar{f}_c = 0.000195$$

To estimate the necessary vacuum level so that the air drag is no longer the dominant loss, we can calculate the power loss due to air drag using the fluid dynamics theory[10], with the molecular mean free path under a pressure of 10^{-5} torr being $\lambda=71$ cm, which is larger than a pertinent dimension of the flow field, so that the rotation occurs in free molecular flow. With some empirical constants applied, the power loss (watt) due to air drag can be simplified as[11]

$$Pcy = 3.665 \times 10^{-8} p\left(\frac{M}{T}\right)^{0.5} L\omega^2 r^3 \text{ (rotor's cylindrical surfaces)} \tag{3a}$$

$$Pdi = 1.832 \times 10^{-8} p\left(\frac{M}{T}\right)^{0.5} \omega^2 r^4 \text{ (rotor's disk surfaces)} \tag{3b}$$

where
p = pressure (torr);
T = temperature (K);
L = surface length (cm);
ω = rotating speed (rad/sec);
r = radius (cm);
M = molecular weight (Mair=28.97).

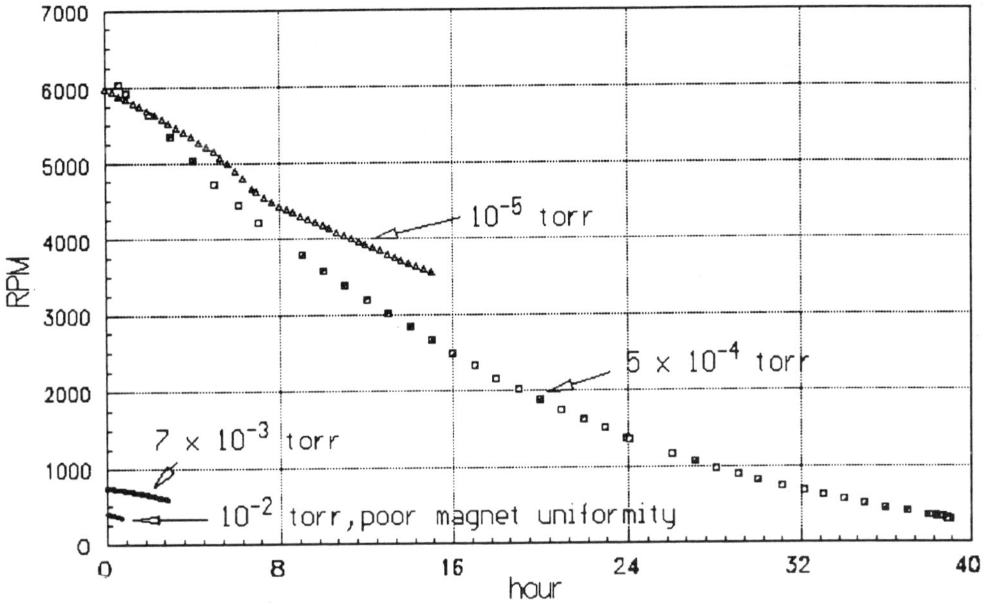

Figure 4. Rotor spin-down tests in different vacuums.

Using (3a) and (3b), we can calculate the power loss from air drag by dividing the rotor geometry into disk and cylindrical surfaces. From Table 1, which compares the air drag loss predicted by theory with the average loss from the measurements, we find that, first, although the loss from the measurements is much higher than that predicted by theory, the difference of the average loss from the two measurements agrees well with theory and secondly, for the current design, the air drag will be a minor factor in the total loss for vacuum levels better than 10^{-5} torr.

2. Friction loss caused by eddy current and hysteresis effects

The total friction-related energy loss in the flywheel rotation may be expressed as:

$$Eloss = Emag + Eair \qquad (4)$$

where E_{air} represents the loss by the residual air drag inside the vacuum chamber, which will slow down the rotation more or less according to the molecular density of the residual gas. Based on the analysis above, we believe that the main contributions to the friction loss

Table 1. Power loss of air drag predicted by theory and average total loss measured from experiments

	10^{-5} torr	5×10^{-4} torr	ΔP
P_{theory}	0.0016 W(6,000 RPM)	0.08 W(6,000 RPM)	0.078 W
P_{measure}	0.285 W(average)	0.356 W(average)	0.071 W

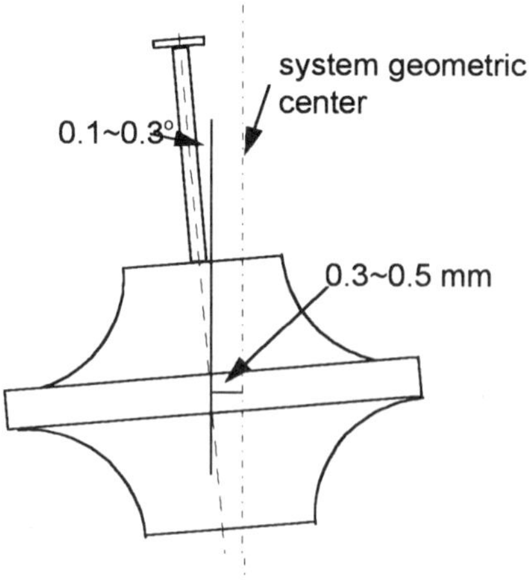

Figure 5. Deviation of levitated rotor from system geometric center.

for this prototype design, where the air pressure is lower than 10^{-5} torr, is from eddy current and hysteresis effects induced by the deviation of magnetic fields from the axial symmetry. Fig. 5 shows a flywheel offset and tilted from the system geometric axis, detected by the position sensors when the flywheel is levitated. This asymmetry may result from: (1), a misalignment of the magnets caused by poor mechanical assembly tolerance; (2), the non-uniformity of the magnets or (3), static inertial imbalance on the rotor which will produce an unbalanced force in rotation. All of these cause the flywheel to be tilted as well as laterally shifted away from the system geometric axis even when the levitated flywheel is at rest. The eddy current and hysteresis loss could happen in superconductors as well as in conducting materials which are located in the vicinity of the magnetic field. Hysteresis effects in superconductors have been studied by Hull[6], by changing the geometry between the magnet and the HTS. Since our current cryostats are made of stainless steel, we can expect considerable energy loss in the top surfaces of the cryostats.

SUMMARY AND FUTURE WORK

We have designed a hybrid superconducting bearing and tested its performance by constructing a flywheel prototype. This hybrid design has high specific load capacity, compared with HTS bearings which use HTS for levitation. The flywheel of *19* kg (*42* lb.) implemented with the HSMB can rotate up to *6,000* RPM. Our rotor spin-down tests show that the frictional energy loss, with the current system parameters, can be <*2%* per hour. Our experiments also indicate that the dominant factor for further energy loss reduction is to reduce the eddy current and magnetic hysteresis losses. Our future work includes: re-design of cryostats; improvement of the system alignment; re-design of clutch mechanism; rotor balance and also a study on the system dynamics.

ACKNOWLEDGMENT

We acknowledge the technical support and HTS material from Ruling Meng & Paul C.W. Chu. This work was supported by the State of Texas through Texas Center for Superconductivity at the University of Houston.

REFERENCES

1. J. A. Kirk. "Flywheel Energy Storage-I, II," *Int. J. mech. Sci.*, Vol.19, pp 223-245, 1977.
2. C. R. Keckler et al. "Integrated Flywheel Technology 1983," *NASA Conference Publication 2290*, Proceedings of a workshop held at NASA Goddard Space Flight Center Greenbelt, Maryland, Aug. 2-3, 1983.
3. B. R. Weinberger et al. "Magnetic bearings using high-temperature superconductors: some practical considerations," *Supercond. Sci. Technol.*, 3(1990): 381-388.
4. R. Takahata, and T. Yotsuya. "Rotation Decay Torques of Superconducting Magnetic Bearing," *IEEE Transactions on Magnetics,* Vol. 27, No.2, March 1991, pp. 2423-2426.
5. K. B. Ma et al. "Spontaneous and persistent rotation of cylindrical magnets levitated by Y-Ba-Cu-O superconductors," *J. Appl. Phys.* 70(7), Oct. 1991, pp. 3961-3963.
6. J.R. Hull et al. "Flywheel energy storage using superconducting magnetic bearings," *Applied Superconductivity,* Vol.2, No. 7/8, July/August 1994, pp. 449-455.
7. E. H. Brandt. "Rigid Levitation and Suspension of High-Temperature Superconductors by Magnets," *Am. J. Phys.* 58, pp. 43-49.
8. K. B. Ma, C. K. McMichael and W.K. Chu. 1992. "Applications of High Temperature Superconductors in Hybrid Magnetic Bearings," *Proceedings of the 1992 TCSUH Workshop on HTS Materials, Bulk Processing, and Bulk Applications,* Feb. 1992.
9. Q. Y. Chen et al. "Hybrid high Tc Superconducting Magnetic Bearings for Flywheel Energy Storage Systems," *Applied Superconductivity*, Vol. 2, No.78, July/Aug. 1994, pp.457-464.
10. S. A. Schaaf and P. L. Chambre. "Flow of Rarefied Gasses," Princeton University Press, 1961.
11. H. P. Liu. Private communication with Dr. Hsing-Pang Liu, Center for Electromechanics, The University of Texas at Austin.

1995 CEC/ICMC Conference: Lance Cooley presenting the Best Poster Awards

ELECTRONIC HIGH VOLTAGE GENERATOR WITH A HIGH TEMPERATURE SUPERCONDUCTING COIL

J. X. Jin,[1] C. Grantham,[2] H. K. Liu,[1] M. Apperley,[3] and S. X. Dou[1]

[1]Center for Superconducting and Electronic Materials
University of Wollongong, Wollongong, NSW 2522, Australia
[2]Dept. of Electric Power
University of NSW, Sydney 2052, Australia
[3]MM Cables, NSW 2170, Australia

ABSTRACT

A novel method for generating high voltages from a low voltage DC source, by using a capacitor and inductor in a R, L, C resonant circuit has been developed with the consideration of using a high temperature superconducting (HTS) coil. To generate high voltages the polarity of a low voltage battery source is reversed each half resonant cycle, the control being achieved by an electronic switch. Resistance in the circuit limits the voltages that can be built up. By replacing a copper winding inductor with another inductor which has a HTS winding, the magnitude of achievable voltages is substantially increased. A $(Bi,Pb)_2Sr_2Ca_2Cu_3O_{10+x}$ multifilament HTS wire is considered in this work to make the superconducting inductor. The high voltages generated are not capable of supplying low impedance loads, however, possible applications of the generator include electrical partial discharge testing and insulation resistance testing. It could also be used as a testing method for the HTS itself with respect to the critical current and AC loss measurement.

INTRODUCTION

Based on the potential use a high temperature superconducting (HTS) coil made by a $(Bi,Pb)_2Sr_2Ca_2Cu_3O_{10+x}$ Ag clad wire, a R, L, C resonant circuit is employed to generate high voltage using a novel method.[1]

This generator mainly consists of a HTS inductor with liquid nitrogen cooling system, a capacitor, a battery DC source, and an electronic switch. For a practical applicable device, as a high voltage and high impedance source, it can be built in a compact size with a fairly simple power electronic circuit.

In addition to the possibility of generating very high voltages with the use of a HTS inductor, the HTS coil or wire itself can be tested in this circuit, for its critical current and

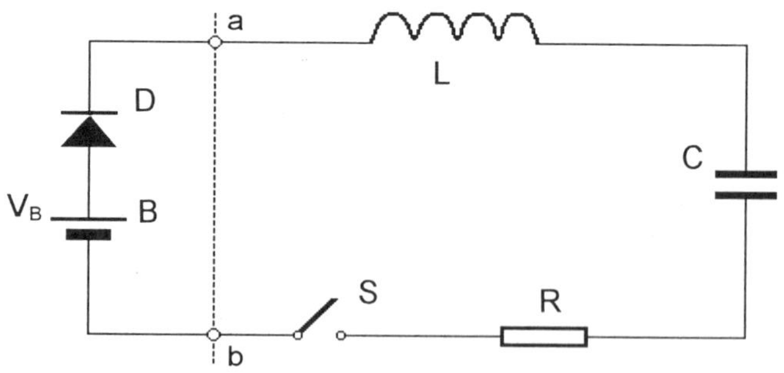

Figure 1. Basic R, L, C resonant circuit for high voltage generation.

AC loss. Therefore the high voltage generator could be used as a method to test the superconductors themselves.

PRINCIPLE OF OPERATION

Figure 1 shows a R, L, C resonant circuit. When a HTS inductor is used, the circuit can be simplified by setting the circuit resistance R to be zero with resonant frequency $\omega = 1/\sqrt{(LC)}$ rad s^{-1}. A diode linked with a DC source in the figure is used to illustrate the circuit current and voltage in the first half resonant cycle.

When the switch S in Figure 1 is closed, the voltage across the capacitor C will be built up to an equilibrium value of $2V_B$ after $t = \pi\sqrt{(LC)}$, as illustrated in Figure 2, where V_C is the voltage across the capacitor C.

If at the point $t = \pi\sqrt{(LC)}$, the battery source a-b in Figure 1 is disconnected, and then reconnected in the opposite polarity for the next half cycle, then the capacitor voltage is increased by a further $2V_B$. If the battery polarity is continued to be reversed each half resonant cycle, the voltage magnitude across the capacitor will increase by $2V_B$ after each reversal. Consequently the voltage across the capacitor C after n cycles will be:

$$V_C(n) = (-1)^{n+1}\,(V_{CO} + 2nV_B) \qquad (1)$$

where n is the iteration number, and V_{CO} is the initial capacitor voltage.

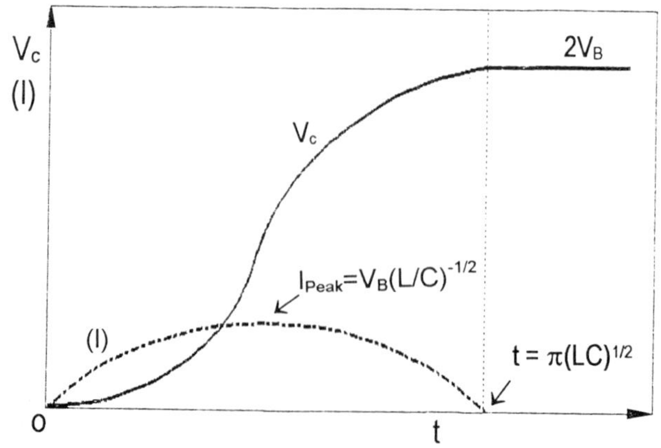

Figure 2. Voltage and current curves for the circuit of Figure 1 after switch S is closed.

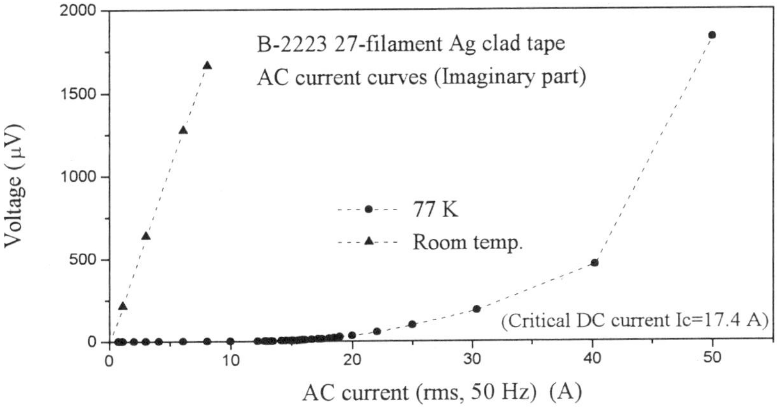

Figure 3. AC I-V curve of a short sample of the superconducting coil.

If the circuit of Figure 1 has some resistance, then some of the energy from the battery is absorbed by the resistance, the corresponding increase in voltage is less than $2V_B$ at each cycle and the magnitude of this voltage increase becomes smaller with each iteration. In this case the maximum voltage which can be generated is related to the Q (= X/R) of the circuit.

In circuits with resistance, no simple theoretical maximum voltage is reached but it can be shown that the lower the resistance, the higher will be the voltage achieved.

HTS INDUCTOR

The superconductor for the HTS inductor is being prepared using a 27 filament $(Bi,Pb)_2Sr_2Ca_2Cu_3O_{10+x}$ Ag clad ribbon. This Bi-based Ag clad HTS wire shows that it can be used to make a compact HTS coil.[2] The best short sample from the produced 27 filament ribbon was tested and shown to have a critical current of 21 A (1 μV criterion) in liquid nitrogen. The cross-sectional dimensions of the tape are 4 mm × 0.2 mm, and the final coil is 306 turns with ϕ_{in} = 30 mm, ϕ_{out} = 55 mm, and L = 65 mm. The final HTS coil critical current is about 5 A (1 μV/cm criterion) at 77 K, and 45 A at 4.2 K.

The inductance of the HTS inductor is about 2 mH with air core and this value can be simply increased by using a magnetic core. To achieve a higher inductance value a couple of such inductors can be used in series.

Since the HTS coil is required to work in an AC circuit, a short sample I-V curve at 77 K is tested with a lock-in amplifier and with a 50 Hz AC power supply. The resistive real part χ' of the short sample from the long HTS ribbon is shown in Figure 3, in which the original DC critical current of this short sample is 17.4 A (1 μV criterion, with 59 mm V-V

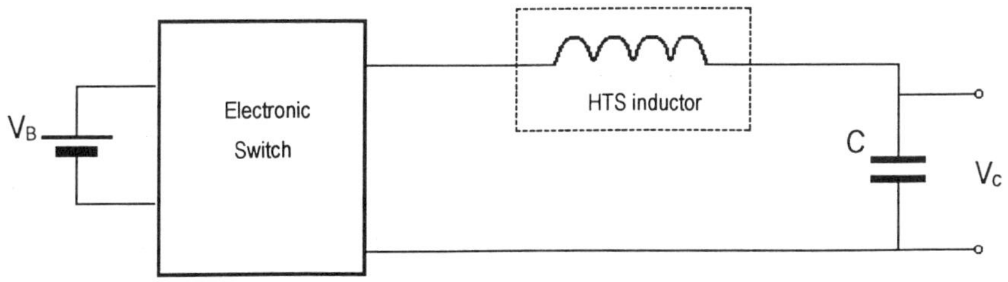

Figure 4. Principle schematic of the superconductor high voltage generator.

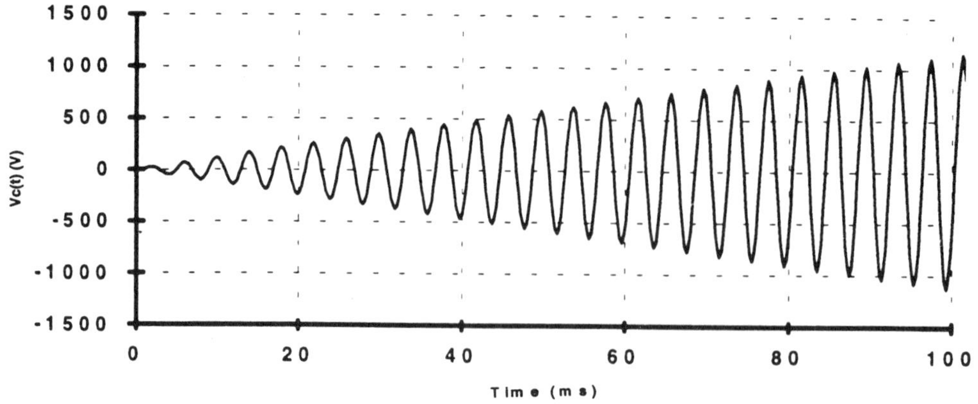

Figure 5. Voltage $V_C(t)$ build-up with the HTS inductor.

interval). The I-V curve of the HTS short sample at room temperature is also plotted. Comparing with the room temperature Ag resistance, the resistance of the HTS sample is dramatically reduced in its superconducting state.

THE GENERATOR

Figure 4 is the principle schematic of the generator. The DC voltage source used is a 12 V lead-acid battery with a low source impedance. The HTS inductor is operated in a liquid nitrogen Dewar vessel. The high voltage is built up across the capacitor C.

In Figure 4, the electronic controlled switch is to reverse the battery polarity, and this can be achieved by using an electronic controlled thyristor bridge or alternatively a simple relay. The signal generator frequency, for supplying the thyristor firing pulses, is set at the natural frequency of the high voltage generator or lower. By firing the thyristor pairs at the natural frequency, the voltage generated is maximized. If the thyristors are fired at a slower frequency, then some of the capacitor charge will leak away between the voltage peak and the firing of the next thyristor pair.

The value of the capacitor is chosen so that the current flowing in the high voltage generator is limited to a value less than the critical current of the HTS inductor at the natural frequency of the circuit, for the range of voltages to be generated.

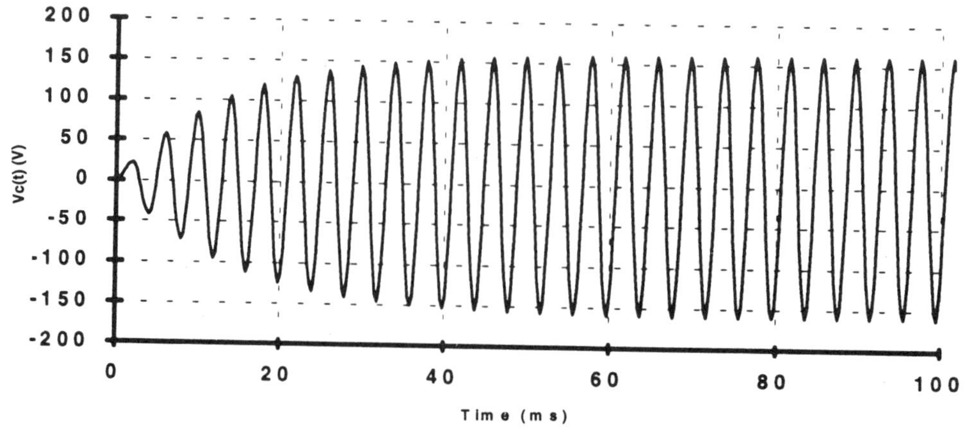

Figure 6. Voltage $V_C(t)$ build-up with normal Cu inductor.

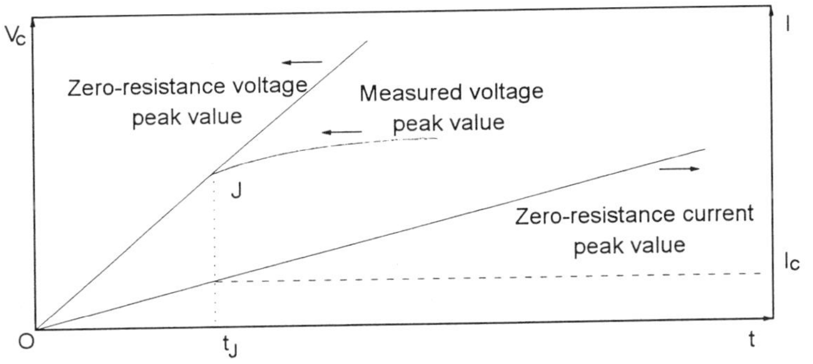

Figure 7. Principle of testing a superconductor with the generator.

To maintain a low current magnitude the inductance should be large. Also since the superconductor critical current reduces with increasing AC frequency, it is advantageous to keep the natural frequency low, by having a large capacitance and large inductance.

LABORATORY SIMULATIONS

During the preparation of the HTS coil for the generator, simulation studies were carried out to predict the voltage build up. To illustrate the operation of the high voltage generator, the values of the circuit components have been chosen to be of a similar size to what will be built into a prototype. The simulation models assume that the bridge thyristors commutate precisely at the natural half cycle, i.e. the DC source polarity is switched at the circuit resonant frequency $[2\pi\sqrt{(LC)}]^{-1}$, and without any voltage drop and capacitor leakage.

To model a circuit containing a HTS inductor, the values chosen were L = 20 mH, C = 20 μF, R = 0.05 Ω, V_B = 12 V and V_{CO} = 0. Using this model the voltage across the capacitor has increased steadily from zero to 700 V in 60 ms , as can be seen in Figure 5, and then to 1149 V in 100 ms (50 iterations). After 6 sec. (3020 iterations) the voltage achieved is 9.66 kV. There is only a 15 V increase of V_C between t = 5 s and t = 6 s.

To model a circuit containing an inductor with copper windings, the values chosen were L = 20 mH, C = 20 μF, R = 3 Ω, V_B = 12 V and V_{CO} = 0. Figure 6 shows the voltage across the capacitor for this model. The voltage achieved after 30 iterations (~60 ms) is 159 V, and after 50 iterations (~100 ms) is 161 V.

By comparing the build-up voltages in the circuit with same L (20 mH) and C (20 μF) values but different R, it can be seen that the build-up voltage with HTS inductor at 100 ms is more than 7 times higher than the copper inductor at the same period. This is achieved by reducing the circuit resistance from 3 Ω to 0.05 Ω. After 2 s, the build-up voltage with the HTS inductor is about 45 times higher than the value with a Cu inductor.

In practice the final voltage across the capacitor would reach a finite maximum value. This would occur as a result of the small voltage increase per cycle being balanced by the same magnitude voltage loss per cycle due to leakage effects.

APPLICATIONS

The high voltage generator can generate voltages of almost any magnitude with the appropriate choice of circuit components. By limiting the number of iterations, or by employing some form of voltage sensing, it is possible to generate a pre-determined set voltage.

The voltage generator can only generate high voltages at low load current. When the resistance load across the capacitor C is low, the maximum build up voltage is considerably reduced. Consequently the device is not capable of supplying a low resistance or low impedance load.

The possible applications in electrical engineering include partial discharge testing and pressure testing of electrical insulation systems, which do not require low impedance loads to be driven.

Since the circuit current in the generator builds up with the build up voltage, the generator can be used as a method to test a superconductor critical current at variable frequency. Figure 7 shows this principle of testing superconductors. When the superconducting coil becomes resistive, the build-up voltage will drop and its curve will shift away from the ideal zero resistance build-up voltage curve. Correspondingly the circuit current at that shifting point, as the point J in Figure 7, is related to the HTS coil critical current I_c.

CONCLUSIONS

A simple circuit for generating high voltages has been described and investigated. By switching the battery polarity in a R, L, C, resonant circuit, it is possible to generate high voltages of many orders of magnitude greater than the DC battery source. This voltage can be generated at a very high value by using a superconducting inductor.

With the prospect of manufacturing large value HTS inductors, this method of voltage generation has the potential to be utilized in industry as a compact device to be used for partial discharge testing and pressure testing of electrical insulation systems. Built with the HTS, the device could be more flexible, and also at much lower cost due to operation with liquid nitrogen rather than liquid helium as with conventional low temperature superconductors.

To achieve a certain value of high generated voltage, a large inductance is selected, resulting in a lower circuit current. Consequently the HTS coil required is not necessary to sustain a very high magnetic field. The presently produced $(Bi,Pb)_2Sr_2Ca_2Cu_3O_{10+x}$ Ag clad wire is a good candidate for this application.

In addition the voltage generator described can be used for testing the critical current and AC loss of HTS coils and wires, by comparing the voltage measured and the theoretical zero-resistance build-up voltage. For the AC loss investigation this testing can be carried out with variable circuit frequencies.

ACKNOWLEDGMENTS

The authors would like to thank R. Bhasale, B. Blears, A.C.G. Secrett, X.P. Gu, K.J.S. Holder, J.N. Li, Y.C. Guo, and A. Bult for providing help in this work.

REFERENCES

1. C. Grantham, F.N. Kong, and H. Jiansheng, Novel electronic high voltage generator, *J. Electr. & Electronics Eng., Australia,* 1991, **2** 117.
2. H.K. Liu, J. Yau, Y.C. Guo, Q.Y. Hu, J.X. Jin, N. Savvides, and S.X. Dou, Long Ag-clad Bi-based superconducting tapes and coils, *Physica B*, 1994, **194-196** 2213.

THERMALLY STRESSED ELEMENTS OF CRYOALTERNATORS: DEVELOPMENT AND INVESTIGATIONS

I.S.Ganzhinov and V.A.Tutaev

VNIIElectromash
St.-Petersburg, Moscovsky av., 100, 196084, Russia

ABSTRACT

Optimization of structures to minimize heat flow to the cold zone and, as a result, improve the on cryostability of field windings is of primary importance in development of promising cryogenic electric machines of middle and high power. Thermal bridges and current leads are the most thermally stressed elements of a superconducting generator connecting the cold zone with the normal one. The present paper outlines the results of development of a calculation procedure for intricately shaped elements made of materials with different thermal and physical properties and operating at significant differences of temperatures. Thermal bridges and current leads of cryogenic machines may be considered as typical representatives of these elements. The results of experimental investigations of a model of the current lead having an insert made of lowered conductivity material for the superconducting generator rotor are given. The results of the work will be used when designing a 220 MW superconducting turbogenerator and other cryogenic electric machines of high power.

INTRODUCTION

During more than 15 years VNIIElectromash has been developing new structures of thermally stressed elements for cryogenic electric machines and modern calculation procedure for these designs which allow the variability of a great number of parameters to be taken into account such as thermal and physical properties of materials, composite selection, and the cooling agent, as well as a great variety of heat exchangers. The first results have been realized in the 20 MVA[1-2] superconducting generator.

Thermal bridges provide minimum heat flows to the rotor cold zone. Current leads are intended to supply the field winding from retaining rings and for extraction of power from the field winding in case the field winding changes from the superconducting state to the normal one.

The current lead is required to provide minimum thermal flows into the cold zone at the rated current. Efficiency of heat flows is characterized by the ratio of the value of heat flow into the cold zone to the current in the current lead. Development of the current lead with lower heat flow into the cold zone offers problems because superconducting generators may require high excitation current. For example, in generators with fast-response excitation under rated conditions the current may reach 5-10 thousand amperes[2]. Composite current leads with inserts made of lowered heat conductivity materials give the possibility to decrease heat flows into the rotor cold zone, while simultaneously properly warming up the cooling gas at the rotor outlet.

1. SETTING UP A PROBLEM

The following assumptions are made when solving the calculation problem of thermally stressed elements of constructions for cryogenic electric machines:

- variations of temperatures across the thermal bridge body and friction losses in the helium flow along cooling channels are not taken into account;
- self-regulating thermal bridge is considered where cooling gas rate corresponds to heat flow into liquid helium at the cold side;
- the effect of the rotor rotation on the heat exchange in cooling channels is neglected.

The temperature distribution in structural elements and cooling helium gas has been deduced by solving the following equation:

$$\frac{d}{dx}\left[\lambda(T)S(x)\frac{dT}{dx}\right]-q(x)\prod_1(x)-\alpha(T-\Theta)\prod(x)+j^2(x)S(x)=0,$$

$$\frac{d}{dx}\left[\lambda_g(T)S_g(x)\frac{d\theta}{dx}\right]+\alpha(T-\Theta)\prod(x)-Gc_p\frac{dQ}{dx}=0, \tag{1}$$

where

T, Θ	- temperatures of an element and helium gas,
$S(x), S_g(x)$	- cross-section areas of an element and channels with helium gas,
$q(x)$	- specific external thermal flux,
α	- coefficient of heat removal from an element to helium,
$j(x)$	- current density,
$\lambda(T), \lambda_g(T)$	- coefficients of heat conductivity of an element and helium,
$\prod(x)$	- periphery of cooling,
$\Pi_1(x)$	- periphery of the place of thermal shield mounting,
G	- mass helium rate,
c_p	- coefficient of heat capacity.

The boundary conditions for solving (1) are:

1. Temperature of the cold end of the constructive element (x=0) is equal to liquid helium temperature

$$T_{/x=0} = 4.2\ K. \tag{2}$$

2. Temperature of the rotor elements at the helium gas outlet is equal to the ambient temperature

$$T|_{x=L_1+L_2+L_3} = 300\ K. \tag{3}$$

3. At the border of the air-cooled part ($x = L1 + L2$), the conditions of equality of temperatures and thermal flows are met, for helium the condition is set by

$$\left.\frac{d\Theta}{dx}\right|_{x=L_1+L_2} = 0. \tag{4}$$

4. Helium temperature at the inlet of helical channels

$$\Theta = 4{,}2 + \frac{Q_{r.h.f}}{G \cdot c_p}, \tag{5}$$

where $Q_{r.h.f}$ is the radiant heat flow through the evacuated hollow of the heat exchanger.

Equations (1) using boundary conditions (2) -(5) have been solved numerically with the help of the finite-difference discreditation method. The obtained system of algebraic equations has been solved with the scalar trial method[3]. The equation system has been solved by iteration. Thermal and physical properties of helium and materials of the calculated assembly has been approximated using cubic spline interpolation of tabular data[4-5].

2. CURRENT LEADS OF SUPERCONDUCTING GENERATORS

A current lead is a complex assembly for a superconducting generator in regard to electrical, thermal and physical processes, as well as to design. As temperatures of current-carrying parts of the current lead and the coolant vary from the cold end to the warm one. During design, one is forced to take into account the change of specific resistance and thermal and physical parameters of structural materials and the cooling media due to the temperature distribution along the current lead. This in turn, depends on the heat exchange between the surfaces and the cooling agent, as well as on the coolant rate and nature of its flow.

A schematic diagram of the current lead for the superconducting turbogenerator rotor is given in Fig.1.

Previous calculations[6] of different design versions of low (LTSC) and high-temperature (HTSC) superconductor current leads have been made (with a low thermally conductive insert and without it).

Next geometrical parameters were determined by solving the problem:

- relation of length of air-cooled and non-air-cooled parts;
- relation of length of the air-cooled low thermally conductive insert and the air-cooled part of the copper conductor:
- coordinates of the insert location within the limits of the air-cooled part.

Figure 2 represents calculated relationships of the variation of the current lead temperature T along the length for different design versions according to data in Table 1.

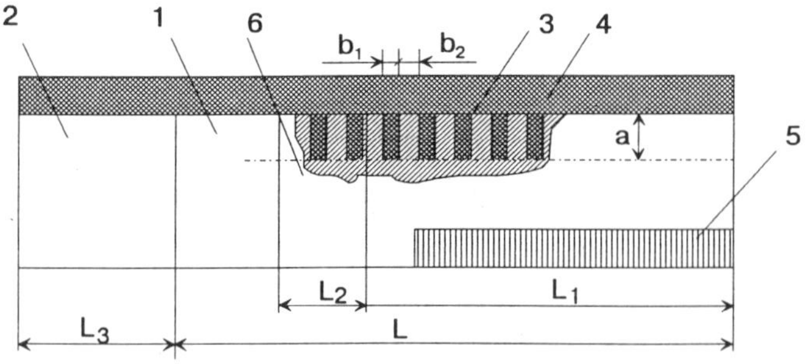

Figure 1. Calculated diagram of the current lead:
1 - air-cooled part, 2 - non-air-cooled part, 3- spiral channel, 4 - insulating tube,
5 - superconductor (LTSC or HTSC), 6 - insert made of lowered heat conductivity material.

Table 1. Parameters of current lead design versions

Design	Helium temperature at outlet, K	Current lead efficiency, W/kA
Current lead :		
copper	161	2.45
with a brass insert and a LTSC-shunt	227	1.86
with a nickel insert	290	0.68
with a nickel insert and a LTSC-shunt	290	0.53
with a nickel insert and a HTSC-shunt	285	0.49

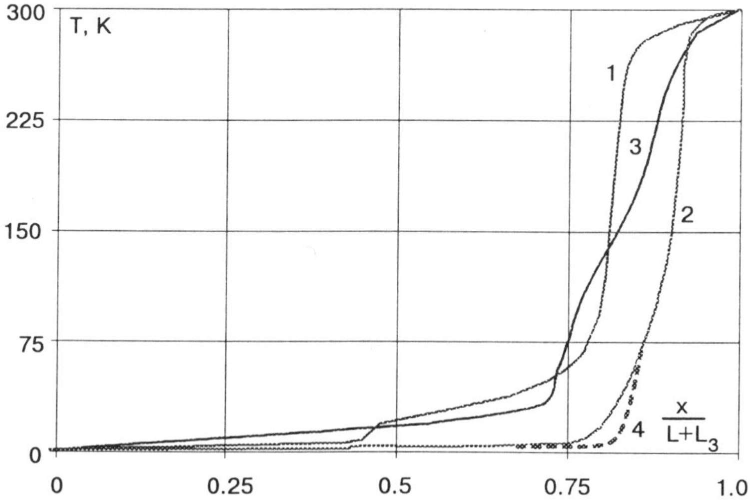

Figure 2. Calculated temperature distributions along the length of different design versions of current leads:
1 - with a nickel insert, 2 - with a nickel insert and LTSC -shunt, 3 - with a brass insert, 4 - with a nickel insert and HTSC-shunt.

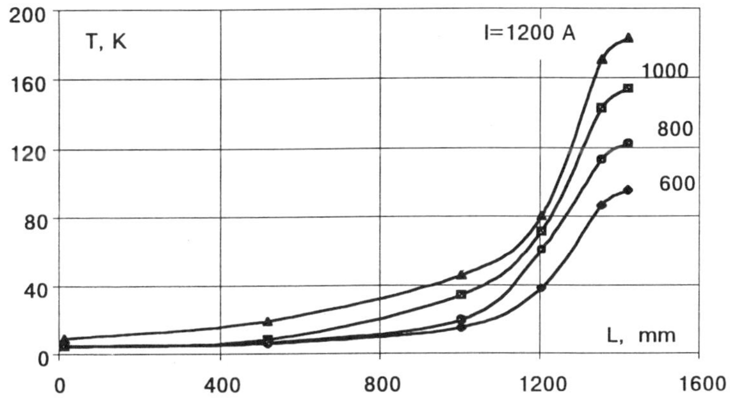

Figure 3. Results of experimental investigations of the LTSC-current lead with a nickel.

3. THERMAL BRIDGES

At present the most generally employed construction of the thermal bridge assembly is that where heat flux from the normal temperature zone is removed with helium gas flow moving along the spiral channels of the evacuated heat exchanger[7]. Schematic diagram of the thermal bridge assembly is shown in Fig.4.

The thermal bridge consists of the air-cooled (1) and non-air-cooled (2 and 3) parts. Inside of the air-cooled part of the bridge there is a heat exchanger with spiral channels (4) through which helium gas evaporated in the rotor bath containing the superconducting field winding flows countercurrent to the heat flow from the warm end of the thermal bridge. The rotor thermal shield is rested on the thermal bridge that results in producing an additional heat flow to the bridge which is removed with helium gas.

A further additional heat flow not taken into account in the previous investi - gations and, importantly, is a radiate heat flow through the evacuated hollow of the thermal bridge heat exchanger.

Figure 5 shows the calculated dependence of dimensionless helium flow rate in the thermal bridge on the location of the thermal shield mounting. Calculations have

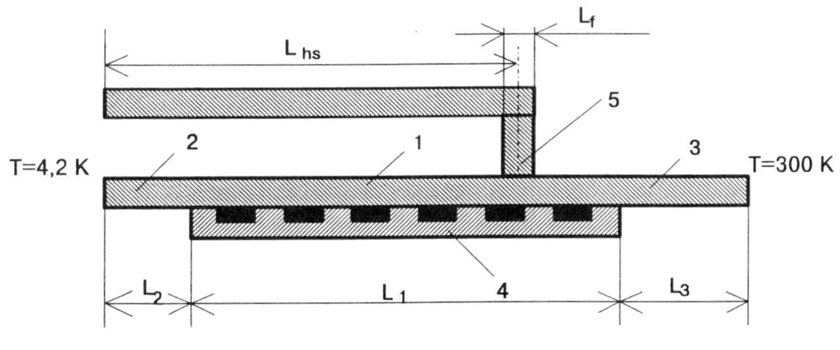

Figure 4. Calculated diagram of the thermal bridge assembly.
L_{ts} - the place of thermal shield mounting, L_f - length of the place of the thermal shield fitting, L_1 - length of the air-cooled part of the thermal bridge, L_2 and L_3 lengths of non-air-cooled parts of the thermal bridge.

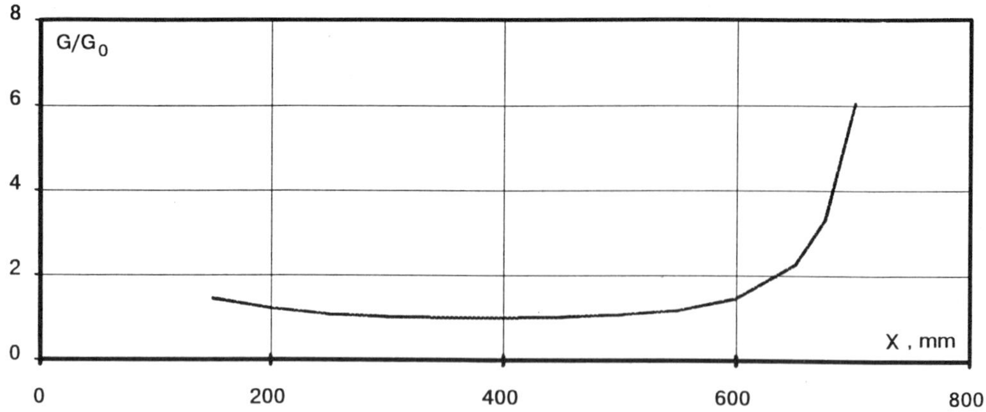

Figure 5. Dependence of dimensionless helium flow rate through the thermal bridge on the place coordinate of the thermal shield mounting X: G_o - minimum value of helium rate through the thermal shield.

been carried out for the thermal bridge with the length of the air-cooled part $L = 510$ mm. For the heat exchanger of the thermal bridge, the spiral channel was chosen to have 4 rotations, with cross-section 1×8 mm. As is evident from the results, the helium flow rate is little dependent on the location of the thermal shield mounting of the thermal bridge. It starts to grow noticeably only when approaching the warm zone. Minimum flow rate corresponds to the thermal shield mounting at the distance from the cold end equal to 0.7 of the length of the air-cooled part of the thermal bridge.

Calculations of dependence of helium gas flow rate through the rotor thermal bridge on the length of the air-cooled part at different fixed places of thermal shield mounting have been carried out. For examining the construction of the thermal bridge there has been used an automatic selection of the thermal bridge parameters based on minimization of heat flow into the cold part of the rotor as geometrical parameters are varied.

Results of these calculations are given in Figure 6. As it is seen from the dependencies, increase in the length of the air-cooled part of the thermal bridge leads to a slight decrease of helium flow rate, but allows helium temperature at outlet of the

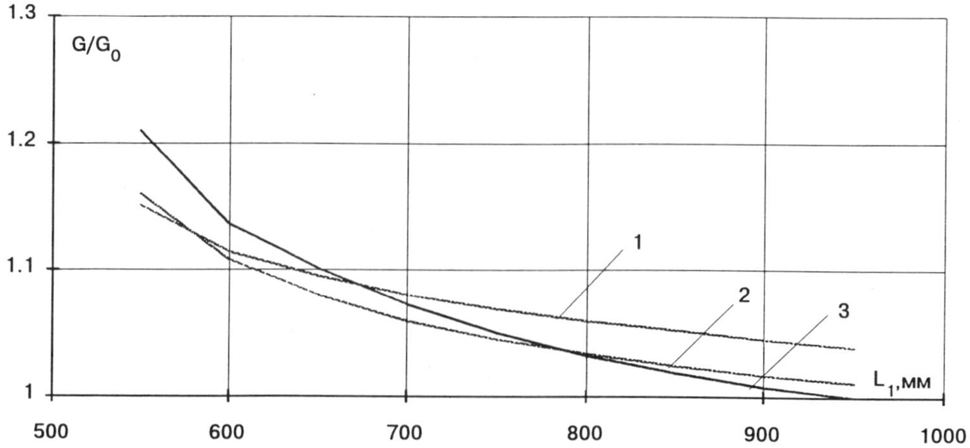

Figure 6. Dependencies of dimensionless helium rate through the thermal bridge on the length of the air-cooled part at the different places of the thermal shield mounting.
Coordinate of the thermal shield mounting: 1 - X_{ts}= 0,7 L_1; 2 - 0,6 L_1; 3 - 0,5 L_1.

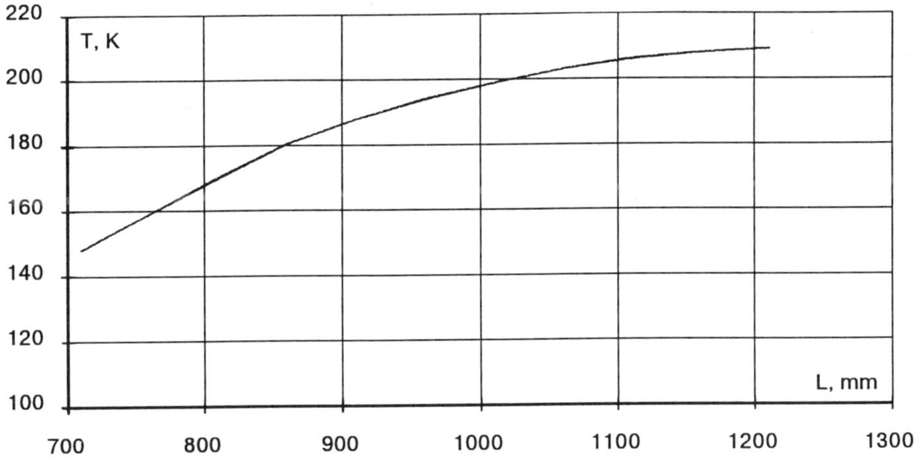

Figure 7. Dependence of helium temperature at the outlet from the heat exchanger on the total length of the thermal bridge: $L=L_1+L_2+L_3$; $L_{hs}=0.5\ L_1$.

thermal bridge to increase (Figure 7) and, in its turn, it results in increasing the temperature of the rotor flanges and avoiding their possible freezing.

Figure 8 shows the curves of temperature distribution along the thermal bridge and in helium gas for the chosen design version. Calculations have been carried out for the thermal bridge having the length of the air-cooled part L_1 = 510 mm and L_{ts} = 0.7 L_1. The total length of the thermal bridge with non-air-cooled parts is 760 mm.

High coefficient of heat exchange in the thermal bridge channels is associated with a slight difference of temperatures between the bridge and cooling gas (near 1 K). Helium gas temperature at outlet from the heat exchanger channels is near 190 K.

The performed analysis has demonstrated that the design procedure may be used for construction of the automatic design system for the thermal bridge assemblies of superconducting devices.

Designing thermal bridges made of materials having low heat conductivity presents the possibility to vary geometrical sizes of spiral channels of the heat exchanger and coordinates of the location of the thermal shield fitting within rather wide limits without noticeable variation of equilibrium helium rate.

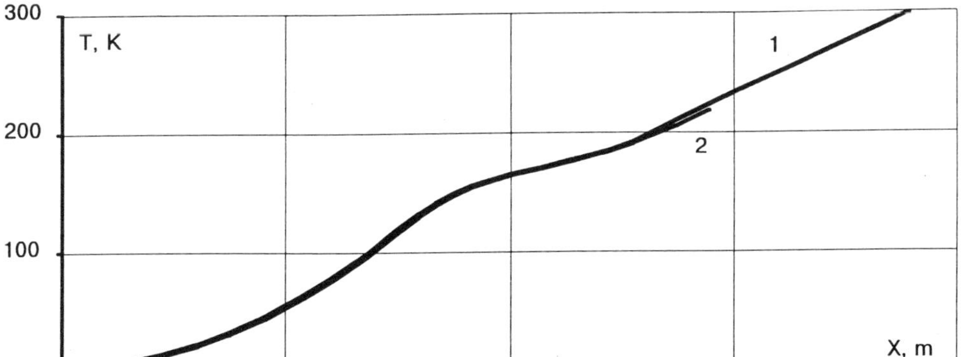

Figure 8. Distribution of temperatures in the thermal bridge assembly of the 220 MW rotor and in helium gas: L_1 = 0,5 m; 1 - body of the thermal bridge; 2 - helium gas.

It is worth noting that radiant heat flow into the cold zone through the evacuated hollow of the heat exchanger has an important role and due to this fact, mounting a thermal shield is effective.

CONCLUSIONS

1. Calculation methods of thermally stressed elements give a possibility to analyze design versions of complex assemblies of cryogenic electric machines at many variable parameters and to optimize the design. These methods may be used in the systems of automatic design.

2. Construction of the composite current lead with a nickel insert and shunting the current-carrying parts with a superconductor enables high values of efficiency. Application of HTSC-current lead with an insert is considered to be the most effective way to solve the problem.

ACKNOWLEDGMENT

The research is supported by the Ministry of Science and Technical Policy of Russia and of the Russian Fund for Fundamental Researches.

REFERENCES

1. I.A.Glebov, Ja.B.Danilevitch, and V.N.Shakhtarin." Turbogenerartors with application of superconductivity", Nauka, Leningrad (1981).
2. Ja.B.Danilevitch and L.I.Chubraeva."New designs of generators and problems of their development", Nauka, S.-Petersburg (1993).
3. A.A.Samarsky."The theory off difference schemes", Nauka, Moscow (1977).
4. I.A.Kozhevnikov and A.A.Novitsky. "Thermal and physical properties of materials at low temperatures", Maschinostroenie, Moscow (1983).
5. R.D.Mc Carty. "Thermodynamic properties of helium 4 from 2 to 1500 K at pressure to 10^8 Pa", Phys. Chem. Ref. Data.. Vol. 2, N 4. 923-1024 (1973).
6. I.S.Ganzhinov, Ja.B.Danilevitch and S.A.Ivanov, Automatic selection of design parameters of the superconducting turbogenerator, in:"Automatization of investigations and calculation of generators",VNIIElectromash,S.-Petersburg (1986), p.89-94.
7. Eu.A.Brynsky, Ja.B.Danilevitch et al., Investigation of thermal bridge assembly of superconducting turbogenerator, Electrotechnika.N1,51:56 (1980).

SUPERCONDUCTING TURBOGENERATORS AS A NEW GENERATION OF HIGH-RATED ELECTRICAL MACHINES

I.A.Glebov and L.I.Chubraeva

VNIIElectromash
St.-Petersburg, Moscovsky av., 100, 196084, Russia

ABSTRACT

Development of turbogenerators is associated with improvement of their cooling system, efficiency, size and weight parameters and operating characteristics. Application of cryogenic cooling and superconducting windings helps to achieve the best results, but there arises the problem of economical efficiency. Superconducting turbogenerator represents, on one hand, just a new type of turbogenerators, incorporating a number of serial units and technologies, and on the other it demands an introduction of new technological processes aimed at achievement of better parameters and higher reliability. Superconducting turbogenerators may be subdivided into two groups: alternators with slow-response excitation for the combined-cycle power plants with the unit rating of 200-250 MW; and, generators with unit ratings starting from 500-600 MW and with high response excitation intended mainly for the nuclear power plants. There exists a possibility to make the first ones relatively cheap while the second will be substantially more expensive. VNIIElectromash is now developing a design of a slow-response generator with the unit rating of 220 MW. Availability of a developed test facility for the generator investigation will help to decrease the investment in a prototype machine.

INTRODUCTION

Turbogenerators are one of the main sources of electrical energy both in Russia and all over the world. They represent the most complicated specimen among the products of the electrical engineering industry. Therefore much attention has been always paid to turbogenerator development and investigations to improve the technological processes involved.

Limitations imposed on the rotor diameter by the mechanical strength of forging and on its length by the permissible vibration level have resulted in the development of more efficient cooling systems together with increased generator unit ratings. It was always important not only to sustain the alternator efficiency, but to improve it. Step-by- step air, hydrogen and water have been introduced as coolants.

Advances in Cryogenic Engineering, Vol. 41
Edited by P. Kittel, Plenum Press, New York, 1996

The guest for enhanced efficiency and reduced specific weight (kg/kVA) of generators resulted in attempts to apply the most efficient coolants, i.e. cryogenic liquids. The design and technology of superconducting turbogenerator manufacturing are to be based on the experience gained when developing previous generations of turbine-type synchronous machines.

During the last decades, combined-cycle technology has been coming into use at thermal power plans, since combined steam-gas plants enable increasing efficiency from 40% up to 53-55%. Gas turbine output power is relatively low (up to 150 MW at present and 200-250 MW feasible in the near future), and it determines the upper level of the turbogenerator rating.

Because gasification of coal presents a considerable problem, future steam turbines are likely to be kept at the power plants utilizing solid fuel and at nuclear power plants. At present, in our country nuclear reactors of high reliability intended for 600-800 MW turbogenerators are being developed. These reactors are expected to be put into commercial use by the year 2000.

At present in Russia there are being manufactured mainly air-cooled and fully water-cooled turbogenerators with the unit rating up to 220 MW and either hydrogen-water (500 and 1000 MW) or fully water-cooled units of 800 MW for the nuclear power plants.

An effective cooling of the stator and rotor windings, stator core and some structural elements finds growing support for the complete water cooling of turbogenerators. Of major interest is a new concept based on the Russian self-pumping design for a water cooling system, which offers the possibility to abandon a unit with poor reliability, i.e. water supply via the rotor shaft. Of course, implementation of the new design demanded a great amount of research work and constant improvement of the design elements. Though the fully water-cooled machine is relatively expensive its advantages are significant for utilities.

Development and application of every new generation of turbogenerators in Russia covers at least two decades: it took 20 years to develop the hydrogen-water cooling system and 25 years to introduce the full water-cooling system. Therefore it is possible to discuss application of superconducting turbogenerators only about the year 2005 and later. It may start with the units intended for combined-cycle power plants having the rating of 200-250 MW. Later generators for the nuclear power plants may be manufactured.

COMBINATION OF CONVENTIONAL AND NEW DESIGNS AND TECHNOLOGIES

The major advantages of superconducting turbogenerators were at first associated mainly with improved size and weight parameters and higher efficiency. Later other advantages were introduced, such as: smooth variation of efficiency with the load increase; higher static stability; possibility to improve transient stability; good performance in the under-excited mode; higher reliability; etc.[1-3] But for application in power plants economic aspects are substantial as well. There exists the opinion that the new alternators are economically feasible only at unit ratings of 500 MW and higher.

It is worth mentioning that superconducting turbogenerators represent just a new step in the development of high-rated electrical machines. Therefore they incorporate many design elements and units of conventional generators. These are: frame; end shields; bearings and oil circulation system; brush-contact unit and sliprings; HV leads; monitoring and diagnostics facilities employing computer techniques; automatic voltage regulators; control and protection systems; etc. Ferromagnetic stator shield is manufactured of the cold-rolled electrotechnical steel, laminated across the direction of rolling, end stacks are stepped. The methods of fastening the shield in the frame to the similar to the conventional technology as well: fastening can be either rigid or elastic. Development of superconducting machines is based on the application of numerous technological processes that find wide use in manufacturing of conventional turbogenerators.

The parameters of superconducting turbogenerator operation in a power system should be compared to those of conventional machines in order to reach equal or better results. From the requirements of transient stability, the ultimate rating of superconducting turbogenerator should be not less that those of a conventional-type generator equipped with a modern automatic voltage regulator and a fast-response thyristor-based excitation system. It refers as well to static stability limits and under-excited operation. Application of superconducting generators in the network implies all the experience obtained for several decades with the operation of conventional turbogenerators.

It is usual now to subdivide superconducting turbogenerators into two groups: slow- and high-response excitation[2]. Until now the majority of developed and tested superconducting turbogenerators are the machines with slow-response excitation. This is due to the properties of superconducting materials applied for the field winding, and to the design of electromagnetic shields which provide effective protection of the excitation winding against varying magnetic fields produced by the stator. The possibility of relatively slow field current variation is retained.

Generators of the high-response type are characterized by the increased value of excitation current and reduced inductance of the superconducting winding, rotor damper system resistance is increased to provide fast regulation of the excitation current. Therefore superconductor which can withstand the varying stator fields is used. As a result the rotor will be more expensive than the previous one.

Stator winding geometry may be either conventional or non-conventional. To decrease the manufacturing costs of the superconducting turbogenerator it is necessary to simplify the armature substantially. The slotless armature design permits new solutions. The experience of VNIIElectromash shows that a non-conventional helical winding is preferable because it opens the way for more simple technological processes and decreases the hand labour involved. The winding support structure incorporates a number of fiber-glass cylinders. The technological process for their manufacturing was developed in Russia for the turbogenerators with oil cooled armatures. Moreover, the arrangement of the armature cooling system in these alternators is the simplest one and allows a decrease in the armature size and weight, improving the design substantially. The cooling system of the armature may be equipped with a heat pump. In terms of equivalent fuel savings, heat pump application is equal to the generator efficiency increase of 0.2 - 0.4 %.

Manufacturing of AC superconductors permits to develop fully superconductive generators[3]. Though the losses in the superconducting armature winding will be relatively high even if compared with a cooper winding at ambient temperature, our studies suggest that the advantages of the alternator with superconducting field and armature windings will be achieved due to improved design of the cooling system. Preference should be given to the design with a single cryostat, where the low-temperature zones of rotor and stator are separated at an intermediate temperature level. The task to develop a rotating sealing for the cold gap between rotor and stator seems to be one of the main problems of creating such a machine.

As mentioned above, the major point to be pursued in the course of developing superconducting turbogenerator is its operation in power system[4].

The opinion about the high stability of superconducting machines of the slow-response type is valid only in case of relatively short power transmission lines (about 300 km for the 220 MW unit). In this case, adequate transient and static stability is provided by a reduced value of synchronous inductance, and hence by low rotor angle relatively to the receiving power system in the initial mode, as well as by the constant value of transient EMF. The ability for stable operation with no additional appliances at short distances between a power plant and a consumer is a prominent advantage of a superconducting turbogenerator.

In other cases, to achieve high transient stability of a generator with slow-response excitation it is necessary to apply a SMES coupled to a semiconductive converter. The main advantage of applying a SMES with a converter for the sake of enhanced stability is that during transients it directly affects the generator output and voltage, whereas in alternators with high-response excitation this influence is accomplished through an inertia unit, i.e. superconducting field winding.

To ensure the required level of transient stability in high-response turbogenerators, one should follow a different approach for the superconducting field winding parameters[4]. Their values should be determined on the basis of equivalent forcing duties of a conventional and superconductive generator, as well as on the voltage level defined by an emergency energy discharge during quench.

DEVELOPMENT OF THE 220 MW SUPERCONDUCTING TURBO-GENERATOR

Development of superconducting turbogenerators is to be based on thorough experimental investigations. VNIIElectromash possesses a unique cryogenic test facility, which provides the possibility to study 20 MWA turbogenerator operation in a power system, to carry out investigation on model installations and comprises all the major equipment necessary for the testing of superconducting turbogenerator rated at 200 - 300 MW. A block diagram of the test facility is shown in Fig.1. The main equipment comprises a 20 MVA superconducting generator (1) with a drive motor (2), excitation system (7) and control panel (10). The automatic system for information pickup and recording is located indoors. To carry out the low-temperature experiments the test facility is equipped with two cryogenic plants, capable of producing up to 100 l/h He each (3) with compressor equipment (4), gas holders (5), receivers (6) and liquid nitrogen vessels (22). Gaseous helium is being stored in room (21). The facility incorporates a helium cryostat (8) of 4600 l capacity

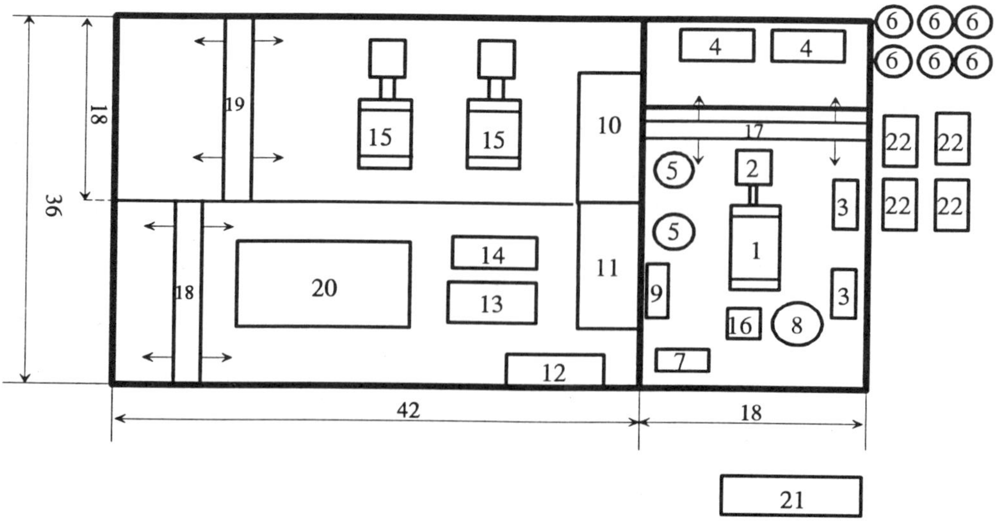

Figure 1. Schematic view of the cryogenic test facilities.

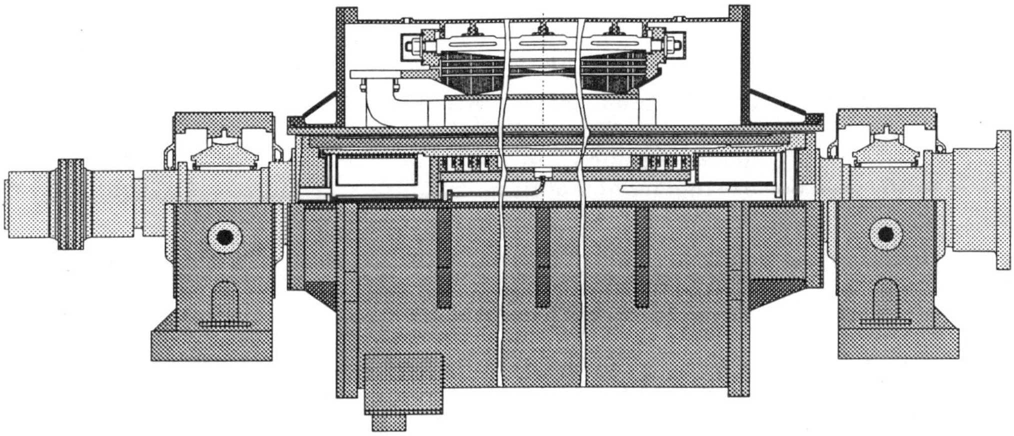

Figure 2. Cross-section of the 220 MW superconducting turbogenerator, n=3000 rpm (L=6.1 m, D=2.4 m).

and hydrodynamic model (16) with a 850 mm outer diameter helium pool. Room (11) houses the stator cooling system with a heat pump. The test-bed is equipped with HV switch gear (9) and reactors for the stator circuit (12). The synchronous generator with a helical winding (13), driving motor and excitation system (14) can be used as a drive motor for the superconducting turbogenerator. The test field (20) foundation can handle the high-rated alternator. An electrical machine-based frequency converter (15) can serve as either a DC power source (up to 7 kA) or an AC power source (up to 0.5 kA) and provides frequency regulation in the range from 0 to 50 Hz. Cranes (17,18 and 19) with carrying capacity up to 32 t are available for equipment transportation.

In previous years the Institute developed the design concept of a 1200 MW superconducting turbogenerator. At present, a 220 MW turbogenerator is being designed. Fig.2 presents the longitudinal cross-section of the latter. The generator data is: voltage (line-to-line) = 15.75 kV; stator current = 9.49 kA; cos φ = 0.85; efficiency = 99.2 %; and, a specific mass less than 0.6 kg/VA. Specific mass is decreased not only by application of a superconducting field winding, but is also due to the reduced mass and dimensions of the stator incorporating helical winding and applying direct cooling of all the stator elements by an insulating liquid.

The experience with the 20 MVA alternator and a number of models is adequate to start manufacturing the 220 MW unit without any intermediate steps. It will reduce investment in the development of the prototype superconducting turbogenerator substantially.

The merits of superconducting machines will be even more obvious with transition from LTSC to HTSC, extending the range of operating temperatures up to 77 K and higher. This will involve simplified rotor design due to simpler thermal insulation, higher critical temperature and the benefits of applying nitrogen as a cooling and insulating medium. It will be possible to apply a coolant inlet system similar to the one used in the fully water-cooled turbogenerators mentioned above. It means the cryoalternator design will be simplified but not totally changed. Development of HTSC turbogenerators will be based on the experience obtained with the LTSC machines. But in case that the HTSC will have appropriate critical parameters only at 20-30 K the alternator design will be practically identical to the one used in a helium system.

CONCLUSIONS

1. Development of superconducting turbogenerators should be based on the best achievements of conventional generators, including modern design elements, modern technology and vast experience of operation in the network.

2. New technologies applied in superconducting turbogenerators will substantially improve existing manufacturing processes and minimize the hand labour involved.

3. There exists a possibility of introducing slow-response type superconducting turbogenerators rated at 220 MW for the combined-cycle power plants alongside with air-cooled and fully water-cooled ones. The investments in the development of the prototype machine may be relatively low because Russia possesses good practical experience and the necessary experimental facilities with contemporary equipment.

4. Application of units 500 MW and higher will be on nuclear power plants. The experience will these type of alternators is inadequate and their development will demand serious financial investment.

5. Development of a fully superconductive turbogenerator may be of interest due not only to lower losses in the armature winding, but simpler design of the alternator in a single cryostat.

6. Application of the HTSC will simplify substantially the alternator design only in case where operating temperatures will be 77 K and higher.

ACKNOWLEDGMENT

Research is being supported by the Ministry of Science and Technical Policy of Russia.

REFERENCES

1. I.A. Glebov, L.I. Chubraeva, J.C. Edmonds et al., Superconducting generators: Current situation and prospects: CIGRE. Sept. 1982. Paper 11-14. 10 p.
2. Research of technical aspects of superconductive generation equipment and materials. Japan: Technova Inc., 1987. 332 p.
3. J.-L. Sabrie, Feasibility of large AC superconducting equipment, J. Phys. Colloque C1. 1984. Vol. 45, N1. P. 717-720.
4. L.I. Chubraeva, Ya. B. Danilevich, I.A. Glebov,and L.G. Mamikoniants, Experience with superconductive turbogenerator in the power system and new developments, CIGRE, Session 1994, Rep. 11-103.

DEVELOPMENT AND EXPERIMENTAL INVESTIGATION OF NEW SEALS FOR CRYOALTERNATORS

I.S.Ganzhinov, V.A.Sapozhnikov, and V.A.Tutaev

VNIIElectromash
St.-Petersburg, Moscovsky av., 100, 196084, Russia

ABSTRACT

One of the vital problems in developing high-speed cryogenically cooled electrical machines with a cryostat common to rotor and stator is availability of reliable shaft seals. Depending on the electric machine design, reliable rotary seals are required both for working media sealing and for maintaining vacuum in rotating cryostats. Newly designed rotary seals employ the principles of radial hydrodynamic sealing. Tentative experimental studies proved the advantages of the new design over the traditional radial hydrodynamic seals. Pressure difference withstood by the new seals with no obvious sealed media leaks is about 1.5 times greater than that provided by a conventional seal.

INTRODUCTION

Development of cryogenically-cooled high-speed electric machinery involves a number of technical problems, one of which is designing the reliable seals for high-rpm shafts. Elaboration of reliable rotary seals for the rotor shaft is a vitally important point in developing the cryogenic machines having a cryostat common to rotor and stator. Application of widely used contact seals and magnetic-liquid seals is impossible in the high-rpm machines because of high linear speeds on the sealed shaft.

This challenging problem was approached by making improvements to hydrodynamic radial seals. This seal is the disk with radial blades made on one or both sides, which rotates together with the sealed shaft and is located inside the circular cavity of the seal frame. The working liquid introduced inside the circular cavity gets curved by the seal blades and develops radial pressure difference. Seals of this type have found intensive use in industry and are advantageous compared to rotary seals of other designs. Radial hydrodynamic seals are most efficient for the shafts with high revolutions per minute. For a fixed difference between the pressure of working liquid and that of the sealed medium, seal dimensions and its power consumption decrease substantially with growing shaft angular speed. Noteworthy also are the small axial dimensions of this type of seals. The major drawback of these seals is that during shut-down or at extremely low rpm, shaft sealing should be provided by a change-over to seals of some other types (say, collar-type or end-face seals), which will be disconnected at higher speeds of rotation. Application of conventional radial hydrodynamic seals in high-speed cryogenic electrical machines is hardly appropriate due to their small margin of sustainable pressure difference and possible leaks of sealed medium. This is most essential if the sealed medium is explosion-hazardous, which can be the case with cryogenic machines.

DEVELOPMENT OF NEW SEAL DESIGN

The working liquid located inside the inter-blade cavity is exposed to the action of several forces (Fig.1). Firstly, the centrifugal force equal to $dm\omega^2 R$, and secondly, Coriolis force equal to $dm2\omega V_R$. The resultant force is balanced by the pressure of adjacent liquid layers, and in consequence the liquid surface in the inter-blade cavity is perpendicular to this resultant force. The slope of working liquid surface has negative effect on hydrodynamic seal performances. First, pressure of seal out-break is brought down. And next, vortices with the radial-directed axes of rotation are formed on the thrust side of blades, which in combination with the slope of working liquid surface give rise to crests whose stalling results in splashing of the working liquid and is the major reason of its leakage. Liquid drops formed at the boundary of phase separation are carried further away to the seal shaft by the vortex arising in the seal gaseous cavity.

The hypothesis has been proposed that the effects of the above-mentioned phenomena could be brought down by filling the inter-blade cavities with capillary-porous filler. Springing up of vortices in liquid and gas are to be suppressed by the capillary-porous structure of the inserts, and transfer of interface into this structure prevents the liquid from splashing. Capillary forces acting of the working liquid are expected to decrease the slope of working liquid level in the seal, increasing thereby the value of seal out-break pressure.

To verify the suggested hypothesis, a model of radial hydrodynamic seal was designed and manufactured (Fig.2). The model was made in a vertical version in compliance with the driving motor design and for the sake of simplification.

The model is mounted on a base attached to a foundation plate. Driving motor is fastened to the base from beneath. On the base surface the seal frame is secured, inside which there is the seal impeller vacuum-fitted onto the shaft. The impeller

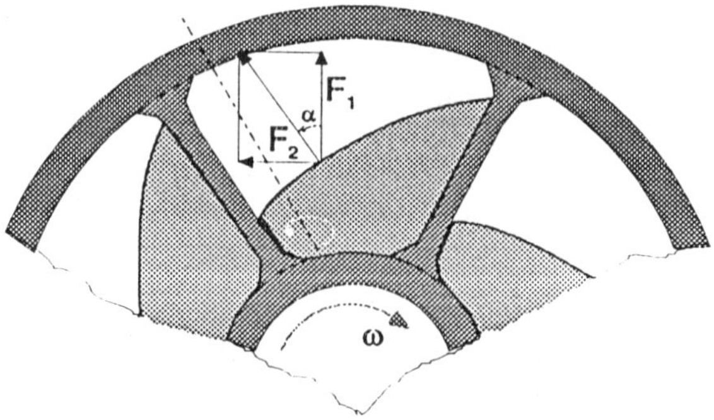

Fig.1. Forces acting on the interface inside impeller: $F_1=dm\omega^2R$, $F_2=dm2\omega V_R$,
were dm - mass of elementary volume of a liquid, ω - angular velocity, R- radius, V_R - projection of velocity liquid stream on radial direction

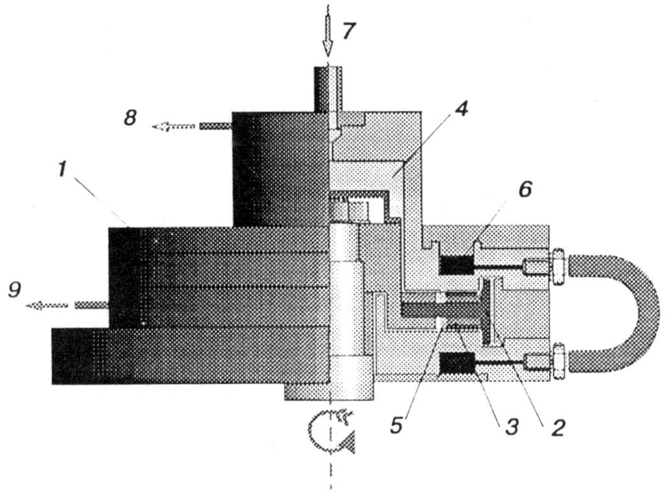

Fig.2.Hydrodynamic radial seal for the cryoalternator.
1 - seal frame, 2 - impeller, 3 - capillary-porous filler, 4 - volume to be sealed, 5 - working liquid, 6 - cooling water, 7 - pipe for maintaining pressure level and for evacuation, 8 - measuring of the pressure in volume to be sealed, 9 - measuring of the pressure in working liquid.

together with the frame are embraced by a cover. Inside the cover there is a cavity acting as a volume to be sealed. All the joints are made air-tight. The ducts for circulating water cooling the seal active zone are made inside the base and the cover. In the seal frame there are apertures for connecting pressure gauges to measure working liquid pressure at the seal bottom, and the valve intended for working liquid discharge. The aperture in the volume to be sealed is made for connecting gauges to measure pressure in the seal cavity, and the pipe connection is used for maintaining pressure level and for evacuation. Circuit pattern of motor connection enables smooth speed regulation within the range up to 8000 rpm. Speed of rotation is measured with the help of an stroboscope.

RESULTS OF MODEL INVESTIGATIONS

Prior to tests, the seal model was filled with working liquid. Equal initial level of working liquid in the seal with no inserts of capillary-porous material and in the seals with those inserts was checked by comparing pressure at the cavity bottom at fixed low value of rpm.

The results of tentative experiments suggest that at increased shaft speed of rotation the seal containing capillary-porous filler provides greater pressure difference than the seal with no porous inserts. The values of seal out-break pressure for the traditional seal and for the seal with capillary-porous inserts start to diverge considerably at the speed of 5000 rpm. Seal design with porous inserts made it possible to reach out-break pressure about $(1.4 - 1.6)\cdot 10^5$ Pa at speeds of 5500-6000 rpm, whereas the maximum out-break pressure encountered with conventional seals was $1.1\cdot 10^5$ Pa.

CONCLUSION

1. Provisional study of radial hydrodynamic seals with capillary-porous filler proves that the design holds much promise for application in high-speed cryogenic

electric machines having the cryostat common to rotor and stator, which enable considerable reduction of machine mass and dimensions.

2. The conducted tentative experiments were not aimed at determining the maximal pressure difference and comparing it with the parameter reached by traditional hydrodynamic seals, but were intended to verify our notions of the physical processes taking place in the rotary seals.

3. Further investigations should define the reasonable application limits of this type of seals and their power characteristics, and will serve as a basis for parameter optimization of the capillary-porous material used for filling up.

ACKNOWLEDGMENT

The research is supported by the Ministry of Science and Technical Policy of Russia and of the Russian Fund for Fundamental Researches.

REFERENCES

1. M.B. Krav, B.V.Ovsiannikov, A.S.Shapiro."The radial hydrodynamic seals of high-speed shafts",Mashinostroenie,Moskow (1976). (In Russian)